## Quadratic Formula

If $ax^2 + bx + c = 0$, where $a \neq 0$, then

$$x = \frac{-b \pm \sqrt{b^2 - 4ac}}{2a}.$$

## Straight Lines

$$m = \frac{y_2 - y_1}{x_2 - x_1} \qquad \text{(slope formula)}$$
$$y - y_1 = m(x - x_1) \qquad \text{(point-slope form)}$$
$$y = mx + b \qquad \text{(slope-intercept form)}$$
$$x = \text{constant} \qquad \text{(vertical line)}$$
$$y = \text{constant} \qquad \text{(horizontal line)}$$

## Inequalities

If $a < b$, then $a + c < b + c$.
If $a < b$ and $c > 0$, then $ac < bc$.
If $a < b$ and $c > 0$, then $a(-c) > b(-c)$.

## Logarithms

$$\log_b x = y \text{ where } x = b^y$$
$$\log_b (mn) = \log_b m + \log_b n$$
$$\log_b \frac{m}{n} = \log_b m - \log_b n$$
$$\log_b m^r = r \log_b m$$
$$\log_b 1 = 0$$
$$\log_b b = 1$$
$$\log_b b^r = r$$
$$b^{\log_b m} = m$$
$$\log_b m = \frac{\log_a m}{\log_a b}$$

## Counting

$$_nP_r = \frac{n!}{(n - r)!}$$

$$_nC_r = \frac{n!}{r!(n - r)!}$$

## Greek Alphabet

| | | | | | |
|---|---|---|---|---|---|
| alpha | A | $\alpha$ | nu | N | $\nu$ |
| beta | B | $\beta$ | xi | $\Xi$ | $\xi$ |
| gamma | $\Gamma$ | $\gamma$ | omicron | O | $o$ |
| delta | $\Delta$ | $\delta$ | pi | $\Pi$ | $\pi$ |
| epsilon | E | $\epsilon$ | rho | P | $\rho$ |
| zeta | Z | $\zeta$ | sigma | $\Sigma$ | $\sigma$ |
| eta | H | $\eta$ | tau | T | $\tau$ |
| theta | $\Theta$ | $\theta$ | upsilon | $\Upsilon$ | $\upsilon$ |
| iota | I | $\iota$ | phi | $\Phi$ | $\phi, \varphi$ |
| kappa | K | $\kappa$ | chi | X | $\chi$ |
| lambda | $\Lambda$ | $\lambda$ | psi | $\Psi$ | $\psi$ |
| mu | M | $\mu$ | omega | $\Omega$ | $\omega$ |

# Introductory Mathematical Analysis

NINTH EDITION

# Introductory Mathematical Analysis

## FOR BUSINESS, ECONOMICS, AND THE LIFE AND SOCIAL SCIENCES

**Ernest F. Haeussler, Jr.**
The Pennsylvania State University

**Richard S. Paul**
The Pennsylvania State University

*with Contributions by Laurel Technical Services*

Prentice Hall
Upper Saddle River, New Jersey 07458

**Library of Congress Cataloging-in-Publication Data**

Haeussler, Ernest F.
  Introductory mathematical analysis for business, economics, and
the life and social sciences / Ernest F. Haeussler, Jr., Richard S.
Paul.—9th ed.
    p.    cm.
  "With contributions by Laurel Technical Services."
  Includes index.
  ISBN 0-13-915760-3
  1. Mathematical analysis.    2. Economics, Mathematical.
3. Business mathematics.    I. Paul, Richard S.    II. Title
QA300.H328 1998                              98–7399
515'.1—dc21                                   CIP

Executive Editor: Sally Simpson
Sponsoring Editor/Supplements Editor: Gina M. Huck
Assistant Editor: Sara Beth Newell
Editorial Director: Tim Bozik
Editor-in-Chief: Jerome Grant
Assistant Vice President of Production and Manufacturing: David W. Riccardi
Editorial/Production Supervision: Richard DeLorenzo
Managing Editor: Linda Mihatov Behrens
Executive Managing Editor: Kathleen Schiaparelli
Manufacturing Buyer: Alan Fischer
Manufacturing Manager: Trudy Pisciotti
Marketing Manager: Patrice Lumumba Jones
Marketing Assistant: Amy Lysik
Creative Director: Paula Maylahn
Associate Creative Director: Amy Rosen
Art Director: Maureen Eide
Assistant to the Art Director: John Christiana
Art Manager: Gus Vibal
Art Editor: Grace Hazeldine
Cover Image: Tokyo Forum; Tokyo Japan—Timothy Hursley/Superstock, Inc.
Cover Designer: Bruce Kenselaar
Interior Design and Layout: Maureen Eide
Illustrator: Monotype Composition
Photo Research: Elaine Estrada

Printed in the United States of America
10 9 8 7 6 5 4 3 2 1

ISBN 0-13-915760-3

Prentice-Hall International (UK) Limited, *London*
Prentice-Hall of Australia Pty. Limited, *Sydney*
Prentice-Hall Canada, Inc., *Toronto*
Prentice-Hall Hispanoamericano, S.A., *Mexico*
Prentice-Hall of India Private Limited, *New Delhi*
Prentice-Hall of Japan, Inc., *Tokyo*
Simon & Schuster Asia Pte. Ltd., *Singapore*
Editora Prentice-Hall do Brasil, Ltda., Rio de Janiero

# Contents

# 4 Lines, Parabolas, and Systems  119

# 5 Exponential and Logarithmic Functions  171

# 6 Matrix Algebra  209

# 7 Linear Programming  285

# Preface

This ninth edition of *Introductory Mathematical Analysis* continues to provide a mathematical foundation for students in business, economics, and the life and social sciences. It begins with noncalculus topics such as equations, functions, matrix algebra, linear programming, mathematics of finance, and probability. Then it progresses through both single-variable and multivariable calculus, including continuous random variables. Technical proofs, conditions, and the like, are sufficiently described, but are not overdone. At times, informal intuitive arguments are given to preserve clarity.

## Applications

An abundance and variety of applications for the intended audience appear throughout the book; students continually see how the mathematics they are learning can be used. These applications cover such diverse areas as business, economics, biology, medicine, sociology, psychology, ecology, statistics, earth science, and archaeology. Many of these real-world situations are drawn from literature and are documented by references. In some, the background and context are given in order to stimulate interest. However, the text is virtually self-contained, in the sense that it assumes no prior exposure to the concepts on which the applications are based.

## Changes to the Ninth Edition

### Principles in Practice

This new element provides students with even more applications. Located in the margin of the text, these additional exercises give students real-world applications and more opportunities to see the chapter material put into practice. *Principles in Practice* applications that can be solved using a graphing calculator are indicated by an icon . Answers to Principles in Practice applications appear at the end of the text.

### Concepts for Calculus Appendix

New to the ninth edition, this useful end-of-text appendix features additional calculus concepts for student review. Such topics include: Slopes and Equations of Lines, Secant Lines and Average Rate of Change, and Slope of a Curve and Derivative.

### Updated Mathematical Snapshots

Included at the end of many chapters, this popular feature has been revised for the ninth edition. Each Snapshot provides an interesting, and at times, novel application involving the mathematics of the chapter in which it occurs. Many of the Snapshots also include exercises, reinforcing the text's strong emphasis on hands-on practice.

## RETAINED FEATURES

Interspersed throughout the text are many warnings to the student that point out commonly made errors. These warnings are indicated under the heading **Pitfall**. Definitions are clearly stated and displayed. Key concepts, as well as important rules and formulas, are boxed to emphasize their importance. Throughout the text, notes to the student are placed in the margin. They reflect passing comments which supplement discussions.

More than 850 examples are worked out in detail. Some include a **strategy** that is specifically designed to guide the student through the logistics of the solution before the solution is obtained.

An abundant number of diagrams (almost 500) and exercises (more than 5,000) are included. In each exercise set, grouped problems are given in increasing order of difficulty. In many exercise sets the problems progress from the basic mechanical-drill type to more interesting thought-provoking problems. Many real-world type problems with real data are included. Considerable effort has been made to produce a proper balance between the drill-type exercises and the problems requiring the integration of the concepts learned.

In order that a student appreciates the value of current **technology**, optional graphics-calculator material appears throughout the text both in the exposition and exercises. It appears for a variety of reasons: as a mathematical tool, to visualize a concept, as a computing aid, and to reinforce concepts. Although calculator displays (see below) for a TI-82 accompany the corresponding technology discussion, our approach is general enough so that it can be applied to other fine graphics calculators.

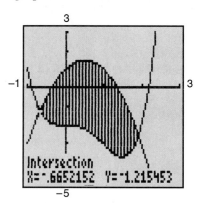

In the exercise sets, graphics-calculator problems are indicated by an icon. To provide flexibility for an instructor in planning assignments, these problems are placed at the end of an exercise set.

Each chapter (except Chapter 0) has a review section that contains a list of important terms and symbols, a chapter summary, and numerous review problems.

Answers to odd-numbered problems appear at the end of the book. For many of the differentiation problems, the answers appear in both unsimplified and simplified forms. This allows students to readily check their work.

## COURSE PLANNING

Because instructors plan a course outline to serve the individual needs of a particular class and curriculum, we shall not attempt to provide sample outlines. However, depending on the background of the students, some instructors will choose to omit Chapter 0, *Algebra Refresher,* or Chapter 1, *Equations.* Others may exclude the topics of matrix algebra and linear programming. Certainly there are other sections that may be omitted at the discretion of the instructor. As an aid to planning a course outline, perhaps a few comments may be helpful. Section 2.1 introduces some business terms, such as total revenue, fixed cost, variable cost and profit. Section 4.2 introduces the notion of supply and demand equations, and Section 4.6 discusses the equilibrium point. Optional sections, which will not cause problems if they are omitted, are: 7.3, 7.5, 15.4, 17.1, 17.2, 19.4, 19.6, 19.9 and 19.10. Section 17.8 may be omitted if Chapter 18 is not covered.

## Supplements

### For Instructors

*Instructor's Solution Manual.* Worked out solutions to all exercises and Principles-in-Practice applications.

*Test Item File.* Provides over 1,700 test questions, keyed to chapter and section.

*Prentice Hall Custom Test.* Allows the instructor to access from the computerized Test Item File and personally prepare and print out tests. Includes an editing feature which allows questions to be added or changed.

### For Students

*Student Solutions Manual with Visual Calculus and Explorations in Finite Mathematics Software.* Worked out solutions for every odd-numbered exercise and all Principles-in-Practice applications. Software includes unique programs which enhance the fundamental concepts of calculus and finite mathematics visually, and include exercises taken directly from the text.

### For Instructors and Students

*PH Companion Website.* Designed to complement and expand upon the text, the PH Companion Website offers a variety of interactive learning tools, including: links to related websites, practice work for students, and the ability for instructors to monitor and evaluate students' work on the website. For more information, contact your local Prentice Hall representative.
*www.prenhall.com/Haeussler*

## Acknowledgments

We express our appreciation to the following colleagues who contributed comments and suggestions that were valuable to us in the evolution of this text:

R. M. Alliston *(Pennsylvania State University)*, R. A. Alo *(University of Houston)*, K. T. Andrews *(Oakland University)*, M. N. de Arce *(University of Puerto Rico)*, G. R. Bates *(Western Illinois University)*, D. E. Bennett *(Murray State University)*, C. Bernett *(Harper College)*, A. Bishop *(Western Illinois University)*, S. A. Book *(California State University)*, A. Brink *(St. Cloud State University)*, R. Brown *(York University)*, R. W. Brown *(University of Alaska)*, S. D. Bulman-Fleming *(Wilfrid Laurier University)*, D. Calvetti *(National College)*, D. Cameron *(University of Akron)*, K. S. Chung *(Kapiolani Community College)*, D. N. Clark *(University of Georgia)*, E. L. Cohen *(University of Ottawa)*, J. Dawson *(Pennsylvania State University)*, A. Dollins *(Pennsylvania State University)*, G. A. Earles *(St. Cloud State University)*, B. H. Edwards *(University of Florida)*, J. R. Elliott *(Wilfrid Laurier University)*, J. Fitzpatrick *(University of Texas at El*

*Paso)*, M. J. Flynn *(Rhode Island Junior College)*, G. J. Fuentes *(University of Maine)*, S. K. Goel *(Valdosta State University)*, G. Goff *(Oklahoma State University)*, J. Goldman *(DePaul University)*, J. T. Gresser *(Bowling Green State University)*, L. Griff *(Pennsylvania State University)*, F. H. Hall *(Pennsylvania State University)*, V. E. Hanks *(Western Kentucky University)*, R. C. Heitmann *(The University of Texas at Austin)*, J. N. Henry *(California State University)*, W. U. Hodgson *(West Chester State College)*, B. C. Horne, Jr. *(Virginia Polytechnic Institute and State University)*, J. Hradnansky *(Pennsylvania State University)*, C. Hurd *(Pennsylvania State University)*, J. A. Jiminez *(Pennsylvania State University)*, W. C. Jones *(Western Kentucky University)*, R. M. King *(Gettysburg College)*, M. M. Kostreva *(University of Maine)*, G. A. Kraus *(Gannon University)*, J. Kucera *(Washington State University)*, M. R. Latina *(Rhode Island Junior College)*, J. F. Longman *(Villanova University)*, I. Marshak *(Loyola University of Chicago)*, D. Mason *(Elmhurst College)*, F. B. Mayer *(Mt. San Antonio College)*, P. McDougle *(University of Miami)*, F. Miles *(California State University)*, E. Mohnike *(Mt. San Antonio College)*, C. Monk *(University of Richmond)*, R. A. Moreland *(Texas Tech University)*, J. G. Morris *(University of Wisconsin-Madison)*, J. C. Moss *(Paducah Community College)*, D. Mullin *(Pennsylvania State University)*, E. Nelson *(Pennsylvania State University)*, S. A. Nett *(Western Illinois University)*, R. H. Oehmke *(University of Iowa)*, Y. Y. Oh *(Pennsylvania State University)*, N. B. Patterson *(Pennsylvania State University)*, V. Pedwaydon *(Lawrence Technical University)*, E. Pemberton *(Wilfrid Laurier University)*, M. Perkel *(Wright State University)*, D. B. Priest *(Harding College)*, J. R. Provencio *(University of Texas)*, L. R. Pulsinelli *(Western Kentucky University)*, M. Racine *(University of Ottawa)*, N. M. Rice *(Queen's University)*, A. Santiago *(University of Puerto Rico)*, J. R. Schaefer *(University of Wisconsin-Milwaukee)*, S. Sehgal *(The Ohio State University)*, W. H. Seybold, Jr. *(West Chester State College)*, G. Shilling *(The University of Texas at Arlington)*, S. Singh *(Pennsylvania State University)*, L. Small *(Los Angeles Pierce College)*, E. Smet *(Huron College)*, M. Stoll *(University of South Carolina)*, A. Tierman *(Saginaw Valley State University)*, B. Toole *(University of Maine)*, J. W. Toole *(University of Maine)*, D. H. Trahan *(Naval Postgraduate School)*, J. P. Tull *(Ohio State University)*, L. O. Vaughan, Jr. *(University of Alabama in Birmingham)*, L. A. Vercoe *(Pennsylvania State University)*, M. Vuilleumier *(Ohio State University)*, B. K. Waits *(Ohio State University)*, A. Walton *(Virginia Polytechnic Institute and State University)*, H. Walum *(Ohio State University)*, E. T. H. Wang *(Wilfrid Laurier University)*, A. J. Weidner *(Pennsylvania State University)*, L. Weiss *(Pennsylvania State University)*, N. A. Weigmann *(California State University)*, G. Woods *(Ohio State University)*, C. R. B. Wright *(University of Oregon)*, C. Wu *(University of Wisconsin-Milwaukee)*.

Some exercises are taken from problem supplements used by students at Wilfrid Laurier University. We wish to extend special thanks to the Department of Mathematics of Wilfrid Laurier University for granting Prentice Hall permission to use and publish this material, and also to thank Prentice Hall, who in turn allowed us to make use of this material.

We also thank Laurel Technical Services for their contribution of the Principles-in-Practice applications, for their input to the Concepts for Calculus appendix, and for their proofreading efforts in the revision process.

Finally, we express our sincere gratitude to the faculty and course coordinators of Ohio State University and Columbus State University who took a keen interest in the ninth edition, offering a number of invaluable suggestions.

Ernest F. Haeussler, Jr.
Richard S. Paul

# Algebra Refresher

## 0.1 Purpose

This chapter is designed to give you a brief review of some terms and methods of manipulative mathematics. No doubt you have been exposed to much of this material before. However, because these topics are important in handling the mathematics that comes later, perhaps an immediate second exposure to them would be beneficial. Devote whatever time is necessary to the sections in which you need review.

## 0.2 Sets and Real Numbers

In simplest terms, a *set* is a collection of objects. For example, we can speak of the set of even numbers between 5 and 11, namely, 6, 8, and 10. An object in a set is called an *element* or *member* of that set.

One way to specify a set is by listing its elements, in any order, inside braces. For example, the previous set is $\{6, 8, 10\}$, which we can denote by a letter such as $A$. A set $A$ is said to be a subset of a set $B$ if and only if every element of $A$ is also an element of $B$. For example, if $A = \{6, 8, 10\}$ and $B = \{6, 8, 10, 12\}$, then $A$ is a subset of $B$.

Certain sets of numbers have special names. The numbers 1, 2, 3, and so on form the set of **positive integers** (or **natural numbers**):

$$\text{set of positive integers} = \{1, 2, 3, \dots\}.$$

The three dots mean that the listing of elements is unending, although we know what the elements are.

The positive integers, together with 0 and the **negative integers** $-1, -2, -3, \dots$, form the set of **integers:**

$$\text{set of integers} = \{\dots, -3, -2, -1, 0, 1, 2, 3, \dots\}.$$

The reason for $q \neq 0$ is that we cannot divide by zero.

The set of **rational numbers** consists of numbers, such as $\frac{1}{2}$ and $\frac{5}{3}$ that can be written as a ratio (quotient) of two integers. That is, a rational number is a number that can be written as $p/q$, where $p$ and $q$ are integers and $q \neq 0$. (The symbol "$\neq$" is read "is not equal to.") For example, the numbers $\frac{19}{20}, \frac{-2}{7}$, and $\frac{-6}{-2}$, are rational. We remark that $\frac{2}{4}, \frac{1}{2}, \frac{3}{6}, \frac{-4}{-8}$, and 0.5 all represent the same

Every integer is a rational number.

rational number. The integer 2 is rational, since $2 = \frac{2}{1}$. In fact, every integer is rational.

All rational numbers can be represented by decimal numbers that *terminate*, such as $\frac{3}{4} = 0.75$ and $\frac{3}{2} = 1.5$, or by *nonterminating repeating decimal numbers* (composed of a group of digits that repeats without end), such as $\frac{2}{3} = 0.666\ldots, \frac{-4}{11} = -0.3636\ldots$, and $\frac{2}{15} = 0.1333\ldots$. Numbers represented by *nonterminating nonrepeating* decimals are called **irrational numbers.** An irrational number cannot be written as an integer divided by an integer. The numbers $\pi$ (pi) and $\sqrt{2}$ are irrational.

The real numbers consist of all decimal numbers.

Together, the rational numbers and irrational numbers form the set of **real numbers.** Real numbers can be represented by points on a line. First we choose a point on the line to represent zero. This point is called the *origin.* (See Fig. 0.1.) Then a standard measure of distance, called a "unit distance," is chosen and is successively marked off both to the right and to the left of the origin. With each point on the line we associate a directed distance, or *signed number,* which depends on the position of the point with respect to the origin. Positions to the right of the origin are considered positive $(+)$ and positions to the left are negative $(-)$. For example, with the point $\frac{1}{2}$ unit to the right of the origin there corresponds the signed number $\frac{1}{2}$, which is called the **coordinate** of that point. Similarly, the coordinate of the point 1.5 units to the left of the origin is $-1.5$. In Figure 0.1, the coordinates of some points are marked. The arrowhead indicates that the direction to the right along the line is considered the positive direction.

Some Points and Their Coordinates

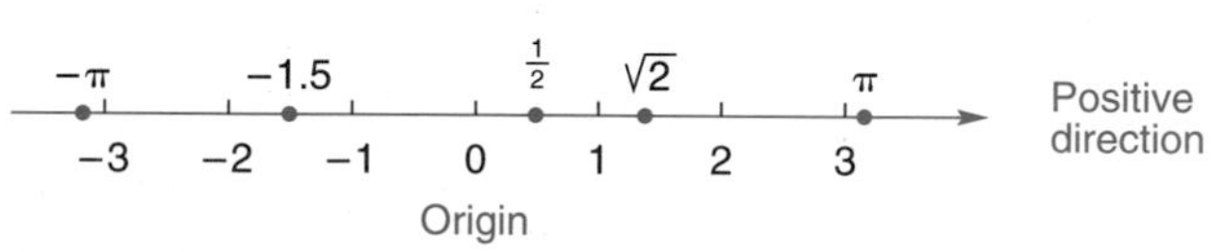

**FIGURE 0.1**  The real-number line.

To each point on the line there corresponds a unique real number, and to each real number there corresponds a unique point on the line. For this reason, we say that there is a *one-to-one correspondence* between points on the line and real numbers. We call this line a **coordinate line** or the **real-number line.** We feel free to treat real numbers as points on a real-number line and vice versa.

■ **Exercise 0.2**

*In Problems **1–12**, classify the statement as either true or false. If false, give a reason.*

**1.** $-7$ is an integer.

**2.** $\frac{1}{6}$ is rational.

**3.** $-3$ is a natural number.

**4.** 0 is not rational.

**5.** 5 is rational.

**6.** $\frac{7}{0}$ is a rational number.

**7.** $\frac{4}{2}$ is not a positive integer.

**8.** $\pi$ is a real number.

**9.** $\frac{0}{6}$ is rational.

**10.** 0 is a natural number.

**11.** $-3$ is to the right of $-4$ on the real number line.

**12.** Every integer is positive or negative.

---

**OBJECTIVE**

To state and illustrate the following properties of real numbers: transitive, commutative, associative, inverse, and distributive. To define subtraction and division in terms of addition and multiplication, respectively.

## 0.3  SOME PROPERTIES OF REAL NUMBERS

We now state a few important properties of the real numbers. Let $a$, $b$, and $c$ be real numbers.

**1. The Transitive Property of Equality**

$$\text{If } a = b \text{ and } b = c, \text{ then } a = c.$$

Thus, two numbers that are both equal to a third number are equal to each other. For example, if $x = y$ and $y = 7$, then $x = 7$.

**2. The Commutative Properties of Addition and Multiplication**

$$a + b = b + a \qquad \text{and} \qquad ab = ba.$$

This means that two numbers can be added or multiplied in any order. For example, $3 + 4 = 4 + 3$ and $7(-4) = (-4)(7)$.

**3. The Associative Properties of Addition and Multiplication**

$$a + (b + c) = (a + b) + c \qquad \text{and} \qquad a(bc) = (ab)c.$$

This means that in addition or multiplication, numbers can be grouped in any order. For example, $2 + (3 + 4) = (2 + 3) + 4$; in both cases, the sum is 9. Similarly, $2x + (x + y) = (2x + x) + y$ and $6(\frac{1}{3} \cdot 5) = (6 \cdot \frac{1}{3}) \cdot 5$.

**4. The Inverse Properties**

For each real number $a$, there is a unique real number denoted $-a$ such that

$$a + (-a) = 0.$$

The number $-a$ is called the **additive inverse**, or **negative**, of $a$.

For example, since $6 + (-6) = 0$, the additive inverse of 6 is $-6$. The additive inverse of a number is not necessarily a negative number. For example, the additive inverse of $-6$ is 6, since $(-6) + (6) = 0$. That is, the negative of $-6$ is 6, so we can write $-(-6) = 6$.

Zero does not have a multiplicative inverse because there is no number that, when multiplied by 0, gives 1.

For each real number $a$, except 0, there is a unique real number denoted $a^{-1}$ such that
$$a \cdot a^{-1} = 1.$$
The number $a^{-1}$ is called the **multiplicative inverse** of $a$.

Thus, all numbers except 0 have a multiplicative inverse. You may recall that $a^{-1}$ can be written $\dfrac{1}{a}$ and is also called the *reciprocal* of $a$. For example, the multiplicative inverse of 3 is $\frac{1}{3}$, since $3(\frac{1}{3}) = 1$. Hence, $\frac{1}{3}$ is the reciprocal of 3. The reciprocal of $\frac{1}{3}$, is 3, since $(\frac{1}{3})(3) = 1$. **The reciprocal of 0 is not defined.**

**5. The Distributive Properties**
$$a(b + c) = ab + ac \qquad \text{and} \qquad (b + c)a = ba + ca.$$

For example, although $2(3 + 4) = 2(7) = 14$, we can write
$$2(3 + 4) = 2(3) + 2(4) = 6 + 8 = 14.$$
Similarly,
$$(2 + 3)(4) = 2(4) + 3(4) = 8 + 12 = 20,$$
$$\text{and} \qquad x(z + 4) = x(z) + x(4) = xz + 4x.$$

The distributive property can be extended to the form
$$a(b + c + d) = ab + ac + ad.$$

In fact, it can be extended to sums involving any number of terms.
   **Subtraction** is defined in terms of addition:
$$a - b \qquad \text{means} \qquad a + (-b),$$
where $-b$ is the additive inverse of $b$. Thus, $6 - 8$ means $6 + (-8)$.
   In a similar way, we define **division** in terms of multiplication. If $b \neq 0$, then $a \div b$, or $\dfrac{a}{b}$, is defined by
$$\frac{a}{b} = a(b^{-1}).$$

Since $b^{-1} = \dfrac{1}{b}$,
$$\frac{a}{b} = a(b^{-1}) = a\left(\frac{1}{b}\right).$$

$\dfrac{a}{b}$ means $a$ times the reciprocal of $b$.

Thus, $\frac{3}{5}$ means 3 times $\frac{1}{5}$, where $\frac{1}{5}$ is the multiplicative inverse of 5. Sometimes we refer to $a \div b$ or $\dfrac{a}{b}$ as the *ratio* of $a$ to $b$. We remark that since 0 does not have a multiplicative inverse, **division by 0 is not defined.**
   The following examples show some manipulations involving the preceding properties.

**EXAMPLE 1   Applying Properties of Real Numbers**

**a.** $x(y - 3z + 2w) = (y - 3z + 2w)x$, by the commutative property of multiplication.

**b.** By the associative property of multiplication, $3(4 \cdot 5) = (3 \cdot 4)5$. Thus, the result of multiplying 3 by the product of 4 and 5 is the same as the result of multiplying the product of 3 and 4 by 5. In either case, the result is 60. ▪

**EXAMPLE 2   Applying Properties of Real Numbers**

**a.** *Show that* $2 - \sqrt{2} = -\sqrt{2} + 2$.

*Solution:* By the definition of subtraction, $2 - \sqrt{2} = 2 + (-\sqrt{2})$.

However, by the commutative property of addition, $2 + (-\sqrt{2}) = -\sqrt{2} + 2$. Hence, by the transitive property, $2 - \sqrt{2} = -\sqrt{2} + 2$. More concisely, we omit intermediate steps and directly write

$$2 - \sqrt{2} = -\sqrt{2} + 2.$$

**b.** *Show that* $(8 + x) - y = 8 + (x - y)$.

*Solution:* Beginning with the left side, we have

$$
\begin{aligned}
(8 + x) - y &= (8 + x) + (-) &&\text{(definition of subtraction)}\\
&= 8 + [x + (-y)] &&\text{(associative property)}\\
&= 8 + (x - y) &&\text{(definition of subtraction).}
\end{aligned}
$$

Hence, by the transitive property,

$$(8 + x) - y = 8 + (x - y).$$

**c.** *Show that* $3(4x + 2y + 8) = 12x + 6y + 24$.

*Solution:* By the distributive property,

$$3(4x + 2y + 8) = 3(4x) + 3(2y) + 3(8).$$

But by the associative property of multiplication,

$$3(4x) = (3 \cdot 4)x = 12x \quad \text{and similarly} \quad 3(2y) = 6y.$$

Thus, $3(4x + 2y + 8) = 12x + 6y + 24$. ▪

**EXAMPLE 3   Applying Properties of Real Numbers**

**a.** *Show that* $\dfrac{ab}{c} = a\left(\dfrac{b}{c}\right)$ *for* $c \neq 0$.

*Solution:* By the definition of division,

$$\frac{ab}{c} = (ab) \cdot \frac{1}{c} \quad \text{for } c \neq 0.$$

But by the associative property,

$$(ab) \cdot \frac{1}{c} = a\left(b \cdot \frac{1}{c}\right).$$

However, by the definition of division, $b \cdot \dfrac{1}{c} = \dfrac{b}{c}$. Thus,

$$\frac{ab}{c} = a\left(\frac{b}{c}\right).$$

We can also show that $\dfrac{ab}{c} = \left(\dfrac{a}{c}\right)b$.

**b.** *Show that* $\dfrac{a + b}{c} = \dfrac{a}{c} + \dfrac{b}{c}$ *for* $c \neq 0$.

**Solution:** By the definition of division and the distributive property,

$$\frac{a + b}{c} = (a + b)\frac{1}{c} = a \cdot \frac{1}{c} + b \cdot \frac{1}{c}.$$

However,

$$a \cdot \frac{1}{c} + b \cdot \frac{1}{c} = \frac{a}{c} + \frac{b}{c}.$$

Hence,

$$\frac{a + b}{c} = \frac{a}{c} + \frac{b}{c}.$$

We remark that $\dfrac{a}{b + c} \neq \dfrac{a}{b} + \dfrac{a}{c}$. For example,

$$\frac{3}{2 + 1} \neq \frac{3}{2} + \frac{3}{1}. \qquad \blacksquare$$

Finding the product of several numbers can be done only by considering products of numbers taken two at a time. For example, to find the product of $x$, $y$, and $z$, we could first multiply $x$ by $y$ and then multiply that product by $z$; that is, we find $(xy)z$. Or alternatively, we could multiply $x$ by the product of $y$ and $z$; that is, we find $x(yz)$. The associative property of multiplication guarantees that both results are identical, regardless of how the numbers are grouped. Thus, it is not ambiguous to write $xyz$. This concept can be extended to more than three numbers and applies equally well to addition.

One final comment is appropriate before we end this section: Not only should you be aware of the manipulative aspects of the properties of the real numbers; you should also be aware of, and familiar with, the terminology involved.

## ■ Exercise 0.3

*In Problems* **1–10,** *classify the statements as either true or false.*

**1.** Every real number has a reciprocal.

**2.** The reciprocal of $\frac{2}{5}$ is $\frac{5}{2}$.

**3.** The additive inverse of 5 is $\frac{1}{5}$.

**4.** $2(3 \cdot 4) = (2 \cdot 3)(2 \cdot 4)$.

**5.** $-x + y = y - x$.

**6.** $(x + 2)(4) = 4x + 8$.

**7.** $\dfrac{x + 2}{2} = \dfrac{x}{2} + 1$.

**8.** $3\left(\dfrac{x}{4}\right) = \dfrac{3x}{4}$.

**9.** $x + (y + 5) = (x + y) + (x + 5)$.

**10.** $8(9x) = 72x$.

*In Problems* **11–20,** *state which properties of the real numbers are being used.*

**11.** $2(x + y) = 2x + 2y$.

**12.** $(x + 5) + y = y + (x + 5)$.

**13.** $2(3y) = (2 \cdot 3)y$.

**14.** $\frac{6}{7} = 6 \cdot \frac{1}{7}$.

**15.** $2(x - y) = (x - y)(2)$.

**16.** $x + (x + y) = (x + x) + y$.

**17.** $8 - y = 8 + (-y)$.

**18.** $5(4 + 7) = 5(7 + 4)$.

**19.** $(7 + x)y = 7y + xy$.

**20.** $(-1)[-3 + 4] = (-1)(-3) + (-1)(4)$.

*In Problems 21–26, show that the statements are true by using properties of the real numbers.*

**21.** $5a(x + 3) = 5ax + 15a$.

**22.** $(2 - x) + y = 2 + (y - x)$.

**23.** $(x + y)(2) = 2x + 2y$.

**24.** $2[27 + (x + y)] = 2[(y + 27) + x]$.

**25.** $x[(2y + 1) + 3] = 2xy + 4x$.

**26.** $(x + 1)(y + 1) = xy + x + y + 1$.

**27.** Show that $a(b + c + d) = ab + ac + ad$.
[*Hint*: $b + c + d = (b + c) + d$.]

---

## OBJECTIVE

**To list and illustrate the most common properties of real numbers.**

## 0.4 OPERATIONS WITH REAL NUMBERS

The following list states important properties of real numbers that you should study thoroughly. Being able to manipulate real numbers is essential to your success in mathematics. A numerical example follows each property. All denominators are different from zero. A knowledge of addition and subtraction of real numbers is assumed.

| *Property* | *Example* |
|---|---|
| **1.** $a - b = a + (-b)$. | $2 - 7 = 2 + (-7) = -5$. |
| **2.** $a - (-b) = a + b$. | $2 - (-7) = 2 + 7 = 9$. |
| **3.** $-a = (-1)(a)$. | $-7 = (-1)(7)$. |
| **4.** $a(b + c) = ab + ac$. | $6(7 + 2) = 6 \cdot 7 + 6 \cdot 2 = 54$. |
| **5.** $a(b - c) = ab - ac$. | $6(7 - 2) = 6 \cdot 7 - 6 \cdot 2 = 30$. |
| **6.** $-(a + b) = -a - b$. | $-(7 + 2) = -7 - 2 = -9$. |
| **7.** $-(a - b) = -a + b$. | $-(2 - 7) = -2 + 7 = 5$. |
| **8.** $-(-a) = a$. | $-(-2) = 2$. |
| **9.** $a(0) = 0$. | $2(0) = 0$. |
| **10.** $(-a)(b) = -(ab) = a(-b)$. | $(-2)(7) = -(2 \cdot 7) = 2(-7) = -14$. |
| **11.** $(-a)(-b) = ab$. | $(-2)(-7) = 2 \cdot 7 = 14$. |
| **12.** $\dfrac{a}{1} = a$. | $\dfrac{7}{1} = 7, \dfrac{-2}{1} = -2$. |
| **13.** $\dfrac{a}{b} = a\left(\dfrac{1}{b}\right)$. | $\dfrac{2}{7} = 2\left(\dfrac{1}{7}\right)$. |
| **14.** $\dfrac{a}{-b} = -\dfrac{a}{b} = \dfrac{-a}{b}$. | $\dfrac{2}{-7} = -\dfrac{2}{7} = \dfrac{-2}{7}$. |
| **15.** $\dfrac{-a}{-b} = \dfrac{a}{b}$. | $\dfrac{-2}{-7} = \dfrac{2}{7}$. |
| **16.** $\dfrac{0}{a} = 0$ when $a \neq 0$. | $\dfrac{0}{7} = 0$. |
| **17.** $\dfrac{a}{a} = 1$ when $a \neq 0$. | $\dfrac{2}{2} = 1, \quad \dfrac{-5}{-5} = 1$. |
| **18.** $a\left(\dfrac{b}{a}\right) = b$. | $2\left(\dfrac{7}{2}\right) = 7$. |
| **19.** $a \cdot \dfrac{1}{a} = 1$ when $a \neq 0$. | $2 \cdot \dfrac{1}{2} = 1$. |
| **20.** $\dfrac{a}{b} \cdot \dfrac{c}{d} = \dfrac{ac}{bd}$ | $\dfrac{2}{3} \cdot \dfrac{4}{5} = \dfrac{2 \cdot 4}{3 \cdot 5} = \dfrac{8}{15}$. |
| **21.** $\dfrac{ab}{c} = \left(\dfrac{a}{c}\right)b = a\left(\dfrac{b}{c}\right)$ | $\dfrac{2 \cdot 7}{3} = \dfrac{2}{3} \cdot 7 = 2 \cdot \dfrac{7}{3}$. |

|  | *Property* | *Example* |
|---|---|---|

**22.** $\dfrac{a}{bc} = \left(\dfrac{a}{b}\right)\left(\dfrac{1}{c}\right) = \left(\dfrac{1}{b}\right)\left(\dfrac{a}{c}\right)$ $\qquad$ $\dfrac{2}{3 \cdot 7} = \dfrac{2}{3} \cdot \dfrac{1}{7} = \dfrac{1}{3} \cdot \dfrac{2}{7}.$

**23.** $\dfrac{a}{b} = \left(\dfrac{a}{b}\right)\left(\dfrac{c}{c}\right) = \dfrac{ac}{bc}$ $\qquad$ $\dfrac{2}{7} = \left(\dfrac{2}{7}\right)\left(\dfrac{5}{5}\right) = \dfrac{2 \cdot 5}{7 \cdot 5}.$
when $c \neq 0$.

**24.** $\dfrac{a}{b(-c)} = \dfrac{a}{(-b)(c)} = \dfrac{-a}{bc} =$ $\qquad$ $\dfrac{2}{3(-5)} = \dfrac{2}{(-3)(5)} = \dfrac{-2}{3(5)} =$

$\dfrac{-a}{(-b)(-c)} = -\dfrac{a}{bc}.$ $\qquad$ $\dfrac{-2}{(-3)(-5)} = -\dfrac{2}{3(5)} = -\dfrac{2}{15}.$

**25.** $\dfrac{a(-b)}{c} = \dfrac{(-a)b}{c} = \dfrac{ab}{-c} =$ $\qquad$ $\dfrac{2(-3)}{5} = \dfrac{(-2)(3)}{5} = \dfrac{2(3)}{-5} =$

$\dfrac{(-a)(-b)}{-c} = -\dfrac{ab}{c}.$ $\qquad$ $\dfrac{(-2)(-3)}{-5} = -\dfrac{2(3)}{5} = -\dfrac{6}{5}.$

**26.** $\dfrac{a}{c} + \dfrac{b}{c} = \dfrac{a + b}{c}.$ $\qquad$ $\dfrac{2}{9} + \dfrac{3}{9} = \dfrac{2 + 3}{9} = \dfrac{5}{9}.$

**27.** $\dfrac{a}{c} - \dfrac{b}{c} = \dfrac{a - b}{c}.$ $\qquad$ $\dfrac{2}{9} - \dfrac{3}{9} = \dfrac{2 - 3}{9} = \dfrac{-1}{9}.$

**28.** $\dfrac{a}{b} + \dfrac{c}{d} = \dfrac{ad + bc}{bd}.$ $\qquad$ $\dfrac{4}{5} + \dfrac{2}{3} = \dfrac{4 \cdot 3 + 5 \cdot 2}{5 \cdot 3} = \dfrac{22}{15}.$

**29.** $\dfrac{a}{b} - \dfrac{c}{d} = \dfrac{ad - bc}{bd}.$ $\qquad$ $\dfrac{4}{5} - \dfrac{2}{3} = \dfrac{4 \cdot 3 - 5 \cdot 2}{5 \cdot 3} = \dfrac{2}{15}.$

**30.** $\dfrac{\frac{a}{b}}{\frac{c}{d}} = \dfrac{a}{b} \div \dfrac{c}{d} = \dfrac{a}{b} \cdot \dfrac{d}{c} = \dfrac{ad}{bc}.$ $\qquad$ $\dfrac{\frac{2}{3}}{\frac{7}{5}} = \dfrac{2}{3} \div \dfrac{7}{5} = \dfrac{2}{3} \cdot \dfrac{5}{7} = \dfrac{10}{21}.$

**31.** $\dfrac{a}{\frac{b}{c}} = a \div \dfrac{b}{c} = a \cdot \dfrac{c}{b} = \dfrac{ac}{b}.$ $\qquad$ $\dfrac{2}{\frac{3}{5}} = 2 \div \dfrac{3}{5} = 2 \cdot \dfrac{5}{3} = \dfrac{2 \cdot 5}{3} = \dfrac{10}{3}.$

**32.** $\dfrac{\frac{a}{b}}{c} = \dfrac{a}{b} \div c = \dfrac{a}{b} \cdot \dfrac{1}{c} = \dfrac{a}{bc}.$ $\qquad$ $\dfrac{\frac{2}{3}}{5} = \dfrac{2}{3} \div 5 = \dfrac{2}{3} \cdot \dfrac{1}{5} = \dfrac{2}{3 \cdot 5} = \dfrac{2}{15}.$

Property 23 is essentially the **fundamental principle of fractions,** which states that *multiplying or dividing both the numerator and denominator of a fraction by the same number, except* 0, *results in a fraction that is equivalent to (that is, has the same value as) the original fraction.* Thus,

$$\frac{7}{\frac{1}{8}} = \frac{7 \cdot 8}{\frac{1}{8} \cdot 8} = \frac{56}{1} = 56.$$

By Properties 28 and 23, we have

$$\frac{2}{5} + \frac{4}{15} = \frac{2 \cdot 15 + 5 \cdot 4}{5 \cdot 15} = \frac{50}{75} = \frac{2 \cdot 25}{3 \cdot 25} = \frac{2}{3}.$$

We can also do this problem by converting $\frac{2}{5}$ and $\frac{4}{15}$ into equivalent fractions that have the same denominators and then using Property 26. The fractions $\frac{2}{5}$ and $\frac{4}{15}$ can be written with a common denominator of $5 \cdot 15$

$$\frac{2}{5} = \frac{2 \cdot 15}{5 \cdot 15} \qquad \text{and} \qquad \frac{4}{15} = \frac{4 \cdot 5}{15 \cdot 5}.$$

However, 15 is the *least* such common denominator and is called the *least common denominator* (L.C.D.) of $\frac{2}{5}$ and $\frac{4}{15}$. Thus,

$$\frac{2}{5} + \frac{4}{15} = \frac{2 \cdot 3}{5 \cdot 3} + \frac{4}{15} = \frac{6}{15} + \frac{4}{15} = \frac{6+4}{15} = \frac{10}{15} = \frac{2}{3}.$$

Similarly,

$$\frac{3}{8} - \frac{5}{12} = \frac{3 \cdot 3}{8 \cdot 3} - \frac{5 \cdot 2}{12 \cdot 2} \qquad \text{(L.C.D. = 24)}$$

$$= \frac{9}{24} - \frac{10}{24} = \frac{9-10}{24}$$

$$= \frac{1}{24}.$$

## ■ Exercise 0.4

*Simplify each of the following if possible.*

**1.** $-2 + (-4)$.     **2.** $-6 + 2$.     **3.** $6 + (-4)$.     **4.** $7 - 2$.

**5.** $7 - (-4)$.     **6.** $-7 - (-4)$.     **7.** $-8 - (-6)$.     **8.** $(-2)(9)$.

**9.** $7(-9)$.     **10.** $(-2)(-12)$.     **11.** $(-1)6$.     **12.** $-(-9)$.

**13.** $-(-6 + x)$.     **14.** $-7(x)$.     **15.** $-12(x - y)$.     **16.** $-[-6 + (-y)]$.

**17.** $-2 \div 6$.     **18.** $-2 \div (-4)$.     **19.** $4 \div (-2)$.     **20.** $2(-6 + 2)$.

**21.** $3[-2(3) + 6(2)]$.     **22.** $(-2)(-4)(-1)$.     **23.** $(-5)(-5)$.     **24.** $x(0)$

**25.** $3(x - 4)$.     **26.** $4(5 + x)$.     **27.** $-(x - 2)$.     **28.** $0(-x)$.

**29.** $8\left(\dfrac{1}{11}\right)$.     **30.** $\dfrac{7}{1}$.     **31.** $\dfrac{-5x}{7y}$.     **32.** $\dfrac{3}{-2x}$.

**33.** $\dfrac{2}{3} \cdot \dfrac{1}{x}$.     **34.** $\dfrac{x}{y}(2z)$.     **35.** $(2x)\left(\dfrac{3}{2x}\right)$.     **36.** $\dfrac{-15x}{-3y}$.

**37.** $\dfrac{7}{y} \cdot \dfrac{1}{x}$.     **38.** $\dfrac{2}{x} \cdot \dfrac{5}{y}$.     **39.** $\dfrac{1}{2} + \dfrac{1}{3}$.     **40.** $\dfrac{5}{12} + \dfrac{3}{4}$.

**41.** $\dfrac{3}{10} - \dfrac{7}{15}$.     **42.** $\dfrac{2}{3} + \dfrac{7}{3}$.     **43.** $\dfrac{x}{9} - \dfrac{y}{9}$.     **44.** $\dfrac{3}{2} - \dfrac{1}{4} + \dfrac{1}{6}$.

**45.** $\dfrac{2}{3} - \dfrac{5}{8}$.     **46.** $\dfrac{6}{\dfrac{x}{y}}$.     **47.** $\dfrac{\dfrac{x}{6}}{y}$.     **48.** $\dfrac{\dfrac{-7}{2}}{\dfrac{5}{8}}$.

**49.** $\dfrac{7}{0}$.     **50.** $\dfrac{0}{7}$.     **51.** $\dfrac{0}{0}$.     **52.** $0 \cdot 0$.

| OBJECTIVE | **0.5 Exponents and Radicals** |
|---|---|
| **To review positive integral exponents, the zero exponent, negative integral exponents, rational exponents, principal roots, radicals, and the procedure of rationalizing the denominator.** | The product $x \cdot x \cdot x$ is abbreviated $x^3$. In general, for $n$ a positive integer, $x^n$ is the abbreviation for the product of $n$ $x$'s. The letter $n$ in $x^n$ is called the *exponent,* and $x$ is called the *base.* More specifically, if $n$ is a positive integer, we have: |

1. $x^n = \underbrace{x \cdot x \cdot x \cdot \ldots \cdot x}_{n \text{ factors}}.$

2. $x^{-n} = \dfrac{1}{x^n} = \dfrac{1}{\underbrace{x \cdot x \cdot x \cdot \ldots \cdot x}_{n \text{ factors}}}.$

3. $\dfrac{1}{x^{-n}} = x^n.$

4. $x^0 = 1$ if $x \neq 0$. $0^0$ is not defined.

**EXAMPLE 1  Exponents**

a. $\left(\dfrac{1}{2}\right)^4 = \left(\dfrac{1}{2}\right)\left(\dfrac{1}{2}\right)\left(\dfrac{1}{2}\right)\left(\dfrac{1}{2}\right) = \dfrac{1}{16}.$

b. $3^{-5} = \dfrac{1}{3^5} = \dfrac{1}{3 \cdot 3 \cdot 3 \cdot 3 \cdot 3} = \dfrac{1}{243}.$

c. $\dfrac{1}{3^{-5}} = 3^5 = 243.$

d. $2^0 = 1,\ \pi^0 = 1,\ (-5)^0 = 1.$

e. $x^1 = x.$  ∎

If $r^n = x$, where $n$ is a positive integer, then $r$ is an *n*th *root* of $x$. For example, $3^2 = 9$, so 3 is a second root (usually called a *square root*) of 9. Since $(-3)^2 = 9$, $-3$ is also a square root of 9. Similarly, $-2$ is a *cube root* of $-8$, since $(-2)^3 = -8$.

Some numbers do not have an *n*th root that is a real number. For example, since the square of any real number is nonnegative, there is no real number that is a square root of $-4$.

The **principal *n*th root** of $x$ is the *n*th root of $x$ that is positive if $x$ is positive and is negative if $x$ is negative and $n$ is odd. We denote the principal *n*th root of $x$ by $\sqrt[n]{x}$. Thus,

$$\sqrt[n]{x} \ \text{ is } \begin{cases} \text{positive if } x \text{ is positive,} \\ \text{negative if } x \text{ is negative and } n \text{ is odd.} \end{cases}$$

For example, $\sqrt[2]{9} = 3$, $\sqrt[3]{-8} = -2$, and $\sqrt[3]{\frac{1}{27}} = \frac{1}{3}$. We define $\sqrt[n]{0} = 0$.

The symbol $\sqrt[n]{x}$ is called a **radical.** Here $n$ is the *index*, $x$ is the *radicand*, and $\sqrt{\phantom{x}}$ is the *radical sign*. With principal square roots, we usually omit the index and write $\sqrt{x}$ instead of $\sqrt[2]{x}$. Thus, $\sqrt{9} = 3$.

***Pitfall*** ▼ Although 2 and $-2$ are square roots of 4, the **principal** square root of 4 is 2, not $-2$. Hence, $\sqrt{4} = 2$.

If $x$ is positive, the expression $x^{p/q}$, where $p$ and $q$ are integers and $q$ is positive, is defined to be $\sqrt[q]{x^p}$. Hence,

$$x^{3/4} = \sqrt[4]{x^3}; \qquad 8^{2/3} = \sqrt[3]{8^2} = \sqrt[3]{64} = 4;$$

$$4^{-1/2} = \sqrt[2]{4^{-1}} = \sqrt{\tfrac{1}{4}} = \tfrac{1}{2}.$$

Here are the basic laws of exponents and radicals.[1]

| | *Law* | *Example* |
|---|---|---|
| **1.** | $x^m \cdot x^n = x^{m+n}$. | $2^3 \cdot 2^5 = 2^8 = 256$;  $x^2 \cdot x^3 = x^5$. |
| **2.** | $x^0 = 1$ if $x \neq 0$. | $2^0 = 1$. |
| **3.** | $x^{-n} = \dfrac{1}{x^n}$. | $2^{-3} = \dfrac{1}{2^3} = \dfrac{1}{8}$. |
| **4.** | $\dfrac{1}{x^{-n}} = x^n$. | $\dfrac{1}{2^{-3}} = 2^3 = 8$;  $\dfrac{1}{x^{-5}} = x^5$. |
| **5.** | $\dfrac{x^m}{x^n} = x^{m-n} = \dfrac{1}{x^{n-m}}$. | $\dfrac{2^{12}}{2^8} = 2^4 = 16$;  $\dfrac{x^8}{x^{12}} = \dfrac{1}{x^4}$. |
| **6.** | $\dfrac{x^m}{x^m} = 1$. | $\dfrac{2^4}{2^4} = 1$. |
| **7.** | $(x^m)^n = x^{mn}$. | $(2^3)^5 = 2^{15}$;  $(x^2)^3 = x^6$. |
| **8.** | $(xy)^n = x^n y^n$. | $(2 \cdot 4)^3 = 2^3 \cdot 4^3 = 8 \cdot 64 = 512$. |
| **9.** | $\left(\dfrac{x}{y}\right)^n = \dfrac{x^n}{y^n}$. | $\left(\dfrac{2}{3}\right)^3 = \dfrac{2^3}{3^3}$;  $\left(\dfrac{1}{3}\right)^5 = \dfrac{1^5}{3^5} = \dfrac{1}{3^5} = 3^{-5}$. |
| **10.** | $\left(\dfrac{x}{y}\right)^{-n} = \left(\dfrac{y}{x}\right)^n$. | $\left(\dfrac{3}{4}\right)^{-2} = \left(\dfrac{4}{3}\right)^2 = \dfrac{16}{9}$. |
| **11.** | $x^{1/n} = \sqrt[n]{x}$. | $3^{1/5} = \sqrt[5]{3}$. |
| **12.** | $x^{-1/n} = \dfrac{1}{x^{1/n}} = \dfrac{1}{\sqrt[n]{x}}$. | $4^{-1/2} = \dfrac{1}{4^{1/2}} = \dfrac{1}{\sqrt{4}} = \dfrac{1}{2}$. |
| **13.** | $\sqrt[n]{x}\,\sqrt[n]{y} = \sqrt[n]{xy}$. | $\sqrt[3]{9}\,\sqrt[3]{2} = \sqrt[3]{18}$. |
| **14.** | $\dfrac{\sqrt[n]{x}}{\sqrt[n]{y}} = \sqrt[n]{\dfrac{x}{y}}$. | $\dfrac{\sqrt[3]{90}}{\sqrt[3]{10}} = \sqrt[3]{\dfrac{90}{10}} = \sqrt[3]{9}$. |
| **15.** | $\sqrt[m]{\sqrt[n]{x}} = \sqrt[mn]{x}$. | $\sqrt[3]{\sqrt[4]{2}} = \sqrt[12]{2}$. |
| **16.** | $x^{m/n} = \sqrt[n]{x^m} = (\sqrt[n]{x})^m$. | $8^{2/3} = \sqrt[3]{8^2} = (\sqrt[3]{8})^2 = 2^2 = 4$. |
| **17.** | $(\sqrt[m]{x})^m = x$. | $(\sqrt[8]{7})^8 = 7$. |

When computing $x^{m/n}$, it is often easier to first find $\sqrt[n]{x}$ and then raise the result to the $m$th power. Thus, $(-27)^{4/3} = (\sqrt[3]{-27})^4 = (-3)^4 = 81$.

**EXAMPLE 2   Exponents and Radicals**

**a.** By Law 1,

$$x^6 x^8 = x^{6+8} = x^{14},$$
$$a^3 b^2 a^5 b = a^3 a^5 b^2 b^1 = a^8 b^3,$$
$$x^{11} x^{-5} = x^{11-5} = x^6,$$
$$z^{2/5} z^{3/5} = z^1 = z,$$
$$x x^{1/2} = x^1 x^{1/2} = x^{3/2}.$$

**b.** By Law 16,

$$\left(\frac{1}{4}\right)^{3/2} = \left(\sqrt{\frac{1}{4}}\right)^3 = \left(\frac{1}{2}\right)^3 = \frac{1}{8}.$$

**c.** $\left(-\dfrac{8}{27}\right)^{4/3} = \left(\sqrt[3]{\dfrac{-8}{27}}\right)^4 = \left(\dfrac{\sqrt[3]{-8}}{\sqrt[3]{27}}\right)^4$   (Laws 16 and 14)

$$= \left(\frac{-2}{3}\right)^4$$

$$= \frac{(-2)^4}{3^4} = \frac{16}{81}$$   (Law 9).

---

[1]Although some laws involve restrictions, they are not vital to our discussion.

**d.** $(64a^3)^{2/3} = 64^{2/3}(a^3)^{2/3}$     (Law 8)

$$= (\sqrt[3]{64})^2 a^2 \quad \text{(Laws 16 and 7)}$$

$$= (4)^2 a^2 = 16a^2.$$    ■

*Rationalizing the denominator* of a fraction is a procedure in which a fraction having a radical in its denominator is expressed as an equivalent fraction without a radical in its denominator. We use the fundamental principle of fractions, as Example 3 shows.

**EXAMPLE 3**   **Rationalizing Denominators**

**a.** $\dfrac{2}{\sqrt{5}} = \dfrac{2}{5^{1/2}} = \dfrac{2 \cdot 5^{1/2}}{5^{1/2} \cdot 5^{1/2}} = \dfrac{2 \cdot 5^{1/2}}{5^1} = \dfrac{2\sqrt{5}}{5}.$

**b.** $\dfrac{2}{\sqrt[6]{3x^5}} = \dfrac{2}{\sqrt[6]{3} \cdot \sqrt[6]{x^5}} = \dfrac{2}{3^{1/6} x^{5/6}} = \dfrac{2 \cdot 3^{5/6} x^{1/6}}{3^{1/6} x^{5/6} \cdot 3^{5/6} x^{1/6}}$

$$= \dfrac{2(3^5 x)^{1/6}}{3x} = \dfrac{2\sqrt[6]{3^5 x}}{3x}.$$    ■

The following examples illustrate various applications of the laws of exponents and radicals.

**EXAMPLE 4**   **Exponents**

**a.** *Eliminate negative exponents in* $\dfrac{x^{-2}y^3}{z^{-2}}$.

*Solution:*

$$\dfrac{x^{-2}y^3}{z^{-2}} = x^{-2} \cdot y^3 \cdot \dfrac{1}{z^{-2}} = \dfrac{1}{x^2} \cdot y^3 \cdot z^2 = \dfrac{y^3 z^2}{x^2}.$$

By comparing our answer with the original expression, we conclude that we can bring a factor of the numerator down to the denominator, and vice versa, by changing the sign of the exponent.

**b.** *Simplify* $\dfrac{x^2 y^7}{x^3 y^5}$.

*Solution:*

$$\dfrac{x^2 y^7}{x^3 \, y^5} = \dfrac{y^{7-5}}{x^{3-2}} = \dfrac{y^2}{x}.$$

**c.** *Simplify* $(x^5 y^8)^5$.

*Solution:*

$$(x^5 y^8)^5 = (x^5)^5 (y^8)^5 = x^{25} y^{40}.$$

**d.** *Simplify* $(x^{5/9} y^{4/3})^{18}$.

*Solution:*

$$(x^{5/9} y^{4/3})^{18} = (x^{5/9})^{18} (y^{4/3})^{18} = x^{10} y^{24}.$$

**e.** *Simplify $\left(\dfrac{x^{1/5}y^{6/5}}{z^{2/5}}\right)^5$.*

*Solution:*

$$\left(\frac{x^{1/5}y^{6/5}}{z^{2/5}}\right)^5 = \frac{(x^{1/5}y^{6/5})^5}{(z^{2/5})^5} = \frac{xy^6}{z^2}.$$

**f.** *Simplify $\dfrac{x^3}{y^2} \div \dfrac{x^6}{y^5}$.*

*Solution:*

$$\frac{x^3}{y^2} \div \frac{x^6}{y^5} = \frac{x^3}{y^2} \cdot \frac{y^5}{x^6} = \frac{y^3}{x^3}.$$

### EXAMPLE 5 Exponents

**a.** *Eliminate negative exponents in $x^{-1} + y^{-1}$ and simplify.*

*Solution:*

$$x^{-1} + y^{-1} = \frac{1}{x} + \frac{1}{y} = \frac{y+x}{xy}. \quad \left(\text{Note: } x^{-1} + y^{-1} \neq \frac{1}{x+y}.\right)$$

**b.** *Simplify $x^{3/2} - x^{1/2}$ by using the distributive law.*

*Solution:*

$$x^{3/2} - x^{1/2} = x^{1/2}(x - 1).$$

**c.** *Eliminate negative exponents in $7x^{-2} + (7x)^{-2}$.*

*Solution:*

$$7x^{-2} + (7x)^{-2} = \frac{7}{x^2} + \frac{1}{(7x)^2} = \frac{7}{x^2} + \frac{1}{49x^2}.$$

**d.** *Eliminate negative exponents in $(x^{-1} - y^{-1})^{-2}$.*

*Solution:*

$$(x^{-1} - y^{-1})^{-2} = \left(\frac{1}{x} - \frac{1}{y}\right)^{-2} = \left(\frac{y-x}{xy}\right)^{-2}$$

$$= \left(\frac{xy}{y-x}\right)^2 = \frac{x^2y^2}{(y-x)^2}.$$

**e.** *Apply the distributive law to $x^{2/5}(y^{1/2} + 2x^{6/5})$.*

*Solution:*

$$x^{2/5}(y^{1/2} + 2x^{6/5}) = x^{2/5}y^{1/2} + 2x^{8/5}.$$

### EXAMPLE 6 Radicals

**a.** *Simplify $\sqrt[4]{48}$*

*Solution:*

$$\sqrt[4]{48} = \sqrt[4]{16 \cdot 3} = \sqrt[4]{16}\,\sqrt[4]{3} = 2\sqrt[4]{3}.$$

**b.** *Rewrite $\sqrt{2 + 5x}$ without using a radical sign.*

*Solution:*

$$\sqrt{2 + 5x} = (2 + 5x)^{1/2}.$$

c. *Rationalize the denominator of* $\dfrac{\sqrt[5]{2}}{\sqrt[3]{6}}$ *and simplify.*

*Solution:*

$$\frac{\sqrt[5]{2}}{\sqrt[3]{6}} = \frac{2^{1/5} \cdot 6^{2/3}}{6^{1/3} \cdot 6^{2/3}} = \frac{2^{3/15}6^{10/15}}{6} = \frac{(2^3 6^{10})^{1/15}}{6} = \frac{\sqrt[15]{2^3 6^{10}}}{6}.$$

d. *Simplify* $\dfrac{\sqrt{20}}{\sqrt{5}}.$

*Solution:*

$$\frac{\sqrt{20}}{\sqrt{5}} = \sqrt{\frac{20}{5}} = \sqrt{4} = 2. \qquad \blacksquare$$

**EXAMPLE 7**   Radicals

a. *Simplify* $\sqrt[3]{x^6 y^4}$

*Solution:*

$$\sqrt[3]{x^6 y^4} = \sqrt[3]{(x^2)^3 y^3 y} = \sqrt[3]{(x^2)^3} \cdot \sqrt[3]{y^3} \cdot \sqrt[3]{y}$$
$$= x^2 y \sqrt[3]{y} \quad \text{(Law 17)}.$$

b. *Simplify* $\sqrt{\dfrac{2}{7}}.$

*Solution:*

$$\sqrt{\frac{2}{7}} = \sqrt{\frac{2 \cdot 7}{7 \cdot 7}} = \sqrt{\frac{14}{7^2}} = \frac{\sqrt{14}}{\sqrt{7^2}} = \frac{\sqrt{14}}{7}.$$

c. *Simplify* $\sqrt{250} - \sqrt{50} + 15\sqrt{2}.$

*Solution:*

$$\sqrt{250} - \sqrt{50} + 15\sqrt{2} = \sqrt{25 \cdot 10} - \sqrt{25 \cdot 2} + 15\sqrt{2}$$
$$= 5\sqrt{10} - 5\sqrt{2} + 15\sqrt{2}$$
$$= 5\sqrt{10} + 10\sqrt{2}.$$

d. *If x is any real number, simplify* $\sqrt{x^2}.$

*Solution:*

Note: $\sqrt{x^2} \neq x.$

$$\sqrt{x^2} = \begin{cases} x, & \text{if } x \text{ is positive,} \\ -x, & \text{if } x \text{ is negative,} \\ 0, & \text{if } x = 0. \end{cases}$$

Thus, $\sqrt{2^2} = 2$ and $\sqrt{(-3)^2} = -(-3) = 3.$ $\qquad \blacksquare$

■ **Exercise 0.5**

*In Problems 1–14, simplify and express all answers in terms of positive exponents.*

**1.** $(2^3)(2^2).$  **2.** $x^6 x^9.$  **3.** $w^4 w^8.$  **4.** $x^6 x^4 x^3.$

**5.** $\dfrac{x^2x^6}{y^7y^{10}}$.

**6.** $(x^{12})^4$.

**7.** $\dfrac{(x^2)^5}{(y^5)^{10}}$.

**8.** $\left(\dfrac{x^2}{y^3}\right)^5$.

**9.** $(2x^2y^3)^3$.

**10.** $\left(\dfrac{w^2s^3}{y^2}\right)^2$.

**11.** $\dfrac{x^8}{x^2}$.

**12.** $\left(\dfrac{2x^2}{4x^4}\right)^3$.

**13.** $\dfrac{(x^3)^6}{x(x^3)}$.

**14.** $\dfrac{(x^2)^3(x^3)^2}{(x^3)^4}$.

*In Problems 15–28, evaluate the expressions.*

**15.** $\sqrt{25}$.

**16.** $\sqrt[3]{64}$.

**17.** $\sqrt[5]{-32}$.

**18.** $\sqrt{0.04}$.

**19.** $\sqrt[4]{\tfrac{1}{16}}$.

**20.** $\sqrt[3]{-\tfrac{8}{27}}$.

**21.** $(100)^{1/2}$.

**22.** $(64)^{1/3}$.

**23.** $4^{3/2}$.

**24.** $(25)^{-3/2}$.

**25.** $(32)^{-2/5}$.

**26.** $(0.09)^{-1/2}$.

**27.** $\left(\dfrac{1}{16}\right)^{5/4}$.

**28.** $\left(-\dfrac{27}{64}\right)^{2/3}$.

*In Problems 29–40, simplify the expressions.*

**29.** $\sqrt{32}$.

**30.** $\sqrt[3]{24}$.

**31.** $\sqrt[3]{2x^3}$.

**32.** $\sqrt{4x}$.

**33.** $\sqrt{16x^4}$.

**34.** $\sqrt[4]{\dfrac{x}{16}}$.

**35.** $2\sqrt{75}-4\sqrt{27}+\sqrt[3]{128}$.

**36.** $\sqrt{\tfrac{5}{11}}$.

**37.** $(9z^4)^{1/2}$.

**38.** $(16y^8)^{3/4}$.

**39.** $\left(\dfrac{27t^3}{8}\right)^{2/3}$.

**40.** $\left(\dfrac{1000}{a^9}\right)^{-2/3}$.

*In Problems 41–52, write the expressions in terms of positive exponents only. Avoid all radicals in the final form. For example,*

$$y^{-1}\sqrt{x}=\dfrac{x^{1/2}}{y}.$$

**41.** $\dfrac{x^3y^{-2}}{z^2}$.

**42.** $\sqrt[5]{x^2y^3z^{-10}}$.

**43.** $2x^{-1}x^{-3}$.

**44.** $x+y^{-1}$.

**45.** $(3t)^{-2}$.

**46.** $(3-z)^{-4}$.

**47.** $\sqrt[3]{7s^2}$.

**48.** $(x^{-2}y^2)^{-2}$.

**49.** $\sqrt{x}-\sqrt{y}$.

**50.** $\dfrac{x^{-2}y^{-6}z^2}{xy^{-1}}$.

**51.** $x^2\sqrt[4]{xy^{-2}z^3}$.

**52.** $(\sqrt[5]{xy^{-3}})x^{-1}y^{-2}$.

*In Problems 53–58, write the exponential forms in equivalent forms involving radicals.*

**53.** $(8x-y)^{4/5}$.

**54.** $(ab^2c^3)^{3/4}$.

**55.** $x^{-4/5}$.

**56.** $2x^{1/2}-(2y)^{1/2}$.

**57.** $2x^{-2/5}-(2x)^{-2/5}$.

**58.** $[(x^{-4})^{1/5}]^{1/6}$.

*In Problems 59–68, rationalize the denominators.*

**59.** $\dfrac{3}{\sqrt{7}}$.

**60.** $\dfrac{5}{\sqrt{11}}$.

**61.** $\dfrac{4}{\sqrt{2x}}$.

**62.** $\dfrac{y}{\sqrt{2y}}$.

**63.** $\dfrac{1}{\sqrt[3]{3x}}$.

**64.** $\dfrac{4}{3\sqrt[3]{x^2}}$.

**65.** $\dfrac{\sqrt{32}}{\sqrt{2}}$.

**66.** $\dfrac{\sqrt{18}}{\sqrt{2}}$.

**67.** $\dfrac{\sqrt[4]{2}}{\sqrt[3]{xy^2}}$.

**68.** $\dfrac{\sqrt{2}}{\sqrt[3]{3}}$.

*In Problems 69–90, simplify. Express all answers in terms of positive exponents. Rationalize the denominator where necessary to avoid fractional exponents in the denominator.*

**69.** $2x^2y^{-3}x^4$.

**70.** $\dfrac{2}{x^{3/2}y^{1/3}}$.

**71.** $\sqrt{\sqrt[3]{t^4}}$.

**72.** $\{[(2x^2)^3]^{-4}\}^{-1}$.

**73.** $\dfrac{2^0}{(2^{-2}x^{1/2}y^{-2})^3}.$

**74.** $\dfrac{\sqrt{s^5}}{\sqrt[3]{s^2}}.$

**75.** $\sqrt[3]{x^2yz^3}\,\sqrt[3]{xy^2}.$

**76.** $(\sqrt[5]{2})^{10}.$

**77.** $3^2(27)^{-4/3}.$

**78.** $(\sqrt[5]{x^2y})^{2/5}.$

**79.** $(2x^{-1}y^2)^2.$

**80.** $\dfrac{3}{\sqrt[3]{y}\,\sqrt[4]{x}}.$

**81.** $\sqrt{x}\sqrt{x^2y^3}\sqrt{xy^2}.$

**82.** $\sqrt{75k^4}.$

**83.** $\dfrac{(x^2y^{-1}z)^{-2}}{(xy^2)^{-4}}.$

**84.** $\sqrt{6(6)}.$

**85.** $\dfrac{(x^2)^3}{x^4}\div\left[\dfrac{x^3}{(x^3)^2}\right]^2.$

**86.** $\sqrt{(-6)(-6)}.$

**87.** $-\dfrac{8s^{-2}}{2s^3}.$

**88.** $(x^{-1}y^{-2}\sqrt{z})^4.$

**89.** $(2x^2y\div 3y^3z^{-2})^2.$

**90.** $\dfrac{1}{\left(\dfrac{\sqrt{2}x^{-2}}{\sqrt{16x^3}}\right)^2}.$

---

OBJECTIVE

To add, subtract, multiply, and divide algebraic expressions. To define a polynomial, to use special products, and to use long division to divide polynomials.

## 0.6   Operations with Algebraic Expressions

If numbers, represented by symbols, are combined by any or all of the operations of addition, subtraction, multiplication, division, and extraction of roots, then the resulting expression is called an *algebraic expression*.

**EXAMPLE 1   Algebraic Expressions**

**a.** $\sqrt[3]{\dfrac{3x^3-5x-2}{10-x}}$ is an algebraic expression in the variable $x$.

**b.** $10-3\sqrt{y}+\dfrac{5}{7+y^2}$ is an algebraic expression in the variable $y$.

**c.** $\dfrac{(x+y)^3-xy}{y}+2$ is an algebraic expression in the variables $x$ and $y$.   ■

The algebraic expression $5ax^3-2bx+3$ consists of three *terms:* $+5ax^3$, $-2bx$, and $+3$. Some of the *factors* of the first term, $5ax^3$, are $5, a, x, x^2$, $x^3, 5ax$, and $ax^2$. Also, $5a$ is the *coefficient* of $x^3$, and 5 is the *numerical coefficient* of $ax^3$. If $a$ and $b$ represent fixed numbers throughout a discussion, then $a$ and $b$ are called *constants*.

Algebraic expressions with exactly one term are called *monomials*. Those having exactly two terms are *binomials*, and those with exactly three terms are *trinomials*. Algebraic expressions with more than one term are called *multinomials*. Thus, the multinomial $2x-5$ is a binomial; the multinomial $3\sqrt{y}+2y-4y^2$ is a trinomial.

A *polynomial in* $x$ is an algebraic expression of the form[2]

$$c_nx^n+c_{n-1}x^{n-1}+\cdots+c_1x+c_0,$$

The words "polynomial" and "multinomial" should not be used interchangeably. For example, $\sqrt{x}+2$ is a multinomial, but not a polynomial. On the other hand, $x+2$ is a multinomial and a polynomial.

where $n$ is a nonnegative integer and the coefficients $c_0, c_1, \ldots, c_n$ are constants with $c_n \neq 0$. We call $n$ the *degree* of the polynomial. Hence, $4x^3-5x^2+x-2$ is a polynomial in $x$ of degree 3, and $y^5-2$ is a polynomial in $y$ of degree 5. A nonzero constant is a polynomial of degree zero; thus, 5 is a polynomial of degree zero. The constant 0 is considered to be a polynomial; however, no degree is assigned to it.

In the following examples, we illustrate operations with algebraic expressions.

---

[2]The three dots indicate all other terms that are understood to be included in the sum.

**EXAMPLE 2   Adding Algebraic Expressions**

*Simplify* $(3x^2y - 2x + 1) + (4x^2y + 6x - 3)$.

*Solution:* We first remove the parentheses. Next, using the commutative property of addition, we gather all similar terms together. *Similar terms* are terms that differ only by their numerical coefficients. In this example, $3x^2y$ and $4x^2y$ are similar, as are the pairs $-2x$ and $6x$, and $1$ and $-3$. Thus,

$$(3x^2y - 2x + 1) + (4x^2y + 6x - 3)$$
$$= 3x^2y - 2x + 1 + 4x^2y + 6x - 3$$
$$= 3x^2y + 4x^2y - 2x + 6x + 1 - 3.$$

By the distributive property,

$$3x^2y + 4x^2y = (3 + 4)x^2y = 7x^2y$$

and   $-2x + 6x = (-2 + 6)x = 4x.$

Hence,   $(3x^2y - 2x + 1) + (4x^2y + 6x - 3) = 7x^2y + 4x - 2.$   ∎

**EXAMPLE 3   Subtracting Algebraic Expressions**

*Simplify* $(3x^2y - 2x + 1) - (4x^2y + 6x - 3)$.

*Solution:* Here we apply the definition of subtraction and the distributive property:

$$(3x^2y - 2x + 1) - (4x^2y + 6x - 3)$$
$$= (3x^2y - 2x + 1) + (-1)(4x^2y + 6x - 3)$$
$$= (3x^2y - 2x + 1) + (-4x^2y - 6x + 3)$$
$$= 3x^2y - 2x + 1 - 4x^2y - 6x + 3$$
$$= 3x^2y - 4x^2y - 2x - 6x + 1 + 3$$
$$= (3 - 4)x^2y + (-2 - 6)x + 1 + 3$$
$$= -x^2y - 8x + 4.$$   ∎

**EXAMPLE 4   Removing Grouping Symbols**

*Simplify* $3\{2x[2x + 3] + 5[4x^2 - (3 - 4x)]\}$.

*Solution:* We first eliminate the innermost grouping symbols (the parentheses). Then we repeat the process until all grouping symbols are removed—combining similar terms whenever possible. We have

$$3\{2x[2x + 3] + 5[4x^2 - (3 - 4x)]\}$$
$$= 3\{2x[2x + 3] + 5[4x^2 - 3 + 4x]\}$$
$$= 3\{4x^2 + 6x + 20x^2 - 15 + 20x\}$$
$$= 3\{24x^2 + 26x - 15\}$$
$$= 72x^2 + 78x - 45.$$   ∎

The distributive property is the key tool in multiplying expressions. For example, to multiply $ax + c$ by $bx + d$ we can consider $ax + c$ to be a single number and then use the distributive property:

$$(ax + c)(bx + d) = (ax + c)bx + (ax + c)d.$$

Using the distributive property again, we have

$$(ax + c)bx + (ax + c)d = abx^2 + cbx + adx + cd$$
$$= abx^2 + (ad + cb)x + cd.$$

Thus, $(ax + c)(bx + d) = abx^2 + (ad + cb)x + cd$. In particular, if $a = 2$, $b = 1$, $c = 3$, and $d = -2$, then

$$(2x + 3)(x - 2) = 2(1)x^2 + [2(-2) + 3(1)]x + 3(-2)$$
$$= 2x^2 - x - 6.$$

We now give a list of special products that may be obtained from the distributive property and are useful in multiplying algebraic expressions.

**Special Products**

**1.** $x(y + z) = xy + xz$ (distributive property).

**2.** $(x + a)(x + b) = x^2 + (a + b)x + ab$.

**3.** $(ax + c)(bx + d) = abx^2 + (ad + cb)x + cd$.

**4.** $(x + a)^2 = x^2 + 2ax + a^2$ (square of a binomial).

**5.** $(x - a)^2 = x^2 - 2ax + a^2$ (square of a binomial).

**6.** $(x + a)(x - a) = x^2 - a^2$ (product of sum and difference).

**7.** $(x + a)^3 = x^3 + 3ax^2 + 3a^2x + a^3$ (cube of a binomial).

**8.** $(x - a)^3 = x^3 - 3ax^2 + 3a^2x - a^3$ (cube of a binomial).

**EXAMPLE 5** **Special Products**

**a.** By Rule 2,

$$(x + 2)(x - 5) = [x + 2][x + (-5)]$$
$$= x^2 + (2 - 5)x + 2(-5)$$
$$= x^2 - 3x - 10.$$

**b.** By Rule 3,

$$(3z + 5)(7z + 4) = 3 \cdot 7z^2 + (3 \cdot 4 + 5 \cdot 7)z + 5 \cdot 4$$
$$= 21z^2 + 47z + 20.$$

**c.** By Rule 5,

$$(x - 4)^2 = x^2 - 2(4)x + 4^2$$
$$= x^2 - 8x + 16.$$

**d.** By Rule 6,

$$(\sqrt{y^2 + 1} + 3)(\sqrt{y^2 + 1} - 3) = (\sqrt{y^2 + 1})^2 - 3^2$$
$$= (y^2 + 1) - 9$$
$$= y^2 - 8.$$

**e.** By Rule 7,

$$(3x + 2)^3 = (3x)^3 + 3(2)(3x)^2 + 3(2)^2(3x) + (2)^3$$
$$= 27x^3 + 54x^2 + 36x + 8.$$

**EXAMPLE 6  Multiplying Multinomials**

*Find the product* $(2t - 3)(5t^2 + 3t - 1)$.

*Solution:* We treat $2t - 3$ as a single number and apply the distributive property twice:

$$(2t - 3)(5t^2 + 3t - 1) = (2t - 3)5t^2 + (2t - 3)3t - (2t - 3)1$$
$$= 10t^3 - 15t^2 + 6t^2 - 9t - 2t + 3$$
$$= 10t^3 - 9t^2 - 11t + 3.$$

In Example 3(b) of Sec. 0.3, we showed that $\dfrac{a + b}{c} = \dfrac{a}{c} + \dfrac{b}{c}$. Similarly, $\dfrac{a - b}{c} = \dfrac{a}{c} - \dfrac{b}{c}$. Using these results, we can divide a multinomial by a monomial by dividing each term in the multinomial by the monomial.

**EXAMPLE 7  Dividing a Multinomial by a Monomial**

**a.** $\dfrac{x^3 + 3x}{x} = \dfrac{x^3}{x} + \dfrac{3x}{x} = x^2 + 3.$

**b.** $\dfrac{4z^3 - 8z^2 + 3z - 6}{2z} = \dfrac{4z^3}{2z} - \dfrac{8z^2}{2z} + \dfrac{3z}{2z} - \dfrac{6}{2z}$
$$= 2z^2 - 4z + \dfrac{3}{2} - \dfrac{3}{z}.$$

## Long Division

To divide a polynomial by a polynomial, we use so-called long division when the degree of the divisor is less than or equal to the degree of the dividend, as the next example shows.

**EXAMPLE 8  Long Division**

*Divide* $2x^3 - 14x - 5$ *by* $x - 3$.

*Solution:* Here $2x^3 - 14x - 5$ is the *dividend* and $x - 3$ is the *divisor*. To avoid errors, it is best to write the dividend as $2x^3 + 0x^2 - 14x - 5$. Note that the powers of $x$ are in decreasing order. We have

$$
\begin{array}{r}
2x^2 + 6x + 4 \leftarrow \text{quotient} \\
x - 3 \overline{)\, 2x^3 + 0x^2 - 14x - 5} \\
\underline{2x^3 - 6x^2} \phantom{xxxxxxxxxx} \\
6x^2 - 14x \phantom{xxx} \\
\underline{6x^2 - 18x} \phantom{xx} \\
4x - 5 \\
\underline{4x - 12} \\
7 \leftarrow \text{remainder.}
\end{array}
$$

Note that we divided $x$ (the first term of the divisor) into $2x^3$ and got $2x^2$. Then we multiplied $2x^2$ by $x - 3$, getting $2x^3 - 6x^2$. After subtracting $2x^3 - 6x^2$ from $2x^3 + 0x^2$, we obtained $6x^2$ and then "brought down" the term $-14x$. This process is continued until we arrive at 7, the *remainder*. We always stop when the remainder is 0 or is a polynomial whose degree is less than the degree of the divisor. Our answer may be written as

$$2x^2 + 6x + 4 + \frac{7}{x - 3}.$$

That is, the answer has the form

$$\text{quotient} + \frac{\text{remainder}}{\text{divisor}}.$$

A way of checking a division is to verify that

$$(\text{quotient})(\text{divisor}) + \text{remainder} = \text{dividend}.$$

By using this equation, you should be able to verify the result of the example. ■

## ■ Exercise 0.6

*Perform the indicated operations and simplify.*

**1.** $(8x - 4y + 2) + (3x + 2y - 5)$.

**2.** $(6x^2 - 10xy + 2) + (2z - xy + 4)$.

**3.** $(8t^2 - 6s^2) + (4s^2 - 2t^2 + 6)$.

**4.** $(\sqrt{x} + 2\sqrt{x}) + (\sqrt{x} + 3\sqrt{x})$.

**5.** $(\sqrt{x} + \sqrt{2y}) + (\sqrt{x} + \sqrt{3z})$.

**6.** $(3x + 2y - 5) - (8x - 4y + 2)$.

**7.** $(6x^2 - 10xy + \sqrt{2}) - (2z - xy + 4)$.

**8.** $(\sqrt{x} + 2\sqrt{x}) - (\sqrt{x} + 3\sqrt{x})$.

**9.** $(\sqrt{x} + \sqrt{2y}) - (\sqrt{x} + \sqrt{3z})$.

**10.** $4(2z - w) - 3(w - 2z)$.

**11.** $3(3x + 2y - 5) - 2(8x - 4y + 2)$.

**12.** $(2s + t) - 3(s - 6) + 4(1 - t)$.

**13.** $3(x^2 + y^2) - x(y + 2x) + 2y(x + 3y)$.

**14.** $2 - [3 + 4(s - 3)]$.

**15.** $2\{3[3(x^2 + 2) - 2(x^2 - 5)]\}$.

**16.** $4\{3(t + 5) - t[1 - (t + 1)]\}$.

**17.** $-3\{4x(x + 2) - 2[x^2 - (3 - x)]\}$.

**18.** $-\{-2[2a + 3b - 1] + 4[a - 2b] - a[2(b - 3)]\}$.

**19.** $(x + 4)(x + 5)$.

**20.** $(x + 3)(x + 2)$.

**21.** $(x + 3)(x - 2)$.

**22.** $(z - 7)(z - 3)$.

**23.** $(2x + 3)(5x + 2)$.

**24.** $(y - 4)(2y + 3)$.

**25.** $(x + 3)^2$.

**26.** $(2x - 1)^2$.

**27.** $(x - 5)^2$.

**28.** $(\sqrt{x} - 1)(2\sqrt{x} + 5)$.

**29.** $(\sqrt{2y} + 3)^2$.

**30.** $(y - 3)(y + 3)$.

**31.** $(2s - 1)(2s + 1)$.

**32.** $(z^2 - 3w)(z^2 + 3w)$.

**33.** $(x^2 - 3)(x + 4)$.

**34.** $(x + 1)(x^2 + x + 3)$.

**35.** $(x^2 - 1)(2x^2 + 2x - 3)$.

**36.** $(2x - 1)(3x^3 + 7x^2 - 5)$.

**37.** $x\{3(x - 1)(x - 2) + 2[x(x + 7)]\}$.

**38.** $[(2z + 1)(2z - 1)](4z^2 + 1)$.

**39.** $(x + y + 2)(3x + 2y - 4)$.

**40.** $(x^2 + x + 1)^2$.

**41.** $(x + 5)^3$.

**42.** $(x - 2)^3$.

**43.** $(2x - 3)^3$.

**44.** $(x + 2y)^3$.

**45.** $\dfrac{z^2 - 4z}{z}$.

**46.** $\dfrac{2x^3 - 7x + 4}{x}$.

**47.** $\dfrac{6x^5 + 4x^3 - 1}{2x^2}$.

**48.** $\dfrac{(3x - 4) - (x + 8)}{4x}$.

**49.** $(x^2 + 3x - 1) \div (x + 3)$.

**50.** $(x^2 - 5x + 4) \div (x - 4)$.

**51.** $(3x^3 - 2x^2 + x - 3) \div (x + 2)$.

**52.** $(x^4 + 2x^2 + 1) \div (x - 1)$.

**53.** $t^2 \div (t - 8)$.

**54.** $(4x^2 + 6x + 1) \div (2x - 1)$.

**55.** $(3x^2 - 4x + 3) \div (3x + 2)$.

**56.** $(z^3 + z^2 + z) \div (z^2 - z + 1)$.

**To state the basic rules for factoring and apply them to factoring expressions.**

# 0.7 Factoring

If two or more expressions are multiplied together, the expressions are called *factors* of the product. Thus, if $c = ab$ then $a$ and $b$ are both factors of the product $c$. The process by which an expression is written as a product of its factors is called *factoring*.

Listed next are rules for factoring expressions, most of which arise from the special products discussed in Sec. 0.6. The right side of each identity is the factored form of the left side.

**Rules for Factoring**

1. $xy + xz = x(y + z)$                                (common factor).
2. $x^2 + (a + b)x + ab = (x + a)(x + b)$.
3. $abx^2 + (ad + cb)x + cd = (ax + c)(bx + d)$.
4. $x^2 + 2ax + a^2 = (x + a)^2$                (perfect-square trinomial).
5. $x^2 - 2ax + a^2 = (x - a)^2$                (perfect-square trinomial).
6. $x^2 - a^2 = (x + a)(x - a)$          (difference of two squares).
7. $x^3 + a^3 = (x + a)(x^2 - ax + a^2)$         (sum of two cubes).
8. $x^3 - a^3 = (x - a)(x^2 + ax + a^2)$     (difference of two cubes).

When factoring a polynomial, we usually choose factors that themselves are polynomials. For example, $x^2 - 4 = (x + 2)(x - 2)$. We shall not write $x - 4$ as $(\sqrt{x} + 2)(\sqrt{x} - 2)$.

Always factor completely. For example,

$$2x^2 - 8 = 2(x^2 - 4) = 2(x + 2)(x - 2).$$

**EXAMPLE 1   Common Factors**

**a.** *Factor $3k^2x^2 + 9k^3x$ completely.*

**Solution:** Since $3k^2x^2 = (3k^2x)(x)$ and $9k^3x = (3k^2x)(3k)$, each term of the original expression contains the common factor $3k^2x$. Thus, by Rule 1,

$$3k^2x^2 + 9k^3x = 3k^2x(x + 3k).$$

Note that although $3k^2x^2 + 9k^3x = 3(k^2x^2 + 3k^3x)$, we do not say that the expression is completely factored, since $k^2x^2 + 3k^3x$ can yet be factored.

**b.** *Factor $8a^5x^2y^3 - 6a^2b^3yz - 2a^4b^4xy^2z^2$ completely.*

**Solution:**

$$8a^5x^2y^3 - 6a^2b^3yz - 2a^4b^4xy^2z^2$$
$$= 2a^2y(4a^3x^2y^2 - 3b^3z - a^2b^4xyz^2).$$ ∎

**EXAMPLE 2   Factoring Trinomials**

**a.** *Factor $3x^2 + 6x + 3$ completely.*

**Solution:** First we remove a common factor. Then we factor the resulting expression completely. Thus, we have

$$3x^2 + 6x + 3 = 3(x^2 + 2x + 1)$$
$$= 3(x + 1)^2 \quad \text{(Rule 4)}.$$

**b.** *Factor $x^2 - x - 6$ completely.*

*Solution:* If this trinomial factors into the form $(x + a)(x + b)$, which is a product of two binomials, then we must determine the values of $a$ and $b$. Since $(x + a)(x + b) = x^2 + (a + b)x + ab$, it follows that

$$x^2 + (-1)x + (-6) = x^2 + (a + b)x + ab.$$

By equating corresponding coefficients, we want

$$a + b = -1 \quad \text{and} \quad ab = -6.$$

If $a = -3$ and $b = 2$, then both conditions are met, and hence,

$$x^2 - x - 6 = (x - 3)(x + 2).$$

As a check, it is wise to multiply the right side to see if it agrees with the left side.

**c.** *Factor $x^2 - 7x + 12$ completely.*

*Solution:*

$$x^2 - 7x + 12 = (x - 3)(x - 4). \quad \blacksquare$$

**EXAMPLE 3** **Factoring**

The following is an assortment of expressions that are completely factored. The numbers in parentheses refer to the rules used.

**a.** $x^2 + 8x + 16 = (x + 4)^2$      (4).

**b.** $9x^2 + 9x + 2 = (3x + 1)(3x + 2)$      (3).

**c.** $6y^3 + 3y^2 - 18y = 3y(2y^2 + y - 6)$      (1)
$$= 3y(2y - 3)(y + 2)$$      (3).

**d.** $x^2 - 6x + 9 = (x - 3)^2$      (5).

**e.** $z^{1/4} + z^{5/4} = z^{1/4}(1 + z)$      (1).

**f.** $x^4 - 1 = (x^2 + 1)(x^2 - 1)$      (6)
$$= (x^2 + 1)(x + 1)(x - 1)$$      (6).

**g.** $x^{2/3} - 5x^{1/3} + 4 = (x^{1/3} - 1)(x^{1/3} - 4)$      (2).

**h.** $ax^2 - ay^2 + bx^2 - by^2 = (ax^2 - ay^2) + (bx^2 - by^2)$
$$= a(x^2 - y^2) + b(x^2 - y^2)$$      (1)
$$= (x^2 - y^2)(a + b)$$      (1)
$$= (x + y)(x - y)(a + b)$$      (6).

**i.** $8 - x^3 = (2)^3 - (x)^3 = (2 - x)(4 + 2x + x^2)$      (8).

**j.** $x^6 - y^6 = (x^3)^2 - (y^3)^2 = (x^3 + y^3)(x^3 - y^3)$      (6)
$$= (x + y)(x^2 - xy + y^2)(x - y)(x^2 + xy + y^2)$$      (7), (8).
$\blacksquare$

Note in Example 3(f) that $x^2 - 1$ is factorable, but $x^2 + 1$ is not. In Example 3(h), we factored by making use of grouping.

## ▪ Exercise 0.7

*Factor the following expressions completely.*

**1.** $6x + 4$.

**2.** $6y^2 - 4y$.

**3.** $10xy + 5xz$.

**4.** $3x^2y - 9x^3y^3$.

**5.** $8a^3bc - 12ab^3cd + 4b^4c^2d^2$.

**6.** $6z^2t^3 + 3zst^4 - 12z^2t^3$.

**7.** $x^2 - 25$.

**8.** $x^2 + 3x - 4$.

**9.** $p^2 + 4p + 3$.

**10.** $s^2 - 6s + 8$.

**11.** $16x^2 - 9$.

**12.** $x^2 + 5x - 24$.

**13.** $z^2 + 6z + 8$.

**14.** $4t^2 - 9s^2$.

**15.** $x^2 + 6x + 9$.

**16.** $y^2 - 15y + 50$.

**17.** $2x^2 + 12x + 16$.

**18.** $2x^2 + 7x - 15$.

**19.** $3x^2 - 3$.

**20.** $4y^2 - 8y + 3$.

**21.** $6y^2 + 13y + 2$.

**22.** $4x^2 - x - 3$.

**23.** $12s^3 + 10s^2 - 8s$.

**24.** $9z^2 + 24z + 16$.

**25.** $x^{2/3}y - 4x^{8/3}y^3$.

**26.** $9x^{4/7} - 1$.

**27.** $2x^3 + 2x^2 - 12x$.

**28.** $x^2y^2 - 4xy + 4$.

**29.** $(4x + 2)^2$.

**30.** $3s^2(3s - 9s^2)^2$.

**31.** $x^3y^2 - 10x^2y + 25x$.

**32.** $(3x^2 + x) + (6x + 2)$.

**33.** $(x^3 - 4x) + (8 - 2x^2)$.

**34.** $(x^2 - 1) + (x^2 - x - 2)$.

**35.** $(y^4 + 8y^3 + 16y^2) - (y^2 + 8y + 16)$.

**36.** $x^3y - xy + z^2x^2 - z^2$.

**37.** $x^3 + 8$.

**38.** $x^3 - 1$.

**39.** $x^6 - 1$.

**40.** $27 + 8x^3$.

**41.** $(x + 3)^3(x - 1) + (x + 3)^2(x - 1)^2$.

**42.** $(x + 5)^2(x + 1)^3 + (x + 5)^3(x + 1)^2$.

**43.** $P(1 + r) + P(1 + r)r$.

**44.** $(x - 3)(2x + 3) - (2x + 3)(x + 5)$.

**45.** $x^4 - 16$.

**46.** $81x^4 - y^4$.

**47.** $y^8 - 1$.

**48.** $t^4 - 4$.

**49.** $x^4 + x^2 - 2$.

**50.** $x^4 - 5x^2 + 4$.

**51.** $x^5 - 2x^3 + x$.

**52.** $4x^3 - 6x^2 - 4x$.

---

### OBJECTIVE

To simplify fractions and to add, subtract, multiply, and divide fractions. To rationalize the denominator of a fraction.

## 0.8 Fractions

### Simplifying Fractions

By using the fundamental principle of fractions (Sec. 0.4), we may be able to simplify fractions. That principle allows us to multiply or divide both the numerator and the denominator of a fraction by the same nonzero quantity. The resulting fraction will be equivalent to the original one. The fractions that we shall consider are assumed to have nonzero denominators.

**EXAMPLE 1   Simplifying Fractions**

**a.** *Simplify* $\dfrac{x^2 - x - 6}{x^2 - 7x + 12}$.

   *Solution:*  First, we completely factor the numerator and denominator:

$$\frac{x^2 - x - 6}{x^2 - 7x + 12} = \frac{(x - 3)(x + 2)}{(x - 3)(x - 4)}.$$

Dividing both numerator and denominator by the common factor $x - 3$ we have

$$\frac{(x-3)(x+2)}{(x-3)(x-4)} = \frac{1(x+2)}{1(x-4)} = \frac{x+2}{x-4}.$$

Usually, we just write

$$\frac{x^2 - x - 6}{x^2 - 7x + 12} = \frac{\overset{1}{(\cancel{x-3})}(x+2)}{\underset{1}{(\cancel{x-3})}(x-4)} = \frac{x+2}{x-4}$$

or

$$\frac{x^2 - x - 6}{x^2 - 7x + 12} = \frac{(x-3)(x+2)}{(x-3)(x-4)} = \frac{x+2}{x-4}.$$

The process of eliminating the common factor $x - 3$ is commonly referred to as "cancellation."

**b.** *Simplify* $\dfrac{2x^2 + 6x - 8}{8 - 4x - 4x^2}.$

*Solution:*

$$\frac{2x^2 + 6x - 8}{8 - 4x - 4x^2} = \frac{2(x^2 + 3x - 4)}{4(2 - x - x^2)} = \frac{2(x-1)(x+4)}{4(1-x)(2+x)}$$

Note how $1 - x$ is written as $(-1)(x - 1)$ to allow cancellation.

$$= \frac{2(x-1)(x+4)}{2(2)[(-1)(x-1)](2+x)}$$

$$= \frac{x+4}{-2(2+x)} = -\frac{x+4}{2(x+2)}. \qquad \blacksquare$$

## Multiplication and Division of Fractions

The rule for multiplying $\dfrac{a}{b}$ by $\dfrac{c}{d}$ is

$$\frac{a}{b} \cdot \frac{c}{d} = \frac{ac}{bd}.$$

### EXAMPLE 2  Multiplying Fractions

**a.** $\dfrac{x}{x+2} \cdot \dfrac{x+3}{x-5} = \dfrac{x(x+3)}{(x+2)(x-5)}.$

**b.** $\dfrac{x^2 - 4x + 4}{x^2 + 2x - 3} \cdot \dfrac{6x^2 - 6}{x^2 + 2x - 8} = \dfrac{[(x-2)^2][6(x+1)(x-1)]}{[(x+3)(x-1)][(x+4)(x-2)]}$

$$= \frac{6(x-2)(x+1)}{(x+3)(x+4)}. \qquad \blacksquare$$

To divide $\dfrac{a}{b}$ by $\dfrac{c}{d}$, where $c \neq 0$, we have

In short, we invert the divisor and multiply.

$$\frac{a}{b} \div \frac{c}{d} = \frac{\dfrac{a}{b}}{\dfrac{c}{d}} = \frac{a}{b} \cdot \frac{d}{c}.$$

**EXAMPLE 3   Dividing Fractions**

a. $\dfrac{x}{x+2} \div \dfrac{x+3}{x-5} = \dfrac{x}{x+2} \cdot \dfrac{x-5}{x+3} = \dfrac{x(x-5)}{(x+2)(x+3)}.$

b. $\dfrac{\dfrac{x-5}{x-3}}{2x} = \dfrac{\dfrac{x-5}{x-3}}{\dfrac{2x}{1}} = \dfrac{x-5}{x-3} \cdot \dfrac{1}{2x} = \dfrac{x-5}{2x(x-3)}.$

c. $\dfrac{\dfrac{4x}{x^2-1}}{\dfrac{2x^2+8x}{x-1}} = \dfrac{4x}{x^2-1} \cdot \dfrac{x-1}{2x^2+8x} = \dfrac{4x(x-1)}{[(x+1)(x-1)][2x(x+4)]}$

$$= \dfrac{2}{(x+1)(x+4)}. \qquad \blacksquare$$

### Rationalizing the Denominator

Sometimes the denominator of a fraction has two terms and involves square roots, such as $2 - \sqrt{3}$ or $\sqrt{5} + \sqrt{2}$. The denominator may then be rationalized by multiplying by an expression that makes the denominator a difference of two squares. For example,

$$\dfrac{4}{\sqrt{5}+\sqrt{2}} = \dfrac{4}{\sqrt{5}+\sqrt{2}} \cdot \dfrac{\sqrt{5}-\sqrt{2}}{\sqrt{5}-\sqrt{2}}$$

$$= \dfrac{4(\sqrt{5}-\sqrt{2})}{(\sqrt{5})^2-(\sqrt{2})^2} = \dfrac{4(\sqrt{5}-\sqrt{2})}{5-2}$$

$$= \dfrac{4\sqrt{5}-\sqrt{2})}{3}.$$

Rationalizing the *numerator* is a similar procedure.

**EXAMPLE 4   Rationalizing Denominators**

a. $\dfrac{x}{\sqrt{2}-6} = \dfrac{x}{\sqrt{2}-6} \cdot \dfrac{\sqrt{2}+6}{\sqrt{2}+6} = \dfrac{x(\sqrt{2}+6)}{(\sqrt{2})^2-6^2}$

$$= \dfrac{x(\sqrt{2}+6)}{2-36} = -\dfrac{x(\sqrt{2}+6)}{34}.$$

b. $\dfrac{\sqrt{5}-\sqrt{2}}{\sqrt{5}+\sqrt{2}} = \dfrac{\sqrt{5}-\sqrt{2}}{\sqrt{5}+\sqrt{2}} \cdot \dfrac{\sqrt{5}-\sqrt{2}}{\sqrt{5}-\sqrt{2}}$

$$= \dfrac{(\sqrt{5}-\sqrt{2})^2}{5-2} = \dfrac{5-2\sqrt{5}\sqrt{2}+2}{3} = \dfrac{7-2\sqrt{10}}{3}. \qquad \blacksquare$$

### Addition and Subtraction of Fractions

In Example 3(b) of Sec 0.3, it was shown that $\dfrac{a}{c} + \dfrac{b}{c} = \dfrac{a+b}{c}$. That is, if we add two fractions having a common denominator, then the result is a fraction whose denominator is the common denominator. The numerator is the sum of the numerators of the original fractions. Similarly, $\dfrac{a}{c} - \dfrac{b}{c} = \dfrac{a-b}{c}$.

**EXAMPLE 5   Adding and Subtracting Fractions**

**a.** $\dfrac{p^2 - 5}{p - 2} + \dfrac{3p + 2}{p - 2} = \dfrac{(p^2 - 5) + (3p + 2)}{p - 2}$

$\qquad\qquad\qquad = \dfrac{p^2 + 3p - 3}{p - 2}.$

**b.** $\dfrac{x^2 - 5x + 4}{x^2 + 2x - 3} - \dfrac{x^2 + 2x}{x^2 + 5x + 6} = \dfrac{(x - 1)(x - 4)}{(x - 1)(x + 3)} - \dfrac{x(x + 2)}{(x + 2)(x + 3)}$

$\qquad\qquad\qquad = \dfrac{x - 4}{x + 3} - \dfrac{x}{x + 3} = \dfrac{(x - 4) - x}{x + 3} = -\dfrac{4}{x + 3}.$

**c.** $\dfrac{x^2 + x - 5}{x - 7} - \dfrac{x^2 - 2}{x - 7} + \dfrac{-4x + 8}{x^2 - 9x + 14}$

$\qquad\qquad\qquad = \dfrac{x^2 + x - 5}{x - 7} - \dfrac{x^2 - 2}{x - 7} + \dfrac{-4(x - 2)}{(x - 2)(x - 7)}$

$\qquad\qquad\qquad = \dfrac{(x^2 + x - 5) - (x^2 - 2) + (-4)}{x - 7}$

$\qquad\qquad\qquad = \dfrac{x - 7}{x - 7} = 1.$   ∎

To add (or subtract) two fractions with different denominators, use the fundamental principle of fractions to rewrite the fractions as equivalent fractions that have the same denominator. Then proceed with the addition (or subtraction) by the method just described.

For example, to find

$$\frac{2}{x^3(x - 3)} + \frac{3}{x(x - 3)^2},$$

we can convert the first fraction into an equivalent fraction by multiplying the numerator and denominator by $x - 3$:

$$\frac{2(x - 3)}{x^3(x - 3)^2};$$

we can convert the second fraction by multiplying the numerator and denominator by $x^2$:

$$\frac{3x^2}{x^3(x - 3)^2}.$$

These fractions have the same denominator. Hence,

$$\frac{2}{x^3(x - 3)} + \frac{3}{x(x - 3)^2} = \frac{2(x - 3)}{x^3(x - 3)^2} + \frac{3x^2}{x^3(x - 3)^2}$$

$$= \frac{3x^2 + 2x - 6}{x^3(x - 3)^2}.$$

We could have converted the original fractions into equivalent fractions with *any* common denominator. However, we chose to convert them into fractions with the denominator $x^3(x - 3)^2$. This is the **least common denominator (L.C.D.)** of the fractions $2/[x^3(x - 3)]$ and $3/[x(x - 3)^2]$.

In general, to find the L.C.D. of two or more fractions, first factor each denominator completely. *The L.C.D. is the product of each of the distinct factors appearing in the denominators, each raised to the highest power to which it occurs in any single denominator.*

**EXAMPLE 6** **Adding and Subtracting Fractions**

**a.** *Subtract:* $\dfrac{t}{3t + 2} - \dfrac{4}{t - 1}$.

*Solution:* The L.C.D. is $(3t + 2)(t - 1)$. Thus, we have

$$\frac{t}{(3t + 2)} - \frac{4}{t - 1} = \frac{t(t - 1)}{(3t + 2)(t - 1)} - \frac{4(3t + 2)}{(3t + 2)(t - 1)}$$

$$= \frac{t(t - 1) - 4(3t + 2)}{(3t + 2)(t - 1)}$$

$$= \frac{t^2 - t - 12t - 8}{(3t + 2)(t - 1)} = \frac{t^2 - 13t - 8}{(3t + 2)(t - 1)}.$$

**b.** *Add:* $\dfrac{4}{q - 1} + 3$.

*Solution:* The L.C.D. is $q - 1$.

$$\frac{4}{q - 1} + 3 = \frac{4}{q - 1} + \frac{3(q - 1)}{q - 1}$$

$$= \frac{4 + 3(q - 1)}{q - 1} = \frac{3q + 1}{q - 1}.$$

**EXAMPLE 7** **Subtracting Fractions**

$$\frac{x - 2}{x^2 + 6x + 9} - \frac{x + 2}{2(x^2 - 9)}$$

$$= \frac{x - 2}{(x + 3)^2} - \frac{x + 2}{2(x + 3)(x - 3)} \qquad [\text{L.C.D.} = 2(x + 3)^2(x - 3)]$$

$$= \frac{(x - 2)(2)(x - 3)}{(x + 3)^2(2)(x - 3)} - \frac{(x + 2)(x + 3)}{2(x + 3)(x - 3)(x + 3)}$$

$$= \frac{(x - 2)(2)(x - 3) - (x + 2)(x + 3)}{2(x + 3)^2(x - 3)}$$

$$= \frac{2(x^2 - 5x + 6) - (x^2 + 5x + 6)}{2(x + 3)^2(x - 3)}$$

$$= \frac{2x^2 - 10x + 12 - x^2 - 5x - 6}{2(x + 3)^2(x - 3)}$$

$$= \frac{x^2 - 15x + 6}{2(x + 3)^2(x - 3)}.$$

Example 8 shows two methods of simplifying a "complex" fraction.

**EXAMPLE 8   Combined Operations with Fractions**

Simplify $\dfrac{\dfrac{1}{x+h} - \dfrac{1}{x}}{h}$.

**Solution:** First we combine the fractions in the numerator and obtain

$$\frac{\dfrac{1}{x+h} - \dfrac{1}{x}}{h} = \frac{\dfrac{x}{x(x+h)} - \dfrac{x+h}{x(x+h)}}{h} = \frac{\dfrac{x-(x+h)}{x(x+h)}}{h}$$

$$= \frac{\dfrac{-h}{x(x+h)}}{\dfrac{h}{1}} = \frac{-h}{x(x+h)h} = -\frac{1}{x(x+h)}.$$

The original fraction can also be simplified by multiplying the numerator and denominator by the L.C.D. of the fractions involved in the numerator (and denominator), namely, $x(x+h)$:

$$\frac{\dfrac{1}{x+h} - \dfrac{1}{x}}{h} = \frac{\left[\dfrac{1}{x+h} - \dfrac{1}{x}\right]x(x+h)}{h[x(x+h)]}$$

$$= \frac{x-(x+h)}{x(x+h)h} = \frac{-h}{x(x+h)h} = -\frac{1}{x(x+h)}. \quad ∎$$

## ▪ Exercise 0.8

*In Problems 1–6, simplify.*

1. $\dfrac{x^2 - 4}{x^2 - 2x}$.

2. $\dfrac{x^2 - 5x - 6}{x^2 - 2x - 3}$.

3. $\dfrac{x^2 - 9x + 20}{x^2 + x - 20}$.

4. $\dfrac{3x^2 - 27x + 24}{2x^3 - 16x^2 + 14x}$.

5. $\dfrac{6x^2 + x - 2}{2x^2 + 3x - 2}$.

6. $\dfrac{12x^2 - 19x + 4}{6x^2 - 17x + 12}$.

*In Problems 7–48, perform the operations and simplify as much as possible.*

7. $\dfrac{y^2}{y-3} \cdot \dfrac{-1}{y+2}$.

8. $\dfrac{z^2 - 4}{z^2 + 2z} \cdot \dfrac{z^2}{z-2}$.

9. $\dfrac{2x-3}{x-2} \cdot \dfrac{2-x}{2x+3}$.

10. $\dfrac{x^2 - y^2}{x+y} \cdot \dfrac{x^2 + 2xy + y^2}{y - x}$.

11. $\dfrac{2x-2}{x^2 - 2x - 8} \div \dfrac{x^2 - 1}{x^2 + 5x + 4}$.

12. $\dfrac{x^2 + 2x}{3x^2 - 18x + 24} \div \dfrac{x^2 - x - 6}{x^2 - 4x + 4}$.

13. $\dfrac{\dfrac{x^2}{6}}{\dfrac{x}{3}}$.

14. $\dfrac{\dfrac{4x^3}{9x}}{\dfrac{x}{18}}$.

15. $\dfrac{\dfrac{2m}{n^3}}{\dfrac{4m}{n^2}}$.

16. $\dfrac{\dfrac{c+d}{c}}{\dfrac{c-d}{2c}}$.

17. $\dfrac{\dfrac{4x}{3}}{2x}$.

18. $\dfrac{4x}{\dfrac{3}{2x}}$.

19. $\dfrac{-9x^3}{\dfrac{x}{3}}$.

20. $\dfrac{\dfrac{-9x^3}{x}}{3}$.

**21.** $\dfrac{\dfrac{x-5}{x^2-7x+10}}{x-2}$.

**22.** $\dfrac{\dfrac{x^2+6x+9}{x}}{x+3}$.

**23.** $\dfrac{\dfrac{10x^3}{x^2-1}}{\dfrac{5x}{x+1}}$.

**24.** $\dfrac{\dfrac{x^2-4}{x^2+2x-3}}{\dfrac{x^2-x-6}{x^2-9}}$.

**25.** $\dfrac{\dfrac{x^2+7x+10}{x^2-2x-8}}{\dfrac{x^2+6x+5}{x^2-3x-4}}$.

**26.** $\dfrac{\dfrac{(x+2)^2}{3x-2}}{\dfrac{9x+18}{4-9x^2}}$.

**27.** $\dfrac{\dfrac{4x^2-9}{x^2+3x-4}}{\dfrac{2x-3}{1-x^2}}$.

**28.** $\dfrac{\dfrac{6x^2y+7xy-3y}{xy-x+5y-5}}{\dfrac{x^3y+4x^2y}{xy-x+4y-4}}$.

**29.** $\dfrac{x^2}{x+3}+\dfrac{5x+6}{x+3}$.

**30.** $\dfrac{2}{x+2}+\dfrac{x}{x+2}$.

**31.** $\dfrac{1}{t}+\dfrac{2}{3t}$.

**32.** $\dfrac{4}{x^2}-\dfrac{1}{x}$.

**33.** $1-\dfrac{p^2}{p^2-1}$.

**34.** $\dfrac{4}{s+4}+s$.

**35.** $\dfrac{4}{2x-1}+\dfrac{x}{x+3}$.

**36.** $\dfrac{x+1}{x-1}-\dfrac{x-1}{x+1}$.

**37.** $\dfrac{1}{x^2-x-2}+\dfrac{1}{x^2-1}$.

**38.** $\dfrac{y}{3y^2-5y-2}-\dfrac{2}{3y^2-7y+2}$.

**39.** $\dfrac{4}{x-1}-3+\dfrac{-3x^2}{5-4x-x^2}$.

**40.** $\dfrac{2x-3}{2x^2+11x-6}-\dfrac{3x+1}{3x^2+16x-12}+\dfrac{1}{3x-2}$.

**41.** $(1+x^{-1})^2$.

**42.** $(x^{-1}+y^{-1})^2$.

**43.** $(x^{-1}-y)^{-1}$.

**44.** $(x-y^{-1})^2$.

**45.** $\dfrac{1+\dfrac{1}{x}}{3}$.

**46.** $\dfrac{\dfrac{x+3}{x}}{x-\dfrac{9}{x}}$.

**47.** $\dfrac{3-\dfrac{1}{2x}}{x+\dfrac{x}{x+2}}$.

**48.** $\dfrac{\dfrac{x-1}{x^2+5x+6}-\dfrac{1}{x+2}}{3+\dfrac{x-7}{3}}$.

*In Problems 49 and 50, perform the indicated operations, but do not rationalize the denominators.*

**49.** $\dfrac{2}{\sqrt{x+h}}-\dfrac{2}{\sqrt{x}}$.

**50.** $\dfrac{x\sqrt{x}}{\sqrt{1+x}}+\dfrac{1}{\sqrt{x}}$.

*In Problems 51–60, simplify, and express your answer in a form that is free of radicals in the denominator.*

**51.** $\dfrac{1}{2+\sqrt{3}}$.

**52.** $\dfrac{1}{1-\sqrt{2}}$.

**53.** $\dfrac{\sqrt{2}}{\sqrt{3}-\sqrt{6}}$.

**54.** $\dfrac{5}{\sqrt{6}+\sqrt{7}}$.

**55.** $\dfrac{2\sqrt{2}}{\sqrt{2}-\sqrt{3}}$.

**56.** $\dfrac{2\sqrt{3}}{\sqrt{5}-\sqrt{2}}$.

**57.** $\dfrac{1}{x+\sqrt{5}}$.

**58.** $\dfrac{x-3}{\sqrt{x}-1}+\dfrac{4}{\sqrt{x}-1}$.

**59.** $\dfrac{5}{1+\sqrt{3}}-\dfrac{4}{2-\sqrt{2}}$.

**60.** $\dfrac{4}{\sqrt{x}+2}\cdot\dfrac{x^2}{3}$.

# Equations

Even beginning students in many areas of study are soon faced with solving elementary equations. In this chapter, we shall develop techniques to accomplish this task. These methods will be applied in the next chapter to some practical situations.

## 1.1 LINEAR EQUATIONS

### Equations

An **equation** is a statement that two expressions are equal. The two expressions that make up an equation are called its **sides** or **members.** They are separated by the **equality sign, "=."**

**EXAMPLE 1**  **Examples of Equations**

**a.** $x + 2 = 3$.

**b.** $x^2 + 3x + 2 = 0$.

**c.** $\dfrac{y}{y - 4} = 6$.

**d.** $w = 7 - z$.  ∎

In Example 1, each equation contains at least one variable. A **variable** is a symbol that can be replaced by any one of a set of different numbers. The most popular symbols for variables are letters from the latter part of the alphabet, such as $x, y, z, w,$ and $t$. Hence, equations (a) and (c) are said to be in the variables $x$ and $y$, respectively. Equation (d) is in the variables $w$ and $z$. In the equation $x + 2 = 3$, the numbers 2 and 3 are called *constants*. They are fixed numbers.

We *never* allow a variable in an equation to have a value for which any expression in that equation is undefined. Thus, in

$$\frac{y}{y - 4} = 6,$$

$y$ cannot be 4, because this would make the denominator zero. (We cannot divide by zero.) In some equations, the allowable values of a variable are restricted for physical reasons. For example, if the variable $t$ represents time, negative values of $t$ may not make sense. Hence, we should assume that $t \geq 0$.

To *solve* an equation means to find all values of its variables for which the equation is true. These values are called *solutions* of the equation and are said to *satisfy* the equation. When only one variable is involved, a solution is also called a **root.** The set of all solutions is called the **solution set** of the equation. Sometimes a letter representing an unknown quantity in an equation is simply called an *unknown.* Let us illustrate these terms.

**EXAMPLE 2   Terminology for Equations**

a. In the equation $x + 2 = 3$, the variable $x$ is the unknown. The only value of $x$ that satisfies the equation is obviously 1. Hence, 1 is a root and the solution set is {1}.

b. $-2$ is a root of $x^2 + 3x + 2 = 0$ because substituting $-2$ for $x$ makes the equation true: $(-2)^2 + 3(-2) + 2 = 0$.

c. $w = 7 - z$ is an equation in two unknowns. One solution is the pair of values $w = 4$ and $z = 3$. However, there are infinitely many solutions. Can you think of another?   ∎

### Equivalent Equations

Solving an equation may involve performing operations on it. We prefer that any such operation result in another equation having exactly the same solutions as the given equation. When this occurs, the equations are said to be **equivalent.** There are three operations that guarantee equivalence:

1. Adding (subtracting) the same polynomial[1] to (from) both sides of an equation, where the polynomial is in the same variable as that occurring in the equation.

For example, if $-5x = 5 - 6x$, then adding $6x$ to both sides gives the equivalent equation $-5x + 6x = 5 - 6x + 6x$, or $x = 5$.

2. Multiplying (dividing) both sides of an equation by the same constant, except by zero.

For example, if $10x = 5$, then dividing both sides by 10 gives the equivalent equation $\dfrac{10x}{10} = \dfrac{5}{10}$, or $x = \dfrac{1}{2}$.

3. Replacing either side of an equation by an equal expression.

For example, if $x(x + 2) = 3$, then replacing the left side by the equal expression $x^2 + 2x$ gives the equivalent equation $x^2 + 2x = 3$.

We repeat: Applying Operations 1–3 guarantees that the resulting equation is equivalent to the given one. However, sometimes in solving an equation we have to apply operations other than 1–3. These operations may *not* necessarily result in equivalent equations. They include the following.

*Equivalence is not guaranteed if both sides are multiplied or divided by an expression involving a variable.*

---

[1]See Sec. 0.6 for a definition of a polynomial.

**Operations That May Not Produce Equivalent Equations**

**4.** Multiplying both sides of an equation by an expression involving the variable;

**5.** Dividing both sides of an equation by an expression involving the variable;

Operation 6 includes taking roots of both sides.

**6.** Raising both sides of an equation to equal powers.

Let us illustrate the last three operations. For example, by inspection, the only root of $x - 1 = 0$ is 1. Multiplying each side by $x$ (Operation 4) gives $x^2 - x = 0$, which is satisfied if $x$ is 0 or 1. (Check this by substitution.) But 0 *does not* satisfy the *original* equation. Thus, the equations are not equivalent.

Continuing, you may check that the equation $(x - 4)(x - 3) = 0$ is satisfied when $x$ is 4 or 3. Dividing both sides by $x - 4$ (Operation 5) gives $x - 3 = 0$, whose only root is 3. Again, we do not have equivalence, since in this case a root has been "lost." Note that when $x$ is 4, division by $x - 4$ implies division by 0, an invalid operation.

Finally, squaring each side of the equation $x = 2$ (Operation 6) gives $x^2 = 4$, which is true if $x = 2$ or $-2$. But $-2$ is not a root of the given equation.

From our discussion, it is clear that when Operations 4–6 are performed, we must be careful about drawing conclusions concerning the roots of a given equation. Operations 4 and 6 *can* produce an equation with more roots. Thus, you should check whether or not each "solution" obtained by the these operations satisfies the *original* equation. Operation 5 *can* produce an equation with fewer roots. In this case, any "lost" root may never be determined. Thus, avoid Operation 5 whenever possible.

In summary, an equation may be thought of as a set of restrictions on any variable in the equation. Operations 4–6 may increase or decrease the number of restrictions, giving solutions different from those of the original equation. However, Operations 1–3 never affect the restrictions.

## TECHNOLOGY

A graphics calculator can be used to test for a root. For example, suppose we want to determine whether 3/2 is a root of the equation

$$2x^3 + 7x^2 = 19x + 60.$$

First, we rewrite the equation so that one side is 0. Subtracting $19x + 60$ from both sides gives the equivalent equation

$$2x^3 + 7x^2 - 19x - 60 = 0.$$

For a TI-82 graphics calculator, we enter the expression $2x^3 + 7x^2 - 19x - 60$ as $Y_1$ and then evaluate $Y_1$ at $x = 3/2$. Figure 1.1 shows that the result is $-66$, which is not 0. Thus, 3/2 is not a root. However, $Y_1$ evaluated at $x = -5/2$ *is* 0. So $-5/2$ is a root of the original equation.

It is worth noting that if the original equation had been in terms of the variable $t$, that is,

$$2t^3 + 7t^2 = 19t + 60,$$

then we must replace $t$ by $x$ because the calculator evaluates $Y_1$ at a specified value of $x$, not $t$.

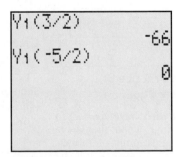

**FIGURE 1.1**  For $2x^3 + 7x^2 - 19x - 60 = 0$, $^3/_2$ is not a root, but $^{-5}/_2$ is a root.

### Linear Equations

The principles presented so far will now be demonstrated in the solution of a *linear equation*.

### DEFINITION

*A **linear equation** in the variable x is an equation that can be written in the form*

$$ax + b = 0, \tag{1}$$

*where a and b are constants and $a \neq 0$.*

A linear equation is also called a first-degree equation or an equation of degree one, since the highest power of the variable that occurs in Eq. (1) is the first.

To solve a linear equation, we perform operations on it until we have an equivalent equation whose solutions are *obvious*. This means an equation in which the variable is isolated on one side, as the following examples show.

**Principles in Practice 2**
Solving a Linear Equation

A trendy coffeehouse's total daily revenue from the sale of $x$ specialty coffees is given by $r = 2.25x$, and its total daily costs are given by $c = 0.75x + 300$. How many specialty coffees need to be sold each day to break even? In other words, when will revenue equal costs?

### EXAMPLE 3   Solving a Linear Equation

*Solve* $5x - 6 = 3x$.

*Solution:* We begin by getting the terms involving $x$ on one side and the constant on the other. Then we solve for $x$ by the appropriate mathematical operation. We have

$$5x - 6 = 3x,$$

$$5x - 6 + (-3x) = 3x + (-3x) \quad \text{(adding } -3x \text{ to both sides),}$$

$$2x - 6 = 0 \quad \text{(simplifying, that is, Operation 3),}$$

$$2x - 6 + 6 = 0 + 6 \quad \text{(adding 6 to both sides),}$$

$$2x = 6 \quad \text{(simplifying),}$$

$$\frac{2x}{2} = \frac{6}{2} \quad \text{(dividing both sides by 2)}$$

$$x = 3.$$

Clearly, 3 is the only root of the last equation. Since each equation is equivalent to the one before it, we conclude that 3 must be the only root of $5x - 6 = 3x$. That is, the solution set is $\{3\}$. We can describe the first step in the solution as moving a term from one side of an equation to the other while changing its sign; this is commonly called *transposing*. Note that since the original equation can be put in the form $2x + (-6) = 0$, it is a linear equation. ■

**Principles in Practice 3**
Solving a Linear Equation

Monica and Pedro have agreed to pool their savings when they have saved the same amount of money. Monica can save $40 a week, but she must use her first $125 to pay her credit card debt. Pedro has been saving $35 a week for three weeks already. When will they pool their savings? How much will they each have saved?

### EXAMPLE 4   Solving a Linear Equation

*Solve* $2(p + 4) = 7p + 2$.

*Solution:* First, we remove parentheses. Then we collect similar terms and solve. We have

$$2(p + 4) = 7p + 2,$$

$$2p + 8 = 7p + 2 \qquad \text{(distributive property)},$$

$$2p = 7p - 6 \qquad \text{(subtracting 8 from both sides)},$$

$$-5p = -6 \qquad \text{(subtracting } 7p \text{ from both sides)},$$

$$p = \frac{-6}{-5} \qquad \text{(dividing both sides by } -5\text{)},$$

$$p = \frac{6}{5}. \qquad \blacksquare$$

### EXAMPLE 5   Solving a Linear Equation

*Solve* $\dfrac{7x + 3}{2} - \dfrac{9x - 8}{4} = 6.$

*Solution:* We first clear the equation of fractions by multiplying *both* sides by the least common denominator (L.C.D.),[2] which is 4. Then we use various algebraic operations to obtain a solution. Thus,

$$4\left(\frac{7x + 3}{2} - \frac{9x - 8}{4}\right) = 4(6),$$

The distributive property requires that *both* terms within the parentheses be multiplied by 4.

$$4 \cdot \frac{7x + 3}{2} - 4 \cdot \frac{9x - 8}{4} = 24 \qquad \text{(distributive property)},$$

$$2(7x + 3) - (9x - 8) = 24 \qquad \text{(simplifying)},$$

$$14x + 6 - 9x + 8 = 24 \qquad \text{(distributive property)},$$

$$5x + 14 = 24 \qquad \text{(simplifying)},$$

$$5x = 10 \qquad \text{(subtracting 14 from both sides)},$$

$$x = 2 \qquad \text{(dividing both sides by 5)}. \qquad \blacksquare$$

Every linear equation has exactly one root.

Each equation in Examples 3–5 has one and only one root. This is true of every linear equation in one variable.

### Literal Equations

Equations in which some of the constants are not specified, but are represented by letters, such as $a, b, c,$ or $d$, are called **literal equations,** and the letters are called **literal constants** or **arbitrary constants.** For example, in the literal equation $x + a = 4b$, we may consider $a$ and $b$ to be arbitrary constants. Formulas, such as $I = Prt$, that express a relationship between certain quantities may be regarded as literal equations. If we want to express a particular letter in a formula in terms of the others, this letter is considered the unknown.

**Principles in Practice 4**
Solving a Literal Equation

The formula $d = rt$ gives the distance $d$ an object travels going a rate $r$ for a time $t$. What is the rate $r$ for a train that travels $d$ miles in $t$ hours?

### EXAMPLE 6   Solving Literal Equations

**a.** *The equation* $I = Prt$ *is the formula for the simple interest* $I$ *on a principal of* $P$ *dollars at the annual interest rate of* $r$ *for a period of* $t$ *years. Express* $r$ *in terms of* $I$, $P$, *and* $t$.

[2]The least common denominator of two or more fractions is the smallest number with all the denominators as factors. That is, the L.C.D. is the least common multiple of all the denominators.

*Solution:* Here we consider $r$ to be the unknown. To isolate $r$, we divide both sides by $Pt$. We have

$$I = Prt,$$

$$\frac{I}{Pt} = \frac{Prt}{Pt},$$

$$\frac{I}{Pt} = r \quad \text{or} \quad r = \frac{I}{Pt}.$$

When we divided both sides by $Pt$, we assumed that $Pt \neq 0$, since we cannot divide by 0. Similar assumptions will be made in solving other literal equations.

**b.** *The equation $S = P + Prt$ is the formula for the value $S$ of an investment of a principal of $P$ dollars at a simple annual interest rate of $r$ for a period of $t$ years. Solve for $P$.*

*Solution:*

$$S = P + Prt,$$

$$S = P(1 + rt) \qquad \text{(factoring)},$$

$$\frac{S}{1 + rt} = P \qquad \text{(dividing both sides by } 1 + rt\text{)}. \qquad \blacksquare$$

---

**Principles in Practice 5**
**Solving a Literal Equation**

The formula $S = 4\pi\left(\dfrac{d}{2}\right)^2$ gives the surface area $S$ of a sphere with diameter $d$. What is the length of the side of the smallest box that will hold a ball with surface area $S$?

**EXAMPLE 7    Solving a Literal Equation**

*Solve $(a + c)x + x^2 = (x + a)^2$ for $x$.*

*Solution:* We first simplify the equation and then get all terms involving $x$ on one side:

$$(a + c)x + x^2 = (x + a)^2,$$

$$ax + cx + x^2 = x^2 + 2ax + a^2,$$

$$ax + cx = 2ax + a^2,$$

$$cx - ax = a^2,$$

$$x(c - a) = a^2,$$

$$x = \frac{a^2}{c - a}. \qquad \blacksquare$$

---

**■ Exercise 1.1**

*In Problems 1–6, determine by substitution which of the given numbers, if any, satisfy the given equation.*

**1.** $9x - x^2 = 0; \quad 1, 0.$

**2.** $20 - 9x = -x^2; \quad 5, 4.$

**3.** $y + 2(y - 3) = 4; \quad \frac{10}{3}, 1.$

**4.** $2x + x^2 - 8 = 0; \quad 2, -4.$

**5.** $x(7 + x) - 2(x + 1) - 3x = -2; \quad -3, 0.$

**6.** $x(x + 1)^2(x + 2) = 0; \quad 0, -1, 2.$

*In Problems 7–16, determine what operations were applied to the first equation to obtain the second. State whether or not the operations guarantee that the equations are equivalent. Do not solve the equations.*

**7.** $x - 5 = 4x + 10; \quad x = 4x + 15.$

**8.** $8x - 4 = 16; \quad x - \frac{1}{2} = 2.$

**9.** $x = 4$;   $x^2 = 16$.

**10.** $\frac{1}{2}x^2 + 3 = x - 9$;   $x^2 + 6 = 2x - 18$.

**11.** $x^2 - 2x = 0$;   $x - 2 = 0$.

**12.** $\frac{2}{x - 2} + x = x^2$;   $2 + x(x - 2) = x^2(x - 2)$.

**13.** $\frac{x^2 - 1}{x - 1} = 3$;   $x^2 - 1 = 3(x - 1)$.

**14.** $x(x + 5)(x + 9) = x(x + 1)$;
$(x + 5)(x + 9) = x + 1$.

**15.** $\frac{x(x + 1)}{x - 5} = x(x + 9)$;   $x + 1 = (x + 9)(x - 5)$.

**16.** $2x^2 - 9 = x$;   $x^2 - \frac{1}{2}x = \frac{9}{2}$.

*In Problems 17–46, solve the equations.*

**17.** $4x = 10$.

**18.** $0.2x = 5$.

**19.** $3y = 0$.

**20.** $2x - 4x = -5$.

**21.** $-5x = 10 - 15$.

**22.** $3 - 2x = 4$.

**23.** $5x - 3 = 9$.

**24.** $\sqrt{2}x + 3 = 8$.

**25.** $7x + 7 = 2(x + 1)$.

**26.** $6z + 5z - 3 = 41$.

**27.** $2(p - 1) - 3(p - 4) = 4p$.

**28.** $t = 2 - 2[2t - 3(1 - t)]$.

**29.** $\frac{x}{5} = 2x - 6$.

**30.** $\frac{5y}{7} - \frac{6}{7} = 2 - 4y$.

**31.** $5 + \frac{4x}{9} = \frac{x}{2}$.

**32.** $\frac{x}{3} - 4 = \frac{x}{5}$.

**33.** $q = \frac{3}{2}q - 4$.

**34.** $\frac{x}{2} + \frac{x}{3} = 7$.

**35.** $3x + \frac{x}{5} - 5 = \frac{1}{5} + 5x$.

**36.** $y - \frac{y}{2} + \frac{y}{3} - \frac{y}{4} = \frac{y}{5}$.

**37.** $\frac{2y - 3}{4} = \frac{6y + 7}{3}$.

**38.** $\frac{p}{3} + \frac{3}{4}p = \frac{9}{2}(p - 1)$.

**39.** $w + \frac{w}{2} - \frac{w}{3} + \frac{w}{4} = 5$.

**40.** $\frac{7 + 2(x + 1)}{3} = \frac{8x}{5}$.

**41.** $\frac{x + 2}{3} - \frac{2 - x}{6} = x - 2$.

**42.** $\frac{x}{5} + \frac{2(x - 4)}{10} = 7$.

**43.** $\frac{9}{5}(3 - x) = \frac{3}{4}(x - 3)$.

**44.** $\frac{2y - 7}{3} + \frac{8y - 9}{14} = \frac{3y - 5}{21}$.

**45.** $\frac{3}{2}(4x - 3) = 2[x - (4x - 3)]$.

**46.** $(3x - 1)^2 - (5x - 3)^2 = -(4x - 2)^2$.

*In Problems 47–54, express the indicated symbol in terms of the remaining symbols.*

**47.** $I = Prt$;   $P$.

**48.** $ax + b = 0$;   $x$.

**49.** $p = 8q - 1$;   $q$.

**50.** $p = -3q + 6$;   $q$.

**51.** $S = P(1 + rt)$;   $r$.

**52.** $r = \frac{2mI}{B(n + 1)}$;   $m$.

**53.** $S = \frac{n}{2}(a_1 + a_n)$;   $a_1$.

**54.** $S = \frac{R[(1 + i)^n - 1]}{i}$;   $R$.

**55. Geometry**   Use the formula $P = 2l + 2w$ to find the width $w$ of a rectangle whose perimeter $P$ is 960 m and whose length $l$ is 360 m.

**56. Geometry**   Use the formula $A = \frac{1}{2}bh$ to find the height $h$ of a triangle whose area $A$ is 75 cm² and whose base $b$ is 15 cm.

**57. Sales Tax**   A salesperson needs to calculate the cost of an item with a sales tax of 8.25%. Write an equation that represents the total cost $c$ of an item costing $x$ dollars.

**58. Revenue**   A day care center's total monthly revenue from the care of $x$ toddlers is given by $r = 650x$, and its total monthly costs are given by $c = 400x + 2000$. How many toddlers need to be enrolled each month to break even? In other words, when will revenue equal costs?

**59. Straight-line Depreciation**   If you purchase an item for business use, in preparing your income tax you may be able to spread out its expense over the life of the item. This is called *depreciation*. One method of depreciation is *straight-line depreciation*, in which the annual depreciation is computed by dividing the cost of the item, less its estimated salvage value, by its useful life. Suppose the cost is $C$ dollars, the useful life is $N$ years, and there is no salvage value. Then the value $V$ (in dollars) of the item at the end of $n$ years is given by

$$V = C\left(1 - \frac{n}{N}\right).$$

If new office furniture is purchased for $1600, has a useful life of 8 years, and has no salvage value, after how many years will it have a value of $1000?

**60. Radar Beam**   When radar is used on a highway to determine the speed of a car, a radar beam is sent out and reflected from the moving car. The difference $F$ (in cycles per second) in frequency between the original and reflected beams is given by

$$F = \frac{vf}{334.8},$$

where $v$ is the speed of the car in miles per hour and $f$ is the frequency of the original beam (in megacycles per second).

Suppose you are driving along a highway with a speed limit of 55 mi/h. A police officer aims a radar

beam with a frequency of 2450 megacycles per second at your car, and the officer observes the difference in frequency to be 420 cycles per second. Can the officer claim that you were speeding?

61. **Savings**    Paula and Sam want to buy a house, so they have decided to save one-fourth of each of their salaries. Paula earns $22.00 per hour and receives an extra $8.00 a week because she declined company benefits, and Sam earns $26.00 per hour plus benefits. They want to save at least $350.00 each week. How many hours must they each work each week?

62. **Gravity**    The equation $h = -4.9t^2 + m$ is the formula for the height $h$ of an object in meters $t$ seconds after is it dropped from a starting point of $m$ meters. What is the time $t$ an object has been falling if it has fallen from a height $m$ and is now at a height $h$?

63. **Linear Expansion**    When solid objects are heated, they expand in length—which is why expansion joints are placed in bridges and pavements. Generally, when the temperature of a solid body of length $I_0$ is increased from $T_0$ to $T$, the body's length $I$ is given by

$$I = I_0[1 + \alpha(T - T_0)],$$

where $\alpha$ (the Greek letter *alpha*) is called the *coefficient of linear expansion*. Suppose a metal rod 1 m long at 0°C expands 0.001 m when it is heated from 0°C to 100°C. Find the coefficient of linear expansion.

64. **Predator-Prey Relation**    To study a predator–prey relationship, an experiment[3] was conducted in which a blindfolded subject, the "predator," stood in front of a 3-ft-square table on which uniform sandpaper discs, the "prey," were placed. For 1 minute the "predator" searched for the discs by tapping with a finger. Whenever a disc was found, it was removed and searching resumed. The experiment was repeated for various disc densities (number of discs per 9 ft²). It was estimated that if $y$ is the number of discs picked up in 1 minute when $x$ discs are on the table, then

$$y = a(1 - by)x,$$

where $a$ and $b$ are constants. Solve this equation for $y$.

*In Problems 65–68, use a graphics calculator to determine which of the given numbers, if any, are roots of the given equation.*

65. $112x^2 = 6x + 1$;    $\frac{1}{8}, -\frac{2}{5}, -\frac{1}{14}$.

66. $8x^3 + 11x + 21 = 58x^2$;    $7, -\frac{1}{2}, \frac{4}{3}$.

67. $\frac{3.1t - 7}{4.8t - 2} = 7$;    $\sqrt{6}, -\frac{47}{52}, \frac{14}{61}$.

68. $\left(\frac{v}{v + 3}\right)^2 = v$;    $0, \frac{27}{4}, \frac{13}{3}$.

---

OBJECTIVE

**To solve fractional and radical equations that lead to linear equations.**

# 1.2 EQUATIONS LEADING TO LINEAR EQUATIONS

## Fractional Equations

In this section, we illustrate that solving a nonlinear equation may lead to a linear equation. We begin with a **fractional equation,** which is an equation in which an unknown is in a denominator.

### Principles in Practice 1
**Solving a Fractional Equation**

A boat with a speed $r$ travels 10 miles downstream in a 2-mile-per-hour current in the same time a boat going the same speed travels 6 miles upstream in the current. Write an equation describing this situation, and find the speed of the boats.

### EXAMPLE 1    Solving a Fractional Equation

*Solve* $\dfrac{5}{x - 4} = \dfrac{6}{x - 3}.$

---

[3]C. S. Holling, "Some Characteristics of Simple Types of Predation and Parasitism," *The Canadian Entomologist,* XCI, no. 7 (1959), 385–98.

*Solution:*

> *Strategy* We first write the equation in a form that is free of fractions. Then we use standard algebraic techniques to solve the resulting linear equation.

Multiplying both sides by the L.C.D., $(x - 4)(x - 3)$, we have

$$(x - 4)(x - 3)\left(\frac{5}{x - 4}\right) = (x - 4)(x - 3)\left(\frac{6}{x - 3}\right),$$

$$5(x - 3) = 6(x - 4) \qquad \text{(linear equation)},$$

$$5x - 15 = 6x - 24,$$

$$9 = x.$$

An alternative solution that avoids multiplying both sides by the L.C.D. is as follows:

$$\frac{5}{x - 4} - \frac{6}{x - 3} = 0.$$

Assuming that $x$ is neither 3 nor 4 and combining fractions gives

$$\frac{9 - x}{(x - 4)(x - 3)} = 0.$$

A fraction can be 0 only when its numerator is 0 and its denominator is not. Hence, $x = 9$.

In the first step, we multiplied each side by an expression involving the *variable x*. As we mentioned in Sec. 1.1, this means that we are not guaranteed that the last equation is equivalent to the *original* equation. Thus, we must check whether or not 9 satisfies the *original* equation. Substituting 9 for $x$ in that equation, we get

$$\frac{5}{9 - 4} = \frac{6}{9 - 3},$$

$$1 = 1,$$

which is a true statement. Hence, 9 is a root. ∎

Some equations that are not linear do not have any solutions. In that case, we say that the solution set is the **empty set** or **null set,** which we denote by { } or ∅. Example 2 will illustrate.

**EXAMPLE 2** **Solving Fractional Equations**

a. *Solve* $\dfrac{3x + 4}{x + 2} - \dfrac{3x - 5}{x - 4} = \dfrac{12}{x^2 - 2x - 8}.$

*Solution:* Observing the denominators and noting that

$$x^2 - 2x - 8 = (x + 2)(x - 4),$$

we conclude that the L.C.D. is $(x + 2)(x - 4)$. Multiplying both sides by the L.C.D., we have

$$(x + 2)(x - 4)\left(\frac{3x + 4}{x + 2} - \frac{3x - 5}{x - 4}\right) = (x + 2)(x - 4) \cdot \frac{12}{(x + 2)(x - 4)},$$

$$(x - 4)(3x + 4) - (x + 2)(3x - 5) = 12,$$

$$3x^2 - 8x - 16 - (3x^2 + x - 10) = 12,$$

$$3x^2 - 8x - 16 - 3x^2 - x + 10 = 12,$$

$$-9x - 6 = 12,$$

$$-9x = 18,$$

$$x = -2. \tag{1}$$

However, the *original* equation is not defined for $x = -2$ (we cannot divide by zero), so there are no roots. Thus, the solution set is ∅. Although

$-2$ is a solution of Eq. (1), it is not a solution of the *original* equation and is called an **extraneous solution** of that equation.

**b.** *Solve* $\dfrac{4}{x-5} = 0$.

*Solution:* The only way a fraction can equal zero is if the numerator is 0 but its denominator is not. Since the numerator, 4, is never 0, the solution set is $\varnothing$. ∎

---

### Principles in Practice 2
**Literal Equation**

The time it takes an airplane to travel a given distance with a tailwind ("with the wind") can be calculated by dividing the distance by the sum of the speed of the airplane and the speed of the wind. Write an equation that calculates the time $t$ it takes an airplane traveling at speed $r$ with a wind $w$ to cover a distance $d$. Solve your equation for $w$.

### EXAMPLE 3 Literal Equation

*If* $s = \dfrac{u}{au+v}$, *express u in terms of the remaining letters; that is, solve for u.*

**Solution:**

> *Strategy:* Since the unknown, $u$, is in the denominator, we first clear fractions and then solve for $u$.

$$s = \frac{u}{au+v},$$

$$s(au+v) = u \qquad \text{(multiplying both sides by } au+v\text{)},$$

$$sau + sv = u,$$

$$sau - u = -sv,$$

$$u(sa-1) = -sv,$$

$$u = \frac{-sv}{sa-1} = \frac{sv}{1-sa}.$$
∎

## Radical Equations

A **radical equation** is one in which an unknown occurs in a radicand. The next two examples illustrate the techniques employed to solve such equations.

---

### Principles in Practice 3
**Solving a Radical Equation**

The difference between the length of a ramp and the length of the horizontal distance it covers is 2 feet. The square of the vertical distance the ramp covers is 16 square feet. Write an equation for the difference and solve. What is the length of the ramp?

### EXAMPLE 4 Solving a Radical Equation

*Solve* $\sqrt{x^2 + 33} - x = 3$.

*Solution:* To solve this radical equation, we raise both sides to the same power to eliminate the radical. This operation does *not* guarantee equivalence, so we must check any resulting "solutions." We begin by isolating the radical on one side. Then we square both sides and solve using standard techniques. Thus,

$$\sqrt{x^2 + 33} = x + 3,$$

$$x^2 + 33 = (x+3)^2 \qquad \text{(squaring both sides)},$$

$$x^2 + 33 = x^2 + 6x + 9,$$

$$24 = 6x,$$

$$4 = x.$$

You should show by substitution that 4 is indeed a root. ∎

With some radical equations, you may have to raise both sides to the same power more than once, as Example 5 shows.

**EXAMPLE 5  Solving a Radical Equation**

*Solve* $\sqrt{y - 3} - \sqrt{y} = -3$.

*Solution:* When an equation has two terms involving radicals, first write the equation so that one radical is on each side, if possible. Then square and solve. We have

$$\sqrt{y - 3} = \sqrt{y} - 3,$$

$$y - 3 = y - 6\sqrt{y} + 9 \qquad \text{(squaring both sides)},$$

$$6\sqrt{y} = 12,$$

$$\sqrt{y} = 2,$$

$$y = 4 \qquad\qquad\qquad \text{(squaring both sides)}.$$

The reason why we desire one radical on each side is to eliminate squaring a binomial with two different radicals.

Substituting 4 into the left side of the *original* equation gives $\sqrt{1} - \sqrt{4}$, which is $-1$. Since this does not equal the right side, $-3$, there is no solution. That is, the solution set is $\varnothing$. Here 4 is an extraneous solution. ∎

## ▪ Exercise 1.2

*In Problems 1–34, solve the equations.*

**1.** $\dfrac{5}{x} = 25$.

**2.** $\dfrac{4}{x - 1} = 2$.

**3.** $\dfrac{3}{7 - x} = 0$.

**4.** $\dfrac{5x - 2}{x + 1} = 0$.

**5.** $\dfrac{4}{8 - x} = \dfrac{3}{4}$.

**6.** $\dfrac{x + 3}{x} = \dfrac{2}{5}$.

**7.** $\dfrac{q}{3q - 4} = 3$.

**8.** $\dfrac{4p}{7 - p} = 1$.

**9.** $\dfrac{1}{p - 1} = \dfrac{2}{p - 2}$.

**10.** $\dfrac{2x - 3}{4x - 5} = 6$.

**11.** $\dfrac{1}{x} + \dfrac{1}{5} = \dfrac{4}{5}$.

**12.** $\dfrac{4}{t - 3} = \dfrac{3}{t - 4}$.

**13.** $\dfrac{3x - 2}{2x + 3} = \dfrac{3x - 1}{2x + 1}$.

**14.** $\dfrac{x + 2}{x - 1} + \dfrac{x + 1}{2 - x} = 0$.

**15.** $\dfrac{y - 6}{y} - \dfrac{6}{y} = \dfrac{y + 6}{y - 6}$.

**16.** $\dfrac{y - 3}{y + 3} = \dfrac{y - 3}{y + 2}$.

**17.** $\dfrac{-4}{x - 1} = \dfrac{7}{2 - x} + \dfrac{3}{x + 1}$.

**18.** $\dfrac{1}{x - 3} - \dfrac{3}{x - 2} = \dfrac{4}{1 - 2x}$.

**19.** $\dfrac{9}{x - 3} = \dfrac{3x}{x - 3}$.

**20.** $\dfrac{x}{x + 3} - \dfrac{x}{x - 3} = \dfrac{3x - 4}{x^2 - 9}$.

**21.** $\sqrt{x + 6} = 3$.

**22.** $\sqrt{z - 2} = 3$.

**23.** $\sqrt{5x - 6} - 16 = 0$.

**24.** $6 - \sqrt{2x + 5} = 0$.

**25.** $\sqrt{\dfrac{x}{2} + 1} = \dfrac{2}{3}$.

**26.** $(x + 6)^{1/2} = 7$.

**27.** $\sqrt{4x - 6} = \sqrt{x}$.

**28.** $\sqrt{7 - 2x} = \sqrt{x - 1}$.

**29.** $(x - 3)^{3/2} = 8$.

**30.** $\sqrt{y^2 - 9} = 9 - y$.

**31.** $\sqrt{y} + \sqrt{y + 2} = 3$.

**32.** $\sqrt{x} - \sqrt{x + 1} = 1$.

**33.** $\sqrt{z^2 + 2z} = 3 + z$.

**34.** $\sqrt{\dfrac{1}{w}} - \sqrt{\dfrac{2}{5w - 2}} = 0$.

*In Problems 35–38, express the indicated letter in terms of the remaining letters.*

**35.** $r = \dfrac{d}{1 - dt}$;  $d$.

**36.** $\dfrac{x - a}{b - x} = \dfrac{x - b}{a - x}$;  $x$.

**37.** $r = \dfrac{2ml}{B(n + 1)}$;  $n$.

**38.** $\dfrac{1}{p} + \dfrac{1}{q} = \dfrac{1}{f}$;  $q$.

**39. Prey Density** In a certain wildlife preserve, the number $y$ of prey consumed by an individual predator over a given period of time is given by

$$y = \frac{10x}{1 + 0.1x},$$

where $x$ is the *prey density* (the number of prey per unit of area). What prey density would allow a predator to survive if it needs to consume 50 prey over the given period?

**40. Store Hours** Suppose the ratio of the number of hours a video store is open to the number of daily customers is constant. When the store is open 8 hours, the number of customers is 92 less than the maximum number of customers. When the store is open 10 hours, the number of customers is 46 less than the maximum number of customers. Write an equation describing this situation, and find the maximum number of daily customers.

**41. Travel Time** The time it takes a boat to travel a given distance upstream (against the current) can be calculated by dividing the distance by the difference of the speed of the boat and the speed of the current. Write an equation that calculates the time $t$ it takes a boat moving at a speed $r$ against a current $c$ to travel a distance $d$. Solve your equation for $r$.

**42. Ramp Length** The difference between the length of a ramp and the length of the horizontal distance it covers is 5 feet. The square of the vertical distance the ramp covers is 45 square feet. Write an equation for the difference and solve. What is the length of the ramp?

**43. Sight Distance** The number of miles a person can see to the horizon from a point above the surface of the Earth is 0.85 of the square root of the person's distance in feet above the surface. Jack is 85 feet higher and sees 4.25 miles farther than Jill. How high are Jack and Jill above the surface?

**44. Automobile Skidding** Police have used the formula $s = \sqrt{30 fd}$ to estimate the speed $s$ (in miles per hour) of a car if it skidded $d$ feet when stopping. The literal number $f$ is the coefficient of friction, determined by the kind of road (such as concrete, asphalt, gravel, or tar) and whether the road is wet or dry. Some values of $f$ are given in Table 1.1. At 40 mi/h, about how many feet will a car skid on a dry concrete road? Give your answer to the nearest foot.

**TABLE 1.1**

|      | Concrete | Tar |
|------|----------|-----|
| Wet  | 0.4      | 0.5 |
| Dry  | 0.8      | 1.0 |

---

**To solve quadratic equations by factoring or by using the quadratic formula.**

# 1.3 QUADRATIC EQUATIONS

To learn how to solve more complicated problems, we turn to methods of solving *quadratic equations.*

**DEFINITION**

*A **quadratic equation** in the variable $x$ is an equation that can be written in the form*

$$ax^2 + bx + c = 0, \tag{1}$$

*where $a, b,$ and $c$ are constants and $a \neq 0$.*

A quadratic equation is also called a *second-degree equation* or an *equation of degree two,* since the highest power of the variable that occurs is the second. Whereas a linear equation has only one root, a quadratic equation may have two different roots.

**Solution by Factoring**

A useful method of solving quadratic equations is based on factoring, as the following example shows.

**Principles in Practice 1**

Solving a Quadratic Equation by Factoring

A number squared is 30 more than the number. What is the number?

**EXAMPLE 1** Solving a Quadratic Equation by Factoring

**a.** *Solve $x^2 + x - 12 = 0$.*

*Solution:* The left side factors easily:

$$(x - 3)(x + 4) = 0.$$

Think of this as two quantities, $x - 3$ and $x + 4$, whose product is zero. **Whenever the product of two or more quantities is *zero*, at least one of the quantities *must* be zero.** This means that either

$$x - 3 = 0 \quad \text{or} \quad x + 4 = 0.$$

Solving these gives $x = 3$ and $x = -4$. Thus, the roots of the original equation are 3 and $-4$, and the solution set is $\{3, -4\}$.

**b.** *Solve $6w^2 = 5w$.*

We do not divide both sides by $w$ (a variable) since equivalence is not guaranteed and we may 'lose' a root.

*Solution:* We write the equation as

$$6w^2 - 5w = 0,$$

so that one side is 0. Factoring gives

$$w(6w - 5) = 0.$$

Setting each factor equal to 0, we have

$$w = 0 \quad \text{or} \quad 6w - 5 = 0,$$
$$6w = 5.$$

Thus, the roots are $w = 0$ and $w = \frac{5}{6}$. Note that if we had divided both sides of $6w^2 = 5w$ by $w$ and obtained $6w = 5$, our only solution would be $w = \frac{5}{6}$. That is, we would lose the root $w = 0$. This confirms our discussion of Operation 5 in Sec. 1.1. ∎

---

**Principles in Practice 2**

Solving a Quadratic Equation by Factoring

The area of a rectangular mural that has a width 10 feet less than its length is 3000 square feet. What are the dimensions of the mural?

**EXAMPLE 2  Solving a Quadratic Equation by Factoring**

*Solve $(3x - 4)(x + 1) = -2$.*

*Pitfall* ▼ You should approach a problem like this with caution. If the product of two quantities is equal to $-2$, it is not true that at least one of the quantities must be $-2$. Why? You should **not** set each factor equal to $-2$; doing so will not provide solutions to the given equation.

*Solution:* We first multiply the factors on the left side:

$$3x^2 - x - 4 = -2.$$

Rewriting this equation so that 0 appears on one side, we have

$$3x^2 - x - 2 = 0,$$
$$(3x + 2)(x - 1) = 0,$$
$$x = -\frac{2}{3}, 1. \qquad ∎$$

Some equations that are not quadratic may be solved by factoring, as Example 3 shows.

**EXAMPLE 3  Solving a Higher Degree Equation by Factoring**

**a.** *Solve $4x - 4x^3 = 0$.*

*Solution:* This is called a *third-degree equation.* We proceed to solve it as follows:

$$4x - 4x^3 = 0,$$
$$4x(1 - x^2) = 0 \quad \text{(factoring)},$$
$$4x(1 - x)(1 + x) = 0 \quad \text{(factoring)}.$$

Do not neglect the fact that the factor $x$ gives rise to a root.

Setting each factor equal to 0 gives $4 = 0$ (impossible), $x = 0, 1 - x = 0$, or $1 + x = 0$. Thus,

$$x = 0, 1, -1,$$

which we can write as $x = 0, \pm 1$.

**b.** *Solve* $x(x + 2)^2(x + 5) + x(x + 2)^3 = 0$.

*Solution:* Factoring $x(x + 2)^2$ from both terms on the left side, we have

$$x(x + 2)^2[(x + 5) + (x + 2)] = 0,$$
$$x(x + 2)^2(2x + 7) = 0.$$

Hence, $x = 0, x + 2 = 0$, or $2x + 7 = 0$, from which it follows that $x = 0, -2, -\frac{7}{2}$.   ■

---

**Principles in Practice 3**

**Solving a Higher Degree Equation by Factoring**

The volume of a rectangular prism with a square base and height that is 5 times as long as its width is 5 times its width. What are the dimensions of the rectangular prism?

---

### EXAMPLE 4   A Fractional Equation Leading to a Quadratic Equation

*Solve*

$$\frac{y + 1}{y + 3} + \frac{y + 5}{y - 2} = \frac{7(2y + 1)}{y^2 + y - 6}. \tag{2}$$

*Solution:* Multiplying both sides by the L.C.D., $(y + 3)(y - 2)$, we get

$$(y - 2)(y + 1) + (y + 3)(y + 5) = 7(2y + 1). \tag{3}$$

Since Eq. (2) was multiplied by an expression involving the variable $y$, remember (from Sec. 1.1) that Eq. (3) is not necessarily equivalent to Eq. (2). After simplifying Eq. (3), we have

$$2y^2 - 7y + 6 = 0 \quad \text{(quadratic equation)},$$
$$(2y - 3)(y - 2) = 0 \quad \text{(factoring)}.$$

Thus, $\frac{3}{2}$ and 2 are *possible* roots of the given equation. But 2 cannot be a root of Eq. (2), since substitution leads to a denominator of 0. However, you should check that $\frac{3}{2}$ does indeed satisfy the *original* equation. Hence, its only root is $\frac{3}{2}$.   ■

Do not hastily conclude that the solution of $x^2 = 3$ consists of $x = \sqrt{3}$ only.

### EXAMPLE 5   Solution by Factoring

*Solve* $x^2 = 3$.

*Solution:*

$$x^2 = 3,$$
$$x^2 - 3 = 0.$$

---

**Principles in Practice 4**

**Solution by Factoring**

If you earned $225 for selling $x$ items for $x$ dollars each, how many items did you sell, and at what price did you sell them?

---

Factoring, we obtain

$$(x - \sqrt{3})(x + \sqrt{3}) = 0.$$

Thus $x - \sqrt{3} = 0$ or $x + \sqrt{3} = 0$, so $x = \pm\sqrt{3}$.

A more general form of the equation $x^2 - 3$ is $u^2 - k$. In the same manner as the preceding we can show that

If $u^2 = k$, then $u = \pm\sqrt{k}$. (4)

## Quadratic Formula

Solving quadratic equations by factoring can be quite difficult, as is evident by trying that method on $0.7x^2 - \sqrt{2}x - 8\sqrt{5} = 0$. However, there is a formula called the *quadratic formula*[4] that gives the roots of any quadratic equation.

> **Quadratic Formula**
>
> The roots of the quadratic equation $ax^2 + bx + c = 0$, where $a$, $b$, and $c$ are constants and $a \neq 0$, are given by
> $$x = \frac{-b \pm \sqrt{b^2 - 4ac}}{2a}.$$

*Pitfall* ▼ Be certain that you use the quadratic formula correctly.
$$x \neq -b \pm \frac{\sqrt{b^2 - 4ac}}{2a}.$$

**Principles in Practice 5**

**A Quadratic Equation with Two Real Roots**

Suppose the height $h$ in feet of fireworks fired straight upward from the ground is given by $h = 160t - 16t^2$, where $t$ is in seconds. When will the fireworks be 300 feet off the ground?

**EXAMPLE 6    A Quadratic Equation with Two Real Roots**

*Solve $4x^2 - 17x + 15 = 0$ by the quadratic formula.*

*Solution:* Here $a = 4$, $b = -17$, and $c = 15$. Thus,

$$x = \frac{-b \pm \sqrt{b^2 - 4ac}}{2a} = \frac{-(-17) \pm \sqrt{(-17)^2 - 4(4)(15)}}{2(4)}$$

$$= \frac{17 \pm \sqrt{49}}{8} = \frac{17 \pm 7}{8}.$$

The roots are $\dfrac{17 + 7}{8} = \dfrac{24}{8} = 3$ and $\dfrac{17 - 7}{8} = \dfrac{10}{8} = \dfrac{5}{4}$.

**Principles in Practice 6**

**A Quadratic Equation with One Real Root**

Suppose the weekly revenue $r$ of a company is given by the equation $r = -2p^2 + 400p$, where $p$ is the price of the company's product. What is the price of the product if the weekly revenue is $20,000?

**EXAMPLE 7    A Quadratic Equation with One Real Root**

*Solve $2 + 6\sqrt{2}y + 9y^2 = 0$ by the quadratic formula.*

*Solution:* Look at the arrangement of the terms. Here $a = 9$, $b = 6\sqrt{2}$, and $c = 2$. Hence,

$$y = \frac{-b \pm \sqrt{b^2 - 4ac}}{2a} = \frac{-6\sqrt{2} \pm \sqrt{0}}{2(9)}.$$

Thus,

$$y = \frac{-6\sqrt{2} + 0}{18} = -\frac{\sqrt{2}}{3} \quad \text{or} \quad y = \frac{-6\sqrt{2} - 0}{18} = -\frac{\sqrt{2}}{3}.$$

Therefore, the only root is $-\dfrac{\sqrt{2}}{3}$.

[4] A derivation of the quadratic formula appears in Sec. 1.4 as a supplement.

**EXAMPLE 8    A Quadratic Equation with No Real Solution**

*Solve $z^2 + z + 1 = 0$ by the quadratic formula.*

*Solution:* Here $a = 1$, $b = 1$, and $c = 1$. The roots are

$$\frac{-b \pm \sqrt{b^2 - 4ac}}{2a} = \frac{-1 \pm \sqrt{-3}}{2}.$$

Now, $\sqrt{-3}$ denotes a number whose square is $-3$. However, no such real number exists, since the square of any real number is nonnegative. Thus, the equation has no real roots.[5]    ■

From Examples 6–8, you can see that a quadratic equation has either two different real roots, exactly one real root, or no real roots, depending on whether $b^2 - 4ac > 0$, $= 0$, or $< 0$, respectively.

*This describes the nature of the roots of a quadratic equation.*

---

**TECHNOLOGY**

```
PROGRAM:QUADROOT
:Prompt A,B,C
:If B²-4AC<0
:Then
:Disp "NOREALROO
T"
:Stop
:End
:Disp (-B+√(B²-4
AC))/(2A)
:Disp (-B-√(B²-4
AC))/(2A)
```

**FIGURE 1.2**   Program to find real roots of $Ax^2 + Bx + C = 0$.

Using the program feature of a graphics calculator, one can create a program that gives the real roots of the quadratic equation $Ax^2 + Bx^2 + C = 0$. Figure

1.2 shows such a program for the TI-82 graphics calculator. To execute it for

$$20x^2 - 33x + 10 = 0,$$

you are prompted to enter the values of A, B, and C. (See Fig. 1.3.) The resulting roots are $x = 1.25$ and $x = 0.4$.

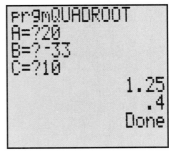

**FIGURE 1.3**   Roots of $20x^2 - 33x + 10 = 0$.

---

**Quadratic Forms**

Sometimes an equation that is not quadratic can be transformed into a quadratic equation by an appropriate substitution. In this case, the given equation is said to have **quadratic form.** The next example will illustrate.

**Principles in Practice 7**

**A Quadratic Equation With No Real Solution**

Suppose the height $h$ in feet of fireworks fired straight upward from the ground is given by $h = 160t - 16t^2$, where $t$ is in seconds. When will the fireworks be 500 feet off the ground?

**EXAMPLE 9    Solving a Quadratic-Form Equation**

*Solve $\dfrac{1}{x^6} + \dfrac{9}{x^3} + 8 = 0$.*

*Solution:* This equation can be written as

$$\left(\frac{1}{x^3}\right)^2 + 9\left(\frac{1}{x^3}\right) + 8 = 0,$$

---

[5] $\dfrac{-1 \pm \sqrt{-3}}{2}$ can be expressed as $\dfrac{-1 \pm i\sqrt{3}}{2}$, where $i = \sqrt{-1}$ is called the *imaginary unit*.

so it is quadratic in $1/x^3$ and hence has quadratic form. Substituting the variable $w$ for $1/x^3$ gives a quadratic equation in the variable $w$, which we can then solve:

$$w^2 + 9w + 8 = 0,$$
$$(w + 8)(w + 1) = 0,$$
$$w = -8 \quad \text{or} \quad w = -1.$$

Returning to the variable $x$, we have

$$\frac{1}{x^3} = -8 \quad \text{or} \quad \frac{1}{x^3} = -1.$$

Do not assume that $-8$ and $-1$ are solutions of the *original* equation.

Thus,

$$x^3 = -\frac{1}{8} \quad \text{or} \quad x^3 = -1,$$

from which it follows that

$$x = -\frac{1}{2}, \quad -1.$$

Checking, we find that these values of $x$ satisfy the original equation.  ∎

## ▪ Exercise 1.3

*In Problems 1–30, solve by factoring.*

**1.** $x^2 - 4x + 4 = 0.$

**2.** $t^2 + 3t + 2 = 0.$

**3.** $y^2 - 7y + 12 = 0.$

**4.** $x^2 + x - 12 = 0.$

**5.** $x^2 - 2x - 3 = 0.$

**6.** $x^2 - 16 = 0.$

**7.** $x^2 - 12x = -36.$

**8.** $3w^2 - 12w + 12 = 0.$

**9.** $x^2 - 4 = 0.$

**10.** $2x^2 + 4x = 0.$

**11.** $z^2 - 8z = 0.$

**12.** $x^2 + 9x = -14.$

**13.** $4x^2 + 1 = 4x.$

**14.** $2z^2 + 7z = 4.$

**15.** $y(2y + 3) = 5.$

**16.** $8 + 2x - 3x^2 = 0.$

**17.** $-x^2 + 3x + 10 = 0.$

**18.** $\frac{1}{7}y^2 = \frac{3}{7}y.$

**19.** $2p^2 = 3p.$

**20.** $-r^2 - r + 12 = 0.$

**21.** $x(x - 1)(x + 2) = 0.$

**22.** $(x - 2)^2(x + 1)^2 = 0.$

**23.** $x^3 - 64x = 0.$

**24.** $x^3 - 4x^2 - 5x = 0.$

**25.** $6x^3 + 5x^2 - 4x = 0.$

**26.** $(x + 1)^2 - 5x + 1 = 0.$

**27.** $(x + 3)(x^2 - x - 2) = 0.$

**28.** $3(x^2 + 2x - 8)(x - 5) = 0.$

**29.** $p(p - 3)^2 - 4(p - 3)^3 = 0.$

**30.** $x^4 - 3x^2 + 2 = 0.$

*In Problems 31–44, find all real roots by using the quadratic formula.*

**31.** $x^2 + 2x - 24 = 0.$

**32.** $x^2 - 2x - 15 = 0.$

**33.** $4x^2 - 12x + 9 = 0.$

**34.** $p^2 + 2p = 0.$

**35.** $p^2 - 5p + 3 = 0.$

**36.** $2 - 2x + x^2 = 0.$

**37.** $4 - 2n + n^2 = 0.$

**38.** $2x^2 + x = 5.$

**39.** $6x^2 + 7x - 5 = 0.$

**40.** $w^2 - 2\sqrt{2}w + 2 = 0.$

**41.** $2x^2 - 3x = 20.$

**42.** $0.01x^2 + 0.2x - 0.6 = 0.$

**43.** $2x^2 + 4x = 5.$

**44.** $-2x^2 - 6x + 5 = 0.$

*In Problems 45–54, solve the given quadratic-form equation.*

**45.** $x^4 - 5x^2 + 6 = 0.$

**46.** $x^4 - 3x^2 - 10 = 0.$

**47.** $\dfrac{2}{x^2} + \dfrac{3}{x} - 2 = 0.$

**48.** $x^{-2} + x^{-1} - 12 = 0.$

**49.** $x^{-4} - 9x^{-2} + 14 = 0.$

**50.** $\dfrac{1}{x^4} - \dfrac{9}{x^2} + 8 = 0.$

**51.** $(x - 3)^2 + 9(x - 3) + 14 = 0.$

**52.** $(x + 5)^2 - 8(x + 5) = 0.$

**53.** $\dfrac{1}{(x - 2)^2} - \dfrac{12}{x - 2} + 35 = 0.$

**54.** $\dfrac{2}{(x + 4)^2} + \dfrac{7}{x + 4} + 3 = 0.$

*In Problems 55–76, solve by any method.*

**55.** $x^2 = \dfrac{x + 3}{2}.$

**56.** $\dfrac{x}{3} = \dfrac{6}{x} - 1.$

**57.** $\dfrac{3}{x - 4} + \dfrac{x - 3}{x} = 2.$

**58.** $\dfrac{2}{x - 1} - \dfrac{6}{2x + 1} = 5.$

**59.** $\dfrac{6x + 7}{2x + 1} - \dfrac{6x + 1}{2x} = 1.$

**60.** $\dfrac{6(w + 1)}{2 - w} + \dfrac{w}{w - 1} = 3.$

**61.** $\dfrac{2}{r - 2} - \dfrac{r + 1}{r + 4} = 0.$

**62.** $\dfrac{2x - 3}{2x + 5} + \dfrac{2x}{3x + 1} = 1.$

**63.** $\dfrac{y + 1}{y + 3} + \dfrac{y + 5}{y - 2} = \dfrac{14y + 7}{y^2 + y - 6}.$

**64.** $\dfrac{3}{t + 1} + \dfrac{4}{t} = \dfrac{12}{t + 2}.$

**65.** $\dfrac{2}{x^2 - 1} - \dfrac{1}{x(x - 1)} = \dfrac{2}{x^2}.$

**66.** $5 - \dfrac{3(x + 3)}{x^2 + 3x} = \dfrac{1 - x}{x}.$

**67.** $\sqrt{x + 2} = x - 4.$

**68.** $3\sqrt{x + 4} = x - 6.$

**69.** $q + 2 = 2\sqrt{4q - 7}.$

**70.** $x + \sqrt{x - 2} = 0.$

**71.** $\sqrt{x + 7} - \sqrt{2x - 1} = 0.$

**72.** $\sqrt{x} - \sqrt{2x - 8} - 2 = 0.$

**73.** $\sqrt{x} - \sqrt{2x + 1} + 1 = 0.$

**74.** $\sqrt{y - 2} + 2 = \sqrt{2y + 3}.$

**75.** $\sqrt{x + 5} + 1 = 2\sqrt{x}.$

**76.** $\sqrt{\sqrt{x} + 2} = \sqrt{2x - 4}.$

*In Problems 77 and 78, find the roots, rounded to two decimal places.*

**77.** $0.02x^2 - 4.7x + 8.6 = 0.$

**78.** $0.01x^2 + 0.2x - 0.6 = 0.$

**79. Geometry** The area of a rectangular picture with a width 2 inches less than its length is 48 square inches. What are the dimensions of the picture?

**80. Temperature** The temperature has been rising $X$ degrees per day for $X$ days. $X$ days ago it was 15 degrees. Today it is 51 degrees. How much has the temperature been rising each day? How many days has it been rising?

**81. Economics** One root of the economics equation

$$\overline{M} = \dfrac{Q(Q + 10)}{44}$$

is $-5 + \sqrt{25 + 44\overline{M}}$. Verify this by using the quadratic formula to solve for $Q$ in terms of $\overline{M}$. Here $Q$ is real income and $\overline{M}$ is the level of money supply.

**82. Diet for Rats** A group of biologists studied the nutritional effects on rats that were fed a diet containing 10%

protein.[6] The protein was made up of yeast and corn flour. By changing the percentage $P$ (expressed as a decimal) of yeast in the protein mix, the group estimated that the average weight gain $g$ (in grams) of a rat over a period of time was given by

$$g = -200P^2 + 200P + 20.$$

What percentage of yeast gave an average weight gain of 70 grams?

**83. Drug Dosage** There are several rules for determining doses of medicine for children when the adult dose has been specified. Such rules may be based on weight,

[6] Adapted from R. Bressani, "The Use of Yeast in Human Foods," in R. I. Mateles and S. R. Tannenbaum (eds.), *Single-Cell Protein* (Cambridge, MA: MIT Press, 1968).

height, and so on. If $A$ is the age of the child, $d$ is the adult dose, and $c$ is the child's dose, then here are two rules:

$$\text{Young's rule:} \quad c = \frac{A}{A + 12}d.$$

$$\text{Cowling's rule:} \quad c = \frac{A + 1}{24}d.$$

At what age are the children's doses the same under both rules? Round your answer to the nearest year.

84. **Delivered Price of a Good**   In a discussion of the delivered price of a good from a mill to a customer, DeCanio[7] arrives at and solves the two quadratic equations

$$(2n - 1)v^2 - 2nv + 1 = 0,$$

and

$$nv^2 - (2n + 1)v + 1 = 0,$$

where $n \geq 1$.

**a.** Solve the first equation for $v$.

**b.** Solve the second equation for $v$ if $v < 1$.

85. **Optics**   An object is 120 cm from a wall. In order to focus the image of the object on the wall, a converging lens with a focal length of 24 cm is used. The lens is placed between the object and the wall at a distance of $p$ centimeters from the object, where

$$\frac{1}{p} + \frac{1}{120 - p} = \frac{1}{24}.$$

Find $p$, rounded to one decimal place.

86. **Physics**   A platinum resistance thermometer of certain specifications operates according to the equation

$$R = 10{,}000 + (4.124 \times 10^{-2})T - (1.779 \times 10^{-5})T^2,$$

where $R$ is the resistance of the thermometer (in ohms) at temperature $T$ (in degrees Celsius). If $R = 13.946$, find the corresponding value of $T$. Round your answer to the nearest degree. Assume that such a thermometer is usable only if $T < 600°C$.

87. **Motion**   Suppose the height $h$ of an object thrown straight upward from the ground is given by

$$h = 44.1t - 4.9t^2,$$

where $h$ is in meters and $t$ is the elapsed time in seconds.

**a.** After how many seconds does the object strike the ground?

**b.** When is the object at a height of 88.2 m?

■ *In Problems 88–93, use a program to determine any real roots of the equation. Round answers to three decimal places. For Problems 88 and 89, confirm your result algebraically.*

 **88.** $2x^2 - 3x - 27 = 0$.

**89.** $8x^2 - 18x + 9 = 0$.

**90.** $14x^2 + 7x - 3 = 0$.

**91.** $23x^2 - \frac{11}{3}x + 5 = 0$.

 **92.** $\frac{9}{2}z^2 - 6.3 = \frac{z}{3}(1.1 - 7z)$.

**93.** $(\pi t - 4)^2 = 4.1t - 3$.

[7]S. J. DeCanio, "Delivered Pricing and Multiple Basing Point Equilibria: A Revolution," *Quarterly Journal of Economics*, XCIX, no. 2 (1984), 329–49.

## 1.4 SUPPLEMENT

Following is a derivation of the quadratic formula. Suppose $ax^2 + bx + c = 0$ is a quadratic equation. Since $a \neq 0$, we can divide both sides by $a$:

$$x^2 + \frac{b}{a}x + \frac{c}{a} = 0,$$

$$x^2 + \frac{b}{a}x = -\frac{c}{a}.$$

If we add $\left(\dfrac{b}{2a}\right)^2$ to both sides, then the left side factors as the square of a binomial:

$$x^2 + \frac{b}{a}x + \left(\frac{b}{2a}\right)^2 = \left(\frac{b}{2a}\right)^2 - \frac{c}{a},$$

$$\left(x + \frac{b}{2a}\right)^2 = \frac{b^2 - 4ac}{4a^2}.$$

This equation has the form $u^2 = k$, so, from Eq. (4) of Sec. 1.3,

$$x + \frac{b}{2a} = \pm\sqrt{\frac{b^2 - 4ac}{4a^2}} = \pm\frac{\sqrt{b^2 - 4ac}}{2a}.$$

Solving for $x$ gives

$$x = -\frac{b}{2a} \pm \frac{\sqrt{b^2 - 4ac}}{2a} = \frac{-b \pm \sqrt{b^2 - 4ac}}{2a}.$$

In summary, the roots of the quadratic equation $ax^2 + bx + c = 0$ are given by the **quadratic formula:**

$$x = \frac{-b \pm \sqrt{b^2 - 4ac}}{2a}.$$

## 1.5 REVIEW

### IMPORTANT TERMS AND SYMBOLS

**Section 1.1**   equation   side (member) of equation   variable   root of equation   solution set
equivalent equations   linear (first-degree) equation   literal equation   arbitrary constant

**Section 1.2**   fractional equation   empty set, $\varnothing$   extraneous solution   radical equation

**Section 1.3**   quadratic (second-degree) equation   quadratic formula

### SUMMARY

When solving an equation, we may apply rules to it that give equivalent equations—that is, equations with exactly the same solutions as the given equation. These rules include adding (or subtracting) the same polynomial to (from) both sides, as well as multiplying (or dividing) both sides by the same constant, except 0.

A linear equation (in $x$) is of the first degree and has the form $ax + b = 0$, where $a \neq 0$. Every linear equation has exactly one root. To solve a linear equation, we apply operations on it until we obtain an equivalent equation in which the unknown is isolated on one side.

A quadratic equation (in $x$) is of the second degree and has the form $ax^2 + bx + c = 0$, where $a \neq 0$. Every quadratic equation has either two different real roots, exactly one real root, or no real roots. A

quadratic equation may be solved either by factoring or by the quadratic formula:

$$x = \frac{-b \pm \sqrt{b^2 - 4ac}}{2a}.$$

When solving a fractional equation or a radical equation, one often applies operations to it that do not guarantee that the resulting equation is equivalent to the given equation. These operations include multiplying both sides by an expression containing the variable and raising both sides to the same power. In these cases, all solutions obtained at the end of such procedures must be checked by substituting them into the given equation. In this way, extraneous solutions may be found.

### REVIEW PROBLEMS

*In Problems 1–44, solve the equations.*

**1.** $4 - 3x = 2 + 5x$.

**2.** $\frac{5}{7}x - \frac{2}{3}x = \frac{3}{21}x$.

**3.** $3[2 - 4(1 + x)] = 5 - 3(3 - x)$.

**4.** $3(x + 4)^2 + 6x = 3x^2 + 7$.

**5.** $2 - w = 3 + w$.

**6.** $x = 2x$.

**7.** $x = 2x - (7 + x)$.

**8.** $3x - 8 = 4(x - 2)$.

**9.** $2(4 - \frac{3}{5}p) = 5$.

**10.** $\frac{5}{7}x - \frac{2}{3}x = \frac{3}{21}$.

**11.** $\dfrac{3x - 1}{x + 4} = 0$.

**12.** $\dfrac{5}{p + 3} - \dfrac{2}{p + 3} = 0$.

**13.** $\dfrac{2x}{x - 3} - \dfrac{x + 1}{x + 2} = 1$.

**14.** $\dfrac{t + 3t + 4}{7 - t} = 14$.

**15.** $3x^2 + 2x - 5 = 0$.

**16.** $x^2 - 2x - 2 = 0$.

**17.** $5q^2 = 7q$.

**18.** $2x^2 - x = 0$.

**19.** $x^2 - 10x + 25 = 0$.

**20.** $r^2 + 10r - 25 = 0$.

**21.** $3x^2 - 5 = 0$.

**22.** $x(x - 9) = 0$.

**23.** $(8t - 5)(2t + 6) = 0$.

**24.** $2(x^2 - 1) + 2x = x^2 - 6x + 1$.

**25.** $-3x^2 + 5x - 1 = 0$.

**26.** $y^2 = 6$.

**27.** $x(x^2 - 9) = 4(x^2 - 9)$.

**28.** $4x^2(x - 5) - 9(x - 5) = 0$.

**29.** $\dfrac{6w + 7}{2w + 1} - \dfrac{6w + 1}{2w} = 1$.

**30.** $\dfrac{3}{x + 1} + \dfrac{4}{x} - \dfrac{12}{x + 2} = 0$.

**31.** $\dfrac{2}{x^2 - 9} - \dfrac{3x}{x + 3} = \dfrac{1}{x - 3}$.

**32.** $\dfrac{3}{x^2 - 4} + \dfrac{2}{x^2 + 4x + 4} - \dfrac{4}{x + 2} = 0$.

**33.** $\sqrt{2x + 5} = 5$.

**34.** $\sqrt{3x - 4} = \sqrt{2x + 5}$.

**35.** $\sqrt[3]{11x + 9} = 4$.

**36.** $\sqrt{x^2 + 5x + 25} = x + 4$.

**37.** $\sqrt{y} + 6 = 5$.

**38.** $\sqrt{z^2 + 9} = 5$.

**39.** $\sqrt{x - 1} + \sqrt{x + 6} = 7$.

**40.** $\sqrt{2x + 1} = x - 7$.

**41.** $x + 2 = 2\sqrt{4x - 7}$.

**42.** $\sqrt{3z} - \sqrt{5z + 1} + 1 = 0$.

**43.** $y^{2/3} + y^{1/3} - 2 = 0$.

**44.** $2y^{-2/3} - 5y^{-1/3} - 3 = 0$.

*In Problems 45–52, solve the given equation for the indicated letter.*

**45.** $E = 4\pi k\dfrac{Q}{A}$;   $Q$.

**46.** $E_1 = i_2 R_1 + i_3 R_1 + i_2 R_2$;   $i_3$.

**47.** $n - 1 = C + \dfrac{C'}{\lambda^2}$;   $C'$.

**48.** $\sigma = \dfrac{n_0 - n_e}{\lambda} L$;   $n_0$.

**49.** $T^2 = 4\pi^2 \left(\dfrac{L}{g}\right)$;   $T$.

**50.** $s = \dfrac{1}{2} at^2$;   $t$.

**51.** $mgh = \dfrac{1}{2}mv^2 + \dfrac{1}{2}I\omega^2$;   $\omega$.

**52.** $P = \dfrac{E^2}{R + r} - \dfrac{E^2 r}{(R + r)^2}$;   $E$.

**53. Electricity**   In studies of electrical networks, the following equation occurs:

$$S^2 + \frac{R}{L}S + \frac{1}{LC} = 0.$$

Show that

$$S = -\frac{R}{2L} \pm \sqrt{\left(\frac{R}{2L}\right)^2 - \frac{1}{LC}}.$$

**54. Electricity**   In an electrical circuit, resonance occurs when

$$2\pi f_r L = \frac{1}{2\pi f_r C},$$

where $f_r$ is a resonant frequency, $L$ is inductance, and $C$ is capacitance. Solve for $f_r$ if $f_r > 0$.

*In Problems 55 and 56, use a graphics calculator to determine which of the given numbers, if any, are roots of the given equation.*

**55.** $12x^3 + 61x = 83x^2 - 30$;   $4, 6, \frac{5}{4}$.

**56.** $\sqrt{t^2 + 4} = t + 1$;   $\frac{2}{3}, \frac{14}{3}, \frac{3}{2}$.

*In Problems 57 and 58, use a program to determine any real roots of the equation. Round answers to three decimal places.*

**57.** $4.6 - 7.2x - 19.3x^2 = 0$.

**58.** $(9x - 3)^2 - \dfrac{7}{6}(x - 2) = 15$.

# MATHEMATICAL *SNAPSHOT*

## REAL GROWTH OF AN INVESTMENT[8]

When we speak of the real growth of an investment, we are referring to the growth in its purchasing power, that is, the growth of the quantity of goods that the investment can buy. The real growth depends on the influence of both interest and inflation. Interest raises the value of the investment, whereas inflation slows down the growth of the investment by increasing prices and, hence, decreasing its purchasing power. The rate of real growth does not usually equal the difference between the rate of interest and the rate of inflation but is described by a different formula called "Fisher's effect."

You can understand Fisher's effect by carefully examining the question posed by Edward P. Foldessy in the *Wall Street Journal* (27 May 1986, 37):

> For the year ending with May 1986, the annual rate of interest was 11% and the annual rate of inflation was 3.6%. Under these conditions, what was the annual rate of real growth of an investment?

You might think that the answer is simply obtained by subtracting percentages: $11\% - 3.6\% = 7.4\%$. However, 7.4% is not the answer.

Suppose you analyze the situation in more specific terms. Consider strawberries selling at $1.00 per pound and assume, because of inflation, that this price increases at the rate of 3.6% a year. From June 1985 to June 1986, the price per pound of $1.00 rose to

$$\$1.00 + (3.6\% \text{ of } \$1.00) = \$1.036.$$

On the other hand, consider $100 invested in June 1985 at an 11% annual interest rate. In June 1986 the interest earned is $100(0.11), so the accumulated amount is

$$\$100 + \$100(0.11) = \$111.$$

Now, compare the purchasing power of $100 in June 1985 to that of $111 in June 1986. In 1985 the $100 bought 100 lb of strawberries at $1.00 per pound. In 1986 strawberries were $1.036 per pound, so the accumulated amount of $111 bought $111/1.036 \approx 107.14$ lb of strawberries (the symbol $\approx$ means *approximately equals*).

What change occurred in the purchasing power of the investment? It increased from 100 lb to 107.14 lb, an increase of 7.14%. That is,

$$\frac{\text{new quantity} - \text{initial quantity}}{\text{initial quantity}} = \frac{107.14 - 100}{100}$$

$$= 0.0714 = 7.14\%.$$

Thus 7.14% is the real rate of growth, which is less than the difference $11\% - 3.6\% = 7.4\%$. Actually, this difference has no meaning, because the three percentages refer to three different quantities: (a) interest (a fraction of the investment—11% of $100), (b) inflation (a fraction of the unit price of goods—3.6% of $1.00), and (c) the rate of real growth (a percentage of the purchasing power—7.14% of the initial quantity of strawberries).

To derive a formula for the rate of real growth, $g$, let $y$ be the annual rate of interest (the yield) and let $i$

........................................
[8] Adapted from Yves Nievergelt, "Fisher's Effect: Real Growth Is Not Interest Less Inflation," *Mathematics Teacher,* 81 (October 1988), 546–47. By permission of the National Council of Teachers of Mathematics.

be the annual rate of inflation. In one year an investment of $P$ dollars (the principal) earns interest of $y \cdot P$ dollars, so it produces an accumulated amount (in dollars) of

$$P + y \cdot P = P(1 + y) \qquad \text{(factoring)}.$$

In one year the price of goods, say $p$ dollars per unit, rises by $i \cdot p$ dollars to a new price of

$$p + i \cdot p = p(1 + i)$$

dollars per unit. The initial purchasing power represents the initial quantity of goods:

$$\text{initial quantity} = \frac{\text{principal}}{\text{initial price}} = \frac{P}{p}.$$

One year later, the new quantity of goods that the accumulated amount of the investment will buy at the new price is given by

$$\text{new quantity} = \frac{\text{new balance}}{\text{new price}} = \frac{P(1 + y)}{p(1 + i)}.$$

Consequently, the rate of growth, or relative change, of the purchasing power is given by

$$g = \frac{\text{new quantity} - \text{initial quantity}}{\text{initial quantity}}$$

$$= \frac{\dfrac{P(1 + y)}{p(1 + i)} - \dfrac{P}{p}}{\dfrac{P}{p}}.$$

Multiplying the numerator and denominator by $p/P$ gives

$$g = \frac{1 + y}{1 + i} - 1$$

$$= \frac{(1 + y) - (1 + i)}{1 + i} = \frac{y - i}{1 + i}.$$

Thus the rate of real growth is given by the literal equation

$$g = \frac{y - i}{1 + i}. \tag{1}$$

The relationship in Eq. (1) is Fisher's effect.[9] To illustrate its use, apply it to the previous example, where $y = 11\%$ and $i = 3.6\%$. Fisher's formula gives

$$g = \frac{0.11 - 0.036}{1 + 0.036} \approx 0.0714 = 7.14\%.$$

### ■ Exercises

1. According to Foldessy (1986), during 1980 the rate of interest averaged 11.4% at a time when inflation advanced by 13.5%.

   a. Compute the accumulated amount of a $100 investment after one year at 11.4%.

   b. If a pound of apples cost $1 in January 1980, how much did it cost a year later?

   c. If a pound of apples cost $1 in January 1980, what quantity of apples did $100 buy in 1980?

   d. What quantity of apples did the accumulated amount [see part (a)] buy a year later?

   e. Use the results of parts (c) and (d) to calculate the rate of real growth by means of the equation

   $$g = \frac{\text{new quantity} - \text{initial quantity}}{\text{initial quantity}}.$$

   f. Verify your answer to part (e) by means of Fisher's formula.

2. Find the rate of real growth, given an interest rate of 10% and an inflation rate of 5%.

---

[9]Irving Fisher, "Appreciation and Interest," *Publications of the American Economic Association,* Third Series, 11 (August 1986), 331–442.

# Applications of Equations and Inequalities

CHAPTER

2

## OBJECTIVE

**To model situations described by linear or quadratic equations.**

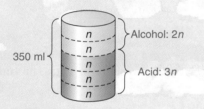

**FIGURE 2.1** Chemical solution (Example 1).

Note that the solution to an equation is not necessarily the solution to the problem posed.

## 2.1 APPLICATIONS OF EQUATIONS

In most cases, to solve practical problems you must translate the relationships stated in the problems into mathematical symbols. This is called *modeling*. The following examples illustrate basic techniques and concepts. Examine each of them carefully before going to the exercises.

**EXAMPLE 1  Mixture**

*A chemist must prepare* 350 ml *of a chemical solution made up of two parts alcohol and three parts acid. How much of each should be used?*

***Solution:*** Let $n$ be the number of milliliters in each part. Figure 2.1 shows the situation. From the diagram, we have

$$2n + 3n = 350,$$

$$5n = 350,$$

$$n = \frac{350}{5} = 70.$$

But $n = 70$ is *not* the answer to the original problem. Each *part* has 70 ml. The amount of alcohol is $2n = 2(70) = 140$, and the amount of acid is $3n = 3(70) = 210$. Thus, the chemist should use 140 ml of alcohol and 210 ml of acid. This example shows how helpful a diagram can be in setting up a word problem. ■

**EXAMPLE 2  Observation Deck**

*A rectangular observation deck overlooking a scenic valley is to be built. [See Fig. 2.2(a).] The deck is to have dimensions* 6 m *by* 12 m. *A rectangular shelter of area* 40 m$^2$ *is to be centered over the deck. The uncovered part of the deck is to be a walkway of uniform width. How wide should this walkway be?*

***Solution:*** A diagram of the deck is shown in Fig. 2.2(b). Let $w$ be the width (in meters) of the walkway. Then the part of the deck for the shelter has dimensions $12 - 2w$ by $6 - 2w$. Since its area must be 40 m$^2$, where area = (length)(width), we have

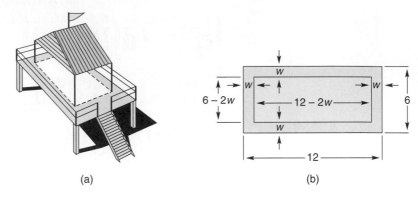

(a)

(b)

**FIGURE 2.2** Deck walkway (Example 2).

$$(12 - 2w)(6 - 2w) = 40,$$
$$72 - 36w + 4w^2 = 40 \qquad \text{(multiplying)},$$
$$4w^2 - 36w + 32 = 0,$$
$$w^2 - 9w + 8 = 0 \qquad \text{(dividing both sides by 4)},$$
$$(w - 8)(w - 1) = 0,$$
$$w = 8, 1.$$

Although 8 is a solution to the equation, it is *not* a solution to our problem, because one of the dimensions of the deck itself is only 6 m. Thus, the only possible solution is that the walkway be 1 m wide. ■

The key words introduced here are *fixed cost, variable cost, total cost, total revenue,* and *profit.* This is the time for you to gain familiarity with these terms because they recur throughout the book.

In the next example, we shall refer to some business terms relative to a manufacturing firm. **Fixed cost** (or *overhead*) is the sum of all costs that are independent of the level of production, such as rent, insurance, and so on. This cost must be paid whether or not output is produced. **Variable cost** is the sum of all costs that are dependent on the level of output, such as labor and material. **Total cost** is the sum of variable cost and fixed cost:

$$\text{total cost} = \text{variable cost} + \text{fixed cost}.$$

**Total revenue** is the money that the manufacturer receives for selling the output:

$$\text{total revenue} = (\text{price per unit})(\text{number of units sold}).$$

**Profit** is total revenue minus total cost:

$$\text{profit} = \text{total revenue} - \text{total cost}.$$

### EXAMPLE 3  Profit

*The Anderson Company produces a product for which the variable cost per unit is $6 and fixed cost is $80,000. Each unit has a selling price of $10. Determine the number of units that must be sold for the company to earn a profit of $60,000.*

*Solution:* Let $q$ be the number of units that must be sold. (In many business problems, $q$ represents quantity.) Then the variable cost (in dollars) is $6q$. The *total* cost for the business is therefore $6q + 80,000$. The total revenue from the sale of $q$ units is $10q$. Since

$$\text{profit} = \text{total revenue} - \text{total cost},$$

our model for this problem is

$$60,000 = 10q - (6q + 80,000).$$

Solving gives

$$60,000 = 10q - 6q - 80,000,$$
$$140,000 = 4q,$$
$$35,000 = q.$$

Thus, 35,000 units must be sold to earn a profit of $60,000. ∎

### EXAMPLE 4  Pricing

*Sportcraft manufactures women's sportswear and is planning to sell its new line of slacks to retail outlets. The cost to the retailer will be $33 per pair of slacks. As a convenience to the retailer, Sportcraft will attach a price tag to each pair. What amount should be marked on the price tag so that the retailer may reduce this price by 20% during a sale and still make a profit of 15% on the cost?*

*Solution:* Here we use the fact that

Note that *price = cost + profit*.

$$\text{selling price} = \text{cost per pair} + \text{profit per pair}.$$

Let $p$ be the tag price per pair, in dollars. During the sale, the retailer actually receives $p - 0.2p$. This must equal the cost, 33, plus the profit, $(0.15)(33)$. Hence,

$$\text{selling price} = \text{cost} + \text{profit},$$
$$p - 0.2p = 33 + (0.15)(33),$$
$$0.8p = 37.95,$$
$$p = 47.4375.$$

From a practical point of view, Sportcraft should mark the price tag at $47.44. ∎

### EXAMPLE 5  Investment

*A total of $10,000 was invested in two business ventures, A and B. At the end of the first year, A and B yielded returns of 6% and $5\frac{3}{4}\%$, respectively, on the original investments. How was the original amount allocated if the total amount earned was $588.75?*

*Solution:* Let $x$ be the amount (in dollars) invested at 6%. Then $10,000 - x$ was invested at $5\frac{3}{4}\%$. The interest earned in $A$ was $(0.06)(x)$, and in $B$ it was $(0.0575)(10,000 - x)$, which total 588.75. Hence,

$$(0.06)x + (0.0575)(10,000 - x) = 588.75,$$
$$0.06x + 575 - 0.0575x = 588.75,$$
$$0.0025x = 13.75,$$
$$x = 5500.$$

Thus, $5500 was invested at 6%, and $10,000 − $5500 = $4500 was invested at $5\frac{3}{4}\%$.   ■

### EXAMPLE 6   Bond Redemption

*The board of directors of Maven Corporation agrees to redeem some of its bonds in two years. At that time, $1,102,500 will be required. Suppose the firm presently sets aside $1,000,000. At what annual rate of interest, compounded annually, will this money have to be invested in order that its future value be sufficient to redeem the bonds?*

*Solution:* Let $r$ be the required annual rate of interest. At the end of the first year, the accumulated amount will be $1,000,000 plus the interest, 1,000,000$r$, for a total of

$$1{,}000{,}000 + 1{,}000{,}000r = 1{,}000{,}000(1 + r).$$

Under compound interest, at the end of the second year the accumulated amount will be $1,000,000(1 + r)$ plus the interest on this, which is $1,000,000(1 + r)r$. Thus, the total value at the end of the second year will be

$$1{,}000{,}000(1 + r) + 1{,}000{,}000(1 + r)r.$$

This must equal $1,102,500:

$$1{,}000{,}000(1 + r) + 1{,}000{,}000(1 + r)r = 1{,}102{,}500. \qquad (1)$$

Since $1,000,000(1 + r)$ is a common factor of both terms on the left side, we have

$$1{,}000{,}000(1 + r)(1 + r) = 1{,}102{,}500,$$

$$1{,}000{,}000(1 + r)^2 = 1{,}102{,}500,$$

$$(1 + r)^2 = \frac{1{,}102{,}500}{1{,}000{,}000} = \frac{11{,}025}{10{,}000} = \frac{441}{400},$$

$$1 + r = \pm\sqrt{\frac{441}{400}} = \pm\frac{21}{20},$$

$$r = -1 \pm \frac{21}{20}.$$

Thus, $r = -1 + (21/20) = 0.05$, or $r = -1 - (21/20) = -2.05$. Although 0.05 and −2.05 are roots of Eq. (1), we reject −2.05, since we want $r$ to be positive. Hence, $r = 0.05$, so the desired rate is 5%.   ■

At times there may be more than one way to model a word problem, as Example 7 shows.

### EXAMPLE 7   Apartment Rent

*A real-estate firm owns the Parklane Garden Apartments, which consist of 90 apartments. At $350 per month, every apartment can be rented. However, for each $10 per month increase, there will be two vacancies with no possibility of filling them. The firm wants to receive $31,980 per month from rents. What rent should be charged for each apartment?*

*Solution:*
**Method I.**  Suppose $r$ is the rent (in dollars) to be charged per apartment. Then the increase over the $350 level is $r − 350$. Thus, the number of $10

increases is $\dfrac{r-350}{10}$. Because each \$10 increase results in two vacancies, the total number of vacancies will be $2\left(\dfrac{r-350}{10}\right)$. Hence, the total number of apartments rented will be $90 - 2\left(\dfrac{r-350}{10}\right)$. Since

$$\text{total rent} = (\text{rent per apartment})(\text{number of apartments rented}),$$

we have

$$31{,}980 = r\left[90 - \frac{2(r-350)}{10}\right],$$

$$31{,}980 = r\left[90 - \frac{r-350}{5}\right],$$

$$31{,}980 = r\left[\frac{450 - r + 350}{5}\right],$$

$$159{,}900 = r[800 - r].$$

Thus,

$$r^2 - 800r + 159{,}900 = 0.$$

By the quadratic formula,

$$r = \frac{800 \pm \sqrt{(-800)^2 - 4(1)(159{,}900)}}{2(1)}$$

$$= \frac{800 \pm \sqrt{400}}{2} = \frac{800 \pm 20}{2} = 400 \pm 10.$$

Hence, the rent for each apartment should be either \$410 or \$390.

**Method II.** Suppose $n$ is the number of \$10 increases. Then the increase in rent per apartment will be $10n$ and there will be $2n$ vacancies. Since

$$\text{total rent} = (\text{rent per apartment})(\text{number of apartments rented}),$$

we have

$$31{,}980 = (350 + 10n)(90 - 2n),$$

$$31{,}980 = 31{,}500 + 200n - 20n^2,$$

$$20n^2 - 200n + 480 = 0,$$

$$n^2 - 10n + 24 = 0.$$

$$(n - 6)(n - 4) = 0.$$

Thus, $n = 6$ or $n = 4$. The rent charged should be either $350 + 10(6) = \$410$ or $350 + 10(4) = \$390$. ◼

## ▪ Exercise 2.1

**1. Fencing**   A fence is to be placed around a rectangular plot so that the enclosed area is 800 ft$^3$ and the length of the plot is twice the width. How many feet of fencing must be used?

**2. Geometry**   The perimeter of a rectangle is 200 ft, and the length of the rectangle is three times the width. Find the dimensions of the rectangle.

3. **Gypsy Moth**    One of the most important defoliating insects is the gypsy moth caterpillar, which feeds on foliage of shade, forest, and fruit trees. A homeowner lives in an area in which the gypsy moth has become a problem. She wishes to spray the trees on her property before more defoliation occurs. She needs 128 oz of a solution made up of 3 parts of insecticide $A$ and 5 parts of insecticide $B$. The solution is then mixed with water. How many ounces of each insecticide should be used?

4. **Concrete Mix**    A builder makes a certain type of concrete by mixing together 1 part cement, 3 parts sand, and 5 parts stone (by volume). If 585 ft$^3$ of concrete are needed, how many cubic feet of each ingredient does he need?

5. **Furniture Finish**    According to *The Consumer's Handbook* [Paul Fargis, ed. (New York: Hawthorn, 1974)], a good oiled furniture finish contains two parts boiled linseed oil and one part turpentine. If you need a pint (16 fluid oz) of this furniture finish, how many fluid ounces of turpentine are needed?

6. **Forest Management**    A lumber company owns a forest that is of rectangular shape, 1 mi by 2 mi. If the company cuts a uniform strip of trees along the outer edges of this forest, how wide should the strip be if $\frac{3}{4}$ sq mi of forest is to remain?

7. **Garden Pavement**    A rectangular plot, 4 m by 8 m, is to be used for a garden. It is decided to put a pavement inside the entire border so that 12 m$^2$ of the plot is left for flowers. How wide should the pavement be?

8. **Ventilating Duct**    The diameter of a circular ventilating duct is 140 mm. This duct is joined to a square duct system as shown in Fig. 2.3. To ensure smooth airflow, the areas of the circle and square sections must be equal. To the nearest millimeter, what should the length $x$ of a side of the square section be?

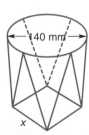

**FIGURE 2.3**  Ventilating duct (Problem 8).

9. **Profit**    The Geometric Products Company produces a product at a variable cost per unit of $2.20. If fixed costs are $95,000 and each unit sells for $3, how many units

must be sold for the company to have a profit of $50,000?

10. **Sales**    The Clark Company management would like to know the total sales units that are required for the company to earn a profit of $100,000. The following data are available: unit selling price of $20; variable cost per unit of $15; total fixed cost of $600,000. From these data, determine the required sales units.

11. **Investment**    A person wishes to invest $20,000 in two enterprises so that the total income per year will be $1440. One enterprise pays 6% annually; the other has more risk and pays $7\frac{1}{2}$% annually. How much must be invested in each?

12. **Investment**    A person invested $20,000, part at an interest rate of 6% annually and the remainder at 7% annually. The total interest at the end of 1 year was equivalent to an annual $6\frac{3}{4}$% rate on the entire $20,000. How much was invested at each rate?

13. **Pricing**    The cost of a product to a retailer is $3.40. If the retailer wishes to make a profit of 20% on the selling price, at what price should the product be sold?

14. **Bond Retirement**    In two years, a company will require $1,123,600 in order to retire some bonds. If the company now invests $1,000,000 for this purpose, what annual rate of interest, compounded annually, must it receive on that amount in order to retire the bonds?

15. **Expansion Program**    In two years, a company will begin an expansion program. It has decided to invest $2,000,000 now so that in two years the total value of the investment will be $2,163,200, the amount required for the expansion. What is the annual rate of interest, compounded annually, that the company must receive to achieve its purpose?

16. **Business**    A company finds that if it produces and sells $q$ units of a product, its total sales revenue in dollars is $100\sqrt{q}$. If the variable cost per unit is $2 and the fixed cost is $1200, find the values of $q$ for which

$$\text{total sales revenue} = \text{variable cost} + \text{fixed cost}.$$

(That is, profit is zero.)

17. **Dormitory Housing**    A college dormitory houses 210 students. This fall, rooms are available for 76 freshmen. On the average, 95% of those freshmen who request room applications actually reserve a room. How many room applications should the college send out if it wants to receive 76 reservations?

18. **Poll**    A group of people were polled, and 20%, or 700, of them favored a new product over the best-selling brand. How many people were polled?

**19. Prison Guard Salary**   It was reported that in a certain women's jail, female prison guards, called matrons, received 30% (or $200) a month less than their male counterparts, deputy sheriffs. Find the yearly salary of a deputy sheriff. Give your answer to the nearest dollar.

**20. Striking Drivers**   A few years ago, cement drivers were on strike for 46 days. Before the strike, these drivers earned $7.50 per hour and worked 260 eight-hour days a year. What percentage increase is needed in yearly income to make up for the lost time within 1 year?

**21. Break Even**   A manufacturer of video-game cartridges sells each cartridge for $19.95. The manufacturing cost of each cartridge is $14.95. Monthly fixed costs are $8000. During the first month of sales of a new game, how many cartridges must be sold in order for the manufacturer to break even (that is, in order that total revenue equal total cost)?

**22. Investment Club**   An investment club bought a bond of an oil corporation for $5000. The bond yields 8% per year. The club now wants to buy shares of stock in a hospital supply company. The stock sells at $20 per share and earns a dividend of $0.50 per share per year. How many shares should the club buy so that its total investment in stocks and bonds yields 5% per year?

**23. Vision Care**   As a fringe benefit for its employees, a company established a vision-care plan. Under this plan, each year the company will pay the first $35 of an employee's vision-care expenses and 80% of all additional vision-care expenses, up to a maximum *total* benefit payment of $100. For an employee, find the total annual vision-care expenses covered by this program.

**24. Quality Control**   Over a period of time, the manufacturer of a caramel-center candy bar found that 2% of the bars were rejected for imperfections.

**a.** If $c$ candy bars are made in a year, how many would the manufacturer expect to be rejected?

**b.** This year, annual consumption of the candy is projected to be 2,000,000 bars. Approximately how many bars will have to be made if rejections are taken into consideration?

**25. Business**   Suppose that consumers will purchase $q$ units of a product when the price is $(80 - q)/4$ dollars *each*. How many units must be sold in order that sales revenue be $400?

**26. Investment**   How long would it take to double an investment at simple interest with a rate of 5% per year? [*Hint:* See Example 6(a) of Sec. 1.1, and express 5% as 0.05.]

**27. Business Alternatives**   The inventor of a new toy offers the Kiddy Toy Company exclusive rights to manufacture and sell the toy for a lump-sum payment of $25,000. After estimating that future sales possibilities beyond one year are nonexistent, the Company management is reviewing an alternative proposal to give a lump-sum payment of $2000 plus a royalty of $0.50 for each unit sold. How many units must be sold the first year to make this alternative as economically attractive to the inventor as the original request? [*Hint:* Determine when the incomes under both proposals are the same.]

**28. Parking Lot**   A company parking lot is 120 ft long and 80 ft wide. Due to an increase in personnel, it is decided to double the area of the lot by adding strips of equal width to one end and one side. Find the width of one such strip.

**29. Rentals**   You are the chief financial advisor to a corporation that owns an office complex consisting of 50 units. At $400 per month, every unit can be rented. However, for each $20 per month increase, there will be two vacancies with no possibility of filling them. The corporation wants to receive a total of $20,240 per month from rents in the complex. You are asked to determine the rent that should be charged for each unit. What is your reply?

**30. Investment**   Six months ago, an investment company had a $3,000,000 portfolio consisting of blue-chip and glamor stocks. Since then, the value of the blue-chip investment increased by $\frac{1}{10}$, whereas the value of the glamor stocks decreased by $\frac{1}{10}$. The current value of the portfolio is $3,140,000. What is the *current* value of the blue-chip investment?

**31. Revenue**   The monthly revenue of a certain company is given by $R = 800p - 7p^2$, where $p$ is the price in dollars of the product the company manufactures. At what price will the revenue be $10,000 if the price must be greater than $50?

**32. Price–earnings Ratio**   The *price–earnings ratio* (*P/E*) of a company is the ratio of the market value of one share of the company's outstanding common stock to the earnings per share. If *P/E* increases by 10% and the earnings per share increase by 20%, determine the percentage increase in the market value per share of the common stock.

33. **Market Equilibrium** When the price of a product is *p* dollars each, suppose that a manufacturer will supply $2p - 8$ units of the product to the market and that consumers will demand to buy $300 - 2p$ units. At the value of *p* for which supply equals demand, the market is said to be in equilibrium. Find this value of *p*.

34. **Market Equilibrium** Repeat Problem 33 for the following conditions: At a price of *p* dollars each, the supply is $3p^2 - 4p$ and the demand is $24 - p^2$.

35. **Security Fence** For security reasons, a company will enclose a rectangular area of 11,200 ft$^2$ in the rear of its plant. One side will be bounded by the building and the other three sides by fencing. (See Fig. 2.4.) If 300 ft of fencing will be used, what will be the dimensions of the rectangular area?

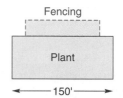

**FIGURE 2.4** Security fencing (Problem 35).

36. **Package Design** A company is designing a package for its product. One part of the package is to be an open box made from a square piece of aluminum by cutting out a 3-in square from each corner and folding up the sides. (See Fig. 2.5.) The box is to contain 75 in$^3$. What are the dimensions of the square piece of aluminum that must be used?

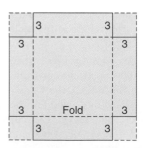

**FIGURE 2.5** Box construction (Problem 36).

37. **Product Design** A candy company makes the popular Dandy Bar. The rectangular-shaped bar is 10 centimeters (cm) long, 5 cm wide, and 2 cm thick. (See Fig. 2.6.) Because of increasing costs, the company has decided to

cut the volume of the bar by a drastic 28%. The thickness will be the same, but the length and width will be reduced by equal amounts. What will be the length and width of the new bar?

**FIGURE 2.6** Candy bar (Problem 37).

38. **Product Design** A candy company makes a washer-shaped candy (a candy with a hole in it); see Fig. 2.7. Because of increasing costs, the company will cut the volume of candy in each piece by 20%. To do this, the firm will keep the same thickness and outer radius, but will make the inner radius larger. At present the thickness is 2 millimeters (mm), the inner radius is 2 mm, and the outer radius is 7 mm. Find the inner radius of the new-style candy. [*Hint:* The volume *V* of a solid disk is $\pi r^2 h$, where *r* is the radius and *h* is the thickness of the disk.)

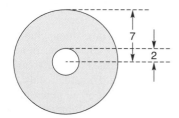

**FIGURE 2.7** Washer-shaped candy (Problem 38).

39. **Compensating Balance** *Compensating balance* refers to that practice wherein a bank requires a borrower to maintain on deposit a certain portion of a loan during the term of the loan. For example, if a firm takes out a $100,000 loan that requires a compensating balance of 20%, it would have to leave $20,000 on deposit and would have the use of $80,000. To meet the expenses of retooling, the Victor Manufacturing Company must borrow $95,000. The Third National Bank, with whom the firm has had no prior association, requires a compensating balance of 15%. To the nearest thousand dollars, what amount of the loan is required to obtain the needed funds?

**40. Incentive Plan** A machine company has an incentive plan for its salespeople. For each machine that a salesperson sells, the commission is $40. The commission for *every* machine sold will increase by $0.04 for each machine sold over 600. For example, the commission on each of 602 machines sold is $40.08. How many machines must a salesperson sell in order to earn $30,800?

**41. Real Estate** A land investment company purchased a parcel of land for $7200. After having sold all but 20 acres at a profit of $30 per acre over the original cost per acre, the company regained the entire cost of the parcel. How many acres were sold?

**42. Margin of Profit** The *margin of profit* of a company is the net income divided by the total sales. A company's margin of profit increased by 0.02 from last year. Last year the company sold its product at $3.00 each and had a net income of $4500. This year it increased the price of its product by $0.50 each, sold 2000 more, and had a net income of $7140. The company never has had a margin of profit greater than 0.15. How many of its product were sold last year and how many were sold this year?

**43. Business** A company manufactures products $A$ and $B$. The cost of producing each unit of $A$ is $2 more than that of $B$. The costs of production of $A$ and $B$ are $1500 and $1000, respectively, and 25 more units of $A$ are produced than of $B$. How many of each are produced?

---

OBJECTIVE

**To solve linear inequalities in one variable and to introduce interval notation.**

## 2.2 LINEAR INEQUALITIES

Suppose $a$ and $b$ are two points on the real-number line. Then either $a$ and $b$ coincide, or $a$ lies to the left of $b$, or $a$ lies to the right of $b$. (See Fig. 2.8.)

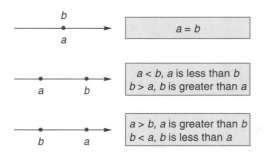

**FIGURE 2.8** Relative position of two points.

If $a$ and $b$ coincide, then $a = b$. If $a$ lies to the left of $b$, we say that $a$ is less than $b$ and write $a < b$, where the *inequality symbol* " $<$ " is read "is less than." On the other hand, if $a$ lies to the right of $b$, we say that $a$ is greater than $b$, written $a > b$. The statements $a > b$ and $b < a$ are equivalent.

Another inequality symbol "$\leq$" is read "is less than or equal to" and is defined as follows: $a \leq b$ if and only if $a < b$ or $a = b$. Similarly, the symbol "$\geq$" is defined as follows: $a \geq b$ if and only if $a > b$ or $a = b$. In this case, we say that $a$ is greater than or equal to $b$.

We shall use the words *real numbers* and *points* interchangeably, since there is a one-to-one correspondence between real numbers and points on a line. Thus, we can speak of the points $-5, -2, 0, 7$, and $9$ and can write $7 < 9, -2 > -5, 7 \leq 7$, and $7 \geq 0$. (See Fig. 2.9.) Clearly, if $a > 0$, then $a$ is positive; if $a < 0$, then $a$ is negative.

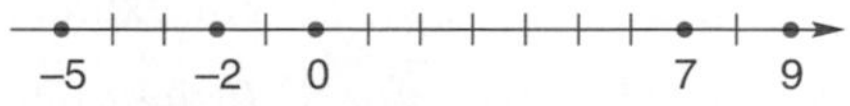

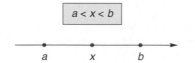

**Figure 2.10** $a < x$ and $x < b$.

**FIGURE 2.9** Points on number line.

Suppose that $a < b$ and $x$ is between $a$ and $b$. (See Fig. 2.10.) Then not only is $a < x$, but also, $x < b$. We indicate this by writing $a < x < b$, which can be thought of as a dual inequality. For example, $0 < 7 < 9$. (Refer back to Fig. 2.9.)

In defining an inequality next, we shall use the less-than relation ($<$), but the others ($>, \leq, \geq$) would also apply.

**DEFINITION**

*An **inequality** is a statement that one number is less than another number.*

Of course, we represent inequalities by means of inequality symbols. If two inequalities have their inequality symbols pointing in the same direction, then the inequalities are said to have the *same sense*. If not, they are said to be *opposite in sense*, or one is said to have the *reverse sense* of the other. Hence, $a < b$ and $c < d$ have the same sense, but $a < b$ has the reverse sense of $c > d$.

Solving an inequality, such as $2(x - 3) < 4$, means finding all values of the variable for which the inequality is true. This involves the application of certain rules, which we now state.

*Rules for Inequalities*

Keep in mind that the rules also apply to $\leq, >$, and $\geq$.

1. *If the same number is added to or subtracted from both sides of an inequality, the resulting inequality has the same sense as the original inequality.* Symbolically,

$$\text{if } a < b, \text{ then } a + c < b + c \text{ and } a - c < b - c.$$

For example, $7 < 10$, so $7 + 3 < 10 + 3$.

2. *If both sides of an inequality are multiplied or divided by the same **positive** number, the resulting inequality has the same sense as the original inequality.* Symbolically,

$$\text{if } a < b \text{ and } c > 0, \text{ then } ac < bc \text{ and } \frac{a}{c} < \frac{b}{c}.$$

For example, $3 < 7$ and $2 > 0$, so $3(2) < 7(2)$ and $\frac{3}{2} < \frac{7}{2}$.

The sense of an inequality must be reversed when multiplying or dividing both sides by a negative number.

3. *If both sides of an inequality are multiplied or divided by the same **negative** number, then the resulting inequality has the **reverse** sense of the original inequality.* Symbolically,

$$\text{if } a < b \text{ and } c > 0, \text{ then } a(-c) > b(-c) \text{ and } \frac{a}{-c} > \frac{b}{-c}.$$

For example, $4 < 7$ but $4(-2) > 7(-2)$ and $\frac{4}{-2} > \frac{7}{-2}$.

4. *Any side of an inequality can be replaced by an expression equal to it.* Symbolically,

$$\text{if } a < b \text{ and } a = c, \text{ then } c < b.$$

For example, if $x < 2$ and $x = y + 4$, then $y + 4 < 2$.

5. *If the sides of an inequality are either both positive or both negative, then their respective reciprocals[1] are unequal in the **reverse** sense.* For example, $2 < 4$, but $\frac{1}{2} > \frac{1}{4}$.

---

[1] The *reciprocal* of a nonzero number $a$ is defined to be $\frac{1}{a}$.

**6.** *If both sides of an inequality are positive and we raise each side to the same positive power, then the resulting inequality has the same sense as the original inequality. Thus, if $0 < a < b$ and $n > 0$, then*

$$a^n < b^n \quad \text{and} \quad \sqrt[n]{a} < \sqrt[n]{b},$$

*where we assume that $n$ is a positive integer in the latter inequality. For example, $4 < 9$, so $4^2 < 9^2$ and $\sqrt{4} < \sqrt{9}$.*

The result of applying Rules 1–4 to an inequality is called an *equivalent inequality*. This is an inequality whose solution is exactly the same as that of the original inequality. We shall apply these rules to a *linear inequality*.

### DEFINITION

*A **linear inequality** in the variable x is an inequality that can be written in the form*

The definition also applies to $\le$, $>$, and $\ge$.

$$ax + b < 0,$$

*where a and b are constants and $a \ne 0$.*

---

**Principles in Practice 1**

**Solving a Linear Inequality**

A salesman has a monthly income given by $I = 200 + 0.8S$, where $S$ is the number of products sold in a month. How many products must he sell to make at least \$4500 a month?

---

**EXAMPLE 1   Solving a Linear Inequality**

*Solve $2(x - 3) < 4$.*

*Solution:*

> *Strategy:*   We shall replace the given inequality by equivalent inequalities until the solution is evident.

$$2(x - 3) < 4,$$
$$2x - 6 < 4 \qquad \text{(Rule 4)},$$
$$2x - 6 + 6 < 4 + 6 \qquad \text{(Rule 1)},$$
$$2x < 10 \qquad \text{(Rule 4)},$$
$$\frac{2x}{2} < \frac{10}{2} \qquad \text{(Rule 2)},$$
$$x < 5.$$

All of the foregoing inequalities are equivalent. Thus, the original inequality is true for *all* real numbers $x$ such that $x < 5$. For example, the inequality is true for $x = -10, -0.1, 0, \frac{1}{2}$, and 4.9. We may write our solution simply as $x < 5$ and can geometrically represent it by the bold line segment in Fig. 2.11. The parenthesis indicates that 5 is *not included* in the solution.  ∎

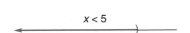

**Figure 2.11**   All real numbers less than 5.

In Example 1 the solution consisted of a set of numbers, namely, all numbers less than 5. In general, it is common to use the term **interval** to describe such a set. In the case of Example 1, the set of all $x$ such that $x < 5$ can be denoted by the *interval notation* $(-\infty, 5)$. The symbol $-\infty$ is not a number, but is merely a convenience for indicating that the interval extends indefinitely to the left.

There are other types of intervals. For example, the set of all numbers $x$ for which $a \le x \le b$ is called a **closed interval** and includes the numbers $a$ and $b$, which are called *endpoints* of the interval. This interval is denoted by $[a, b]$ and is shown in  Fig. 2.12(a). The square brackets indicated that $a$ and $b$ *are included* in the interval. On the other hand, the set of all $x$ for which

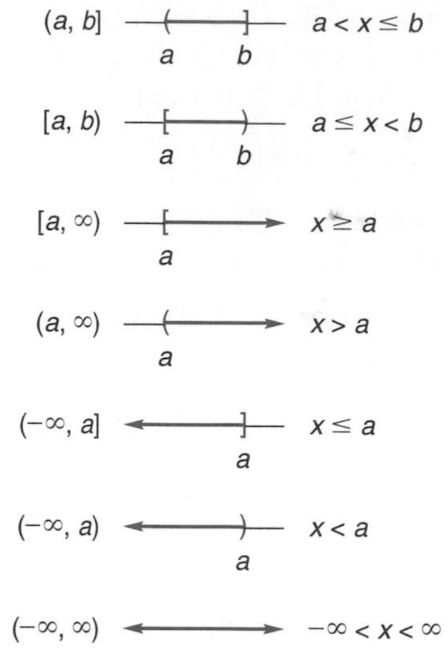

$(a, b]$    $a < x \le b$

$[a, b)$    $a \le x < b$

$[a, \infty)$    $x \ge a$

$(a, \infty)$    $x > a$

$(-\infty, a]$    $x \le a$

$(-\infty, a)$    $x < a$

$(-\infty, \infty)$    $-\infty < x < \infty$

**FIGURE 2.13**   Intervals.

Dividing both sides by $-2$ reverses the sense of the inequality.

Closed interval $[a, b]$     Open interval $(a, b)$

(a)          (b)

**FIGURE 2.12**   Closed and open intervals.

$a < x < b$ is called an **open interval** and is denoted by $(a, b)$. The endpoints are *not* part of this set. [See Fig. 2.12(b).] Extending these concepts, we have the intervals shown in Fig. 2.13.

**EXAMPLE 2**   **Solving a Linear Inequality**

*Solve* $3 - 2x \le 6$.

*Solution:*

$$3 - 2x \le 6,$$
$$-2x \le 3 \qquad \text{(Rule 1)},$$
$$x \ge -\frac{3}{2} \qquad \text{(Rule 3)}.$$

The solution is $x \ge -\frac{3}{2}$, or, in interval notation, $[-\frac{3}{2}, \infty)$. This is represented geometrically in Fig. 2.14.

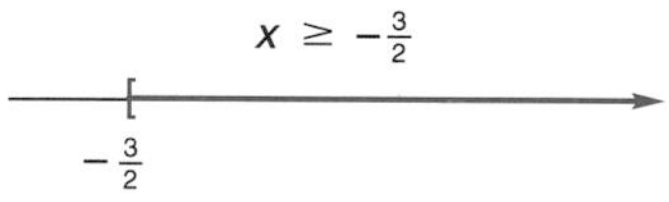

$x \ge -\frac{3}{2}$

$-\frac{3}{2}$

**FIGURE 2.14**   The interval $[-\frac{3}{2}, \infty)$.

## Principles in Practice 2
### Solving a Linear Inequality

A zoo veterinarian can purchase four different animal foods with various nutrient values for the zoo's grazing animals. Let $x_1$ represent the number of bags of food 1, $x_2$ represent the number of bags of food 2, and so on. The number of bags of each food needed can be described by the following equations:

$$x_1 = 150 - x_4$$
$$x_2 = 3x_4 - 210$$
$$x_3 = x_4 + 60$$

Develop four inequalities from these equations, assuming that each variable must be nonnegative.

**FIGURE 2.15**   The interval $(\frac{20}{7}, \infty)$.

**EXAMPLE 3**   **Solving a Linear Inequality**

*Solve* $\frac{3}{2}(s - 2) + 1 > -2(s - 4)$.

*Solution:*

$$\frac{3}{2}(s - 2) + 1 > -2(s - 4),$$
$$2[\tfrac{3}{2}(s - 2) + 1] > 2[-2(s - 4)] \qquad \text{(Rule 2)},$$
$$3(s - 2) + 2 > -4(s - 4),$$
$$3s - 4 > -4s + 16,$$
$$7s > 20 \qquad \text{(Rule 1)},$$
$$s > \frac{20}{7} \qquad \text{(Rule 2)}.$$

The solution is $(\frac{20}{7}, \infty)$; see Fig. 2.15.

**EXAMPLE 4**   **Solving Linear Inequalities**

**a.** *Solve* $2(x - 4) - 3 > 2x - 1$.

*Solution:*

$$2(x - 4) - 3 > 2x - 1,$$
$$2x - 8 - 3 > 2x - 1,$$
$$-11 > -1.$$

Since it is never true that $-11 > -1$, there is no solution, and the solution set is $\varnothing$.

$-\infty < x < \infty$

**FIGURE 2.16** The interval $(-\infty, \infty)$.

**b.** *Solve* $2(x - 4) - 3 < 2x - 1$.

*Solution:* Proceeding as in part (a), we obtain $-11 < -1$. This is true for all real numbers $x$, so the solution is $(-\infty, \infty)$; see Fig. 2.16. ∎

## ▪ Exercise 2.2

*In Problems 1–34, solve the inequalities. Give your answer in interval notation, and indicate the answer geometrically on the real-number line.*

**1.** $3x > 12$.

**2.** $4x < -2$.

**3.** $4x - 13 \le 7$.

**4.** $3x \ge 0$.

**5.** $-4x \ge 2$.

**6.** $2y + 1 > 0$.

**7.** $3 - 5s > 5$.

**8.** $4s - 1 < -5$.

**9.** $3 < 2y + 3$.

**10.** $6 \le 5 - 3y$.

**11.** $2x - 3 \le 4 + 7x$.

**12.** $-3 \ge 8(2 - x)$.

**13.** $3(2 - 3x) > 4(1 - 4x)$.

**14.** $8(x + 1) + 1 < 3(2x) + 1$.

**15.** $2(3x - 2) > 3(2x - 1)$.

**16.** $3 - 2(x - 1) \le 2(4 + x)$.

**17.** $x + 2 < \sqrt{3} - x$.

**18.** $\sqrt{2}(x + 2) > \sqrt{8}(3 - x)$.

**19.** $\dfrac{5}{3}x < 10$.

**20.** $-\dfrac{1}{2}x > 6$.

**21.** $\dfrac{9y + 1}{4} \le 2y - 1$.

**22.** $\dfrac{4y - 3}{2} \ge \dfrac{1}{3}$.

**23.** $4x - 1 \ge 4(x - 2) + 7$.

**24.** $0x \le 0$.

**25.** $\dfrac{1 - t}{2} < \dfrac{3t - 7}{3}$.

**26.** $\dfrac{3(2t - 2)}{2} > \dfrac{6t - 3}{5} + \dfrac{t}{10}$.

**27.** $2x + 3 \ge \dfrac{1}{2}x - 4$.

**28.** $4x - \dfrac{1}{2} \le \dfrac{3}{2}x$.

**29.** $\dfrac{2}{3}r < \dfrac{5}{6}r$.

**30.** $\dfrac{7}{4}t > -\dfrac{2}{3}t$.

**31.** $\dfrac{y}{2} + \dfrac{y}{3} > y + \dfrac{y}{5}$.

**32.** $9 - 0.1x \le \dfrac{2 - 0.01x}{0.2}$.

**33.** $0.1(0.03x + 4) \ge 0.02x + 0.434$.

**34.** $\dfrac{5y - 1}{-3} < \dfrac{7(y + 1)}{-2}$.

**35. Earnings** Each month last year, a company had earnings that were greater than \$37,000, but less than \$53,000. If $S$ represents the total earnings for the year, describe $S$ by using inequalities.

**36.** Using inequalities, symbolize the following statement: The number of labor hours $x$ to produce a product is not less than $2\frac{1}{2}$ nor more than 4.

**37. Geometry** In a right triangle, one of the acute angles $x$ is less than 3 times the other acute angle plus 10 degrees. Solve for $x$.

**38. Spending** A student has \$330 to spend on a stereo system and some compact disks. If she buys a stereo that costs \$198 and the disks are \$9.50 each, find the greatest number of disks she can buy.

**To model situations in terms of inequalities.**

## 2.3 APPLICATIONS OF INEQUALITIES

Solving word problems may sometimes involve inequalities, as the following examples illustrate.

### EXAMPLE 1   Profit

*For a company that manufactures thermostats, the combined cost for labor and material is $5 per thermostat. Fixed costs (costs incurred in a given period, regardless of output) are $60,000. If the selling price of a thermostat is $7, how many must be sold for the company to earn a profit?*

*Solution:*

> *Strategy:* Recall that
>
> $$\text{profit} = \text{total revenue} - \text{total cost}.$$
>
> We shall find total revenue and total cost and then determine when their difference is positive.

Let $q$ be the number of thermostats that must be sold. Then their cost is $5q$. The total cost for the company is therefore $5q + 60,000$. The total revenue from the sale of $q$ thermostats will be $7q$. Now,

$$\text{profit} = \text{total revenue} - \text{total cost},$$

and we want profit $> 0$. Thus,

$$\text{total revenue} - \text{total cost} > 0.$$
$$7q - (5q + 60,000) > 0,$$
$$2q > 60,000,$$
$$q > 30,000.$$

Therefore, at least 30,001 thermostats must be sold for the company to earn a profit. ■

### EXAMPLE 2   Renting vs. Purchasing

*A builder must decide whether to rent or buy an excavating machine. If he were to rent the machine, the rental fee would be $600 per month (on a yearly basis), and the daily cost (gas, oil, and driver) would be $60 for each day the machine is used. If he were to buy it, his fixed annual cost would be $4000, and daily operating and maintenance costs would be $80 for each day the machine is used. What is the least number of days each year that the builder would have to use the machine to justify renting it rather than buying it?*

*Solution:*

> *Strategy:*   We shall determine expressions for the annual cost of renting and the annual cost of purchasing. We then find when the cost of renting is less than that of purchasing.

Let $d$ be the number of days each year that the machine is used. If the machine is rented, the total yearly cost consists of rental fees, which are $(12)(600)$, and daily charges of $60d$. If the machine is purchased, the cost per year is $4000 + 80d$. We want

$$\text{cost}_{\text{rent}} < \text{cost}_{\text{purchase}},$$

$$12(600) + 60d < 4000 + 80d,$$

$$7200 + 60d < 4000 + 80d,$$

$$3200 < 20d,$$

$$160 < d.$$

Thus, the builder must use the machine at least 161 days to justify renting it.  ■

### EXAMPLE 3  Current Ratio

The *current ratio* of a business is the ratio of its current assets (such as cash, merchandise inventory, and accounts receivable) to its current liabilities (such as short-term loans and taxes payable).

*After consulting with the comptroller, the president of the Ace Sports Equipment Company decides to make a short-term loan to build up inventory. The company has current assets of $350,000 and current liabilities of $80,000. How much can the company borrow if the current ratio is to be no less than 2.5? (Note: The funds received are considered as current assets and the loan as a current liability.)*

*Solution:* Let $x$ denote the amount the company can borrow. Then current assets will be $350,000 + x$, and current liabilities will be $80,000 + x$. Thus,

$$\text{current ratio} = \frac{\text{current assets}}{\text{current liabilities}} = \frac{350{,}000 + x}{80{,}000 + x}.$$

We want

Although the inequality that must be solved is not linear, it leads to a linear inequality.

$$\frac{350{,}000 + x}{80{,}000 + x} \geq 2.5.$$

Since $x$ is positive, so is $80,000 + x$. Hence, we can multiply both sides of the inequality by $80,000 + x$ and the sense of the inequality will remain the same. We have

$$350{,}000 + x \geq 2.5(80{,}000 + x),$$

$$150{,}000 \geq 1.5x,$$

$$100{,}000 \geq x.$$

Consequently, the company may borrow as much as $100,000 and yet maintain a current ratio of no less than 2.5.  ■

### EXAMPLE 4  Publishing

*A publishing company finds that the cost of publishing each copy of a certain magazine is $1.50. The revenue from dealers is $1.40 per copy. The advertising revenue is 10% of the revenue received from dealers for all copies sold beyond 10,000. What is the least number of copies that must be sold so as to have a profit for the company?*

*Solution:*

*Strategy:*  We have profit = total revenue − total cost, so we find an expression for profit and then set it greater than 0.

Let $q$ be the number of copies that are sold. The revenue from dealers is $1.40q$, and the revenue from advertising is $(0.10)[(1.40)(q - 10,000)]$. The total cost of publication is $1.50q$. Thus,

$$\text{total revenue} - \text{total cost} > 0.$$

$$1.40q + (0.10)[(1.40)(q - 10,000)] - 1.50q > 0,$$

$$1.4q + 0.14q - 1400 - 1.5q > 0,$$

$$0.04q - 1400 > 0,$$

$$0.04q > 1400,$$

$$q > 35,000.$$

Therefore, the total number of copies must be greater than 35,000. That is, at least 35,001 copies must be sold to guarantee a profit. ∎

## ■ Exercise 2.3

1. **Profit** The Davis Company manufactures a product that has a unit selling price of $20 and a unit cost of $15. If fixed costs are $600,000, determine the least number of units that must be sold for the company to have a profit.

2. **Profit** To produce 1 unit of a new product, a company determines that the cost for material is $2.50 and the cost of labor is $4. The constant overhead, regardless of sales volume, is $5000. If the cost to a wholesaler is $7.40 per unit, determine the least number of units that must be sold by the company to realize a profit.

3. **Renting vs. Purchasing** A businesswoman wants to determine the difference between the costs of owning and renting an automobile. She can rent a car for $400 per month (on an annual basis). Under this plan, the cost per mile (gas and oil) is $0.10. If she were to purchase the car, the fixed annual expense would be $3000, and other costs would amount to $0.18 per mile. What is the least number of miles she would have to drive per year to make renting no more expensive than purchasing?

4. **Shirt Manufacturer** A T-shirt manufacturer produces $N$ shirts at a total labor cost (in dollars) of $1.2N$ and a total material cost of $0.3N$. The constant overhead for

the plant is $6000. If each shirt sells for $3, how many must be sold by the company to realize a profit?

5. **Publishing** The cost of publication of each copy of a magazine is $0.65. It is sold to dealers for $0.60 each, and the amount received for advertising is 10% of the amount received for all magazines issued beyond 10,000. Find the least number of magazines that can be published without loss—that is, such that profit $\geq 0$. (Assume that all issues will be sold.)

6. **Production Allocation** A company produces alarm clocks. During the regular workweek, the labor cost for producing one clock is $2.00. However, if a clock is produced on overtime, the labor cost is $3.00. Management has decided to spend no more than a total of $25,000 per week for labor. The company must produce 11,000 clocks this week. What is the minimum number of clocks that must be produced during the regular workweek?

7. **Investment** A company invests a total of $30,000 of surplus funds at two annual rates of interest: 5% and $6\frac{3}{4}\%$. It wishes an annual yield of no less than $6\frac{1}{2}\%$. What is the least amount of money that the company must invest at the $6\frac{3}{4}\%$ rate?

8. **Current Ratio** The current ratio of Precision Machine Products is 3.8. If the firm's current assets are $570,000, what are its current liabilities? To raise additional funds, what is the maximum amount the company can borrow on a short-term basis if the current ratio is to be no less than 2.6? (See Example 3 for an explanation of current ratio.)

9. **Sales Allocation** At present, a manufacturer has 2500 units of product in stock. The product is now selling at $4 per unit. Next month the unit price will increase by $0.50. The manufacturer wants the total revenue received from the sale of the 2500 units to be no less than $10,750. What is the maximum number of units that can be sold this month?

10. **Revenue** Suppose consumers will purchase $q$ units of a product at a price of $\dfrac{100}{q} + 1$ dollars per unit. What is the minimum number of units that must be sold in order that sales revenue be greater than $5000$?

**11. Hourly Rate** Painters arc oftcn paid cither by the hour or on a per-job basis. The rate they receive can affect their working speed. For example, suppose they can work either for $8.50 per hour or for $300 plus $3 for each hour less than 40 if they complete the job in less than 40 hours. Suppose the job will take $t$ hours. If $t \geq 40$, clearly the hourly rate is better. If $t < 40$, for what values of $t$ is the hourly rate the better pay scale?

**12. Compensation** Suppose a company offers you a sales position with your choice of two methods of determining your yearly salary. One method pays $12,600 plus a bonus of 2% of your yearly sales. The other method pays a straight 8% commission on your sales. For what yearly sales amount is it better to choose the first method?

**13. Campus Concert** The dean of student affairs of a college is arranging for a rock group to perform a concert on campus. The group charges a flat fee of $2440 or, instead, a fee of $1000 plus 40% of the gate. It is likely that 800 students will attend. At most, how much could the dean charge for a ticket so that the second arrangement is no more costly than the flat fee? If this maximum is charged, how much money will be left over to pay for publicity, guards, and other concert expenses?

---

OBJECTIVE

To solve equations and inequalities involving absolute values.

Basically, the absolute value of a real number is its value when its sign is ignored.

## 2.4 ABSOLUTE VALUE

### Absolute-Value Equations

On the real-number line, the distance of a number $x$ from 0 is called the **absolute value** of $x$ and is denoted by $|x|$. For example, $|5| = 5$ and $|-5| = 5$ because both 5 and $-5$ are 5 units from 0. (See Fig. 2.17.) Similarly, $|0| = 0$. Notice that $|x|$ can never be negative; that is, $|x| \geq 0$.

If $x$ is positive or zero, then $|x|$ is simply $x$ itself, so we can omit the vertical bars and write $|x| = x$. On the other hand, consider the absolute value of a negative number, like $x = -5$.

$$|x| = |-5| = 5 = -(-5) = -x.$$

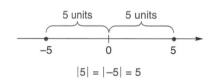

**FIGURE 2.17** Absolute value.

Thus, if $x$ is negative, then $|x|$ is the positive number $-x$. The minus sign indicates that we have changed the sign of $x$. Accordingly, aside from its geometrical interpretation, the absolute value can be defined as follows.

**DEFINITION**
The **absolute value** of a real number $x$, written $|x|$, is defined as

$$|x| = \begin{cases} x, & \text{if } x \geq 0, \\ -x, & \text{if } x < 0. \end{cases}$$

Applying the definition, we have $|3| = 3$, $|-8| = -(-8)$, and $|\frac{1}{2}| = \frac{1}{2}$. Also, $-|2| = -2$ and $-|-2| = -2$.

*Pitfall ▼* $\sqrt{x^2}$ is not necessarily $x$, but

$$\sqrt{x^2} = |x|.$$

For example, $\sqrt{(-2)^2} = |-2| = 2$, not $-2$. This agrees with the fact that

$$\sqrt{(-2)^2} = \sqrt{4} = 2.$$

Also, $|-x| \neq x$ and

$$|-x - 1| \neq x + 1.$$

For example, if we let $x = -3$, then $|-(-3)| \neq -3$, and

$$|-(-3) - 1| \neq -3 + 1.$$

### EXAMPLE 1   Solving Absolute-Value Equations

**a.** *Solve* $|x - 3| = 2$.

  *Solution:* This equation states that $x - 3$ is a number 2 units from 0. Thus, either

$$x - 3 = 2 \quad \text{or} \quad x - 3 = -2.$$

  Solving these equations gives $x = 5$ or $x = 1$.

**b.** *Solve* $|7 - 3x| = 5$.

  *Solution:* The equation is true if $7 - 3x = 5$ or if $7 - 3x = -5$. Solving these equations gives $x = \frac{2}{3}$ or $x = 4$.

**c.** *Solve* $|x - 4| = -3$.

  *Solution:* The absolute value of a number is never negative, so the solution set is $\varnothing$. ■

We may interpret $|a - b|$ or $|b - a|$ as the distance between $a$ and $b$. For example, the distance between 5 and 9 is either

$$|9 - 5| = |4| = 4,$$
$$\text{or} \quad |5 - 9| = |-4| = 4.$$

Similarly, the equation $|x - 3| = 2$ states that the distance between $x$ and 3 is 2 units. Thus, $x$ can be 1 or 5, as shown in Example 1(a) and Fig. 2.18.

### Absolute-Value Inequalities

Let us turn now to inequalities involving absolute values. If $|x| < 3$, then $x$ is less than 3 units from 0. Hence, $x$ must lie between $-3$ and 3, that is, on the in-

2 units    2 units

**FIGURE 2.18**   The solution of $|x - 3| = 2$ is 1 or 5.

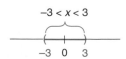

$-3 < x < 3$

(a) Solution of $|x| < 3$

$x < -3$      $x > 3$

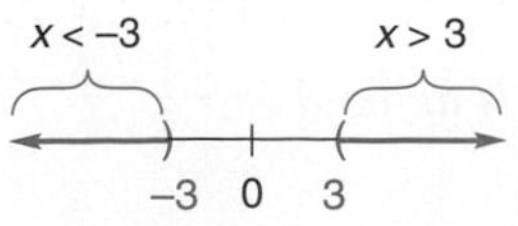

(b) Solution of $|x| > 3$

**FIGURE 2.19**   Solution of $|x| < 3$ and $|x| > 3$.

**TABLE 2.1**

| Inequality ($d > 0$) | Solution |
|---|---|
| $|x| < d$ | $-d < x < d$ |
| $|x| \leq d$ | $-d \leq x \leq d$ |
| $|x| > d$ | $x < -d$  or  $x > d$ |
| $|x| \geq d$ | $x \leq -d$  or  $x \geq d$ |

terval $-3 < x < 3$. [See Fig. 2.19(a).] On the other hand, if $|x| > 3$, then $x$ must be greater than 3 units from 0. Hence, there are two intervals in the solution: Either $x < -3$ or $x > 3$. [See Fig. 2.19(b).] We can extend these ideas as follows: If $|x| \le 3$, then $-3 \le x \le 3$; if $|x| \ge 3$, then $x \le -3$ or $x \ge 3$. Table 2.1 gives a summary of the solutions to absolute-value inequalities.

### EXAMPLE 2   Solving Absolute-Value Inequalities

**a.** *Solve* $|x - 2| < 4$.

*Solution:* The number $x - 2$ must be less than 4 units from 0. From the preceding discussion, this means that $-4 < x - 2 < 4$. We may set up the procedure for solving this inequality as follows:

$$-4 < x - 2 < 4,$$
$$-4 + 2 < x < 4 + 2 \qquad \text{(adding 2 to each member)},$$
$$-2 < x < 6.$$

Thus, the solution is the open interval $(-2, 6)$. This means that all numbers between $-2$ and 6 satisfy the original inequality. (See Fig. 2.20.)

**b.** *Solve* $|3 - 2x| \le 5$.

*Solution:*

$$-5 \le 3 - 2x \le 5,$$
$$-5 - 3 \le -2x \le 5 - 3 \qquad \text{(subtracting 3 from each member)},$$
$$-8 \le -2x \le 2,$$
$$4 \ge x \ge -1 \qquad \text{(dividing each member by } -2),$$
$$-1 \le x \le 4 \qquad \text{(rewriting)}.$$

Note that the sense of the original inequality was *reversed* when we divided by a negative number. The solution is the closed interval $[-1, 4]$. ■

### EXAMPLE 3   Solving Absolute-Value Inequalities

**a.** *Solve* $|x + 5| \ge 7$.

*Solution:* Here $x + 5$ must be *at least* 7 units from 0. Thus, either $x + 5 \le -7$ or $x + 5 \ge 7$. This means that either $x \le -12$ or $x \ge 2$. Thus, the solution consists of two intervals: $(-\infty, -12]$ and $[2, \infty)$. We can abbreviate this collection of numbers by writing

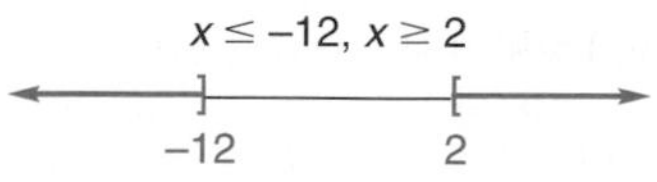

$$(-\infty, -12] \cup [2, \infty).$$

where the connecting symbol $\cup$ is called the *union* symbol. (See Fig. 2.21.) More formally, the **union** of sets $A$ and $B$ is the set consisting of all elements that are in either $A$ or $B$ (or in both $A$ and $B$).

**b.** *Solve* $|3x - 4| > 1$.

*Solution:* Either $3x - 4 < -1$ or $3x - 4 > 1$. Thus, either $3x < 3$ or $3x > 5$. Therefore, $x < 1$ or $x > \frac{5}{3}$, so the solution consists of all numbers in the set $(-\infty, 1) \cup \left(\frac{5}{3}, \infty\right)$. ■

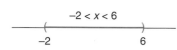

$-2 < x < 6$

**FIGURE 2.20**   The solution of $|x - 2| < 4$ is the interval $(-2, 6)$.

$x \le -12,\ x \ge 2$

**FIGURE 2.21**   The union $(-\infty, -12] \cup [2, \infty)$.

The inequalities $x < 1$ and $x > \frac{5}{3}$ cannot be combined into a single inequality, although you might like to do so. It is incorrect to combine $\frac{5}{3} < x$ and $x < 1$ as $\frac{5}{3} < x < 1$, since this implies that $\frac{5}{3} < 1$.

**Principles in Practice 1**
**Absolute-Value Notation**

Express the following statement using absolute-value notation:

The actual weight $w$ of a box of cereal must be within 0.3 oz of the weight stated on the box, which is 22 oz.

**EXAMPLE 4  Absolute-Value Notation**

*Using absolute-value notation, express the following statements:*

**a.** $x$ is less than 3 units from 5.

*Solution:*
$$|x - 5| < 3.$$

**b.** $x$ differs from 6 by at least 7.

*Solution:*
$$|x - 6| \geq 7.$$

**c.** $x < 3$ and $x > -3$ simultaneously.

*Solution:*
$$|x| < 3.$$

**d.** $x$ is more than 1 unit from $-2$.

*Solution:*
$$|x - (-2)| > 1,$$
$$|x + 2| > 1.$$

**e.** $x$ is less than $\sigma$ (a Greek letter read "sigma") units from $\mu$ (a Greek letter read "mu").

*Solution:*
$$|x - \mu| < \sigma. \qquad \blacksquare$$

**Properties of the Absolute Value**

Four basic properties of the absolute value are as follows:

> **1.** $|ab| = |a| \cdot |b|$.
>
> **2.** $\left|\dfrac{a}{b}\right| = \dfrac{|a|}{|b|}$.
>
> **3.** $|a - b| = |b - a|$.
>
> **4.** $-|a| \leq a \leq |a|$.

For example, Property 1 states that the absolute value of the product of two numbers is equal to the product of the absolute values of the numbers.

**EXAMPLE 5  Properties of Absolute Value**

**a.** $|(-7) \cdot 3| = |-7| \cdot |3| = 21$.

**b.** $|4 - 2| = |2 - 4| = 2$.

**c.** $|7 - x| = |x - 7|$.

**d.** $\left|\dfrac{-7}{3}\right| = \dfrac{|-7|}{|3|} = \dfrac{7}{3}$;    $\left|\dfrac{-7}{-3}\right| = \dfrac{|-7|}{|-3|} = \dfrac{7}{3}$.

**e.** $\left|\dfrac{x - 3}{-5}\right| = \dfrac{|x - 3|}{|-5|} = \dfrac{|x - 3|}{5}$.

**f.** $-|2| \leq 2 \leq |2|$. $\qquad \blacksquare$

# ■ Exercise 2.4

*In Problems 1–10, write an equivalent form without the absolute-value symbol.*

**1.** $|-13|$.

**2.** $|2^{-1}|$.

**3.** $|8 - 2|$.

**4.** $|(-4 - 6)/2|$.

**5.** $|3(-\frac{5}{3})|$.

**6.** $|2 - 7| - |7 - 2|$.

**7.** $|x| < 3$.

**8.** $|x| < 10$.

**9.** $|2 - \sqrt{5}|$.

**10.** $|\sqrt{5} - 2|$.

**11.** Using the absolute value symbol, express each fact.

  **a.** $x$ is less than 3 units from 7.

  **b.** $x$ differs from 2 by less than 3.

  **c.** $x$ is no more than 5 units from 7.

  **d.** The distance between 7 and $x$ is 4.

  **e.** $x + 4$ is less than 2 units from 0.

  **f.** $x$ is between $-3$ and 3, but is not equal to 3 or $-3$.

  **g.** $x < -6$ or $x > 6$.

  **h.** $x - 6 > 4$ or $x - 6 < -4$.

  **i.** The number $x$ of hours that a machine will operate efficiently differs from 105 by less than 3.

  **j.** The average monthly income $x$ (in dollars) of a family differs from 850 by less than 100.

**12.** Use absolute-value notation to indicate that $x$ and $\mu$ differ by no more than $\sigma$.

**13.** Use absolute-value notation to indicate that the prices $p_1$ and $p_2$ of two products may differ by no more than 2 (dollars).

**14.** Find all values of $x$ such that $|x - \mu| \le 2\sigma$.

*In Problems 15–36, solve the given equation or inequality.*

**15.** $|x| = 7$.

**16.** $|-x| = 2$.

**17.** $\left|\dfrac{x}{3}\right| = 2$.

**18.** $\left|\dfrac{4}{x}\right| = 8$.

**19.** $|x - 5| = 8$.

**20.** $|4 + 3x| = 2$.

**21.** $|5x - 2| = 0$.

**22.** $|7x + 3| = x$.

**23.** $|7 - 4x| = 5$.

**24.** $|1 - 2x| = 1$.

**25.** $|x| < 4$.

**26.** $|-x| < 3$.

**27.** $\left|\dfrac{x}{4}\right| > 2$.

**28.** $\left|\dfrac{x}{3}\right| > \dfrac{1}{2}$.

**29.** $|x + 7| < 2$.

**30.** $|5x - 1| < -6$.

**31.** $|x - \frac{1}{2}| > \frac{1}{2}$.

**32.** $|1 - 3x| > 2$.

**33.** $|5 - 2x| \le 1$.

**34.** $|4x - 1| \ge 0$.

**35.** $\left|\dfrac{3x - 8}{2}\right| \ge 4$.

**36.** $\left|\dfrac{x - 8}{4}\right| \le 2$.

*In Problems 37–38, express the statement using absolute value notation.*

**37.** In a science experiment, the measurement of a distance $d$ is 10 m, and is accurate to $\pm 1$ cm.

**38.** The difference in temperature between two chemicals which are to be mixed must be no less than 5 degrees and no more than 10 degrees.

**39. Statistics**   In statistical analysis, the Chebyshev inequality asserts that if $x$ is a random variable, $\mu$ is its mean, and $\sigma$ is its standard deviation, then

$$\text{(probability that } |x - \mu| > h\sigma) \ge \frac{1}{h^2}.$$

Find those values of $x$ such that $|x - \mu| > h\sigma$.

**40. Manufacturing Tolerance**   In the manufacture of widgets, the average dimension of a part is 0.01 cm. Using the absolute-value symbol, express the fact that an individual measurement $x$ of a part does not differ from the average by more than 0.005 cm.

# 2.5 REVIEW

## IMPORTANT TERMS AND SYMBOLS

| Section 2.1 | fixed cost | overhead | variable cost | total cost | total revenue | profit |
|---|---|---|---|---|---|---|

| Section 2.2 | $a < b$ | $a \le b$ | $a > b$ | $a \ge b$ | $a < x < b$ | inequality |
|---|---|---|---|---|---|---|
| | sense of inequality | | equivalent inequality | | linear inequality | $-\infty < x < \infty$ |
| | open interval | | closed interval | endpoints | interval notation | |

| Section 2.4 | absolute value, $|x|$ | union, $\cup$ |
|---|---|---|

## SUMMARY

With a word problem, an equation is not handed to you. Instead, you must set it up by translating verbal statements into an equation (or inequality). This is *mathematical modeling.* It is important that you first read the problem more than once so that you clearly understand what facts are given and what you are asked to find. Then choose a letter to represent the unknown quantity that you want to find. Use the relationships and facts given in the problem and translate them into an equation involving the letter. Finally, solve the equation, and see if your solution answers what was asked. Sometimes the solution to the *equation* will not be the answer to the *problem,* but it may be useful in obtaining that answer.

Some basic relationships that are used in solving business problems are:

> total cost = variable cost + fixed cost,
>
> total revenue = (price per unit)(number of units sold),
>
> profit = total revenue − total cost.

The inequality symbols $<, \leq, >,$ and $\geq$ are used to represent an inequality, which is a statement that one number is, for example, less than another number. Three basic operations that, when applied to an inequality, guarantee an equivalent inequality are:

**1.** Adding (or subtracting) the same number to (or from) both sides.

**2.** Multiplying (or dividing) both sides by the same positive number.

**3.** Multiplying (or dividing) both sides by the same negative number and reversing the sense of the inequality.

These operations are useful in solving a linear inequality (an inequality that can be put in the form $ax + b < 0$ or $ax + b \leq 0$, where $a \neq 0$).

> $|x| = x,$ if $x \geq 0$   and   $|x| = -x,$ if $x < 0.$

An algebraic definition of absolute value is We interpret $|a - b|$ or $|b - a|$ as the distance between $a$ and $b$. If $d > 0$, then the solution to the inequality $|x| < d$ is the interval $(-d, d)$. The solution to $|x| > d$ consists of two intervals and is given by $(-\infty, -d) \cup (d, \infty)$. Some basic properties of the absolute value are:

> **1.** $|ab| = |a| \cdot |b|,$
>
> **2.** $\left|\dfrac{a}{b}\right| = \dfrac{|a|}{|b|},$
>
> **3.** $|a - b| = |b - a|.$
>
> **4.** $-|a| \leq a \leq |a|.$

## REVIEW PROBLEMS

*In Problems 1–15, solve the equation or inequality.*

**1.** $3x - 8 \geq 4(x - 2).$

**2.** $2x - (7 + x) \leq x.$

**3.** $-(5x + 2) < -(2x + 4).$

**4.** $-2(x + 6) > x + 4.$

**5.** $3p(1 - p) > 3(2 + p) - 3p^2.$

**6.** $2(4 - \frac{3}{5}q) < 5.$

**7.** $\dfrac{x + 1}{3} - \dfrac{1}{2} \leq 2.$

**8.** $\dfrac{x}{2} + \dfrac{x}{3} > \dfrac{x}{4}.$

**9.** $\dfrac{1}{4}s - 3 \leq \dfrac{1}{8}(3 + 2s).$

**10.** $\dfrac{1}{3}(t + 2) \geq \dfrac{1}{4}t + 4.$

**11.** $|3 - 2x| = 7.$

**12.** $\left|\dfrac{5x - 8}{13}\right| = 0.$

**13.** $|4t - 1| < 1.$

**14.** $4 < \left|\dfrac{2}{3}x + 5\right|.$

**15.** $|3 - 2x| \geq 4.$

**16. Profit**   A profit of 40% on the selling price of a product is equivalent to what percent profit on the cost?

**17. Stock Exchange**   On a certain day, there were 1132 different issues traded on the New York Stock Exchange. There were 48 more issues showing an increase than showing a decline, and no issues remained the same. How many issues suffered a decline?

**18. Sales Tax**   The sales tax in a certain state is 6%. If a total of $3017.29 in purchases, including tax, is made in the course of a year, how much of it is tax?

**19. Production Allocation**   A company will manufacture a total of 10,000 units of its product at plants A and B. Available data are as follows:

| | Plant A | Plant B |
|---|---|---|
| Unit cost for labor and material | $5 | $5.50 |
| Fixed cost | $30,000 | $35,000 |

Between the two plants the company has decided to allot no more than $117,000 for total costs. What is the minimum number of units that must be produced at plant A?

**20. Storage Tanks**   A company is replacing two cylindrical oil-storage tanks with one new tank. The old tanks are each 16 ft high. One has a radius of 15 ft and the other a radius of 20 ft. The new tank will also be 16 ft high. Find its radius if it is to have the same volume as the old tanks combined. [*Hint:* The volume $V$ of a cylindrical tank is $V = \pi r^2 h$, where $r$ is the radius of the circular base and $h$ is the height of the tank.]

# MATHEMATICAL *SNAPSHOT*

## QUALITY VCR RECORDING[2]

*The following is an entertaining discussion of VCR recording and illustrates mathematical concepts of this chapter.*

If you are like the millions of others who have a VCR, you have seen how convenient it is to record television shows for future viewing. Here you will learn how to get the highest quality recording from this marvel of home entertainment hardware.

With a VHS format, you may have your choice of standard play speed (SP), long play speed (LP), or extended play speed (EP). SP is the fastest speed and provides the best picture quality. LP, a slower speed, gives lower quality recording, and EP, which is the slowest speed, gives the lowest picture quality.

With the common T-120 videotape, the maximum recording time at SP is 2 hours. At LP it is 4 hours, and at EP it is 6 hours. In the following discussion you may assume that these recording times are exact and that the amount of tape used changes uniformly with recording time.

If you wish to record a movie that is no more than 2 hours long, obviously SP should be used to get the best picture quality. However, for recording a 3-hour movie on a single T-120 tape, using only SP speed would cause the tape to be filled 1 hour before the movie is over. You can overcome this difficulty by using SP in conjunction with another speed, making sure that the time at SP is maximized.

For example, you can begin at LP and then complete the recording at SP. Obviously your problem is to determine when the changeover to SP should be made. Let $t$ be the time, in hours, that LP is used. Then $3 - t$ hours of the movie remain to be recorded at SP.

Since the tape speed in the LP mode is $\frac{1}{4}$ tape per hour and the speed in the SP mode is $\frac{1}{2}$ tape per hour, the portion of the tape used at LP is $t/4$ and the portion at SP is $(3 - t)/2$. The sum of these portions must be 1 because the whole tape would be used up. Therefore, you need to solve a linear equation.

$$\frac{t}{4} + \frac{3 - t}{2} = 1,$$
$$t + 2(3 - t) = 4,$$
$$6 - t = 4,$$
$$t = 2.$$

Thus you should record at LP for 2 hours and then switch to SP for the remaining $3 - t = 3 - 2 = 1$ hour. This means that one-third of the movie will be recorded with the best picture quality.

Instead of restricting yourself to a 3-hour movie, you can generalize the above problem to handle a movie of length $l$ hours, where $2 < l \le 4$. This situation gives

$$\frac{t}{4} + \frac{l - t}{2} = 1,$$

whose solution is

$$t = 2l - 4.$$

Alternatively, you might feel that there is not much difference between the picture qualities at LP

[2]Adapted from Gregory N. Fiore, "An Application of Linear Equations to the VCR," *Mathematics Teacher,* 81 (October 1988), 570–72. By permission of the National Council of Teachers of Mathematics.

and EP speeds. If you desire to begin at EP and end with SP, you could handle a movie of length $l$, where $2 < l \le 6$. Let $t$ be the time, in hours, that EP is used. Then

$$\frac{t}{6} + \frac{l-t}{2} = 1,$$

$$t + 3(l - t) = 6,$$

$$-2t + 3l = 6,$$

$$3l - 6 = 2t,$$

$$t = \frac{3}{2}l - 3.$$

For example, with a 3-hour movie you would record at EP for $t = \frac{3}{2}(3) - 3 = 1\frac{1}{2}$ hours and then at SP for $3 - 1\frac{1}{2} = 1\frac{1}{2}$ hours. This shows that using EP instead of LP gives you $\frac{1}{2}$ hour more of quality recording at SP. As a second example, consider recording a 4-hour-and-20-minute movie. Here $l = 4\frac{1}{3}$ hours, so you would use EP for

$$t = \frac{3}{2}\left(\frac{13}{3}\right) - 3 = 3\frac{1}{2} \text{ hours}$$

and use SP for the rest of the movie.

Finally, suppose that a television show has commercial interruptions, but you want to record it with the commercials eliminated. You can handle this by stopping the VCR at the beginning of a commercial and restarting when the commercial is over. Assume you estimate that every hour of an $l$-hour show contains $c$ minutes of commercials. Then the total commercial time, in hours, is $lc/60$. Thus the noncommercial length of the show, in hours, is

$$l - \frac{lc}{60}.$$

If you begin recording at EP and then switch to SP, you will have

$$\frac{t}{6} + \frac{l - \dfrac{lc}{60} - t}{2} = 1,$$

where $t$ is the time, in hours, at EP. Solving gives

$$t + 3\left(l - \frac{lc}{60} - t\right) = 6,$$

$$t + 3l - \frac{lc}{20} - 3t = 6,$$

$$-2t = -3l + \frac{lc}{20} + 6,$$

$$t = \frac{3}{2}l - \frac{c}{40}l - 3.$$

For example, take the case of a 3-hour show that has 6 minutes of commercials each hour. Then $l = 3$ and $c = 6$, so

$$t = \frac{3}{2}(3) - \frac{6}{40}(3) - 3$$

$$= \frac{21}{20} \text{ hour.}$$

$$= 1 \text{ hour and 3 minutes.}$$

This means that you should record 1 hour and 3 minutes of the noncommercial part of the show at EP and the remainder of the noncommercial time at SP.

## ▪ Exercises

1. If LP and SP modes are used to record a $2\frac{1}{2}$-hour movie, how long after the start of the movie should the switch from LP to SP be made?

2. If EP and SP modes are used to record a $2\frac{1}{2}$-hour show, how many minutes after the start of the show should the switch from EP to SP be made?

3. If EP and SP modes are used to record a movie of length 2 hours and 40 minutes, how long after the start of the movie should the switch from EP to SP be made?

4. EP and SP modes are used to record a 3-hour movie. How long after the start of the movie should the switch from EP to SP be made if the viewer eliminates 8 minutes of commercials each hour?

5. Repeat Problem 4 if LP and SP modes are to be used.

# Functions and Graphs

**To understand what a function is and to determine domains and function values.**

## 3.1 FUNCTIONS

In the 17th century, Gottfried Wilhelm Leibniz, one of the inventors of calculus, introduced the term *function* into the mathematical vocabulary. The concept of a function is one of the most basic in all of mathematics, and it is essential to the study of calculus.

Briefly, a function is a special type of relation that expresses how one quantity (the *output*) depends on another quantity (the *input*). For example, when money is invested at some interest rate, the interest $I$ (output) depends on the length of time $t$ (input) that the money is invested. To express this dependence, we say that $I$ is a "function of" $t$. Functional relations like this are usually specified by a formula that shows what must be done to the input to find the output.

To illustrate, suppose $100 earns simple interest at an annual rate of 6%. Then it can be shown that interest and time are related by the formula

$$I = 100(0.06)t, \tag{1}$$

where $I$ is in dollars and $t$ is in years. For example,

$$\text{if } t = \tfrac{1}{2}, \quad \text{then} \quad I = 100(0.06)(\tfrac{1}{2}) = 3. \tag{2}$$

Thus, Formula (1) assigns the output 3 to the input $\tfrac{1}{2}$. We can think of Formula (1) as defining a *rule:* Multiply $t$ by $100(0.06)$. The rule assigns to each input number $t$ exactly one output number $I$, which we symbolize by the following arrow notation:

$$t \to I \quad \text{or} \quad t \to 100(0.06)t.$$

This rule is an example of a *function* in the following sense:

**DEFINITION**

*A **function** is a rule that assigns to each input number exactly one output number. The set of all input numbers to which the rule applies is called the **domain** of the function. The set of all output numbers is called the **range**.*

For the interest function defined by Formula (1), the input number $t$ cannot be negative, because negative time makes no sense. Thus, the domain consists of all nonnegative numbers—that is, all $t \geq 0$. From (2), we see that when the input is $\tfrac{1}{2}$, the output is 3. Thus, 3 is in the range.

We have been using the term *function* in a restricted sense because, in general, the inputs or outputs do not have to be numbers. For example, a list of states and their capitals assigns to each state its capital (exactly one output). Hence, a function is implied. However, for the time being, we shall consider only functions whose domains and ranges consist of real numbers.

A variable that represents input numbers for a function is called an **independent variable.** A variable that represents output numbers is called a **dependent variable** because its value *depends* on the value of the independent variable. We say that the dependent variable is a *function of* the independent variable. That is, output is a function of input. Thus, for the interest formula $I = 100(0.06)t$, the independent variable is $t$, the dependent variable is $I$, and $I$ is a function of $t$.

As another example, the equation (or formula)

$$y = x + 2 \qquad\qquad (3)$$

defines $y$ as a function of $x$. The equation gives the rule, "Add 2 to $x$." This rule assigns to each input $x$ exactly one output $x + 2$, which is $y$. If $x = 1$, then $y = 3$; if $x = -4$, then $y = -2$. The independent variable is $x$ and the dependent variable is $y$.

Not all equations in $x$ and $y$ define $y$ as a function of $x$. For example, let $y^2 = x$. If $x$ is 9, then $y^2 = 9$, so $y = \pm 3$. Hence, to the input 9, there are assigned not one, but *two*, output numbers: 3 and $-3$. This violates the definition of a function, so $y$ is **not** a function of $x$.

In $y^2 = x$, $x$ and $y$ are related, but the relationship is not a function of $x$.

On the other hand, some equations in two variables define either variable as a function of the other variable. For example, if $y = 2x$, then for each input $x$, there is exactly one output, $2x$. Thus, $y$ is a function of $x$. However, solving the equation for $x$ gives $x = y/2$. For each input $y$, there is exactly one output, $y/2$. Consequently, $x$ is a function of $y$.

Usually, the letters $f$, $g$, $h$, $F$, $G$, and so on are used to represent function rules. For example, Eq. (3), $y = x + 2$, defines $y$ as a function of $x$, where the rule is "Add 2 to the input." Suppose we let $f$ represent this rule. Then we say that $f$ is the function. To indicate that $f$ assigns the output 3 to the input 1, we write $f(1) = 3$, which is read "$f$ of 1 equals 3." Similarly, $f(-4) = -2$. More generally, if $x$ is any input, we have the following notation:

*$f(x)$ is an output number.*

$f(x)$, which is read "$f$ of $x$," means the output number in the range of $f$ that corresponds to the input number $x$ in the domain.

input
↓
$f(x)$
‿
↑
output

Thus, the output $f(x)$ is the same as $y$. But since $y = x + 2$, we may write $y = f(x) = x + 2$, or simply,

$$f(x) = x + 2.$$

For example, to find $f(3)$, which is the output corresponding to the input 3, we replace each $x$ in $f(x) = x + 2$ by 3:

$$f(3) = 3 + 2 = 5.$$

Likewise,

$$f(8) = 8 + 2 = 10,$$
$$f(-4) = -4 + 2 = -2.$$

Output numbers such as $f(-4)$ are called **function values** (or functional values). Keep in mind that they are in the range of $f$.

*Pitfall* ▼ $f(x)$ does **not** mean $f$ times $x$. $f(x)$ is the output that corresponds to the input $x$.

Functional notation is used extensively in calculus.

Quite often, functions are defined by "functional notation." For example, the equation $g(x) = x^3 + x^2$ defines the function $g$ that assigns the output number $x^3 + x^2$ to an input number $x$:

$$g: x \rightarrow x^3 + x^2.$$

In other words, $g$ adds the cube and the square of an input number. Some function values are

$$g(2) = 2^3 + 2^2 = 12,$$
$$g(-1) = (-1)^3 + (-1)^2 = -1 + 1 = 0,$$
$$g(t) = t^3 + t^2,$$
$$g(x + 1) = (x + 1)^3 + (x + 1)^2.$$

The idea of *replacement* is very important in determining function values.

Note that $g(x + 1)$ was found by replacing each $x$ in $x^3 + x^2$ by the input $x + 1$.

When we refer to the function $g$ defined by $g(x) = x^3 + x^2$, we shall feel free to call the equation itself a function. Thus, we speak of "the function $g(x) = x^3 + x^2$," and, similarly, "the function $y = x + 2$."

Let's be specific about the domain of a function. Unless otherwise stated, the domain consists of all real numbers for which the rule of the function makes sense; that is, the rule gives function values that are real numbers.

For example, suppose

$$h(x) = \frac{1}{x - 6}.$$

Here any real number can be used for $x$ except 6, because the denominator is 0 when $x$ is 6. So the domain of $h$ is understood to be all real numbers except 6.

---

**Principles in Practice 1**

**Finding Domains**

The area of a circle depends on the length of the radius of the circle.

**a.** Write a function $a(r)$ for the area of a circle when the length of the radius is $r$.

**b.** What is the domain of this function out of context?

**c.** What is the domain of this function in the given context?

---

**EXAMPLE 1  Finding Domains**

*Find the domain of each function.*

**a.** $f(x) = \dfrac{x}{x^2 - x - 2}.$

*Solution:* We cannot divide by zero, so we must find any values of $x$ that make the denominator 0. These *cannot* be input numbers. Thus, we set the denominator equal to 0 and solve for $x$:

$$x^2 - x - 2 = 0 \quad \text{(quadratic equation),}$$
$$(x - 2)(x + 1) = 0 \quad \text{(factoring),}$$
$$x = 2, -1.$$

Therefore, the domain of $f$ is all real numbers *except* 2 and $-1$.

**b.** $g(t) = \sqrt{2t - 1}.$

*Solution:* $\sqrt{2t - 1}$ is a real number if $2t - 1$ is greater than or equal to 0. If $2t - 1$ is negative, then $\sqrt{2t - 1}$ is not a real number. (It is an *imaginary number*.) Since function values must be real numbers, we must assume that

$$2t - 1 \geq 0,$$

$$2t \geq 1 \qquad \text{(adding 1 to both sides)},$$

$$t \geq \frac{1}{2} \qquad \text{(dividing both sides by 2)}.$$

Thus, the domain is the interval $[\frac{1}{2}, \infty)$.

---

**Principles in Practice 2**

**Finding Domain and Function Values**

The time it takes to go a given distance depends on the speed at which one is traveling.

a. Write a function $t(r)$ for the time it takes if the distance is 300 miles and the speed is $r$.

b. What is the domain of this function out of context?

c. What is the domain of this function in the given context?

d. Find $t(x)$, $t\left(\dfrac{x}{2}\right)$, and $t\left(\dfrac{x}{4}\right)$.

e. What happens to the time if the speed is reduced (divided) by a constant $c$? Describe this situation using an equation.

---

**EXAMPLE 2  Finding Domain and Function Values**

Let $g(x) = 3x^2 - x + 5$. Any real number can be used for $x$, so the domain of $g$ is all real numbers.

**a.** *Find $g(z)$.*

*Solution:* Replacing each $x$ in $g(x) = 3x^2 - x + 5$ by $z$ gives

$$g(z) = 3(z)^2 - z + 5 = 3z^2 - z + 5.$$

**b.** *Find $g(r^2)$.*

*Solution:* Replacing each $x$ in $g(x) = 3x^2 - x + 5$ by $r^2$ gives

$$g(r^2) = 3(r^2)^2 - r^2 + 5 = 3r^4 - r^2 + 5.$$

**c.** *Find $g(x + h)$.*

*Solution:*

$$g(x + h) = 3(x + h)^2 - (x + h) + 5$$
$$= 3(x^2 + 2hx + h^2) - x - h + 5$$
$$= 3x^2 + 6hx + 3h^2 - x - h + 5.$$

---

*Pitfall* ▼ Don't be confused by notation. In Example 2(c), we found $g(x + h)$ by replacing each $x$ in $g(x) = 3x^2 - x + 5$ by the input $x + h$. **Don't** write the function and then add $h$. That is, $g(x + h) \neq g(x) + h$:

$$g(x + h) \neq 3x^2 - x + 5 + h.$$

Also, **don't** use the distributive law on $g(x + h)$. The parentheses do **not** stand for multiplication. That is,

$$g(x + h) \neq g(x) + g(h).$$

---

**EXAMPLE 3  Finding a Difference Quotient**

*If $f(x) = x^2$, find $\dfrac{f(x + h) - f(x)}{h}$.*

The difference quotient of a function is an important mathematical concept.

*Solution:* The expression $\dfrac{f(x + h) - f(x)}{h}$ is referred to as a **difference quotient.** Here the numerator is a difference of function values. We have

$$\frac{f(x + h) - f(x)}{h} = \frac{(x + h)^2 - x^2}{h}$$

$$= \frac{x^2 + 2hx + h^2 - x^2}{h} = \frac{2hx + h^2}{h}$$

$$= \frac{h(2x + h)}{h} = 2x + h.$$

In some cases, the domain of a function is restricted for physical or economic reasons. For example, the previous interest function $I = 100(0.06)t$ has $t \geq 0$ because $t$ represents time. Example 4 will give another illustration.

### EXAMPLE 4   Demand Function

Suppose that the equation $p = 100/q$ describes the relationship between the price per unit $p$ of a certain product and the number of units $q$ of the product that consumers will buy (that is, demand) per week at the stated price. This equation is called a *demand equation* for the product. If $q$ is an input number, then to each value of $q$, there is assigned exactly one output number $p$:

$$q \to \frac{100}{q} = p.$$

For example,

$$20 \to \frac{100}{20} = 5;$$

that is, when $q$ is 20, $p$ is 5. Thus, price $p$ is a function of quantity demanded, $q$. This function is called a **demand function.** The independent variable is $q$, and $p$ is the dependent variable. Since $q$ cannot be 0 (division by 0 is not defined) and cannot be negative ($q$ represents quantity), the domain is all values of $q$ such that $q > 0$. ∎

We have seen that a function is essentially a *correspondence* whereby to each input number in the domain there is assigned exactly one output number in the range. For the correspondence given by $f(x) = x^2$, some sample assignments are shown by the arrows in Fig. 3.1. The next example discusses a functional correspondence that is not given by an algebraic formula.

## Principles in Practice 3
### Demand Function

Suppose the weekly demand function for large pizzas at a local pizza parlor is $p = 26 - \dfrac{q}{40}$.

**a.** If the current price is $18.50 per pizza, how many pizzas are sold each week?

**b.** If 200 pizzas are sold each week, what is the current price?

**c.** If the owner wants to double the number of large pizzas sold each week (to 400), what should the price be?

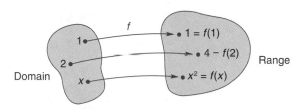

**FIGURE 3.1**    Functional correspondence for $f(x) = x^2$.

### EXAMPLE 5   Supply Schedule

The table in Fig. 3.2 is a *supply schedule*. Such a table gives a correspondence between the price $p$ of a certain product and the quantity $q$ the producers will supply per week at that price. For each price, there corresponds exactly one quantity, and vice versa.

If $p$ is the independent variable, then $q$ is a function of $p$, say, $q = f(p)$, and

$$f(500) = 11, \quad f(600) = 14, \quad f(700) = 17, \quad \text{and} \quad f(800) = 20.$$

Notice that as price per unit increases, the producers are willing to supply more units per week.

On the other hand, if $q$ is the independent variable, then $p$ is a function of $q$, say $p = g(q)$, and

**Supply Schedule**

| $p$ Price per Unit in Dollars | $q$ Quantity Supplied per Week |
|---|---|
| 500 | 11 |
| 600 | 14 |
| 700 | 17 |
| 800 | 20 |

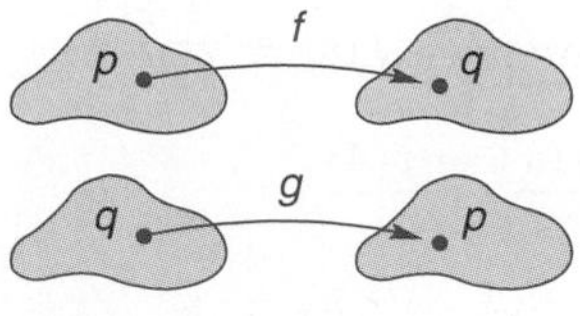

**FIGURE 3.2**    Supply schedule and supply functions.

$$g(11) = 500, \quad g(14) = 600, \quad g(17) = 700, \quad \text{and} \quad g(20) = 800.$$

We speak of $f$ and $g$ as **supply functions.**  ■

## TECHNOLOGY

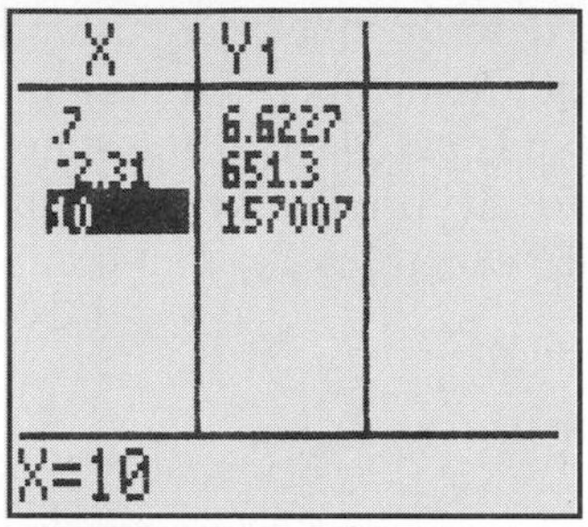

**FIGURE 3.3** Table of function values for $f(x) = 17x^4 - 13x^3 + 7$.

Function values are easily computed with a graphics calculator. For example, suppose

$$f(x) = 17x^4 - 13x^3 + 7,$$

and we wish to find $f(0.7)$, $f(-2.31)$, and $f(10)$. With a TI-82, we first enter the function as $Y_1$:

$$Y_1 = 17X^\wedge 4 - 13X^\wedge 3 + 7.$$

After pressing the TABLE key, we successively enter the $x$-values .7, $-2.31$, and 10. The results are shown in Fig. 3.3. We remark that there are other methods of determining function values with the TI-82.

## ■ Exercise 3.1

*In Problems 1–12, give the domain of each function.*

**1.** $f(x) = \dfrac{8}{x}$.

**2.** $g(x) = \dfrac{x}{5}$.

**3.** $h(x) = \sqrt{x - 3}$.

**4.** $H(z) = \dfrac{1}{\sqrt{z}}$.

**5.** $F(t) = 4t^2 - 6$.

**6.** $H(x) = \dfrac{x}{x + 8}$.

**7.** $f(x) = \dfrac{3x - 1}{2x + 5}$.

**8.** $g(x) = \sqrt{4x + 3}$.

**9.** $G(y) = \dfrac{4}{y^2 - y}$.

**10.** $f(x) = \dfrac{x + 1}{x^2 + 6x + 5}$.

**11.** $h(s) = \dfrac{4 - s^2}{2s^2 - 7s - 4}$.

**12.** $G(r) = \dfrac{2}{r^2 + 1}$.

*In Problems 13–24, find the function values for each function.*

**13.** $f(x) = 2x + 1$; $f(0), f(3), f(-4)$.

**14.** $H(s) = s^2 - 3$; $H(4), H(\sqrt{2}), H(\frac{2}{3})$.

**15.** $G(x) = 2 - x^2$; $G(-8), G(u), G(u^2)$.

**16.** $f(x) = 7x$; $f(s), f(t + 1), f(x + 3)$.

**17.** $g(u) = u^2 + u$; $g(-2), g(2v), g(-x^2)$.

**18.** $h(v) = \dfrac{1}{\sqrt{v}}$; $h(16), h\left(\dfrac{1}{4}\right), h(1 - x)$.

**19.** $f(x) = x^2 + 2x + 1$; $f(1), f(-1), f(x + h)$.

**20.** $H(x) = (x + 4)^2$; $H(0), H(2), H(t - 4)$.

**21.** $g(x) = \dfrac{x - 5}{x^2 + 4}$; $g(5), g(3x), g(x + h)$.

**22.** $H(x) = \sqrt{4 + x}$; $H(-4), H(-3), H(x + 1) - H(x)$.

**23.** $f(x) = x^{4/3}$; $f(0), f(64), f(\frac{1}{8})$.

**24.** $g(x) = x^{2/5}$; $g(32), g(-64), g(t^{10})$.

*In Problems 25–32, find (a) $f(x + h)$ and (b) $\dfrac{f(x + h) - f(x)}{h}$; simplify your answers.*

**25.** $f(x) = 4x - 5$.

**26.** $f(x) = \dfrac{x}{2}$.

**27.** $f(x) = x^2 + 2x$.

**28.** $f(x) = 2x^2 - 3x - 5$.

**29.** $f(x) = 2 - 4x - 3x^2$.

**30.** $f(x) = x^3$.

**31.** $f(x) = \dfrac{1}{x}$.

**32.** $f(x) = \dfrac{x + 1}{x}$.

**33.** If $f(x) = 9x + 7$, find $\dfrac{f(2 + h) - f(2)}{h}$.

**34.** If $f(x) = x^2 - x$, find $\dfrac{f(x) - f(4)}{x - 4}$.

*In Problems 35–38, is y a function of x? Is x a function of y?*

**35.** $y - 3x - 4 = 0$.

**36.** $x^2 + y = 0$.

**37.** $y = 7x^2$.

**38.** $x^2 + y^2 = 1$.

**39.** The formula for the area of a circle of radius $r$ is $A = \pi r^2$. Is the area a function of the radius?

**40.** Suppose $f(b) = ab^2 + a^2b$. (a) Find $f(a)$. (b) Find $f(ab)$.

**41. Value of Business**   A business with an original capital of $10,000 has income and expenses each week of $2000 and $1600, respectively. If all profits are retained in the business, express the value $V$ of the business at the end of $t$ weeks as a function of $t$.

**42. Depreciation**   If a $30,000 machine depreciates 2% of its original value each year, find a function $f$ that expresses the machine's value $V$ after $t$ years have elapsed.

**43. Profit Function**   If $q$ units of a certain product are sold ($q$ is nonnegative), the profit $P$ is given by the equation $P = 1.25q$. Is $P$ a function of $q$? What is the dependent variable? the independent variable?

**44. Demand Function**   Suppose the yearly demand function for a particular actor to star in a film is $p = \dfrac{600,000}{q}$, where $q$ is the number of films he stars in during the year. If the actor currently charges $300,000 per film, how many films does he star in each year? If he wants to star in four films per year, what should his price be?

**45. Supply Function**   Suppose the weekly supply function for a pound of your house-blend coffee at a local coffee shop is $p = \dfrac{q}{50}$, where $q$ is the number of pounds of coffee supplied per week. How many pounds of coffee per week will be supplied if the price is $8.00 a pound? How many pounds of coffee per week will be supplied if the price is $20.00 a pound? How does the amount supplied change as the price increases?

**46. Hospital Discharges**   An insurance company examined the records of a group of individuals hospitalized for a particular illness. It was found that the total proportion discharged at the end of $t$ days of hospitalization is given by

$$f(t) = 1 - \left(\frac{300}{300 + t}\right)^3.$$

Evaluate (a) $f(0)$, (b) $f(100)$, and (c) $f(300)$. (d) At the end of how many days was 99.9% (0.999) of the group discharged?

**47. Psychology**   A psychophysical experiment was conducted to analyze human response to electrical shocks.[1] The subjects received a shock of a certain intensity. They were told to assign a magnitude of 10 to this particular shock, called the standard stimulus. Then other shocks (stimuli) of various intensities were given. For each one, the response $R$ was to be a number that indicated the perceived magnitude of the shock relative to that of the standard stimulus. It was found that $R$ was a function of the intensity $I$ of the shock ($I$ in microamperes) and was estimated by

$$R = f(I) = \frac{I^{4/3}}{2500}, \qquad 500 \le I \le 3500.$$

Evaluate (a) $f(1000)$ and (b) $f(2000)$. (c) Suppose that $I_0$ and $2I_0$ are in the domain of $f$. Express $f(2I_0)$ in terms of $f(I_0)$. What effect does the doubling of intensity have on response?

**48. Psychology**   In a paired-associate learning experiment,[2] the probability of a correct response as a function of the number $n$ of trials has the form

$$P(n) = 1 - \frac{1}{2}(1 - c)^{n-1}, \qquad n \ge 1,$$

where the estimated value of $c$ is 0.344. Find $P(1)$ and $P(2)$ by using this value of $c$.

**49. Demand Schedule**   The following table is called a *demand schedule*. It gives a correspondence between the price $p$ of a product and the quantity $q$ that consumers will demand (that is, purchase) at that price. (a) If $p = f(q)$, list the numbers in the domain of $f$. Find $f(2900)$ and $f(3000)$, (b) If $q = g(p)$, list the numbers in the domain of $g$. Find $g(10)$ and $g(17)$.

| Price per Unit, $p$ | Quantity Demanded per Week, $q$ |
|---|---|
| $10 | 3000 |
| 12 | 2900 |
| 17 | 2300 |
| 20 | 2000 |

*In Problems 50–53, use your calculator to find the indicated values for the given function. Round answers to two decimal places.*

 **50.** $f(x) = 2.03x^3 - 5.27x^2 - 13.71$;  (a) $f(1.73)$, (b) $f(-5.78)$, (c) $f(\sqrt{2})$.

 **52.** $f(x) = (20 - 3x)(2.25x^2 - 17.1x - 13)^4$; (a) $f(0.1)$,  (b) $f(-0.01)$,  (c) $f(6.6)$.

**51.** $f(x) = \dfrac{14.7x^2 - 3.95x - 15.76}{24.3 - x^3}$;  (a) $f(4)$, (b) $f(-17/4)$,  (c) $f(\pi)$.

 **53.** $f(x) = \sqrt{\dfrac{\sqrt{2}x^2 + 47.62(x + 1)}{9.07}}$;  (a) $f(15.93)$, (b) $f(-146)$, (c) $f(0)$.

[1] Adapted from H. Babkoff, "Magnitude Estimation of Short Electrocutaneous Pulses," *Psychological Research*, 39, no. 1 (1976), 39–49.

[2] D. Laming, *Mathematical Psychology* (New York: Academic Press, 1983).

**To introduce constant functions, polynomial functions, rational functions, compound functions, the absolute-value function, and factorial notation.**

---

### Principles in Practice 1
#### Constant Function

Suppose the monthly health insurance premiums for an individual are $125.00.
a. Write the monthly health insurance premiums as a function of the number of visits the individual makes to the doctor.
b. How do the health insurance premiums change as the number of visits to the doctor increases?
c. What kind of function is this?

Each term in a polynomial function is either a constant or a constant times a positive integral power of $x$.

---

### Principles in Practice 2
#### Polynomial Functions

The function $d(t) = 3t^2$ represents the distance in meters a car will go in $t$ seconds when it is has a constant acceleration of 6 meters per second.
a. What kind of function is this?
b. What is its degree?
c. What is its leading coefficient?

---

## 3.2 SPECIAL FUNCTIONS

In this section, we shall look at functions having special forms and representations. We begin with perhaps the simplest type of function there is: a *constant function.*

**EXAMPLE 1 Constant Function**

Let $h(x) = 2$. The domain of $h$ is all real numbers. All function values are 2. For example,

$$h(10) = 2, \qquad h(-387) = 2, \qquad h(x + 3) = 2.$$

We call $h$ a *constant function* because all the function values are the same. More generally, we have this definition:

A function of the form $h(x) = c$, where $c$ is a *constant,* is called a **constant function.**

∎

A constant function belongs to a broader class of functions, called *polynomial functions.* In general, a function of the form

$$f(x) = c_n x^n + c_{n-1} x^{n-1} + \cdots + c_1 x + c_0,$$

where $n$ is a nonnegative integer and $c_n, c_{n-1}, \ldots, c_0$ are constants with $c_n \neq 0$, is called a **polynomial function** (in $x$). The number $n$ is called the **degree** of the polynomial, and $c_n$ is the **leading coefficient.** Thus,

$$f(x) = 3x^2 - 8x + 9$$

is a polynomial function of degree 2 with leading coefficient 3. Likewise, $g(x) = 4 - 2x$ has degree 1 and leading coefficient $-2$. Polynomial functions of degree 1 or 2 are called **linear** or **quadratic functions,** respectively. For example, $g(x) = 4 - 2x$ is linear and $f(x) = 3x^2 - 8x + 9$ is quadratic. Note that a nonzero constant function, such as $f(x) = 5$ [which can be written as $f(x) = 5x^0$], is a polynomial function of degree 0. The constant function $f(x) = 0$ is also considered a polynomial function, but has no degree assigned to it. The domain of any polynomial function is all real numbers.

**EXAMPLE 2 Polynomial Functions**

a. $f(x) = x^3 - 6x^2 + 7$ is a polynomial (function) of degree 3 with leading coefficient 1.

b. $g(x) = \dfrac{2x}{3}$ is a linear function with leading coefficient $\dfrac{2}{3}$.

c. $f(x) = \dfrac{2}{x^3}$ is *not* a polynomial function. Because $f(x) = 2x^{-3}$ and the exponent for $x$ is not a nonnegative integer, this function does not have the proper form for a polynomial. Similarly, $g(x) = \sqrt{x}$ is not a polynomial, because $g(x) = x^{1/2}$.

∎

Another type of function is a *rational function,* which involves polynomials.

A function that is a quotient of polynomial functions is called a **rational function.**

### EXAMPLE 3   Rational Functions

**a.** $f(x) = \dfrac{x^2 - 6x}{x + 5}$ is a rational function, since the numerator and denominator are each polynomials. Note that this rational function is not defined for $x = -5$.

**b.** $g(x) = 2x + 3$ is a rational function, since $2x + 3 = \dfrac{2x + 3}{1}$. In fact, every polynomial function is also a rational function. ■

*Every polynomial function is a rational function.*

Sometimes more than one expression is needed to define a function, as Example 4 shows.

### EXAMPLE 4   Compound Function

Let

$$F(s) = \begin{cases} 1, & \text{if } -1 \le s < 1, \\ 0, & \text{if } 1 \le s \le 2, \\ s - 3 & \text{if } 2 < s \le 8. \end{cases}$$

This is called a **compound function** because the rule for specifying it is given by more than one expression. Here $s$ is the independent variable, and the domain of $F$ is all $s$ such that $-1 \le s \le 8$. The value of $s$ determines which expression to use.

Find $F(0)$:   Since $-1 \le 0 < 1$, we have $F(0) = 1$.

Find $F(2)$:   Since $1 \le 2 \le 2$, we have $F(2) = 0$.

Find $F(7)$:   Since $2 < 7 \le 8$, we substitute 7 for $s$ in $s - 3$.

$$F(7) = 7 - 3 = 4.$$
■

---

## TECHNOLOGY

**FIGURE 3.4**   Entering a compound function.

To illustrate how to enter a compound function with a TI-82, Fig. 3.4 shows a key sequence for the function

$$f(x) = \begin{cases} 2x, & \text{if } x < 0, \\ x^2, & \text{if } 0 \le x < 10, \\ -x, & \text{if } x \ge 10. \end{cases}$$

---

### EXAMPLE 5   Absolute-Value Function

The function $f(x) = |x|$ is called the *absolute-value function.* Recall that the **absolute value,** or **magnitude,** of a real number $x$ is denoted $|x|$ and is defined by

The absolute-value function can be considered a compound function.

$$|x| = \begin{cases} x, & \text{if } x \geq 0, \\ -x, & \text{if } x < 0. \end{cases}$$

Thus, the domain of $f$ is all real numbers. Some function values are

$$f(16) = |16| = 16,$$
$$f(-\tfrac{4}{3}) = |-\tfrac{4}{3}| = -(-\tfrac{4}{3}) = \tfrac{4}{3},$$
$$f(0) = |0| = 0.$$

In our next examples, we make use of *factorial notation*.

> The symbol $r!$, with $r$ a positive integer, is read "**$r$ factorial.**" It represents the product of the first $r$ positive integers:
>
> $$r! = 1 \cdot 2 \cdot 3 \cdots r.$$
>
> We define $0!$ to be $1$.

**Principles in Practice 4**

**Factorials**

Seven different books are to be placed on a shelf. How many ways can they be arranged? Represent the question as a factorial problem and give the solution.

**EXAMPLE 6   Factorials**

**a.** $5! = 1 \cdot 2 \cdot 3 \cdot 4 \cdot 5 = 120$.

**b.** $3!(6 - 5)! = 3! \cdot 1! = (3 \cdot 2 \cdot 1)(1) = (6)(1) = 6$.

**c.** $\dfrac{4!}{0!} = \dfrac{1 \cdot 2 \cdot 3 \cdot 4}{1} = \dfrac{24}{1} = 24$.

**EXAMPLE 7   Genetics**

*Suppose two black guinea pigs are bred and produce exactly five offspring. Under certain conditions, it can be shown that the probability $P$ that exactly $r$ of the offspring will be brown and the others black is a function of $r$, say, $P = P(r)$, where*

$$P(r) = \frac{5!(\tfrac{1}{4})^r(\tfrac{3}{4})^{5-r}}{r!(5-r)!}, \qquad r = 0, 1, 2, \ldots, 5.$$

Factorials occur frequently in probability theory.

*The letter $P$ in $P = P(r)$ is used in two ways. On the right side, $P$ represents the function rule. On the left side, $P$ represents the dependent variable. The domain of $P$ is all integers from 0 to 5, inclusive. Find the probability that exactly three guinea pigs will be brown.*

*Solution:* We want to find $P(3)$. We have

$$P(3) = \frac{5!(\tfrac{1}{4})^3(\tfrac{3}{4})^2}{3!2!} = \frac{120(\tfrac{1}{64})(\tfrac{9}{16})}{6(2)} = \frac{45}{512}.$$

## ■ Exercise 3.2

*In Problems 1–4, determine whether the given function is a polynomial function.*

**1.** $f(x) = x^2 - x^4 + 4$.     **2.** $f(x) = \dfrac{x^2 + 7}{3}$.     **3.** $g(x) = \dfrac{3}{x^2 + 7}$.     **4.** $g(x) = 3^{-2}x^2$.

*In Problems 5–8, determine whether the given function is a rational function.*

**5.** $f(x) = \dfrac{x^2 + x}{x^3 + 4}$.     **6.** $f(x) = \dfrac{3}{2x + 1}$.     **7.** $g(x) = 8x$.     **8.** $g(x) = 4x^{-4}$.

*In Problems 9–12, find the domain of each function.*

**9.** $H(z) = 16.$

**10.** $f(t) = \pi.$

**11.** $f(x) = \begin{cases} 5x, & \text{if } x > 1, \\ 4, & \text{if } x \le 1. \end{cases}$

**12.** $f(x) = \begin{cases} 4, & \text{if } x = 3, \\ x^2, & \text{if } 1 \le x < 3. \end{cases}$

*In Problems 13–16, state (a) the degree and (b) the leading coefficient of the given polynomial function.*

**13.** $F(x) = 7x^3 - 2x^2 + 6.$  **14.** $f(x) = x.$  **15.** $f(x) = 2 - 3x^4 + 2x.$  **16.** $f(x) = 9.$

*In Problems 17–22, find the function values for each function.*

**17.** $f(x) = 8;$  $f(2),$  $f(t + 8),$  $f(-\sqrt{17}).$

**18.** $g(x) = |x - 3|;$  $g(10),$  $g(3),$  $g(-3).$

**19.** $F(t) = \begin{cases} 1, & \text{if } t > 0 \\ 0, & \text{if } t = 0; \\ -1, & \text{if } t < 0 \end{cases}$
$F(10),$  $F(-\sqrt{3}),$  $F(0),$  $F(-\frac{18}{5}).$

**20.** $f(x) = \begin{cases} 4, & \text{if } x \ge 0 \\ 3, & \text{if } x < 0 \end{cases};$
$f(3),$  $f(-4),$  $f(0).$

**21.** $G(x) = \begin{cases} x, & \text{if } x \ge 3 \\ 2 - x, & \text{if } x < 3 \end{cases};$
$G(8),$  $G(3),$  $G(-1),$  $G(1).$

**22.** $h(r) = \begin{cases} 3r - 1, & \text{if } r > 2 \\ r^2 - 4r + 7, & \text{if } r < 2 \end{cases};$
$h(3),$  $h(-3),$  $h(2).$

*In Problems 23–28, determine the value of each expression.*

**23.** $6!.$

**24.** $0!.$

**25.** $(4 - 2)!.$

**26.** $5! \cdot 3!.$

**27.** $\dfrac{5!}{4!}.$

**28.** $\dfrac{8!}{5!(8 - 5)!}.$

**29. Train Trip**  A daily round-trip train ticket to the city costs $3.50. Write the cost of a daily round-trip ticket as a function of a passenger's income. What kind of function is this?

**30. Geometry**  A rectangular prism has length three more than its width and height one less than twice the width. Write the volume of the rectangular prism as a function of the width. What kind of function is this?

**31. Cost Function**  In manufacturing a component for a machine, the initial cost of a die is $850 and all other additional costs are $3 per unit produced. (a) Express the total cost $C$ (in dollars) as a linear function of the number $q$ of units produced. (b) How many units are produced if the total cost is $1600?

**32. Investment**  If a principal of $P$ dollars is invested at a simple annual interest rate of $r$ for $t$ years, express the total accumulated amount of the principal and interest as a function of $t$. Is your result a linear function of $t$?

**33. Sales**  To encourage large group sales, a theater charges two rates. If your group is less than 10, each ticket costs $7.50. If your group is 10 or more, each ticket costs $7.00. Write a compound function to represent the cost of buying $n$ tickets.

**34. Factorials**  At an amusement park, a group of friends wants to ride the log ride in every possible order. How many rides will a group of three have to take? How about a group of four? Five?

**35. Genetics**  Under certain conditions, if two brown-eyed parents have exactly three children, the probability that there will be exactly $r$ blue-eyed children is given by the function $P = P(r)$, where

$$P(r) = \frac{3!\left(\frac{1}{4}\right)^r\left(\frac{3}{4}\right)^{3-r}}{r!(3 - r)!}, \qquad r = 0, 1, 2, 3.$$

Find the probability that exactly two of the children will be blue-eyed.

**36. Genetics**  In Example 7, find the probability that all five offspring will be brown.

**37. Bacteria Growth**  Bacteria are growing in a culture. The time $t$ (in hours) for the bacteria to double in number (the generation time) is a function of the temperature $T$ (in °C) of the culture. If this function is given by[3]

$$t = f(T) = \begin{cases} \dfrac{1}{24}T + \dfrac{11}{4}, & \text{if } 30 \le T \le 36, \\[2mm] \dfrac{4}{3}T - \dfrac{175}{4}, & \text{if } 36 < T \le 39, \end{cases}$$

(a) determine the domain of $f$, and (b) find $f(30), f(36)$, and $f(39)$.

[3]Adapted from F. K. E. Imrie and A. J. Vlitos, "Production of Fungal Protein from Carob," in *Single-Cell Protein II*, ed. S. R. Tannenbaum and D. I. C. Wang (Cambridge, MA. MIT Press, 1975).

*In Problems 38–41, use your calculator to find the indicated function values for the given function. Round answers to two decimal places.*

**38.** $f(x) = \begin{cases} 0.08x^5 - 47.98, & \text{if } x \geq 7.98 \\ 0.67x^6 - 37.41, & \text{if } x < 7.98 \end{cases}$.

(a) $f(7.98)$, (b) $f(2.26)$, (c) $f(9)$.

**39.** $f(x) = \begin{cases} 47.1x^5 + 30.4, & \text{if } x > 0 \\ 9.4x^3 - x, & \text{if } x \leq 0 \end{cases}$

(a) $f(5.5)$, (b) $f(-3.6)$, (c) $f(6/7)$.

**40.** $f(x) = \begin{cases} 4.07x - 2.3 & \text{if } x < -8 \\ 19.12, & \text{if } -8 \leq x < -2; \\ x^2 - 4x^{-2}, & \text{if } x \geq -2 \end{cases}$

(a) $f(-6.3)$, (b) $f(-13.2)$, (c) $f(13.2)$.

**41.** $f(x) = \begin{cases} x/(x+3), & \text{if } x < -5 \\ x(x-4)^2, & \text{if } -5 \leq x < 0; \\ \sqrt{2.1x + 3}, & \text{if } x \geq 0 \end{cases}$

(a) $f(-\sqrt{30})$, (b) $f(46)$, (c) $f(-2/3)$.

---

OBJECTIVE

**To combine functions by means of addition, subtraction, multiplication, division, and composition.**

## 3.3 COMBINATIONS OF FUNCTIONS

There are different ways of combining two functions to create a new function. Suppose $f$ and $g$ are the functions given by

$$f(x) = x^2 \quad \text{and} \quad g(x) = 3x.$$

Adding $f(x)$ and $g(x)$ gives

$$f(x) + g(x) = x^2 + 3x.$$

This operation defines a new function called the *sum* of $f$ and $g$, denoted $f + g$. Its function value at $x$ is $f(x) + g(x)$. That is,

$$(f + g)(x) = f(x) + g(x) = x^2 + 3x.$$

For example,

$$(f + g)(2) = 2^2 + 3(2) = 10.$$

In general, for any functions $f$ and $g$, we define the **sum $f + g$**, the **difference $f - g$**, the **product $fg$**, and the **quotient $\dfrac{f}{g}$** as follows:[4]

$$(f + g)(x) = f(x) + g(x),$$
$$(f - g)(x) = f(x) - g(x),$$
$$(fg)(x) = f(x) \cdot g(x),$$
$$\frac{f}{g}(x) = \frac{f(x)}{g(x)}.$$

Thus, for $f(x) = x^2$ and $g(x) = 3x$, we have

$$(f + g)(x) = f(x) + g(x) = x^2 + 3x,$$
$$(f - g)(x) = f(x) - g(x) = x^2 - 3x,$$
$$(fg)(x) = f(x) \cdot g(x) = x^2(3x) = 3x^3,$$
$$\frac{f}{g}(x) = \frac{f(x)}{g(x)} = \frac{x^2}{3x} = \frac{x}{3}, \quad \text{for } x \neq 0.$$

---

[4]In each of the four combinations, we assume that $x$ is in the domains of both $f$ and $g$. In the quotient, we also do not allow any value of $x$ for which $g(x)$ is 0.

**EXAMPLE 1  Combining Functions**

If $f(x) = 3x - 1$ and $g(x) = x^2 + 3x$, find

**a.** $(f + g)(x)$,          **b.** $(f - g)(x)$,

**c.** $(fg)(x)$,          **d.** $\dfrac{f}{g}(x)$.

*Solution:*

**a.** $(f + g)(x) = f(x) + g(x) = (3x - 1) + (x^2 + 3x) = x^2 + 6x - 1$.

**b.** $(f - g)(x) = f(x) - g(x) = (3x - 1) - (x^2 + 3x) = -1 - x^2$.

**c.** $(fg)(x) = f(x)g(x) = (3x - 1)(x^2 + 3x) = 3x^3 + 8x^2 - 3x$.

**d.** $\dfrac{f}{g}(x) = \dfrac{f(x)}{g(x)} = \dfrac{3x - 1}{x^2 + 3x}$. ∎

## Composition

We may also combine two functions by first applying one function to a number and then applying the other function to the result. For example, suppose $g(x) = 3x$, $f(x) = x^2$, and $x = 2$. Then $g(2) = 3(2) = 6$. Thus, $g$ sends the input 2 into the output 6:

$$2 \xrightarrow{\;g\;} 6.$$

Next, we let the output 6 become input to $f$:

$$f(6) = 6^2 = 36.$$

So $f$ sends 6 into 36:

$$6 \xrightarrow{\;f\;} 36.$$

By first applying $g$ and then $f$, we send 2 into 36:

$$2 \xrightarrow{\;g\;} 6 \xrightarrow{\;f\;} 36.$$

To be more general, let's replace the 2 by $x$, where $x$ is in the domain of $g$. (See Fig. 3.5). Applying $g$ to $x$, we get the number $g(x)$, which we shall assume is in the domain of $f$. By applying $f$ to $g(x)$, we get $f(g(x))$, read "$f$ of $g$ of $x$," which is in the range of $f$. The operation of applying $g$ and then applying $f$ to the result defines a so-called composite function denoted $f \circ g$. This function assigns the output number $f(g(x))$ to the input number $x$. (See the bottom arrow in Fig. 3.5.) Thus, $(f \circ g)(x) = f(g(x))$. We can think of $f(g(x))$ as a function of a function.

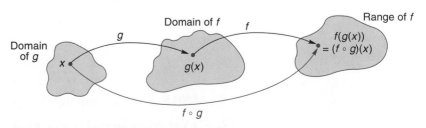

**FIGURE 3.5**  Composition of $f$ with $g$.

**DEFINITION**

*If f and g are functions, the **composition of f with g** is the function f ∘ g defined by*

$$(f \circ g)(x) = f(g(x)),$$

*where the domain of f ∘ g is the set of all x in the domain of g such that g(x) is in the domain of f.*

For $f(x) = x^2$ and $g(x) = 3x$, we can get a simple form for $f \circ g$ :

$$(f \circ g)(x) = f(g(x)) = f(3x) = (3x)^2 = 9x^2.$$

For example, $(f \circ g)(2) = 9(2)^2 = 36$, as we saw before.

---

**Principles in Practice 1**

**Composition**

A CD costs $x$ dollars wholesale. The price the store pays is given by the function $s(x) = x + 3$. The price the customer pays is $c(x) = 2x$, where $x$ is the price the store pays. Write a composite function to find the customer's price as a function of the wholesale price.

---

**EXAMPLE 2   Composition**

*Let $f(x) = \sqrt{x}$ and $g(x) = x + 1$. Find*

**a.** $(f \circ g)(x)$,          **b.** $(g \circ f)(x)$.

*Solution:*

**a.** $(f \circ g)(x)$ is $f(g(x))$. Now $g$ adds 1 to $x$, and $f$ takes the square root of the result. Thus,

$$(f \circ g)(x) = f(g(x)) = f(x + 1) = \sqrt{x + 1}.$$

The domain of $g$ is all real numbers $x$, and the domain of $f$ is all nonnegative reals. Hence, the domain of the composition is all $x$ for which $g(x) = x + 1$ is nonnegative. That is, the domain is all $x \geq -1$, or equivalently, the interval $[-1, \infty)$.

**b.** $(g \circ f)(x)$ is $g(f(x))$. Now $f$ takes the square root of $x$, and $g$ adds 1 to the result. Thus, $g$ adds 1 to $\sqrt{x}$, and we have

$$(g \circ f)(x) = g(f(x)) = g(\sqrt{x}) = \sqrt{x} + 1.$$

The domain of $f$ is all $x \geq 0$, and the domain of $g$ is all reals. Hence, the domain of the composition is all $x \geq 0$ for which $f(x) = \sqrt{x}$ is real, namely, all $x \geq 0$. ∎

***Pitfall*** ▼ Generally, $f \circ g \neq g \circ f$. In Example 2,

$$(f \circ g)(x) = \sqrt{x + 1},$$

but we have

$$(g \circ f)(x) = \sqrt{x} + 1.$$

Also, do not confuse $f(g(x))$ with $(fg)(x)$, which is the product $f(x)g(x)$. Here

$$f(g(x)) = \sqrt{x + 1},$$

but

$$f(x)g(x) = \sqrt{x}(x + 1).$$

**EXAMPLE 3   Composition**

*If $F(p) = p^2 + 4p - 3$ and $G(p) = 2p + 1$, find*

**a.** $F(G(p))$,          **b.** $G(F(1))$.

*Solution:*

**a.** $F(G(p)) = F(2p + 1) = (2p + 1)^2 + 4(2p + 1) - 3$

$$= 4p^2 + 12p + 2.$$

**b.** $G(F(1)) = G(1^2 + 4 \cdot 1 - 3) = G(2) = 2 \cdot 2 + 1 = 5.$ ∎

In calculus, it is necessary at times to think of a particular function as a composition of two simpler functions, as the next example shows.

**Principles in Practice 2**

**Expressing a Function as a Composition**

Suppose the area of a square garden is $g(x) = (x + 3)^2$. Express $g$ as a composition of two functions, and explain what each function represents.

**EXAMPLE 4  Expressing a Function as a Composition**

*Express $h(x) = (2x - 1)^3$ as a composition.*

*Solution:*

We note that $h(x)$ is obtained by finding $2x - 1$ and cubing the result. Suppose we let $g(x) = 2x - 1$ and $f(x) = x^3$. Then

$$h(x) = (2x - 1)^3 = [g(x)]^3 = f(g(x)) = (f \circ g)(x),$$

which gives $h$ as a composition of two functions. ∎

**TECHNOLOGY**

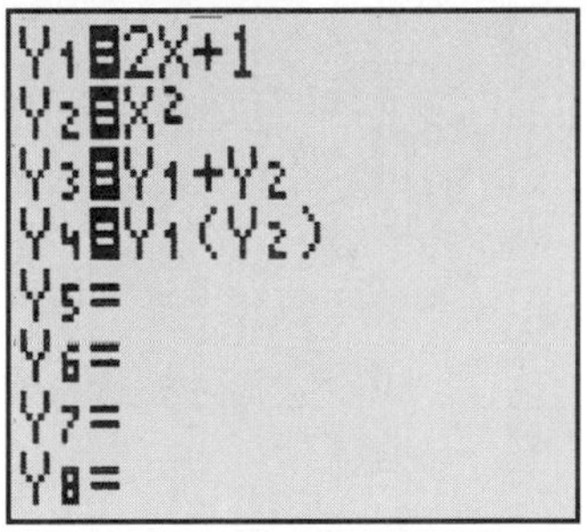

```
Y₁◻2X+1
Y₂◻X²
Y₃◻Y₁+Y₂
Y₄◻Y₁(Y₂)
Y₅=
Y₆=
Y₇=
Y₈=
```

**FIGURE 3.6**  $Y_3$ and $Y_4$ are combinations of $Y_1$ and $Y_2$.

Two functions can be combined by using a graphics calculator. Consider the functions

$$f(x) = 2x + 1 \quad \text{and} \quad g(x) = x^2,$$

which we enter as $Y_1$ and $Y_2$, as shown in Fig. 3.6. The sum of $f$ and $g$ is given by $Y_3 = Y_1 + Y_2$ and the composition $f \circ g$ by $Y_4 = Y_1(Y_2)$. For example, $f(g(3))$ is obtained by evaluating $Y_4$ at 3.

■ **Exercise 3.3**

**1.** If $f(x) = x + 3$ and $g(x) = x + 5$, find the following.
   **a.** $(f + g)(x)$.    **b.** $(f + g)(0)$.    **c.** $(f - g)(x)$.

   **d.** $(fg)(x)$.    **e.** $(fg)(-2)$.    **f.** $\dfrac{f}{g}(x)$.

   **g.** $(f \circ g)(x)$.    **h.** $(f \circ g)(3)$.    **i.** $(g \circ f)(x)$.

**3.** If $f(x) = x^2$ and $g(x) = x^2 + x$, find the following.
   **a.** $(f + g)(x)$.    **b.** $(f - g)(x)$.    **c.** $(f - g)(-\frac{1}{2})$.

   **d.** $(fg)(x)$.    **e.** $\dfrac{f}{g}(x)$.    **f.** $\dfrac{f}{g}(-\frac{1}{2})$.

   **g.** $(f \circ g)(x)$.    **h.** $(g \circ f)(x)$.    **i.** $(g \circ f)(-3)$.

**5.** If $f(x) = 3x^2 + 6$ and $g(x) = 4 - 2x$, find $f(g(2))$ and $g(f(2))$.

**2.** If $f(x) = 2x$ and $g(x) = 4 + x$, find the following.
   **a.** $(f + g)(x)$.    **b.** $(f - g)(x)$.    **c.** $(f - g)(4)$.

   **d.** $(fg)(x)$.    **e.** $\dfrac{f}{g}(x)$.    **f.** $\dfrac{f}{g}(2)$.

   **g.** $(f \circ g)(x)$.    **h.** $(g \circ f)(x)$.    **i.** $(g \circ f)(2)$.

**4.** If $f(x) = x^2 - 1$ and $g(x) = 4$, find the following.
   **a.** $(f + g)(x)$.    **b.** $(f + g)(\frac{1}{2})$.    **c.** $(f - g)(x)$.

   **d.** $(fg)(x)$.    **e.** $(fg)(4)$.    **f.** $\dfrac{f}{g}(x)$.

   **g.** $(f \circ g)(x)$.    **h.** $(f \circ g)(100)$.    **i.** $(g \circ f)(x)$.

**6.** If $f(p) = \dfrac{4}{p}$ and $g(p) = \dfrac{p - 2}{3}$, find both $(f \circ g)(p)$ and $(g \circ f)(p)$.

**7.** If $F(t) = t^2 + 3t + 1$ and $G(t) = \dfrac{2}{t-1}$, find $(F \circ G)(t)$ and $(G \circ F)(t)$.

**8.** If $F(s) = \sqrt{s}$ and $G(t) = 3t^2 + 4t + 2$, find $(F \circ G)(t)$ and $(G \circ F)(t)$.

**9.** If $f(w) = \dfrac{1}{w^2 + 1}$ and $g(v) = \sqrt{v + 2}$, find $(f \circ g)(v)$ and $(g \circ f)(w)$.

**10.** If $f(x) = x^2 + 3$, find $(f \circ f)(x)$.

*In Problems 11–16, find functions f and g such that $h(x) = f(g(x))$.*

**11.** $h(x) = (4x - 3)^5$.

**12.** $h(x) = \sqrt{x^2 - 2}$.

**13.** $h(x) = \dfrac{1}{x^2 - 2}$.

**14.** $h(x) = (3x^3 - 2x)^3 - (3x^3 - 2x)^2 + 7$.

**15.** $h(x) = \sqrt[5]{\dfrac{x+1}{3}}$.

**16.** $h(x) = \dfrac{x+1}{(x+1)^2 + 2}$.

**17. Profit**  A coffeehouse sells a pound of coffee for \$9.75. Expenses are \$4500 each month, plus \$4.25 for each pound of coffee sold.

   (a) Write a function $r(x)$ for the total monthly revenue as a function of the number of pounds of coffee sold.

   (b) Write a function $e(x)$ for the total monthly expenses as a function of the number of pounds of coffee sold.

   (c) Write a function $(r - e)(x)$ for the total monthly profit as a function of the number of pounds of coffee sold.

**18. Geometry**  Suppose the volume of a cube is $v(x) = (4x - 2)^3$. Express $v$ is a composition of two functions, and explain what each function represents.

**19. Business**  A manufacturer determines that the total number of units of output per day, $q$, is a function of the number of employees, $m$, where

$$q = f(m) = \frac{(40m - m^2)}{4}.$$

The total revenue $r$ that is received for selling $q$ units is given by the function $g$, where $r = g(q) = 40q$. Find $(g \circ f)(m)$. What does this composition function describe?

**20. Sociology**  Studies have been conducted concerning the statistical relations between a person's status, education, and income.[5] Let $S$ denote a numerical value of status based on annual income $I$. For a certain population, suppose

$$S = f(I) = 0.45(I - 1000)^{0.53}.$$

Furthermore, suppose a person's income $I$ is a function of the number of years of education $E$, where

$$I = g(E) = 7202 + 0.29E^{3.68}.$$

Find $(f \circ g)(E)$. What does this function describe?

*In Problems 21–24, for the given functions f and g, find the indicated function values. Round answers to two decimal places.*

**21.** $f(x) = (4x - 13)^2$, $g(x) = 0.2x^2 - 4x + 3$;
(a) $(f + g)(4.1)$,   (b) $(f \circ g)(3)$.

**22.** $f(x) = \sqrt{\dfrac{x+2}{x}}$, $g(x) = 13.4x + 7.31$;
(a) $\dfrac{f}{g}(10)$,   (b) $(g \circ f)(-6)$.

**23.** $f(x) = x^{2/3}$, $g(x) = x^3 - 7$;   (a) $(fg)(5)$,
(b) $(g \circ f)(2.25)$.

**24.** $f(x) = \dfrac{5}{x+3}$, $g(x) = \dfrac{2}{x^2}$;   (a) $(f - g)(7.3)$,
(b) $(f \circ g)(-4.17)$.

[5]R. K. Leik and B. F. Meeker, *Mathematical Sociology* (Englewood Cliffs, NJ; Prentice-Hall, 1975)

---

**OBJECTIVE**

To graph equations and functions in rectangular coordinates, to determine intercepts, to apply the vertical-line test, and to determine the domain and range of a function from a graph.

## 3.4  GRAPHS IN RECTANGULAR COORDINATES

A **rectangular** (or **Cartesian**) **coordinate system** allows us to specify and locate points in a plane. It also provides a geometric way to represent equations in two variables as well as functions.

   In a plane, two real-number lines, called *coordinate axes,* are constructed perpendicular to each other so that their origins coincide, as in Fig. 3.7. Their point of intersection is called the *origin* of the coordinate system. For now, we shall call the horizontal line the *x-axis* and the vertical line the *y-axis.* The unit distance on the *x*-axis need not be the same as on the *y*-axis.

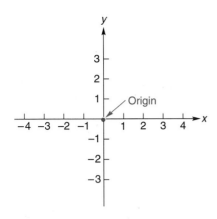

**FIGURE 3.7**   Coordinate axes.

The plane on which the coordinate axes are placed is called a *rectangular coordinate plane* or, more simply, an *x,y-plane*. Every point in the *x,y*-plane can be labeled to indicate its position. To label point *P* in Fig. 3.8(a), we draw perpendiculars from *P* to the *x*-axis and *y*-axis. They meet these axes at 4 and 2, respectively. Thus, *P* determines two numbers, 4 and 2. We say that the **rectangular coordinates** of *P* are given by the **ordered pair** (4, 2). The word "ordered" is important. In Fig. 3.8(b), the point corresponding to (4, 2) is not the same as that for (2, 4):

$$(4, 2) \neq (2, 4).$$

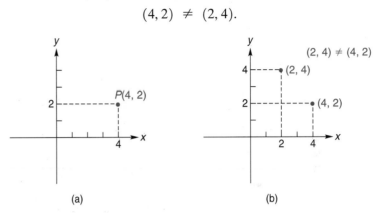

(a)                    (b)

**FIGURE 3.8**   Rectangular coordinates.

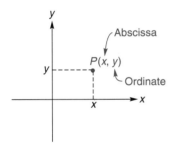

**FIGURE 3.9**   Coordinates of *P*.

In general, if *P* is any point, then its rectangular coordinates will be given by an ordered pair of the form $(x, y)$. (See Fig. 3.9.) We call *x* the *abscissa* or *x-coordinate* of *P*, and *y* the *ordinate* or *y-coordinate* of *P*.

Accordingly, with each point in a given coordinate plane, we can associate exactly one ordered pair $(x, y)$ of real numbers. Also, it should be clear that with each ordered pair $(x, y)$ of real numbers, we can associate exactly one point in that plane. Since there is a *one-to-one correspondence* between the points in the plane and all ordered pairs of real numbers, we shall refer to a point *P* with abscissa *x* and ordinate *y* simply as the point $(x, y)$, or as $P(x, y)$. Moreover, we shall use the words "point" and "ordered pair" interchangeably.

In Fig. 3.10, the coordinates of various points are indicated. For example, the point $(1, -4)$ is located one unit to the right of the *y*-axis and four units below the *x*-axis. The origin is $(0, 0)$. The *x*-coordinate of every point on the *y*-axis is 0, and the *y*-coordinate of every point on the *x*-axis is 0.

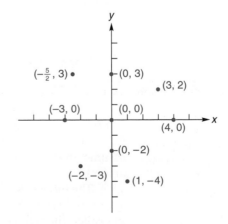

**FIGURE 3.10**   Coordinates of points.

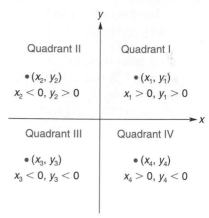

**FIGURE 3.11** Quadrants.

The coordinate axes divide the plane into four regions called *quadrants* (Fig. 3.11). For example, quadrant I consists of all points $(x_1, y_1)$ with $x_1 > 0$ and $y_1 > 0$. The points on the axes do not lie in any quadrant.

Using a rectangular coordinate system, we can geometrically represent equations in two variables. For example, let us consider

$$y = x^2 + 2x - 3. \qquad (1)$$

A solution of this equation is a value of $x$ and a value of $y$ that make the equation true. For example, if $x = 1$, substituting into Eq. (1) gives

$$y = 1^2 + 2(1) - 3 = 0.$$

Thus, $x = 1$, $y = 0$ is a solution. Similarly,

$$\text{if } x = -2, \quad \text{then} \quad y = (-2)^2 + 2(-2) - 3 = -3,$$

and so $x = -2$, $y = -3$ is also a solution. By choosing other values for $x$, we can get more solutions. [See Fig. 3.12(a).] It should be clear that there are infinitely many solutions of Eq. (1).

Each solution gives rise to a point $(x, y)$. For example, to $x = 1$ and $y = 0$ corresponds $(1, 0)$. The **graph** of $y = x^2 + 2x - 3$ is the geometric representation of all its solutions. In Fig. 3.12(b), we have plotted the points corresponding to the solutions in the table.

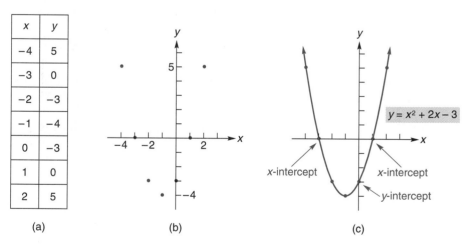

| $x$ | $y$ |
|-----|-----|
| $-4$ | 5 |
| $-3$ | 0 |
| $-2$ | $-3$ |
| $-1$ | $-4$ |
| 0 | $-3$ |
| 1 | 0 |
| 2 | 5 |

(a)  (b)  (c)

**FIGURE 3.12** Graphing $y = x^2 + 2x - 3$.

Since the equation has infinitely many solutions, it seems impossible to determine its graph precisely. However, we are concerned only with the graph's general shape. For this reason, we plot enough points so that we may intelligently guess its proper shape. Then we join these points by a smooth curve wherever conditions permit. This gives the curve in Fig. 3.12(c). Of course, the more points we plot, the better our graph is. Here we assume that the graph extends indefinitely upward, which is indicated by arrows.

Often, we simply say that the $y$-intercept is 3 and the $x$-intercepts are $-3$ and 1.

The point $(0, -3)$ where the curve intersects the $y$-axis is called the *y-intercept*. The points $(-3, 0)$ and $(1, 0)$ where the curve intersects the $x$-axis are called the *x-intercepts*. In general, we have the following definition.

**DEFINITION**

*An **x-intercept** of the graph of an equation in x and y is a point where the graph intersects the x-axis. A **y-intercept** is a point where the graph intersects the y-axis.*

To find the *x*-intercepts of the graph of an equation in *x* and *y*, we first set *y* = 0 and solve the resulting equation for *x*. To find the *y*-intercepts, we first set *x* = 0 and solve for *y*. For example, let us find the *x*-intercepts for the graph of $y = x^2 + 2x - 3$. Setting *y* = 0 and solving for *x* gives

$$0 = x^2 + 2x - 3,$$

$$0 = (x + 3)(x - 1),$$

$$x = -3, 1.$$

Thus, the *x*-intercepts are $(-3, 0)$ and $(1, 0)$, as we saw before. If *x* = 0, then

$$y = 0^2 + 2(0) - 3 = -3.$$

So $(0, -3)$ is the *y*-intercept. Keep in mind that an *x*-intercept has its *y*-coordinate 0, and a *y*-intercept has its *x*-coordinate 0. Intercepts are useful because they indicate precisely where the graph intersects the axes.

**EXAMPLE 1   Intercepts and Graph**

*Find the x- and y-intercepts of the graph of y = 2x + 3, and sketch the graph.*

*Solution:* If *y* = 0, then

$$0 = 2x + 3 \qquad \text{or} \qquad x = -\frac{3}{2}.$$

Thus, the *x*-intercept is $(-\frac{3}{2}, 0)$. If *x* = 0, then

$$y = 2(0) + 3 = 3.$$

So the *y*-intercept is $(0, 3)$. Figure 3.13 shows a table of other points on the graph and a sketch of the graph.   ∎

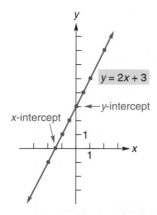

| x | 0 | $-\frac{3}{2}$ | $\frac{1}{2}$ | $-\frac{1}{2}$ | 1 | −1 | 2 | −2 |
|---|---|---|---|---|---|---|---|---|
| y | 3 | 0 | 4 | 2 | 5 | 1 | 7 | −1 |

**FIGURE 3.13**   Graph of $y = 2x + 3$.

**EXAMPLE 2   Intercepts and Graph**

*Determine the intercepts, if any, of the graph of $s = \dfrac{100}{t}$, and sketch the graph.*

**Principles in Practice 1**

**Intercepts and Graph**

Rachel has saved $7250 for college expenses. She plans to spend $600 a month from this account. Write an equation to represent the situation, and identify the intercepts.

## Principles in Practice 2

### Intercepts and Graph

The price of admission to an amusement park is $24.95. This fee allows the customer to ride all the rides at the park as often as he or she likes. Write an equation that represents the relationship between the number of rides, *x*, that a customer takes and the cost per ride, *y*, to that customer. Describe the graph of this equation, and identify the intercepts. Assume $x > 0$.

*Solution:* For the graph, we shall label the horizontal axis *t* and the vertical axis *s* (Fig. 3.14). Because *t* cannot equal 0 (division by 0 is not defined), there is no *s*-intercept. Thus, the graph has no point corresponding to $t = 0$. Moreover, there is no *t*-intercept, because if $s = 0$, then the equation

$$0 = \frac{100}{t}$$

has no solution. Figure 3.14 shows the graph. In general, the graph of $s = k/t$, where *k* is a nonzero constant, is called a *hyperbola*. ∎

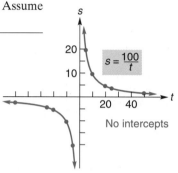

| *t* | 5 | −5 | 10 | −10 | 20 | −20 | 25 | −25 | 50 | −50 |
|-----|---|----|----|----|----|----|----|----|----|----|
| *s* | 20 | −20 | 10 | −10 | 5 | −5 | 4 | −4 | 2 | −2 |

**FIGURE 3.14** Graph of $s = \dfrac{100}{t}$.

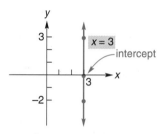

| *x* | 3 | 3 | 3 |
|-----|---|---|---|
| *y* | 0 | 3 | −2 |

**FIGURE 3.15** Graph of $x = 3$.

### EXAMPLE 3 Intercepts and Graph

*Determine the intercepts of the graph of $x = 3$, and sketch the graph.*

*Solution:* We can think of $x = 3$ as an equation in the variables *x* and *y* if we write it as $x = 3 + 0y$. Here *y* can be any value, but *x* must be 3. Because $x = 3$ when $y = 0$, the *x*-intercept is $(3, 0)$. There is no *y*-intercept, because *x* cannot be 0. (See Fig. 3.15.) The graph is a vertical line. ∎

In addition to equations, we can also represent functions in a coordinate plane. If *f* is a function with independent variable *x* and dependent variable *y*, then the graph of *f* is simply the graph of the equation $y = f(x)$. This graph consists of all points $(x, y)$ or $(x, f(x))$, where *x* is in the domain of *f*. The vertical axis can be labeled either *y* or $f(x)$ and is referred to as the **function-value axis.** *We always label the horizontal axis with the independent variable.*

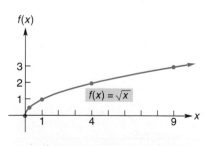

| *x* | 0 | $\frac{1}{4}$ | 1 | 4 | 9 |
|-----|---|---------------|---|---|---|
| $f(x)$ | 0 | $\frac{1}{2}$ | 1 | 2 | 3 |

**FIGURE 3.16** Graph of $f(x) = \sqrt{x}$.

### EXAMPLE 4 Graph of the Square-Root Function

*Graph $f(x) = \sqrt{x}$.*

*Solution:* The graph is shown in Fig. 3.16. We label the vertical axis as $f(x)$. Recall that $\sqrt{x}$ denotes the *principal* square root of *x*. Thus, $f(9) = \sqrt{9} = 3$, not $\pm 3$. Also, we cannot choose negative values for *x*, because we don't want imaginary numbers for $\sqrt{x}$. That is, we must have $x \geq 0$. Let us now consider intercepts. If $f(x) = 0$, then $\sqrt{x} = 0$, or $x = 0$. Also, if $x = 0$, then $f(x) = 0$. Thus, the *x*-intercept and the vertical-axis intercept are the same, namely, $(0, 0)$. ∎

## Principles in Practice 3
### Graph of the Absolute-Value Function

Brett rented a bike from a rental shop, rode at a constant rate of 12 mph for 2.5 hours along a bike path, and then returned along the same path. Graph the absolute-value function to represent Brett's distance from the rental shop as a function of time over the appropriate domain.

**EXAMPLE 5   Graph of the Absolute-Value Function**

*Graph* $p = G(q) = |q|$.

*Solution:* We use the independent variable $q$ to label the horizontal axis. The function-value axis can be labeled either $G(q)$ or $p$. (See Fig. 3.17.) Notice that the $q$- and $p$-intercepts are the same point, $(0, 0)$. ■

| $q$ | 0 | 1 | −1 | 3 | −3 | 5 | −5 |
|---|---|---|---|---|---|---|---|
| $p$ | 0 | 1 | 1 | 3 | 3 | 5 | 5 |

**FIGURE 3.17**   Graph of $p = |q|$.

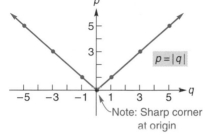

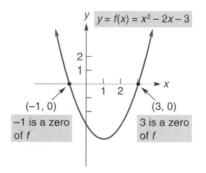

**FIGURE 3.18**   Zeros of function.

In general, the real solutions to an equation $f(x) = 0$ are the real zeros of $f$.

The notion of a *zero* is important in the study of functions.

**DEFINITION**

*A **zero** of a function $f$ is any value of $x$ for which $f(x) = 0$.*

For example, a zero of the function $f(x) = 2x - 6$ is 3 because $f(3) = 2(3) - 6 = 0$. Here we call 3 a *real* zero because it is a real number. We note that zeros of $f$ can be found by setting $f(x) = 0$ and solving for $x$. Thus, the real zeros of a function are precisely the $x$-intercepts of its graph, because it is at these points that $f(x) = 0$.

To further illustrate, Fig. 3.18 shows the graph of the function $y = f(x) = x^2 - 2x - 3$. The $x$-intercepts of the graph are $-1$ and 3. Hence, $-1$ and 3 are zeros of $f$, or equivalently, $-1$ and 3 are solutions to the equation $x^2 - 2x - 3 = 0$.

## TECHNOLOGY

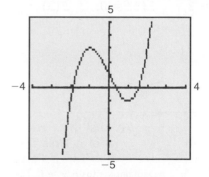

**FIGURE 3.19**   The roots of $x^3 - 3x + 1 = 0$ are approximately $-1.88$, $0.35$, and $1.53$.

To solve the equation $x^3 = 3x - 1$ with a graphics calculator, we first express the equation in the form $f(x) = 0$

$$f(x) = x^3 - 3x + 1 = 0.$$

Next we graph $f$ and then estimate the $x$-intercepts, either by using zoom and trace or by using the root operation. (See Fig. 3.19.) Note that we defined our window for $-4 \le x \le 4$ and $-5 \le y \le 5$.

Figure 3.20 shows the graph of some function $y = f(x)$. The point $(x, f(x))$ implies that corresponding to the input number $x$ on the horizontal axis is the output number $f(x)$ on the vertical axis, as indicated by the arrow. For example,

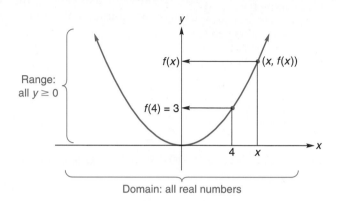

**FIGURE 3.20**   Domain, range, function values.

corresponding to the input 4 is the output 3, so $f(4) = 3$. From the shape of the graph, it seems reasonable to assume that, for any value of $x$, there is an output number, so the domain of $f$ is all real numbers. Notice that the set of all $y$-coordinates of points on the graph is the set of all nonnegative numbers. Thus, the range of $f$ is all $y \geq 0$. This shows that we can make an "educated" guess about the domain and range of a function by looking at its graph. *In general, the domain consists of all x-values that are included in the graph, and the range is all y-values that are included.* For example, Fig. 3.16 implies that both the domain and range of $f(x) = \sqrt{x}$ are all nonnegative numbers. From Fig. 3.17, it is clear that the domain of $p = G(q) = |q|$ is all real numbers and the range is all $p \geq 0$.

**EXAMPLE 6**   **Domain, Range, and Function Values**

Figure 3.21 shows the graph of a function $F$. To the right of 4, assume that the graph repeats itself indefinitely. Then the domain of $F$ is all $t \geq 0$. The range is $-1 \leq s \leq 1$. Some function values are

$$F(0) = 0, \qquad F(1) = 1, \qquad F(2) = 0, \qquad F(3) = -1. \qquad \blacksquare$$

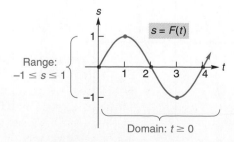

**FIGURE 3.21**   Domain, range, function values.

## TECHNOLOGY

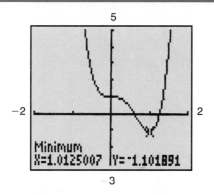

**FIGURE 3.22** The range of $f(x) = 6x^4 - 8.1x^3 + 1$ is approximately $[-1.10, \infty)$.

Using a graphics calculator, we can estimate the range of a function. The graph of

$$f(x) = 6x^4 - 8.1x^3 + 1$$

is shown in Fig. 3.22. The lowest point on the graph corresponds to the minimum value of $f(x)$, and the range is all reals greater than or equal to this minimum. We can estimate this minimum $y$-value either by using trace and zoom or by selecting the "minimum" operation.

---

**Principles in Practice 4**

Graph of a Compound Function

To encourage conservation, a gas company charges two rates. You pay $0.53 per therm for 0–70 therms and $0.74 for each therm over 70. Graph the compound function which represents the monthly cost of $t$ therms of gas.

**EXAMPLE 7    Graph of a Compound Function**

*Graph the compound function*

$$f(x) = \begin{cases} x, & \text{if } 0 \le x < 3, \\ x - 1, & \text{if } 3 \le x \le 5, \\ 4, & \text{if } 5 < x \le 7. \end{cases}$$

*Solution:* The domain of $f$ is $0 \le x \le 7$. The graph is given in Fig. 3.23, where the *hollow dot* means that the point is *not* included in the graph. Notice that the range of $f$ is all real numbers $y$ such that $0 \le y \le 4$. ■

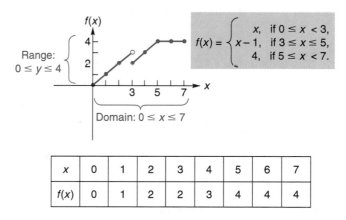

| $x$ | 0 | 1 | 2 | 3 | 4 | 5 | 6 | 7 |
|-----|---|---|---|---|---|---|---|---|
| $f(x)$ | 0 | 1 | 2 | 2 | 3 | 4 | 4 | 4 |

**FIGURE 3.23**    Graph of compound function.

There is an easy way to tell whether a curve is the graph of a function. In Fig. 3.24(a), notice that with the given $x$ there are associated *two* values of $y$: $y_1$ and $y_2$. Thus, the curve is *not* the graph of a function of $x$. Looking at it another way, we have the following general rule, called the **vertical-line test.** If a *vertical* line $L$ can be drawn that intersects a curve in at least two points, then the curve is *not* the graph of a function of $x$. When no such vertical line can be drawn, the curve *is* the graph of a function of $x$. Consequently, the curves in Fig. 3.24 do not represent functions of $x$, but those in Fig. 3.25 do.

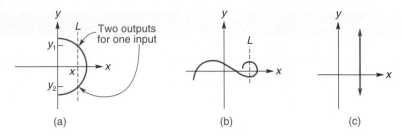

**FIGURE 3.24**   *y* is not a function of *x*.

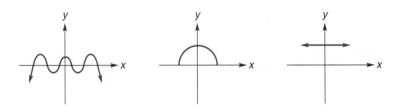

**FIGURE 3.25**   Functions of *x*.

**EXAMPLE 8   A Graph That Does Not Represent a Function of *x***

*Graph* $x = 2y^2$.

*Solution:* Here it is easier to choose values of *y* and then find the corresponding values of *x*. Figure 3.26 shows the graph. By the vertical-line test, the equation $x = 2y^2$ does not define a function of *x*.

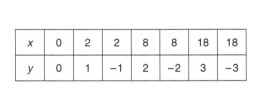

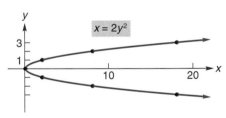

| *x* | 0 | 2 | 2 | 8 | 8 | 18 | 18 |
|---|---|---|---|---|---|---|---|
| *y* | 0 | 1 | −1 | 2 | −2 | 3 | −3 |

**FIGURE 3.26**   Graph of $x = 2y^2$.

## ▪ Exercise 3.4

*In Problems **1** and **2**, locate and label each of the points, and give the quadrant, if possible, in which each point lies.*

**1.** $(2, 7), (8, -3), (-\frac{1}{2}, -2), (0, 0)$.

**2.** $(-4, 5), (3, 0), (1, 1), (0, -6)$.

**3.** Figure 3.27(a) shows the graph of $y = f(x)$.
   **a.** Estimate $f(0), f(2), f(4)$, and $f(2)$.
   **b.** What is the domain of *f*?
   **c.** What is the range of *f*?
   **d.** What is a real zero of *f*?

**4.** Figure 3.27(b) shows the graph of $y = f(x)$.
   **a.** Estimate $f(0)$ and $f(2)$.
   **b.** What is the domain of *f*?
   **c.** What is the range of *f*?
   **d.** What is a real zero of *f*?

**5.** Figure 3.28(a) shows the graph of $y = f(x)$.
   **a.** Estimate $f(0), f(1)$, and $f(-1)$.
   **b.** What is the domain of *f*?
   **c.** What is the range of *f*?
   **d.** What is a real zero of *f*?

**6.** Figure 3.28(b) shows the graph of $y = f(x)$.
   **a.** Estimate $f(0), f(2), f(3)$, and $f(4)$.
   **b.** What is the domain of *f*?
   **c.** What is the range of *f*?
   **d.** What is a real zero of *f*?

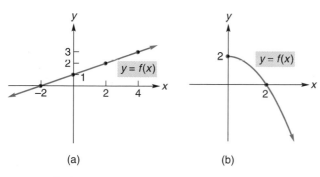

(a)　　　　　　　　　　　(b)

**FIGURE 3.27**　Diagram for Problems 3 and 4.

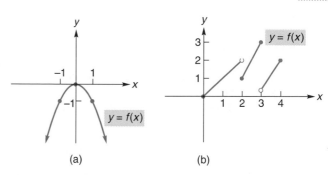

(a)　　　　　　　　　　　(b)

**FIGURE 3.28**　Diagram for Problems 5 and 6.

*In Problems 7–20, determine the intercepts of the graph of each equation, and sketch the graph. Based on your graph, is y a function of x, and if so, what are the domain and range?*

**7.** $y = x$.

**8.** $y = x + 1$.

**9.** $y = 3x - 5$.

**10.** $y = 3 - 2x$.

**11.** $y = x^2$.

**12.** $y = \dfrac{3}{x}$.

**13.** $x = 0$.

**14.** $y = x^2 - 9$.

**15.** $y = x^3$.

**16.** $x = -4$.

**17.** $x = -3y^2$.

**18.** $x^2 = y^2$.

**19.** $2x + y - 2 = 0$.

**20.** $x + y = 1$.

*In Problems 21–34, graph each function and give the domain and range. Also, determine the intercepts.*

**21.** $s = f(t) = 4 - t^2$.

**22.** $f(x) = 5 - 2x^2$.

**23.** $y = g(x) = 2$.

**24.** $G(s) = -8$.

**25.** $y = h(x) = x^2 - 4x + 1$.

**26.** $y = f(x) = x^2 + 2x - 8$.

**27.** $f(t) = -t^3$.

**28.** $p = h(q) = q(2 - q)$.

**29.** $s = F(r) = \sqrt{r - 5}$.

**30.** $F(r) = -\dfrac{1}{r}$.

**31.** $f(x) = |2x - 1|$.

**32.** $v = H(u) = |u - 3|$.

**33.** $F(t) = \dfrac{16}{t^2}$.

**34.** $y = f(x) = \dfrac{2}{x - 4}$.

*In Problems 35–38, graph each compound function and give the domain and range.*

**35.** $c = g(p) = \begin{cases} p, & \text{if } 0 \le p < 2, \\ 2, & \text{if } p \ge 2. \end{cases}$

**36.** $f(x) = \begin{cases} 2x + 1, & \text{if } -1 \le x < 2, \\ 9 - x^2, & \text{if } x \ge 2. \end{cases}$

**37.** $g(x) = \begin{cases} x + 6, & \text{if } x \ge 3, \\ x^2, & \text{if } x < 3. \end{cases}$

**38.** $f(x) = \begin{cases} x + 1, & \text{if } 0 < x \le 3, \\ 4, & \text{if } 3 < x \le 5. \\ x - 1, & \text{if } x > 5. \end{cases}$

**39.** Which of the graphs in Fig. 3.29 represent functions of $x$?

**40. Debt Payments**　Jane has charged $800 on her credit cards. She plans to pay them off at the rate of $75 per month. Write an equation to represent the amount she owes, excluding any finance charges, and identify the intercepts.

**41. Pricing**　To encourage an even flow of customers, a restaurant varies the price of an item throughout the day. From 6:00 P.M. to 8:00 P.M., customers pay full price. At lunch, from 10:30 A.M. until 2:30 P.M., customers pay half price. From 2:30 P.M. until 4:30 P.M., customers get a

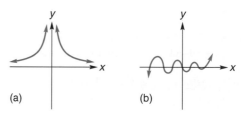

(a)　　　　　　　　　　　(b)

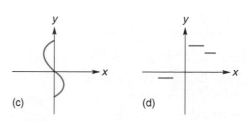

(c)　　　　　　　　　　　(d)

**FIGURE 3.29**　Diagram for Problem 39.

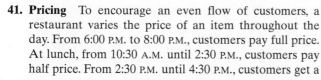

dollar off the lunch price. From 4:30 P.M. until 6:00 P.M., customers get $5.00 off the dinner price. From 8:00 P.M. until closing time at 10:00 P.M., customers get $5.00 off the dinner price. Graph the compound function to represent the cost of an item throughout the day for a dinner price of $18.

**42. Supply Schedule** Given the following supply schedule (see Example 5 of Sec. 3.1), plot each quantity–price pair by choosing the horizontal axis for the possible quantities. Approximate the points in between the data by connecting the data points with a smooth curve. The

| Quantity Supplied per Week, $q$ | Price per Unit, $p$ |
|---|---|
| 30 | $10 |
| 100 | 20 |
| 150 | 30 |
| 190 | 40 |
| 210 | 50 |

result is a *supply curve*. From the graph, determine the relationship between price and supply. (That is, as price increases, what happens to the quantity supplied?) Is price per unit a function of quantity supplied?

**43. Demand Schedule** The following table is called a *demand schedule*. It indicates the quantities of brand X that consumers will demand (that is, purchase) each week at certain prices per unit (in dollars). Plot each quantity–price pair by choosing the vertical axis for the possible prices. Connect the points with a smooth curve. In this way, we approximate points in between the given

| Quantity Demanded, $q$ | Price per Unit, $p$ |
|---|---|
| 5 | $20 |
| 10 | 10 |
| 20 | 5 |
| 25 | 4 |

data. The result is called a *demand curve*. From the graph, determine the relationship between the price of brand X and the amount that will be demanded. (That is, as price decreases, what happens to the quantity demanded?) Is price per unit a function of quantity demanded?

**44. Inventory** Sketch the graph of

$$y = f(x) = \begin{cases} -100x + 600, & \text{if } 0 \le x < 5, \\ -100x + 1100, & \text{if } 5 \le x < 10, \\ -100x + 1600, & \text{if } 10 \le x < 15. \end{cases}$$

A function such as this might describe the inventory $y$ of a company at time $x$.

**45. Psychology** In a psychological experiment on visual information, a subject briefly viewed an array of letters and was then asked to recall as many letters as possible from the array. The procedure was repeated several times. Suppose that $y$ is the average number of letters recalled from arrays with $x$ letters. The graph of the results approximately fits the graph of

$$y = f(x) = \begin{cases} x, & \text{if } 0 \le x \le 4, \\ \frac{1}{2}x + 2, & \text{if } 4 < x \le 5, \\ 4.5, & \text{if } 5 < x \le 12. \end{cases}$$

Plot this function.[6]

 *In Problems 46–49, use a graphics calculator to find all real roots of the given equation. Round answers to two decimal places.*

**46.** $7x^3 + 2x = 3$. **47.** $x(x^2 - 3) = x^4 + 1$. **48.** $(9x + 3.1)^2 = 7.4 - 4x^2$. **49.** $(x - 3)^3 = x^2 - 8$.

 *In Problems 50–53, use a graphics calculator to find all real zeros of the given function. Round answers to two decimal places.*

**50.** $f(x) = x^3 + 5x + 7$. **51.** $f(x) = x^4 - 2.5x^3 - 2$.

**52.** $g(x) = x^4 - 2.5x^3 + x$. **53.** $g(x) = \sqrt{3}x^5 - 4x^2 + 1$.

 *In Problems 54–56, use a graphics calculator to find (a) the maximum value of $f(x)$ and (b) the minimum value of $f(x)$ for the indicated values of $x$. Round answers to two decimal places.*

**54.** $f(x) = x^4 - 4.1x^3 + x^2 + 10$, $1 \le x \le 4$. **55.** $f(x) = x(2.1x^2 - 3)^2 - x^3 + 1$, $-1 \le x \le 1$.

**56.** $f(x) = \dfrac{x^2 - 3}{2x - 5}$, $3 \le x \le 5$.

[6]Adapted from G. R. Loftus and E. F. Loftus, *Human Memory: The Processing of Information* (New York: Lawrence Erlbaum Associates, Inc., distributed by the Halsted Press, Division of John Wiley & Sons, Inc., 1976).

**57.** From the graph of $f(x) = \sqrt{2}x^3 + 1.1x^2 + 4$, find (a) the range and (b) the intercepts. Round values to two decimal places.

 **58.** From the graph of $f(x) = 2 - 3x^3 - x^4$, find (a) the maximum value of $f(x)$, (b) the range of $f$, and (c) the real zeros of $f$. Round values to two decimal places.

**59.** From the graph of $f(x) = \dfrac{x^2 + 5.2}{2.1 + \sqrt{x}}$, find (a) the minimum value of $f(x)$, (b) the range of $f$, and (c) the intercepts. (d) Does $f$ have any real zeros? Round values to two decimal places.

 **60.** Graph $f(x) = \dfrac{4.1x^3 + \sqrt{2}}{x^2 - 3}$ for $2 \le x \le 5$. Determine (a) the maximum value of $f(x)$, (b) the minimum value of $f(x)$, (c) the range of $f$, and (d) all intercepts. Round values to two decimal places.

---

## OBJECTIVE

**To study symmetry about the *x*-axis, the *y*-axis, and the origin, and to apply symmetry to curve sketching.**

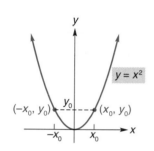

**FIGURE 3.30** Symmetry about *y*-axis.

# 3.5 SYMMETRY

Examining the graphical behavior of equations is a basic part of mathematics. In this section, we shall examine equations to determine whether their graphs have *symmetry*. In a later chapter, you will see that calculus is a *great* aid in graphing because it helps determine the shape of a graph. It provides powerful techniques for determining whether or not a curve "wiggles" between points.

Consider the graph of $y = x^2$ in Fig. 3.30. The portion to the left of the *y*-axis is the reflection (or mirror image) through the *y*-axis of that portion to the right of the *y*-axis, and vice versa. More precisely, if $(x_0, y_0)$ is any point on this graph, then the point $(-x_0, y_0)$ must also lie on the graph. We say that this graph is *symmetric about the y-axis*.

**DEFINITION**

*A graph is **symmetric about the y-axis** if and only if $(-x_0, y_0)$ lies on the graph when $(x_0, y_0)$ does.*

**EXAMPLE 1** *y*-Axis Symmetry

*Use the preceding definition to show that the graph of $y = x^2$ is symmetric about the y-axis.*

***Solution:*** Suppose $(x_0, y_0)$ is *any* point on the graph of $y = x^2$. Then

$$y_0 = x_0^2 \,.$$

We must show that the coordinates of $(-x_0, y_0)$ satisfy $y = x^2$:

$$y_0 = (-x_0)^2?$$
$$y_0 = x_0^2?$$

But we already know that $y_0 = x_0^2$. Thus, the graph *is* symmetric about the *y*-axis. ∎

When one is testing for symmetry in Example 1, $(x_0, y_0)$ can be any point on the graph. In the future, for convenience, we shall omit the subscripts. This means that a graph is symmetric about the *y*-axis if replacing $x$ by $-x$ in its equation results in an equivalent equation.

Another type of symmetry is shown by the graph of $x = y^2$ in Fig. 3.31. Here the portion below the *x*-axis is the reflection through the *x*-axis of that portion above the *x*-axis, and vice versa. If the point $(x, y)$ lies on the graph, then $(x, -y)$ also lies on it. This graph is said to be *symmetric about the x-axis*.

**DEFINITION**

*A graph is **symmetric about the x-axis** if and only if $(x, -y)$ lies on the graph when $(x, y)$ does.*

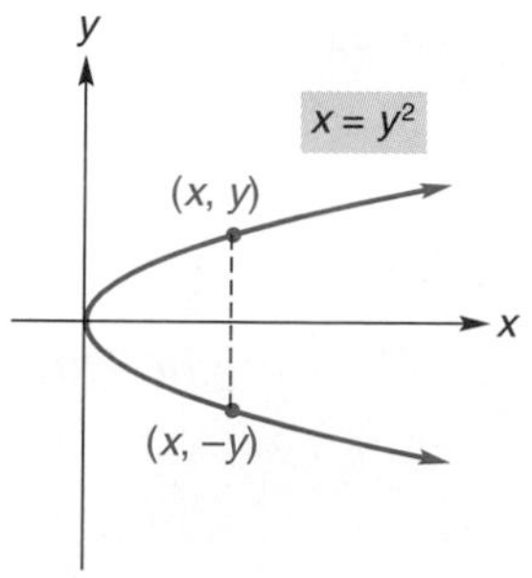

**FIGURE 3.31** Symmetry about *x*-axis.

Thus, the graph of an equation in $x$ and $y$ has $x$-axis symmetry if replacing $y$ by $-y$ results in an equivalent equation. For example, applying this test to the graph of $x = y^2$ shown in Fig. 3.31 gives

$$x = (-y)^2,$$
$$x = y^2,$$

which is equivalent to the original equation. Hence, the graph is symmetric about the $x$-axis.

A third type of symmetry, *symmetry about the origin,* is illustrated by the graph of $y = x^3$ (Fig. 3.32). Whenever the point $(x, y)$ lies on the graph, $(-x, -y)$ also lies on it. As a result, the line segment joining points $(x, y)$ and $(-x, -y)$ is bisected by the origin.

**DEFINITION**

*A graph is **symmetric about the origin** if and only if $(-x, -y)$ lies on the graph when $(x, y)$ does.*

Thus, the graph of an equation in $x$ and $y$ has symmetry about the origin if replacing $x$ by $-x$ and $-y$ by $y$ results in an equivalent equation. For example, applying this test to the graph of $y = x^3$ shown in Fig. 3.32 gives

$$-y = (-x)^3,$$
$$-y = -x^3,$$
$$y = x^3,$$

which is equivalent to the original equation. Accordingly, the graph is symmetric about the origin.

Table 3.1 summarizes the tests for symmetry. When we know that a graph has symmetry, we can sketch it by plotting fewer points than would otherwise be needed.

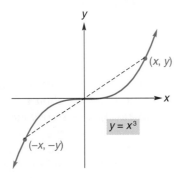

**FIGURE 3.32** Symmetry about origin.

**TABLE 3.1**  Tests for Symmetry

| | |
|---|---|
| Symmetry about $x$-axis | Replace $y$ by $-y$ in given equation. Symmetric if equivalent equation is obtained. |
| Symmetry about $y$-axis | Replace $x$ by $-x$ in given equation. Symmetric if equivalent equation is obtained. |
| Symmetry about origin | Replace $x$ by $-x$ and $y$ by $-y$ in given equation. Symmetric if equivalent equation is obtained. |

**EXAMPLE 2**  **Graphing with Intercepts and Symmetry**

*Test $y = \dfrac{1}{x}$ for symmetry about the x-axis, the y-axis, and the origin. Then find the intercepts and sketch the graph.*

*Solution:*
***Symmetry** x-axis:* Replacing $y$ by $-y$ in $y = 1/x$ gives

$$-y = \frac{1}{x} \qquad \text{or} \qquad y = -\frac{1}{x},$$

which is not equivalent to the given equation. Thus, the graph is *not* symmetric about the *x*-axis.

*y-axis:* Replacing *x* by $-x$ in $y = 1/x$ gives

$$y = \frac{1}{-x} \quad \text{or} \quad y = -\frac{1}{x},$$

which is not equivalent to the given equation. Hence, the graph is *not* symmetric about the *y*-axis.

*Origin:* Replacing *x* by $-x$ and *y* by $-y$ in $y = 1/x$ gives

$$-y = \frac{1}{-x} \quad \text{or} \quad y = \frac{1}{x},$$

which is equivalent to the given equation. Consequently, the graph *is* symmetric about the origin.

***Intercepts*** Since *x* cannot be 0, the graph has no *y*-intercept. If *y* is 0, then $0 = 1/x$, and this equation has no solution. Thus, no *x*-intercept exists.

***Discussion*** Because no intercepts exist, the graph cannot intersect either axis. If $x > 0$, we obtain points only in quadrant I. Fig. 3.33 shows the portion of the graph in quadrant I. By symmetry, we reflect that portion through the origin to obtain the entire graph.

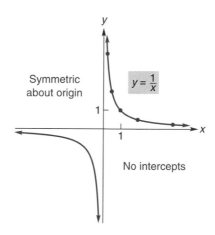

Symmetric about origin

$y = \dfrac{1}{x}$

No intercepts

| *x* | $\frac{1}{4}$ | $\frac{1}{2}$ | 1 | 2 | 4 |
|---|---|---|---|---|---|
| *y* | 4 | 2 | 1 | $\frac{1}{2}$ | $\frac{1}{4}$ |

**FIGURE 3.33**  Graph of $y = \dfrac{1}{x}$.

### EXAMPLE 3  Graphing with Intercepts and Symmetry

*Test $y = f(x) = 1 - x^4$ for symmetry about the x-axis, the y-axis, and the origin. Then find the intercepts and sketch the graph.*

***Solution:***
***Symmetry*** *x-axis:* Replacing *y* by $-y$ in $y = 1 - x^4$ gives

$$-y = 1 - x^4 \quad \text{or} \quad y = -1 + x^4,$$

which is not equivalent to the given equation. Thus, the graph is *not* symmetric about the *x*-axis.

*y-axis:* Replacing *x* by $-x$ in $y = 1 - x^4$ gives

$$y = 1 - (-x)^4 \quad \text{or} \quad y = 1 - x^4,$$

which is equivalent to the given equation. Hence, the graph *is* symmetric about the *y*-axis.

*Origin:* Replacing *x* by $-x$ and *y* by $-y$ in $y = 1 - x^4$ gives

$$-y = 1 - (-x)^4, \quad -y = 1 - x^4, \quad y = -1 + x^4,$$

which is not equivalent to the given equation. Thus, the graph is *not* symmetric about the origin.

***Intercepts*** Testing for *x*-intercepts, we set $y = 0$ in $y = 1 - x^4$. Then

$$1 - x^4 = 0,$$
$$(1 - x^2)(1 + x^2) = 0,$$
$$(1 - x)(1 + x)(1 + x^2) = 0,$$
$$x = 1 \quad \text{or} \quad x = -1.$$

The $x$-intercepts are therefore $(1, 0)$ and $(-1, 0)$. Testing for $y$-intercepts, we set $x = 0$. Then $y = 1$, so $(0, 1)$ is the only $y$-intercept.

***Discussion*** If the intercepts and some points $(x, y)$ to the right of the $y$-axis are plotted, we can sketch the *entire* graph by using symmetry about the $y$-axis (Fig. 3.34). ■

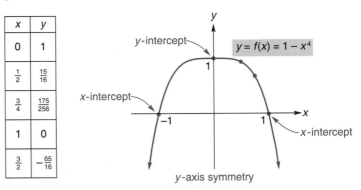

| $x$ | $y$ |
|-----|-----|
| $0$ | $1$ |
| $\frac{1}{2}$ | $\frac{15}{16}$ |
| $\frac{3}{4}$ | $\frac{175}{256}$ |
| $1$ | $0$ |
| $\frac{3}{2}$ | $-\frac{65}{16}$ |

**FIGURE 3.34**   Graph of $y = 1 - x^4$.

In Example 3, we showed that the graph of $y = f(x) = 1 - x^4$ does not have $x$-axis symmetry. With the exception of the constant function $f(x) = 0$, *the graph of any **function** $y = f(x)$ cannot be symmetric about the x-axis*, because such symmetry implies two $y$-values with the same $x$-value, which violates the definition of a function.

**EXAMPLE 4   Graphing with Intercepts and Symmetry**

*Test the graph of $4x^2 + 9y^2 = 36$ for intercepts and symmetry. Sketch the graph.*

***Solution:***
***Intercepts*** If $y = 0$, then $4x^2 = 36$, so $x = \pm 3$. Thus, the $x$-intercepts are $(3, 0)$ and $(-3, 0)$. If $x = 0$, then $9y^2 = 36$, so $y = \pm 2$. Hence, the $y$-intercepts are $(0, 2)$ and $(0, -2)$.

***Symmetry*** *x-axis:* Replacing $y$ by $-y$ in $4x^2 + 9y^2 = 36$ gives

$$4x^2 + 9(-y)^2 = 36, \quad \text{or} \quad 4x^2 + 9y^2 = 36.$$

Since we obtain the original equation, there is symmetry about the $x$-axis.

*y-axis:* Replacing $x$ by $-x$ in $4x^2 + 9y^2 = 36$ gives

$$4(-x)^2 + 9y^2 = 36, \quad \text{or} \quad 4x^2 + 9y^2 = 36.$$

Again we have the original equation, so there is also symmetry about the $y$-axis.

*Origin:* Replacing $x$ by $-x$ and $y$ by $-y$ in $4x^2 + 9y^2 = 36$ gives

$$4(-x)^2 + 9(-y)^2 = 36, \quad \text{or} \quad 4x^2 + 9y^2 = 36.$$

Since this is the original equation, the graph is also symmetric about the origin.

***Discussion*** In Fig. 3.35, the intercepts and some points in the first quadrant are plotted. The points in that quadrant are then connected by a smooth curve. By symmetry about the $x$-axis, the points in the fourth quadrant are obtained. Then, by symmetry about the $y$-axis, the complete graph is found. There are other ways of graphing the equation by using symmetry. For example, after plotting the intercepts and some points in the first quadrant, we can obtain the

| $x$ | $y$ |
|-----|-----|
| $\pm 3$ | 0 |
| 0 | $\pm 2$ |
| 1 | $\frac{4\sqrt{2}}{3}$ |
| 2 | $\frac{2\sqrt{5}}{3}$ |
| $\frac{5}{2}$ | $\frac{\sqrt{11}}{3}$ |

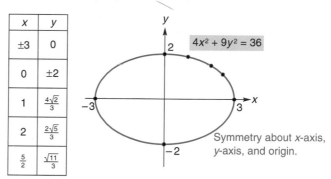

Symmetry about $x$-axis, $y$-axis, and origin.

**FIGURE 3.35**   Graph of $4x^2 + 9y^2 = 36$.

points in the third quadrant by symmetry about the origin. By symmetry about the $x$-axis (or $y$-axis), we can then obtain the entire graph. ■

*This fact can be a timesaving device in checking for symmetry.*

In Example 4, the graph is symmetric about the $x$-axis, the $y$-axis, and the origin. It can be shown that **for any graph, if any two of the three types of symmetry exist, then the remaining type must also exist.**

## ■ Exercise 3.5

*In Problems 1–16, find the x- and y-intercepts of the graphs of the equations. Also, test for symmetry about the x-axis, the y-axis, and the origin. Do not sketch the graphs.*

**1.** $y = 5x$.

**2.** $y = f(x) = x^2 - 4$.

**3.** $2x^2 + y^2 x^4 = 8 - y$.

**4.** $x = y^3$.

**5.** $4x^2 - 9y^2 = 36$.

**6.** $y = 7$.

**7.** $x = -2$.

**8.** $y = |2x| - 2$.

**9.** $x = -y^{-4}$.

**10.** $y = \sqrt{x^2 - 4}$.

**11.** $x - 4y - y^2 + 21 = 0$.

**12.** $x^3 - xy + y^2 = 0$.

**13.** $y = f(x) = \dfrac{x^3}{x^2 + 5}$.

**14.** $x^2 + xy + y^2 = 0$.

**15.** $y = \dfrac{1}{x^3 + 1}$.

**16.** $y = \dfrac{x^4}{x + y}$.

*In Problems 17–24, find the x- and y-intercepts of the graphs of the equations. Also, test for symmetry about the x-axis, the y-axis, and the origin. Then sketch the graphs.*

**17.** $2x + y^2 = 4$.

**18.** $x = y^4$.

**19.** $y = f(x) = x^3 - 4x$.

**20.** $y = x - x^3$.

**21.** $|x| - |y| = 0$.

**22.** $x^2 + y^2 = 16$.

**23.** $4x^2 + y^2 = 16$.

**24.** $x^2 - y^2 = 1$.

**25.** Prove that the graph of $y = f(x) = 2 - 0.03x^2 - x^4$ is symmetric about the $y$-axis, and then graph the function. (a) Make use of symmetry, where possible, to find all intercepts. Determine (b) the maximum value of $f(x)$ and (c) the range of $f$. Round all values to two decimal places.

**26.** Prove that the graph of $y = f(x) = 2x^4 - 7x^2 + 5$ is symmetric about the $y$-axis, and then graph the function. Find all real zeros of $f$. Round your answers to two decimal places.

## 3.6 TRANSLATIONS AND REFLECTIONS

**To be familiar with the shapes of the graphs of six basic functions and to consider translation, reflection, and vertical stretching or shrinking of the graph of a function.**

Up to now, our approach to graphing has been based on plotting points and making use of any symmetry that exists. But this technique is not necessarily the preferred way. Later in this book, we shall analyze graphs by using other techniques. However, some functions and their associated graphs occur so frequently—for example, for illustrative purposes—that we find it worthwhile to memorize them. Figure 3.36 shows six such functions.

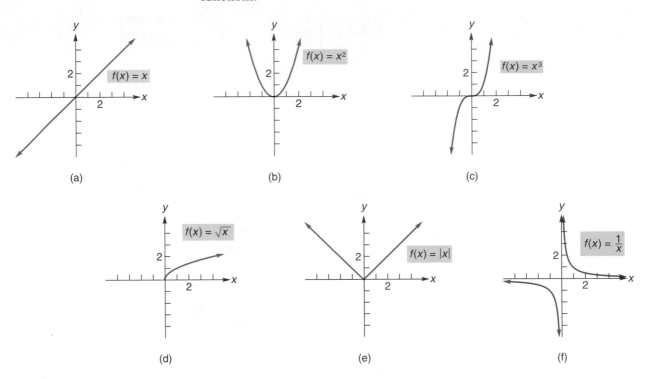

(a)  (b)  (c)

(d)  (e)  (f)

**FIGURE 3.36** Functions frequently used.

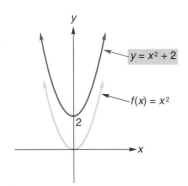

**FIGURE 3.37** Graph of $y = x^2 + 2$.

At times, by altering a function through an *algebraic* manipulation, the graph of the new function can be obtained from the graph of the original function by performing a *geometric* manipulation. For example, we can use the graph of $f(x) = x^2$ to graph $y = x^2 + 2$. Note that $y = f(x) + 2$. Thus, for each $x$, the corresponding ordinate for the graph of $y = x^2 + 2$ is 2 more than the ordinate for the graph of $f(x) = x^2$. This means that the graph of $y = x^2 + 2$ is simply the graph of $f(x) = x^2$, shifted, or *translated*, 2 units upward. (See Fig. 3.37.) We say that the graph of $y = x^2 + 2$ is a *transformation* of the graph of $f(x) = x^2$. Table 3.2 gives a list of basic types of transformations.

### EXAMPLE 1 Horizontal Translation

*Sketch the graph of $y = (x - 1)^3$.*

*Solution:* We observe that $(x - 1)^3$ is $x^3$ with $x$ replaced by $x - 1$. Thus, if $f(x) = x^3$, then $y = (x - 1)^3 = f(x - 1)$, which has the form $f(x - c)$, where $c = 1$. From Table 3.2, the graph of $y = (x - 1)^3$ is the graph of $f(x) = x^3$, shifted 1 unit to the right. (See Fig. 3.38.) ∎

### EXAMPLE 2 Shrinking and Reflection

*Sketch the graph of $y = -\frac{1}{2}\sqrt{x}$.*

*Solution:* We can do this problem in two steps. First, observe that $\frac{1}{2}\sqrt{x}$ is $\sqrt{x}$ multiplied by $\frac{1}{2}$. Thus, if $f(x) = \sqrt{x}$, then $\frac{1}{2}\sqrt{x} = \frac{1}{2}f(x)$, which has the form

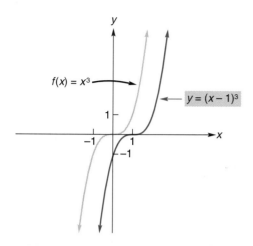

**FIGURE 3.38**   Graph of
$y = (x - 1)^3$.

**TABLE 3.2**   Transformations, $c > 0$

| Equation | How to transform graph of $y = f(x)$ to obtain graph of equation |
|---|---|
| **1.** $y = f(x) + c$ | shift $c$ units upward |
| **2.** $y = f(x) - c$ | shift $c$ units downward |
| **3.** $y = f(x - c)$ | shift $c$ units to right |
| **4.** $y = f(x + c)$ | shift $c$ units to left |
| **5.** $y = -f(x)$ | reflect about $x$-axis |
| **6.** $y = f(-x)$ | reflect about $y$-axis |
| **7.** $y = cf(x),\ c > 1$ | vertically stretch away from $x$-axis by a factor of $c$ |
| **8.** $y = cf(x),\ c < 1$ | vertically shrink toward $x$-axis by a factor of $c$ |

$cf(x)$, where $c = \frac{1}{2}$. So the graph of $y = \frac{1}{2}\sqrt{x}$ is the graph of $f$ shrunk vertically toward the $x$-axis by a factor of $\frac{1}{2}$ (transformation 8, Table 3.2; see Fig. 3.39). Second, the minus sign in $y = -\frac{1}{2}\sqrt{x}$ causes a reflection in the graph of $y = \frac{1}{2}\sqrt{x}$ about the $x$-axis (transformation 5, Table 3.2; see Fig. 3.39). ■

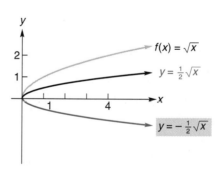

**FIGURE 3.39**   To graph $y = -\frac{1}{2}\sqrt{x}$, shrink $y = \sqrt{x}$ and reflect result about $x$-axis.

■ **Exercise 3.6**

*In Problems 1–12, use the graphs of the functions in Fig. 3.36 and transformation techniques to plot the given functions.*

**1.** $y = x^2 - 2$.  
**2.** $y = -x^2$.  
**3.** $y = \dfrac{1}{x - 2}$.  
**4.** $y = \sqrt{x + 2}$.

**5.** $y = \dfrac{2}{x}$.  
**6.** $y = |x| + 1$.  
**7.** $y = |x + 1| - 2$.  
**8.** $y = -\frac{1}{2}x^3$.

**9.** $y = 1 - (x - 1)^2$.  
**10.** $y = (x - 1)^2 + 1$.  
**11.** $y = \sqrt{-x}$.  
**12.** $y = \dfrac{1}{2 - x}$.

*In Problems 13–16, describe what must be done to the graph of $y = f(x)$ to obtain the graph of the given equation.*

**13.** $y = f(x - 4) + 3$.  
**14.** $y = f(x + 3) - 4$.  
**15.** $y = f(-x) + 3$.  
**16.** $y = -f(x + 3)$.

**17.** Graph the function $y = \sqrt[3]{x} + k$ for $k = 0, 1, 2, 3,$ $-1, -2,$ and $-3$. Observe the vertical translations compared to the first graph.

**18.** Graph the function $y = \sqrt[3]{x + k}$ for $k = 0, 1, 2, 3,$ $-1, -2,$ and $-3$. Observe the horizontal translations compared to the first graph.

**19.** Graph the function $y = k\sqrt[3]{x}$ for $k = 1, 2, \frac{1}{2},$ and $3$. Observe the vertical stretching and shrinking compared to the first graph. Graph the function for $k = -2$. Observe that the graph is the same as that obtained by stretching the reflection of $y = \sqrt[3]{x}$ about the $x$-axis by a factor of 2.

## 3.7 REVIEW

### IMPORTANT TERMS AND SYMBOLS

**Section 3.1**    function    domain    range    independent variable    dependent variable    $f(x)$

        function value    difference quotient, $\dfrac{f(x + h) - f(x)}{h}$    demand function    supply function

**Section 3.2**    constant function    polynomial function    linear function    quadratic function
        compound function    absolute value, $|x|$    factorial, $r!$    rational function

**Section 3.3**    $f + g$    $f - g$    $fg$    $f/g$    $f \circ g$    composition function

**Section 3.4**    rectangular coordinate system    coordinate axes    origin    $x, y$-plane    ordered pair, $(x, y)$
        coordinates of a point    $x$-coordinate    $y$-coordinate    abscissa    ordinate    quadrant
        graph of equation    $y$-intercept    $x$-intercept    graph of function    function-value axis
        zeros of function    vertical-line test

**Section 3.5**    $x$-axis symmetry    $y$-axis symmetry    symmetry about origin

### SUMMARY

A function $f$ is a rule of correspondence that assigns exactly one output number $f(x)$ to each input number $x$. Usually, a function is specified by an equation that indicates what must be done to an input $x$ to obtain $f(x)$. To obtain a particular function value $f(a)$, we replace each $x$ in the equation by $a$.

The domain of a function consists of all input numbers, and the range consists of all output numbers. Unless otherwise specified, the domain of $f$ consists of all real numbers $x$ for which $f(x)$ is also a real number.

Some special types of functions are constant functions, polynomial functions, and rational functions. A function that is defined by more than one expression is called a compound function.

In economics, supply functions and demand functions give a correspondence between the price $p$ of a product and the number of units $q$ of the product that producers (or consumers) will supply (or buy) at that price.

Two functions $f$ and $g$ can be combined to form a sum, difference, product, quotient, or composition as follows:

$$(f + g)(x) = f(x) + g(x),$$
$$(f - g)(x) = f(x) - g(x),$$
$$(fg)(x) = f(x)g(x),$$
$$\frac{f}{g}(x) = \frac{f(x)}{g(x)},$$
$$(f \circ g)(x) = f(g(x)).$$

A rectangular coordinate system allows us to geometrically represent equations in two variables, as well as functions. The graph of an equation in $x$ and $y$ consists of all points $(x, y)$ that correspond to the solutions of the equation. We plot a sufficient number of points and connect them (where appropriate) so that the basic shape of the graph is apparent. Points where the graph intersects the $x$- and $y$-axes are called $x$-intercepts and $y$-intercepts, respectively. An $x$-intercept is found by letting $y$ be 0 and solving for $x$; a $y$-intercept is found by letting $x$ be 0 and solving for $y$.

The graph of a function $f$ is the graph of the equation $y = f(x)$ and consists of all points $(x, f(x))$ such that $x$ is in the domain of $f$. The zeros of $f$ are the values of $x$ for which $f(x) = 0$. From the graph of a function, it is easy to determine the domain and range.

The fact that a graph represents a function can be determined by using the vertical-line test. A vertical line cannot cut the graph of a function at more than one point.

When the graph of an equation has symmetry, the mirror-image effect allows us to sketch the graph by plotting fewer points than would otherwise be needed. The tests for symmetry are as follows:

| Symmetry about $x$-axis | Replace $y$ by $-y$ in given equation. Symmetric if equivalent equation is obtained. |
|---|---|

| Symmetry about $y$-axis | Replace $x$ by $-x$ in given equation. Symmetric if equivalent equation is obtained. |
|---|---|
| Symmetry about origin | Replace $x$ by $-x$ and $y$ by $-y$ in given equation. Symmetric if equivalent equation is obtained. |

Sometimes the graph of a function can be obtained from that of a familiar function by means of a vertical shift upward or downward, a horizontal shift to the right or left, a reflection about the $x$-axis or $y$-axis, or a vertical stretching or shrinking away from or toward the $x$-axis. Such transformations are indicated in Table 3.2 in Sec. 3.6.

## REVIEW PROBLEMS

*In Problems 1–6, give the domain of each function.*

**1.** $f(x) = \dfrac{x}{x^2 - 3x + 2}$.

**2.** $g(x) = x^2 + 3x$.

**3.** $F(t) = 7t + 4t^2$.

**4.** $G(x) = 18$.

**5.** $h(x) = \dfrac{\sqrt{x}}{x - 1}$.

**6.** $H(s) = \dfrac{\sqrt{s - 5}}{4}$.

*In Problems 7–14, find the function values for the given function.*

**7.** $f(x) = 3x^2 - 4x + 7$; $f(0), f(-3), f(5), f(t)$.

**8.** $g(x) = 4$; $g(4), g(\frac{1}{100}), g(-156), g(x + 4)$.

**9.** $G(x) = \sqrt{x - 1}$; $G(1), G(10), G(t + 1), G(x^2)$.

**10.** $F(x) = \dfrac{x - 3}{x + 4}$; $F(-1), F(0), F(5), F(x + 3)$.

**11.** $h(u) = \dfrac{\sqrt{u + 4}}{u}$; $h(5), h(-4), h(x), h(u - 4)$.

**12.** $H(s) = \dfrac{(s - 4)^2}{3}$; $H(-2), H(7), H(\frac{1}{2}), H(x^2)$.

**13.** $f(x) = \begin{cases} 4, & \text{if } x < 2 \\ 8 - x^2, & \text{if } x > 2 \end{cases}$;

$f(4), f(-2), f(0), f(10)$.

**14.** $h(q) = \begin{cases} q, & \text{if } -1 \le q < 0 \\ 3 - q, & \text{if } 0 \le q < 3; \\ 2q^2, & \text{if } 3 \le g \le 5 \end{cases}$

$h(0), h(4), h(-\frac{1}{2}), h(\frac{1}{2})$.

*In Problems 15–18, find (a)* $f(x + h)$ *and (b)* $\dfrac{f(x + h) - f(x)}{h}$, *and simplify your answers.*

**15.** $f(x) = 3 - 7x$.

**16.** $f(x) = x^2 + 4$.

**17.** $f(x) = 4x^2 + 2x - 5$.

**18.** $f(x) = \dfrac{7}{x + 1}$.

**19.** If $f(x) = 3x - 1$ and $g(x) = 2x + 3$, find the following.

  **a.** $(f + g)(x)$.  **b.** $(f + g)(4)$.  **c.** $(f - g)(x)$.

  **d.** $(fg)(x)$.  **e.** $(fg)(1)$.  **f.** $\dfrac{f}{g}(x)$.

  **g.** $(f \circ g)(x)$.  **h.** $(f \circ g)(5)$.  **i.** $(g \circ f)(x)$.

**20.** If $f(x) = x^2$ and $g(x) = 2x + 1$, find the following.

  **a.** $(f + g)(x)$.  **b.** $(f - g)(x)$.  **c.** $(f - g)(-3)$.

  **d.** $(fg)(x)$.  **e.** $\dfrac{f}{g}(x)$.  **f.** $\dfrac{f}{g}(2)$.

  **g.** $(f \circ g)(x)$.  **h.** $(g \circ f)(x)$.  **i.** $(g \circ f)(-4)$.

*In Problems 21–24, find $(f \circ g)(x)$ and $(g \circ f)(x)$.*

**21.** $f(x) = \dfrac{1}{x}, \quad g(x) = x - 1.$

**22.** $f(x) = \dfrac{x + 1}{4}, \quad g(x) = \sqrt{x}.$

**23.** $f(x) = x + 2, \quad g(x) = x^3.$

**24.** $f(x) = 2, \quad g(x) = 3.$

*In Problems 25 and 26, find the intercepts of the graph of each equation, and test for symmetry about the x-axis, the y-axis, and the origin. Do not sketch the graph.*

**25.** $y = 2x - 3x^3.$

**26.** $\dfrac{xy^2}{x^2 + 1} = 4.$

*In Problems 27 and 28, find the x- and y-intercepts of the graphs of the equations. Also, test for symmetry about the x-axis, the y-axis, and the origin. Then sketch the graphs.*

**27.** $y = 9 - x^2.$

**28.** $y = 3x - 7.$

*In Problems 29–32, graph each function and give its domain and range. Also, determine the intercepts.*

**29.** $G(u) = \sqrt{u + 4}.$

**30.** $f(x) = |x| + 1.$

**31.** $y = g(t) = \dfrac{2}{t - 4}.$

**32.** $g(t) = \sqrt{4t}.$

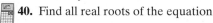

**33.** Graph the following compound function, and give its domain and range:

$$y = f(x) = \begin{cases} 1 - x, & \text{if } x \le 0, \\ 1, & \text{if } x > 0. \end{cases}$$

**34.** Use the graph of $f(x) = \sqrt{x}$ to sketch the graph of $y = \sqrt{x - 2} - 1.$

**35.** Use the graph of $f(x) = x^2$ to sketch the graph of $y = -\frac{1}{2}x^2 + 2.$

**36. Trend Equation** The projected annual sales (in dollars) of a new product is given by the equation $S = 150,000 + 3000t$, where $t$ is the time in years from 1996. Such an equation is called a *trend equation*. Find the projected annual sales for 2001. Is $S$ a function of $t$?

**37.** In Fig. 3.40, which graphs represent functions of $x$?

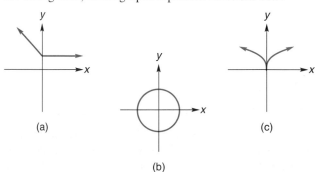

(a)

(b)

(c)

**FIGURE 3.40** Diagram for Problem 37.

**38.** If $f(x) = (4x^{2.3} - 3x^3 + 7)^5$, find (a) $f(2)$ and (b) $f(2.3)$. Round your answers to two decimal places.

**39.** Find all real roots of the equation

$$5x^3 - 7x^2 = 4x - 2.$$

Round your answers to two decimal places.

**40.** Find all real roots of the equation

$$x^4 - 4x^3 = (2x - 1)^2.$$

Round your answers to two decimal places.

**41.** Find all real zeros of

$$f(x) = x(2.1x^2 - 3)^2 - x^3 + 1.$$

Round your answers to two decimal places.

**42.** Determine the range of

$$f(x) = \begin{cases} -2.5x - 4, & \text{if } x < 0, \\ 6 + 4.1x - x^2, & \text{if } x \ge 0. \end{cases}$$

**43.** From the graph of $f(x) = -x^3 + 0.04x + 7$, find (a) the range and (b) the intercepts. Round values to two decimal places.

**44.** From the graph of $f(x) = \sqrt{x + 1}(x^2 - 3)$, find (a) the minimum value of $f(x)$, (b) the range of $f$, and (c) all real zeros of $f$. Round values to two decimal places.

**45.** Graph $y = f(x) = x^3 + x^k$ for $k = 0, 1, 2, 3,$ and $4$. For which values of $k$ does the graph have (a) symmetry about the y-axis? (b) symmetry about the origin?

# MATHEMATICAL *SNAPSHOT*

## A TAXING EXPERIENCE!

You probably have heard the old saying; "There are only two things that are certain in life: death and taxes." Here you will see how functions can be applied to one of these "certainties," namely, taxes.

You will be using the 1997 federal tax rates for a married couple filing a joint return. Suppose that you want to determine a formula for the function $f$ such that $f(x)$ is the tax in dollars on a taxable income of $x$ dollars. The tax is based on various ranges of taxable income. According to Schedule Y-1 of the Internal Revenue Service (see Fig. 3.41):

- If $x$ is $0 or less, the tax is $0.
- If $x$ is over $0, but not over $41,200, the tax is 15% of $x$.
- If $x$ is over $41,200, but not over $99,600, the tax is $6,180, plus 28% of the amount over $41,200.
- If $x$ is over $99,600, but not over $151,750, the tax is $22,532, plus 31% of the amount over $99,600.
- If $x$ is over $151,750, but not over $271,050, the tax is $38,698.50, plus 36% of the amount over $151,750.
- If $x$ is over $271,050, the tax is $81,646.50, plus 39.6% of the amount over $271,050.

Clearly, if $x \leq 0$, then

$$f(x) = 0.$$

If $0 < x \leq 41,200$, then

$$f(x) = 0.15x.$$

Notice that the tax on a taxable income of $41,200 is

$$f(41,200) = 0.15(41,200) = \$6,180.$$

**Schedule Y-1**—Use if your filing status is **Married filing jointly or Qualifying widow(er)**

| If the amount on Form 1040 line 38, is Over– | But not over– | Enter on Form 1040, line 39 | | of the amount over– |
|---|---|---|---|---|
| $0 | $41,200 | | 15% | $0 |
| 41,200 | 99,600 | **$6,180.00 +** | 28% | 41,200 |
| 99,600 | 151,750 | **22,532.00 +** | 31% | 99,600 |
| 151,750 | 271,050 | **38,698.50 +** | 36% | 151,750 |
| 271,050 | ---------- | **81,646.50 +** | 39.6% | 271,050 |

**FIGURE 3.41**   Internal Revenue Service 1997 Schedule Y-1

If $41,200 < x \leq 99,600$, then the amount over 41,200 is $x - 41,200$, so

$$f(x) = 6180 + 0.28(x - 41,200).$$

Because 6180 is 15% of 41,200, for a taxable income between $41,200 and $99,600 you are essentially taxed at the rate of 15% for the first $41,200 of income and at the rate of 28% for the remaining income. Notice that the tax on $99,600 is

$$f(99,600) = 6180 + 0.28(99,600 - 41,200$$
$$= 6180 + 0.28(58,400)$$
$$= 6180 + 16,352 = \$22,532.$$

**117**

If $99,600 < x \le 151,750$, then the amount over 99,600 is $x - 99,600$, so

$$f(x) = 22,532 + 0.31(x - 99,600).$$

Because $22,532 is the tax on $99,600, for a taxable income between $99,600 and $151,750 you are taxed at the rate of 15% for the first $41,200 of income, at the rate of 28% for the next $58,400 of income $(99,600 - 41,200 = 58,400)$, and at the rate of 31% for the remaining income. Notice that the tax on $151,750 is

$$f(151,750) = 22,532 + 0.31(151,750 - 99,600)$$

$$= 22,532 + 0.31(52,150)$$

$$= 22,532 + 16,166.50$$

$$= \$38,698.50.$$

If $151,750 < x \le 271,050$, then the amount over 151,750 is $x - 151,750$, so

$$f(x) = 38,698.50 + 0.36(x - 151,750).$$

Because $38,698.50 is the tax on $151,750, for a taxable income between $151,750 and $271,050 you are taxed at the rate of 15% for the first $41,200 of income, at the rate of 28% for the next $58,400 of income, at the rate of 31% for the next $52,150 of income $(151,750 - 99,600 = 52,150)$, and at the rate of 36% for the remaining income. Notice that the tax on $271,050 is

$$f(271,050) = 38,698.50 + 0.36(271,050 - 151,750)$$

$$= 38,698.50 + 0.36(119,300)$$

$$= 38,698.50 + 42,948 = \$81,646.50.$$

If $x > 271,050$, then the amount over 271,050 is $x - 271,050$, so

$$f(x) = 81,646.50 + 0.396(x - 271,050).$$

Because $81,646.50 is the tax on $271,050, for a taxable income over $271,050 you are taxed at the rate of 15% for the first $41,200 of income, at the rate of 28% for the next $58,400 of income, at the rate of 31% for

the next $52,150 of income, at the rate of 36% for the next $119,300 in income, and at the rate of 39.6% for the remaining income.

Summarizing all of these results gives the compound function

$$f(x) = \begin{cases} 0, & \text{if } x \le 0, \\ 0.15x, & \text{if } 0 < x \le 41,200, \\ 6,180 + 0.28(x - 41,200), \\ \quad \text{if } 41,200 < x \le 99,600, \\ 22,532 + 0.31(x - 99,600), \\ \quad \text{if } 99,600 < x \le 151,750, \\ 38,698.50 + 0.36(x - 151,750), \\ \quad \text{if } 151,750 < x \le 271,050, \\ 81,646.50 + 0.396(x - 271,050), \\ \quad \text{if } x > 271,050. \end{cases}$$

For example, the tax on a taxable income of $60,000 is (using the third expression)

$$f(60,000) = 6,180 + 0.28(60,000 - 41,200)$$

$$= 6,180 + 0.28(18,800)$$

$$= 6,180 + 5,264 = \$11,444.$$

With these formulas, you can geometrically depict the income tax function, as in Fig. 3.42.

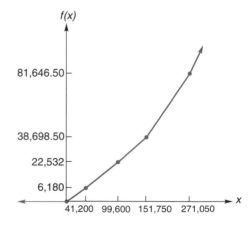

**FIGURE 3.42**   Income tax function.

# ■ Exercises

*Use the preceding income tax function f to determine the tax on the given taxable income.*

**1.** $100,000.        **2.** $25,350.        **3.** $280,000.        **4.** $162,700.

# Lines, Parabolas, and Systems

**To develop the notion of slope and different forms of equations of lines.**

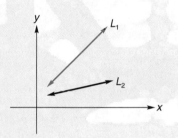

**FIGURE 4.1** Line $L_1$ is "steeper" than $L_2$.

## 4.1 LINES

### Slope of a Line

Many relationships between quantities can be represented conveniently by straight lines. One feature of a straight line is its "steepness." For example, in Fig. 4.1, line $L_1$ rises faster as it goes from left to right than does line $L_2$. In this sense, $L_1$ is steeper.

To measure the steepness of a line, we use the notion of *slope*. In Fig. 4.2, as we move along line $L$ from $(1, 3)$ to $(3, 7)$, the $x$-coordinate increases

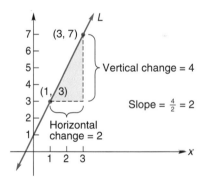

**FIGURE 4.2** Slope of a line.

from 1 to 3, and the $y$-coordinate increases from 3 to 7. The average rate of change of $y$ with respect to $x$ is the ratio

$$\frac{\text{change in } y}{\text{change in } x} = \frac{\text{vertical change}}{\text{horizontal change}} = \frac{7 - 3}{3 - 1} = \frac{4}{2} = 2.$$

The ratio of 2 means that for each 1-unit increase in $x$, there is a 2-unit *increase* in $y$. Due to the increase, the line *rises* from left to right. It can be shown that, regardless of which two points on $L$ are chosen to compute the ratio of the change in $y$ to the change in $x$, the result is always 2, which we call the *slope* of the line.

## DEFINITION

*Let $(x_1, y_1)$ and $(x_2, y_2)$ be two different points on a nonvertical line. The slope of the line is*

$$m = \frac{y_2 - y_1}{x_2 - x_1} \quad \left( = \frac{vertical\ change}{horizontal\ change} \right). \qquad (1)$$

Having no slope does not mean having a slope of zero.

A vertical line does not have a slope, because any two points on it must have $x_1 = x_2$ [see Fig. 4.3(a)], which gives a denominator of zero in Eq. (1). For a horizontal line, any two points must have $y_1 = y_2$. [See Fig. 4.3(b).] This gives a numerator of zero in Eq. (1), and hence, the slope of the line is zero.

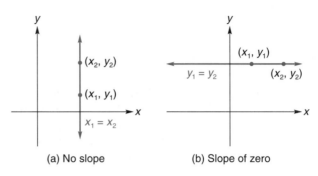

(a) No slope    (b) Slope of zero

**FIGURE 4.3** Vertical and horizontal lines.

This example shows how the slope can be interpreted.

---

### Principles in Practice 1

**Price–Time Relationship**

A doctor purchased a new car in 1991 for $32,000. In 1994, he sold it to a friend for $26,000. Draw a line showing the relationship between the selling price of the car and the year in which it was sold. Find and interpret the slope.

---

### EXAMPLE 1   Price–Quantity Relationship

*The line in Fig. 4.4 shows the relationship between the price p of a widget (in dollars) and the quantity q of widgets (in thousands) that consumers will buy at that price. Find and interpret the slope.*

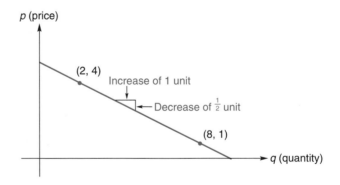

**FIGURE 4.4** Price–quantity line.

*Solution:* In the slope formula (1), we replace the $x$'s by $q$'s and the $y$'s by $p$'s. Either point in Fig. 4.4 may be chosen as $(q_1, p_1)$. Letting $(2, 4) = (q_1, p_1)$ and $(8, 1) = (q_2, p_2)$, we have

$$m = \frac{p_2 - p_1}{q_2 - q_1} = \frac{1 - 4}{8 - 2} = \frac{-3}{6} = -\frac{1}{2}.$$

The slope is negative, $-\frac{1}{2}$. This means that, for each 1-unit increase in quantity (one thousand widgets), there corresponds a **decrease** in price of $\frac{1}{2}$ (dollar per widget). Due to this decrease, the line **falls** from left to right. ∎

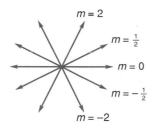

**FIGURE 4.5** Slopes of lines.

In summary, we can characterize the orientation of a line by its slope:

| | |
|---|---|
| Zero slope: | horizontal line. |
| Undefined slope: | vertical line. |
| Positive slope: | line rises from left to right. |
| Negative slope: | line falls from left to right. |

Lines with different slopes are shown in Fig. 4.5. Notice that *the closer the slope is to 0, the more nearly horizontal is the line. The greater the absolute value of the slope, the more nearly vertical is the line.* We remark that two lines are parallel if and only if they have the same slope or are vertical.

### Equations of Lines

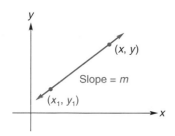

**FIGURE 4.6** Line through $(x_1, y_1)$ with slope $m$.

If we know a point on a line and the slope of the line, we can find an equation whose graph is that line. Suppose that line $L$ has slope $m$ and passes through the point $(x_1, y_1)$. If $(x, y)$ is *any* other point on $L$ (see Fig. 4.6), we can find an algebraic relationship between $x$ and $y$. Using the slope formula on the points $(x_1, y_1)$ and $(x, y)$ gives

$$\frac{y - y_1}{x - x_1} = m,$$

$$y - y_1 = m(x - x_1). \tag{2}$$

Every point on $L$ satisfies Eq. (2). It is also true that every point satisfying Eq. (2) must lie on $L$. Thus, Eq. (2) is an equation for $L$ and is given a special name:

$$y - y_1 = m(x - x_1)$$

is a **point–slope form** of an equation of the line through $(x_1, y_1)$ with slope $m$.

**EXAMPLE 2    Point–Slope Form**

*Find an equation of the line that has slope 2 and passes through $(1, -3)$.*

*Solution:* Using a point–slope form with $m = 2$ and $(x_1, y_1) = (1, -3)$ gives

$$y - y_1 = m(x - x_1),$$
$$y - (-3) = 2(x - 1),$$
$$y + 3 = 2x - 2,$$

which can be rewritten as

$$2x - y - 5 = 0. \quad \blacksquare$$

An equation of the line passing through two given points can be found easily, as Example 3 shows.

**EXAMPLE 3    Determining a Line from Two Points**

*Find an equation of the line passing through $(-3, 8)$ and $(4, -2)$.*

### Principles in Practice 3
**Determining a Line from Two Points**

Find an equation of the line passing through the given points. A temperature of 41°F is equivalent to 5°C, and a temperature of 77°F is equivalent to 25°C.

*Solution:*

*Strategy:*   First we find the slope of the line from the given points. Then we substitute the slope and one of the points into a point–slope form.

The line has slope

$$m = \frac{-2 - 8}{4 - (-3)} = -\frac{10}{7}.$$

Using a point–slope form with $(-3, 8)$ as $(x_1, y_1)$ gives

$$y - 8 = -\tfrac{10}{7}[x - (-3)],$$
$$y - 8 = -\tfrac{10}{7}(x + 3),$$
$$7y - 56 = -10x - 30,$$

or

$$10x + 7y - 26 = 0. \qquad \blacksquare$$

Choosing $(4, -2)$ as $(x_1, y_1)$ would give an equivalent result.

Recall that a point $(0, b)$ where a graph intersects the $y$-axis is called a $y$-intercept (Fig. 4.7). If the slope $m$ and $y$-intercept $b$ of a line are known, an equation for the line is [by using a point–slope form with $(x_1, y_1) = (0, b)$]

$$y - b = m(x - 0).$$

Solving for $y$ gives $y = mx + b$, called the *slope–intercept form* of an equation of the line:

$$y = mx + b$$

is the **slope–intercept form** of an equation of the line with slope $m$ and $y$-intercept $b$.

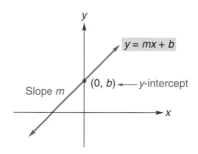

**FIGURE 4.7**   Line with slope $m$ and $y$-intercept $b$.

### EXAMPLE 4   Slope–Intercept Form

*Find an equation of the line with slope 3 and y-intercept* $-4$.

*Solution:* Using the slope–intercept form $y = mx + b$ with $m = 3$ and $b = -4$ gives

$$y = 3x + (-4),$$
$$y = 3x - 4. \qquad \blacksquare$$

### EXAMPLE 5   Find the Slope and *y*-Intercept of a Line

*Find the slope and y-intercept of the line with equation* $y = 5(3 - 2x)$.

*Solution:*

*Strategy:*   We shall rewrite the equation so it has the slope–intercept form $y = mx + b$. Then the slope is the coefficient of $x$ and the $y$-intercept is the constant term.

### Principles in Practice 4
**Finding the Slope and y-Intercept of a Line**

One formula for the recommended dosage (in milligrams) of medication for a child $t$ years old is $y = \dfrac{1}{24}(t + 1)a$, where $a$ is the adult dosage. For an over-the-counter pain reliever, $a = 1000$. Find the slope and $y$-intercept of this equation.

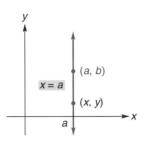

**FIGURE 4.8** Vertical line through $(a, b)$.

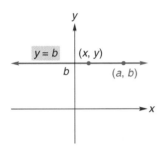

**FIGURE 4.9** Horizontal line through $(a, b)$.

Table 4.1 provides a good summary for you.

Do not confuse the forms of equations of horizontal and vertical lines. Remember which one has the form $x = $ constant and which one has the form $y = $ constant.

We have

$$y = 5(3 - 2x),$$
$$y = 15 - 10x,$$
$$y = -10x + 15.$$

Thus, $m = -10$ and $b = 15$, so the slope is $-10$ and the $y$-intercept is 15.

If a *vertical* line passes through $(a, b)$ (see Fig. 4.8), then any other point $(x, y)$ lies on the line if and only if $x = a$. The $y$-coordinate can have any value. Hence, an equation of the line is $x = a$. Similarly, an equation of the *horizontal* line passing through $(a, b)$ is $y = b$. (See Fig. 4.9.) Here the $x$-coordinate can have any value.

**EXAMPLE 6** Equations of Horizontal and Vertical Lines

**a.** An equation of the vertical line through $(-2, 3)$ is $x = -2$. An equation of the horizontal line through $(-2, 3)$ is $y = 3$.

**b.** The $x$- and $y$-axes are horizontal and vertical lines, respectively. Because $(0, 0)$ lies on both axes, an equation of the $x$-axis is $y = 0$, and an equation of the $y$-axis is $x = 0$.

From our discussions, we can show that every straight line is the graph of an equation of the form $Ax + By + C = 0$, where $A$, $B$, and $C$ are constants and $A$ and $B$ are not both zero. We call this a **general linear equation** (or an *equation of the first degree*) **in the variables $x$ and $y$,** and $x$ and $y$ are said to be **linearly related.** For example, a general linear equation for $y = 7x - 2$ is $(-7)x + (1)y + (2) = 0$. Conversely, the graph of a general linear equation is a straight line. Table 4.1 gives the various forms of equations of straight lines.

**TABLE 4.1** Forms of Equations of Straight Lines

| | |
|---|---|
| Point–slope form | $y - y_1 = m(x - x_1)$ |
| Slope–intercept form | $y = mx + b$ |
| General linear form | $Ax + By + C = 0$ |
| Vertical line | $x = a$ |
| Horizontal line | $y = b$ |

**Principles in Practice 5**
**Converting Forms of Equations of Lines**

Find a general linear form of the Fahrenheit–Celsius conversion equation whose slope–intercept form is $F = \dfrac{9}{5}C + 32.$

**EXAMPLE 7** Converting Forms of Equations of Lines

**a.** *Find a general linear form of the line whose slope–intercept form is*

$$y = -\tfrac{2}{3}x + 4.$$

*Solution:* Getting one side to be 0, we obtain

$$\tfrac{2}{3}x + y - 4 = 0,$$

which is a general linear form with $A = \frac{2}{3}$, $B = 1$, and $C = -4$. An alternative general form can be obtained by clearing fractions:

$$2x + 3y - 12 = 0.$$

This illustrates that a general linear form of a line is not unique.

**b.** *Find the slope–intercept form of the line having a general linear form $3x + 4y - 2 = 0$.*

**Solution:** We want the form $y = mx + b$, so we solve the given equation for $y$. We have

$$3x + 4y - 2 = 0,$$
$$4y = -3x + 2,$$
$$y = -\frac{3}{4}x + \frac{1}{2},$$

which is the slope–intercept form. Note that the line has slope $-\frac{3}{4}$ and $y$-intercept $\frac{1}{2}$.  ∎

---

**Principles in Practice 6**
**Graphing a General Linear Equation**

Sketch the graph of the Fahrenheit–Celsius conversion equation that you found in Principles in Practice 5. How could you use this graph to convert a Celsius temperature to Fahrenheit?

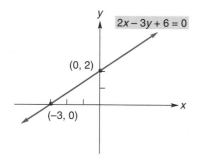

**FIGURE 4.10**  Graph of $2x - 3y + 6 = 0$.

**EXAMPLE 8  Graphing a General Linear Equation**

*Sketch the graph of $2x - 3y + 6 = 0$.*

**Solution:**

> *Strategy:*  Since this is a general linear equation, its graph is a straight line. Thus, we need only determine two different points on the graph in order to sketch it. We shall find the intercepts.

If $x = 0$, then $-3y + 6 = 0$, so the $y$-intercept is 2. If $y = 0$, then $2x + 6 = 0$, so the $x$-intercept is $-3$. We now draw the line passing through $(0, 2)$ and $(-3, 0)$. (See Fig. 4.10.)  ∎

---

**TECHNOLOGY**

To graph the equation of Example 8 with a graphics calculator, we first express $y$ in terms of $x$:

$$2x - 3y + 6 = 0,$$
$$3y = 2x + 6,$$
$$y = \tfrac{1}{3}(2x + 6).$$

Essentially, $y$ is expressed as a function of $x$; the graph is shown in Fig. 4.11.

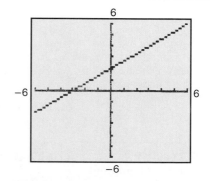

**FIGURE 4.11**  Calculator graph of $2x - 3y + 6 = 0$.

---

**Parallel and Perpendicular Lines**

As stated previously, there is a rule for parallel lines:

### Parallel Lines

Two lines are parallel if and only if they have the same slope or are vertical.

There is also a rule for perpendicular lines. Look back to Fig. 4.5 and observe that the line with slope $-\frac{1}{2}$ is perpendicular to the line with slope 2. The fact that the slope of either of these lines is the negative reciprocal of the slope of the other line is not a coincidence, as the following rule states.

### Perpendicular Lines

Two lines with slopes $m_1$ and $m_2$ are perpendicular to each other if, and only if,

$$m_1 = -\frac{1}{m_2}.$$

Moreover, a horizontal line and a vertical line are perpendicular to each other.

**Principles in Practice 7**
Parallel and Perpendicular Lines

Show that a triangle with vertices at $A(0, 0)$, $B(6, 0)$, and $C(7, 7)$ is not a right triangle.

### EXAMPLE 9   Parallel and Perpendicular Lines

*Figure* 4.12 *shows two lines passing through* $(3, -2)$. *One is parallel to the line* $y = 3x + 1$, *and the other is perpendicular to it. Find equations of these lines.*

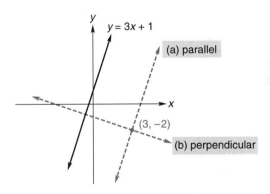

**FIGURE 4.12**   Lines parallel and perpendicular to $y = 3x + 1$ (Example 9).

*Solution:* The slope of $y = 3x + 1$ is 3. Thus, the line through $(3, -2)$ that is *parallel* to $y = 3x + 1$ also has slope 3. Using a point–slope form, we get

$$y - (-2) = 3(x - 3),$$
$$y + 2 = 3x - 9,$$
$$y = 3x - 11.$$

The slope of a line *perpendicular* to $y = 3x + 1$ must be $-\frac{1}{3}$ (the negative reciprocal of 3). Using a point–slope form, we get

$$y - (-2) = -\frac{1}{3}(x - 3),$$
$$y + 2 = -\frac{1}{3}x + 1,$$
$$y = -\frac{1}{3}x - 1.$$

## ▪ Exercise 4.1

*In Problems 1–8, find the slope of the straight line that passes through the given points.*

**1.** $(4, 1), (7, 10)$.  
**2.** $(-3, 11), (2, 1)$.  
**3.** $(4, -2), (-6, 3)$.  
**4.** $(2, -4), (3, -4)$.  
**5.** $(5, 3), (5, -8)$.  
**6.** $(0, -6), (3, 0)$.  
**7.** $(5, -2), (4, -2)$.  
**8.** $(1, -6), (1, 0)$.

*In Problems 9–24, find a general linear equation $(Ax + By + C = 0)$ of the straight line that has the indicated properties, and sketch each line.*

**9.** Passes through $(2, 8)$ and has slope 6.  
**10.** Passes through origin and has slope $-5$.  
**11.** Passes through $(-2, 5)$ and has slope $-\frac{1}{4}$.  
**12.** Passes through $(-\frac{5}{2}, 5)$ and has slope $\frac{1}{3}$.  
**13.** Passes through $(-6, 1)$ and $(1, 4)$.  
**14.** Passes through $(7, 1)$ and $(7, -5)$.  
**15.** Passes through $(3, -1)$ and $(-2, -9)$.  
**16.** Passes through $(0, 0)$ and $(2, 3)$.  
**17.** Has slope 2 and $y$-intercept 4.  
**18.** Has slope 7 and $y$-intercept $-5$.  
**19.** Has slope $-\frac{1}{2}$ and $y$-intercept $-3$.  
**20.** Has slope 0 and $y$-intercept $-\frac{1}{2}$.  
**21.** Is horizontal and passes through $(-3, -2)$.  
**22.** Is vertical and passes through $(-1, 4)$.  
**23.** Passes through $(2, -3)$ and is vertical.  
**24.** Passes through the origin and is horizontal.

*In Problems 25–34, find, if possible, the slope and $y$-intercept of the straight line determined by the equation, and sketch the graph.*

**25.** $y = 2x - 1$.  
**26.** $x - 1 = 5$.  
**27.** $x + 2y - 3 = 0$.  
**28.** $y + 4 = 7$.  
**29.** $x = -5$.  
**30.** $x - 1 = 5y + 3$.  
**31.** $y = 3x$.  
**32.** $y - 7 = 3(x - 4)$.  
**33.** $y = 1$.  
**34.** $2y - 3 = 0$.

*In Problems 35–40, find a general linear form and the slope–intercept form of the given equation.*

**35.** $2x = 5 - 3y$.  
**36.** $3x + 2y = 6$.  
**37.** $4x + 9y - 5 = 0$.  
**38.** $2(x - 3) - 4(y + 2) = 8$.  
**39.** $\dfrac{x}{2} - \dfrac{y}{3} = -4$.  
**40.** $y = \dfrac{1}{300}x + 8$.

*In Problems 41–50, determine whether the lines are parallel, perpendicular, or neither.*

**41.** $y = 7x + 2$,  $y = 7x - 3$.  
**42.** $y = 4x + 3$,  $y = 5 + 4x$.  
**43.** $y = 5x + 2$,  $-5x + y - 3 = 0$.  
**44.** $y = x$,  $y = -x$.  
**45.** $x + 2y + 1 = 0$,  $y = -2x$.  
**46.** $x + 2y = 0$,  $x + y - 4 = 0$.  
**47.** $y = 3$,  $x = -\frac{1}{3}$.  
**48.** $x = 3$,  $x = -4$.  
**49.** $3x + y = 4$,  $x - 3y + 1 = 0$.  
**50.** $x - 1 = 0$,  $y = 0$.

*In Problems 51–60, find an equation of the line satisfying the given conditions. Give the answer in slope–intercept form if possible.*

**51.** Passing through $(-1, 3)$ and parallel to $y = 4x - 5$.  
**52.** Passing through $(2, -8)$ and parallel to $x = -4$.  
**53.** Passing through $(2, 1)$ and parallel to $y = 2$.  
**54.** Passing through $(3, -4)$ and parallel to $y = 3 + 2x$.  
**55.** Perpendicular to $y = 3x - 5$ and passing through $(3, 4)$.  
**56.** Perpendicular to $y = -4$ and passing through $(1, 1)$.  
**57.** Passing through $(7, 4)$ and perpendicular to $y = -4$.  
**58.** Passing through $(-5, 4)$ and perpendicular to the line $2y = -x + 1$.  
**59.** Passing through $(-7, -5)$ and parallel to the line $2x + 3y + 6 = 0$.  
**60.** Passing through $(-2, 1)$ and parallel to the $y$-axis.

**61.** A straight line passes through $(1, 2)$ and $(-3, 8)$. Find the point on it that has an $x$-coordinate of 5.  
**62.** A straight line has slope 2 and $y$-intercept $(0, 1)$. Does the point $(-1, -1)$ lie on the line?

**63. Stock** In 1986, the stock in a biotechnology company traded for $30 per share. In 1996 the company started having trouble, and the stock price dropped to $10 per share. Draw a line showing the relationship between the price per share and the year in which it traded, with years on the $x$-axis and price on the $y$-axis. Find and interpret the slope.

**64. Speed of Sound** A graph of the speed of sound, $S$ (in meters per second), at sea level versus the temperature $T$ of the air (in Celsius) has a slope of 0.61. The equation that describes the relationship between speed of sound and air temperature is $S = 0.61T + b$. When the temperature is $15°C$, a researcher measures the speed of sound as 340.55 meters per second. Determine $b$ to complete the equation.

*In Problems 65–66, find an equation of the line describing the following information.*

**65. Home Runs** In one season, a major league baseball player has hit 14 home runs by the end of the third month and 20 home runs by the end of the fifth month.

**66. Business** A delicatessen owner starts her business with debts of $100,000. After operating for five years, she has accumulated a profit of $40,000.

**67. Due Date** The length, $L$, of a human fetus more than 12 weeks old can be estimated by the formula $L = 1.53t - 6.7$, where $L$ is in centimeters and $t$ is in weeks from conception. An obstetrician uses the length of a fetus, measured by ultrasound, to determine the approximate age of the fetus and establish a due date for the mother. The formula must be rewritten to result in an age $t$, given a fetal length $L$. Find the slope and $L$-intercept of the equation.

**68. Discus Throw** A mathematical model can approximate the winning distance for the Olympic discus throw by the formula $d = 175 + 1.75t$, where $d$ is in feet and $t = 0$ corresponds to the year 1948. Find a general linear form of this equation.

**69. Campus Map** A coordinate map of a college campus gives the coordinates $(x, y)$ of three major buildings as follows: computer center, $(3.5, -1)$; engineering lab, $(0.5, 0)$; and library $(-1, -4.5)$. Find the equations (in slope–intercept form) of the straight–line paths connecting (a) the engineering lab with the computer center and (b) the engineering lab with the library. Show that these two paths are perpendicular to each other.

**70. Geometry** Show that the points $A(0, 0)$, $B(0, 4)$, $C(2, 3)$, and $D(2, 7)$ are the vertices of a parallelogram. (Opposite sides of a parallelogram are parallel.)

**71. Approach Angle** A small plane is landing at an airport with an approach angle of 45 degrees, or slope of $-1$. The plane begins its descent when it has an elevation of 3300 feet. Find the equation that describes the relationship between the craft's altitude and distance traveled, assuming that at distance 0 it starts the approach angle. Graph your equation on a graphics calculator. What does the graph tell you about the approach if the airport is 4000 feet from where the plane starts its landing.

**72. Cost Equation** The average daily cost $C$ for a room at a city hospital has risen by $42.50 per year for the years 1980 through 1990. If the average cost in 1987 was $542.50, what is an equation which describes the average cost during this decade, as a function of the number of years $T$ since 1980.

**73. Revenue Equation** A small business predicts its revenue growth by a straight-line method with a slope of $50,000 per year. In its fifth year it had revenues of $330,000. Find an equation that describes the relationship between revenues $R$ and the number of years $T$ since it opened for business.

**74.** Graph $y = 1.3x + 7$ and verify that the $y$-intercept is 7.

**75.** Graph the lines whose equations are

$$y = 1.5x + 1,$$
$$y = 1.5x - 1,$$

and

$$y = 1.5x + 2.5.$$

What do you observe about the orientation of these lines? Why would you expect this result from the equations of the lines themselves?

**76.** Graph the line $y = 3.4x - 2.3$. Find the coordinates of any two points on the line, and use them to estimate the slope. What is the actual slope of the line?

**77.** Using the standard window, graph the lines with equations

$$0.1875x - 0.3y + 0.94 = 0$$

and

$$0.32x + 0.2y + 1.01 = 0$$

on the same viewing rectangle. Now, change the window to a square window (for example, on a TI-82, use ZOOM, ZSquare). Notice that the lines appear to be perpendicular to each other. Prove that this is indeed the case.

---

**OBJECTIVE**

To develop the notion of demand and supply curves and to introduce linear functions.

## 4.2 APPLICATIONS AND LINEAR FUNCTIONS

Many situations in economics can be described by using straight lines, as Example 1 shows.

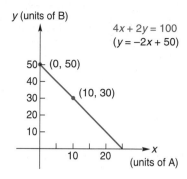

y (units of B)

$4x + 2y = 100$
$(y = -2x + 50)$

50 — (0, 50)
40 —
30 — (10, 30)
20 —
10 —

10    20
(units of A)

**FIGURE 4.13** Linearly related production levels.

### EXAMPLE 1  Production Levels

Suppose that a manufacturer uses 100 lb of material to produce products A and B, which require 4 lb and 2 lb of material per unit, respectively. If $x$ and $y$ denote the number of units produced of A and B, respectively, then all levels of production are given by the combinations of $x$ and $y$ that satisfy the equation

$$4x + 2y = 100, \quad \text{where } x, y \geq 0.$$

Thus, the levels of production of A and B are linearly related. Solving for $y$ gives

$$y = -2x + 50 \quad \text{(slope–intercept form)},$$

so the slope is $-2$. The slope reflects the rate of change of the level of production of B with respect to the level of production of A. For example, if 1 more unit of A is to be produced, it will require 4 more pounds of material, resulting in $\frac{4}{2} = 2$ *fewer* units of B. Accordingly, as $x$ increases by 1 unit, the corresponding value of $y$ decreases by 2 units. To sketch the graph of $y = -2x + 50$, we can use the $y$-intercept $(0, 50)$ and the fact that when $x = 10$, $y = 30$. (See Fig. 4.13.) ∎

### Demand and Supply Curves

**Principles in Practice 1**
**Production Levels**

A sporting-goods manufacturer allocates 1000 units of time per day to make skis and ski boots. If it takes 8 units of time to make a ski, and 14 units of time to make a boot, find an equation to describe all possible production levels of the two products.

For each price level of a product, there is a corresponding quantity of that product which consumers will demand (that is, purchase) during some time period. Usually, the higher the price, the smaller is the quantity demanded; as the price falls, the quantity demanded increases. If the price per unit of the product is given by $p$ and the corresponding quantity (in units) is given by $q$, then an equation relating $p$ and $q$ is called a **demand equation.** Its graph is called a **demand curve.** Figure 4.14(a) shows a demand curve. In keeping with the practice of most economists, the horizontal axis is the $q$-axis and the vertical axis is the $p$-axis. We shall assume that the price per unit is given in dollars and the period is one week. Thus, the point $(a, b)$ in Fig. 4.14(a) indicates that, at a price of $b$ dollars per unit, consumers will demand $a$ units per week. Since negative prices or quantities are not meaningful, both $a$ and $b$ must be nonnegative. For most products, an increase in the quantity demanded corresponds to a decrease in price. Thus, a demand curve typically falls from left to right, as in Fig. 4.14(a).

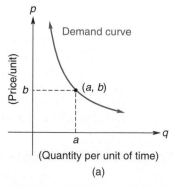

p

Demand curve

(Price/unit)

b ------ (a, b)

a    q

(Quantity per unit of time)

(a)

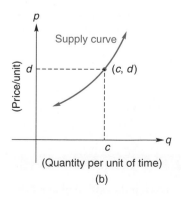

p

Supply curve

(Price/unit)

d ------ (c, d)

c    q

(Quantity per unit of time)

(b)

**FIGURE 4.14** Demand and supply curves.

In response to various prices, there is a corresponding quantity of product that *producers* are willing to supply to the market during some time period. Usually, the higher the price per unit, the larger is the quantity that producers are willing to supply; as the price falls, so will the quantity supplied. If $p$ denotes the price per unit and $q$ denotes the corresponding quantity, then an equation relating $p$ and $q$ is called a **supply equation,** and its graph is called a **supply curve.** Figure 4.14(b) shows a supply curve. If $p$ is in dollars and the period is one week, then the point $(c, d)$ indicates that, at a price of $d$ dollars each, producers will supply $c$ units per week. As before, $c$ and $d$ are nonnegative. A supply curve usually rises from left to right, as in Fig. 4.14(b). This indicates that a producer will supply more of a product at higher prices.

We shall now focus on demand and supply curves that are straight lines (Fig. 4.15). They are called *linear* demand and *linear* supply curves. Such curves have equations in which $p$ and $q$ are linearly related. Because a demand curve typically falls from left to right, a linear demand curve has a negative slope. [See Fig. 4.15(a).] However, the slope of a linear supply curve is positive, because the curve rises from left to right. [See Fig. 4.15(b).]

> Typically, a demand curve falls from left to right and a supply curve rises from left to right. However, there are exceptions. For example, the demand for insulin could be represented by a vertical line, since this demand can remain constant regardless of price.

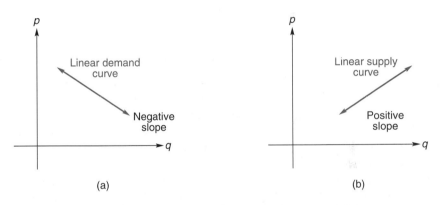

(a)                                    (b)

**FIGURE 4.15**   Linear demand and supply curves.

## EXAMPLE 2   Finding a Demand Equation

*Suppose the demand per week for a product is* 100 *units when the price is* $58 *per unit and* 200 *units at* $51 *each. Determine the demand equation, assuming that it is linear.*

*Solution:*

> *Strategy:*   Since the demand equation is linear, the demand curve must be a straight line. We are given that quantity $q$ and price $p$ are linearly related such that $p = 58$ when $q = 100$ and $p = 51$ when $q = 200$. Thus, the given data can be represented in a $q, p$-coordinate plane [see Fig. 4.15(a)] by points $(100, 58)$ and $(200, 51)$. With these points, we can find an equation of the line—that is, the demand equation.

The slope of the line passing through $(100, 58)$ and $(200, 51)$ is

$$m = \frac{51 - 58}{200 - 100} = -\frac{7}{100}.$$

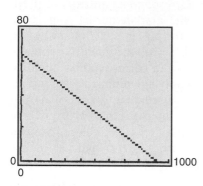

**FIGURE 4.16**    Graph of demand function $p = -\frac{7}{100}q + 65$.

An equation of the line (point–slope form) is

$$p - p_1 = m(q - q_1),$$

$$p - 58 = -\frac{7}{100}(q - 100).$$

Simplifying gives the demand equation

$$p = -\frac{7}{100}q + 65. \tag{1}$$

Customarily, a demand equation (as well as a supply equation) expresses $p$ in terms of $q$ and actually defines a function of $q$. For example, Eq. (1) defines $p$ as a function of $q$ and is called the *demand function* for the product. (See Fig. 4.16).    ∎

### Linear Functions

A *linear function* was described in Sec. 3.2. Here is a formal definition.

**DEFINITION**

*A function $f$ is a **linear function** if and only if $f(x)$ can be written in the form $f(x) = ax + b$, where $a$ and $b$ are constants and $a \neq 0$.*

Suppose that $f(x) = ax + b$ is a linear function, and let $y = f(x)$. Then $y = ax + b$, which is an equation of a straight line with slope $a$ and $y$-intercept $b$. Thus, **the graph of a linear function is a straight line.** We say that the function $f(x) = ax + b$ has slope $a$.

**Principles in Practice 3**
Graphing Linear Functions

A computer repair company charges a fixed amount plus an hourly rate for a service call. If $x$ is the number of hours needed for a service call, the total cost of a call is described by the function $f(x) = 40x + 60$. Graph the function by finding and plotting two points.

**EXAMPLE 3    Graphing Linear Functions**

a. *Graph $f(x) = 2x - 1$.*

*Solution:* Here $f$ is a linear function (with slope 2), so its graph is a straight line. Since two points determine a straight line, we need only plot two points and then draw a line through them. [See Fig. 4.17(a).] Note that one of the points plotted is the vertical-axis intercept, $-1$, which occurs when $x = 0$.

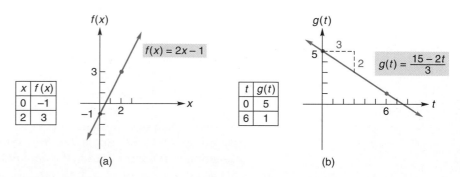

**FIGURE 4.17**    Graphs of linear functions.

b. *Graph $g(t) = \dfrac{15 - 2t}{3}$.*

*Solution:* Notice that $g$ is a linear function, because we can express it in the form $g(t) = at + b$.

$$g(t) = \frac{15 - 2t}{3} = \frac{15}{3} - \frac{2t}{3} = -\frac{2}{3}t + 5.$$

The graph of $g$ is shown in Fig. 4.17(b). Since the slope is $-\frac{2}{3}$, observe that as $t$ increases by 3 units, $g(t)$ *decreases* by 2. ∎

### EXAMPLE 4 Determining a Linear Function

*Suppose f is a linear function with slope 2 and $f(4) = 8$. Find $f(x)$.*

*Solution:* Since $f$ is linear, it has the form $f(x) = ax + b$. The slope is 2, so $a = 2$, and we have

$$f(x) = 2x + b. \qquad (2)$$

Now we determine $b$. Since $f(4) = 8$, we replace $x$ by 4 in Eq. (2) and solve for $b$:

$$f(4) = 2(4) + b,$$
$$8 = 8 + b,$$
$$0 = b.$$

Hence, $f(x) = 2x$. ∎

### EXAMPLE 5 Determining a Linear Function

*If $y = f(x)$ is a linear function such that $f(-2) = 6$ and $f(1) = -3$, find $f(x)$.*

*Solution:*

> *Strategy:* The function values correspond to points on the graph of $f$. With these points we can determine an equation of the line and hence the linear function.

The condition that $f(-2) = 6$ means that when $x = -2$, then $y = 6$. Thus, $(-2, 6)$ lies on the graph of $f$, which is a straight line. Similarly, $f(1) = -3$ implies that $(1, -3)$ also lies on the line. If we set $(x_1, y_1) = (-2, 6)$ and $(x_2, y_2) = (1, -3)$, the slope of the line is given by

$$m = \frac{y_2 - y_1}{x_2 - x_1} = \frac{-3 - 6}{1 - (-2)} = \frac{-9}{3} = -3.$$

We can find an equation of the line by using a point–slope form:

$$y - y_1 = m(x - x_1),$$
$$y - 6 = -3[x - (-2)],$$
$$y - 6 = -3x - 6,$$
$$y = -3x.$$

Because $y = f(x)$, $f(x) = -3x$. Of course, the same result is obtained if we set $(x_1, y_1) = (1, -3)$. ∎

In many studies, data are collected and plotted on a coordinate system. An analysis of the results may indicate a functional relationship between the

---

### Principles in Practice 4
**Determining a Linear Function**

The height of children between the ages of 6 years and 10 years can be modeled by a linear function of age $t$ in years. The height of one child changes by 2.3 inches per year, and she is 50.6 inches tall at age 8. Find a function that describes the height of this child at age $t$.

### Principles in Practice 5
**Determining a Linear Function**

An antique necklace is expected to be worth \$360 after 3 years and \$640 after 7 years. Find a function that describes the value of the necklace after $x$ years.

variables involved. For example, the data points may be approximated by points on a straight line. This would indicate a linear functional relationship, such as the one in the next example.

### EXAMPLE 6  Diet for Hens

*In testing an experimental diet for hens, it was determined that the average live weight w (in grams) of a hen was statistically a linear function of the number of days d after the diet began, where $0 \le d \le 50$. Suppose the average weight of a hen beginning the diet was 40 grams and 25 days later it was 675 grams.*

**a.** *Determine w as a linear function of d.*

> **Solution:** Since $w$ is a linear function of $d$, its graph is a straight line. When $d = 0$ (the beginning of the diet), $w = 40$. Thus, $(0, 40)$ lies on the graph. (See Fig. 4.18.) Similarly, $(25, 675)$ lies on the graph. If we set $(d_1, w_1) = (0, 40)$ and $(d_2, w_2) = (25, 675)$, the slope of the line is

$$m = \frac{w_2 - w_1}{d_2 - d_1} = \frac{675 - 40}{25 - 0} = \frac{635}{25} = \frac{127}{5}.$$

Using a point–slope form, we have

$$w - w_1 = m(d - d_1),$$

$$w - 40 = \frac{127}{5}(d - 0),$$

$$w - 40 = \frac{127}{5}d,$$

$$w = \frac{127}{5}d + 40,$$

which expresses $w$ as a linear function of $d$.

**b.** *Find the average weight of a hen when $d = 10$.*

> **Solution:** When $d = 10$, $w = \frac{127}{5}(10) + 40 = 254 + 40 = 294$. Thus, the average weight of a hen 10 days after the beginning of the diet is 294 grams.

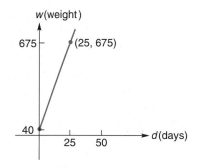

**FIGURE 4.18**  Linear function describing diet for hens.

### ■ Exercise 4.2

*In Problems 1–6, find the slope and vertical-axis intercept of the linear function, and sketch the graph.*

**1.** $y = f(x) = -4x$.

**2.** $y = f(x) = x + 1$.

**3.** $g(t) = 2t - 4$.

**4.** $g(t) = 2(4 - t)$.

**5.** $h(q) = \dfrac{7 - q}{2}$.

**6.** $h(q) = 0.5q + 0.25$.

*In Problems 7–14, find f(x) if f is a linear function that has the given properties.*

**7.** slope $= 4$,  $f(2) = 8$.

**8.** $f(0) = 3$,  $f(4) = -5$.

**9.** $f(1) = 2$,  $f(-2) = 8$.

**10.** slope $= -6$,  $f(\frac{1}{2}) = -2$.

**11.** slope $= -\frac{1}{2}, f(-\frac{1}{2}) = 4$.

**12.** $f(1) = 1$,  $f(2) = 2$.

**13.** $f(-2) = -1$,  $f(-4) = -3$.

**14.** slope $= 0.01$,  $f(0.1) = 0.01$.

**15. Demand Equation**  Suppose consumers will demand 40 units of a product when the price is $12 per unit and 25 units when the price is $18 each. Find the demand equation, assuming that it is linear. Find the price per unit when 30 units are demanded.

**16. Demand Equation** The demand per week for a best-selling book is 26,000 books when the price is $16 each, and 10,000 books when the price is $24 each. Find the demand equation for the book, assuming that it is linear.

**17. Supply Equation** A refrigerator manufacturer will produce 300 units when the price is $840, and 220 units when the price is $640. Assume that price $p$ and quantity $q$ produced are linearly related and find the supply equation.

**18. Supply Equation** Suppose a manufacturer of shoes will place on the market 50 (thousand pairs) when the price is 35 (dollars per pair) and 35 when the price is 30. Find the supply equation, assuming that price $p$ and quantity $q$ are linearly related.

**19. Cost Equation** Suppose the cost to produce 10 units of a product is $40 and the cost of 20 units is $70. If cost $c$ is linearly related to output $q$, find a linear equation relating $c$ and $q$. Find the cost to produce 35 units.

**20. Cost Equation** An advertiser goes to a printer and is charged $73 for 100 copies of one flyer and $82 for 400 copies of another flyer. This printer charges a fixed setup cost plus a charge for every copy of single-page flyers. Find a function that describes the cost of a printing job, if $x$ is the number of copies made.

**21. Electric Rates** An electric utility company charges residential customers 11.5 cents per kilowatt hour plus a base charge each month. One customer's monthly bill comes to $51.65 for 380 kilowatt hours. Find a linear function that describes the total monthly charges for electricity if $x$ is the number of kilowatt hours used in a month.

**22. Radiation Therapy** A cancer patient is to receive drug and radiation therapies. Each cubic centimeter of the drug to be used contain 200 curative units, and each minute of radiation exposure gives 300 curative units. The patient requires 2400 curative units. If $d$ cubic centimeters of the drug and $r$ minutes of radiation are administered, determine an equation relating $d$ and $r$. Graph the equation for $d \geq 0$, and $r \geq 0$; label the horizontal axis as $d$.

**23. Depreciation** Suppose the value of a piece of machinery decreases each year by 10% of is original value. If the original value is $8000, find an equation that expresses the value $v$ of the machinery $t$ years after purchase, where $0 \leq t \leq 10$. Sketch the equation, choosing $t$ as the horizontal axis and $v$ as the vertical axis. What is the slope of the resulting line? This method of considering the value of equipment is called *straight-line depreciation.*

**24. Depreciation** A new television depreciates $120 per year, and it is worth $340 after four years. Find a function that describes the value of this television, if $x$ is the age of the television in years.

**25. Appreciation** A new apartment building was sold for $760,000 five years after it was purchased. The original owners calculated that the building appreciated $40,000 per year while they owned it. Find a linear function that describes the appreciation of the building, if $x$ is the number of years since the original purchase.

**26. Appreciation** A house purchased for $198,000 is expected to double in value in 18 years. Find a linear equation that describes the house's value after $x$ years.

**27. Repair Charges** A business-copier repair company charges a fixed amount plus an hourly rate for a service call. If a customer is billed $150 for a one-hour service call and $280 for a three-hour service call, find a linear function that describes the price of a service call, where $x$ is the number of hours of service.

**28. Sheep Wool Length** For sheep maintained at high environmental temperatures, respiratory rate $r$ (per minute) increases as wool length $l$ (in centimeters) decreases.[1] Suppose sheep with a wool length of 2 cm have an (average) respiratory rate of 160, and those with a wool length of 4 cm have a respiratory rate of 125. Assume that $r$ and $l$ are linearly related. (a) Find an equation that gives $r$ in terms of $l$. (b) Find the respiratory rate of sheep with a wool length of 1 cm.

**29. Isocost Line** In production analysis, an *isocost line* is a line whose points represent all combinations of two factors of production that can be purchased for the same amount. Suppose a farmer has allocated $20,000 for the purchase of $x$ tons of fertilizer (costing $200 per ton) and $y$ acres of land (costing $2000 per acre). Find an equation of the isocost line which describes the various combinations that can be purchased for $20,000. Observe that neither $x$ nor $y$ can be negative.

**30. Isoprofit Line** A manufacturer produces products $X$ and $Y$ for which the profits per unit are $4 and $6, respectively. If $x$ units of $X$ and $y$ units of $Y$ are sold, then the total profit $P$ is given by $P = 4x + 6y$, where $x, y \geq 0$. (a) Sketch the graph of this equation for $P = 240$. The result is called an *isoprofit line,* and its points represent all combinations of sales that produce a profit of $240. (b) Determine the slope for $P = 240$. (c) If $P = 600$, determine the slope. (d) Are isoprofit lines for products $X$ and $Y$ parallel?

**31. Grade Scaling** For reasons of comparison, a professor wants to rescale the scores on a set of test papers so that the maximum score is still 100, but the mean (average) is 80 instead of 56. (a) Find a linear equation that will do

[1]Adapted from G. E. Folk, Jr., *Textbook of Environmental Physiology,* 2d ed. (Philadelphia: Lea & Febiger, 1974).

this. [*Hint:* You want 56 to become an 80 and 100 to remain 100. Consider the points (56, 80) and (100, 100) and, more generally, $(x, y)$, where $x$ is the old score and $y$ is the new score. Find the slope and use a point–slope form. Express $y$ in terms of $x$.] (b) If 60 on the new scale is the lowest passing score, what was the lowest passing score on the original scale?

32. **Psychology**   The result of Sternberg's psychological experiment[2] on information retrieval is that a person's reaction time $R$, in milliseconds, is statistically a linear function of memory set size $N$ as follows:

$$R = 38N + 397.$$

Sketch the graph for $1 \le N \le 5$. What is the slope?

33. **Psychology**   In a certain learning experiment involving repetition and memory,[3] the proportion $p$ of items recalled was estimated to be linearly related to the effective study time $t$ (in seconds), where $t$ is between 5 and 9. For an effective study time of 5 seconds, the proportion of items recalled was 0.32. For each 1-second increase in study time, the proportion recalled increased by 0.059. (a) Find an equation that gives $p$ in terms of $t$. (b) What proportion of items was recalled with 9 seconds of effective study time?

34. **Diet for Pigs**   In testing an experimental diet for pigs, it was determined that the (average) live weight $w$ (in kilograms) of a pig was statistically a linear function of the number of days $d$ after the diet was initiated, where $0 \le d \le 100$. If the weight of a pig beginning the diet was 20 kg, and thereafter the pig gained 6.6 kg every 10 days, determine $w$ as a function of $d$, and find the weight of a pig 50 days after the beginning of the diet.

35. **Cricket Chirps**   Biologists have found that the number of chirps made per minute by crickets of a certain species is related to the temperature. The relationship is very close to being linear. At 68°F, those crickets chirp about 124 times a minute. At 80°F, they chirp about 172 times a minute. (a) Find an equation that gives Fahrenheit temperature $t$ in terms of the number of chirps, $c$, per minute. (b) If you count chirps for only 15 seconds, how can you quickly estimate the temperature?

36. **Electrical Circuit**   In a circuit, the voltage $V$ (in volts) and current $i$ (in amperes) are linearly related. When $i = 4, V = 2$; when $i = 12, V = 6$.
   a. Determine $V$ as a function of $i$.
   b. Find the voltage when the current is 10.

37. **Physics**   The pressure $P$ of a fixed volume of gas, in centimeters of mercury, is linearly related to the temperature $T$, in degrees Celsius. In an experiment with dry air, it was found that $P = 90$ when $T = 40$ and that $P = 100$ when $T = 80$. Express $P$ as a function of $T$.

38. **Electrical Theory**   When a graph of the terminal potential difference $V$, in volts, of a Daniell cell is plotted as a function of the current $i$, in amperes, delivered to an external resistor, a straight line is obtained. The slope of this line is the negative of the internal resistance of the cell. For a particular cell with an internal resistance of 0.06 ohms it was found that $V = 0.6$ V when $i = 0.12$ A. Express $V$ as a function of $i$.

39. **Hydraulics**   A formula used in hydraulics is

$$Q = 3.340b^3 + 1.8704b^2 x,$$

where $b$ is a constant.
   a. Is the graph of this equation a straight line?
   b. If so, what is the slope of the line when $b = 1$?

[2]G. R. Loftus and E. F. Loftus, *Human Memory: The Processing of Information* (New York: Lawrence Erlbaum Associates, Inc., distributed by the Halsted Press, Division of John Wiley & Sons, Inc., 1976).

[3]D. L. Hintzman, "Repetition and Learning," in *The Psychology of Learning*, Vol. 10, ed. G. H. Bower (New York: Academic Press, Inc., 1976), p. 77.

---

**OBJECTIVE**

To sketch parabolas arising from quadratic functions.

## 4.3 QUADRATIC FUNCTIONS

In Sec. 3.2, a *quadratic function* was described as a polynomial function of degree 2. Here is a formal definition.

**DEFINITION**

*A function f is a **quadratic function** if and only if $f(x)$ can be written in the form $f(x) = ax^2 + bx + c$, where a, b, and c are constants and $a \neq 0$.*

For example, the functions $f(x) = x^2 - 3x + 2$ and $F(t) = -3t^2$ are quadratic. However, $g(x) = \dfrac{1}{x^2}$ is *not* quadratic, because it cannot be written in the form $g(x) = ax^2 + bx + c$.

The graph of the quadratic function $y = f(x) = ax^2 + bx + c$ is called a **parabola** and has a shape like the curves in Fig. 4.19. If $a > 0$, the graph extends upward indefinitely, and we say that the parabola *opens upward* [Fig. 4.19(a)]. If $a < 0$, the parabola *opens downward* [Fig. 4.19(b)].

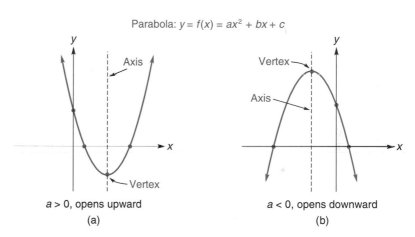

Parabola: $y = f(x) = ax^2 + bx + c$

| | |
|:---:|:---:|
| $a > 0$, opens upward | $a < 0$, opens downward |
| (a) | (b) |

**FIGURE 4.19** Parabolas.

Each parabola in Fig. 4.19 is *symmetric* about a vertical line, called the **axis of symmetry** of the parabola. That is, if the page were folded on one of these lines, then the two halves of the corresponding parabola would coincide. The axis (of symmetry) is *not* part of the parabola, but is a useful aid in sketching the parabola.

Figure 4.19 also shows points labeled **vertex,** where the axis cuts the parabola. If $a > 0$, the vertex is the "lowest" point on the parabola. This means that $f(x)$ has a minimum value at this point. By performing algebraic manipulations on $ax^2 + bx + c$ (referred to as *completing the square*), we can determine not only this minimum value, but also where it occurs. We have

$$f(x) = ax^2 + bx + c = (ax^2 + bx) + c.$$

Adding and subtracting $\dfrac{b^2}{4a}$ gives

$$f(x) = \left(ax^2 + bx + \frac{b^2}{4a}\right) + c - \frac{b^2}{4a}$$

$$= a\left(x^2 + \frac{b}{a}x + \frac{b^2}{4a^2}\right) + c - \frac{b^2}{4a},$$

so that

$$f(x) = a\left(x + \frac{b}{2a}\right)^2 + c - \frac{b^2}{4a}.$$

Since $\left(x + \dfrac{b}{2a}\right)^2 \geq 0$ and $a > 0$, it follows that $f(x)$ has a minimum value when $x + \dfrac{b}{2a} = 0$, that is, when $x = -\dfrac{b}{2a}$. The $y$-coordinate corresponding to this value of $x$ is $f\left(-\dfrac{b}{2a}\right)$. Thus, the vertex is given by

$$\text{vertex} = \left(-\frac{b}{2a}, f\left(-\frac{b}{2a}\right)\right).$$

This is also the vertex of a parabola that opens downward ($a < 0$), but in this case $f\left(-\dfrac{b}{2a}\right)$ is the maximum value of $f(x)$. [See Fig. 4.19(b).]

The point where the parabola $y = ax^2 + bx + c$ intersects the $y$-axis (that is, the $y$-intercept) occurs when $x = 0$. The $y$-coordinate of this point is $c$, so the $y$-intercept is $(0, c)$, or, more simply, $c$. In summary, we have the following.

> **Graph of Quadratic Function**
>
> The graph of the quadratic function $y = f(x) = ax^2 + bx + c$ is a parabola.
>
> **1.** If $a > 0$, the parabola opens upward. If $a < 0$, it opens downward.
>
> **2.** The vertex is $\left(-\dfrac{b}{2a}, f\left(-\dfrac{b}{2a}\right)\right)$.
>
> **3.** The $y$-intercept is $c$.

We can quickly sketch the graph of a quadratic function by first locating the vertex, the $y$-intercept, and a few other points, such as those where the parabola intersects the $x$-axis. These $x$-*intercepts* are found by setting $y = 0$ and solving for $x$. Once the intercepts and vertex are found, it is then relatively easy to pass the appropriate parabola through these points. In the event that the $x$-intercepts are very close to the vertex or that no $x$-intercepts exist, we find a point on each side of the vertex, so that we can give a reasonable sketch of the parabola. Keep in mind that passing a (dashed) vertical line through the vertex gives the axis of symmetry. By plotting points to one side of the axis, we can use symmetry and obtain corresponding points on the other side.

**Principles in Practice 1**

**Graphing a Quadratic Function**

The daily profit for a car dealership from the sale of a minivan is given by $P(x) = -x^2 + 2x + 399$, where $x$ is the number of minivans sold. Find the function's vertex and intercepts, and graph the function.

**EXAMPLE 1  Graphing a Quadratic Function**

*Graph the quadratic function* $y = f(x) = -x^2 - 4x + 12$.

*Solution:* Here $a = -1, b = -4$, and $c = 12$. Since $a < 0$, the parabola opens downward and thus has a highest point. The $x$-coordinate of the vertex is

$$-\frac{b}{2a} = -\frac{-4}{2(-1)} = -2.$$

The $y$-coordinate is $f(-2) = -(-2)^2 - 4(-2) + 12 = 16$. Thus, the vertex is $(-2, 16)$, so the maximum value of $f(x)$ is 16. Since $c = 12$, the $y$-intercept is 12. To find the $x$-intercepts, we let $y$ be 0 in $y = -x^2 - 4x + 12$ and solve for $x$:

$$0 = -x^2 - 4x + 12,$$

$$0 = -(x^2 + 4x - 12),$$

$$0 = -(x + 6)(x - 2).$$

Hence, $x = -6$ or $x = 2$, so the $x$-intercepts are $-6$ and 2. Now we plot the vertex, axis of symmetry, and intercepts. [See Fig. 4.20(a).] Since $(0, 12)$ is *two* units to the *right* of the axis, there is a corresponding point *two* units to the *left* of the axis with the same *y*-coordinate. Thus, we get the point $(-4, 12)$. Through all points, we draw a parabola opening downward. [See Fig. 4.20(b).] ∎

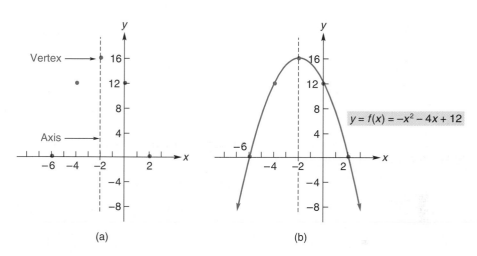

(a)  (b)

**FIGURE 4.20** Graph of parabola $y = f(x) = -x^2 - 4x + 12$.

### EXAMPLE 2 Graphing a Quadratic Function

*Graph $p = 2q^2$.*

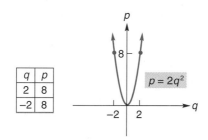

**FIGURE 4.21** Graph of parabola $p = 2q^2$.

| q | p |
|---|---|
| 2 | 8 |
| -2 | 8 |

*Solution:* Here *p* is a quadratic function of *q*, where $a = 2$, $b = 0$, and $c = 0$. Since $a > 0$, the parabola opens upward and thus has a lowest point. The *q*-coordinate of the vertex is

$$-\frac{b}{2a} = -\frac{0}{2(2)} = 0,$$

and the *p*-coordinate is $2(0)^2 = 0$. Consequently, the *minimum* value of *p* is 0 and the vertex is $(0, 0)$. In this case, the *p*-axis is the axis of symmetry. A parabola opening upward with vertex at $(0, 0)$ cannot have any other intercepts. Hence, to draw a reasonable graph, we plot a point on each side of the vertex. If $q = 2$, then $p = 8$. This gives the point $(2, 8)$ and, by symmetry, the point $(-2, 8)$. (See Fig. 4.21.) ∎

Example 3 illustrates that finding intercepts may require use of the quadratic formula.

### Principles in Practice 2
#### Graphing a Quadratic Function

A man standing on a pitcher's mound throws a ball straight up with an initial velocity of 32 feet per second. The height *h* of the ball in feet *t* seconds after it was thrown is described by the function $h(t) = -16t^2 + 32t + 8$, for $t \geq 0$. Find the function's vertex and intercepts, and graph the function.

### EXAMPLE 3 Graphing a Quadratic Function

*Graph $g(x) = x^2 - 6x + 7$.*

*Solution:* Here *g* is a quadratic function, where $a = 1$, $b = -6$, and $c = 7$. The parabola opens upward, because $a > 0$. The *x*-coordinate of the vertex (lowest point) is

$$-\frac{b}{2a} = -\frac{-6}{2(1)} = 3.$$

and $g(3) = 3^2 - 6(3) + 7 = -2$, which is the minimum value of $g(x)$. Thus, the vertex is $(3, -2)$. Since $c = 7$, the vertical-axis intercept is 7. To find *x*-intercepts, we set $g(x) = 0$.

$$0 = x^2 - 6x + 7.$$

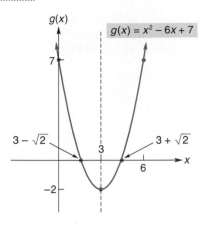

**FIGURE 4.22** Graph of parabola $g(x) = x^2 - 6x + 7$.

The right side does not factor easily, so we shall use the quadratic formula to solve for $x$:

$$x = \frac{-b \pm \sqrt{b^2 - 4ac}}{2a} = \frac{-(-6) \pm \sqrt{(-6)^2 - 4(1)(7)}}{2(1)}$$

$$= \frac{6 \pm \sqrt{8}}{2} = \frac{6 \pm \sqrt{4 \cdot 2}}{2} = \frac{6 \pm 2\sqrt{2}}{2}$$

$$= \frac{6}{2} \pm \frac{2\sqrt{2}}{2} = 3 \pm \sqrt{2}.$$

Therefore, the $x$-intercepts are $3 + \sqrt{2}$ and $3 - \sqrt{2}$. After plotting the vertex, intercepts, and (by symmetry) the point $(6, 7)$, we draw a parabola opening upward in Fig. 4.22. ∎

**EXAMPLE 4 Graphing a Quadratic Function**

*Graph $y = f(x) = 2x^2 + 2x + 3$ and find the range of $f$.*

*Solution:* This function is quadratic with $a = 2, b = 2$, and $c = 3$. Since $a > 0$, the graph is a parabola opening upward. The $x$-coordinate of the vertex is

$$-\frac{b}{2a} = -\frac{2}{2(2)} = -\frac{1}{2}.$$

and the $y$-coordinate is $2(-\frac{1}{2})^2 + 2(-\frac{1}{2}) + 3 = \frac{5}{2}$. Thus, the vertex is $(-\frac{1}{2}, \frac{5}{2})$. Since $c = 3$, the $y$-intercept is 3. A parabola opening upward with its vertex above the $x$-axis has no $x$-intercepts. In Fig. 4.23 we plotted the $y$-intercept, the

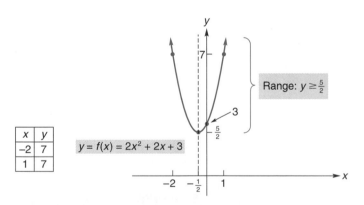

| $x$ | $y$ |
|-----|-----|
| $-2$ | $7$ |
| $1$ | $7$ |

**FIGURE 4.23** Graph of $y = f(x) = 2x^2 + 2x + 3$.

vertex, and an additional point $(-2, 7)$ to the left of the vertex. By symmetry, we also get the point $(1, 7)$. Passing a parabola through these points gives the desired graph. From the figure, we see that the range of $f$ is all $y \geq \frac{5}{2}$, that is, the interval $[\frac{5}{2}, \infty)$.

**EXAMPLE 5 Maximum Revenue**

*The demand function for a manufacturer's product is $p = 1000 - 2q$, where $p$ is the price (in dollars) per unit when $q$ units are demanded (per week) by consumers. Find the level of production that will maximize the manufacturer's total revenue, and determine this revenue.*

*Solution:*

> *Strategy:* To maximize revenue, we must determine the revenue function, $r = f(q)$. Using the relation
>
> $$\textbf{total revenue} = \textbf{(price)}\textbf{(quantity)},$$
>
> we have
>
> $$r = pq.$$
>
> Using the demand equation, we can express $p$ in terms of $q$, so $r$ will be strictly a function of $q$.

The formula for total revenue should be added to your repertoire of relationships in business and economics.

We have

$$r = pq$$
$$= (1000 - 2q)q.$$
$$r = 1000q - 2q^2.$$

Note that $r$ is a quadratic function of $q$, with $a = -2$, $b = 1000$, and $c = 0$. Since $a < 0$ (the parabola opens downward), $r$ is maximum at the vertex $(q, r)$, where

$$q = -\frac{b}{2a} = -\frac{1000}{2(-2)} = 250.$$

The maximum value of $r$ is given by

$$r = 1000(250) - 2(250)^2$$
$$= 250{,}000 - 125{,}000 = 125{,}000.$$

Thus, the maximum revenue that the manufacturer can receive is $125,000, which occurs at a production level of 250 units. Figure 4.24(a) shows the graph of the revenue function. Only that portion for which $q \geq 0$ and $r \geq 0$ is drawn, since quantity and revenue cannot be negative. ∎

### Principles in Practice 3
#### Maximum Revenue

The demand function for a publisher's line of cookbooks is $p = 6 - 0.003q$, where $p$ is the price (in dollars) per unit when $q$ units are demanded (per day) by consumers. Find the level of production that will maximize the manufacturer's total revenue, and determine this revenue.

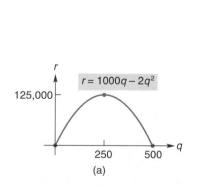

(a)

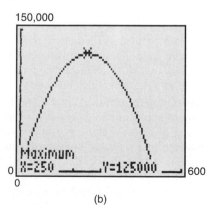

(b)

**FIGURE 4.24**   Graph of revenue function.

## TECHNOLOGY

The maximum (or minimum) value of a function can be conveniently found with a graphics calculator either by using trace and zoom or by using the "maximum" (or "minimum") feature. Figure 4.24(b) shows the display for the revenue function of Example 5, namely, the graph of $y = 1000x - 2x^2$. Note that we replaced $r$ by $y$ and $q$ by $x$.

## ■ Exercise 4.3

*In Problems 1–8, state whether the function is quadratic.*

**1.** $f(x) = 5x^2.$

**2.** $g(x) = \dfrac{1}{2x^2 - 4}.$

**3.** $g(x) = 7 - 6x.$

**4.** $h(s) = 2s^2\,(s^2 + 1).$

**5.** $h(q) = (q + 4)^2.$

**6.** $f(t) = 2t\,(3 - t) + 4t.$

**7.** $f(s) = \dfrac{s^2 - 4}{2}.$

**8.** $g(t) = (t^2 - 1)^2.$

*In Problems 9–12, do not include a graph.*

**9.** (a) For the parabola $y = f(x) = -4x^2 + 8x + 7$, find the vertex. (b) Does the vertex correspond to the highest point or the lowest point on the graph?

**10.** Repeat Problem 9 if $y = f(x) = 8x^2 + 4x - 1.$

**11.** For the parabola $y = f(x) = x^2 + 2x - 8$, find (a) the $y$-intercept, (b) the $x$-intercepts, and (c) the vertex.

**12.** Repeat Problem 11 if $y = f(x) = 3 + x - 2x^2.$

*In Problems 13–22, graph each function. Give the vertex and intercepts, and state the range.*

**13.** $y = f(x) = x^2 - 6x + 5.$

**14.** $y = f(x) = -3x^2.$

**15.** $y = g(x) = -2x^2 - 6x.$

**16.** $y = f(x) = x^2 - 1.$

**17.** $s = h(t) = t^2 + 2t + 1.$

**18.** $s = h(t) = 2t^2 + 3t - 2.$

**19.** $y = f(x) = -9 + 8x - 2x^2.$

**20.** $y = H(x) = 1 - x - x^2.$

**21.** $t = f(s) = s^2 - 8s + 13.$

**22.** $t = f(s) = s^2 + 6s + 11.$

*In Problems 23–26, state whether $f(x)$ has a maximum value or a minimum value, and find that value.*

**23.** $f(x) = 100x^2 - 20x + 25.$

**24.** $f(x) = -2x^2 - 16x + 3.$

**25.** $f(x) = 4x - 50 - 0.1x^2.$

**26.** $f(x) = x(x + 3) - 12.$

**27. Revenue** The demand function for a manufacturer's product is $p = f(q) = 1200 - 3q$, where $p$ is the price (in dollars) per unit when $q$ units are demanded (per week). Find the level of production that maximizes the manufacturer's total revenue and determine this revenue.

**28. Revenue** The demand function for an office supply company's line of plastic rulers is $p = 0.8 - 0.0004q$, where $p$ is the price (in dollars) per unit when $q$ units are demanded (per day) by consumers. Find the level of production that will maximize the manufacturer's total revenue, and determine this revenue.

**29. Revenue** The demand function for an electronics company's laptop computer line is $p = 2400 - 6q$, where $p$ is the price (in dollars) per unit when $q$ units are demanded (per week) by consumers. Find the level of production that will maximize the manufacturer's total revenue, and determine this revenue.

**30. Marketing** A marketing firm estimates that $n$ months after the introduction of a client's new product, $f(n)$ thousand households will use it, where

$$f(n) = \tfrac{10}{9}n(12 - n), \qquad 0 \le n \le 12.$$

Estimate the maximum number of households that will use the product.

**31. Profit** The daily profit for the garden department of a store from the sale of trees is given by $P(x) = -x^2 + 18x + 144$, where $x$ is the number of trees sold. Find the function's vertex and intercepts, and graph the function.

**32. Psychology** A prediction made by early psychology relating the magnitude of a stimulus $x$ to the magnitude of a response $y$ is expressed by the equation $y = kx^2$, where $k$ is a constant of the experiment. In an experiment on pattern recognition, $k = 2$. Find the function's vertex and graph the equation. (Assume no restriction on $x$.)

**33. Biology** Biologists studied the nutritional effects on rats that were fed a diet containing 10% protein.[4] The protein consisted of yeast and corn flour. By varying the percentage $P$ of yeast in the protein mix, the group estimated that the average weight gain (in grams) of a rat over a period of time was

$$f(P) = -\tfrac{1}{50}P^2 + 2P + 20, \qquad 0 \le P \le 100.$$

Find the maximum weight gain.

**34. Height of Ball** Suppose that the height $s$ of a ball thrown vertically upward from the ground is given by

$$s = -4.9t^2 + 58.8t,$$

where $s$ is in meters and $t$ is elapsed time in seconds. (See Fig. 4.25.) After how many seconds will the ball reach its maximum height? What is the maximum height?

[4]Adapted from R. Bressani, "The Use of Yeast in Human Foods," in *Single-Cell Protein*, ed. R. I. Mateles and S. R. Tannenbaum (Cambridge, MA: MIT Press, 1968).

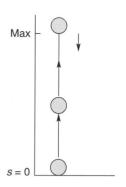

**FIGURE 4.25**
Ball thrown upward
(Problem 34).

**35. Archery**   A boy standing on a hill shoots an arrow straight up with an initial velocity of 64 feet per second. The height $h$ of the arrow in feet $t$ seconds after it was released is described by the function $h(t) = -16t^2 + 64t + 32$. What is the maximum height reached by the arrow? How many seconds after release does it take to reach this height?

**36. Toy Toss**   A 6 year old girl standing on a toy chest throws a doll straight up with an initial velocity of 16 feet per second. The height $h$ of the doll in feet $t$ seconds after it was released is described by the function $h(t) = -16t^2 + 16t + 4$. How long does it take the doll to reach its maximum height? What is the maximum height?

**37. Rocket Launch**   A toy rocket is launched straight up from the roof of a garage with an initial velocity of 80 feet per second. The height $h$ of the rocket in feet $t$ seconds after it was released is described by the function $h(t) = -16t^2 + 80t + 16$. Find the function's vertex and intercepts, and graph the function.

**38. Suspension Cable**   The shape of the main cable on a suspension bridge can be described by the function

$$y = f(x) = \frac{1}{500}x^2 + \frac{1}{250}x + 10, \quad -100 \le x \le 100,$$

where $f(x)$ is the height of the cable (in feet) above the roadbed and $x$ is the horizontal distance (in feet) from the center of the bridge. Graph the function and find its range.

**39. Physics**   The displacement of an object from a reference point at time $t$ is given by

$$s = 3.2t^2 - 16t + 28.7,$$

where $s$ is in meters and $t$ is in seconds.
  **a.** For what value of $t$ does the minimum displacement occur?
  **b.** What is the minimum displacement of the object from the reference point?

**40. Force**   During a collision, the force $F$ (in newtons) that acted on an object varied with time $t$ according to the equation $F = 87t - 21t^2$, where $t$ is in seconds.

  **a.** For what value of $t$ was the force a maximum?
  **b.** What was the maximum value of the force?

**41. Loaded Beam**   When a horizontal beam of length $l$ is uniformly loaded, the moment equation is

$$M = \frac{wlx}{2} - \frac{wx^2}{2},$$

where $w$ is related to the load and $x$ is measured from the left end of the beam.

  **a.** For what value of $x$ is $M$ a maximum? (Assume that $w > 0$.)
  **b.** What is the maximum value of $M$?
  **c.** For what values of $x$ does $M = 0$?

**42. Area**   Express the area of the rectangle shown in Fig. 4.26 as a quadratic function of $x$. For what value of $x$ will the area be a maximum?

**FIGURE 4.26**   Diagram for Problem 42.

**43. Enclosing Plot**   A building contractor wants to fence in a rectangular plot adjacent to a straight stream by using the stream for one side of the enclosed area. (See Fig. 4.27.) If the contractor has 200 feet of fence, find the dimensions of the maximum enclosed area.

**FIGURE 4.27**   Diagram for Problem 43.

**44.** Find two numbers whose sum is 40 and whose product is a maximum.

**45.** From the graph of $y = 1.4x^2 - 3.1x + 4.6$, determine the coordinates of the vertex. Round values to two decimal places. Verify your answer by using the vertex formula.

**46.** Find the zeros of $f(x) = -\sqrt{2}x^2 + 3x + 8.5$ by examining the graph of $f$. Round values to two decimal places.

**47.** Determine the number of real zeros for each of the following quadratic functions:
  **a.** $f(x) = 4.2x^2 - 8.1x + 10.4$.
  **b.** $f(x) = 5x^2 - 2\sqrt{35}x + 7$.
  **c.** $f(x) = \dfrac{5.1 - 7.2x - x^2}{4.8}$.

**48.** Find the maximum value (rounded to two decimal places) of the function $f(x) = 5.4 + 12x - 4.1x^2$ from its graph.

**49.** Find the minimum value (rounded to two decimal places) of the function $f(x) = 20x^2 - 3x + 7$ from its graph.

**To solve systems of linear equations in both two and three variables by using the technique of elimination by addition or by substitution. (In Chapter 6, other methods are shown.)**

## 4.4  SYSTEMS OF LINEAR EQUATIONS

### Two-Variable Systems

When a situation must be described mathematically, it is not unusual for a *set* of equations to arise. For example, suppose that the manager of a factory is setting up a production schedule for two models of a new product. Model A requires 4 widgets and 9 klunkers. Model B requires 5 widgets and 14 klunkers. From its suppliers, the factory gets 335 widgets and 850 klunkers each day. How many of each model should the manager plan to make each day so that all the widgets and klunkers are used?

It's a good idea to construct a table that summarizes the important information. Table 4.2 shows the number of widgets and klunkers required for each model, as well as the total number available.

**TABLE 4.2**

|          | Model A | Model B | Total Available |
|----------|:-------:|:-------:|:---------------:|
| Widgets  | 4       | 5       | 335             |
| Klunkers | 9       | 14      | 850             |

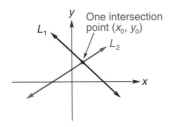

**FIGURE 4.28**   Linear system (one solution).

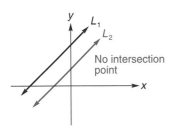

**FIGURE 4.29**   Linear system (no solution).

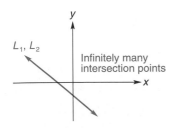

**FIGURE 4.30**   Linear system (infinitely many solutions).

Suppose we let $x$ be the number of model A made each day and $y$ be the number of model B. Then these require a total of $4x + 5y$ widgets and $9x + 14y$ klunkers. Since 335 widgets and 850 klunkers are available, we have

$$\begin{cases} 4x + 5y = 335, & (1) \\ 9x + 14y = 850. & (2) \end{cases}$$

We call this set of equations a **system** of two linear equations in the variables (or unknowns) $x$ and $y$. The problem is to find values of $x$ and $y$ for which *both* equations are true *simultaneously*. These are called *solutions* of the system.

Since Eqs. (1) and (2) are linear, their graphs are straight lines; call these lines $L_1$ and $L_2$. Now, the coordinates of any point on a line satisfy the equation of that line; that is, they make the equation true. Thus, the coordinates of any point of intersection of $L_1$ and $L_2$ will satisfy both equations. This means that a point of intersection gives a solution of the system.

If $L_1$ and $L_2$ are drawn on the same plane, there are three situations that could occur:

1. $L_1$ and $L_2$ may intersect at exactly one point, say, $(x_0, y_0)$. (See Fig., 4.28.) Thus, the system has the solution $x = x_0$ and $y = y_0$.

2. $L_1$ and $L_2$ may be parallel and have no points in common. (See Fig. 4.29.) In this case, there is no solution.

3. $L_1$ and $L_2$ may be the same line. (See Fig. 4.30.) Here the coordinates of any point on the line are a solution of the system. Consequently, there are infinitely many solutions.

Our main concern in this section is algebraic methods of solving a system of linear equations. Essentially, we successively replace the system by other systems that have the same solution (that is, we replace the original system by *equivalent systems*), but whose equations have a progressively more desirable form for determining the solution. More precisely, we seek an equivalent system containing an equation in which one of the variables does not appear. (That is, one of the variables is *eliminated*.) We shall illustrate this procedure for the system in the problem originally posed:

$$\begin{cases} 4x + 5y = 335, & (3) \\ 9x + 14y = 850. & (4) \end{cases}$$

To begin, we shall obtain an equivalent system in which $x$ does not appear in one equation. First we find an equivalent system in which the coefficients of the $x$-terms in each equation are the same except for their sign. Multiplying Eq. (3) by 9 [that is, multiplying both sides of Eq. (3) by 9] and multiplying Eq. (4) by $-4$ gives

$$\begin{cases} 36x + 45y = 3015, & (5) \\ -36x - 56y = -3400. & (6) \end{cases}$$

The left and right sides of Eq. (6) are equal, so each side can be *added* to the corresponding side of Eq. (5). This results in

$$-11y = -385,$$

which has only one variable, as planned. Solving gives

$$y = 35,$$

so we obtain the equivalent system

$$\begin{cases} \qquad\quad y = 35, & (7) \\ -36x - 56y = -3400. & (8) \end{cases}$$

Replacing $y$ in Eq. (8) by 35, we get

$$-36x - 56(35) = -3400,$$

$$-36x - 1960 = -3400,$$

$$-36x = -1440,$$

$$x = 40.$$

Thus, the original system is equivalent to

$$\begin{cases} y = 35, \\ x = 40. \end{cases}$$

We can check our answer by substituting $x = 40$ and $y = 35$ into *both* of the original equations. In Eq. (3), we get $4(40) + 5(35) = 335$, or $335 = 335$. In Eq. (4), we get $9(40) + 14(35) = 850$, or $850 = 850$. Hence, the solution is

$$x = 40 \quad \text{and} \quad y = 35.$$

Each day the manager should plan to make 40 of model A and 35 of model B. Our procedure is referred to as **elimination by addition.** Although we chose to eliminate $x$ first, we could have done the same for $y$ by a similar procedure.

### EXAMPLE 1   Elimination-by-Addition Method

*Use elimination by addition to solve the system.*

$$\begin{cases} 3x - 4y = 13, \\ 3y + 2x = 3. \end{cases}$$

## Principles in Practice 1
### Elimination-by-Addition Method

A computer consultant has \$200,000 invested for retirement, part at 9% and part at 8%. If the total yearly income from the investments is \$17,200, how much is invested at each rate?

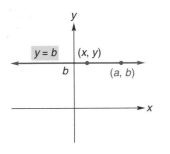

**FIGURE 4.31** Linear system of Example 1: one solution.

*Solution:* Aligning the $x$- and $y$-terms for convenience gives

$$\begin{cases} 3x - 4y = 13, & (9) \\ 2x + 3y = 3. & (10) \end{cases}$$

To eliminate $y$, we multiply Eq. (9) by 3 and Eq. (10) by 4:

$$\begin{cases} 9x - 12y = 39, & (11) \\ 8x + 12y = 12. & (12) \end{cases}$$

Adding Eq. (11) to Eq. (12) gives $17x = 51$, from which $x = 3$. We have the equivalent system

$$\begin{cases} 9x - 12y = 39, & (13) \\ x = 3. & (14) \end{cases}$$

Replacing $x$ by 3 in Eq. (13) results in

$$9(3) - 12y = 39,$$
$$-12y = 12,$$
$$y = -1,$$

so the original system is equivalent to

$$\begin{cases} y = -1, \\ x = 3. \end{cases}$$

The solution is $x = 3$ and $y = -1$. Figure 4.31 shows a graph of the system. ∎

The system in Example 1,

$$\begin{cases} 3x - 4y = 13, & (15) \\ 2x + 3y = 3, & (16) \end{cases}$$

can be solved another way. We first choose one of the equations—for example, Eq. (15)—and solve it for one unknown in terms of the other, say $x$ in terms of $y$. Hence Eq. (15) is equivalent to $3x = 4y + 13$, or

$$x = \frac{4}{3}y + \frac{13}{3},$$

and we obtain

$$\begin{cases} x = \frac{4}{3}y + \frac{13}{3}, & (17) \\ 2x + 3y = 3. & (18) \end{cases}$$

*Substituting* the right side of Eq. (17) for $x$ in Eq. (18) gives

$$2\left(\frac{4}{3}y + \frac{13}{3}\right) + 3y = 3. \qquad (19)$$

Thus, $x$ has been eliminated. Solving Eq. (19), we have

$$\frac{8}{3}y + \frac{26}{3} + 3y = 3,$$
$$8y + 26 + 9y = 9 \quad \text{(clearing fractions)},$$
$$17y = -17,$$
$$y = -1.$$

Replacing $y$ in Eq. (17) by $-1$ gives $x = 3$, and the original system is equivalent to

$$\begin{cases} x = 3, \\ y = -1, \end{cases}$$

as before, This method is called **elimination by substitution.**

**Principles in Practice 2**

**Method of Elimination by Substitution**

Two species of deer, A and B, living in a wildlife refuge are given extra food in the winter. Each week, they receive 2 tons of food pellets and 4.75 tons of hay. Each deer of species A requires 4 pounds of the pellets and 5 pounds of hay. Each deer of species B requires 2 pounds of the pellets and 7 pounds of hay. How many of each species of deer will the food support so that all of the food is consumed each week?

**EXAMPLE 2** **Method of Elimination by Substitution**

*Use elimination by substitution to solve the system*

$$\begin{cases} x + 2y - 8 = 0, \\ 2x + 4y + 4 = 0. \end{cases}$$

*Solution:* It is easy to solve the first equation for $x$. Doing so gives the equivalent system

$$\begin{cases} x = -2y + 8, & (20) \\ 2x + 4y + 4 = 0. & (21) \end{cases}$$

Substituting $-2y + 8$ for $x$ in Eq. (21) yields

$$2(-2y + 8) + 4y + 4 = 0,$$

$$-4y + 16 + 4y + 4 = 0.$$

The latter equation simplifies to $20 = 0$. Thus, we have the system

$$\begin{cases} x = -2y + 8, & (22) \\ 20 = 0. & (23) \end{cases}$$

Since Eq. (23) is *never* true, there is **no solution** of the original system. The reason is clear if we observe that the original equations can be written in slope–intercept form as

$$y = -\frac{1}{2}x + 4$$

and

$$y = -\frac{1}{2}x - 1.$$

These equations represent straight lines having slopes of $-\frac{1}{2}$, but different $y$-intercepts, namely, 4 and $-1$. That is, they determine different parallel lines. (See Fig. 4.32.) ∎

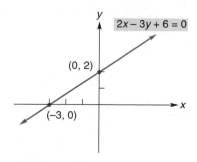

**FIGURE 4.32** Linear system of Example 2: no solution.

**Principles in Practice 3**

**A Linear System with Infinitely Many Solutions**

Two species of fish, A and B, are raised in one pond at a fish farm where they are fed two vitamin supplements. Each day, they receive 100 grams of the first supplement and 200 grams of the second supplement. Each fish of species A requires 15 mg of the first supplement and 30 mg of the second supplement. Each fish of species B requires 20 mg of the first supplement and 40 mg of the second supplement. How many of each species of fish will the pond support so that all of the supplements are consumed each day?

### EXAMPLE 3   A Linear System with Infinitely Many Solutions

*Solve*

$$\begin{cases} x + 5y = 2, & (24) \\ \dfrac{1}{2}x + \dfrac{5}{2}y = 1. & (25) \end{cases}$$

*Solution:* We begin by eliminating $x$ from the second equation. Multiplying Eq. (25) by $-2$, we have

$$\begin{cases} x + 5y = 2, & (26) \\ -x - 5y = -2. & (27) \end{cases}$$

Adding Eq. (26) to Eq. (27) gives

$$\begin{cases} x + 5y = 2, & (28) \\ \phantom{x + 5y} 0 = 0. & (29) \end{cases}$$

Because Eq. (29) is *always* true, any solution of Eq. (28) is a solution of the system. Now let us see how we can express our answer. From Eq. (28), we have $x = 2 - 5y$, where $y$ can be any real number, say, $r$. Thus, we can write $x = 2 - 5r$. The complete solution is

$$x = 2 - 5r,$$

$$y = r,$$

where $r$ is any real number. In this situation $r$ is called a **parameter,** and we say that we have a one-parameter family of solutions. Each value of $r$ determines a particular solution. For example, if $r = 0$, then $x = 2$ and $y = 0$ is a solution; if $r = 5$, then $x = -23$ and $y = 5$ is another solution. Clearly, the given system has infinitely many solutions.

It is worthwhile to note that by writing Eqs. (24) and (25) in their slope–intercept forms, we get the equivalent system

$$\begin{cases} y = -\dfrac{1}{5}x + \dfrac{2}{5}, \\ y = -\dfrac{1}{5}x + \dfrac{2}{5}, \end{cases}$$

in which both equations represent the same line. Hence, the lines coincide (Fig. 4.33), and Eqs. (24) and (25) are equivalent. The solution of the system consists of the coordinate pairs of all points on the line $x + 5y = 2$, and these points are given by our parametric solution. ∎

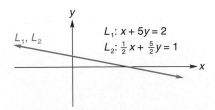

**FIGURE 4.33**   Linear system of Example 3: infinitely many solutions.

## TECHNOLOGY

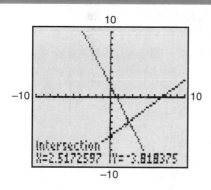

**FIGURE 4.34**  Graphical solution of system.

*Graphically solve the system*

$$\begin{cases} 9x + 4.1y = 7, \\ 2.6x - 3y = 18. \end{cases}$$

*Solution:* First we solve each equation for $y$, so that each equation has the form $y = f(x)$:

$$y = \frac{1}{4.1}(7 - 9x),$$

$$y = -\frac{1}{3}(18 - 2.6x).$$

Next we enter these functions as $Y_1$ and $Y_2$ and display them on the same viewing rectangle. (See Fig. 4.34.) Finally, either using trace and zoom or using the intersection feature, we estimate the solution to be $x = 2.52$, $y = -3.82$.

### EXAMPLE 4  Mixture

*A chemical manufacturer wishes to fill an order for 500 liters of a 25% acid solution. (Twenty-five percent by volume is acid.) If solutions of 30% and 18% are available in stock, how many liters of each must be mixed to fill the order?*

*Solution:*  Let $x$ and $y$ respectively be the number of liters of the 30% and 18% solutions that should be mixed. Then

$$x + y = 500.$$

To help visualize the situation, we draw the diagram in Fig. 4.35. In 500 liters of a 25% solution, there will be $0.25(500) = 125$ liters of acid. This acid comes

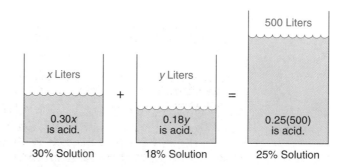

**FIGURE 4.35**  Mixture problem.

from two sources: $0.30x$ liters of it come from the 30% solution, and $0.18y$ liters of it come from the 18% solution. Hence,

$$0.30x + 0.18y = 125.$$

These two equations form a system of two linear equations in two unknowns. Solving the first for $x$ gives $x = 500 - y$. Substituting in the second gives

$$0.30(500 - y) + 0.18y = 125.$$

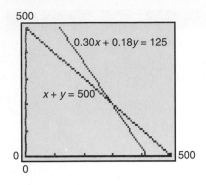

**FIGURE 4.36** Graph for Example 4.

---

### Principles in Practice 4

Solving a Three-Variable Linear System

A coffee shop specializes in blending gourmet coffees. From type A, type B, and type C coffees, the owner wants to prepare a blend that will sell for $8.50 for a 1-pound bag. The cost per pound of these coffees is $12, $9, and $7, respectively. The amount of type B is to be twice the amount of type A. How much of each type of coffee will be in the final blend?

---

Solving this equation for $y$, we find that $y = 208\frac{1}{3}$ liters. Thus, $x = 500 - 208\frac{1}{3} = 291\frac{2}{3}$ liters. (See Fig. 4.36.)

## Three-Variable Systems

The methods used in solving a two-variable system of linear equations can be used to solve a three-variable system of linear equations. A **general linear equation in the three variables $x$, $y$, and $z$** is an equation having the form

$$Ax + By + Cz = D,$$

where $A$, $B$, $C$, and $D$ are constants and $A$, $B$, and $C$ are not all zero. For example, $2x - 4y + z = 2$ is such an equation. Geometrically, a general linear equation in three variables represents a *plane* in space, and a solution to a system of such equations is the intersection of planes. Example 5 shows how to solve a system of three linear equations in three variables.

### EXAMPLE 5 Solving a Three-Variable Linear System

*Solve*

$$\begin{cases} 2x + y + z = 3, & (30) \\ -x + 2y + 2z = 1, & (31) \\ x - y - 3z = -6. & (32) \end{cases}$$

*Solution:* This system consists of three linear equations in three variables. From Eq. (32), $x = y + 3z - 6$. By substituting for $x$ in Eqs. (30) and (31), we obtain

$$\begin{cases} 2(y + 3z - 6) + y + z = 3, \\ -(y + 3z - 6) + 2y + 2z = 1, \\ \phantom{xxxxxxxxxx} x = y + 3z - 6. \end{cases}$$

Simplifying gives

$$\begin{cases} 3y + 7z = 15, & (33) \\ y - z = -5, & (34) \\ x = y + 3z - 6. & (35) \end{cases}$$

Note that $x$ does not appear in Eqs. (33) and (34). Since any solution of the original system must satisfy Eqs. (33) and (34), we shall consider their solution first:

$$\begin{cases} 3y + 7z = 15, & (33) \\ y - z = -5. & (34) \end{cases}$$

From Eq. (34), $y = z - 5$. This means that we can replace Eq. (33) by

$$3(z - 5) + 7z = 15, \quad \text{or} \quad z = 3.$$

Since $z$ is 3, we can replace Eq. (34) with $y = -2$. Hence, the previous system is equivalent to

$$\begin{cases} z = 3, \\ y = -2. \end{cases}$$

The original system becomes

$$\begin{cases} z = 3, \\ y = -2, \\ x = y + 3z - 6, \end{cases}$$

from which $x = 1$. The solution is $x = 1$, $y = -2$, and $z = 3$, which you may verify. ∎

Just as a two-variable system may have a one-parameter family of solutions, a three-variable system may have a one- or two-parameter family of solutions.[5] The next two examples illustrate.

**EXAMPLE 6  One-Parameter Family of Solutions**

*Solve*

$$\begin{cases} x - 2y = 4, & (35) \\ 2x - 3y + 2z = -2, & (36) \\ 4x - 7y + 2z = 6. & (37) \end{cases}$$

*Solution:* Note that since Eq. (35) can be written $x - 2y + 0z = 4$, we can view Eqs. (35) to (37) as a system of three linear equations in the variables $x$, $y$, and $z$. From Eq. (35), we have $x = 2y + 4$. Using this equation and substitution, we can eliminate $x$ from Eqs. (36) and (37):

$$\begin{cases} x = 2y + 4, \\ 2(2y + 4) - 3y + 2z = -2, \\ 4(2y + 4) - 7y + 2z = 6. \end{cases}$$

Or, more simply,

$$\begin{cases} x = 2y + 4, & (38) \\ y + 2z = -10, & (39) \\ y + 2z = -10. & (40) \end{cases}$$

Multiplying Eq. (40) by $-1$ gives

$$\begin{cases} x = 2y + 4, \\ y + 2z = -10, \\ -y - 2z = 10. \end{cases}$$

Adding the second equation to the third yields

$$\begin{cases} x = 2y + 4, \\ y + 2z = -10, \\ 0 = 0. \end{cases}$$

Since the equation $0 = 0$ is always true, we can essentially deal with the system

$$\begin{cases} x = 2y + 4, & (41) \\ y + 2z = -10. & (42) \end{cases}$$

Solving Eq. (42) for $y$, we have

$$y = -10 - 2z,$$

which expresses $y$ in terms of $z$. We can also express $x$ in terms of $z$. From Eq. (41),

$$\begin{aligned} x &= 2y + 4 \\ &= 2(-10 - 2z) + 4 \\ &= -16 - 4z. \end{aligned}$$

---

[5]Note to instructor: Examples 6 and 7 may be omitted without loss of continuity.

Thus, we have

$$\begin{cases} x = -16 - 4z, \\ y = -10 - 2z. \end{cases}$$

Since no restriction is placed on $z$, this suggests a parametric family of solutions. Setting $z = r$, we have the following family of solutions of the given system:

$$x = -16 - 4r,$$
$$y = -10 - 2r,$$
$$z = r,$$

Other parametric representations of the solution are possible.

where $r$ can be any real number. We see, then, that the given system has infinitely many solutions. For example, setting $r = 1$ gives the particular solution $x = -20$, $y = -12$, and $z = 1$. ∎

**EXAMPLE 7  Two-Parameter Family of Solutions**

*Solve the system*

$$\begin{cases} x + 2y + z = 4, \\ 2x + 4y + 2z = 8. \end{cases}$$

*Solution:* This is a system of two linear equations in three variables. We shall eliminate $x$ from the second equation by first multiplying that equation by $-\frac{1}{2}$:

$$\begin{cases} x + 2y + z = 4, \\ -x - 2y - z = -4. \end{cases}$$

Adding the first equation to the second gives

$$\begin{cases} x + 2y + z = 4, \\ \qquad\qquad 0 = 0. \end{cases}$$

From the first equation, we obtain

$$x = 4 - 2y - z.$$

Since no restriction is placed on either $y$ or $z$, they can be arbitrary real numbers, giving us a two-parameter family of solutions. Setting $y = r$ and $z = s$, we find that the solution of the given system is

$$x = 4 - 2r - s,$$
$$y = r,$$
$$z = s,$$

where $r$ and $s$ can be any real numbers. Each assignment of values to $r$ and $s$ results in a solution of the given system, so there are infinitely many solutions. For example, letting $r = 1$ and $s = 2$ gives the particular solution $x = 0$, $y = 1$, and $z = 2$. ∎

## ■ Exercise 4.4

*In Problems 1–24, solve the systems algebraically.*

1. $\begin{cases} x + 4y = 3, \\ 3x - 2y = -5. \end{cases}$
2. $\begin{cases} 4x + 2y = 9, \\ 5y - 4x = 5. \end{cases}$
3. $\begin{cases} 3x - 4y = 13, \\ 2x + 3y = 3. \end{cases}$

**4.** $\begin{cases} 2x - y = 1, \\ -x + 2y = 7. \end{cases}$

**5.** $\begin{cases} 5v + 2w = 36, \\ 8v - 3w = -54. \end{cases}$

**6.** $\begin{cases} p + q = 3, \\ 3p + 2q = 19. \end{cases}$

**7.** $\begin{cases} x - 2y = 8, \\ 5x + 3y = 1. \end{cases}$

**8.** $\begin{cases} 3x + 5y = 7, \\ 5x + 9y = 7. \end{cases}$

**9.** $\begin{cases} 4x - 3y - 2 = 3x - 7y, \\ x + 5y - 2 = y + 4. \end{cases}$

**10.** $\begin{cases} 5x + 7y + 2 = 9y - 4x + 6, \\ \frac{21}{2}x - \frac{4}{3}y - \frac{11}{4} = \frac{3}{2}x + \frac{2}{3}y + \frac{5}{4}. \end{cases}$

**11.** $\begin{cases} \frac{2}{3}x + \frac{1}{2}y = 2, \\ \frac{3}{8}x + \frac{5}{6}y = -\frac{11}{2}. \end{cases}$

**12.** $\begin{cases} \frac{1}{2}z - \frac{1}{4}w = \frac{1}{6}, \\ z + \frac{1}{2}w = \frac{2}{3}. \end{cases}$

**13.** $\begin{cases} 4p + 12q = 6, \\ 2p + 6q = 3. \end{cases}$

**14.** $\begin{cases} 5x - 3y = 2, \\ -10x + 6y = 4. \end{cases}$

**15.** $\begin{cases} 2x + y + 6z = 3, \\ x - y + 4z = 1, \\ 3x + 2y - 2z = 2. \end{cases}$

**16.** $\begin{cases} x + y + z = -1, \\ 3x + y + z = 1, \\ 4x - 2y + 2z = 0. \end{cases}$

**17.** $\begin{cases} 5x - 7y + 4z = 2, \\ 3x + 2y - 2z = 3, \\ 2x - y + 3z = 4. \end{cases}$

**18.** $\begin{cases} 3x - 2y + z = 0, \\ -2x + y - 3z = 15, \\ \frac{3}{2}x + \frac{4}{5}y + 4z = 10. \end{cases}$

**[6]19.** $\begin{cases} x - 2z = 1, \\ y + z = 3. \end{cases}$

**[6]20.** $\begin{cases} 2y + 3z = 1, \\ 3x - 4z = 0. \end{cases}$

**[6]21.** $\begin{cases} x - y + 2z = 0, \\ 2x + y - z = 0, \\ x + 2y - 3z = 0. \end{cases}$

**[6] 22.** $\begin{cases} x - 2y - z = 0, \\ 2x - 4y - 2z = 0, \\ -x + 2y + z = 0. \end{cases}$

**[6]23.** $\begin{cases} 2x + 2y - z = 3, \\ 4x + 4y - 2z = 6. \end{cases}$

**[6]24.** $\begin{cases} x + 2y - 3z = -4, \\ 2x + y - 3z = 4. \end{cases}$

---

**25. Mixture** A chemical manufacturer wishes to fill an order for 700 gallons of a 24% acid solution. Solutions of 20% and 30% are in stock. How many gallons of each solution must be mixed to fill the order?

**26. Mixture** A gardener has two fertilizers that contain different concentrations of nitrogen. One is 3% nitrogen and the other is 11% nitrogen. How many pounds of each should she mix to obtain 20 pounds of a 9% concentration?

**27. Fabric** A textile mill produces fabric made from different fibers. From cotton, polyester, and nylon, the owners want to produce a fabric blend that will cost $3.25 per pound to make. The cost per pound of these fibers is $4.00, $3.00, and $2.00, respectively. The amount of nylon is to be the same as the amount of polyester. How much of each fiber will be in the final fabric?

**28. Taxes** A company has taxable income of $312,000. The federal tax is 25% of that portion left after the state tax has been paid. The state tax is 10% of that portion left after the federal tax has been paid. Find the federal and state taxes.

**29. Airplane Speed** An airplane travels 900 mi in 3 h with the aid of a tail wind. It takes 3 h, 36 min, for the return trip, flying against the same wind. Find the speed of the airplane in still air and the speed of the wind.

**30. Speed of Raft** On a trip on a raft, it took $\frac{3}{4}$ h to travel 12 mi downstream. The return trip took $1\frac{1}{2}$ h. Find the speed of the raft in still water and the speed of the current.

**31. Furniture Sales** A manufacturer of dining-room sets produces two styles: early American and contemporary. From past experience, management has determined that 20% more of the early American styles can be sold than the contemporary styles. A profit of $250 is made on each early American set sold, whereas a profit of $350 is made on each contemporary set. If, in the forthcoming year, management desires a total profit of $130,000, how many units of each style must be sold?

**32. Survey** National Surveys was awarded a contract to perform a product-rating survey for Crispy Crackers. A total of 250 people were interviewed. National Surveys reported that 62.5% more people like Crispy Crackers than disliked them. However, the report did not indicate that 16% of those interviewed had no comment. How many of those surveyed liked Crispy Crackers? How many disliked them? How many had no comment?

**33. Equalizing Cost** United Products Co. manufactures calculators and has plants in the cities of Exton and

[6]Refers to concepts in Examples 6 and 7.

Whyton. At the Exton plant, fixed costs are $7000 per month, and the cost of producing each calculator is $7.50. At the Whyton plant, fixed costs are $8800 per month, and each calculator costs $6.00 to produce. Next month, United Products must produce 1500 calculators. How many must be made at each plant if the total cost at each plant is to be the same?

**34. Coffee Blending**  A coffee wholesaler blends together three types of coffee that sell for $2.20, $2.30, and $2.60 per pound, so as to obtain 100 lb of coffee worth $2.40 per pound. If the wholesaler uses the same amount of the two higher priced coffees, how much of each type must be used in the blend?

**35. Commissions**  A company pays its salespeople on a basis of a certain percentage of the first $100,000 in sales, plus a certain percentage of any amount over $100,000 in sales. If one salesperson earned $8500 on sales of $175,000 and another salesperson earned $14,800 on sales of $280,000, find the two percentages.

**36. Yearly Profits**  In news reports, profits of a company this year $(T)$ are often compared with those of last year $(L)$, but actual values of $T$ and $L$ are not always given. This year, a company had profits of $20 million more than last year. The profits were up 25%. Determine $T$ and $L$ from these data.

**37. Production**  Universal Control Co. makes industrial control units. Their new models are the Argon I and the Argon II. To make each Argon I unit, they use 6 doodles and 3 skeeters. To make each Argon II unit, they use 10 doodles and 8 skeeters. The company receives a total of 760 doodles and 500 skeeters each day from its supplier. How many units of each model of the Argon can the company make each day? Assume that all the parts are used.

**38. Investments**  A person made two investments, and the percentage return per year on each was the same. Of the total amount invested, $\frac{3}{10}$ of it plus $600 was invested in one venture, and at the end of 1 year the person received a return of $384 from that venture. If the total return after 1 year was $1120, find the total amount invested.

**39. Production Run**  A company makes three types of patio furniture: chairs, rockers, and chaise lounges. Each requires wood, plastic, and aluminum, in the amounts shown in the following table. The company has in stock 400 units of wood, 600 units of plastic, and 1500 units of aluminum. For its end-of-the-season production run, the company wants to use up all the stock. To do this, how many chairs, rockers, and chaise lounges should it make?

|              | Wood   | Plastic | Aluminum |
|--------------|--------|---------|----------|
| Chair        | 1 unit | 1 unit  | 2 units  |
| Rocker       | 1 unit | 1 unit  | 3 units  |
| Chaise lounge| 1 unit | 2 units | 5 units  |

**40. Investments**  A total of $35,000 was invested at three interest rates: 7, 8, and 9%. The interest for the first year was $2830, which was not reinvested. The second year the amount originally invested at 9% earned 10% instead, and the other rates remained the same. The total interest the second year was $2960. How much was invested at each rate?

**41. Hiring Workers**  A company pays skilled workers in its assembly department $15 per hour. Semiskilled workers in that department are paid $9 per hour. Shipping clerks are paid $10 per hour. Because of an increase in orders, the company needs to hire a total of 70 workers in the assembly and shipping departments. It will pay a total of $760 per hour to these employees. Because of a union contract, twice as many semiskilled workers as skilled workers must be employed. How many semiskilled workers, skilled workers, and shipping clerks should the company hire?

**42. Solvent Storage**  A 10,000-gallon railroad tank car is to be filled with solvent from two storage tanks, $A$ and $B$. Solvent from $A$ is pumped at the rate of 20 gal/min. Solvent from $B$ is pumped at 30 gal/min. Usually, both pumps operate at the same time. However, because of a blown fuse, the pump on $A$ is delayed 10 minutes. How many gallons from each storage tank will be used to fill the car?

**43.** Verify your answer to Problem 1 by using your graphics calculator.

**44.** Verify your answer to Problem 11 by using your graphics calculator.

**45.** Graphically solve the system.

$$\begin{cases} 0.24x - 0.34y = 0.28, \\ 0.11x + 0.21y = 0.86. \end{cases}$$

**46.** Graphically solve the system

$$\begin{cases} x + y = 2, \\ \frac{1}{4}x + \frac{2}{5}y = \frac{3}{5}. \end{cases}$$

Round the $x$- and $y$-values to two decimal places.

**47.** Graphically solve the system

$$\begin{cases} 0.5736x - 0.3420y = 0, \\ 0.8192x + 0.9397y = 20. \end{cases}$$

Round the $x$- and $y$-values to one decimal place

O B J E C T I V E

**To use substitution to solve nonlinear systems.**

# 4.5 NONLINEAR SYSTEMS

A system of equations in which at least one equation is not linear is called a **nonlinear system.** We can often solve a nonlinear system by substitution, as was done with linear systems. The following examples illustrate.

**EXAMPLE 1   Solving a Nonlinear System**

*Solve*

$$\begin{cases} x^2 - 2x + y - 7 = 0, & (1) \\ 3x - y + 1 = 0. & (2) \end{cases}$$

*Solution:*

> *Strategy:* If a nonlinear system contains a linear equation, we usually solve the linear equation for one variable and substitute for that variable in the other equation.

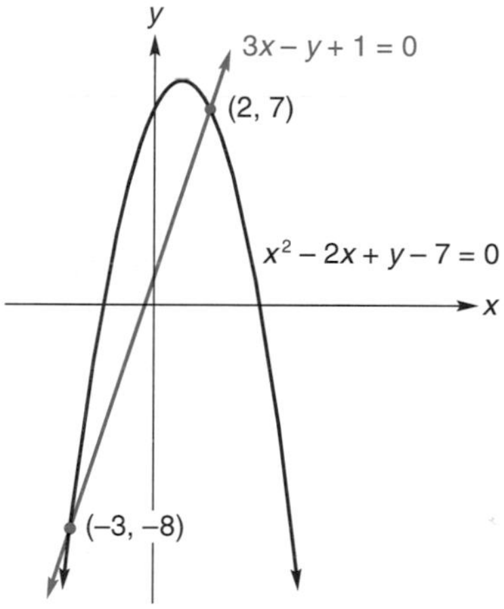

Solving Eq. (2) for $y$ gives

$$y = 3x + 1. \qquad (3)$$

Substituting into Eq. (1) and simplifying, we have

$$x^2 - 2x + (3x + 1) - 7 = 0,$$
$$x^2 + x - 6 = 0,$$
$$(x + 3)(x - 2) = 0,$$
$$x = -3 \quad \text{or} \quad x = 2.$$

If $x = -3$, then Eq. (3) implies that $y = -8$; if $x = 2$, then $y = 7$. You should verify that each pair of values satisfies the given system. Hence, the solutions are $x = -3$, $y = -8$ and $x = 2$, $y = 7$. These solutions can be seen geometrically in the graph of the system in Fig. 4.37. Notice that the graph of Eq. (1) is a parabola and the graph of Eq. (2) is a line. The solutions correspond to the intersection points $(-3, -8)$ and $(2, 7)$.  ■

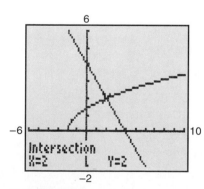

**FIGURE 4.37**   Nonlinear system of equations.

This example illustrates the need for checking all "solutions."

**EXAMPLE 2   Solving a Nonlinear System**

*Solve*

$$\begin{cases} y = \sqrt{x + 2}, \\ x + y = 4. \end{cases}$$

*Solution:* Solving the second equation, which is linear, for $y$ gives

$$y = 4 - x. \qquad (4)$$

Substituting into the first equation yields

$$4 - x = \sqrt{x + 2},$$
$$16 - 8x + x^2 = x + 2 \qquad \text{(squaring both sides)},$$
$$x^2 - 9x + 14 = 0,$$
$$(x - 2)(x - 7) = 0.$$

**FIGURE 4.38**   Nonlinear system of Example 2.

Thus, $x = 2$ or $x = 7$. From Eq. (4), if $x = 2$, then $y = 2$; if $x = 7$, then $y = -3$. Since we performed the operation of squaring both sides, we must check our results. Although the pair $x = 2$, $y = 2$ satisfies both of the original equations, this is not the case for $x = 7$ and $y = -3$. Thus, the solution is $x = 2$, $y = 2$. (See Fig. 4.38.) ▬

## TECHNOLOGY

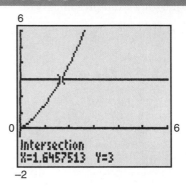

**FIGURE 4.39**   Solution of $0.5x^2 + x = 3$.

*Graphically solve the equation $0.5x^2 + x = 3$, where $x \geq 0$.*

**Solution:** To solve the equation, we could find zeros of the function $f(x) = 0.5x^2 + x - 3$. Alternatively, we can think of this problem as solving the nonlinear system

$$y = 0.5x^2 + x,$$

$$y = 3.$$

In Fig. 4.39, the intersection point is estimated to be $x = 1.65$, $y = 3$. Note that the graph of $y = 3$ is a horizontal line. The solution of the given equation is $x = 1.65$.

## ▪ EXERCISE 4.5

*In Problems **1–14**, solve the given nonlinear system.*

**1.** $\begin{cases} y = 4 - x^2, \\ 3x + y = 0. \end{cases}$

**2.** $\begin{cases} y = x^3, \\ x - y = 0. \end{cases}$

**3.** $\begin{cases} p^2 = 4 - q, \\ p = q + 2. \end{cases}$

**4.** $\begin{cases} y^2 - x^2 = 28, \\ x - y = 14. \end{cases}$

**5.** $\begin{cases} x = y^2, \\ y = x^2. \end{cases}$

**6.** $\begin{cases} p^2 - q = 0, \\ 3q - 2p - 1 = 0. \end{cases}$

**7.** $\begin{cases} y = 4x - x^2 + 8, \\ y = x^2 - 2x. \end{cases}$

**8.** $\begin{cases} x^2 - y = 8, \\ y - x^2 = 0. \end{cases}$

**9.** $\begin{cases} p = \sqrt{q}, \\ p = q^2. \end{cases}$

**10.** $\begin{cases} z = 4/w, \\ 3z = 2w + 2. \end{cases}$

**11.** $\begin{cases} x^2 = y^2 + 14, \\ y = x^2 - 16. \end{cases}$

**12.** $\begin{cases} x^2 + y^2 - 2xy = 1, \\ 3x - y = 5. \end{cases}$

**13.** $\begin{cases} x = y + 6, \\ y = 3\sqrt{x} + 4. \end{cases}$

**14.** $\begin{cases} y = \dfrac{x^2}{x - 1} + 1, \\ y = \dfrac{1}{x - 1}. \end{cases}$

**15. Decorations**  The shape of a paper streamer suspended above a dance floor can be described by the function $y = 0.01x^2 + 0.01x + 7$, where $y$ is the height of the streamer (in feet) above the floor and $x$ is the horizontal distance (in feet) from the center of the room. A rope holding up other decorations touches the streamer and is described by the function $y = 0.01x + 8.0$. Where does the rope touch the paper streamer?

**16. Awning**  The shape of a decorative awning over a storefront can be described by the function $y = 0.06x^2 + 0.012x + 8$, where $y$ is the height of the edge of the awning (in feet) above the sidewalk and $x$ is the distance (in feet) from the center of the store's doorway. A vandal pokes a stick through the awning, piercing it in two places. The position of the stick can be described by the function $y = 0.912x + 5$. Where are the holes in the awning caused by the vandal?

**17.** 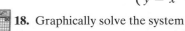 Graphically determine how many solutions there are to the system

$$\begin{cases} y = \dfrac{1}{x}, \\ y = x^2 - 4. \end{cases}$$

**18.** 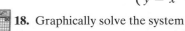 Graphically solve the system

$$\begin{cases} y = x^3, \\ y = 6 - x^2 \end{cases}$$

to one-decimal-place accuracy.

**19.**  Graphically solve the system

$$\begin{cases} y = x^2 - 2x + 1, \\ y = x^3 + x^2 - 2x + 3 \end{cases}$$

to one-decimal-place accuracy.

**20.** Graphically solve the system

$$\begin{cases} y = x^3 + x, \\ y = 4x \end{cases}$$

to one-decimal-place accuracy.

*In Problems 21–23, graphically solve the equation by treating it as a system. Round answers to two decimal places.*

**21.** $0.8x^2 + 2x = 6$, where $x \geq 0$.  **22.** $\sqrt{x + 2} = 5 - x$.

**23.** $x^3 - 3x^2 = x - 8$.

---

| OBJECTIVE | 4.6  APPLICATIONS OF SYSTEMS OF EQUATIONS |
|---|---|

**To solve systems describing equilibrium and break-even points.**

### Equilibrium

Recall from Sec. 4.2 that an equation that relates price per unit and quantity demanded (supplied) is called a *demand equation (supply equation)*. Suppose that, for product Z, the demand equation is

$$p = -\frac{1}{180}q + 12 \tag{1}$$

and the supply equation is

$$p = \frac{1}{300}q + 8, \tag{2}$$

where $q, p \geq 0$. The corresponding demand and supply curves are the lines in Figs. 4.40 and 4.41, respectively. In analyzing Fig. 4.40, we see that consumers will purchase 540 units per week when the price is \$9 per unit, 1080 units when the price is \$6, and so on. Figure 4.41 shows that when the price is \$9 per unit

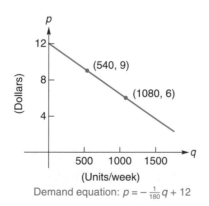

**FIGURE 4.40**  Demand curve.

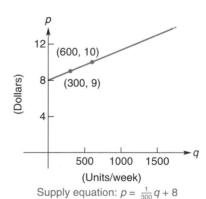

**FIGURE 4.41**  Supply curve.

producers will place 300 units per week on the market, at \$10 they will supply 600 units, and so on.

When the demand and supply curves of a product are represented on the same coordinate plane, the point $(m, n)$ where the curves intersect is called the **point of equilibrium.** (See Fig. 4.42.) The price $n$, called the **equilibrium price,** is the price at which consumers will purchase the same quantity of a product that producers wish to sell at that price. In short, $n$ is the price at which

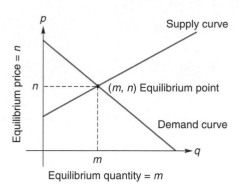

**FIGURE 4.42** Equilibrium.

stability in the producer–consumer relationship occurs. The quantity $m$ is called the **equilibrium quantity.**

To determine precisely the equilibrium point, we solve the system formed by the supply and demand equations. Let us do this for our previous data, namely, the system

$$
\begin{cases}
p = -\dfrac{1}{180}q + 12 & \text{(demand equation)}, \\[2mm]
p = \dfrac{1}{300}q + 8 & \text{(supply equation)}.
\end{cases}
$$

By substituting $\dfrac{1}{300}q + 8$ for $p$ in the demand equation, we get

$$
\frac{1}{300}q + 8 = -\frac{1}{180}q + 12,
$$

$$
\left(\frac{1}{300} + \frac{1}{180}\right)q = 4,
$$

$$
q = 450 \qquad \text{(equilibrium quantity)}.
$$

Thus,

$$
p = \frac{1}{300}(450) + 8
$$

$$
= 9.50 \qquad \text{(equilibrium price)},
$$

and the equilibrium point is (450, 9.50). Therefore, at the price of \$9.50 per unit, manufacturers will produce exactly the quantity (450) of units per week that consumers will purchase at that price. (See Fig. 4.43.)

### EXAMPLE 1   Tax Effect on Equilibrium

*Let $p = \frac{8}{100}q + 50$ be the supply equation for a manufacturer's product, and suppose the demand equation is $p = -\frac{7}{100}q + 65$.*

**a.** *If a tax of \$1.50 per unit is to be imposed on the manufacturer, how will the original equilibrium price be affected if the demand remains the same?*

*Solution:* Before the tax, the equilibrium price is obtained by solving the system

$$
\begin{cases}
p = \dfrac{8}{100}q + 50, \\[2mm]
p = -\dfrac{7}{100}q + 65.
\end{cases}
$$

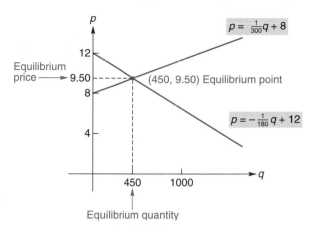

**FIGURE 4.43**  Equilibrium.

By substitution,

$$-\frac{7}{100}q + 65 = \frac{8}{100}q + 50,$$

$$15 = \frac{15}{100}q,$$

$$100 = q,$$

and

$$p = \frac{8}{100}(100) + 50 = 58.$$

Thus, \$58 is the original equilibrium price. Before the tax, the manufacturer supplies $q$ units at a price of $p = \frac{8}{100}q + 50$ per unit. After the tax, he will sell the same $q$ units for an additional \$1.50 per unit. The price per unit will be $\left(\frac{8}{100}q + 50\right) + 1.50$, so the new supply equation is

$$p = \frac{8}{100}q + 51.50.$$

Solving the system

$$\begin{cases} p = \dfrac{8}{100}q + 51.50, \\[2mm] p = -\dfrac{7}{100}q + 65 \end{cases}$$

will give the new equilibrium price:

$$\frac{8}{100}q + 51.50 = -\frac{7}{100}q + 65,$$

$$\frac{15}{100}q = 13.50,$$

$$q = 90,$$

$$p = \frac{8}{100}(90) + 51.50 = 58.70.$$

The tax of $1.50 per unit increases the equilibrium price by $0.70. (See Fig. 4.44.) Note that there is also a decrease in the equilibrium quantity from $q = 100$ to $q = 90$, because of the change in the equilibrium price. (In the exercises, you are asked to find the effect of a subsidy given to the manufacturer, which will reduce the price of the product.)

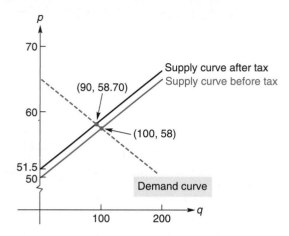

**FIGURE 4.44**    Equilibrium before and after tax.

**b.** *Determine the total revenue obtained by the manufacturer at the equilibrium point both before and after the tax.*

*Solution:* If $q$ units of a product are sold at a price of $p$ dollars each, then the total revenue is given by

$$y_{TR} = pq.$$

Before the tax, the revenue at $(100, 58)$ is (in dollars)

$$y_{TR} = (58)(100) = 5800.$$

After the tax, it is

$$y_{TR} = (58.70)(90) = 5283,$$

which is a decrease.                                                             ▀

**EXAMPLE 2    Equilibrium with Nonlinear Demand**

*Find the equilibrium point if the supply and demand equations of a product are* $p = \dfrac{q}{40} + 10$ *and* $p = \dfrac{8000}{q}$, *respectively.*

*Solution:* Here the demand equation is not linear. Solving the system

$$\begin{cases} p = \dfrac{q}{40} + 10, \\[2mm] p = \dfrac{8000}{q} \end{cases}$$

by substitution gives

$$\frac{8000}{q} = \frac{q}{40} + 10,$$

$$320{,}000 = q^2 + 400q \qquad \text{(multiplying both sides by } 40q\text{)},$$

$$q^2 + 400q - 320{,}000 = 0,$$

$$(q + 800)(q - 400) = 0,$$

$$q = -800 \quad \text{or} \quad q = 400.$$

We disregard $q = -800$, since $q$ represents quantity. Choosing $q = 400$, we have $p = (8000/400) = 20$, so the equilibrium point is $(400, 20)$. (See Fig. 4.45.) ∎

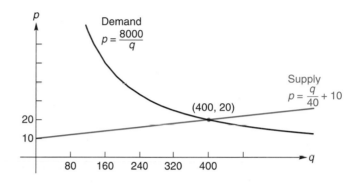

**FIGURE 4.45** Equilibrium with nonlinear demand.

## Break-even Points

Suppose a manufacturer produces product A and sells it at $8 per unit. Then the total revenue $y_{TR}$ received (in dollars) from selling $q$ units is

$$y_{TR} = 8q \qquad \text{(total revenue)}.$$

The difference between the total revenue received for $q$ units and the total cost of $q$ units is the manufacturer's profit (or loss if the difference is negative):

**profit (or loss) = total revenue − total cost.**

**Total cost,** $y_{TC}$, is the sum of total variable costs $y_{VC}$ and total fixed costs $y_{FC}$:

$$y_{TC} = y_{VC} + y_{FC}.$$

**Fixed costs** are those costs that, under normal conditions, do not depend on the level of production; that is, over some period of time they remain constant at all levels of output. (Examples are rent, officers' salaries, and normal maintenance.) **Variable costs** are those costs that vary with the level of production (such as the cost of materials, labor, maintenance due to wear and tear, etc.). For $q$ units of product A, suppose that

$$y_{FC} = 5000 \qquad \text{(fixed cost)}$$

$$\text{and} \quad y_{VC} = \frac{22}{9}q \qquad \text{(variable cost)}.$$

Then

$$y_{TC} = \frac{22}{9}q + 5000 \qquad \text{(total cost)}.$$

The graphs of total cost and total revenue appear in Fig. 4.46. The horizontal axis represents the level of production, $q$, and the vertical axis represents the total dollar value, be it revenue or cost. The **break-even point** is the

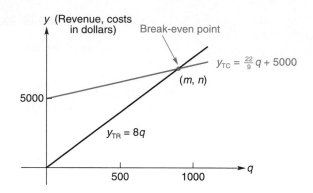

**FIGURE 4.46** Break-even chart.

point at which total revenue equals total cost (TR = TC). It occurs when the levels of production and sales result in neither a profit nor a loss to the manufacturer. In the diagram, called a *break-even chart*, the break-even point is the point $(m, n)$ at which the graphs of $y_{TR} = 8q$ and $y_{TC} = \frac{22}{9}q + 5000$ intersect. We call $m$ the **break-even quantity** and $n$ the **break-even revenue.** When total cost and revenue are linearly related to output, as in this case, for any production level greater than $m$, total revenue is greater than total cost, resulting in a profit. However, at any level less than $m$ units, total revenue is less than total cost, resulting in a loss. At an output of $m$ units, the profit is zero. In the following example, we shall examine our data in more detail.

**EXAMPLE 3**   **Break-even Point, Profit, and Loss**

*A manufacturer sells a product at $8 per unit, selling all that is produced. Fixed cost is $5000 and variable cost per unit is $\frac{22}{9}$ (dollars).*

**a.** *Find the total output and revenue at the break-even point.*

> *Solution:* At an output level of $q$ units, the variable cost is $y_{VC} = \frac{22}{9}q$ and the total revenue is $y_{TR} = 8q$. Hence,
>
> $$y_{TR} = 8q,$$
>
> $$y_{TC} = y_{VC} + y_{FC} = \frac{22}{9}q + 5000.$$
>
> At the break-even point, total revenue equals total cost. Thus, we solve the system formed by the foregoing equations. Since

$$y_{TR} = y_{TC},$$

we have

$$8q = \frac{22}{9}q + 5000,$$

$$\frac{50}{9}q = 5000,$$

$$q = 900.$$

Hence, the desired output is 900 units, resulting in a total revenue (in dollars) of

$$y_{TR} = 8(900) = 7200.$$

(See Fig. 4.47.)

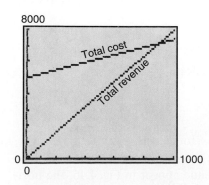

**FIGURE 4.47**   Equilibrium point (900, 7200).

**b.** *Find the profit when* 1800 *units are produced.*

*Solution:* Since profit = total revenue − total cost, when $q = 1800$ we have

$$y_{TR} - y_{TC} = 8(1800) - \left[ \frac{22}{9}(1800) + 5000 \right]$$

$$= 5000.$$

The profit when 1800 units are produced and sold is $5000.

**c.** *Find the loss when* 450 *units are produced.*

*Solution:* When $q = 450$,

$$y_{TR} - y_{TC} = 8(450) - \left[ \frac{22}{9}(450) + 5000 \right] = -2500.$$

A loss of $2500 occurs when the level of production is 450 units.

**d.** *Find the output required to obtain a profit of* $10,000.

*Solution:* In order to obtain a profit of $10,000, we have

$$\text{profit} = \text{total revenue} - \text{total cost},$$

$$10,000 = 8q - \left( \frac{22}{9}q + 5000 \right),$$

$$15,000 = \frac{50}{9}q,$$

$$q = 2700.$$

Thus, 2700 units must be produced.

**EXAMPLE 4  Break-even Quantity**

*Determine the break-even quantity of XYZ Manufacturing Co., given the following data: total fixed cost, $1200; variable cost per unit, $2; total revenue for selling q units, $y_{TR} = 100\sqrt{q}$.*

*Solution:* For $q$ units of output,

$$y_{TR} = 100\sqrt{q},$$

$$y_{TC} = 2q + 1200.$$

Equating total revenue to total cost gives

$$100\sqrt{q} = 2q + 1200,$$

$$50\sqrt{q} = q + 600 \qquad \text{(dividing both sides by 2)}.$$

Squaring both sides, we have

$$2500q = q^2 + 1200q + (600)^2,$$

$$0 = q^2 - 1300q + 360,000.$$

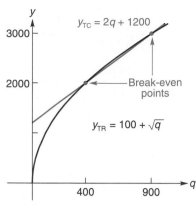

**FIGURE 4.48**  Two break-even points.

By the quadratic formula,

$$q = \frac{1300 \pm \sqrt{250{,}000}}{2},$$

$$q = \frac{1300 \pm 500}{2},$$

$$q = 400 \quad \text{or} \quad q = 900.$$

Although both $q = 400$ and $q = 900$ are break-even quantities, observe in Fig. 4.48 that when $q > 900$, total cost is greater than total revenue, so there will always be a loss. This occurs because here total revenue is not linearly related to output. Thus, producing more than the break-even quantity does not necessarily guarantee a profit. ■

## ■ Exercise 4.6

*In Problems 1–8, you are given a supply equation and a demand equation for a product. If p represents price per unit in dollars and q represents the number of units per unit of time, find the equilibrium point. In Problems 1 and 2, sketch the system.*

**1.** Supply: $p = \frac{3}{100}q + 2$,
Demand: $p = -\frac{7}{100}q + 12$.

**2.** Supply: $p = \frac{1}{2000}q + 3$,
Demand: $p = -\frac{1}{2500}q + \frac{42}{5}$.

**3.** Supply: $35q - 2p + 250 = 0$,
Demand: $65q + p - 537.5 = 0$.

**4.** Supply: $246p - 3.25q - 2460 = 0$,
Demand: $410p + 3q - 14{,}452.5 = 0$.

**5.** Supply: $p = 2q + 20$,
Demand: $p = 200 - 2q^2$.

**6.** Supply: $p = (q + 10)^2$,
Demand: $p = 388 - 16q - q^2$.

**7.** Supply: $p = \sqrt{q + 10}$,
Demand: $p = 20 - q$.

**8.** Supply: $p = \frac{1}{5}q + 5$,
Demand: $p = \frac{3000}{q + 200}$.

*In Problems 9–14, $y_{TR}$ represents total revenue in dollars and $y_{TC}$ represents total cost in dollars for a manufacturer. If q represents both the number of units produced and the number of units sold, find the break-even quantity. Sketch a break-even chart in Problems 9 and 10.*

**9.** $y_{TR} = 3q$,
$y_{TC} = 2q + 4500$,

**10.** $y_{TR} = 14q$,
$y_{TC} = \frac{40}{3}q + 1200$.

**11.** $y_{TR} = 0.05q$,
$y_{TC} = 0.85q + 600$.

**12.** $y_{TR} = 0.25q$,
$y_{TC} = 0.16q + 360$.

**13.** $y_{TR} = 100 - \frac{1000}{q + 10}$,
$y_{TC} = q + 40$.

**14.** $y_{TR} = 0.1q^2 + 7q$,
$y_{TC} = 2q + 500$.

**15. Business**  Supply and demand equations for a certain product are

$$3q - 200p + 1800 = 0$$

and

$$3q + 100p - 1800 = 0,$$

respectively, where $p$ represents the price per unit in dollars and $q$ represents the number of units sold per time period.

**a.** Find the equilibrium price algebraically, and derive it graphically.

**b.** Find the equilibrium price when a tax of 27 cents per unit is imposed on the supplier.

**16. Business**  A manufacturer of a product sells all that is produced. The total revenue is given by $y_{TR} = 7q$, and the total cost is given by $y_{TC} = 6q + 800$, where $q$ represents the number of units produced and sold.

**a.** Find the level of production at the break-even point, and draw the break-even chart.

**b.** Find the level of production at the break-even point if the total cost increases by 5%.

**17. Business**  A manufacturer sells a product at $8.35 per unit, selling all produced. The fixed cost is $2116 and the variable cost is $7.20 per unit. At what level of production will there be a profit of $4600? At what level of production will there be a loss of $1150? At what level of production will the break-even point occur?

**18. Business** The market equilibrium point for a product occurs when 13,500 units are produced at a price of $4.50 per unit. The producer will supply no units at $1, and the consumers will demand no units at $20. Find the supply and demand equations if they are both linear.

**19. Business** A manufacturer of widgets will break even at a sales volume of $200,000. Fixed costs are $40,000, and each unit of output sells for $5. Determine the variable cost per unit.

**20. Business** The Footsie Sandal Co. manufactures sandals for which the material cost is $0.80 per pair and the labor cost is $0.90 per pair. Additional variable costs amount to $0.30 per pair. Fixed costs are $70,000. If each pair sells for $2.50, how many pairs must be sold for the company to break even?

**21. Business** Find the break-even point for company Z, which sells all it produces, if the variable cost per unit is $2, fixed costs are $1050, and $y_{TR} = 50\sqrt{q}$, where $q$ is the number of units of output produced.

**22. Business** A company has determined that the demand equation for its product is $p = 1000/q$, where $p$ is the price per unit for $q$ units produced and sold in some period. Determine the quantity demanded when the price per unit is (a) $4, (b) $2, and (c) $0.50. For each of these prices, determine the total revenue that the company will receive. What will be the revenue regardless of the price? [*Hint:* Find the revenue when the price is $p$ dollars.]

**23. Business** Using the data in Example 1, determine how the original equilibrium price will be affected if the company is given a government subsidy of $1.50 per unit.

**24. Business** The Monroe Forging Company sells a corrugated steel product to the Standard Manufacturing Company and is in competition on such sales with other suppliers of the Standard Manufacturing Co. The vice president of sales of Monroe Forging Co. believes that by reducing the price of the product, a 40% increase in the volume of units sold to the Standard Manufacturing Co. could be secured. As the manager of the cost and analysis department, you have been asked to analyze the proposal of the vice president and submit your recommendations as to whether it is financially beneficial to the Monroe Forging Co. You are specifically requested to determine the following.

**a.** Net profit or loss based on the pricing proposal.

**b.** Unit sales volume under the proposed price that is required to make the same $40,000 profit which is now earned at the current price and unit sales volume.

Use the following data in your analysis:

| | Current Operations | Proposal of Vice President of Sales |
|---|---|---|
| Unit price | $2.50 | $2.00 |
| Unit sales volume | 200,000 units | 280,000 units |
| Variable cost | | |
| Total | $350,000 | $490,000 |
| Per unit | $1.75 | $1.75 |
| Fixed cost | $110,000 | $110,000 |
| Profit | $40,000 | ? |

**25. Business** Suppose products A and B have demand and supply equations that are related to each other. If $q_A$ and $q_B$ are the quantities produced and sold of A and B, respectively, and $p_A$ and $p_B$ are their respective prices, the demand equations are

$$q_A = 8 - p_A + p_B$$

and

$$q_B = 26 + p_A - p_B,$$

and the supply equations are

$$q_A = -2 + 5p_A - p_B$$

and

$$q_B = -4 - p_A + 3p_B.$$

Eliminate $q_A$ and $q_B$ to get the equilibrium prices.

**26. Business** The supply equation for a product is

$$p = 0.3q^2 + 14.6,$$

and the demand equation is

$$p = \frac{34.2}{1 + 0.3q}.$$

Here $p$ represents price per unit in dollars and $q$ represents number of units (in thousands) per unit time. Graph both equations, and from your graph, determine the equilibrium price and equilibrium quantity to one decimal place.

**27. Business** For a manufacturer, the total-revenue equation is

$$y_{TR} = 20.5\sqrt{q + 4} - 41$$

and the total-cost equation is

$$y_{TC} = 0.02q^3 + 10.4,$$

where $q$ represents (in thousands) both the number of units produced and the number of units sold. Graph a break-even chart and find the break-even quantity.

# 4.7 REVIEW

## IMPORTANT TERMS AND SYMBOLS

| | |
|---|---|
| **Section 4.1** | slope of a line    point–slope form    slope–intercept form    general linear equation in $x$ and $y$    linearly related |
| **Section 4.2** | demand equation    demand curve    supply equation    supply curve    linear function |
| **Section 4.3** | quadratic function    parabola    axis of symmetry    vertex |
| **Section 4.4** | system of equations    equivalent systems    elimination by addition    elimination by substitution    parameter    general linear equation in $x, y,$ and $z$ |
| **Section 4.5** | nonlinear system |
| **Section 4.6** | point of equilibrium    equilibrium price    equilibrium quantity    profit    total cost    fixed cost    variable cost    break-even point    break-even quantity    break-even revenue |

## SUMMARY

The orientation of a nonvertical line is characterized by the slope of the line given by

$$m = \frac{y_2 - y_1}{x_2 - x_1},$$

where $(x_1, y_1)$ and $(x_2, y_2)$ are two different points on the line. The slope of a vertical line is not defined, and the slope of a horizontal line is zero. Rising lines have positive slopes; falling lines have negative slopes. Two lines are parallel if and only if they have the same slope or are vertical. Two lines with slopes $m_1$ and $m_2$ are perpendicular to each other if and only if $m_1 = -\dfrac{1}{m_2}$. A horizontal line and a vertical line are perpendicular to each other.

Basic forms of equations of lines are:

| | |
|---|---|
| $y - y_1 = m(x - x_1)$ | (point–slope form) |
| $y = mx + b$ | (slope-intercept form) |
| $x = a$ | (vertical line) |
| $y = b$ | (horizontal line) |
| $Ax + By + C = 0$ | (general) |

The linear function $f(x) = ax + b \ (a \neq 0)$ has a straight line for its graph.

In economics, supply functions and demand functions have the form $p = f(q)$ and play an impor-tant role. Each gives a correspondence between the price $p$ of a product and the number of units $q$ of the product that manufacturers (or consumers) will supply (or purchase) at that price during some time period.

A quadratic function has the form

$$f(x) = ax^2 + bx + c \qquad (a \neq 0).$$

The graph of $f$ is a parabola that opens upward if $a > 0$ and downward if $a < 0$. The vertex is

$$\left( -\frac{b}{2a}, f\left( -\frac{b}{2a} \right) \right),$$

and the $y$-intercept is $c$. The axis of symmetry, as well as the $x$- and $y$-intercepts, are useful in sketching the graph.

A system of linear equations may be solved with the method of elimination by addition or elimination by substitution. A solution may involve one or more parameters. Substitution is also useful in solving non-linear systems.

Solving a system formed by the supply and de-mand equations for a product gives the equilibrium point, which indicates the price at which consumers will purchase the same quantity of a product that pro-ducers wish to sell at that price.

Profit is total revenue minus total cost, where total cost is the sum of fixed costs and variable costs. The break-even point is the point where total revenue equals total cost.

## REVIEW PROBLEMS

**1.** The slope of the line through $(2, 5)$ and $(3, k)$ is 4. Find $k$.

**2.** The slope of the line through $(2, 3)$ and $(k, 3)$ is 0. Find $k$.

*In Problems 3–9, determine the slope–intercept form and a general linear form of an equation of the straight line that has the indicated properties.*

**3.** Passes through $(3, -2)$ and has $y$-intercept 1.

**5.** Passes through $(10, 4)$ and has slope $\frac{1}{2}$.

**7.** Passes through $(-2, 4)$ and is horizontal.

**9.** Has $y$-intercept 2 and is perpendicular to $y + 3x = 2$.

**10.** Determine whether the point $(0, -7)$ lies on the line through $(1, -3)$ and $(4, 9)$.

**4.** Passes through $(-1, -1)$ and is parallel to the line $y = 3x - 4$.

**6.** Passes through $(3, 5)$ and is vertical.

**8.** Passes through $(1, 2)$ and is perpendicular to the line $-3y + 5x = 7$.

*In Problems 11–16, determine whether the lines are parallel, perpendicular, or neither.*

**11.** $x + 4y + 2 = 0$,   $8x - 2y - 2 = 0$.

**13.** $x - 3 = 2(y + 4)$,   $y = 4x + 2$.

**15.** $y = \frac{1}{2}x + 5$,   $2x = 4y - 3$.

**12.** $y - 2 = 2(x - 1)$,   $2x + 4y - 3 = 0$.

**14.** $3x + 5y + 4 = 0$,   $6x + 10y = 0$.

**16.** $y = 7x$,   $y = 7$.

*In Problems 17–20, write each line in slope–intercept form, and sketch. What is the slope of the line?*

**17.** $3x - 2y = 4$.       **18.** $x = -3y + 4$.       **19.** $4 - 3y = 0$.       **20.** $y = 2x$.

*In Problems 21–30, graph each function. For those that are linear, give the slope and the vertical-axis intercept. For those that are quadratic, give all intercepts and the vertex.*

**21.** $y = f(x) = 4 - 2x$.

**23.** $y = f(x) = 9 - x^2$.

**25.** $y = h(t) = t^2 - 4t - 5$.

**27.** $p = g(t) = 3t$.

**29.** $y = F(x) = -(x^2 + 2x + 3)$.

**22.** $s = g(t) = 8 - 2t - t^2$.

**24.** $y = f(x) = 3x - 7$.

**26.** $y = h(t) = 1 + 3t$.

**28.** $y = F(x) = (2x - 1)^2$.

**30.** $y = f(x) = \dfrac{x}{3} - 2$.

*In Problems 31–44, solve the given system.*

**31.** $\begin{cases} 2x - y = 6, \\ 3x + 2y = 5. \end{cases}$

**32.** $\begin{cases} 8x - 4y = 7, \\ y = 2x - 4. \end{cases}$

**33.** $\begin{cases} 4x + 5y = 3, \\ 3x + 4y = 2. \end{cases}$

**34.** $\begin{cases} 3x + 6y = 9, \\ 4x + 8y = 12. \end{cases}$

**35.** $\begin{cases} \dfrac{1}{4}x - \dfrac{3}{2}y = -4, \\ \dfrac{3}{4}x + \dfrac{1}{2}y = 8. \end{cases}$

**36.** $\begin{cases} \dfrac{1}{3}x - \dfrac{1}{4}y = \dfrac{1}{12}, \\ \dfrac{4}{3}x + 3y = \dfrac{5}{3}. \end{cases}$

**37.** $\begin{cases} 3x - 2y + z = -2, \\ 2x + y + z = 1, \\ x + 3y - z = 3. \end{cases}$

**38.** $\begin{cases} x + \dfrac{2y + x}{6} = 14, \\ y + \dfrac{3x + y}{4} = 20. \end{cases}$

**39.** $\begin{cases} x^2 - y + 2x = 7, \\ x^2 + y = 5. \end{cases}$

**40.** $\begin{cases} y = \dfrac{18}{x + 4}, \\ x - y + 7 = 0. \end{cases}$

**[7]41.** $\begin{cases} x + 2z = -2, \\ x + y + z = 5. \end{cases}$

**[7]42.** $\begin{cases} x + y + z = 0, \\ x - y + z = 0, \\ x + z = 0. \end{cases}$

**[7]43.** $\begin{cases} x - y - z = 0, \\ 2x - 2y + 3z = 0. \end{cases}$

**[7]44.** $\begin{cases} 2x - 5y + 6z = 1, \\ 4x - 10y + 12z = 2. \end{cases}$

**45.** Suppose $a$ and $b$ are linearly related so that $a = 1$ when $b = 2$ and $a = 2$ when $b = 1$. Find a general linear form of an equation that relates $a$ and $b$. Also, find $a$ when $b = 3$.

**46. Temperature and Heart Rate**   When the temperature $T$ (in degrees Celsius) of a cat is reduced, the cat's heart rate $r$ (in beats per minute) decreases. Under laboratory conditions, a cat at a temperature of 37°C had a heart rate of 220, and at a temperature of 32°C its heart rate was 150. If $r$ is linearly related to $T$, where $T$ is between 26 and 38, (a) determine an equation for $r$ in terms of $T$,

[7]Refer to concepts in Examples 6 and 7 in Sec. 4.4.

and (b) determine the cat's heart rate at a temperature of 28°C.

47. Suppose $f$ is a linear function such that $f(1) = 5$ and $f(x)$ decreases by four units for every three-unit increase in $x$. Find $f(x)$.

48. If $f$ is a linear function such that $f(-1) = 8$ and $f(2) = 5$, find $f(x)$.

49. **Maximum Revenue**  The demand function for a manufacturer's product is $p = f(q) = 200 - 2q$, where $p$ is the price (in dollars) per unit when $q$ units are demanded. Find the level of production that maximizes the manufacturer's total revenue, and determine this revenue.

50. **Sales Tax**  The difference in price of two items before a 5% sales tax is imposed is $4. The difference in price after the sales tax is imposed is $4.20. Find the price of each item before the sales tax.

51. **Equilibrium Price**  If the supply and demand equations of a certain product are $125p - q - 250 = 0$ and $100p + q - 1100 = 0$, respectively, find the equilibrium price.

52. **Break-even Point**  A manufacturer of a certain product sells all that is produced. Determine the break-even point if the product is sold at $16 per unit, fixed cost is $10,000, and variable cost is given by $y_{VC} = 8q$, where $q$ is the number of units produced ($y_{VC}$ expressed in dollars).

53. **Psychology**  In psychology, the term *semantic memory* refers to our knowledge of the meaning and relationships of words, as well as the means by which we store and retrieve such information.[8] In a network model of semantic memory, there is a hierarchy of levels at which information is stored. In an experiment by Collins and Quillian based on a network model, data were obtained on the reaction time to respond to simple questions about nouns. The graph of the results shows that, on the average, the reaction time $R$ (in milliseconds) is a linear function of the level $L$ at which a characterizing property of the noun is stored. At level 0, the reaction time is 1310; at level 2, the reaction time is 1460. (a) Find the linear function. (b) Find the reaction time at level 1. (c) Find the slope and determine its significance.

54. **Temperature Conversion**  Celsius temperature C is a linear function of Fahrenheit temperature F. Use the facts that 32°F is the same as 0°C and 212°F is the same as 100°C to find this function. Also, find C when F = 50.

55. **Enclosing Land**  A company has set aside $3000 to fence a rectangular portion of land adjacent to a straight stream by using the stream for one side of the enclosed area. The cost of the fencing parallel to the stream is $5 per foot installed, and the fencing for the remaining two sides is $3 per foot installed. Find the dimensions of the maximum enclosed area. [*Hint:* If the side opposite the stream has length $x$, first show that each of the other two sides has length $(3000 - 5x)/6$.]

56. Graphically solve the linear system

$$\begin{cases} 3x + 4y = 20, \\ 7x + 5y = 64. \end{cases}$$

57. Graphically solve the linear system

$$\begin{cases} 0.2x + 0.3y = 7, \\ 0.3x + 0.5y = 4. \end{cases}$$

Round $x$ and $y$ to two decimal places.

58. Graphically solve the nonlinear system

$$\begin{cases} y = \dfrac{2}{x}, \text{ where } x > 0, \\ y = x^2 - 6. \end{cases}$$

Round $x$ and $y$ to two decimal places.

59. Graphically solve the nonlinear system

$$\begin{cases} y = x^3 + 1, \\ y = 2 - x^2. \end{cases}$$

Round $x$ and $y$ to two decimal places.

60. Graphically solve the equation

$$x^2 + 4 = x^3 - 3x$$

by treating it as a system. Round $x$ to two decimal places.

[8]G. R. Loftus and E. F. Loftus, *Human Memory: The Processing of Information* (New York: Lawrence Erlbaum Associates, Inc., distributed by the Halsted Press, Division of John Wiley & Sons, Inc., 1976).

# MATHEMATICAL *SNAPSHOT*

Perhaps at some time you have made an appointment to see someone or to get something done, only to be met with a long wait. For example, people with appointments complain frequently about long waits to see physicians or to have their cars serviced. Even in sports, there are complaints about waiting. Tennis players may have to wait until 10:00 to begin their 9:00 match. Similarly for golfers, and so on. It appears that scheduling is at the heart of all these problems. Here you will learn one way to devise a time schedule for matches at a tennis tournament.

Suppose that 11 courts are available for a tennis tournament that begins at 8:00 A.M. Using an average match time of 1 hour and 30 minutes, a tournament director would typically schedule 11 matches at 8:00 A.M., 11 matches at 9:30 A.M., 11 matches at 11:00 A.M., and so on. However, the actual time length of a match varies. Some matches are completed after 30 minutes; others take more than 2 hours. Because of this, matches scheduled near the end of the tournament may run several hours late. Perhaps a better scheduling method is to have 11 matches begin at 8:00 A.M., then schedule some matches at 8:30 A.M., some at 9:00 A.M., some at 9:30 A.M., and continuing in this way, scheduling some matches every 30 minutes later. For convenience, refer to these 30-minute time intervals as periods 1, 2, 3, and so on, with period 1 beginning at 8:00 A.M. The problem is to determine the number of matches to be scheduled for each period.

Based on records of this tournament over the past several years, suppose that you estimate the average match time to be 1 hour and 37 minutes. These records also give rise to Table 4.3, which gives the average number of matches, as well as the cumulative total, played on the courts during each 30-minute interval. From Table 4.3 it appears that a schedule with 11 matches at 8:00 A.M., 1 match at 8:30 A.M., 4 matches at 9:00 A.M., and so on, is more reasonable than the typical schedule of 11 matches every $1\frac{1}{2}$ hours. In this situation, no one would have to wait more than 30 minutes. Denoting the period by $x$ and the corresponding cumulative total number of matches by $y$, Fig. 4.49 gives a geometrical representation of the points $(x, y)$. For example, the point $(2, 12)$ indicates that during period 2 the total number of matches played on the courts since the beginning of the tournament was 12.

It is obvious from Fig. 4.49 that the points almost lie on a straight line. Knowing the equation of such a line would allow you to predict the total number of matches to be scheduled by a given period. Because 11 matches must be scheduled for period 1, and the last match, the eighty-fifth, must be scheduled for period 23, it is not unreasonable to choose as your "predictor" line the one through the points $(1, 11)$ and $(23, 85)$, as shown in Fig. 4.49. The slope, $m$, is given by

$$m = \frac{85 - 11}{23 - 1} = \frac{74}{22} = \frac{37}{11}.$$

[9]Adapted from Brian Garman, "Applying a Linear Function to Schedule Tennis Matches," *The Mathematics Teacher*, 77, no. 7 (October 1984), 544–47. By permission of the National Council of Teachers of Mathematics.

**TABLE 4.3**

| Period | Time | Matches per Period | Total Matches | Period | Time | Matches per Period | Total Matches |
|--------|------|--------------------|---------------|--------|------|--------------------|---------------|
| 1 | 8:00 | 11 | 11 | 13 | 2:00 | 3 | 51 |
| 2 | 8:30 | 1 | 12 | 14 | 2:30 | 4 | 55 |
| 3 | 9:00 | 4 | 16 | 15 | 3:00 | 3 | 58 |
| 4 | 9:30 | 5 | 21 | 16 | 3:30 | 2 | 60 |
| 5 | 10:00 | 4 | 25 | 17 | 4:00 | 3 | 63 |
| 6 | 10:30 | 3 | 28 | 18 | 4:30 | 4 | 67 |
| 7 | 11:00 | 3 | 31 | 19 | 5:00 | 4 | 71 |
| 8 | 11:30 | 3 | 34 | 20 | 5:30 | 3 | 74 |
| 9 | 12:00 | 3 | 37 | 21 | 6:00 | 4 | 78 |
| 10 | 12:30 | 3 | 40 | 22 | 6:30 | 4 | 82 |
| 11 | 1:00 | 4 | 44 | 23 | 7:00 | 3 | 85 |
| 12 | 1:30 | 4 | 48 | | | | |

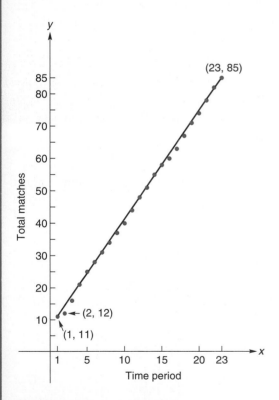

**FIGURE 4.49**   Total tennis matches as a function of time

A point–slope form of an equation of the line is

$$y - 11 = \frac{37}{11}(x - 1),$$

which can be rewritten as

$$y = \frac{37}{11}(x - 1) + 11. \qquad (1)$$

For example, if $x = 2$, then Eq. (1) gives $y = \frac{158}{11} \approx$ 14.36. Since fractional matches make no sense, the function in Eq. (1) must be refined. One method is to round the $y$-values in Eq. (1) to the nearest integer by making use of a function called the *greatest-integer function* and denoted by $[x]$. The notation $[x]$ means the greatest integer less than or equal to $x$. For example,

$$[4] = 4, \quad [4.1] = 4, \quad [5.9] = 5.$$

If you add 0.5 to any given number and then find the greatest integer in the sum, the result will be the given number rounded to the nearest integer. For instance, $[4.1 + 0.5] = [4.6] = 4$, which is 4.1 rounded to the nearest integer; $[5.9 + 0.5] = [6.4] = 6$, which is 5.9 rounded to the nearest integer. Thus the refined function, say $f$, that predicts the total number of matches up to and including period $x$ is given by

$$f(x) = \left[ \frac{37}{11}(x - 1) + 11.5 \right].$$

In particular, if $x = 2$, then

$$\frac{37}{11}(x - 1) + 11.5 = \frac{37}{11}(2 - 1) + 11.5 \approx 14.86$$

and $[14.86] = 14$. Thus $f(2) = 14$. Table 4.4 gives the predicted totals, $f(x)$, for $x = 1, 2, 3, \ldots, 23$, as well as

---

**TABLE 4.4**

| Period, $x$ | Time | Actual Total | Predicted Total, $f(x)$ | Adjusted Total, $F(x)$ |
|---|---|---|---|---|
| 1 | 8:00 | 11 | 11 | 11 |
| 2 | 8:30 | 12 | 14 | 12 |
| 3 | 9:00 | 16 | 18 | 16 |
| 4 | 9:30 | 21 | 21 | 21 |
| 5 | 10:00 | 25 | 24 | 24 |
| 6 | 10:30 | 28 | 28 | 28 |
| 7 | 11:00 | 31 | 31 | 31 |
| 8 | 11:30 | 34 | 35 | 35 |
| 9 | 12:00 | 37 | 38 | 38 |
| 10 | 12:30 | 40 | 41 | 41 |
| 11 | 1:00 | 44 | 45 | 45 |
| 12 | 1:30 | 48 | 48 | 48 |
| 13 | 2:00 | 51 | 51 | 51 |
| 14 | 2:30 | 55 | 55 | 55 |
| 15 | 3:00 | 58 | 58 | 58 |
| 16 | 3:30 | 60 | 61 | 61 |
| 17 | 4:00 | 63 | 65 | 65 |
| 18 | 4:30 | 67 | 68 | 68 |
| 19 | 5:00 | 71 | 72 | 72 |
| 20 | 5:30 | 74 | 75 | 75 |
| 21 | 6:00 | 78 | 78 | 78 |
| 22 | 6:30 | 82 | 82 | 82 |
| 23 | 7:00 | 85 | 85 | 85 |

---

the actual totals. Notice that $f$ is a fairly good predictor of the actual totals. However, $f$ predicts more matches for periods 2 and 3 than experience warrants. There are minor discrepancies for other periods. Since proper scheduling at the beginning of the tournament is of utmost importance, you can adjust the predicted totals to reflect 12 matches in period 2 and 16 matches in period 3. The adjusted totals are also shown in Table 4.4. It appears that a schedule based on adjusted totals is reasonable. The function, say, $F$, that describes this schedule is thus given by

$$F(x) = \begin{cases} 12, & \text{if } x = 2, \\ 16, & \text{if } x = 3, \\ \left[\frac{37}{11}(x - 1) + 11.5\right] & \text{otherwise,} \end{cases}$$

where $F(x)$ is the total number of matches scheduled through period $x$. Thus the number of matches assigned a starting time of period $x$ is $F(x) - F(x - 1)$, where $x > 1$.

Of course not all tennis tournaments involve 85 matches on 11 courts. Also, the average match time varies and depends, for example, on the class of players or the type of court. Thus to handle other situations, suppose that you generalize the above scheduling function to a tournament involving $m$ matches on $c$ courts based on an average match time of $h$ hours and $t$ minutes.

First, you need to determine the number of 30-minute periods involved. Suppose that the tournament is to start at time $T = 0$ and $E$ is the average time (in minutes) that a court is used throughout the day. Then

$$E = \frac{m(60h + t)}{c}.$$

The last match would end approximately at time $T = E$, so it would begin approximately at time $T = E - (60h + t)$. The number of 30-minute intervals from $T = 0$ to the starting time of the last match is

$$\frac{E - (60h + t)}{30} = \frac{\dfrac{m(60h + t)}{c} - (60h + t)}{30}$$

$$= \frac{\dfrac{m(60h + t)}{c} - \dfrac{(60h + t)c}{c}}{30}$$

$$= \frac{(m - c)(60h + t)}{30c}.$$

This number does not include the final period. The number of periods to be scheduled is the rounded-off value of

$$\frac{(m - c)(60h + t)}{30c} + 1.$$

Denoting the number of periods by $z$ gives

$$z = \left[\frac{(m - c)(60h + t)}{30c} + 1.5\right]. \qquad (2)$$

Assume that the total number of matches, $y$, scheduled through period $x$ is a linear function of $x$. Because $c$ matches are scheduled for period 1, and by period $z$ a total of $m$ matches are scheduled, the points

$(1, c)$ and $(z, m)$ lie on the graph of this function. The slope is

$$\frac{m - c}{z - 1},$$

so a point–slope form of an equation of the line is

$$y - c = \frac{m - c}{z - 1}(x - 1).$$

Simplifying gives

$$y = \frac{m - c}{z - 1}(x - 1) + c.$$

This function must be refined so that the $y$ values are rounded to the nearest integer. Furthermore, experience indicates that for periods 2 and 3, the total number of matches assigned should reflect the same proportions as in the original data based on 11 courts. That is, for period 2 there should be a total of $[\frac{12}{11}c + 0.5]$ matches assigned, and for period 3 there should be $[\frac{16}{11}c + 0.5]$. The function, say $F$, that describes this schedule is

$$F(x) = \begin{cases} [\frac{12}{11}c + 0.5], & \text{if } x = 2, \\ [\frac{16}{11}c + 0.5], & \text{if } x = 3, \\ \left[\dfrac{m - c}{z - 1}(x - 1) + c + 0.5\right] & \text{otherwise}, \end{cases}$$

where $F(x)$ is the total number of matches scheduled through period $x$, $m$ is the number of matches in the tournament, $c$ is the number of courts, and $z$ is the number of periods, where $z$ is given by Eq. (2).

This function describes the tennis scheduling system known as the "Garman system," which is now used in many championship tennis tournaments.

■ **Exercises**

*In the following problems, use the Garman system of scheduling for a tennis tournament that involves* **16 matches with an average match time of** **1 hour and** **45 minutes and four courts.**

1. Determine (a) the value of $z$, the number of 30-minute periods, (b) $F(x)$, and (c) the values of $F(x)$ for $x = 1, 2, 3, \ldots, z$.

2. If the tournament is to begin at 10:00 A.M., for each period indicate the starting time and the number of matches to be played during the period.

# Exponential and Logarithmic Functions

---

CHAPTER

5

## OBJECTIVE

**To study exponential functions and their applications to such areas as compound interest, population growth, and radioactive decay.**

Do not confuse the exponential function $y = 2^x$ with the *power function* $y = x^2$, which has a variable base and a constant exponent.

If you wish to review exponents, refer to Sec. 0.5.

## 5.1 EXPONENTIAL FUNCTIONS

There is a function that has an important role not only in mathematics, but also in business, economics, and other areas of study. It involves a constant raised to a variable power, such as $f(x) = 2^x$. We call such functions *exponential functions*.

**DEFINITION**

*The function f defined by*

$$f(x) = b^x,$$

*where $b > 0$, $b \neq 1$, and the exponent $x$ is any real number, is called an* **exponential function** *with base $b$.*[1]

Since the exponent in $b^x$ can be any real number, you may wonder how we assign a value to something like $2^{\sqrt{2}}$, where the exponent is an irrational number. Stated simply, we use approximations. Because $\sqrt{2} = 1.41421\ldots$, $2^{\sqrt{2}}$ is approximately $2^{1.4} = 2^{7/5} = \sqrt[5]{2^7}$, which *is* defined. Better approximations are $2^{1.41} = 2^{141/100} = \sqrt[100]{2^{141}}$, and so on. In this way, the meaning of $2^{\sqrt{2}}$ becomes clear. A calculator value of $2^{\sqrt{2}}$ is (approximately) 2.66514.

When you work with exponential functions, it may be necessary to apply rules for exponents. These rules are as follows, where $m$ and $n$ are real numbers and $a$ and $b$ are positive.

| **Rules for Exponents** | |
|---|---|
| **1.** $a^m a^n = a^{m+n}$. | **2.** $\dfrac{a^m}{a^n} = a^{m-n}$. |
| **3.** $(a^m)^n = a^{mn}$. | **4.** $(ab)^n = a^n b^n$. |
| **5.** $\left(\dfrac{a}{b}\right)^n = \dfrac{a^n}{b^n}$. | **6.** $a^1 = a$. |
| **7.** $a^0 = 1$. | **8.** $a^{-n} = \dfrac{1}{a^n}$. |

---

[1] If $b = 1$, then $f(x) = 1^x = 1$. This function is so uninteresting that we do not call it an exponential function.

Some functions that do not appear to have the exponential form $b^x$ can be put in that form by applying the preceding rules. For example, $2^{-x} = 1/(2^x) = (\frac{1}{2})^x$ and $3^{2x} = (3^2)^x = 9^x$.

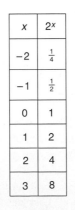

### Principles in Practice 1

**Bacteria Growth**

The number of bacteria in a culture that doubles every hour is given by $N(t) = A \cdot 2^t$, where $A$ is the number originally present and $t$ is the number of hours the bacteria have been doubling. Use a graphics calculator to plot this function for various values of $A > 1$. How are the graphs similar? How does the value of $A$ alter the graph?

### Principles in Practice 2

**Graphing Exponential Functions with $b > 1$**

Suppose an investment increases by 10% every year. Make a table of the factor by which the investment increases from the original amount for 0 to 4 years. For each year, write an expression for the increase as a power of some base. What base did you use? How does that base relate to the problem? Use your table to graph the multiplicative increase as a function of the number of years. Use your graph to determine when the investment will double.

## EXAMPLE 1  Bacteria Growth

*The number of bacteria present in a culture after t minutes is given by*

$$N(t) = 300\left(\frac{4}{3}\right)^t.$$

*Note that $N(t)$ is a constant multiple of the exponential function $\left(\frac{4}{3}\right)^t$.*

**a.** *How many bacteria are present initially?*

*Solution:* Here we want to find $N(t)$ when $t = 0$. We have

$$N(0) = 300\left(\frac{4}{3}\right)^0 = 300(1) = 300.$$

Thus, 300 bacteria are initially present.

**b.** *Approximately how many bacteria are present after 3 minutes?*

*Solution:*

$$N(3) = 300\left(\frac{4}{3}\right)^3 = 300\left(\frac{64}{27}\right) = \frac{6400}{9} \approx 711.$$

Hence, approximately 711 bacteria are present after 3 minutes. ∎

## Graphs of Exponential Functions

## EXAMPLE 2  Graphing Exponential Functions with $b > 1$

*Graph the exponential functions $f(x) = 2^x$ and $f(x) = 5^x$.*

*Solution:* By plotting points and connecting them, we obtain the graphs in Fig. 5.1. For the graph of $f(x) = 5^x$, because of the unit distance chosen on the $y$-axis, the points $(-2, \frac{1}{25})$, $(2, 25)$, and $(3, 125)$ are not shown.

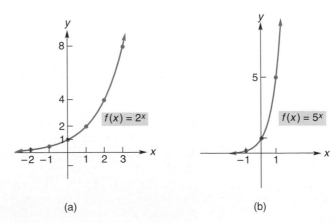

| $x$ | $2^x$ |
|---|---|
| $-2$ | $\frac{1}{4}$ |
| $-1$ | $\frac{1}{2}$ |
| $0$ | $1$ |
| $1$ | $2$ |
| $2$ | $4$ |
| $3$ | $8$ |

(a)

(b)

| $x$ | $5^x$ |
|---|---|
| $-2$ | $\frac{1}{25}$ |
| $-1$ | $\frac{1}{5}$ |
| $0$ | $1$ |
| $1$ | $5$ |
| $2$ | $25$ |
| $3$ | $125$ |

**FIGURE 5.1**  Graphs of $f(x) = 2^x$ and $f(x) = 5^x$.

We can make some observations about these graphs. The domain of each function consists of all real numbers, and the range consists of all positive real numbers. Each graph has $y$-intercept $(0, 1)$. Moreover, the graphs have the same general shape. Each *rises* from left to right. As $x$ increases, $f(x)$ also increases. In fact, $f(x)$ increases without bound. However, in quadrant I, the graph of $f(x) = 5^x$ rises more quickly than that of $f(x) = 2^x$, because the base in $5^x$ is *greater* than the base in $2^x$ (that is, $5 > 2$). Looking at quadrant II, we see that as $x$ becomes very negative, the graphs of both functions approach the $x$-axis.[2] This implies that the function values get very close to 0. ■

The observations made in Example 2 are true for all exponential functions whose base $b$ is greater than 1. Example 3 will examine the case for a base between 0 and 1 $(0 < b < 1)$.

## Principles in Practice 3

**Graphing Exponential Functions with $0 < b < 1$**

Suppose the value of a car depreciates by 15% every year. Make a table of the factor by which the value decreases from the original amount for 0 to 3 years. For each year, write an expression for the decrease as a power of some base. What base did you use? How does that base relate to the problem? Use your table to graph the multiplicative decrease as a function of the number of years. Use your graph to determine when your car will be worth half as much as its original price.

## EXAMPLE 3   Graphing Exponential Functions with $0 < b < 1$

*Graph the exponential function $f(x) = \left(\dfrac{1}{2}\right)^x$.*

*Solution:* By plotting points and connecting them, we obtain the graph in Fig. 5.2. Notice that the domain consists of all real numbers, and the range consists of all positive real numbers. The graph has $y$-intercept $(0, 1)$. Compared to the graphs in Example 2, the graph here *falls* from left to right. That is, as $x$

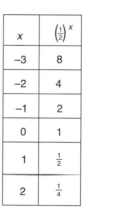

| $x$ | $\left(\frac{1}{2}\right)^x$ |
|---|---|
| $-3$ | $8$ |
| $-2$ | $4$ |
| $-1$ | $2$ |
| $0$ | $1$ |
| $1$ | $\frac{1}{2}$ |
| $2$ | $\frac{1}{4}$ |

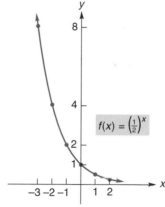

**FIGURE 5.2**   Graph of $f(x) = \left(\frac{1}{2}\right)^x$.

increases, $f(x)$ decreases. Notice that as $x$ becomes very positive, $f(x)$ takes on values close to 0 and the graph approaches the $x$-axis. However, as $x$ becomes very negative, the function values are unbounded. ■

In general, the graph of an exponential function has one of two shapes, depending on the value of the base $b$. This is illustrated in Fig. 5.3. The basic properties of an exponential function and its graph are summarized in Table 5.1.

Recall from Sec. 3.6 that the graph of one function may be related to that of another by means of a certain transformation. Our next example has to do with this concept.

[2]We say that the $x$-axis is an *asymptote* for each graph.

There are two basic shapes for the graphs of exponential functions, and they depend on the base involved.

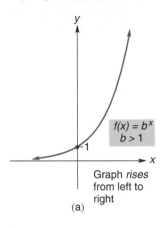

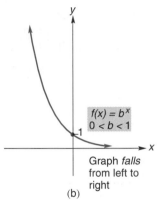

**FIGURE 5.3** General shapes of $f(x) = b^x$.

Example 4 makes use of transformations from Table 3.2.

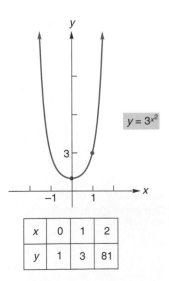

| x | 0 | 1 | 2 |
|---|---|---|---|
| y | 1 | 3 | 81 |

**FIGURE 5.6** Graph of $y = 3^{x^2}$.

**TABLE 5.1** Properties of the Exponential Function $f(x) = b^x$

1. The domain of an exponential function consists of all real numbers. The range consists of all positive numbers.

2. The graph of $f(x) = b^x$ has $y$-intercept $(0, 1)$. There is no $x$-intercept.

3. If $b > 1$, the graph *rises* from left to right. If $0 < b < 1$, the graph *falls* from left to right.

4. If $b > 1$, the graph approaches the $x$-axis as $x$ becomes more and more negative. If $0 < b < 1$, the graph approaches the $x$-axis as $x$ becomes more and more positive.

**EXAMPLE 4** **Transformations of Exponential Functions**

**a.** *Use the graph of $y = 2^x$ to plot $y = 2^x - 3$.*

**Solution:** The function has the form $f(x) - c$, where $f(x) = 2^x$ and $c = 3$. Thus its graph is obtained by shifting the graph of $f(x) = 2^x$ three units downward. (See Fig. 5.4.)

**b.** *Use the graph of $y = (\frac{1}{2})^x$ to graph $y = (\frac{1}{2})^{x-4}$.*

**Solution:** The function has the form $f(x - c)$, where $f(x) = (\frac{1}{2})^x$ and $c = 4$. Hence, its graph is obtained by shifting the graph of $f(x) = (\frac{1}{2})^x$ four units to the right. (See Fig. 5.5.)  ∎

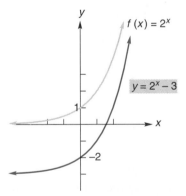

**FIGURE 5.4** Graph of $y = 2^x - 3$.

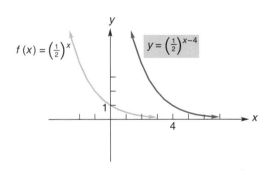

**FIGURE 5.5** Graph of $y = (\frac{1}{2})^{x-4}$.

**EXAMPLE 5** **Graph of a Function with a Constant Base**

*Graph $y = 3^{x^2}$.*

**Solution:** Although this is not an exponential function, it does have a constant base. We see that replacing $x$ by $-x$ results in the same equation. Thus, the graph is symmetric about the $y$-axis. Plotting some points and using symmetry gives the graph in Fig. 5.6.  ∎

## TECHNOLOGY

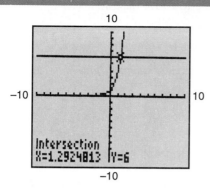

**FIGURE 5.7**   Solving the equation
$6 = 4^x$.

If $y = 4^x$, consider the problem of finding $x$ when
$y = 6$. One way to solve this equation is to find the intersection of the graphs of $y = 6$ and $y = 4^x$. Figure
5.7 shows that $x$ is approximately 1.29.

### Compound Interest

Exponential functions are involved in **compound interest,** whereby the interest earned by an invested amount of money (or **principal**) is reinvested so that
it, too, earns interest. That is, the interest is converted (or *compounded*) into
principal, and hence, there is "interest on interest."

For example, suppose that \$100 is invested at the rate of 5% compound-
ed annually. At the end of the first year, the value of the investment is the original principal (\$100), plus the interest on the principal [100(0.05)]:

$$100 + 100(0.05) = \$105.$$

This is the amount on which interest is earned for the second year. At the end
of the second year, the value of the investment is the principal at the end of the
first year (\$105), plus the interest on that sum [105(0.05)]:

$$105 + 105(0.05) = \$110.25.$$

Thus, each year the principal increases by 5%. The \$110.25 represents the orig-
inal principal, plus all accrued interest; it is called the **accumulated amount** or
**compound amount.** The difference between the compound amount and the
original principal is called the **compound interest.** Here the compound interest
is $110.25 - 100 = \$10.25$.

More generally, if a principal of $P$ dollars is invested at a rate of $100r$ percent compounded annually (for example, at 5%, $r$ is 0.05), the compound
amount after 1 year is $P + Pr$, or, by factoring, $P(1 + r)$. At the end of the
second year, the compound amount is

$$P(1 + r) + [P(1 + r)]r$$

$$= P(1 + r)[1 + r] \qquad \text{(factoring)}$$

$$= P(1 + r)^2.$$

This pattern continues. After three years, the compound amount is $P(1 + r)^3$.
In general, **the compound amount $S$ of the principal $P$ at the end of $n$ years at
the rate of $r$ compounded annually** is given by

$$S = P(1 + r)^n. \qquad (1)$$

**Principles in Practice 4**

**Transformations of Exponential
Functions**

After watching his sister's money
grow for three years in a plan with an
8% yearly return, George started a
savings account with the same plan.
If $y = 1.08^t$ represents the multiplicative increase in his sister's account, write an equation that will
represent the multiplicative increase
in George's account, using the same
time reference. If George has a graph
of the multiplicative increase in his
sister's money at time $t$ years since
she started saving, how could he use
the graph to project the increase in
his money?

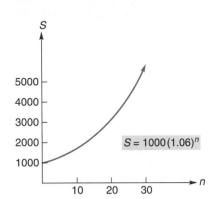

**FIGURE 5.8** Graph of $S = 1000(1.06)^n$.

Notice from Eq. (1) that, for a given principal and rate, $S$ is a function of $n$. In fact, $S$ involves an exponential function with base $1 + r$.

**EXAMPLE 6**   **Compound Amount and Compound Interest**

*Suppose* $1000 *is invested for* 10 *years at* 6% *compounded annually.*

**a.** *Find the compound amount.*

   *Solution:*  We use Eq. (1) with $P = 1000$, $r = 0.06$, and $n = 10$:

$$S = 1000(1 + 0.06)^{10} = 1000(1.06)^{10} \approx \$1790.85.$$

   Figure 5.8 shows the graph of $S = 1000(1.06)^n$. Notice that as time goes on, the compound amount grows dramatically.

**b.** *Find the compound interest.*

   *Solution:*  Using the results from part (a), we have

$$\text{compound interest} = S - P$$
$$= 1790.85 - 1000 = \$790.85. \quad \blacksquare$$

**Principles in Practice 5**
**Compound Amount and Compound Interest**

Suppose $2000 is invested at 13% compounded annually. Find the value of the investment after five years. Find the interest earned over the first five years.

The abbreviation A.P.R. is a common one and is found on credit card statements and in advertising.

Suppose the principal of $1000 in Example 6 is invested for 10 years as before, but this time the compounding takes place every three months (that is, *quarterly*) at the rate of $1\frac{1}{2}$% *per quarter*. Then there are four **interest periods** or **conversion periods** per year, and in 10 years there are $10(4) = 40$ interest periods. Thus, the compound amount with $r = 0.015$ is now

$$1000(1.015)^{40} \approx \$1814.02,$$

and the compound interest is $814.02. Usually, the interest rate per conversion period is stated as an annual rate. Here we would speak of an annual rate of 6% compounded quarterly, so that the rate per interest period, or the **periodic rate,** is $6\%/4 = 1.5\%$. This *quoted* annual rate of 6% is called the **nominal rate** or the **annual percentage rate (A.P.R.).** Unless otherwise stated, all interest rates will be assumed to be annual (nominal) rates. Thus a rate of 15% compounded monthly corresponds to a periodic rate of $15\%/12 = 1.25\%$.

On the basis of our discussion, we can generalize Eq. (1). The formula

$$S = P(1 + r)^n \tag{2}$$

gives **the compound amount $S$ of a principal $P$ at the end of $n$ interest periods at the periodic rate of $r$.**

We have seen that for a principal of $1000 at a nominal rate of 6% over a period of 10 years, annual compounding results in a compound interest of $790.85, and with quarterly compounding the compound interest is $814.02. It is typical that for a given nominal rate, the more frequent the compounding, the greater is the compound interest. However, as the number of interest periods increases, the effect tends to be less meaningful. For example, with weekly compounding the compound interest is

$$1000\left(1 + \frac{0.06}{52}\right)^{10(52)} - 1000 \approx \$821.49,$$

and with daily compounding it is

$$1000\left(1 + \frac{0.06}{365}\right)^{10(365)} - 1000 \approx \$822.03.$$

The difference is not too significant here.

***Pitfall*** ▼ A nominal rate of 6% does not necessarily mean that an investment increases in value by 6% in a year's time. The increase depends on the frequency of compounding.

Sometimes the phrase "money is worth" is used to express an annual interest rate. Thus, saying that money is worth 6% compounded quarterly refers to an annual (nominal) rate of 6% compounded quarterly.

**Principles in Practice 6**

**Semiannual Compounding**

Suppose $2000 is invested at a nominal rate of 6.5% compounded semiannually. Find the value of the investment after five years. Find the interest earned over the first five years.

**EXAMPLE 7**   **Semiannual Compounding**

*Suppose $3000 is placed in a savings account. If money is worth 6% compounded semiannually, what is the balance in the account after seven years? (Assume that no other deposits and no withdrawals are made.)*

***Solution:*** Here $P = 3000$. With two interest periods per year, we have a total of $n = 7(2) = 14$ interest periods. The periodic rate $r$ is $0.06/2 = 0.03$. By Eq. (2), we have

$$S = 3000(1.03)^{14} \approx \$4537.77.$$   ∎

A more detailed discussion of compound interest and the mathematics of finance appears in Chapter 8.

**Population Growth**

Equation (2) can be applied not only to the growth of money, but also to other types of growth, such as population growth. For example, suppose the population $P$ of a town of 10,000 is increasing at the rate of 2% per year. Then $P$ is a function of time $t$, in years. It is common to indicate this functional dependence by writing

$$P = P(t).$$

Here the letter $P$ is used in two ways: On the right side, $P$ represents the function; on the left side, $P$ represents the dependent variable. From Eq. (2), we have

$$P(t) = 10{,}000(1 + 0.02)^t = 10{,}000(1.02)^t.$$

**EXAMPLE 8**   **Population Growth**

*The population of a town of 10,000 grows at the rate of 2% per year. Find the population three years from now.*

***Solution:*** From the preceding discussion,

$$P(t) = 10{,}000(1.02)^t.$$

For $t = 3$, we have

$$P(3) = 10{,}000(1.02)^3 \approx 10{,}612.$$

Thus, the population three years from now will be 10,612. (See Fig. 5.9.)   ∎

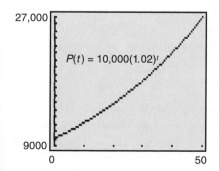

**FIGURE 5.9**   Graph of population function $P(t) = 10{,}000(1.02)^t$.

## Exponential Function with Base *e*

One of the numbers most useful for a base in an exponential function is a certain irrational number denoted by the letter *e* in honor of the Swiss mathematician Leonhard Euler (1707–1783):

$$e = 2.71828. \ldots$$

The exponential function with base *e* is called the **natural exponential function.**

Although *e* may seem to be a strange base, it arises quite naturally in calculus (as you will see in a later chapter). It also occurs in economic analysis and problems involving natural growth or decay, such as population studies, compound interest, and radioactive decay. Approximate values of $e^x$ can be found with calculators. The graph of $y = e^x$ is shown in Fig. 5.10. The accompanying table indicates *y*-values to two decimal places. Of course, the graph has the general shape of an exponential function with base greater than 1.

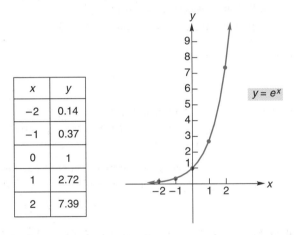

| x | y |
|----|------|
| −2 | 0.14 |
| −1 | 0.37 |
| 0 | 1 |
| 1 | 2.72 |
| 2 | 7.39 |

**FIGURE 5.10**   Graph of natural exponential function.

You should gain familiarity with the graph of the natural exponential function in Fig. 5.10.

**EXAMPLE 9**   **Graphs of Functions Involving *e***

a. *Graph $y = e^{-x}$.*

*Solution:* Since $e^{-x} = \left(\dfrac{1}{e}\right)^x$ and $0 < \dfrac{1}{e} < 1$, the graph is that of an exponential function falling from left to right. (See Fig. 5.11.) Alternatively, we can consider the graph of $y = e^{-x}$ as a transformation of the graph of $f(x) = e^x$. Because $e^{-x} = f(-x)$, the graph of $y = e^{-x}$ is simply the reflection of the graph of *f* about the *y*-axis. (Compare the graphs in Figs. 5.10 and 5.11.)

b. *Graph $y = e^{x+2}$.*

*Solution:* The graph of $y = e^{x+2}$ is related to that of $f(x) = e^x$. Since $e^{x+2}$ is $f(x + 2)$, we can obtain the graph of $y = e^{x+2}$ by horizontally shifting the graph of $f(x) = e^x$ two units to the left. (See Fig. 5.12.)   ■

**EXAMPLE 10**   **Population Growth**

*The projected population P of a city is given by*

$$P = 100{,}000e^{0.05t},$$

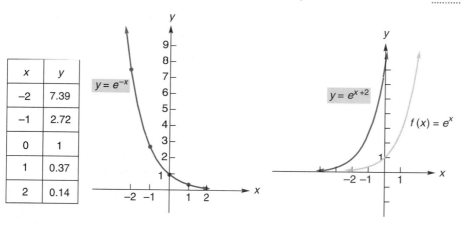

**FIGURE 5.11**  Graph of $y = e^{-x}$

**FIGURE 5.12**  Graph of $y = e^{x+2}$.

*where t is the number of years after* 1990. *Predict the population for the year* 2010.

*Solution:*  The number of years from 1990 to 2010 is 20, so let $t = 20$. Then

$$P = 100{,}000e^{0.05(20)} = 100{,}000e^1 = 100{,}000e \approx 271{,}828.$$

Many forecasts are based on population studies.  ■

In statistics, an important function used as a model in describing events occurring in nature is the **Poisson distribution function:**

$$f(x) = \frac{e^{-\mu}\mu^x}{x!}, \qquad x = 0, 1, 2, \ldots.$$

The symbol $\mu$ (read "mu") is a Greek letter. In certain situations, $f(x)$ gives the probability that exactly $x$ events will occur in an interval of time or space. The constant $\mu$ is the mean, or average, number of occurrences in the interval. The next example illustrates the Poisson distribution.

### EXAMPLE 11  Hemocytometer and Cells

*A hemocytometer is a counting chamber divided into squares and is used in studying the number of microscopic structures in a liquid. In a well-known experiment,[3] yeast cells were diluted and thoroughly mixed in a liquid, and the mixture was placed in a hemocytometer. With a microscope, the number of yeast cells on each square were counted. The probability that there were exactly x yeast cells on a hemocytometer square was found to fit a Poisson distribution with $\mu = 1.8$. Find the probability that there were exactly four cells on a particular square.*

*Solution:*  We use the Poisson distribution function with $\mu = 1.8$ and $x = 4$:

$$f(x) = \frac{e^{-\mu}\mu^x}{x!},$$

$$f(4) = \frac{e^{-1.8}(1.8)^4}{4!} \approx 0.072.$$

[3]R. R. Sokal and F. J. Rohlf, *Introduction to Biostatistics* (San Francisco: W. H. Freeman and Company, Publishers, 1973).

For example, this means that in 400 squares we would *expect* $400(0.072) \approx 29$ squares to contain exactly 4 cells. (In the experiment, in 400 squares the actual number observed was 30.) ▪

### Radioactive Decay

Radioactive elements are such that the amount of the element decreases with respect to time. We say that the element *decays*. It can be shown that, if $N$ is the amount at time $t$, then

$$N = N_0 e^{-\lambda t},\tag{3}$$

where $N_0$ and $\lambda$ (a Greek letter read "lambda") are positive constants. Notice that $N$ involves an exponential function of $t$. We say that $N$ follows an **exponential law of decay.** If $t = 0$, then $N = N_0 e^0 = N_0 \cdot 1 = N_0$. Thus, the constant $N_0$ represents the amount of the element present at time $t = 0$ and is called the **initial amount.** The constant $\lambda$ depends on the particular element involved and is called the **decay constant.**

Because $N$ decreases as time progresses, suppose we let $T$ be the length of time it takes for the element to decrease to half of the initial amount. Then at time $t = T$, we have $N = N_0/2$. Equation (3) implies that

$$\frac{N_0}{2} = N_0 e^{-\lambda T}.$$

We shall now use this fact to show that over *any* time interval of length $T$, half of the amount of the element decays. Consider the interval from time $t$ to $t + T$, which has length $T$. At time $t$, the amount of the element is $N_0 e^{-\lambda t}$, and at time $t + T$ it is

$$N_0 e^{-\lambda(t+T)} = N_0 e^{-\lambda t} e^{-\lambda T} = (N_0 e^{-\lambda T}) e^{-\lambda t}$$

$$= \frac{N_0}{2} e^{-\lambda t} = \frac{1}{2}(N_0 e^{-\lambda t}),$$

which is half of the amount at time $t$. This means that if the initial amount present, $N_0$, were 1 gram, then at time $T$, $\frac{1}{2}$ gram would remain; at time $2T$, $\frac{1}{4}$ gram would remain; and so on. The value of $T$ is called the **half-life** of the radioactive element. Figure 5.13 shows a graph of radioactive decay.

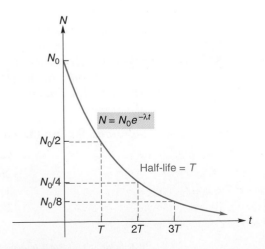

**FIGURE 5.13** Radioactive decay.

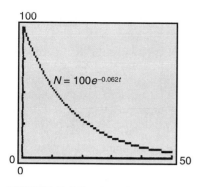

**FIGURE 5.14**   Graph of radioactive decay function $N = 100e^{-0.062t}$.

**EXAMPLE 12   Radioactive Decay**

*A radioactive element decays such that after t days the number of milligrams present is given by*

$$N = 100e^{-0.062t}.$$

a. *How many milligrams are initially present?*

> *Solution:* This equation has the form of Eq. (3), $N = N_0 e^{-\lambda t}$, where $N_0 = 100$ and $\lambda = 0.062$. $N_0$ is the initial amount and corresponds to $t = 0$. Thus, 100 milligrams are initially present. (See Fig. 5.14.)

b. *How many milligrams are present after 10 days?*

> *Solution:* When $t = 10$,
>
> $$N = 100e^{-0.062(10)} = 100e^{-0.62} \approx 53.8.$$

Therefore, approximately 53.8 milligrams are present after 10 days.   ■

# ▪ Exercise 5.1

*In Problems 1–12, graph each function.*

**1.** $y = f(x) = 4^x$.    **2.** $y = f(x) = 3^x$.    **3.** $y = f(x) = (\frac{1}{3})^x$.    **4.** $y = f(x) = (\frac{1}{4})^x$.

**5.** $y = f(x) = 2(\frac{1}{4})^x$.    **6.** $y = f(x) = 3(2)^x$.    **7.** $y = f(x) = 3^{x+2}$.    **8.** $y = f(x) = 2^{x-1}$.

**9.** $y = f(x) = 2^x - 1$.    **10.** $y = f(x) = 3^{x-1} - 1$.    **11.** $y = f(x) = 2^{-x}$.    **12.** $y = f(x) = \frac{1}{2}(3^{x/2})$.

*Problems 13 and 14 refer to Fig. 5.15, which shows the graphs of $y = 0.4^x$, $y = 2^x$, and $y = 5^x$.*

**13.** Of the curves *A*, *B*, and *C*, which is the graph of $y = 5^x$?

**14.** Of the curves *A*, *B*, and *C*, which is the graph of $y = 0.4^x$?

**15. Population**   The projected population of a city is given by $P = 125{,}000(1.12)^{t/20}$, where *t* is the number of years after 1995. What is the projected population in 2015?

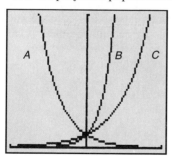

**FIGURE 5.15**   Diagram for Problems 13 and 14.

**16. Population**   For a certain city, the population *P* grows at the rate of 2% per year. The formula $P = 1{,}000{,}000(1.02)^t$ gives the population *t* years after 1998. Find the population in (a) 1999 and (b) 2000.

**17. Paired-Associate Learning**   In a psychological experiment involving learning,[4] subjects were asked to give particular responses after being shown certain stimuli. Each stimulus was a pair of letters, and each response was either the digit 1 or 2. After each response, the subject was told the correct answer. In this so-called *paired-associate* learning experiment, the theoretical probability *P* that a subject makes a correct response on the *n*th trial is given by

$$P = 1 - \tfrac{1}{2}(1 - c)^{n-1}, \qquad n \geq 1, \qquad 0 < c < 1.$$

Find *P* when $n = 1$.

**18.** Express $y = 2^{3x}$ as an exponential function in base 8.

*In Problems 19–27, find (a) the compound amount and (b) the compound interest for the given investment and annual rate.*

**19.** $4000 for 7 years at 6% compounded annually.

**20.** $5000 for 20 years at 5% compounded annually.

**21.** $700 for 15 years at 7% compounded semiannually.

**22.** $4000 for 12 years at 6% compounded semiannually.

**23.** $4000 for 15 years at $8\frac{1}{2}$% compounded quarterly.

**24.** $900 for 11 years at 10% compounded quarterly.

**25.** $5000 for $2\frac{1}{2}$ years at 9% compounded monthly.

**26.** $500 for 5 years at 11% compounded semiannually.

**27.** $8000 for 3 years at $6\frac{1}{4}$% compounded daily. (Assume that there are 365 days in a year.)

---

[4]D. Laming, *Mathematical Psychology* (New York: Academic Press, Inc., 1973).

**28. Investment**  Suppose $1000 is placed in a savings account that earns interest at the rate of 5% compounded semiannually. (a) What is the value of the account at the end of four years? (b) If the account had earned interest at the rate of 5% compounded annually, what would be the value after four years?

**29. Investment**  A $6000 certificate of deposit is purchased for $6000 and is held seven years. If the certificate earns 8% compounded quarterly, what is it worth at the end of seven years?

**30. Population Growth**  The population of a town of 5000 grows at the rate of 3% per year. (a) Determine an equation that gives the population after $t$ years from now. (b) Find the population three years from now. Give your answer to (b) to the nearest integer.

**31. Bacteria Growth**  Bacteria are growing in a culture, and their number is increasing at the rate of 5% an hour. Initially, 400 bacteria are present. (a) Determine an equation that gives the number $N$ of bacteria present after $t$ hours. (b) How many bacteria are present after

one hour? (c) After four hours? Give your answers to (b) and (c) to the nearest integer.

**32. Bacteria Reduction**  A certain medicine reduces the bacteria present in a person by 10% each hour. Currently, 100,000 bacteria are present. Make a table of values for the number of bacteria present each hour for 0 to 4 hours. For each hour, write an expression for the number of bacteria as a product of 100,000 and a power of $\frac{9}{10}$. Use the expressions to make an entry in your table for the number of bacteria after $t$ hours. Write a function $N$ for the number of bacteria after $t$ hours.

**33. Recycling**  Suppose the amount of plastic being recycled increases by 30% every year. Make a table of the factor by which recycling increases over the original amount for 0 to 3 years. For each year, write an expression for the increase as a power of some base. What base did you use? How does that base relate to the problem? Use your table to graph the multiplicative increase as a function of years. Use your graph to determine when the recycling will triple.

**34. Population Growth**  Cities A and B presently have populations of 70,000 and 60,000, respectively. City A grows at the rate of 4% per year, and B grows at the rate of 5% per year. Determine the difference in the populations of the cities at the end of five years. Give your answer to the nearest integer.

*Problems 35 and 36 involve a declining population. If a population declines at the rate of r per time period, then the population after t time periods is given by*

$$P = P_0(1 - r)^t.$$

*where $P_0$ is the initial population (the population when $t = 0$).*

**35. Population**  Because of an economic downturn, the population of a certain urban area declines at the rate of 1% per year. Initially, the population is 100,000. To the nearest person, what is the population after three years?

**36. Workforce**  In a dramatic cost-cutting effort, a company will reduce its workforce at the rate of 2% per month for 12 months. If the company presently employs 500 workers, to the nearest worker, how many workers will it employ 12 months from now?

*In Problems 37–40, use a calculator to find the value (rounded to four decimal places) of each expression.*

**37.** $e^{1.5}$          **38.** $e^{3.4}$.          **39.** $e^{-0.4}$          **40.** $e^{-3/4}$.

*In Problems 41 and 42, graph the functions.*

**41.** $y = -e^x$.          **42.** $y = 2e^x$.

**43. Telephone Calls**  The probability that a telephone operator will receive exactly $x$ calls during a certain period is given by

$$P = \frac{e^{-3}3^x}{x!}.$$

Find the probability that the operator will receive exactly three calls. Round your answer to four decimal places.

**44. Normal Distribution**  An important function used in economic and business decisions is the *normal distribution density function,* which, in standard form, is

$$f(x) = \frac{1}{\sqrt{2\pi}}e^{-\left(\frac{1}{2}\right)x^2}.$$

Evaluate $f(0), f(-1)$, and $f(1)$. Round your answers to three decimal places.

**45.** Express $e^{kt}$ in the form $b^t$.

**46.** Express $\frac{1}{e^x}$ in the form $b^x$.

**47. Radioactive Decay**  A radioactive element is such that $N$ grams remain after $t$ hours, where

$$N = 10e^{-0.028t}.$$

(a) How many grams are initially present? To the nearest tenth of a gram, how many grams remain after (b) 10 hours? (c) 50 hours? (d) Based on your answer to part (c), what is your estimate of the half-life of this element?

**48. Radioactive Decay** At a certain time, there are 100 milligrams of a radioactive substance. The substance decays so that after $t$ years the number of milligrams present, $N$, is given by

$$N = 100e^{-0.035t}.$$

How many milligrams are present after 20 years? Give your answer to the nearest milligram.

**49. Radioactive Decay** If a radioactive substance has a half-life of 8 years, how long does it take for 1 gram of the substance to decay to $\frac{1}{16}$ gram?

**50. Marketing** A mail-order company advertises in a national magazine. The company finds that, of all small towns, the percentage (given as a decimal) in which exactly $x$ people respond to an ad fits a Poisson distribution with $\mu = 0.5$. From what percentage of small towns can the company expect exactly two people to respond? Round your answer to four decimal places.

**51. Emergency-Room Admissions** Suppose the number of patients admitted into a hospital emergency room during a certain hour of the day has a Poisson distribution with mean 4. Find the probability that during that hour there will be exactly two emergency patients. Round your answer to four decimal places.

**52.** Graph $y = 10^x$ and $y = (\frac{1}{10})^x$ on the same screen. Determine the intersection point.

**53.** Graph $y = 2^x$ and $y = 4 \cdot 2^x$ on the same screen. It appears that the graph of $y = 4 \cdot 2^x$ is the graph of $y = 2^x$ shifted two units to the left. Prove algebraically that this is indeed true.

**54.** For $y = 7^x$, find $x$ if $y = 4$. Round your answer to two decimal places.

**55.** For $y = 2^x$, find $x$ if $y = 3$. Round your answer to two decimal places.

**56. Cell Growth** Cells are growing in a culture, and their number is increasing at the rate of 7% per hour. Initially, 1000 cells are present. After how many full hours will there be at least 3000 cells?

**57. Bacteria Growth** Refer to Example 1. How long will it take for 1000 bacteria to be present? Round your answer to the nearest tenth of a minute.

**58. Demand Equation** The demand equation for a new toy is

$$q = 10,000(0.95123)^p.$$

**a.** Evaluate $q$ to the nearest integer when $p = 10$.

**b.** Convert the demand equation to the form

$$q = 10,000e^{-0.05p}.$$

[*Hint:* Find a number $x$ such that $0.95123 \approx e^{-x}$.]

**c.** Use the equation in part **b** to evaluate $q$ to the nearest integer when $p = 10$. Your answers in parts **a** and **c** should be the same.

**59. Investment** If $1000 is invested in a savings account that earns interest at the rate of 5% compounded annually, after how many full years will the amount at least double?

---

**To introduce logarithmic functions and their graphs. Properties of logarithms will be discussed in Sec. 5.3.**

## 5.2 LOGARITHMIC FUNCTIONS

In this section the functions of interest to us are *logarithmic functions,* which are related to exponential functions. Figure 5.16(a) shows the graph of the exponential function $s = f(t) = 2^t$. Here $f$ sends an input number $t$ into a *positive* output number $s$:

$$f: t \to s \quad \text{where} \quad s = 2^t.$$

For example, $f$ sends 2 into 4.

Looking at the same curve in Fig. 5.16(b), you can see from the small arrows that, with each positive number $s$ on the vertical axis, we can associate exactly one value of $t$. For example, with $s = 4$, we associate $t = 2$. By thinking of $s$ as an input and $t$ as an output, we have a function that sends $s$'s into $t$'s. We shall denote this function by $f^{-1}$ (read "$f$ inverse"):[5]

$$f^{-1}: s \to t \quad \text{where} \quad s = 2^t.$$

[5]The $-1$ in $f^{-1}$ is not an exponent, so $f^{-1}$ does *not* mean $\frac{1}{f}$.

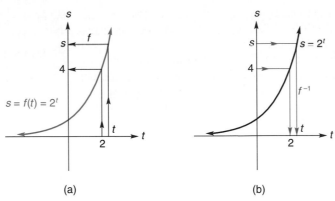

**FIGURE 5.16**   Graph of $s = 2^t$.

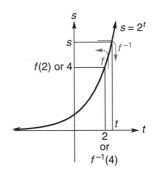

**FIGURE 5.17**   Actions of $f$ and $f^{-1}$.

Thus, $f^{-1}(s) = t$. The domain of $f^{-1}$ is the range of $f$ (all positive real numbers), and the range of $f^{-1}$ is the domain of $f$ (all real numbers).

The functions $f$ and $f^{-1}$ are related. Figure 5.17 shows that $f^{-1}$ *reverses* the action of $f$, and vice versa. For example.

$$f \text{ sends 2 into 4} \quad \text{and} \quad f^{-1} \text{ sends 4 into 2.}$$

More generally, $f(t) = s$ and $f^{-1}(s) = t$. In terms of composition, when either $f^{-1} \circ f$ or $f \circ f^{-1}$ is applied to an input number, that number is obtained for output because of the reversing effects of $f$ and $f^{-1}$. That is,

$$(f^{-1} \circ f)(t) = f^{-1}(f(t)) = f^{-1}(s) = t$$

and

$$(f \circ f^{-1})(s) = f(f^{-1}(s)) = f(t) = s.$$

We give a special name to $f^{-1}$: the **logarithmic function with base 2,** written as $\log_2$ [read "logarithm (or log) base 2"]. Thus, $f^{-1}(4) = \log_2 4 = 2$, and we say that the *logarithm* base 2 of 4 is 2.

In summary,

$$\text{if } s = 2^t, \quad \text{then} \quad t = \log_2 s. \tag{1}$$

We now generalize our discussion to other bases. Replacing 2 by $b$, $s$ by $x$, and $t$ by $y$ in Eq. (1) gives the following definition.

**DEFINITION**

*The **logarithmic function** with base $b$, where $b > 0$ and $b \neq 1$, is denoted by $\log_b$ and is defined by*

$$y = \log_b x \quad \text{if and only if} \quad b^y = x.$$

*The domain of $\log_b$ is all positive real numbers, and the range is all real numbers.*

Because a logarithmic function reverses the action of the corresponding exponential function and vice versa, each logarithmic function is called the *inverse* of its corresponding exponential function, and that exponential function is the inverse of its corresponding logarithmic function.

Remember, when we say that $y$ is the log base $b$ of $x$, we mean that $b$ raised to the $y$ power is $x$. That is,

$$y = \log_b x \quad \text{means} \quad b^y = x.$$

Logarithmic and
exponential forms

**FIGURE 5.18** A logarithm can be considered an exponent.

In this sense, *a logarithm of a number is an exponent:* $\log_b x$ is the power to which we must raise $b$ to get $x$. For example,

$$\log_2 8 = 3 \quad \text{because} \quad 2^3 = 8.$$

We say that $\log_2 8 = 3$ is the **logarithmic form** of the **exponential form** $2^3 = 8$. (See Fig. 5.18.)

**EXAMPLE 1** Converting from Exponential to Logarithmic Form

|   |       | *Exponential Form* |                 | *Logarithmic Form* |
|---|-------|--------------------|-----------------|--------------------|
| a. | Since | $5^2 = 25,$ | it follows that | $\log_5 25 = 2.$ |
| b. | Since | $3^4 = 81,$ | it follows that | $\log_3 81 = 4.$ |
| c. | Since | $10^0 = 1,$ | it follows that | $\log_{10} 1 = 0.$ |

■

---

**Principles in Practice 1**

**Converting from Exponential to Logarithmic Form**

If bacteria have been doubling every hour and the current amount is 16 times the amount first measured, then the situation can be represented by $16 = 2^t$. Represent this equation in logarithmic form. What does $t$ represent?

---

**EXAMPLE 2** Converting from Logarithmic to Exponential Form

|   | *Logarithmic Form* |       | *Exponential Form* |
|---|--------------------|-------|--------------------|
| a. | $\log_{10} 1000 = 3$ | means | $10^3 = 1000.$ |
| b. | $\log_{64} 8 = \dfrac{1}{2}$ | means | $64^{1/2} = 8.$ |
| c. | $\log_2 \dfrac{1}{16} = -4$ | means | $2^{-4} = \dfrac{1}{16}.$ |

■

---

**Principles in Practice 2**

**Converting from Logarithmic to Exponential Form**

An earthquake measuring 8.3 on the Richter scale can be represented by $8.3 = \log_{10}\left(\dfrac{I}{I_0}\right)$, where $I$ is the intensity of the earthquake and $I_0$ is the intensity of a zero-level earthquake. Represent this equation in exponential form.

---

**EXAMPLE 3** Graph of a Logarithmic Function with $b > 1$

*Graph the function $y = \log_2 x$.*

*Solution:* It can be awkward to substitute values of $x$ and then find corresponding values of $y$. For example, if $x = 3$, then $y = \log_2 3$, which is not easily determined. An easier way to plot points is to use the equivalent exponential form $x = 2^y$. We choose values of $y$ and find the corresponding values of $x$. For example, if $y = 0$, then $x = 1$. This gives the point $(1, 0)$. Other points are shown in Fig. 5.19.

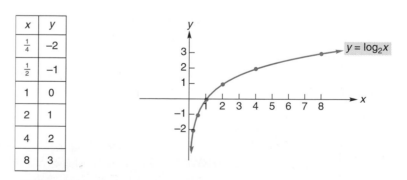

| $x$ | $y$ |
|-----|-----|
| $\frac{1}{4}$ | $-2$ |
| $\frac{1}{2}$ | $-1$ |
| $1$ | $0$ |
| $2$ | $1$ |
| $4$ | $2$ |
| $8$ | $3$ |

**FIGURE 5.19** Graph of $y = \log_2 x$.

From the graph, we can see that the domain is all positive real numbers. Thus, *negative numbers and 0 do not have logarithms.* The range is all real numbers. Notice that the graph rises from left to right. Numbers between 0 and

1 have negative logarithms, and the closer a number is to 0, the more negative is its logarithm. Numbers greater than 1 have positive logarithms. The log of 1 is 0, which corresponds to the $x$-intercept $(1, 0)$. There is no $y$-intercept. This graph is typical for a logarithmic function with $b > 1$. ∎

---

**Principles in Practice 3**

**Graph of a Logarithmic Function with $b > 1$**

Suppose a recycling plant has found that the amount of material being recycled has increased by 50% every year since the plant's first year of operation. Graph each year as a function of the multiplicative increase in recycling since the first year. Label the graph with the name of the function.

---

**EXAMPLE 4  Graph of a Logarithmic Function with $0 < b < 1$**

*Graph $y = \log_{1/2} x$.*

*Solution:* To plot points, we use the equivalent exponential form $x = \left(\frac{1}{2}\right)^y$. (See Fig. 5.20.)

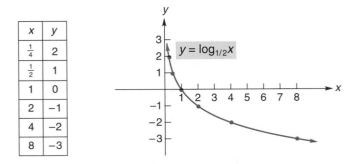

| $x$ | $y$ |
|-----|-----|
| $\frac{1}{4}$ | 2 |
| $\frac{1}{2}$ | 1 |
| 1 | 0 |
| 2 | −1 |
| 4 | −2 |
| 8 | −3 |

**FIGURE 5.20**  Graph of $y = \log_{1/2} x$.

From the graph, we can see that the domain is all positive reals and the range is all real numbers. The graph falls from left to right. Numbers between 0 and 1 have positive logarithms, and the closer a number is to 0, the larger is its logarithm. Numbers greater than 1 have negative logarithms. The logarithm of 1 is 0 and corresponds to the $x$-intercept $(1, 0)$. This graph is typical for a logarithmic function with $0 < b < 1$. ∎

Summarizing the results of Examples 3 and 4, we can say that the graph of a logarithmic function has one of two general shapes, depending on whether $b > 1$ or $0 < b < 1$. (See Fig. 5.21.) For $b > 1$, the graph rises from left to right; as $x$ gets closer and closer to 0, the function values decrease without

---

**Principles in Practice 4**

**Graph of a Logarithmic Function with $0 < b < 1$**

Suppose a boat depreciates 20% every year. Graph the number of years the boat is owned as a function of the multiplicative decrease in its original value. Label the graph with the name of the function.

---

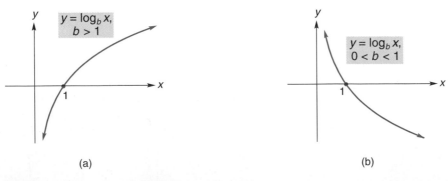

(a)  (b)

**FIGURE 5.21**  General shapes of $y = \log_b x$.

bound, and the graph gets closer and closer to the $y$-axis. For $0 < b < 1$, the graph falls from left to right; as $x$ gets closer and closer to 0, the function values increase without bound, and the graph gets closer and closer to the $y$-axis. In each case, note that:

1. The domain of a logarithmic function is the interval $(0, \infty)$. Thus, the logarithm of either a negative number or 0 does not exist.

2. The range is the interval $(-\infty, \infty)$.

3. The logarithm of 1 is 0, which corresponds to the $x$-intercept $(1, 0)$.

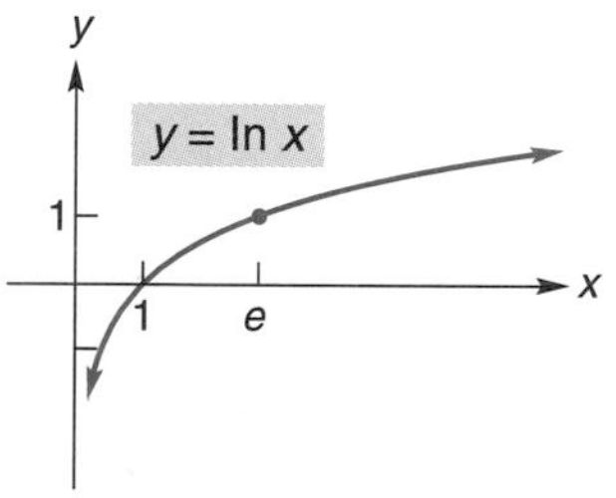

Logarithms to the base 10 are called **common logarithms.** They were frequently used for computational purposes before the calculator age. The subscript 10 is usually omitted from the notation:

$$\log x \quad \text{means} \quad \log_{10} x.$$

Important in calculus are logarithms to the base $e$, called **natural logarithms.** We use the notation "ln" for such logarithms:

$$\ln x \quad \text{means} \quad \log_e x.$$

The symbol $\ln x$ may be read "ell-en of $x$." Your calculator gives approximate values for natural and common logarithms. For example, verify that $\ln 2 \approx 0.69315$. This means that $e^{0.69315} \approx 2$. Figure 5.22 shows the graph of $y = \ln x$. Because $e > 1$, the graph has the general shape of that of a logarithmic function with $b > 1$ [see Fig. 5.21(a)] and rises from left to right.

$y = \ln x$

**FIGURE 5.22**  Graph of natural logarithmic function.

You should gain familiarity with the graph of the natural logarithmic function in Fig. 5.22

Remember: A logarithm (in a sense) is an exponent.

### Principles in Practice 5
#### Finding Logarithms

The number of years it takes for an amount invested at an annual rate of $r$ and compounded continuously to quadruple is a function of the annual rate $r$ given by $t(r) = \dfrac{\ln 4}{r}$. Use a calculator to find the rate needed to quadruple an investment in 10 years.

### EXAMPLE 5  Finding Logarithms

**a.** *Find* log 100.

**Solution:** Here the base is 10. Thus, log 100 is the power to which we must raise 10 to get 100. Since $10^2 = 100$, $\log 100 = 2$.

**b.** *Find* ln 1.

**Solution:** Here the base is $e$. Because $e^0 = 1$, $\ln 1 = 0$.

**c.** *Find* log 0.1.

**Solution:** Since $0.1 = \frac{1}{10} = 10^{-1}$, $\log 0.1 = -1$.

**d.** *Find* $\ln e^{-1}$.

**Solution:** Since $\ln e^{-1}$ is the power to which $e$ must be raised to obtain $e^{-1}$, clearly $\ln e^{-1} = -1$.

**e.** *Find* $\log_{36} 6$.

**Solution:** Because $36^{1/2}$ (or $\sqrt{36}$) is 6, $\log_{36} 6 = \frac{1}{2}$.  ∎

Many equations involving logarithmic or exponential forms can be solved for an unknown quantity by first transforming from logarithmic form to exponential form or vice versa. Example 6 will illustrate.

### EXAMPLE 6  Solving Logarithmic and Exponential Equations

**a.** *Solve* $\log_2 x = 4$.

**Solution:** We can get an explicit expression for $x$ by writing the equation in exponential form. This gives

$$2^4 = x,$$

so $x = 16$.

The multiplicative increase $m$ of an amount invested at an annual rate of $r$ compounded continuously for a time $t$ is given by $m = e^{rt}$. What annual percentage rate is needed to triple the investment in 12 years?

**b.** *Solve* $\ln(x + 1) = 7$.

*Solution:* The exponential form yields $e^7 = x + 1$. Thus, $x = e^7 - 1$.

**c.** *Solve* $\log_x 49 = 2$.

*Solution:* In exponential form, $x^2 = 49$, so $x = 7$. We reject $x = -7$ because a negative number cannot be a base of a logarithmic function.

**d.** *Solve* $e^{5x} = 4$.

*Solution:* We can get an explicit expression for $x$ by writing the equation in logarithmic form. We have

$$\ln 4 = 5x,$$

$$x = \frac{\ln 4}{5}.$$

**Radioactive Decay and Half-Life**

From our discussion of the decay of a radioactive element in Sec. 5.1, we know that the amount of the element present at time $t$ is given by

$$N = N_0 e^{-\lambda t}, \tag{2}$$

where $N_0$ is the initial amount (the amount at time $t = 0$) and $\lambda$ is the decay constant. Let us now determine the half-life $T$ of the element. At time $T$, half of the initial amount is present. That is, when $t = T$, $N = N_0/2$. Thus, from Eq. (2), we have

$$\frac{N_0}{2} = N_0 e^{-\lambda T}.$$

Solving for $T$ gives

$$\frac{1}{2} = e^{-\lambda T},$$

$$2 = e^{\lambda T} \qquad \text{(taking reciprocals of both sides)}.$$

To get an explicit expression for $T$, we shall convert to logarithmic form. This results in

$$\lambda T = \ln 2,$$

$$T = \frac{\ln 2}{\lambda}.$$

Summarizing, we have the following:

If a radioactive element has decay constant $\lambda$, then the half-life of the element is given by

$$T = \frac{\ln 2}{\lambda}. \tag{3}$$

**EXAMPLE 7   Finding Half-Life**

*A 10-milligram sample of radioactive polonium* 210 ($^{210}$Po) *decays according to the equation*

$$N = 10e^{-0.00501t},$$

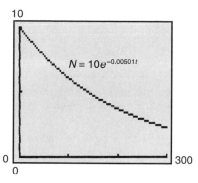

**FIGURE 5.23**   Radioactive decay function $N = 10e^{-0.00501t}$.

*where N is the number of milligrams present after t days. (See Fig. 5.23) Determine the half-life of* $^{210}$*Po.*

**Solution:** Here the decay constant $\lambda$ is 0.00501. By Eq. (3), the half-life is given by

$$T = \frac{\ln 2}{\lambda} = \frac{\ln 2}{0.00501} \approx 138.4 \text{ days.}$$

## ▪ Exercise 5.2

*In Problems 1–8, express each logarithmic form exponentially and each exponential form logarithmically.*

**1.** $10^4 = 10{,}000$.

**2.** $2 = \log_{12} 144$.

**3.** $\log_2 64 = 6$.

**4.** $8^{2/3} = 4$.

**5.** $e^2 = 7.3891$.

**6.** $e^{0.33647} = 1.4$.

**7.** $\ln 3 = 1.09861$.

**8.** $\log 5 = 0.6990$.

*In Problems 9–16, graph the functions.*

**9.** $y = f(x) = \log_3 x$.

**10.** $y = f(x) = \log_4 x$.

**11.** $y = f(x) = \log_{1/4} x$.

**12.** $y = f(x) = \log_{1/3} x$.

**13.** $y = f(x) = \log_2(x - 4)$.

**14.** $y = f(x) = \log_2(-x)$.

**15.** $y = f(x) = -2 \ln x$.

**16.** $y = f(x) = \ln(x + 2)$.

*In Problems 17–28, evaluate the expression.*

**17.** $\log_6 36$.

**18.** $\log_2 32$.

**19.** $\log_3 27$.

**20.** $\log_{16} 4$.

**21.** $\log_7 7$.

**22.** $\log 10{,}000$.

**23.** $\log 0.01$.

**24.** $\log_2 \sqrt{2}$.

**25.** $\log_5 1$.

**26.** $\log_5 \frac{1}{25}$.

**27.** $\log_2 \frac{1}{8}$.

**28.** $\log_4 \sqrt[5]{4}$.

*In Problems 29–48, find x.*

**29.** $\log_3 x = 2$.

**30.** $\log_2 x = 4$.

**31.** $\log_5 x = 3$.

**32.** $\log_4 x = 0$.

**33.** $\log x = -1$.

**34.** $\ln x = 1$.

**35.** $\ln x = 2$.

**36.** $\log_x 100 = 2$.

**37.** $\log_x 8 = 3$.

**38.** $\log_x 3 = \frac{1}{2}$.

**39.** $\log_x \frac{1}{6} = -1$.

**40.** $\log_x y = 1$.

**41.** $\log_3 x = -4$.

**42.** $\log_x(2x - 3) = 1$.

**43.** $\log_x(6 - x) = 2$.

**44.** $\log_8 64 = x - 1$.

**45.** $2 + \log_2 4 = 3x - 1$.

**46.** $\log_3(x + 2) = -2$.

**47.** $\log_x(2x + 8) = 2$.

**48.** $\log_x(30 - 4x - x^2) = 2$.

*In Problems 49–52, find x and express your answer in terms of natural logarithms.*

**49.** $e^{3x} = 2$.

**50.** $0.1e^{0.1x} = 0.5$.

**51.** $e^{2x-5} + 1 = 4$.

**52.** $3e^{2x} - 1 = \frac{1}{2}$.

*In Problems 53–56, use your calculator to find the approximate value of each expression. Round your answer to five decimal places.*

**53.** $\ln 5$.

**54.** $\ln 3.12$.

**55.** $\ln 7.39$.

**56.** $\ln 9.98$.

**57. Chemistry**   If the pH of a substance is 5.5, then the concentration of hydrogen ions, $h$, in gram atoms per liter can be represented by $5.5 = \log \frac{1}{h}$. Represent this equation in exponential form.

**58. Depreciation**   Suppose a car depreciates 25% every year. Graph the number of years it is owned as a function of the multiplicative decrease in its original value. Label the graph with the name of the function.

**59. Cost Equation**   The cost for a firm producing $q$ units of a product is given by the cost equation

$$c = (2q \ln q) + 20.$$

Evaluate the cost when $q = 6$. (Round your answer to two decimal places.)

**60. Supply Equation**   A manufacturer's supply equation is

$$p = \log\left(10 + \frac{q}{2}\right),$$

where $q$ is the number of units supplied at a price $p$ per unit. At what price will the manufacturer supply 1980 units?

**61. Earthquake** The magnitude $M$ of an earthquake and its energy $E$ are related by the equation[6]

$$1.5M = \log\left(\frac{E}{2.5 \times 10^{11}}\right).$$

where $M$ is given in terms of Richter's preferred scale of 1958 and $E$ is in ergs. Solve the equation for $E$.

**62. Biology** For a certain population of cells, the number of cells at time $t$ is given by $N = N_0(2^{t/k})$, where $N_0$ is the number of cells at $t = 0$ and $k$ is a positive constant. (a) Find $N$ when $t = k$. (b) What is the significance of $k$? (c) Show that the time it takes to have population $N_1$ can be written

$$t = k \log_2 \frac{N_1}{N_0}.$$

**63. Earth Science** Atmospheric pressure $p$ varies with the altitude $h$ above the earth's surface. For altitudes up to about 10 kilometers, the pressure (in millimeters of mercury) is given approximately by

$$p = 760e^{-0.125h},$$

where $h$ is in kilometers. (a) Find $p$ at an altitude of 7.3 km. (b) At what altitude will the pressure be 400 mm of mercury?

**64. Work** The work, in joules, done by a 1-kilogram sample of nitrogen gas as its volume changes from an initial value $V_i$ to a final value $V_f$ during a constant-temperature process is given by

$$W = 8.1 \times 10^4 \ln \frac{V_f}{V_i}.$$

If such a sample expands from a volume of 3 liters to a volume of 7 liters, determine the work done by the gas to the nearest hundred joules.

**65. Inferior Good** In a discussion of an inferior good, Persky[7] solves an equation of the form

$$u_0 = A \ln(x_1) + \frac{x_2^2}{2}$$

for $x_1$, where $x_1$ and $x_2$ are quantities of two products, $u_0$ is a measure of utility, and $A$ is a positive constant. Determine $x_1$.

**66. Radioactive Decay** A 1-gram sample of radioactive lead 211 ($^{211}$Pb) decays according to the equation $N = e^{-0.01920t}$, where $N$ is the number of grams present after $t$ minutes. Find the half-life of $^{211}$Pb to the nearest tenth of a minute.

**67. Radioactive Decay** A 100-milligram sample of radioactive actinium 227 ($^{227}$Ac) decays according to the equation

$$N = 100e^{-0.03194t},$$

where $N$ is the number of milligrams present after $t$ years. Find the half-life of $^{227}Ac$ to the nearest tenth of a year.

**68.** If $\log_y x = 3$ and $\log_z x = 2$, find a formula for $z$ as an explicit function of $y$ only.

**69.** Solve for $y$ as an explicit function of $x$ if

$$x + 2e^{3y} - 10 = 0.$$

**70.** Suppose $y = f(x) = x \ln x$. (a) For what values of $x$ is $y < 0$? [*Hint:* Determine when the graph is below the $x$-axis.] (b) Determine the range of $f$.

**71.** Find the $x$-intercept of $y = x^2 \ln x$.

**72.** Use the graph of $y = e^x$ to estimate $\ln 2$. Round your answer to two decimal places.

**73.** Use the graph of $y = \ln x$ to estimate $e^2$. Round your answer to two decimal places.

**74.** Determine the $x$-values of points of intersection of the graphs of $y = (x - 2)^2$ and $y = \ln x$. Round your answers to two decimal places.

[6]K. E. Bullen, *An Introduction to the Theory of Seismology* (Cambridge, U.K.: Cambridge at the University Press, 1963).

[7]A. L. Persky, "An Inferior Good and a Novel Indifference Map," *The American Economist,* XXIX, no. 1 (Spring 1985).

---

**OBJECTIVE**

To study basic properties of logarithmic functions.

## 5.3 PROPERTIES OF LOGARITHMS

The logarithmic function has many important properties. For example, the logarithm of a product of two numbers is the sum of the logarithms of the numbers. Symbolically, we have $\log_b(mn) = \log_b m + \log_b n$. To prove this, we let $x = \log_b m$ and $y = \log_b n$. Then $b^x = m, b^y = n$, and

$$mn = b^x b^y = b^{x+y}.$$

Thus, $mn = b^{x+y}$. In logarithmic form, this means that $\log_b(mn) = x + y$. Therefore, $\log_b(mn) = \log_b m + \log_b n$.

**1.** $\log_b(mn) = \log_b m + \log_b n$.

*That is, the logarithm of a product is the sum of the logarithms of the factors.*

We shall not prove the next two properties, since their proofs are similar to that of Property 1.

**2.** $\log_b \dfrac{m}{n} = \log_b m - \log_b n$.

*That is, the logarithm of a quotient is the difference of the logarithm of the numerator and the logarithm of the denominator.*

**3.** $\log_b m^r = r \log_b m$.

*That is, the logarithm of a power of a number is the exponent times the logarithm of the number.*

*Pitfall* ▼ Make sure that you clearly understand properties 1–3. They do not apply to the log of a sum $[\log_b(m + n)]$, to the log of a difference $[\log_b(m - n)]$, or to the quotient of two logs $\left[\dfrac{\log_b m}{\log_b n}\right]$. For example,

$$\log_b(m + n) \neq \log_b m + \log_b n,$$

$$\log_b(m - n) \neq \log_b m - \log_b n,$$

$$\frac{\log_b m}{\log_b n} \neq \log_b(m - n),$$

and

$$\frac{\log_b m}{\log_b n} \neq \log_b\left(\frac{m}{n}\right).$$

Table 5.2 gives the values of a few common logarithms. Most entries are approximate. For example, $\log 4 \approx 0.6021$, which means $10^{0.6021} \approx 4$. To illustrate the use of properties of logarithms, we shall use this table in some of the examples that follow.

**TABLE 5.2**  Common Logarithms

| $x$ | $\log x$ | $x$ | $\log x$ |
|---|---|---|---|
| 2 | 0.3010 | 7 | 0.8451 |
| 3 | 0.4771 | 8 | 0.9031 |
| 4 | 0.6021 | 9 | 0.9542 |
| 5 | 0.6990 | 10 | 1.0000 |
| 6 | 0.7782 | $e$ | 0.4343 |

Although the logarithms in Example 1 can be found with a calculator, we shall make use of properties of logarithms.

**EXAMPLE 1   Finding Logarithms by Using Table 5.2**

a. *Find* log 56.

*Solution:* Log 56 is not in the table. But we can write 56 as the product $8 \cdot 7$. Thus, by Property 1,

$$\log 56 = \log(8 \cdot 7) = \log 8 + \log 7 \approx 0.9031 + 0.8451 = 1.7482.$$

b. *Find* $\log \frac{9}{2}$.

*Solution:* By Property 2,

$$\log \tfrac{9}{2} = \log 9 - \log 2 \approx 0.9542 - 0.3010 = 0.6532.$$

c. *Find* log 64.

*Solution:* Since $64 = 8^2$, by Property 3.

$$\log 64 = \log 8^2 = 2 \log 8 \approx 2(0.9031) = 1.8062.$$

d. *Find* $\log \sqrt{5}$.

*Solution:* By Property 3, we have

$$\log \sqrt{5} = \log 5^{1/2} = \tfrac{1}{2} \log 5 \approx \tfrac{1}{2}(0.6990) = 0.3495.$$

e. *Find* $\log \dfrac{16}{21}$.

*Solution:*

$$\log \frac{16}{21} = \log 16 - \log 21 = \log(4^2) - \log(3 \cdot 7)$$
$$= 2 \log 4 - [\log 3 + \log 7]$$
$$\approx 2(0.6021) - [0.4771 + 0.8451] = -0.1180.$$

Note the use of brackets in the second line. It is wrong to write $2 \log 4 - \log 3 + \log 7$. ∎

**EXAMPLE 2   Rewriting Logarithmic Expressions**

a. *Express* $\log \dfrac{1}{x^2}$ *in terms of* log $x$.

*Solution:*

$$\log \frac{1}{x^2} = \log x^{-2} = -2 \log x \qquad \text{(Property 3)}.$$

Here we have assumed that $x > 0$. Although $\log(1/x^2)$ is defined for $x \neq 0$, the expression $-2 \log x$ is defined only if $x > 0$.

b. *Express* $\log \dfrac{1}{x}$ *in terms of* log $x$.

*Solution:* By Property 3,

$$\log \frac{1}{x} = \log x^{-1} = -1 \log x = -\log x. \qquad ∎$$

From Example 2(b), we see that $\log(1/x) = -\log x$. Generalizing gives the following property:

> **4.** $\log_b \dfrac{1}{m} = -\log_b m.$
>
> *That is, the logarithm of the reciprocal of a number is minus the logarithm of the number.*

For example, $\log \dfrac{2}{3} = -\log \dfrac{3}{2}.$

Manipulations such as those in Example 3 are frequently used in calculus.

### EXAMPLE 3   Writing Logarithms in Terms of Simpler Logarithms

**a.** *Write* $\ln \dfrac{x}{zw}$ *in terms of* $\ln x, \ln z,$ *and* $\ln w.$

*Solution:*

$$\ln \frac{x}{zw} = \ln x - \ln(zw) \qquad \text{(Property 2)}$$

$$= \ln x - (\ln z + \ln w) \qquad \text{(Property 1)}$$

$$= \ln x - \ln z - \ln w.$$

**b.** *Write* $\ln \sqrt[3]{\dfrac{x^5(x-2)^8}{x-3}}$ *in terms of* $\ln x, \ln(x-2),$ *and* $\ln(x-3).$

*Solution:*

$$\ln \sqrt[3]{\frac{x^5(x-2)^8}{x-3}} = \ln\left[\frac{x^5(x-2)^8}{x-3}\right]^{1/3} = \frac{1}{3}\ln\frac{x^5(x-2)^8}{x-3}$$

$$= \frac{1}{3}\{\ln[x^5(x-2)^8] - \ln(x-3)\}$$

$$= \frac{1}{3}[\ln x^5 + \ln(x-2)^8 - \ln(x-3)]$$

$$= \frac{1}{3}[5\ln x + 8\ln(x-2) - \ln(x-3)]. \qquad \blacksquare$$

### Principles in Practice 1
#### Combining Logarithms

The Richter scale measure of an earthquake is given by $R = \log\left(\dfrac{I}{I_0}\right)$, where $I$ is the intensi-ty of the earthquake and $I_0$, is the intensity of a zero-level earthquake. How much more on the Richter scale is an earthquake with intensity 900,000 times the intensity of a zero-level earthquake than an earthquake with intensity 9000 times the intensity of a zero-level earthquake? Write the answer as an expression involving logarithms. Simplify the expression by combining logarithms, and then to evaluate the resulting expression.

### EXAMPLE 4   Combining Logarithms

**a.** *Write* $\ln x - \ln(x + 3)$ *as a single logarithm.*

*Solution:*

$$\ln x - \ln(x+3) = \ln\frac{x}{x+3} \qquad \text{(Property 2).}$$

**b.** *Write* $\ln 3 + \ln 7 - \ln 2 - 2\ln 4$ *as a single logarithm.*

*Solution:*

$$\ln 3 + \ln 7 - \ln 2 - 2\ln 4$$

$$= \ln 3 + \ln 7 - \ln 2 - \ln(4^2) \qquad \text{(Property 3)}$$

$$= \ln 3 + \ln 7 - [\ln 2 + \ln(4^2)]$$

$$= \ln(3 \cdot 7) - \ln(2 \cdot 4^2) \qquad \text{(Property 1)}$$

$$= \ln 21 - \ln 32$$

$$= \ln \frac{21}{23} \qquad \text{(Property 2).} \qquad \blacksquare$$

Since $b^0 = 1$ and $b^1 = b$, by converting to logarithmic forms we have the following properties:

> **5.** $\log_b 1 = 0$.
>
> **6.** $\log_b b = 1$.

By Property 3, $\log_b b^r = r \log_b b$. But by Property 6, $\log_b b = 1$. Thus, we have the next property:

> **7.** $\log_b b^r = r$.

**Principles in Practice 2**
**Simplifying Logarithmic Expressions**

If an earthquake is 10,000 times as intense as a zero-level earthquake, what is its measurement on the Richter scale? Write the answer as a logarithmic expression and simplify it. (See previous page for formula.)

**EXAMPLE 5    Simplifying Logarithmic Expressions**

**a.** *Find* $\ln e^{3x}$.

*Solution:* By Property 7 with $b = e$, we have $\ln e^{3x} = 3x$. Alternatively, by Properties 3 and 6,

$$\ln e^{3x} = 3x \ln e = 3x(1) = 3x.$$

**b.** *Find* $\log 1 + \log 1000$.

*Solution:* By Property 5, $\log 1 = 0$. Thus,

$$\log 1 + \log 1000 = 0 + \log 10^3$$

$$= 0 + 3 \qquad \text{(Property 7 with } b = 10)$$

$$= 3.$$

**c.** *Find* $\log_7 \sqrt[9]{7^8}$.

*Solution:*

$$\log_7 \sqrt[9]{7^8} = \log_7 7^{8/9} = \tfrac{8}{9}.$$

**d.** *Find* $\log_3 \left( \dfrac{27}{81} \right)$.

*Solution:*

$$\log_3 \left( \frac{27}{81} \right) = \log_3 \left( \frac{3^3}{3^4} \right) = \log_3(3^{-1}) = -1.$$

**e.** *Find* $\ln e + \log \frac{1}{10}$.

*Solution:*

$$\ln e + \log \tfrac{1}{10} = \ln e + \log 10^{-1}$$

$$= 1 + (-1) = 0. \qquad \blacksquare$$

Do not confuse $\ln x^2$ with $(\ln x)^2$. We have

$$\ln x^2 = \ln(x \cdot x),$$

but

$$(\ln x)^2 = (\ln x)(\ln x),$$

which can be written as $\ln^2 x$. That is, in $\ln x^2$ we square $x$; in $(\ln x)^2$, or $\ln^2 x$, we square $\ln x$.

Our next property is the following:

**8.** $b^{\log_b m} = m$

and, in particular,

$$10^{\log x} = x \qquad \text{and} \qquad e^{\ln x} = x.$$

Property 8 is true because it states, in logarithmic form, that $\log_b m = \log_b m$.

**EXAMPLE 6** Use of Property 8

**a.** *Find* $e^{\ln x^2}$.

*Solution:* By Property 8, $e^{\ln x^2} = x^2$.

**b.** *Solve* $10^{\log x^2} = 25$ *for* $x$.

*Solution:*

$$10^{\log x^2} = 25.$$
$$x^2 = 25 \qquad \text{(Property 8)},$$
$$x = \pm 5.$$

**EXAMPLE 7** Evaluating a Logarithm Base 5

*Use a calculator to find* $\log_5 2$.

*Solution:* Calculators typically have keys for logarithms in base 10 and base $e$, but not for base 5. However, we can convert logarithms in one base to logarithms in another base. Let us convert from base 5 to base 10. First, let $x = \log_5 2$. Then $5^x = 2$. Taking the common logarithms of both sides of $5^x = 2$ gives

$$\log 5^x = \log 2,$$
$$x \log 5 = \log 2,$$
$$x = \frac{\log 2}{\log 5} \approx 0.4307.$$

If we had taken natural logarithms of both sides, the result would be $x = (\ln 2)/(\ln 5) \approx 0.4307$, the same as before.

Generalizing the method used in Example 7, we obtain the so-called *change-of-base* formula:

**Change-of-Base Formula**

**9.** $\log_b m = \dfrac{\log_a m}{\log_a b}.$

The change-of-base formula allows logarithms to be converted from one base $(b)$ to another $(a)$.

**EXAMPLE 8** Change-of-Base Formula

*Express* log *x* *in terms of natural logarithms.*

*Solution:* We must transform from base 10 into base *e*. Thus, we use the change-of-base formula (Property 9) with $b = 10$, $m = x$, and $a = e$:

$$\log x = \log_{10} x = \frac{\log_e x}{\log_e 10} = \frac{\ln x}{\ln 10}.$$ ∎

## TECHNOLOGY

*Problem:* Display the graph of $y = \log_2 x$.

*Solution:* To enter the function, we must first convert it to base *e* or base 10. We shall choose base *e*. By Property 9,

$$y = \log_2 x = \frac{\log_e x}{\log_e 2} = \frac{\ln x}{\ln 2}.$$

Now we graph $y = (\ln x)/(\ln 2)$, which is shown in Fig. 5.24.

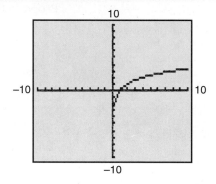

**FIGURE 5.24** Graph of $y = \log_2 x$.

## ■ Exercise 5.3

*In Problems 1–10, let* $\log 2 = a$, $\log 3 = b$, *and* $\log 5 = c$. *Express the indicated logarithm in terms of a, b, or c.*

**1.** $\log 15$.     **2.** $\log 16$.     **3.** $\log \frac{2}{3}$.     **4.** $\log \frac{5}{2}$.

**5.** $\log \frac{8}{3}$.     **6.** $\log \frac{3}{10}$.     **7.** $\log 36$.     **8.** $\log 0.0002$.

**9.** $\log_2 3$.     **10.** $\log_3 5$.

*In Problems 11–20, determine the value of the expression without the use of a calculator.*

**11.** $\log_7 7^{48}$.     **12.** $\log_5 (5\sqrt{5})^5$.     **13.** $\log 0.0001$.     **14.** $10^{\log 3.4}$.

**15.** $\ln e^{5.01}$.     **16.** $\ln e$.     **17.** $\ln \frac{1}{e}$.     **18.** $\log_5 25$.

**19.** $\log 10 + \ln e^3$.     **20.** $e^{\ln 6}$.

*In Problems 21–32, write the expression in terms of* $\ln x$, $\ln(x + 1)$, *and/or* $\ln(x + 2)$.

**21.** $\ln[x(x + 1)^2]$.     **22.** $\ln \frac{\sqrt{x}}{x + 1}$.     **23.** $\ln \frac{x^2}{(x + 1)^3}$.     **24.** $\ln[x(x + 1)]^3$.

**25.** $\ln \left( \frac{x}{x + 1} \right)^3$.     **26.** $\ln \sqrt{x(x + 1)}$.     **27.** $\ln \frac{x}{(x + 1)(x + 2)}$.     **28.** $\ln \frac{x^2(x + 1)}{x + 2}$.

**29.** $\ln \frac{\sqrt{x}}{(x + 1)^2(x + 2)^3}$.     **30.** $\ln \frac{1}{x(x + 1)(x + 2)}$.     **31.** $\ln \left[ \frac{1}{x + 2} \sqrt[5]{\frac{x^2}{x + 1}} \right]$.     **32.** $\ln \sqrt{\frac{x^4(x + 1)^3}{x + 2}}$.

*In Problems 33–40, express each of the given forms as a single logarithm.*

**33.** $\log 7 + \log 4$.     **34.** $\log_3 10 - \log_3 5$.     **35.** $\log_2(2x) - \log_2(x + 1)$.

**36.** $2 \log x - \frac{1}{2} \log(x - 2)$.     **37.** $9 \log 7 + 5 \log 23$.     **38.** $3(\log x + \log y - 2 \log z)$.

**39.** $2 + 10 \log 1.05$.     **40.** $\frac{1}{2}(\log 215 + 8 \log 6 - 3 \log 121)$.

*In Problems 41–44, determine the values of the expressions without using a calculator.*

**41.** $e^{4 \ln 3 - 3 \ln 4}$.     **42.** $\log_2 [\ln(\sqrt{7 + e^2} + \sqrt{7}) + \ln(\sqrt{7 + e^2} - \sqrt{7})]$.

**43.** $\log_6 54 - \log_6 9$.     **44.** $\log_2 \sqrt{2} + \log_3 \sqrt[3]{3} - \log_4 \sqrt[4]{4}$.

*In Problems 45–48, find x.*

**45.** $e^{\ln(2x)} = 5$.

**46.** $4^{\log_4 x + \log_4 2} = 3$.

**47.** $10^{\log x^2} = 4$.

**48.** $e^{3 \ln x} = 8$.

*In Problems 49–52, write each expression in terms of natural logarithms.*

**49.** $\log(x + 8)$.

**50.** $\log_2 x$.

**51.** $\log_3(x^2 + 1)$.

**52.** $\log_5 (9 - x^2)$.

**53.** If $e^{\ln z} = 7e^y$, solve for $y$ in terms of $z$.

**54. Statistics**   In statistics, the sample regression equation $y = ab^x$ is reduced to a linear form by taking logarithms of both sides. Express $\log y$ in terms of $x, \log a$, and $\log b$.

**55. Military Compensation**   In a study of military enlistments, Brown[8] considers total military compensation $C$ as the sum of basic military compensation $B$ (which includes the value of allowances, tax advantages, and base pay) and educational benefits $E$. Thus, $C = B + E$. Brown states that

$$\ln C = \ln B + \ln\left(1 + \frac{E}{B}\right).$$

Verify this.

**56. Sound Intensity**   The intensity level of a sound wave of intensity $I$ is given by

$$\beta = 10 \log \frac{I}{I_0},$$

where $\beta$ is the Greek letter "beta" and $I_0$ is a standard reference intensity taken to be $10^{-12}$, which approximately corresponds to the faintest sound a person can hear. The intensity level is measured in decibels (db). For example, the intensity level of an ordinary conversation is 40 db, and that for a subway train is 100 db. Find the intensity level of the sound of rustling leaves, which has an intensity of $10^{-11}$.

**57. Earthquake**   According to Richter,[9] the magnitude $M$ of an earthquake occurring 100 km from a certain type of seismometer is given by $M = \log(A) + 3$, where $A$ is the recorded trace amplitude (in millimeters) of the quake. (a) Find the magnitude of an earthquake that records a trace amplitude of 1 mm. (b) If a particular earthquake has amplitude $A_1$ and magnitude $M_1$, determine the magnitude of a quake with amplitude $100A_1$. Express your answer to (b) in terms of $M_1$.

[8]C. Brown, "Military Enlistments: What Can We Learn from Geographic Variation?" *The American Economic Review,* 75, no. 1 (1985), 228–34.

[9]C. F. Richter, *Elementary Seismology* (San Francisco: W. H. Freeman and Company, Publisher, 1958).

**Chemistry**   *A chemist can determine the acidity or basicity of an aqueous solution at room temperature by finding the* pH *of the solution. To do this, the chemist can first find the hydrogen-ion concentration (in moles per liter). The symbol* $[H^+]$ *stands for this concentration. The* pH *is then given by*

$$pH = -\log[H^+].$$

*If* pH $< 7$, *the solution is acidic. If* pH $= 7$, *we say that the solution is neutral. Use this information in Problems* **58** *and* **59.**

**58.** A cleaning solution has a pH of 8. What is $[H^+]$ of this solution?

**59.** What is the pH of vinegar with $[H^+]$ equal to $3 \times 10^{-4}$?

**Chemistry**   *For an aqueous solution at room temperature, the product of the hydrogen-ion concentration,* $[H^+]$, *and the hydroxide-ion concentration,* $[OH^-]$, *is* $10^{-14}$ *(where the concentrations are in moles per liter).*

$$[H^+][OH^-] = 10^{-14}.$$

*In Problems* **60** *and* **61,** *find the pH of a solution (see explanation just before Problem 58) with the given* $[OH^-]$.

**60.** $[OH^-] = 10^{-4}$.

**61.** $[OH^-] = 3 \times 10^{-2}$.

**62.**  Display the graph of $y = \log_6 x$.

**63.**  Display the graph of $y = \log_4(x + 2)$.

**64.**  Display the graphs of $y = \log x$ and $y = \dfrac{\ln x}{\ln 10}$ on the same screen. The graphs appear to be identical. Why?

**65.**  On the same screen, display the graphs of $y = \ln x$ and $y = \ln(4x)$. It appears that the graph of $y = \ln(4x)$ is the graph of $y = \ln x$ shifted upward. Determine algebraically the value of this shift.

**66.** On the same screen, display the graphs of $y = \ln x$ and $y = \ln(x/3)$. It appears that the graph of $y = \ln(x/3)$ is the graph of $y = \ln x$ shifted downward. Determine algebraically the value of this shift.

---

| OBJECTIVE | |
|---|---|
| **To develop techniques for solving logarithmic and exponential equations.** | **5.4  LOGARITHMIC AND EXPONENTIAL EQUATIONS** |

Here we shall solve *logarithmic* and *exponential equations.* A **logarithmic equation** is an equation that involves the logarithm of an expression containing an unknown. For example, $2 \ln(x + 4) = 5$ is a logarithmic equation. On

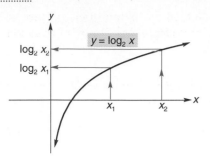

**FIGURE 5.25** If $x_1 \neq x_2$, then $\log_2 x_1 \neq \log_2 x_2$.

the other hand, an **exponential equation** has the unknown appearing in an exponent, as in $2^{3x} = 7$.

To solve some logarithmic equations, we use a property of logarithms that we shall now develop.

For many functions $f$, if $f(m) = f(n)$, this does not imply that $m = n$. For example, if $f(x) = x^2$ then $f(2) = f(-2)$, but $2 \neq -2$. This is not the case for the logarithmic function. You can see in Fig. 5.25 that the graph of $y = \log_2 x$ rises from left to right. Thus, if $x_1$ and $x_2$ are different, then their logarithms ($y$-values) are different. This means that if $\log_2 m = \log_2 n$, then $m = n$. Generalizing to base $b$, we have the following property:

$$\text{If } \log_b m = \log_b n, \text{ then } m = n.$$

There is a similar property for exponentials:

$$\text{If } b^m = b^n, \text{ then } m = n.$$

### EXAMPLE 1 Oxygen Composition

*An experiment was conducted with a particular type of small animal.*[10] *The logarithm of the amount of oxygen consumed per hour was determined for a number of the animals and was plotted against the logarithms of the weights of the animals. It was found that*

$$\log y = \log 5.934 + 0.885 \log x,$$

*where $y$ is the number of microliters of oxygen consumed per hour and $x$ is the weight of the animal (in grams). Solve for $y$.*

*Solution:* We first combine the terms on the right side into a single logarithm:

$$\log y = \log 5.934 + 0.885 \log x$$
$$= \log 5.934 + \log x^{0.885} \qquad \text{(Property 3 of Sec. 5.3)}.$$
$$\log y = \log(5.934 x^{0.885}) \qquad \text{(Property 1 of Sec. 5.3)}.$$

By the property of equality of logarithms, we have

$$y = 5.934 x^{0.885}. \qquad ■$$

### Principles in Practice 1
### Solving an Exponential Equation

Greg took a number and multiplied it by a power of 32. Jean started with the same number and got the same result when she multiplied it by 4 raised to a number that was nine less than three times the exponent that Greg used. What power of 32 did Greg use?

### EXAMPLE 2 Solving an Exponential Equation

*Find $x$ if $(25)^{x+2} = 5^{3x-4}$.*

*Solution:* Since $25 = 5^2$, we can express both sides of the equation as powers of 5:

$$(25)^{x+2} = 5^{3x-4},$$
$$(5^2)^{x+2} = 5^{3x-4},$$
$$5^{2x+4} = 5^{3x-4}.$$

By the property of equality of exponentials,

$$2x + 4 = 3x - 4,$$
$$x = 8. \qquad ■$$

[10]R. W. Poole, *An Introduction to Quantitative Ecology* (New York: McGraw-Hill Book Company, 1974).

Some exponential equations can be solved by taking the logarithm of both sides after the equation is put in a desirable form. The following example illustrates.

**Principles in Practice 2**

**Using Logarithms to Solve an Exponential Equation**

The sales manager at a fast-food chain finds that breakfast sales begin to fall after the end of a promotional campaign. The sales in dollars as a function of the number of days $d$ after the campaign's end is given by $S = 800\left(\dfrac{4}{3}\right)^{-0.1d}$. If the manager does not want sales to drop below 450 per day before starting a new campaign, when should he start such a campaign?

**EXAMPLE 3**  **Using Logarithms to Solve an Exponential Equation**

*Solve* $5 + (3)4^{x-1} = 12$.

*Solution:* We first isolate the exponential expression $4^{x-1}$ on one side of the equation:

$$5 + (3)4^{x-1} = 12,$$
$$(3)4^{x-1} = 7,$$
$$4^{x-1} = \frac{7}{3}.$$

Now we take the natural logarithm of both sides:

$$\ln 4^{x-1} = \ln \frac{7}{3}.$$

Simplifying gives

$$(x-1)\ln 4 = \ln \frac{7}{3},$$
$$x - 1 = \frac{\ln \frac{7}{3}}{\ln 4},$$
$$x = \frac{\ln \frac{7}{3}}{\ln 4} + 1 \approx 1.61120.$$

In Example 3, we used natural logarithms to solve the given equation. However, logarithms in any base can be employed. Generally, natural or common logarithms are used if a decimal form of the solution is desired. If we used common logarithms, we would obtain

$$x = \frac{\log \frac{7}{3}}{\log 4} + 1 \approx 1.61120.$$

---

**TECHNOLOGY**

Figure 5.26 shows a graphical solution of the equation $5 + (3)4^{x-1} = 12$ of Example 3. This solution occurs at the intersection of the graphs of $y = 5 + (3)4^{x-1}$ and $y = 12$.

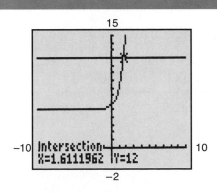

**FIGURE 5.26**  The solution of $5 + (3)4^{x-1} = 12$ is approximately 1.61120.

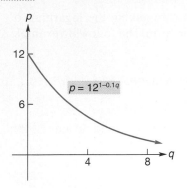

**FIGURE 5.27** Graph of the demand equation $p = 12^{1-0.1q}$.

### EXAMPLE 4  Demand Equation

*The demand equation for a product is $p = 12^{1-0.1q}$. Use common logarithms to express q in terms of p.*

*Solution:* Figure 5.27 shows the graph of this demand equation for $q \geq 0$. As is typical of a demand equation, the graph falls from left to right. We want to solve the equation for $q$. Taking the common logarithms of both sides of $p = 12^{1-0.1q}$ gives

$$\log p = \log(12^{1-0.1q}),$$

$$\log p = (1 - 0.1q)\log 12,$$

$$\frac{\log p}{\log 12} = 1 - 0.1q,$$

$$0.1q = 1 - \frac{\log p}{\log 12},$$

$$q = 10\left(1 - \frac{\log p}{\log 12}\right).$$ ■

To solve some exponential equations involving base $e$ or base 10, such as $10^{2x} = 3$, rather than taking logarithms of both sides, it may be easier to first transform the equation into an equivalent logarithmic form. In this case, we have

$$10^{2x} = 3,$$

$$2x = \log 3 \quad \text{(logarithmic form)},$$

$$x = \frac{\log 3}{2} \approx 0.2386.$$

### EXAMPLE 5  Predator–Prey Relation

*In an article concerning predators and prey, Holling[11] refers to an equation of the form*

$$y = K(1 - e^{-ax}),$$

*where x is the prey density, y is the number of prey attacked, and K and a are constants. Verify his claim that*

$$\ln \frac{K}{K - y} = ax.$$

*Solution:* To find $ax$, we shall first solve the given equation for $e^{-ax}$:

$$y = K(1 - e^{-ax}),$$

$$\frac{y}{K} = 1 - e^{-ax},$$

$$e^{-ax} = 1 - \frac{y}{K},$$

$$e^{-ax} = \frac{K - y}{K}.$$

---

[11]C. S. Holling, "Some Characteristics of Simple Types of Predation and Parasitism," *The Canadian Entomologist*, 91, no. 7 (1959), 385–98.

## Principles in Practice 3
### Solving a Logarithmic Equation

The Richter scale measure of an earthquake is given by $R = \log\left(\dfrac{I}{I_0}\right)$, where $I$ is the intensity of the earthquake, and $I_0$ is the intensity of a zero-level earthquake. An earthquake that is 675,000 times as intense as a zero-level earthquake has a magnitude on the Richter scale which is 4 more than another earthquake. What is the intensity of the other earthquake?

Now we convert to logarithmic form:

$$\ln \frac{K - y}{K} = -ax,$$

$$-\ln \frac{K - y}{K} = ax,$$

$$\ln \frac{K}{K - y} = ax \qquad \text{(Property 4 of Sec. 5.3),}$$

as was to be shown. ∎

Some logarithmic equations can be solved by rewriting them in exponential forms.

**EXAMPLE 6   Solving a Logarithmic Equation**

*Solve* $\log_2 x = 5 - \log_2(x + 4)$.

*Solution:* Here we must assume that both $x$ and $x + 4$ are positive, so that their logarithms are defined. Both conditions are satisfied if $x > 0$. To solve the equation, we first place all logarithms on one side so that we can combine them:

$$\log_2 x + \log_2(x + 4) = 5,$$
$$\log_2[x(x + 4)] = 5.$$

In exponential form, we have

$$x(x + 4) = 2^5,$$
$$x^2 + 4x = 32,$$
$$x^2 + 4x - 32 = 0 \qquad \text{(quadratic equation),}$$
$$(x - 4)(x + 8) = 0,$$
$$x = 4 \quad \text{or} \quad x = -8.$$

Because we must have $x > 0$, the only solution is 4, as can be verified by substituting into the original equation:

$$\log_2 4 \overset{?}{=} 5 - \log_2 8,$$
$$2 \overset{?}{=} 5 - 3,$$
$$2 = 2.$$

In solving a logarithmic equation, it is a good idea to check for extraneous solutions. ∎

## ■ EXERCISE 5.4

*In Problems 1–36, find x. Round your answers to three decimal places.*

**1.** $\log(2x + 1) = \log(x + 6)$.

**2.** $\log x + \log 3 = \log 5$.

**3.** $\log x - \log(x - 1) = \log 4$.

**4.** $\log_2 x + 3 \log_2 2 = \log_2 \dfrac{2}{x}$.

**5.** $\ln(-x) = \ln(x^2 - 6)$.

**6.** $\ln(4 - x) + \ln 2 = 2 \ln x$.

**7.** $e^{2x} \cdot e^{5x} = e^{14}$.

**8.** $(e^{5x+1})^2 = e$.

**9.** $(16)^{3x} = 2$.

**10.** $(27)^{2x+1} = \frac{1}{3}$.

**11.** $e^{2x} = 5$.

**12.** $e^{4x} = \frac{3}{4}$.

**13.** $3e^{3x+1} = 15$.

**14.** $6e^{1-x} + 1 = 25$.

**15.** $10^{4/x} = 6$.

**16.** $\frac{4(10)^{0.2x}}{5} = 3$.

**17.** $\frac{5}{10^{2x}} = 7$.

**18.** $2(10)^x + (10)^{x+1} = 4$.

**19.** $2^x = 5$.

**20.** $4^{x+3} = 7$.

**21.** $5^{2x-5} = 9$.

**22.** $4^{x/2} = 20$.

**23.** $2^{-2x/3} = \frac{4}{5}$.

**24.** $5(3^x - 6) = 10$.

**25.** $(4)5^{3-x} - 7 = 2$.

**26.** $\frac{8}{3^x} = 4$.

**27.** $\log(x - 3) = 3$.

**28.** $\log_2(x + 1) = 4$.

**29.** $\log_4(3x - 4) = 2$.

**30.** $\log_4(2x + 4) - 3 = \log_4 3$.

**31.** $\log(3x - 1) - \log(x - 3) = 2$.

**32.** $\log x + \log(x - 15) = 2$.

**33.** $\log_3(2x + 3) = 4 - \log_3(x + 6)$.

**34.** $\log(x + 2)^2 = 2$, where $x > 0$.

**35.** $\log_2\left(\frac{2}{x}\right) = 3 + \log_2 x$.

**36.** $\ln x = \ln(3x + 1) + 1$.

---

**37. Rooted Plants** In a study of rooted plants in a certain geographic region,[12] it was determined that on plots of size $A$ (in square meters), the average number of species that occurred was $S$. When $\log S$ was graphed as a function of $\log A$, the result was a straight line given by

$$\log S = \log 12.4 + 0.26 \log A.$$

Solve for $S$.

**38. Gross National Product** In an article, Taagepera and Hayes refer to an equation of the form

$$\log T = 1.7 + 0.2068 \log P - 0.1334 \log^2 P.$$

Here $T$ is the percentage of a country's gross national product (GNP) that corresponds to foreign trade (exports plus imports), and $P$ is the country's population (in units of 100,000).[13] Verify the claim that

$$T = 50P^{(0.2068 - 0.1334 \log P)}.$$

You may assume that $\log 50 = 1.7$.

**39. Radioactivity** The number of milligrams of a radioactive substance present after $t$ years is given by

$$Q = 100e^{-0.035t}.$$

**a.** How many milligrams are present after 0 years?

**b.** After how many years will there be 20 milligrams present?

Give your answer to the nearest year.

**40. Blood Sample** On the surface of a glass slide is a grid that divides the surface into 225 equal squares. Suppose a blood sample containing $N$ red cells is spread on the slide and the cells are randomly distributed. Then the number of squares containing no cells is (approximately) given by $225e^{-N/225}$. If 100 of the squares contain no cells, estimate the number of cells the blood sample contained.

**41. Population** In one city, the population $P$ grows at the rate of 2% per year. The equation $P = 1{,}000{,}000(1.02)^t$ gives the population $t$ years after 1995. Find the value of $t$ for which the population is 1,500,000. Give your answer to the nearest tenth.

**42. Market Penetration** In a discussion of market penetration by new products, Hurter and Rubenstein[14] refer to the function

$$F(t) = \frac{q - pe^{-(t+C)(p+q)}}{q[1 + e^{(t+C)(p+q)}]},$$

where $p$, $q$, and $C$ are constants. They claim that if $F(0) = 0$, then

$$C = -\frac{1}{p + q} \ln \frac{q}{p}.$$

Show that their claim is true.

[12]R. W. Poole, *An Introduction to Quantitative Ecology* (New York: McGraw-Hill Book Company, 1974).

[13]R. Taagepera and J. P. Hayes, "How Trade/GNP Ratio Decreases with Country Size," *Social Science Research*, 6 (1977), 108–32.

[14]A. P. Hurter, Jr., A. H. Rubenstein, et al., "Market Penetration by New Innovations: The Technological Literature," *Technological Forecasting and Social Change*, 11 (1978), 197–221.

**43. Demand Equation**   The demand equation for a consumer product is $q = 80 - 2^p$. Solve for $p$ and express your answer in terms of common logarithms, as in Example 4. Evaluate $p$ to two decimal places when $q = 60$.

**44. Investment**   The equation $A = P(1.1)^t$ gives the value $A$ at the end of $t$ years of an investment of $P$ dollars compounded annually at an annual interest rate of 10%. How many years will it take for an investment to double? Give your answer to the nearest year.

**45. Light Intensity**   A translucent material has the property that, although light passes through the material, its intensity is reduced. A particular translucent plastic has the property that a sheet 1 mm thick reduces the intensity of light by 10%. How many such sheets are necessary to reduce the intensity of a beam of light to about 50% of its original value?

**46. Sales**   After $t$ years the number of units of a product sold per year is given by $q = 1000(\frac{1}{2})^{0.8^t}$. Such an equation is called a *Gompertz equation* and describes natural growth in many areas of study. Solve this equation for $t$ in the same manner as in Example 4, and show that

$$t = \frac{\log \dfrac{3 - \log q}{\log 2}}{(3 \log 2) - 1}.$$

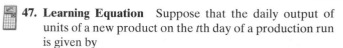 **47. Learning Equation**   Suppose that the daily output of units of a new product on the $t$th day of a production run is given by

$$q = 500(1 - e^{-0.2t}).$$

Such an equation is called a *learning equation* and indicates that as time progresses, output per day will increase. This may be due to a gain in a worker's proficiency at his or her job. Determine, to the nearest complete unit, the output on (a) the first day and (b) the tenth day after the start of a production run. (c) After how many days will a daily production run of 400 units be reached? Give your answer to the nearest day.

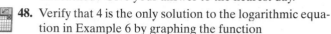 **48.** Verify that 4 is the only solution to the logarithmic equation in Example 6 by graphing the function

$$y = 5 - \log_2(x + 4) - \log_2 x$$

and observing when $y = 0$.

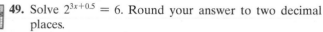 **49.** Solve $2^{3x+0.5} = 6$. Round your answer to two decimal places.

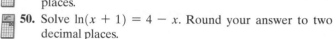 **50.** Solve $\ln(x + 1) = 4 - x$. Round your answer to two decimal places.

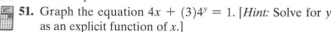 **51.** Graph the equation $4x + (3)4^y = 1$. [*Hint:* Solve for $y$ as an explicit function of $x$.]

# 5.5  REVIEW

## IMPORTANT TERMS AND SYMBOLS

**Section 5.1**   exponential function, $b^x$   compound interest   principal   compound amount
interest period   periodic rate   nominal rate   $e$   natural exponential function, $e^x$
exponential law of decay   initial amount   decay constant   half-life

**Section 5.2**   logarithmic function, $\log_b x$   common logarithm, $\log x$   natural logarithm, $\ln x$

**Section 5.3**   change-of-base formula

**Section 5.4**   logarithmic equation   exponential equation

## SUMMARY

An exponential function has the form $f(x) = b^x$. The graph of $f(x) = b^x$ has one of two general shapes, depending on the value of the base $b$. (See Fig. 5.3.) An exponential function is involved in the compound interest formula

$$S = P(1 + r)^n,$$

where $S$ is the compound amount of a principal of $P$ at the end of $n$ interest periods at the periodic rate $r$.

A frequently used base in an exponential function is the irrational number $e \approx 2.71828$. This base occurs in economic analysis and many situations involving growth or decay, such as population studies

and radioactive decay. Radioactive elements follow the exponential law of decay,

$$N = N_0 e^{-\lambda t},$$

where $N$ is the amount of an element present at time $t$, $N_0$ is the initial amount, and $\lambda$ is the decay constant. The time required for half of the amount of the element to decay is called the half-life.

The logarithmic function is the inverse function of the exponential function, and vice versa. The logarithmic function with base $b$ is denoted $\log_b$, and $y = \log_b x$ if and only if $b^y = x$. The graph of $y = \log_b x$ has one of two general shapes, depending on the value of the base $b$. (See Fig. 5.21.) Logarithms with base $e$ are called natural logarithms and are denoted ln; those with base 10 are called common logarithms and are denoted log. The half-life $T$ of a radioactive element can be given in terms of a natural logarithm and the decay constant: $T = (\ln 2)/\lambda$.

Some important properties of logarithms are the following:

$$\log_b(mn) = \log_b m + \log_b n,$$

$$\log_b \frac{m}{n} = \log_b m - \log_b n,$$

$$\log_b m^r = r \log_b m,$$

$$\log_b \frac{1}{m} = -\log_b m,$$

$$\log_b 1 = 0,$$

$$\log_b b = 1,$$

$$\log_b b^r = r,$$

$$b^{\log_b m} = m,$$

$$\log_b m = \frac{\log_a m}{\log_a b}.$$

Moreover, if $\log_b m = \log_b n$, then $m = n$. Similarly, if $b^m = b^n$, then $m = n$. Many of these properties are used in solving logarithmic and exponential equations.

## REVIEW PROBLEMS

*In Problems 1–6, write each exponential form logarithmically and each logarithmic form exponentially.*

**1.** $3^5 = 243$.  
**4.** $10^5 = 100,000$.

**2.** $\log_7 343 = 3$.  
**5.** $e^4 = 54.598$.

**3.** $\log_{16} 2 = \frac{1}{4}$.  
**6.** $\log_9 9 = 1$.

*In Problems 7–12, find the value of the expression without using a calculator.*

**7.** $\log_5 125$.  
**10.** $\log_{1/3} \frac{1}{9}$.

**8.** $\log_4 16$.  
**11.** $\log_{1/3} 9$.

**9.** $\log_2 \frac{1}{16}$.  
**12.** $\log_4 2$.

*In Problems 13–18, find x without using a calculator.*

**13.** $\log_5 125 = x$.  
**16.** $\ln \frac{1}{e} = x$.

**14.** $\log_x \frac{1}{8} = -3$.  
**17.** $\log_x(2x + 3) = 2$.

**15.** $\log x = -2$.  
**18.** $e^{\ln(x+4)} = 7$.

*In Problems 19 and 20, let log 2 = a and log 3 = b. Express the given logarithm in terms of a and b.*

**19.** $\log 8000$.  

**20.** $\log \frac{9}{\sqrt{2}}$.

*In Problems 21–26, write each expression as a single logarithm.*

**21.** $2 \log 5 - 3 \log 3$.  
**23.** $2 \ln x + \ln y - 3 \ln z$.  
**25.** $\frac{1}{2} \log_2 x + 2 \log_2(x^2) - 3 \log_2(x + 1) - 4 \log_2(x + 2)$.

**22.** $6 \ln x + 4 \ln y$.  
**24.** $\log_6 2 - \log_6 4 - 2 \log_6 3$.  
**26.** $3 \log x + \log y - 2(\log z + \log w)$.

*In Problems 27–32, write the expression in terms of ln x, ln y, and ln z.*

**27.** $\ln \frac{x^2 y}{z^3}$.  

**28.** $\ln \frac{\sqrt{x}}{(yz)^2}$.  

**29.** $\ln \sqrt[3]{xyz}$.

**30.** $\ln \left[ \frac{xy^3}{z^2} \right]^4$.  

**31.** $\ln \left[ \frac{1}{x} \sqrt{\frac{y}{z}} \right]$.  

**32.** $\ln \left[ \left( \frac{x}{y} \right)^2 \left( \frac{x}{z} \right)^3 \right]$.

**33.** Write $\log_3(x + 5)$ in terms of natural logarithms.

**34.** Write $\log_5(2x^2 + 1)$ in terms of common logarithms.

**35.** Suppose that $\log_2 19 = 4.2479$ and $\log_2 5 = 2.3219$. Find $\log_5 19$.

**36.** Use natural logarithms to determine the value of $\log_4 5$.

**37.** If $\ln 3 = x$ and $\ln 4 = y$, express $\ln(16\sqrt{3})$ in terms of $x$ and $y$.

**38.** Express $\log \frac{x^2 \sqrt{x + 1}}{\sqrt[3]{x^2 + 2}}$ in terms of $\log x$, $\log(x + 1)$, and $\log(x^2 + 2)$.

**39.** Simplify $e^{\ln x} + \ln e^x + \ln 1$.

**40.** Simplify $\log 10^2 + \log 1000 - 5$.

**41.** If $\ln y = x^2 + 2$, find $y$.

**42.** Sketch the graphs of $y = 3^x$ and $y = \log_3 x$.

**43.** Sketch the graph of $y = 2^{x+3}$.

**44.** Sketch the graph of $y = -2 \log_2 x$.

### In Problems 45–52, find x.

**45.** $\log(4x + 1) = \log(x + 2)$.

**46.** $\log x + \log 2 = 1$.

**47.** $3^{4x} = 9^{x+1}$.

**48.** $4^{3-x} = \dfrac{1}{16}$.

**49.** $\log x + \log(10x) = 3$.

**50.** $\log_3(x + 1) = \log_3(x - 1) + 1$.

**51.** $\ln(\log_x 2) = -1$.

**52.** $\log_2 x + \log_4 x = 3$.

### In Problems 53–58, find x. Round your answers to three decimal places.

**53.** $e^{3x} = 2$.

**54.** $10^{3x/2} = 5$.

**55.** $3(10^{x+4} - 3) = 9$.

**56.** $7e^{3x-1} - 2 = 1$.

**57.** $4^{x+3} = 7$.

**58.** $5^{2/x} = 2$.

**59. Investment** If \$2600 is invested for $6\frac{1}{2}$ years at 6% compounded quarterly, find (a) the compound amount and (b) the compound interest.

**60.** Find the nominal rate that corresponds to a periodic rate of $1\frac{1}{6}\%$ per month.

**61. Investment** Find the compound amount of an investment of \$4000 for five years at the rate of 11% compounded monthly.

**62. Bacteria Growth** Bacteria are growing in a culture, and their number is increasing at the rate of 4% an hour. Initially, 500 bacteria are present. (a) Determine an equation that gives the number $N$ of bacteria present after $t$ hours. (b) How many bacteria are present after one hour? (c) After three hours? Give your answer to (c) to the nearest integer.

**63. Population Growth** The population of a town of 8000 grows at the rate of 2% per year. (a) Determine an equation that gives the population $P$ after $t$ years from now. (b) Find the population two years from now. Give your answer to (b) to the nearest integer.

**64. Revenue** Due to ineffective advertising, the Kleer-Kut Razor Company finds that its annual revenues have been cut sharply. Moreover, the annual revenue $R$ at the end of $t$ years of business satisfies the equation $R = 200{,}000e^{-0.2t}$. Find the annual revenue at the end of two years and at the end of three years.

**65. Radioactivity** A radioactive substance decays according to the formula

$$N = 10e^{-0.41t},$$

where $N$ is the number of milligrams present after $t$ hours. (a) Determine the initial amount of the substance present. (b) To the nearest tenth of a milligram, determine the amount present after 2 hours. (c) After 10 hours. (d) To the nearest tenth of an hour, determine the

half-life of the substance, and (e) determine the number of hours for 1 milligram to remain.

**66. Radioactivity** If a radioactive substance has a half-life of 10 days, in how many days will $\frac{1}{8}$ of the initial amount be present?

**67. Marketing** A marketing-research company needs to determine how people adapt to the taste of a new cough drop. In one experiment, a person was given a cough drop and was asked periodically to assign a number, on a scale from 0 to 10, to the perceived taste. This number was called the *response magnitude*. The number 10 was assigned to the initial taste. After conducting the experiment several times, the company estimated that the response magnitude is given by

$$R = 10e^{-t/40},$$

where $t$ is the number of seconds after the person is given the cough drop. (a) Find the response magnitude after 20 seconds. Give your answer to the nearest integer. (b) After how many seconds does a person have a response magnitude of 5? Give your answer to the nearest second.

**68. Sediment in Water** The water in a midwestern lake contains sediment, and the presence of the sediment reduces the transmission of light through the water. Experiments indicate that the intensity of light is reduced by 10% by passage through 20 cm of water. Suppose that the lake is uniform with respect to the amount of sediment contained by the water. A measuring instrument can detect light at the intensity of 0.17% of full sunlight. This measuring instrument is lowered into the lake. At what depth will it first cease to record the presence of light? Give your answer to the nearest 10 cm.

69. **Body Cooling** In a discussion of the rate of cooling of isolated portions of the body when they are exposed to low temperatures, there occurs the equation[15]

$$T_t - T_e = (T_t - T_e)_o e^{-at},$$

where $T_t$ is the temperature of the portion at time $t$, $T_e$ is the environmental temperature, the subscript $o$ refers to the initial temperature difference, and $a$ is a constant. Show that

$$a = \frac{1}{t} \ln \frac{(T_t - T_e)_o}{T_t - T_e}.$$

[15]R. W. Stacy et al., *Essentials of Biological and Medical Physics* (New York: McGraw-Hill Book Company, 1955).

 70. Graph $y = 3^x$ and $y = \dfrac{3^x}{9}$ on the same screen. It appears that the graph of $y = \dfrac{3^x}{9}$ is the graph of $y = 3^x$ shifted two units to the right. Prove algebraically that this is indeed true.

 71. If $y = f(x) = \dfrac{\ln x}{x}$, determine the range of $f$. Round values to two decimal places.

 72. Determine the points of intersection of the graphs of $y = \ln(x + 2)$ and $y = x^2 - 4$. Round your answers to two decimal places.

 73. Solve $\ln x = 4 - x$. Round your answer to two decimal places.

 74. Solve $6^{3-4x} = 15$. Round your answer to two decimal places.

 75. Display the graph of $y = \log_3(x^2 + 1)$.

 76. Display the graph of the equation $(6)5^y + x = 2$. [*Hint:* Solve for $y$ as an explicit function of $x$.]

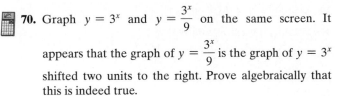

# MATHEMATICAL *SNAPSHOT*

## DRUG DOSAGES[16]

Determining and prescribing drug dosages are extremely important aspects of the medical profession. Quite often, caution must be taken because of possible adverse side or toxic effects of drugs.

Many drugs are used up by the human body in such a way that the amount present follows an exponential law of decay. That is, if $N$ is the amount of the drug present in the body at time $t$, then

$$N = N_0 e^{-kt}, \tag{1}$$

where $k$ is a positive constant and $N_0$ is the amount present at time $t = 0$. If $H$ is the half-life of such a drug, then from Sec. 5.2, $H = (\ln 2)/k$ or, equivalently, $k = (\ln 2)/H$.

Suppose that you want to analyze the situation whereby equal doses of such a drug are introduced into a patient's system every $I$ units of time until a therapeutic level is attained, and then the dosage is reduced to maintain the therapeutic level. The reason for *reduced* maintenance doses is frequently related to the toxic effects of drugs.

In particular, assume that there are $d$ doses of $P$ units each, a dose is given at times $t = 0, I, 2I, \ldots,$ and $(d - 1)/I$, and that the therapeutic level, $T$, is attained at $t = dI$, which occurs one time interval after the last dose is administered. You will now see how to determine a formula that gives the therapeutic level.

At time $t = 0$ the patient receives the first $P$ units, so the amount of drug in the body is $P$. At time $t = I$ the amount present from the first dose is [from Eq. (1)] $Pe^{-kI}$. In addition, at $t = I$ the second $P$ units are given. Thus the total amount of the drug present is

$$P + Pe^{-kI}.$$

At time $t = 2I$, the amount remaining from the first dose is $Pe^{-2kI}$; from the second dose, which has been in the system for only one time interval, the amount present is $Pe^{-kI}$. Also, at time $t = 2I$ the third dose of $P$ units is given, so the total amount of the drug present is

$$P + Pe^{-kI} + Pe^{-2kI}.$$

Continuing in this way, the amount $T$ of drug present in the system at time $dI$, one time interval after the last dose, is given by

$$T = Pe^{-kI} + Pe^{-2kI} + \cdots + Pe^{-dkI}. \tag{2}$$

You can express the right side of Eq. (2) in a different form. First, multiply both sides of Eq. (2) by $e^{-kI}$.

$$e^{-kI}T = e^{-kI}(Pe^{-kI} + Pe^{-2kI} + \cdots + Pe^{-dkI}),$$

$$e^{-kI}T = Pe^{-2kI} + Pe^{-3kI} + \cdots + Pe^{-(d+1)kI}. \tag{3}$$

Subtracting the sides of Eq. (3) from the corresponding sides of Eq. (2), you have

$$T - e^{-kI}T = Pe^{-kI} - Pe^{-(d+1)kI}.$$

[16]This discussion is adapted from Gerald M. Armstrong and Calvin P. Midgley, "The Exponential-Decay Law Applied to Medical Dosages," *The Mathematics Teacher,* 80, no. 3 (February 1987), 110–13. By permission of the National Council of Teachers of Mathematics.

Simplifying and solving for $T$ gives

$$(1 - e^{-kI})T = Pe^{-kI}(1 - e^{-dkI}),$$

$$T = \frac{Pe^{-kI}(1 - e^{-dkI})}{1 - e^{-kI}}, \qquad (4)$$

$$T = \frac{P(1 - e^{-dkI})}{e^{kI}(1 - e^{-kI})},$$

$$T = \frac{P(1 - e^{-dkI})}{e^{kI} - 1}. \qquad (5)$$

Equation (5) allows you to determine the therapeutic level, $T$, in terms of the dose, $P$, the length of the time intervals, $I$, the number of doses, $d$, and the half-life, $H$, of the drug [since $k = (\ln 2)/H$]. Among other possibilities, you can determine the dose $P$ if $T$, $H$, $I$, and $d$ are known.

The objective now is to maintain the therapeutic level in the patient's system. To do this, a reduced dose, $R$, is given at times $t = dI, (d + 1)I, (d + 2)I$, and so on. You can determine a formula for $R$ in the following way.

At time $t = (d + 1)I$, but before the second reduced dose is given, the amount of drug in the system from the first reduced dose is $Re^{-kI}$ and the amount that remains from the therapeutic level is $Te^{-kI}$. Suppose you require that the sum of these amounts be the therapeutic level, $T$; that is,

$$T = Re^{-kI} + Te^{-kI}.$$

Solving for $R$ gives

$$Re^{-kI} = T - Te^{-kI},$$

$$R = T(1 - e^{-kI})e^{kI}.$$

Replacing $T$ by the right side of Eq. (4) gives

$$R = \frac{Pe^{-kI}(1 - e^{-dkI})}{1 - e^{-kI}}(1 - e^{-kI})e^{kI},$$

or, more simply,

$$R = P(1 - e^{-dkI}). \qquad (6)$$

By continuing the reduced doses at time intervals of length $I$, you are assured that the drug level in the system never falls below $T$. Furthermore, note that because $-dkI < 0$, then $0 < e^{-dkI} < 1$. Consequently the factor $1 - e^{-dkI}$ in Eq. (6) is between 0 and 1. This ensures that $R$ is less than $P$; hence $R$ is indeed a *reduced* dose.

It is interesting to note that Armstrong and Midgley state that "the therapeutic amount $T$ must be chosen from a range of empirically determined values. Medical discretion and experience are needed to select proper intervals and durations of time to administer the drug. Even the half-life of a drug can vary somewhat among different patients. Many other factors play a role, such as absorption levels of drugs, their distribution in the system, drug interactions, age of patients, their general health, and the health of such vital organs as the liver and kidneys."

### ■ Exercises

1. Solve Eq. (5) above for (a) $P$ and (b) $d$.

2. Show that if $I$ is equal to the half-life of the drug, Eq. (5) can be written as

$$T = \left(1 - \frac{1}{2^d}\right)P.$$

Note that $0 < 1 - (1/2^d) < 1$ for $d > 0$. Hence, the foregoing equation implies that when doses of $P$ units are administered at time intervals equal to the half-life of the drug, it follows that at one time interval after any dose is given, but before the next dose is given, the total level of the drug in a patient's system is less than $P$.

3. Theophylline is a drug used to treat bronchial asthma and has a half-life of 8 hours in the system of a relatively healthy nonsmoking patient. Suppose that such a patient achieves the desired therapeutic level of this drug in 12 hours when 100 milligrams is administered every 4 hours. Here $d = 3$. Because of toxicity, the dose must be reduced thereafter. To the nearest milligram, determine (a) the therapeutic level and (b) the reduced dose.

4. A main component of thyroid hormone is thyroxine, denoted $T_4$. A person with a certain thyroid deficiency is said to have hypothyroidism or myxedema. Suppose that such a person must have his or her $T_4$ level increased by 500 micrograms. It is medically recommended that the increase occur slowly. To do this, a daily dose of $P$ micrograms of $T_4$ is administered for 28 days ($d = 28$), and then the dose is reduced to $R$ micrograms per day. Assume that the half-life of $T_4$ is 9 days. Determine $P$ and $R$ to the nearest microgram.

# Matrix Algebra

To introduce the concept of a matrix and to consider special types of matrices.

## 6.1 MATRICES

Finding ways to describe many situations in mathematics and economics leads to the study of rectangular arrays of numbers. Consider, for example, the system of linear equations

$$\begin{cases} 3x + 4y + 3z = 0, \\ 2x + y - z = 0, \\ 9x - 6y + 2z = 0. \end{cases}$$

The features that characterize this system are the numerical coefficients in the equations, together with their relative positions. For this reason, the system can be described by the rectangular array

$$\begin{bmatrix} 3 & 4 & 3 \\ 2 & 1 & -1 \\ 9 & -6 & 2 \end{bmatrix},$$

which is called a *matrix* (plural: *matrices,* pronounced may′ tri sees). We shall consider such rectangular arrays to be objects in themselves, and our custom, as just shown, will be to enclose them by brackets. Parentheses are also commonly used. In symbolically representing matrices, we shall use bold capital letters such as **A, B, C,** and so on.

*Pitfall* ▼ Do not use vertical bars, | |, instead of brackets or parentheses, for they have a different meaning.

In economics it is often convenient to use matrices in formulating problems and displaying data. For example, a manufacturer who produces products A, B, and C could represent the units of labor and material involved in one week's production of these items as in Table 6.1. More simply, the data can be represented by the matrix

$$\mathbf{A} = \begin{bmatrix} 10 & 12 & 16 \\ 5 & 9 & 7 \end{bmatrix}.$$

**TABLE 6.1**

|          | Product |    |    |
|----------|---------|----|----|
|          | **A**   | **B** | **C** |
| Labor    | 10      | 12 | 16 |
| Material | 5       | 9  | 7  |

The horizontal rows of a matrix are numbered consecutively from top to bottom, and the vertical columns are numbered from left to right. For the foregoing matrix **A**, we have

$$
\begin{array}{cc}
& \text{column 1} \quad \text{column 2} \quad \text{column 3} \\
\begin{array}{c} \text{row 1} \\ \text{row 2} \end{array} &
\left[\begin{array}{ccc} 10 & 12 & 16 \\ 5 & 9 & 7 \end{array}\right] = \mathbf{A}.
\end{array}
$$

Since **A** has two rows and three columns, we say that **A** has *order*, or *size*, $2 \times 3$ (read "2 by 3"), where the number of rows is specified first. Similarly, the matrices

$$
\mathbf{B} = \left[\begin{array}{ccc} 1 & 6 & -2 \\ 5 & 1 & -4 \\ -3 & 5 & 0 \end{array}\right]
\quad \text{and} \quad
\mathbf{C} = \left[\begin{array}{cc} 1 & 2 \\ -3 & 4 \\ 5 & 6 \\ 7 & -8 \end{array}\right]
$$

have orders $3 \times 3$ and $4 \times 2$, respectively.

The numbers in a matrix are called its **entries** or **elements.** To denote arbitrary entries in a matrix, say, of order $2 \times 3$, there are two common methods. First, we may use different letters:

$$
\left[\begin{array}{ccc} a & b & c \\ d & e & f \end{array}\right].
$$

Second, a single letter may be used, say, *a*, along with appropriate *double* subscripts to indicate position:

$$
\left[\begin{array}{ccc} a_{11} & a_{12} & a_{13} \\ a_{21} & a_{22} & a_{23} \end{array}\right].
$$

The row subscript appears to the left of the column subscript. In general, $a_{ij} \neq a_{ji}$.

For the entry $a_{12}$ (read "*a* sub one-two"), the first subscript, 1, specifies the row and the second subscript, 2, the column in which the entry appears. Similarly, the entry $a_{23}$ (read "*a* sub two-three") is the entry in the second row and the third column. Generalizing, we say that the symbol $a_{ij}$ denotes the entry in the *i*th row and *j*th column.

Our concern in this chapter is the manipulation and application of various types of matrices. For completeness, we now give a formal definition of a matrix.

**DEFINITION**

*A rectangular array of numbers consisting of m horizontal rows and n vertical columns,*

$$
\left[\begin{array}{cccc}
a_{11} & a_{12} & \cdots & a_{1n} \\
a_{21} & a_{22} & \cdots & a_{2n} \\
\cdot & \cdot & \cdots & \cdot \\
\cdot & \cdot & \cdots & \cdot \\
\cdot & \cdot & \cdots & \cdot \\
a_{m1} & a_{m2} & \cdots & a_{mn}
\end{array}\right],
$$

*is called an **m $\times$ n matrix** or a **matrix of order m $\times$ n**. For the entry $a_{ij}$, we call i the row subscript and j the column subscript.*

The number of entries in an $m \times n$ matrix is $mn$. For brevity, an $m \times n$ matrix can be denoted by the symbol $[a_{ij}]_{m \times n}$ or, more simply, $[a_{ij}]$, where the order is understood to be that which is appropriate for the given context. This notation merely indicates what types of symbols we are using to denote the general entry.

*Pitfall* ▼ Do not confuse the general entry $a_{ij}$ with the matrix $[a_{ij}]$.

A matrix that has exactly one row, such as the $1 \times 4$ matrix

$$\mathbf{A} = [1 \quad 7 \quad 12 \quad 3],$$

is called a **row matrix** or a **row vector.** A matrix consisting of a single column, such as the $5 \times 1$ matrix

$$\begin{bmatrix} 1 \\ -2 \\ 15 \\ 9 \\ 16 \end{bmatrix},$$

is called a **column matrix** or a **column vector.**

## Principles in Practice 1
### Order (or Size) of a Matrix

A manufacturer who uses raw materials A and B is interested in tracking the costs of these materials from three different sources. What is the order of the matrix she would use?

## EXAMPLE 1 Order (or Size) of a Matrix

**a.** The matrix $[1 \quad 2 \quad 0]$ has order $1 \times 3$.

**b.** The matrix $\begin{bmatrix} 1 & -6 \\ 5 & 1 \\ 9 & 4 \end{bmatrix}$ has size $3 \times 2$.

**c.** The matrix $[7]$ has order $1 \times 1$.

**d.** The matrix $\begin{bmatrix} 1 & 3 & 7 & -2 & 4 \\ 9 & 11 & 5 & 6 & 8 \\ 6 & -2 & -1 & 1 & 1 \end{bmatrix}$ has order $3 \times 5$ and $3(5) = 15$ entries. ■

## Principles in Practice 2
### Constructing Matrices

An analysis of a workplace uses a $3 \times 5$ matrix to describe the time spent on each of three phases of five different projects. Project 1 requires 1 hour for each phase, project 2 requires twice as much time as project 1, project 3 requires twice as much time as project 2, . . ., and so on. Construct this time-analysis matrix.

## EXAMPLE 2 Constructing Matrices

**a.** *Construct a three-entry column matrix such that $a_{21} = 6$ and $a_{i1} = 0$ otherwise.*

*Solution:* Since $a_{11} = a_{31} = 0$, the matrix is

$$\begin{bmatrix} 0 \\ 6 \\ 0 \end{bmatrix}.$$

**b.** *If $\mathbf{A} = [a_{ij}]$ has order $3 \times 4$ and $a_{ij} = i + j$, find $\mathbf{A}$.*

*Solution:* Here $i = 1, 2, 3$ and $j = 1, 2, 3, 4$, and $\mathbf{A}$ has $(3)(4) = 12$ entries. Since $a_{ij} = i + j$, the entry in row $i$ and column $j$ is obtained by adding the numbers $i$ and $j$. Hence, $a_{11} = 1 + 1 = 2$, $a_{12} = 1 + 2 = 3$, $a_{13} = 1 + 3 = 4$, and so on. Thus,

$$\mathbf{A} = \begin{bmatrix} 1+1 & 1+2 & 1+3 & 1+4 \\ 2+1 & 2+2 & 2+3 & 2+4 \\ 3+1 & 3+2 & 3+3 & 3+4 \end{bmatrix} = \begin{bmatrix} 2 & 3 & 4 & 5 \\ 3 & 4 & 5 & 6 \\ 4 & 5 & 6 & 7 \end{bmatrix}.$$

**c.** *Construct the $3 \times 3$ matrix $\mathbf{I}$, given that $a_{11} = a_{22} = a_{33} = 1$ and $a_{ij} = 0$ otherwise.*

*Solution:* The matrix is given by

$$\mathbf{I} = \begin{bmatrix} 1 & 0 & 0 \\ 0 & 1 & 0 \\ 0 & 0 & 1 \end{bmatrix}.$$ ■

## Equality of Matrices

We now define what is meant by saying that two matrices are *equal*.

**DEFINITION**

*Matrices* $\mathbf{A} = [a_{ij}]$ *and* $\mathbf{B} = [b_{ij}]$ *are* **equal** *if and only if they have the same order and* $a_{ij} = b_{ij}$ *for each i and j (that is, corresponding entries are equal).*

Thus,

$$\begin{bmatrix} 1+1 & \frac{2}{2} \\ 2 \cdot 3 & 0 \end{bmatrix} = \begin{bmatrix} 2 & 1 \\ 6 & 0 \end{bmatrix},$$

but

$$[1 \quad 1] \neq \begin{bmatrix} 1 \\ 1 \end{bmatrix} \quad \text{and} \quad [1 \quad 1] \neq [1 \quad 1 \quad 1] \quad \text{(different sizes).}$$

A matrix equation can define a system of equations. For example, suppose that

$$\begin{bmatrix} x & y+1 \\ 2z & 5w \end{bmatrix} = \begin{bmatrix} 2 & 7 \\ 4 & 2 \end{bmatrix}.$$

By equating corresponding entries, we must have

$$\begin{cases} x = 2, \\ y + 1 = 7, \\ 2z = 4, \\ 5w = 2. \end{cases}$$

Solving gives $x = 2$, $y = 6$, $z = 2$, and $w = \frac{2}{5}$. It is a significant fact that a matrix equation can define a system of linear equations.

## Transpose of a Matrix

If $\mathbf{A}$ is a matrix, the matrix formed from $\mathbf{A}$ by interchanging its rows with its columns is called the *transpose* of $\mathbf{A}$.

**DEFINITION**

*The* **transpose** *of an* $m \times n$ *matrix* $\mathbf{A}$, *denoted* $\mathbf{A}^{\mathrm{T}}$, *is the* $n \times m$ *matrix whose ith row is the ith column of* $\mathbf{A}$.

**EXAMPLE 3** **Transpose of a Matrix**

*If* $\mathbf{A} = \begin{bmatrix} 1 & 2 & 3 \\ 4 & 5 & 6 \end{bmatrix}$, *find* $\mathbf{A}^{\mathrm{T}}$.

*Solution:* Matrix $\mathbf{A}$ is $2 \times 3$, so $\mathbf{A}^{\mathrm{T}}$ is $3 \times 2$. Column 1 of $\mathbf{A}$ becomes row 1 of $\mathbf{A}^{\mathrm{T}}$, column 2 becomes row 2, and column 3 becomes row 3. Thus,

$$\mathbf{A}^{\mathrm{T}} = \begin{bmatrix} 1 & 4 \\ 2 & 5 \\ 3 & 6 \end{bmatrix}.$$

Observe that the columns of $\mathbf{A}^{\mathrm{T}}$ are the rows of $\mathbf{A}$. Also, if we take the transpose of our answer, the original matrix $\mathbf{A}$ is obtained. That is, the transpose operation has the property that

$$(\mathbf{A}^{\mathrm{T}})^{\mathrm{T}} = \mathbf{A}.$$

## TECHNOLOGY

Graphics calculators have the ability to manipulate matrices. For example, Fig. 6.1 shows the result of applying the transpose operation to matrix $\mathbf{A}$.

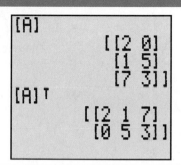

**FIGURE 6.1**    $\mathbf{A}$ and $\mathbf{A}^{\mathrm{T}}$

### Special Matrices

Certain types of matrices play important roles in matrix theory. We now consider some of these special types.

An $m \times n$ matrix whose entries are all 0 is called the $m \times n$ **zero matrix** and is denoted by $\mathbf{O}_{m \times n}$ or, more simply, by $\mathbf{O}$ if its size is understood. Thus, the $2 \times 3$ zero matrix is

$$\mathbf{O} = \begin{bmatrix} 0 & 0 & 0 \\ 0 & 0 & 0 \end{bmatrix},$$

and in general,

$$\mathbf{O} = \begin{bmatrix} 0 & 0 & \dots & 0 \\ 0 & 0 & \dots & 0 \\ \cdot & \cdot & \dots & \cdot \\ \cdot & \cdot & \dots & \cdot \\ \cdot & \cdot & \dots & \cdot \\ 0 & 0 & \dots & 0 \end{bmatrix}.$$

*Pitfall* ▼ Do not confuse the matrix $\mathbf{O}$ with the real number 0.

A matrix having the same number of columns as rows—for example, $n$ rows and $n$ columns—is called a **square matrix** of order $n$. That is, an $m \times n$ matrix is square if and only if $m = n$. For example, matrices

$$\begin{bmatrix} 2 & 7 & 4 \\ 6 & 2 & 0 \\ 4 & 6 & 1 \end{bmatrix} \qquad \text{and} \qquad [3]$$

are square with orders 3 and 1, respectively.

In a square matrix of order $n$, the entries $a_{11}, a_{22}, a_{33}, \dots, a_{nn}$ that lie on the "main" diagonal extending from the upper left corner to the lower right corner are called the *main-diagonal* entries, or more simply the **main diagonal.** Thus, in the matrix

$$\begin{bmatrix} 1 & 2 & 3 \\ 4 & 5 & 6 \\ 7 & 8 & 9 \end{bmatrix},$$

the main diagonal (see the shaded region) consists of $a_{11} = 1$, $a_{22} = 5$, and $a_{33} = 9$.

A square matrix **A** is called a **diagonal matrix** if all the entries that are off the main diagonal are zero—that is, if $a_{ij} = 0$ for $i \neq j$. Examples of diagonal matrices are

$$\begin{bmatrix} 1 & 0 \\ 0 & 1 \end{bmatrix} \quad \text{and} \quad \begin{bmatrix} 3 & 0 & 0 \\ 0 & 6 & 0 \\ 0 & 0 & 9 \end{bmatrix}.$$

A square matrix **A** is said to be an **upper triangular matrix** if all entries *below* the main diagonal are zero—that is, if $a_{ij} = 0$ for $i > j$. Similarly, a matrix **A** is said to be a **lower triangular matrix** if all entries *above* the main diagonal are zero—that is, if $a_{ij} = 0$ for $i < j$. When a matrix is either upper triangular or lower triangular, it is called a **triangular matrix.** Thus, the matrices

$$\begin{bmatrix} 5 & 1 & 1 \\ 0 & -3 & 7 \\ 0 & 0 & 4 \end{bmatrix} \quad \text{and} \quad \begin{bmatrix} 7 & 0 & 0 & 0 \\ 3 & 2 & 0 & 0 \\ 6 & 5 & -4 & 0 \\ 1 & 6 & 0 & 1 \end{bmatrix}$$

A diagonal matrix is considered to be both upper triangular and lower triangular.

are upper and lower triangular matrices, respectively, and are therefore triangular matrices.

## ▪ Exercise 6.1

**1.** Let

$$\mathbf{A} = \begin{bmatrix} 1 & -6 & 2 \\ -4 & 2 & 1 \end{bmatrix}, \qquad \mathbf{B} = \begin{bmatrix} 1 & 2 & 3 \\ 4 & 5 & 6 \\ 7 & 8 & 9 \end{bmatrix}, \qquad \mathbf{C} = \begin{bmatrix} 1 & 1 \\ 2 & 2 \\ 3 & 3 \end{bmatrix}, \qquad \mathbf{D} = \begin{bmatrix} 1 & 0 \\ 2 & 3 \end{bmatrix}, \qquad \mathbf{E} = \begin{bmatrix} 1 & 2 & 3 & 4 \\ 0 & 1 & 6 & 0 \\ 0 & 0 & 2 & 0 \\ 0 & 0 & 6 & 1 \end{bmatrix},$$

$$\mathbf{F} = [6 \quad 2], \qquad \mathbf{G} = \begin{bmatrix} 5 \\ 6 \\ 1 \end{bmatrix}, \qquad \mathbf{H} = \begin{bmatrix} 1 & 6 & 2 \\ 0 & 0 & 0 \\ 0 & 0 & 0 \end{bmatrix}, \qquad \mathbf{J} = [4].$$

**a.** State the order of each matrix.
**b.** Which matrices are square?
**c.** Which matrices are upper triangular? lower triangular?

**d.** Which are row vectors?
**e.** Which are column vectors?

*In Problems 2–9, let*

$$\mathbf{A} = [a_{ij}] = \begin{bmatrix} 7 & -2 & 14 & 6 \\ 6 & 2 & 3 & -2 \\ 5 & 4 & 1 & 0 \\ 8 & 0 & 2 & 0 \end{bmatrix}.$$

**2.** What is the order of **A**?

*Find the following entries.*

**3.** $a_{43}$.

**4.** $a_{12}$.

**5.** $a_{32}$.

**6.** $a_{34}$.

**7.** $a_{14}$.

**8.** $a_{55}$.

**9.** What are the main-diagonal entries?

**10.** Write the upper triangular matrix of order 5, given that all entries which are not required to be 0 are equal to 1.

**11.** Construct the matrix $\mathbf{A} = [a_{ij}]$ if $\mathbf{A}$ is $3 \times 4$ and $a_{ij} = 2i + 3j$.

**12.** Construct the matrix $\mathbf{B} = [b_{ij}]$ if $\mathbf{B}$ is $2 \times 2$ and $b_{ij} = (-1)^{i+j}(i^2 + j^2)$.

**13.** If $\mathbf{A} = [a_{ij}]$ is $12 \times 10$, how many entries does $\mathbf{A}$ have? If $a_{ij} = 1$ for $i = j$ and $a_{ij} = 0$ for $i \neq j$, find $a_{33}$, $a_{52}$, $a_{10,10}$, and $a_{12,10}$.

**14.** List the main diagonal of

**a.** $\begin{bmatrix} 1 & 4 & -2 & 0 \\ 7 & 0 & 4 & -1 \\ -6 & 6 & -5 & 1 \\ 2 & 1 & 7 & 2 \end{bmatrix}$

**b.** $\begin{bmatrix} x & 1 & y \\ 9 & y & 7 \\ y & 0 & z \end{bmatrix}$.

**15.** Write the zero matrix of order (a) 4 and (b) 6.

**16.** If $\mathbf{A}$ is a $4 \times 5$ matrix, what is the order of $A^T$?

*In Problems 17–20, find $\mathbf{A}^\mathsf{T}$.*

**17.** $\mathbf{A} = \begin{bmatrix} 6 & -3 \\ 2 & 4 \end{bmatrix}$.

**18.** $\mathbf{A} = [2 \quad 4 \quad 6 \quad 8]$.

**19.** $\mathbf{A} = \begin{bmatrix} 1 & 3 & 2 & 3 \\ 3 & 2 & -2 & 0 \\ -4 & 2 & 0 & 1 \end{bmatrix}$.

**20.** $\mathbf{A} = \begin{bmatrix} 2 & -1 & 0 \\ -1 & 5 & 1 \\ 0 & 1 & 3 \end{bmatrix}$.

**21.** Let

$$\mathbf{A} = \begin{bmatrix} 7 & 0 \\ 0 & 6 \end{bmatrix}, \qquad \mathbf{B} = \begin{bmatrix} 1 & 0 & 0 \\ 0 & 2 & 0 \\ 0 & 10 & -3 \end{bmatrix},$$

$$\mathbf{C} = \begin{bmatrix} 0 & 0 & 0 \\ 0 & 0 & 0 \\ 0 & 0 & 0 \end{bmatrix}, \qquad \mathbf{D} = \begin{bmatrix} 2 & 0 & -1 \\ 0 & 4 & 0 \\ 0 & 0 & 6 \end{bmatrix}.$$

**a.** Which are diagonal matrices?

**b.** Which are triangular matrices?

**22.** A matrix is *symmetric* if $\mathbf{A}^\mathsf{T} = \mathbf{A}$. Is the matrix of Problem 20 symmetric?

**23.** If

$$\mathbf{A} = \begin{bmatrix} 1 & 2 & 3 \\ 4 & 5 & 6 \\ 7 & 8 & 9 \end{bmatrix}.$$

verify the general property that $(\mathbf{A}^\mathsf{T})^\mathsf{T} = \mathbf{A}$ by finding $\mathbf{A}^\mathsf{T}$ and then $(\mathbf{A}^\mathsf{T})^\mathsf{T}$.

*In Problems 24–27, solve the matrix equation.*

**24.** $\begin{bmatrix} 2x & y \\ z & 3w \end{bmatrix} = \begin{bmatrix} 4 & 6 \\ 0 & 7 \end{bmatrix}$.

**25.** $\begin{bmatrix} 6 & 2 \\ x & 7 \\ 3y & 2z \end{bmatrix} = \begin{bmatrix} 6 & 2 \\ 6 & 7 \\ 2 & 7 \end{bmatrix}$.

**26.** $\begin{bmatrix} 4 & 2 & 1 \\ 3x & y & 3z \\ 0 & w & 7 \end{bmatrix} = \begin{bmatrix} 4 & 2 & 1 \\ 6 & 7 & 9 \\ 0 & 9 & 8 \end{bmatrix}$.

**27.** $\begin{bmatrix} 2x & 7 \\ 7 & 2y \end{bmatrix} = \begin{bmatrix} y & 7 \\ 7 & y \end{bmatrix}$.

**28. Stocks** A stockbroker sold a customer 200 shares of stock $A$, 300 shares of stock $B$, 500 shares of stock $C$, and 300 shares of stock $D$. Write a row vector that gives the number of shares of each stock sold. If the stocks sell for $20, $30, $45, and $100 per share, respectively, write this information as a column vector.

**29. Sales Analysis** The Widget Company has its monthly sales reports given by means of matrices whose rows, in order, represent the number of regular, deluxe, and super-duper models sold, and the columns, in order, give the number of red, white, blue, and purple units sold. The matrices for January (**J**) and February (**F**) are

$$\mathbf{J} = \begin{bmatrix} 2 & 6 & 1 & 2 \\ 0 & 1 & 3 & 5 \\ 2 & 7 & 6 & 0 \end{bmatrix}, \qquad \mathbf{F} = \begin{bmatrix} 0 & 2 & 4 & 4 \\ 2 & 3 & 3 & 2 \\ 4 & 0 & 2 & 6 \end{bmatrix}.$$

(a) How many white super-duper models were sold in January? (b) How many blue deluxe models were sold in February? (c) In which month were more purple regular models sold? (d) Which model and color sold the same number of units in both months? (e) In which month were more deluxe models sold? (f) In which month were more red widgets sold? (g) How many widgets were sold in January?

**30. Input-Output Matrix** Input-output matrices, which were developed by W. W. Leontief, indicate the interrelationships that exist among the various sectors of an economy during some period of time. A hypothetical example for a simplified economy is given by matrix **M** at the end of this problem. The consuming sectors are the same as the producing sectors and can be thought of as manufacturers, government, steel, agriculture, households, and so on. Each row shows how the output of a given sector is consumed by the four sectors. For example, of the total output of industry A, 50 went to industry A itself, 70 to B, 200 to C, and 360 to all others. The sum of the entries in row 1, namely, 680, gives the total output of A for a given period. Each column gives the output of each sector that is consumed by a given sector. For example, in producing 680 units, industry A consumed 50 units of A, 90 of B, 120 of C, and 420 from all other producers. For each column, find the sum of the entries. Do the same for each row. What do you observe in comparing these totals? Suppose sector A increases its output by 20%, namely, by 136 units. Assuming that this results in a uniform 20% increase of all its inputs, by how many units will sector B have to increase its output? Answer the same question for C and for all other producers.

CONSUMERS

|  | Industry A | Industry B | Industry C | All Other Consumers |
|---|---|---|---|---|
| PRODUCERS | | | | |
| Industry A | 50 | 70 | 200 | 360 |
| **M** = Industry B | 90 | 30 | 270 | 320 |
| Industry C | 120 | 240 | 100 | 1050 |
| All Other Producers | 420 | 370 | 940 | 4960 |

**31.** Find all the values of $x$ for which

$$\begin{bmatrix} x^2 + 1993x & \sqrt{x^2} \\ x^2 & \ln(e^x) \end{bmatrix} = \begin{bmatrix} 1994 & -x \\ 1994 - 1993x & x \end{bmatrix}.$$

*In Problems 32 and 33, find* $\mathbf{A}^{\mathrm{T}}$.

 **32.** $\mathbf{A} = \begin{bmatrix} 3 & -4 & 5 \\ -2 & 1 & 6 \end{bmatrix}$.

 **33.** $\mathbf{A} = \begin{bmatrix} 3 & 1 & 4 & 2 \\ 1 & 7 & 3 & 6 \\ 1 & 4 & 1 & 2 \end{bmatrix}$.

---

**OBJECTIVE**

To define matrix addition and scalar multiplication and to consider properties related to these operations.

## 6.2 MATRIX ADDITION AND SCALAR MULTIPLICATION

### Matrix Addition

Consider a snowmobile dealer who sells two models, Deluxe and Super. Each is available in one of two colors, red and blue. Suppose that the sales for January and February are represented by the matrices

$$\mathbf{J} = \begin{array}{c} \text{red} \\ \text{blue} \end{array} \begin{bmatrix} 1 & 2 \\ 3 & 5 \end{bmatrix}, \qquad \mathbf{F} = \begin{bmatrix} 3 & 1 \\ 4 & 2 \end{bmatrix}.$$

Each row of **J** and **F** gives the number of each model sold for a given color. Each column gives the number of each color sold for a given model. A matrix representing total sales for each model and color over the two months can be obtained by adding the corresponding entries in **J** and **F**:

$$\begin{bmatrix} 4 & 3 \\ 7 & 7 \end{bmatrix}.$$

This situation provides some motivation for introducing the operation of matrix addition for two matrices of the same order.

**DEFINITION**

*If* $\mathbf{A} = [a_{ij}]$ *and* $\mathbf{B} = [b_{ij}]$ *are both* $m \times n$ *matrices, then the **sum** $\mathbf{A} + \mathbf{B}$ is the* $m \times n$ *matrix obtained by adding corresponding entries of* $\mathbf{A}$ *and* $\mathbf{B}$; *that is,* $\mathbf{A} + \mathbf{B} = [a_{ij} + b_{ij}]$.

For example, let

$$\mathbf{A} = \begin{bmatrix} 3 & 0 & -2 \\ 2 & -1 & 4 \end{bmatrix} \quad \text{and} \quad \mathbf{B} = \begin{bmatrix} 5 & -3 & 6 \\ 1 & 2 & -5 \end{bmatrix}.$$

## Principles in Practice 1
### Matrix Addition

An office furniture company manufactures desks and tables at two plants, A and B. Matrix **J** represents the production of the two plants in January, and matrix **F** represents the production of the two plants in February. Write a matrix that represents the total production at the two plants for the two months. **J** and **F** are as follows:

$$\mathbf{J} = \begin{matrix} \text{desks} \\ \text{tables} \end{matrix} \begin{matrix} A & B \\ \begin{bmatrix} 120 & 80 \\ 105 & 130 \end{bmatrix} \end{matrix};$$

$$\mathbf{F} = \begin{matrix} \text{desks} \\ \text{tables} \end{matrix} \begin{bmatrix} 110 & 140 \\ 85 & 125 \end{bmatrix}.$$

[[M 217]]These properties of matrix addition are similar to the corresponding properties of real numbers.
[[E 217]]

Since **A** and **B** are the same size ($2 \times 3$), their sum is defined. We have

$$\mathbf{A} + \mathbf{B} = \begin{bmatrix} 3+5 & 0+(-3) & -2+6 \\ 2+1 & -1+2 & 4+(-5) \end{bmatrix} = \begin{bmatrix} 8 & -3 & 4 \\ 3 & 1 & -1 \end{bmatrix}.$$

### EXAMPLE 1   Matrix Addition

a.
$$\begin{bmatrix} 1 & 2 \\ 3 & 4 \\ 5 & 6 \end{bmatrix} + \begin{bmatrix} 7 & -2 \\ -6 & 4 \\ 3 & 0 \end{bmatrix} = \begin{bmatrix} 1+7 & 2-2 \\ 3-6 & 4+4 \\ 5+3 & 6+0 \end{bmatrix} = \begin{bmatrix} 8 & 0 \\ -3 & 8 \\ 8 & 6 \end{bmatrix}.$$

b. $\begin{bmatrix} 1 & 2 \\ 3 & 4 \end{bmatrix} + \begin{bmatrix} 2 \\ 1 \end{bmatrix}$ is not defined, since the matrices are not the same size.  ∎

If **A, B, C,** and **O** have the same order, then the following properties hold for matrix addition:

| Properties of Matrix Addition[[MR 217]] | |
|---|---|
| **1. A + B = B + A** | (commutative property), |
| **2. A + (B + C) = (A + B) + C** | (associative property), |
| **3. A + O = O + A = A** | (identity property). |

Property 1 states that matrices can be added in any order, and Property 2 allows matrices to be grouped for the addition operation. Property 3 states that the zero matrix plays the same role in matrix addition as does the number 0 in the addition of real numbers. These properties are illustrated in the next example.

### EXAMPLE 2   Properties of Matrix Addition

*Let*

$$\mathbf{A} = \begin{bmatrix} 1 & 2 & 1 \\ -2 & 0 & 1 \end{bmatrix}, \qquad \mathbf{B} = \begin{bmatrix} 0 & 1 & 2 \\ 1 & -3 & 1 \end{bmatrix},$$

$$\mathbf{C} = \begin{bmatrix} -2 & 1 & -1 \\ 0 & -2 & 1 \end{bmatrix}, \qquad \mathbf{O} = \begin{bmatrix} 0 & 0 & 0 \\ 0 & 0 & 0 \end{bmatrix}.$$

a. *Show that* **A + B = B + A**.

*Solution:*

$$\mathbf{A} + \mathbf{B} = \begin{bmatrix} 1 & 3 & 3 \\ -1 & -3 & 2 \end{bmatrix}; \qquad \mathbf{B} + \mathbf{A} = \begin{bmatrix} 1 & 3 & 3 \\ -1 & -3 & 2 \end{bmatrix}.$$

Thus, **A + B = B + A.**

b. *Show that* **A + (B + C) = (A + B) + C.**

*Solution:*

$$\mathbf{A} + (\mathbf{B} + \mathbf{C}) = \mathbf{A} + \begin{bmatrix} -2 & 2 & 1 \\ 1 & -5 & 2 \end{bmatrix} = \begin{bmatrix} -1 & 4 & 2 \\ -1 & -5 & 3 \end{bmatrix},$$

$$(\mathbf{A} + \mathbf{B}) + \mathbf{C} = \begin{bmatrix} 1 & 3 & 3 \\ -1 & -3 & 2 \end{bmatrix} + \mathbf{C} = \begin{bmatrix} -1 & 4 & 2 \\ -1 & -5 & 3 \end{bmatrix}.$$

**c.** *Show that* $\mathbf{A} + \mathbf{O} = \mathbf{A}.$

*Solution:*

$$\mathbf{A} + \mathbf{O} = \begin{bmatrix} 1 & 2 & 1 \\ -2 & 0 & 1 \end{bmatrix} + \begin{bmatrix} 0 & 0 & 0 \\ 0 & 0 & 0 \end{bmatrix} = \begin{bmatrix} 1 & 2 & 1 \\ -2 & 0 & 1 \end{bmatrix} = \mathbf{A}. \quad \blacksquare$$

### EXAMPLE 3  Demand Vectors for an Economy

Consider a simplified hypothetical economy having three industries, say, coal, electricity, and steel, and three consumers, 1, 2, and 3. Suppose that each consumer may use some of the output of each industry and also that each industry uses some of the output of each other industry. Then the needs of each consumer and industry can be represented by a (row) demand vector whose entries, in order, give the amount of coal, electricity, and steel needed by the consumer or industry in some convenient units. For example, the demand vectors for the consumers might be

$$\mathbf{D}_1 = [3 \quad 2 \quad 5], \quad \mathbf{D}_2 = [0 \quad 17 \quad 1], \quad \mathbf{D}_3 = [4 \quad 6 \quad 12],$$

and for the industries they might be

$$\mathbf{D}_C = [0 \quad 1 \quad 4], \quad \mathbf{D}_E = [20 \quad 0 \quad 8], \quad \mathbf{D}_S = [30 \quad 5 \quad 0],$$

where the subscripts C, E, and S stand for coal, electricity, and steel, respectively. The total demand for these goods by the consumers is given by the sum

$$\mathbf{D}_1 + \mathbf{D}_2 + \mathbf{D}_3 = [3 \quad 2 \quad 5] + [0 \quad 17 \quad 1] + [4 \quad 6 \quad 12] = [7 \quad 25 \quad 18].$$

The total industrial demand is given by the sum

$$\mathbf{D}_C + \mathbf{D}_E + \mathbf{D}_S = [0 \quad 1 \quad 4] + [20 \quad 0 \quad 8] + [30 \quad 5 \quad 0] = [50 \quad 6 \quad 12].$$

Therefore, the total overall demand is given by

$$[7 \quad 25 \quad 18] + [50 \quad 6 \quad 12] = [57 \quad 31 \quad 30].$$

Thus, the coal industry sells a total of 57 units, the total units of electricity sold is 31, and the total units of steel that are sold is 30.[1] $\quad \blacksquare$

### Scalar Multiplication

Returning to the snowmobile dealer, recall that February sales were given by the matrix

$$\mathbf{F} = \begin{bmatrix} 3 & 1 \\ 4 & 2 \end{bmatrix}.$$

If, in March, the dealer doubles February's sales of each model and color of snowmobile, the sales matrix for March could be obtained by multiplying each entry in $\mathbf{F}$ by 2, yielding

$$\mathbf{M} = \begin{bmatrix} 2(3) & 2(1) \\ 2(4) & 2(2) \end{bmatrix}.$$

It seems reasonable to write this operation as

$$\mathbf{M} = 2\mathbf{F} = 2\begin{bmatrix} 3 & 1 \\ 4 & 2 \end{bmatrix} = \begin{bmatrix} 2 \cdot 3 & 2 \cdot 1 \\ 2 \cdot 4 & 2 \cdot 2 \end{bmatrix} = \begin{bmatrix} 6 & 2 \\ 8 & 4 \end{bmatrix},$$

---

[1]This example, as well as some others in this chapter, are from John G. Kemeny, J. Laurie Snell, and Gerald L. Thompson, *Introduction to Finite Mathematics,* 3d ed. © 1974. Reprinted by permission of Prentice-Hall, Inc., Englewood Cliffs, New Jersey.

which is thought of as multiplying a matrix by a real number. Indeed, we have the following definition.

**DEFINITION**

*If **A** is an m × n matrix and k is a real number (also called a scalar), then, by k**A**, we denote the m × n matrix obtained by multiplying each entry in **A** by k. This operation is called **scalar multiplication**, and k**A** is called a **scalar multiple** of **A**.*

For example,

$$-3\begin{bmatrix} 1 & 0 & -2 \\ 2 & -1 & 4 \end{bmatrix} = \begin{bmatrix} -3(1) & -3(0) & -3(-2) \\ -3(2) & -3(-1) & -3(4) \end{bmatrix} = \begin{bmatrix} -3 & 0 & 6 \\ -6 & 3 & -12 \end{bmatrix}.$$

**EXAMPLE 4  Scalar Multiplication**

Let

$$\mathbf{A} = \begin{bmatrix} 1 & 2 \\ 4 & -2 \end{bmatrix}, \quad \mathbf{B} = \begin{bmatrix} 3 & -4 \\ 7 & 1 \end{bmatrix}, \quad \mathbf{O} = \begin{bmatrix} 0 & 0 \\ 0 & 0 \end{bmatrix}.$$

*Compute the following.*

**a.** 4**A**.

*Solution:*

$$4\mathbf{A} = 4\begin{bmatrix} 1 & 2 \\ 4 & -2 \end{bmatrix} = \begin{bmatrix} 4(1) & 4(2) \\ 4(4) & 4(-2) \end{bmatrix} = \begin{bmatrix} 4 & 8 \\ 16 & -8 \end{bmatrix}.$$

**b.** $-\dfrac{2}{3}$ **B**.

*Solution:*

$$-\frac{2}{3}\mathbf{B} = \begin{bmatrix} -\frac{2}{3}(3) & -\frac{2}{3}(-4) \\ -\frac{2}{3}(7) & -\frac{2}{3}(1) \end{bmatrix} = \begin{bmatrix} -2 & \frac{8}{3} \\ -\frac{14}{3} & -\frac{2}{3} \end{bmatrix}.$$

**c.** $\dfrac{1}{2}$**A** + 3**B**.

*Solution:*

$$\frac{1}{2}\mathbf{A} + 3\mathbf{B} = \frac{1}{2}\begin{bmatrix} 1 & 2 \\ 4 & -2 \end{bmatrix} + 3\begin{bmatrix} 3 & -4 \\ 7 & 1 \end{bmatrix}$$

$$= \begin{bmatrix} \frac{1}{2} & 1 \\ 2 & -1 \end{bmatrix} + \begin{bmatrix} 9 & -12 \\ 21 & 3 \end{bmatrix} = \begin{bmatrix} \frac{19}{2} & -11 \\ 23 & 2 \end{bmatrix}.$$

**d.** 0**A**.

*Solution:*

$$0\mathbf{A} = 0\begin{bmatrix} 1 & 2 \\ 4 & -2 \end{bmatrix} = \begin{bmatrix} 0 & 0 \\ 0 & 0 \end{bmatrix} = \mathbf{O}.$$

**e.** k**O**.

*Solution:*

$$k\mathbf{O} = k\begin{bmatrix} 0 & 0 \\ 0 & 0 \end{bmatrix} = \begin{bmatrix} 0 & 0 \\ 0 & 0 \end{bmatrix} = \mathbf{O}.$$

If **A**, **B**, and **O** are the same size, then, for any scalars $k$, $k_1$, and $k_2$, we have the following properties of scalar multiplication:

**Properties of Scalar Multiplication**

1. $k(\mathbf{A} + \mathbf{B}) = k\mathbf{A} + k\mathbf{B}.$
2. $(k_1 + k_2)\mathbf{A} = k_1\mathbf{A} + k_2\mathbf{A}.$
3. $k_1(k_2\mathbf{A}) = (k_1 k_2)\mathbf{A}.$
4. $0\mathbf{A} = \mathbf{O}.$
5. $k\mathbf{O} = \mathbf{O}.$

Remember that $\mathbf{O} \neq 0$, for 0 is a *scalar* and **O** is a zero *matrix*.

Properties 4 and 5 were illustrated in Examples 4(d) and (e); the others will be illustrated in the exercises.

We also have the following properties of the transpose operation, where **A** and **B** are of the same size and $k$ is any scalar:

$$(\mathbf{A} + \mathbf{B})^{\mathrm{T}} = \mathbf{A}^{\mathrm{T}} + \mathbf{B}^{\mathrm{T}}.$$
$$(k\mathbf{A})^{\mathrm{T}} = k\mathbf{A}^{\mathrm{T}}.$$

The first property states that *the transpose of a sum is the sum of the transposes.*

**Subtraction of Matrices**

If **A** is any matrix, then the scalar multiple $(-1)\mathbf{A}$ is simply written as $-\mathbf{A}$ and is called the **negative of A**:

$$-\mathbf{A} = (-1)\mathbf{A}.$$

Thus, if

$$\mathbf{A} = \begin{bmatrix} 3 & 1 \\ -4 & 5 \end{bmatrix},$$

then

$$-\mathbf{A} = (-1)\begin{bmatrix} 3 & 1 \\ -4 & 5 \end{bmatrix} = \begin{bmatrix} -3 & -1 \\ 4 & -5 \end{bmatrix}.$$

Note that $-\mathbf{A}$ is the matrix obtained by multiplying each entry of **A** by $-1$.

Subtraction of matrices is defined in terms of matrix addition:

**DEFINITION**

*If **A** and **B** are the same size, then, by **A** − **B**, we mean **A** + (−**B**).*

**EXAMPLE 5   Matrix Subtraction**

a.
$$\begin{bmatrix} 2 & 6 \\ -4 & 1 \\ 3 & 2 \end{bmatrix} - \begin{bmatrix} 6 & -2 \\ 4 & 1 \\ 0 & 3 \end{bmatrix} = \begin{bmatrix} 2 & 6 \\ -4 & 1 \\ 3 & 2 \end{bmatrix} + (-1)\begin{bmatrix} 6 & -2 \\ 4 & 1 \\ 0 & 3 \end{bmatrix}$$

$$= \begin{bmatrix} 2 & 6 \\ -4 & 1 \\ 3 & 2 \end{bmatrix} + \begin{bmatrix} -6 & 2 \\ -4 & -1 \\ 0 & -3 \end{bmatrix}$$

$$= \begin{bmatrix} 2-6 & 6+2 \\ -4-4 & 1-1 \\ 3+0 & 2-3 \end{bmatrix} = \begin{bmatrix} -4 & 8 \\ -8 & 0 \\ 3 & -1 \end{bmatrix}.$$

More simply, to find **A** − **B**, we can subtract each entry in **B** from the corresponding entry in **A**.

## Principles in Practice 2
### Matrix Equation

A manufacturer of doors, windows, and cabinets writes her yearly profit (in thousands of dollars) for each category in a vector as $\mathbf{P} = \begin{bmatrix} 248 \\ 319 \\ 532 \end{bmatrix}$. Her fixed costs of production can be described by the vector $\mathbf{C} = \begin{bmatrix} 40 \\ 30 \\ 60 \end{bmatrix}$. She calculates that, with a new pricing structure that generates an income that is 80% of her competitor's income, she can double her profit, assuming that her fixed costs remain the same. This calculation can be represented by

$$0.8 \begin{bmatrix} x_1 \\ x_2 \\ x_3 \end{bmatrix} - \begin{bmatrix} 40 \\ 30 \\ 60 \end{bmatrix} = 2 \begin{bmatrix} 248 \\ 319 \\ 532 \end{bmatrix}.$$

Solve for $x_1$, $x_2$, and $x_3$, which represent her competitor's income from each category.

**b.** If $\mathbf{A} = \begin{bmatrix} 6 & 0 \\ 2 & -1 \end{bmatrix}$ and $\mathbf{B} = \begin{bmatrix} 3 & -3 \\ 1 & 2 \end{bmatrix}$, then

$$\mathbf{A}^{\mathrm{T}} - 2\mathbf{B} = \begin{bmatrix} 6 & 2 \\ 0 & -1 \end{bmatrix} - \begin{bmatrix} 6 & -6 \\ 2 & 4 \end{bmatrix} = \begin{bmatrix} 0 & 8 \\ -2 & -5 \end{bmatrix}.$$ ∎

### EXAMPLE 6   Matrix Equation

*Solve the equation* $2 \begin{bmatrix} x_1 \\ x_2 \end{bmatrix} - \begin{bmatrix} 3 \\ 4 \end{bmatrix} = 5 \begin{bmatrix} 5 \\ -4 \end{bmatrix}.$

*Solution:*

> *Strategy:* We first simplify each side into one matrix. Then, by equality of matrices, we equate corresponding entries.

We have

$$2 \begin{bmatrix} x_1 \\ x_2 \end{bmatrix} - \begin{bmatrix} 3 \\ 4 \end{bmatrix} = 5 \begin{bmatrix} 5 \\ -4 \end{bmatrix},$$

$$\begin{bmatrix} 2x_1 \\ 2x_2 \end{bmatrix} - \begin{bmatrix} 3 \\ 4 \end{bmatrix} = \begin{bmatrix} 25 \\ -20 \end{bmatrix},$$

$$\begin{bmatrix} 2x_1 - 3 \\ 2x_2 - 4 \end{bmatrix} = \begin{bmatrix} 25 \\ -20 \end{bmatrix}.$$

By equality of matrices, we must have $2x_1 - 3 = 25$, which gives $x_1 = 14$; from $2x_2 - 4 = -20$, we get $x_2 = -8$. ∎

## TECHNOLOGY

The matrix operations of addition, subtraction, and scalar multiplication can be performed on a graphics calculator. For example, Fig. 6.2 shows 2**A**   3**B**, where

$$\mathbf{A} = \begin{bmatrix} -2 & 0 \\ 1 & 3 \end{bmatrix} \quad \text{and} \quad \mathbf{B} = \begin{bmatrix} 1 & 2 \\ 4 & 1 \end{bmatrix}.$$

```
2[A]-3[B]
        [[-7  -6]
         [-10 3 ]]
```

**FIGURE 6.2**   Matrix operations with graphics calculator.

## ■ Exercise 6.2

*In Problems 1–12, perform the indicated operations.*

**1.** $\begin{bmatrix} 2 & 0 & -3 \\ -1 & 4 & 0 \\ 1 & -6 & 5 \end{bmatrix} + \begin{bmatrix} 2 & -3 & 4 \\ -1 & 6 & 5 \\ 9 & 11 & -2 \end{bmatrix}.$

**2.** $\begin{bmatrix} 2 & -7 \\ -6 & 4 \end{bmatrix} + \begin{bmatrix} 7 & -4 \\ -2 & 1 \end{bmatrix} + \begin{bmatrix} 2 & 7 \\ 7 & 2 \end{bmatrix}.$

**3.** $\begin{bmatrix} 1 & 4 \\ -2 & 7 \\ 6 & 9 \end{bmatrix} - \begin{bmatrix} 6 & -1 \\ 7 & 2 \\ 1 & 0 \end{bmatrix}.$

**4.** $2 \begin{bmatrix} 3 & -1 & 4 \\ 2 & 1 & -1 \\ 0 & 0 & 2 \end{bmatrix}.$

**5.** $3[1 \quad -3 \quad 1] + 2[-6 \quad 1 \quad 4] - 0[-2 \quad 7 \quad 4].$

**6.** $[7 \quad 7] + 66.$

**7.** $\begin{bmatrix} 1 & 2 \\ 3 & 4 \end{bmatrix} + \begin{bmatrix} 5 \\ 6 \end{bmatrix}$.

**8.** $\begin{bmatrix} 2 & -1 \\ 7 & 4 \end{bmatrix} + 3\begin{bmatrix} 0 & 0 \\ 0 & 0 \end{bmatrix}$.

**9.** $-6\begin{bmatrix} 2 & -6 & 7 & 1 \\ 7 & 1 & 6 & -2 \end{bmatrix}$.

**10.** $\begin{bmatrix} 1 & -1 \\ 2 & 0 \\ 3 & -6 \\ 4 & 9 \end{bmatrix} - 3\begin{bmatrix} -6 & 9 \\ 2 & 6 \\ 1 & -2 \\ 4 & 5 \end{bmatrix}$.

**11.** $\begin{bmatrix} 2 & -4 & 0 \\ 0 & 6 & -2 \\ -4 & 0 & 10 \end{bmatrix} + \frac{1}{3}\begin{bmatrix} 9 & 0 & 3 \\ 0 & 3 & 0 \\ 3 & 9 & 9 \end{bmatrix}$.

**12.** $2\begin{bmatrix} 1 & 0 & 0 \\ 0 & 1 & 0 \\ 0 & 0 & 1 \end{bmatrix} - 3\left(\begin{bmatrix} 2 & 1 & 0 \\ 1 & -2 & 3 \\ 1 & 0 & 0 \end{bmatrix} - \begin{bmatrix} 6 & -2 & 1 \\ -5 & 1 & -2 \\ 0 & 1 & 3 \end{bmatrix}\right)$.

*In Problems 13–24, compute the required matrices if*

$$\mathbf{A} = \begin{bmatrix} 2 & 1 \\ 3 & -3 \end{bmatrix}, \qquad \mathbf{B} = \begin{bmatrix} -6 & -5 \\ 2 & -3 \end{bmatrix}, \qquad \mathbf{C} = \begin{bmatrix} -2 & -1 \\ -3 & 3 \end{bmatrix}, \qquad \mathbf{O} = \begin{bmatrix} 0 & 0 \\ 0 & 0 \end{bmatrix}.$$

**13.** $-\mathbf{B}$.

**14.** $-(\mathbf{A} - \mathbf{B})$.

**15.** $2\mathbf{O}$.

**16.** $\mathbf{A} + \mathbf{B} - \mathbf{C}$.

**17.** $2(\mathbf{A} - 2\mathbf{B})$.

**18.** $0(\mathbf{A} + \mathbf{B})$.

**19.** $3(\mathbf{A} - \mathbf{C}) + 6$.

**20.** $\mathbf{A} + (\mathbf{C} + 2\mathbf{O})$.

**21.** $2\mathbf{B} - 3\mathbf{A} + 2\mathbf{C}$.

**22.** $3\mathbf{C} - 2\mathbf{B}$.

**23.** $\frac{1}{2}\mathbf{A} - 2(\mathbf{B} + 2\mathbf{C})$.

**24.** $2\mathbf{A} - \frac{1}{2}(\mathbf{B} - \mathbf{C})$.

*In Problems 25–28, verify the equations for the preceding matrices* **A, B,** *and* **C.**

**25.** $3(\mathbf{A} + \mathbf{B}) = 3\mathbf{A} + 3\mathbf{B}$.

**26.** $(2 + 3)\mathbf{A} = 2\mathbf{A} + 3\mathbf{A}$.

**27.** $k_1(k_2\mathbf{A}) = (k_1k_2)\mathbf{A}$.

**28.** $k(\mathbf{A} + \mathbf{B} + \mathbf{C}) = k\mathbf{A} + k\mathbf{B} + k\mathbf{C}$.

*In Problems 29–34, let*

$$\mathbf{A} = \begin{bmatrix} 1 & 2 \\ 0 & -1 \\ 2 & 0 \end{bmatrix}, \qquad \mathbf{B} = \begin{bmatrix} 1 & 3 \\ 4 & -1 \end{bmatrix}, \qquad \mathbf{C} = \begin{bmatrix} 1 & 0 \\ 1 & 2 \end{bmatrix}, \qquad \mathbf{D} = \begin{bmatrix} 1 & 2 & -1 \\ 1 & 0 & 2 \end{bmatrix}.$$

*Compute the indicated matrices if possible.*

**29.** $3\mathbf{A}^{\mathrm{T}} + \mathbf{D}$.

**30.** $(\mathbf{B} - \mathbf{C})^{\mathrm{T}}$.

**31.** $2\mathbf{B}^{\mathrm{T}} - 3\mathbf{C}^{\mathrm{T}}$.

**32.** $2\mathbf{B} + \mathbf{B}^{\mathrm{T}}$.

**33.** $\mathbf{C}^{\mathrm{T}} - \mathbf{D}$.

**34.** $(\mathbf{D} - 2\mathbf{A}^{\mathrm{T}})^{\mathrm{T}}$.

**35.** Express the matrix equation

$$x\begin{bmatrix} 2 \\ 1 \end{bmatrix} - y\begin{bmatrix} -3 \\ 5 \end{bmatrix} = 2\begin{bmatrix} 8 \\ 11 \end{bmatrix}$$

as a system of linear equations and solve.

**36.** In the reverse of the manner used in Problem 35, write the system

$$\begin{cases} 3x + 5y = 16 \\ 2x - 6y = -4 \end{cases}$$

as a matrix equation.

*In Problems 37–40, solve the matrix equations.*

**37.** $3\begin{bmatrix} x \\ y \end{bmatrix} - 3\begin{bmatrix} -2 \\ 4 \end{bmatrix} = 4\begin{bmatrix} 6 \\ -2 \end{bmatrix}$.

**38.** $3\begin{bmatrix} x \\ 2 \end{bmatrix} - 4\begin{bmatrix} 7 \\ -y \end{bmatrix} = \begin{bmatrix} -x \\ 2y \end{bmatrix}$.

**39.** $\begin{bmatrix} 2 \\ 4 \\ 6 \end{bmatrix} + 2\begin{bmatrix} x \\ y \\ 4z \end{bmatrix} = \begin{bmatrix} -10 \\ -24 \\ 14 \end{bmatrix}$.

**40.** $x\begin{bmatrix} 2 \\ 0 \\ 3 \end{bmatrix} + 2\begin{bmatrix} -1 \\ 0 \\ 6 \end{bmatrix} + y\begin{bmatrix} 0 \\ 2 \\ -3 \end{bmatrix} = \begin{bmatrix} 8 \\ 4 \\ 3x + 12 - 3y \end{bmatrix}$.

41. **Production**   An electronics company manufactures televisions, VCRs, and CD players at two plants, A and B. Matrix **X** represents the production of the two plants for retailer X, and matrix **Y** represents the production of the two plants for retailer Y. Write a matrix that represents the total production at the two plants for both retailers. Matrices **X** and **Y** are as follows:

$$\mathbf{X} = \begin{array}{c} \\ \text{TV} \\ \text{VCR} \\ \text{CD} \end{array} \begin{array}{cc} A & B \\ \begin{bmatrix} 20 & 40 \\ 45 & 30 \\ 15 & 10 \end{bmatrix} \end{array}; \mathbf{Y} = \begin{array}{c} \\ \text{TV} \\ \text{VCR} \\ \text{CD} \end{array} \begin{array}{cc} A & B \\ \begin{bmatrix} 15 & 25 \\ 30 & 25 \\ 10 & 5 \end{bmatrix} \end{array}.$$

42. **Sales**   Let matrix **A** represent the sales (in thousands of dollars) of a toy company in 1994 in three cities, and let **B** represent the sales in the same cities in 1996, where **A** and **B** are given by

$$\mathbf{A} = \begin{array}{c} \text{Action} \\ \text{Educational} \end{array} \begin{bmatrix} 400 & 350 & 150 \\ 450 & 280 & 850 \end{bmatrix},$$

$$\mathbf{B} = \begin{array}{c} \text{Action} \\ \text{Educational} \end{array} \begin{bmatrix} 380 & 330 & 220 \\ 460 & 320 & 750 \end{bmatrix}.$$

If the company buys a competitor and doubles its 1996 sales in 1997, what is the change in sales between 1994 and 1997?

43. Suppose the prices of products A, B, and C are given, in that order, by the price vector

$$\mathbf{P} = [p_1 \quad p_2 \quad p_3].$$

If the prices are to be increased by 10%, the vector for the new prices can be obtained by multiplying **P** by what scalar?

44. Prove that $(\mathbf{A} - \mathbf{B})^{\mathrm{T}} = \mathbf{A}^{\mathrm{T}} - \mathbf{B}^{\mathrm{T}}$. [*Hint:* Use the definition of subtraction and properties of the transpose operation.]

 *In Problems 45–47, compute the given matrices if*

$$\mathbf{A} = \begin{bmatrix} 3 & -4 & 5 \\ -2 & 1 & 6 \end{bmatrix}, \quad \mathbf{B} = \begin{bmatrix} 1 & 4 & 2 \\ 4 & 1 & 2 \end{bmatrix}, \quad and \quad \mathbf{C} = \begin{bmatrix} -1 & 1 & 3 \\ 2 & 6 & -6 \end{bmatrix}.$$

45. $4\mathbf{A} + 3\mathbf{B}$.

46. $-2(\mathbf{A} + \mathbf{B}) - \mathbf{C}$.

47. $2(3\mathbf{C} - \mathbf{A}) + 2\mathbf{B}$.

---

OBJECTIVE

**To define multiplication of matrices and to consider associated properties. To express a system as a single matrix equation by using matrix multiplication.**

## 6.3 MATRIX MULTIPLICATION

Besides the operations of matrix addition and scalar multiplication, the product **AB** of matrices **A** and **B** can be defined under certain conditions, namely, that the number of columns of **A** is equal to the number of rows of **B**. Although the following definition of matrix multiplication may not appear to you to be a natural one (it would seem more natural to simply multiply corresponding entries), a more thorough study of matrices would convince you that the definition is appropriate and extremely practical for applications.

### DEFINITION

*Let **A** be an m × n matrix and **B** be an n × p matrix. Then the product **AB** is the m × p matrix **C** whose entry $c_{ij}$ in row i and column j is obtained as follows: Sum the products formed by multiplying, in order, each entry (that is, first, second, etc.) in row i of **A** by the "corresponding" entry (that is, first, second, etc.) in column j of **B**.*

Three points must be completely understood concerning this definition of **AB**. First, the condition that **A** be m × n and **B** be n × p is equivalent to saying that the number of columns of **A** must be equal to the number of rows of **B.** Second, the product will be a matrix of order $m \times p$ it will have as many rows as **A** and as many columns as **B.**

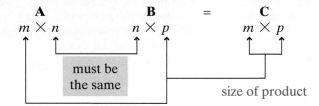

Third, the definition refers to the product **AB,** *in that order;* **A** is the left factor and **B** is the right factor. For **AB,** we say that **B** is *premultiplied* by **A** or **A** is *postmultiplied* by **B.**

To apply the definition, let us find the product

$$\mathbf{AB} = \begin{bmatrix} 2 & 1 & -6 \\ 1 & -3 & 2 \end{bmatrix} \begin{bmatrix} 1 & 0 & -3 \\ 0 & 4 & 2 \\ -2 & 1 & 1 \end{bmatrix}.$$

Matrix **A** has size $2 \times 3$ $(m \times n)$ and matrix **B** has size $3 \times 3$ $(n \times p)$. The number of columns of **A** is equal to the number of rows of **B** $(n = 3)$, so the product **C** is defined and will be a $2 \times 3$ $(m \times p)$ matrix; that is,

$$\mathbf{C} = \begin{bmatrix} c_{11} & c_{12} & c_{13} \\ c_{21} & c_{22} & c_{23} \end{bmatrix}.$$

The entry $c_{11}$ is obtained by summing the products of each entry in row 1 of **A** by the "corresponding" entry in column 1 of **B.** Thus,

row 1 entries of **A**

$$c_{11} = (2)(1) + (1)(0) + (-6)(-2) = 14.$$

column 1 entries of **B**

At this stage, we have

$$\begin{bmatrix} 2 & 1 & -6 \\ 1 & -3 & 2 \end{bmatrix} \begin{bmatrix} 1 & 0 & -3 \\ 0 & 4 & 2 \\ -2 & 1 & 1 \end{bmatrix} = \begin{bmatrix} 14 & c_{12} & c_{13} \\ c_{21} & c_{22} & c_{23} \end{bmatrix}.$$

Similarly, for $c_{12}$, we use the entries in row 1 of **A** and those in column 2 of **B:**

row 1 entries of **A**

$$c_{12} = (2)(0) + (1)(4) + (-6)(1) = -2.$$

column 2 entries of **B**

We now have

$$\begin{bmatrix} 2 & 1 & -6 \\ 1 & -3 & 2 \end{bmatrix} \begin{bmatrix} 1 & 0 & -3 \\ 0 & 4 & 2 \\ -2 & 1 & 1 \end{bmatrix} = \begin{bmatrix} 14 & -2 & c_{13} \\ c_{21} & c_{22} & c_{23} \end{bmatrix}.$$

For the remaining entries of **AB,** we obtain

$$c_{13} = (2)(-3) + (1)(2) + (-6)(1) = -10,$$
$$c_{21} = (1)(1) + (-3)(0) + (2)(-2) = -3,$$

$$c_{22} = (1)(0) + (-3)(4) + (2)(1) = -10,$$

$$c_{23} = (1)(-3) + (-3)(2) + (2)(1) = -7.$$

Thus,

$$\mathbf{AB} = \begin{bmatrix} 2 & 1 & -6 \\ 1 & -3 & 2 \end{bmatrix} \begin{bmatrix} 1 & 0 & -3 \\ 0 & 4 & 2 \\ -2 & 1 & 1 \end{bmatrix} = \begin{bmatrix} 14 & -2 & -10 \\ -3 & -10 & -7 \end{bmatrix}.$$

Note that if we reverse the order of the factors, then the product

$$\mathbf{BA} = \begin{bmatrix} 1 & 0 & -3 \\ 0 & 4 & 2 \\ -2 & 1 & 1 \end{bmatrix} \begin{bmatrix} 2 & 1 & -6 \\ 1 & -3 & 2 \end{bmatrix}$$

Matrix multiplication is not commutative.

is *not* defined, because the number of columns of **B** does *not* equal the number of rows of **A.** This shows that matrix multiplication is not commutative. That is, for any matrices **A** and **B**, it is usually the case that $\mathbf{AB} \neq \mathbf{BA}$ (even if both products are defined), so the order in which the matrices in a product are written is extremely important.

### EXAMPLE 1   Sizes of Matrices and Their Product

Let **A** be a $3 \times 5$ matrix and **B** be a $5 \times 3$ matrix. Then **AB** is defined and is a $3 \times 3$ matrix. Moreover, **BA** is also defined and is a $5 \times 5$ matrix.

    If **C** is a $3 \times 5$ matrix and **D** is a $7 \times 3$ matrix, then **CD** is undefined, but **DC** is defined and is a $7 \times 5$ matrix. ∎

### EXAMPLE 2   Matrix Product

*Compute the matrix product*

$$\mathbf{AB} = \begin{bmatrix} 2 & -4 & 2 \\ 0 & 1 & -3 \end{bmatrix} \begin{bmatrix} 2 & 1 \\ 0 & 4 \\ 2 & 2 \end{bmatrix}.$$

*Solution:* Since **A** is $2 \times 3$ and **B** is $3 \times 2$, the product **AB** is defined and will have order $2 \times 2$. By simultaneously moving the index finger of your left hand along the rows of **A** and the index finger of your right hand along the columns of **B,** it should not be difficult for you to find mentally the entries of the product. We obtain

$$\begin{bmatrix} 2 & -4 & 2 \\ 0 & 1 & -3 \end{bmatrix} \begin{bmatrix} 2 & 1 \\ 0 & 4 \\ 2 & 2 \end{bmatrix} = \begin{bmatrix} 8 & -10 \\ -6 & -2 \end{bmatrix}.$$
∎

### Principles in Practice 1
**Matrix Products**

A bookstore has 100 dictionaries, 70 cookbooks, and 90 thesauruses in stock. If the value of each dictionary is \$28, each cookbook is \$22, and each thesaurus is \$16, use a matrix product to find the total value of the bookstore's inventory.

### EXAMPLE 3   Matrix Products

**a.** *Compute* $\begin{bmatrix} 1 & 2 & 3 \end{bmatrix} \begin{bmatrix} 4 \\ 5 \\ 6 \end{bmatrix}$.

*Solution:* The product has order $1 \times 1$:

$$\begin{bmatrix} 1 & 2 & 3 \end{bmatrix} \begin{bmatrix} 4 \\ 5 \\ 6 \end{bmatrix} = \begin{bmatrix} 32 \end{bmatrix}.$$

**b.** *Compute* $\begin{bmatrix} 1 \\ 2 \\ 3 \end{bmatrix} \begin{bmatrix} 1 & 6 \end{bmatrix}$.

*Solution:* The product has order $3 \times 2$:

$$\begin{bmatrix} 1 \\ 2 \\ 3 \end{bmatrix} \begin{bmatrix} 1 & 6 \end{bmatrix} = \begin{bmatrix} 1 & 6 \\ 2 & 12 \\ 3 & 18 \end{bmatrix}.$$

**c.** $\begin{bmatrix} 1 & 3 & 0 \\ -2 & 2 & 1 \\ 1 & 0 & -4 \end{bmatrix} \begin{bmatrix} 1 & 0 & 2 \\ 5 & -1 & 3 \\ 2 & 1 & -2 \end{bmatrix} = \begin{bmatrix} 16 & -3 & 11 \\ 10 & -1 & 0 \\ -7 & -4 & 10 \end{bmatrix}.$

**d.** $\begin{bmatrix} a_{11} & a_{12} \\ a_{21} & a_{22} \end{bmatrix} \begin{bmatrix} b_{11} & b_{12} \\ b_{21} & b_{22} \end{bmatrix} = \begin{bmatrix} a_{11}b_{11} + a_{12}b_{21} & a_{11}b_{12} + a_{12}b_{22} \\ a_{21}b_{11} + a_{22}b_{21} & a_{21}b_{12} + a_{22}b_{22} \end{bmatrix}.$ ▬

### EXAMPLE 4 Matrix Products

*Compute* **AB** *and* **BA** *if*

$$\mathbf{A} = \begin{bmatrix} 2 & -1 \\ 3 & 1 \end{bmatrix} \quad and \quad \mathbf{B} = \begin{bmatrix} -2 & 1 \\ 1 & 4 \end{bmatrix}.$$

Example 4 shows that even when the matrix products **AB** and **BA** are defined, they are not necessarily equal.

*Solution:* We have

$$\mathbf{AB} = \begin{bmatrix} 2 & -1 \\ 3 & 1 \end{bmatrix} \begin{bmatrix} -2 & 1 \\ 1 & 4 \end{bmatrix} = \begin{bmatrix} -5 & -2 \\ -5 & 7 \end{bmatrix},$$

$$\mathbf{BA} = \begin{bmatrix} -2 & 1 \\ 1 & 4 \end{bmatrix} \begin{bmatrix} 2 & -1 \\ 3 & 1 \end{bmatrix} = \begin{bmatrix} -1 & 3 \\ 14 & 3 \end{bmatrix}.$$

Note that although both **AB** and **BA** are defined, **AB** $\neq$ **BA.** ▬

## TECHNOLOGY

Figure 6.3 shows the result of using a graphics calculator to find the product **AB** in Example 4.

```
[A][B]
         [[-5 -2]
          [-5  7]]
```

**FIGURE 6.3** Calculator solution of matrix product of Example 4.

### EXAMPLE 5 Cost Vector

Suppose that the prices (in dollars per unit) for products A, B, and C are represented by the price vector

Price of
A  B  C

$$\mathbf{P} = \begin{bmatrix} 2 & 3 & 4 \end{bmatrix}.$$

If the quantities (in units) of A, B, and C that are purchased are given by the column vector

$$\mathbf{Q} = \begin{bmatrix} 7 \\ 5 \\ 11 \end{bmatrix} \begin{matrix} \text{units of A} \\ \text{units of B} \\ \text{units of C,} \end{matrix}$$

then the total cost (in dollars) of the purchases is given by the entry in the cost vector

$$\mathbf{PQ} = \begin{bmatrix} 2 & 3 & 4 \end{bmatrix} \begin{bmatrix} 7 \\ 5 \\ 11 \end{bmatrix} = [(2 \cdot 7) + (3 \cdot 5) + (4 \cdot 11)] = [73]. \quad \blacksquare$$

## Principles in Practice 2

### Cost Vector

The prices (in dollars per unit) for three textbooks are represented by the price vector $\mathbf{P} = [26.25 \ 34.75 \ 28.50]$. A university bookstore orders these books in the quantities given by the column vector $\mathbf{Q} = \begin{bmatrix} 250 \\ 325 \\ 175 \end{bmatrix}$. Find the total cost (in dollars) of the purchase.

**EXAMPLE 6  Profit for an Economy**

In Example 3 of Sec. 6.2, suppose that in the hypothetical economy the price of coal is \$10,000 per unit, the price of electricity is \$20,000 per unit, and the price of steel is \$40,000 per unit. These prices can be represented by the (column) price vector

$$\mathbf{P} = \begin{bmatrix} 10{,}000 \\ 20{,}000 \\ 40{,}000 \end{bmatrix}.$$

Consider the steel industry. It sells a total of 30 units of steel at \$40,000 per unit, and its total income is therefore \$1,200,000. Its costs for the various goods are given by the matrix product

$$\mathbf{D_S P} = \begin{bmatrix} 30 & 5 & 0 \end{bmatrix} \begin{bmatrix} 10{,}000 \\ 20{,}000 \\ 40{,}000 \end{bmatrix} = [400{,}000].$$

Hence, the profit for the steel industry is \$1,200,000 − \$400,000 = \$800,000.

$\blacksquare$

Matrix multiplication satisfies the following properties, provided that all sums and products are defined:

**Properties of Matrix Multiplication**

1. $\mathbf{A(BC)} = \mathbf{(AB)C}$ \qquad (associative property),

2. $\mathbf{A(B + C)} = \mathbf{AB} + \mathbf{AC},$ \quad (distributive properties).
   $\mathbf{(A + B)C} = \mathbf{AC} + \mathbf{BC}$

**EXAMPLE 7  Associative Property**

*If*

$$\mathbf{A} = \begin{bmatrix} 1 & -2 \\ -3 & 4 \end{bmatrix}, \quad \mathbf{B} = \begin{bmatrix} 3 & 0 & -1 \\ 1 & 1 & 2 \end{bmatrix}, \quad and \quad \mathbf{C} = \begin{bmatrix} 1 & 0 \\ 0 & 2 \\ 1 & 1 \end{bmatrix},$$

*compute* **ABC** *in two ways.*

*Solution:* Grouping **BC** gives

$$\mathbf{A(BC)} = \begin{bmatrix} 1 & -2 \\ -3 & 4 \end{bmatrix} \left( \begin{bmatrix} 3 & 0 & -1 \\ 1 & 1 & 2 \end{bmatrix} \begin{bmatrix} 1 & 0 \\ 0 & 2 \\ 1 & 1 \end{bmatrix} \right)$$

$$= \begin{bmatrix} 1 & -2 \\ -3 & 4 \end{bmatrix} \begin{bmatrix} 2 & -1 \\ 3 & 4 \end{bmatrix} = \begin{bmatrix} -4 & -9 \\ 6 & 19 \end{bmatrix}.$$

Alternatively, grouping **AB** gives

$$(\mathbf{AB})\mathbf{C} = \left( \begin{bmatrix} 1 & -2 \\ -3 & 4 \end{bmatrix} \begin{bmatrix} 3 & 0 & -1 \\ 1 & 1 & 2 \end{bmatrix} \right) \begin{bmatrix} 1 & 0 \\ 0 & 2 \\ 1 & 1 \end{bmatrix}$$

$$= \begin{bmatrix} 1 & -2 & -5 \\ -5 & 4 & 11 \end{bmatrix} \begin{bmatrix} 1 & 0 \\ 0 & 2 \\ 1 & 1 \end{bmatrix}$$

$$= \begin{bmatrix} -4 & -9 \\ 6 & 19 \end{bmatrix}.$$

Note that $\mathbf{A}(\mathbf{BC}) = (\mathbf{AB})\mathbf{C}.$ ■

### EXAMPLE 8   Distributive Property

*Verify that* $\mathbf{A}(\mathbf{B} + \mathbf{C}) = \mathbf{AB} + \mathbf{AC}$ *if*

$$\mathbf{A} = \begin{bmatrix} 1 & 0 \\ 2 & 3 \end{bmatrix}, \quad \mathbf{B} = \begin{bmatrix} -2 & 0 \\ 1 & 3 \end{bmatrix}, \quad and \quad \mathbf{C} = \begin{bmatrix} -2 & 1 \\ 0 & 2 \end{bmatrix}.$$

*Solution:* On the left side, we have

$$\mathbf{A}(\mathbf{B} + \mathbf{C}) = \begin{bmatrix} 1 & 0 \\ 2 & 3 \end{bmatrix} \left( \begin{bmatrix} -2 & 0 \\ 1 & 3 \end{bmatrix} + \begin{bmatrix} -2 & 1 \\ 0 & 2 \end{bmatrix} \right)$$

$$= \begin{bmatrix} 1 & 0 \\ 2 & 3 \end{bmatrix} \begin{bmatrix} -4 & 1 \\ 1 & 5 \end{bmatrix} = \begin{bmatrix} -4 & 1 \\ -5 & 17 \end{bmatrix}.$$

On the right side,

$$\mathbf{AB} + \mathbf{AC} = \begin{bmatrix} 1 & 0 \\ 2 & 3 \end{bmatrix} \begin{bmatrix} -2 & 0 \\ 1 & 3 \end{bmatrix} + \begin{bmatrix} 1 & 0 \\ 2 & 3 \end{bmatrix} \begin{bmatrix} -2 & 1 \\ 0 & 2 \end{bmatrix}$$

$$= \begin{bmatrix} -2 & 0 \\ -1 & 9 \end{bmatrix} + \begin{bmatrix} -2 & 1 \\ -4 & 8 \end{bmatrix} = \begin{bmatrix} -4 & 1 \\ -5 & 17 \end{bmatrix}.$$

Thus, $\mathbf{A}(\mathbf{B} + \mathbf{C}) = \mathbf{AB} + \mathbf{AC}.$ ■

### EXAMPLE 9   Raw Materials and Cost

Suppose that a building contractor has accepted orders for five ranch-style houses, seven Cape Cod-style houses, and 12 colonial-style houses. Then his orders can be represented by the row vector

$$\mathbf{Q} = [5 \quad 7 \quad 12].$$

Furthermore, suppose that the "raw materials" which go into each type of house are steel, wood, glass, paint, and labor. The entries in the following matrix **R** give the number of units of each raw material going into each type of house (the entries are not necessarily realistic, but are chosen for convenience):

| | Steel | Wood | Glass | Paint | Labor | |
|---|---|---|---|---|---|---|
| Ranch | 5 | 20 | 16 | 7 | 17 | |
| Cape Cod | 7 | 18 | 12 | 9 | 21 | $= \mathbf{R}.$ |
| Colonial | 6 | 25 | 8 | 5 | 13 | |

Each row indicates the amount of each raw material needed for a given type of house; each column indicates the amount of a given raw material needed for each type of house. Suppose now that the contractor wishes to compute the amount of each raw material needed to fulfill his orders. Then such information is given by the matrix

$$\mathbf{QR} = \begin{bmatrix} 5 & 7 & 12 \end{bmatrix} \begin{bmatrix} 5 & 20 & 16 & 7 & 17 \\ 7 & 18 & 12 & 9 & 21 \\ 6 & 25 & 8 & 5 & 13 \end{bmatrix}$$

$$= \begin{bmatrix} 146 & 526 & 260 & 158 & 388 \end{bmatrix}.$$

Thus, the contractor should order 146 units of steel, 526 units of wood, 260 units of glass, and so on.

The contractor is also interested in the costs he will have to pay for these materials. Suppose steel costs $1500 per unit, wood costs $800 per unit, and glass, paint, and labor cost $500, $100, and $1000 per unit, respectively. These data can be written as the column cost vector

$$\mathbf{C} = \begin{bmatrix} 1500 \\ 800 \\ 500 \\ 100 \\ 1000 \end{bmatrix}.$$

Then the cost of each type of house is given by the matrix

$$\mathbf{RC} = \begin{bmatrix} 5 & 20 & 16 & 7 & 17 \\ 7 & 18 & 12 & 9 & 21 \\ 6 & 25 & 8 & 5 & 13 \end{bmatrix} \begin{bmatrix} 1500 \\ 800 \\ 500 \\ 100 \\ 1000 \end{bmatrix} = \begin{bmatrix} 49{,}200 \\ 52{,}800 \\ 46{,}500 \end{bmatrix}.$$

Consequently, the cost of materials for the ranch-style house is $49,200, for the Cape Code house $52,800, and for the colonial house $46,500.

The total cost of raw materials for all the houses is given by

$$\mathbf{QRC} = \mathbf{Q}(\mathbf{RC}) = \begin{bmatrix} 5 & 7 & 12 \end{bmatrix} \begin{bmatrix} 49{,}200 \\ 52{,}800 \\ 46{,}500 \end{bmatrix} = \begin{bmatrix} 1{,}173{,}600 \end{bmatrix}.$$

The total cost is $1,173,600.    ■

Another property of matrices involves scalar and matrix multiplications. If $k$ is a scalar and the product $\mathbf{AB}$ is defined, then

$$k(\mathbf{AB}) = (k\mathbf{A})\mathbf{B} = \mathbf{A}(k\mathbf{B}).$$

The product $k(\mathbf{AB})$ may be written simply as $k\mathbf{AB}$. Thus,

$$k\mathbf{AB} = k(\mathbf{AB}) = (k\mathbf{A})\mathbf{B} = \mathbf{A}(k\mathbf{B}).$$

For example,

$$3\begin{bmatrix} 2 & 1 \\ 0 & -1 \end{bmatrix}\begin{bmatrix} 1 & 3 \\ 2 & 0 \end{bmatrix} = \left(3\begin{bmatrix} 2 & 1 \\ 0 & -1 \end{bmatrix}\right)\begin{bmatrix} 1 & 3 \\ 2 & 0 \end{bmatrix}$$

$$= \begin{bmatrix} 6 & 3 \\ 0 & -3 \end{bmatrix}\begin{bmatrix} 1 & 3 \\ 2 & 0 \end{bmatrix}$$

$$= \begin{bmatrix} 12 & 18 \\ -6 & 0 \end{bmatrix}.$$

There is an interesting property concerning the transpose of a matrix product:

$$(\mathbf{AB})^{\mathrm{T}} = \mathbf{B}^{\mathrm{T}}\mathbf{A}^{\mathrm{T}}.$$

In words, the transpose of a product of matrices is equal to the product of their transposes in the *reverse* order.

This property can be extended to the case of more than two factors. For example,

$$(\mathbf{A}^{\mathrm{T}}\mathbf{B}\mathbf{C})^{\mathrm{T}} = \mathbf{C}^{\mathrm{T}}\mathbf{B}^{\mathrm{T}}(\mathbf{A}^{\mathrm{T}})^{\mathrm{T}} = \mathbf{C}^{\mathrm{T}}\mathbf{B}^{\mathrm{T}}\mathbf{A}.$$

Here we used the fact that $(\mathbf{A}^{\mathrm{T}})^{\mathrm{T}} = \mathbf{A}.$

**EXAMPLE 10    Transpose of a Product**

*Let*

$$\mathbf{A} = \begin{bmatrix} 1 & 0 \\ 1 & 2 \end{bmatrix} \quad and \quad \mathbf{B} = \begin{bmatrix} 1 & 2 \\ 1 & 0 \end{bmatrix}.$$

*Show that* $(\mathbf{AB})^{\mathrm{T}} = \mathbf{B}^{\mathrm{T}}\mathbf{A}^{\mathrm{T}}.$

*Solution:* We have

$$\mathbf{AB} = \begin{bmatrix} 1 & 2 \\ 3 & 2 \end{bmatrix}, \quad so \quad (\mathbf{AB})^{\mathrm{T}} = \begin{bmatrix} 1 & 3 \\ 2 & 2 \end{bmatrix}.$$

Now,

$$\mathbf{A}^{\mathrm{T}} = \begin{bmatrix} 1 & 1 \\ 0 & 2 \end{bmatrix} \quad and \quad \mathbf{B}^{\mathrm{T}} = \begin{bmatrix} 1 & 1 \\ 2 & 0 \end{bmatrix}.$$

Thus,

$$\mathbf{B}^{\mathrm{T}}\mathbf{A}^{\mathrm{T}} = \begin{bmatrix} 1 & 1 \\ 2 & 0 \end{bmatrix}\begin{bmatrix} 1 & 1 \\ 0 & 2 \end{bmatrix} = \begin{bmatrix} 1 & 3 \\ 2 & 2 \end{bmatrix} = (\mathbf{AB})^{\mathrm{T}},$$

so $(\mathbf{AB})^{\mathrm{T}} = \mathbf{B}^{\mathrm{T}}\mathbf{A}^{\mathrm{T}}.$  ∎

Just as the zero matrix plays an important role as the identity in matrix addition, there is a special matrix, called the *identity matrix,* that plays a corresponding role in matrix multiplication:

The $n \times n$ **identity matrix,** denoted $\mathbf{I}_n$, is the diagonal matrix whose main diagonal entries are 1's.

For example, the identity matrices $I_3$ and $I_4$ are

$$I_3 = \begin{bmatrix} 1 & 0 & 0 \\ 0 & 1 & 0 \\ 0 & 0 & 1 \end{bmatrix} \quad \text{and} \quad I_4 = \begin{bmatrix} 1 & 0 & 0 & 0 \\ 0 & 1 & 0 & 0 \\ 0 & 0 & 1 & 0 \\ 0 & 0 & 0 & 1 \end{bmatrix}.$$

When the size of an identity matrix is understood to be appropriate for an operation to be defined, we shall omit the subscript and simply denote the matrix by **I**. It should be clear that

$$I^T = I.$$

The identity matrix plays the same role in matrix multiplication as does the number 1 in the multiplication of real numbers. That is, just as the product of a real number and 1 is the number itself, the product of a matrix and the identity matrix is the matrix itself. For example,

$$\begin{bmatrix} 2 & 4 \\ 1 & 5 \end{bmatrix} I = \begin{bmatrix} 2 & 4 \\ 1 & 5 \end{bmatrix}\begin{bmatrix} 1 & 0 \\ 0 & 1 \end{bmatrix} = \begin{bmatrix} 2 & 4 \\ 1 & 5 \end{bmatrix}$$

and

$$I\begin{bmatrix} 2 & 4 \\ 1 & 5 \end{bmatrix} = \begin{bmatrix} 1 & 0 \\ 0 & 1 \end{bmatrix}\begin{bmatrix} 2 & 4 \\ 1 & 5 \end{bmatrix} = \begin{bmatrix} 2 & 4 \\ 1 & 5 \end{bmatrix}.$$

In general, if **I** is $n \times n$ and **A** has $n$ columns, then $AI = A$. If **B** has $n$ rows, then $IB = B$. Moreover, if **A** is $n \times n$, then

$$AI = IA = A.$$

**EXAMPLE 11** Matrix Operations Involving I and O

*If*

$$A = \begin{bmatrix} 3 & 2 \\ 1 & 4 \end{bmatrix}, \quad B = \begin{bmatrix} \frac{2}{5} & -\frac{1}{5} \\ -\frac{1}{10} & \frac{3}{10} \end{bmatrix},$$

$$I = \begin{bmatrix} 1 & 0 \\ 0 & 1 \end{bmatrix}, \quad \text{and} \quad O = \begin{bmatrix} 0 & 0 \\ 0 & 0 \end{bmatrix}.$$

*compute each of the following.*

**a. I − A.**

*Solution:*

$$I - A = \begin{bmatrix} 1 & 0 \\ 0 & 1 \end{bmatrix} - \begin{bmatrix} 3 & 2 \\ 1 & 4 \end{bmatrix} = \begin{bmatrix} -2 & -2 \\ -1 & -3 \end{bmatrix}.$$

**b.** $3(A − 2I)$.

*Solution:*

$$3(A - 2I) = 3\left(\begin{bmatrix} 3 & 2 \\ 1 & 4 \end{bmatrix} - 2\begin{bmatrix} 1 & 0 \\ 0 & 1 \end{bmatrix}\right)$$
$$= 3\left(\begin{bmatrix} 3 & 2 \\ 1 & 4 \end{bmatrix} - \begin{bmatrix} 2 & 0 \\ 0 & 2 \end{bmatrix}\right)$$
$$= 3\begin{bmatrix} 1 & 2 \\ 1 & 2 \end{bmatrix} = \begin{bmatrix} 3 & 6 \\ 3 & 6 \end{bmatrix}.$$

c. **AO.**

*Solution:*

$$\mathbf{AO} = \begin{bmatrix} 3 & 2 \\ 1 & 4 \end{bmatrix} \begin{bmatrix} 0 & 0 \\ 0 & 0 \end{bmatrix} = \begin{bmatrix} 0 & 0 \\ 0 & 0 \end{bmatrix} = \mathbf{O}.$$

In general, if **AO** and **OA** are defined, then

$$\mathbf{AO} = \mathbf{OA} = \mathbf{O}.$$

d. **AB.**

*Solution:*

$$\mathbf{AB} = \begin{bmatrix} 3 & 2 \\ 1 & 4 \end{bmatrix} \begin{bmatrix} \frac{2}{5} & -\frac{1}{5} \\ -\frac{1}{10} & \frac{3}{10} \end{bmatrix} = \begin{bmatrix} 1 & 0 \\ 0 & 1 \end{bmatrix} = \mathbf{I}.$$

If **A** is a square matrix, we can speak of a *power* of **A**:

If **A** is a square matrix and $p$ is a positive integer, then the **$p$th power** of **A,** written $\mathbf{A}^p$, is the product of $p$ factors of **A**:

$$\mathbf{A}^p = \underbrace{\mathbf{A} \cdot \mathbf{A} \cdots \mathbf{A}}_{p \text{ factors}}.$$

If **A** is $n \times n$, we define $\mathbf{A}^0 = \mathbf{I}_n$.

We remark that $\mathbf{I}^p = \mathbf{I}$.

**EXAMPLE 12** Power of a Matrix

*If* $\mathbf{A} = \begin{bmatrix} 1 & 0 \\ 1 & 2 \end{bmatrix}$, *compute* $\mathbf{A}^3$.

*Solution:* Since $\mathbf{A}^3 = (\mathbf{A}^2)\mathbf{A}$ and

$$\mathbf{A}^2 = \begin{bmatrix} 1 & 0 \\ 1 & 2 \end{bmatrix} \begin{bmatrix} 1 & 0 \\ 1 & 2 \end{bmatrix} = \begin{bmatrix} 1 & 0 \\ 3 & 4 \end{bmatrix},$$

we have

$$\mathbf{A}^3 = \mathbf{A}^2\mathbf{A} = \begin{bmatrix} 1 & 0 \\ 3 & 4 \end{bmatrix} \begin{bmatrix} 1 & 0 \\ 1 & 2 \end{bmatrix} = \begin{bmatrix} 1 & 0 \\ 7 & 8 \end{bmatrix}.$$

## TECHNOLOGY

Use of the graphics calculator to compute $\mathbf{A}^4$, where $\mathbf{A} = \begin{bmatrix} 2 & -3 \\ 1 & 4 \end{bmatrix}$, is shown in Fig. 6.4.

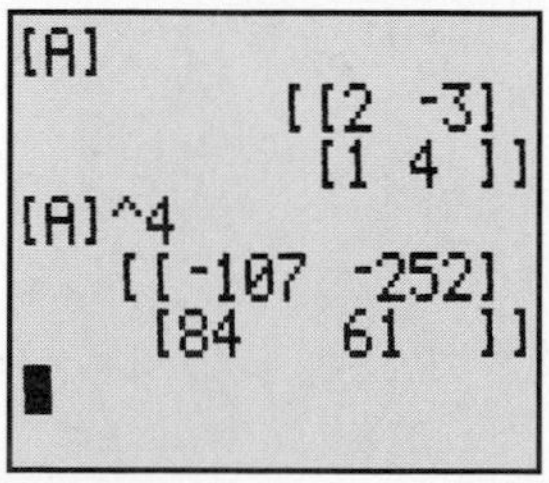

```
[A]
        [[2 -3]
         [1 4 ]]
[A]^4
    [[-107 -252]
     [84    61 ]]
```

**FIGURE 6.4** Power of a matrix.

## Matrix Equations

Systems of linear equations can be represented by using matrix multiplication. For example, consider the matrix equation

$$\begin{bmatrix} 1 & 4 & -2 \\ 2 & -3 & 1 \end{bmatrix} \begin{bmatrix} x_1 \\ x_2 \\ x_3 \end{bmatrix} = \begin{bmatrix} 4 \\ -3 \end{bmatrix}. \tag{1}$$

The product on the left side has order $2 \times 1$ and hence is a column matrix. Thus,

$$\begin{bmatrix} x_1 + 4x_2 - 2x_3 \\ 2x_1 - 3x_2 + x_3 \end{bmatrix} = \begin{bmatrix} 4 \\ -3 \end{bmatrix}.$$

By equality of matrices, corresponding entries must be equal, so we obtain the system

$$\begin{cases} x_1 + 4x_2 - 2x_3 = 4 \\ 2x_1 - 3x_2 + x_3 = -3. \end{cases}$$

Hence, this system of linear equations can be defined by matrix equation (1). We usually describe Eq. (1) by saying that it has the form

$$\mathbf{AX = B},$$

where $\mathbf{A}$ is the matrix obtained from the coefficients of the variables, $\mathbf{X}$ is a column matrix obtained from the variables, and $\mathbf{B}$ is a column matrix obtained from the constants. Matrix $\mathbf{A}$ is called the *coefficient matrix* for the system.

### Principles in Practice 3
**Matrix Form of a System Using Matrix Multiplication**

Write the following pair of lines in matrix form, using matrix multiplication.

$$y = -\frac{8}{5}x + \frac{8}{5}, y = -\frac{1}{3}x + \frac{5}{3}.$$

**EXAMPLE 13** Matrix Form of a System Using Matrix Multiplication

*Write the system*

$$\begin{cases} 2x_1 + 5x_2 = 4, \\ 8x_1 + 3x_2 = 7, \end{cases}$$

*in matrix form by using matrix multiplication.*

**Solution:** If

$$\mathbf{A} = \begin{bmatrix} 2 & 5 \\ 8 & 3 \end{bmatrix}, \quad \mathbf{X} = \begin{bmatrix} x_1 \\ x_2 \end{bmatrix}, \quad \text{and} \quad \mathbf{B} = \begin{bmatrix} 4 \\ 7 \end{bmatrix}.$$

then the given system is equivalent to the single matrix equation

$$\mathbf{AX = B},$$

or

$$\begin{bmatrix} 2 & 5 \\ 8 & 3 \end{bmatrix} \begin{bmatrix} x_1 \\ x_2 \end{bmatrix} = \begin{bmatrix} 4 \\ 7 \end{bmatrix}.$$

## ■ Exercise 6.3

If $\mathbf{A} = \begin{bmatrix} 1 & 3 & -2 \\ -2 & 1 & -1 \\ 0 & 4 & 3 \end{bmatrix}$, $\mathbf{B} = \begin{bmatrix} 0 & -2 & 3 \\ -2 & 4 & -2 \\ 3 & 1 & -1 \end{bmatrix}$, and $\mathbf{AB = C}$, *find each of the following.*

**1.** $c_{11}$.

**2.** $c_{23}$.

**3.** $c_{32}$.

**4.** $c_{33}$.

**5.** $c_{22}$.

**6.** $c_{13}$.

*If A is 2 × 3, B is 3 × 1, C is 2 × 5, D is 4 × 3, E is 3 × 2, and F is 2 × 3, find the order and number of entries of each of the following.*

7. **AE.**          8. **DE.**          9. **EC.**          10. **DB.**

11. **FB.**         12. **BA.**        13. **EA.**        14. **E(AE).**

15. **E(FB).**      16. **(F + A)B.**

*Write the identity matrix that has the following order.*

17. 4               18. 6.

*In Problems 19–36, perform the indicated operations.*

19. $\begin{bmatrix} 2 & -4 \\ 3 & 2 \end{bmatrix}\begin{bmatrix} 3 & 0 \\ -1 & 4 \end{bmatrix}$.

20. $\begin{bmatrix} -1 & 1 \\ 0 & 4 \\ 2 & 1 \end{bmatrix}\begin{bmatrix} 1 & -2 \\ 3 & 4 \end{bmatrix}$.

21. $\begin{bmatrix} 2 & 0 & 3 \\ -1 & 4 & 5 \end{bmatrix}\begin{bmatrix} 1 \\ 4 \\ 7 \end{bmatrix}$.

22. $\begin{bmatrix} 1 & 0 & 6 & 2 \end{bmatrix}\begin{bmatrix} 0 \\ 1 \\ 2 \\ 3 \end{bmatrix}$.

23. $\begin{bmatrix} 1 & 4 & -1 \\ 0 & 0 & 2 \\ -2 & 1 & 1 \end{bmatrix}\begin{bmatrix} -2 & 1 & 0 \\ 0 & 1 & 1 \\ 1 & 1 & 2 \end{bmatrix}$.

24. $\begin{bmatrix} 3 & 2 & -1 \\ 4 & 10 & 0 \\ 0 & 1 & 2 \end{bmatrix}\begin{bmatrix} 2 & 0 & 1 & 0 \\ 0 & 1 & 0 & 0 \\ 0 & 1 & 0 & 1 \end{bmatrix}$.

25. $\begin{bmatrix} -1 & 2 & 3 \end{bmatrix}\begin{bmatrix} 3 & 1 & -1 & 2 \\ 0 & 4 & 3 & 1 \\ -1 & 3 & 1 & -2 \end{bmatrix}$.

26. $\begin{bmatrix} 1 & -4 \end{bmatrix}\begin{bmatrix} -2 & 1 \\ 0 & 5 \\ 1 & 0 \end{bmatrix}$.

27. $\begin{bmatrix} 2 \\ 3 \\ -4 \\ 1 \end{bmatrix}\begin{bmatrix} 2 & 3 & -2 & 3 \end{bmatrix}$.

28. $\begin{bmatrix} 0 & 1 \\ 2 & 3 \end{bmatrix}\left(\begin{bmatrix} 1 & 0 & 1 \\ 0 & 1 & 0 \end{bmatrix} + \begin{bmatrix} 0 & 1 & 0 \\ 0 & 0 & 1 \end{bmatrix}\right)$.

29. $3\left(\begin{bmatrix} -2 & 0 & 2 \\ 3 & -1 & 1 \end{bmatrix} + 2\begin{bmatrix} -1 & 0 & 2 \\ 1 & 1 & -2 \end{bmatrix}\right)\begin{bmatrix} 1 & 2 \\ 3 & 4 \\ 5 & 6 \end{bmatrix}$.

30. $\begin{bmatrix} -1 & 3 \\ -1 & 0 \end{bmatrix}\begin{bmatrix} -1 & 0 & 2 & -1 \\ 2 & 1 & -3 & -2 \end{bmatrix}$.

31. $\begin{bmatrix} 1 & 2 \\ 3 & 4 \end{bmatrix}\left(\begin{bmatrix} 2 & 0 & 1 \\ 1 & 0 & -2 \end{bmatrix}\begin{bmatrix} 1 & -2 \\ 2 & 1 \\ 3 & 0 \end{bmatrix}\right)$.

32. $3\begin{bmatrix} 1 & 2 \\ -1 & 4 \end{bmatrix} - 4\left(\begin{bmatrix} 1 & 0 \\ 0 & 1 \end{bmatrix}\begin{bmatrix} -2 & 4 \\ 6 & 1 \end{bmatrix}\right)$.

33. $\begin{bmatrix} 1 & 0 & 0 \\ 0 & 1 & 0 \\ 0 & 0 & 1 \end{bmatrix}\begin{bmatrix} x \\ y \\ z \end{bmatrix}$.

34. $\begin{bmatrix} a_{11} & a_{12} \\ a_{21} & a_{22} \end{bmatrix}\begin{bmatrix} x_1 \\ x_2 \end{bmatrix}$.

35. $\begin{bmatrix} 2 & 1 & 3 \\ 4 & 9 & 7 \end{bmatrix}\begin{bmatrix} x_1 \\ x_2 \\ x_3 \end{bmatrix}$.

36. $\begin{bmatrix} 1 & -2 \\ 0 & 1 \\ 3 & 2 \end{bmatrix}\begin{bmatrix} x_1 \\ x_2 \end{bmatrix}$.

*In Problems 37–44, compute the required matrices if*

$$\mathbf{A} = \begin{bmatrix} 1 & -2 \\ 0 & 3 \end{bmatrix}, \quad \mathbf{B} = \begin{bmatrix} -2 & 3 & 0 \\ 1 & -4 & 1 \end{bmatrix}, \quad \mathbf{C} = \begin{bmatrix} -1 & 1 \\ 0 & 3 \\ 2 & 4 \end{bmatrix}, \quad \mathbf{D} = \begin{bmatrix} 1 & 0 & 0 \\ 0 & 1 & 1 \\ 1 & 2 & 1 \end{bmatrix},$$

$$\mathbf{E} = \begin{bmatrix} 3 & 0 & 0 \\ 0 & 6 & 0 \\ 0 & 0 & 3 \end{bmatrix}, \quad \mathbf{F} = \begin{bmatrix} \frac{1}{3} & 0 & 0 \\ 0 & \frac{1}{6} & 0 \\ 0 & 0 & \frac{1}{3} \end{bmatrix}, \quad \mathbf{I} = \begin{bmatrix} 1 & 0 & 0 \\ 0 & 1 & 0 \\ 0 & 0 & 1 \end{bmatrix}.$$

37. **DI** − $\frac{1}{3}$**E.**          38. **DD.**          39. **3A** − **2BC.**          40. **B(D** + **E).**

**41.** $2\mathbf{I} - \frac{1}{2}\mathbf{EF}.$          **42.** $\mathbf{F}(2\mathbf{D} - 3\mathbf{I}).$          **43.** $(\mathbf{DC})\mathbf{A}.$          **44.** $\mathbf{A}(\mathbf{BC}).$

*In each of Problems 45–58, compute the required matrix, if it exists, given that*

$$\mathbf{A} = \begin{bmatrix} 1 & -1 & 0 \\ 0 & 1 & 1 \end{bmatrix}, \quad \mathbf{B} = \begin{bmatrix} 0 & 0 & -1 \\ 2 & -1 & 0 \\ 0 & 0 & 2 \end{bmatrix}, \quad \mathbf{C} = \begin{bmatrix} 1 & 0 \\ 2 & -1 \\ 0 & 1 \end{bmatrix},$$

$$\mathbf{I} = \begin{bmatrix} 1 & 0 & 0 \\ 0 & 1 & 0 \\ 0 & 0 & 1 \end{bmatrix}, \quad \mathbf{O} = \begin{bmatrix} 0 & 0 & 0 \\ 0 & 0 & 0 \\ 0 & 0 & 0 \end{bmatrix}.$$

**45.** $\mathbf{A}^2.$

**46.** $\mathbf{A}^{\mathsf{T}}\mathbf{A}.$

**47.** $\mathbf{B}^3.$

**48.** $\mathbf{A}(\mathbf{B}^{\mathsf{T}})^2.$

**49.** $(\mathbf{AC})^2.$

**50.** $\mathbf{A}^{\mathsf{T}}(2\mathbf{C}^{\mathsf{T}}).$

**51.** $(\mathbf{BA}^{\mathsf{T}})^{\mathsf{T}}.$

**52.** $(2\mathbf{B})^{\mathsf{T}}.$

**53.** $(2\mathbf{I})^2 - 2\mathbf{I}^2.$

**54.** $\mathbf{IA}^0.$

**55.** $\mathbf{A}(\mathbf{I} - \mathbf{O}).$

**56.** $\mathbf{I}^{\mathsf{T}}\mathbf{O}.$

**57.** $(\mathbf{AB})(\mathbf{AB})^{\mathsf{T}}.$

**58.** $\mathbf{B}^2 - 3\mathbf{B} + 2\mathbf{I}.$

*In Problems 59–61, represent the given system by using matrix multiplication.*

**59.** $\begin{cases} 3x + y = 6, \\ 7x - 2y = 5. \end{cases}$

**60.** $\begin{cases} x + y + z = 6, \\ x - y + z = 2, \\ 2x - y + 3z = 6. \end{cases}$

**61.** $\begin{cases} 4r - s + 3t = 9, \\ 3r - t = 7, \\ 3s + 2t = 15. \end{cases}$

**62. Secret Messages** Secret messages can be encoded by using a code and an encoding matrix. Suppose we have the following code:

| a | b | c | d | e | f | g | h | i | j | k | l | m |
|---|---|---|---|---|---|---|---|---|---|---|---|---|
| 1 | 2 | 3 | 4 | 5 | 6 | 7 | 8 | 9 | 10 | 11 | 12 | 13 |

| n | o | p | q | r | s | t | u | v | w | x | y | z |
|---|---|---|---|---|---|---|---|---|---|---|---|---|
| 14 | 15 | 16 | 17 | 18 | 19 | 20 | 21 | 22 | 23 | 24 | 25 | 26 |

Let the encoding matrix be $\mathbf{E} = \begin{bmatrix} 1 & 3 \\ 2 & 4 \end{bmatrix}$. Then we can encode a message by taking every two letters of the message, converting them to their corresponding numbers, creating a $2 \times 1$ matrix, and then multiplying each matrix by $\mathbf{E}$. Use this code and matrix to encode the message "meet/at/noon/Friday," leaving the slashes to separate words.

**63. Inventory** A pet store has 6 kittens, 10 puppies, and 7 parrots in stock. If the value of each kitten is \$55, each puppy is \$150, and each parrot is \$35, using matrix multiplication, find the total value of the pet store's inventory.

**64. Stocks** A stockbroker sold a customer 200 shares of stock A, 300 shares of stock B, 500 shares of stock C, and 250 shares of stock D. The prices per share of A, B, C, and D are \$100, \$150, \$200, and \$300, respectively. Write a row vector representing the number of shares of each stock bought. Write a column vector representing the price per share of each stock. Using matrix multiplication, find the total cost of the stocks.

**65. Construction Cost** In Example 9, assume that the contractor is to build seven ranch-style, three Cape Cod-style, and five colonial-style houses. Using matrix multiplication, compute the total cost of raw materials.

**66. Costs** In Example 9, assume that the contractor wishes to take into account the cost of transporting raw materials to the building site, as well as the purchasing cost. Suppose the costs are given in the following matrix:

$$\mathbf{C} = \begin{bmatrix} 1500 & 45 \\ 800 & 20 \\ 500 & 30 \\ 100 & 5 \\ 1000 & 0 \end{bmatrix} \begin{matrix} \text{Steel} \\ \text{Wood} \\ \text{Glass} \\ \text{Paint} \\ \text{Labor.} \end{matrix}$$

with column headings Purchase and Transport.

**a.** By computing $\mathbf{RC}$, find a matrix whose entries give the purchase and transportation costs of the materials for each type of house.

**b.** Find the matrix $\mathbf{QRC}$ whose first entry gives the total purchase price and whose second entry gives the total transportation cost.

**c.** Let $\mathbf{Z} = \begin{bmatrix} 1 \\ 1 \end{bmatrix}$, and then compute $\mathbf{QRCZ}$, which gives the total cost of materials and transportation for all houses being built.

**67.** Perform the following calculations for Example 6.

**a.** Compute the amount that each industry and each consumer have to pay for the goods they receive.

**b.** Compute the profit earned by each industry.

**c.** Find the total amount of money that is paid out by all the industries and consumers.

**d.** Find the proportion of the total amount of money found in (c) paid out by the industries. Find the proportion of the total amount of money found in (c) that is paid out by the consumers.

**68.** Prove that if $\mathbf{AB} = \mathbf{BA}$, then $(\mathbf{A} + \mathbf{B})(\mathbf{A} - \mathbf{B}) = \mathbf{A}^2 - \mathbf{B}^2$.

**69.** Show that if

$$\mathbf{A} = \begin{bmatrix} 1 & 2 \\ 1 & 2 \end{bmatrix} \quad \text{and} \quad \begin{bmatrix} 2 & -3 \\ -1 & \frac{3}{2} \end{bmatrix},$$

then $\mathbf{AB} = \mathbf{O}$. Observe that since neither $\mathbf{A}$ nor $\mathbf{B}$ is the zero matrix, the algebraic rule for real numbers, "if $ab = 0$, then either $a$ or $b$ is zero," does not hold for ma-

trices. It can also be shown that the cancellation law is not true for matrices; that is, if $\mathbf{AB} = \mathbf{AC}$, then it is not necessarily true that $\mathbf{B} = \mathbf{C}$.

**70.** Let $\mathbf{D}_1$ and $\mathbf{D}_2$ be two arbitrary $3 \times 3$ diagonal matrices. By computing $\mathbf{D}_1\mathbf{D}_2$ and $\mathbf{D}_2\mathbf{D}_1$, show that

**a.** Both $\mathbf{D}_1\mathbf{D}_2$ and $\mathbf{D}_2\mathbf{D}_1$ are diagonal matrices.

**b.** $\mathbf{D}_1$ and $\mathbf{D}_2$ commute.

 *In Problems 71–74, compute the required matrices, given that*

$$\mathbf{A} = \begin{bmatrix} 3.2 & -4.1 & 5.1 \\ -2.6 & 1.2 & 6.8 \end{bmatrix}, \qquad \mathbf{B} = \begin{bmatrix} 1.1 & 4.8 \\ -2.3 & 3.2 \\ 4.6 & -1.4 \end{bmatrix}, \qquad \textit{and} \qquad \mathbf{C} = \begin{bmatrix} -1.2 & 1.5 \\ 2.4 & 6.2 \end{bmatrix}.$$

**71.** $\mathbf{A}(2\mathbf{B})$.　　　　**72.** $-2(\mathbf{BC})$.　　　　**73.** $(-\mathbf{C})(3\mathbf{A})\mathbf{B}$.　　　　**74.** $\mathbf{C}^3$.

---

O B J E C T I V E

To show how to reduce a matrix and to use matrix reduction to solve a linear system.

## 6.4 METHOD OF REDUCTION

In this section, we shall illustrate a method by which matrices can be used to solve a system of linear equations. In developing this *method of reduction,* we shall first solve a system by the usual method of elimination. Then we shall obtain the same solution by using matrices.

Let us consider the system

$$\begin{cases} 3x - y = 1, & (1) \\ x + 2y = 5, & (2) \end{cases}$$

consisting of two linear equations in two unknowns, $x$ and $y$. Although this system can be solved by various algebraic methods, we shall solve it by a method that is readily adapted to matrices.

For reasons that will be obvious later, we begin by replacing Eq. (1) by Eq. (2), and Eq. (2) by Eq. (1), thus obtaining the equivalent system,[2]

$$\begin{cases} x + 2y = 5, & (3) \\ 3x - y = 1. & (4) \end{cases}$$

Multiplying both sides of Eq. (3) by $-3$ gives $-3x - 6y = -15$. Adding the left and right sides of this equation to the corresponding sides of Eq. (4) produces an equivalent system in which $x$ is eliminated from the second equation:

$$\begin{cases} x + 2y = 5, & (5) \\ 0x - 7y = -14. & (6) \end{cases}$$

Now we shall eliminate $y$ from the first equation. Multiplying both sides of Eq. (6) by $-\frac{1}{7}$ gives the equivalent system,

$$\begin{cases} x + 2y = 5, & (7) \\ 0x + y = 2. & (8) \end{cases}$$

---

[2]Recall from Sec. 4.4 that two or more systems are equivalent if they have the same solution.

From Eq. (8), $y = 2$, and hence, $-2y = -4$. Adding the sides of $-2y = -4$ to the corresponding sides of Eq. (7), we get the equivalent system,

$$\begin{cases} x + 0y = 1, \\ 0x + \phantom{2}y = 2. \end{cases}$$

Therefore, $x = 1$ and $y = 2$, so the original system is solved.

Note that in solving the original system, we successively replaced it by an equivalent system that was obtained by performing one of the following three operations (called *elementary operations*) which leave the solution unchanged:

1. Interchanging two equations.

2. Adding a constant multiple of the sides of one equation to the corresponding sides of another equation.

3. Multiplying one equation by a nonzero constant.

Before showing a matrix method of solving the original system,

$$\begin{cases} 3x - \phantom{2}y = 1, \\ \phantom{3}x + 2y = 5, \end{cases}$$

we first need to define some terms. Recall from Sec. 6.3 that the matrix

$$\begin{bmatrix} 3 & -1 \\ 1 & 2 \end{bmatrix}$$

is the **coefficient matrix** of this system. The entries in the first column correspond to the coefficients of the $x$'s in the equations. For example, the entry in the first row and first column corresponds to the coefficient of $x$ in the first equation; and the entry in the second row and first column corresponds to the coefficient of $x$ in the second equation. Similarly, the entries in the second column correspond to the coefficients of the $y$'s.

Another matrix associated with this system is called the **augmented coefficient matrix** and is given by

$$\begin{bmatrix} 3 & -1 & 1 \\ 1 & 2 & 5 \end{bmatrix}$$

The first and second columns are the first and second columns, respectively, of the coefficient matrix. The entries in the third column correspond to the constant terms in the system: the entry in the first row of this column is the constant term of the first equation, whereas the entry in the second row is the constant term of the second equation. Although it is not necessary to include the vertical line in the augmented coefficient matrix, it serves to remind us that the 1 and the 5 are the constant terms that appear on the right sides of the equations. The augmented coefficient matrix itself completely describes the system of equations.

The procedure that was used to solve the original system involved a number of equivalent systems. With each of these systems, we can associate its augmented coefficient matrix. Following are the systems that were involved, together with their corresponding augmented coefficient matrices, which we have labeled **A, B, C, D,** and **E**:

$$\begin{cases} 3x - y = 1, \\ x + 2y = 5. \end{cases} \qquad \begin{bmatrix} 3 & -1 & | & 1 \\ 1 & 2 & | & 5 \end{bmatrix} = \mathbf{A}.$$

$$\begin{cases} x + 2y = 5, \\ 3x - y = 1. \end{cases} \qquad \begin{bmatrix} 1 & 2 & | & 5 \\ 3 & -1 & | & 1 \end{bmatrix} = \mathbf{B}.$$

$$\begin{cases} x + 2y = 5, \\ 0x - 7y = -14. \end{cases} \qquad \begin{bmatrix} 1 & 2 & | & 5 \\ 0 & -7 & | & -14 \end{bmatrix} = \mathbf{C}.$$

$$\begin{cases} x + 2y = 5, \\ 0x + y = 2. \end{cases} \qquad \begin{bmatrix} 1 & 2 & | & 5 \\ 0 & 1 & | & 2 \end{bmatrix} = \mathbf{D}.$$

$$\begin{cases} x + 0y = 1, \\ 0x + y = 2. \end{cases} \qquad \begin{bmatrix} 1 & 0 & | & 1 \\ 0 & 1 & | & 2 \end{bmatrix} = \mathbf{E}.$$

Let us see how these matrices are related.

**B** can be obtained from **A** by interchanging the first and second rows of **A.** This operation corresponds to interchanging the two equations in the original system.

**C** can be obtained from **B** by adding to each entry in the second row of **B** $-3$ times the corresponding entry in the first row of **B**:

$$\mathbf{C} = \begin{bmatrix} 1 & 2 & | & 5 \\ 3 + (-3)(1) & -1 + (-3)(2) & | & 1 + (-3)(5) \end{bmatrix}$$

$$= \begin{bmatrix} 1 & 2 & | & 5 \\ 0 & -7 & | & -14 \end{bmatrix}.$$

This operation is described as the addition of $-3$ times the first row of **B** to the second row of **B.**

**D** can be obtained from **C** by multiplying each entry in the second row of **C** by $-\frac{1}{7}$. This operation is referred to as multiplying the second row of **C** by $-\frac{1}{7}$.

**E** can be obtained from **D** by adding $-2$ times the second row of **D** to the first row of **D.**

Observe that **E,** which essentially gives the solution, was obtained from **A** by successively performing one of three matrix operations, called **elementary row operations:**

---

**Elementary Row Operation**

1. Interchanging two rows of a matrix;

2. Adding a multiple of one row of a matrix to a different row of that matrix;

3. Multiplying a row of a matrix by a nonzero scalar.

---

These elementary row operations correspond to the three elementary operations used in the algebraic method of elimination. Whenever a matrix can be obtained from another by one or more elementary row operations, we say that the matrices are **equivalent.** Thus, **A** and **E** are equivalent. (We could also obtain **A** from **E** by performing similar row operations in the reverse order, so the term *equivalent* is appropriate.) When describing particular elementary row operations, we will use the following notation for convenience:

| Notation | Corresponding Row Operation |
|---|---|
| $R_i \leftrightarrow R_j$ | interchange rows $R_i$ and $R_j$ |
| $kR_i$ | Multiply row $R_i$ by the constant $k$ |
| $kR_i + R_j$ | Add $k$ times row $R_i$ to row $R_j$ (but row $R_i$ remains the same) |

For example, writing

$$\begin{bmatrix} 1 & 0 & -2 \\ 4 & -2 & 1 \\ 5 & 0 & 3 \end{bmatrix} \xrightarrow{-4R_1 + R_2} \begin{bmatrix} 1 & 0 & -2 \\ 0 & -2 & 9 \\ 5 & 0 & 3 \end{bmatrix}$$

means that the second matrix was obtained from the first by adding $-4$ times row 1 to row 2. Note that we may write $(-k)R_i$ as $-kR_i$.

We are now ready to describe a matrix procedure for solving a system of linear equations. First, we form the augmented coefficient matrix of the system; then, by means of elementary row operations, we determine an equivalent matrix that clearly indicates the solution. Let us be quite specific as to what we mean by a matrix that clearly indicates the solution. This is a matrix, called a *reduced matrix*, defined as follows:

**Reduced Matrix**

A matrix is said to be a **reduced matrix,**[3] provided that all of the following are true:

1. If a row does not consist entirely of zeros, then the first nonzero entry in the row, called the **leading entry,** is 1, whereas all other entries in the column in which the 1 appears are zeros.

2. The first nonzero entry in each row is to the right of the first nonzero entry in each row above it.

3. Any rows that consist entirely of zeros are at the bottom of the matrix.

In other words, to solve the system, we must find the reduced matrix such that the augmented coefficient matrix is equivalent to it. In our previous discussion of elementary row operations, the matrix

$$\mathbf{E} = \left[\begin{array}{cc|c} 1 & 0 & 1 \\ 0 & 1 & 2 \end{array}\right]$$

is a reduced matrix.

**EXAMPLE 1   Reduced Matrices**

*For each of the following matrices, determine whether it is reduced or not reduced.*

**a.** $\begin{bmatrix} 1 & 0 \\ 0 & 3 \end{bmatrix}$.    **b.** $\begin{bmatrix} 1 & 0 & 0 \\ 0 & 1 & 0 \end{bmatrix}$.    **c.** $\begin{bmatrix} 0 & 1 \\ 1 & 0 \end{bmatrix}$.

[3]Or in *reduced row echelon form.*

**d.** $\begin{bmatrix} 0 & 0 & 0 \\ 0 & 0 & 0 \end{bmatrix}$.   **e.** $\begin{bmatrix} 1 & 0 & 0 \\ 0 & 0 & 0 \\ 0 & 1 & 0 \end{bmatrix}$.   **f.** $\begin{bmatrix} 0 & 1 & 0 & 3 \\ 0 & 0 & 1 & 2 \\ 0 & 0 & 0 & 0 \end{bmatrix}$.

*Solution:*

**a.** Not a reduced matrix, because the leading entry in the second row is not 1.
**b.** Reduced matrix.
**c.** Not a reduced matrix, because the leading entry in the second row is not to the right of the first nonzero entry in the first row.
**d.** Reduced matrix.
**e.** Not a reduced matrix, because the second row, consisting entirely of zeros, is not at the bottom of the matrix.
**f.** Reduced matrix.

**EXAMPLE 2   Reducing a Matrix**

*Reduce the matrix*

$$\begin{bmatrix} 0 & 0 & 1 & 2 \\ 3 & -6 & -3 & 0 \\ 6 & -12 & 2 & 11 \end{bmatrix}.$$

> *Strategy:* To reduce the matrix, we must get the leading entry to be a 1 in the first row, a 1 in the second row, and so on, until we arrive at a zero row, if there is any. Moreover, we must work from left to right, because the leading 1 in each row must be to the *left* of all other leading 1's in the rows below it.

*Solution:* Since there are no zero rows to move to the bottom, we proceed to find the first column that contains a nonzero entry; this turns out to be column 1. Accordingly, in the reduced matrix, the leading 1 in the first row must be in column 1. To accomplish this, we shall begin by interchanging the first two rows so that a nonzero entry is in row 1 of column 1:

$$\begin{bmatrix} 0 & 0 & 1 & 2 \\ 3 & -6 & -3 & 0 \\ 6 & -12 & 2 & 11 \end{bmatrix}$$

$$\xrightarrow{\;R_1 \leftrightarrow R_2\;} \begin{bmatrix} 3 & -6 & -3 & 0 \\ 0 & 0 & 1 & 2 \\ 6 & -12 & 2 & 11 \end{bmatrix}.$$

Next, we multiply row 1 by $\frac{1}{3}$ so that the leading entry is a 1:

$$\xrightarrow{\;\frac{1}{3}R_1\;} \begin{bmatrix} 1 & -2 & -1 & 0 \\ 0 & 0 & 1 & 2 \\ 6 & -12 & 2 & 11 \end{bmatrix}.$$

Now, because we must have zeros below (and above) each leading 1, we add $-6$ times row 1 to row 3:

$$\xrightarrow{\;-6R_1 + R_3\;} \begin{bmatrix} 1 & -2 & -1 & 0 \\ 0 & 0 & 1 & 2 \\ 0 & 0 & 8 & 11 \end{bmatrix}.$$

Next, we move to the right of column 1 to find the first column that has a nonzero entry in row 2 or below; this is column 3. Consequently, in the reduced matrix, the leading 1 in the second row must be in column 3. The foregoing matrix already does have a leading 1 there. Thus, all we need do to get zeros below and above the leading 1 is add 1 times row 2 to row 1 and add $-8$ times row 2 to row 3:

$$\xrightarrow[\,-8R_2 + R_3\,]{(1)R_2 + R_1} \begin{bmatrix} 1 & -2 & 0 & 2 \\ 0 & 0 & 1 & 2 \\ 0 & 0 & 0 & -5 \end{bmatrix}.$$

Again, we move to the right to find the first column that has a nonzero entry in row 3; namely, column 4. To make the leading entry a 1, we multiply row 3 by $-\frac{1}{5}$:

$$\xrightarrow{\,-\frac{1}{5}R_3\,} \begin{bmatrix} 1 & -2 & 0 & 2 \\ 0 & 0 & 1 & 2 \\ 0 & 0 & 0 & 1 \end{bmatrix}.$$

Finally, to get all other entries in column 4 to be zeros, we add $-2$ times row 3 to both row 1 and row 2:

The sequence of steps that is used to reduce a matrix is not unique; however, the reduced form *is* unique.

$$\xrightarrow[\,-2R_3 + R_2\,]{-2R_3 + R_1} \begin{bmatrix} 1 & -2 & 0 & 0 \\ 0 & 0 & 1 & 0 \\ 0 & 0 & 0 & 1 \end{bmatrix}.$$

The last matrix is in reduced form.   ■

## TECHNOLOGY

Although elementary row operations can be performed on a graphics calculator, the procedure is rather awkward.

The method of reduction described for solving our original system can be generalized to systems consisting of $m$ linear equations in $n$ unknowns. To solve such a system as

$$\begin{cases} a_{11}x_1 + a_{12}x_2 + \cdots + a_{1n}x_n = c_1, \\ a_{21}x_1 + a_{22}x_2 + \cdots + a_{2n}x_n = c_2, \\ \quad \vdots \qquad\quad \vdots \qquad\qquad \vdots \qquad \vdots \\ a_{m1}x_1 + a_{m2}x_2 + \cdots + a_{mn}x_n = c_m \end{cases}$$

involves

1. determining the augmented coefficient matrix of the system, which is

$$\begin{bmatrix} a_{11} & a_{12} & \cdots & a_{1n} & c_1 \\ a_{21} & a_{22} & \cdots & a_{2n} & c_2 \\ \vdots & \vdots & & \vdots & \vdots \\ a_{m1} & a_{m2} & \cdots & a_{mn} & c_m \end{bmatrix},$$

and

**2.** determining a reduced matrix such that the augmented coefficient matrix is equivalent to it.

Frequently, step 2 is called *reducing the augmented coefficient matrix.*

**Principles in Practice 1**
Solving a System by Reduction

An investment firm offers three stock portfolios: A, B, and C. The number of blocks of each type of stock in each of these portfolios is summarized in the following table:

|  |  | Portfolio | | |
|---|---|---|---|---|
|  |  | A | B | C |
|  | High | 6 | 1 | 3 |
| Risk: | Moderate | 3 | 2 | 3 |
|  | Low | 1 | 5 | 3 |

A client wants 35 blocks of high-risk stock, 22 blocks of moderate-risk stock, and 18 blocks of low-risk stock. How many of each portfolio should be suggested?

**EXAMPLE 3**   Solving a System by Reduction

*By using matrix reduction, solve the system*

$$\begin{cases} 2x + 3y = -1, \\ 2x + y = 5, \\ x + y = 1. \end{cases}$$

*Solution:* Reducing the augmented coefficient matrix of the system, we have

$$\begin{bmatrix} 2 & 3 & | & -1 \\ 2 & 1 & | & 5 \\ 1 & 1 & | & 1 \end{bmatrix} \xrightarrow{R_1 \leftrightarrow R_3} \begin{bmatrix} 1 & 1 & | & 1 \\ 2 & 1 & | & 5 \\ 2 & 3 & | & -1 \end{bmatrix}$$

$$\xrightarrow{-2R_1 + R_2} \begin{bmatrix} 1 & 1 & | & 1 \\ 0 & -1 & | & 3 \\ 2 & 3 & | & -1 \end{bmatrix}$$

$$\xrightarrow{-2R_1 + R_3} \begin{bmatrix} 1 & 1 & | & 1 \\ 0 & -1 & | & 3 \\ 0 & 1 & | & -3 \end{bmatrix}$$

$$\xrightarrow{(-1)R_2} \begin{bmatrix} 1 & 1 & | & 1 \\ 0 & 1 & | & -3 \\ 0 & 1 & | & -3 \end{bmatrix}$$

$$\xrightarrow{-R_2 + R_1} \begin{bmatrix} 1 & 0 & | & 4 \\ 0 & 1 & | & -3 \\ 0 & 1 & | & -3 \end{bmatrix}$$

$$\xrightarrow{-R_2 + R_3} \begin{bmatrix} 1 & 0 & | & 4 \\ 0 & 1 & | & -3 \\ 0 & 0 & | & 0 \end{bmatrix}.$$

The last matrix is reduced and corresponds to the system

$$\begin{cases} x + 0y = 4, \\ 0x + y = -3, \\ 0x + 0y = 0. \end{cases}$$

Since the original system is equivalent to this system, it has a unique solution, namely,

$$x = 4,$$

$$y = -3.$$

## Principles in Practice 2

### Solving a System by Reduction

A health spa customizes the diet and vitamin supplements of each of its clients. The spa offers three different vitamin supplements, each containing different percentages of the recommended daily allowance (RDA) of vitamins A, C, and D. One tablet of supplement X provides 40% of the RDA of A, 20% of the RDA of C, and 10% of the RDA of D. One tablet of supplement Y provides 10% of the RDA of A, 10% of the RDA of C, and 30% of the RDA of D. One tablet of supplement Z provides 10% of the RDA of A, 50% of the RDA of C, and 20% of the RDA of D. The spa staff determines that one client should take 180% of the RDA of vitamin A, 200% of the RDA of vitamin C, and 190% of the RDA of vitamin D each day. How many tablets of each supplement should she take each day?

**EXAMPLE 4    Solving a System by Reduction**

*Using matrix reduction, solve*

$$\begin{cases} x + 2y + 4z - 6 = 0, \\ \qquad 2z + y - 3 = 0, \\ x + y + 2z - 1 = 0. \end{cases}$$

*Solution:* Rewriting the system so that the variables are aligned and the constant terms appear on the right sides of the equations, we have

$$\begin{cases} x + 2y + 4z = 6, \\ \qquad y + 2z = 3, \\ x + y + 2z = 1. \end{cases}$$

Reducing the augmented coefficient matrix, we obtain

$$\begin{bmatrix} 1 & 2 & 4 & 6 \\ 0 & 1 & 2 & 3 \\ 1 & 1 & 2 & 1 \end{bmatrix}$$

$$\xrightarrow{-R_1 + R_3} \begin{bmatrix} 1 & 2 & 4 & 6 \\ 0 & 1 & 2 & 3 \\ 0 & -1 & -2 & -5 \end{bmatrix}$$

$$\xrightarrow[\;(1)R_2 + R_3\;]{-2R_2 + R_1} \begin{bmatrix} 1 & 0 & 0 & 0 \\ 0 & 1 & 2 & 3 \\ 0 & 0 & 0 & -2 \end{bmatrix}$$

$$\xrightarrow{-\frac{1}{2}R_3} \begin{bmatrix} 1 & 0 & 0 & 0 \\ 0 & 1 & 2 & 3 \\ 0 & 0 & 0 & 1 \end{bmatrix}$$

$$\xrightarrow{-3R_3 + R_2} \begin{bmatrix} 1 & 0 & 0 & 0 \\ 0 & 1 & 2 & 0 \\ 0 & 0 & 0 & 1 \end{bmatrix}.$$

The last matrix is reduced and corresponds to

$$\begin{cases} x = 0, \\ y + 2z = 0, \\ \qquad 0 = 1. \end{cases}$$

Whenever we get a row with all 0's to the left side of the vertical rule and a nonzero entry to the right, no solution exists.

Since $0 \neq 1$, there are no values of $x$, $y$, and $z$ for which all equations are satisfied simultaneously. Thus, the original system has no solution.    ■

**EXAMPLE 5    Parametric Form of a Solution**

*Using matrix reduction, solve*

$$\begin{cases} 2x_1 + 3x_2 + 2x_3 + 6x_4 = 10, \\ \qquad x_2 + 2x_3 + x_4 = 2, \\ 3x_1 \qquad - 3x_3 + 6x_4 = 9. \end{cases}$$

## Principles in Practice 3
### Parametric Form of a Solution

A zoo veterinarian can purchase animal food of four different types: A, B, C, and D. Each food comes in the same size bag, and the number of grams of each of three nutrients in each bag are summarized in the following table:

Food

| | | A | B | C | D |
|---|---|---|---|---|---|
| | $N_1$ | 5 | 5 | 10 | 5 |
| Nutrient | $N_2$ | 10 | 5 | 30 | 10 |
| | $N_3$ | 5 | 15 | 10 | 25 |

For one animal, the veterinarian determines that she needs to combine the bags to get 10,000 g of $N_1$, 20,000 g of $N_2$, and 20,000 g of $N_3$. How many bags of each type of food should she order?

*Solution:* Reducing the augmented coefficient matrix, we have

$$\begin{bmatrix} 2 & 3 & 2 & 6 & | & 10 \\ 0 & 1 & 2 & 1 & | & 2 \\ 3 & 0 & -3 & 6 & | & 9 \end{bmatrix}$$

$$\xrightarrow{\frac{1}{2}R_1} \begin{bmatrix} 1 & \frac{3}{2} & 1 & 3 & | & 5 \\ 0 & 1 & 2 & 1 & | & 2 \\ 3 & 0 & -3 & 6 & | & 9 \end{bmatrix}$$

$$\xrightarrow{-3R_1 + R_3} \begin{bmatrix} 1 & \frac{3}{2} & 1 & 3 & | & 5 \\ 0 & 1 & 2 & 1 & | & 2 \\ 0 & -\frac{9}{2} & -6 & -3 & | & -6 \end{bmatrix}$$

$$\xrightarrow[\frac{9}{2}R_2 + R_3]{-\frac{3}{2}R_2 + R_1} \begin{bmatrix} 1 & 0 & -2 & \frac{3}{2} & | & 2 \\ 0 & 1 & 2 & 1 & | & 2 \\ 0 & 0 & 3 & \frac{3}{2} & | & 3 \end{bmatrix}$$

$$\xrightarrow{\frac{1}{3}R_3} \begin{bmatrix} 1 & 0 & -2 & \frac{3}{2} & | & 2 \\ 0 & 1 & 2 & 1 & | & 2 \\ 0 & 0 & 1 & \frac{1}{2} & | & 1 \end{bmatrix}$$

$$\xrightarrow[-2R_3 + R_2]{2R_3 + R_1} \begin{bmatrix} 1 & 0 & 0 & \frac{5}{2} & | & 4 \\ 0 & 1 & 0 & 0 & | & 0 \\ 0 & 0 & 1 & \frac{1}{2} & | & 1 \end{bmatrix}.$$

This matrix is reduced and corresponds to the system

$$\begin{cases} x_1 + \frac{5}{2}x_4 = 4, \\ \qquad x_2 = 0, \\ x_3 + \frac{1}{2}x_4 = 1. \end{cases}$$

Thus,

$$x_1 = -\frac{5}{2}x_4 + 4, \tag{9}$$

$$x_2 = 0, \tag{10}$$

$$x_3 = -\frac{1}{2}x_4 + 1, \tag{11}$$

$$x_4 = x_4. \tag{12}$$

If $x_4$ is any real number, $r$, then Eqs. (9)–(12) determine a particular solution of the original system. For example, if $r = 0$ (that is, $x_4 = 0$), then a *particular* solution is

$$x_1 = 4, \quad x_2 = 0, \quad x_3 = 1, \quad \text{and} \quad x_4 = 0.$$

If $r = 2$, then

$$x_1 = -1, \quad x_2 = 0, \quad x_3 = 0, \quad \text{and} \quad x_4 = 2.$$

Recall [see Example 3 of Sec. 4.4] that the variable $r$, on which $x_1$, $x_3$, and $x_4$ depend, is called a **parameter.** There are infinitely many solutions of the system—one corresponding to each value of the parameter. We say that the *general* solution of the original system is given by

$$x_1 = -\tfrac{5}{2}r + 4,$$

$$x_2 = 0,$$

$$x_3 = -\tfrac{1}{2}r + 1,$$

$$x_4 = r,$$

where $r$ is any real number, and we speak of having a *one-parameter family* of solutions. ∎

Examples 3–5 illustrate the fact that a system of linear equations may have a unique solution, no solution, or infinitely many solutions.

## ▪ Exercise 6.4

*In each of Problems 1–6, determine whether the matrix is reduced or not reduced.*

1. $\begin{bmatrix} 1 & 2 \\ 3 & 0 \end{bmatrix}.$

2. $\begin{bmatrix} 1 & 0 & 0 & 3 \\ 0 & 0 & 1 & 2 \end{bmatrix}.$

3. $\begin{bmatrix} 1 & 0 & 0 \\ 0 & 1 & 0 \\ 0 & 0 & 1 \end{bmatrix}.$

4. $\begin{bmatrix} 1 & 1 \\ 0 & 1 \\ 0 & 0 \\ 0 & 0 \end{bmatrix}.$

5. $\begin{bmatrix} 0 & 0 & 0 & 0 \\ 0 & 1 & 0 & 0 \\ 0 & 0 & 1 & 0 \\ 0 & 0 & 0 & 0 \end{bmatrix}.$

6. $\begin{bmatrix} 0 & 0 & 5 \\ 1 & 0 & 4 \\ 0 & 1 & 2 \\ 0 & 0 & 0 \end{bmatrix}.$

*In each of Problems 7–12, reduce the given matrix.*

7. $\begin{bmatrix} 1 & 3 \\ 4 & 0 \end{bmatrix}.$

8. $\begin{bmatrix} 0 & -2 & 0 & 1 \\ 1 & 2 & 0 & 4 \end{bmatrix}.$

9. $\begin{bmatrix} 2 & 4 & 6 \\ 1 & 2 & 3 \\ 1 & 2 & 3 \end{bmatrix}.$

10. $\begin{bmatrix} 2 & 3 \\ 1 & -6 \\ 4 & 8 \\ 1 & 7 \end{bmatrix}.$

11. $\begin{bmatrix} 2 & 0 & 3 & 1 \\ 1 & 4 & 2 & 2 \\ -1 & 3 & 1 & 4 \\ 0 & 2 & 1 & 0 \end{bmatrix}.$

12. $\begin{bmatrix} 0 & 0 & 2 \\ 2 & 0 & 3 \\ 0 & -1 & 0 \\ 0 & 4 & 1 \end{bmatrix}.$

*Solve the systems in Problems 13–26, by the method of reduction.*

13. $\begin{cases} 2x + 3y = 5, \\ x - 2y = -1. \end{cases}$

14. $\begin{cases} x - 3y = -11, \\ 4x + 3y = 9. \end{cases}$

15. $\begin{cases} 3x + y = 4, \\ 12x + 4y = 2. \end{cases}$

16. $\begin{cases} x + 2y - 3z = 0, \\ -2x - 4y + 6z = 1. \end{cases}$

17. $\begin{cases} x + 2y + z - 4 = 0, \\ 3x + 2z - 5 = 0. \end{cases}$

18. $\begin{cases} x + 2y + 5z - 1 = 0, \\ x + y + 3z - 2 = 0. \end{cases}$

19. $\begin{cases} x_1 - 3x_2 = 0, \\ 2x_1 + 2x_2 = 3, \\ 5x_1 - x_2 = 1. \end{cases}$

20. $\begin{cases} x_1 + 3x_2 = 5, \\ 2x_1 + x_2 = 5, \\ x_1 + x_2 = 3. \end{cases}$

21. $\begin{cases} x - y - 3z = -4, \\ 2x - y - 4z = -7, \\ x + y - z = -2. \end{cases}$

22. $\begin{cases} x + y - z = 6, \\ 2x - 3y - 2z = 2, \\ x - y - 5z = 18. \end{cases}$

23. $\begin{cases} 2x - 4z = 8, \\ x - 2y - 2z = 14, \\ x + y - 2z = -1, \\ 3x + y + z = 0. \end{cases}$

24. $\begin{cases} x + 3z = -1, \\ 3x + 2y + 11z = 1, \\ x + y + 4z = 1, \\ 2x - 3y + 3z = -8. \end{cases}$

25. $\begin{cases} x_1 + x_2 - x_3 + x_4 + x_5 = 0, \\ x_1 + x_2 + x_3 - x_4 + x_5 = 0, \\ x_1 - x_2 - x_3 + x_4 - x_5 = 0, \\ x_1 + x_2 - x_3 - x_4 - x_5 = 0. \end{cases}$

26. $\begin{cases} x_1 + x_2 - x_3 + x_4 = 0, \\ x_1 + x_2 + x_3 - x_4 = 0, \\ x_1 - x_2 - x_3 + x_4 = 0, \\ x_1 + x_2 - x_3 - x_4 = 0. \end{cases}$

*Solve Problems 27–33 by using matrix reduction.*

27. **Taxes**   A company has taxable income of $312,000. The federal tax is 25% of that portion which is left after the state tax has been paid. The state tax is 10% of that portion which is left after the federal tax has been paid. Find the company's federal and state taxes.

28. **Decision Making**   A manufacturer produces two products, A and B. For each unit of A sold, the profit is $8, and for each unit of B sold, the profit is $11. From past experience, it has been found that 25% more of A can be sold than of B. Next year the manufacturer desires a total profit of $42,000. How many units of each product must be sold?

29. **Production Scheduling**   A manufacturer produces three products: A, B, and C. The profits for each unit sold of A, B, and C are $1, $2, and $3, respectively. Fixed costs are $17,000 per year, and the costs of producing each unit of A, B, and C are $4, $5, and $7, respectively. Next year, a total of 11,000 units of all three products is to be produced and sold, and a total profit of $25,000 is to be realized. If total cost is to be $80,000, how many units of each of the products should be produced next year?

30. **Production Allocation**   National Desk Co. has plants for producing desks on both the east and west coasts. At the east coast plant, fixed costs are $16,000 per year and the cost of producing each desk is $90. At the west coast plant, fixed costs are $20,000 per year and the cost of

producing each desk is $80. Next year the company wants to produce a total of 800 desks. Determine the production order for each plant for the forthcoming year if the total cost for each plant is to be the same.

31. **Vitamins**   A person is ordered by a doctor to take 10 units of vitamin A, 9 units of vitamin D, and 19 units of vitamin E each day. The person can choose from three brands of vitamin pills. Brand X contains 2 units of vitamin A, 3 units of vitamin D, and 5 units of vitamin E;

brand Y has 1, 3, and 4 units, respectively; and brand Z has 1 unit of vitamin A, none of vitamin D, and 1 of vitamin E.

a. Find all possible combinations of pills that will provide exactly the required amounts of vitamins.

b. If brand X costs 1 cent a pill, brand Y 6 cents, and brand Z 3 cents, are there any combinations in part (a) costing exactly 15 cents a day?

c. What is the least expensive combination in part (a)? The most expensive?

32. **Production**   A firm produces three products, A, B, and C, that require processing by three machines, I, II, and III. The time in hours required for processing one unit of each product by the three machines is given by the following table:

|   | I | II | III |
|---|---|----|-----|
| A | 3 | 1 | 2 |
| B | 1 | 2 | 1 |
| C | 2 | 4 | 1 |

Machine I is available for 850 hours, machine II for 1200 hours, and machine III for 550 hours. Find how many units of each product should be produced to make use of all the available time on the machines.

33. **Investments**   An investment company sells three types of pooled funds, Standard (S), Deluxe (D), and Gold Star (G).

Each unit of S contains 12 shares of stock A, 16 of stock B, and 8 of stock C.

Each unit of D contains 20 shares of stock A, 12 of stock B, and 28 of stock C.

Each unit of G contains 32 shares of stock A, 28 of stock B, and 36 of stock C.

Suppose an investor wishes to purchase exactly 220 shares of stock A, 176 shares of stock B, and 264 shares of stock C by buying units of the three funds.

a. Determine those combinations of units of S, D, and G which will meet the investor's requirements exactly.

b. Suppose each unit of S (respectively, D, G) costs the investor $300 (respectively, $400, $600). Which of the combinations from (a) will minimize the total cost to the investor?

OBJECTIVE

**To focus our attention on nonhomogeneous systems that involve more than one parameter in their general solution and to solve, and consider the theory of, homogeneous systems.**

# 6.5 METHOD OF REDUCTION (CONTINUED)[4]

As we saw in Sec. 6.4, a system of linear equations may have a unique solution, no solution, or infinitely many solutions. When there are infinitely many, the general solution is expressed in terms of at least one parameter. For example, the general solution in Example 5 was given in terms of the parameter $r$:

$$x_1 = -\tfrac{5}{2}r + 4,$$
$$x_2 = 0,$$
$$x_3 = -\tfrac{1}{2}r + 1,$$
$$x_4 = r.$$

At times, more than one parameter is necessary,[5] as the next example shows.

### EXAMPLE 1  Two-Parameter Family of Solutions

*Using matrix reduction, solve*

$$\begin{cases} x_1 + 2x_2 + 5x_3 + 5x_4 = -3, \\ x_1 + x_2 + 3x_3 + 4x_4 = -1, \\ x_1 - x_2 - x_3 + 2x_4 = 3. \end{cases}$$

*Solution:* The augmented coefficient matrix is

$$\begin{bmatrix} 1 & 2 & 5 & 5 & | & -3 \\ 1 & 1 & 3 & 4 & | & -1 \\ 1 & -1 & -1 & 2 & | & 3 \end{bmatrix},$$

whose reduced form is

$$\begin{bmatrix} 1 & 0 & 1 & 3 & | & 1 \\ 0 & 1 & 2 & 1 & | & -2 \\ 0 & 0 & 0 & 0 & | & 0 \end{bmatrix}.$$

Hence,

$$\begin{cases} x_1 + x_3 + 3x_4 = 1, \\ x_2 + 2x_3 + x_4 = -2, \end{cases}$$

from which it follows that

$$x_1 = 1 - x_3 - 3x_4,$$
$$x_2 = -2 - 2x_3 - x_4.$$

Since no restriction is placed on either $x_3$ or $x_4$, they can be arbitrary real numbers, giving us a parametric family of solutions. Setting $x_3 = r$ and $x_4 = s$, we can give the solution of the given system as

$$x_1 = 1 - r - 3s,$$
$$x_2 = -2 - 2r - s,$$
$$x_3 = r,$$
$$x_4 = s,$$

[4]This section may be omitted.

[5]See Example 7 of Sec. 4.4.

where the parameters $r$ and $s$ can be any real numbers. By assigning specific values to $r$ and $s$, we get particular solutions. For example, if $r = 1$ and $s = 2$, then the corresponding particular solution is $x_1 = -6$, $x_2 = -6$, $x_3 = 1$, and $x_4 = 2$. ∎

It is customary to classify a system of linear equations as being either *homogeneous* or *nonhomogeneous*, depending on whether the constant terms are all zero.

**DEFINITION**

*The system*

$$\begin{cases} a_{11}x_1 + a_{12}x_2 + \cdots + a_{1n}x_n = c_1, \\ a_{21}x_1 + a_{22}x_2 + \cdots + a_{2n}x_n = c_2, \\ \qquad \cdot \qquad\qquad \cdot \qquad\qquad\qquad \cdot \qquad\qquad \cdot \\ \qquad \cdot \qquad\qquad \cdot \qquad\qquad\qquad \cdot \qquad\qquad \cdot \\ \qquad \cdot \qquad\qquad \cdot \qquad\qquad\qquad \cdot \qquad\qquad \cdot \\ a_{m1}x_1 + a_{m2}x_2 + \cdots + a_{mn}x_n = c_m \end{cases}$$

*is called a **homogeneous system** if $c_1 = c_2 = \cdots = c_m = 0$. The system is a **nonhomogeneous system** if at least one of the $c$'s is not equal to 0.*

**EXAMPLE 2**   **Nonhomogeneous and Homogeneous Systems**

The system

$$\begin{cases} 2x + 3y = 4, \\ 3x - 4y = 0, \end{cases}$$

is nonhomogeneous because of the 4 in the top equation. The system

$$\begin{cases} 2x + 3y = 0, \\ 3x - 4y = 0, \end{cases}$$

is homogeneous. ∎

If the homogeneous system

$$\begin{cases} 2x + 3y = 0, \\ 3x - 4y = 0, \end{cases}$$

were solved by the method of reduction, first the augmented coefficient matrix would be written

$$\begin{bmatrix} 2 & 3 & | & 0 \\ 3 & -4 & | & 0 \end{bmatrix}.$$

Observe that the last column consists entirely of zeros. This is typical of the augmented coefficient matrix of any homogeneous system. We would then reduce this matrix by using elementary row operations:

$$\begin{bmatrix} 2 & 3 & | & 0 \\ 3 & -4 & | & 0 \end{bmatrix} \rightarrow \cdots \rightarrow \begin{bmatrix} 1 & 0 & | & 0 \\ 0 & 1 & | & 0 \end{bmatrix}.$$

The last column of the reduced matrix also consists only of zeros. This does not occur by chance. When any elementary row operation is performed on a matrix that has a column consisting entirely of zeros, the corresponding column of the resulting matrix will also be all zeros. For convenience, it will be our custom when solving a homogeneous system by matrix reduction to delete the last column of the matrices involved. That is, we shall reduce only the *coefficient matrix* of the system. For the preceding system, we would have

$$\begin{bmatrix} 2 & 3 \\ 3 & -4 \end{bmatrix} \quad \rightarrow \cdots \rightarrow \quad \begin{bmatrix} 1 & 0 \\ 0 & 1 \end{bmatrix}.$$

Here the reduced matrix, called the *reduced coefficient matrix,* corresponds to the system

$$\begin{cases} x + 0y = 0, \\ 0x + y = 0. \end{cases}$$

so the solution is $x = 0$ and $y = 0$.

Let us now consider the number of solutions of the homogeneous system

$$\begin{cases} a_{11}x_1 + a_{12}x_2 + \cdots + a_{1n}x_n = 0, \\ a_{21}x_1 + a_{22}x_2 + \cdots + a_{2n}x_n = 0, \\ \quad\cdot\quad\quad\quad\cdot\quad\quad\quad\quad\cdot\quad\quad\cdot \\ \quad\cdot\quad\quad\quad\cdot\quad\quad\quad\quad\cdot\quad\quad\cdot \\ \quad\cdot\quad\quad\quad\cdot\quad\quad\quad\quad\cdot\quad\quad\cdot \\ a_{m1}x_1 + a_{m2}x_2 + \cdots + a_{mn}x_n = 0. \end{cases}$$

One solution always occurs when $x_1 = 0$, $x_2 = 0, \cdots$, and $x_n = 0$, since each equation is satisfied for these values. This solution, called the **trivial solution,** is a solution of *every* homogeneous system.

There is a theorem that allows us to determine whether a homogeneous system has a unique solution (the trivial solution only) or infinitely many solutions. The theorem is based on the number of nonzero rows that appear in the reduced coefficient matrix of the system. A *nonzero row* is a row that does not consist entirely of zeros.

---

**Theorem**

Let **A** be the *reduced* coefficient matrix of a homogeneous system of $m$ linear equations in $n$ unknowns. If **A** has exactly $k$ nonzero rows, then $k \le n$. Moreover,

**a.** if $k < n$, the system has infinitely many solutions,

and

**b.** if $k = n$, the system has a unique solution (the trivial solution).

---

If a homogeneous system consists of $m$ equations in $n$ unknowns, then the coefficient matrix of the system has order $m \times n$. Thus, if $m < n$ and $k$ is the number of nonzero rows in the reduced coefficient matrix, then $k \le m$, and hence, $k < n$. By the foregoing theorem, the system must have infinitely many solutions. Consequently, we have the following corollary.

---

**Corollary**

A homogeneous system of linear equations with fewer equations than unknowns has infinitely many solutions.

*Pitfall* ▼ The preceding theorem and corollary apply only to **homogeneous** systems of linear equations. For example, consider the system

$$\begin{cases} x + y - 2z = 3, \\ 2x + 2y - 4z = 4, \end{cases}$$

which consists of two linear equations in three unknowns. We **cannot** conclude that this system has infinitely many solutions, since it is not homogeneous. Indeed, you should verify that it has no solution.

### EXAMPLE 3   Number of Solutions of a Homogeneous System

*Determine whether the system*

$$\begin{cases} x + y - 2z = 0, \\ 2x + 2y - 4z = 0, \end{cases}$$

*has a unique solution or infinitely many solutions.*

*Solution:* There are two equations in this homogeneous system, and this number is less than the number of unknowns (three). Thus, by the previous corollary, the system has infinitely many solutions. ■

### Principles in Practice 1
#### Solving Homogeneous Systems

A plane in three-dimensional space can be written as $ax + by + cz = d$. We can find the possible intersections of planes in this form by writing them as systems of linear equations and using reduction to solve them. If $d = 0$ in each equation, then we have a homogeneous system with either a unique solution or infinitely many solutions. Determine whether the intersection of the planes

$$5x + 3y + 4z = 0,$$

$$6x + 8y + 7z = 0,$$

$$3x + 1y + 2z = 0$$

has a unique solution or infinitely many solutions; then solve the system.

### EXAMPLE 4   Solving Homogeneous Systems

*Determine whether the following homogeneous systems have a unique solution or infinitely many solutions; then solve the systems.*

**a.** $\begin{cases} x - 2y + z = 0, \\ 2x - y + 5z = 0, \\ x + y + 4z = 0. \end{cases}$

*Solution:* Reducing the coefficient matrix, we have

$$\begin{bmatrix} 1 & -2 & 1 \\ 2 & -1 & 5 \\ 1 & 1 & 4 \end{bmatrix} \rightarrow \cdots \rightarrow \begin{bmatrix} 1 & 0 & 3 \\ 0 & 1 & 1 \\ 0 & 0 & 0 \end{bmatrix}.$$

The number of nonzero rows (2) in the reduced coefficient matrix is less than the number of unknowns (3) in the system. By the previous theorem, there are infinitely many solutions.

Since the reduced coefficient matrix corresponds to

$$\begin{cases} x + 3z = 0, \\ y + z = 0, \end{cases}$$

the solution may be given in parametric form by

$$x = -3r,$$

$$y = -r,$$

$$z = r,$$

where $r$ is any real number.

**b.** $\begin{cases} 3x + 4y = 0, \\ x - 2y = 0, \\ 2x + y = 0, \\ 2x + 3y = 0. \end{cases}$

*Solution:* Reducing the coefficient matrix, we have

$$\begin{bmatrix} 3 & 4 \\ 1 & -2 \\ 2 & 1 \\ 2 & 3 \end{bmatrix} \rightarrow \cdots \rightarrow \begin{bmatrix} 1 & 0 \\ 0 & 1 \\ 0 & 0 \\ 0 & 0 \end{bmatrix}.$$

The number of nonzero rows (2) in the reduced coefficient matrix equals the number of unknowns in the system. By the theorem, the system must have a unique solution, namely, the trivial solution $x = 0$, $y = 0$.   ■

## ■ Exercise 6.5

*In Problems 1–8, solve the systems by using matrix reduction.*

**1.** $\begin{cases} w - x - y + 4z = 5, \\ 2w - 3x - 4y + 9z = 13, \\ 2w + x + 4y + 5z = 1. \end{cases}$

**2.** $\begin{cases} 3w - x + 12y + 18z = -4, \\ w - 2x + 4y + 11z = -13, \\ w + x + 4y + 2z = 8. \end{cases}$

**3.** $\begin{cases} 3w - x - 3y - z = -2, \\ 2w - 2x - 6y - 6z = -4, \\ 2w - x - 3y - 2z = -2, \\ 3w + x + 3y + 7z = 2. \end{cases}$

**4.** $\begin{cases} w + x + 5z = 1, \\ w + y + 2z = 1, \\ w - 3x + 4y - 7z = 1, \\ x - y + 3z = 0. \end{cases}$

**5.** $\begin{cases} w + x + 3y - z = 2, \\ 2w + x + 5y - 2z = 0, \\ 2w - x + 3y - 2z = -8, \\ 3w + 2x + 8y - 3z = 2, \\ w + 2y - z = -2. \end{cases}$

**6.** $\begin{cases} w + x + y + 2z = 4, \\ 2w + x + 2y + 2z = 7, \\ w + 2x + y + 4z = 5, \\ 3w - 2x + 3y - 4z = 7, \\ 4w - 3x + 4y - 6z = 9. \end{cases}$

**7.** $\begin{cases} 4x_1 - 3x_2 + 5x_3 - 10x_4 + 11x_5 = -8, \\ 2x_1 + x_2 + 5x_3 + 3x_5 = 6. \end{cases}$

**8.** $\begin{cases} x_1 + 2x_3 + x_4 + 4x_5 = 1, \\ x_2 + x_3 - 3x_4 = -2, \\ 4x_1 - 3x_2 + 5x_3 + 13x_4 + 16x_5 = 10, \\ x_1 + 2x_2 + 4x_3 - 5x_4 + 4x_5 = -3. \end{cases}$

*For each of Problems 9–14, determine whether the system has infinitely many solutions or only the trivial solution. Do not solve the systems.*

**9.** $\begin{cases} 0.07x + 0.3y + 0.02z = 0, \\ 0.053x - 0.4y + 0.08z = 0. \end{cases}$

**10.** $\begin{cases} 3w + 5x - 4y + 2z = 0, \\ 7w - 2x + 9y + 3z = 0. \end{cases}$

**11.** $\begin{cases} 3x - 4y = 0, \\ x + 5y = 0, \\ 4x - y = 0. \end{cases}$

**12.** $\begin{cases} 2x + 3y + 12z = 0, \\ 3x - 2y + 5z = 0, \\ 4x + y + 14z = 0. \end{cases}$

**13.** $\begin{cases} x + y + z = 0, \\ x - z = 0, \\ x - 2y - 5z = 0. \end{cases}$

**14.** $\begin{cases} 2x + 5y = 0, \\ x + 4y = 0, \\ 3x - 2y = 0. \end{cases}$

*Solve each of the following systems.*

**15.** $\begin{cases} x + y = 0, \\ 3x - 4y = 0. \end{cases}$

**16.** $\begin{cases} 2x - 5y = 0, \\ 8x - 20y = 0. \end{cases}$

**17.** $\begin{cases} x + 6y - 2z = 0, \\ 2x - 3y + 4z = 0. \end{cases}$

**18.** $\begin{cases} 4x + 7y = 0, \\ 2x + 3y = 0. \end{cases}$

**19.** $\begin{cases} x + y = 0, \\ 3x - 4y = 0, \\ 5x - 8y = 0. \end{cases}$

**20.** $\begin{cases} 4x - 3y + 2z = 0, \\ x + 2y + 3z = 0, \\ x + y + z = 0. \end{cases}$

**21.** $\begin{cases} x + y + z = 0, \\ 5x - 2y - 9z = 0, \\ 3x + y - z = 0, \\ 3x - 2y - 7z = 0. \end{cases}$

**22.** $\begin{cases} x + y + 7z = 0, \\ x - y - z = 0, \\ 2x - 3y - 6z = 0, \\ 3x + y + 13z = 0. \end{cases}$

**23.** $\begin{cases} w + x + y + 4z = 0, \\ w + x + 5z = 0, \\ 2w + x + 3y + 4z = 0, \\ w - 3x + 2y - 9z = 0. \end{cases}$

**24.** $\begin{cases} w + x + 2y + 7z = 0, \\ w - 2x - y + z = 0, \\ w + 2x + 3y + 9z = 0, \\ 2w - 3x - y + 4z = 0. \end{cases}$

O B J E C T I V E

**To determine the inverse of an invertible matrix and to use inverses to solve systems.**

# 6.6 Inverses

We have seen how useful the method of reduction is for solving systems of linear equations. But it is by no means the only method that uses matrices. In this section, we shall discuss a different method that applies to certain systems of $n$ linear equations in $n$ unknowns.

In Sec. 6.3, we showed how a system of linear equations can be written in matrix form as the single matrix equation $\mathbf{AX} = \mathbf{B}$, where $\mathbf{A}$ is the coefficient matrix. For example, the system

$$\begin{cases} x_1 + 2x_2 = 3, \\ x_1 - x_2 = 1 \end{cases}$$

can be written in the matrix form $\mathbf{AX} = \mathbf{B}$, where

$$\mathbf{A} = \begin{bmatrix} 1 & 2 \\ 1 & -1 \end{bmatrix}, \quad \mathbf{X} = \begin{bmatrix} x_1 \\ x_2 \end{bmatrix}, \quad \text{and} \quad \mathbf{B} = \begin{bmatrix} 3 \\ 1 \end{bmatrix}.$$

If we can determine the values of the entries in the unknown matrix $\mathbf{X}$, we have a solution of the system. Thus, we would like to find a method to solve the matrix equation $\mathbf{AX} = \mathbf{B}$ for $\mathbf{X}$. Some motivation is provided by looking at the procedure for solving the algebraic equation $ax = b$. The latter equation is solved by simply multiplying both sides by the multiplicative inverse of $a$. [Recall that the multiplicative inverse of a nonzero number $a$ is denoted $a^{-1}$ (which is $1/a$) and has the property that $a^{-1}a = 1$.] For example, if $3x = 11$, then

$$3^{-1}(3x) = 3^{-1}(11), \quad \text{so} \quad x = \frac{11}{3}.$$

If we can apply a similar procedure to the *matrix* equation

$$\mathbf{AX} = \mathbf{B}, \tag{1}$$

then we need a multiplicative inverse of $\mathbf{A}$—that is, a matrix $\mathbf{C}$ such that $\mathbf{CA} = \mathbf{I}$. Then we can simply multiply both sides of Eq. (1) by $\mathbf{C}$:

$$\mathbf{C}(\mathbf{AX}) = \mathbf{CB},$$

$$(\mathbf{CA})\mathbf{X} = \mathbf{CB},$$

$$\mathbf{IX} = \mathbf{CB},$$

$$\mathbf{X} = \mathbf{CB}. \tag{2}$$

Thus, the solution is $\mathbf{X} = \mathbf{CB}$. Of course, this method is based on the existence of a matrix $\mathbf{C}$ such that $\mathbf{CA} = \mathbf{I}$. When such a matrix does exist, we say that it is an *inverse matrix* (or simply an *inverse*) of $\mathbf{A}$.

## DEFINITION

*If* **A** *is a square matrix and there exists a matrix* **C** *such that* **CA** $=$ **I** *then* **C** *is called an **inverse** of* **A,** *and* **A** *is said to be **invertible** (or nonsingular).*

### Principles in Practice 1
#### Inverse of a Matrix

Secret messages can be encoded by using a code and an encoding matrix. Suppose we have the following code:

| a | b | c | d | e | f | g | h | i | j | k | l | m |
|---|---|---|---|---|---|---|---|---|---|---|---|---|
| 1 | 2 | 3 | 4 | 5 | 6 | 7 | 8 | 9 | 10 | 11 | 12 | 13 |

| n | o | p | q | r | s | t | u | v | w | x | y | z |
|---|---|---|---|---|---|---|---|---|---|---|---|---|
| 14 | 15 | 16 | 17 | 18 | 19 | 20 | 21 | 22 | 23 | 24 | 25 | 26 |

Let the encoding matrix be **E.** Then we can encode a message by taking every two letters of the message, converting them to their corresponding numbers, creating a $2 \times 1$ matrix, and then multiply each matrix by **E.** The message may be unscrambled with a decoding matrix that is the inverse of the coding matrix—that is, $\mathbf{E}^{-1}$. Determine whether the encoding matrices

$$\begin{bmatrix} 1 & 3 \\ 2 & 4 \end{bmatrix} \quad \text{and} \quad \begin{bmatrix} -2 & 1.5 \\ 1 & -0.5 \end{bmatrix}$$

are inverses of each other.

### EXAMPLE 1   Inverse of a Matrix

Let $\mathbf{A} = \begin{bmatrix} 1 & 2 \\ 3 & 7 \end{bmatrix}$ and $\mathbf{C} = \begin{bmatrix} 7 & -2 \\ -3 & 1 \end{bmatrix}$.   Since

$$\mathbf{CA} = \begin{bmatrix} 7 & -2 \\ -3 & 1 \end{bmatrix} \begin{bmatrix} 1 & 2 \\ 3 & 7 \end{bmatrix} = \begin{bmatrix} 1 & 0 \\ 0 & 1 \end{bmatrix} = \mathbf{I},$$

matrix **C** is an inverse of **A.**   ∎

It can be shown that an invertible matrix has one and only one inverse; that is, an inverse is unique. Thus, in Example 1, matrix **C** is the *only* matrix such that **CA** $=$ **I.** For this reason, we can speak of *the* inverse of an invertible matrix **A,** which we denote by the symbol $\mathbf{A}^{-1}$. Accordingly, $\mathbf{A}^{-1}\mathbf{A} = \mathbf{I}$. Moreover, although matrix multiplication is not generally commutative, it is a fact that $\mathbf{A}^{-1}$ commutes with **A:**

$$\mathbf{A}^{-1}\mathbf{A} = \mathbf{A}\mathbf{A}^{-1} = \mathbf{I}.$$

Returning to the matrix equation **AX** $=$ **B,** from Eq. (2) we can now state the following:

> If **A** is an invertible matrix, then the matrix equation **AX** $=$ **B** has the unique solution $\mathbf{X} = \mathbf{A}^{-1}\mathbf{B}$.

### Principles in Practice 2
#### Using the Inverse to Solve a System

Suppose the encoding matrix $\mathbf{E} = \begin{bmatrix} 1 & 3 \\ 2 & 4 \end{bmatrix}$ was used to encode a message. Use the code from Principles in Practice 1 and the inverse $\mathbf{E}^{-1} = \begin{bmatrix} -2 & 1.5 \\ 1 & -0.5 \end{bmatrix}$ to decode the message, broken into the following pieces:

28, 46, 65, 90

61, 82

59, 88, 57, 86

60, 84, 21, 34, 76, 102

### EXAMPLE 2   Using the Inverse to Solve a System

*Solve the system*

$$\begin{cases} x_1 + 2x_2 = 5, \\ 3x_1 + 7x_2 = 18. \end{cases}$$

*Solution:* In matrix form, we have **AX** $=$ **B,** where

$$\mathbf{A} = \begin{bmatrix} 1 & 2 \\ 3 & 7 \end{bmatrix}, \quad \mathbf{X} = \begin{bmatrix} x_1 \\ x_2 \end{bmatrix}, \quad \text{and} \quad \mathbf{B} = \begin{bmatrix} 5 \\ 18 \end{bmatrix}.$$

In Example 1, we showed that

$$\mathbf{A}^{-1} = \begin{bmatrix} 7 & -2 \\ -3 & 1 \end{bmatrix}.$$

Therefore,

$$\mathbf{X} = \mathbf{A}^{-1}\mathbf{B} = \begin{bmatrix} 7 & -2 \\ -3 & 1 \end{bmatrix} \begin{bmatrix} 5 \\ 18 \end{bmatrix} = \begin{bmatrix} -1 \\ 3 \end{bmatrix},$$

so $x_1 = -1$ and $x_2 = 3$.   ∎

In order that the method of Example 2 apply to a system, two conditions must be met:

1. The system must have the same number of equations as there are unknowns.

2. The coefficient matrix must be invertible.

As far as condition 2 is concerned, we caution that not all square matrices are invertible. For example, if

$$\mathbf{A} = \begin{bmatrix} 0 & 1 \\ 0 & 1 \end{bmatrix},$$

then

$$\begin{bmatrix} a & b \\ c & d \end{bmatrix} \begin{bmatrix} 0 & 1 \\ 0 & 1 \end{bmatrix} = \begin{bmatrix} 0 & a + b \\ 0 & c + d \end{bmatrix} \neq \begin{bmatrix} 1 & 0 \\ 0 & 1 \end{bmatrix}.$$

Hence, there is no matrix which, when postmultiplied by **A,** yields the identity matrix. Thus, **A** is not invertible.

Before discussing a procedure for finding the inverse of an invertible matrix, we introduce the concept of *elementary matrices*. An $n \times n$ **elementary matrix** is a matrix obtained from the $n \times n$ identity matrix **I** by an elementary row operation. There are three basic types of elementary matrices:

---

**Elementary Matrices**

1. One obtained by interchanging two rows of **I;**

2. One obtained by multiplying any row of **I** by a nonzero scalar;

3. One obtained by adding a constant multiple of one row of **I** to another row.

---

**EXAMPLE 3**   **Elementary Matrices**

The matrices

$$\mathbf{E}_1 = \begin{bmatrix} 1 & 0 & 0 \\ 0 & 0 & 1 \\ 0 & 1 & 0 \end{bmatrix}, \quad \mathbf{E}_2 = \begin{bmatrix} -4 & 0 \\ 0 & 1 \end{bmatrix}, \quad \text{and} \quad \mathbf{E}_3 = \begin{bmatrix} 1 & 0 \\ 3 & 1 \end{bmatrix}$$

are elementary matrices. $\mathbf{E}_1$ is obtained from the $3 \times 3$ identity matrix by interchanging the second and third rows. $\mathbf{E}_2$ is obtained from the $2 \times 2$ identity matrix by multiplying the first row by $-4$. $\mathbf{E}_3$ is obtained from the $2 \times 2$ identity matrix by adding 3 times the first row to the second. ■

Suppose that **E** is an $n \times n$ elementary matrix obtained from **I** by a certain elementary row operation and **A** is an $n \times n$ matrix. Then it can be shown that the product **EA** is equal to the matrix that is obtained from **A** by applying the same elementary row operation to **A.** For example, let

$$\mathbf{A} = \begin{bmatrix} 1 & 2 \\ 3 & 4 \end{bmatrix}, \quad \mathbf{E}_1 = \begin{bmatrix} 0 & 1 \\ 1 & 0 \end{bmatrix},$$

$$\mathbf{E}_2 = \begin{bmatrix} 1 & 0 \\ 0 & 2 \end{bmatrix}, \quad \text{and} \quad \mathbf{E}_3 = \begin{bmatrix} 1 & -2 \\ 0 & 1 \end{bmatrix}.$$

Observe that $\mathbf{E}_1$, $\mathbf{E}_2$, and $\mathbf{E}_3$ are elementary matrices. $\mathbf{E}_1$ is obtained by interchanging the first and second rows of $\mathbf{I}$. Likewise, the product

$$\mathbf{E}_1\mathbf{A} = \begin{bmatrix} 0 & 1 \\ 1 & 0 \end{bmatrix}\begin{bmatrix} 1 & 2 \\ 3 & 4 \end{bmatrix} = \begin{bmatrix} 3 & 4 \\ 1 & 2 \end{bmatrix}$$

is the matrix obtained from $\mathbf{A}$ by interchanging the first and second rows of $\mathbf{A}$. Matrix $\mathbf{E}_2$ is obtained by multiplying the second row of $\mathbf{I}$ by 2. Accordingly, the product

$$\mathbf{E}_2\mathbf{A} = \begin{bmatrix} 1 & 0 \\ 0 & 2 \end{bmatrix}\begin{bmatrix} 1 & 2 \\ 3 & 4 \end{bmatrix} = \begin{bmatrix} 1 & 2 \\ 6 & 8 \end{bmatrix}$$

is the matrix obtained by multiplying the second row of $\mathbf{A}$ by 2. Matrix $\mathbf{E}_3$ is obtained by adding $-2$ times the second row of $\mathbf{I}$ to the first row. The product

$$\mathbf{E}_3\mathbf{A} = \begin{bmatrix} 1 & -2 \\ 0 & 1 \end{bmatrix}\begin{bmatrix} 1 & 2 \\ 3 & 4 \end{bmatrix} = \begin{bmatrix} -5 & -6 \\ 3 & 4 \end{bmatrix}$$

is the matrix obtained from $\mathbf{A}$ by the same elementary row operation.

If we wanted to reduce the matrix

$$\mathbf{A} = \begin{bmatrix} 1 & 0 \\ 2 & 2 \end{bmatrix},$$

we might proceed through a sequence of steps as follows:

$$\begin{bmatrix} 1 & 0 \\ 2 & 2 \end{bmatrix} \xrightarrow{-2\mathbf{R}_1 + \mathbf{R}_2} \begin{bmatrix} 1 & 0 \\ 0 & 2 \end{bmatrix}$$

$$\xrightarrow{\frac{1}{2}\mathbf{R}_2} \begin{bmatrix} 1 & 0 \\ 0 & 1 \end{bmatrix}.$$

Note that $\mathbf{A}$ reduces to $\mathbf{I}$. Since our reduction procedure involves elementary row operations, it seems natural that elementary matrices can be used to reduce $\mathbf{A}$. If $\mathbf{A}$ is premultiplied by the elementary matrix $\mathbf{E}_1 = \begin{bmatrix} 1 & 0 \\ -2 & 1 \end{bmatrix}$, then $\mathbf{E}_1\mathbf{A}$ is the matrix obtained from $\mathbf{A}$ by adding $-2$ times the first row to the second row:

$$\mathbf{E}_1\mathbf{A} = \begin{bmatrix} 1 & 0 \\ -2 & 1 \end{bmatrix}\begin{bmatrix} 1 & 0 \\ 2 & 2 \end{bmatrix} = \begin{bmatrix} 1 & 0 \\ 0 & 2 \end{bmatrix}.$$

Premultiplying $\mathbf{E}_1\mathbf{A}$ by the elementary matrix $\mathbf{E}_2 = \begin{bmatrix} 1 & 0 \\ 0 & \frac{1}{2} \end{bmatrix}$ gives the matrix obtained by multiplying the second row of $\mathbf{E}_1\mathbf{A}$ by $\frac{1}{2}$:

$$\mathbf{E}_2(\mathbf{E}_1\mathbf{A}) = \begin{bmatrix} 1 & 0 \\ 0 & \frac{1}{2} \end{bmatrix}\begin{bmatrix} 1 & 0 \\ 0 & 2 \end{bmatrix} = \begin{bmatrix} 1 & 0 \\ 0 & 1 \end{bmatrix} = \mathbf{I}.$$

Thus, we have reduced $\mathbf{A}$ by multiplying it by a product of elementary matrices.

Since $(\mathbf{E}_2\mathbf{E}_1)\mathbf{A} = \mathbf{E}_2(\mathbf{E}_1\mathbf{A}) = \mathbf{I}$, the product $\mathbf{E}_2\mathbf{E}_1$ is $\mathbf{A}^{-1}$. However,

$$\mathbf{A}^{-1} = \mathbf{E}_2\mathbf{E}_1 = (\mathbf{E}_2\mathbf{E}_1)\mathbf{I} = \mathbf{E}_2(\mathbf{E}_1\mathbf{I}).$$

Consequently, $\mathbf{A}^{-1}$ can be obtained by applying the same elementary row operations, beginning with $\mathbf{I}$, that were used to reduce $\mathbf{A}$ to $\mathbf{I}$:

$$\begin{bmatrix} 1 & 0 \\ 0 & 1 \end{bmatrix} \xrightarrow{\;-2R_1 + R_2\;} \begin{bmatrix} 1 & 0 \\ -2 & 1 \end{bmatrix}$$

$$\xrightarrow{\;\frac{1}{2}R_2\;} \begin{bmatrix} 1 & 0 \\ -1 & \frac{1}{2} \end{bmatrix}.$$

Therefore,

$$\mathbf{A}^{-1} = \begin{bmatrix} 1 & 0 \\ -1 & \frac{1}{2} \end{bmatrix}.$$

Our result can be verified by showing that $\mathbf{A}^{-1}\mathbf{A} = \mathbf{I}$:

$$\mathbf{A}^{-1}\mathbf{A} = \begin{bmatrix} 1 & 0 \\ -1 & \frac{1}{2} \end{bmatrix}\begin{bmatrix} 1 & 0 \\ 2 & 2 \end{bmatrix} = \begin{bmatrix} 1 & 0 \\ 0 & 1 \end{bmatrix} = \mathbf{I}.$$

In summary, to find $\mathbf{A}^{-1}$, we apply the identical elementary row operations, beginning with $\mathbf{I}$ and proceeding in the same order, as those that were used to reduce $\mathbf{A}$ to $\mathbf{I}$. Finding $\mathbf{A}^{-1}$ by this technique can be done conveniently by using the following format. First, we write the matrix

$$[\mathbf{A} \mid \mathbf{I}] = \begin{bmatrix} 1 & 0 & 1 & 0 \\ 2 & 2 & 0 & 1 \end{bmatrix}.$$

Then we apply elementary row operations until $[\mathbf{A} \mid \mathbf{I}]$ is equivalent to a matrix that has $\mathbf{I}$ as its first two columns. The last two columns of this matrix will be $\mathbf{A}^{-1}$. Thus,

$$[\mathbf{A} \mid \mathbf{I}] = \begin{bmatrix} 1 & 0 & 1 & 0 \\ 2 & 2 & 0 & 1 \end{bmatrix} \rightarrow \begin{bmatrix} 1 & 0 & 1 & 0 \\ 0 & 2 & -2 & 1 \end{bmatrix}$$

$$\rightarrow \begin{bmatrix} 1 & 0 & 1 & 0 \\ 0 & 1 & -1 & \frac{1}{2} \end{bmatrix} = [\mathbf{I} \mid \mathbf{A}^{-1}].$$

Note that the first two columns of $[\mathbf{I} \mid \mathbf{A}^{-1}]$ form a reduced matrix.

This procedure can be extended to find the inverse of *any* invertible matrix:

A matrix is invertible if and only if it is equivalent to the identity matrix.

---

**Method to Find the Inverse of a Matrix**

If $\mathbf{M}$ is an invertible $n \times n$ matrix, form the $n \times (2n)$ matrix $[\mathbf{M} \mid \mathbf{I}]$. Then perform elementary row operations until the first $n$ columns form a reduced matrix equal to $\mathbf{I}$. The last $n$ columns will be $\mathbf{M}^{-1}$. Symbolically,

$$[\mathbf{M} \mid \mathbf{I}] \rightarrow \cdots \rightarrow [\mathbf{I} \mid \mathbf{M}^{-1}].$$

If a matrix $\mathbf{M}$ does not reduce to $\mathbf{I}$, then $\mathbf{M}^{-1}$ does not exist.

---

**EXAMPLE 4   Finding the Inverse of a Matrix**

*Determine $\mathbf{A}^{-1}$ if $\mathbf{A}$ is invertible.*

a. $\mathbf{A} = \begin{bmatrix} 1 & 0 & -2 \\ 4 & -2 & 1 \\ 1 & 2 & -10 \end{bmatrix}.$

*Solution:* Following the foregoing procedure, we have

## Principles in Practice 3
### Finding the Inverse of a Matrix

We could extend the encoding scheme used in Principles in Practice 1 to a 3 × 3 matrix, encoding three letters of a message at a time. Find the inverses of the following 3 × 3 encoding matrices:

$$\mathbf{E} = \begin{bmatrix} 3 & 1 & 2 \\ 2 & 2 & 2 \\ 2 & 1 & 3 \end{bmatrix}, \quad \mathbf{F} = \begin{bmatrix} 2 & 1 & 2 \\ 3 & 2 & 3 \\ 4 & 3 & 4 \end{bmatrix}.$$

$$[\mathbf{A} \mid \mathbf{I}] = \left[ \begin{array}{ccc|ccc} 1 & 0 & -2 & 1 & 0 & 0 \\ 4 & -2 & 1 & 0 & 1 & 0 \\ 1 & 2 & -10 & 0 & 0 & 1 \end{array} \right]$$

$$\xrightarrow[-1R_1 + R_3]{-4R_1 + R_2} \left[ \begin{array}{ccc|ccc} 1 & 0 & -2 & 1 & 0 & 0 \\ 0 & -2 & 9 & -4 & 1 & 0 \\ 0 & 2 & -8 & -1 & 0 & 1 \end{array} \right]$$

$$\xrightarrow{-\frac{1}{2}R_2} \left[ \begin{array}{ccc|ccc} 1 & 0 & -2 & 1 & 0 & 0 \\ 0 & 1 & -\frac{9}{2} & 2 & -\frac{1}{2} & 0 \\ 0 & 2 & -8 & -1 & 0 & 1 \end{array} \right]$$

$$\xrightarrow{-2R_2 + R_3} \left[ \begin{array}{ccc|ccc} 1 & 0 & -2 & 1 & 0 & 0 \\ 0 & 1 & -\frac{9}{2} & 2 & -\frac{1}{2} & 0 \\ 0 & 0 & 1 & -5 & 1 & 1 \end{array} \right]$$

$$\xrightarrow[\frac{9}{2}R_3 + R_2]{2R_3 + R_1} \left[ \begin{array}{ccc|ccc} 1 & 0 & 0 & -9 & 2 & 2 \\ 0 & 1 & 0 & -\frac{41}{2} & 4 & \frac{9}{2} \\ 0 & 0 & 1 & -5 & 1 & 1 \end{array} \right].$$

The first three columns of the last matrix form **I**. Thus, **A** is invertible and

$$\mathbf{A}^{-1} = \begin{bmatrix} -9 & 2 & 2 \\ -\frac{41}{2} & 4 & \frac{9}{2} \\ -5 & 1 & 1 \end{bmatrix}.$$

**b.** $\mathbf{A} = \begin{bmatrix} 3 & 2 \\ 6 & 4 \end{bmatrix}.$

*Solution:* We have

$$[\mathbf{A} \mid \mathbf{I}] = \left[ \begin{array}{cc|cc} 3 & 2 & 1 & 0 \\ 6 & 4 & 0 & 1 \end{array} \right] \xrightarrow{-2R_1 + R_2} \left[ \begin{array}{cc|cc} 3 & 2 & 1 & 0 \\ 0 & 0 & -2 & 1 \end{array} \right]$$

$$\xrightarrow{\frac{1}{3}R_1} \left[ \begin{array}{cc|cc} 1 & \frac{2}{3} & \frac{1}{3} & 0 \\ 0 & 0 & -2 & 1 \end{array} \right].$$

The first two columns of the last matrix form a reduced matrix different from **I**. Thus, **A** is not invertible. ■

## TECHNOLOGY

Finding the inverse of an invertible matrix with a graphics calculator can be a real timesaver. Figure 6.5 shows the inverse of

$$\mathbf{A} = \begin{bmatrix} 3 & 2 \\ 1 & 4 \end{bmatrix}.$$

Moreover, on a TI-82, we can display our answer with fractional entries.

```
[A]⁻¹
      [[.4  -.2]
       [-.1 .3 ]]
Ans►Frac
      [[2/5   -1/5]
       [-1/10 3/10]]
```

**FIGURE 6.5** Inverse of **A** with decimal entries and with fractional entries.

Now we shall solve a system by using the inverse.

**Principles in Practice 4**
**Using the Inverse to Solve a System**

A group of investors has \$500,000 to invest in the stocks of three companies. Company A sells for \$50 a share and has an expected growth of 13% per year. Company B sells for \$20 per share and has an expected growth of 15% per year. Company C sells for \$80 a share and has an expected growth of 10% per year. The group plans to buy twice as many shares of Company A as of Company C. If the group's goal is 12% growth per year, how many shares of each stock should the investors buy?

**EXAMPLE 5**   **Using the Inverse to Solve a System**

*Solve the system*

$$\begin{cases} x_1 \quad\quad - 2x_3 = \quad 1, \\ 4x_1 - 2x_2 + \quad x_3 = \quad 2, \\ x_1 + 2x_2 - 10x_3 = -1, \end{cases}$$

*by finding the inverse of the coefficient matrix.*

**Solution:**  In matrix form the system is $\mathbf{AX} = \mathbf{B}$, where

$$\mathbf{A} = \begin{bmatrix} 1 & 0 & -2 \\ 4 & -2 & 1 \\ 1 & 2 & -10 \end{bmatrix}$$

is the coefficient matrix. From Example 4(a),

$$\mathbf{A}^{-1} = \begin{bmatrix} -9 & 2 & 2 \\ -\frac{41}{2} & 4 & \frac{9}{2} \\ -5 & 1 & 1 \end{bmatrix}.$$

The solution is given by $\mathbf{X} = \mathbf{A}^{-1}\mathbf{B}$:

$$\begin{bmatrix} x_1 \\ x_2 \\ x_3 \end{bmatrix} = \begin{bmatrix} -9 & 2 & 2 \\ -\frac{41}{2} & 4 & \frac{9}{2} \\ -5 & 1 & 1 \end{bmatrix} \begin{bmatrix} 1 \\ 2 \\ -1 \end{bmatrix} = \begin{bmatrix} -7 \\ -17 \\ -4 \end{bmatrix},$$

so $x_1 = -7$, $x_2 = -17$, and $x_3 = -4$.  ∎

It can be shown that a system of $n$ linear equations in $n$ unknowns has a unique solution if and only if the coefficient matrix is invertible. Indeed, in the previous example the coefficient matrix is invertible, and a unique solution does in fact exist. When the coefficient matrix is not invertible, the system will have either no solution or infinitely many solutions.

**EXAMPLE 6**   **A Coefficient Matrix That Is Not Invertible**

*Solve the system*

$$\begin{cases} x - 2y + \quad z = 0, \\ 2x - \quad y + 5z = 0, \\ x + \quad y + 4z = 0. \end{cases}$$

**Solution:**  The coefficient matrix is

$$\begin{bmatrix} 1 & -2 & 1 \\ 2 & -1 & 5 \\ 1 & 1 & 4 \end{bmatrix}.$$

Since

$$\left[\begin{array}{ccc|ccc} 1 & -2 & 1 & 1 & 0 & 0 \\ 2 & -1 & 5 & 0 & 1 & 0 \\ 1 & 1 & 4 & 0 & 0 & 1 \end{array}\right] \rightarrow \cdots \rightarrow \left[\begin{array}{ccc|ccc} 1 & 0 & 3 & -\frac{1}{3} & \frac{2}{3} & 0 \\ 0 & 1 & 1 & -\frac{2}{3} & \frac{1}{3} & 0 \\ 0 & 0 & 0 & 1 & -1 & 1 \end{array}\right],$$

the coefficient matrix is not invertible. Hence, the system *cannot* be solved by inverses. Instead, another method must be used. In Example 4(a) of Sec. 6.5, the solution was found to be $x = -3r$, $y = -r$, and $z = r$.   ■

## TECHNOLOGY

To solve the system

$$\begin{cases} 3x + 2y = \phantom{-}6, \\ \phantom{3}x + 4y = -8, \end{cases}$$

with a graphics calculator, we enter the coefficient matrix as [A] and the column matrix of constants as [B]. The product $[A]^{-1}[B]$ in Fig. 6.6 gives the solution $x = 4$, $y = -3$.

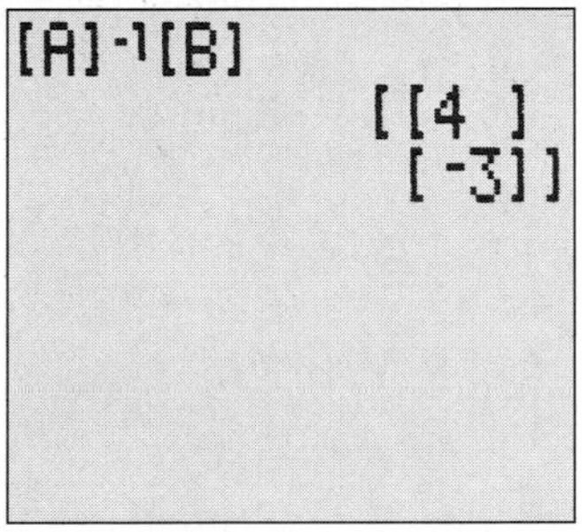

```
[A]⁻¹[B]
              [[4 ]
               [-3]]
```

**FIGURE 6.6**   $[A]^{-1}[B]$
gives solution $x = 4$, $y = -3$
the to system of equations.

## ■ Exercise 6.6

*In each of Problems **1–18**, if the given matrix is invertible, find its inverse.*

**1.** $\begin{bmatrix} 6 & 1 \\ 5 & 1 \end{bmatrix}$.

**2.** $\begin{bmatrix} 2 & 8 \\ 3 & 12 \end{bmatrix}$.

**3.** $\begin{bmatrix} 1 & 1 \\ 1 & 1 \end{bmatrix}$.

**4.** $\begin{bmatrix} 4 & 9 \\ 0 & -6 \end{bmatrix}$.

**5.** $\begin{bmatrix} 1 & 0 & 0 \\ 0 & -3 & 0 \\ 0 & 0 & 4 \end{bmatrix}$.

**6.** $\begin{bmatrix} 2 & 0 & 8 \\ -1 & 4 & 0 \\ 2 & 1 & 0 \end{bmatrix}$.

**7.** $\begin{bmatrix} 1 & 2 & 3 \\ 0 & 0 & 4 \\ 0 & 0 & 5 \end{bmatrix}$.

**8.** $\begin{bmatrix} 2 & 0 & 0 \\ 0 & 0 & 0 \\ 0 & 0 & -4 \end{bmatrix}$.

**9.** $\begin{bmatrix} 2 & 4 \\ 8 & 1 \\ 6 & 3 \end{bmatrix}$.

**10.** $\begin{bmatrix} 0 & 0 & 0 \\ 0 & 0 & 0 \\ 0 & 0 & 0 \end{bmatrix}$.

**11.** $\begin{bmatrix} 1 & 1 & 1 \\ 0 & 1 & 1 \\ 0 & 0 & 1 \end{bmatrix}$.

**12.** $\begin{bmatrix} 1 & 2 & -1 \\ 0 & 1 & 4 \\ 1 & -1 & 2 \end{bmatrix}$.

**13.** $\begin{bmatrix} 7 & 0 & -2 \\ 0 & 1 & 0 \\ -3 & 0 & 1 \end{bmatrix}$.

**14.** $\begin{bmatrix} 7 & -8 & 5 \\ -4 & 5 & -3 \\ 1 & -1 & 1 \end{bmatrix}$.

**15.** $\begin{bmatrix} 2 & 1 & 0 \\ 4 & -1 & 5 \\ 1 & -1 & 2 \end{bmatrix}$.

**16.** $\begin{bmatrix} -5 & 4 & -3 \\ 10 & -7 & 6 \\ 8 & -6 & 5 \end{bmatrix}$.

**17.** $\begin{bmatrix} 1 & 2 & 3 \\ 1 & 3 & 5 \\ 1 & 5 & 12 \end{bmatrix}$.

**18.** $\begin{bmatrix} 2 & -1 & 3 \\ 0 & 2 & 0 \\ 2 & 1 & 1 \end{bmatrix}$.

**19.** Solve $\mathbf{AX} = \mathbf{B}$ if

$$\mathbf{A}^{-1} = \begin{bmatrix} 1 & 2 \\ 1 & 1 \end{bmatrix} \quad \text{and} \quad \mathbf{B} = \begin{bmatrix} 2 \\ 4 \end{bmatrix}.$$

**20.** Solve $\mathbf{AX} = \mathbf{B}$ if

$$\mathbf{A}^{-1} = \begin{bmatrix} 1 & 0 & 1 \\ 0 & 3 & 0 \\ 2 & 0 & 4 \end{bmatrix} \quad \text{and} \quad \mathbf{B} = \begin{bmatrix} 1 \\ 2 \\ -1 \end{bmatrix}.$$

*For each of Problems **21–34**, if the coefficient matrix of the system is invertible, solve the system by using the inverse. If not, solve the system by the method of reduction.*

**21.** $\begin{cases} 6x + 5y = \phantom{-}2, \\ \phantom{6}x + \phantom{5}y = -3. \end{cases}$

**22.** $\begin{cases} 2x + 3y = \phantom{-}4, \\ -x + 5y = -2. \end{cases}$

**23.** $\begin{cases} 2x + y = 5, \\ 3x - y = 0. \end{cases}$

**24.** $\begin{cases} 3x + 2y = 26, \\ 4x + 3y = 37. \end{cases}$

**25.** $\begin{cases} 2x + 6y = 2, \\ 3x + 9y = 3. \end{cases}$

**26.** $\begin{cases} 2x + 8y = 3, \\ 3x + 12y = 6. \end{cases}$

**27.** $\begin{cases} x + 2y + z = 4, \\ 3x \quad\;\; + z = 2, \\ x - y + z = 1. \end{cases}$

**28.** $\begin{cases} x + y + z = 2, \\ x - y + z = -2, \\ x - y - z = 0. \end{cases}$

**29.** $\begin{cases} x + y + z = 2, \\ x - y + z = 1, \\ x - y - z = 0 \end{cases}$

**30.** $\begin{cases} 2x \quad\;\; + 8z = 8, \\ -x + 4y \quad\;\; = 36, \\ 2x + y \quad\;\; = 9. \end{cases}$

**31.** $\begin{cases} x + 3y + 3z = 7, \\ 2x + y + z = 4, \\ x + y + z = 4. \end{cases}$

**32.** $\begin{cases} x + 3y + 3z = 7, \\ 2x + y + z = 4, \\ x + y + z = 3. \end{cases}$

**33.** $\begin{cases} w \quad\quad + 2y + z = 4, \\ w - x \quad\quad + 2z = 12, \\ 2w + x \quad\quad + z = 12, \\ w + 2x + y + z = 12. \end{cases}$

**34.** $\begin{cases} w + x \quad\quad + z = 2, \\ w \quad\quad + y \quad\quad = 0, \\ x + y + z = 4, \\ y + z = 1. \end{cases}$

*For each of Problems 35 and 36, find $(\mathbf{I} - \mathbf{A})^{-1}$ for the given matrix $\mathbf{A}$.*

**35.** $\mathbf{A} = \begin{bmatrix} 2 & -1 \\ 1 & 3 \end{bmatrix}$.

**36.** $\mathbf{A} = \begin{bmatrix} -3 & 2 \\ 4 & 3 \end{bmatrix}$.

---

**37. Auto Production** Solve the following problems by using the inverse of the matrix involved.

**a.** An automobile factory produces two models, A and B. Model A requires 1 labor hour to paint and $\frac{1}{2}$ labor

hour to polish; model B requires 1 labor hour for each process. During each hour that the assembly line is operating, there are 100 labor hours available for painting and 80 labor hours for polishing. How many of each model can be produced each hour if all the labor hours available are to be utilized?

**b.** Suppose each model A requires 10 widgets and 14 shims and each model B requires 7 widgets and 10 shims. The factory can obtain 800 widgets and 1130 shims each hour. How many cars of each model can it produce while utilizing all the parts available?

**38.** If $\mathbf{A} = \begin{bmatrix} a & 0 & 0 \\ 0 & b & 0 \\ 0 & 0 & c \end{bmatrix}$, where $a, b, c \neq 0$, show that

$$\mathbf{A}^{-1} = \begin{bmatrix} 1/a & 0 & 0 \\ 0 & 1/b & 0 \\ 0 & 0 & 1/c \end{bmatrix}.$$

**39. a.** If $\mathbf{A}$ and $\mathbf{B}$ are invertible matrices with the same order, show that $(\mathbf{AB})^{-1} = \mathbf{B}^{-1}\mathbf{A}^{-1}$. [*Hint:* Show that

$$(\mathbf{B}^{-1}\mathbf{A}^{-1})(\mathbf{AB}) = \mathbf{I},$$

and use the fact that the inverse is unique.]

**b.** If

$$\mathbf{A}^{-1} = \begin{bmatrix} 1 & 2 \\ 3 & 4 \end{bmatrix} \quad \text{and} \quad \mathbf{B}^{-1} = \begin{bmatrix} 1 & 1 \\ 1 & 2 \end{bmatrix},$$

find $(\mathbf{AB})^{-1}$.

**40.** If $\mathbf{A}$ is invertible, it can be shown that $(\mathbf{A}^{\mathsf{T}})^{-1} = (\mathbf{A}^{-1})^{\mathsf{T}}$. Verify this relationship if

$$\mathbf{A} = \begin{bmatrix} 1 & 2 \\ 3 & 4 \end{bmatrix}.$$

**41.** A matrix $\mathbf{P}$ is said to be *orthogonal* if $\mathbf{P}^{-1} = \mathbf{P}^{\mathsf{T}}$. Is the matrix $\mathbf{P} = \frac{1}{5}\begin{bmatrix} 3 & -4 \\ 4 & 3 \end{bmatrix}$ orthogonal?

**42. Secret Message** A friend has sent you a secret message that consists of three row matrices of numbers as follows:

$$\mathbf{R}_1 = \begin{bmatrix} -5 & -9 & 29 \end{bmatrix}, \quad \mathbf{R}_2 = \begin{bmatrix} 7 & 23 & 48 \end{bmatrix},$$

$$\mathbf{R}_3 = \begin{bmatrix} 34 & 89 & 64 \end{bmatrix}.$$

Both you and your friend have committed the following matrix to memory (your friend used it to code the message):

$$\mathbf{A} = \begin{bmatrix} 1 & 2 & -1 \\ 2 & 5 & 2 \\ -1 & -2 & 2 \end{bmatrix}.$$

Decipher the message by proceeding as follows:

**a.** Calculate the three matrix products $\mathbf{R}_1\mathbf{A}^{-1}$, $\mathbf{R}_2\mathbf{A}^{-1}$, and $\mathbf{R}_3\mathbf{A}^{-1}$.

**b.** Assume that the letters of the alphabet correspond to the numbers 1 through 26, replace the numbers in the preceding three matrices by letters, and determine the message.

**43. Investing**  A group of investors decides to invest $500,000 in the stocks of three companies. Company D sells for $60 a share and has an expected growth of 16% per year. Company E sells for $80 per share and has an expected growth of 12% per year. Company F sells for $30 a share and has an expected growth of 9% per year. The group plans to buy four times as many shares of company F as of company E. If the group's goal is 13.68% growth per year, how many shares of each stock should the investors buy?

**44. Investing**  The investors in Problem 43 decide to try a new investment strategy with the same companies. They wish to buy twice as many shares of company F as of company E, and they have a goal of 14.52% growth per year. How many shares of each stock should they buy?

*In Problems 45 and 46 use a graphics calculator to: (a) find $\mathbf{A}^{-1}$, and express its entries in decimal form rounded to two decimal places. (b) Express the entries of $\mathbf{A}^{-1}$ in fractional form if your calculator has such capability. [Caution: For part (b), use the calculator matrix $\mathbf{A}^{-1}$ to convert to fractional entries; do not use the matrix of rounded values from part (a).]*

 **45.** $\mathbf{A} = \begin{bmatrix} \frac{4}{5} & -\frac{1}{3} \\ -\frac{3}{10} & \frac{13}{15} \end{bmatrix}.$

 **46.** $\mathbf{A} = \begin{bmatrix} 2 & 6 & -3 \\ 4 & 8 & 9 \\ -7 & 2 & 5 \end{bmatrix}.$

 **47.** If $\mathbf{A} = \begin{bmatrix} 0.4 & 0.6 & -0.3 \\ 0.2 & 0.1 & -0.1 \\ 0.3 & 0.2 & -0.4 \end{bmatrix}$, find $(\mathbf{I} - \mathbf{A})^{-1}$, where $\mathbf{I}$ is the identity matrix of order 3. Round entries to two decimal places.

*In Problems 48 and 49, use a graphics calculator to solve the system by using the inverse of the coefficient matrix.*

**48.** $\begin{cases} 0.6x + 3y - 4.7z = 13, \\ 2x - 0.4y + 2z = 4.7, \\ x - 0.8y - 0.5z = 7.2. \end{cases}$

**49.** $\begin{cases} \frac{2}{5}w + 4x + \frac{1}{2}y - \frac{3}{7}z = \frac{14}{13}, \\ \frac{5}{9}w - \frac{2}{3}x - 4y - z = \frac{7}{8}, \\ x - \frac{4}{9}y + \frac{5}{6}z = 9, \\ \frac{1}{2}w + 4y - \frac{1}{3}z = \frac{4}{7}. \end{cases}$

---

### OBJECTIVE

To find the determinant of a square matrix by using minors and cofactors and to consider properties that simplify the evaluation of a determinant.

The determinant of **A** is also denoted det **A**.

## 6.7 DETERMINANTS

We now introduce a new function, the *determinant function.* Here our inputs will be *square* matrices, but our outputs will be real numbers. If **A** is a square matrix, then the determinant function associates with **A** exactly one real number called the *determinant* of **A**. Denoting the determinant of **A** by $|\mathbf{A}|$ (that is, using vertical bars), we can think of the determinant function as a correspondence:

$$\mathbf{A} \rightarrow |\mathbf{A}|$$

$$\frac{\text{square} \quad \text{real}}{\text{matrix} \quad \text{number}} = \frac{\text{determinant}}{\text{of } \mathbf{A}}$$

The use of determinants in solving systems of linear equations will be discussed later. Turning to how a real number is assigned to a square matrix, we shall first consider the special cases of matrices of orders 1 and 2. Then we shall extend the definition to matrices of order *n*.

**DEFINITION**

*If* $\mathbf{A} = \begin{bmatrix} a_{11} \end{bmatrix}$ *is a square matrix of order 1, then* $|\mathbf{A}| = a_{11}$

That is, the determinant function assigns to the one-entry matrix $\begin{bmatrix} a_{11} \end{bmatrix}$ the number $a_{11}$. Hence, if $\mathbf{A} = \begin{bmatrix} 6 \end{bmatrix}$, then $|\mathbf{A}| = 6$.

**DEFINITION**

*If* $\mathbf{A} = \begin{bmatrix} a_{11} & a_{12} \\ a_{21} & a_{22} \end{bmatrix}$ *is a square matrix of order 2, then*

$$|\mathbf{A}| = a_{11}a_{22} - a_{12}a_{21}.$$

That is, the determinant of a $2 \times 2$ matrix is obtained by taking the product of the entries in the main diagonal and subtracting from it the product of the entries in the other diagonal:

$$\begin{vmatrix} a_{11} & a_{12} \\ a_{21} & a_{22} \end{vmatrix} = a_{11}a_{22} - a_{12}a_{21}.$$

We speak of the determinant of a $2 \times 2$ matrix as a *determinant of order* 2.

**EXAMPLE 1**   **Evaluating Determinants of Order 2**

a. $\begin{vmatrix} 2 & 1 \\ 3 & -4 \end{vmatrix} = (2)(-4) - (1)(3) = -8 - 3 = -11.$

b. $\begin{vmatrix} -3 & -2 \\ 0 & 1 \end{vmatrix} = (-3)(1) - (-2)(0) = -3 - 0 = -3.$

c. $\begin{vmatrix} 1 & 0 \\ 0 & 1 \end{vmatrix} = (1)(1) - (0)(0) = 1.$

d. $\begin{vmatrix} x & 0 \\ y & 1 \end{vmatrix} = (x)(1) - (0)(y) = x.$ ∎

The determinant of a square matrix $\mathbf{A}$ of order $n > 2$ is defined in the following manner. With a given entry of $\mathbf{A}$, we associate the square matrix of order $n - 1$ obtained by deleting the entries in the row and column in which the given entry lies. For example, given the matrix

$$\begin{bmatrix} a_{11} & a_{12} & a_{13} \\ a_{21} & a_{22} & a_{23} \\ a_{31} & a_{32} & a_{33} \end{bmatrix},$$

for entry $a_{21}$, we delete the entries in row 2 and column 1, shown by shading as follows:

$$\begin{bmatrix} a_{11} & a_{12} & a_{13} \\ a_{21} & a_{22} & a_{23} \\ a_{31} & a_{32} & a_{33} \end{bmatrix}.$$

This leaves the matrix

$$\begin{bmatrix} a_{12} & a_{13} \\ a_{32} & a_{33} \end{bmatrix},$$

of order 2. The *determinant* of this matrix is called the **minor** of $a_{21}$. Similarly, the minor of $a_{22}$ is

$$\begin{vmatrix} a_{11} & a_{13} \\ a_{31} & a_{33} \end{vmatrix},$$

and for $a_{23}$, it is

$$\begin{vmatrix} a_{11} & a_{12} \\ a_{31} & a_{32} \end{vmatrix}.$$

With each entry $a_{ij}$ we also associate a number determined by the subscript of the entry, namely,

$$(-1)^{i+j},$$

where $i + j$ is the sum of the row number $i$ and column number $j$ in which the entry lies. With entry $a_{21}$ we associate $(-1)^{2+1} = -1$, with $a_{22}$ the number $(-1)^{2+2} = 1$, and with $a_{23}$ the number $(-1)^{2+3} = -1$. The **cofactor** $c_{ij}$ of the entry $a_{ij}$ is the product of $(-1)^{i+j}$ and the minor of $a_{ij}$. For example, the cofactor of $a_{21}$ is

$$c_{21} = (-1)^{2+1}\begin{vmatrix} a_{12} & a_{13} \\ a_{32} & a_{33} \end{vmatrix}.$$

The only difference between a cofactor and a minor is the factor $(-1)^{i+j}$.

> **Determinant of a Square Matrix**
>
> To find the determinant of any square matrix **A** of order $n > 2$, select *any* row (or column) of **A**, and multiply each entry in the row (column) by its cofactor. The sum of these products is defined to be the determinant of **A** and is called a **determinant of order $n$.**

For example, we shall find the determinant of

$$\begin{bmatrix} 2 & -1 & 3 \\ 3 & 0 & -5 \\ 2 & 1 & 1 \end{bmatrix}$$

by applying the preceding rule to the first row (sometimes referred to as "expanding along the first row"). For entry $a_{11}$, we obtain

$$(2)(-1)^{1+1}\begin{vmatrix} 0 & -5 \\ 1 & 1 \end{vmatrix} = (2)(1)(5) = 10.$$

For $a_{12}$ we obtain

$$(-1)(-1)^{1+2}\begin{vmatrix} 3 & -5 \\ 2 & 1 \end{vmatrix} = (-1)(-1)(13) = 13,$$

and for $a_{13}$, we obtain

$$(3)(-1)^{1+3}\begin{vmatrix} 3 & 0 \\ 2 & 1 \end{vmatrix} = 3(1)(3) = 9.$$

Hence,

$$\begin{vmatrix} 2 & -1 & 3 \\ 3 & 0 & -5 \\ 2 & 1 & 1 \end{vmatrix} = 10 + 13 + 9 = 32.$$

Alternatively, if we had expanded along the second column, then

$$\begin{vmatrix} 2 & -1 & 3 \\ 3 & 0 & -5 \\ 2 & 1 & 1 \end{vmatrix} = (-1)(-1)^{1+2}\begin{vmatrix} 3 & -5 \\ 2 & 1 \end{vmatrix} + 0 + (1)(-1)^{3+2}\begin{vmatrix} 2 & 3 \\ 3 & -5 \end{vmatrix}$$

$$= 13 + 0 + 19 = 32,$$

as before.

It can be shown that the determinant of a matrix is unique and does not depend on the row or column chosen for its evaluation. In the foregoing problem, the second expansion is preferable, since the 0 in column 2 contributed nothing to the sum, thus simplifying the calculation.

---

**Principles in Practice 1**

**Evaluating a Determinant of Order 3 by Using Cofactors**

A botanist is growing three different types of algae in the same environment in her laboratory. She gives the algae a nutrient mixture each day that contains three different nutrients (1, 2, and 3). The requirements of each cell of the three types of algae (A, B, and C) can be represented by the following matrix:

$$\begin{array}{c} \\ 1 \\ 2 \\ 3 \end{array}\begin{array}{ccc} A & B & C \\ \begin{bmatrix} 1 & 1 & 2 \\ 3 & 4 & 10 \\ 4 & 2 & 6 \end{bmatrix} \end{array}.$$

Find the determinant of this matrix.

---

**EXAMPLE 2   Evaluating a Determinant of Order 3 by Using Cofactors**

*Find* $|\mathbf{A}|$ *if*

**a.** $\mathbf{A} = \begin{bmatrix} 12 & -1 & 3 \\ -3 & 1 & -1 \\ -10 & 2 & -3 \end{bmatrix}.$

**Solution:** Expanding along the first row, we have

$$|\mathbf{A}| = 12(-1)^{1+1}\begin{vmatrix} 1 & -1 \\ 2 & -3 \end{vmatrix} + (-1)(-1)^{1+2}\begin{vmatrix} -3 & -1 \\ -10 & -3 \end{vmatrix} + 3(-1)^{1+3}\begin{vmatrix} -3 & 1 \\ -10 & 2 \end{vmatrix}$$

$$= 12(1)(-1) + (-1)(-1)(-1) + 3(1)(4) = -1.$$

**b.** $\mathbf{A} = \begin{bmatrix} 0 & 1 & 1 \\ 2 & 3 & 2 \\ 0 & -1 & 3 \end{bmatrix}.$

**Solution:** Expanding along column 1 for convenience, we have

$$|\mathbf{A}| = 0 + 2(-1)^{2+1}\begin{vmatrix} 1 & 1 \\ -1 & 3 \end{vmatrix} + 0 = 2(-1)(4) = -8. \quad\blacksquare$$

**EXAMPLE 3   Evaluating a Determinant of Order 4**

*Evaluate* $|\mathbf{A}| = \begin{vmatrix} 2 & 0 & 0 & 1 \\ 0 & 1 & 0 & 3 \\ 0 & 0 & 1 & 2 \\ 1 & 2 & 3 & 0 \end{vmatrix}$ *by expanding along the first row.*

**Solution:**

$$|\mathbf{A}| = 2(-1)^{1+1}\begin{vmatrix} 1 & 0 & 3 \\ 0 & 1 & 2 \\ 2 & 3 & 0 \end{vmatrix} + 1(-1)^{1+4}\begin{vmatrix} 0 & 1 & 0 \\ 0 & 0 & 1 \\ 1 & 2 & 3 \end{vmatrix}.$$

We have now expressed $|\mathbf{A}|$ in terms of determinants of order 3. Expanding each of these along the first row, we have

$$|\mathbf{A}| = 2(1)\left[1(-1)^{1+1}\begin{vmatrix} 1 & 2 \\ 3 & 0 \end{vmatrix} + 3(-1)^{1+3}\begin{vmatrix} 0 & 1 \\ 2 & 3 \end{vmatrix}\right] + 1(-1)\left[1(-1)^{1+2}\begin{vmatrix} 0 & 1 \\ 1 & 3 \end{vmatrix}\right]$$

$$= 2[1(1)(-6) + 3(1)(-2)] + (-1)[(1)(-1)(-1)] = -25. \quad\blacksquare$$

We can also evaluate a determinant of order 3 as follows. Copy the first and second columns of the determinant to its right, thus giving:

$$\begin{vmatrix} a_{11} & a_{12} & a_{13} \\ a_{21} & a_{22} & a_{23} \\ a_{31} & a_{32} & a_{33} \end{vmatrix}\begin{matrix} a_{11} & a_{12} \\ a_{21} & a_{22} \\ a_{31} & a_{32} \end{matrix}.$$

Then take the sum of the three products of the entries on the arrows extending to the right, and subtract from this the sum of the three products of the entries on the arrows extending to the left. The result is

$$a_{11}a_{22}a_{33} + a_{12}a_{23}a_{31} + a_{13}a_{21}a_{32} - (a_{12}a_{21}a_{33} + a_{11}a_{23}a_{32} + a_{13}a_{22}a_{31}).$$

You should verify this method for the determinants in Example 2. We emphasize that there is no similar device for evaluating determinants of order greater than 3.

The evaluation of determinants is often simplified by the use of various properties, some of which appear in the following list (in each case, **A** denotes a square matrix):

**1. If each of the entries in a row (or column) of A is 0, then $|\mathbf{A}| = 0$.**

Thus,

$$\begin{vmatrix} 6 & 2 & 5 \\ 7 & 1 & 4 \\ 0 & 0 & 0 \end{vmatrix} = 0.$$

**2. If two rows (or columns) of A are identical, $|\mathbf{A}| = 0$.**

For example,

$$\begin{vmatrix} 2 & 5 & 2 & 1 \\ 2 & 6 & 2 & 3 \\ 2 & 4 & 2 & 1 \\ 6 & 5 & 6 & 1 \end{vmatrix} = 0, \quad \text{since column 1} = \text{column 3.}$$

**3. If A is upper (or lower) triangular, then $|\mathbf{A}|$ is equal to the product of the main diagonal entries.**

Hence,

$$\begin{vmatrix} 2 & 6 & 1 & 0 \\ 0 & 5 & 7 & 6 \\ 0 & 0 & -2 & 5 \\ 0 & 0 & 0 & 1 \end{vmatrix} = (2)(5)(-2)(1) = -20.$$

From this property, we conclude that the determinant of an identity matrix is 1.

**4. If B is the matrix obtained by adding a multiple of one row (or column) of A to another row (column), then $|\mathbf{B}| = |\mathbf{A}|$.**

Thus, if

$$\mathbf{A} = \begin{bmatrix} 2 & 4 & 2 & 6 \\ 1 & 3 & 5 & 2 \\ 1 & 2 & 1 & 3 \\ 0 & 5 & 6 & 2 \end{bmatrix}$$

and **B** is the matrix obtained from **A** by adding $-2$ times row 3 to row 1, then

$$|\mathbf{A}| = \begin{vmatrix} 2 & 4 & 2 & 6 \\ 1 & 3 & 5 & 2 \\ 1 & 2 & 1 & 3 \\ 0 & 5 & 6 & 2 \end{vmatrix} = \begin{vmatrix} 0 & 0 & 0 & 0 \\ 1 & 3 & 5 & 2 \\ 1 & 2 & 1 & 3 \\ 0 & 5 & 6 & 2 \end{vmatrix} = |\mathbf{B}|.$$

By property 1, $|\mathbf{B}| = 0$, and hence, $|\mathbf{A}| = 0$.

**5. If B is the matrix obtained by interchanging two rows (or columns) of A, then $|\mathbf{B}| = -|\mathbf{A}|$, or equivalently, $|\mathbf{A}| = -|\mathbf{B}|$.**

For example, if

$$\mathbf{A} = \begin{bmatrix} 2 & 2 & 1 & 6 \\ 0 & 0 & 0 & 1 \\ 0 & 0 & 2 & 0 \\ 0 & 1 & -3 & 4 \end{bmatrix},$$

then, by interchanging rows 2 and 4, we have

$$|\mathbf{A}| = \begin{vmatrix} 2 & 2 & 1 & 6 \\ 0 & 0 & 0 & 1 \\ 0 & 0 & 2 & 0 \\ 0 & 1 & -3 & 4 \end{vmatrix} = -\begin{vmatrix} 2 & 2 & 1 & 6 \\ 0 & 1 & -3 & 4 \\ 0 & 0 & 2 & 0 \\ 0 & 0 & 0 & 1 \end{vmatrix} = -(2)(1)(2)(1) = -4,$$

by property 3.

**6. If B is the matrix obtained by multiplying each entry of a row (or column) of A by the same number $k$, then $|\mathbf{B}| = k|\mathbf{A}|$.**

Essentially, with this property, a number can be "factored out" of one row or column. For example,

$$\begin{vmatrix} 6 & 10 & 14 \\ 5 & 2 & 1 \\ 9 & 15 & 21 \end{vmatrix} = \begin{vmatrix} 2(3) & 2(5) & 2(7) \\ 5 & 2 & 1 \\ 9 & 15 & 21 \end{vmatrix} = 2\begin{vmatrix} 3 & 5 & 7 \\ 5 & 2 & 1 \\ 9 & 15 & 21 \end{vmatrix}.$$

Thus,

$$\begin{vmatrix} 6 & 10 & 14 \\ 5 & 2 & 1 \\ 9 & 15 & 21 \end{vmatrix} \overset{\frac{1}{2}\mathbf{R}_1}{=} 2\begin{vmatrix} 3 & 5 & 7 \\ 5 & 2 & 1 \\ 9 & 15 & 21 \end{vmatrix},$$

where the notation $\frac{1}{2}\mathbf{R}_1$ indicates that we multiplied row 1 by $\frac{1}{2}$ and inserted a factor of 2 in front. Continuing, we have

$$2\begin{vmatrix} 3 & 5 & 7 \\ 5 & 2 & 1 \\ 9 & 15 & 21 \end{vmatrix} \overset{\frac{1}{3}\mathbf{R}_3}{=} 2(3)\begin{vmatrix} 3 & 5 & 7 \\ 5 & 2 & 1 \\ 3 & 5 & 7 \end{vmatrix} = 2(3)(0) = 0,$$

because rows 1 and 3 are the same.

> **7. If $k$ is a constant and A has order $n$, then $|k\mathbf{A}| = k^n|\mathbf{A}|$. This follows from property 6, since each of the $n$ rows of $k\mathbf{A}$ has a common factor of $k$.**

For example, if

$$\mathbf{A} = \begin{bmatrix} 1 & 2 \\ 3 & 4 \end{bmatrix},$$

then $|\mathbf{A}| = -2$, so $|4\mathbf{A}| = 4^2|\mathbf{A}| = 16(-2) = -32$.

> **8. The determinant of the product of two matrices of order $n$ is the product of their determinants. That is, $|\mathbf{AB}| = |\mathbf{A}||\mathbf{B}|$.**

Thus, if

$$\mathbf{A} = \begin{bmatrix} 1 & 2 \\ 3 & 4 \end{bmatrix} \quad \text{and} \quad \mathbf{B} = \begin{bmatrix} 1 & 2 \\ 0 & 3 \end{bmatrix},$$

then

$$|\mathbf{AB}| = |\mathbf{A}| \cdot |\mathbf{B}| = \begin{vmatrix} 1 & 2 \\ 3 & 4 \end{vmatrix} \cdot \begin{vmatrix} 1 & 2 \\ 0 & 3 \end{vmatrix} = (-2)(3) = -6.$$

The fact that properties 1–6 hold for columns as well as rows is a result of another property: *The determinant of a square matrix and the determinant of its transpose are equal;* symbolically,

$$|\mathbf{A}| = |\mathbf{A}^{\mathrm{T}}|.$$

For example,

$$\begin{vmatrix} 1 & 2 \\ 3 & 4 \end{vmatrix} = -2 \quad \text{and} \quad \det\begin{bmatrix} 1 & 2 \\ 3 & 4 \end{bmatrix}^{\mathrm{T}} = \begin{vmatrix} 1 & 3 \\ 2 & 4 \end{vmatrix} = -2.$$

Properties 1–6 are useful in evaluating $|\mathbf{A}|$ because they give us a way of expressing $\mathbf{A}$ in triangular form (we say that we "triangulate"); then, by property 3, we take the product of the main diagonal.

### EXAMPLE 4   Evaluating a Determinant by Triangulation

*Evaluate*

$$\begin{vmatrix} 2 & -3 & 0 \\ 3 & 6 & 9 \\ 4 & 8 & 1 \end{vmatrix}.$$

***Solution:*** We have

$$\begin{vmatrix} 2 & -3 & 0 \\ 3 & 6 & 9 \\ 4 & 8 & 1 \end{vmatrix} \overset{\frac{1}{3}R_2}{=} 3\begin{vmatrix} 2 & -3 & 0 \\ 1 & 2 & 3 \\ 4 & 8 & 1 \end{vmatrix} \overset{R_1 \leftrightarrow R_2}{=} -3\begin{vmatrix} 1 & 2 & 3 \\ 2 & -3 & 0 \\ 4 & 8 & 1 \end{vmatrix}$$

$$\overset{-2R_1 + R_2}{=} -3\begin{vmatrix} 1 & 2 & 3 \\ 0 & -7 & -6 \\ 4 & 8 & 1 \end{vmatrix} \overset{-4R_1 + R_3}{=} -3\begin{vmatrix} 1 & 2 & 3 \\ 0 & -7 & -6 \\ 0 & 0 & -11 \end{vmatrix}$$

$$= -3(1)(-7)(-11) = -231. \qquad \blacksquare$$

**EXAMPLE 5**  **Evaluating a Determinant by Triangulation**

*Evaluate*

$$\begin{vmatrix} 1 & 1 & 0 & 5 \\ 1 & 2 & 1 & 0 \\ 0 & 2 & 1 & 1 \\ 3 & 0 & 0 & -4 \end{vmatrix}.$$

*Solution:*  We have

$$\begin{vmatrix} 1 & 1 & 0 & 5 \\ 1 & 2 & 1 & 0 \\ 0 & 2 & 1 & 1 \\ 3 & 0 & 0 & -4 \end{vmatrix} \underset{=}{-R_1 + R_2} \begin{vmatrix} 1 & 1 & 0 & 5 \\ 0 & 1 & 1 & -5 \\ 0 & 2 & 1 & 1 \\ 3 & 0 & 0 & -4 \end{vmatrix}$$

$$\underset{=}{-3R_1 + R_4} \begin{vmatrix} 1 & 1 & 0 & 5 \\ 0 & 1 & 1 & -5 \\ 0 & 2 & 1 & 1 \\ 0 & -3 & 0 & -19 \end{vmatrix} \underset{=}{-2R_2 + R_3} \begin{vmatrix} 1 & 1 & 0 & 5 \\ 0 & 1 & 1 & -5 \\ 0 & 0 & -1 & 11 \\ 0 & -3 & 0 & -19 \end{vmatrix}$$

$$\underset{=}{3R_2 + R_4} \begin{vmatrix} 1 & 1 & 0 & 5 \\ 0 & 1 & 1 & -5 \\ 0 & 0 & -1 & 11 \\ 0 & 0 & 3 & -34 \end{vmatrix} \underset{=}{3R_3 + R_4} \begin{vmatrix} 1 & 1 & 0 & 5 \\ 0 & 1 & 1 & -5 \\ 0 & 0 & -1 & 11 \\ 0 & 0 & 0 & -1 \end{vmatrix}$$

$$= (1)(1)(-1)(-1) = 1.$$

## TECHNOLOGY

Figure 6.7 shows the result of evaluating $|\mathbf{A}|$ with a graphics calculator, where

$$\mathbf{A} = \begin{bmatrix} 0.2 & 0 & 0.1 \\ 0.8 & 1 & -0.3 \\ 0.4 & 2 & 0.5 \end{bmatrix}.$$

The evaluation yields $|\mathbf{A}| = 0.34$.

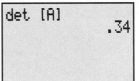

**FIGURE 6.7**  Evaluating det $\mathbf{A}$ yields 0.34.

■ **Exercise 6.7**

*In Problems 1–6, evaluate the determinants.*

1. $\begin{vmatrix} 2 & 1 \\ 3 & 2 \end{vmatrix}.$

2. $\begin{vmatrix} 3 & 2 \\ -5 & -4 \end{vmatrix}.$

3. $\begin{vmatrix} -2 & -3 \\ -4 & -6 \end{vmatrix}.$

4. $\begin{vmatrix} -3 & 1 \\ -a & b \end{vmatrix}.$

5. $\begin{vmatrix} 1 & x \\ 0 & y \end{vmatrix}.$

6. $\begin{vmatrix} -2 & -a \\ -a & 2 \end{vmatrix}.$

*In Problems 7 and 8, evaluate the given expressions.*

7. $\dfrac{\begin{vmatrix} 1 & 2 \\ 3 & 4 \end{vmatrix}}{\begin{vmatrix} 2 & 1 \\ 5 & 6 \end{vmatrix}}.$

8. $\dfrac{\begin{vmatrix} 6 & 2 \\ 1 & 5 \end{vmatrix}}{\begin{vmatrix} 2 & -6 \\ 5 & 3 \end{vmatrix}}.$

9. Solve for $k$ if $\begin{vmatrix} 2 & 3 \\ 4 & k \end{vmatrix} = 12.$

*In Problems 10–13, if*

$$A = \begin{bmatrix} 1 & 2 & 3 \\ 4 & 5 & 6 \\ 7 & 8 & 9 \end{bmatrix},$$

*determine each expression.*

**10.** The minor of $a_{31}$.

**11.** The minor of $a_{22}$.

**12.** The cofactor of $a_{23}$.

**13.** The cofactor of $a_{32}$.

**14.** If $A = [a_{ij}]$ is $50 \times 50$ and the minor of $a_{43,47}$ equals 20, what is the value of the cofactor of $a_{43,47}$?

*In Problems 15–18, if*

$$A = \begin{bmatrix} a_{11} & a_{12} & a_{13} & a_{14} \\ a_{21} & a_{22} & a_{23} & a_{24} \\ a_{31} & a_{32} & a_{33} & a_{34} \\ a_{41} & a_{42} & a_{43} & a_{44} \end{bmatrix},$$

*write each expression.*

**15.** The minor of $a_{32}$.

**16.** The minor of $a_{24}$.

**17.** The cofactor of $a_{13}$.

**18.** The cofactor of $a_{43}$.

*In Problems 19–38, evaluate the determinant. Use properties of determinants if possible.*

**19.** $\begin{vmatrix} 2 & 1 & 3 \\ 2 & 0 & 1 \\ -4 & 0 & 6 \end{vmatrix}.$

**20.** $\begin{vmatrix} 3 & 2 & 1 \\ 1 & -2 & 3 \\ -1 & 3 & 2 \end{vmatrix}.$

**21.** $\begin{vmatrix} 1 & 2 & -3 \\ 4 & 5 & 4 \\ 3 & -2 & 1 \end{vmatrix}.$

**22.** $\begin{vmatrix} 1 & 0 & -1 \\ 0 & 1 & 0 \\ 1 & -1 & 1 \end{vmatrix}.$

**23.** $\begin{vmatrix} 2 & 1 & 5 \\ -3 & 4 & -1 \\ 0 & 6 & -1 \end{vmatrix}.$

**24.** $\begin{vmatrix} 1 & 2 & 3 \\ 4 & 5 & 4 \\ 3 & 2 & 1 \end{vmatrix}.$

**25.** $\begin{vmatrix} 2 & -1 & 3 \\ 1 & 1 & -1 \\ 1 & 2 & -3 \end{vmatrix}.$

**26.** $\begin{vmatrix} 1 & 2 & 3 \\ 4 & 5 & 6 \\ 7 & 8 & 9 \end{vmatrix}.$

**27.** $\begin{vmatrix} \frac{1}{2} & \frac{2}{3} & -\frac{1}{2} \\ -1 & \frac{1}{3} & \frac{2}{3} \\ 3 & -4 & 1 \end{vmatrix}.$

**28.** $\begin{vmatrix} -\frac{1}{3} & \frac{1}{4} & 4 \\ \frac{3}{2} & \frac{3}{8} & -2 \\ -\frac{1}{8} & \frac{9}{2} & 1 \end{vmatrix}.$

**29.** $\begin{vmatrix} 1 & 0 & 3 & 2 \\ 4 & -1 & 0 & 1 \\ 2 & 1 & 0 & 3 \\ -1 & 2 & 3 & -1 \end{vmatrix}.$

**30.** $\begin{vmatrix} 7 & 6 & 0 & 5 \\ -3 & 2 & 0 & 1 \\ 4 & -3 & 0 & 2 \\ 1 & 0 & 0 & 6 \end{vmatrix}.$

**31.** $\begin{vmatrix} 1 & 7 & -3 & 8 \\ 0 & 1 & -5 & 4 \\ 0 & 0 & 1 & 7 \\ 0 & 0 & 0 & 1 \end{vmatrix}.$

**32.** $\begin{vmatrix} 1 & 2 & -3 & 4 \\ 3 & -1 & 2 & 4 \\ -2 & -4 & 6 & -8 \\ 0 & 3 & -1 & 2 \end{vmatrix}.$

**33.** $\begin{vmatrix} 1 & 0 & 0 & 0 \\ 0 & -2 & 0 & 0 \\ 0 & 0 & 4 & 0 \\ 0 & 0 & 0 & -3 \end{vmatrix}.$

**34.** $\begin{vmatrix} 1 & -3 & 2 & 6 & 4 \\ 0 & 13 & 0 & 1 & 5 \\ -2 & 1 & 2 & 3 & 4 \\ 1 & 1 & 4 & 5 & 9 \end{vmatrix}.$

**35.** $\begin{vmatrix} 1 & -1 & 2 & -1 \\ 2 & 3 & -1 & 2 \\ 1 & 4 & -3 & 3 \\ 4 & 1 & 3 & 0 \end{vmatrix}.$

**36.** $\begin{vmatrix} 1 & 0 & -2 & 3 \\ 4 & 2 & 1 & 4 \\ -5 & 3 & -1 & -7 \\ 1 & 5 & -4 & 3 \end{vmatrix}.$

**37.** $\begin{vmatrix} 5 & -1 & 4 & -2 & 2 \\ 0 & 2 & 2 & 3 & -4 \\ 2 & 0 & 3 & 5 & 0 \\ 7 & 1 & 1 & 2 & -2 \\ 3 & 0 & 1 & 0 & 0 \end{vmatrix}.$

**38.** $\begin{vmatrix} 3 & 6 & 9 & 12 & 15 \\ 1 & 2 & 3 & 4 & 3 \\ 2 & 3 & 1 & 6 & 3 \\ 4 & 1 & 3 & 3 & 5 \\ 2 & 2 & 1 & 4 & 3 \end{vmatrix}.$

*In Problems 39 and 40, solve for x.*

**39.** $\begin{vmatrix} x & -2 \\ 7 & 7-x \end{vmatrix} = 26.$

**40.** $\begin{vmatrix} 3 & x & 2x \\ 0 & x & 99 \\ 0 & 0 & x-1 \end{vmatrix} = 60.$

**41.** If $\mathbf{A}$ is of order $4 \times 4$ and $|\mathbf{A}| = 12$, what is the value of the determinant of the matrix obtained by multiplying every entry in $\mathbf{A}$ by 2?

**42.** Suppose that $\mathbf{A}$ is a square matrix of order 5 and $|\mathbf{A}| = \frac{1}{2}$. Let $\mathbf{B}$ be the matrix obtained by multiplying the third row of $\mathbf{A}$ by 7. (The other rows are unchanged.) Find $|2\mathbf{B}|$.

**43.** It can be shown that a square matrix $\mathbf{A}$ is invertible if and only if $|\mathbf{A}| \neq 0$.

  **a.** If $\mathbf{A}$ is invertible, prove that
  $$|\mathbf{A}^{-1}| = \frac{1}{|\mathbf{A}|}.$$

  **b.** If $|\mathbf{A}| = 3$, find $|\mathbf{A}^{-1}|$.

**44.** If matrix $\mathbf{A}$ has order $4 \times 4$ and $|\mathbf{A}| = 2$, find the values of (a) $|3\mathbf{A}|$, (b) $|-\mathbf{A}|$, and (c) $|\mathbf{A}^{-1}|$. [*Hint:* For (c), refer to Problem 43.]

**45.** Determine the value(s) of the constant $c$ for which the following system has infinitely many solutions:
$$\begin{cases} x = -2z - 3y, \\ cy + x = -4z, \\ 2y + cz = 0. \end{cases}$$

[*Hint:* See both the paragraph immediately preceding Example 6 of Sec. 6.6 and the beginning statement of Problem 43 here.]

*In Problems 46–48, use a graphics calculator to evaluate the determinant.*

**46.** $\begin{vmatrix} 40 & 80 & 7 \\ -23 & 46 & 18 \\ 15 & 10 & -9 \end{vmatrix}.$

**47.** $\begin{vmatrix} 2 & -3 & 4 & 6 \\ 0 & 7 & 2 & -3 \\ 5 & -1 & 2 & -4 \\ 3 & 1 & 0 & 6 \end{vmatrix}.$

**48.** $\begin{vmatrix} 0.3 & -9.1 & 7.4 & 4.7 \\ -6.2 & 3.4 & 9.6 & 3.2 \\ 5.2 & 0.2 & 7.8 & 1.6 \\ 5.1 & 7.2 & 9.6 & -0.4 \end{vmatrix}.$

**49.** If $\mathbf{A} = \begin{bmatrix} 1 & 2 & -3 \\ 0 & 4 & -8 \\ 7 & 2 & 1 \end{bmatrix}$ and $\mathbf{B} = \begin{bmatrix} 2 & 0 & -1 \\ 0 & 4 & 3 \\ -1 & 2 & 4 \end{bmatrix}$, find $|2\mathbf{A} - \mathbf{B}^2|$.

---

**OBJECTIVE**

To motivate a formula, called Cramer's rule, for the solution of a system of two linear equations in two unknowns and to generalize the rule to $n$ linear equations in $n$ unknowns.

## 6.8 CRAMER'S RULE

Determinants can be applied to solving certain systems of $n$ linear equations in $n$ unknowns. In fact, it is from the analysis of such systems that the study of determinants took its origin. We shall first consider a system of two linear equations in two unknowns. Then the results will be extended to include more general situations.

Let us solve
$$\begin{cases} a_{11}x + a_{12}y = c_1, \\ a_{21}x + a_{22}y = c_2. \end{cases} \tag{1}$$

To find an explicit formula for $x$, we look at $x\begin{vmatrix} a_{11} & a_{12} \\ a_{21} & a_{22} \end{vmatrix}$:

$$x\begin{vmatrix} a_{11} & a_{12} \\ a_{21} & a_{22} \end{vmatrix} = \begin{vmatrix} a_{11}x & a_{12} \\ a_{21}x & a_{22} \end{vmatrix} \qquad \text{(property 6 of Sec. 6.7)}$$

$$= \begin{vmatrix} a_{11}x + a_{12}y & a_{12} \\ a_{21}x + a_{22}y & a_{22} \end{vmatrix} \qquad \text{(adding } y \text{ times column 2 to column 1)}$$

$$= \begin{vmatrix} c_1 & a_{12} \\ c_2 & a_{22} \end{vmatrix} \qquad \text{[from Eq. (1)].}$$

Thus,

$$x \begin{vmatrix} a_{11} & a_{12} \\ a_{21} & a_{22} \end{vmatrix} = \begin{vmatrix} c_1 & a_{12} \\ c_2 & a_{22} \end{vmatrix},$$

so

$$x = \frac{\begin{vmatrix} c_1 & a_{12} \\ c_2 & a_{22} \end{vmatrix}}{\begin{vmatrix} a_{11} & a_{12} \\ a_{21} & a_{22} \end{vmatrix}}. \tag{2}$$

To find a formula for $y$, we look at $y \begin{vmatrix} a_{11} & a_{12} \\ a_{21} & a_{22} \end{vmatrix}$:

$$y \begin{vmatrix} a_{11} & a_{12} \\ a_{21} & a_{22} \end{vmatrix} = \begin{vmatrix} a_{11} & a_{12}y \\ a_{21} & a_{22}y \end{vmatrix} \qquad \text{(property 6 of Sec. 6.7)}$$

$$= \begin{vmatrix} a_{11} & a_{11}x + a_{12}y \\ a_{21} & a_{21}x + a_{22}y \end{vmatrix} \qquad \text{(adding } x \text{ times column 1 to column 2)}$$

$$= \begin{vmatrix} a_{11} & c_1 \\ a_{21} & c_2 \end{vmatrix} \qquad \text{[from Eq. (1)].}$$

Therefore,

$$y \begin{vmatrix} a_{11} & a_{12} \\ a_{21} & a_{22} \end{vmatrix} = \begin{vmatrix} a_{11} & c_1 \\ a_{21} & c_2 \end{vmatrix},$$

so

$$y = \frac{\begin{vmatrix} a_{11} & c_1 \\ a_{21} & c_2 \end{vmatrix}}{\begin{vmatrix} a_{11} & a_{12} \\ a_{21} & a_{22} \end{vmatrix}}. \tag{3}$$

Note that in Eqs. (2) and (3) the denominators are the same, namely, the determinant of the coefficient matrix of the given system. In finding $x$, the numerator in Eq. (2) is the determinant of the matrix obtained from the coefficient matrix by replacing the "$x$-column" (that is, column 1) by the column of constants, $\begin{smallmatrix} c_1 \\ c_2 \end{smallmatrix}$. Similarly, the numerator in Eq. (3) is the determinant of the matrix obtained from the coefficient matrix when the "$y$-column" (that is, column 2) is replaced by $\begin{smallmatrix} c_1 \\ c_2 \end{smallmatrix}$. Provided that the determinant of the coefficient matrix is not zero, the original system will have a unique solution. However, if this determinant is zero, the procedure is not applicable, and the system may have either no solution or infinitely many solutions. In such cases, previous methods should be used to solve the system.

We shall illustrate the preceding results by solving the system

$$\begin{cases} 2x + y + 5 = 0, \\ \quad\ 3y + x = 6. \end{cases}$$

First, the system is written in the appropriate form:

$$\begin{cases} 2x + y = -5, \\ x + 3y = \quad 6. \end{cases}$$

The determinant $\Delta$ of the coefficient matrix is

$$\Delta = \begin{vmatrix} 2 & 1 \\ 1 & 3 \end{vmatrix} = 2(3) - 1(1) = 5.$$

Since $\Delta \neq 0$, there is a unique solution. Solving for $x$, we have

$$x = \frac{\begin{vmatrix} -5 & 1 \\ 6 & 3 \end{vmatrix}}{\Delta} = \frac{-21}{5} = -\frac{21}{5}.$$

Solving for $y$, we obtain

$$y = \frac{\begin{vmatrix} 2 & -5 \\ 1 & 6 \end{vmatrix}}{\Delta} = \frac{17}{5}.$$

Thus, the solution is $x = -\frac{21}{5}$ and $y = \frac{17}{5}$.

The method just described can be extended to systems of $n$ linear equations in $n$ unknowns and is referred to as *Cramer's rule*.

---

**Cramer's Rule**

Let a system of $n$ linear equations in $n$ unknowns be given by

$$\begin{cases} a_{11}x_1 + a_{12}x_2 + \cdots + a_{1n}x_n = c_1, \\ a_{21}x_1 + a_{22}x_2 + \cdots + a_{2n}x_n = c_2, \\ \quad \vdots \qquad\quad \vdots \qquad\qquad\quad \vdots \\ a_{n1}x_1 + a_{n2}x_2 + \cdots + a_{nn}x_n = c_n. \end{cases}$$

If the determinant $\Delta$ of the coefficient matrix **A** is different from 0, then the system has a unique solution. Moreover, the solution is given by

$$x_1 = \frac{\Delta_1}{\Delta}, \quad x_2 = \frac{\Delta_2}{\Delta}, \quad \cdots, \quad x_n = \frac{\Delta_n}{\Delta},$$

where $\Delta_k$, the numerator of $x_k$, is the determinant of the matrix obtained by replacing the $k$th column of **A** by the column of constants.

---

**EXAMPLE 1    Applying Cramer's Rule**

*Solve the following system by Cramer's rule:*

$$\begin{cases} 2x + y + z = 0, \\ 4x + 3y + 2z = 2, \\ 2x - y - 3z = 0. \end{cases}$$

*Solution:* The determinant of the coefficient matrix is

$$\Delta = \begin{vmatrix} 2 & 1 & 1 \\ 4 & 3 & 2 \\ 2 & -1 & -3 \end{vmatrix} = -8.$$

Since $\Delta \neq 0$, there is a unique solution. Solving for $x$, we replace the first column of the coefficient matrix by the column of constants and obtain

$-1(-6) + 1(-2)$
$= 4$

$$x = \frac{\begin{vmatrix} 0 & 1 & 1 \\ 2 & 3 & 2 \\ 0 & -1 & -3 \end{vmatrix}}{\Delta} = \frac{4}{-8} = -\frac{1}{2},$$

$$\begin{matrix} 2 & 3 & 2 \\ 0 & 1 & 1 \\ 0 & 0 & -2 \end{matrix}$$

Similarly,

$$y = \frac{\begin{vmatrix} 2 & 0 & 1 \\ 4 & 2 & 2 \\ 2 & 0 & -3 \end{vmatrix}}{\Delta} = \frac{-16}{-8} = 2,$$

$2(-6) - 0 + 1(-4)$
$= -16$

$$z = \frac{\begin{vmatrix} 2 & 1 & 0 \\ 4 & 3 & 2 \\ 2 & -1 & 0 \end{vmatrix}}{\Delta} = \frac{8}{-8} = -1.$$

The solution is $x = -\frac{1}{2}$, $y = 2$, and $z = -1$. ■

**EXAMPLE 2   Applying Cramer's Rule**

*Solve the following system for z by using Cramer's rule:*

$$\begin{cases} x + y \quad\quad\; + 5w = 6, \\ x + 2y + z \quad\quad = 4, \\ \quad\quad 2y + z + w = 6, \\ 3x \quad\quad\quad - 4w = 2. \end{cases}$$

*Solution:*  We have

$$\Delta = \begin{vmatrix} 1 & 1 & 0 & 5 \\ 1 & 2 & 1 & 0 \\ 0 & 2 & 1 & 1 \\ 3 & 0 & 0 & -4 \end{vmatrix} = \begin{vmatrix} 1 & 1 & 0 & 5 \\ 0 & 1 & 1 & -5 \\ 0 & 0 & -1 & 11 \\ 0 & 0 & 0 & -1 \end{vmatrix} = 1.$$

Here we transformed into upper-triangular form and found the product of the main diagonal entries (Sec. 6.7, Example 5). In a similar fashion, we obtain

$$\Delta_z = \begin{vmatrix} 1 & 1 & 6 & 5 \\ 1 & 2 & 4 & 0 \\ 0 & 2 & 6 & 1 \\ 3 & 0 & 2 & -4 \end{vmatrix} = \begin{vmatrix} 1 & 1 & 6 & 5 \\ 0 & 1 & -2 & -5 \\ 0 & 0 & 10 & 11 \\ 0 & 0 & 0 & -\frac{49}{5} \end{vmatrix} = -98.$$

Cramer's rule allows us to solve for one unknown without having to solve for the others.

Hence, $z = \Delta_z/\Delta = -98/1 = -98$. ■

## ■ Exercise 6.8

*In Problems 1–16, solve. Use Cramer's rule if possible.*

**1.** $\begin{cases} 2x - y = 4, \\ 3x + y = 5. \end{cases}$

**2.** $\begin{cases} 3x + y = 6, \\ 7x - 2y = 5. \end{cases}$

**3.** $\begin{cases} -2x = 4 - 3y, \\ \quad y = 6x - 1. \end{cases}$

**4.** $\begin{cases} x + 2y - 6 = 0, \\ y - 1 = 3x. \end{cases}$

**5.** $\begin{cases} 3(x + 2) = 5, \\ 6(x + y) = -8. \end{cases}$

**6.** $\begin{cases} w - 2z = 4, \\ 3w - 4z = 6. \end{cases}$

**7.** $\begin{cases} \frac{3}{2}x - \frac{1}{4}z = 1, \\ \frac{1}{3}x + \frac{1}{2}z = 2. \end{cases}$

**8.** $\begin{cases} 0.6x - 0.7y = 0.33, \\ 2.1x - 0.9y = 0.69. \end{cases}$

**9.** $\begin{cases} x + y + z = 6, \\ x - y + z = 2, \\ 2x - y + 3z = 6. \end{cases}$

**10.** $\begin{cases} 2x - y + 3z = 12, \\ x + y - z = -3, \\ x + 2y - 3z = -10. \end{cases}$

**11.** $\begin{cases} 2x - 3y + 4z = 0, \\ x + y - 3z = 4, \\ 3x + 2y - z = 0. \end{cases}$

**12.** $\begin{cases} 3r \quad\;\; - t = 7, \\ 4r - s + 3t = 9, \\ \quad\;\; 3s + 2t = 15. \end{cases}$

**13.** $\begin{cases} x - 2y + z = 3, \\ 2x + y + 2z = 6, \\ x + 8y + z = 3. \end{cases}$

**14.** $\begin{cases} 2x + y + z = 1, \\ x - y + z = 4, \\ 5x + y + 3z = 5. \end{cases}$

**15.** $\begin{cases} 2x - 3y + z = -2, \\ x - 6y + 3z = -2, \\ 3x + 3y - 2z = 2. \end{cases}$

**16.** $\begin{cases} x \quad\;\; - z = 14, \\ \quad\; y + z = 21, \\ x - y + z = -10. \end{cases}$

*In Problems* **17** *and* **18,** *use Cramer's rule to solve for the indicated unknowns.*

**17.** $\begin{cases} x - y + 3z + w = -14, \\ x + 2y \quad\;\; - 3w = 12, \\ 2x + 3y + 6z + w = 1, \\ x + y + z + w = 6. \end{cases}$ ; $y, w$.

**18.** $\begin{cases} x + y + 5z \quad\quad = 6, \\ x + 2y \quad\;\; + w = 4, \\ \quad\; 2y + z + w = 6, \\ 3x \quad\quad - 4z \quad\;\; = 2. \end{cases}$ ; $x, y$.

**19.** Show that Cramer's rule does *not* apply to

$$\begin{cases} 2 - y = x, \\ 3 + x = -y, \end{cases}$$

but that, from geometrical considerations, there is no solution.

**20.** Determine all values of $c$ such that Cramer's rule cannot be used to solve the following system:

$$\begin{cases} x + cy + 8z = -4, \\ cx \quad\quad - z = 1, \\ -53x - 6y + z = 2. \end{cases}$$

**21. Balancing Attendance at Special Events** A student determined that she has sufficient spare time to attend 24 special events during the school year. Among the events being offered are concerts, hockey games, and theater productions. She feels that an ideal balance would be achieved if she attended twice as many concerts as hockey games and if the number of concerts she

attended was equal to the *average* of the number of hockey games and the number of theater productions attended. Use Cramer's rule to determine the number of hockey games she should attend to achieve this ideal balance.

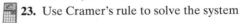 **22.** Use Cramer's rule to solve the following system for $y$:

$$\begin{cases} 3w + 2x - 7y + z = 12, \\ \quad\quad\; -5x - 6y + 3z = 8, \\ 4w \quad\quad\; + 2y + 9z = 3, \\ 7w - 2x + 4y + 5z = 9. \end{cases}$$

Round your answer to two decimal places.

**23.** Use Cramer's rule to solve the system

$$\begin{cases} \frac{4}{7}x - \frac{7}{3}y + \frac{2}{5}z = \frac{14}{9}, \\ -8x + \frac{5}{8}y - 6z = \frac{13}{9}, \\ 2x + \frac{3}{5}y + \frac{4}{3}z = \frac{10}{3}. \end{cases}$$

Round your answers to two decimal places.

**To use the methods of this chapter to analyze the production of industrial sectors of an economy.**

## 6.9 INPUT-OUTPUT ANALYSIS WITH A GRAPHICS CALCULATOR

Input-output matrices, which were developed by Wassily W. Leontief,[6] indicate the supply and demand interrelationships that exist between the various sec-

[6]Leontief won the 1973 Nobel prize in economic science for the development of the "input-output" method and its applications to economic problems.

tors of an economy during some time period. The phrase "input-output" is used because the matrices show the values of outputs of each industry that are sold as inputs to each industry and for final use by consumers.

A hypothetical example for an oversimplified two-industry economy is given by the input-output matrix to be presented next. Before we present and explain the matrix, however, let us say that the *industrial* sectors can be thought of as manufacturing, steel, agriculture, coal, and so on. The *other production factors* sector consists of costs to the respective industries, such as labor, profits, and so on. The *final-demand* sector could be consumption by households, government, and so on. The matrix is as follows:

$$
\begin{array}{c}
& Consumers\ (input) \\[4pt]
\begin{array}{r}
Producers\ (output):\\
Industry\ A\\
Industry\ B\\[18pt]
Other\ Production\ Factors\\
Totals
\end{array}
&
\begin{array}{ccccc}
Industry & Industry & Final & & \\
A & B & Demand & & Totals\\[4pt]
\left[\begin{array}{cc|c} 240 & 500 & 460 \\ 360 & 200 & 940 \\ \hline 600 & 800 & - \end{array}\right] & & & 1200 \\
& & & & 1500 \\[10pt]
1200 & 1500 & & &
\end{array}
\end{array}
$$

Each industry appears in a row and column. The row shows the purchases of an industry's output by the industrial sectors and by consumers for final use (hence the term "final demand"). The entries represent the value of the products and might be in units of millions of dollars of product. For example, of the total output of industry $A$, 240 went as input to industry $A$ itself (for internal use), 500 went to industry $B$, and 460 went directly to the final-demand sector. The total output of $A$ is the sum of industrial demand and final demand $(240 + 500 + 460 = 1200)$.

Each industry column gives the value of what the industry purchased for input from each industry, as well as what it spent for other costs. For example, in order to produce its 1200 units, $A$ purchased 240 units of output from itself, bought 360 of $B$'s output, and had labor and other costs of 600 units.

Note that for each industry, the sum of the entries in its row is equal to the sum of the entries in its column. That is, the value of the total output of $A$ is equal to the value of the total input to $A$.

Input-output analysis allows us to estimate the total production of each *industrial* sector if there is a change in final demand, *as long as the basic structure of the economy remains the same.* This important assumption means that for each industry, the amount spent on each input for each dollar's worth of output must remain fixed.

For example, in producing 1200 units' worth of product, industry $A$ purchases 240 units' worth from industry $A$, purchases 360 units' worth from $B$, and spends 600 units on other costs. Thus, for each dollar's worth of output, industry $A$ spends $\frac{240}{1200} = \frac{1}{5} (= \$0.20)$ on $A$, $\frac{360}{1200} = \frac{3}{10} (= \$0.30)$ on $B$, and $\frac{600}{1200} = \frac{1}{2} (= \$0.50)$ on other costs. Combining these fixed ratios of industry $A$ with those of industry $B$, we can give the input requirements per dollar of output for each industry:

$$
\begin{array}{c}
\begin{array}{cc} A & B \end{array} \\
\begin{array}{c} A \\ B \\[18pt] Other \end{array}
\left[\begin{array}{cc} \frac{240}{1200} & \frac{500}{1500} \\[4pt] \frac{360}{1200} & \frac{200}{1500} \\[4pt] \hline \frac{600}{1200} & \frac{800}{1500} \end{array}\right]
\end{array}
=
\begin{array}{c}
\begin{array}{cc} A & B \end{array} \\
\left[\begin{array}{cc} \frac{1}{5} & \frac{1}{3} \\[4pt] \frac{3}{10} & \frac{2}{15} \\[4pt] \hline \frac{1}{2} & \frac{8}{15} \end{array}\right]
\begin{array}{c} A \\ B \\[18pt] Other \end{array}
\end{array}
$$

The entries in the matrix are called **input-output coefficients.** The sum of each column is 1.

Now, suppose the value of final demand changes from 460 to 500 for industry $A$ and from 940 to 1200 for industry $B$. We would like to estimate the value of *total* output that $A$ and $B$ must produce for both industry and final demand to meet this goal, provided that the structure in the preceding matrix remains the same.

Let the new values of total outputs for industries $A$ and $B$ be $X_A$ and $X_B$, respectively. Now, for $A$,

$$\frac{\text{total value of}}{\text{output of } A} = \frac{\text{value consumed}}{\text{by } A} + \frac{\text{value consumed}}{\text{by } B} + \frac{\text{value consumed}}{\text{by final demand}},$$

so we have

$$X_A = \tfrac{1}{5}X_A + \tfrac{1}{3}X_B + 500.$$

Similarly, for $B$,

$$X_B = \tfrac{3}{10}X_A + \tfrac{2}{15}X_B + 1200.$$

Using matrix notation, we can write

$$\begin{bmatrix} X_A \\ X_B \end{bmatrix} = \begin{bmatrix} \tfrac{1}{5} & \tfrac{1}{3} \\ \tfrac{3}{10} & \tfrac{2}{15} \end{bmatrix} \begin{bmatrix} X_A \\ X_B \end{bmatrix} + \begin{bmatrix} 500 \\ 1200 \end{bmatrix}. \tag{1}$$

In this matrix equation, let

$$\mathbf{X} = \begin{bmatrix} X_A \\ X_B \end{bmatrix}, \qquad \mathbf{A} = \begin{bmatrix} \tfrac{1}{5} & \tfrac{1}{3} \\ \tfrac{3}{10} & \tfrac{2}{15} \end{bmatrix}, \qquad \text{and} \qquad \mathbf{C} = \begin{bmatrix} 500 \\ 1200 \end{bmatrix}.$$

We call $\mathbf{X}$ the **output matrix, A** the **coefficient matrix,** and $\mathbf{C}$ the **final-demand matrix.** From Eq. (1),

$$\mathbf{X} = \mathbf{AX} + \mathbf{C},$$

$$\mathbf{X} - \mathbf{AX} = \mathbf{C}.$$

If $\mathbf{I}$ is the $2 \times 2$ identity matrix, then

$$\mathbf{IX} - \mathbf{AX} = \mathbf{C},$$

$$(\mathbf{I} - \mathbf{A})\mathbf{X} = \mathbf{C}.$$

If $(\mathbf{I} - \mathbf{A})^{-1}$ exists, then

$$\mathbf{X} = (\mathbf{I} - \mathbf{A})^{-1}\mathbf{C}.$$

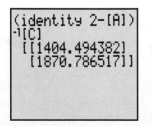

**FIGURE 6.8** Evaluating an output matrix.

The matrix $\mathbf{I} - \mathbf{A}$ is called the **Leontief matrix.** We enter the matrices $\mathbf{A}$ and $\mathbf{C}$ into a graphics calculator. With a TI-82, the identity matrix of order 2 is obtained with the command "identity 2." Evaluating $(\mathbf{I} - \mathbf{A})^{-1}\mathbf{C}$ as shown in Fig. 6.8 results in the output matrix

$$\mathbf{X} = (\mathbf{I} - \mathbf{A})^{-1}\mathbf{C} = \begin{bmatrix} 1404.49 \\ 1870.79 \end{bmatrix}.$$

Here we rounded the entries in $\mathbf{X}$ to two decimal places. Thus, to meet the goal, industry $A$ must produce 1404.49 units of value, and industry $B$ must produce 1870.79. If we were interested in the value of other production factors for $A$, say, $P_A$, then

$$P_A = \tfrac{1}{2}X_A = 702.25.$$

**EXAMPLE 1   Input-Output Analysis**

*Given the input-output matrix*

|  |  | Industry | | | Final |
|---|---|---|---|---|---|
|  |  | A | B | C | Demand |
| Industry: | A | 240 | 180 | 144 | 36 |
|  | B | 120 | 36 | 48 | 156 |
|  | C | 120 | 72 | 48 | 240 |
| Other |  | 120 | 72 | 240 | — |

*suppose final demand changes to 77 for A, 154 for B, and 231 for C. Find the output matrix for the economy. (The entries are in millions of dollars.)*

*Solution:* We separately add the entries in the first three rows. The total values of output for industries $A$, $B$, and $C$ are 600, 360, and 480, respectively. To get the coefficient matrix $\mathbf{A}$, we divide the industry entries in each industry column by the total value of output for that industry:

$$\mathbf{A} = \begin{bmatrix} \frac{240}{600} & \frac{180}{360} & \frac{144}{480} \\ \frac{120}{600} & \frac{36}{360} & \frac{48}{480} \\ \frac{120}{600} & \frac{72}{360} & \frac{48}{480} \end{bmatrix}.$$

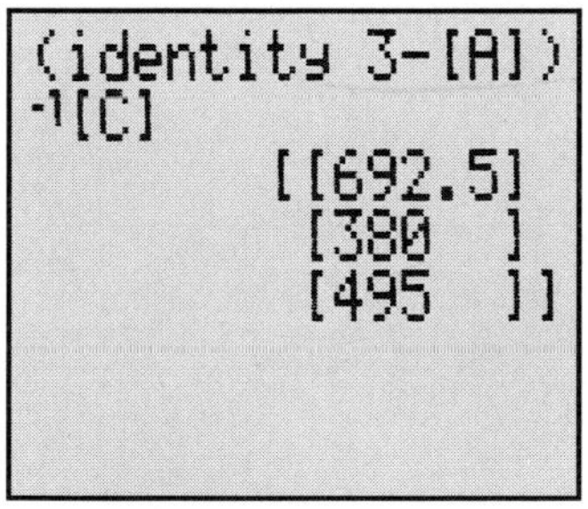

The final-demand matrix is

$$\mathbf{C} = \begin{bmatrix} 77 \\ 154 \\ 231 \end{bmatrix}.$$

Figure 6.9 shows the result of evaluating $(\mathbf{I} - \mathbf{A})^{-1}\mathbf{C}$. Thus, the output matrix is

$$\mathbf{X} = (\mathbf{I} - \mathbf{A})^{-1}\mathbf{C} = \begin{bmatrix} 692.5 \\ 380 \\ 495 \end{bmatrix}.$$

**FIGURE 6.9**   Evaluating the output matrix of Example 1.

## ■ Exercise 6.9

**1.** Given the input-output matrix

|  |  | Industry | | Final |
|---|---|---|---|---|
|  |  | Steel | Coal | Demand |
| Industry: | Steel | 200 | 500 | 500 |
|  | Coal | 400 | 200 | 900 |
| Other |  | 600 | 800 | — |

find the output matrix if final demand changes to 600 for Steel and 805 for Coal. Find the total value of the other production costs that this involves.

**2.** Given the input-output matrix

|  |  | Industry | | Final |
|---|---|---|---|---|
|  |  | Education | Government | Demand |
| Industry: | Education | 40 | 120 | 40 |
|  | Government | 120 | 90 | 90 |
| Other |  | 40 | 90 | — |

find the output matrix if final demand changes to (a) 200 for Education and 300 for Government; (b) 64 for Education and 64 for Government.

**3.** Given the input-output matrix

|  | Industry | | | Final Demand |
|---|---|---|---|---|
|  | Grain | Fertilizer | Cattle |  |
| Industry: Grain | 18 | 30 | 45 | 15 |
| Fertilizer | 27 | 30 | 60 | 3 |
| Cattle | 54 | 40 | 60 | 26 |
| Other | 9 | 20 | 15 | — |

find the output matrix (with entries rounded to two decimal places) if final demand changes to (a) 50 for Grain, 40 for Fertilizer, and 30 for Cattle; (b) 10 for Grain, 10 for Fertilizer, and 24 for Cattle.

**4.** Given the input-output matrix

|  | Industry | | | Final Demand |
|---|---|---|---|---|
|  | Water | Electric Power | Agriculture |  |
| Industry: Water | 100 | 400 | 240 | 260 |
| Electric Power | 100 | 80 | 480 | 140 |
| Agriculture | 300 | 160 | 240 | 500 |
| Other | 500 | 160 | 240 | — |

find the output matrix if final demand changes to 300 for Water, 200 for Electric Power, and 400 for Agriculture. Round your entries to two decimal places.

**5.** Given the input-output matrix

|  | Industry | | | Final Demand |
|---|---|---|---|---|
|  | Government | Agriculture | Manufacturing |  |
| Industry: Government | 400 | 200 | 200 | 200 |
| Agriculture | 200 | 400 | 100 | 300 |
| Manufacturing | 200 | 100 | 300 | 400 |
| Other | 200 | 300 | 400 | — |

with entries in billions of dollars, find the output matrix for the economy if the final demand changes to 300 for Government, 350 for Agriculture, and 450 for Manufacturing. Round your entries to the nearest billion dollars.

**6.** Given the input-output matrix in Problem 5, find the output matrix for the economy if the final demand changes to 150 for Government, 200 for Agriculture, and 300 for Manufacturing. Round your entries to the nearest billion dollars.

**7.** Given the input-output matrix in Problem 5, find the output matrix for the economy if the final demand changes to 250 for Government, 300 for Agriculture, and 350 for Manufacturing. Round your entries to the nearest billion dollars.

**8.** Given the input-output matrix in Problem 5, find the output matrix for the economy if the final demand changes to 400 for Government, 500 for Agriculture, and 300 for Manufacturing. Round your entries to the nearest billion dollars.

## 6.10 REVIEW

### IMPORTANT TERMS AND SYMBOLS

**Section 6.1**   matrix    order (or size)    entry    $a_{ij}$    $[a_{ij}]$    row matrix (or vector)
column matrix (or vector)    equality of matrices    transpose of matrix, $\mathbf{A}^T$    zero matrix, $\mathbf{O}$
square matrix    main diagonal    diagonal matrix    upper (lower) triangular matrix

**Section 6.2**   scalar multiplication    addition and subtraction of matrices

**Section 6.3**   matrix multiplication    identity matrix, $\mathbf{I}$    power of a matrix    matrix equation, $\mathbf{AX} = \mathbf{B}$

## SUMMARY

A matrix is a rectangular array of numbers enclosed within brackets. Three special types are the zero matrix $\mathbf{O}$, a square matrix, and the identity matrix $\mathbf{I}$. Besides the basic operation of scalar multiplication, there are the operations of matrix addition and subtraction, which apply to matrices of the same order. The product $\mathbf{AB}$ is defined when the number of columns of $\mathbf{A}$ is equal to the number of rows of $\mathbf{B}$. Although matrix addition is commutative, matrix multiplication is not. By using matrix multiplication, we can express a system of linear equations as the matrix equation $\mathbf{AX} = \mathbf{B}$.

A system of linear equations may have a unique solution, no solution, or infinitely many solutions. Three methods of solving a system of linear equations with matrices are (1) by using the three elementary row operations, (2) by using an inverse matrix, and (3) with determinants. The first method involves applying elementary row operations to the augmented coefficient matrix of the system until an equivalent reduced matrix is obtained. The reduced matrix makes the solution(s) to the system obvious (assuming that any exist). If there are infinitely many solutions, the general solution involves at least one parameter.

The second method of solving a system of linear equations involves inverses. The inverse (if it exists) of a square matrix $\mathbf{A}$ is a matrix $\mathbf{A}^{-1}$ such that $\mathbf{A}^{-1}\mathbf{A} = \mathbf{I}$. If $\mathbf{A}$ is invertible, we can find $\mathbf{A}^{-1}$ by augmenting $\mathbf{A}$ with $\mathbf{I}$ and applying elementary row operations until $\mathbf{A}$ is reduced to $\mathbf{I}$. The result of applying the same elementary row operations to $\mathbf{I}$ is $\mathbf{A}^{-1}$. The inverse of a matrix can be used to solve a system of $n$ equations in $n$ unknowns given by $\mathbf{AX} = \mathbf{B}$ provided that the coefficient matrix $\mathbf{A}$ is invertible. The unique solution is given by $\mathbf{X} = \mathbf{A}^{-1}\mathbf{B}$. If $\mathbf{A}$ is not invertible, the system has either no solution or infinitely many solutions.

The third method of solving a system of linear equations makes use of determinants and is known as Cramer's rule. It applies to a system of $n$ equations in $n$ unknowns when the determinant of the coefficient matrix is not zero.

Our final application of matrices dealt with the interrelationships that exist between the various sectors of an economy and is known as input-output analysis.

## REVIEW PROBLEMS

*In Problems 1–8, simplify.*

**1.** $3\begin{bmatrix} 3 & 4 \\ -5 & 1 \end{bmatrix} - 2\begin{bmatrix} 1 & 0 \\ 2 & 4 \end{bmatrix}.$

**2.** $5\begin{bmatrix} 1 & 2 \\ 7 & 0 \end{bmatrix} - 6\begin{bmatrix} 1 & 0 \\ 0 & 1 \end{bmatrix}.$

**3.** $\begin{bmatrix} 1 & 7 \\ 2 & -3 \\ 1 & 0 \end{bmatrix} \begin{bmatrix} 1 & 0 & -2 \\ 0 & 5 & 1 \end{bmatrix}.$

**4.** $\begin{bmatrix} 1 & 4 & 5 \end{bmatrix} \begin{bmatrix} 2 & 1 \\ 0 & -1 \\ 8 & 1 \end{bmatrix}.$

**5.** $\begin{bmatrix} 1 & 0 \\ -1 & 4 \end{bmatrix}\left(\begin{bmatrix} 1 & 4 \\ 6 & 5 \end{bmatrix} - \begin{bmatrix} 2 & 6 \\ 5 & 0 \end{bmatrix}\right).$

**6.** $-\left(\begin{bmatrix} 2 & 0 \\ 7 & 8 \end{bmatrix} + 2\begin{bmatrix} 0 & -5 \\ 6 & -4 \end{bmatrix}\right).$

**7.** $2\begin{bmatrix} 1 & -2 \\ 3 & 1 \end{bmatrix}^2 [1 \, -2]^{\mathrm{T}}.$

**8.** $\dfrac{1}{3}\begin{bmatrix} 3 & 0 \\ 3 & 6 \end{bmatrix}\left(\begin{bmatrix} 1 & 0 \\ 1 & 3 \end{bmatrix}^{\mathrm{T}}\right)^2.$

*In Problems 9–12, compute the required matrix if*

$$\mathbf{A} = \begin{bmatrix} 1 & 1 \\ -1 & 2 \end{bmatrix}, \qquad \mathbf{B} = \begin{bmatrix} 1 & 0 \\ 0 & 2 \end{bmatrix}.$$

**9.** $(2\mathbf{A})^{\mathrm{T}} - 3\mathbf{I}^2$.

**10.** $\mathbf{A}(2\mathbf{I}) - \mathbf{A}\mathbf{O}^{\mathrm{T}}$.

**11.** $\mathbf{B}^4 + \mathbf{I}^4$.

**12.** $(\mathbf{AB})^{\mathrm{T}} - \mathbf{B}^{\mathrm{T}}\mathbf{A}^{\mathrm{T}}$.

*In Problems 13 and 14, solve for x and y.*

**13.** $\begin{bmatrix} 5 \\ 2 \end{bmatrix}[x] = \begin{bmatrix} 15 \\ y \end{bmatrix}$.

**14.** $\begin{bmatrix} 1 & x \\ 2 & y \end{bmatrix}\begin{bmatrix} 2 & 1 \\ x & 3 \end{bmatrix} = \begin{bmatrix} 3 & 4 \\ 3 & y \end{bmatrix}$.

*In Problems 15–18, reduce the given matrices.*

**15.** $\begin{bmatrix} 1 & 4 \\ 5 & 8 \end{bmatrix}$.

**16.** $\begin{bmatrix} 0 & 0 & 4 \\ 0 & 3 & 5 \end{bmatrix}$.

**17.** $\begin{bmatrix} 2 & 4 & 3 \\ 1 & 2 & 3 \\ 4 & 8 & 6 \end{bmatrix}$.

**18.** $\begin{bmatrix} 0 & 0 & 0 & 1 \\ 0 & 0 & 0 & 0 \\ 1 & 0 & 0 & 0 \end{bmatrix}$.

*In Problems 19–22, solve each of the systems by the method of reduction.*

**19.** $\begin{cases} 2x - 5y = 0, \\ 4x + 3y = 0. \end{cases}$

**20.** $\begin{cases} x - y + 2z = 3, \\ 3x + y + z = 5. \end{cases}$

**21.** $\begin{cases} x + y + 2z = 1, \\ 3x - 2y - 4z = -7, \\ 2x - y - 2z = 2. \end{cases}$

**22.** $\begin{cases} x - y - z - 2 = 0, \\ x + y + 2z + 5 = 0, \\ 2x + z + 3 = 0. \end{cases}$

*In Problems 23–26, find the inverses of the matrices.*

**23.** $\begin{bmatrix} 1 & 5 \\ 3 & 9 \end{bmatrix}$.

**24.** $\begin{bmatrix} 0 & 1 \\ 1 & 0 \end{bmatrix}$.

**25.** $\begin{bmatrix} 1 & 3 & -2 \\ 4 & 1 & 0 \\ 3 & -2 & 2 \end{bmatrix}$.

**26.** $\begin{bmatrix} 1 & 0 & 0 \\ 2 & 3 & -1 \\ 1 & -1 & 2 \end{bmatrix}$.

*In Problems 27 and 28, solve the given system by using the inverse of the coefficient matrix.*

**27.** $\begin{cases} 3x + y + 4z = 1, \\ x + z = 0, \\ 2y + z = 2. \end{cases}$

**28.** $\begin{cases} 2x + y - z = 0, \\ 3x + z = 0, \\ x - y + z = 0. \end{cases}$

*In Problems 29–34, evaluate the determinants.*

**29.** $\begin{vmatrix} 2 & -1 \\ 4 & 7 \end{vmatrix}$.

**30.** $\begin{vmatrix} 5 & 8 \\ 3 & 0 \end{vmatrix}$.

**31.** $\begin{vmatrix} 1 & 2 & -1 \\ 0 & 1 & 4 \\ 1 & 2 & 2 \end{vmatrix}$.

**32.** $\begin{vmatrix} 2 & 0 & 3 \\ 1 & 4 & 6 \\ -1 & 2 & -1 \end{vmatrix}$.

**33.** $\begin{vmatrix} r & p & q & a \\ 0 & i & j & m \\ 0 & 0 & c & n \\ 0 & 0 & 0 & h \end{vmatrix}$.

**34.** $\begin{vmatrix} e & 0 & 0 & 0 \\ a & r & 0 & 0 \\ p & j & n & 0 \\ s & k & t & i \end{vmatrix}$.

*Solve the systems in Problems 35 and 36 by using Cramer's rule.*

**35.** $\begin{cases} 3x - y = 1, \\ 2x + 3y = 8. \end{cases}$

**36.** $\begin{cases} x + 2y - z = 0, \\ y + 4z = 0, \\ x + 2y + 2z = 0. \end{cases}$

**37.** Given that $|\mathbf{A}| = -2$, $|\mathbf{B}| = 4$, and $|\mathbf{A}^{-1}| = \dfrac{1}{|\mathbf{A}|}$, find $|\mathbf{A}^{-1}\mathbf{B}^{\mathrm{T}}|$.

**38.** Construct matrix $\mathbf{A} = [a_{ij}]_{3\times 3}$ if $a_{ij} = |i - j|$.

**39.** Let $\mathbf{A} = \begin{bmatrix} 0 & 0 & 1 \\ 0 & 1 & 0 \\ 1 & 0 & 0 \end{bmatrix}$. Find the matrices $\mathbf{A}^2$, $\mathbf{A}^{-1}$, and $\mathbf{A}^{1994}$.

**40.** $\mathbf{A} = \begin{bmatrix} 2 & 0 \\ 0 & 4 \end{bmatrix}$, show that $(\mathbf{A}^{\mathrm{T}})^{-1} = (\mathbf{A}^{-1})^{\mathrm{T}}$.

**41.** Suppose $a, b$, and $c$ are nonzero constants. Use Cramer's rule to solve the following system of linear equations:
$$\begin{cases} ax + cz = a, \\ bx + by = b, \\ ax + ay + cz = c. \end{cases}$$

**42.** Show that Cramer's rule can be used to solve the following system, and then use it to find the value of $x$ that satisfies the system:
$$\begin{cases} x + 3y + 2w + 1 = z, \\ 2w + z = 2x + 4y + 1, \\ x + 2y + 3z + 3w - 3 = 0, \\ 2x + 7y + 6z + 2w = 6. \end{cases}$$

**43.** A consumer wishes to supplement his vitamin intake by *exactly* 13 units of vitamin A, 22 units of vitamin B, and 31 units of vitamin C per week. There are three brands of vitamin capsules available. Brand I contains 1 unit each of vitamins A, B, and C per capsule; brand II contains 1 unit of vitamin A, 2 of B, and 3 of C; and brand III contains 4 units of A, 7 of B, and 10 of C.

  **a.** What combinations of capsules of brands I, II, and III will produce *exactly* the desired amounts?

  **b.** If brand I capsules cost 5 cents each, brand II 7 cents each, and brand III 20 cents each, what combination will minimize the consumer's weekly cost?

**44.** Suppose that $\mathbf{A}$ is an invertible $n \times n$ matrix.

  **a.** Prove that $\mathbf{A}^3$ is invertible.

  **b.** Prove that $\mathbf{B}$ and $\mathbf{C}$ are $n \times n$ matrices such that $\mathbf{AB} = \mathbf{AC}$, then $\mathbf{B} = \mathbf{C}$.

  **c.** If $\mathbf{A}^2 = \mathbf{A}$ (we say that $\mathbf{A}$ is *idempotent*), find $\mathbf{A}$.

**45.** If $\mathbf{A} = \begin{bmatrix} 10 & -3 \\ 4 & 7 \end{bmatrix}$ and $\mathbf{B} = \begin{bmatrix} 8 & 6 \\ -7 & -3 \end{bmatrix}$ find $3\mathbf{AB} - 4\mathbf{B}^2$.

**46.** Solve the system
$$\begin{cases} 9.7x - 3.4y + 7.2z = 18.2, \\ 4.3x + 8.5y - 6.7z = 20.8, \\ 5.4x - 2.6y - 4.7z = 30.9, \end{cases}$$
by using the inverse of the coefficient matrix. Round your answers to two decimal places.

**47.** Given the input-output matrix

| | Industry | | Final |
| --- | A | B | Demand |
| Industry: A | 10 | 20 | 4 |
| B | 15 | 14 | 10 |
| Other | 9 | 5 | — |

find the output matrix if final demand changes to 8 for $A$ and 8 for $B$. (Data are in tens of billions of dollars.)

# MATHEMATICAL *SNAPSHOT*

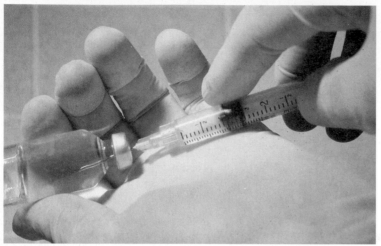

A vacation lodge in the mountains of Washington State has a well-deserved reputation for attending to the special health needs of its guests. Next week the manager of the lodge is expecting four guests, each of whom has insulin-dependent diabetes. These guests plan to stay at the lodge for 7, 14, 21, and 28 days, respectively.

The lodge is quite a distance from the nearest drugstore, so before the arrival of the guests the manager plans to obtain the total amount of insulin that will be needed. Three different types of insulin are required: lente, semi-lente, and ultra-lente. The manager will store the insulin, and then lodge personnel will administer the daily dose of the three different types of insulin to each of the guests.

The daily requirements of the four guests are:

Guest 1  20 insulin units of semi-lente, 30 units of lente, 10 units of ultra-lente;

Guest 2  40 insulin units of semi-lente, 0 units of lente, 0 units of ultra-lente;

Guest 3  30 insulin units of semi-lente, 10 units of lente, 30 units of ultra-lente;

Guest 4  10 insulin units of semi-lente, 10 units of lente, 50 units of ultra-lente.

This information will be represented by the following "requirement" matrix **A**:

$$\mathbf{A} = [a_{ij}]_{3\times4}, \quad \text{where } \mathbf{A} \text{ is given by}$$

|  | Guest 1 | Guest 2 | Guest 3 | Guest 4 |
|---|---|---|---|---|
| semi-lente insulin | 20 | 40 | 30 | 10 |
| lente insulin | 30 | 0 | 10 | 10 |
| ultra-lente insulin | 10 | 0 | 30 | 50 |

Recall that Guest 1 will stay for 7 days, Guest 2 for 14 days, Guest 3 for 21 days, and Guest 4 for 28 days. You can let the following column vector **T** represent the time, in days, that each guest is staying at the lodge:

$$\mathbf{T} = \begin{bmatrix} 7 \\ 14 \\ 21 \\ 28 \end{bmatrix}.$$

To determine the total amounts of the different types of insulin needed by the four guests, you compute the matrix product **AT**.

$$\mathbf{AT} = \begin{bmatrix} 20 & 40 & 30 & 10 \\ 30 & 0 & 10 & 10 \\ 10 & 0 & 30 & 50 \end{bmatrix} \begin{bmatrix} 7 \\ 14 \\ 21 \\ 28 \end{bmatrix}$$

$$= 10(7) \begin{bmatrix} 2 & 4 & 3 & 1 \\ 3 & 0 & 1 & 1 \\ 1 & 0 & 3 & 5 \end{bmatrix} \begin{bmatrix} 1 \\ 2 \\ 3 \\ 4 \end{bmatrix}$$

[7]Adapted from Richard F. Baum, "Insulin Requirements as a Linear Process," in R. M. Thrall, J. A. Mortimer, K. R. Rebman, and R. F. Baum, (eds.), *Some Mathematical Models in Biology,* rev. ed. Report 40241-R-7. Prepared at the University of Michigan, 1967.

$$= 70 \begin{bmatrix} 23 \\ 10 \\ 30 \end{bmatrix} = \begin{bmatrix} 1610 \\ 700 \\ 2100 \end{bmatrix} - \mathbf{B}.$$

Vector **B** (or **AT**) indicates that a total of 1610 insulin units of semi-lente, 700 insulin units of lente, and 2100 insulin units of ultra-lente are required by the four guests.

Now, change the problem a bit. Suppose that each guest decided to double the original length of stay. The resulting vector that gives the total amount needed of semi-lente, lente, and ultra-lente insulin is

$$\mathbf{A}(2\mathbf{T}) = 2(\mathbf{AT}) = 2\mathbf{B} = \begin{bmatrix} 3220 \\ 1400 \\ 4200 \end{bmatrix}.$$

In fact, if each guest planned to spend a factor $k$ ($k \geq 0$) of the original time at the lodge (that is, Guest 1 planned to stay for $k \cdot 7$ days, Guest 2 for $k \cdot 14$ days, and so on), then the insulin requirements would be

$$\mathbf{A}(k\mathbf{T}) = k(\mathbf{AT}) = k\mathbf{B} = \begin{bmatrix} k \cdot 1610 \\ k \cdot 700 \\ k \cdot 2100 \end{bmatrix}.$$

Similarly, if the guests decided to add 1, 3, 4, and 6 days, respectively, to the times they originally intended to stay, then the amounts of insulin required would be

$$\mathbf{A}(\mathbf{T} + \mathbf{T}_1) = \mathbf{AT} + \mathbf{AT}_1, \quad \text{where } \mathbf{T}_1 = \begin{bmatrix} 1 \\ 3 \\ 4 \\ 6 \end{bmatrix}.$$

Based on the results thus far, it is obvious that the following matrix equation generalizes the situation.

$$\mathbf{AX} = \mathbf{B}$$

or

$$\begin{bmatrix} 20 & 40 & 30 & 10 \\ 30 & 0 & 10 & 10 \\ 10 & 0 & 30 & 50 \end{bmatrix} \begin{bmatrix} x_1 \\ x_2 \\ x_3 \\ x_4 \end{bmatrix} = \begin{bmatrix} b_1 \\ b_2 \\ b_3 \end{bmatrix},$$

which represents the linear system

$$\begin{cases} 20x_1 + 40x_2 + 30x_3 + 10x_4 = b_1, \\ 30x_1 \qquad\quad + 10x_3 + 10x_4 = b_2, \\ 10x_1 \qquad\quad + 30x_3 + 50x_4 = b_3, \end{cases}$$

where $x_i$ is the number of days that Guest $i$ stays at the lodge, and $b_1, b_2, b_3$ give, respectively, the total number of units of semi-lente, lente, and ultra-lente insulin needed by the four guests for their entire stay at the lodge.

Finally suppose once again that vector **T** represents the number of days that each guest originally planned to stay at the lodge. Furthermore, suppose vector **C** gives the cost (in cents) per insulin unit of the three types of insulin, where

$$\mathbf{C} = \begin{bmatrix} 9 \\ 8 \\ 10 \end{bmatrix} = \text{cost matrix}.$$

That is, one unit of semi-lente costs 9¢, one unit of lente costs 8¢, and one unit of ultra-lente costs 10¢. Then the total amount paid by the lodge for all the insulin required by the four guests is

$$\mathbf{C}^{\mathrm{T}}(\mathbf{AT}) = \mathbf{C}^{\mathrm{T}}\mathbf{B} = \begin{bmatrix} 9 & 8 & 10 \end{bmatrix} \begin{bmatrix} 1610 \\ 700 \\ 2100 \end{bmatrix} = [41{,}090],$$

that is, 41,090¢ or $410.90.

## ■ Exercises

1. Suppose that Guest 1 will stay at the lodge for 7 days, Guest 2 for 10 days, Guest 3 for 7 days, and Guest 4 for 5 days. Assume that the daily requirements of the four guests and the cost matrix are the same as given in the discussion. Find the total amount (in dollars) that the lodge must pay for all the insulin required by the guests.

2. Suppose that Guest 1 will stay at the lodge for 4 days, Guest 2 for 7 days, and Guest 3 for 10 days. Guest 4 changes plans and will not be staying at the lodge. Assume that the daily requirements of the three guests and the cost matrix are the same as given in the discussion. Find the total amount (in dollars) that the lodge must pay for all the insulin required by the guests.

# Linear Programming

**To geometrically represent the solution of a linear inequality in two variables and to extend this representation to a system of linear inequalities.**

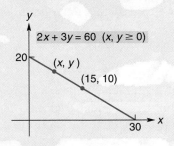

**FIGURE 7.1** Budget line.

## 7.1 LINEAR INEQUALITIES IN TWO VARIABLES

Suppose a consumer receives a fixed income of $60 per week and uses it all to purchase products A and B. If $x$ kilograms of A cost $2 per kilogram and $y$ kilograms of B cost $3 per kilogram, then

$$2x + 3y = 60, \qquad \text{where } x, y \geq 0.$$

The solutions of this equation, called a *budget equation*, give the possible combinations of A and B that can be purchased for $60. The graph of the equation is the *budget line* in Fig. 7.1. Note that $(15, 10)$ lies on the line. This means that if 15 kg of A are purchased, then 10 kg of B must be bought, for a total cost of $60.

On the other hand, suppose the consumer does not necessarily wish to spend all of the $60. In this case, the possible combinations are described by the inequality

$$2x + 3y \leq 60, \qquad \text{where } x, y \geq 0. \tag{1}$$

When inequalities in one variable were discussed in Chapter 2, their solutions were represented geometrically by *intervals* on the real number line. However, for an inequality in two variables, like the inequality (1), the solution is usually represented by a *region* in the coordinate plane. We shall find the region corresponding to (1) after considering such inequalities in general.

### DEFINITION

*A **linear inequality** in the variables $x$ and $y$ is an inequality that can be written in the form*

$$ax + by + c < 0 \qquad (or \leq 0, \ \geq 0, \ > 0),$$

*where $a$, $b$, and $c$ are constants and $a$ and $b$ are not both zero.*

Geometrically, the solution (or graph) of a linear inequality in $x$ and $y$ consists of all points $(x, y)$ in the plane whose coordinates satisfy the inequality. For example, a solution of $x + 3y < 20$ is the point $(-2, 4)$, because substitution gives

$$-2 + 3(4) < 20,$$

$$10 < 20, \qquad \text{which is true.}$$

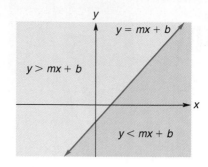

**FIGURE 7.2** A nonvertical line determines two half planes.

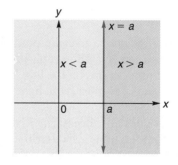

**FIGURE 7.3** A vertical line determines two half planes.

Clearly, there are infinitely many solutions, which is typical of every linear inequality.

To consider linear inequalities in general, we first note that the graph of a nonvertical line $y = mx + b$ separates the plane into three distinct parts (see Fig. 7.2):

1. the line itself, consisting of all points $(x, y)$ whose coordinates satisfy the equation $y = mx + b$;

2. the region *above* the line, consisting of all points $(x, y)$ whose coordinates satisfy the inequality $y > mx + b$ (this region is called an *open* **half plane**);

3. the open half plane *below* the line, consisting of all points $(x, y)$ whose coordinates satisfy the inequality $y < mx + b$.

In the situation where the strict inequality "$<$" is replaced by "$\leq$", the solution of $y \leq mx + b$ consists of the line $y = mx + b$, as well as the half plane below it. In this case, we say that the solution is a *closed* half plane. A similar statement can be made when "$>$" is replaced by "$\geq$". For a vertical line $x = a$ (see Fig. 7.3), we speak of a half plane to the right $(x > a)$ of the line or to its left $(x < a)$. Since any linear inequality (in two variables) can be put into one of the forms we have discussed, we can say that *the solution of a linear inequality must be a half plane.*

To apply these facts, we shall solve the linear inequality

$$2x + y < 5.$$

From our previous discussion, we know that the solution is a half plane. To find it, we begin by replacing the inequality symbol by an equals sign and then graphing the resulting *line*, $2x + y = 5$. This is easily done by choosing two points on the line—for instance, the intercepts $(\frac{5}{2}, 0)$ and $(0, 5)$. [See Fig. 7.4.] Because points on the line do not satisfy the "$<$" inequality, we used a *dashed* line to indicate that the line is not part of the solution. We must now determine whether the solution is the half plane *above* the line or the one *below* it. This can be done by solving the inequality for $y$. Once $y$ is isolated, the appropriate half plane will be apparent. We have

$$y < 5 - 2x.$$

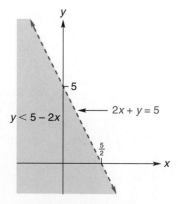

**FIGURE 7.4** Graph of $2x + y < 5$.

Geometrically, the solution of a linear inequality in two variables is a *region* in the plane, not an interval on the real-number line.

From the aforementioned statement 3, we conclude that the solution consists of the half plane *below* the line. Part of this region is shaded in Fig. 7.4. Thus, if $(x_0, y_0)$ is *any* point in this region, then its ordinate $y_0$ is less than the number $5 - 2x_0$. (See Fig. 7.5.) For example, $(-2, -1)$ is in the region, and

$$-1 < 5 - 2(-2),$$
$$-1 < 9.$$

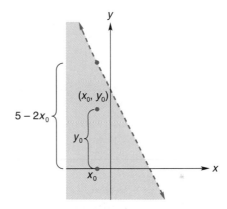

**FIGURE 7.5**  Analysis of point satisfying $y < 5 - 2x$.

If, instead, the original inequality had been $y \leq 5 - 2x$, then the line $y = 5 - 2x$ would have been included in the solution. We would indicate its inclusion by using a solid line rather than a dashed line. This solution, which is a closed half plane, is shown in Fig. 7.6. Keep in mind that **a solid line *is* included in the solution, and a dashed line *is not*.**

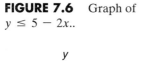

**FIGURE 7.6**  Graph of $y \leq 5 - 2x$..

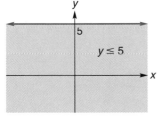

**FIGURE 7.7**  Graph of $y \leq 5$.

**EXAMPLE 1**  Solving a Linear Inequality

*Find the region defined by the inequality $y \leq 5$.*

*Solution:* Since $x$ does not appear, the inequality is assumed to be true for all values of $x$. The region consists of the line $y = 5$, together with the half plane below it. (See Fig. 7.7.)  ■

**EXAMPLE 2**  Solving a Linear Inequality

*Solve the inequality $2(2x - y) < 2(x + y) - 4$.*

*Solution:* We first solve the inequality for $y$, so that the appropriate half plane is obvious. The inequality is equivalent to

$$4x - 2y < 2x + 2y - 4,$$
$$4x - 4y < 2x - 4,$$
$$-4y < -2x - 4,$$
$$y > \frac{x}{2} + 1 \qquad \text{(dividing both sides by } -4 \text{ and reversing the sense).}$$

Using a dashed line, we now sketch $y = (x/2) + 1$ by noting that its intercepts are $(0, 1)$ and $(-2, 0)$. Because the inequality symbol is $>$, we shade the

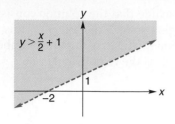

**FIGURE 7.8** Graph of $y > \frac{x}{2} + 1$

---

### Principles in Practice 1

**Solving a Linear Inequality**

To earn some extra money, you make two types of refrigerator magnets, type A and type B, for sale. You have an initial start-up expense of $50. The production cost for type A is $0.90 per magnet, and the production cost for type B is $0.70 per magnet. The price for type A is $2.00 per magnet, and the price for type B is $1.50 per magnet. Let $x$ be the number of type A and $y$ be the number of type B produced and sold. Write an inequality describing revenue greater than cost. Solve the inequality and describe the region. Also, describe what this result means in terms of magnets.

---

The point where the graphs of $y = x$ and $y = -2x + 3$ intersect is not included in the solution.

---

### Principles in Practice 2

**Solving a System of Linear Inequalities**

A store sells two types of cameras. In order to cover overhead, it must sell at least 50 cameras per week, and in order to satisfy distribution requirements, it must sell at least twice as many of type I as type II. Write a system of inequalities to describe the situation. Let $x$ be the number of type I that the store sells in a week and $y$ be the number of type II that it sells in a week. Find the region described by the system of linear inequalities.

---

half plane above the line. (See Fig. 7.8.) Each point in this region is a solution. ∎

### Systems of Inequalities

The solution of a *system* of inequalities consists of all points whose coordinates simultaneously satisfy all of the given inequalities. Geometrically, it is the region that is common to all the regions determined by the given inequalities. For example, let us solve the system

$$\begin{cases} 2x + y > 3, \\ x \geq y, \\ 2y - 1 > 0. \end{cases}$$

We first rewrite each inequality so that $y$ is isolated. This gives the equivalent system

$$\begin{cases} y > -2x + 3, \\ y \leq x, \\ y > \frac{1}{2}. \end{cases}$$

Now we sketch the corresponding lines $y = -2x + 3$, $y = x$, and $y = \frac{1}{2}$. Noting that the first and third inequalities define *open* half planes, but the second one defines a *closed* half plane, we then shade the region that is simultaneously above the first line, on or below the second line, and above the third line. (See Fig. 7.9.) When you are sketching the lines, **it is best to draw dashed lines everywhere until it is clear which portions are to be included in the solution.**

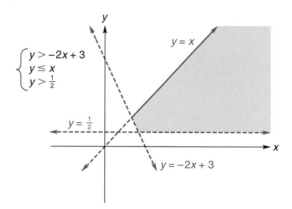

**FIGURE 7.9** Solution of system of linear inequalities.

### EXAMPLE 3 Solving a System of Linear Inequalities

*Solve the system*

$$\begin{cases} y \geq -2x + 10, \\ y \geq x - 2. \end{cases}$$

*Solution:* The solution consists of all points that are simultaneously on or above the line $y = -2x + 10$ and on or above the line $y = x - 2$. It is the shaded region in Fig. 7.10. ∎

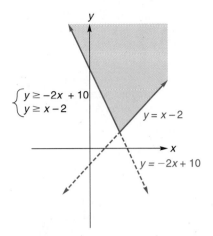

**FIGURE 7.10**   Solution of system of linear inequalities.

**EXAMPLE 4**   **Solving a System of Linear Inequalities**

*Find the region described by*

$$\begin{cases} 2x + 3y \le 60, \\ \quad\ x \ge 0, \\ \quad\ y \ge 0. \end{cases}$$

*Solution:* This system relates to inequality (1) at the beginning of the section. The first inequality is equivalent to $y \le -\frac{2}{3}x + 20$. The last two inequalities restrict the solution to points that are both on or to the right of the *y*-axis *and* also on or above the *x*-axis. The desired region is shaded in Fig. 7.11. ∎

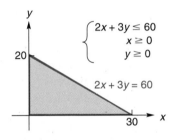

**FIGURE 7.11**   Solution of system of linear inequalities

# ▪ Exercise 7.1

*In Problems 1–24, solve the inequalities.*

**1.** $2x + 3y > 6$.

**2.** $3x - 2y \ge 12$.

**3.** $x + 2y \le 7$.

**4.** $y > 6 - 2x$.

**5.** $-x \le 2y - 4$.

**6.** $2x + y \ge 10$.

**7.** $3x + y < 0$.

**8.** $x + 5y < -5$.

**9.** $\begin{cases} 3x - 2y < 6, \\ \quad x - 3y > 9. \end{cases}$

**10.** $\begin{cases} 2x + 3y > -6, \\ 3x - \ \ y < 6. \end{cases}$

**11.** $\begin{cases} 2x + 3y \le 6, \\ \qquad x \ge 0. \end{cases}$

**12.** $\begin{cases} 2y - 3x < 6, \\ \qquad x < 0. \end{cases}$

**13.** $\begin{cases} y - 3x < 6, \\ x - \ \ y > -3. \end{cases}$

**14.** $\begin{cases} x - y < 1, \\ y - x < 1. \end{cases}$

**15.** $\begin{cases} 2x - 2 \ge y, \\ \qquad 2x \le 3 - 2y. \end{cases}$

**16.** $\begin{cases} 2y < 4x + 2, \\ \ \ y < 2x + 1. \end{cases}$

**17.** $\begin{cases} x - y > 4, \\ \quad\ x < 2, \\ \quad\ y > -5. \end{cases}$

**18.** $\begin{cases} 2x + y < -1, \\ \qquad y > -x, \\ 2x + 6 < 0. \end{cases}$

**19.** $\begin{cases} y < 2x + 4, \\ x \ge -2, \\ y < 1. \end{cases}$

**20.** $\begin{cases} 4x + 3y \ge 12, \\ \qquad y \ge x, \\ \qquad 2y \le 3x + 6. \end{cases}$

**21.** $\begin{cases} \quad\ x + y > 1, \\ 3x - 5 \le y, \\ \qquad y < 2x. \end{cases}$

**22.** $\begin{cases} 2x - 3y > -12, \\ \ \ 3x + y > -6, \\ \qquad\ y > x. \end{cases}$

**23.** $\begin{cases} 3x + y > -6, \\ \ \ x - y > -5, \\ \qquad x \ge 0. \end{cases}$

**24.** $\begin{cases} 5y - 2x \le 10, \\ 4x - 6y \le 12, \\ \qquad y \ge 0. \end{cases}$

*If a consumer wants to spend no more than P dollars to purchase quantities x and y of two products having prices of $p_1$ and $p_2$ dollars per unit, respectively, then $p_1 x + p_2 y \leq P$, where $x, y \geq 0$. In Problems 25 and 26, find geometrically the possible combinations of purchases by determining the solution of this system for the given values of $p_1, p_2$, and P.*

**25.** $p_1 = 5, p_2 = 3, P = 15$.

**26.** $p_1 = 6, p_2 = 4, P = 24$.

**27.** If a manufacturer wishes to purchase a *total* of no more than 100 lb of product Z from suppliers A and B, set up a system of inequalities which describes the possible combinations of quantities that can be purchased from each supplier. Sketch the solution in the plane.

**28. Manufacturing** The XYZ Corporation produces two models of home computers: the Alpha model and the Beta model. Let x be the number of Alpha models and y the number of Beta models produced at the San Antonio factory per week. If the factory can produce at most 200 Alpha and Beta models combined in a week, write inequalities to describe this situation.

**29. Manufacturing** A chair company produces two models of chairs: the Sequoia and the Saratoga. The Sequoia model takes 3 hours to assemble and $\frac{1}{2}$ hour to paint. The Saratoga model takes 2 hours to assemble and 1 hour to paint. The maximum number of hours available to assemble chairs is 24 per day, and the maximum number of hours available to paint chairs is 8 per day. Write a system of linear inequalities to describe the situation. Let x represent the number of Sequoia models produced in a day and y represent the number of Saratoga models produced in a day. Find the region described by this system of linear inequalities.

---

**To state the nature of a linear programming problem, to introduce terminology associated with it, and to solve it geometrically.**

A great deal of terminology is used in discussing linear programming, and you are advised to learn this terminology as it is introduced.

## 7.2 LINEAR PROGRAMMING

Sometimes it is desired to maximize or minimize a function, subject to certain restrictions (or *constraints*). For example, a manufacturer may want to maximize a profit function, subject to production restrictions imposed by limitations on the use of machinery and labor.

We shall now consider how to solve such problems when the function to be maximized or minimized is *linear*. A **linear function in x and y** has the form

$$Z = ax + by,$$

where $a$ and $b$ are constants. We shall also require that the corresponding constraints be represented by a system of linear inequalities (involving "$\leq$" or "$\geq$") or linear equations in $x$ and $y$, and that all variables be non negative. A problem involving all of these conditions is called a *linear programming problem*.

Linear programming was developed by George B. Danzig in the late 1940s and was first used by the U.S. Air Force as an aid in decision making. Today it has wide application in industrial and economic analysis.

In a linear programming problem, the function to be maximized or minimized is called the **objective function.** Although there are usually infinitely many solutions to the system of constraints (these are called **feasible solutions** or **feasible points**), the aim is to find one such solution that is an **optimum solution** (that is, one that gives the maximum or minimum value of the objective function).

We shall now give a geometrical approach to linear programming. In Sec. 7.4, a matrix approach will be discussed that will enable us to work with more than two variables and, hence, a wider range of problems.

We consider the following problem. A company produces two types of widgets: manual and electric. Each requires in its manufacture the use of three machines: A, B, and C. Table 7.1 gives data relating to the manufacture of these widgets. Each manual widget requires the use of machine A for 2 hours, machine B for 1 hour, and machine C for 1 hour. An electric widget requires 1 hour on A, 2 hours on B, and 1 hour on C. Furthermore, suppose the maximum numbers of hours available per month for the use of machines A, B, and C are

**TABLE 7.1**

|             | A      | B      | C      | Profit/Unit |
|-------------|--------|--------|--------|-------------|
| Manual      | 2 hr   | 1 hr   | 1 hr   | $4          |
| Electric    | 1 hr   | 2 hr   | 1 hr   | 6           |
| Hours available | 180 | 160   | 100    |             |

180, 160, and 100, respectively. The profit on a manual widget is $4, and on an electric widget it is $6. If the company can sell all the widgets it can produce, how many of each type should it make in order to maximize the monthly profit?

To solve the problem, let $x$ and $y$ denote the number of manual and electric widgets, respectively, that are made in a month. Since the number of widgets made is not negative,

$$x \geq 0 \quad \text{and} \quad y \geq 0.$$

For machine A, the time needed for working on $x$ manual widgets is $2x$ hours, and the time needed for working on $y$ electric widgets is $1y$ hours. The sum of these times cannot be greater than 180, so

$$2x + y \leq 180.$$

Similarly, the restrictions for machines B and C give

$$x + 2y \leq 160 \quad \text{and} \quad x + y \leq 100.$$

The profit is a function of $x$ and $y$ and is given by the *profit function*

$$P = 4x + 6y.$$

Summarizing, we want to maximize the *objective function*

$$P = 4x + 6y, \tag{1}$$

subject to the condition that $x$ and $y$ must be a solution of the system of constraints

$$
\begin{cases}
x \geq 0, & (2) \\
y \geq 0, & (3) \\
2x + y \leq 180, & (4) \\
x + 2y \leq 160, & (5) \\
x + y \leq 100. & (6)
\end{cases}
$$

Thus, we have a linear programming problem. Constraints (2) and (3) are called **nonnegativity conditions.** The region simultaneously satisfying constraints (2)–(6) is shaded in Fig. 7.12. Each point in this region represents a feasible solution, and the region is called the **feasible region.** Although there are infinitely many feasible solutions, we must find one that maximizes the profit function.

Since the objective function, $P = 4x + 6y$, is equivalent to

$$y = -\frac{2}{3}x + \frac{P}{6},$$

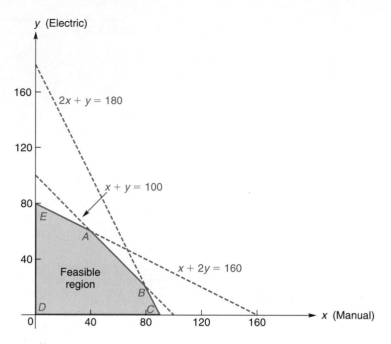

**FIGURE 7.12** Feasible region.

it defines a so-called family of parallel lines, each having a slope of $-2/3$ and $y$-intercept $(0, P/6)$. For example, if $P = 600$, then we obtain the line

$$y = -\frac{2}{3}x + 100$$

shown in Fig. 7.13. This line, called an **isoprofit line,** gives all possible combinations of $x$ and $y$ that yield the same profit, \$600. Note that this isoprofit line has no point in common with the feasible region, whereas the isoprofit line for $P = 300$ has infinitely many such points. Let us look for the member of the

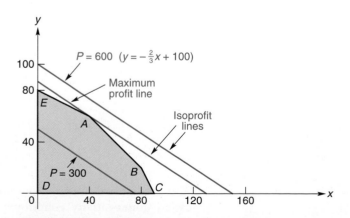

**FIGURE 7.13** Isoprofit lines and feasible region.

family that contains a feasible point and whose $P$-value is maximum. *This will be the line whose y-intercept is furthest from the origin (giving a maximum value of P) and that has at least one point in common with the feasible region.* It is not difficult to observe that such a line will contain the *corner point A.* Any isoprofit line with a greater profit will contain no points of the feasible region.

From Fig. 7.12, we see that $A$ lies on both the line $x + y = 100$ and the line $x + 2y = 160$. Thus, its coordinates may be found by solving the system

$$\begin{cases} x + y = 100, \\ x + 2y = 160. \end{cases}$$

This gives $x = 40$ and $y = 60$. Substituting these values into the equation $P = 4x + 6y$, we find that the maximum profit subject to the constraints is $520, which is obtained by producing 40 manual widgets and 60 electric widgets per month.

If a feasible region can be contained within a circle, such as the region in Fig. 7.13, it is called a **bounded feasible region.** Otherwise, it is **unbounded.** When a feasible region contains at least one point, it is said to be **nonempty.** Otherwise, it is **empty.** The region in Fig. 7.13 is a nonempty bounded feasible region.

It can be shown that:

> A linear function defined on a nonempty bounded feasible region has a maximum (minimum) value, and this value can be found at a corner point.

This statement gives us a way of finding an optimum solution without drawing isoprofit lines, as we did above: We simply evaluate the objective function at each of the corner points of the feasible region and then choose a corner point at which the function is optimum.

For example, in Fig. 7.13 the corner points are $A$, $B$, $C$, $D$, and $E$. We found $A$ before to be $(40, 60)$. To find $B$, we see from Fig. 7.12 that we must solve $2x + y = 180$ and $x + y = 100$ simultaneously. This gives the point $B = (80, 20)$. In a similar way, we obtain all the corner points:

$$A = (40, 60), \qquad B = (80, 20), \qquad C = (90, 0),$$
$$D = (0, 0), \qquad E = (0, 80).$$

We now evaluate the objective function $P = 4x + 6y$ at each point:

$$P(A) = 4(40) + 6(60) = 520,$$
$$P(B) = 4(80) + 6(20) = 440,$$
$$P(C) = 4(90) + 6(0) = 360,$$
$$P(D) = 4(0) + 6(0) = 0,$$
$$P(E) = 4(0) + 6(80) = 480.$$

Thus, $P$ has a maximum value of 520 at $A$, where $x = 40$ and $y = 60$.

The optimum solution to a linear programming problem is given by the optimum value of the objective function and the point where the optimum value of the objective function occurs.

**EXAMPLE 1** Solving a Linear Programming Problem

*Maximize the objective function $Z = 3x + y$, subject to the constraints*

$$2x + y \leq 8,$$
$$2x + 3y \leq 12,$$
$$x \geq 0,$$
$$y \geq 0.$$

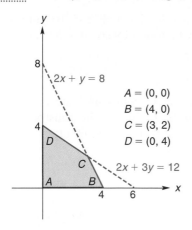

**FIGURE 7.14** $A, B, C,$ and $D$ are corner points of feasible region.

**Solution:** In Fig. 7.14, the feasible region is nonempty and bounded. Thus, $Z$ is maximum at one of the four corner points. The coordinates of $A$, $B$, and $D$ are obvious on inspection. To find the coordinates of $C$, we solve the equations $2x + y = 8$ and $2x + 3y = 12$ simultaneously, which gives $x = 3$, $y = 2$. Thus,

$$A = (0, 0), \quad B = (4, 0), \quad C = (3, 2), \quad D = (0, 4).$$

Evaluating $Z$ at these points, we obtain

$$Z(A) = 3(0) + 0 = 0,$$
$$Z(B) = 3(4) + 0 = 12,$$
$$Z(C) = 3(3) + 2 = 11,$$
$$Z(D) = 3(0) + 4 = 4.$$

Hence, the maximum value of $Z$, subject to the constraints, is 12, and it occurs when $x = 4$ and $y = 0$. ∎

### Empty Feasible Region

The next example illustrates a situation where no optimum solution exists.

**EXAMPLE 2  Empty Feasible Region**

*Minimize the objective function* $Z = 8x - 3y$, *subject to the constraints*

$$-x + 3y = 21,$$
$$x + y \leq 5,$$
$$x \geq 0,$$
$$y \geq 0.$$

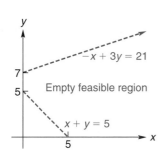

**FIGURE 7.15** Empty feasible region.

**Solution:** Notice that the first constraint, $-x + 3y = 21$, is an *equality*. The portions of the lines $-x + 3y = 21$ and $x + y = 5$ for which $x \geq 0$ and $y \geq 0$ are shown in Fig. 7.15. They will remain dashed lines until we determine whether or not they are to be included in the feasible region. A feasible point $(x, y)$ must have $x \geq 0$ and $y \geq 0$, and must lie both on the top dashed line and on or below the bottom dashed line (since $y \leq 5 - x$). However, no such point exists. Hence, the feasible region is *empty,* and the problem has *no* optimum solution. ∎

The situation in Example 2 can be made more general:

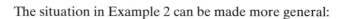

Whenever the feasible region of a linear programming problem is empty, no optimum solution exists.

### Unbounded Feasible Region

Suppose a feasible region is defined by

$$y = 2, \quad x \geq 0, \quad \text{and} \quad y \geq 0.$$

This region is the portion of the horizontal line $y = 2$ indicated in Fig. 7.16. Since the region cannot be contained within a circle, it is *unbounded.* Let us consider maximizing

$$Z = x + y,$$

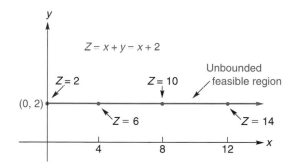

**FIGURE 7.16** Unbounded feasible region on which $Z$ has no maximum.

subject to the foregoing constraints. Since $y = 2$, $Z = x + 2$. Clearly, as $x$ increases without bound, so does $Z$. Thus, no feasible point maximizes $Z$, so no optimum solution exists. In this case, we say that the solution is "unbounded." On the other hand, suppose we want to *minimize* $Z = x + y$ over the same region. Since $Z = x + 2$, then $Z$ is minimum when $x$ is as small as possible, namely, when $x = 0$. This gives a minimum value of $Z = x + y = 0 + 2 = 2$, and the optimum solution is the corner point $(0, 2)$.

In general, it can be shown that:

> If a feasible region is unbounded, and if the objective function has a maximum (or minimum) value, then that value occurs at a corner point.

**EXAMPLE 3   Unbounded Feasible Region**

*A produce grower is purchasing fertilizer containing three nutrients, A, B, and C. The minimum needs are 160 units of A, 200 units of B, and 80 units of C. There are two popular brands of fertilizer on the market. Fast Grow, costing $4 a bag, contains 3 units of A, 5 units of B, and 1 unit of C. Easy Grow, costing $3 a bag, contains 2 units of each nutrient. If the grower wishes to minimize cost while still maintaining the nutrients required, how many bags of each brand should be bought? The information is summarized as follows:*

|  | **A** | **B** | **C** | **Cost/Bag** |
|---|---|---|---|---|
| Fast Grow | 3 units | 5 units | 1 unit | $4 |
| Easy Grow | 2 units | 2 units | 2 units | 3 |
| Units required | 160 | 200 | 80 | |

***Solution:*** Let $x$ be the number of bags of Fast Grow that are bought and $y$ the number of bags of Easy Grow that are bought. Then we wish to *minimize* the cost function

$$C = 4x + 3y \tag{7}$$

subject to the constraints

$$x \geq 0, \tag{8}$$

$$y \geq 0, \tag{9}$$

$$3x + 2y \geq 160, \tag{10}$$

$$5x + 2y \geq 200, \tag{11}$$

$$x + 2y \geq 80. \tag{12}$$

The feasible region satisfying constraints (8)–(12) is shaded in Fig. 7.17, along with *isocost lines* for $C = 200$ and $C = 300$. The feasible region is unbounded. The member of the family of lines $C = 4x + 3y$ that gives a minimum cost,

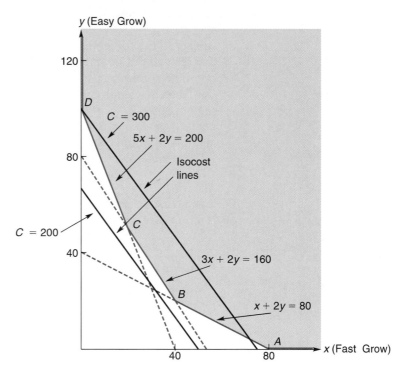

**FIGURE 7.17**  Minimum cost at corner point $B$ of unbounded feasible region.

subject to the constraints, intersects the feasible region at the corner point $B$. Here we chose the isocost line whose $y$-intercept was *closest* to the origin and that had at least one point in common with the feasible region. The coordinates of $B$ are found by solving the system

$$\begin{cases} 3x + 2y = 160, \\ x + 2y = 80. \end{cases}$$

Thus, $x = 40$ and $y = 20$, which gives a minimum cost of \$220. The produce grower should buy 40 bags of Fast Grow and 20 bags of Easy Grow.  ∎

In Example 3, we found that the function $C = 4x + 3y$ has a minimum value at a corner point of the unbounded feasible region. On the other hand, suppose we want to *maximize* $C$ over that region and take the approach of evaluating $C$ at all corner points. These points are

$$A = (80, 0), \qquad B = (40, 20), \qquad C = (20, 50), \qquad D = (0, 100),$$

from which we obtain

$$C(A) = 4(80) + 3(0) = 320,$$
$$C(B) = 4(40) + 3(20) = 220,$$
$$C(C) = 4(20) + 3(50) = 230,$$
$$C(D) = 4(0) + 3(100) = 300.$$

A hasty conclusion is that the maximum value of $C$ is 320. This is *false!* There is *no* maximum value, since isocost lines with arbitrarily large values of $C$ intersect the feasible region.

*Pitfall* ▼ When working with an unbounded feasible region, do not simply conclude that an optimum solution exists at a corner point, since there may not be an optimum solution.

## TECHNOLOGY

**Problem:** Maximize $Z = 4.1x - 3.2y$, subject to the constraints

$$y \leq 8 - x, \tag{13}$$
$$y \leq 6 - 0.2x, \tag{14}$$
$$y \geq 2 + 0.3x, \tag{15}$$

and

$$x \geq 0, y \geq 0.$$

**Solution:** As shown in Fig. 7.18, we enter the objective function as $Y_1$, where $y$ is entered as "alpha Y." Next, the equations corresponding to constraints (13)–(15) are entered as $Y_2$, $Y_3$, and $Y_4$. To begin, the $Y_1$ function is "turned off" (that is, the symbol "=" is not highlighted), and we obtain the graphs of $Y_2$, $Y_3$, and $Y_4$. (See Fig. 7.19.) It is a good idea to make a pencil sketch of the graphs and label the lines. From the sketch, we determine the feasible region and label

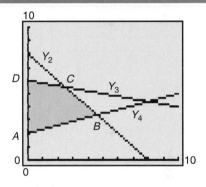

**FIGURE 7.19**  Determining feasible region and labeling corner points.

any corner points. In Fig. 7.19, the feasible region is shaded and the corner points are $A$, $B$, $C$, and $D$. Because the feasible region is nonempty and bounded, the maximum value of $Z$ will occur at one of the corner points.

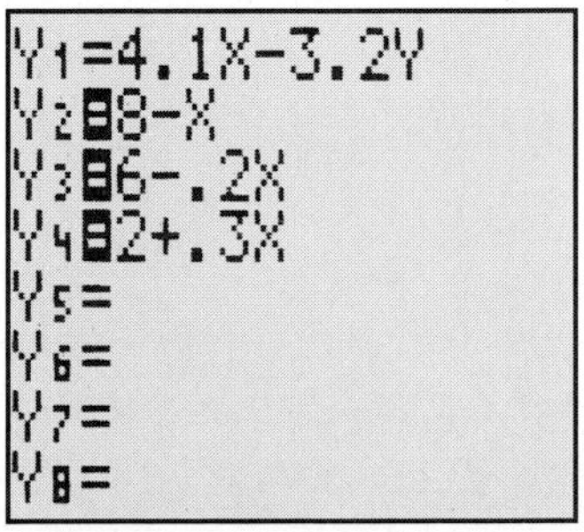

Point $A$ is the $y$-intercept of $Y_4$ and is easily found to be $(0, 2)$. The value of $Z$ at $A$ is found by storing 0 in X, and 2 in Y and then evaluating $Y_1$. (See Fig. 7.20.) Thus, $Z = -6.4$ at this corner point.

Point $B$ is the intersection of $Y_2$ and $Y_4$. To find the coordinates of $B$, we first turn off the $Y_3$ function and highlight only $Y_2$ and $Y_4$. After displaying the graphs of $Y_2$ and $Y_4$, we find their intersection point. On the TI-82 it is convenient to use the "intersection" feature. (See Fig. 7.21.) The values of X and Y at the intersection are automatically stored in the X and Y registers. Returning to the home screen, we evaluate $Y_1$ and obtain 8.09 (rounded to two decimal places). Thus, $Z = 8.09$ at corner point $B$.

```
Y₁=4.1X-3.2Y
Y₂■8-X
Y₃■6-.2X
Y₄■2+.3X
Y₅=
Y₆=
Y₇=
Y₈=
```

**FIGURE 7.18**  Entering objective function and equations corresponding to constraints and "turning off" objective function.

Continuing in this fashion, we find the coordinates of $C$ and $D$ and evaluate $Y_1$ (or $Z$) there:

$$C = (2.5, 5.5) \qquad Z = -7.35,$$
$$D = (0, 6), \qquad Z = -19.2.$$

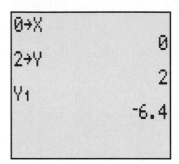

**FIGURE 7.20** Evaluating objective function at corner point $A = (0, 2)$.

Hence, the maximum value of $Z$ is 8.09 and occurs at the corner point $B$, where $x \approx 4.62$ and $y \approx 3.38$.

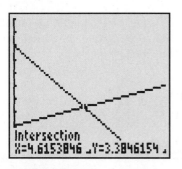

**FIGURE 7.21** Determining corner point $B$.

## ■ Exercise 7.2

**1.** Maximize
$$P = 10x + 12y,$$
subject to
$$x + y \leq 60,$$
$$x - 2y \geq 0,$$
$$x, y \geq 0.$$

**2.** Maximize
$$P = 5x + 6y,$$
subject to
$$x + y \leq 80,$$
$$3x + 2y \leq 220,$$
$$2x + 3y \leq 210,$$
$$x, y \geq 0.$$

**3.** Maximize
$$Z = 4x - 6y,$$
subject to
$$y \leq 7,$$
$$3x - y \leq 3,$$
$$x + y \geq 5,$$
$$x, y \geq 0.$$

**4.** Minimize
$$Z = x + y,$$
subject to
$$x - y \geq 0,$$
$$4x + 3y \geq 12,$$
$$9x + 11y \leq 99,$$
$$x \leq 8,$$
$$x, y \geq 0.$$

**5.** Maximize
$$Z = 4x - 10y,$$
subject to
$$x - 4y \geq 4,$$
$$2x - y \leq 2,$$
$$x, y \geq 0.$$

**6.** Minimize
$$Z = 20x + 30y,$$
subject to
$$2x + y \leq 10,$$
$$3x + 4y \leq 24,$$
$$8x + 7y \geq 56,$$
$$x, y \geq 0.$$

**7.** Minimize
$$Z = 7x + 3y,$$
subject to
$$3x - y \geq -2,$$
$$x + y \leq 9,$$
$$x - y = -1,$$
$$x, y \geq 0.$$

**8.** Maximize
$$Z = 0.5x - 0.3y,$$
subject to
$$x - y \geq -2,$$
$$2x - y \leq 4,$$
$$2x + y = 8,$$
$$x, y \geq 0.$$

**9.** Minimize
$$C = 2x + y,$$
subject to
$$3x + y \geq 3,$$
$$4x + 3y \geq 6,$$
$$x + 2y \geq 2,$$
$$x, y \geq 0.$$

**10.** Minimize
$$C = 2x + 2y,$$
subject to
$$x + 2y \geq 80,$$
$$3x + 2y \geq 160,$$
$$5x + 2y \geq 200,$$
$$x, y \geq 0.$$

**11.** Maximize
$$Z = 10x + 2y,$$
subject to
$$x + 2y \geq 4,$$
$$x - 2y \geq 0,$$
$$x, y \geq 0.$$

**12.** Minimize
$$Z = y - x,$$
subject to
$$x \geq 3,$$
$$x + 3y \geq 6,$$
$$x - 3y \geq -6,$$
$$x, y \geq 0.$$

**13. Production for Maximum Profit** A toy manufacturer preparing a production schedule for two new toys, widgets and wadgits, must use the information concerning their construction times given in the following table:

|  | Machine A | Machine B | Finishing |
|---|---|---|---|
| Widget | 2 hr | 1 hr | 1 hr |
| Wadgit | 1 hr | 1 hr | 3 hr |

For example, each widget requires 2 hours on machine A. The available employee hours per week are as follows: for operating machine A, 70 hours; for B, 40 hours; for finishing, 90 hours. If the profits on each widget and wadgit are $4 and $6, respectively, how many of each toy should be made per week in order to maximize profit? What would the maximum profit be?

14. **Production for Maximum Profit** A manufacturer produces two types of barbecue grills: Old Smokey and Blaze Away. During production, the grills require the use of two machines, A and B. The number of hours needed on both machines are indicated in the following table:

| | Machine A | Machine B |
|---|---|---|
| Old Smokey | 2 hr | 4 hr |
| Blaze Away | 4 hr | 2 hr |

If each machine can be used 24 hours a day, and the profits on the Old Smokey and Blaze Away models are $4 and $6, respectively, how many of each type of grill should be made per day to obtain maximum profit? What is the maximum profit?

15. **Diet Formulation** A diet is to contain at least 16 units of carbohydrates and 20 units of protein. Food A contains 2 units of carbohydrates and 4 of protein; food B contains 2 units of carbohydrates and 1 of protein. If food A costs $1.20 per unit and food B costs $0.80 per unit, how many units of each food should be purchased in order to minimize cost? What is the minimum cost?

16. **Fertilizer Nutrients** A produce grower is purchasing fertilizer containing three nutrients: A, B, and C. The minimum weekly requirements are 80 units of A, 120 of B, and 240 of C. There are two popular blends of fertilizer on the market. Blend I, costing $4 a bag, contains 2 units of A, 6 of B, and 4 of C. Blend II, costing $5 a bag, contains 2 units of A, 2 of B, and 12 of C. How many bags of each blend should the grower buy each week to minimize the cost of meeting the nutrient requirements?

17. **Mineral Extraction** A company extracts minerals from ore. The numbers of pounds of minerals A and B that can be extracted from each ton of ores I and II are given in the following table, together with the costs per ton of the ores:

| | Ore I | Ore II |
|---|---|---|
| Mineral A | 100 lb | 200 lb |
| Mineral B | 200 lb | 50 lb |
| Cost per ton | $50 | $60 |

If the company must produce at least 3000 lb of A and 2500 lb of B, how many tons of each ore should be processed in order to minimize cost? What is the minimum cost?

18. **Production Scheduling** An oil company that has two refineries needs at least 800, 1400, and 500 barrels of low-, medium-, and high-grade oil, respectively. Each day, Refinery I produces 200 barrels of low-, 300 barrels of medium-, and 100 barrels of high-grade oil, whereas Refinery II produces 100 barrels each of low- and high- and 200 barrels of medium-grade oil. If it costs $2500 per day to operate Refinery I and $2000 per day to operate Refinery II, how many days should each refinery be operated to satisfy the production requirements at minimum cost? What is the minimum cost? (Assume that a minimum cost exists.)

19. **Construction Cost** A chemical company is designing a plant for producing two types of polymers, $P_1$ and $P_2$. The plant must be capable of producing at least 100 units of $P_1$ and 420 units of $P_2$ each day. There are two possible designs for the basic reaction chambers that are to be included in the plant. Each chamber of type A costs $600,000 and is capable of producing 10 units of $P_1$ and 20 units of $P_2$ per day; type B is of cheaper design, costing $300,000, and is capable of producing 4 units of $P_1$ and 30 units of $P_2$ per day. Because of operating costs, it is necessary to have at least four chambers of each type in the plant. How many chambers of each type should be included to minimize the cost of construction and still meet the required production schedule? (Assume that a minimum cost exists.)

20. **Pollution Control** Because of new federal regulations on pollution, a chemical company has introduced into its plant a new, more expensive process to supplement or replace an older process in the production of a particular chemical. The older process discharges 15 grams of sulfur dioxide and 40 grams of particulate matter into the atmosphere for each liter of chemical produced. The new process discharges 5 grams of sulfur dioxide and 20 grams

of particulate matter into the atmosphere for each liter produced. The company makes a profit of 30 cents per liter and 20 cents per liter on the old and new processes, respectively. If the government allows the plant to discharge no more than 10,500 grams of sulfur dioxide and no more than 30,000 grams of particulate matter into the atmosphere each day; how many liters of chemical should be produced daily, by each process, to maximize daily profit? What is the maximum daily profit?

21. **Construction Discount**   The highway department has decided to add exactly 200 km of highway and exactly 100 km of expressway to its road system this year. The standard price for road construction is \$1 million per kilometer of highway and \$5 million per kilometer of expressway. Only two contractors, company A and company B, can do this kind of construction, so the entire 300 km of road must be built by these two companies. However, company A can construct at most 200 km of roadway (highway and expressway), and company B can construct at most 150 km. For political reasons, each company must be awarded a contract with a standard price of at least \$250 million (before discounts). Company A offers a discount of \$1000 per kilometer of highway and \$6000 per kilometer of expressway; company B offers a discount of \$2000 for each kilometer of highway and \$5000 for each kilometer of expressway.

a. Let $x$ and $y$ represent the number of kilometers of highway and expressway, respectively, awarded to company A. Show that the total discount received from both companies is given by

$$D = 900 - x + y,$$

where $D$ is in thousands of dollars.

b. The highway department wishes to maximize the total discount $D$. Show that this problem is equivalent to the following linear programming problem, by showing exactly how the first four constraints arise:

Maximize $D = 900 - x + y$,

subject to

$$
\begin{aligned}
x + y &\le 200, \\
x + y &\ge 150, \\
x + 5y &\ge 250, \\
x + 5y &\le 450, \\
x &\ge 0, \\
y &\ge 0.
\end{aligned}
$$

c. Find the values of $x$ and $y$ that maximize $D$.

*In Problems 22–25, round answers to two decimal places.*

22. Maximize

   $Z = 2x + 0.3y$,

   subject to

   $$
   \begin{aligned}
   y &\le 6 - 4x, \\
   y &\ge 2 - 0.5x, \\
   x, y &\ge 0.
   \end{aligned}
   $$

23. Maximize

   $Z = 14x - 3y$,

   subject to

   $$
   \begin{aligned}
   y &\ge 12.5 - 4x, \\
   y &\le 9.3 - x, \\
   y &\ge 4.7 + 0.8x, \\
   x, y &\ge 0.
   \end{aligned}
   $$

24. Minimize

   $Z = 6.3y - 2.7x$,

   subject to

   $$
   \begin{aligned}
   7.4x + y &\ge 25, \\
   1.2x + y &\le 10.4, \\
   0.4x - y &\ge -0.6, \\
   x, y &\ge 0.
   \end{aligned}
   $$

25. Minimize

   $Z = 17.3x - 14.4y$,

   subject to

   $$
   \begin{aligned}
   0.73x - y &\le -2.4, \\
   1.22x - y &\ge -5.1, \\
   0.45x - y &\ge -12.4, \\
   x, y &\ge 0.
   \end{aligned}
   $$

OBJECTIVE

**To consider situations in which a linear programming problem has more than one optimum solution.**

# 7.3 MULTIPLE OPTIMUM SOLUTIONS[1]

Sometimes an objective function attains its optimum value at more than one feasible point, in which case **multiple optimum solutions** are said to exist. Example 1 will illustrate.

**EXAMPLE 1   Multiple Optimum Solutions**

*Maximize Z = 2x + 4y, subject to the constraints*

$$
\begin{aligned}
x - 4y &\le -8, \\
x + 2y &\le 16, \\
x \ge 0, \quad y &\ge 0.
\end{aligned}
$$

*Solution:* The feasible region appears in Fig. 7.22. Since the region is non-empty and bounded, $Z$ has a maximum value at a corner point. The corner points are

$$A = (0, 2), \quad B = (8, 4), \quad C = (0, 8).$$

---

[1]This section can be omitted.

Evaluating the objective function at $A$, $B$, and $C$ gives

$$Z(A) = 2(0) + 4(2) = 8,$$
$$Z(B) = 2(8) + 4(4) = 32,$$
$$Z(C) = 2(0) + 4(8) = 32.$$

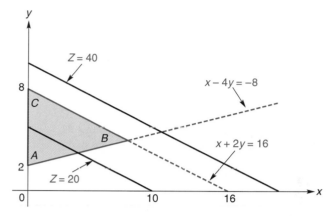

**FIGURE 7.22**    $Z = 2x + 4y$ is maximized at each point on the line segment $\overline{BC}$.

Thus, the maximum value of $Z$ over the region is 32, and it occurs at *two* corner points, $B$ and $C$. In fact, this maximum value also occurs at *all* points on the line segment *joining* $B$ and $C$, for the following reason. Each member of the family of lines $Z = 2x + 4y$ has slope $-\frac{1}{2}$. Moreover, the constraint line $x + 2y = 16$, which contains $B$ and $C$, also has slope $-\frac{1}{2}$ and hence is parallel to each member of $Z = 2x + 4y$. Figure 7.22 shows lines for $Z = 20$ and $Z = 40$. Note that the member of the family that maximizes $Z$ contains not only $B$ and $C$, but also all points on the line segment $\overline{BC}$. It thus has infinitely many points in common with the feasible region. Hence, this linear programming problem has infinitely many optimum solutions. In fact, it can be shown that:

> If $(x_1, y_1)$ and $(x_2, y_2)$ are two corner points at which an objective function is optimum, then the function will also be optimum at all points $(x, y)$ where
>
> $$x = (1 - t)x_1 + tx_2,$$
> $$y = (1 - t)y_1 + ty_2,$$
>
> and
>
> $$0 \le t \le 1.$$

In our case, if $(x_1, y_1) = B = (8, 4)$ and $(x_2, y_2) = C = (0, 8)$, then $Z$ is maximum at any point $(x, y)$ where

$$x = (1 - t)8 + t \cdot 0 = 8(1 - t),$$
$$y = (1 - t)4 + t \cdot 8 = 4(1 + t),$$

$$\text{and} \quad 0 \le t \le 1.$$

These equations give the coordinates of any point on the line segment $\overline{BC}$. In particular, if $t = 0$, then $x = 8$ and $y = 4$, which gives the corner point $B = (8, 4)$. If $t = 1$, we get the corner point $C = (0, 8)$. The value $t = \frac{1}{2}$ gives the point $(4, 6)$. Notice that at $(4, 6)$, $Z = 2(4) + 4(6) = 32$, which is the maximum value of $Z$.

■ **Exercise 7.3**

**1.** Minimize

$$Z = 3x + 9y,$$

subject to

$$y \geq -\tfrac{3}{2}x + 6,$$
$$y \geq -\tfrac{1}{3}x + \tfrac{11}{3},$$
$$y \geq x - 3,$$
$$x, y \geq 0.$$

**2.** Maximize

$$Z = 3x + 6y,$$

subject to

$$x - y \geq -3,$$
$$2x - y \leq 4,$$
$$x + 2y = 12,$$
$$x, y \geq 0.$$

**3.** Maximize

$$Z = 18x + 9y,$$

subject to

$$2x + 3y \leq 12,$$
$$2x + y \leq 8,$$
$$x, y \geq 0.$$

**4. Minimize Cost**   Suppose a car dealer has showrooms in Atherton and Berkeley and warehouses in Concord and Dublin. The cost of delivering a car is $60 from Concord to Atherton, $45 from Concord to Berkeley, $50 from Dublin to Atherton, and $35 from Dublin to Berkeley. Suppose that the showroom in Atherton orders seven cars and the showroom in Berkeley orders four cars. Suppose also that the warehouse in Concord has six cars and the warehouse in Dublin has eight cars available. Find the best way to minimize cost, and find the minimum cost. [*Hint:* Let $x$ be the number of cars delivered from Concord to Atherton and $y$ be the number of cars delivered from Concord to Berkeley. Then $7 - x$ is the number of cars delivered from Dublin to Atherton and $4 - y$ the number of cars shipped from Dublin to Berkeley.]

---

### OBJECTIVE

To show how the simplex method is used to solve a standard linear programming problem. This method will allow you to solve problems that cannot be solved geometrically.

## 7.4 THE SIMPLEX METHOD

Up to now, we have solved linear programming problems by a geometric method. This method is not practical when the number of variables increases to three, and is not possible beyond that. Now we shall look at a different technique—the **simplex method,** whose name is linked in more advanced discussions to a geometrical object called a simplex.

The simplex method begins with a feasible solution and tests whether it is optimum. If it is not, the method proceeds to a *better* solution. We say "better" in the sense that the new solution brings you closer to optimization of the objective function.[2] Should this new solution not be optimum, we repeat the procedure. Eventually, the simplex method leads to an optimum solution, if one exists.

Besides being efficient, the simplex method has other advantages. For one, it is completely mechanical. (We use matrices, elementary row operations, and basic arithmetic.) Moreover, no geometry is involved; this allows us to solve linear programming problems having any number of constraints and variables.

In this section, we shall consider only so-called **standard linear programming problems.** These can be put in the following form.

**Standard Linear Programming Problem**

Maximize the linear function $Z = c_1 x_1 + c_2 x_2 + \cdots + c_n x_n$, subject to the constraints

$$
\left.
\begin{aligned}
a_{11}x_1 + a_{12}x_2 + \cdots + a_{1n}x_n &\leq b_1, \\
a_{21}x_1 + a_{22}x_2 + \cdots + a_{2n}x_n &\leq b_2, \\
&\ \ \vdots \\
a_{m1}x_1 + a_{m2}x_2 + \cdots + a_{mn}x_n &\leq b_m,
\end{aligned}
\right\} \tag{1}
$$

where $x_1, x_2, \ldots, x_n$ and $b_1, b_2, \ldots, b_m$ are nonnegative.

---

[2]In most cases this is true. In some situations, however, the new solution may be just as good as the previous one. Example 2 will illustrate this.

Note that one feasible solution to a standard linear programming problem is always $x_1 = 0, x_2 = 0, \ldots, x_n = 0$. Other types of linear programming problems will be discussed in Secs. 7.6 and 7.7.

*The procedure that we follow here will be outlined later in this section.*

We shall now apply the simplex method to the problem in Example 1 of Sec. 7.2, which has the form

$$\text{maximize } Z = 3x_1 + x_2,$$

subject to the constraints

$$2x_1 + x_2 \leq 8 \tag{2}$$

and

$$2x_1 + 3x_2 \leq 12, \tag{3}$$

where $x_1 \geq 0$ and $x_2 \geq 0$. This problem is of standard form. We begin by expressing constraints (2) and (3) as equations. In (2), $2x_1 + x_2$ will *equal* 8 if we add some nonnegative number $s_1$ to $2x_1 + x_2$, so that

$$2x_1 + x_2 + s_1 = 8, \qquad \text{where } s_1 \geq 0.$$

We call $s_1$ a **slack variable,** since it makes up for the "slack" on the left side of (2) so that we have equality. Similarly, inequality (3) can be written as an equation by using the slack variable $s_2$; we have

$$2x_1 + 3x_2 + s_2 = 12, \qquad \text{where } s_2 \geq 0.$$

The variables $x_1$ and $x_2$ are called **structural variables.**

Now we can restate the problem in terms of equations:

$$\text{Maximize } Z = 3x_1 + x_2 \tag{4}$$

such that

$$2x_1 + x_2 + s_1 = 8 \tag{5}$$

and

$$2x_1 + 3x_2 + s_2 = 12, \tag{6}$$

where $x_1, x_2, s_1$, and $s_2$ are nonnegative.

From Sec. 7.2, we know that the optimum solution occurs at a corner point of the feasible region in Fig. 7.23. At each of these points, at least *two* of the variables $x_1, x_2, s_1$, and $s_2$ are 0, as the following listing indicates:

1. At $A$, we have $x_1 = 0$ and $x_2 = 0$.

2. At $B$, $x_1 = 4$ and $x_2 = 0$. But from Eq. (5), $2(4) + 0 + s_1 = 8$. Thus, $s_1 = 0$.

3. At $C$, $x_1 = 3$ and $x_2 = 2$. But from Eq. (5), $2(3) + 2 + s_1 = 8$. Hence, $s_1 = 0$. From Eq. (6), $2(3) + 3(2) + s_2 = 12$. Therefore, $s_2 = 0$.

4. At $D$, $x_1 = 0$ and $x_2 = 4$. From Eq. (6), $2(0) + 3(4) + s_2 = 12$. Thus, $s_2 = 0$.

It can also be shown that any solution to Eqs. (5) and (6), such that at least *two* of the four variables $x_1, x_2, s_1$, and $s_2$ are zero, corresponds to a corner point. Any such solution where at least two of these variables are zero is called a **basic feasible solution** (abbreviated B.F.S.). This number, 2, is determined by the expression $n - m$, where $m$ is the number of constraints (excluding the nonnegativity conditions) and $n$ is the number of variables that occur after these

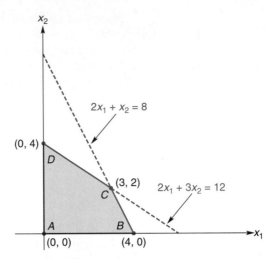

**FIGURE 7.23** Optimum solution must occur at corner point of feasible region.

constraints are converted to equations. In our case, $n = 4$ and $m = 2$. For any particular B.F.S., the two variables held at zero value are called **nonbasic variables,** whereas the others are called **basic variables** for that B.F.S. Thus, for the B.F.S. corresponding to item 3 in the preceding list, $s_1$ and $s_2$ are the nonbasic variables, but for the B.F.S. corresponding to item 4, the nonbasic variables are $x_1$ and $s_2$. We eventually want to find a B.F.S. that maximizes $Z$.

We shall first find an initial B.F.S. and then determine whether the corresponding value of $Z$ can be increased by a different B.F.S. Since $x_1 = 0$ and $x_2 = 0$ is a feasible solution to this standard linear programming problem, let us initially find the B.F.S. such that the structural variables $x_1$ and $x_2$ are nonbasic. That is, we choose $x_1 = 0$ and $x_2 = 0$ and find the corresponding values of $s_1, s_2$, and $Z$. This can be done most conveniently by matrix techniques, based on the methods developed in Chapter 6.

If we write Eq. (4) as $-3x_1 - x_2 + Z = 0$, then Eqs. (5), (6), and (4) form the system

$$\begin{cases} 2x_1 + \phantom{3}x_2 + s_1 \phantom{{}+ s_2} \phantom{{}+ Z} = \phantom{0}8, \\ 2x_1 + 3x_2 \phantom{{}+ s_1} + s_2 \phantom{{}+ Z} = 12, \\ -3x_1 - \phantom{3}x_2 \phantom{{}+ s_1 + s_2} + Z = \phantom{0}0. \end{cases}$$

In terms of an augmented coefficient matrix, called the **initial simplex tableau,** we have

$$\begin{array}{c} \phantom{s_1} \\ s_1 \\ s_2 \\ Z \end{array} \begin{array}{cccccc} x_1 & x_2 & s_1 & s_2 & Z & \\ \left[ \begin{array}{ccccc|c} 2 & 1 & 1 & 0 & 0 & 8 \\ 2 & 3 & 0 & 1 & 0 & 12 \\ \hdashline -3 & -1 & 0 & 0 & 1 & 0 \end{array} \right]. \end{array}$$

The first two rows correspond to the constraints, and the last row, called the **objective row,** corresponds to the objective equation—thus the dashed horizontal separating line. Notice that if $x_1 = 0$ and $x_2 = 0$, then, from rows 1, 2, and 3, we can directly read off the values of $s_1, s_2$, and $Z$: $s_1 = 8$, $s_2 = 12$, and $Z = 0$. That is why we placed the letters $s_1, s_2$, and $Z$ to the left of the rows.

(We remind you that $s_1$ and $s_2$ are the basic variables.) Thus, our initial basic feasible solution is

$$x_1 = 0, \qquad x_2 = 0, \qquad s_1 = 8, \qquad s_2 = 12,$$

at which $Z = 0$. Let us see if we can find a B.F.S. that gives a larger value of $Z$.

The variables $x_1$ and $x_2$ are nonbasic in the preceding B.F.S. We shall now look for a B.F.S. in which one of these variables is basic while the other remains nonbasic. Which one should we choose as the basic variable? Let us examine the possibilities. From the $Z$-row of the matrix above, $Z = 3x_1 + x_2$. If $x_1$ is allowed to become basic, then $x_2$ remains at 0 and $Z = 3x_1$; thus, for each one-unit increase in $x_1$, $Z$ increases by three units. On the other hand, if $x_2$ is allowed to become basic, then $x_1$ remains at 0 and $Z = x_2$; hence, for each one-unit increase in $x_2$, $Z$ increases by one unit. Consequently, we get a *greater* increase in the value of $Z$ if $x_1$, rather than $x_2$, enters the basic-variable category. In this case, we call $x_1$ an **entering variable.** Thus, in terms of the simplex tableau below (which is the same as the matrix presented earlier, except for some additional labeling), the entering variable can be found by looking at the "most negative" of the numbers enclosed by the brace in the $Z$-row. (By *most negative*, we mean the negative indicator having the greatest magnitude.) Since that number is $-3$ and appears in the $x_1$-column, $x_1$ is the entering variable. The numbers in the brace are called **indicators.**

$$
\begin{array}{c}
\begin{array}{ccccc}
x_1 & x_2 & s_1 & s_2 & Z
\end{array} \\
\begin{array}{c}
s_1 \\
s_2 \\
Z
\end{array}
\left[
\begin{array}{ccccc|c}
2 & 1 & 1 & 0 & 0 & 8 \\
2 & 3 & 0 & 1 & 0 & 12 \\
\hline
-3 & -1 & 0 & 0 & 1 & 0
\end{array}
\right]
\end{array}
$$

$$\underbrace{\phantom{-3 \quad -1}}$$
$$\uparrow \qquad \text{indicators}$$
$$\text{entering}$$
$$\text{variable}$$

Let us summarize the information that can be obtained from this tableau. It gives a B.F.S. where $s_1$ and $s_2$ are the basic variables and $x_1$ and $x_2$ are nonbasic. The B.F.S. is $s_1 = 8$ (the right-hand side of the $s_1$-row), $s_2 = 12$ (the right-hand side of the $s_2$-row), $x_1 = 0$, and $x_2 = 0$. The $-3$ in the $x_1$-column of the $Z$-row indicates that if $x_2$ remains 0, then $Z$ increases three units for each one-unit increase in $x_1$. The $-1$ in the $x_2$-column of the $Z$-row indicates that if $x_1$ remains 0, then $Z$ increases one unit for each one-unit increase in $x_2$. The column in which the most negative indicator, $-3$, lies gives the entering variable $x_1$—that is, the variable that should become basic in the next B.F.S.

In our new B.F.S., the larger the increase in $x_1$ (from $x_1 = 0$), the larger is the increase in $Z$. Now, by how much can we increase $x_1$? Since $x_2$ is still held at 0, from rows 1 and 2 of the simplex tableau, it follows that

$$s_1 = 8 - 2x_1$$

and

$$s_2 = 12 - 2x_1.$$

Since $s_1$ and $s_2$ are nonnegative, we have

$$8 - 2x_1 \geq 0$$

and

$$12 - 2x_1 \geq 0.$$

From the first inequality, $x_1 \leq \frac{8}{2} = 4$; from the second, $x_1 \leq \frac{12}{2} = 6$. Thus, $x_1$ must be less than or equal to the smaller of the quotients $\frac{8}{2}$ and $\frac{12}{2}$, which is $\frac{8}{2}$. Hence, $x_1$ can increase at most by 4. However, in a B.F.S., two variables must be 0. We already have $x_2 = 0$. Since $s_1 = 8 - 2x_1$, $s_1$ must be 0 for $x_1 = 4$. Therefore, we have a new B.F.S. with $x_1$ replacing $s_1$ as a basic variable. That is, $s_1$ will *depart* from the category of basic variables in the previous B.F.S. and will be nonbasic in the new B.F.S. We say that $s_1$ is the **departing variable** for the previous B.F.S. In summary, for our new B.F.S., we want $x_1$ and $s_2$ as basic variables with $x_1 = 4$, and $x_2$ and $s_1$ as nonbasic variables ($x_2 = 0$, $s_1 = 0$).

Before proceeding, let us update our tableau. To the right of the following tableau, the quotients $\frac{8}{2}$ and $\frac{12}{2}$ are indicated:

$$
\begin{array}{c}
\text{departing} \\
\text{variable}
\end{array} \rightarrow
\begin{array}{c}
s_1 \\
s_2 \\
Z
\end{array}
\begin{bmatrix}
\begin{array}{ccccc|c}
x_1 & x_2 & s_1 & s_2 & Z & b \\
2 & 1 & 1 & 0 & 0 & 8 \\
2 & 3 & 0 & 1 & 0 & 12 \\
\hline
-3 & -1 & 0 & 0 & 1 & 0
\end{array}
\end{bmatrix}
\qquad
\begin{array}{l}
\textit{Quotients} \\
8 \div 2 = 4 \text{ (smaller).} \\
12 \div 2 = 6.
\end{array}
$$

$\uparrow$

entering variable (most negative indicator)

These quotients are obtained by dividing each entry in the first two rows of the $b$-column by the entry in the corresponding row of the entering-variable column. Notice that the departing variable is in the same row as the *smaller* quotient, $8 \div 2$.

Since $x_1$ and $s_2$ will be basic variables in our new B.F.S., it would be convenient to change our previous tableau by elementary row operations into a form in which the values of $x_1$, $s_2$, and $Z$ can be read off with ease (just as we were able to do with the solution corresponding to $x_1 = 0$ and $x_2 = 0$). To do this, we want to find a matrix that is equivalent to the preceding tableau, but that has the form

$$
\begin{array}{ccccc}
x_1 & x_2 & s_1 & s_2 & Z
\end{array}
$$
$$
\begin{bmatrix}
\begin{array}{ccccc|c}
1 & ? & ? & 0 & 0 & ? \\
0 & ? & ? & 1 & 0 & ? \\
\hline
0 & ? & ? & 0 & 1 & ?
\end{array}
\end{bmatrix},
$$

where the question marks represent numbers to be determined. Notice here that if $x_2 = 0$ and $s_1 = 0$, then $x_1$ equals the number in row 1 of the last column, $s_2$ equals the number in row 2, and $Z$ is the number in row 3. Thus, we must transform the tableau

$$
\begin{array}{c}
\text{departing} \\
\text{variable}
\end{array} \rightarrow
\begin{array}{c}
s_1 \\
s_2 \\
Z
\end{array}
\begin{bmatrix}
\begin{array}{ccccc|c}
x_1 & x_2 & s_1 & s_2 & Z & \\
\boxed{2} & 1 & 1 & 0 & 0 & 8 \\
2 & 3 & 0 & 1 & 0 & 12 \\
\hline
-3 & -1 & 0 & 0 & 1 & 0
\end{array}
\end{bmatrix}
\qquad (7)
$$

$\uparrow$

entering variable

into an equivalent matrix that has a 1 where the shaded entry appears and 0's elsewhere in the $x_1$-column. The shaded entry is called the **pivot entry**—it is in the column of the entering variable (called the *pivot column*) and the row of

*Circling a pivot entry can be used in place of shading.*

the departing variable (called the *pivot row*). By elementary row operations, we have

$$
\begin{array}{ccccc}
x_1 & x_2 & s_1 & s_2 & Z
\end{array}
$$
$$
\left[\begin{array}{ccccc|c}
2 & 1 & 1 & 0 & 0 & 8 \\
2 & 3 & 0 & 1 & 0 & 12 \\
\hline
-3 & -1 & 0 & 0 & 1 & 0
\end{array}\right]
$$

$$\xrightarrow{\frac{1}{2}R_1}
\left[\begin{array}{ccccc|c}
1 & \frac{1}{2} & \frac{1}{2} & 0 & 0 & 4 \\
2 & 3 & 0 & 1 & 0 & 12 \\
\hline
-3 & -1 & 0 & 0 & 1 & 0
\end{array}\right]
$$

$$\xrightarrow[3R_1 + R_3]{-2R_1 + R_2}
\left[\begin{array}{ccccc|c}
1 & \frac{1}{2} & \frac{1}{2} & 0 & 0 & 4 \\
0 & 2 & -1 & 1 & 0 & 4 \\
0 & \frac{1}{2} & \frac{3}{2} & 0 & 1 & 12
\end{array}\right].
$$

Thus, we have a new simplex tableau:

$$
\begin{array}{c}
x_1 \\ s_2 \\ Z
\end{array}
\left[\begin{array}{ccccc|c}
1 & \frac{1}{2} & \frac{1}{2} & 0 & 0 & 4 \\
0 & 2 & -1 & 1 & 0 & 4 \\
\hline
0 & \frac{1}{2} & \frac{3}{2} & 0 & 1 & 12
\end{array}\right]. \tag{8}
$$

indicators

For $x_2 = 0$ and $s_1 = 0$, from the first row, we have $x_1 = 4$; from the second, we obtain $s_2 = 4$. These values give us the new B.F.S. Note that we replaced the $s_1$ located to the left of the initial tableau (7) by $x_1$ in our new tableau (8), so that $s_1$ *departed* and $x_1$ *entered*. From row 3, for $x_2 = 0$ and $s_1 = 0$, we get $Z = 12$, which is a larger value than we had before ($Z = 0$).

In our present B.F.S., $x_2$ and $s_1$ are nonbasic variables ($x_2 = 0$, $s_1 = 0$). Suppose we look for another B.F.S. that gives a larger value of $Z$ and is such that one of $x_2$ or $s_1$ is basic. The equation corresponding to the $Z$-row is given by $\frac{1}{2}x_2 + \frac{3}{2}s_1 + Z = 12$, or

$$ Z = 12 - \tfrac{1}{2}x_2 - \tfrac{3}{2}s_1. \tag{9}$$

If $x_2$ becomes basic and therefore $s_1$ remains nonbasic, then

$$ Z = 12 - \tfrac{1}{2}x_2 \quad (\text{since } s_1 = 0).$$

Here, each one-unit increase in $x_2$ *decreases* $Z$ by $\frac{1}{2}$ unit. Thus, any increase in $x_2$ would make $Z$ smaller than before. On the other hand, if $s_1$ becomes basic and $x_2$ remains nonbasic, then, from Eq. (9),

$$ Z = 12 - \tfrac{3}{2}s_1 \quad (\text{since } x_2 = 0).$$

Here each one-unit increase in $s_1$ *decreases* $Z$ by $\frac{3}{2}$ units. Hence, any increase in $s_1$ would make $Z$ smaller than before. Consequently, we cannot move to a better B.F.S. In short, no B.F.S. gives a larger value of $Z$ than the B.F.S. $x_1 = 4$, $s_2 = 4$, $x_2 = 0$, and $s_1 = 0$ (which gives $Z = 12$).

In fact, since $x_2 \geq 0$ and $s_1 \geq 0$, and since the coefficients of $x_2$ and $s_1$ in Eq. (9) are negative, $Z$ is maximum when $x_2 = 0$ and $s_1 = 0$. That is, in (8), *having all nonnegative indicators means that we have an optimum solution.*

In terms of our original problem, if

$$ Z = 3x_1 + x_2,$$

such that

$$2x_1 + x_2 \leq 8, \qquad 2x_1 + 3x_2 \leq 12, \qquad x_1 \geq 0, \qquad \text{and} \qquad x_2 \geq 0,$$

then $Z$ is maximum when $x_1 = 4$ and $x_2 = 0$, and the maximum value of $Z$ is 12. (This confirms our result in Example 1 of Sec. 7.2.) Note that the values of $s_1$ and $s_2$ do not have to appear here.

Let us outline the simplex method for a standard linear-programming problem with three structural variables and four constraints, not counting non-negativity conditions. The outline implies how the simplex method works for any number of structural variables and constraints.

**Simplex Method**

*Problem:*

$$\text{Maximize } Z = c_1x_1 + c_2x_2 + c_3x_3$$

such that

$$a_{11}x_1 + a_{12}x_2 + a_{13}x_3 \leq b_1,$$

$$a_{21}x_1 + a_{22}x_2 + a_{23}x_3 \leq b_2,$$

$$a_{31}x_1 + a_{32}x_2 + a_{33}x_3 \leq b_3,$$

$$a_{41}x_1 + a_{42}x_2 + a_{43}x_3 \leq b_4,$$

where $x_1$, $x_2$, $x_3$ and $b_1$, $b_2$, $b_3$, $b_4$ are nonnegative.

*Method:*

1. Set up the initial simplex tableau:

|       | $x_1$    | $x_2$    | $x_3$    | $s_1$ | $s_2$ | $s_3$ | $s_4$ | $Z$ | $b$   |
|-------|----------|----------|----------|-------|-------|-------|-------|-----|-------|
| $s_1$ | $a_{11}$ | $a_{12}$ | $a_{13}$ | 1     | 0     | 0     | 0     | 0   | $b_1$ |
| $s_2$ | $a_{21}$ | $a_{22}$ | $a_{23}$ | 0     | 1     | 0     | 0     | 0   | $b_2$ |
| $s_3$ | $a_{31}$ | $a_{32}$ | $a_{33}$ | 0     | 0     | 1     | 0     | 0   | $b_3$ |
| $s_4$ | $a_{41}$ | $a_{42}$ | $a_{43}$ | 0     | 0     | 0     | 1     | 0   | $b_4$ |
| $Z$   | $-c_1$   | $-c_2$   | $-c_3$   | 0     | 0     | 0     | 0     | 1   | 0     |

indicators

There are four slack variables: $s_1$, $s_2$, $s_3$, and $s_4$—one for each constraint.

2. If all the indicators in the last row are nonnegative, then $Z$ has a maximum when $x_1 = 0$, $x_2 = 0$, and $x_3 = 0$. The maximum value is 0. If there are any negative indicators, locate the column in which the most negative indicator appears. This pivot column gives the entering variable. (If more than one column contains the most negative indicator, the choice of pivot column is arbitrary.)

3. Divide each *positive*[3] entry above the dashed line in the entering-variable column *into* the corresponding value of $b$.

4. Mark the entry in the pivot column that corresponds to the smallest quotient in step 3. This is the pivot entry. The departing variable is the one to the left of the pivot row.

[3]This will be discussed after Example 1.

5. Use elementary row operations to transform the tableau into a new equivalent tableau that has a 1 where the pivot entry was and 0's elsewhere in that column.

6. On the left side of this tableau, the entering variable replaces the departing variable.

7. If the indicators of the new tableau are all nonnegative, we have an optimum solution. The maximum value of $Z$ is the entry in the last row and last column. It occurs when the variables to the left of the tableau are equal to the corresponding entries in the last column. All other variables are 0. If at least one of the indicators is negative, repeat the process, beginning with step 2 applied to the new tableau.

As an aid in understanding the simplex method, we should be able to interpret certain entries in a tableau. Suppose that we obtain a tableau in which the last row is as shown in the following array:

$$
\begin{array}{c}
\begin{array}{ccccccc}
x_1 & x_2 & x_3 & s_1 & s_2 & s_3 & s_4 & Z
\end{array} \\
\begin{bmatrix}
\cdot & \cdot & \cdot & \cdot & \cdot & \cdot & \cdot & \cdot & \vline & \cdot \\
\cdot & \cdot & \cdot & \cdot & \cdot & \cdot & \cdot & \cdot & \vline & \cdot \\
\cdot & \cdot & \cdot & \cdot & \cdot & \cdot & \cdot & \cdot & \vline & \cdot \\
\hdashline
a & b & c & d & e & f & g & 1 & \vline & h
\end{bmatrix}
\end{array}
\begin{array}{c} \\ \\ \\ Z \end{array}
$$

We can interpret the entry $b$, for example, as follows: If $x_2$ is nonbasic and were to become basic, then, for each one-unit increase in $x_2$,

$$\text{if } b < 0, Z \text{ } \textit{increases} \text{ by } |b| \text{ units;}$$

$$\text{if } b > 0, Z \text{ } \textit{decreases} \text{ by } b \text{ units;}$$

$$\text{if } b = 0, \text{ there is no change in } Z.$$

**EXAMPLE 1** **The Simplex Method**

*Maximize $Z = 5x_1 + 4x_2$, subject to*

$$x_1 + x_2 \leq 20,$$
$$2x_1 + x_2 \leq 35,$$
$$-3x_1 + x_2 \leq 12,$$

*and $x_1 \geq 0, x_2 \geq 0$.*

*Solution:* This linear programming problem fits the standard form. The initial simplex tableau is

| | $x_1$ | $x_2$ | $s_1$ | $s_2$ | $s_3$ | $Z$ | $b$ | Quotients |
|---|---|---|---|---|---|---|---|---|
| $s_1$ | 1 | 1 | 1 | 0 | 0 | 0 | 20 | $20 \div 1 = 20.$ |
| $s_2$ | 2 | 1 | 0 | 1 | 0 | 0 | 35 | $35 \div 2 = \frac{35}{2}.$ |
| $s_3$ | $-3$ | 1 | 0 | 0 | 1 | 0 | 12 | no quotient, since |
| $Z$ | $-5$ | $-4$ | 0 | 0 | 0 | 1 | 0 | $-3$ is not positive. |

departing variable → $s_2$

↑ indicators

entering variable

The most negative indicator, $-5$, occurs in the $x_1$-column. Thus, $x_1$ is the entering variable. The smaller quotient is $\frac{35}{2}$, so $s_2$ is the departing variable. The pivot entry is 2. Using elementary row operations to get a 1 in the pivot position and 0's elsewhere in its column, we have

$$
\begin{array}{c}
\begin{array}{ccccccc}
x_1 & x_2 & s_1 & s_2 & s_3 & Z & b
\end{array}\\
\left[\begin{array}{ccccccc|c}
1 & 1 & 1 & 0 & 0 & 0 & 20\\
2 & 1 & 0 & 1 & 0 & 0 & 35\\
-3 & 1 & 0 & 0 & 1 & 0 & 12\\
\hline
-5 & -4 & 0 & 0 & 0 & 1 & 0
\end{array}\right]
\end{array}
$$

$$
\xrightarrow{\frac{1}{2}R_2}
\left[\begin{array}{ccccccc|c}
1 & 1 & 1 & 0 & 0 & 0 & 20\\
1 & \frac{1}{2} & 0 & \frac{1}{2} & 0 & 0 & \frac{35}{2}\\
-3 & 1 & 0 & 0 & 1 & 0 & 12\\
\hline
-5 & -4 & 0 & 0 & 0 & 1 & 0
\end{array}\right]
$$

$$
\begin{array}{c}
\xrightarrow[\substack{3R_2+R_3\\5R_2+R_4}]{-1R_2+R_1}
\end{array}
\left[\begin{array}{ccccccc|c}
0 & \frac{1}{2} & 1 & -\frac{1}{2} & 0 & 0 & \frac{5}{2}\\
1 & \frac{1}{2} & 0 & \frac{1}{2} & 0 & 0 & \frac{35}{2}\\
0 & \frac{5}{2} & 0 & \frac{3}{2} & 1 & 0 & \frac{129}{2}\\
\hline
0 & -\frac{3}{2} & 0 & \frac{5}{2} & 0 & 1 & \frac{175}{2}
\end{array}\right].
$$

Our new tableau is

$$
\begin{array}{c}
\text{departing} \rightarrow \\
\text{variable}
\end{array}
\begin{array}{c}
\\
s_1\\
x_1\\
s_3\\
Z
\end{array}
\begin{array}{c}
\begin{array}{ccccccc}
x_1 & x_2 & s_1 & s_2 & s_3 & Z & b
\end{array}\\
\left[\begin{array}{ccccccc|c}
0 & \frac{1}{2} & 1 & -\frac{1}{2} & 0 & 0 & \frac{5}{2}\\
1 & \frac{1}{2} & 0 & \frac{1}{2} & 0 & 0 & \frac{35}{2}\\
0 & \frac{5}{2} & 0 & \frac{3}{2} & 1 & 0 & \frac{129}{2}\\
\hline
0 & -\frac{3}{2} & 0 & \frac{5}{2} & 0 & 1 & \frac{175}{2}
\end{array}\right]
\end{array}
\begin{array}{l}
\textit{Quotients}\\
\frac{5}{2}\div\frac{1}{2}=5.\\
\frac{35}{2}\div\frac{1}{2}=35.\\
\frac{129}{2}\div\frac{5}{2}=25\frac{4}{5}.
\end{array}
$$

$$\uparrow \quad \text{indicators}$$

entering
variable

Note that on the left side, $x_1$ replaced $s_2$. Since $-\frac{3}{2}$ is the most negative indicator, we must continue our process. The entering variable is now $x_2$. The smallest quotient is 5. Hence, $s_1$ is the departing variable and $\frac{1}{2}$ is the pivot entry. Using elementary row operations, we have

$$
\begin{array}{c}
\begin{array}{ccccccc}
x_1 & x_2 & s_1 & s_2 & s_3 & Z & b
\end{array}\\
\left[\begin{array}{ccccccc|c}
0 & \frac{1}{2} & 1 & -\frac{1}{2} & 0 & 0 & \frac{5}{2}\\
1 & \frac{1}{2} & 0 & \frac{1}{2} & 0 & 0 & \frac{35}{2}\\
0 & \frac{5}{2} & 0 & \frac{3}{2} & 1 & 0 & \frac{129}{2}\\
\hline
0 & -\frac{3}{2} & 0 & \frac{5}{2} & 0 & 1 & \frac{175}{2}
\end{array}\right]
\end{array}
$$

$$
\xrightarrow[\substack{-5R_1+R_3\\3R_1+R_4}]{-1R_1+R_2}
\left[\begin{array}{ccccccc|c}
0 & \frac{1}{2} & 1 & -\frac{1}{2} & 0 & 0 & \frac{5}{2}\\
1 & 0 & -1 & 1 & 0 & 0 & 15\\
0 & 0 & -5 & 4 & 1 & 0 & 52\\
\hline
0 & 0 & 3 & 1 & 0 & 1 & 95
\end{array}\right]
$$

$$
\xrightarrow{2R_1}
\left[\begin{array}{ccccccc|c}
0 & 1 & 2 & -1 & 0 & 0 & 5\\
1 & 0 & -1 & 1 & 0 & 0 & 15\\
0 & 0 & -5 & 4 & 1 & 0 & 52\\
\hline
0 & 0 & 3 & 1 & 0 & 1 & 95
\end{array}\right].
$$

Our new tableau is

$$
\begin{array}{c}
\\
x_2 \\
x_1 \\
s_3 \\
Z
\end{array}
\begin{array}{c}
\begin{array}{ccccccc}
x_1 & x_2 & s_1 & s_2 & s_3 & Z & b
\end{array} \\
\left[
\begin{array}{cccccc|c}
0 & 1 & 2 & -1 & 0 & 0 & 5 \\
1 & 0 & -1 & 1 & 0 & 0 & 15 \\
0 & 0 & -5 & 4 & 1 & 0 & 52 \\
\hline
0 & 0 & 3 & 1 & 0 & 1 & 95
\end{array}
\right]
\end{array},
$$

indicators

where $x_2$ replaced $s_1$ on the left side. Since all indicators are nonnegative, the maximum value of $Z$ is 95 and occurs when $x_2 = 5$ and $x_1 = 15$ (and $s_3 = 52, s_1 = 0$, and $s_2 = 0$).  ■

It is interesting to see how the values of $Z$ got progressively "better" in successive tableaus in Example 1. These are the entries in the last row and column of each tableau. In the initial tableau, we had $Z = 0$. From then on, we obtained $Z = \frac{175}{2} = 87\frac{1}{2}$ and then $Z = 95$, the maximum.

In Example 1, you may wonder why no quotient is considered in the third row of the initial tableau. The B.F.S. for this tableau is

$$ s_1 = 20, \qquad s_2 = 35, \qquad s_3 = 12, \qquad x_1 = 0, \qquad x_2 = 0, $$

where $x_1$ is the entering variable. The quotients 20 and $\frac{35}{2}$ reflect that, for the next B.F.S., we have $x_1 \leq 20$ and $x_1 \leq \frac{35}{2}$. Since the third row represents the equation $s_3 = 12 + 3x_1 - x_2$, and $x_2 = 0$, it follows that $s_3 = 12 + 3x_1$. But $s_3 \geq 0$, so $12 + 3x_1 \geq 0$, which implies that $x_1 \geq -\frac{12}{3} = -4$. Thus, we have

$$ x_1 \leq 20, \qquad x_1 \leq \frac{35}{2}, \qquad \text{and} \qquad x_1 \geq -4. $$

Hence, $x_1$ can increase at most by $\frac{35}{2}$. The condition $x_1 \geq -4$ has no influence in determining the maximum increase in $x_1$. That is why the quotient $12/(-3) = -4$ is not considered in row 3. In general, *no quotient is considered for a row if the entry in the entering-variable column is negative (or, of course, 0)*.

Although the simplex procedure that has been developed in this section applies only to linear programming problems of standard form, other forms may be adapted to fit this form. Suppose that a constraint has the form

$$ a_1 x_1 + a_2 x_2 + \cdots + a_n x_n \geq -b, $$

where $b > 0$. Here the inequality symbol is "$\geq$", and the constant on the right side is *negative*. Thus, the constraint is not in standard form. However, multiplying both sides by $-1$ gives

$$ -a_1 x_1 - a_2 x_2 - \cdots - a_n x_n \leq b, $$

which *does* have the proper form. Accordingly, it may be necessary to rewrite a constraint before proceeding with the simplex method.

In a simplex tableau, several indicators may "tie" for being most negative. In this case, we choose any one of these indicators to give the column for the entering variable. Likewise, there may be several quotients that "tie" for being the smallest. We may then choose any one of these quotients to determine the departing variable and pivot entry. Example 2 will illustrate this situation. When a tie for the smallest quotient exists, then, along with the nonbasic variables, a B.F.S. will have a basic variable that is 0. In this case we say that the B.F.S. is *degenerate* or that the linear programming problem has a *degeneracy*. More will be said about degeneracies in Sec. 7.5.

## Principles in Practice 1
### The Simplex Method

The What If Company has $30,000 for the purchase of materials to make three types of gadgets. The company has allocated a total of 1200 hours of assembly time and 180 hours of packaging time for the gadgets. The following table gives the cost per gadget, the number of hours per gadget, and the profit per gadget for each type:

|  | Type 1 | Type 2 | Type 3 |
|---|---|---|---|
| Cost/gadget | $300 | $300 | $400 |
| Assembly Hours/gadget | 15 | 15 | 10 |
| Packaging Hours/gadget | 2 | 2 | 3 |
| Profit | $150 | $250 | $200 |

Find the number of gadgets of each type the company should produce to maximize profit.

### EXAMPLE 2  The Simplex Method

*Maximize* $Z = 3x_1 + 4x_2 + \frac{3}{2}x_3$, *subject to*

$$-x_1 - 2x_2 \qquad\ \geq -10, \qquad (10)$$
$$2x_1 + 2x_2 + x_3 \leq 10,$$

*and* $x_1, x_2, x_3 \geq 0$.

**Solution:** Constraint (10) does not fit the standard form. However, multiplying both sides of inequality (10) by $-1$ gives

$$x_1 + 2x_2 \leq 10,$$

which *does* have the proper form. Thus, our initial simplex tableau is tableau I:

### SIMPLEX TABLEAU I

$$
\begin{array}{c}
\text{departing} \\
\text{variable}
\end{array} \rightarrow
\begin{array}{c}
\\ s_1 \\ s_2 \\ Z
\end{array}
\begin{array}{c}
x_1\ \ x_2\ \ x_3\ \ s_1\ \ s_2\ \ Z \ \ \ b \\
\left[\begin{array}{cccccc|c}
1 & 2 & 0 & 1 & 0 & 0 & 10 \\
2 & 2 & 1 & 0 & 1 & 0 & 10 \\
\hline
-3 & -4 & -\frac{3}{2} & 0 & 0 & 1 & 0
\end{array}\right]
\end{array}
\qquad
\begin{array}{c}
\textit{Quotients} \\
10 \div 2 = 5. \\
10 \div 2 = 5.
\end{array}
$$

indicators ↗

entering variable

The entering variable is $x_2$. Since there is a tie for the smallest quotient, we can choose either $s_1$ or $s_2$ as the departing variable. Let us choose $s_1$. The pivot entry is shaded. Using elementary row operations, we get tableau II:

### SIMPLEX TABLEAU II

$$
\begin{array}{c}
\\ \\ \text{departing} \\ \text{variable}
\end{array} \rightarrow
\begin{array}{c}
x_2 \\ \\ s_2 \\ Z
\end{array}
\begin{array}{c}
x_1\ \ x_2\ \ x_3\ \ s_1\ \ s_2\ \ Z\ \ \ b \\
\left[\begin{array}{cccccc|c}
\frac{1}{2} & 1 & 0 & \frac{1}{2} & 0 & 0 & 5 \\
1 & 0 & 1 & -1 & 1 & 0 & 0 \\
\hline
-1 & 0 & -\frac{3}{2} & 2 & 0 & 1 & 20
\end{array}\right]
\end{array}
\qquad
\begin{array}{c}
\textit{Quotients} \\
\text{no quotient, since 0 is} \\
\text{not positive.} \\
0 \div 1 = 0.
\end{array}
$$

indicators ↗

entering variable

Tableau II corresponds to a B.F.S. in which a basic variable, $s_2$, is zero. Thus, the B.F.S. is degenerate. Since there are negative indicators, we continue. The entering variable is now $x_3$, the departing variable is $s_2$, and the pivot is shaded. Using elementary row operations, we get tableau III:

### SIMPLEX TABLEAU III

$$
\begin{array}{c}
x_2 \\ x_3 \\ Z
\end{array}
\begin{array}{c}
x_1\ \ x_2\ \ x_3\ \ \ s_1\ \ s_2\ \ Z\ \ \ b \\
\left[\begin{array}{cccccc|c}
\frac{1}{2} & 1 & 0 & \frac{1}{2} & 0 & 0 & 5 \\
1 & 0 & 1 & -1 & 1 & 0 & 0 \\
\hline
\frac{1}{2} & 0 & 0 & \frac{1}{2} & \frac{3}{2} & 1 & 20
\end{array}\right]
\end{array}
$$

indicators

Since all indicators are nonnegative, $Z$ is maximum when $x_2 = 5$, $x_3 = 0$, and $x_1 = s_1 = s_2 = 0$. The maximum value is $Z = 20$. Note that this value is the same as that value of $Z$ corresponding to tableau II. In degenerate problems, it is possible to arrive at the same value of $Z$ at various stages of the simplex process. In Exercise 7.4 you are asked to solve this example by using $s_2$ as the departing variable in the initial tableau. ∎

Because of its mechanical nature, the simplex procedure is readily adaptable to computers to solve linear programming problems involving many variables and constraints.

## ■ Exercise 7.4

*Use the simplex method to solve the following problems.*

**1.** Maximize
$$Z = x_1 + 2x_2,$$
subject to
$$2x_1 + x_2 \le 8,$$
$$2x_1 + 3x_2 \le 12,$$
$$x_1, x_2 \ge 0.$$

**2.** Maximize
$$Z = 2x_1 + x_2,$$
subject to
$$-x_1 + x_2 \le 4,$$
$$x_1 + x_2 \le 6,$$
$$x_1, x_2 \ge 0.$$

**3.** Maximize
$$Z = -x_1 + 3x_2,$$
subject to
$$x_1 + x_2 \le 6,$$
$$-x_1 + x_2 \le 4,$$
$$x_1, x_2 \ge 0.$$

**4.** Maximize
$$Z = 3x_1 + 8x_2,$$
subject to
$$x_1 + 2x_2 \le 8,$$
$$x_1 + 6x_2 \le 12,$$
$$x_1, x_2 \ge 0.$$

**5.** Maximize
$$Z = 8x_1 + 2x_2,$$
subject to
$$x_1 - x_2 \le 1,$$
$$x_1 + 2x_2 \le 8,$$
$$x_1 + x_2 \le 5,$$
$$x_1, x_2 \ge 0.$$

**6.** Maximize
$$Z = 2x_1 - 6x_2,$$
subject to
$$x_1 - x_2 \le 4,$$
$$-x_1 + x_2 \le 4,$$
$$x_1 + x_2 \le 6,$$
$$x_1, x_2 \ge 0.$$

**7.** Solve the problem in Example 2 by choosing $s_2$ as the departing variable in tableau I.

**8.** Maximize
$$Z = 2x_1 - x_2 + x_3,$$
subject to
$$2x_1 + x_2 - x_3 \le 4,$$
$$x_1 + x_2 + x_3 \le 2,$$
$$x_1, x_2, x_3 \ge 0.$$

**9.** Maximize
$$Z = 2x_1 + x_2 - x_3,$$
subject to
$$x_1 + x_2 \le 1,$$
$$x_1 - 2x_2 - x_3 \ge -2,$$
$$x_1, x_2, x_3 \ge 0.$$

**10.** Maximize
$$Z = -x_1 + 2x_2,$$
subject to
$$x_1 + x_2 \le 1,$$
$$x_1 - x_2 \le 1,$$
$$x_1 - x_2 \ge -2,$$
$$x_1 \le 2,$$
$$x_1, x_2 \ge 0.$$

**11.** Maximize
$$Z = x_1 + x_2,$$
subject to
$$x_1 - x_2 \le 4,$$
$$-x_1 + x_2 \le 4,$$
$$8x_1 + 5x_2 \le 40,$$
$$2x_1 + x_2 \le 6,$$
$$x_1, x_2 \ge 0.$$

**12.** Maximize
$$W = 2x_1 + x_2 - 2x_3,$$
subject to
$$-2x_1 + x_2 + x_3 \ge -2,$$
$$x_1 - x_2 + x_3 \le 4,$$
$$x_1 + x_2 + 2x_3 \le 6,$$
$$x_1, x_2, x_3 \ge 0.$$

**13.** Maximize
$$W = x_1 - 12x_2 + 4x_3,$$
subject to
$$4x_1 + 3x_2 - x_3 \le 1,$$
$$x_1 + x_2 - x_3 \ge -2,$$
$$-x_1 + x_2 + x_3 \ge -1,$$
$$x_1, x_2, x_3 \ge 0.$$

**14.** Maximize
$$W = 4x_1 + 0x_2 - x_3,$$
subject to
$$x_1 + x_2 + x_3 \le 6,$$
$$x_1 - x_2 + x_3 \le 10,$$
$$x_1 - x_2 - x_3 \le 4,$$
$$x_1, x_2, x_3 \ge 0.$$

**15.** Maximize
$$Z = 60x_1 + 0x_2 + 90x_3 + 0x_4,$$
subject to
$$x_1 - 2x_2 \le 2,$$
$$x_1 + x_2 \le 5,$$
$$x_3 + x_4 \le 4,$$
$$x_3 - 2x_4 \le 7,$$
$$x_1, x_2, x_3, x_4 \ge 0.$$

**16.** Maximize
$$Z = 4x_1 + 10x_2 - 6x_3 - x_4,$$
subject to
$$x_1 + x_3 - x_4 \le 1,$$
$$x_1 - x_2 + x_4 \le 2,$$
$$x_1 + x_2 - x_3 + x_4 \le 4,$$
$$x_1, x_2, x_3, x_4 \ge 0.$$

17. **Freight Shipments** A freight company handles shipments by two corporations, A and B, that are located in the same city. Corporation A ships boxes that weigh 3 lb each and have a volume of 2 ft³; B ships 1 ft³ boxes that weigh 5 lb each. Both A and B ship to the same destination. The transportation cost for each box from A is $0.75, and from B it is $0.50. The freight company has a truck with 2400 ft³ of cargo space and a maximum capacity of 9200 lb. In one haul, how many boxes from each corporation should be transported by this truck so that the freight company receives maximum revenue? What is the maximum revenue?

18. **Production** A company manufactures three products: X, Y, and Z. Each product requires machine time and finishing time as shown in the following table:

|   | Machine Time | Finishing Time |
|---|---|---|
| X | 1 hr | 4 hr |
| Y | 2 hr | 4 hr |
| Z | 3 hr | 8 hr |

The numbers of hours of machine time and finishing time available per month are 900 and 5000, respectively. The unit profit on X, Y, and Z is $3, $4, and $6, respectively. What is the maximum profit per month that can be obtained?

19. **Production** A company manufactures three types of patio furniture: chairs, rockers, and chaise lounges. Each requires wood, plastic, and aluminum as shown in the following table:

|   | Wood | Plastic | Aluminum |
|---|---|---|---|
| Chair | 1 unit | 1 unit | 2 units |
| Rocker | 1 unit | 1 unit | 3 units |
| Chaise lounge | 1 unit | 2 units | 5 units |

The company has available 400 units of wood, 500 units of plastic, and 1450 units of aluminum. Each chair, rocker, and chaise lounge sells at $7, $8, and $12, respectively. Assuming that all furniture can be sold, determine a production order so that total revenue will be maximum. What is the maximum revenue?

---

OBJECTIVE

To consider the simplex method in relation to degeneracy, unbounded solutions, and multiple optimum solutions.

## 7.5 DEGENERACY, UNBOUNDED SOLUTIONS, AND MULTIPLE OPTIMUM SOLUTIONS[4]

### Degeneracy

In the preceding section, we stated that a B.F.S. is **degenerate** if, along with one of the nonbasic variables, one of the basic variables is 0. Suppose $x_1$, $x_2$, $x_3$, and $x_4$ are the variables in a degenerate B.F.S., where $x_1$ and $x_2$ are basic with $x_1 = 0$, $x_3$ and $x_4$ are nonbasic, and $x_3$ is the entering variable. The corresponding simplex tableau has the form

$$\begin{array}{c} \text{departing} \to \\ \text{variable} \end{array} \begin{array}{c} \\ x_1 \\ x_2 \\ Z \end{array} \begin{bmatrix} s & x_1 & x_2 & x_3 & x_4 & Z & b \\ 1 & 0 & a_{13} & a_{14} & 0 & 0 \\ 0 & 1 & a_{23} & a_{24} & 0 & a \\ \hline 0 & 0 & d_1 & d_2 & 1 & d_3 \end{bmatrix} \quad 0 \div a_{13} = 0.$$

indicators ↖

entering variable

Thus, the B.F.S. is

$$x_1 = 0, \qquad x_2 = a, \qquad x_3 = 0, \qquad x_4 = 0.$$

Suppose $a_{13} > 0$. Then the smaller quotient is 0, and we can choose $a_{13}$ as the pivot entry. Therefore, $x_1$ is the departing variable. Elementary row operations

---

[4]This section may be omitted.

give the following tableau, where the question marks represent numbers to be determined:

$$
\begin{array}{c}
\phantom{x_3} \\
x_3 \\
x_2 \\
Z
\end{array}
\begin{array}{c}
\begin{array}{cccccc}
x_1 & x_2 & x_3 & x_4 & Z & b
\end{array} \\
\left[
\begin{array}{ccccc|c}
? & 0 & 1 & ? & 0 & 0 \\
? & 1 & 0 & ? & 0 & a \\
\hline
? & 0 & 0 & ? & 1 & d_3
\end{array}
\right].
\end{array}
$$

For the B.F.S. corresponding to this tableau, $x_3$ and $x_2$ are basic variables, and $x_1$ and $x_4$ are nonbasic. The B.F.S. is

$$
x_3 = 0, \qquad x_2 = a, \qquad x_1 = 0, \qquad x_4 = 0,
$$

which is the same B.F.S. as before. Actually, these are usually considered different B.F.S.'s, with the only distinction being that $x_1$ is basic in the first B.F.S., whereas in the second it is nonbasic. The value of $Z$ for both B.F.S.'s is the same: $d_3$. Thus, no "improvement" in $Z$ is obtained.

In a degenerate situation, some problems may develop in the simplex procedure. It is possible to obtain a sequence of tableaus that correspond to B.F.S.'s which give the same $Z$-value. Moreover, we may eventually return to the first tableau in the sequence. In Fig. 7.24, we arrive at B.F.S.$_1$, proceed to B.F.S.$_2$, go on to B.F.S.$_3$, and finally return to B.F.S.$_1$. This is called *cycling*. When cycling occurs, we may never obtain the optimum value of $Z$. This phenomenon rarely is encountered in practical linear programming problems; however, there are techniques (which will not be considered in this text) for resolving such difficulties.

A degenerate B.F.S. will occur when two quotients in a simplex tableau tie for being the smallest. For example, consider the following (partial) tableau:

$$
\begin{array}{c}
\phantom{x_1} \\
x_1 \\
x_2
\end{array}
\begin{array}{c}
\begin{array}{cc}
x_3 & \quad \textit{Quotients}
\end{array} \\
\left[
\begin{array}{c|cc}
q_1 & p_1 & p_1/q_1, \\
q_2 & p_2 & p_2/q_2.
\end{array}
\right]
\end{array}
$$

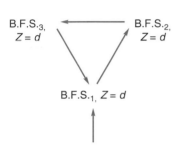

B.F.S.$_3$, $Z = d$    B.F.S.$_2$, $Z = d$

B.F.S.$_1$, $Z = d$

**FIGURE 7.24**  Cycling.

Here $x_1$ and $x_2$ are basic variables. Suppose $x_3$ is nonbasic and entering, and $p_1/q_1$ and $p_2/q_2$ are equal and also the smallest quotients involved. Choosing $q_1$ as the pivot entry, we obtain, by elementary row operations,

$$
\begin{array}{c}
\phantom{x_3} \\
x_3 \\
\\
x_2
\end{array}
\begin{array}{c}
\begin{array}{c}
x_3
\end{array} \\
\left[
\begin{array}{c|c}
1 & p_1/q_1 \\
\\
0 & p_2 - q_2 \dfrac{p_1}{q_1}
\end{array}
\right].
\end{array}
$$

Since $p_1/q_1 = p_2/q_2$, $p_2 - q_2(p_1/q_1) = 0$. Thus, the B.F.S. corresponding to this tableau has $x_2 = 0$, which gives a *degenerate* B.F.S. Although such a B.F.S. may produce cycling, we shall not encounter such situations in this book.

### Unbounded Solutions

We now turn our attention to "unbounded problems." In Sec. 7.2, we saw that a linear programming problem may have no maximum value because the feasible region is such that the objective function may become arbitrarily large therein. In this case, the problem is said to have an **unbounded solution.** This is a way of saying specifically that no optimum solution exists. Such a situation occurs when no quotients are possible in a simplex tableau for an entering variable. For example, consider the following tableau:

$$
\begin{array}{c}
\phantom{x_1} \\
x_1 \\
x_3 \\
Z
\end{array}
\begin{array}{c}
x_1 \quad x_2 \quad x_3 \quad x_4 \quad Z \quad\; b \\
\left[\begin{array}{ccccc|c}
1 & -3 & 0 & 2 & 0 & 5 \\
0 & 0 & 1 & 4 & 0 & 1 \\
\hline
0 & -5 & 0 & -2 & 1 & 10
\end{array}\right]
\end{array}
\begin{array}{l}
\text{no quotient.} \\
\text{no quotient.}
\end{array}
$$

$\uparrow$ indicators

entering
variable

Here $x_2$ is the entering variable, and for each one-unit increase in $x_2$, $Z$ increases by 5. Since there are no positive entries in the first two rows of the $x_2$ column, no quotients exist. From rows 1 and 2 we get

$$x_1 = 5 + 3x_2 - 2x_4$$

and

$$x_3 = 1 - 4x_4.$$

If we try to proceed to the next B.F.S., what is an upper bound on $x_2$? In that B.F.S., $x_4$ will remain nonbasic ($x_4 = 0$). Thus, $x_1 = 5 + 3x_2$ and $x_3 = 1$. Since $x_1 \geq 0$, $x_2 \geq -\frac{5}{3}$. Therefore, there is no upper bound on $x_2$. Hence, $Z$ can be arbitrarily large, and we have an unbounded solution. In general:

> If no quotients exist in a simplex tableau, then the linear programming problem has an unbounded solution.

### EXAMPLE 1   Unbounded Solution

*Maximize $Z = x_1 + 4x_2 - x_3$, subject to*

$$-5x_1 + 6x_2 - 2x_3 \leq 30,$$
$$-x_1 + 3x_2 + 6x_3 \leq 12,$$

*and $x_1, x_2, x_3 \geq 0$.*

**Solution:** The initial simplex tableau is

$$
\begin{array}{c}
\phantom{s_1} \\
s_1 \\
s_2 \\
Z
\end{array}
\begin{array}{c}
x_1 \quad x_2 \quad x_3 \quad s_1 \quad s_2 \quad Z \quad\; b \\
\left[\begin{array}{cccccc|c}
-5 & 6 & -2 & 1 & 0 & 0 & 30 \\
-1 & 3 & 6 & 0 & 1 & 0 & 12 \\
\hline
-1 & -4 & 1 & 0 & 0 & 1 & 0
\end{array}\right]
\end{array}
\begin{array}{l}
Quotients \\
30 \div 6 = 5. \\
12 \div 3 = 4.
\end{array}
$$

departing $\rightarrow$
variable

$\nearrow$ indicators

entering
variable

The second tableau is

$$
\begin{array}{c c}
& \begin{array}{c c c c c c c}
x_1 & x_2 & x_3 & s_1 & s_2 & Z & b
\end{array} \\
\begin{array}{c}
s_1 \\ x_2 \\ Z
\end{array} &
\left[
\begin{array}{c c c c c c | c}
-3 & 0 & -14 & 1 & -2 & 0 & 6 \\
-\frac{1}{3} & 1 & 2 & 0 & \frac{1}{3} & 0 & 4 \\
\hline
-\frac{7}{3} & 0 & 9 & 0 & \frac{4}{3} & 1 & 16
\end{array}
\right]
\end{array}
\begin{array}{l}
\text{no quotient.} \\
\text{no quotient.}
\end{array}
$$

$\underbrace{\phantom{xxxxxxxxxxxx}}$

$\uparrow$  indicators

entering
variable

Here the entering variable is $x_1$. Since the entries in the first two rows of the $x_1$-column are negative, no quotients exist. Hence, the problem has an unbounded solution.    ■

## Multiple Optimum Solutions

We conclude this section with a discussion of "multiple optimum solutions." Suppose that

$$x_1 = a_1, \quad x_2 = a_2, \quad \ldots, \quad x_n = a_n$$

and

$$x_1 = b_1, \quad x_2 = b_2, \quad \ldots, \quad x_n = b_n$$

are two *different* B.F.S.'s for which a linear programming problem is optimum. By "different B.F.S.'s," we mean that $a_i \neq b_i$ for some $i$, where $1 \leq i \leq n$. It can be shown that the values

$$
\begin{aligned}
x_1 &= (1-t)a_1 + tb_1, \\
x_2 &= (1-t)a_2 + tb_2, \\
&\;\;\vdots \\
x_n &= (1-t)a_n + tb_n,
\end{aligned}
\tag{1}
$$

for any $t$ such that $0 \leq t \leq 1$,

also give an optimum solution (although it may not necessarily be a B.F.S.). Thus, there are *multiple (optimum) solutions* to the problem.

We can determine the possibility of multiple optimum solutions from a simplex tableau that gives an optimum solution, such as the following (partial) tableau:

$$
\begin{array}{c c}
& \begin{array}{c c c c c}
x_1 & x_2 & x_3 & x_4 & Z
\end{array} \\
\begin{array}{c}
x_1 \\ x_2 \\ Z
\end{array} &
\left[
\begin{array}{c c c c c | c}
& & & & & p_1 \\
& & & & & q_1 \\
\hline
0 & 0 & a & 0 & 1 & r
\end{array}
\right]
\end{array}.
$$

$\underbrace{\phantom{xxxxxxxxxxxx}}$

indicators

Here $a$ must be nonnegative. The corresponding B.F.S. is

$$x_1 = p_1, \quad x_2 = q_1, \quad x_3 = 0, \quad x_4 = 0,$$

and the maximum value of $Z$ is $r$. If $x_4$ were to become basic, the indicator 0 in the $x_4$-column means that for each one-unit increase in $x_4$, $Z$ does not change.

Thus, we can find a B.F.S. in which $x_4$ is basic and the corresponding $Z$-value is the same as before. This is done by treating $x_4$ as an entering variable in the preceding tableau. If, for instance, $x_1$ is the departing variable, the new B.F.S. has the form

$$x_1 = 0, \qquad x_2 = q_2, \qquad x_3 = 0, \qquad x_4 = p_2.$$

If this B.F.S. is different from the previous one, multiple solutions exist. In fact, from Eqs. (1), an optimum solution is given by any values of $x_1, x_2, x_3,$ and $x_4,$ such that

$$x_1 = (1 - t)p_1 + t \cdot 0 = (1 - t)p_1,$$
$$x_2 = (1 - t)q_1 + tq_2,$$
$$x_3 = (1 - t) \cdot 0 + t \cdot 0 = 0,$$
$$x_4 = (1 - t) \cdot 0 + tp_2 = tp_2,$$

where $0 \le t \le 1$.

Note that when $t = 0$, we get the first optimum B.F.S.; when $t = 1$, we get the second. Of course, it may be possible to repeat the procedure by using the tableau corresponding to the last B.F.S. and obtain more optimum solutions by using Eqs. (1).

In general:

> In a tableau that gives an optimum solution, a zero indicator for a non-basic variable suggests the possibility of multiple optimum solutions.

### Principles in Practice 1
**Multiple Solutions**

A company produces three kinds of devices requiring three different production procedures. The company has allocated a total of 190 hours for procedure 1, 180 hours for procedure 2, and 165 hours for procedure 3. The following table gives the number of hours per device for

|  | Device 1 | Device 2 | Device 3 |
|---|---|---|---|
| Procedure 1 | 5.5 | 5.5 | 6.5 |
| Procedure 2 | 3.5 | 6.5 | 7.5 |
| Procedure 3 | 4.5 | 6.0 | 6.5 |

each procedure. If the profit is $50 per device 1, $50 per device 2, and $50 per device 3, find the number of devices of each kind the company should produce to maximize profit. Enter the initial tableau into a matrix on your graphics calculator, and perform the necessary row operations to find the answer. Round answers to the nearest whole number.

**EXAMPLE 2**  **Multiple Solutions**

*Maximize $Z = -x_1 + 4x_2 + 6x_3,$ subject to*

$$x_1 + 2x_2 + 3x_3 \le 6,$$
$$-2x_1 - 5x_2 + x_3 \le 10,$$

*and $x_1, x_2, x_3 \ge 0$.*

*Solution:* Our initial simplex tableau is

|  | $x_1$ | $x_2$ | $x_3$ | $s_1$ | $s_2$ | $Z$ | $b$ | Quotients |
|---|---|---|---|---|---|---|---|---|
| departing variable → $s_1$ | 1 | 2 | 3 | 1 | 0 | 0 | 6 | $6 \div 3 = 2.$ |
| $s_2$ | $-2$ | $-5$ | 1 | 0 | 1 | 0 | 10 | $10 \div 1 = 10.$ |
| $Z$ | 1 | $-4$ | $-6$ | 0 | 0 | 1 | 0 | |

↑ indicators
entering variable

Since there is a negative indicator, we continue and obtain

|  | $x_1$ | $x_2$ | $x_3$ | $s_1$ | $s_2$ | $Z$ | $b$ | Quotients |
|---|---|---|---|---|---|---|---|---|
| departing variable → $x_3$ | $\frac{1}{3}$ | $\frac{2}{3}$ | 1 | $\frac{1}{3}$ | 0 | 0 | 2 | $2 \div \frac{2}{3} = 3.$ |
| $s_2$ | $-\frac{7}{3}$ | $-\frac{17}{3}$ | 0 | $-\frac{1}{3}$ | 1 | 0 | 8 | no quotient. |
| $Z$ | 3 | 0 | 0 | 2 | 0 | 1 | 12 | |

↑ indicators
entering variable

All indicators are nonnegative: hence, an optimum solution occurs for the B.F.S.

$$x_3 = 2, \qquad s_2 = 8, \qquad x_1 = 0, \qquad x_2 = 0, \qquad s_1 = 0,$$

and the maximum value of $Z$ is 12. However, since $x_2$ is a nonbasic variable and its indicator is 0, we shall check for multiple solutions. Treating $x_2$ as an entering variable, we obtain the following tableau:

$$
\begin{array}{c}
\begin{array}{ccccccc} & x_1 & x_2 & x_3 & s_1 & s_2 & Z & b \end{array} \\
\begin{array}{c} x_2 \\ s_2 \\ \\ Z \end{array}
\left[
\begin{array}{cccccc|c}
\frac{1}{2} & 1 & \frac{3}{2} & \frac{1}{2} & 0 & 0 & 3 \\
\frac{1}{2} & 0 & \frac{17}{2} & \frac{5}{2} & 1 & 0 & 25 \\
\hline
3 & 0 & 0 & 2 & 0 & 1 & 12
\end{array}
\right].
\end{array}
$$

The B.F.S. here is

$$x_2 = 3, \qquad s_2 = 25, \qquad x_1 = 0, \qquad x_3 = 0, \qquad s_1 = 0$$

(for which $Z = 12$, as before) and is different from the previous one. Thus, multiple solutions exist. Since we are concerned only with values of the structural variables, we have an optimum solution,

$$x_1 = (1 - t) \cdot 0 + t \cdot 0 = 0,$$
$$x_2 = (1 - t) \cdot 0 + t \cdot 3 = 3t,$$
$$x_3 = (1 - t) \cdot 2 + t \cdot 0 = 2(1 - t),$$

for each value of $t$ such that $0 \le t \le 1$. (For example, if $t = \frac{1}{2}$, then $x_1 = 0$, $x_2 = \frac{3}{2}$, and $x_3 = 1$ is an optimum solution.)

In the last B.F.S., $x_3$ is nonbasic and its indicator is 0. However, if we repeated the process for determining other optimum solutions, we would return to the second tableau. Therefore, our procedure gives no other optimum solutions. ∎

## ■ Exercise 7.5

*In each of Problems 1 and 2, does the linear programming problem associated with the given tableau have a degeneracy? If so, why?*

**1.**
$$
\begin{array}{c}
\begin{array}{ccccc} & x_1 & x_2 & s_1 & s_2 & Z \end{array} \\
\begin{array}{c} x_1 \\ s_2 \\ \\ Z \end{array}
\left[
\begin{array}{cccc|c}
1 & 2 & 4 & 0 & 0 & 6 \\
0 & 1 & 1 & 1 & 0 & 3 \\
\hline
0 & -3 & -2 & 0 & 1 & 10
\end{array}
\right].
\end{array}
$$
indicators

**2.**
$$
\begin{array}{c}
\begin{array}{cccccc} & x_1 & x_2 & x_3 & s_1 & s_2 & Z \end{array} \\
\begin{array}{c} s_1 \\ x_2 \\ \\ Z \end{array}
\left[
\begin{array}{ccccc|c}
2 & 0 & 2 & 1 & 1 & 0 & 4 \\
3 & 1 & 1 & 0 & 1 & 0 & 0 \\
\hline
-5 & 0 & 1 & 0 & -3 & 1 & 2
\end{array}
\right].
\end{array}
$$
indicators

*In Problems 3–11, use the simplex method.*

**3.** Maximize

$$Z = 2x_1 + 7x_2,$$

subject to
$$4x_1 - 3x_2 \le 4,$$
$$3x_1 - x_2 \le 6,$$
$$5x_1 \le 8,$$
$$x_1, x_2 \ge 0.$$

**4.** Maximize

$$Z = x_1 + x_2,$$

subject to
$$x_1 - x_2 \le 4,$$
$$-x_1 + x_2 \le 4,$$
$$8x_1 + 5x_2 \le 40,$$
$$x_1 + x_2 \le 6,$$
$$x_1, x_2 \ge 0.$$

**5.** Maximize

$$Z = 3x_1 - 3x_2,$$

subject to
$$x_1 - x_2 \le 4,$$
$$-x_1 + x_2 \le 4,$$
$$x_1 + x_2 \le 6,$$
$$x_1, x_2 \ge 0.$$

**6.** Maximize

$$Z = 4x_1 + x_2 + 2x_3,$$

subject to
$$x_1 - x_2 + 4x_3 \le 6,$$
$$x_1 - x_2 - x_3 \ge -4,$$
$$x_1 - 6x_2 + x_3 \le 8,$$
$$x_1, x_2, x_3 \ge 0.$$

**7.** Maximize

$$Z = 5x_1 + 6x_2 + x_3,$$

subject to

$$\begin{aligned} 9x_1 + 3x_2 - 2x_3 &\leq 5, \\ 4x_1 + 2x_2 - x_3 &\leq 2, \\ x_1 - 4x_2 + x_3 &\leq 3, \\ x_1, x_2, x_3 &\geq 0. \end{aligned}$$

**8.** Maximize

$$Z = 2x_1 + x_2 - 4x_3,$$

subject to

$$\begin{aligned} 6x_1 + 3x_2 - 3x_3 &\leq 10, \\ x_1 - x_2 + x_3 &\leq 1, \\ 2x_1 - x_2 + 2x_3 &\leq 12, \\ x_1, x_2, x_3 &\geq 0. \end{aligned}$$

**9.** Maximize

$$Z = 6x_1 + 2x_2 + x_3,$$

subject to

$$\begin{aligned} 2x_1 + x_2 + x_3 &\leq 7, \\ -4x_1 - x_2 &\geq -6, \\ x_1, x_2, x_3 &\geq 0. \end{aligned}$$

**10.** Maximize

$$P = 4x_1 + 3x_2 + 2x_3 + x_4,$$

subject to

$$\begin{aligned} x_1 - x_2 &\leq 5, \\ x_2 - x_3 &\leq 2, \\ x_2 - 2x_3 + x_4 &\leq 4, \\ x_1, x_2, x_3, x_4 &\geq 0. \end{aligned}$$

**11. Production** A company manufactures three types of patio furniture: chairs, rockers, and chaise lounges. Each requires wood, plastic, and aluminum as given in the following table:

|  | **Wood** | **Plastic** | **Aluminum** |
|---|---|---|---|
| Chair | 1 unit | 1 unit | 2 units |
| Rocker | 1 unit | 1 unit | 3 units |
| Chaise lounge | 1 unit | 2 units | 5 units |

The company has available 400 units of wood, 600 units of plastic, and 1500 units of aluminum. Each chair, rocker, and chaise lounge sells at \$6, \$8, and \$12, respectively. Assuming that all furniture can be sold, what is the maximum total revenue that can be obtained? Determine the possible production orders that will generate this revenue.

---

**To handle maximization problems that are not of standard form by introducing artificial variables.**

## 7.6 Artificial Variables

To initiate the simplex method, a B.F.S. is required. For a standard linear programming problem, we begin with the B.F.S. in which all structural variables are zero. However, for a maximization problem that is not of standard form, such a B.F.S. may not exist. In this section, we will learn how the simplex method is used in such situations.

Let us consider the following problem:

$$\text{Maximize } Z = x_1 + 2x_2,$$

subject to

$$x_1 + x_2 \leq 9, \tag{1}$$

$$x_1 - x_2 \geq 1, \tag{2}$$

and $x_1, x_2 \geq 0$. Since constraint (2) cannot be written as $a_1x_1 + a_2x_2 \leq b$, where $b$ is nonnegative, this problem cannot be put into standard form. Note that $(0, 0)$ is not a feasible point. To solve the problem, we begin by writing constraints (1) and (2) as equations. Constraint (1) becomes

$$x_1 + x_2 + s_1 = 9, \tag{3}$$

where $s_1 \geq 0$ is a slack variable. For constraint (2), $x_1 - x_2$ will equal 1 if we *subtract* a nonnegative slack variable $s_2$ from $x_1 - x_2$. That is, by subtracting $s_2$, we are making up for the "surplus" on the left side of (2), so that we have equality. Thus,

$$x_1 - x_2 - s_2 = 1, \tag{4}$$

where $s_2 \geq 0$. We can now restate the problem:

$$\text{Maximize } Z = x_1 + 2x_2, \tag{5}$$

subject to

$$x_1 + x_2 + s_1 = 9, \tag{6}$$

$$x_1 - x_2 - s_2 = 1, \tag{7}$$

and $x_1, x_2, s_1, s_2 \geq 0$.

Since $(0, 0)$ is not in the feasible region, we do not have a B.F.S. in which $x_1 = x_2 = 0$. In fact, if $x_1 = 0$ and $x_2 = 0$ are substituted into Eq. (7), then $0 - 0 - s_2 = 1$, which gives $s_2 = -1$. But this contradicts the condition that $s_2 \geq 0$.

To get the simplex method started, we need an initial B.F.S. Although none is obvious, there is an ingenious method to arrive at one *artificially*. It requires that we consider a related linear programming problem called the *artificial problem*. First, a new equation is formed by adding a nonnegative variable $t$ to the left side of the equation in which the coefficient of the slack variable is $-1$. The variable $t$ is called an **artificial variable.** In our case, we replace Eq. (7) by $x_1 - x_2 - s_2 + t = 1$. Thus, Eqs. (6) and (7) become

$$x_1 + x_2 + s_1 = 9, \tag{8}$$

$$x_1 - x_2 - s_2 + t = 1, \tag{9}$$

where $x_1, x_2, s_1, s_2, t \geq 0$.

An obvious solution to Eqs. (8) and (9) is found by setting $x_1$, $x_2$, and $s_2$ equal to 0. This gives

$$x_1 = x_2 = s_2 = 0, \qquad s_1 = 9, \qquad t = 1,$$

Note that these values do not satisfy Eqs. (6) and (7). However, it is clear that any solution of Eqs. (8) and (9) for which $t = 0$ will give a solution to Eqs. (6) and (7), and conversely.

We can eventually force $t$ to be 0 if we alter the original objective function. We define the **artificial objective function** to be

$$W = Z - Mt = x_1 + 2x_2 - Mt, \tag{10}$$

where the constant $M$ is a large positive number. We shall not worry about the particular value of $M$ and shall proceed to maximize $W$ by the simplex method. Since there are $m = 2$ constraints (excluding the nonnegativity conditions) and $n = 5$ variables in Eqs. (8) and (9), any B.F.S. must have at least $n - m = 3$ variables equal to zero. We start with the following B.F.S.:

$$x_1 = x_2 = s_2 = 0, \qquad s_1 = 9, \qquad t = 1. \tag{11}$$

In this initial B.F.S., the nonbasic variables are the structural variables and the slack variable with coefficient $-1$ in Eqs. (8) and (9). The corresponding value of $W$ is $W = x_1 + 2x_2 - Mt = -M$, which is "extremely" negative. A significant improvement in $W$ will occur if we can find another B.F.S. for which $t = 0$. Since the simplex method seeks better values of $W$ at each stage, we shall apply it until we reach such a B.F.S. if possible. That solution will be an initial B.F.S. for the original problem.

To apply the simplex method to the artificial problem, we first write Eq. (10) as

$$-x_1 - 2x_2 + Mt + W = 0. \tag{12}$$

The augmented coefficient matrix of Eqs. (8), (9), and (12) is

$$
\begin{array}{c}
\\
s_1 \\
t \\
\\
\end{array}
\begin{array}{c}
\begin{array}{cccccc}
x_1 & x_2 & s_1 & s_2 & t & W \\
\end{array} \\
\left[\begin{array}{cccccc|c}
1 & 1 & 1 & 0 & 0 & 0 & 9 \\
1 & -1 & 0 & -1 & 1 & 0 & 1 \\
\hdashline
-1 & -2 & 0 & 0 & M & 1 & 0
\end{array}\right].
\end{array}
\tag{13}
$$

An initial B.F.S. is given by (11). Notice that, from row 1, when $x_1 = x_2 = s_2 = 0$, we can directly read the value of $s_1$, namely, $s_1 = 9$. From row 2, we get $t = 1$. From row 3, $Mt + W = 0$. Since $t = 1$, $W = -M$. But in a simplex tableau we want the value of $W$ to appear in the last row and last column. This is not so in (13); thus, we modify that matrix.

To do this, we transform (13) into an equivalent matrix whose last row has the form

$$
\begin{array}{cccccc}
x_1 & x_2 & s_1 & s_2 & t & W \\
? & ? & 0 & ? & 0 & 1 \mid ?
\end{array}
$$

That is, the $M$ in the $t$-column is replaced by 0. As a result, if $x_1 = x_2 = s_2 = 0$, then $W$ equals the last entry. Proceeding to obtain such a matrix, we have

$$
\begin{array}{c}
\begin{array}{cccccc}
x_1 & x_2 & s_1 & s_2 & t & W \\
\end{array} \\
\left[\begin{array}{cccccc|c}
1 & 1 & 1 & 0 & 0 & 0 & 9 \\
1 & -1 & 0 & -1 & 1 & 0 & 1 \\
\hdashline
-1 & -2 & 0 & 0 & M & 1 & 0
\end{array}\right]
\end{array}
$$

$$
\xrightarrow{-MR_2 + R_3}
\begin{array}{c}
\begin{array}{cccccc}
x_1 & x_2 & s_1 & s_2 & t & W \\
\end{array} \\
\left[\begin{array}{cccccc|c}
1 & 1 & 1 & 0 & 0 & 0 & 9 \\
1 & -1 & 0 & -1 & 1 & 0 & 1 \\
\hdashline
-1-M & -2+M & 0 & M & 0 & 1 & -M
\end{array}\right].
\end{array}
$$

Let us now check things out. If $x_1 = 0$, $x_2 = 0$, and $s_2 = 0$, then from row 1 we get $s_1 = 9$, from row 2, $t = 1$; and from row 3, $W = -M$. Thus, we now have initial simplex tableau I:

SIMPLEX TABLEAU I

$$
\begin{array}{c}
\\
\\
\text{departing} \rightarrow \\
\text{variable} \\
\\
\end{array}
\begin{array}{c}
\\
s_1 \\
t \\
\\
W \\
\end{array}
\begin{array}{c}
\begin{array}{cccccc}
x_1 & x_2 & s_1 & s_2 & t & W \\
\end{array} \\
\left[\begin{array}{cccccc|c}
1 & 1 & 1 & 0 & 0 & 0 & 9 \\
\boxed{1} & -1 & 0 & -1 & 1 & 0 & 1 \\
\hdashline
-1-M & -2+M & 0 & M & 0 & 1 & -M
\end{array}\right]
\end{array}
\begin{array}{c}
\textit{Quotients} \\
9 \div 1 = 9. \\
1 \div 1 = 1. \\
\\
\end{array}
$$

↑      indicators<br>entering<br>variable

From this point, we can use the procedures of Sec. 7.4. Since $M$ is a large positive number, the most negative indicator is $-1 - M$. Thus, the entering variable is $x_1$. From the quotients, we choose $t$ as the departing variable. The pivot entry is shaded. Using elementary row operations to get 1 in the pivot position and 0's elsewhere in that column, we get tableau II:

### SIMPLEX TABLEAU II

$$
\begin{array}{c}
\text{departing} \\ \text{variable}
\end{array} \rightarrow
\begin{array}{c}
s_1 \\ x_1 \\ W
\end{array}
\left[
\begin{array}{cccccc|c}
x_1 & x_2 & s_1 & s_2 & t & W & \\
0 & 2 & 1 & 1 & -1 & 0 & 8 \\
1 & -1 & 0 & -1 & 1 & 0 & 1 \\
\hline
0 & -3 & 0 & -1 & 1+M & 1 & 1
\end{array}
\right]
\quad
\begin{array}{l}
\textit{Quotients} \\
8 \div 2 = 4. \\
\text{no quotient,} \\
\text{since } -1 \text{ is not} \\
\text{positive.}
\end{array}
$$

$\uparrow$ indicators

entering
variable

From tableau II, we have the following B.F.S.:

$$ s_1 = 8, \qquad x_1 = 1, \qquad x_2 = 0, \qquad s_2 = 0, \qquad t = 0. $$

Since $t = 0$, the values $s_1 = 8, x_1 = 1, x_2 = 0$, and $s_2 = 0$ form an initial B.F.S. for the *original* problem! The artificial variable has served its purpose. For succeeding tableaus, we shall delete the $t$-column (since we want to solve the original problem) and change the $W$'s to $Z$'s (since $W = Z$ for $t = 0$). From tableau II, the entering variable is $x_2$, the departing variable is $s_1$, and the pivot entry is shaded. Using elementary row operations (omitting the $t$-column), we get tableau III:

### SIMPLEX TABLEAU III

$$
\begin{array}{c}
x_2 \\ x_1 \\ Z
\end{array}
\left[
\begin{array}{ccccc|c}
x_1 & x_2 & s_1 & s_2 & Z & \\
0 & 1 & \frac{1}{2} & \frac{1}{2} & 0 & 4 \\
1 & 0 & \frac{1}{2} & -\frac{1}{2} & 0 & 5 \\
\hline
0 & 0 & \frac{3}{2} & \frac{1}{2} & 1 & 13
\end{array}
\right]
$$

indicators

Since all the indicators are nonnegative, the maximum value of $Z$ is 13. It occurs when $x_1 = 5$ and $x_2 = 4$.

*Here is a summary of the procedure involving artificial variables.*

It is worthwhile to review the steps we performed to solve our problem:

$$ \text{Maximize } Z = x_1 + 2x_2, $$

subject to

$$ x_1 + x_2 \le 9, \tag{14} $$

$$ x_1 - x_2 \ge 1, \tag{15} $$

and $x_1 \ge 0, x_2 \ge 0$. We write Eq. (14) as

$$ x_1 + x_2 + s_1 = 9. \tag{16} $$

Since Eq. (15) involves the symbol $\ge$, and the constant on the right side is nonnegative, we write Eq. (15) in a form having both a slack variable (with coefficient $-1$) and an artificial variable:

$$ x_1 - x_2 - s_2 + t = 1. \tag{17} $$

The artificial objective equation to consider is $W = x_1 + 2x_2 - Mt$, or equivalently,

$$ -x_1 - 2x_2 + Mt + W = 0. \tag{18} $$

The augmented coefficient matrix of the system formed by Eqs. (16)–(18) is

$$
\begin{array}{ccccccc}
x_1 & x_2 & s_1 & s_2 & t & W & \\
\end{array}
$$

$$
\left[
\begin{array}{cccccc|c}
1 & 1 & 1 & 0 & 0 & 0 & 9 \\
1 & -1 & 0 & -1 & 1 & 0 & 1 \\
\hline
-1 & -2 & 0 & 0 & M & 1 & 0
\end{array}
\right].
$$

Next, we remove the $M$ from the artificial variable column and replace it by 0 by using elementary row operations. The resulting simplex tableau I corresponds to the initial B.F.S. of the artificial problem in which the structural variables, $x_1$ and $x_2$, and the slack variable, $s_2$ (the one associated with the constraint involving the symbol $\geq$), are each 0:

**SIMPLEX TABLEAU I**

$$
\begin{array}{c}
\\
s_1 \\
t \\
\\
W
\end{array}
\begin{array}{ccccccc}
x_1 & & x_2 & s_1 & s_2 & t & W \\
\end{array}
\left[
\begin{array}{cccccc|c}
1 & & 1 & 1 & 0 & 0 & 0 & 9 \\
1 & & -1 & 0 & -1 & 1 & 0 & 1 \\
\hline
-1-M & & -2+M & 0 & M & 0 & 1 & -M
\end{array}
\right].
$$

The basic variables $s_1$ and $t$ on the left side of the tableau correspond to the non-structural variables in Eqs. (16) and (17) that have positive coefficients. We now apply the simplex method until we obtain a B.F.S. in which the artificial variable $t$ equals 0. Then we can delete the artificial variable column, change the $W$'s to $Z$'s, and continue the procedure until the maximum value of $Z$ is obtained.

### EXAMPLE 1   Artificial Variables

*Use the simplex method to maximize $Z = 2x_1 + x_2$, subject to*

$$x_1 + x_2 \leq 12, \tag{19}$$

$$x_1 + 2x_2 \leq 20, \tag{20}$$

$$-x_1 + x_2 \geq 2, \tag{21}$$

*and $x_1 \geq 0, x_2 \geq 0$.*

**Solution:** The equations for (19)–(21) will involve a total of three slack variables: $s_1$, $s_2$, and $s_3$. Since (21) contains the symbol $\geq$, and the constant on the right side is nonnegative, its equation will also involve an artificial variable $t$, and the coefficient of its slack variable $s_3$ will be $-1$. We thus have

$$x_1 + x_2 + s_1 \qquad\qquad = 12, \tag{22}$$

$$x_1 + 2x_2 \qquad + s_2 \qquad\qquad = 20, \tag{23}$$

$$-x_1 + x_2 \qquad\qquad - s_3 + t = 2. \tag{24}$$

We consider $W = Z - Mt = 2x_1 + x_2 - Mt$ as the artificial objective equation, or equivalently,

$$-2x_1 - x_2 + Mt + W = 0, \tag{25}$$

where $M$ is a large positive number. Now we construct the augmented coefficient matrix of Eqs. (22)–(25):

$$
\begin{array}{ccccccc}
x_1 & x_2 & s_1 & s_2 & s_3 & t & W \\
\end{array}
$$

$$
\left[
\begin{array}{ccccccc|c}
1 & 1 & 1 & 0 & 0 & 0 & 0 & 12 \\
1 & 2 & 0 & 1 & 0 & 0 & 0 & 20 \\
-1 & 1 & 0 & 0 & -1 & 1 & 0 & 2 \\
\hline
-2 & -1 & 0 & 0 & 0 & M & 1 & 0
\end{array}
\right].
$$

To get simplex tableau I, we replace the $M$ in the artificial variable column by zero by adding $-M$ times row 3 to row 4:

### SIMPLEX TABLEAU I

$$
\begin{array}{c}
\\
s_1 \\
s_2 \\
\text{departing} \rightarrow \; t \\
\text{variable} \\
W
\end{array}
\begin{array}{ccccccc|c}
x_1 & x_2 & s_1 & s_2 & s_3 & t & W & \\
1 & 1 & 1 & 0 & 0 & 0 & 0 & 12 \\
1 & 2 & 0 & 1 & 0 & 0 & 0 & 20 \\
-1 & 1 & 0 & 0 & -1 & 1 & 0 & 2 \\
\hline
-2+M & -1-M & 0 & 0 & M & 0 & 1 & -2M
\end{array}
\begin{array}{l}
Quotients \\
12 \div 1 = 12. \\
20 \div 2 = 10. \\
2 \div 1 = 2.
\end{array}
$$

$\uparrow$ indicators
entering
variable

The variables $s_1$, $s_2$, and $t$ on the left side of tableau I are the nonstructural variables with positive coefficients in Eqs. (22)–(24). Since $M$ is a large positive number, $-1-M$ is the most negative indicator. The entering variable is $x_2$, the departing variable is $t$, and the pivot entry is shaded. Proceeding, we get tableau II:

### SIMPLEX TABLEAU II

$$
\begin{array}{c}
\\
\text{departing} \rightarrow \; s_1 \\
\text{variable} \quad s_2 \\
x_2 \\
W
\end{array}
\begin{array}{ccccccc|c}
x_1 & x_2 & s_1 & s_2 & s_3 & t & W & \\
2 & 0 & 1 & 0 & 1 & -1 & 0 & 10 \\
3 & 0 & 0 & 1 & 2 & -2 & 0 & 16 \\
-1 & 1 & 0 & 0 & -1 & 1 & 0 & 2 \\
\hline
-3 & 0 & 0 & 0 & -1 & 1+M & 1 & 2
\end{array}
\begin{array}{l}
Quotients \\
10 \div 2 = 5. \\
16 \div 3 = 5\frac{1}{3}.
\end{array}
$$

$\uparrow$ indicators
entering
variable

The B.F.S. corresponding to tableau II has $t = 0$. Thus, we shall delete the $t$-column and change $W$'s to $Z$'s in succeeding tableaus. Continuing, we obtain tableau III:

### SIMPLEX TABLEAU III

$$
\begin{array}{c}
x_1 \\
s_2 \\
x_2 \\
Z
\end{array}
\begin{array}{cccccc|c}
x_1 & x_2 & s_1 & s_2 & s_3 & Z & \\
1 & 0 & \frac{1}{2} & 0 & \frac{1}{2} & 0 & 5 \\
0 & 0 & -\frac{3}{2} & 1 & \frac{1}{2} & 0 & 1 \\
0 & 1 & \frac{1}{2} & 0 & -\frac{1}{2} & 0 & 7 \\
\hline
0 & 0 & \frac{3}{2} & 0 & \frac{1}{2} & 1 & 17
\end{array}
$$

indicators

All indicators are nonnegative. Hence, the maximum value of $Z$ is 17. It occurs when $x_1 = 5$ and $x_2 = 7$. ∎

## Equality Constraints

When an *equality* constraint of the form

$$a_1 x_1 + a_2 x_2 + \cdots + a_n x_n = b, \qquad \text{where } b \geq 0,$$

occurs in a linear programming problem, artificial variables are used in the simplex method. To illustrate, consider the following problem:

$$\text{Maximize } Z = x_1 + 3x_2 - 2x_3,$$

subject to

$$x_1 + x_2 - x_3 = 6 \tag{26}$$

and $x_1, x_2, x_3 \geq 0$. Constraint (26) is already expressed as an equation, so no slack variable is necessary. Since $x_1 = x_2 = x_3 = 0$ is not a feasible solution, we do not have an obvious starting point for the simplex procedure. Thus, we create an artificial problem by first adding an artificial variable $t$ to the left side of Eq. (26):

$$x_1 + x_2 - x_3 + t = 6.$$

Here an obvious B.F.S. is $x_1 = x_2 = x_3 = 0, t = 6$. The artificial objective function is

$$W = Z - Mt = x_1 + 3x_2 - 2x_3 - Mt,$$

where $M$ is a large positive number. The simplex procedure is applied to this artificial problem until we obtain a B.F.S. in which $t = 0$. This solution will give an initial B.F.S. for the original problem, and we then proceed as before.

In general, the simplex method may be used to

$$\text{maximize } Z = c_1x_1 + c_2x_2 + \cdots + c_nx_n,$$

subject to

$$\left.\begin{array}{l} a_{11}x_1 + a_{12}x_2 + \cdots + a_{1n}x_n \ \{\leq, \geq, =\} \ b_1, \\ a_{21}x_1 + a_{22}x_2 + \cdots + a_{2n}x_n \ \{\leq, \geq, =\} \ b_2, \\ \ \vdots \qquad \ \vdots \qquad \qquad \ \vdots \qquad \qquad \quad \vdots \\ a_{m1}x_1 + a_{m2}x_2 + \cdots + a_{mn}x_n \ \{\leq, \geq, =\} \ b_m, \end{array}\right\} \tag{27}$$

where $x_1, x_2, \ldots, x_n$ and $b_1, b_2, \ldots, b_m$ are nonnegative. The symbolism $\{\leq, \geq, =\}$ means that one of the relations "$\leq$," "$\geq$," or "$=$" exists for a constraint. If all constraints involve "$\leq$", the problem is of standard form and the simplex techniques of the previous sections apply. If any constraint involves "$\geq$" or "$=$", we begin with an artificial problem, which is obtained as follows.

Each constraint that contains "$\leq$" is written as an equation involving a slack variable $s_i$ with coefficient $+1$:

$$a_{i1}x_1 + a_{i2}x_2 + \cdots + a_{in}x_n + s_i = b_i.$$

Each constraint that contains "$\geq$" is written as an equation involving a slack variable $s_j$ with coefficient $-1$ and an artificial variable $t_j$:

$$a_{j1}x_1 + a_{j2}x_2 + \cdots + a_{jn}x_n - s_j + t_j = b_j.$$

A nonnegative artificial variable $t_k$ is inserted into each equality constraint:

$$a_{k1}x_1 + a_{k2}x_2 + \cdots + a_{kn}x_n + t_k = b_k.$$

Should the artificial variables involved in this problem be, for example, $t_1, t_2$, and $t_3$, then the artificial objective function is

$$W = Z - Mt_1 - Mt_2 - Mt_3,$$

where $M$ is a large positive number. An initial B.F.S. occurs when $x_1 = x_2 = \cdots = x_n = 0$ and each slack variable with a coefficient of $-1$ equals 0. After obtaining an initial simplex tableau, we apply the simplex procedure until we arrive at a tableau that corresponds to a B.F.S. in which *all* artificial variables are 0. We then delete the artificial variable columns, change $W$'s to $Z$'s, and continue by using the procedures of the previous sections.

### EXAMPLE 2  An Equality Constraint

*Use the simplex method to maximize $Z = x_1 + 3x_2 - 2x_3$, subject to*

$$-x_1 - 2x_2 - 2x_3 = -6, \tag{28}$$

$$-x_1 - x_2 + x_3 \leq -2, \tag{29}$$

*and $x_1, x_2, x_3 \geq 0$.*

***Solution:*** Constraints (28) and (29) will have the forms indicated in (27) [that is, $b$'s positive] if we multiply both sides of each constraint by $-1$:

$$x_1 + 2x_2 + 2x_3 = 6, \tag{30}$$

$$x_1 + x_2 - x_3 \geq 2. \tag{31}$$

Since constraints (30) and (31) involve "$=$" and "$\geq$", two artificial variables, $t_1$ and $t_2$, will occur. The equations for the artificial problem are

$$x_1 + 2x_2 + 2x_3 \qquad + t_1 \quad = 6, \tag{32}$$

and

$$x_1 + x_2 - x_3 - s_2 \qquad + t_2 = 2. \tag{33}$$

Here the subscript 2 on $s_2$ reflects the order of the equations. The artificial objective function is $W = Z - Mt_1 - Mt_2$, or equivalently,

$$-x_1 - 3x_2 + 2x_3 + Mt_1 + Mt_2 + W = 0, \tag{34}$$

where $M$ is a large positive number. The augmented coefficient matrix of Eqs. (32)–(34) is

$$
\begin{array}{ccccccc}
x_1 & x_2 & x_3 & s_2 & t_1 & t_2 & W \\
\end{array}
$$
$$
\left[
\begin{array}{ccccccc|c}
1 & 2 & 2 & 0 & 1 & 0 & 0 & 6 \\
1 & 1 & -1 & -1 & 0 & 1 & 0 & 2 \\
\hline
-1 & -3 & 2 & 0 & M & M & 1 & 0
\end{array}
\right].
$$

We now use elementary row operations to remove the $M$'s from *all* the artificial variable columns. By adding $-M$ times row 1 to row 3 and adding $-M$ times row 2 to row 3, we get initial simplex tableau I:

### SIMPLEX TABLEAU I

| | $x_1$ | $x_2$ | $x_3$ | $s_2$ | $t_1$ | $t_2$ | $W$ | | Quotients |
|---|---|---|---|---|---|---|---|---|---|
| $t_1$ | 1 | 2 | 2 | 0 | 1 | 0 | 0 | 6 | $6 \div 2 = 3.$ |
| $t_2$ | 1 | 1 | -1 | -1 | 0 | 1 | 0 | 2 | $2 \div 1 = 2.$ |
| $W$ | $-1-2M$ | $-3-3M$ | $2-M$ | $M$ | 0 | 0 | 1 | $-8M$ | |

departing → $t_2$
variable

$\underbrace{\phantom{-1-2M \quad -3-3M \quad 2-M \quad M \quad 0 \quad 0}}$
indicators

$\uparrow$
entering
variable

Proceeding, we obtain simplex tableaus II and III:

SIMPLEX TABLEAU II

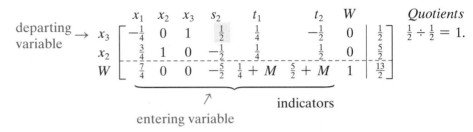

$$
\begin{array}{c}
\text{departing}\rightarrow \\
\text{variable}
\end{array}
\begin{array}{c}
t_1 \\
x_2 \\
W
\end{array}
\left[
\begin{array}{ccccccc|c}
x_1 & x_2 & x_3 & s_2 & t_1 & t_2 & W & \\
-1 & 0 & \boxed{4} & 2 & 1 & -2 & 0 & 2 \\
1 & 1 & -1 & -1 & 0 & 1 & 0 & 2 \\
\hline
2+M & 0 & -1-4M & -3-2M & 0 & 3+3M & 1 & 6-2M
\end{array}
\right]
\quad
\begin{array}{l}
\textit{Quotients} \\
2 \div 4 = \tfrac{1}{2}.
\end{array}
$$

↑ indicators
entering variable

SIMPLEX TABLEAU III

$$
\begin{array}{c}
\text{departing}\rightarrow \\
\text{variable}
\end{array}
\begin{array}{c}
x_3 \\
x_2 \\
W
\end{array}
\left[
\begin{array}{cccccc|c}
x_1 & x_2 & x_3 & s_2 & t_1 & t_2 & W \\
-\tfrac{1}{4} & 0 & 1 & \tfrac{1}{2} & \tfrac{1}{4} & -\tfrac{1}{2} & 0 & \tfrac{1}{2} \\
\tfrac{3}{4} & 1 & 0 & -\tfrac{1}{2} & \tfrac{1}{4} & \tfrac{1}{2} & 0 & \tfrac{5}{2} \\
\hline
\tfrac{7}{4} & 0 & 0 & -\tfrac{5}{2} & \tfrac{1}{4}+M & \tfrac{5}{2}+M & 1 & \tfrac{13}{2}
\end{array}
\right]
\quad
\begin{array}{l}
\textit{Quotients} \\
\tfrac{1}{2} \div \tfrac{1}{2} = 1.
\end{array}
$$

↗ indicators
entering variable

For the B.F.S. corresponding to tableau III, the artificial variables $t_1$ and $t_2$ are both 0. We now can delete the $t_1$- and $t_2$-columns and change $W$'s to $Z$'s. Continuing, we obtain simplex tableau IV:

SIMPLEX TABLEAU IV

$$
\begin{array}{c}
s_2 \\
x_2 \\
Z
\end{array}
\left[
\begin{array}{ccccc|c}
x_1 & x_2 & x_3 & s_2 & Z \\
-\tfrac{1}{2} & 0 & 2 & 1 & 0 & 1 \\
\tfrac{1}{2} & 1 & 1 & 0 & 0 & 3 \\
\hline
\tfrac{1}{2} & 0 & 5 & 0 & 1 & 9
\end{array}
\right].
$$

indicators

Since all indicators are nonnegative, we have reached the final tableau. The maximum value of $Z$ is 9, and it occurs when $x_1 = 0$, $x_2 = 3$, and $x_3 = 0$.   ∎

## Empty Feasible Regions

It is possible that the simplex procedure terminates and not all artificial variables are 0. It can be shown that in this situation *the feasible region of the original problem is empty,* and hence, there is *no optimum solution.* The following example will illustrate.

### EXAMPLE 3   An Empty Feasible Region

*Use the simplex method to maximize $Z = 2x_1 + x_2$, subject to*

$$-x_1 + x_2 \geq 2, \tag{35}$$

$$x_1 + x_2 \leq 1,$$

*and $x_1, x_2 \geq 0$.*

*Solution:* Since constraint (35) is of the form $a_{11}x_1 + a_{12}x_2 \geq b_1$, where $b_1 \geq 0$, an artificial variable will occur. The equations to consider are

$$-x_1 + x_2 - s_1 \qquad + t_1 = 2 \qquad (36)$$

and
$$x_1 + x_2 \qquad + s_2 \qquad = 1, \qquad (37)$$

where $s_1$ and $s_2$ are slack variables and $t_1$ is artificial. The artificial objective function is $W = Z - Mt_1$, or equivalently,

$$-2x_1 - x_2 + Mt_1 + W = 0. \qquad (38)$$

The augmented coefficient matrix of Eqs. (36)–(38) is

$$
\begin{array}{cccccc}
x_1 & x_2 & s_1 & s_2 & t_1 & W \\
\end{array}
$$
$$
\begin{bmatrix}
-1 & 1 & -1 & 0 & 1 & 0 & 2 \\
1 & 1 & 0 & 1 & 0 & 0 & 1 \\
\hdashline
-2 & -1 & 0 & 0 & M & 1 & 0
\end{bmatrix}.
$$

The simplex tableaus are as follows:

### SIMPLEX TABLEAU I

| | $x_1$ | $x_2$ | $s_1$ | $s_2$ | $t_1$ | $W$ | | Quotients |
|---|---|---|---|---|---|---|---|---|
| $t_1$ | $-1$ | $1$ | $-1$ | $0$ | $1$ | $0$ | $2$ | $2 \div 1 = 2.$ |
| departing → $s_2$ | $1$ | $1$ | $0$ | $1$ | $0$ | $0$ | $1$ | $1 \div 1 = 1.$ |
| $W$ | $-2 + M$ | $-1 - M$ | $M$ | $0$ | $0$ | $1$ | $-2M$ | |

departing variable

↑ indicators

entering variable

### SIMPLEX TABLEAU II

| | $x_1$ | $x_2$ | $s_1$ | $s_2$ | $t_1$ | $W$ | |
|---|---|---|---|---|---|---|---|
| $t_1$ | $-2$ | $0$ | $-1$ | $-1$ | $1$ | $0$ | $1$ |
| $x_2$ | $1$ | $1$ | $0$ | $1$ | $0$ | $0$ | $1$ |
| $W$ | $-1 + 2M$ | $0$ | $M$ | $1 + M$ | $0$ | $1$ | $1 - M$ |

indicators

Since $M$ is a large positive number, the indicators in simplex tableau II are nonnegative, so the simplex procedure terminates. The value of the artificial variable $t_1$ is 1. Therefore, as previously stated, the feasible region of the original problem is empty, and hence, no solution exists. This result can be obtained geometrically. Figure 7.25 shows the graphs of $-x_1 + x_2 = 2$ and $x_1 + x_2 = 1$ for $x_1, x_2 \geq 0$. Since there is no point $(x_1, x_2)$ that simultaneously lies above $-x_1 + x_2 = 2$ and below $x_1 + x_2 = 1$ such that $x_1, x_2 \geq 0$, the feasible region is empty, and thus, no solution exists. ∎

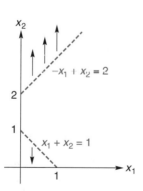

**FIGURE 7.25** Empty feasible region (no solution exists).

In the next section we shall use the simplex method on minimization problems.

## ■ Exercise 7.6

*Use the simplex method to solve the following problems.*

**1.** Maximize

$$Z = 2x_1 + x_2,$$

subject to
$$
\begin{aligned}
x_1 + x_2 &\leq 6, \\
-x_1 + x_2 &\geq 4, \\
x_1, x_2 &\geq 0.
\end{aligned}
$$

**2.** Maximize

$$Z = 3x_1 + 4x_2,$$

subject to
$$
\begin{aligned}
x_1 + 2x_2 &\leq 8, \\
x_1 + 6x_2 &\geq 12, \\
x_1, x_2 &\geq 0.
\end{aligned}
$$

**3.** Maximize

$$Z = 2x_1 + x_2 - x_3,$$

subject to
$$
\begin{aligned}
x_1 + 2x_2 + x_3 &\leq 5, \\
-x_1 + x_2 + x_3 &\geq 1, \\
x_1, x_2, x_3 &\geq 0.
\end{aligned}
$$

**4.** Maximize

$$Z = x_1 - x_2 + 4x_3,$$

subject to
$$
\begin{aligned}
x_1 + x_2 + x_3 &\leq 9, \\
x_1 - 2x_2 + x_3 &\geq 6, \\
x_1, x_2, x_3 &\geq 0.
\end{aligned}
$$

**5.** Maximize

$$Z = 4x_1 + x_2 + 2x_3,$$

subject to

$$2x_1 + x_2 + 3x_3 \leq 10,$$
$$x_1 - x_2 + x_3 = 4,$$
$$x_1, x_2, x_3 \geq 0.$$

**6.** Maximize

$$Z = x_1 + 2x_2 + 3x_3,$$

subject to

$$x_2 - 2x_3 \geq 5,$$
$$x_1 + x_2 + x_3 = 8,$$
$$x_1, x_2, x_3 \geq 0.$$

**7.** Maximize

$$Z = x_1 - 10x_2,$$

subject to

$$x_1 - x_2 \leq 1,$$
$$x_1 + 2x_2 \leq 8,$$
$$x_1 + x_2 \geq 5,$$
$$x_1, x_2 \geq 0.$$

**8.** Maximize

$$Z = x_1 + 4x_2 - x_3,$$

subject to

$$x_1 + x_2 - x_3 \geq 5,$$
$$x_1 + x_2 + x_3 \leq 3,$$
$$x_1 - x_2 + x_3 = 7,$$
$$x_1, x_2, x_3 \geq 0.$$

**9.** Maximize

$$Z = 3x_1 - 2x_2 + x_3,$$

subject to

$$x_1 + x_2 + x_3 \leq 1,$$
$$x_1 - x_2 + x_3 \geq 2,$$
$$x_1 - x_2 - x_3 \leq -6,$$
$$x_1, x_2, x_3 \geq 0.$$

**10.** Maximize

$$Z = x_1 + 4x_2,$$

subject to

$$x_1 + 2x_2 \leq 8,$$
$$x_1 + 6x_2 \geq 12,$$
$$x_2 \geq 2,$$
$$x_1, x_2 \geq 0.$$

**11.** Maximize

$$Z = -3x_1 + 2x_2,$$

subject to

$$x_1 - x_2 \leq 4,$$
$$-x_1 + x_2 = 4,$$
$$x_1 \geq 6,$$
$$x_1, x_2 \geq 0.$$

**12.** Maximize

$$Z = x_1 - 5x_2,$$

subject to

$$x_1 - 2x_2 \geq -13,$$
$$-x_1 + x_2 \geq 3,$$
$$x_1 + x_2 \geq 11,$$
$$x_1, x_2 \geq 0.$$

**13. Production** A company manufactures two types of bookcases: Standard and Executive. Each type requires assembly and finishing times as given in the following table:

| | Assembly Time | Finishing Time | Profit per Unit |
|---|---|---|---|
| Standard | 1 hr | 2 hr | $10 |
| Executive | 2 hr | 3 hr | 12 |

The profit on each unit is also indicated. The number of hours available per week in the assembly department is 400, and in the finishing department it is 510. Because of a union contract, the finishing department is guaranteed at least 240 hours of work per week. How many units of each type should the company produce each week to maximize profit?

**14. Production** A company manufactures three products: X, Y, and Z. Each product requires the use of time on machines A and B as given in the following table:

| | Machine A | Machine B |
|---|---|---|
| Product X | 1 hr | 1 hr |
| Product Y | 2 hr | 1 hr |
| Product Z | 2 hr | 2 hr |

The numbers of hours per week that A and B are available for production are 40 and 30, respectively. The profit per unit on X, Y, and Z is $50, $60, and $75, respectively. At least five units of Z must be produced next week. What should be the production order for that period if maximum profit is to be achieved? What is the maximum profit?

**15. Investments** The prospectus of an investment fund states that all money is invested in bonds that are rated A, AA, and AAA; no more than 30% of the total investment is in A and AA bonds, and at least 50% is in AA and AAA bonds. The A, AA, and AAA bonds respectively yield 8%, 7%, and 6% annually. Determine the percentages of the total investment that should be committed to each type of bond so that the fund maximizes its annual yield. What is this yield?

---

**OBJECTIVE**

**To show how to solve a minimization problem by altering the objective function so that a maximization problem results.**

## 7.7 MINIMIZATION

So far we have used the simplex method to *maximize* objective functions. In general, to *minimize,* a function it suffices to maximize the negative of the function. To understand why, consider the function $f(x) = x^2 - 4$. In Fig. 7.26(a), observe that the minimum value of $f$ is $-4$, and it occurs when $x = 0$. Figure 7.26(b) shows the graph of $g(x) = -f(x) = -(x^2 - 4)$. This graph is the reflection through the $x$-axis of the graph of $f$. Notice that the maximum value of $g$ is 4 and occurs when $x = 0$. Thus the minimum value of $x^2 - 4$ is the negative of the maximum value of $-(x^2 - 4)$. That is,

$$\min f = -\max(-f).$$

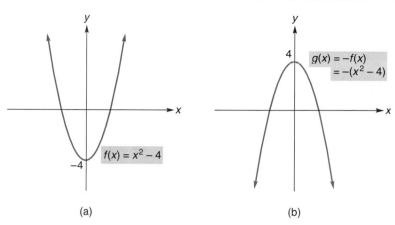

**FIGURE 7.26** Minimum value of $f(x)$ is equal to the negative of the maximum value of $-f(x)$.

The problem in Example 1 will be solved more efficiently in Example 4 of Sec. 7.8.

## EXAMPLE 1 Minimization

*Use the simplex method to minimize $Z = x_1 + 2x_2$, subject to*

$$-2x_1 + x_2 \geq 1, \qquad (1)$$

$$-x_1 + x_2 \geq 2, \qquad (2)$$

*and $x_1, x_2 \geq 0$.*

**Solution:** To minimize $Z$, we can maximize $-Z = -x_1 - 2x_2$. Note that constraints (1) and (2) each have the form $a_1x_1 + a_2x_2 \geq b$, where $b \geq 0$. Thus, their equations involve two slack variables $s_1$ and $s_2$, each with coefficient $-1$, and two artificial variables $t_1$ and $t_2$:

$$-2x_1 + x_2 - s_1 + t_1 = 1, \qquad (3)$$

$$-x_1 + x_2 - s_2 + t_2 = 2. \qquad (4)$$

Since there are *two* artificial variables, we maximize the objective function

$$W = (-Z) - Mt_1 - Mt_2,$$

where $M$ is a large positive number. Equivalently,

$$x_1 + 2x_2 + Mt_1 + Mt_2 + W = 0. \qquad (5)$$

The augmented coefficient matrix of Eqs. (3)–(5) is

$$
\begin{array}{ccccccc}
x_1 & x_2 & s_1 & s_2 & t_1 & t_2 & W \\
\end{array}
$$
$$
\left[
\begin{array}{ccccccc|c}
-2 & 1 & -1 & 0 & 1 & 0 & 0 & 1 \\
-1 & 1 & 0 & -1 & 0 & 1 & 0 & 2 \\
\hline
1 & 2 & 0 & 0 & M & M & 1 & 0 \\
\end{array}
\right].
$$

Proceeding, we obtain tableaus I, II, III:

### SIMPLEX TABLEAU I

|  | $x_1$ | $x_2$ | $s_1$ | $s_2$ | $t_1$ | $t_2$ | $W$ |  | Quotients |
|---|---|---|---|---|---|---|---|---|---|
| departing → $t_1$ | $-2$ | **1** | $-1$ | $0$ | $1$ | $0$ | $0$ | $1$ | $1 \div 1 = 1.$ |
| variable $\quad t_2$ | $-1$ | $1$ | $0$ | $-1$ | $0$ | $1$ | $0$ | $2$ | $2 \div 1 = 2.$ |
| $W$ | $1 + 3M$ | $2 - 2M$ | $M$ | $M$ | $0$ | $0$ | $1$ | $-3M$ |  |

↑    indicators

entering variable

SIMPLEX TABLEAU II

|  | $x_1$ | $x_2$ | $s_1$ | $s_2$ | $t_1$ | $t_2$ | $W$ |  | Quotients |
|---|---|---|---|---|---|---|---|---|---|
| $x_2$ | $-2$ | $1$ | $-1$ | $0$ | $1$ | $0$ | $0$ | $1$ | |
| departing → $t_2$ | $1$ | $0$ | $1$ | $-1$ | $-1$ | $1$ | $0$ | $1$ | $1 \div 1 = 1.$ |
| variable $W$ | $5 - M$ | $0$ | $2 - M$ | $M$ | $-2 + 2M$ | $0$ | $1$ | $-2 - M$ | |

$\underbrace{\phantom{5 - M \quad 0 \quad 2 - M \quad M \quad -2 + 2M}}$

↗  indicators

entering variable

SIMPLEX TABLEAU III

|  | $x_1$ | $x_2$ | $s_1$ | $s_2$ | $t_1$ | $t_2$ | $W$ |  |
|---|---|---|---|---|---|---|---|---|
| $x_2$ | $-1$ | $1$ | $0$ | $-1$ | $0$ | $1$ | $0$ | $2$ |
| $s_1$ | $1$ | $0$ | $1$ | $-1$ | $-1$ | $1$ | $0$ | $1$ |
| $W$ | $3$ | $0$ | $0$ | $2$ | $M$ | $-2 + M$ | $1$ | $-4$ |

$\underbrace{\phantom{3 \quad 0 \quad 0 \quad 2 \quad M \quad -2 + M}}$

indicators

The B.F.S. corresponding to tableau III has both artificial variables equal to 0. Thus, the $t_1$- and $t_2$-columns are no longer needed. However, the indicators in the $x_1$-, $x_2$-, $s_1$-, and $s_2$-columns are nonnegative, and hence, an optimum solution has been reached. Since $W = -Z$ when $t_1 = t_2 = 0$, the maximum value of $-Z$ is $-4$. Consequently, the *minimum* value of $Z$ is $-(-4)$, or 4. It occurs when $x_1 = 0$ and $x_2 = 2$. ∎

### EXAMPLE 2  Reducing Dust Emissions

Here is an interesting example dealing with environmental controls.

*A cement plant produces 2,500,000 barrels of cement per year. The kilns emit 2 lb of dust for each barrel produced. A governmental agency dealing with environmental protection requires that the plant reduce its dust emissions to no more than 800,000 lb per year. There are two emission control devices available, A and B. Device A reduces emissions to $\frac{1}{2}$ lb per barrel, and its cost is \$0.20 per barrel of cement produced. For device B, emissions are reduced to $\frac{1}{5}$ lb per barrel, and the cost is \$0.25 per barrel of cement produced. Determine the most economical course of action that the plant should take so that it complies with the agency's requirement and also maintains its annual production of 2,500,000 barrels of cement.*[5]

*Solution:* We must minimize the annual cost of emission control. Let $x_1$, $x_2$, and $x_3$ be the annual numbers of barrels of cement produced in kilns that use device A, device B, and no device, respectively. Then $x_1, x_2, x_3 \geq 0$, and the annual emission control cost (in dollars) is

$$C = \tfrac{1}{5}x_1 + \tfrac{1}{4}x_2 + 0x_3. \tag{6}$$

Since 2,500,000 barrels of cement are produced each year,

$$x_1 + x_2 + x_3 = 2{,}500{,}000. \tag{7}$$

The numbers of pounds of dust emitted annually by the kilns that use device A, device B, and no device are $\frac{1}{2}x_1, \frac{1}{5}x_2$, and $2x_3$, respectively. Since the total number of pounds of dust emission is to be no more than 800,000,

$$\tfrac{1}{2}x_1 + \tfrac{1}{5}x_2 + 2x_3 \leq 800{,}000. \tag{8}$$

---

[5]This example is adapted from Robert E. Kohn, "A Mathematical Model for Air Pollution Control," *School Science and Mathematics,* 69 (1969), 487–94.

To minimize $C$ subject to constraints (7) and (8), where $x_1, x_2, x_3 \geq 0$, we shall first maximize $-C$ by using the simplex method. The equations to consider are

$$x_1 + x_2 + x_3 + t_1 = 2{,}500{,}000 \qquad (9)$$

and

$$\tfrac{1}{2}x_1 + \tfrac{1}{5}x_2 + 2x_3 + s_2 = 800{,}000, \qquad (10)$$

where $t_1$ and $s_2$ are artificial and slack variables, respectively. The artificial objective equation is $W = (-C) - Mt_1$, or equivalently,

$$\tfrac{1}{5}x_1 + \tfrac{1}{4}x_2 + 0x_3 + Mt_1 + W = 0, \qquad (11)$$

where $M$ is a large positive number. The augmented coefficient matrix of Eqs. (9)–(11) is

$$
\begin{array}{cccccc}
x_1 & x_2 & x_3 & s_2 & t_1 & W \\
\end{array}
$$
$$
\left[
\begin{array}{cccccc|c}
1 & 1 & 1 & 0 & 1 & 0 & 2{,}500{,}000 \\
\tfrac{1}{2} & \tfrac{1}{5} & 2 & 1 & 0 & 0 & 800{,}000 \\
\hline
\tfrac{1}{5} & \tfrac{1}{4} & 0 & 0 & M & 1 & 0
\end{array}
\right].
$$

Once we determine our initial simplex tableau, we proceed and obtain (after three additional tableaus) our final tableau:
Notice that $W$ is replaced by $-C$ when $t_1 = 0$. The maximum value of $-C$

$$
\begin{array}{ccccc}
 & x_1 & x_2 & x_3 & s_2 & -C \\
\end{array}
$$
$$
\begin{array}{c}
x_2 \\
x_1 \\
\\
-C
\end{array}
\left[
\begin{array}{ccccc|c}
0 & 1 & -5 & -\tfrac{10}{3} & 0 & 1{,}500{,}000 \\
1 & 0 & 6 & \tfrac{10}{3} & 0 & 1{,}000{,}000 \\
\hline
0 & 0 & \tfrac{1}{20} & \tfrac{1}{6} & 1 & -575{,}000
\end{array}
\right].
$$

$$\underbrace{\phantom{0 \quad 0 \quad \tfrac{1}{20} \quad \tfrac{1}{6}}}_{\text{indicators}}$$

is $-575{,}000$ and occurs when $x_1 = 1{,}000{,}000$, $x_2 = 1{,}500{,}000$, and $x_3 = 0$. Thus, the *minimum* annual cost of the emission control must be $-(-575{,}000) = \$575{,}000$. Device A should be installed on kilns producing 1,000,000 barrels of cement annually, and device B should be installed on kilns producing 1,500,000 barrels annually. ∎

# ▪ Exercise 7.7

*Use the simplex method to solve the following problems.*

**1.** Minimize

$$Z = 3x_1 + 6x_2,$$

subject to
$$-x_1 + x_2 \geq 6,$$
$$x_1 + x_2 \geq 10,$$
$$x_1, x_2 \geq 0.$$

**2.** Minimize

$$Z = 8x_1 + 12x_2,$$

subject to
$$2x_1 + 2x_2 \geq 1,$$
$$x_1 + 3x_2 \geq 2,$$
$$x_1, x_2 \geq 0.$$

**3.** Minimize

$$Z = 4x_1 + 2x_2 + x_3,$$

subject to
$$x_1 - x_2 - x_3 \geq 9,$$
$$x_1, x_2, x_3 \geq 0.$$

**4.** Minimize

$$Z = x_1 + x_2 + 2x_3,$$

subject to
$$x_1 + 2x_2 - x_3 \geq 4,$$
$$x_1, x_2, x_3 \geq 0.$$

**5.** Minimize

$$Z = 2x_1 + 3x_2 + x_3,$$

subject to

$$x_1 + x_2 + x_3 \leq 6,$$
$$x_1 \qquad - x_3 \leq -4,$$
$$x_2 + x_3 \leq 5,$$
$$x_1, x_2, x_3 \geq 0.$$

**6.** Minimize

$$Z = 4x_1 + x_2 + 2x_3,$$

subject to

$$4x_1 + x_2 - x_3 \leq 3,$$
$$x_1 \qquad + x_3 \leq 4,$$
$$x_1 + x_2 + x_3 \geq 1,$$
$$x_1, x_2, x_3 \geq 0.$$

**7.** Minimize

$$Z = x_1 - x_2 - 3x_3,$$

subject to

$$x_1 + 2x_2 + x_3 = 4,$$
$$x_2 + x_3 = 1,$$
$$x_1 + x_2 \qquad \leq 6,$$
$$x_1, x_2, x_3 \geq 0.$$

**8.** Minimize

$$Z = x_1 + x_2 - 2x_3,$$

subject to

$$x_1 - x_2 + x_3 \leq 4,$$
$$2x_1 + x_2 - 3x_3 \geq 6,$$
$$x_1 - x_2 - 2x_3 = 2,$$
$$x_1, x_2, x_3 \geq 0.$$

**9.** Minimize

$$Z = x_1 + 8x_2 + 5x_3,$$

subject to

$$x_1 + x_2 + x_3 \geq 8,$$
$$-x_1 + 2x_2 + x_3 \geq 2,$$
$$x_1, x_2, x_3 \geq 0.$$

**10.** Minimize

$$Z = 4x_1 + 4x_2 + 6x_3,$$

subject to

$$x_1 - x_2 - x_3 \leq 3,$$
$$x_1 - x_2 + x_3 \geq 3,$$
$$x_1, x_2, x_3 \geq 0.$$

**11. Emission Control** A cement plant produces 3,300,000 barrels of cement per year. The kilns emit 2 lb of dust for each barrel produced. The plant must reduce its dust emissions to no more than 1,000,000 lb per year. There are two devices available, A and B, that will control emissions. Device A will reduce emissions to $\frac{1}{2}$ lb per barrel, and the cost is $0.25 per barrel of cement produced. For device B, emissions are reduced to $\frac{1}{4}$ lb per barrel, and the cost is $0.40 per barrel of cement produced. Determine the most economical course of action the plant should take so that it maintains an annual production of exactly 3,300,00 barrels of cement.

**12. Delivery Truck Scheduling** Because of increased business, a catering service finds that it must rent additional delivery trucks. The minimum needs are 12 units each of refrigerated and nonrefrigerated space. Two standard types of trucks are available in the rental market. Type A has 2 units of refrigerated space and 1 unit of nonrefrigerated space. Type B has 2 units of refrigerated space and 3 units of nonrefrigerated space. The costs per mile are $0.40 for A and $0.60 for B. How many of each type of truck should be rented so as to minimize total cost per mile? What is the minimum total cost per mile?

**13. Transportation Costs** A retailer has stores in Exton and Whyton and has warehouses A and B in two other cities. Each store requires delivery of exactly 30 refrigerators. In warehouse A there are 50 refrigerators, and in

B there are 20. The transportation costs to ship refrigerators from the warehouses to the stores are given in the following table:

| | Exton | Whyton |
|---|---|---|
| Warehouse A | $15 | $13 |
| Warehouse B | 11 | 12 |

For example, the cost to ship a refrigerator from A to the Exton store is $15. How should the retailer order the refrigerators so that the requirements of the stores are met and the total transportation costs are minimized? What is the minimum transportation cost?

**14. Battery Purchasing** An auto manufacturer purchases batteries from two suppliers, X and Y. The manufacturer has two plants, A and B, and requires delivery of exactly 6000 batteries to plant A and exactly 4000 to plant B. Supplier X charges $30 and $32 per battery (including transportation cost) to A and B, respectively. For these prices, X requires that the auto manufacturer order at least a total of 2000 batteries. However, X can supply no more than 4000 batteries. Supplier Y charges $34 and

$28 per battery to A and B, respectively, and requires a minimum order of 6000 batteries. Determine how the auto manufacturer should order the necessary batteries so that their total cost is a minimum. What is this minimum cost?

**15. Producing Wrapping Paper** A paper company stocks its holiday wrapping paper in 48-in.-wide rolls, called stock rolls, and cuts such rolls into smaller widths, depending on customers' orders. Suppose that an order for 50 rolls of 15-in.-wide paper and 60 rolls of 10-in.-wide paper is received. From a stock roll, the company can cut three 15-in.-wide rolls and one 3-in.-wide roll. (See Fig. 7.27.) Since the 3-in.-wide roll cannot be used in the order, 3 in. is called the trim loss for this roll. Similarly, from a stock roll, two 15-in.-wide rolls, one 10-in. wide

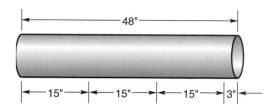

**FIGURE 7.27**   Diagram for Problem 15.

roll, and one 8-in.-wide roll could be cut. Here the trim loss would be 8 in. The following table indicates the number of 15-in. and 10-in. rolls, together with trim loss, that can be cut from a stock roll:

| Roll width | 15 in. | 3 | 2 | 1 | — |
|---|---|---|---|---|---|
|  | 10 in. | 0 | 1 | — | — |
| Trim loss |  | 3 | 8 | — | — |

(a) Complete the last two columns of the table. (b) Assume that the company has a sufficient number of stock rolls to fill the order and that *at least* 50 rolls of 15-in.-wide and *at least* 60 rolls of 10-in.-wide wrapping paper will be cut. If $x_1$, $x_2$, $x_3$, and $x_4$ are the numbers of stock rolls that are cut in a manner described by columns 1–4 of the table, respectively, determine the values of the $x$'s so that the total trim loss is minimized. (c) What is the minimum amount of total trim loss?

**To first motivate and then formally define the dual of a linear programming problem.**

## 7.8 THE DUAL

There is a fundamental principle, called *duality,* that allows us to solve a maximization problem by solving a related minimization problem. Let us illustrate.

Suppose that a company produces two types of widgets, manual and electric, and each requires the use of machines A and B in its production. Table 7.2 indicates that a manual widget requires the use of A for 1 hour and B for

**TABLE 7.2**

|  | Machine A | Machine B | Profit/Unit |
|---|---|---|---|
| Manual | 1 hr | 1 hr | $10 |
| Electric | 2 hr | 4 hr | $24 |
| Hours available | 120 | 180 |  |

1 hour. An electric widget requires A for 2 hours and B for 4 hours. The maximum numbers of hours available per month for machines A and B are 120 and 180, respectively. The profit on a manual widget is $10, and on an electric widget it is $24. Assuming that the company can sell all the widgets it can produce, we shall determine the maximum monthly profit. If $x_1$ and $x_2$ are the numbers of manual and electric widgets produced per month, respectively, then we want to maximize the monthly profit function

$$P = 10x_1 + 24x_2,$$

subject to

$$x_1 + 2x_2 \le 120, \tag{1}$$

$$x_1 + 4x_2 \le 180, \tag{2}$$

and $x_1$, $x_2 \ge 0$. Writing constraints (1) and (2) as equations, we have

$$x_1 + 2x_2 + s_1 = 120 \tag{3}$$

and
$$x_1 + 4x_2 + s_2 = 180,$$

where $s_1$ and $s_2$ are slack variables. In Eq. (3), $x_1 + 2x_2$ is the number of hours that machine A is used. Since 120 hours on A are available, $s_1$ is the number of available hours that are *not* used for production. That is, $s_1$ represents unused capacity (in hours) for A. Similarly, $s_2$ represents unused capacity for B. Solving this problem by the simplex method, we find that the final tableau is

$$
\begin{array}{c}
\begin{array}{ccccc} x_1 & x_2 & s_1 & s_2 & P \end{array} \\
\begin{array}{c} x_1 \\ x_2 \\ P \end{array}
\left[
\begin{array}{ccccc|c}
1 & 0 & 2 & -1 & 0 & 60 \\
0 & 1 & -\frac{1}{2} & \frac{1}{2} & 0 & 30 \\
\hline
0 & 0 & 8 & 2 & 1 & 1320
\end{array}
\right].
\end{array}
\tag{4}
$$

$$\underbrace{\qquad\qquad\qquad}_{\text{indicators}}$$

Thus, the maximum profit per month is \$1320, which occurs when $x_1 = 60$ and $x_2 = 30$.

Now let us look at the situation from a different point of view. Suppose that the company wishes to rent out machines A and B. What is the minimum monthly rental fee they should charge? Certainly, if the charge is too high, no one would rent the machines. On the other hand, if the charge is too low, it may not pay the company to rent them at all. Obviously, the minimum rent should be \$1320. That is, the minimum the company should charge is the profit it could make by using the machines itself. We can arrive at this minimum rental fee directly by solving a linear programming problem.

Let $R$ be the total monthly rental fee. To determine $R$, suppose the company assigns values or "worths" to each hour of capacity on machines A and B. Let these worths be $y_1$ and $y_2$ dollars, respectively, where $y_1, y_2 \geq 0$. Then the monthly worth of machine A is $120y_1$, and for B it is $180y_2$. Thus,

$$R = 120y_1 + 180y_2.$$

The total worth of machine time to produce a manual widget is $1y_1 + 1y_2$. This should be at least equal to the \$10 profit the company can earn by producing that widget. If not, the company would make more money by using the machine time to produce a manual widget. Accordingly,

$$1y_1 + 1y_2 \geq 10.$$

Similarly, the total worth of machine time to produce an electric widget should be at least \$24:

$$2y_1 + 4y_2 \geq 24.$$

Therefore, the company wants to

$$\text{minimize } R = 120y_1 + 180y_2,$$

subject to

$$y_1 + y_2 \geq 10, \tag{5}$$

$$2y_1 + 4y_2 \geq 24, \tag{6}$$

and $y_1, y_2 \geq 0$.

To minimize $R$, we shall maximize $-R$. Since constraints (5) and (6) have the form $a_1 y_1 + a_2 y_2 \geq b$, where $b \geq 0$, we consider an artificial problem. If

$r_1$ and $r_2$ are slack variables and $t_1$ and $t_2$ are artificial variables, then we want to maximize

$$W = (-R) - Mt_1 - Mt_2,$$

where $M$ is a large positive number, such that

$$y_1 + y_2 - r_1 + t_1 = 10,$$
$$2y_1 + 4y_2 - r_2 + t_2 = 24,$$

and the $y$'s, $r$'s, and $t$'s are nonnegative. The final simplex tableau for this problem (with the artificial variable columns deleted and $W$ changed to $-R$) is

$$\begin{array}{c} \\ y_1 \\ y_2 \\ -R \end{array} \begin{array}{c} \begin{array}{cccccc} y_1 & y_2 & r_1 & r_2 & -R & \\ \end{array} \\ \left[\begin{array}{ccccc|c} 1 & 0 & -2 & \frac{1}{2} & 0 & 8 \\ 0 & 1 & 1 & -\frac{1}{2} & 0 & 2 \\ \hline 0 & 0 & 60 & 30 & 1 & -1320 \end{array}\right] \end{array}.$$

$$\underbrace{\phantom{0 \quad 0 \quad 60 \quad 30}}_{\text{indicators}}$$

Since the maximum value of $-R$ is $-1320$, the *minimum* value of $R$ is $-(-1320) = \$1320$ (as anticipated). It occurs when $y_1 = 8$ and $y_2 = 2$. We have therefore determined the optimum value of one linear programming problem (maximizing profit) by finding the optimum value of another problem (minimizing rental fee).

The values $y_1 = 8$ and $y_2 = 2$ could have been anticipated from the final tableau of the maximization problem. In (4), the indicator 8 in the $s_1$-column means that at the optimum level of production, if $s_1$ increases by one unit, then the profit $P$ *decreases* by 8. That is, 1 unused hour of capacity on A decreases the maximum profit by \$8. Thus, 1 hour of capacity on A is worth \$8. We say that the **shadow price** or **accounting price** of 1 hour of capacity on A is \$8. Now, recall that $y_1$ in the rental problem is the worth of 1 hour of capacity on A. Therefore, $y_1$ must equal 8 in the optimum solution for that problem. Similarly, since the indicator in the $s_2$-column is 2, the shadow price of 1 hour of capacity on B is \$2, which is the value of $y_2$ in the optimum solution of the rental problem.

Let us now analyze the structure of our two linear programming problems:

| Maximize | Minimize |
|---|---|
| $P = 10x_1 + 24x_2,$ | $R = 120y_1 + 180y_2,$ |
| subject to | subject to |
| $\left.\begin{array}{l} x_1 + 2x_2 \le 120 \\ x_1 + 4x_2 \le 180 \end{array}\right\} \quad (7)$ | $\left.\begin{array}{l} y_1 + y_2 \ge 10 \\ 2y_1 + 4y_2 \ge 24 \end{array}\right\} \quad (8)$ |
| and $x_1, x_2 \ge 0.$ | and $y_1, y_2 \ge 0.$ |

Note that in (7) the inequalities are all $\le$, but in (8) they are all $\ge$. The coefficients of the objective function in the minimization problem are the constant terms in (7). The constant terms in (8) are the coefficients of the objective function of the maximization problem. The coefficients of the $y_1$'s in (8) are the coefficients of $x_1$ and $x_2$ in the first constraint of (7); the coefficients of the $y_2$'s in (8) are the coefficients of $x_1$ and $x_2$ in the second constraint of (7). The minimization problem is called the *dual* of the maximization problem, and vice versa.

In general, with any given linear programming problem, we can associate another linear programming problem called its **dual.** The given problem is called **primal.** If the primal is a maximization problem, then its dual is a minimization problem. Similarly, if the primal involves minimization, then the dual involves maximization.

Any primal maximization problem can be written in the form indicated in Table 7.3. Note that there are no restrictions on the $b$'s.[6] The corresponding dual minimization problem can be written in the form indicated in Table 7.4. Similarly, any primal minimization problem can be put in the form of Table 7.4, and its dual is the maximization problem in Table 7.3.

Let us compare the primal and its dual in Tables 7.3 and 7.4. For convenience, when we refer to constraints, we shall mean those in (9) or (10); we shall not include the nonnegativity conditions. Observe that if all the con-

---

**TABLE 7.3**  Primal (Dual)

Maximize $Z = c_1x_1 + c_2x_2 + \cdots + c_nx_n,$

subject to

$$
\left.
\begin{aligned}
a_{11}x_1 + a_{12}x_2 + \cdots + a_{1n}x_n &\leq b_1, \\
a_{21}x_1 + a_{22}x_2 + \cdots + a_{2n}x_n &\leq b_2, \\
&\vdots \\
a_{m1}x_1 + a_{m2}x_2 + \cdots + a_{mn}x_n &\leq b_m,
\end{aligned}
\right\} \quad (9)
$$

and $x_1, x_2, \ldots, x_n \geq 0.$

**TABLE 7.4**  Dual (Primal)

Minimize $W = b_1y_1 + b_2y_2 + \cdots + b_my_m,$

subject to

$$
\left.
\begin{aligned}
a_{11}y_1 + a_{21}y_2 + \cdots + a_{m1}y_m &\geq c_1, \\
a_{12}y_1 + a_{22}y_2 + \cdots + a_{m2}y_m &\geq c_2, \\
&\vdots \\
a_{1n}y_1 + a_{2n}y_2 + \cdots + a_{mn}y_m &\geq c_n,
\end{aligned}
\right\} \quad (10)
$$

and $y_1, y_2, \ldots, y_m \geq 0.$

---

straints in the primal involve $\leq (\geq)$, then all the constraints in its dual involve $\geq (\leq)$. The coefficients in the dual's objective function are the constant terms in the primal's constraints. Similarly, the constant terms in the dual's constraints are the coefficients of the primal's objective function. The coefficient matrix of the left sides of the dual's constraints is the *transpose* of the coefficient matrix of the left sides of the primal's constraints. That is,

$$
\begin{bmatrix}
a_{11} & a_{12} & \cdots & a_{1n} \\
a_{21} & a_{22} & \cdots & a_{2n} \\
\vdots & \vdots & & \vdots \\
a_{m1} & a_{m2} & \cdots & a_{mn}
\end{bmatrix}^{\mathrm{T}}
=
\begin{bmatrix}
a_{11} & a_{21} & \cdots & a_{m1} \\
a_{12} & a_{22} & \cdots & a_{m2} \\
\vdots & \vdots & & \vdots \\
a_{1n} & a_{2n} & \cdots & a_{mn}
\end{bmatrix}.
$$

If the primal involves $n$ structural variables and $m$ slack variables, then the dual involves $m$ structural variables and $n$ slack variables. It should be noted that the dual of the *dual* is the primal.

There is an important relationship between the primal and its dual:

If the primal has an optimum solution, then so does the dual, and the optimum value of the primal's objective function is the *same* as that of its dual.

---

[6]If an inequality constraint involves $\geq$, multiplying both sides by $-1$ yields an inequality involving $\leq$. If a constraint is an equality, it can be written in terms of two inequalities, one involving $\leq$ and one involving $\geq$.

Moreover, suppose that the primal's objective function is

$$Z = c_1 x_1 + c_2 x_2 + \cdots + c_n x_n.$$

Then:

> If $s_i$ is the slack variable associated with the $i$th constraint in the dual, then the indicator in the $s_i$-column of the final simplex tableau of the dual is the value of $x_i$ in the optimum solution of the primal.

Thus, we can solve the primal by merely solving its dual. At times this is more convenient than solving the primal directly.

**Principles in Practice 1**

**Finding the Dual of a Maximization Problem**

Find the dual of the following problem:

Suppose that the What If Company has \$60,000 for the purchase of materials to make three types of gadgets. The company has allocated a total of 2000 hours of assembly time and 120 hours of packaging time for the gadgets. The following table gives the cost per gadget, the number of hours per gadget, and the profit per gadget for each type:

|  | Type 1 | Type 2 | Type 3 |
|---|---|---|---|
| Cost/Gadget | \$300 | \$220 | \$180 |
| Assembly Hours/Gadget | 20 | 40 | 20 |
| Packaging Hours/Gadget | 3 | 1 | 2 |
| Profit | \$300 | \$200 | \$200 |

**EXAMPLE 1    Finding the Dual of a Maximization Problem**

*Find the dual of the following:*

$$\text{Maximize } Z = 3x_1 + 4x_2 + 2x_3,$$

*subject to*

$$x_1 + 2x_2 + 0x_3 \le 10,$$
$$2x_1 + 2x_2 + x_3 \le 10,$$

*and $x_1, x_2, x_3 \ge 0$.*

*Solution:* The primal is of the form of Table 7.3. Thus, the dual is

$$\text{minimize } W = 10y_1 + 10y_2,$$

subject to

$$y_1 + 2y_2 \ge 3,$$
$$2y_1 + 2y_2 \ge 4,$$
$$0y_1 + y_2 \ge 2,$$

and $y_1, y_2 \ge 0$.　■

**EXAMPLE 2    Finding the Dual of a Minimization Problem**

*Find the dual of the following:*

$$\text{Minimize } Z = 4x_1 + 3x_2,$$

*subject to*

$$3x_1 - x_2 \ge 2, \tag{11}$$
$$x_1 + x_2 \le 1, \tag{12}$$
$$-4x_1 + x_2 \le 3, \tag{13}$$

*and $x_1, x_2 \ge 0$.*

*Solution:* Since the primal is a minimization problem, we want constraints (12) and (13) to involve $\ge$. (See Table 7.4.) Multiplying both sides of (12) and

(13) by $-1$, we get $-x_1 - x_2 \geq -1$ and $4x_1 - x_2 \geq -3$. Thus, constraints (11)–(13) become

$$3x_1 - x_2 \geq 2,$$
$$-x_1 - x_2 \geq -1,$$
$$4x_1 - x_2 \geq -3.$$

The dual is

$$\text{maximize } W = 2y_1 - y_2 - 3y_3,$$

subject to

$$3y_1 - y_2 + 4y_3 \leq 4,$$
$$-y_1 - y_2 - y_3 \leq 3,$$

and $y_1, y_2, y_3 \geq 0.$ ■

---

## Principles in Practice 2

**Finding the Dual of a Minimization Problem**

Find the dual of the following problem:

A person decides to take two different dietary supplements. Each supplement contains two essential ingredients, A and B, for which there are minimum daily requirements, and each contains a third ingredient, C, that needs to be minimized.

| | Supplement 1 | Supplement 2 | Daily Requirement |
|---|---|---|---|
| A | 20 mg/oz | 6 mg/oz | 98 mg |
| B | 8 mg/oz | 16 mg/oz | 80 mg |
| C | 6 mg/oz | 2 mg/oz | |

## EXAMPLE 3 Applying the Simplex Method to the Dual

*Use the dual and the simplex method to*

$$\text{maximize } Z = 4x_1 - x_2 - x_3,$$

*subject to*

$$3x_1 + x_2 - x_3 \leq 4,$$
$$x_1 + x_2 + x_3 \leq 2,$$

*and* $x_1, x_2, x_3 \geq 0.$

**Solution:** The dual is

$$\text{minimize } W = 4y_1 + 2y_2,$$

subject to

$$3y_1 + y_2 \geq 4, \tag{14}$$
$$y_1 + y_2 \geq -1, \tag{15}$$
$$-y_1 + y_2 \geq -1, \tag{16}$$

and $y_1, y_2 \geq 0.$ To use the simplex method, we must get nonnegative constants in (15) and (16). Multiplying both sides of these equations by $-1$ gives

$$-y_1 - y_2 \leq 1, \tag{17}$$
$$y_1 - y_2 \leq 1. \tag{18}$$

Since (14) involves $\geq$, an artificial variable is required. The corresponding equations of (14), (17), and (18) are, respectively,

$$3y_1 + y_2 - s_1 + t_1 = 4,$$
$$-y_1 - y_2 + s_2 = 1,$$

and

$$y_1 - y_2 + s_3 = 1,$$

where $t_1$ is an artificial variable and $s_1$, and $s_2$, and $s_3$ are slack variables. To minimize $W$, we maximize $-W$. The artificial objective function is $U = (-W) - Mt_1$, where $M$ is a large positive number. After computations, we find that the final simplex tableau is

$$
\begin{array}{c}
\begin{array}{cccccc}
y_1 & y_2 & s_1 & s_2 & s_3 & -W
\end{array} \\
\begin{array}{c}
y_2 \\
s_2 \\
y_1 \\
\\
-W
\end{array}
\left[
\begin{array}{cccccc|c}
0 & 1 & -\frac{1}{4} & 0 & -\frac{3}{4} & 0 & \frac{1}{4} \\
0 & 0 & -\frac{1}{2} & 1 & -\frac{1}{2} & 0 & \frac{5}{2} \\
1 & 0 & -\frac{1}{4} & 0 & \frac{1}{4} & 0 & \frac{5}{4} \\
\hline
0 & 0 & \frac{3}{2} & 0 & \frac{1}{2} & 1 & -\frac{11}{2}
\end{array}
\right].
\end{array}
$$

$$\underbrace{\hphantom{0 \quad 0 \quad \frac{3}{2} \quad 0 \quad \frac{1}{2}}}_{\text{indicators}}$$

The maximum value of $-W$ is $-\frac{11}{2}$, so the *minimum* value of $W$ is $\frac{11}{2}$. Hence, the maximum value of $Z$ is also $\frac{11}{2}$. Note that the indicators in the $s_1$-, $s_2$-, and $s_3$-columns are $\frac{3}{2}$, 0, and $\frac{1}{2}$, respectively. Thus, the maximum value of $Z$ occurs when $x_1 = \frac{3}{2}$, $x_2 = 0$, and $x_3 = \frac{1}{2}$. ∎

In Example 1 of Sec. 7.7 we used the simplex method to

$$\text{minimize } Z = x_1 + 2x_2$$

such that

$$-2x_1 + x_2 \geq 1,$$
$$-x_1 + x_2 \geq 2,$$

and $x_1, x_2 \geq 0$. The initial simplex tableau had 24 entries and involved two artificial variables. The tableau of the dual has only 18 entries and *no artificial variables* and is easier to handle, as Example 4 will show. Thus, there may be a distinct advantage in solving the dual to determine the solution of the primal.

**This discussion shows the advantage of solving the dual problem.**

---

### Principles in Practice 3

#### Applying the Simplex Method to the Dual

A company produces three kinds of devices requiring three different production procedures. The company has allocated a total of 300 hours for procedure 1, 400 hours for procedure 2, and 600 hours for procedure 3. The following table gives the number of hours per device for each procedure:

|            | Device 1 | Device 2 | Device 3 |
|------------|----------|----------|----------|
| Procedure 1 | 30       | 15       | 10       |
| Procedure 2 | 20       | 30       | 20       |
| Procedure 3 | 40       | 30       | 25       |

If the profit is \$30 per device 1, \$20 per device 2, and \$20 per device 3, then, using the dual and the simplex method, find the number of devices of each kind the company should produce to maximize profit.

---

### EXAMPLE 4  Using the Dual and the Simplex Method

*Use the dual and the simplex method to*

$$\text{minimize } Z = x_1 + 2x_2,$$

*subject to*

$$-2x_1 + x_2 \geq 1,$$
$$-x_1 + x_2 \geq 2,$$

*and* $x_1, x_2 \geq 0$.

*Solution:* The dual is

$$\text{maximize } W = y_1 + 2y_2,$$

subject to

$$-2y_1 - y_2 \leq 1,$$
$$y_1 + y_2 \leq 2,$$

and $y_1, y_2 \geq 0$. The initial simplex tableau is tableau I:

## SIMPLEX TABLEAU I

$$
\begin{array}{c}
\begin{array}{ccccc} y_1 & y_2 & s_1 & s_2 & W \end{array} \\
\begin{array}{c} s_1 \\ \text{departing} \rightarrow s_2 \\ \text{variable} \\ W \end{array}
\left[
\begin{array}{ccccc|c}
-2 & -1 & 1 & 0 & 0 & 1 \\
1 & 1 & 0 & 1 & 0 & 2 \\
\hline
-1 & -2 & 0 & 0 & 1 & 0
\end{array}
\right]
\end{array}
\qquad
\begin{array}{l}
\textit{Quotients} \\
\\
2 \div 1 = 2.
\end{array}
$$

$$\underbrace{\phantom{-1 \quad -2 \quad 0 \quad 0}}_{}$$

$\uparrow$ indicators

entering
variable

Continuing, we get tableau II.

## SIMPLEX TABLEAU II

$$
\begin{array}{c}
\begin{array}{ccccc} y_1 & y_2 & s_1 & s_2 & W \end{array} \\
\begin{array}{c} s_1 \\ y_2 \\ W \end{array}
\left[
\begin{array}{ccccc|c}
-1 & 0 & 1 & 1 & 0 & 3 \\
1 & 1 & 0 & 1 & 0 & 2 \\
\hline
1 & 0 & 0 & 2 & 1 & 4
\end{array}
\right]
\end{array}.
$$

$$\underbrace{\phantom{1 \quad 0 \quad 0 \quad 2}}_{}$$

indicators

Since all indicators are nonnegative in tableau II, the maximum value of $W$ is 4. Hence, the minimum value of $Z$ is also 4. The indicators 0 and 2 in the $s_1$- and $s_2$-columns of tableau II mean that the minimum value of $Z$ occurs when $x_1 = 0$ and $x_2 = 2$. ∎

## ■ Exercise 7.8

*In Problems 1–8, find the duals. Do not solve.*

**1.** Maximize

   $Z = 2x_1 + 3x_2,$

   subject to

   $x_1 + x_2 \le 6,$
   $-x_1 + x_2 \le 4,$
   $x_1, x_2 \ge 0.$

**2.** Maximize

   $Z = 2x_1 + x_2 - x_3,$

   subject to

   $x_1 + x_2 \quad \le 1,$
   $-x_1 + 2x_2 + x_3 \le 2,$
   $x_1, x_2, x_3 \ge 0.$

**3.** Minimize

   $Z = x_1 + 8x_2 + 5x_3,$

   subject to

   $x_1 + x_2 + x_3 \ge 8,$
   $-x_1 + 2x_2 + x_3 \ge 2,$
   $x_1, x_2, x_3 \ge 0.$

**4.** Minimize

   $Z = 8x_1 + 12x_2,$

   subject to

   $2x_1 + 2x_2 \ge 1,$
   $x_1 + 3x_2 \ge 2,$
   $x_1, x_2 \ge 0.$

**5.** Maximize

   $Z = x_1 - x_2,$

   subject to

   $-x_1 + 2x_2 \le 13,$
   $-x_1 + x_2 \ge 3,$
   $x_1 + x_2 \ge 11,$
   $x_1, x_2 \ge 0.$

**6.** Maximize

   $Z = x_1 - x_2 + 4x_3,$

   subject to

   $x_1 + x_2 + x_3 \le 9,$
   $x_1 - 2x_2 + x_3 \ge 6,$
   $x_1, x_2, x_3 \ge 0.$

**7.** Minimize

   $Z = 4x_1 + 4x_2 + 6x_3,$

   subject to

   $x_1 - x_2 - x_3 \le 3,$
   $x_1 - x_2 + x_3 \ge 3,$
   $x_1, x_2, x_3 \ge 0.$

**8.** Minimize

   $Z = 6x_1 + 3x_2,$

   subject to

   $-3x_1 + 4x_1 \ge -12,$
   $13x_1 - 8x_2 \le 80,$
   $x_1, x_2 \ge 0.$

*In Problems 9–14, solve by using duals and the simplex method.*

**9.** Minimize

   $Z = 4x_1 + 4x_2 + 6x_3,$

   subject to

   $x_1 - x_2 + x_3 \ge 1,$
   $-x_1 + x_2 + x_3 \ge 2,$
   $x_1, x_2, x_3 \ge 0,$

**10.** Minimize

   $Z = x_1 + x_2,$

   subject to

   $x_1 + 4x_2 \ge 28,$
   $2x_1 - x_2 \ge 2,$
   $-3x_1 + 8x_2 \ge 16,$
   $x_1, x_2 \ge 0.$

**11.** Maximize

   $Z = 3x_1 + 8x_2,$

   subject to

   $x_1 + 2x_2 \le 8,$
   $x_1 + 6x_2 \le 12,$
   $x_1, x_2 \ge 0.$

**12.** Maximize

$$Z = 2x_1 + 6x_2,$$

subject to

$$3x_1 + x_2 \leq 12,$$
$$x_1 + x_2 \leq 8,$$
$$x_1, x_2 \geq 0.$$

**13.** Minimize

$$Z = 6x_1 + 4x_2,$$

subject to

$$-x_1 + x_2 \leq 1,$$
$$x_1 + x_2 \geq 3,$$
$$x_1, x_2 \geq 0.$$

**14.** Minimize

$$Z = x_1 + x_2 + 2x_3,$$

subject to

$$-x_1 - x_2 + x_3 \leq 1,$$
$$x_1 - x_2 + x_3 \geq 2,$$
$$x_1, x_2, x_3 \geq 0.$$

**15. Advertising** A firm is comparing the costs of advertising in two media; newspaper and radio. For every dollar's worth of advertising, the following table gives the number of people by income group, reached by these media:

|  | Under $20,000 | Over $20,000 |
|---|---|---|
| Newspaper | 40 | 100 |
| Radio | 50 | 25 |

The firm wants to reach at least 8000 persons earning under $20,000 and at least 6000 earning over $20,000. Use the dual and the simplex method to find the amounts that the firm should spend on newspaper and radio advertising so as to reach these numbers of people

at a minimum total advertising cost. What is the minimum total advertising cost?

**16.** Use the dual and the simplex method to find the minimum total cost per mile in Problem 12 of Exercise 7.7.

**17. Labor Costs** A company pays skilled and semiskilled workers in its assembly department $7 and $4 per hour, respectively. In the shipping department, shipping clerks are paid $5 per hour and shipping clerk apprentices are paid $2 per hour. The company requires at least 90 workers in the assembly department and at least 60 in the shipping department. Because of union agreements, at least twice as many semiskilled workers must be employed as skilled workers. Also, at least twice as many shipping clerks must be employed as shipping clerk apprentices. Use the dual and the simplex method to find the number of each type of worker that the company must employ so that the total hourly wage paid to these employees is a minimum. What is the minimum total hourly wage?

# 7.9 REVIEW

## IMPORTANT TERMS AND SYMBOLS

**Section 7.1**      linear inequality      half plane (open, closed)      system of inequalities

**Section 7.2**      constraints      linear function in $x$ and $y$      linear programming problem      objective function
     feasible solution      nonnegativity conditions      feasible region      isoprofit line
     corner point      bounded feasible region      unbounded feasible region
     nonempty feasible region      empty feasible region      isocost line      unbounded solution

**Section 7.3**      multiple optimum solution

**Section 7.4**      standard linear programming problem      slack variable      structural variable
     basic feasible solution      nonbasic variable      basic variable      simplex tableau
     objective row      entering variable      indicator      departing variable      pivot entry
     simplex method      degeneracy

**Section 7.6**      artificial problem      artificial variable      artificial objective function

**Section 7.8**      shadow price      dual      primal

## SUMMARY

The solution of a system of linear inequalities consists of all points whose coordinates simultaneously satisfy all of the inequalities. Geometrically, for two variables, it is the region that is common to all of the regions determined by the inequalities.

Linear programming involves maximizing or minimizing a linear function (the objective function), subject to a system of constraints, which are linear inequalities or linear equations. One method for finding an optimum solution for a nonempty feasible region is the corner-point method. The objective function is evaluated at each of the corner points of the feasible region, and we choose a corner point at which the objective function is optimum.

For a problem involving more than two variables, the corner–point method is either impractical or impossible. Instead, we use a matrix method called the simplex method, which is efficient and completely mechanical.

## REVIEW PROBLEMS

*In Problems 1–10, solve the given inequality or system of inequalities.*

**1.** $-3x + 2y > -6$.

**2.** $x - 2y + 6 \geq 0$.

**3.** $2y \leq -3$.

**4.** $-x < 2$.

**5.** $\begin{cases} y - 3x < 6, \\ x - y > -3. \end{cases}$

**6.** $\begin{cases} x - 2y > 4, \\ x + y > 1. \end{cases}$

**7.** $\begin{cases} x - y < 4, \\ y - x < 4. \end{cases}$

**8.** $\begin{cases} x > y, \\ x + y < 0. \end{cases}$

**9.** $\begin{cases} 3x + y > -4, \\ x - y > -5, \\ x \geq 0. \end{cases}$

**10.** $\begin{cases} x - y > 4, \\ x < 2, \\ y < -4. \end{cases}$

*In Problems 11–18, do not use the simplex method.*

**11.** Maximize
$$Z = x - 2y,$$
subject to
$$y - x \leq 2,$$
$$x + y \leq 4,$$
$$x \leq 3,$$
$$x, y \geq 0.$$

**12.** Maximize
$$Z = 4x + 2y,$$
subject to
$$x + 2y \leq 10,$$
$$x \leq 4,$$
$$y \geq 1,$$
$$x, y \geq 0.$$

**13.** Minimize
$$Z = 2x - y,$$
subject to
$$x - y \geq -2,$$
$$x + y \geq 1,$$
$$x - 2y \leq 2,$$
$$x, y \geq 0.$$

**14.** Minimize
$$Z = x + y,$$
subject to
$$x + 3y \leq 15,$$
$$3x + 2y \leq 17,$$
$$x - 5y \leq 0,$$
$$x, y \geq 0.$$

**15.** Minimize
$$Z = 4x - 3y,$$
subject to
$$x + y \leq 3,$$
$$2x + 3y \leq 12,$$
$$5x + 8y \geq 40,$$
$$x, y \geq 0.$$

**[7]16.** Minimize
$$Z = 2x + 2y,$$
subject to
$$x + y \geq 4,$$
$$-x + 3y \leq 18,$$
$$x \leq 6,$$
$$x, y \geq 0.$$

**[7]17.** Maximize
$$Z = 9x + 6y,$$
subject to
$$x + 2y \leq 8,$$
$$3x + 2y \leq 12,$$
$$x, y \geq 0.$$

**18.** Maximize
$$Z = 4x + y,$$
subject to
$$x + 2y \geq 8,$$
$$3x + 2y \geq 12,$$
$$x, y \geq 0.$$

*In Problems 19–28, use the simplex method.*

**19.** Maximize
$$Z = 4x_1 + 5x_2,$$
subject to
$$x_1 + 6x_2 \leq 12,$$
$$x_1 + 2x_2 \leq 8,$$
$$x_1, x_2 \geq 0.$$

**20.** Maximize
$$Z = 18x_1 + 20x_2,$$
subject to
$$2x_1 + 3x_2 \leq 18,$$
$$4x_1 + 3x_2 \leq 24,$$
$$x_2 \leq 5,$$
$$x_1, x_2 \geq 0.$$

**21.** Minimize
$$Z = 2x_1 + 3x_2 + x_3,$$
subject to
$$x_1 + 2x_2 + 3x_3 \geq 6,$$
$$x_1, x_2, x_3 \geq 0.$$

**22.** Minimize
$$Z = x_1 + x_2,$$
subject to
$$3x_1 + 4x_2 \geq 24,$$
$$x_2 \geq 3,$$
$$x_1, x_2 \geq 0.$$

**23.** Maximize
$$Z = x_1 + 2x_2,$$
subject to
$$x_1 + x_2 \leq 12,$$
$$x_1 + x_2 \geq 5,$$
$$x_1 \leq 10,$$
$$x_1, x_2 \geq 0.$$

**24.** Minimize
$$Z = 2x_1 + x_2,$$
subject to
$$x_1 + 2x_2 \leq 6,$$
$$x_1 + x_2 \geq 1,$$
$$x_1, x_2 \geq 0.$$

**25.** Minimize
$$Z = x_1 + 2x_2 + x_3,$$
subject to
$$x_1 - x_2 - x_3 \leq -1,$$
$$6x_1 + 3x_2 + 2x_3 = 12,$$
$$x_1, x_2, x_3 \geq 0.$$

**26.** Maximize
$$Z = x_1 + 3x_2 + 2x_3,$$
subject to
$$x_1 + x_2 + 4x_3 \geq 6,$$
$$2x_1 + x_2 + 3x_3 \leq 4,$$
$$x_1, x_2, x_3 \geq 0,$$

---

[7]Refers to Sec. 7.3.

[8]**27.** Maximize

$$Z = x_1 + 4x_2 + 2x_3,$$

subject to

$$4x_1 - x_2 \quad\quad \le 2,$$
$$-10x_1 + x_2 + 3x_3 \le 1,$$
$$x_1, x_2, x_3 < 0.$$

[8]**28.** Minimize

$$Z = x_1 + x_2,$$

subject to

$$x_1 + x_2 + 2x_3 \le 4,$$
$$x_3 \ge 1,$$
$$x_1, x_2, x_3 \ge 0.$$

*In Problems 29 and 30, solve by using duals and the simplex method.*

**29.** Minimize

$$Z = 2x_1 + 7x_2 + 8x_3,$$

subject to

$$x_1 + 2x_2 + 3x_3 \ge 35,$$
$$x_1 + x_2 + x_3 \ge 25,$$
$$x_1, x_2, x_3 \ge 0.$$

**30.** Maximize

$$Z = x_1 - 2x_2,$$

subject to

$$x_1 - x_2 \le 3,$$
$$x_1 + 2x_2 \le 4,$$
$$4x_1 + x_2 \ge 2,$$
$$x_1, x_2 \ge 0.$$

**31. Production Order** A company manufactures three products: X, Y, and Z. Each product requires the use of time on machines A and B as given in the following table:

|  | Machine A | Machine B |
|---|---|---|
| Product X | 1 hr | 1 hr |
| Product Y | 2 hr | 1 hr |
| Product Z | 2 hr | 2 hr |

The numbers of hours per week that A and B are available for production are 40 and 34, respectively. The profit per unit on X, Y, and Z is $10, $15, and $22, respectively. What should be the weekly production order if maximum profit is to be obtained? What is the maximum profit?

**32.** Repeat Problem 31 if the company must produce at least a total of 24 units per week.

**33. Oil Transportation** An oil company has storage facilities for heating fuel in cities A, B, C, and D. Cities C and D are each in need of exactly 500,000 gal of fuel. The company determines that A and B can each sacrifice at most 600,000 gal to satisfy the needs of C and D. The following table gives the costs per gallon to transport fuel between the cities:

|  | To | |
|---|---|---|
| From | C | D |
| A | $0.01 | $0.02 |
| B | $0.02 | $0.04 |

How should the company distribute the fuel in order to minimize the total transportation cost? What is the minimum transportation cost?

**34. Profit** J. Smith, who operates from a pay-phone booth, sells two electronic devices: the "Zeta" and the "Gamma." These devices are built for Smith by three friends, named A, B, and C, each of whom must do some of the work on each device. The time that each must spend in the manufacture of each device is given in the following table:

|  | Friend A | Friend B | Friend C |
|---|---|---|---|
| Zeta | 2 hr | 1 hr | 1 hr |
| Gamma | 1 hr | 1 hr | 3 hr |

Smith's friends have other work to do, but they find that each month they can spend up to 70, 50, and 90 hours, respectively, for work on Smith's products. Smith makes a profit of $5 on each Zeta and $7 on each Gamma device. How many of each device should Smith have built each month to maximize profit, and what would be this maximum profit?

**35. Diet Formulation** A technician in a zoo must formulate a diet from two commercial products, food A and food B, for a certain group of animals. Each ounce of food A contains 8 units of fat, 16 units of carbohydrate, and 2 units of protein. Each ounce of food B contains 4 units of fat, 32 units of carbohydrate, and 5 units of protein. The minimum daily requirements are 176 units of fat, 1024 units of carbohydrate, and 200 units of protein. If food A costs 4 cents per ounce and food B costs 11 cents per ounce, how many ounces of each food should be used to meet the minimum daily requirements at the least cost? (Assume that a minimum cost exists.)

[8]Refers to Sec. 7.5.

*In Problems 36 and 37, do not use the simplex method. Round your answers to two decimal places.*

 **36.** Minimize

$$Z = 4.2x - 2.1y,$$

subject to

$$y \leq 3.4 + 1.2x,$$
$$y \leq -7.6 + 3.5x,$$
$$y \leq 18.7 - 0.6x,$$
$$x, y \geq 0.$$

 **37.** Maximize

$$Z = 12.4x + 8.3y,$$

subject to

$$1.4x + 1.7y \leq 15.9,$$
$$3.6x - 2.6y \geq -10.7,$$
$$1.3x - 4.3y \leq -5.2,$$
$$x, y \geq 0.$$

# MATHEMATICAL *SNAPSHOT*

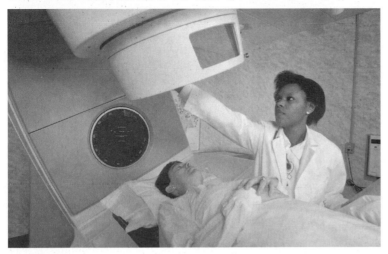

## DRUG AND RADIATION THERAPIES[9]

Frequently there are alternative forms of treatment available for a patient diagnosed to have a particular disease complex. With each treatment there may be not only positive effects on the patient but also negative effects, such as toxicity or discomfort. A physician must make the best choice of these treatments or combination of treatments. This choice will depend not only on curative effects but also on the effects of toxicity and discomfort.

Suppose that you are a physician with a cancer patient under your care and there are two possible treatments available: drug administration and radiation therapy. Let us assume that the efficacies of the treatments are expressed in common units, say, curative units. The drug contains 1000 curative units per ounce, and the radiation gives 1000 curative units per minute. Your analysis indicates that the patient must receive at least 3000 curative units.

However, a degree of toxicity is involved with each treatment. Assume that the toxic effects of each treatment are measured in a common unit of toxicity, say, a toxic unit. The drug contains 400 toxic units per ounce, and the radiation induces 1000 toxic units per minute. Based on your studies, you believe that the patient must receive not more than 2000 toxic units.

In addition, each treatment involves a degree of discomfort to the patient. The drug is three times as noxious per ounce as the radiation per minute.

Table 7.5 summarizes the data. The problem posed to you is to determine the doses of the drug and radiation that will satisfy the curative and toxicity requirements and, at the same time, minimize the discomfort to the patient.

**TABLE 7.5**

|  | Curative Units | Toxic Units | Relative Discomfort |
|---|---|---|---|
| Drug (per ounce) | 1000 | 400 | 3 |
| Radiation (per minute) | 1000 | 1000 | 1 |
| Requirement | $\geq 3000$ | $\leq 2000$ | |

Let $x_1$ be the number of ounces of the drug and $x_2$ be the number of minutes of radiation to be administered. Then you want to minimize the discomfort $D$ given by

$$D = 3x_1 + x_2,$$

subject to the curative condition

$$1000x_1 + 1000x_2 \geq 3000$$

and the toxic condition

$$400x_1 + 1000x_2 \leq 2000,$$

where $x_1 \geq 0$ and $x_2 \geq 0$. You should recognize that this is a linear programming problem. By graphing, the feasible region shown in Fig. 7.28 is obtained. The corner points are $(3, 0)$, $(5, 0)$, and $(\frac{5}{3}, \frac{4}{3})$. Evaluating $D$ at each corner point gives the following:

[9]Adapted from R. S. Ledley and L. B. Lusted, "Medical Diagnosis and Modern Decision Making," *Proceedings of Symposia in Applied Mathematics*, Vol. XIV; *Mathematical Problems in the Biological Sciences* (American Mathematical Society, 1962).

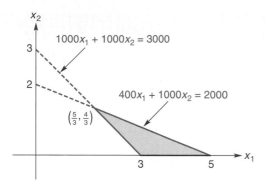

**FIGURE 7.28** Feasible region for drug and radiation problem.

$$\text{at } (3,0), \quad D = 3(3) + 0 = 9,$$

$$\text{at } (5,0), \quad D = 3(5) + 0 = 15,$$

and $\text{at } \left(\frac{5}{3}, \frac{4}{3}\right), \quad D = 3\left(\frac{5}{3}\right) + \frac{4}{3} = \frac{19}{3} \approx 6.3.$

Because $D$ is minimum at $\left(\frac{5}{3}, \frac{4}{3}\right)$, you should prescribe a treatment of $\frac{5}{3}$ ounces of the drug and $\frac{4}{3}$ minutes of radiation. Thus by solving a linear programming problem, you have determined the "best" treatment for the patient.

### ■ Exercises

1. Suppose that drug and radiation treatments are available to a patient. Each ounce of the drug contains 500 curative units and 400 toxic units. Each minute of radiation gives 1000 curative units and 600 toxic units. The patient requires at least 2000 curative units and can tolerate no more than 1400 toxic units. If each ounce of the drug is as noxious as each minute of radiation, determine the doses of the drug and radiation so that the discomfort to the patient is minimized.

2. Suppose that drug A, drug B, and radiation therapy are treatments available to a patient. Each ounce of drug A contains 600 curative units and 500 toxic units. Each ounce of drug B contains 500 curative units and 100 toxic units. Each minute of radiation gives 1000 curative units and 1000 toxic units. The patient requires at least 3000 curative units and can tolerate no more than 2000 toxic units. If each ounce of drug A and each minute of radiation are equally as noxious, and each ounce of drug B is twice as noxious as each ounce of drug A, determine the doses of the drugs and radiation so that the discomfort to the patient is minimized.

# Mathematics of Finance

**To extend the notion of compound interest to include effective rates and to solve interest problems whose solutions require logarithms.**

## 8.1 COMPOUND INTEREST

In this chapter we shall model selected topics in finance that deal with the time value of money, such as investments, loans, and so on. In later chapters, when more mathematics is at our disposal, certain topics will be revisited and expanded.

Let us first review some facts from Sec. 5.1, where the notion of compound interest was introduced. Under compound interest, at the end of each interest period the interest earned for that period is added to the *principal* (the invested amount) so that it, too, earns interest over the next interest period. The basic formula for the value (or *compound amount*) of an investment after *n* interest periods under compound interest is as follows:

**Compound Interest Formula**

For an original principal of $P$, the formula

$$S = P(1 + r)^n, \qquad (1)$$

gives the **compound amount** $S$ at the end of $n$ interest (or *conversion*) *periods* at the *periodic rate* of $r$.

The compound amount is also called the *accumulated amount,* and the difference between the compound amount and the original principal, $S - P$, is called the *compound interest.*

Recall that an interest rate is usually quoted as an *annual* rate, called the *nominal rate* or the *annual percentage rate* (A.P.R.). The periodic rate (or rate per conversion period) is obtained by dividing the nominal rate by the number of conversion periods per year.

For example, let us compute the compound amount when $1000 is invested for 5 years at the nominal rate of 12% compounded quarterly. The rate *per period* is 0.12/4, and the number of interest periods is 5(4). From Eq. (1), we have

$$S = 1000\left(1 + \frac{0.12}{4}\right)^{5(4)}$$

$$= 1000(1 + 0.03)^{20} \approx \$1806.11.$$

Keep a calculator handy as you read this chapter.

**Principles in Practice 1**
**Compound Interest**

Suppose you leave an initial amount of $518 in a savings account for three years. If interest is compounded daily (365 times per year), use a graphics calculator to graph the compound amount $S$ as a function of the nominal rate of interest. From the graph, estimate the nominal rate of interest so that there is $600 after three years.

**EXAMPLE 1  Compound Interest**

*Suppose that $500 amounted to $588.38 in a savings account after three years. If interest was compounded semiannually, find the nominal rate of interest, compounded semiannually, that was earned by the money.*

*Solution:* Let $r$ be the semiannual rate. There are $2(3) = 6$ interest periods. From Eq. (1),

$$500(1 + r)^6 = 588.38,$$

$$(1 + r)^6 = \frac{588.38}{500},$$

$$1 + r = \sqrt[6]{\frac{588.38}{500}},$$

$$r = \sqrt[6]{\frac{588.38}{500}} - 1 \approx 0.0275.$$

Thus, the semiannual rate was 2.75%, so the nominal rate was $5\frac{1}{2}\%$ compounded semiannually. ∎

**EXAMPLE 2  Doubling Money**

*At what nominal rate of interest, compounded yearly, will money double in eight years?*

*Solution:* Let $r$ be the rate at which a principal of $P$ doubles in eight years. Then the compound amount is $2P$. From Eq. (1),

$$P(1 + r)^8 = 2P,$$

$$(1 + r)^8 = 2,$$

$$1 + r = \sqrt[8]{2},$$

$$r = \sqrt[8]{2} - 1 \approx 0.0905.$$

Note that the doubling rate is independent of the principal $P$.

Hence, the desired rate is 9.05%. ∎

We can determine how long it takes for a given principal to accumulate to a particular amount by using logarithms, as Example 3 shows.

**Principles in Practice 2**
**Compound Interest**

Suppose you leave an initial amount of $520 in a savings account at an annual rate of 5.2% compounded daily (365 days per year). Use a graphics calculator to graph the compound amount $S$ as a function of the interest periods. From the graph, estimate how long it takes for the amount to accumulate to $750.

**EXAMPLE 3  Compound Interest**

*How long will it take for $600 to amount to $900 at an annual rate of 8% compounded quarterly?*

*Solution:* The periodic rate is $r = 0.08/4 = 0.02$. Let $n$ be the number of interest periods it takes for a principal of $P = 600$ to amount to $S = 900$. Then, from Eq. (1),

$$900 = 600(1.02)^n, \qquad (2)$$

$$(1.02)^n = \frac{900}{600},$$

$$(1.02)^n = 1.5.$$

To solve for $n$, we first take the natural logarithms of both sides:

$$\ln(1.02)^n = \ln 1.5,$$

$$n \ln 1.02 = \ln 1.5 \qquad \text{since } \ln m^r = r \ln m,$$

$$n = \frac{\ln 1.5}{\ln 1.02} \approx 20.475.$$

The number of years that corresponds to 20.475 quarterly interest periods is $20.475/4 \approx 5.1188$, which is slightly more than 5 years and 1 month. Actually, the principal doesn't amount to $900 until $5\frac{1}{4}$ years pass, because interest is compounded quarterly. ∎

## TECHNOLOGY

We can solve Eq. (2) in Example 3 by graphing

$$Y_1 = 900$$

$$Y_2 = 600(1.02)\text{^}X$$

and finding the intersection. (See Fig. 8.1.)

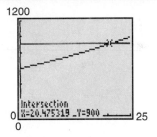

**FIGURE 8.1** Solution of Example 3.

### Effective Rate

If $P$ dollars are invested at a nominal rate of 12% compounded quarterly for one year, the principal will earn more than 12% that year. In fact, the compound interest is

$$S - P = P\left(1 + \frac{0.12}{4}\right)^4 - P = [(1.03)^4 - 1]P$$

$$\approx 0.125509P,$$

which is about 12.55% of $P$. That is, 12.55% is the approximate rate of interest compounded *annually* that is actually earned, and that rate is called the **effective rate** of interest, or the **yield.** The effective rate is independent of $P$. In general, the effective interest rate is just the rate of *simple* interest earned over a period of one year. Thus, we have shown that the nominal rate of 12% compounded quarterly is equivalent to an effective rate of 12.55%. Following the preceding procedure, we can generalize our result:

### Effective Rate

The **effective rate** $r_e$ that is equivalent to a nominal rate of $r$ compounded $n$ times a year is given by

$$r_e = \left(1 + \frac{r}{n}\right)^n - 1. \tag{3}$$

**EXAMPLE 4** **Effective Rate**

*What effective rate is equivalent to a nominal rate of 6% compounded (a) semi-annually and (b) quarterly?*

*Solution:*

**a.** From Eq. (3), the effective rate is

$$r_e = \left(1 + \frac{0.06}{2}\right)^2 - 1 = (1.03)^2 - 1 = 0.0609, \quad \text{or} \quad 6.09\%.$$

**b.** The effective rate is

$$r_e = \left(1 + \frac{0.06}{4}\right)^4 - 1 = (1.015)^4 - 1 \approx 0.061364, \quad \text{or} \quad 6.14\%. \quad \blacksquare$$

Example 4 illustrates that, for a given nominal rate $r$, the effective rate increases as the number of interest periods per year ($n$) increases. However, in Sec. 11.3 it is shown that, regardless of how large $n$ is, the maximum effective rate that can be obtained is $e^r - 1$.

**EXAMPLE 5** **Effective Rate**

*To what amount will $12,000 accumulate in 15 years if it is invested at an effective rate of 5%?*

*Solution:* Since an effective rate is the rate that is compounded annually, we have

$$S = 12,000(1.05)^{15} \approx \$24,947.14. \quad \blacksquare$$

**EXAMPLE 6** **Doubling Money**

*How many years will it take for money to double at the effective rate of r?*

*Solution:* Let $n$ be the number of years it takes for a principal of $P$ to double. Then the compound amount is $2P$. Thus,

$$2P = P(1 + r)^n,$$

$$2 = (1 + r)^n,$$

$$\ln 2 = n \ln(1 + r) \quad \text{(taking logs of both sides)}.$$

Hence,

$$n = \frac{\ln 2}{\ln(1 + r)}.$$

For example, if $r = 0.06$, the number of years it takes to double a principal is

$$\frac{\ln 2}{\ln 1.06} \approx 11.9 \text{ years}. \quad \blacksquare$$

We remark that when alternative interest rates are available to an investor, effective rates are used to compare them—that is, to determine which of them is the "best." The next example illustrates.

## Principles in Practice 4
### Comparing Interest Rates

Suppose you have two investment opportunities. You can invest $10,000 at 11% compounded monthly, or you can invest $9700 at 11.25% compounded quarterly. Which has the better effective rate of interest? Which is the better investment over 20 years?

**EXAMPLE 7    Comparing Interest Rates**

*If an investor has a choice of investing money at 6% compounded daily or $6\frac{1}{8}\%$ compounded quarterly, which is the better choice?*

*Solution:*

> *Strategy:*   We determine the equivalent effective rate of interest for each nominal rate and then compare our results.

The respective effective rates of interest are

$$r_e = \left(1 + \frac{0.06}{365}\right)^{365} - 1 \approx 6.18\%$$

and

$$r_e = \left(1 + \frac{0.06125}{4}\right)^4 - 1 \approx 6.27\%.$$

Since the second choice gives the higher effective rate, it is the better choice (in spite of the fact that daily compounding may be psychologically more appealing).    ■

## ■ Exercise 8.1

*In Problems 1 and 2, find (a) the compound amount and (b) the compound interest for the given investment and rate.*

**1.** $6000 for eight years at an effective rate of 8%.

**2.** $750 for 12 months at an effective rate of 10%.

*In Problems 3–6, find the effective rate that corresponds to the given nominal rate. Round answers to three decimal places.*

**3.** 8% compounded quarterly.

**4.** 12% compounded monthly.

**5.** 8% compounded daily.

**6.** 12% compounded daily.

**7.** Find the effective rate of interest (rounded to three decimal places) that is equivalent to a nominal rate of 10% compounded.
  **a.** yearly,
  **b.** semiannually,
  **c.** quarterly,
  **d.** monthly,
  **e.** daily.

**8.** Find (i) the compound interest (rounded to two decimal places) and (ii) the effective rate (to three decimal places) if $1000 is invested for five years at an annual rate of 7% compounded
  **a.** quarterly,
  **b.** monthly,
  **c.** weekly,
  **d.** daily.

**9.** Over a five-year period, an original principal of $2000 accumulated to $2950 in an account in which interest was compounded quarterly. Determine the effective rate of interest, rounded to two decimal places.

**10.** Suppose that over a six-year period, $1000 accumulated to $1725 in an investment certificate in which interest was compounded quarterly. Find the nominal rate of interest, compounded quarterly, that was earned. Round your answer to two decimal places.

*In Problems 11 and 12, find how many years it would take to double a principal at the given effective rate. Give your answer to one decimal place.*

**11.** 8%.

**12.** 5%.

**13.** A $6000 certificate of deposit is purchased for $6000 and is held seven years. If the certificate earns an effective rate of 8%, what is it worth at the end of that period?

**14.** How many years will it take for money to triple at the effective rate of $r$?

15. **College Costs** Suppose attending a certain college costs $18,500 in the 1996–1997 school year. This price includes tuition, room, board, books, and other expenses. Assuming an effective 6% inflation rate for these costs, determine what the college costs will be in the 2006–2007 school year.

16. **College Costs** Repeat Problem 15 for an inflation rate of 6% compounded semiannually.

17. **Finance Charge** A major credit-card company has a finance charge of $1\frac{1}{2}$% per month on the outstanding indebtedness. (a) What is the nominal rate compounded monthly? (b) What is the effective rate?

18. How long would it take for a principal of $P$ to double if money is worth 12% compounded monthly? Give your answer to the nearest month.

19. To what sum will $2000 amount in eight years if invested at a 6% effective rate for the first four years and at 6% compounded semiannually thereafter?

20. How long will it take for $500 to amount to $700 if invested at 8% compounded quarterly?

21. An investor has a choice of investing a sum of money at 8% compounded annually or at 7.8% compounded semiannually. Which is the better of the two rates?

22. What nominal rate of interest, compounded quarterly, corresponds to an effective rate of 4%?

23. **Savings Account** A bank advertises that it pays interest on savings accounts at the rate of $5\frac{1}{4}$% compounded daily. Find the effective rate if the bank assumes that a year consists of (a) 360 days or (b) 365 days in determining the *daily rate*. Assume that compounding occurs 365 times a year, and round your answer to two decimal places.

24. **Savings Account** Suppose that $700 amounted to $801.06 in a savings account after two years. If interest was compounded quarterly, find the nominal rate of interest, compounded quarterly, that was earned by the money.

25. **Inflation** As a hedge against inflation, an investor purchased a painting in 1980 for $100,000. It was sold in 1990 for $300,000. At what effective rate did the painting appreciate in value?

26. **Inflation** If the rate of inflation for certain goods is $7\frac{1}{4}$% compounded daily, how many years will it take for the average price of such a good to double?

27. **Zero-Coupon Bond** A *zero-coupon bond* is a bond that is sold for less than its face value (that is, it is *discounted*) and has no periodic interest payments. Instead, the bond is redeemed for its face value at maturity. Thus, in this sense, interest is paid at maturity. Suppose that a zero-coupon bond sells for $220 and can be redeemed in 14 years for its face value of $1000. The bond earns interest at what nominal rate, compounded semiannually?

28. **Inflation** Suppose $1000 is hidden under a mattress for safekeeping. Each year, because of inflation, the purchasing power of the money is 95% of what it was the previous year. After five years, what is the purchasing power of the $1000? [*Hint:* Consider Eq. (1) with $r = -0.05$.]

---

OBJECTIVE

To study present value and to solve problems involving the time value of money by using equations of value. To introduce the net present value of cash flows.

# 8.2 Present Value

Suppose that $100 is deposited in a savings account that pays 6% compounded annually. Then at the end of two years, the account is worth

$$100(1.06)^2 = \$112.36.$$

To describe this relationship, we say that the compound amount of $112.36 is the *future value* of the $100, and $100 is the *present value* of the $112.36. In general, there are times when we may know the future value of an investment and wish to find the present value. To obtain a formula for doing this, we solve the equation $S = P(1 + r)^n$ for $P$. The result is $P = S/(1 + r)^n = S(1 + r)^{-n}$.

**Present Value**

The principal $P$ which must be invested at the periodic rate of $r$ for $n$ interest periods so that the compound amount is $S$ is given by

$$P = S(1 + r)^{-n} \qquad (1)$$

and is called the **present value** of $S$.

**EXAMPLE 1    Present Value**

*Find the present value of* $1000 *due after three years if the interest rate is* 9% *compounded monthly.*

***Solution:*** We use Eq. (1) with $S = 1000, r = 0.09/12 = 0.0075$, and $n = 3(12) = 36$:

$$P = 1000(1.0075)^{-36} \approx \$764.15.$$

This means that $764.15 must be invested at 9% compounded monthly to have $1000 in three years. ■

If the interest rate in Example 1 were 10% compounded monthly, the present value would be

$$P = 1000 \left( 1 + \frac{0.1}{12} \right)^{-36} \approx \$741.74,$$

which is less than before. It is typical that the present value for a given future value decreases as the interest rate per conversion period increases.

**EXAMPLE 2    Single–Payment Trust Fund**

*A trust fund for a child's education is being set up by a single payment so that at the end of* 15 *years there will be* $24,000. *If the fund earns interest at the rate of* 7% *compounded semiannually, how much money should be paid into the fund?*

***Solution:*** We want the present value of $24,000, due in 15 years. From Eq. (1) with $S = 24,000, r = 0.07/2 = 0.035$, and $n = 15(2) = 30$, we have

$$P = 24,000(1.035)^{-30} \approx \$8550.68.$$
■

### Equations of Value

Suppose that Mr. Smith owes Mr. Jones two sums of money: $1000, due in two years, and $600, due in five years. If Mr. Smith wishes to pay off the total debt now by a single payment, how much should the payment be? Assume an interest rate of 8% compounded quarterly.

The single payment $x$ due now must be such that it would grow and eventually pay off the debts when they are due. That is, it must equal the sum of the present values of the future payments. As shown in Fig. 8.2, we have

$$x = 1000(1.02)^{-8} + 600(1.02)^{-20}. \tag{2}$$

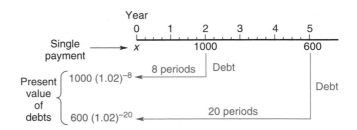

**FIGURE 8.2**    Replacing two future payments by a single payment now.

Figure 8.2 is a useful tool for visualizing the time value of money. It is a great aid in setting up an equation of value.

This equation is called an *equation of value*. We find that

$$x \approx \$1257.27.$$

Thus, the single payment now due is \$1257.27. Let us analyze the situation in more detail. There are two methods of payment of the debt: a single payment now or two payments in the future. Notice that Eq. (2) indicates that the value *now* of all payments under one method must equal the value *now* of all payments under the other method. In general, this is true not just *now,* but at *any time.* For example, if we multiply both sides of Eq. (2) by $(1.02)^{20}$, we get the equation of value

$$x(1.02)^{20} = 1000(1.02)^{12} + 600. \qquad (3)$$

The left side of Eq. (3) gives the value five years from now of the single payment (see Fig. 8.3), while the right side gives the value five years from now of all payments under the other method. Solving Eq. (3) for $x$ gives the same

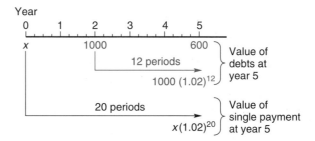

**FIGURE 8.3** Diagram for equation of value.

result, $x \approx \$1257.27$. In general, an **equation of value** illustrates that when one is considering two methods of paying a debt (or of making some other transaction), *at any time* the value of all payments under one method must equal the value of all payments under the other method.

In certain situations, one equation of value may be more convenient to use than another, as Example 3 illustrates.

**EXAMPLE 3** Equation of Value

*A debt of* \$3000 *due six years from now is instead to be paid off by three payments:* \$500 *now,* \$1500 *in three years, and a final payment at the end of five years. What would this payment be if an interest rate of* 6% *compounded annually is assumed?*

*Solution:* Let $x$ be the final payment due in five years. For computational convenience, we shall set up an equation of value to represent the situation at the end of that time, for in that way the coefficient of $x$ will be 1, as seen in Fig. 8.4.

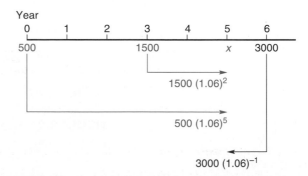

**FIGURE 8.4** Time values of payments for Example 3.

Notice that at year 5 we compute the future values of $500 and $1500, and the present value of $3000. The equation of value is

$$500(1.06)^5 + 1500(1.06)^2 + x = 3000(1.06)^{-1},$$

so

$$x = 3000(1.06)^{-1} - 500(1.06)^5 - 1500(1.06)^2$$

$$\approx \$475.68.$$ ▪

When one is considering a choice of two investments, a comparison should be made of the value of each investment at a certain time, as Example 4 shows.

### EXAMPLE 4   Comparing Investments

*Suppose that you had the opportunity of investing $5000 in a business such that the value of the investment after five years would be $6300. On the other hand, you could instead put the $5000 in a savings account that pays 6% compounded semiannually. Which investment is better?*

*Solution:* Let us consider the value of each investment at the end of five years. At that time the business investment would have a value of $6300, while the savings account would have a value of $5000(1.03)^{10} \approx \$6719.58$. Clearly, the better choice is putting the money in the savings account. ▪

### Net Present Value

If an initial investment will bring in payments at future times, the payments are called **cash flows.** The **net present value,** denoted NPV, of the cash flows is defined to be the sum of the present values of the cash flows, minus the initial investment. If NPV > 0, then the investment is profitable; if NPV < 0, the investment is not profitable.

### EXAMPLE 5   Net Present Value

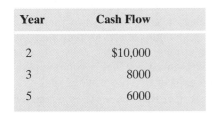

| Year | Cash Flow |
|------|-----------|
| 2 | $10,000 |
| 3 | 8000 |
| 5 | 6000 |

*Suppose that you can invest $20,000 in a business that guarantees you cash flows at the end of years, 2, 3, and 5, as indicated in the table to the left. Assume an interest rate of 7% compounded annually, and find the net present value of the cash flows.*

*Solution:* Subtracting the initial investment from the sum of the present values of the cash flows gives

$$\text{NPV} = 10,000(1.07)^{-2} + 8000(1.07)^{-3} + 6000(1.07)^{-5} - 20,000$$

$$\approx -\$457.31.$$

Since NPV < 0, the business venture is not profitable if one considers the time value of money. It would be better to invest the $20,000 in a bank paying 7%, since the venture is equivalent to investing only $20,000 - 457.31 = \$19,542.69$. ▪

## ■ Exercise 8.2

*In Problems* **1–10,** *find the present value of the given future payment at the specified interest rate.*

**1.** $6000 due in 20 years at 5% compounded annually.

**2.** $3500 due in eight years at 6% effective.

**3.** $4000 due in 12 years at 7% compounded semiannually.

**4.** $750 due in three years at 18% compounded monthly.

5. $8000 due in $7\frac{1}{2}$ years at 6% compounded quarterly.

6. $6000 due in $6\frac{1}{2}$ years at 10% compounded semiannually.

7. $8000 due in five years at 10% compounded monthly.

8. $500 due in three years at $8\frac{3}{4}$% compounded quarterly.

9. $10,000 due in four years at $9\frac{1}{2}$% compounded daily.

10. $1250 due in $1\frac{1}{2}$ years at $13\frac{1}{2}$% compounded weekly.

11. A bank account pays 12% annual interest, compounded monthly. How much must be deposited now so that the account contains exactly $10,000 at the end of one year?

12. Repeat Problem 11 for the nominal rate of 6% compounded quarterly.

13. **Trust Fund** A trust fund for a 10-year-old child is being set up by a single payment so that at age 21 the child will receive $27,000. Find how much the payment is if an interest rate of 6% compounded semiannually is assumed.

14. A debt of $550 due in four years and $550 due in five years is to be repaid by a single payment now. Find how much the payment is if an interest rate of 10% compounded quarterly is assumed.

15. A debt of $600 due in three years and $800 due in four years is to be repaid by a single payment two years from now. If the interest rate is 8% compounded semiannually, how much is the payment?

16. A debt of $5000 due in five years is to be repaid by a payment of $2000 now and a second payment at the end of six years. How much should the second payment be if the interest rate is 6% compounded quarterly?

17. A debt of $5000 due five years from now and $5000 due ten years from now is to be repaid by a payment of $2000 in two years, a payment of $4000 in four years, and a final payment at the end of six years. If the interest rate is 7% compounded annually, how much is the final payment?

18. A debt of $2000 due in three years and $3000 due in seven years is to be repaid by a single payment of $1000 now and two equal payments that are due one year from now and four years from now. If the interest rate is 6% compounded annually, how much are each of the equal payments?

19. **Cash Flows** An initial investment of $25,000 in a business guarantees the following cash flows:

| Year | Cash Flow |
|------|-----------|
| 3 | $ 8000 |
| 4 | $10,000 |
| 6 | $14,000 |

Assume an interest rate of 5% compounded semiannually.

a. Find the net present value of the cash flows.

b. Is the investment profitable?

20. **Cash Flows** Repeat Problem 19 for the interest rate of 6% compounded semiannually.

21. **Decision Making** Suppose that a person has the following choices of investing $10,000:

a. placing the money in a savings account paying 6% compounded semiannually;

b. investing in a business such that the value of the investment after 8 years is $16,000.

Which is the better choice?

22. A owes B two sums of money: $1000 plus interest at 7% compounded annually, which is due in five years, and $2000 plus interest at 8% compounded semiannually, which is due in seven years. If both debts are to be paid off by a single payment at the end of six years, find the amount of the payment if money is worth 6% compounded quarterly.

23. **Purchase Incentive** A jewelry store advertises that for every $1000 spent on diamond jewelry, the purchaser receives a $1000 bond at absolutely no cost. In reality, the $1000 is the full maturity value of a zero-coupon bond (see Problem 27 of Exercise 8.1), which the store purchases at a heavily reduced price. If the bond earns interest at the rate of 11.5% compounded quarterly and matures after 20 years, how much does the bond cost the store?

24. Find the present value of $3000 due in two years at a bank rate of 8% compounded daily. Assume that the bank uses 360 days in determining the daily rate and that there are 365 days in a year; that is, compounding occurs 365 times in a year.

25. **Promissory Note** A *(promissory) note* is a written statement agreeing to pay a sum of money either on demand or at a definite future time. When a note is purchased for its present value at a given interest rate, the note is said to be *discounted,* and the interest rate is called the *discount rate.* Suppose a $10,000 note due eight years from now is sold to a financial institution for $4700. What is the nominal discount rate with quarterly compounding?

**To introduce the notions of ordinary annuities and annuities due. To use geometric series to model the present value and future value of an annuity. To determine payments to be placed in a sinking fund.**

# 8.3 ANNUITIES

## Sequences and Geometric Series

In mathematics we use the word **sequence** to describe a list of numbers, called *terms,* that are arranged in a definite order. For example, the list

$$2, 4, 6, 8$$

is a (finite) sequence. The first term is 2, the second is 4, and so on.

In the sequence

$$3, 6, 12, 24, 48,$$

each term, after the first, can be obtained by multiplying the preceding term by 2:

$$6 = 3(2), \qquad 12 = 6(2), \qquad \text{and so on.}$$

This means that the *ratio* of every two consecutive terms is 2:

$$\frac{6}{3} = 2, \qquad \frac{12}{6} = 2, \qquad \text{and so on.}$$

We call the sequence a *geometric sequence* with *common ratio* 2. Note that it can be written as

$$3, 3(2), 3(2)(2), 3(2)(2)(2), 3(2)(2)(2)(2)$$

or in the form

$$3, 3(2), 3(2^2), 3(2^3), 3(2^4).$$

More generally, if a geometric sequence has $n$ terms such that the first term is $a$ and the common ratio is the constant $r$, then the sequence has the form

$$a, ar, ar^2, ar^3, \ldots, ar^{n-1}.$$

Note that the $n$th term in the sequence is $ar^{n-1}$.

### DEFINITION

*The sequence of n numbers*

$$a, ar, ar^2, \ldots, ar^{n-1}, \qquad \text{where } a \neq 0,\text{[1]}$$

*is called a **geometric sequence** with **first term** a and **common ratio** r.*

---

**Principles in Practice 1**

**Geometric Sequences**

A rubber ball always bounces back $\dfrac{3}{4}$ of its previous height. If the ball is dropped from a height of 64 feet, what are the next five heights of the ball?

---

**EXAMPLE 1    Geometric Sequences**

**a.** The geometric sequence with $a = 3$, common ratio $\frac{1}{2}$, and $n = 5$ is

$$3, 3(\tfrac{1}{2}), 3(\tfrac{1}{2})^2, 3(\tfrac{1}{2})^3, 3(\tfrac{1}{2})^4,$$

or

$$3, \tfrac{3}{2}, \tfrac{3}{4}, \tfrac{3}{8}, \tfrac{3}{16}.$$

**b.** The numbers

$$1, 0.1, 0.01, 0.001$$

---

[1]If $a = 0$, the sequence is $0, 0, 0, \ldots, 0$. We shall not consider this uninteresting case.

form a geometric sequence with $a = 1$, $r = 0.1$, and $n = 4$. ∎

**Principles in Practice 2**
Geometric Sequence

Suppose bacteria increases at a rate of 50% each minute for six minutes. If the starting population is 500, list the population at the end of each minute as a geometric sequence.

**EXAMPLE 2  Geometric Sequence**

If \$100 is invested at the rate of 6% compounded annually, then the list of compound amounts at the end of each year for eight years is

$$100(1.06),\ 100(1.06)^2,\ 100(1.06)^3,\ \ldots,\ 100(1.06)^8.$$

This is a geometric sequence with common ratio 1.06. ∎

The indicated sum of the terms of the geometric sequence $a, ar, ar^2, \ldots, ar^{n-1}$ is called a **geometric series:**

$$a + ar + ar^2 + \cdots + ar^{n-1}. \tag{1}$$

For example,

$$1 + \tfrac{1}{2} + (\tfrac{1}{2})^2 + \cdots + (\tfrac{1}{2})^6$$

is a geometric series with $a = 1$, common ratio $r = \tfrac{1}{2}$, and $n = 7$.

Let us compute the sum $s$ of the geometric series in Eq. (1):

$$s = a + ar + ar^2 + \cdots + ar^{n-1}. \tag{2}$$

We can express $s$ in a more compact form. Multiplying both sides by $r$ gives

$$rs = ar + ar^2 + ar^3 + \cdots + ar^n. \tag{3}$$

Subtracting corresponding sides of Eq. (3) from Eq. (2) yields

$$s - rs = a - ar^n,$$

$$s(1 - r) = a(1 - r^n) \qquad \text{(factoring)}.$$

Dividing both sides by $1 - r$, we have $s = a(1 - r^n)/(1 - r)$.

> **Sum of Geometric Series**
>
> The **sum** $s$ of a geometric series[2] of $n$ terms whose first term is $a$ and common ratio is $r$ is given by
>
> $$s = \frac{a(1 - r^n)}{1 - r}. \tag{4}$$

**Principles in Practice 3**
Sum of Geometric Series

A ball rebounds $\dfrac{2}{3}$ of its previous height after each bounce. If the ball is tossed up to a height of 6 meters, how far has it traveled in the air when it hits the ground for the 12th time?

**EXAMPLE 3  Sum of Geometric Series**

*Find the sum of the geometric series*

$$1 + \tfrac{1}{2} + (\tfrac{1}{2})^2 + \cdots + (\tfrac{1}{2})^6.$$

*Solution:* Here $a = 1$, $r = \tfrac{1}{2}$, and $n = 7$ (not 6). From Eq. (4), we have

$$s = \frac{a(1 - r^n)}{1 - r} = \frac{1[1 - (\tfrac{1}{2})^7]}{1 - \tfrac{1}{2}} = \frac{\tfrac{127}{128}}{\tfrac{1}{2}} = \frac{127}{64}.$$ ∎

---

[2]This formula assumes that $r \neq 1$. However, if $r = 1$, then $s = a + \cdots + a = na$.

**Principles in Practice 4**

Sum of Geometric Series

A company earns a profit of $2000 in its first month. Suppose the profit increases by 10% each month for two years. Find the amount of profit the company earns in its first two years.

**EXAMPLE 4** Sum of Geometric Series

*Find the sum of the geometric series*

$$3^5 + 3^6 + 3^7 + \cdots + 3^{11}.$$

*Solution:* Here $a = 3^5$, $r = 3$, and $n = 7$. From Eq. (4),

$$s = \frac{3^5(1 - 3^7)}{1 - 3} = \frac{243(1 - 2187)}{-2} = 265{,}599.$$

■

### Present Value of an Annuity

The notion of a geometric series is the basis of the mathematical model of an *annuity*. Essentially, an **annuity** is a sequence of payments made at fixed periods of time over a given interval. The fixed period is called the **payment period,** and the given interval is the **term** of the annuity. An example of an annuity is the depositing of $100 in a savings account every three months for a year.

The **present value of an annuity** is the sum of the *present values* of all the payments. It represents the amount that must be invested now to purchase the payments due in the future. Unless otherwise specified, we assume that each payment is made at the *end* of a payment period; such an annuity is called an **ordinary annuity.** We also assume that interest is computed at the end of each payment period.

Let us consider an annuity of $n$ payments of $R$ (dollars) each, where the interest rate *per period* is $r$ (see Fig. 8.5) and the first payment is due one

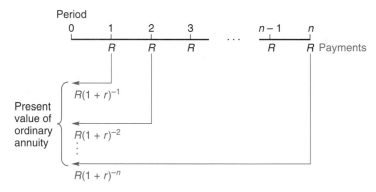

**FIGURE 8.5** Present value of ordinary annuity.

period from now. The present value of the annuity is given by

$$A = R(1 + r)^{-1} + R(1 + r)^{-2} + \cdots + R(1 + r)^{-n}.$$

This is a geometric series of $n$ terms with first term $R(1 + r)^{-1}$ and common ratio $(1 + r)^{-1}$. Hence, from Eq. (4), we obtain the formula

$$A = \frac{R(1 + r)^{-1}[1 - (1 + r)^{-n}]}{1 - (1 + r)^{-1}}$$

$$= \frac{R[1 - (1 + r)^{-n}]}{(1 + r)[1 - (1 + r)^{-1}]} = \frac{R[1 - (1 + r)^{-n}]}{(1 + r) - 1}$$

$$= R \cdot \frac{1 - (1 + r)^{-n}}{r}.$$

**Present Value of Annuity**

The formula

$$A = R \cdot \frac{1 - (1 + r)^{-n}}{r} \qquad (5)$$

gives the **present value** $A$ of an ordinary annuity of $R$ dollars per payment period for $n$ periods at the interest rate of $r$ per period.

In Eq. (5), the expression $[1 - (1 + r)^{-n}]/r$ is denoted $a_{\overline{n}|r}$ and (letting $R = 1$) represents the present value of an annuity of $1 per period. The symbol $a_{\overline{n}|r}$ is read "$a$ angle $n$ at $r$." Thus, Eq. (5) can be written as

$$A = Ra_{\overline{n}|r}. \qquad (6)$$

Selected values of $a_{\overline{n}|r}$ are given in Appendix B. (Most are approximate.)

Whenever a desired value of $a_{\overline{n}|r}$ is not in Appendix B, we shall use a calculator to compute it.

**EXAMPLE 5  Present Value of Annuity**

*Find the present value of an annuity of $100 per month for $3\frac{1}{2}$ years at an interest rate of 6% compounded monthly.*

**Solution:** Substituting in Eq. (6), we set $R = 100$, $r = 0.06/12 = 0.005$, and $n = (3\frac{1}{2})(12) = 42$. Thus,

$$A = 100a_{\overline{42}|0.005}.$$

From Appendix B, $a_{\overline{42}|0.005} \approx 37.798300$. Hence,

$$A \approx 100(37.798300) = \$3779.83.$$

■

**Principles in Practice 5**
**Present Value of Annuity**

Given a payment of $500 per month for six years, use a graphics calculator to graph the present value $A$ as a function of the interest rate per month, $r$. Determine the nominal rate if the present value of the annuity is $30,000.

**Principles in Practice 6**
**Present Value of Annuity**

Suppose a man purchases a house with an initial down payment of $20,000 and then makes quarterly payments: $2000 at the end of each quarter for six years and $3500 at the end of each quarter for eight more years. Given an interest rate of 6% compounded quarterly, find the present value of the payments and the list price of the house.

**EXAMPLE 6  Present Value of Annuity**

*Given an interest rate of 5% compounded annually, find the present value of an annuity of $2000 due at the end of each year for three years and $5000 due thereafter at the end of each year for four years. (See Fig. 8.6.)*

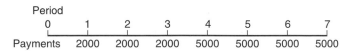

**FIGURE 8.6**  Annuity of Example 6.

**Solution:** The present value is obtained by summing the present values of all payments:

$$2000(1.05)^{-1} + 2000(1.05)^{-2} + 2000(1.05)^{-3} + 5000(1.05)^{-4} +$$

$$5000(1.05)^{-5} + 5000(1.05)^{-6} + 5000(1.05)^{-7}.$$

Rather than evaluating this expression, we can simplify our work by considering the payments to be an annuity of $5000 for seven years, minus an annuity of $3000 for three years, so that the first three payments are $2000 each. Thus, the present value is

$$5000a_{\overline{7}|0.05} - 5000a_{\overline{3}|0.05}$$

$$\approx 5000(5.786373) - 3000(2.723248)$$

$$\approx \$20,762.12.$$

■

**Principles in Practice 7**

Periodic Payment of Annuity

Given an annuity with equal payments at the end of each quarter for six years and an interest rate of 4.8% compounded quarterly, use a graphics calculator to graph the present value $A$ as a function of the monthly payment $R$. Determine the monthly payment if the present value of the annuity is \$15,000.

**EXAMPLE 7**    **Periodic Payment of Annuity**

*If* \$10,000 *is used to purchase an annuity consisting of equal payments at the end of each year for the next four years and the interest rate is 6% compounded annually, find the amount of each payment.*

*Solution:* Here $A = \$10{,}000$, $n = 4$, $r = 0.06$, and we want to find $R$. From Eq. (6),

$$10{,}000 = Ra_{\overline{4}|0.06}.$$

Solving for $R$ gives

$$R = \frac{10{,}000}{a_{\overline{4}|0.06}} \approx \frac{10{,}000}{3.465106} \approx \$2885.91.$$

In general, the formula

$$R = \frac{A}{a_{\overline{n}|r}}$$

gives the periodic payment $R$ of an ordinary annuity whose present value is $A$. ∎

**EXAMPLE 8**    **Annuity Due**

*The premiums on an insurance policy are* \$50 *per quarter, payable at the beginning of each quarter. If the policyholder wishes to pay one year's premiums in advance, how much should be paid, provided that the interest rate is 4% compounded quarterly?*

*Solution:* We want the present value of an annuity of \$50 per period for four periods at a rate of 1% per period. However, each payment is due at the *beginning* of the payment period. Such an annuity is called an **annuity due.** The given annuity can be thought of as an initial payment of \$50, followed by an ordinary annuity of \$50 for three periods. (See Fig. 8.7.) Thus, the present value is

An example of a situation involving an annuity due is an apartment lease for which the first payment is made immediately.

$$50 + 50a_{\overline{3}|0.01} \approx 50 + 50(2.940985) \approx \$197.05.$$

We remark that the general formula for the **present value of an annuity due** is $A = R + Ra_{\overline{n-1}|r}$, or

$$A = R(1 + a_{\overline{n-1}|r}).$$    ∎

**Principles in Practice 8**

Annuity Due

A man makes house payments of \$1000 at the beginning of every month. If the man wishes to pay one year's worth of payments in advance, how much should he pay, provided that the interest rate is 4.8% compounded monthly?

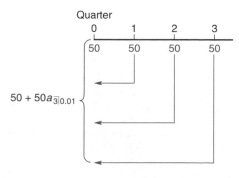

**FIGURE 8.7**    Annuity due (present value).

### Amount of an Annuity

The **amount** (or **future value**) **of an annuity** is the value, at the end of the term, of all payments. That is, it is the sum of the compound amounts of all payments. Let us consider an ordinary annuity of $n$ payments of $R$ (dollars) each, where the interest rate per period is $r$. The compound amount of the last payment is $R$, since it occurs at the end of the last interest period and hence does not accrue interest. (See Fig. 8.8.) The $(n - 1)$th payment earns interest for one

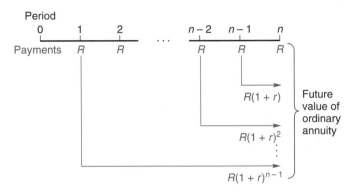

**FIGURE 8.8** Future value of ordinary annuity.

period, and so on, and the first payment earns interest for $n - 1$ periods. Thus, the future value of the annuity is

$$R + R(1 + r) + R(1 + r)^2 + \cdots + R(1 + r)^{n-1}.$$

This is a geometric series of $n$ terms with first term $R$ and common ratio $1 + r$. Consequently, its sum $S$ is [by using Eq. (4)]

$$S = \frac{R[1 - (1 + r)^n]}{1 - (1 + r)} = R \cdot \frac{1 - (1 + r)^n}{-r} = R \cdot \frac{(1 + r)^n - 1}{r}.$$

> ### Amount of an Annuity
>
> The formula
>
> $$S = R \cdot \frac{(1 + r)^n - 1}{r} \qquad (7)$$
>
> gives the **amount** $S$ of an ordinary annuity of $R$ (dollars) per payment period for $n$ periods at the interest rate of $r$ per period.

The expression $[(1 + r)^n - 1]/r$ is abbreviated $s_{\overline{n}|r}$, and some approximate values of $s_{\overline{n}|r}$ are given in Appendix B. Thus,

$$S = Rs_{\overline{n}|r}. \qquad (8)$$

---

**Principles in Practice 9**

**Amount of Annuity**

Suppose you invest in an IRA by depositing $2000 at the end of every tax year for the next 15 years. If the interest rate is 8.7% compounded annually, how much will you have at the end of 15 years?

---

### EXAMPLE 9    Amount of Annuity

*Find the amount of an annuity consisting of payments of $50 at the end of every three months for three years at the rate of 6% compounded quarterly. Also, find the compound interest.*

**Solution:** To find the amount of the annuity, we use Eq. (8) with $R = 50$, $n = 4(3) = 12$, and $r = 0.06/4 = 0.015$

$$S = 50s_{\overline{12}|0.015} \approx 50(13.041211) \approx \$652.06.$$

The compound interest is the difference between the amount of the annuity and the sum of the payments, namely,

$$652.06 - 12(50) = 652.06 - 600 = \$52.06.$$ ■

**Principles in Practice 10**

**Amount of Annuity Due**

Suppose you invest in an IRA by depositing \$2000 at the beginning of every tax year for the next 15 years. If the interest rate is 8.7% compounded annually, how much will you have at the end of 15 years?

**EXAMPLE 10    Amount of Annuity Due**

*At the beginning of each quarter, \$50 is deposited into a savings account that pays 6% compounded quarterly. Find the balance in the account at the end of three years.*

*Solution:* Since the deposits are made at the beginning of a payment period, we want the amount of an *annuity due*, as defined in Example 8. (See Fig. 8.9.)

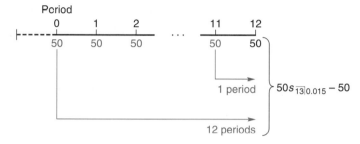

**FIGURE 8.9**    Future value of annuity due.

The given annuity can be thought of as an ordinary annuity of \$50 for 13 periods, minus the final payment of \$50. Thus, the amount is

$$50s_{\overline{13}|0.015} - 50 \approx 50(14.236830) - 50 \approx \$661.84.$$

The formula for the **future value of an annuity due** is $S = Rs_{\overline{n+1}|r} - R$, or

$$S = R(s_{\overline{n+1}|r} - 1).$$ ■

**Sinking Fund**

Our final examples involve the notion of a *sinking fund*.

**EXAMPLE 11    Sinking Fund**

*A **sinking fund** is a fund into which periodic payments are made in order to satisfy a future obligation. Suppose a machine costing \$7000 is to be replaced at the end of eight years, at which time it will have a salvage value of \$700. In order to provide money at that time for a new machine costing the same amount, a sinking fund is set up. The amount in the fund at the end of eight years is to be the difference between the replacement cost and the salvage value. If equal payments are placed in the fund at the end of each quarter and the fund earns 8% compounded quarterly, what should each payment be?*

*Solution:* The amount needed after eight years is $7000 - 700 = \$6300$. Let $R$ be the quarterly payment. The payments into the sinking fund form an annuity with $n = 4(8) = 32$, $r = 0.08/4 = 0.02$, and $S = 6300$. Thus, from Eq. (8), we have

$$6300 = Rs_{\overline{32}|0.02},$$

$$R = \frac{6300}{s_{\overline{32}|0.02}} \approx \frac{6300}{44.227030} \approx \$142.45.$$

In general, the formula

$$R = \frac{S}{s_{\overline{n}|r}}$$

gives the periodic payment $R$ of an annuity that is to amount to $S$. ∎

### EXAMPLE 12   Sinking Fund

*A rental firm estimates that, if purchased, a machine will yield an annual net return of $1000 for six years, after which the machine would be worthless. How much should the firm pay for the machine if it wants to earn 7% annually on its investment and also set up a sinking fund to replace the purchase price? For the fund, assume annual payments and a rate of 5% compounded annually.*

*Solution:* Let $x$ be the purchase price. Each year, the return on the investment is $0.07x$. Since the machine gives a return of $1000 a year, the amount left to be placed into the fund each year is $1000 - 0.07x$. These payments must accumulate to $x$. Hence,

$$(1000 - 0.07x)s_{\overline{6}|0.05} = x,$$

$$1000s_{\overline{6}|0.05} - 0.07x s_{\overline{6}|0.05} = x,$$

$$1000s_{\overline{6}|0.05} = x(1 + 0.07s_{\overline{6}|0.05}),$$

$$\frac{1000s_{\overline{6}|0.05}}{1 + 0.07s_{\overline{6}|0.05}} = x,$$

$$x \approx \frac{1000(6.801913)}{1 + 0.07(6.801913)}$$

$$\approx \$4607.92.$$

Another way to look at the problem is as follows: Each year, the $1000 must account for a return of $0.07x$ and also a payment of $\dfrac{x}{s_{\overline{6}|0.05}}$, into the sinking fund. Thus, we have $1000 = 0.07x + \dfrac{x}{s_{\overline{6}|0.05}}$, which, when solved, gives the same result. ∎

## ■ Exercise 8.3

*In Problems 1–4, write the geometric sequence satisfying the given conditions. Simplify the terms.*

**1.** $a = 64, r = \frac{1}{2}, n = 5$.

**2.** $a = 2, r = -3, n = 4$.

**3.** $a = 100, r = 1.02, n = 3$.

**4.** $a = 81, r = 3^{-1}, n = 4$.

*In Problems 5–8, find the sum of the given geometric series by using Eq. (4) of this section.*

**5.** $\frac{2}{3} + (\frac{2}{3})^2 + \cdots + (\frac{2}{3})^5$.

**6.** $1 + \frac{1}{4} + (\frac{1}{4})^2 + \cdots + (\frac{1}{4})^5$.

**7.** $1 + 0.1 + (0.1)^2 + \cdots + (0.1)^5$.

**8.** $(1.1)^{-1} + (1.1)^{-2} + \cdots + (1.1)^{-6}$.

*In Problems 9–12, use Appendix B and find the value of the given expression.*

**9.** $a_{\overline{35}|0.04}$.

**10.** $a_{\overline{15}|0.07}$.

**11.** $s_{\overline{8}|0.0075}$.

**12.** $s_{\overline{12}|0.005}$.

*In Problems* **13–16,** *find the present value of the given (ordinary) annuity.*

**13.** $500 per year for five years at the rate of 7% compounded annually.

**14.** $1000 every six months for four years at the rate of 10% compounded semiannually.

**15.** $2000 per quarter for $4\frac{1}{2}$ years at the rate of 8% compounded quarterly.

**16.** $1500 per month for 15 months at the rate of 9% compounded monthly.

*In Problems* **17 and 18,** *find the present value of the given annuity due.*

**17.** $800 paid at the beginning of each six-month period for six years at the rate of 7% compounded semiannually.

**18.** $100 paid at the beginning of each quarter for five years at the rate of 6% compounded quarterly.

*In Problems* **19–22,** *find the future value of the given (ordinary) annuity.*

**19.** $2000 per month for three years at the rate of 15% compounded monthly.

**20.** $600 per quarter for four years at the rate of 8% compounded quarterly.

**21.** $5000 per year for 20 years at the rate of 7% compounded annually.

**22.** $2000 every six months for 10 years at the rate of 6% compounded semiannually.

*In Problems* **23 and 24,** *find the future value of the given annuity due.*

**23.** $1200 each year for 12 years at the rate of 8% compounded annually.

**24.** $500 every quarter for $5\frac{3}{4}$ years at the rate of 5% compounded quarterly.

**25.** For an interest rate of 6% compounded monthly, find the present value of an annuity of $50 at the end of each month for six months and $75 thereafter at the end of each month for two years.

**26. Leasing Office Space**  A company wishes to lease temporary office space for a period of six months. The rental fee is $500 a month, payable in advance. Suppose that the company wants to make a lump-sum payment, at the beginning of the rental period, to cover all rental fees due over the six-month period. If money is worth 9% compounded monthly, how much should the payment be?

**27.** An annuity consisting of equal payments at the end of each quarter for three years is to be purchased for $5000. If the interest rate is 6% compounded quarterly, how much is each payment?

**28. Equipment Purchase**  A machine is purchased for $3000 down and payments of $250 at the end of every six months for six years. If interest is at 8% compounded semiannually, find the corresponding cash price of the machine.

**29.** Suppose $50 is placed in a savings account at the end of each month for four years. If no further deposits are made, (a) how much is in the account after six years, and (b) how much of this amount is compound interest? Assume that the savings account pays 6% compounded monthly.

**30. Insurance Settlement Options**  The beneficiary of an insurance policy has the option of receiving a lump-sum payment of $35,000 or 10 equal yearly payments, where the first payment is due at once. If interest is at 4% compounded annually, find the yearly payment.

**31. Sinking Find**  In 10 years, a $40,000 machine will have a salvage value of $4000. A new machine at that time is expected to sell for $52,000. In order to provide funds for the difference between the replacement cost and the salvage value, a sinking fund is set up into which equal payments are placed at the end of each year. If the fund earns 7% compounded annually, how much should each payment be?

**32. Sinking Fund**  A paper company is considering the purchase of a forest that is estimated to yield an annual return of $50,000 for 10 years, after which the forest will have no value. The company wants to earn 8% on its investment and also set up a sinking fund to replace the purchase price. If money is placed in the fund at the end of each year and earns 6% compounded annually, find the price the company should pay for the forest. Round your answer to the nearest hundred dollars.

**33. Sinking Fund**  In order to replace a machine in the future, a company is placing equal payments into a sinking fund at the end of each year so that after 10 years the amount in the fund is $25,000. The fund earns 6% compounded annually. After 6 years, the interest rate increases and the fund pays 7% compounded annually. Because of the higher interest rate, the company decreases the amount of the remaining payments. Find the amount of the new payment. Round your answer to the nearest dollar.

**34.** A owes B the sum of $5000 and agrees to pay B the sum of $1000 at the end of each year for five years and a final payment at the end of the sixth year. How much should the final payment be if interest is at 8% compounded annually?

*In Problems 35–43, use the following formulas:*

$$a_{\overline{n}|r} = \frac{1 - (1 + r)^{-n}}{r},$$

$$s_{\overline{n}|r} = \frac{(1 + r)^n - 1}{r},$$

$$R = \frac{A}{a_{\overline{n}|r}} = \frac{Ar}{1 - (1 + r)^{-n}} = \frac{Ar(1 + r)^n}{(1 + r)^n - 1},$$

$$R = \frac{S}{s_{\overline{n}|r}} = \frac{Sr}{(1 + r)^n - 1}.$$

**35.** Find $s_{\overline{60}|0.017}$ to five decimal places.

**36.** Find $a_{\overline{10}|0.073}$ to five decimal places.

**37.** Find $700a_{\overline{360}|0.0125}$ to two decimal places.

**38.** Find $1000s_{\overline{120}|0.01}$ to two decimal places.

**39.** Equal payments are to be deposited in a savings account at the end of each quarter for five years so that at the end of that time there will be $3000. If interest is at $5\frac{1}{2}\%$ compounded quarterly, find the quarterly payment.

**40. Insurance Proceeds**   Suppose that insurance proceeds of $25,000 are used to purchase an annuity of equal payments at the end of each month for five years. If interest is at the rate of 10% compounded monthly, find the amount of each payment.

**41. Lottery**   Mary Jones won a state $1,000,000 lottery and will receive a check for $50,000 now and a similar one each year for the next 19 years. To provide these 20 payments, the State Lottery Commission purchased an annuity due at the interest rate of 12% compounded annually. How much did the annuity cost the Commission?

**42. Pension Plan Options**   Suppose an employee of a company is retiring and has the choice of two benefit options under the company pension plan. Option A consists of a guaranteed payment of $450 at the end of each month for 10 years. Alternatively, under option B, the employee receives a lump-sum payment equal to the present value of the payments described under option A.

  **a.** Find the sum of the payments under option A.

  **b.** Find the lump-sum payment under option B if it is determined by using an interest rate of 6% compounded monthly. Round your answer to the nearest dollar.

**43. An Early Start to Investing**   An insurance agent offers services to clients who are concerned about their personal financial planning for retirement. To emphasize the advantages of an early start to investing, she points out that a 25-year-old person who saves $2000 a year for 10 years (and makes no more contributions after age 34) will earn more than by waiting 10 years and then saving $2000 a year from age 35 until retirement at age 65 (a total of 30 contributions). Find the net earnings (compound amount minus total contributions) at age 65 for both situations. Assume an effective annual rate of 7%, and suppose that deposits are made at the beginning of each year. Round answers to the nearest dollar.

---

**O B J E C T I V E**

**To learn how to amortize a loan and set up an amortization schedule.**

## 8.4 Amortization of Loans

Suppose that a bank lends a borrower $1500 and charges interest at the nominal rate of 12% compounded monthly. The $1500 plus interest is to be repaid by equal payments of $R$ dollars at the end of each month for three months. Essentially, by paying the borrower $1500, the bank is purchasing an annuity of three payments of $R$ each. Using the formula from Example 7 of the preceding section, we find that the monthly payment is given by

$$R = \frac{A}{a_{\overline{n}|r}} = \frac{1500}{a_{\overline{3}|0.01}} \approx \frac{1500}{2.940985} \approx \$510.0332.$$

We shall round the payment to $510.03, which may result in a slightly higher final payment. However, it is not unusual for a bank to round *up* to the nearest cent, in which case the final payment may be less than the other payments.

  The bank can consider each payment as consisting of two parts: (1) interest on the outstanding loan and (2) repayment of part of the loan. This is called **amortizing.** A loan is **amortized** when part of each payment is used to pay interest and the remaining part is used to reduce the outstanding principal. Since each payment reduces the outstanding principal, the interest portion of a payment decreases as time goes on. Let us analyze the loan just described.

At the end of the first month, the borrower pays $510.03. The interest on the outstanding principal is $0.01(1500) = \$15$. The balance of the payment, $510.03 - 15 = \$495.03$, is then applied to reduce the principal. Hence, the principal outstanding is now $1500 - 495.03 = \$1004.97$. At the end of the second month, the interest is $0.01(1004.97) \approx \$10.05$. Thus, the amount of the loan repaid is $510.03 - 10.05 = \$499.98$, and the outstanding balance is $1004.97 - 499.98 = \$504.99$. The interest due at the end of the third and final month is $0.01(504.99) \approx \$5.05$, so the amount of the loan repaid is $510.03 - 5.05 = \$504.98$. This would leave an outstanding balance of $504.99 - 504.98 = \$0.10$, so we take the final payment to be $510.04, and the debt is paid off. As we said earlier, the final payment is adjusted to offset rounding errors. An analysis of how each payment in the loan is handled can be given in a table called an **amortization schedule.** (See Table 8.1.) The total interest paid is $30.10, which is often called the **finance charge.**

Many end-of-year mortgage statements are issued in the form of an amortization schedule.

**TABLE 8.1** Amortization Schedule

| Period | Principal Outstanding at Beginning of Period | Interest for Period | Payment at End of Period | Principal Repaid at End of Period |
|--------|----------------------------------------------|---------------------|--------------------------|-----------------------------------|
| 1 | $1500 | $15 | 510.03 | $495.03 |
| 2 | 1004.97 | 10.05 | 510.03 | 499.98 |
| 3 | 504.99 | 5.05 | 510.04 | 504.99 |
| Total | | 30.10 | 1530.10 | 1500.00 |

When one is amortizing a loan, at the beginning of any period the principal outstanding is the present value of the remaining payments. Using this fact together with our previous development, we obtain the formulas listed in Table 8.2, which describe the amortization of an interest-bearing loan of $A$ dollars, at a rate $r$ per period, by $n$ equal payments of $R$ dollars each and such that a payment is made at the end of each period. In particular, notice that Formula 1 for the periodic payment $R$ involves $a_{\overline{n}|r}$, which, as you recall, is defined as $[1 - (1 + r)^{-n}]/r$.

**TABLE 8.2** Amortization Formulas

1. Periodic payment: $R = \dfrac{A}{a_{\overline{n}|r}} = A \cdot \dfrac{r}{1 - (1 + r)^{-n}}$

2. Principal outstanding at beginning of $k$th period:

$$Ra_{\overline{n-k+1}|r} = R \cdot \frac{1 - (1 + r)^{-n+k-1}}{r}$$

3. Interest in $k$th payment: $Rra_{\overline{n-k+1}|r}$

4. Principal contained in $k$th payment: $R[1 - ra_{\overline{n-k+1}|r}]$

5. Total interest paid: $R(n - a_{\overline{n}|r})$, or $nR - A$

**EXAMPLE 1** Amortizing a Loan

*A person amortizes a loan of $30,000 for a new home by obtaining a 20-year mortgage at the rate of 9% compounded monthly. Find (a) the monthly payment, (b) the total interest charges, and (c) the principal remaining after 5 years.*

*Solution:*

a. The number of payment periods is $n = 12(20) = 240$, the interest rate per period is $r = 0.09/12 = 0.0075$, and $A = 30,000$. From Formula 1 in Table 8.2, the monthly payment $R$ is $30{,}000/a_{\overline{240}|0.0075}$. Since $a_{\overline{240}|0.0075}$ is not in Appendix B, we use the following equivalent formula and a calculator:

$$R = 30{,}000 \left[ \frac{0.0075}{1 - (1.0075)^{-240}} \right]$$

$$\approx \$269.92.$$

b. From Formula 5, the total interest charges are

$$240(269.92) - 30{,}000 = 64{,}780.80 - 30{,}000$$

$$= \$34{,}780.80.$$

This is more than the loan itself.

c. After 5 years, we are at the beginning of the 61st period. Using Formula 2 with $n - k + 1 = 240 - 61 + 1 = 180$, we find that the principal remaining is

$$269.92 \left[ \frac{1 - (1.0075)^{-180}}{0.0075} \right] \approx \$26{,}612.33. \qquad \blacksquare$$

At one time, a very common type of installment loan involved the "add-on method" of determining the finance charge. With that method, the finance charge is found by applying a quoted annual interest rate [under simple (that is, noncompounded) interest] to the borrowed amount of the loan. The charge is then added to the principal, and the total is divided by the number of *months* of the loan to determine the monthly installment payment. In loans of this type, the borrower may not immediately realize that the true annual rate is significantly higher than the quoted rate, as the following technology example shows.

## TECHNOLOGY

**Problem:** A $1000 loan is taken for one year at 9% interest under the add-on method. Estimate the true annual interest rate if monthly compounding is assumed.

**Solution:** Since the add-on method is used, payments will be made monthly. The finance charge for $1000 at 9% simple interest for one year is $0.09(1000) = \$90$. Adding this to the loan amount gives $1000 + 90 = \$1090$. Thus, the monthly installment payment is $1090/12 \approx \$90.83$. Hence, we have a loan of $1000 with 12 equal payments of $90.83. From Formula 1 in Table 8.2,

$$R = \frac{A}{a_{\overline{n}|r}},$$

$$\frac{1090}{12} = \frac{1000}{a_{\overline{12}|r}},$$

$$a_{\overline{12}|r} = \frac{1000(12)}{1090} \approx 11.009174.$$

We now solve $a_{\overline{12}|r} = 11.009174$ for the monthly rate $r$. We have

$$\frac{1 - (1 + r)^{-12}}{r} = 11.009174.$$

Graphing

$$Y_1 = (1 - (1 + X)^{\wedge} - 12)/X$$

$$Y_2 = 11.009174$$

and finding the intersection (see Fig. 8.10) gives

$$r \approx 0.01351374,$$

which corresponds to an annual rate of

$$12(0.01351374) \approx 0.1622, \text{ or } 16.22\%.$$

Thus, the true annual rate is 16.22%. Federal regulations concerning truth-in-lending laws have made add-

on loans virtually obsolete.

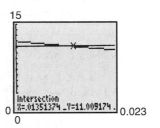

**FIGURE 8.10**  Solution of $a_{\overline{12}|r} = 11.009174$.

The annuity formula

$$A = R \cdot \frac{1 - (1 + r)^{-n}}{r}$$

can be solved for $n$ to give the number of periods of a loan. Multiplying both sides by $\dfrac{r}{R}$ gives

$$\frac{Ar}{R} = 1 - (1 + r)^{-n},$$

$$(1 + r)^{-n} = 1 - \frac{Ar}{R} = \frac{R - Ar}{R},$$

$$-n \ln(1 + r) = \ln\left(\frac{R - Ar}{R}\right) \quad \text{(taking logs of both sides)},$$

$$n = -\frac{\ln\left(\dfrac{R - Ar}{R}\right)}{\ln(1 + r)}.$$

Using properties of logarithms, we eliminate the minus sign by inverting the quotient in the numerator:

$$n = \frac{\ln\left(\dfrac{R}{R - Ar}\right)}{\ln(1 + r)}. \tag{1}$$

**EXAMPLE 2**  **Periods of a Loan**

*Bob Smith recently purchased a computer for $1500 and agreed to pay it off by making monthly payments of $75. If the store charges interest at the rate of 12% compounded monthly, how many months will it take to pay off the debt?*

*Solution:* From Eq. (1),

$$n = \frac{\ln\left[\dfrac{75}{75 - 1500(0.01)}\right]}{\ln(1.01)} \approx 22.4 \text{ months.}$$

In reality, there are 23 payments; however, the final payment will be less than $75. ■

## ■ Exercise 8.4

1. A person borrows $2000 from a bank and agrees to pay it off by equal payments at the end of each month for three years. If interest is at 15% compounded monthly, how much is each payment?

2. A person wishes to make a three-year loan and can afford payments of $50 at the end of each month. If interest is at 12% compounded monthly, how much can the person afford to borrow?

3. **Finance Charge** Determine the finance charge on a 36-month $8000 auto loan with monthly payments if interest is at the rate of 12% compounded monthly.

4. For a one-year loan of $500 at the rate of 15% compounded monthly, find (a) the monthly installment payment and (b) the finance charge.

5. **Car Loan** A person is amortizing a 36-month car loan of $7500 with interest at the rate of 12% compounded monthly. Find (a) the monthly payment, (b) the interest in the first month, and (c) the principal repaid in the first payment.

6. **Real-Estate Loan** A person is amortizing a 48-month loan of $10,000 for a house lot. If interest is at the rate of 9% compounded monthly, find (a) the monthly payment, (b) the interest in the first payment, and (c) the principal repaid in the first payment.

*In Problems 7–10, construct amortization schedules for the indicated debts. Adjust the final payments if necessary.*

7. $5000 repaid by four equal yearly payments with interest at 7% compounded annually.

8. $8000 repaid by six equal semiannual payments with interest at 8% compounded semiannually.

9. $900 repaid by five equal quarterly payments with interest at 10% compounded quarterly.

10. $10,000 repaid by five equal monthly payments with interest at 9% compounded monthly.

11. A loan of $1000 is being paid off by quarterly payments of $100. If interest is at the rate of 8% compounded quarterly, how many *full* payments will be made?

12. A loan of $2000 is being amortized over 48 months at an interest rate of 12% compounded monthly. Find:

   a. the monthly payment;

   b. the principal outstanding at the beginning of the 36th month;

   c. the interest in the 36th payment;

   d. the principal in the 36th payment;

   e. the total interest paid.

13. A debt of $10,000 is being repaid by 10 equal semiannual payments, with the first payment to be made six months from now. Interest is at the rate of 8% compounded semiannually. However, after two years, the interest rate increases to 10% compounded semiannually. If the debt must be paid off on the original date agreed upon, find the new annual payment. Give your answer to the nearest dollar.

14. A person borrows $2000 and will pay off the loan by equal payments at the end of each month for five years. If interest is at the rate of 16.8% compounded monthly, how much is each payment?

15. **Mortgage** A $45,000 mortgage for 25 years for a new home is obtained at the rate of 10.2% compounded monthly. Find (a) the monthly payment, (b) the interest in the first payment, (c) the principal repaid in the first payment, and (d) the finance charge.

16. **Auto Loan** An automobile loan of $8500 is to be amortized over 48 months at an interest rate of 13.2% compounded monthly. Find (a) the monthly payment and (b) the finance charge.

17. **Furniture Loan** A person purchases furniture for $2000 and agrees to pay off this amount by monthly payments of $100. If interest is charged at the rate of 18% compounded monthly, how many *full* payments will there be?

18. Find the monthly payment of a five-year loan for $7000 if interest is at 12.12% compounded monthly.

19. **Mortgage** Bob and Mary Rodgers want to purchase a new house and feel that they can afford a mortgage payment of $600 a month. They are able to obtain a 30-year 12.6% mortgage (compounded monthly), but must put down 25% of the cost of the house. Assuming that they have enough savings for the down payment, how expensive a house can they afford? Give your answer to the nearest dollar.

20. **Mortgage** Suppose you have the choice of taking out an $80,000 mortgage at 12% compounded monthly for either 15 years or 30 years. How much savings is there in the finance charge if you were to choose the 15-year mortgage?

21. On a $25,000 five-year loan, how much less is the monthly payment if the loan were at the rate of 12% compounded monthly rather than at 15% compounded monthly?

22. **Home Loan** The federal government has a program to aid low-income homeowners in urban areas. This program allows certain qualified homeowners to obtain low-interest home improvement loans. Each loan is processed through a commercial bank. The bank makes home improvement loans at an annual rate of $11\frac{1}{4}\%$ compounded monthly. However, the government subsidizes the bank so that the loan to the homeowner is at the annual rate of 4% compounded monthly. If the monthly payment at the 4% rate is $x$ dollars ($x$ dollars is the homeowner's monthly payment) and the monthly payment at the $11\frac{1}{4}\%$ rate is $y$ dollars ($y$ dollars is the monthly payment the bank must receive), then the gov-

ernment makes up the difference $y - x$ to the bank each month. From a practical point of view, the government does not want to bother with *monthly* payments. Instead, at the beginning of the loan, the government pays the present value of all such monthly differences, at

an annual rate of $11\frac{1}{4}\%$ compounded monthly.

If a qualified homeowner takes out a loan for $5000 for five years, determine the government's payment to the bank at the beginning of the loan.

# 8.5 REVIEW

## IMPORTANT TERMS AND SYMBOLS

**Section 8.1**   effective rate   yield

**Section 8.2**   present value   future value   equation of value   cash flows   net present value

**Section 8.3**   geometric sequence   geometric series   common ratio   annuity   ordinary annuity   annuity due   present value of annuity, $a_{\overline{n}|r}$   amount of annuity, $s_{\overline{n}|r}$

**Section 8.4**   amortizing   amortization schedules   finance charge

## SUMMARY

The concept of compound interest lies at the heart of any discussion dealing with the time value of money—that is, the present value of money due in the future or the future value of money presently invested. Under compound interest, interest is converted into principal and earns interest itself. The basic compound-interest formulas are

$$S = P(1 + r)^n \quad \text{(future value)},$$
$$P = S(1 + r)^{-n} \quad \text{(present value)},$$

where   $S$ = compound amount (future value),

$P$ = principal (present value),

$r$ = periodic rate,

$n$ = number of conversion periods.

Interest rates are usually quoted as an annual rate called the nominal rate. The periodic rate is obtained by dividing the nominal rate by the number of conversion periods each year. The effective rate is the annual simple-interest rate that is equivalent to the nominal rate of $r$ compounded $n$ times a year and is given by

$$r_e = \left(1 + \frac{r}{n}\right)^n - 1 \quad \text{(effective rate)}.$$

Effective rates are used to compare different interest rates.

An annuity is a sequence of payments made at fixed periods of time over some interval. The mathematical basis for formulas dealing with annuities is the notion of the sum of a geometric series—that is,

$$s = \frac{a(1 - r^n)}{1 - r} \quad \text{(sum of geometric series)},$$

where

$s$ = sum,

$a$ = first term,

$r$ = common ratio,

$n$ = number of terms.

An ordinary annuity is an annuity in which each payment is made at the *end* of a payment period, whereas an annuity due is an annuity in which each payment is made at the *beginning* of a payment period. The basic formulas dealing with ordinary annuities are

$$A = R \cdot \frac{1 - (1 + r)^{-n}}{r} = R a_{\overline{n}|r}$$
$$\text{(present value)},$$

$$S = R \cdot \frac{(1 + r)^n - 1}{r} = R s_{\overline{n}|r} \quad \text{(future value)},$$

where

$A$ = present value of annuity,

$S$ = amount (future value) of annuity,

$R$ = amount of each payment,

$n$ = number of payment periods,

$r$ = periodic rate.

For an annuity due, the corresponding formulas are

$$A = R(1 + a_{\overline{n-1}|r}) \quad \text{(present value)},$$
$$S = R(s_{\overline{n+1}|r} - 1) \quad \text{(future value)}.$$

A loan, such as a mortgage, is amortized when part of each installment payment is used to pay interest and the remaining part is used to reduce the principal. A complete analysis of each payment is given in an amortization schedule. The following formulas deal with amortizing a loan of $A$ dollars, at the periodic rate of $r$, by $n$ equal payments of $R$ dollars each and such that a payment is made at the end of each period:

Periodic payment:

$$R = \frac{A}{a_{\overline{n}|r}} = A \cdot \frac{r}{1 - (1 + r)^{-n}}.$$

Principal outstanding at beginning of $k$th period:

$$Ra_{\overline{n-k+1}|r} = R \cdot \frac{1 - (1 + r)^{-n+k-1}}{r}.$$

Interest in $k$th payment: $Rra_{\overline{n-k+1}|r}$.

Principal contained in $k$th payment:

$$R[1 - ra_{\overline{n-k+1}|r}].$$

Total interest paid: $R(n - a_{\overline{n}|r})$ or $nR - A$.

## REVIEW PROBLEMS

1. Find the sum of the geometric series

$$2 + 1 + \tfrac{1}{2} + \cdots + 2(\tfrac{1}{2})^5.$$

2. Find the effective rate that corresponds to a nominal rate of 6% compounded quarterly.

3. An investor has a choice of investing a sum of money at either 8.5% compounded annually or 8.2% compounded semiannually. Which is the better choice?

4. **Cash Flows** Find the net present value of the following cash flows, which can be purchased by an initial investment of $7000:

| Year | Cash Flow |
|------|-----------|
| 2 | $3400 |
| 4 | 3500 |

Assume that interest is at 7% compounded semiannually.

5. A debt of $1200 due in four years and $1000 due in six years is to be repaid by a payment of $1000 now and a second payment at the end of two years. How much should the second payment be if interest is at 8% compounded semiannually?

6. Find the present value of an annuity of $250 at the end of each month for four years if interest is at 6% compounded monthly.

7. For an annuity of $200 at the end of every six months for $6\frac{1}{2}$ years, find (a) the present value and (b) the future value at an interest rate of 8% compounded semiannually.

8. Find the amount of an annuity due which consists of 10 yearly payments of $100, provided that the interest rate is 6% compounded annually.

9. Suppose $100 is initially placed in a savings account and $100 is deposited at the end of every six months for the next four years. If interest is at 7% compounded semiannually, how much is in the account at the end of four years?

10. A savings account pays interest at the rate of 5% compounded semiannually. What amount must be deposited now so that $250 can be withdrawn at the end of every six months for the next 10 years?

11. **Sinking Fund** A company borrows $5000 on which it will pay interest at the end of each year at the annual rate of 11%. In addition, a sinking fund is set up so that the $5000 can be repaid at the end of five years. Equal payments are placed in the fund at the end of each year, and the fund earns interest at the effective rate of 6%. Find the annual payment in the *sinking fund*.

12. **Car Loan** A debtor is to amortize a $7000 car loan by making equal payments at the end of each month for 36 months. If interest is at 12% compounded monthly, find (a) the amount of each payment and (b) the finance charge.

13. A person has debts of $500 due in three years with interest at 5% compounded annually and $500 due in four years with interest at 6% compounded semiannually. The debtor wants to pay off these debts by making two payments: the first payment now, and the second, which is double the first payment, at the end of the third year. If money is worth 7% compounded annually, how much is the first payment?

14. Construct an amortization schedule for a loan of $2000 repaid by three monthly payments with interest at 12% compounded monthly.

15. Construct an amortization schedule for a loan of $15,000 repaid by five monthly payments with interest at 9% compounded monthly.

16. Find the present value of an ordinary annuity of $540 every month for seven years at the rate of 10% compounded monthly.

17. **Auto Loan** Determine the finance charge for a 48-month auto loan of $11,000 with monthly payments at the rate of 13.5% compounded monthly.

# MATHEMATICAL *SNAPSHOT*

## THE RULE OF 78's[3]

If you have a loan and decide to pay it in full before the last payment is due, you would certainly expect the lender to refund or credit you with a portion of the total finance charge. One accounting method the lender might use to determine this rebate is called the "rule of 78's" or the "sum of the digits" method. Basically, this method requires you to pay much of the total finance charge at the beginning of the loan period—the reasoning being that you have use of more of the money at the beginning of the period than at the end.

To illustrate, suppose that you have a loan with 12 equal monthly payments. According to the rule of 78's, the integers from 1 to 12 are added (because the number of months in the loan period is 12). This gives

$$1 + 2 + 3 + \cdots + 12 = 78$$

(hence the 78 in the name of the rule). During the first month you have use of $\frac{12}{12}$ of the loan money, so the rule entitles the lender to take $\frac{12}{78}$ of the total finance charge as the interest portion of your first payment. During the second month you are considered to have use of $\frac{11}{12}$ of the money, so the lender is entitled to $\frac{11}{78}$ of the finance charge. The third month the lender takes $\frac{10}{78}$ of it, and so on. Thus the rule of 78's is a way of allocating on a monthly basis the total finance charge.

For example, suppose that you were to pay off the loan one-half of the way through the loan period (after 6 months). To compute your interest rebate you find that the fraction of the total finance charge you have already paid is

$$\frac{12 + 11 + 10 + 9 + 8 + 7}{78} = \frac{57}{78}.$$

Thus the rebate is $\frac{21}{78} \approx 26.9\%$ of the total finance charge.

The rule of 78's applies not only to a 12-month loan. In general, the rule states that for a loan with $n$ equal monthly payments, the portion of the total interest that the lender earns during the $k$th month is

$$\frac{n - k + 1}{1 + 2 + 3 + \cdots + n}.$$

As a specific example, suppose that you go to a bank and make a $3000 installment loan for 3 months with an A.P.R. of 11.96%. From the annuity formula (see Sec. 8.4)

$$R = A \cdot \frac{r}{1 - (1 + r)^{-n}}, \tag{1}$$

with $A = 3000$, $r = 0.1196/12 \approx 0.0099667$, and $n = 3$, the monthly payment $R$ is $1020. The corresponding amortization schedule is given in Table 8.3 where, as you know, monthly interest is computed on the current unpaid balance. For example, the interest charge for the first month is $29.90, that is,

$$3000(0.0099667) = \$29.90.$$

This method of amortizing a loan is called the *actuarial method*. Observe that the total interest charge is $60.

[3]Adapted from Alonzo F. Johnson, "The Rule of 78: A Rule That Outlived Its Useful Life," *The Mathematical Teacher,* 81 (September 1988), 450–53. By permission of the National Council of Teachers of Mathematics.

**375**

**TABLE 8.3** Actuarial Method for a $3000 Loan for 3 Months at 11.96% A.P.R.

| Month | Amount Owed at Beginning of Month | Interest Charged at End of Month | Payment at End of Month | Principal Repaid at End of Month |
|-------|-----------------------------------|----------------------------------|-------------------------|----------------------------------|
| 1 | $3000.00 | $29.90 | $1020 | $ 990.10 |
| 2 | 2009.90 | 20.03 | 1020 | 999.97 |
| 3 | 1009.93 | 10.07 | 1020 | 1009.93 |
| Total | | 60.00 | 3060 | 3000.00 |

Now, suppose that you wish to pay off the loan after one month. Then in addition to the monthly payment, the proper payoff amount, based on the actuarial method, is $2009.90, that is,

$$3000 - (1020 - 29.90) = \$2009.90.$$

But if the bank uses the rule of 78's instead, what is the payoff amount? Consider making the corresponding amortization schedule to see how the total interest charge of $60 is handled. Because you have a 3-month loan, you must add the integers from 1 to 3: $1 + 2 + 3 = 6$. Thus at the end of the first month, the bank is entitled to $\frac{3}{6}$ of the $60, or $30 (which is subtracted from the payment). At the end of the second month the bank earns $\frac{2}{6} \cdot 60 = \$20$, and so on. This is shown in Table 8.4. The schedule shows that the payoff after one month is

$$3000 - (1020 - 30) = \$2010,$$

which is $0.10 more than the payoff under the actuarial method. Note also that the difference in the interest

allocations between the two methods at the end of the second and third months is only $0.03 and $0.07, respectively. Certainly for this situation the rule of 78's gives a reasonable estimate of the interest allocations under the actuarial method. Indeed, for other short-term loans this rule may also give reasonable estimates (the term "reasonable" is subjective, of course).

However, in some situations the interest rate and length of term may be such that allocating interest by the rule of 78's can result in an interest charge that is greater than the monthly payment itself! For example, suppose that you take out a home improvement loan of $15,000 with monthly payments for 15 years (180 months) at an A.P.R. of 15%. Using Eq. (1) you find that the monthly payment is $209.94. Over the term of this loan the total interest charge is

$$180(209.94) - 15,000 = \$22,789.20.$$

**TABLE 8.4** Rule of 78's for a $3000 Loan for 3 Months with Total Interest of $60

| Month | Amount Owed at Beginning of Month | Interest Charged at End of Month | Payment at End of Month | Principal Repaid at End of Month |
|-------|-----------------------------------|----------------------------------|-------------------------|----------------------------------|
| 1 | $3000 | $30 | $1020 | $ 990 |
| 2 | 2010 | 20 | 1020 | 1000 |
| 3 | 1010 | 10 | 1020 | 1010 |
| Total | | 60 | 3060 | 3000 |

**TABLE 8.5**   Rule of 78's for a $15,000 Loan for 180 Months at 15% A.P.R.

| Month | Amount Owed at Beginning of Month | Interest Charged at End of Month | Payment at End of Month | Principal Repaid at End of Month |
|---|---|---|---|---|
| 1 | $15,000 | $251.81 | $209.94 | –$41.87 |

By the rule of 78's, the interest charged the first month is[4]

$$\frac{180}{1 + 2 + \cdots + 180}(22{,}789.20) - \$251.81,$$

which is $41.87 more than the payment (and corresponds to an A.P.R. of 18.94%). Thus the payoff amount after one month is

$$15{,}000 - (209.94 - 251.81) = \$15{,}041.87,$$

which is more than the amount of the loan. The beginning of the amortization schedule has a negative entry as shown in Table 8.5. On the other hand, at the end of one month the proper interest charged by the actuarial method is

$$15{,}000\left(\frac{0.15}{12}\right) = \$187.50,$$

so the payoff amount is

$$15{,}000 - (209.94 - 187.50) = \$14{,}977.56.$$

The difference between the payoff amount by the rule of 78's and the payoff by the actuarial method, namely

$$15{,}041.87 - 14{,}977.56 = \$64.31,$$

may be considered to be the *penalty* for paying the loan off after one month. Table 8.6 gives payoff amounts and corresponding penalties for several months of this loan.

The data in Table 8.6 are indicated by the curves in Fig. 8.11. You can see from the top curve that under

**TABLE 8.6**   Payoff Amounts and Penalties

| Month | Payoff Amount 78's | Payoff Amount Actuarial | Penalty Col. 1–Col. 2 |
|---|---|---|---|
| 1 | $15,041.87 | $14,977.56 | $ 64.31 |
| 12 | 15,410.16 | 14,711.42 | 698.74 |
| 36 | 15,626.13 | 13,987.66 | 1,638.47 |
| 60 | 15,036.28 | 13,012.47 | 2,023.81 |
| 72 | 14,439.19 | 12,404.39 | 2,034.80 |
| 84 | 13,640.54 | 11,698.55 | 1,941.99 |
| 96 | 12,640.64 | 10,879.25 | 1,761.39 |
| 108 | 11,439.18 | 9,928.22 | 1,510.96 |
| 120 | 10,036.29 | 8,824.30 | 1,211.99 |
| 144 | 6,626.13 | 6,055.60 | 570.53 |
| 168 | 2,410.16 | 2,325.22 | 84.94 |
| 179 | 208.54 | 206.47 | 2.07 |

[4]Here we shall use the fact that $1 + 2 + \cdots + n = \dfrac{n(n + 1)}{2}$. Thus $1 + 2 + \cdots + 180 = \dfrac{180(181)}{2} = 16{,}290.$

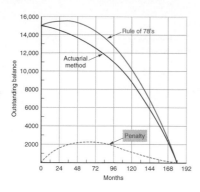

**FIGURE 8.11** Payoff amounts for loan.

the rule of 78's the payoff amount increases each month for 30 months, and for 60 months the payoff amount is greater than the $15,000 borrowed. Approximately one-third through the term of the loan (67 months) you pay a maximum penalty of $2043. This time period for the maximum penalty (one third through the term) is typical under the rule of 78's.

Other maximum prepayment penalties that occur when the rule of 78's is applied to a $10,000 loan at different interest rates for different terms of the loan are given in Table 8.7. The number in parentheses refers to the number of months into the loan at which the maximum penalty occurs. For example, on a $10,000, 16%, 10-year loan the maximum penalty is $688; it occurs if the loan is paid off after 44 months. Increasing the interest rate or the term of a loan increases the maximum penalty.

It is worthwhile to point out that since the rule of 78's allows the lender to earn most of the interest at the beginning of the term of the loan, it is to the advantage of the lender to have the consumer refinance the loan after several payments have been made. According to Johnson, "The National Consumer Law Center, Inc., has estimated that the rule of 78's costs consumers several hundred million dollars a year above what actuarial interest would cost them (Senate Hearings 1980, 144)."

Johnson also notes that "at one time every state except Arkansas either allowed or required the use of the rule of 78's in allocating interest to the payment on a loan. Since 1970 at least sixteen states have passed laws limiting its use. Of these sixteen, Iowa does not allow its use."

## ■ Exercises

1. Suppose that the total interest charge for a 12-month installment loan is $150. If the rule of 78's is used, how much interest has been charged by the lender during the first 4 months?

2. Suppose that the finance charge on a 24-month installment loan is $775. Find the amount of interest charged by the lender during each of the first 2 months if the lender uses the rule of 78's.

3. A person has a 36-month installment loan of $7500 with an A.P.R. of 12%. If the person wants to pay it off after 2 months, find the payoff amount if the lender computes it by using the rule of 78's.

4. A person has a five-month installment loan of $10,000 at a 9% A.P.R. Find the prepayment penalty after one month if the lender uses the rule of 78's to determine the payoff amount.

**TABLE 8.7** Maximum Prepayment Penalty Caused by Rule of 78's for a $10,000 Loan (with Month Occurring in Parentheses)

| Length of Loan (in Years) | 12% | 14% | 16% | 18% | 20% |
|---|---|---|---|---|---|
| 1 | $ 4 (4) | $ 5 (4) | 7 (4) | $ 9 (4) | $ 11 (4) |
| 2 | 15 (8) | 21 (8) | 27 (8) | 35 (8) | 43 (8) |
| 3 | 34 (12) | 46 (12) | 61 (12) | 77 (12) | 95 (13) |
| 4 | 60 (17) | 82 (17) | 108 (17) | 137 (17) | 170 (17) |
| 5 | 94 (21) | 129 (21) | 169 (21) | 215 (21) | 266 (21) |
| 7 | 186 (29) | 255 (30) | 334 (30) | 424 (30) | 526 (30) |
| 10 | 384 (43) | 525 (43) | 688 (44) | 872 (44) | 1077 (44) |
| 12 | 555 (52) | 758 (52) | 993 (53) | 1256 (54) | 1548 (54) |
| 15 | 871 (66) | 1187 (67) | 1549 (68) | 1951 (69) | 2392 (69) |
| 20 | 1548 (90) | 2094 (91) | 2708 (93) | 3381 (94) | 4105 (95) |

# Introduction to Probability and Statistics

The term *probability* is familiar to most of us. It is not uncommon to hear such phrases as "the probability of precipitation," "the probability of flooding," and "the probability of receiving an A in a course." Loosely speaking, probability refers to a number that indicates the degree of likelihood that some future event will have a particular outcome. For example, before tossing a well-balanced coin, you do not know with certainty whether the outcome will be a head or a tail. However, no doubt you consider these outcomes as being equally likely to occur. This means that if the coin were tossed a large number of times, you would expect that approximately half of the tosses would give heads. Thus, we say that the probability of a head occurring on any toss is $\frac{1}{2}$, or 50%.

The field of probability forms the basis of the study of statistics. In statistics, we are concerned about making an inference—that is, a prediction or decision—about a population (a large set of objects under consideration) by using a sample of data drawn from that population. In other words, in statistics, we make an inference about a population based on a known sample. For example, by drawing a sample of units from an assembly line, we can statistically make an inference about *all* the units in a production run. However, in the study of probability, we work with a known population and consider the likelihood (or probability) of drawing a particular sample from it. For example, if we select a card from a deck, we may be interested in the probability that it will be the ace of hearts.

## 9.1 Basic Counting Principle and Permutations

### Basic Counting Principle

Later on, you will find that computing a probability may require you to calculate the number of elements in a set. Because counting the elements individually may be extremely tedious (or even prohibitive), we shall spend some time developing efficient counting techniques. We begin by motivating the *basic counting principle,* which is useful in solving a wide variety of problems.

Suppose a manufacturer wants to produce coffee brewers in 2-, 8-, and 10-cup capacities, with each capacity available in colors of white, beige, red, and green. How many types of brewers must the manufacturer produce? To answer the question, it is not necessary that we count the capacity-color pairs one by one (such as 2-white and 8-beige). Since there are three capacities, and for each capacity there are four colors, the number of types is the product $3 \cdot 4$, or 12. We can systematically list the different types by using the **tree diagram** of Fig. 9.1. From the starting point, there are three branches that indicate the possible capacities. From each of these branches are four more branches that indicate the possible colors. This tree determines 12 paths, each beginning at the starting point and ending at a tip. Each path determines a different type of coffee brewer. We refer to the diagram as being a *two-level* tree: there is a level for capacity and a level for color.

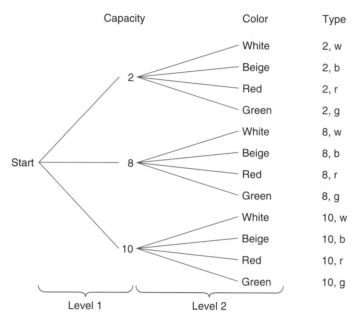

**FIGURE 9.1** Two-level tree diagram for types of coffee brewers.

We can consider the listing of the types of coffee brewers as a two-stage procedure. In the first stage we indicate a capacity and in the second a color. The number of types of coffee brewers is the number of ways the first stage can occur (3), times the number of ways the second stage can occur (4), which yields $3 \cdot 4 = 12$. This multiplication procedure can be generalized into a basic counting principle:

**Basic Counting Principle**

Suppose that a procedure involves a sequence of $k$ stages. Let $n_1$ be the number of ways the first can occur and $n_2$ be the number of ways the second can occur after the first stage has occurred. Continuing in this way, let $n_k$ be the number of ways the $k$th stage can occur after the first $k - 1$ stages have occurred. Then the total number of different ways the procedure can occur is

$$n_1 \cdot n_2 \cdots n_k.$$

**EXAMPLE 1  Travel Routes**

*Two roads connect cities* A *and* B, *four connect* B *and* C, *and five connect* C *and* D. *(See Fig. 9.2.) To drive from* A, *to* B, *to* C, *and then to city* D, *how many different routes are possible?*

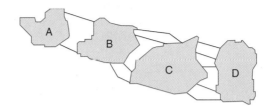

**FIGURE 9.2**  Roads connecting cities A, B, C, D.

*Solution:* Here we have a three-stage procedure. The first $(A \rightarrow B)$ has two possibilities, the second $(B \rightarrow C)$ has four, and the third $(C \rightarrow D)$ has five. By the basic counting principle, the total number of routes is $2 \cdot 4 \cdot 5$, or 40. ∎

**EXAMPLE 2  Coin Tosses and Roll of a Die**

*When a coin is tossed, a head* (H) *or a tail* (T) *may show. If a die is rolled, a* 1, 2, 3, 4, 5, *or* 6 *may show. Suppose a coin is tossed twice and then a die is rolled, and the result is noted (such as* H *on first toss,* T *on second, and* 4 *on roll of die). How many different results can occur?*

*Solution:* Tossing a coin twice and then rolling a die can be considered a three-stage procedure. Each of the first two stages (the coin toss) has two possible outcomes. The third stage (rolling the die) has six possible outcomes. By the basic counting principle, the number of different results for the procedure is

$$2 \cdot 2 \cdot 6 = 24.$$ ∎

**EXAMPLE 3  Answering a Quiz**

*In how many different ways can a quiz be answered under each of the following conditions?*

**a.** *The quiz consists of three multiple-choice questions with four choices for each.*

*Solution:* Successively answering the three questions is a three-stage procedure. The first question can be answered in any of four ways. Likewise, each of the other two questions can be answered in four ways. By the basic counting principle, the number of ways to answer the quiz is

$$4 \cdot 4 \cdot 4 = 4^3 = 64.$$

**b.** *The quiz consists of three multiple-choice questions (with four choices for each) and five true-false questions.*

*Solution:* Answering the quiz can be considered a two-stage procedure. First we can answer the multiple-choice questions (the first stage), and then we can answer the true-false questions (the second stage). From part (a), the three multiple-choice questions can be answered in $4 \cdot 4 \cdot 4$ ways. Each of the true-false questions has two choices ("true" or "false"), so the total number of ways of answering all five of them is $2 \cdot 2 \cdot 2 \cdot 2 \cdot 2$. By the basic counting principle, the number of ways the entire quiz can be answered is

$$\underbrace{(4 \cdot 4 \cdot 4)}_{\substack{\text{multiple-}\\\text{choice}}} \underbrace{(2 \cdot 2 \cdot 2 \cdot 2 \cdot 2)}_{\substack{\text{true-}\\\text{false}}} = 4^3 \cdot 2^5 = 2048.$$

∎

### EXAMPLE 4  Letter Arrangements

*From the five letters* A, B, C, D, *and* E, *how many three-letter horizontal arrangements (called "words") are possible if no letter may be repeated? (A "word" need not make sense.) For example,* BDE *and* DEB *are two acceptable words, but* CAC *is not.*

*Solution:* To form a word, we must successively fill the positions __ __ __ with different letters. Thus, we have a three-stage procedure. For the first position, we can choose any of the five letters. After filling that position with some letter, we can fill the second position with any of the remaining four letters. After that position is filled, the third position can be filled with any of the three letters that have not yet been used. By the basic counting principle, the total number of three-letter words is

$$5 \cdot 4 \cdot 3 = 60.$$

∎

If repetitions are allowed, the number of words is $5 \cdot 5 \cdot 5 = 125$.

## Permutations

In Example 4, we selected three different letters from five letters and arranged them in an *order*. Each result is called a *permutation of five letters taken three at a time*. More generally, we have the following definition.

### DEFINITION

*An ordered arrangement of r objects, without repetition, selected from n distinct objects is called a* **permutation of n objects taken r at a time.** *The number of such permutations is denoted* $_nP_r$.

Thus, in Example 4, we found that

$$_5P_3 = 5 \cdot 4 \cdot 3 = 60.$$

By a similar analysis, we shall now find a general formula for $_nP_r$. In making an ordered arrangement of $r$ objects from $n$ objects, for the first position we may choose any one of the $n$ objects. (See Fig. 9.3.) After the first position is filled, there remain $n - 1$ objects that may be chosen for the second position.

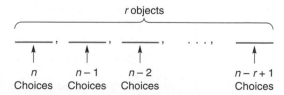

**FIGURE 9.3**  An ordered arrangement of $r$ objects selected from $n$ objects.

After that position is filled, there are $n - 2$ objects that may be chosen for the third position. Continuing in this way and using the basic counting principle, we arrive at the following formula:

> The number of permutations of $n$ objects taken $r$ at a time is given by
> $$_nP_r = \underbrace{n(n-1)(n-2)\cdots(n-r+1)}_{r \text{ factors}}. \qquad (1)$$

The formula for $_nP_r$ can be expressed in terms of factorials.[1] Multiplying the right side of Eq. (1) by

$$\frac{(n-r)(n-r-1)\cdots(2)(1)}{(n-r)(n-r-1)\cdots(2)(1)}$$

gives

$$_nP_r = \frac{n(n-1)(n-2)\cdots(n-r+1)\cdot(n-r)(n-r-1)\cdots(2)(1)}{(n-r)(n-r-1)\cdots(2)(1)}.$$

The numerator is simply $n!$, and the denominator is $(n-r)!$. Thus, we have the following result:

> The number of permutations of $n$ objects taken $r$ at a time is given by
> $$_nP_r = \frac{n!}{(n-r)!}. \qquad (2)$$

For example, from Eq. (2), we have

$$_7P_3 = \frac{7!}{(7-3)!} = \frac{7!}{4!} = \frac{7\cdot6\cdot5\cdot4\cdot3\cdot2\cdot1}{4\cdot3\cdot2\cdot1} = 210.$$

This calculation can be obtained easily with a calculator by using the factorial key. Alternatively, we can conveniently write

$$\frac{7!}{4!} = \frac{7\cdot6\cdot5\cdot4!}{4!} = 7\cdot6\cdot5 = 210.$$

Notice how 7! was written so that the 4!'s would cancel.

Many calculators can directly calculate $_nP_r$.

### EXAMPLE 5  Club Officers

*A club has* 20 *members. The offices of president, vice president, secretary, and treasurer are to be filled, and no member may serve in more than one office. How many different slates of candidates are possible?*

*Solution:* We shall consider a slate in the order of president, vice president, secretary, and treasurer. Each ordering of four members constitutes a slate, so the number of possible slates is $_{20}P_4$. By Eq. (1),

$$_{20}P_4 = 20\cdot19\cdot18\cdot17 = 116{,}280.$$

Alternatively, using Eq. (2) gives

$$_{20}P_4 = \frac{20!}{(20-4)!} = \frac{20!}{16!} = \frac{20\cdot19\cdot18\cdot17\cdot16!}{16!}$$

$$= 20\cdot19\cdot18\cdot17 = 116{,}280.$$

Note the large number of slates that are possible!  ▬

[1]Factorials are discussed in Sec. 3.2.

**EXAMPLE 6** Political Questionnaire

*A politician sends a questionnaire to her constituents to determine their concerns about six important national issues: unemployment, the environment, taxes, interest rates, national defense, and social security. A respondent is to select four issues of personal concern and rank them by placing the number 1, 2, 3, or 4 after each issue to indicate the degree of concern, with 1 indicating the greatest concern and 4 the least. In how many ways can a respondent reply to the questionnaire?*

*Solution:* A respondent is to rank four of the six issues. Thus, we can consider a reply as an ordered arrangement of six items taken four at a time, where the first item is the issue with rank 1, the second is the issue with rank 2, and so on. Hence, we have a permutation problem, and the number of possible replies is

$$_6P_4 = \frac{6!}{(6-4)!} = \frac{6!}{2!} = \frac{6 \cdot 5 \cdot 4 \cdot 3 \cdot 2!}{2!} = 6 \cdot 5 \cdot 4 \cdot 3 = 360. \qquad \blacksquare$$

In case you want to find the number of permutations of $n$ objects taken all at a time, setting $r = n$ in Eq. (2) gives

$$_nP_n = \frac{n!}{(n-n)!} = \frac{n!}{0!} = \frac{n!}{1} = n!.$$

Each of these permutations is simply called a **permutation of $n$ objects.**

> The number of permutations of $n$ objects is $n!$

For example, the number of permutations of the letters in the word SET is 3!, or 6. These permutations are

$$\text{SET,} \qquad \text{STE,} \qquad \text{EST,} \qquad \text{ETS,} \qquad \text{TES,} \qquad \text{TSE.}$$

**EXAMPLE 7** Name of Legal Firm

*Lawyers Smith, Jones, Jacobs, and Bell want to form a legal firm and will name it by using all four of their last names. How many possible names are there?*

*Solution:* Since order is important, we must find the number of permutations of four names, which is

$$4! = 4 \cdot 3 \cdot 2 \cdot 1 = 24.$$

Thus, there are 24 possible names for the firm. $\qquad \blacksquare$

## ▪ Exercise 9.1

1. **Production Process** In a production process, a product goes through one of the assembly lines A, B, or C and then goes through one of the finishing lines D or E. Draw a tree diagram that indicates the possible production routes for a unit of the product. How many production routes are possible?

2. **Air Conditioner Models** A manufacturer produces air conditioners having 6000-, 8000-, and 10,000-BTU capacities. Each capacity is available with one- or two-speed fans. Draw a tree diagram that represents all types of models. How many types are there?

3. **Die Roll and Coin Toss** A die is rolled and then a coin is tossed. Draw a tree diagram to indi-

cate the possible results. How many results are possible?

4. **Coin Toss** A coin is tossed three times. Draw a tree diagram to indicate the possible results. How many results are possible?

*In Problems 5–10, use the basic counting principle.*

5. **Course Selection**   A student must take a science course and a humanities course. The available science courses are biology, chemistry, physics, computer science, and mathematics. In humanities, the available courses are English, history, speech communications, and classics. How many two-course selections can the student make?

6. **Auto Routes**   A person lives in city A and commutes by automobile to city B. There are four roads connecting A and B. (a) How many routes are possible for a round-trip? (b) How many round-trip routes are possible if a different road is to be used for the return trip?

7. **Dinner Choices**   At a restaurant, a complete dinner consists of an appetizer, an entree, a dessert, and a beverage. The choices for the appetizer are soup and salad; for the entree, the choices are chicken, fish, steak, and

lamb; for the dessert, the choices are cherries jubilee, fresh peach cobbler, chocolate truffle cake, and blueberry roly-poly; for the beverage, the choices are coffee, tea, and milk. How many complete dinners are possible?

8. **Multiple-Choice Exam**   In how many ways is it possible to answer a six-question multiple-choice examination if each question has four choices (and one choice is selected for each question)?

9. **True-False Exam**   In how many ways is it possible to answer a 10-question true-false examination?

10. **Product Codes**   A manufacturer place a five-symbol code on each unit of product. The code consists of a letter followed by four digits, the first of which is not 0. How many codes are possible?

*In Problems 11–16, determine the values.*

11. $_5P_2$.

12. $_{100}P_1$.

13. $_6P_6$.

14. $_9P_4$.

15. $_4P_2 \cdot _5P_3$.

16. $\dfrac{_6P_3}{_6P_2}$.

17. Compute $1000!/999!$ without using a calculator. Now try it with your calculator, using the factorial feature.

18. Determine $\dfrac{_nP_r}{n!}$.

*In Problems 19–42, use any appropriate counting method.*

19. **Name of Firm**   Flynn, Peters, and Walters are forming an advertising firm and agree to name it by their three last names. How many names for the firm are possible?

20. **Softball**   If a softball league has seven teams, how many different end-of-the-season rankings are possible? Assume that there are no ties.

21. **Contest**   In how many ways can a judge award first, second, and third prizes in a contest having eight contestants?

22. **Matching-Type Exam**   On a history exam, each of four items in one column is to be matched with exactly one of six items in another column. No item in the second column can be selected more than once. In how many ways can the matching be done?

23. **Die Roll**   A die is rolled three times and the outcome of each roll is noted. How many results are possible?

24. **Coin Toss**   A coin is tossed four times. How many results are possible if the order of the tosses is considered?

25. **Problem Assignment**   In a mathematics class with 12 students, the instructor wants homework problems 1, 3, and 5 put on the board by three different students. In how many ways can the instructor assign the problems?

26. **Combination Lock**   A combination lock has 10 different letters, and a sequence of three different letters must be selected for the lock to open. How many combinations are possible?

27. **Employee Questionnaire**   A company issues a questionnaire whereby each employee must rank the three items with which he or she is most dissatisfied. The items are

| | |
|---|---|
| wages, | supervisors, |
| work environment, | health insurance, |
| vacation time, | break time, |
| job security, | retirement plan. |

The ranking is to be indicated by the numbers 1, 2, and 3, where 1 indicates the item involving the greatest dissatisfaction and 3 the least. In how many ways can an employee answer the questionnaire?

28. **Die Roll**   A die is rolled four times. How many results are possible if the order of the rolls is considered and the second roll is even?

29. **Letter Arrangements**   How many six-letter words from the letters in the word MEADOW are possible if no letter is repeated?

30. **Letter Arrangements**   Using the letters in the word DISC, how many four-letter words are possible if no letter is repeated?

31. **Book Arrangements**   In how many ways can five of seven books be arranged on a bookshelf? In how many ways can all seven books be arranged on the shelf?

32. **Lecture Hall**   A lecture hall has five doors.
   **a.** In how many ways can a student enter the hall by one door and exit by a different door?
   **b.** Exit by any door?

**33. Poker Hand** A poker hand consists of 5 cards drawn from a deck of 52 playing cards. The hand is said to be "four of a kind" if four of the cards have the same face value. For example, hands with four 10's or four jacks or four 2's are four-of-a-kind hands. How many such hands are possible?

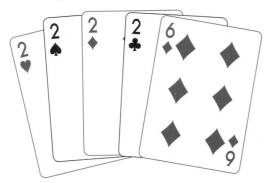

**34. Merchandise Choice** In a merchandise catalog, a blanket is available in the colors of blue, pink, yellow, and beige. When placing an order for one, customers must indicate their first and second color choices. In how many ways can this be done?

**35. Diner Order** Four students go to a diner and order a hamburger, a cheeseburger, a fish sandwich, and a roast beef sandwich (one sandwich for each). When the waitress returns with the food, she forgets which student ordered which item and simply places a sandwich before each student. In how many ways can the waitress do this?

**36. Group Photograph** In how many ways can three men and two women line up for a group picture? In how many ways can they line up if a woman is to be at each end?

**37. Club Officers** A club has 12 members.
   **a.** In how many ways can the offices of president, vice president, secretary, and treasurer be filled if no member can serve in more than one office?
   **b.** In how many ways can the four offices be filled if the president and vice president must be different members?

**38. Fraternity Names** Suppose a fraternity is named by three Greek letters. (There are 24 letters in the Greek alphabet.)
   **a.** How many names are possible?
   **b.** How many names are possible if no letter can be used more than one time?

**39. Basketball** In how many ways can a basketball coach assign positions to her five-member team if two of the members are qualified for the center position and all five are qualified for all the other positions?

**40. Call Letters** Suppose the call letters of a radio station consist of four letters, of which the first must be a K or a W. How many such identifications are possible?

**41. Baseball** A baseball manager determines that, of his nine team members, four are strong hitters and five are weak. If the manager wants the strong hitters to be the first four batters in a batting order, how many batting orders are possible?

**42. Signal Flags** When at least one of four flags colored red, green, yellow, and blue are arranged vertically on a flagpole, the result indicates a signal (or message). Different arrangements give different signals.
   **a.** How many different signals are possible if all four flags are used?
   **b.** How many different signals are possible if at least one flag is used?

O B J E C T I V E

**To discuss combinations, permutations with repeated objects, and assignments to cells.**

# 9.2 COMBINATIONS AND OTHER COUNTING PRINCIPLES

## Combinations

We continue our discussion of counting methods by considering the following. In a 20-member club the offices of president, vice president, secretary, and treasurer are to be filled, and no member may serve in more than one office. If these offices, in the order given, are filled by members A, B, C, and D, respectively, then we can represent this slate by

$$ABCD.$$

A different slate is

$$BACD.$$

These two slates represent different permutations of 20 members taken four at a time. Now, as a different situation, let us consider four-person *committees* that may be formed from the 20 members. In that case, the two arrangements

$$ABCD \quad \text{and} \quad BACD$$

represent the *same* committee. Here *the order of listing the members is of no concern*. These two arrangements are considered to give the same *combination* of A, B, C, and D.

**DEFINITION**

The important phrase here is "without regard to order," since order implies a permutation rather than a combination.

*An arrangement of r objects, without regard to order and without repetition, selected from n distinct objects is called a* **combination of n objects taken r at a time.** *The number of such combinations is denoted $_nC_r$.*

**EXAMPLE 1   Comparing Combinations and Permutations**

*List all combinations and all permutations of the four letters*

$$A, \quad B, \quad C, \quad and \quad D$$

*when they are taken three at a time.*

*Solution:* The combinations are

$$ABC, \quad ABD, \quad ACD, \quad BCD.$$

There are four combinations, so $_4C_3 = 4$. The permutations are

| | | | |
|---|---|---|---|
| ABC, | ABD, | ACD, | BCD, |
| ACB, | ADB, | ADC, | BDC, |
| BAC, | BAD, | CAD, | CBD, |
| BCA, | BDA, | CDA, | CDB, |
| CAB, | DAB, | DAC, | DBC, |
| CBA, | DBA, | DCA, | DCB. |

There are 24 permutations. ■

In Example 1, notice that each column consists of all the permutations for the same combination of letters. With this observation, we can determine a formula for $_nC_r$—the number of combinations of $n$ objects taken $r$ at a time. Suppose one such combination is

$$x_1 x_2 \cdots x_r.$$

The number of permutations of these $r$ objects is $r!$. If we listed all other such combinations and then listed all permutations of these combinations, we would obtain a complete list of permutations of the $n$ objects taken $r$ at a time. Thus,

$$_nC_r \cdot r! = {}_nP_r.$$

Solving for $_nC_r$ gives

$$_nC_r = \frac{_nP_r}{r!} = \frac{\dfrac{n!}{(n-r)!}}{r!} = \frac{n!}{r!(n-r)!}.$$

The number of combinations of $n$ objects taken $r$ at a time is given by

$$_nC_r = \frac{n!}{r!(n-r)!}.$$

Many calculators can directly compute $_nC_r$.

**EXAMPLE 2**  **Committee Selection**

*If a club has 20 members, how many different four-member committees are possible?*

*Solution:* Order is not important because, no matter how the members of a committee are arranged, we have the same committee. Thus, we simply have to compute the number of combinations of 20 objects taken four at a time, $_{20}C_4$:

$$_{20}C_4 = \frac{20!}{4!(20-4)!} = \frac{20!}{4!16!}$$

$$= \frac{20 \cdot 19 \cdot 18 \cdot 17 \cdot \cancel{16!}}{4 \cdot 3 \cdot 2 \cdot 1 \cdot \cancel{16!}} = 4845.$$

Observe how 20! was written so that the 16!'s would cancel.

There are 4845 possible committees. ∎

It is important to remember that if a selection of objects is made and *order is important,* then *permutations* should be considered. If *order is not important,* consider *combinations.*

**EXAMPLE 3**  **Poker Hand**

*A **poker hand** consists of five cards dealt from an ordinary deck of 52 cards. How many different poker hands are there?*

*Solution:* One possible hand is

2 of hearts, 3 of diamonds, 6 of clubs,

4 of spades, king of hearts,

which we can abbreviate as

2H, 3D, 6C, 4S, KH.

The order in which the cards are dealt does not matter, so this hand is the same as

KH, 4S, 6C, 3D, 2H.

Thus, the number of possible hands is the number of ways that five objects can be selected from 52, without regard to order. This is a combination problem. We have

$$_{52}C_5 = \frac{52!}{5!(52-5)!} = \frac{52!}{5!47!}$$

$$= \frac{52 \cdot 51 \cdot 50 \cdot 49 \cdot 48 \cdot 47!}{5 \cdot 4 \cdot 3 \cdot 2 \cdot 1 \cdot 47!}$$

$$= \frac{52 \cdot 51 \cdot 50 \cdot 49 \cdot 48}{5 \cdot 4 \cdot 3 \cdot 2} = 2{,}598{,}960.$$

∎

**EXAMPLE 4**  **Majority Decision and Sum of Combinations**

*A college promotion committee consists of five members. In how many ways can the committee reach a majority decision in favor of a promotion?*

*Strategy:* A favorable majority decision is reached if, and only if,

exactly three members vote favorably,

or exact four members vote favorably,

or all five members vote favorably.

To determine the total number of ways to reach a favorable majority decision, we *add* the number of ways that each of the preceding votes can occur.

*Solution:* Suppose exactly three members vote favorably. The order of the members is of no concern, and thus, we can think of these members as forming a combination. Hence, the number of ways three of the five members can vote favorably is $_5C_3$. Similarly, the number of ways exactly four can vote favorably is $_5C_4$, and the number of ways all five can vote favorably is $_5C_5$ (which, of course, is 1). Thus, the number of ways to reach a majority decision in favor of a promotion is

$$_5C_3 + {_5C_4} + {_5C_5} = \frac{5!}{3!(5-3)!} + \frac{5!}{4!(5-4)!} + \frac{5!}{5!(5-5)!}$$

$$= \frac{5!}{3!2!} + \frac{5!}{4!1!} + \frac{5!}{5!0!}$$

$$= \frac{5 \cdot 4 \cdot 3!}{3! \cdot 2 \cdot 1} + \frac{5 \cdot 4!}{4! \cdot 1} + 1$$

$$= 10 + 5 + 1 = 16.$$ ∎

## Permutations with Repeated Objects

In Sec. 9.1, we discussed permutations of objects that were all different. Now we examine the case where some of the objects are alike (or *repeated*). For example, consider determining the number of different permutations of the seven letters in the word

SUCCESS.

Here the letters C and S are repeated. If the two C's were interchanged, the resulting permutation would be indistinguishable from SUCCESS. Thus, the number of distinct permutations is not 7!, as it would be with 7 different objects. To determine the number of distinct permutations, we use an approach that involves combinations.

Figure 9.4(a) shows boxes representing the different letters in the word SUCCESS. In these boxes we place the integers from 1 through 7. We place three integers in the S's box (because there are three S's), one in the U box, two in the C's box, and one in the E box. A typical placement is indicated in

(a)                                (b)

**FIGURE 9.4** Permutations with repeated objects.

Fig. 9.4(b). That placement can be thought of as indicating a permutation of the seven letters in SUCCESS, namely, the permutation in which (going from left to right) the S's are in the second, third, and sixth positions, the U is in the first position, and so on. Thus, Fig. 9.4(b) corresponds to the permutation

$$U\ S\ S\ E\ C\ S\ C.$$
$$\uparrow\ \uparrow\ \uparrow\ \uparrow\ \uparrow\ \uparrow\ \uparrow$$
$$1\ 2\ 3\ 4\ 5\ 6\ 7$$

To count the number of distinct permutations, it suffices to determine the number of ways the integers from 1 to 7 can be placed in the boxes. Since the order in which they are placed into a box is not important, the S's box can be filled in $_7C_3$ ways. Then the U box can be filled with one of the remaining four integers in $_4C_1$ ways. Finally, the C's and E boxes can be successively filled in $_3C_2$ and $_1C_1$ ways, respectively. Since we have a four-stage procedure, by the basic counting principle the total number of ways to fill the boxes or, equivalently, the number of distinguishable permutations of the letters in SUCCESS is

$$_7C_3 \cdot {}_4C_1 \cdot {}_3C_2 \cdot {}_1C_1 = \frac{7!}{3!\,4!} \cdot \frac{4!}{1!\,3!} \cdot \frac{3!}{2!\,1!} \cdot \frac{1!}{1!\,0!}$$

$$= \frac{7!}{3!\,1!\,2!\,1!}$$

$$= 420.$$

In summary, the word SUCCESS has four types of letters: S, U, C, and E. There are three S's, one U, two C's, and one E, and the number of distinguishable permutations of the 7 letters is

$$\frac{7!}{3!\,1!\,2!\,1!}.$$

Observing the forms of the numerator and denominator, we can make the following generalization:

---

**Permutations with Repeated Objects**

The number of distinguishable permutations of $n$ objects such that $n_1$ are of one type, $n_2$ are of a second type, . . . , and $n_k$ are of a $k$th type, where $n_1 + n_2 + \cdots + n_k = n$, is

$$\frac{n!}{n_1!\,n_2! \cdots n_k!} \qquad (1)$$

---

**EXAMPLE 5**   **Letter Arrangements with and without Repetition**

*For each of the following words, how many distinguishable permutations of the letters are possible?*

a. *APOLLO*

   *Solution:* The word APOLLO has six letters with repetition. We have one A, one P, two O's, and two L's. Using Eq. (1), we find that the number of permutations is

$$\frac{6!}{1!\,1!\,2!\,2!} = 180.$$

**b.** *GERM*

*Solution:* None of the four letters in GERM is repeated, so the number of permutations is

$$_4P_4 = 4! = 24.$$ ∎

### EXAMPLE 6  Name of Legal Firm

*A group of four lawyers, Smith, Jones, Smith, and Bell (the Smiths are cousins), want to form a legal firm and will name it by using all of their last names. How many possible names exist?*

*Solution:* Each different permutation of the last four names is a name for the firm. There are two Smiths, one Jones, and one Bell. From Eq. (1), the number of distinguishable names is

$$\frac{4!}{2!1!1!} = 12.$$ ∎

### Cells

At times, we want to find the number of ways in which objects can be placed into "compartments," or *cells*. For example, suppose that from a group of five people, three are to be assigned to room A and two to room B. In how many ways can this be done? Figure 9.5 shows one such assignment, where the numbers $1, 2, \ldots, 5$ represent the people. Obviously, the order in which people are placed into the rooms is of no concern. The boxes (or cells) remind us of those in Fig. 9.4(b), and, by an analysis similar to the discussion of permutations with repeated objects, the number of ways to assign the people is

$$\frac{5!}{3!2!} = \frac{5 \cdot 4 \cdot 3!}{3!2!} = 10.$$

In general, we have the following principle:

**FIGURE 9.5**
Assignment of people to rooms.

> **Assignment to Cells**
>
> Suppose $n$ distinct objects are assigned to $k$ ordered cells with $n_i$ objects in cell $i$ $(i = 1, 2, \ldots, k)$ and the order in which the objects are assigned to cell $i$ is of no concern. The number of all such assignment is
>
> $$\frac{n!}{n_1!n_2! \cdots n_k!},$$ (2)
>
> where $n_1 + n_2 + \cdots + n_k = n$.

### EXAMPLE 7  Assigning Mourners to Limousines

*A funeral director must assign 15 mourners to three limousines: 6 in the first limousine, 5 in the second, and 4 in the third. In how many ways can this be done?*

*Solution:* Here 15 people are placed into three cells (limousines): 6 in cell 1, 5 in cell 2, and 4 in cell 3. By Eq. (2), the number of ways this can be done is

$$\frac{15!}{6!5!4!} = 630,630.$$ ∎

Example 8 will show three different approaches to a counting problem. Indeed, many counting problems have alternative methods of solution.

**EXAMPLE 8**  **Art Exhibit**

*An artist has created 20 original paintings, and she will exhibit some of them in three galleries. Four paintings will be sent to gallery* A, *four to gallery* B, *and three to gallery* C. *In how many ways can this be done?*

*Solution:*

*Method 1*  The artist must send $4 + 4 + 3 = 11$ paintings to the galleries, and the 9 that are not sent can be thought of as staying in her studio. Thus, we can think of this situation as placing 20 paintings into four cells:

4 in gallery A,

4 in gallery B,

3 in gallery C,

9 in the artist's studio.

From Eq. (2), the number of ways this can be done is

$$\frac{20!}{4!4!3!9!} = 1{,}939{,}938{,}000.$$

*Method 2*  We can handle the problem in terms of a two-stage procedure and use the basic counting principle. First, 11 paintings are selected for exhibit. Then, these are split into three groups (cells) corresponding to the three galleries. We proceed as follows.

Selecting 11 of the 20 paintings for exhibit (order is of no concern) can be done in $_{20}C_{11}$ ways. Once a selection is made, four of the paintings go into one cell (gallery A), four go to a second cell (gallery B), and three go to a third cell (gallery C). By Eq. (2), this can be done in $\dfrac{11!}{4!4!3!}$ ways. Applying the basic counting principle gives the number of ways the artist can send the paintings to the galleries:

$$_{20}C_{11} \cdot \frac{11!}{4!4!3!} = \frac{20!}{11!9!} \cdot \frac{11!}{4!4!3!} = 1{,}939{,}938{,}000.$$

*Method 3*  Another approach to this problem is in terms of a three-stage procedure. First, 4 of the 20 paintings are selected for shipment to gallery A. This can be done in $_{20}C_4$ ways. Then, from the remaining 16 paintings, the number of ways 4 can be selected for gallery B is $_{16}C_4$. Finally, the number of ways 3 can be sent to gallery C from the 12 paintings that have not yet been selected is $_{12}C_3$. By the basic counting principle, the entire procedure can be done in

$$_{20}C_4 \cdot {}_{16}C_4 \cdot {}_{12}C_3$$

ways, which gives the previous answer, as expected!  ■

## ■ Exercise 9.2

*In Problems 1–6, determine the values.*

**1.** $_6C_4$.

**2.** $_6C_2$.

**3.** $_{100}C_{100}$.

**4.** $_{100}C_1$.

**5.** $_3P_2 \cdot {}_3C_2$.

**6.** $_4P_2 \cdot {}_5C_3$.

**7.** Verify that $_nC_r = {}_nC_{n-r}$.

**8.** Determine $_nC_n$.

**9.** **Committee**  In how many ways can a five-member committee be formed from a group of 15 people?

**10.** **Horse Race**  In a horse race, a horse is said to *finish in the money* if it finishes in first, second, or third place. For an eight-horse race, in how many ways can the horses finish in the money? Assume no ties.

**11. Math Exam** On a 12-question mathematics examination, a student must answer any 10 questions. In how many ways can the 10 questions be chosen (without regard to order)?

**12. Cards** From a deck of 52 playing cards, how many 3-card hands are there?

**13. Quality Control** A quality-control technician must select a sample of 10 dresses from a production lot of 74 couture dresses. How many different samples are possible? Express your answer in terms of factorials.

**14. Packaging** A jelly producer makes seven types of jelly. The producer packages gift boxes containing four jars of jelly, no two of which are of the same type. To reflect the three national chains through which the jelly is distributed, the producer uses three types of boxes. How many different gift boxes are possible?

**15. Scoring on Exam** In a 10-question examination, each question is worth 10 points and is graded right or wrong. Considering the individual questions, in how many ways can a student score 80 or better?

**16. Team Results** A sports team plays nine games. In how many ways can the outcomes of the games result in four wins, three losses, and two ties?

**17. Letter Arrangements** How many distinguishable horizontal arrangements of all the letters in the word REMEMBER are possible?

**18. Letter Arrangements** How many distinguishable horizontal arrangements of all the letters in the word ALABAMA are possible?

**19. Coin Toss** If a coin is tossed six times and the outcome of each toss is noted, in how many ways can four heads and two tails occur?

**20. Die Roll** A die is rolled six times and the order of the rolls is considered. In how many ways can two 2's, three 3's, and one 4 occur?

**21. Repair Scheduling** An appliance repairman must go out on six service calls. In how many ways can he arrange his schedule?

**22. Baseball** A Little League baseball team has 12 members and must play an away game. Three cars will be used for transportation. In how many ways can the manager assign the members to specific cars if each car can accommodate four members?

**23. Project Assignment** The director of research and development for a company has nine scientists who are equally qualified to work on projects A, B, and C. In how many ways can the director assign three scientists to each project?

**24. Begonias** A gardener buys a container of six begonia plants. The label indicates that three of the plants will have red flowers and three will have pink. However, none of the plants are in bloom. If the gardener plants the begonias in a horizontal row, how many different color arrangements are possible when all the begonias are in bloom?

**25. True-False Exam** A biology instructor includes several true-false questions on quizzes. From past experience, a student believes that half of the questions are true and half are false. If there are eight true-false questions on the next quiz, in how many ways can the student answer half of them "true" and the other half "false"?

**26. Food Order** A waiter takes the following order from a table with seven people: three hamburgers, two cheeseburgers, and two steak sandwiches. Upon returning with the food, he forgets who ordered what item and simply places an item in front of each person. In how many ways can the waiter do this?

**27. Caseworker Assignment** A social services office has 15 new clients. The supervisor wants to assign 5 clients to each of three specific caseworkers. In how many ways can this be done?

**28. Basketball** There are nine members on a basketball team, and all are qualified for the five positions. In how many ways can the coach form a starting lineup?

**29. Flag Signals** Colored flags arranged vertically on a flagpole indicate a signal (or message). How many different signals are possible if

**a.** two red, two green, and two yellow flags are used?

**b.** two red, two green, and three yellow flags are available, and all the red and green flags and at least one yellow flag are used?

**30. Hiring** A company personnel director must hire five people: three for the assembly department and two for the shipping department. There are 10 applicants who are equally qualified to work in each department. In how many ways can the personnel director fill the positions?

**31. Financial Portfolio** A financial advisor wants to create a portfolio consisting of eight stocks and four bonds. If twelve stocks and seven bonds are acceptable for the portfolio, in how many ways can the portfolio be created?

**32. World Series** A baseball team wins the World Series if it is the first team in the series to win four games. Thus, a series could range from four to seven games. For example, a team winning the first four games would be the champion. Likewise, a team losing the first three games and winning the last four would be champion. In how many ways can a team win the World Series?

**33. Subcommittee** A committee has seven members, three of whom are male and four female. In how many ways can a subcommittee be selected if it is to consist exactly of

**a.** three males?

**b.** four females?

**c.** two males and two females?

**34. Subcommittee** A committee has four male and four female members. In how many ways can a subcommittee of four be selected if at least two females are to serve on it?

**35. Poker Hand** A poker hand consists of 5 cards from a deck of 52 playing cards. The hand is a "full house" if there are 3 cards of one denomination and 2 cards of another. For example, three 10's and two jacks form a full house. How many full-house hands are possible?

**36. Poker Hand** In poker, 2 cards of the same denomination form a "pair." For example, two 8's form a pair. A poker hand (5 cards from a 52-card deck) is said to be a "two-pair" hand if it contains two pairs and there are three different face values involved in the five cards. For example, a pair of 3's, a pair of 8's, and a 10 constitute a two-pair hand. How many two-pair hands are possible?

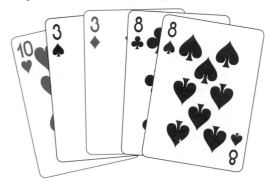

**37. Tram Loading** At a tourist attraction, two trams carry sightseers up a picturesque mountain. One tram can accommodate six people and the other eight. A busload of 18 tourists arrives, and both trams are at the bottom of the mountain. Obviously, only 14 tourists can initially go up the mountain. In how many ways can the attendant load 14 tourists onto the two trams?

**38. Discussion Groups** A history instructor wants to split a class of 10 students into three discussion groups. One group will consist of four students and discuss topic A. The second and third groups will discuss topics B and C, respectively, and consist of three students each.

    **a.** In how many ways can the instructor form the groups?

    **b.** If the instructor designates a group leader and a secretary (different students) for each group, in how many ways can the class be split?

---

**To determine a sample space and to consider events associated with it. (These notions involve sets and subsets.) To represent a sample space and events by means of a Venn diagram. To introduce the notions of complement, union, and intersection.**

## 9.3 SAMPLE SPACES AND EVENTS

### Sample Spaces

Inherent in any discussion of probability is the performance of an experiment (a procedure) in which a particular result, or *outcome,* involves chance. For example, consider the experiment of tossing a coin. There are only two ways the coin can fall, a head (H) or a tail (T), but the actual outcome is determined by chance. (We assume that the coin does not land on its edge.) The set of possible outcomes,

$$\{H, T\},$$

is called a *sample space* for the experiment, and H and T are called *sample points.*

**DEFINITION**

*A **sample space** S for an experiment is the set of all possible outcomes of the experiment such that each outcome corresponds to exactly one element in S. The elements of S are called **sample points.** If there is a finite number of sample points, that number is denoted n(S), and S is said to be a **finite sample space.***

When determining "possible outcomes" of an experiment, we must be sure that they reflect the situation about which we are concerned. For example, consider the experiment of rolling a die and observing the top face. We could say that a sample space is

$$S_1 = \{1, 2, 3, 4, 5, 6\},$$

The order in which sample points are listed in a sample space is of no concern.

where the possible outcomes are the number of dots on the top face. However, other possible outcomes are

        odd number of dots appear    (odd)

and    even number of dots appear    (even).

Thus, the set

$$S_2 = \{\text{odd, even}\}$$

is also a sample space for the experiment, so you can see that an experiment may have more than one sample space.

If an outcome in $S_1$ occurred, then we know which outcome in $S_2$ occurred, but the reverse is not true. To describe this asymmetry, we say that $S_1$ is a **more primitive** sample space than $S_2$. Usually, the more primitive a sample space is, the more questions pertinent to the experiment it allows us to answer. For example, with $S_1$, we can answer such questions as

"Did a three occur?"

"Did a number greater than two occur?"

"Did a number less than four occur?"

But with $S_2$, we cannot answer these questions. As a rule of thumb, the more primitive a sample space is, the more elements it has and the more detail it indicates. Unless otherwise stated, when an experiment has more than one sample space, it will be our practice to consider only a sample space that gives sufficient detail to answer all pertinent questions relative to the experiment. For example, for the experiment of rolling a die and observing the top face, it will be tacitly understood that we are observing the number of dots. Thus, we shall consider the sample space to be

$$S_1 = \{1, 2, 3, 4, 5, 6\}$$

and shall refer to it as the *usual* sample space for the experiment.

### EXAMPLE 1  Sample Space: Toss of Two Coins

*Two different coins are tossed, and the result* (H *or* T) *for each coin is observed. Determine a sample space.*

*Solution:* One possible outcome is a head on the first coin and a head on the second, which we can indicate by the ordered pair (H, H) or, more simply, HH. Similarly, we indicate a head on the first coin and a tail on the second by HT, and so on. A sample space is

$$S = \{\text{HH, HT, TH, TT}\}.$$

A tree diagram is given in Fig. 9.6 and, in a sense, indicates the sample space. We remark that $S$ is also a sample space for the experiment of tossing a single

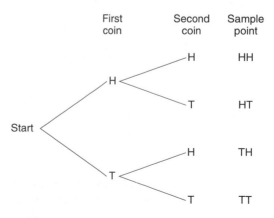

**FIGURE 9.6**  Tree diagram for toss of two coins.

coin twice in succession. In fact, these two experiments can be considered one and the same. Although other sample spaces can be contemplated, we take $S$ to be the *usual* sample space for these experiments.  ■

---

**Principles in Practice 1**

**Sample Space**

A video store has 400 different movies to rent. A customer wants to rent 3 movies. If she chooses the videos at random, how many 3-movie choices (sample points) does she have?

---

**EXAMPLE 2   Sample Space: Three Tosses of Coin**

*A coin is tossed three times, and the result of each toss is observed. Describe a sample space and determine the number of sample points.*

**Solution:** Because there are three tosses, we shall choose a sample point to be an ordered *triple,* such as HHT, where each component is either H or T. By the basic counting principle, the total number of sample points is $2 \cdot 2 \cdot 2$, or 8. A sample space (the *usual* one) is

$$S = \{HHH, HHT, HTH, HTT, THH, THT, TTH, TTT\},$$

and a tree diagram appears in Fig. 9.7. Note that it is not necessary to list the entire sample space to determine the number of sample points in it.

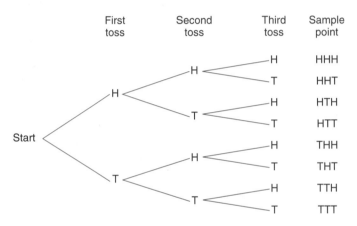

**FIGURE 9.7**   Tree diagram for three tosses of a coin.  ■

**FIGURE 9.8**   Four colored marbles in urn.

**EXAMPLE 3   Sample Space: Marbles in Urn**

*An urn contains four marbles: one red, one pink, one black, and one white. (See Fig. 9.8.)*

a. *A marble is drawn at random, its color is noted, and it is placed back into the urn. Then a marble is again randomly drawn and its color noted. Describe a sample space and determine the number of sample points.*

**Solution:** In this experiment we say that the two marbles are drawn **with replacement.** Let R, P, B, and W denote drawing a red, pink, black, and white marble, respectively. Then our sample space consists of the sample points RW, PB, RB, WW, and so on, where (for example) RW represents the outcome that the first marble drawn is red and the second is white. There are four possibilities for the first draw and, since that marble is placed back, four possibilities for the second draw. By the basic counting principle, the number of sample points is $4 \cdot 4$, or 16.

b. *Determine the number of sample points in the sample space if two marbles are selected in succession* **without replacement** *and the colors are noted.*

**Solution:** The first marble drawn can have any of four colors. Since it is *not* returned to the urn, the second marble drawn can have any of the *three* remaining colors. Thus, the number of sample points is $4 \cdot 3$, or 12. Alternatively, there are $_4P_2 = 12$ sample points. ■

**EXAMPLE 4   Sample Space: Poker Hand**

*From an ordinary deck of 52 playing cards, a poker hand is dealt. Describe a sample space and determine the number of sample points.*

*Solution:* A sample space consists of all combinations of 52 cards taken 5 at a time. From Example 3 of Sec. 9.2, the number of sample points is $_{52}C_5 = 2{,}598{,}960$. ■

**EXAMPLE 5   Sample Space: Roll of Two Dice**

*A pair of dice is rolled once, and for each die, the number that turns up is observed. Describe a sample space.*

*Solution:* Think of the dice as being distinguishable, as if one were red and the other green. Each die can turn up in six ways, so we can take a sample point to be an ordered pair in which each component is an integer between 1 and 6, inclusive. For example, $(4, 6), (3, 2)$, and $(2, 3)$ are three different sample points. By the basic counting principle, the number of sample points is $6 \cdot 6$, or 36. ■

## Events

At times, we are concerned with the outcomes of an experiment that satisfy a particular relationship. For example, we may be interested in whether the outcome of rolling a single die is an even number. This relationship can be considered as the set of outcomes $\{2, 4, 6\}$, which is a subset of the sample space

$$S = \{1, 2, 3, 4, 5, 6\}.$$

In general, any subset of a sample space is called an *event* for the experiment. Thus,

$$\{2, 4, 6\}$$

is the event that an even number turns up, which can also be described by

$$\{\text{an even number}\}.$$

Note that although an event is a set, it can be described verbally. Usually, an event is denoted by $E$. When several events are involved in a discussion, they may be denoted by $E, F, G, H$, and so on, or by $E_1, E_2, E_3$, and so on.

**DEFINITION**

*Any subset $E$ of a sample space for an experiment is called an **event** for the experiment. If the outcome of the experiment is a sample point in $E$, then event $E$ is said to **occur**.*

In the previous experiment of rolling a die, we saw that $\{2, 4, 6\}$ is an event. Thus, if the outcome is a 2, that event occurs. Some other events are

$$E = \{1, 3, 5\} = \{\text{an odd number}\},$$

$$F = \{3, 4, 5, 6\} = \{\text{a number} \geq 3\},$$

$$G = \{1\}.$$

A sample space is a subset of itself, so it, too, is an event, called a **certain event;** it must occur no matter what the outcome. An event, such as $\{1\}$, that consists of a single sample point is called a **simple event.** We can also consider an event such as "7 occurs." This event contains no sample points, so it is the empty set $\varnothing$ (the set with no elements in it). In fact, $\varnothing$ is called an **impossible event,** because it can never occur.

### EXAMPLE 6   Events

*A coin is tossed three times, and the result of each toss is noted. The usual sample space (from Example 2) is*

$$\{HHH, HHT, HTH, HTT, THH, THT, TTH, TTT\}.$$

*Determine the following events.*

**a.** $E = \{\text{one head and two tails}\}.$

*Solution:*

$$E = \{HTT, THT, TTH\}.$$

**b.** $F = \{\text{at least two heads}\}.$

*Solution:*

$$F = \{HHH, HHT, HTH, THH\}.$$

**c.** $G = \{\text{all heads}\}.$

*Solution:*

$$G = \{HHH\}.$$

**d.** $I = \{\text{head on first toss}\}.$

*Solution:*

$$I = \{HHH, HHT, HTH, HTT\}. \qquad \blacksquare$$

Sometimes it is convenient to represent a sample space $S$ and an event $E$ by a *Venn diagram,* as in Fig. 9.9. The region inside the rectangle represents the sample points in $S$. (The sample points are not specifically shown.) The sample points in $E$ are represented by the points inside the circle. Because $E$ is a subset of $S$, the circular region cannot extend outside the rectangle.

With Venn diagrams, it is easy to see how events for an experiment can be used to form other events. Figure 9.10 shows sample space $S$ and event $E$. The shaded region inside the rectangle, but outside the circle, represents the set of all sample points in $S$ that are not in $E$. This set is an event called the *complement of E* and is denoted by $E'$. Figure 9.11(a) shows two events, $E$ and $F$. The shaded region represents the set of all sample points either in $E$, or in $F$, or in both $E$ and $F$. This set is an event called the *union* of $E$ and $F$ and is denoted by $E \cup F$. The shaded region in Fig. 9.11(b) represents the event consisting of all sample points that are common to both $E$ and $F$. This event is called the *intersection* of $E$ and $F$ and is denoted by $E \cap F$. In summary, we have the following definitions.

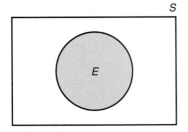

**FIGURE 9.9**   Venn diagram for sample space $S$ and event $E$.

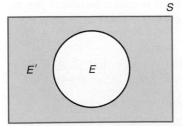

$E'$ is the shaded region

**FIGURE 9.10**   Venn diagram for the complement of $E$.

### DEFINITION

*Let $S$ be a sample space for an experiment with events $E$ and $F$. The* **complement** *of $E$, denoted by $E'$, is the event consisting of all sample points in $S$ that are not in $E$. The* **union** *of $E$ and $F$, denoted by $E \cup F$, is the event consisting of all*

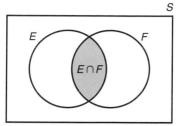

$E \cup F$, union of $E$ and $F$

(a)

$E \cap F$, intersection of $E$ and $F$

(b)

**FIGURE 9.11**  Representation of $E \cup F$ and $E \cap F$.

*sample points that are either in E, or in F, or in both E and F. The **intersection** of E and F, denoted by $E \cap F$, is the event consisting of all samples points that are common to both E and F.*

Note that if a sample point is in the event $E \cup F$, then the point is in at least one of the sets $E$ and $F$. Thus, for the event $E \cup F$ to occur, *at least one* of the events $E$ and $F$ must occur, and conversely. On the other hand, if event $E \cap F$ occurs, then *both* $E$ and $F$ must occur, and conversely. If event $E'$ occurs, then $E$ *does not* occur, and conversely.

**EXAMPLE 7**  Complement, Union, Intersection

*Given the usual sample space*

$$S = \{1, 2, 3, 4, 5, 6\}$$

*for the rolling of a die, let E, F, and G be the events*

$$E = \{1, 3, 5\}, \qquad F = \{3, 4, 5, 6\}, \qquad G = \{1\}.$$

*Determine each of the following events.*

**a.** $E'$.

*Solution:* Event $E'$ consists of those sample points in $S$ that are not in $E$, so

$$E' = \{2, 4, 6\}.$$

We note that $E'$ is the event that an even number appears.

**b.** $E \cup F$.

*Solution:* We want the sample points in $E$, or $F$, or both. Thus,

$$E \cup F = \{1, 3, 4, 5, 6\}.$$

**c.** $E \cap F$.

*Solution:* The sample points common to both $E$ and $F$ are 3 and 5, so

$$E \cap F = \{3, 5\}.$$

**d.** $F \cap G$.

*Solution:* Since $F$ and $G$ have no sample point in common,

$$F \cap G = \varnothing.$$

**e.** $E \cup E'$.

**Solution:** Using the result of part (a), we have

$$E \cup E' = \{1, 3, 5\} \cup \{2, 4, 6\} = (1, 2, 3, 4, 5, 6) = S.$$

**f.** $E \cap E'$.

**Solution:**

$$E \cap E' = \{1, 3, 5\} \cap \{2, 4, 6\} = \varnothing.$$

The results of Examples 7(e) and (f) can be generalized as follows:

> If $E$ is any event for an experiment with sample space $S$, then
>
> $$E \cup E' = S \qquad \text{and} \qquad E \cap E' = \varnothing.$$

Thus, the union of an event and its complement is the sample space; the intersection of an event and its complement is the empty set. Other properties of events are listed in Table 9.1.

---

**TABLE 9.1**    Properties of Events

If $E$ and $F$ are any events for an experiment with sample space $S$, then

| | |
|---|---|
| **1.** $E \cup E = E$. | |
| **2.** $E \cap E = E$. | |
| **3.** $(E')' = E$ | (the complement of the complement of an event is the event). |
| **4.** $E \cup E' = S$. | |
| **5.** $E \cap E' = \varnothing$. | |
| **6.** $E \cup S = S$. | |
| **7.** $E \cap S = E$. | |
| **8.** $E \cup \varnothing = E$. | |
| **9.** $E \cap \varnothing = \varnothing$. | |
| **10.** $E \cup F = F \cup E$ | (commutative property of union). |
| **11.** $E \cap F = F \cap E$ | (commutative property of intersection). |
| **12.** $(E \cup F)' = E' \cap F'$ | (the complement of a union is the intersection of complements). |
| **13.** $(E \cap F)' = E' \cup F'$ | (the complement of an intersection is the union of complements). |
| **14.** $E \cup (F \cup G) = (E \cup F) \cup G$ | (associative property of union). |
| **15.** $E \cap (F \cap G) = (E \cap F) \cap G$ | (associative property of intersection). |
| **16.** $E \cap (F \cup G) = (E \cap F) \cup (E \cap G)$ | (distributive property of intersection through union). |
| **17.** $E \cup (F \cap G) = (E \cup F) \cap (E \cup G)$ | (distributive property of union through intersection). |

---

When two events have no sample point in common, they are called *mutually exclusive* events. For example, in the rolling of a die, the events

$$E = \{2, 4, 6\} \qquad \text{and} \qquad F = \{1\}$$

are mutually exclusive (see Fig. 9.12); here, $E \cap F = \varnothing$, which, of course, is always true for two mutually exclusive events.

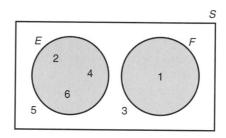

**FIGURE 9.12**   Mutually exclusive events.

**DEFINITION**

*Events E and F are said to be* **mutually exclusive** *events if and only if* $E \cap F = \varnothing$.

When two events are mutually exclusive, the occurrence of one event means that the other event cannot occur; that is, the two events cannot occur simultaneously. An event and its complement are mutually exclusive, since $E \cap E' = \varnothing$.

**EXAMPLE 8**   **Mutually Exclusive Events**

*If E, F, and G are events for an experiment and F and G are mutually exclusive, show that events* $E \cap F$ *and* $E \cap G$ *are also mutually exclusive.*

*Solution:* Given that $F \cap G = \varnothing$, we must show that the intersection of $E \cap F$ and $E \cap G$ is the empty set. Using the properties in Table 9.1, we have

$$
\begin{aligned}
(E \cap F) \cap (E \cap G) &= (E \cap F \cap E) \cap G && \text{(property 15)} \\
&= (E \cap E \cap F) \cap G && \text{(property 11)} \\
&= (E \cap F) \cap G && \text{(property 2)} \\
&= E \cap (F \cap G) && \text{(property 15)} \\
&= E \cap \varnothing && \text{(given)} \\
&= \varnothing && \text{(property 9)} \quad \blacksquare
\end{aligned}
$$

## ■ Exercise 9.3

*In Problems **1–6**, determine a sample space for the given experiment.*

**1. Card Selection**   A card is drawn from a four-card deck consisting of the 9 of diamonds, 9 of hearts, 9 of clubs, and 9 of spades.

**2. Coin Toss**   A coin is tossed four times in succession, and the faces showing are observed.

**3. Die Roll and Coin Toss**   A die is rolled and then a coin is tossed.

**4. Dice Roll**   Two dice are rolled, and the sum of the numbers that turn up is observed.

**5. Letter Selection**   Two different letters are selected in succession from the letters in the word "love."

**6. Sexes of Children**   The sexes of the first, second, and third children of a three-child family are noted. (Let, for example, BGB denote that the first, second, and third children are boy, girl, boy, respectively.)

**7. Marble Selection**   An urn contains three colored marbles: one red, one white, and one blue. Determine a sample space if (a) two marbles are selected with replacement and (b) two marbles are selected without replacement.

**8. Manufacturing Process**   A company makes a product that goes through three processes during its manufacture. The first is an assembly line, the second is a finishing line, and the third is an inspection line. There are three assembly lines (A, B, and C), two finishing lines (D and E), and two inspection lines (F and G). For each process, the company chooses a line at random. Determine a sample space.

*In Problems 9–14, describe the nature of a sample space for the given experiment, and determine the number of sample points.*

**9. Coin Toss**   A coin is tossed six times in succession, and the faces showing are observed.

**10. Dice Roll**   Four dice are rolled, and the numbers that turn up are observed.

**11. Card and Die**   A card is drawn from an ordinary deck of 52 cards, and then a die is rolled.

**12. Ball Selection**   From an urn containing eight different balls, four balls are drawn successively without replacement.

**13. Card Deal**   A 13-card hand is dealt from a deck of 52 cards. Do not simplify your answer.

**14. Letter Selection**   A four-letter "word" is formed by successively choosing any four letters from the alphabet with replacement.

*Suppose that S = {1, 2, 3, 4, 5, 6, 7, 8, 9, 10} is the sample space for an experiment with events*

$$E = \{1, 3, 5\}, \quad F = \{3, 5, 7, 9\} \quad and \quad G = \{2, 4, 6, 8\}.$$

*In Problems 15–22, determine the indicated events.*

**15.** $E \cup F$.

**16.** $G'$.

**17.** $E \cap F$.

**18.** $E \cap G$.

**19.** $F'$.

**20.** $(E \cup F)'$.

**21.** $(F \cap G)'$.

**22.** $(E \cup G) \cap F'$.

**23.** Of the following events, which pairs are mutually exclusive?

$$E_1 = \{1, 2, 3\}, \quad E_2 = \{3, 4, 5\},$$
$$E_3 = \{1, 2\}, \quad E_4 = \{5, 6, 7\}.$$

**24. Card Selection**   From a standard deck of 52 playing cards, 2 cards are drawn without replacement. Suppose $E_A$ is the event that both cards are aces, $E_H$ is the event that both cards are hearts, and $E_2$ is the event that both cards are 2's. Which pairs of these events are mutually exclusive?

**25. Card Selection**   From a standard deck of 52 playing cards, 1 card is selected. Which pairs of the following events are mutually exclusive?

$$E = \{\text{diamond}\},$$
$$F = \{\text{ace}\},$$
$$G = \{\text{red}\},$$
$$H = \{\text{club}\},$$
$$I = \{\text{ace of diamonds}\}.$$

**26. Dice**   A red and a green die are thrown, and the numbers on each are noted. Which pairs of the following events are mutually exclusive?

$$E = \{\text{both are even}\},$$
$$F = \{\text{both are odd}\},$$
$$G = \{\text{sum is 2}\},$$
$$H = \{\text{sum is 4}\},$$
$$I = \{\text{sum is greater than 10}\}.$$

**27. Coin Toss**   A coin is tossed three times in succession, and the results are observed. Determine each of the following.

**a.** The usual sample space $S$.

**b.** The event $E_1$ that at least one head occurs.

**c.** The event $E_2$ that at least one tail occurs.

**d.** $E_1 \cup E_2$.

**e.** $E_1 \cap E_2$.

**f.** $(E_1 \cup E_2)'$.

**g.** $(E_1 \cap E_2)'$.

**28. Sexes of Children**   A husband and wife have two children. The outcome of the first child being a boy and the second a girl can be represented by BG. Determine each of the following.

**a.** Sample space that describes all the orders of the possible sexes of the children.

**b.** The event that at least one child is a girl.

**c.** The event that at least one child is a boy.

**d.** Is the event in part (c) the complement of the event in part (b)?

**29. Arrivals**   Persons A, B, and C enter a building at different times. The outcome of A arriving first, B second, and C third can be indicated by ABC. Determine each of the following.

**a.** The sample space involved for the arrivals.

**b.** The event that A arrives first.

**c.** The event that A does not arrive first.

**30. Supplier Selection**   A manufacturer can order electronic components from suppliers U, V, W, and X and mechanical components from suppliers U, V, Y, and Z. The manufacturer selects one supplier for each type of component. The outcome of U being selected for electronic components and V for mechanical components can be represented by UV.

**a.** Determine a sample space.

**b.** Determine the event $E$ that the suppliers are different.

**c.** Determine $E'$ and give a verbal description of this event.

**31.** If $E$ and $F$ are events for an experiment, prove that events $E \cap F$ and $E \cap F'$ are mutually exclusive.

**32.** If $E$ and $F$ are events for an experiment, show that

$$(E \cap F) \cup (E \cap F') = E.$$

Note that from Problem 31, $E \cap F$ and $E \cap F'$ are mutually exclusive events. Thus, the foregoing equation expresses $E$ as a union of mutually exclusive events. [*Hint:* Make use of a distributive property.]

---

**To define what is meant by the probability of an event. To develop formulas that are used in computing probabilities. Emphasis is placed on equiprobable spaces.**

## 9.4 PROBABILITY

### Equiprobable Spaces

We now introduce the basic concepts underlying the study of probability. Consider tossing a well-balanced die and observing the number that turns up. The usual sample space for the experiment is

$$S = \{1, 2, 3, 4, 5, 6\}.$$

Before the experiment is performed, we cannot predict with certainty which of the six possible outcomes (sample points) will occur. But it does seem reasonable that each outcome has the same chance of occurring; that is, the outcomes are *equally likely*. This does not mean that in six tosses each number must turn up once. Rather, it means that if the experiment were performed a large number of times, each outcome would occur about $\frac{1}{6}$ of the time.

To be more specific, let the experiment be performed $n$ times. Each performance of an experiment is called a **trial.** Suppose that we are interested in the event of obtaining a 1 (that is, the simple event consisting of the sample point 1). If a 1 occurs in $k$ of these $n$ trials, then the proportion of times that 1 occurs is $k/n$. This ratio is called the **relative frequency** of the event. Because getting a 1 is just one of six possible equally likely outcomes, we expect that in the long run a 1 will occur $\frac{1}{6}$ of the time. That is, as $n$ becomes very large, we expect the relative frequency $k/n$ to approach $\frac{1}{6}$. The number $\frac{1}{6}$ is taken to be the probability of getting a 1 on the toss of a well-balanced die, which is denoted $P(1)$. Thus, $P(1) = \frac{1}{6}$. Similarly, $P(2) = \frac{1}{6}$, $P(3) = \frac{1}{6}$, and so on.

In this experiment, all of the simple events in the sample space were understood to be equally likely to occur. To describe this equal likelihood, we say that $S$ is an *equiprobable space*.

### DEFINITION

*A sample space $S$ is called an **equiprobable space** if and only if all the simple events are equally likely to occur.*

We remark that besides the phrase "equally likely," other words and phrases used in the context of an equiprobable space are "well balanced," "fair," "unbiased," and "at random." For example, we may have a *well-balanced* die (as above), a *fair* coin, or *unbiased* dice, or we may select a marble *at random* from an urn.

We now generalize our discussion of the die experiment to other (finite) equiprobable spaces.

**DEFINITION**

*If S is an equiprobable sample space with N sample points (or outcomes)*
$s_1, s_2, \ldots, s_N$, *then the **probability of the simple event** $\{s_i\}$ is given by*

$$P(s_i) = \frac{1}{N},$$

*for $i = 1, 2, \ldots, N$.*

We remark that $P(s_i)$ can be interpreted as the relative frequency of $\{s_i\}$ occurring in the long run.

We can also assign probabilities to events that are not simple. For example, in the die experiment, consider the event $E$ of a 1 or a 2 turning up:

$$E = \{1, 2\}.$$

Because the die is well balanced, in $n$ trials (where $n$ is large) we expect that a 1 should turn up approximately $\frac{1}{6}$ of the time and a 2 should turn up approximately $\frac{1}{6}$ of the time. Thus, a 1 or 2 should turn up approximately $\frac{1}{6} + \frac{1}{6}$ of the time, or $\frac{2}{6}$ of the time. Hence, it is reasonable to assume that the long-run relative frequency of $E$ is $\frac{2}{6}$. For this reason, we define $\frac{2}{6}$ to be the probability of $E$ and denote it $P(E)$.

$$P(E) = \frac{1}{6} + \frac{1}{6} = \frac{2}{6}.$$

Note that $P(E)$ is simply the sum of the probabilities of the simple events that form $E$. Equivalently, $P(E)$ is the ratio of the number of sample points in $E$ (two) to the number of sample points in the sample space (six).

**DEFINITION**

*If S is a finite equiprobable space for an experiment and $E = \{s_1, s_2, \cdots, s_j\}$ is an event, then the **probability of E** is given by*

$$P(E) = P(s_1) + P(s_2) + \cdots + P(s_j).$$

*Equivalently,*

$$P(E) = \frac{n(E)}{n(S)},$$

*where $n(E)$ is the number of outcomes in E and $n(S)$ is the number of outcomes in S.*

Note that we can think of $P$ as a function that associates with each event $E$ the probability of $E$, namely, $P(E)$. The probability of $E$ can be interpreted as the relative frequency of $E$ occurring in the long run. Thus, in $k$ trials, we would expect $E$ to occur approximately $k \cdot P(E)$ times, provided that $k$ is large.

**EXAMPLE 1    Coin Tossing**

*Two fair coins are tossed. Determine the probability that*

**a.** *two heads occur,*

**b.** *at least one head occurs.*

*Solution:* The usual sample space is

$$S = \{\text{HH}, \text{HT}, \text{TH}, \text{TT}\}.$$

Since the four outcomes are equally likely, $S$ is equiprobable and $n(S) = 4$.

**a.** If $E = \{HH\}$, then $E$ is a simple event, so

$$P(E) = \frac{n(E)}{n(S)} = \frac{1}{4}.$$

**b.** Let $F = \{\text{at least one head}\}$. Then

$$F = \{HH, HT, TH\},$$

which has three outcomes. Thus,

$$P(F) = \frac{n(F)}{n(S)} = \frac{3}{4}.$$

Alternatively,

$$P(F) = P(HH) + P(HT) + P(TH)$$

$$= \frac{1}{4} + \frac{1}{4} + \frac{1}{4} = \frac{3}{4}.$$

Consequently, in 1000 trials of this experiment, we would expect $F$ to occur approximately $1000 \cdot \frac{3}{4}$, or 750, times. ∎

### EXAMPLE 2  Cards

*From an ordinary deck of 52 playing cards, 2 cards are randomly drawn without replacement. If E is the event that one card is a 2 and the other a 3, find P(E).*

*Solution:* We can disregard the order in which the 2 cards are drawn. As our sample space $S$, we choose the set of all combinations of the 52 cards taken 2 at a time. Thus, $S$ is equiprobable and $n(S) = {}_{52}C_2$. To find $n(E)$, we note that since there are four suits, a 2 can be drawn in four ways and a 3 in four ways. Hence, a 2 and a 3 can be drawn in $4 \cdot 4$ ways, so

$$P(E) = \frac{n(E)}{n(S)} = \frac{4 \cdot 4}{{}_{52}C_2} = \frac{16}{1326} = \frac{8}{663}. \quad ∎$$

### EXAMPLE 3  Four-of-a-Kind Poker Hand

*Find the probability of drawing four of a kind in a poker hand (for example, four 10's and one 4).*

*Solution:* The set of all combinations of 52 cards taken 5 at a time is an equiprobable sample space. (The order in which the cards are dealt is of no concern.) Thus, $n(S) = {}_{52}C_5$. We now must find $n(E)$, where $E$ is the event of drawing four of a kind. Each of the four suits has 13 denominations, so four cards of one denomination can be drawn in 13 ways. There are $52 - 4 = 48$ possible selections for the fifth card. Hence, four of a kind can be drawn in $13 \cdot 48$ ways, and we have

$$P(\text{four of a kind}) = \frac{n(E)}{n(S)} = \frac{13 \cdot 48}{{}_{52}C_5} = \frac{13 \cdot 48}{2,598,960} \approx 0.00024. \quad ∎$$

### EXAMPLE 4  Selecting a Subcommittee

*From a committee of three males and four females, a subcommittee of four is to be randomly selected. Find the probability that it consists of two males and two females.*

*Solution:* Since order of selection is not important, the number of subcommittees of four that can be selected from the seven members is $_7C_4$. The two males can be selected in $_3C_2$ ways and the two females in $_4C_2$ ways. By the basic counting principle, the number of subcommittees of two males and two females is $_3C_2 \cdot _4C_2$. Thus,

$$P(\text{two males and two females}) = \frac{_3C_2 \cdot _4C_2}{_7C_4}.$$

$$= \frac{\dfrac{3!}{2!1!} \cdot \dfrac{4!}{2!2!}}{\dfrac{7!}{4!3!}} = \frac{18}{35}. \qquad \blacksquare$$

### Properties of Probability

We now develop some properties of probability. Let $S$ be an equiprobable sample space with $N$ outcomes; that is, $n(S) = N$. (We assume a finite sample space throughout this section.) If $E$ is an event, then $0 \le n(E) \le N$. Dividing each member by $n(S)$, or $N$, gives

$$0 \le \frac{n(E)}{n(S)} \le \frac{N}{N}.$$

But $\dfrac{n(E)}{n(S)} = P(E)$, so we have the following property:

$$0 \le P(E) \le 1.$$

That is, the probability of an event is a number between 0 and 1, inclusive. Moreover, $P(\varnothing) = \dfrac{n(\varnothing)}{n(S)} = \dfrac{0}{N} = 0$. Thus,

$$P(\varnothing) = 0.$$

Also, $P(S) = \dfrac{n(S)}{n(S)} = \dfrac{N}{N} = 1$, so

$$P(S) = 1.$$

Accordingly, the probability of an impossible event is 0 and the probability of a certain event is 1.

Since $P(S)$ is the sum of the probabilities of the outcomes in the sample space, we conclude that the sum of the probabilities of all the simple events for a sample space is 1.

Now let us focus on the probability of the union of two events $E$ and $F$. The event $E \cup F$ occurs if and only if at least one of the events ($E$ or $F$) occurs. Thus, $P(E \cup F)$ is the probability that *at least one* of the events $E$ and $F$ occurs. We know that

$$P(E \cup F) = \frac{n(E \cup F)}{n(S)}.$$

Although you might think that $n(E \cup F) = n(E) + n(F)$, this is not necessarily true. Since event $E \cap F$ is contained in both $E$ and $F$ (see Fig. 9.13), the sum $n(E) + n(F)$ includes $n(E \cap F)$ twice. Thus, we must subtract $n(E \cap F)$ from the sum to obtain $n(E \cup F)$:

$$n(E \cup F) = n(E) + n(F) - n(E \cap F).$$

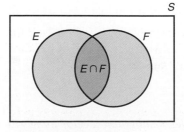

**FIGURE 9.13**   $E \cap F$ is contained in both $E$ and $F$.

Dividing both sides by $n(S)$ gives the following result:

> **Probability of a Union of Events**
>
> If $E$ and $F$ are events, then
> $$P(E \cup F) = P(E) + P(F) - P(E \cap F). \qquad (1)$$

*The probability of a union is not, in general, the sum of the probabilities.*

For example, let a fair die be rolled, and let $E = \{1, 3, 5\}$ and $F = \{1, 2, 3\}$. Then $E \cap F = \{1, 3\}$, so

$$P(E \cup F) = P(E) + P(F) - P(E \cap F)$$

$$= \frac{3}{6} + \frac{3}{6} - \frac{2}{6} = \frac{2}{3}.$$

Alternatively, $E \cup F = \{1, 2, 3, 5\}$, so $P(E \cup F) = \frac{4}{6} = \frac{2}{3}$.

If $E$ and $F$ are mutually exclusive events, then $E \cap F = \varnothing$, so $P(E \cap F) = P(\varnothing) = 0$. Hence, from Eq. (1), we obtain the following law:

> **Addition Law for Mutually Exclusive Events**
>
> If $E$ and $F$ are *mutually exclusive* events, then
> $$P(E \cup F) = P(E) + P(F).$$

For example, let a fair die be rolled, and let $E = \{2, 3\}$ and $F = \{1, 5\}$. Then $E \cap F = \varnothing$, so

$$P(E \cup F) = P(E) + P(F) = \frac{2}{6} + \frac{2}{6} = \frac{2}{3}.$$

The addition law can be extended to more than two mutually exclusive events.[2] For example, if events, $E$, $F$, and $G$ are mutually exclusive, then

$$P(E \cup F \cup G) = P(E) + P(F) + P(G).$$

An event and its complement are mutually exclusive, so by the addition law,

$$P(E \cup E') = P(E) + P(E').$$

But $P(E \cup E') = P(S) = 1$. Thus,

$$1 = P(E) + P(E'),$$

$$P(E') = 1 - P(E),$$

or, equivalently,

$$P(E) = 1 - P(E').$$

*In order to find the probability of an event, sometimes it may be more convenient to first find the probability of its complement and then subtract the result from 1. See, especially, Example 6(c).*

Accordingly, if we know the probability of an event, then we can easily find the probability of its complement, and vice versa. For example, if $P(E) = \frac{1}{4}$, then $P(E') = 1 - \frac{1}{4} = \frac{3}{4}$. $P(E')$ is the probability that $E$ does not occur.

---

[2]Two or more events are **mutually exclusive** if, and only if, no two of them can occur at the same time. That is, given any two of them, their intersection must be empty. For example, to say the events $E$, $F$, and $G$ are mutually exclusive means that

$$E \cap F = E \cap G = F \cap G = \varnothing.$$

### EXAMPLE 5    Quality Control

*From a production run of 5000 light bulbs, 2% of which are defective, 1 bulb is selected at random. What is the probability that the bulb is defective? What is the probability that it is not defective?*

**Solution:**  The sample space $S$ consists of the 5000 bulbs. Since a bulb is selected at random, the possible outcomes are equally likely. Let $E$ be the event of selecting a defective bulb. The number of outcomes in $E$ is $0.02 \cdot 5000$, or 100. Thus,

$$P(E) = \frac{n(E)}{n(S)} = \frac{100}{5000} = \frac{1}{50} = 0.02.$$

Alternatively, since the probability of selecting a particular bulb is $\frac{1}{5000}$ and $E$ contains 100 sample points, by summing probabilities we have

$$P(E) = 100 \cdot \frac{1}{5000} = 0.02.$$

The event that the bulb selected is *not* defective is $E'$. Hence,

$$P(E') = 1 - P(E) = 1 - 0.02 = 0.98.$$ ∎

### EXAMPLE 6    Dice

*A pair of well-balanced dice is rolled, and the number on each die is noted. Determine the probability that the sum of the numbers that turn up is (a) 7, (b) 7 or 11, and (c) greater than 3.*

**Solution:**  Since each die can turn up in any of six ways, by the basic counting principle the number of possible outcomes is $6 \cdot 6$, or 36. Our sample space consists of the following ordered pairs:

$$
\begin{array}{cccccc}
(1,1), & (1,2), & (1,3), & (1,4), & (1,5), & (1,6), \\
(2,1), & (2,2), & (2,3), & (2,4), & (2,5), & (2,6), \\
(3,1), & (3,2), & (3,3), & (3,4), & (3,5), & (3,6), \\
(4,1), & (4,2), & (4,3), & (4,4), & (4,5), & (4,6), \\
(5,1), & (5,2), & (5,3), & (5,4), & (5,5), & (5,6), \\
(6,1), & (6,2), & (6,3), & (6,4), & (6,5), & (6,6).
\end{array}
$$

The outcomes are equally likely, so the probability of each outcome is $\frac{1}{36}$.

**a.**  Let $E_7$ be the event that the sum of the numbers appearing is 7. Then

$$E_7 = \{(1,6), (2,5), (3,4), (4,3), (5,2), (6,1)\},$$

which has six outcomes. Thus,

$$P(E_7) = \frac{6}{36} = \frac{1}{6}.$$

**b.**  Let $E_{7\,\text{or}\,11}$ be the event that the sum is 7 or 11. If $E_{11}$ is the event that the sum is 11, then

$$E_{11} = \{(5,6), (6,5)\},$$

which has two outcomes. Since $E_{7\,\text{or}\,11} = E_7 \cup E_{11}$ and $E_7$ and $E_{11}$ are mutually exclusive, we have

$$P(E_{7\,\text{or}\,11}) = P(E_7) + P(E_{11}) = \frac{6}{36} + \frac{2}{36} = \frac{8}{36} = \frac{2}{9}.$$

Alternatively, we can determine $P(E_{7 \text{ or } 11})$ by counting the number of outcomes in $E_{7 \text{ or } 11}$. We obtain

$$E_{7 \text{ or } 11} = \{(1, 6), (2, 5), (3, 4), (4, 3), (5, 2), (6, 1), (5, 6), (6, 5)\},$$

which has eight outcomes. Thus,

$$P(E_{7 \text{ or } 11}) = \frac{8}{36} = \frac{2}{9}.$$

c.  Let $E$ be the event that the sum is greater than 3. The number of outcomes in $E$ is relatively large. Thus, to determine $P(E)$, it is easier to find $E'$, rather than $E$, and then use the formula $P(E) = 1 - P(E')$. Here $E'$ is the event that the sum is 2 or 3. We have

$$E' = \{(1, 1), (1, 2), (2, 1)\},$$

which has three outcomes. Hence,

$$P(E) = 1 - P(E') = 1 - \frac{3}{36} = \frac{11}{12}.$$

∎

## Probability Functions in General

Many of the properties of equiprobable spaces carry over to sample spaces that are not equiprobable. To illustrate, consider the experiment of tossing two fair coins and observing the number of heads. The coins can fall in one of four ways, namely,

$$HH, HT, TH, TT,$$

which correspond to two heads, one head, one head, and zero heads, respectively. Because we are interested in the number of heads, we can choose a sample space to be

$$S = \{0, 1, 2\}.$$

However, the simple events in $S$ are *not* equally likely to occur, because of the four possible ways in which the coins can fall: Two of these ways correspond to the one-head outcome, whereas only one corresponds to the two-head outcome and similarly for the zero-head outcome. In the long run, it is reasonable to expect repeated trials to result in one head about $\frac{2}{4}$ of the time, zero heads about $\frac{1}{4}$ of the time, and two heads about $\frac{1}{4}$ of the time. If we were to assign probabilities to these simple events, it is natural to have

$$P(0) = \frac{1}{4}, \qquad P(1) = \frac{2}{4} = \frac{1}{2}, \qquad P(2) = \frac{1}{4}.$$

Although $S$ is not equiprobable, these probabilities lie between 0 and 1, inclusive, and their sum is 1. This is consistent with what was stated for an equiprobable space.

Based on our discussion, we can consider a *probability function* that relates to sample spaces in general.

## DEFINITION

*Let $S = \{s_1, s_2, \ldots, s_N\}$ be a sample space for an experiment. The function $P$ is called a **probability function** if both of the following are true:*

**1.** $0 \le P(s_i) \le 1$ *for $i = 1$ to $N$.*

**2.** $P(s_1) + P(s_2) + \cdots + P(s_N) = 1.$

*If E is an event, then P(E) is the sum of the probabilities of the sample points in E. We define P(∅) to be 0.*

From a mathematical point of view, any function $P$ that satisfies conditions 1 and 2 is a probability function for a sample space. For example, consider the sample space for the previous experiment of tossing two fair coins and observing the number of heads:

$$S = \{0, 1, 2\}.$$

We could assign probabilities as follows:

$$P(0) = 0.1, \qquad P(1) = 0.2, \qquad P(2) = 0.7.$$

Here $P$ satisfies both conditions 1 and 2 and thus is a legitimate probability function. However, this assignment does not reflect the long-run interpretation of probability and, consequently, would not be acceptable from a practical point of view.

In general, for any probability function defined on a sample space (finite or infinite), the following properties hold:

$$P(E') = 1 - P(E),$$

$$P(S) = 1,$$

$$P(E_1 \cup E_2) = P(E_1) + P(E_2) \qquad \text{if } E_1 \cap E_2 = \varnothing.$$

## Empirical Probability

You have seen how easy it is to assign probabilities to simple events when we have an equiprobable sample space. For example, when a fair coin is tossed, we have $S = \{H, T\}$ and $P(H) = P(T) = \frac{1}{2}$. These probabilities are determined, in essence, by the intrinsic nature of the experiment—namely, that there are two possible outcomes that should have the same probability because the outcomes are equally likely. Such probabilities are called *theoretical* probabilities. However, suppose the coin is not fair. How can probabilities then be assigned? By tossing the coin a number of times, we can determine the relative frequencies of heads and tails occurring. For example, suppose that in 1000 tosses, heads occurs 517 times and tails occurs 483 times. Then the relative frequencies of heads and tails occurring are $\frac{517}{1000}$ and $\frac{483}{1000}$, respectively. In this situation, the assignment $P(H) = 0.517$ and $P(T) = 0.483$ would be quite reasonable. Probabilities assigned in this way are called *empirical probabilities*. In general, probabilities based on sample or historical data are empirical. Now suppose that the coin were tossed 2000 times, and the relative frequencies of heads and tails occurring were $\frac{1023}{2000} = 0.5115$ and $\frac{977}{2000} = 0.4885$, respectively. Then in this case, the assignment $P(H) = 0.5115$ and $P(T) = 0.4885$ would be acceptable. In a certain sense, the latter probabilities may be more indicative of the true nature of the coin than would be the probabilities associated with 1000 tosses.

In the next example, probabilities (empirical) are assigned on the basis of sample data.

### EXAMPLE 7   Opinion Survey

*An opinion survey of a sample of 150 adult residents of a town was conducted. Each person was asked his or her opinion about floating a bond issue to build a community swimming pool. The results are summarized in Table 9.2.*

**TABLE 9.2**

|        | Favor | Oppose | Total |
|--------|-------|--------|-------|
| Male   | 60    | 20     | 80    |
| Female | 40    | 30     | 70    |
| Total  | 100   | 50     | 150   |

  *Suppose an adult resident from the town is randomly selected. Let M be the event "male selected" and F be the event "selected person favors the bond issue." Find each of the following:*

**a.** $P(M)$,   **b.** $P(F)$,   **c.** $P(M \cap F)$,   **d.** $P(M \cup F)$.

> *Strategy:* We shall assume that proportions which apply to the sample also apply to the adult population of the town.

*Solution:*

**a.** Of the 150 persons in the sample, 80 are males. Thus, for the adult population of the town (the sample space), we assume that $\frac{80}{150}$ are male. Hence, the (empirical) probability of selecting a male is

$$P(M) = \frac{80}{150} = \frac{8}{15}.$$

**b.** Of the 150 persons in the sample, 100 favor the bond issue. Therefore,

$$P(F) = \frac{100}{150} = \frac{2}{3}.$$

**c.** Table 9.2 indicates that 60 males favor the bond issue. Hence,

$$P(M \cap F) = \frac{60}{150} = \frac{2}{5}.$$

**d.** To find $P(M \cup F)$, we use Eq. (1):

$$P(M \cup F) = P(M) + P(F) - P(M \cap F)$$

$$= \frac{80}{150} + \frac{100}{150} - \frac{60}{150} = \frac{120}{150} = \frac{4}{5}.$$ ■

## Odds

The probability of an event is sometimes expressed in terms of *odds,* especially in gaming situations.

## DEFINITION

*The **odds** in favor of event E occurring is the ratio*

$$\frac{P(E)}{P(E')},$$

*provided that $P(E') \neq 0$. Odds are usually expressed as the ratio $\dfrac{p}{q}$ (or p:q) of two positive integers, which is read "p to q."*

**EXAMPLE 8   Odds for an "A" in an Exam**

*A student believes that the probability of getting an A on the next mathematics exam is 0.2. What are the odds (in favor) of this occurring?*

*Solution:* If $E =$ "gets an A," then $P(E) = 0.2$ and $P(E') = 1 - 0.2 = 0.8$. Hence, the odds of getting an A are

$$\frac{P(E)}{P(E')} = \frac{0.2}{0.8} = \frac{2}{8} = \frac{1}{4} = 1{:}4.$$

That is, the odds are 1 to 4. (We remark that the odds *against* getting an A are 4 to 1.) ▪

If the odds that event $E$ occurs are $a{:}b$, then the probability of $E$ can be easily determined. We are given that

$$\frac{P(E)}{1 - P(E)} = \frac{a}{b}.$$

Solving for $P(E)$ gives

$$bP(E) = [1 - P(E)]a \qquad \text{(clearing fractions)},$$
$$aP(E) + bP(E) = a,$$
$$(a + b)P(E) = a,$$
$$P(E) = \frac{a}{a + b}.$$

---

**Finding Probability from Odds**

If the odds that event $E$ occurs are $a{:}b$, then

$$P(E) = \frac{a}{a + b}.$$

---

Over the long run, if the odds that $E$ occurs are $a{:}b$, then, on the average, $E$ should occur $a$ times in every $a + b$ trials of the experiment.

**EXAMPLE 9   Probability of Winning a Prize**

*A $1000 savings bond is one of the prizes listed on a contest brochure received in the mail. The odds in favor of winning the bond are stated to be 1:10,000. What is the probability of winning this prize?*

*Solution:* Here $a = 1$ and $b = 10,000$. From the preceding rule,

$$P(\text{winning prize}) = \frac{a}{a + b}$$

$$= \frac{1}{1 + 10,000} = \frac{1}{10,001}. \qquad ▪$$

## ▪ Exercise 9.4

1. In 2000 trials of an experiment, how many times would you expect event $E$ to occur if $P(E) = 0.3$?

2. In 3000 trials of an experiment, how many times would you expect event $E$ to occur if $P(E') = 0.45$?

3. If $P(E) = 0.2$, $P(F) = 0.3$, and $P(E \cap F) = 0.1$, find (a) $P(E')$ and (b) $P(E \cup F)$.

4. If $P(E) = \frac{1}{4}$, $P(F) = \frac{1}{2}$, and $P(E \cap F) = \frac{1}{8}$, find (a) $P(E')$ and (b) $P(E \cup F)$.

5. If $P(E \cap F) = 0.831$, are $E$ and $F$ mutually exclusive?

6. If $P(E) = \frac{1}{4}$, $P(E \cup F) = \frac{13}{20}$, and $P(E \cap F) = \frac{1}{10}$, find $P(F)$.

7. **Dice** A pair of well-balanced dice is tossed. Find the probability that the sum of the numbers is (a) 8, (b) 2 or 3, (c) 3, 4, or 5, (d) 12 or 13, (e) even, (f) odd, and (g) less than 10.

8. **Dice** A pair of fair dice is tossed. Determine the probability that at least one die shows a 2.

9. **Card Selection** A card is randomly selected from a standard deck of 52 playing cards. Determine the probability that the card is (a) the king of hearts, (b) a diamond, (c) a jack, (d) red, (e) a heart or a club, (f) a club and a 4, (g) a club or a 4, (h) red and a king, and (i) a spade and a heart.

10. **Coin and Die** A fair coin and a fair die are tossed. Find the probability that (a) a head and a 5 show, (b) a head shows, (c) a 3 shows, and (d) a head and an even number show.

11. **Coin, Die, and Card** A fair coin and a fair die are tossed, and a card is randomly selected from a standard deck of 52 playing cards. Determine the probability that the coin, die, and card respectively show (a) a tail, a 3, and the queen of hearts, (b) a tail, a 3, and a queen, (c) a head, a 2 or 3, and a queen, and (d) a head, an even number, and a diamond.

12. **Coins** Three fair coins are tossed. Find the probability that (a) three heads show, (b) exactly one tail shows, (c) no more than two heads show, and (d) no more than one tail shows.

13. **Card Selection** Two cards from a standard deck of 52 playing cards are successively drawn at random without replacement. Find the probability that (a) both cards are kings and (b) one card is a diamond and the other is a heart.

14. **Card Selection** Two cards from a standard deck of 52 playing cards are successively drawn at random with replacement. Find the probability that (a) both cards are kings and (b) one card is a king and the other is a heart.

15. **Sexes of Children** Assuming that the sex of a person is determined at random, determine the probability that a family with three children has (a) three girls, (b) exactly one boy, (c) no girls, and (d) at least one girl.

16. **Marble Selection** A marble is randomly drawn from an urn that contains seven red, five white, and eight blue marbles. Find the probability that the marble is (a) blue, (b) not red, (c) red or white, (d) neither red nor blue, (e) yellow, and (f) red or yellow.

17. **Stock Selection** A stock is selected at random from a list of 60 utility stocks, 48 of which have an annual dividend yield of 10% or more. Find the probability that the stock pays an annual dividend that yields (a) 10% or more and (b) less than 10%.

18. **Inventory** A clothing store maintains its inventory of suits so that 25% are 100% pure wool. If a suit is selected at random, what is the probability that it is (a) 100% pure wool and (b) not 100% pure wool?

19. **Examination Grades** On an examination given to 40 students, 10% received an A, 25% a B, 35% a C, 25% a D, and 5% an F. If a student is selected at random, what is the probability that the student (a) received an A, (b) received an A or a B, (c) received neither a D nor an F, and (d) did not receive an F? (e) Answer questions (a)-(d) if the number of students that were given the examination is unknown.

20. **Marble Selection** Two urns contain colored marbles. Urn 1 contains three red and two green marbles, and urn 2 contains four red and five green marbles. A marble is selected at random from each urn. Find the probability that (a) both marbles are red and (b) one marble is red and the other is green.

21. **Committee Selection** From a group of two women and three men, two persons are selected at random to form a committee. Find the probability that the committee consists of women only.

22. **Committee Selection** For the committee selection in Problem 21, find the probability that the committee consists of a man and a woman.

23. **Examination Score** A student answers each question on a 10-question true-false examination in a random fashion. If each question is worth 10 points, what is the probability that the student scores (a) 100 points and (b) 90 or more points?

24. **Multiple-Choice Examination** On a five-question, multiple-choice examination there are four choices for each question, only one of which is correct. If a student answers each question in a random fashion, find the probability that the student answers (a) each question correctly and (b) exactly four questions correctly.

25. **Poker Hand** Find the probability of being dealt a full house in a poker game. A full house is three of one kind and two of another, such as three queens and two 10's. Express your answer using the symbol $_nC_r$.

26. Suppose $P(E) = \frac{1}{3}$, $P(E \cup F) = \frac{8}{15}$, and $P(E \cap F) = \frac{1}{5}$.
   a. Find $P(F)$.
   b. Find $P(E' \cup F)$.
   [*Hint:*
$$F = (E \cap F) \cup (E' \cap F),$$
   where $E \cap F$ and $E' \cap F$ are mutually exclusive.]

27. **Faculty Committee** The classification of faculty at a college is indicated in Table 9.3. If a committee of three faculty members is selected at random, what is the probability that it consists of (a) all females; (b) a professor and two associate professors?

28. **Biased Die** A die is biased such that $P(1) = \frac{3}{10}$, $P(2) = P(5) = \frac{2}{10}$, and $P(3) = P(4) = P(6) = \frac{1}{10}$. If the die is tossed, find $P(\text{even number})$.

29. **Biased Die** When a biased die is tossed, the probabilities of 1, 3, and 5 showing are the same. The probabilities of 2, 4, and 6 showing are also the same, but are twice those of 1, 3, and 5. Determine $P(1)$.

**TABLE 9.3** Faculty Classification

|  | Male | Female | Total |
|---|---|---|---|
| Professor | 12 | 3 | 15 |
| Associate Professor | 15 | 9 | 24 |
| Assistant Professor | 18 | 8 | 26 |
| Instructor | 20 | 15 | 35 |
| Total | 65 | 35 | 100 |

**30.** For the sample space $\{a, b, c, d, e\}$, suppose that the probabilities of $a$, $b$, $c$, and $d$ are the same. Is it possible to determine $P(e)$?

**31. Tax Increase**   A legislative body is considering a tax increase to support education. An opinion survey of 100 registered voters was conducted, and the results are indicated in Table 9.4. Assume that the survey reflects the opinion of the voting population. If a person from that population is selected at random, determine each of the following (empirical) probabilities.

**a.** $P$(favors tax increase).

**b.** $P$(opposes tax increase).

**c.** $P$(is a Republican with no opinion).

**32. Camcorder Sales**   A department store chain has stores in the cities of Exton and Whyton. Each store sells three brands of camcorders, A, B, and C. Over the past year, the average monthly unit sales of the camcorders was determined, and the results are indicated in Table 9.5. Assume that future sales follow the pattern indicated in the table.

**TABLE 9.4** Tax Increase Survey

|  | Favor | Oppose | No Opinion | Total |
|---|---|---|---|---|
| Democrat | 32 | 26 | 2 | 60 |
| Republican | 15 | 17 | 3 | 35 |
| Other | 4 | 1 | 0 | 5 |
| Total | 51 | 44 | 5 | 100 |

**TABLE 9.5** Unit Sales Per Month

|  | A | B | C |
|---|---|---|---|
| Exton | 25 | 40 | 30 |
| Whyton | 20 | 25 | 30 |

**a.** Determine the probability that a sale of a camcorder next month is for brand B.

**b.** Next month, if a sale occurs at the Exton store, find the probability that it is for brand C.

*In Problems 33–36, for the given probability, find the odds that E will occur.*

**33.** $P(E) = \dfrac{4}{5}$.

**34.** $P(E) = \dfrac{1}{3}$.

**35.** $P(E) = 0.3$.

**36.** $P(E) = 0.001$.

*In Problems 37–40, the odds that E will occur are given. Find P(E).*

**37.** 5:4.

**38.** 100:1.

**39.** 4:10.

**40.** $a{:}a$.

**41. Weather Forecast**   A television weather forecaster reported that the odds that rain will occur tomorrow are 3 to 1. What is the probability of rain?

**42.** If the odds of event $E$ occurring are $a{:}b$, what are the odds that $E$ does not occur?

---

**OBJECTIVE**

To discuss conditional probability via a reduced sample space as well as the original space. To analyze a stochastic process with the aid of a probability tree. To develop the general multiplication law for $P(E \cap F)$.

## 9.5 Conditional Probability and Stochastic Processes

### Conditional Probability

The probability of an event could be affected when additional related information about the experiment is known. For example, if you guess at the answer to a multiple-choice question having five choices, the probability of getting the correct answer is $\frac{1}{5}$. However, if you know that answers A and B are wrong and thus can be ignored, the probability of guessing the correct answer increases to $\frac{1}{3}$. In this section, we consider similar situations in which we want the probability of an event $E$ when it is known that some other event $F$ has occurred. This is called a **conditional probability** and is denoted by $P(E|F)$, which is read "the conditional probability of $E$, given $F$." For instance, in the situation involving the multiple-choice question, we have

$$P(\text{guessing correct answer} \mid \text{A and B eliminated}) = \frac{1}{3}.$$

To investigate the notion of conditional probability, we consider the following situation. A fair die is rolled, and we are interested in the probability of the event

$$E = \{\text{even number shows}\}.$$

The usual equiprobable sample space for this experiment is

$$S = \{1, 2, 3, 4, 5, 6\},$$

so

$$E = \{2, 4, 6\}.$$

Thus,

$$P(E) = \frac{n(E)}{n(S)} = \frac{3}{6} = \frac{1}{2}.$$

Now we change the situation a bit. Suppose the die is rolled out of our sight, and then we are told that a number greater than 3 occurred. In light of this additional information, what now is the probability of an even number? To answer that question we reason as follows. The event $F$ of a number greater than 3 is

$$F = \{4, 5, 6\}.$$

Since $F$ already occurred, the set of possible outcomes is no longer $S$; it is $F$. That is, $F$ becomes our new sample space, called a **reduced sample space** or a *subspace* of $S$. The outcomes in $F$ are equally likely, and, of these, only 4 and 6 are favorable to $E$; that is,

$$E \cap F = \{4, 6\}.$$

Since two of the three outcomes in the reduced sample space are favorable to an even number occurring, we say that $\frac{2}{3}$ is *the conditional probability of an even number, given that a number greater than* 3 *occurred:*

$$P(E|F) = \frac{n(E \cap F)}{n(F)} = \frac{2}{3}. \tag{1}$$

The Venn diagram in Fig. 9.14 illustrates the situation.

If we compare the conditional probability $P(E|F) = \frac{2}{3}$ with the "unconditional" probability $P(E) = \frac{1}{2}$, we see that $P(E|F) > P(E)$. This means that knowing that a number greater than 3 occurred *increases* the likelihood that an even number occurred. There are situations, however, in which conditional and unconditional probabilities are the same. These are discussed in the next section.

In summary, we have the following generalization of Eq. (1):

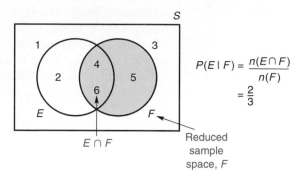

$$P(E \mid F) = \frac{n(E \cap F)}{n(F)}$$

$$= \frac{2}{3}$$

**FIGURE 9.14** Venn diagram for conditional probability.

---

**Formula for a Conditional Probability**

If $E$ and $F$ are events associated with an equiprobable sample space and $F \neq \emptyset$, then

$$P(E|F) = \frac{n(E \cap F)}{n(F)}. \tag{2}$$

It can be shown that

$$P(E'|F) = 1 - P(E|F).$$

**EXAMPLE 1    Marbles in Urn**

*An urn contains two blue marbles (say, $B_1$ and $B_2$) and two white marbles ($W_1$ and $W_2$). If two marbles are randomly drawn without replacement, find the probability that the second marble drawn is white, given that the first marble drawn is blue. (See Fig. 9.15.)*

Draw two marbles without replacement.

**FIGURE 9.15    Two white and two blue marbles in urn.**

***Solution:*** For our equiprobable sample space, we take all ordered pairs, such as $B_1W_2$ and $W_2W_1$, whose components indicate the marbles selected on the first and on the second draw. Let $B$ and $W$ be the events

$$B = (\text{blue on first draw}),$$

$$W = \{\text{white on second draw}\}.$$

We are interested in

$$P(W|B) = \frac{n(W \cap B)}{n(B)}.$$

The reduced sample space $B$ consists of all outcomes in which a blue marble is drawn first:

$$B = \{B_1B_2, B_1W_1, B_1W_2, B_2B_1, B_2W_1, B_2W_2\}.$$

Event $W \cap B$ consists of the outcomes in $B$ for which the second marble is white:

$$W \cap B = \{B_1W_1, B_1W_2, B_2W_1, B_2W_2\}.$$

Since $n(B) = 6$ and $n(W \cap B) = 4$, we have

$$P(W|B) = \frac{4}{6} = \frac{2}{3}.$$ ∎

Example 1 showed how efficient the use of a reduced sample space can be. Note that it was not necessary to list all the outcomes either in the original sample space or in event $W$. Although we listed the outcomes in $B$, we could have found $n(B)$ by using counting methods:

There are two ways in which the first marble can be blue, and three possibilities for the second marble, which can be either the remaining blue marble or one of the two white marbles. Thus, $n(B) = 2 \cdot 3 = 6$.

The number $n(W \cap B)$ could also be found by means of counting methods.

**EXAMPLE 2**   **Survey**

*In a survey of 150 people, each person was asked his or her marital status and opinion about floating a bond issue to build a community swimming pool. The results are summarized in Table 9.6. If one of these persons is randomly selected, find each of the following conditional probabilities.*

**TABLE 9.6**

|  | Favor (F) | Oppose (F') | Total |
|---|---|---|---|
| Married (M) | 60 | 20 | 80 |
| Single (M') | 40 | 30 | 70 |
| Total | 100 | 50 | 150 |

a. *The probability that the person favors the bond issue, given that the person is married.*

*Solution:* We are interested in $P(F|M)$. The reduced sample space ($M$) contains 80 married persons, of which 60 favor the bond issue. Thus,

$$P(F|M) = \frac{n(F \cap M)}{n(M)} = \frac{60}{80} = \frac{3}{4}.$$

b. *The probability that the person is married, given that the person favors the bond issue.*

*Solution:* We want to find $P(M|F)$. The reduced sample space ($F$) contains 100 persons who favor the bond issue. Of these, 60 are married. Hence,

$$P(M|F) = \frac{n(M \cap F)}{n(F)} = \frac{60}{100} = \frac{3}{5}.$$

Note here that $P(M|F) \neq P(F|M)$. [However, in general it is not always true that $P(E|F) \neq P(F|E)$.]  ▬

Another method of computing a conditional probability is by means of a formula involving *probabilities* with respect to the *original* sample space itself. Before stating the formula, we shall provide some motivation so that it seems reasonable to you. (The discussion that follows is oversimplified in the sense that certain assumptions are tacitly made.)

To consider $P(E|F)$, we shall assume that event $F$ has probability $P(F)$ and event $E \cap F$ has probability $P(E \cap F)$. Let the experiment associated with this problem be repeated $n$ times, where $n$ is very large. Then the number of trials in which $F$ occurs is approximately $n \cdot P(F)$. Of these, the number in which event $E$ *also* occurs is approximately $n \cdot P(E \cap F)$. For large $n$, we estimate $P(E|F)$ by the relative frequency of the number of occurrences of $E \cap F$ with respect to the number of occurrences of $F$, which is approximately

$$\frac{n \cdot P(E \cap F)}{n \cdot P(F)}, \quad \text{or} \quad \frac{P(E \cap F)}{P(F)}.$$

This result strongly suggests the formula that appears in the following formal definition of conditional probability. (The definition applies to any sample space, whether or not it is equiprobable.)

**DEFINITION**

*The **conditional probability** of an event E, given that event F has occurred, is denoted $P(E|F)$ and is defined by*

$$P(E|F) = \frac{P(E \cap F)}{P(F)}, \quad \text{if } P(F) \neq 0. \tag{3}$$

Similarly,

$$P(F|E) = \frac{P(F \cap E)}{P(E)}, \quad \text{if } P(E) \neq 0. \tag{4}$$

We emphasize that **the probabilities in Eqs. (3) and (4) are with respect to the original sample space.** Here we do *not* deal directly with a reduced sample space.

**EXAMPLE 3** Quality Control

*After the initial production run of a new style of steel desk, a quality control technician found that 40 percent of the desks had an alignment problem and 10 percent had both a defective paint job and an alignment problem. If a desk is randomly selected from this run, and it has an alignment problem, what is the probability that it also has a defective paint job?*

*Solution:* Let $A$ and $D$ be the events

$$A = \{\text{alignment problem}\},$$

$$D = \{\text{defective paint job}\}.$$

We are interested in $P(D|A)$, the probability of a defective paint job, given an alignment problem. From the given data, we have $P(A) = 0.4$ and $P(D \cap A) = 0.1$. Substituting into Eq. (3) gives

$$P(D|A) = \frac{P(D \cap A)}{P(A)} = \frac{0.1}{0.4} = \frac{1}{4}.$$

Note that we cannot use Eq. (2) to solve this problem, because we are given probabilities rather than information about the sample space. ∎

### EXAMPLE 4  Sexes of Offspring

*If a family has two children, find the probability that both are boys, given that one of the children is a boy. Assume that a child of either sex is equally likely and that, for example, having a girl first and a boy second is just as likely as having a boy first and a girl second.*

*Solution:*  Let $E$ and $F$ be the events

$$E = \{\text{both children are boys}\},$$

$$F = \{\text{at least one of the children is a boy}\}.$$

We are interested in $P(E|F)$. Letting the letter B denote "boy" and G denote "girl," we use the equiprobable sample space

$$S = \{BB, BG, GG, GB\},$$

where, in each outcome, the order of the letters indicates the order in which the children are born. Thus,

$$E = \{BB\}, \quad F = \{BB, BG, GB\}, \quad \text{and} \quad E \cap F = \{BB\}.$$

From Eq. (3),

$$P(E|F) = \frac{P(E \cap F)}{P(F)} = \frac{\frac{1}{4}}{\frac{3}{4}} = \frac{1}{3}.$$

Alternatively, this problem can be solved by using the reduced sample space $F$:

$$P(E|F) = \frac{n(E \cap F)}{n(F)} = \frac{1}{3}. \quad ∎$$

Equations (3) and (4) can be rewritten in terms of products by clearing fractions. This gives

$$P(E \cap F) = P(F)P(E|F)$$

and

$$P(F \cap E) = P(E)P(F|E).$$

By the commutative law, $P(E \cap F) = P(F \cap E)$, so we can combine the preceding equations to get an important law:

**General Multiplication Law**

$$P(E \cap F) = P(E)P(F|E)$$
$$= P(F)P(E|F). \tag{5}$$

The general multiplication law states that the probability that two events *both* occur is equal to the probability that one of them occurs, times the conditional probability that the other one occurs, given that the first has occurred.

**EXAMPLE 5**   Advertising

*A computer hardware company placed an ad for its new modem in a popular computer magazine. The company believes that the ad will be read by 32 percent of the magazine's readers and that 2 percent of those who read the ad will buy the modem. Assume that this is true, and find the probability that a reader of the magazine will read the ad and buy the modem.*

*Solution:* Letting $R$ denote the event "read ad" and $B$ denote "buy modem," we are interested in $P(R \cap B)$. We are given that $P(R) = 0.32$. The fact that 2 percent of the readers of the ad will buy the modem can be written $P(B|R) = 0.02$. By the general multiplication law, Eq. (5),

$$P(R \cap B) = P(R)P(B|R) = (0.32)(0.02) = 0.0064. \qquad \blacksquare$$

## Stochastic Processes

The general multiplication law is also called the **law of compound probability.** The reason is that it is extremely useful when applied to an experiment that can be expressed as a *sequence* (or a compounding) of two or more other experiments, called **trials** or **stages.** The original experiment is called a **compound experiment,** and the sequence of trials is called a **stochastic process.** The probabilities of the events associated with each trial (beyond the first) could depend on what events occurred in the preceding trials, so they are conditional probabilities.

When we analyze a compound experiment, a tree diagram is extremely useful in keeping track of the possible outcomes at each stage. A complete path from the start to a tip of the tree gives an outcome of the experiment.

The notion of a compound experiment is discussed in detail in the next example. Read it carefully! Although the discussion is lengthy for the sake of developing a new idea, the actual computation takes little time.

**EXAMPLE 6**   Cards and Probability Tree

*Two cards are drawn without replacement from a standard deck of cards. Find the probability that the second card is red.*

*Solution:* The experiment of drawing two cards without replacement can be thought of as a compound experiment consisting of a sequence of two trials: The first is drawing a card, and the second is drawing a card after the first card has been drawn. The first trial has two possible outcomes:

$$R_1 = \{\text{red card}\} \qquad \text{or} \qquad B_1 = \{\text{black card}\}.$$

(Here the subscript "1" refers to the first trial.) In Fig. 9.16, these outcomes are represented by the two branches in the first level of the tree. Keep in mind that these outcomes are mutually exclusive, and they are also *exhaustive* in the sense that there are no other possibilities. Since there are 26 cards of each color, we have

$$P(R_1) = \frac{26}{52} \qquad \text{and} \qquad P(B_1) = \frac{26}{52}.$$

These *unconditional* probabilities are written along the corresponding branches. We appropriately call Fig. 9.16 a **probability tree.**

Now, if a red card is obtained in the first trial, then, of the remaining 51 cards, 25 are red and 26 are black. The card drawn in the second trial can be

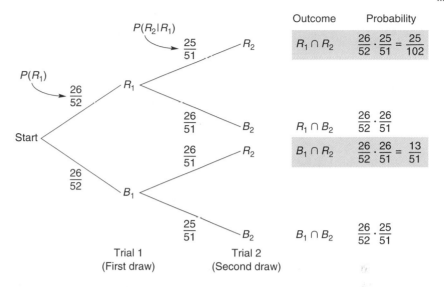

**FIGURE 9.16**   Probability tree for compound experiment.

red $(R_2)$ or black $(B_2)$. Thus, in the tree, the fork at $R_1$ has two branches: red and black. The *conditional* probabilities $P(R_2|R_1) = \dfrac{25}{51}$ and $P(B_2|R_1) = \dfrac{26}{51}$ are placed along these branches. Similarly, if a black card is obtained in the first trial, then, of the remaining 51 cards, 26 are red and 25 are black. Hence, $P(R_2|B_1) = \dfrac{26}{51}$ and $P(B_2|B_1) = \dfrac{25}{51}$, as indicated alongside the two branches emanating from $B_1$. The complete tree has two levels (one for each trial) and four paths (one for each of the four mutually exclusive and exhaustive events of the compound experiment).

Note that the sum of the probabilities along the branches from the vertex "Start" to $R_1$ and $B_1$ is 1:

$$\frac{26}{52} + \frac{26}{52} = 1.$$

In general, the sum of the probabilities along all the branches emanating from a single vertex to an outcome of that trial must be 1. Thus, for the vertex at $R_1$,

$$\frac{25}{51} + \frac{26}{51} = 1,$$

and for the vertex at $B_1$,

$$\frac{26}{51} + \frac{25}{51} = 1.$$

Now, consider the topmost path. It represents the event "red on first draw and red on second draw." By the general multiplication law,

$$P(R_1 \cap R_2) = P(R_1)P(R_2|R_1) = \frac{26}{52} \cdot \frac{25}{51} = \frac{25}{102}.$$

That is, *the probability of an event is obtained by multiplying the probabilities in the branches of the path for that event.* The probabilities for the other three paths are also indicated in the tree.

Returning to the original question, we see that two paths give a red card on the second draw, namely, the paths for $R_1 \cap R_2$ and $B_1 \cap R_2$. Therefore, the event "second card red" is the union of two mutually exclusive events. By the addition law, the probability of the event is the sum of the probabilities for the two paths:

$$P(R_2) = \frac{26}{52} \cdot \frac{25}{51} + \frac{26}{52} \cdot \frac{26}{51} = \frac{25}{102} + \frac{13}{51} = \frac{1}{2}.$$

Note how easy it was to find $P(R_2)$ by using a probability tree.

Here is a summary of what we have done:

$$R_2 = (R_1 \cap R_2) \cup (B_1 \cap R_2),$$

$$P(R_2) = P(R_1 \cap R_2) + P(B_1 \cap R_2)$$

$$= P(R_1)P(R_2|R_1) + P(B_1)P(R_2|B_1)$$

$$= \frac{26}{52} \cdot \frac{25}{51} + \frac{26}{52} \cdot \frac{26}{51} = \frac{25}{102} + \frac{13}{51} = \frac{1}{2}. \qquad \blacksquare$$

### EXAMPLE 7   Cards

*Two cards are drawn without replacement from a standard deck of cards. Find the probability that both cards are red.*

*Solution:* Refer back to the probability tree in Fig. 9.16. Only one path gives a red card on both draws, namely, that for $R_1 \cap R_2$. Thus, multiplying the probabilities along this path gives the desired probability:

$$P(R_1 \cap R_2) = P(R_1)P(R_2|R_1) = \frac{26}{52} \cdot \frac{25}{51} = \frac{25}{102}. \qquad \blacksquare$$

### EXAMPLE 8   Defective Computer Chips

*A company uses one computer chip in assembling each unit of a product. The chips are purchased from suppliers A, B, and C and are randomly picked for assembling a unit. Twenty percent come from A, 30 percent come from B, and the remainder come from C. The company believes that the probability that a chip from A will prove to be defective in the first 24 hours of use is 0.03, and the corresponding probabilities for B and C are 0.04 and 0.01, respectively. If an assembled unit is chosen at random and tested for 24 continuous hours, what is the probability that the chip in it is defective?*

*Solution:* In this problem, there is a sequence of two trials: selecting a chip ($A$, $B$, or $C$) and then testing the selected chip [defective ($D$) or nondefective ($D'$)]. We are given the unconditional probabilities

$$P(A) = 0.2 \quad \text{and} \quad P(B) = 0.3.$$

Since $A$, $B$, and $C$ are mutually exclusive and exhaustive,

$$P(C) = 1 - (0.2 + 0.3) = 0.5.$$

From the statement of the problem, we also have the conditional probabilities

$$P(D|A) = 0.03, \quad P(D|B) = 0.04, \quad \text{and} \quad P(D|C) = 0.01.$$

We want to find $P(D)$. To begin, we construct the two-level probability tree shown in Fig. 9.17. We see that the paths that give a defective chip are those for the events

$$A \cap D, \quad B \cap D, \quad \text{and} \quad C \cap D.$$

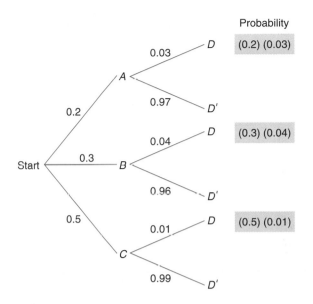

**FIGURE 9.17**   Probability tree for Example 8.

Since these events are mutually exclusive,

$$P(D) = P(A \cap D) + P(B \cap D) + P(C \cap D)$$
$$= P(A)P(D|A) + P(B)P(D|B) + P(C)P(D|C)$$
$$= (0.2)(0.03) + (0.3)(0.04) + (0.5)(0.01) = 0.023. \quad\blacksquare$$

The general multiplication law can be extended so that it applies to more than two events. For $n$ events, we have

$$P(E_1 \cap E_2 \cap \cdots \cap E_n)$$
$$= P(E_1)P(E_2|E_1)P(E_3|E_1 \cap E_2) \cdots P(E_n|E_1 \cap E_2 \cap \cdots \cap E_{n-1}).$$

(We assume that all conditional probabilities are defined.) In words, the probability that two or more events all occur is equal to the probability that one of them occurs, times the conditional probability that a second one occurs given that the first occurred, times the conditional probability that a third occurs given that the first two occurred, and so on. For example, in the manner of Example 7, the probability of drawing three red cards from a deck without replacement is

$$P(R_1 \cap R_2 \cap R_3) = P(R_1)P(R_2|R_1)P(R_3|R_1 \cap R_2) = \frac{26}{52} \cdot \frac{25}{51} \cdot \frac{24}{50}.$$

### EXAMPLE 9   Marbles in Urn

*Urn I contains one black and two red marbles, and Urn II contains one pink marble. (See Fig. 9.18.) An urn is selected at random. Then a marble is randomly drawn from it and placed in the other urn. A marble is then randomly drawn from that urn. Find the probability this marble is pink.*

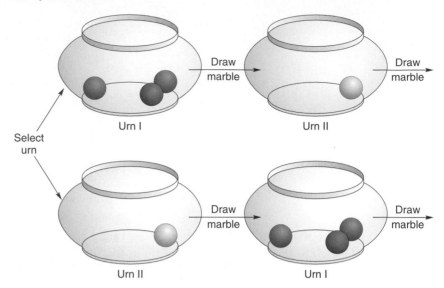

**FIGURE 9.18** Marble selections from urns.

*Solution:* This is a compound experiment with three trials:

1. Selecting an urn.
2. Drawing a marble from the urn.
3. Drawing a marble from the other urn after the marble drawn in trial 2 is placed in it.

We want to find $P$(pink marble on second draw).

We analyze the situation by constructing a three-level probability tree. (See Fig. 9.19.) The first trial has two equally likely possible outcomes, "Urn I" or "Urn II," so each has probability of $\frac{1}{2}$.

If Urn I was selected, the second trial has two possible outcomes, "red" ($R$) or "black" ($B$), with conditional probabilities $P(R|\text{I}) = \frac{2}{3}$ and

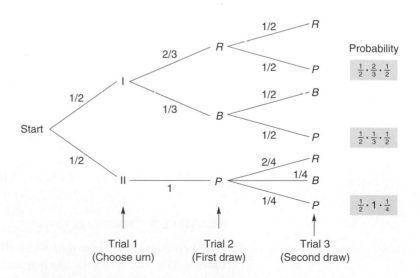

**FIGURE 9.19** Three-level probability tree.

$P(B|\text{I}) = \frac{1}{3}$. If Urn II was selected, there is one possible outcome, "pink" ($P$), so $P(P|\text{II}) = 1$. Thus, the second level of the tree has three branches.

Now we turn to the third trial. If Urn I was selected and a red marble drawn from it and placed in Urn II, then Urn II contains one red and one pink marble. Hence, at the end of the second trial, the fork at vertex $R$ has two branches, $R$ and $P$, with conditional probabilities

$$P(R|\text{I} \cap R) = \frac{1}{2} \quad \text{and} \quad P(P|\text{I} \cap R) = \frac{1}{2}.$$

Similarly, the tree shows the two possibilities if Urn I was initially selected and a black marble was placed into Urn II. Now, if Urn II was selected in the first trial, then the pink marble in it was drawn and placed into Urn I, so Urn I contains two red, one black, and one pink marble. Thus, the fork at $P$ has *three* branches, one with probability $\frac{2}{4}$ and two with probability $\frac{1}{4}$.

We see that three paths give a pink marble on the third trial, so for each, we multiply the probabilities along its branches. For example, the second path from the top represents $\text{I} \to R \to P$; the probability of this event is

$$P(\text{I} \cap R \cap P) = P(\text{I})P(R|\text{I})P(P|\text{I} \cap R)$$

$$= \frac{1}{2} \cdot \frac{2}{3} \cdot \frac{1}{2}.$$

Adding the probabilities for the three paths gives

$$P(\text{pink marble on second draw}) = \frac{1}{2} \cdot \frac{2}{3} \cdot \frac{1}{2} + \frac{1}{2} \cdot \frac{1}{3} \cdot \frac{1}{2} + \frac{1}{2} \cdot 1 \cdot \frac{1}{4}$$

$$= \frac{1}{6} + \frac{1}{12} + \frac{1}{8} = \frac{3}{8}. \quad \blacksquare$$

# ▪ Exercise 9.5

**1.** Given the equiprobable sample space

$$S = \{1, 2, 3, 4, 5, 6, 7, 8\}$$

and events

$$E = \{1, 3, 5\},$$

$$F = \{1, 2, 4, 5, 6\},$$

$$G = \{2, 3, 4, 5\},$$

find each of the following.

**a.** $P(E|F)$.      **b.** $P(E'|F)$.

**c.** $P(E|F')$.      **d.** $P(F|E)$.

**e.** $P(E|F \cap G)$.

**2.** Given the equiprobable sample space

$$S = \{1, 2, 3, 4\}$$

and events

$$E = \{1\},$$

$$F = \{2\},$$

$$G = \{1, 2\},$$

find each of the following.

**a.** $P(E)$.      **b.** $P(E|F)$.

**c.** $P(E|G)$.      **d.** $P(G|E)$.

**e.** $P(G|F')$.      **f.** $P(E'|F')$.

**3.** If $P(E) > 0$, find $P(E|E)$.

**4.** If $P(E) > 0$, find $P(\varnothing|E)$.

**5.** If $P(E \mid F) = 0.63$, find $P(E'|F)$.

**6.** If $F$ and $G$ are mutually exclusive events with positive probabilities, find $P(F|G)$.

**7.** If $P(E) = \frac{1}{4}$, $P(F) = \frac{1}{3}$, and $P(E \cap F) = \frac{1}{6}$, find each of the following.

**a.** $P(E|F)$.      **b.** $P(F|E)$.

**8.** If $P(E) = \frac{1}{2}$, $P(F) = \frac{2}{5}$, and $P(E|F) = \frac{3}{4}$, find $P(E \cup F)$. [*Hint:* Use the addition law to find $P(E \cup F)$.]

**9.** If $P(E) = \frac{1}{3}$, $P(E \cup F) = \frac{8}{15}$, and $P(E \cap F) = \frac{1}{5}$, find each of the following.

**a.** $P(F|E)$.

**b.** $P(F)$. [*Hint:* Use the addition law.]

**c.** $P(E|F)$.

**d.** $P(E|F')$. [*Hint:* Find $P(E \cap F')$ by using the identity $P(E) = P(E \cap F) + P(E \cap F')$.]

**10.** If $P(E) = \frac{3}{5}$, $P(F) = \frac{3}{10}$, and $P(E \cup F) = \frac{7}{10}$, find $P(E|F)$.

**11. Gypsy Moth** Because of gypsy moth infestation of three large areas that are densely populated with trees, consideration is being given to aerial spraying to destroy larvae. A survey was made of the 200 residents of these areas to determine whether or not they favor the spraying. The resulting data are shown in Table 9.7. Suppose that a resident is randomly selected. Let I be the event "the resident is from Area I" and so on. Find each of the following.

**TABLE 9.7**

|  | Area I | Area II | Area III | Total |
|---|---|---|---|---|
| Favor ($F$) | 46 | 35 | 44 | 125 |
| Opposed ($O$) | 22 | 15 | 10 | 47 |
| No opinion ($N$) | 10 | 8 | 10 | 28 |
| Total | 78 | 58 | 64 | 200 |

**a.** $P(F)$.   **b.** $P(F|\text{II})$.
**c.** $P(O|\text{I})$.   **d.** $P(\text{III})$.
**e.** $P(\text{III}|O)$.   **f.** $P(\text{II}|N')$.

**12. College Selection and Family Income** A survey of 175 students resulted in the data shown in Table 9.8. It shows the type of college the student attends and the income level of the student's family. Suppose a student in the survey is randomly selected.

**TABLE 9.8**

| Income | College Private | Public | Total |
|---|---|---|---|
| High | 15 | 10 | 25 |
| Middle | 25 | 55 | 80 |
| Low | 10 | 60 | 70 |
| Total | 50 | 125 | 175 |

**a.** Find the probability that the student attends a public college, given that the student comes from a middle-income family.

**b.** Find the probability that the student is from a high-income family, given that the student attends a private college.

**c.** If the student comes from a high-income family, find the probability that the student attends a private college.

**d.** Find the probability that the student attends a public college or comes from a low-income family.

**13. Cola Preference** A survey was taken among cola drinkers to see which of two popular brands people preferred. It was found that 45% liked brand A, 40% liked brand B, and 20% liked both brands. Suppose that a person in the survey is randomly selected.

**a.** Find the probability that the person liked brand A, given that he or she liked brand B.

**b.** Find the probability that the person liked brand B, given that he or she liked brand A.

**14. Quality Control** A company manufactures product X. Of the units produced, 15 percent have defective assembly and 12 percent have both defective assembly and poor finishing. If a unit of the product is randomly selected from stock and it has defective assembly, what is the probability that it has poor finishing?

*In Problems 15 and 16, assume that a child of either sex is equally likely and that, for example, having a girl first and a boy second is just as likely as having a boy first and a girl second.*

**15. Sexes of Offspring** If a family has two children, what is the probability that one child is a girl, given that at least one child is a boy?

**16. Sexes of Offspring** If a family has three children, find each of the following.

**a.** The probability that it has two girls, given that at least one child is a boy.

**b.** The probability that it has at least two girls, given that the oldest child is a girl.

**17. Coin Toss** If a fair coin is tossed three times in succession, find each of the following.

**a.** The probability of getting exactly two tails, given that the second toss is a tail.

**b.** The probability of getting exactly two tails, given that the second toss is a head.

**18. Coin Toss** If a fair coin is tossed three times in succession, find the odds of getting three tails, given that the first toss is a tail.

**19. Die Roll** If a fair die is rolled, find the probability of getting a number less than 4, given that the number is odd.

**20. Die Roll** If a fair die is rolled, find the probability of getting an even number, given that the number is less than 5.

**21. Dice Roll** If two fair dice are rolled, find the probability that two 6's occur, given that at least one die shows a 6.

**22. Dice Roll** If a fair red die and a fair green die are rolled, find the probability that the sum is greater than 9, given that a 5 shows on the red die.

**23. Dice Roll** If a fair red die and a fair green die are rolled, find the probability of getting a total of 7, given that the green die shows an even number.

**24. Dice Roll** A fair die is tossed two times in succession.

   **a.** Find the probability that the sum is 5, given that the second toss is neither a 2 nor a 4.

   **b.** Find the probability that the sum is 5 and that the second toss is neither a 2 nor a 4.

**25. Dice Roll** If fair die is tossed two times in succession, find the probability of getting a total greater than 5, given that the first toss is less than 4.

**26. Coin and Die** If a fair coin and a fair die are thrown, find the probability that the coin shows heads, given that the number on the die is even.

**27. Cards** If a card is randomly drawn from a deck of 52 cards, find the probability that the card is a king, given that it is a heart.

**28. Cards** If a card is randomly drawn from a deck of 52 cards, find the probability that the card is a heart, given that it is a face card (a jack, queen, or king).

**29. Cards** If two cards are randomly drawn without replacement from a standard deck, find the probability that the second card is not a face card, given that the first card is a face card (a jack, queen, or king).

*In Problems 30–35, consider the experiment to be a compound experiment.*

**30. Cards** If two cards are randomly drawn from a standard deck, find the probability that both cards are aces if:

   **a.** the cards are drawn without replacement.

   **b.** the cards are drawn with replacement.

**31. Cards** If three cards are randomly drawn without replacement from a standard deck, find the probability of getting a king, a queen, and a jack, in that order.

**32. Cards** If three cards are randomly drawn without replacement from a standard deck, find the probability of getting the jack of diamonds, the ace of hearts, and the king of clubs, in that order.

**33. Cards** If three cards are randomly drawn without replacement from a standard deck, find the probability that all three cards are diamonds.

**34. Cards** If two cards are randomly drawn without replacement from a standard deck of cards, find the probability that the second card is a heart.

**35. Cards** If two cards are randomly drawn without replacement from a standard deck, find the probability of getting two diamonds, given that the first card is red.

**36. Wake-Up Call** Barbara Smith, a sales representative, is staying overnight at a hotel and has a breakfast meeting with an important client the following morning. She asked room service to give her a 7 A.M. wake-up call that morning in order that she be prompt for the meeting. The probability that she will get the call is 0.9. If she gets the call, the probability that she will be on time is 0.9. If the call is not given, the probability that she will be on time is 0.6. Find the probability that she will be on time for the meeting.

**37. Taxpayer Survey** In a certain school district, a questionnaire was sent to all property-tax payers concerning whether or not a new high school should be built. Of those that responded, 60% favored its construction, 30% opposed it, and 10% had no opinion. Further analysis of the data concerning the area in which the respondents lived gave the results in Table 9.9.

   **a.** If one of the respondents is selected at random, what is the probability that he or she lives in an urban area?

   **b.** If a respondent is selected at random, use the result of part (a) to find the probability that he or she favors the construction of the school, given that the person lives in an urban area.

**TABLE 9.9**

|            | Urban | Suburban |
|------------|-------|----------|
| Favor      | 45%   | 55%      |
| Oppose     | 55%   | 45%      |
| No opinion | 35%   | 65%      |

**38. Marketing** A travel agency has a computerized telephone that randomly selects telephone numbers for advertising group bus tours. The telephone automatically dials the selected number and plays a prerecorded message to the recipient of the call. Experience has shown that 5% of those called show interest and contact the agency. However, of these, 22% actually agree to purchase a seat on a tour.

   **a.** Find the probability that a person called will contact the agency and purchase a seat on a tour.

   **b.** If 1000 people are called, how many can be expected to contact the agency and purchase a seat on a tour?

**39. Marbles in Urn** An urn contains three yellow and two red marbles.

   **a.** If two marbles are randomly drawn without replacement, find the probability that the second marble drawn is yellow, given that the first marble drawn is red.

   **b.** Repeat part (a), but assume that the first marble is replaced before the second marble is drawn.

**40. Marbles in Urn** Urn 1 contains four green and three red marbles, and Urn 2 contains three green, one white, and two red marbles. A marble is randomly drawn from Urn 1 and placed into Urn 2. If a marble is then randomly drawn from Urn 2, find the probability that the marble is green.

**41. Balls in Urn** Urn 1 contains three red and two white balls. Urn 2 contains two red and two white balls. An urn is chosen at random and then a ball is chosen at random from it. What is the probability that the ball is white?

**42. Balls in Urn** Urn 1 contains two red and three white balls. Urn 2 contains three red and four white balls. Urn 3 contains two red, two white, and two green balls. An urn is chosen at random, and then a ball is chosen at random from it.

   **a.** Find the probability that the ball is white.

   **b.** Find the probability that the ball is red.

   **c.** Find the probability that the ball is green.

**43. Marbles in Urn** Urn 1 contains one green and one red marble, and Urn 2 contains one white and one red marble. An urn is selected at random. A marble is randomly drawn from it and placed in the other urn. A marble is then randomly drawn from that urn. Find the probability that the marble is white.

**44. Defective Fuses** During a thunderstorm, two fuses blew in Mr. Smith's house and caused the lights in the kitchen and garage to go out. The fuse box is located in the garage, along with a box containing 10 fuses. Of the 10, 3 are defective from previous storms (and should have been thrown out). Mr. Smith went to the garage and, in the dark, selected 2 fuses at random from the box. Find the probability that both fuses were defective.

**45. Quality Control** A manufacturing process requires the use of a robotic welder on each of two assembly lines. Assembly line A produces 200 units of product per day, and assembly line B produces 400 units per day. Over a period of time, it has been found that the welder on A produces 2% defective units, whereas the welder on B produces 5% defective units. At the end of a day, a unit was selected at random from the total production. Find the probability that the unit has defective welding.

**46. Game Show** A TV game show host presents the following situation to a contestant. On a table are three identical boxes. One of them contains two identical envelopes. In one is a check for $5000, and in the other is a check for $1. Another box contains two envelopes with a check for $5000 in each and six envelopes with a check for $1 in each. The remaining box contains one envelope with a check for $5000 inside and five envelopes with a check for $1 inside each. If the contestant must select a box at random and then randomly draw an envelope, find the probability that a check for $5000 is inside.

**47. Quality Control** A company uses one computer chip in assembling each unit of a product. The chips are purchased from suppliers A, B, and C and are randomly picked for assembling a unit. Ten percent come from A, 20% come from B, and the remainder come from C. The probability that a chip from A will prove to be defective in the first 24 hours of use is 0.06, and the corresponding probabilities for B and C are 0.04 and 0.05, respectively. If an assembled unit is chosen at random and tested for 24 continuous hours, what is the probability that the chip in it will prove to be defective?

**48. Quality Control** A manufacturer of widgets has four assembly lines: A, B, C, and D. The percentages of output produced by the lines are 30%, 20%, 35%, and 15%, respectively, and the percentages of defective units they produce are 7%, 4%, 2%, and 5%. If a widget is randomly selected from stock, what is the probability that it is defective?

**49. Voting** In a certain town, 40% of eligible voters are registered Democrats, 35% are Republicans, and the remainder are Independents. In the last primary election, 15% of the Democrats, 20% of the Republicans, and 10% of the Independents voted.

   **a.** If an eligible voter is chosen at random, what is the probability that he or she is a Democrat who voted?

   **b.** If an eligible voter is chosen at random, what is the probability that he or she voted?

**50. Job Applicants** A company has three openings in its accounting department. Suppose Frank, Joan, Sam, Barbara, and Doris are the only applicants for these positions, and all are equally qualified. If three are hired at random, find the probability that Frank, Joan, and Doris were chosen, given that Sam was not hired.

**51. Committee Selection** Suppose five female and four male students wish to fill three openings on a campus committee on cultural diversity. If three of the students are chosen at random for the committee, find the probability that all three are female, given that at least one is female.

---

**To develop the notion of independent events and apply the special multiplication law.**

## 9.6 INDEPENDENT EVENTS

In our discussion of conditional probability, you saw that the probability of an event may be affected by the knowledge that another event has occurred. In this section, we consider the situation where the additional information has no effect. That is, the conditional probability $P(E \mid F)$ and the unconditional probability $P(E)$ are the same.

When $P(E|F) = P(E)$, we say that $E$ is *independent* of $F$. If $E$ is independent of $F$, it follows that $F$ is independent of $E$ (and vice versa). To prove this, assume that $P(E|F) = P(E)$. Then

$$P(F|E) = \frac{P(E \cap F)}{P(E)} = \frac{P(F)P(E|F)}{P(E)} = \frac{P(F)P(E)}{P(E)} = P(F),$$

which means that $F$ is independent of $E$. Thus, to prove independence, it suffices to show that either $P(E|F) = P(E)$ or $P(F|E) = P(F)$, and when one of these is true, we simply say that $E$ and $F$ are *independent events*.

Independence of two events is defined by probabilities, not by a causal relationship.

## DEFINITION

*Let E and F be events with positive probabilities. Then E and F are said to be* **independent events** *if either*

$$P(E|F) = P(E) \tag{1}$$

*or*

$$P(F|E) = P(F). \tag{2}$$

*If E and F are not independent, they are said to be* **dependent events.**

Thus, with dependent events, the occurrence of one of the events *does* affect the probability of the other. If $E$ and $F$ are independent events, it can be shown that the events in each of the following pairs are also independent:

$$E \text{ and } F', \qquad E' \text{ and } F, \qquad \text{and} \qquad E' \text{ and } F'.$$

### EXAMPLE 1   Showing that Two Events Are Independent

*A fair coin is tossed twice. Let E and F be the events*

$$E = \{head\ on\ first\ toss\},$$

$$F = \{head\ on\ second\ toss\}.$$

*Determine whether or not E and F are independent events.*

**Solution:** We suspect that they are independent, because one coin toss should not influence the outcome of another toss. To confirm our suspicion, we shall compare $P(E)$ with $P(E|F)$. For the equiprobable sample space $S = \{HH, HT, TH, TT\}$, we have $E = \{HH, HT\}$ and $F = \{HH, TH\}$. Thus,

$$P(E) = \frac{n(E)}{n(S)} = \frac{2}{4} = \frac{1}{2},$$

$$P(E|F) = \frac{n(E \cap F)}{n(F)} = \frac{n(\{HH\})}{n(F)} = \frac{1}{2}.$$

Since $P(E|F) = P(E)$, events $E$ and $F$ are independent.  ■

In Example 1 we suspected the result, and certainly there are other situations where we have an intuitive feeling as to whether or not two events are independent. For example, if a red die and green die are tossed, we expect (and it is indeed true) that the events "3 on red die" and "6 on green die" are independent, because the outcome on one die should not be influenced by the outcome on the other die. Similarly, if two cards are drawn *with replacement* from a deck of cards, we would assume that the events "first card is a jack" and "second card is a jack" are independent. However, suppose the cards are drawn *without replacement.* Because the first card drawn is not put back in the deck, it should have an effect on the outcome of the second draw, so we expect the events to be dependent. In many problems, your intuitive notion of independence or the context of the problem may make it clear whether or not independence can be assumed. In spite of one's intuition (which can prove to be wrong!), the only sure way to determine whether events $E$ and $F$ are independent (or dependent) is by showing that Eq. (1) or Eq. (2) is true (or is not true).

### EXAMPLE 2  Lung Cancer and Smoking

*In a study of smoking and lung cancer, 750 people were studied, with the results as given in Table 9.10. Suppose a person from the study is selected at random. On the basis of the data, determine whether or not the events "having lung cancer" (L) and "smoking" (S) are independent events.*

**TABLE 9.10**  Lung Cancer and Smoking

|  | Smoker | Nonsmoker | Total |
|---|---|---|---|
| Lung cancer | 150 | 100 | 250 |
| No lung cancer | 125 | 375 | 500 |
| Total | 275 | 475 | 750 |

*Solution:*  We shall compare $P(L)$ with $P(L \mid S)$. The number $P(L)$ is the proportion of the people studied that have lung cancer:

$$P(L) = \frac{250}{750} = \frac{1}{3} \approx 0.33.$$

For $P(L \mid S)$, the sample space is reduced to 275 smokers, of which 150 have lung cancer:

$$P(L \mid S) = \frac{150}{275} = \frac{6}{11} \approx 0.55.$$

Since $P(L \mid S) \neq P(L)$, having lung cancer and smoking are dependent.  ∎

The general multiplication law takes on an extremely important form for independent events. Recall that law:

$$P(E \cap F) = P(E)P(F \mid E)$$
$$= P(F)P(E \mid F).$$

If events $E$ and $F$ are independent, then $P(F \mid E) = P(F)$, so substitution in the first equation gives

$$P(E \cap F) = P(E)P(F).$$

The same result is obtained from the second equation. Thus, we have the following law:

**Special Multiplication Law**

If $E$ and $F$ are *independent events,* then

$$P(E \cap F) = P(E)P(F). \tag{3}$$

Equation (3) states that if $E$ and $F$ are independent events, then the probability that $E$ and $F$ both occur is the probability that $E$ occurs times the probability that $F$ occurs. Keep in mind that Eq. (3) is *not* valid when $E$ and $F$ are dependent.

**EXAMPLE 3** Survival Rates

*Suppose the probability of the event "Bob lives 20 more years" (B) is 0.8 and the probability of the event "Doris lives 20 more years" (D) is 0.85. Assume that B and D are independent events.*

**a.** *Find the probability that both Bob and Doris live 20 more years.*

**Solution:** We are interested in $P(B \cap D)$. Since $B$ and $D$ are independent events, the special multiplication law applics:

$$P(B \cap D) = P(B)P(D) = (0.8)(0.85) = 0.68.$$

**b.** *Find the probability that at least one of them lives 20 more years.*

**Solution:** Here we want $P(B \cup D)$. By the addition law,

$$P(B \cup D) = P(B) + P(D) - P(B \cap D).$$

From part a, $P(B \cap D) = 0.68$, so

$$P(B \cup D) = 0.8 + 0.85 - 0.68 = 0.97.$$

**c.** *Find the probability that exactly one of them lives 20 more years.*

**Solution:** We first express the event

$$E = \{\text{exactly one of them lives 20 more years}\}$$

in terms of the given events, $B$ and $D$. Now, event $E$ can occur in one of two *mutually exclusive* ways: Bob lives 20 more years but Doris does not ($B \cap D'$), or Doris lives 20 more years but Bob does not ($B' \cap D$). Thus,

$$E = (B \cap D') \cup (B' \cap D).$$

By the addition law (for mutually exclusive events),

$$P(E) = P(B \cap D') + P(B' \cap D). \tag{4}$$

To compute $P(B \cap D')$, we note that, since $B$ and $D$ are independent, so are $B$ and $D'$ (from the statement preceding Example 1). Accordingly, we can use the multiplication law and the rule for complements:

$$P(B \cap D') = P(B)P(D')$$

$$= P(B)[1 - P(D)] = (0.8)(0.15) = 0.12.$$

Similarly,

$$P(B' \cap D) = P(B')P(D) = (0.2)(0.85) = 0.17.$$

Substituting into Eq. (4) gives

$$P(E) = 0.12 + 0.17 = 0.29. \qquad \blacksquare$$

In Example 3, it was assumed that events $B$ and $D$ are independent. However, if Bob and Doris are related in some way, it is quite possible that the survival of one of them has a bearing on the survival of the other. In that case the assumption of independence is not justified, and we could not use the special multiplication law, Eq. (3).

**EXAMPLE 4** **Cards**

*In a math exam, a student was given the following two-part problem. A card is randomly drawn from a deck of 52 cards. Let H, K, and R be the events*

$$H = \{heart\ drawn\},$$
$$K = \{king\ drawn\},$$
$$R = \{red\ card\ drawn\}.$$

*Find $P(H \cap K)$ and $P(H \cap R)$.*
*For the first part, the student wrote*

$$P(H \cap K) = P(H)P(K) = \frac{13}{52} \cdot \frac{4}{52} = \frac{1}{52},$$

*and for the second part, she wrote*

$$P(H \cap R) = P(H)P(R) = \frac{13}{52} \cdot \frac{26}{52} = \frac{1}{8}.$$

*The answer was correct for $P(H \cap K)$, but not for $P(H \cap R)$. Why?*

**Solution:** The reason is that the student assumed independence in *both* parts by using the special multiplication law to multiply unconditional probabilities when, in fact, that assumption should *not* have been made. Let us examine the first part of the exam problem for independence. We shall see whether $P(H)$ and $P(H|K)$ are the same. We have

$$P(H) = \frac{13}{52} = \frac{1}{4}$$

and

$$P(H|K) = \frac{1}{4} \qquad \text{(one heart out of four kings).}$$

Since $P(H) = P(H|K)$, events $H$ and $K$ are independent, so the student's procedure is valid. (The student was lucky!) For the second part, again we have $P(H) = \frac{1}{4}$, but

$$P(H|R) = \frac{13}{26} = \frac{1}{2} \qquad \text{(13 hearts out of 26 red cards).}$$

Since $P(H|R) \neq P(H)$, events $H$ and $R$ are dependent, so the student should not have multiplied the unconditional probabilities. (The student was unlucky.) However, the student would have been safe by using the *general* multiplication law, that is,

$$P(H \cap R) = P(H)P(R|H) = \frac{13}{52} \cdot 1 = \frac{1}{4}$$

or

$$P(H \cap R) = P(R)P(H|R) = \frac{26}{52} \cdot \frac{13}{26} = \frac{1}{4}.$$

More simply, observe that $H \cap R = H$, so

$$P(H \cap R) = P(H) = \frac{13}{52} = \frac{1}{4}.$$

Equation (3) is often used as an alternative means of defining independent events, and we shall consider it as such:

Events $E$ and $F$ are independent if and only if

$$P(E \cap F) = P(E)P(F) \qquad (3)$$

Putting everything together, we can say that, to prove events $E$ and $F$ are independent, only one of the following relationships has to be shown:

$$P(E|F) = P(E), \qquad (1)$$

or

$$P(F|E) = P(F), \qquad (2)$$

or

$$P(E \cap F) = P(E)P(F). \qquad (3)$$

In other words, if any one of these equations is true, then all of them are true; if any is false, then all of them are false, and $E$ and $F$ are dependent.

**EXAMPLE 5   Dice**

*Two fair dice, one red and the other green, are rolled, and the numbers on the top faces are noted. Let $E$ and $F$ be the events*

$$E = \{number\ on\ red\ die\ is\ even\},$$

$$F = \{sum\ is\ 7\}.$$

*Test whether $P(E \cap F) = P(E)P(F)$ to determine whether $E$ and $F$ are independent.*

**Solution:** Our usual sample space for the roll of two dice has $6 \cdot 6 = 36$ equally likely outcomes. For event $E$, the red die can fall in any of three ways and the green die any of six ways, so $E$ consists of $3 \cdot 6 = 18$ outcomes. Thus, $P(E) = \frac{18}{36} = \frac{1}{2}$. Event $F$ has six outcomes:

$$F = \{(1, 6), (2, 5), (3, 4), (4, 3), (5, 2), (6, 1)\} \qquad (5)$$

where, for example, we take $(1, 6)$ to mean "1" on the red die and "6" on the green die. Therefore, $P(F) = \frac{6}{36} = \frac{1}{6}$, so

$$P(E)P(F) = \frac{1}{2} \cdot \frac{1}{6} = \frac{1}{12}.$$

Now, event $E \cap F$ consists of all outcomes in which the red die is even and the sum is 7. Using Eq. (5) as an aid, we see that

$$E \cap F = \{(2, 5), (4, 3), (6, 1)\}.$$

Thus,

$$P(E \cap F) = \frac{3}{36} = \frac{1}{12}.$$

Since $P(E \cap F) = P(E)P(F)$, events $E$ and $F$ are independent. This fact may not have been obvious before the problem was solved.   ■

**EXAMPLE 6** **Sexes of Offspring**

*For a family with at least two children, let E and F be the events*

$$E = \{\text{at most one boy}\},$$

$$F = \{\text{at least one child of each sex}\}.$$

*Assume that a child of either sex is equally likely and that, for example, having a girl first and a boy second is just as likely as having a boy first and a girl second. Determine whether E and F are independent in each of the following situations.*

**a.** *The family has exactly two children.*

*Solution:* We will use the equiprobable sample space

$$S = \{\text{BB, BG, GG, GB}\}$$

and test whether $P(E \cap F) = P(E)P(F)$. We have

$$E = \{\text{BG, GB, GG}\}, \quad F = \{\text{BG, GB}\}; \quad E \cap F = \{\text{BG, GB}\}.$$

Thus, $P(E) = \dfrac{3}{4}, P(F) = \dfrac{2}{4} = \dfrac{1}{2}$, and $P(E \cap F) = \dfrac{2}{4} = \dfrac{1}{2}$. We ask whether

$$P(E \cap F) \stackrel{?}{=} P(E)P(F)$$

and see that

$$\frac{1}{2} \neq \frac{3}{4} \cdot \frac{1}{2} = \frac{3}{8},$$

so $E$ and $F$ are dependent events.

**b.** *The family has exactly three children.*

*Solution:* Based on the result of part a, you may have an intuitive feeling that $E$ and $F$ are dependent. Nevertheless, we must test this conjecture. For three children, we use the equiprobable sample space

$$S = \{\text{BBB, BBG, BGB, BGG, GBB, GBG, GGB, GGG}\}.$$

Again we test whether $P(E \cap F) = P(E)P(F)$. We have

$$E = \{\text{BGG, GBG, GGB, GGG}\},$$

$$F = \{\text{BBG, BGB, BGG, GBB, GBG, GGB}\},$$

$$E \cap F = \{\text{BGG, GBG, GGB}\}.$$

Hence, $P(E) = \dfrac{4}{8} = \dfrac{1}{2}, P(F) = \dfrac{6}{8} = \dfrac{3}{4}$, and $P(E \cap F) = \dfrac{3}{8}$, so

$$P(E)P(F) = \frac{1}{2} \cdot \frac{3}{4} = \frac{3}{8} = P(E \cap F).$$

Therefore, we have the *unexpected* result that events $E$ and $F$ are independent. (*Moral:* You cannot always trust your intuition.) ∎

We now generalize our discussion of independence to the case of more than two events.

**DEFINITION**

*The events $E_1, E_2, \ldots, E_n$ are said to be **independent** if, and only if, for each set of two or more of the events, the probability of the intersection of the events in the set is equal to the product of the probabilities of the events in that set.*

For instance, let us apply the definition to the case of three events ($n = 3$). We say that $E$, $F$, and $G$ are independent events if the special multiplication law is true for these events, taken two at a time and three at a time. That is, each of the following equations must be true:

$$\left.\begin{array}{l} P(E \cap F) = P(E)P(F), \\ P(E \cap G) = P(E)P(G), \\ P(F \cap G) = P(F)P(G), \end{array}\right\} \text{Two at a time}$$

$$P(E \cap F \cap G) = P(E)P(F)P(G). \quad \} \text{Three at a time}$$

As another example, if events $E$, $F$, $G$, and $H$ are independent, then we can assert such things as

$$P(E \cap F \cap G \cap H) = P(E)P(F)P(G)P(H),$$

$$P(E \cap G \cap H) = P(E)P(G)P(H),$$

and

$$P(F \cap H) = P(F)P(H).$$

Similar conclusions can be made if any of the events are replaced by their complements.

### EXAMPLE 7 Cards

*Four cards are randomly drawn, with replacement, from a deck of 52 cards. Find the probability that the cards chosen, in order, are a king (K), a queen (Q), a jack (J), and a heart (H).*

*Solution:* Since there is replacement, what happens on any draw does not affect the outcome on any other draw, so we can assume independence and multiply the unconditional probabilities. We obtain

$$P(K \cap Q \cap J \cap H) = P(K)P(Q)P(J)P(H)$$

$$= \frac{4}{52} \cdot \frac{4}{52} \cdot \frac{4}{52} \cdot \frac{13}{52} = \frac{1}{8788}. \quad \blacksquare$$

### EXAMPLE 8 Aptitude Test

*Personnel Temps, a temporary-employment agency, requires that each job applicant take the company's aptitude test, which has 90% accuracy.*

**a.** *Find the probability that the test will be accurate for the next three applicants who are tested.*

*Solution:* Let $A$, $B$, and $C$ be the events that the test will be accurate for applicants A, B, and C, respectively. We are interested in

$$P(A \cap B \cap C).$$

Since the accuracy of the test for one applicant should not affect the accuracy for any of the others, it seems reasonable to assume that $A$, $B$, and $C$ are independent. Thus, we can multiply probabilities:

$$P(A \cap B \cap C) = P(A)P(B)P(C)$$

$$= (0.9)(0.9)(0.9) = (0.9)^3 = 0.729.$$

**b.** *Find the probability that the test will be accurate for at least two of the next three applicants who are tested.*

*Solution:* Here, "at least two" means exactly two or three. In the first case, the possible ways of choosing the two tests that are accurate are

$$A \text{ and } B, \quad A \text{ and } C, \quad B \text{ and } C.$$

In each of these three possibilities, the test for the remaining applicant is not accurate. For example, choosing $A$ and $B$ gives the event $A \cap B \cap C'$, whose probability is

$$P(A)P(B)P(C') = (0.9)(0.9)(0.1) = (0.9)^2(0.1).$$

You should verify that the probability for each of the other two possibilities is also $(0.9)^2(0.1)$. Summing the three probabilities gives

$$P(\text{exactly two accurate}) = 3[(0.9)^2(0.1)] = 0.243.$$

Using this result and that of part a, we obtain

$$P(\text{at least two accurate}) = P(\text{exactly two accurate}) + P(\text{three accurate})$$

$$= 0.243 + 0.729 = 0.972.$$

Alternatively, the problem could be solved by computing

$$1 - [P(\text{none accurate}) + P(\text{exactly one accurate})].$$

Why? ∎

We conclude with a note of caution: **Do not confuse independent events with mutually exclusive events.** The concept of independence is defined in terms of probability, whereas mutual exclusiveness is not. When two events are independent, the occurrence of one of them does not affect the probability of the other. However, when two events are mutually exclusive, they cannot occur simultaneously. Although these two concepts are not the same, we can draw some conclusions about their relationship. If $E$ and $F$ are mutually exclusive events with positive probabilities, the occurrence of $F$ means that $E$ cannot occur. Hence,

$$P(E|F) = 0 \neq P(E), \quad \text{since } P(E) > 0.$$

Thus, $E$ and $F$ must be dependent. In short, *mutually exclusive events with positive probabilities must be dependent.* Another way of saying this is that *independent events with positive probabilities are not mutually exclusive.*

## ■ Exercise 9.6

**1.** If events $E$ and $F$ are independent with $P(E) = \frac{1}{3}$ and $P(F) = \frac{3}{4}$, find each of the following.

  **a.** $P(E \cap F)$.       **b.** $P(E \cup F)$.

  **c.** $P(E|F)$.        **d.** $P(E'|F)$.

  **e.** $P(E \cap F')$.     **f.** $P(E \cup F')$.

  **g.** $P(E|F')$.

**2.** If events $E$, $F$, and $G$ are independent with $P(E) = 0.2$, $P(F) = 0.4$, and $P(G) = 0.5$, find each of the following.

  **a.** $P(E \cap F)$.       **b.** $P(F \cap G)$.

  **c.** $P(E \cap F \cap G)$.   **d.** $P(E|(F \cap G))$.

  **e.** $P(E \cap F' \cap G')$.

**3.** If events $E$ and $F$ are independent with $P(E) = \frac{2}{5}$ and $P(E \cap F) = \frac{1}{3}$, find $P(F)$.

**4.** If events $E$ and $F$ are independent with $P(E|F) = \frac{1}{3}$, find $P(E')$.

*In Problems 5 and 6, events E and F satisfy the given conditions. Determine whether E and F are independent or dependent.*

**5.** $P(E) = \frac{3}{4}$, $P(F) = \frac{8}{9}$, $P(E \cap F) = \frac{2}{3}$.

**6.** $P(E) = 0.26$, $P(F) = 0.15$, $P(E \cap F) = 0.038$.

**7. Stockbrokers**  Six hundred investors were surveyed to determine whether a person who uses a full-service stockbroker has better performance in his or her investment portfolio than one who uses a discount broker. In general, discount brokers usually offer no investment advice to their clients, whereas full-service brokers usually offer help in selecting stocks, but charge larger fees. The data, based on the last 12 months, are given in Table 9.11. Determine whether the event of having a full-service broker and the event of having an increase in portfolio value are independent or dependent.

**TABLE 9.11**  Portfolio Value

|              | Increase | Decrease | Total |
|--------------|----------|----------|-------|
| Full service | 320      | 80       | 400   |
| Discount     | 160      | 40       | 200   |
| Total        | 480      | 120      | 600   |

**8. Cinema Offenses**  An observation of 175 patrons in a theater resulted in the data shown in Table 9.12. The table shows three types of cinema offenses committed by male and female patrons. Crunchers include noisy eaters of popcorn and other morsels, as well as cold-drink slurpers. Determine whether the event of being a male and the event of being a cruncher are independent or dependent. (See page 5D of the July 21, 1991, issue of

**TABLE 9.12**  Theatre Patrons

|             | Male | Female | Total |
|-------------|------|--------|-------|
| Talkers     | 60   | 10     | 70    |
| Crunchers   | 55   | 25     | 80    |
| Seat kickers| 15   | 10     | 25    |
| Total       | 130  | 45     | 175   |

*USA TODAY* for the article "Pests now appearing at a theater near you.")

**9. Dice**  Two fair dice are rolled, one red and one green, and the numbers on the top faces are noted. Let event $E$ be "number on red die is neither 4 nor 5" and event $F$ be "sum is 5." Determine whether $E$ and $F$ are independent or dependent.

**10. Cards**  A card is randomly drawn from an ordinary deck of 52 cards. Let $R$ and $F$ be the events "red card drawn" and "a jack, queen, or king drawn," respectively. Determine whether $R$ and $F$ are independent or dependent.

**11. Coins**  If two fair coins are tossed, let $E$ be the event "at most one head" and $F$ be the event "exactly one head." Determine whether $E$ and $F$ are independent or dependent.

**12. Coins**  If three fair coins are tossed, let $E$ be the event "at most one head" and $F$ be the event "at least one head and one tail." Determine whether $E$ and $F$ are independent or dependent.

**13. Chips in Urn**  An urn contains six chips numbered from 1 to 6. Two chips are randomly drawn with replacement. Let $E$, $F$, and $G$ be the events

$$E = \text{"3 on first draw,"}$$

$$F = \text{"3 on second draw,"}$$

$$G = \text{"sum is odd."}$$

**a.** Determine whether $E$ and $F$ are independent or dependent.

**b.** Determine whether $E$ and $G$ are independent or dependent.

**c.** Determine whether $F$ and $G$ are independent or dependent.

**d.** Are $E$, $F$, and $G$ independent?

**14. Chips in Urn**  An urn contains six chips numbered from 1 to 6. A chip is randomly drawn. Let $E$ be the event of drawing a 4 and $F$ be the event of drawing a 6.

**a.** Are $E$ and $F$ mutually exclusive?

**b.** Are $E$ and $F$ independent?

*In Problems 15 and 16, events E and F satisfy the given conditions. Determine whether E and F are independent or dependent.*

**15.** $P(E|F) = 0.6$, $P(E \cap F) = 0.28$, $P(F|E) = 0.4$.

**16.** $P(E|F) = \frac{2}{3}$, $P(E \cup F) = \frac{17}{18}$, $P(E \cap F) = \frac{5}{9}$.

*In Problems 17–37, you may make use of your intuition concerning independent events if nothing to that effect is specified.*

**17. Dice**  Two fair dice are rolled, one red and one green. Find the probability that the red die is a 4 and the green die is a number greater than 4.

**18. Die**  If a fair die is rolled four times, find the probability that a 2 or 3 comes up each time.

**19. Fitness Classes**  At a certain fitness center, the probability that a member regularly attends an aerobics class is $\frac{1}{5}$. If two members are randomly selected, find the probability that both attend the class regularly. Assume independence.

**20. Monopoly** In the game of Monopoly, a player rolls two fair dice. One special situation that can arise is that the numbers on the top faces of the dice are the same (such as two 3's). This result is called a "double," and when it occurs, the player continues his or her turn and rolls the dice again. The pattern continues, unless the player is unfortunate enough to throw doubles three consecutive times. In that case, the player goes to jail. Find the probability that a player goes to jail in this way.

**21. Cards** Three cards are randomly drawn, with replacement, from an ordinary deck of 52 cards. Find the probability that the cards drawn, in order, are an ace, a face card (a jack, queen, or king), and a spade.

**22. Die** If a fair die is rolled four times, find each of the following.

**a.** The probability of getting a 3 each time.

**b.** The probability of getting an odd number each time.

**23. Exam Grades** In a sociology course, the probability that Bill gets an A on the final exam is $\frac{3}{4}$, and for Jim and Linda, the probabilities are $\frac{1}{2}$ and $\frac{4}{5}$, respectively. Assume independence and find each of the following.

**a.** The probability that all three of them get an A on the exam.

**b.** The probability that none of them get an A on the exam.

**c.** The probability that, of the three, only Linda gets an A.

**24. Die** If a fair die is rolled three times, find the probability of getting at least one 6.

**25. Survival Rates** The probability that person A survives 15 more years is $\frac{2}{3}$, and the probability that person B survives 15 more years is $\frac{3}{5}$. Find the probability of each of the following. Assume independence.

**a.** A and B both survive 15 years.

**b.** B survives 15 years, but A does not.

**c.** Exactly one of A and B survives 15 years.

**d.** At least one of A and B survives 15 years.

**e.** Neither A nor B survives 15 years.

**26. Nuts and Bolts** In his workshop, a carpenter has a box containing a mixture of three sizes of carriage bolts (A, B, and C) and another box containing a mixture of the three corresponding nuts. The percentages of each size of bolt and nut in the boxes are given in Table 9.13. If a bolt and a nut are randomly drawn, find the probability that they will fit each other. Round your answer to two decimal places.

**27. Marbles in Urn** An urn contains four red, six white, and five green marbles. If two marbles are randomly drawn with replacement, find each of the following.

**a.** The probability that the first marble is white and the second is green.

**b.** The probability that one marble is white and the other one is green.

**28. Dice** Suppose two fair dice are rolled twice. Find the probability of getting a total of 7 on one of the rolls and a total of 12 on the other one.

**29. Marbles in Urn** An urn contains three red, six white, and nine green marbles. If two marbles are randomly drawn with replacement, find the probability that they have the same color.

**30. Die** Find the probability of rolling the same number in three throws of a fair die.

**31. Tickets in Hat** Twenty tickets numbered from 1 to 20 are placed in a hat. If two tickets are randomly drawn with replacement, find the probability that the sum is 35.

**32. Coins and Dice** Suppose two fair coins are tossed and then two fair dice are rolled. Find each of the following.

**a.** The probability that two tails and two 3's occur.

**b.** The probability that two heads, one 4, and one 6 occur.

**33. Carnival Game** In a carnival game, a well-balanced roulette-type wheel has 12 equally spaced slots that are numbered from 1 to 12. The wheel is spun, and a ball travels along the rim of the wheel. When the wheel stops, the number of the slot in which the ball finally rests is considered the result of the spin. If the wheel is spun three times, find each of the following.

**a.** The probability that the first number will be 4 and the second and third numbers will be 5.

**b.** The probability that there will be one even number and two odd numbers.

**34. Cards** Two cards are randomly drawn, with replacement, from an ordinary deck of 52 cards. Find each of the following.

**a.** The probability of drawing, in order, a jack and a heart.

**b.** The probability of drawing two jacks.

**c.** The probability that one jack and one heart are drawn.

**d.** The probability of drawing exactly one jack.

**TABLE 9.13** Nuts and Bolts

| | Boxes | |
| --- | --- | --- |
| Size | Bolts | Nuts |
| A | 62% | 55% |
| B | 27% | 32% |
| C | 11% | 13% |

35. **Multiple-Choice Exam**   A quiz contains five multiple-choice problems. Each problem has four choices for the answer, but only one of them is correct. Suppose a student randomly guesses the answer to each problem. Find each of the following by assuming that the guesses are independent.

   a. The probability that the student gets exactly four correct answers.

   b. The probability that the student gets at least four correct answers.

   c. The probability that the student gets three or more correct answers.

36. **Shooting Gallery**   At a shooting gallery, suppose Bill, Jim, and Linda each take one shot at a moving target. The probability that Bill hits the target is 0.8, and for Jim and Linda, the probabilities are 0.6 and 0.7, respectively. Assume independence and find each of the following.

   a. The probability that none of them hit the target.

   b. The probability that Linda is the only one of them that hits the target.

   c. The probability that exactly one of them hits the target.

   d. The probability that exactly two of them hit the target.

37. **Decision Making**[3]   The president of Zeta Construction Company must decide which of two actions to take, namely, to rent or to buy expensive excavating equipment. The probability that the vice president makes a faulty analysis and thus recommends the wrong decision to the president is 0.05. To be thorough, the president hires two consultants, who study the problem independently and make their recommendations. After having observed them at work, the president estimates that the first consultant is likely to recommend the wrong decision with probability 0.05, the other with probability 0.1. He decides to take the action recommended by a majority of the three recommendations he receives. What is the probability that he will make the wrong decision?

[3]Samuel Goldberg, *Probability, an Introduction* (Prentice-Hall, Inc., 1960, Dover Publications, Inc., 1986), p. 113. Adapted by permission of the author.

# 9.7 BAYES' FORMULA

In this section, we shall be dealing with a two-stage experiment in which we know the outcome of the second stage and are interested in the probability that a particular outcome has occurred in the first stage.

To illustrate, suppose it is believed that of the total population (our sample space), 8 percent have a particular disease. Imagine also that there is a new blood test for detecting the disease and that researchers have evaluated its effectiveness. Data from extensive testing show that the blood test is not perfect: Not only is it positive for only 95 percent of those who have the disease, but it is also positive for 3 percent of those who do not. Suppose a person from the population is selected at random and given the blood test. If the result is positive, what is the probability that the person has the disease?

To analyze this problem, we consider the following events:

$$D_1 = \{\text{having the disease}\},$$

$$D_2 = \{\text{not having the disease}\},$$

$$T_1 = \{\text{testing positive}\},$$

$$T_2 = \{\text{testing negative}\}.$$

We are given that

$$P(D_1) = 0.08,$$

so

$$P(D_2) = 1 - 0.08 = 0.92,$$

because $D_1$ and $D_2$ are complements. It is reasonable to assume that $T_1$ and $T_2$ are also complements; in that case, we have the conditional probabilities

$$P(T_1|D_1) = 0.95, \qquad P(T_2|D_1) = 1 - 0.95 = 0.05,$$

$$P(T_1|D_2) = 0.03, \qquad P(T_2|D_2) = 1 - 0.03 = 0.97.$$

Figure 9.20 shows a two-stage probability tree that reflects this information. The first stage takes into account either having or not having the disease, and the second stage shows possible test results.

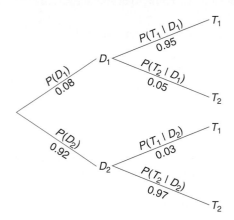

**FIGURE 9.20** Two-stage probability tree.

We are interested in the probability that a person who tests positive has the disease. That is, we want to find the conditional probability that $D_1$ occurred in the first stage, given that $T_1$ occurred in the second stage:

$$P(D_1|T_1).$$

It is important that you understand the difference between the conditional probabilities $P(D_1|T_1)$ and $P(T_1|D_1)$. The probability $P(T_1|D_1)$, which is *given* to us, is a "typical" conditional probability, in that it deals with the probability of an outcome in the second stage *after* an outcome in the first stage has occurred. However, with $P(D_1|T_1)$, we have a "reverse" situation. Here we must find the probability of an outcome in the *first* stage, given that an outcome in the second stage occurred. In a sense, we have the "cart before the horse" in that this probability does not fit the usual (and more natural) pattern of a typical conditional probability. Fortunately, we have all the tools needed to find $P(D_1|T_1)$. We proceed as follows.

From the definition of conditional probability,

$$P(D_1|T_1) = \frac{P(D_1 \cap T_1)}{P(T_1)}. \tag{1}$$

Consider the numerator. Applying the general multiplication law gives

$$P(D_1 \cap T_1) = P(D_1)P(T_1|D_1)$$
$$= (0.08)(0.95) = 0.076,$$

which is indicated in the path through $D_1$ and $T_1$ in Fig. 9.21. The denominator, $P(T_1)$, is the sum of the probabilities for all paths of the tree ending in $T_1$. Thus,

$$P(T_1) = P(D_1 \cap T_1) + P(D_2 \cap T_1)$$
$$= P(D_1)P(T_1|D_1) + P(D_2)P(T_1|D_2)$$
$$= (0.08)(0.95) + (0.92)(0.03) = 0.1036.$$

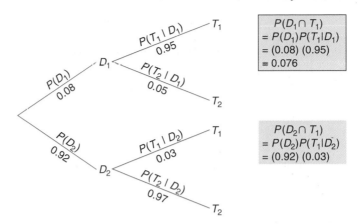

**FIGURE 9.21**   Probability tree to determine $P(D_1|T_1)$.

Hence,

$$P(D_1|T_1) = \frac{P(D_1 \cap T_1)}{P(T_1)}$$

$$= \frac{\text{probability of path through } D_1 \text{ and } T_1}{\text{sum of probabilities of all paths to } T_1}$$

$$= \frac{0.076}{0.1036} = \frac{760}{1036} = \frac{190}{259} \approx 0.734.$$

So the probability that the person has the disease, given that the test is positive, is approximately 0.734. In other words, about 73.4 percent of people who test positive actually have the disease. You would probably agree that this probability was relatively easy to find by using basic principles [Eq. (1)] and a probability tree (Fig. 9.21).

At this point, some terminology should be introduced. The *unconditional* probabilities $P(D_1)$ and $P(D_2)$ are called **prior** (or *a priori*) **probabilities,** because they are given *before* we have any knowledge about the outcome of a blood test. The conditional probability $P(D_1|T_1)$ is called a **posterior** (or *a posteriori*) **probability,** because it is found *after* the outcome ($T_1$) of the test is known.

From our answer for $P(D_1|T_1)$, we can easily find the posterior probability of not having the disease given a positive test result:

$$P(D_2|T_1) = 1 - P(D_1|T_1) = 1 - \frac{190}{259} = \frac{69}{259} \approx 0.266.$$

Of course, this can also be found by using the probability tree:

$$P(D_2|T_1) = \frac{\text{probability of path through } D_2 \text{ and } T_1}{\text{sum of probabilities of all paths to } T_1}$$

$$= \frac{(0.92)(0.03)}{0.1036} = \frac{0.0276}{0.1036} = \frac{276}{1036} = \frac{69}{259} \approx 0.266.$$

It is not really necessary to use a probability tree to find $P(D_1|T_1)$. Instead, a formula can be developed. We know that

$$P(D_1|T_1) = \frac{P(D_1 \cap T_1)}{P(T_1)} = \frac{P(D_1)P(T_1|D_1)}{P(T_1)}. \tag{2}$$

Although we used a probability tree to express $P(T_1)$ conveniently as a sum of probabilities, the sum can be found another way. Take note that events $D_1$ and $D_2$ have two properties: They are mutually exclusive and their union is the sample space $S$. Such events are collectively called a **partition** of $S$. Using this partition, we can break up event $T_1$ into mutually exclusive "pieces":

$$T_1 = T_1 \cap S = T_1 \cap (D_1 \cup D_2).$$

Then, by the distributive and commutative laws,

$$T_1 = (D_1 \cap T_1) \cup (D_2 \cap T_1). \tag{3}$$

Since $D_1$ and $D_2$ are mutually exclusive, so are events $D_1 \cap T_1$ and $D_2 \cap T_1$.[4] Thus, $T_1$ has been expressed as a union of mutually exclusive events. In this form, we can find $P(T_1)$ by adding probabilities. Applying the addition law for mutually exclusive events to Eq. (3) gives

$$P(T_1) = P(D_1 \cap T_1) + P(D_2 \cap T_1)$$
$$= P(D_1)P(T_1|D_1) + P(D_2)P(T_1|D_2).$$

Substituting into Eq. (2), we obtain

$$P(D_1|T_1) = \frac{P(D_1)P(T_1|D_1)}{P(D_1)P(T_1|D_1) + P(D_2)P(T_1|D_2)}, \tag{4}$$

which is a formula for computing $P(D_1|T_1)$.

Equation (4) is a special case (namely, for a partition of $S$ into two events) of the following general formula, called **Bayes' formula,**[5] which has has wide application in decision making:

---

**Bayes' Formula**

Suppose $F_1, F_2, \ldots, F_n$ are $n$ events that partition a sample space $S$. That is, the $F_i$'s are mutually exclusive and their union is $S$. Furthermore, suppose that $E$ is any event in $S$, where $P(E) > 0$. Then the conditional probability of $F_i$ given that event $E$ has occurred is expressed by

$$P(F_i|E) = \frac{P(F_i)P(E|F_i)}{P(F_1)P(E|F_1) + P(F_2)P(E|F_2) + \cdots + P(F_n)P(E|F_n)}$$

for each value of $i$, where $i = 1, 2, \ldots, n$.

---

Rather than memorize the formula, a probability tree can be used to obtain $P(F_i|E)$. Using the tree in Fig. 9.22, we have

$$P(F_i|E) = \frac{\text{probability for path through } F_i \text{ and } E}{\text{sum of all probabilities for paths to } E}.$$

### EXAMPLE 1   Quality Control

*A camcorder manufacturer uses one microchip in assembling each camcorder it produces. The microchips are purchased from suppliers* A, B, *and* C *and are randomly picked for assembling each camcorder. Twenty percent of the*

---

[4]See Example 8 of Sec. 9.3.

[5]After Thomas Bayes (1702–1761), the 18th-century English minister who discovered the formula.

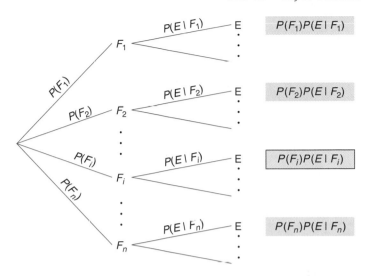

**FIGURE 9.22** Probability tree for $P(F_i|E)$.

*microchips come from* A, 35 *percent come from* B, *and the remainder come from* C. *Based on past experience, the manufacturer believes that the probability that a microchip from* A *is defective is 0.03, and the corresponding probabilities for* B *and* C *are 0.02 and 0.01, respectively. A camcorder is selected at random from a day's production, and its microchip is found to be defective. Find the probability that it was supplied (a) from* A, *(b) from* B, *and (c) from* C. *(d) From what supplier was the microchip most likely supplied?*

*Solution:* We define the following events:

$$S_1 = \{\text{supplier A}\},$$

$$S_2 = \{\text{supplier B}\},$$

$$S_3 = \{\text{supplier C}\},$$

$$D = \{\text{defective microchip}\}.$$

We have

$$P(S_1) = 0.2, \qquad P(S_2) = 0.35, \qquad P(S_3) = 0.45,$$

and the conditional probabilities

$$P(D|S_1) = 0.03, \qquad P(D|S_2) = 0.02, \qquad P(D|S_3) = 0.01,$$

which are reflected in the probability tree in Fig.9.23. Note that the figure shows only the portion of the complete probability tree that relates to event $D$. This is all that actually needs to be drawn, and this abbreviated form is often called a *Bayes' probability tree*.

For part (a), we want to find the probability of $S_1$ given that $D$ has occurred. That is,

$$P(S_1|D) = \frac{\text{probability of path through } S_1 \text{ and } D}{\text{sum of probabilities of all paths to } D}$$

$$= \frac{(0.2)(0.03)}{(0.2)(0.03) + (0.35)(0.02) + (0.45)(0.01)}$$

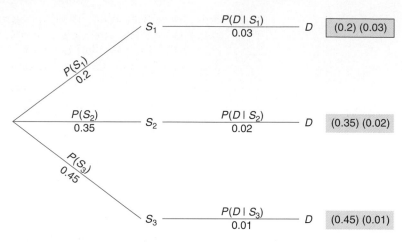

**FIGURE 9.23**   Bayes' probability tree for Example 1.

$$= \frac{0.006}{0.006 + 0.007 + 0.0045}$$

$$= \frac{0.006}{0.0175} = \frac{60}{175} = \frac{12}{35} \approx 0.343.$$

This means that approximately 34.3 percent of the defective microchips come from supplier A.

For part (b), we have

$$P(S_2|D) = \frac{\text{probability of path through } S_2 \text{ and } D}{\text{sum of probabilities of all paths to } D}$$

$$= \frac{(0.35)(0.02)}{0.0175} = \frac{0.007}{0.0175} = \frac{70}{175} = \frac{14}{35}.$$

For part (c),

$$P(S_3|D) = \frac{\text{probability of path through } S_3 \text{ and } D}{\text{sum of probabilities of all paths to } D}$$

$$= \frac{(0.45)(0.01)}{0.0175} = \frac{0.0045}{0.0175} = \frac{45}{175} = \frac{9}{35}.$$

For part (d), the greatest of $P(S_1|D)$, $P(S_2|D)$, and $P(S_3|D)$ is $P(S_2|D)$. Thus, the defective microchip was most likely supplied by B.  ■

### EXAMPLE 2   Marbles in Urn

*Two identical urns, Urn I and Urn II, are on a table. Urn I contains one red and one black marble; Urn II contains two red marbles. (See Fig. 9.24.) An urn is se-lected at random, and then a marble is randomly drawn from it. The marble is red. What is the probability that the other marble in the selected urn is red?*

Urn I          Urn II

**FIGURE 9.24**   Diagram for Example 2.

*Solution:* Because the other marble could be red or black, you might hastily conclude that the answer is $\frac{1}{2}$. Sorry, your intuition fails! The question can be restated as follows: Find the probability that the marble came from Urn II, given that the marble is red. We define the events

$$U_1 = \{\text{Urn I selected}\},$$

$$U_2 = \{\text{Urn II selected}\},$$

$$R = \{\text{red marble selected}\}.$$

We want to find $P(U_2|R)$. Since an urn is selected at random, $P(U_1) = P(U_2) = \frac{1}{2}$. From Fig. 9.24, we conclude that

$$P(R|U_1) = \frac{1}{2} \quad \text{and} \quad P(R|U_2) = 1.$$

We will show two methods of solving this problem, the first with a probability tree and the second with Bayes' formula.

*Method 1: Probability Tree* Figure 9.25 shows a Bayes' probability tree for

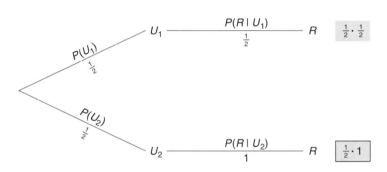

**FIGURE 9.25**    Bayes' probability tree for Example 2.

our problem. Since all paths end at $R$,

$$P(U_2|R) = \frac{\text{probability for path through } U_2 \text{ and } R}{\text{sum of probabilities of all paths}}$$

$$= \frac{\left(\dfrac{1}{2}\right)(1)}{\left(\dfrac{1}{2}\right)\left(\dfrac{1}{2}\right) + \left(\dfrac{1}{2}\right)(1)} = \frac{\dfrac{1}{2}}{\dfrac{3}{4}} = \frac{2}{3}.$$

Note that the unconditional probability of choosing Urn II, namely, $P(U_2) = \frac{1}{2}$, increases to $\frac{2}{3}$, given that a red marble was drawn. An increase is reasonable: Since there are only red marbles in Urn II, choosing a red marble should make it more likely that it came from Urn II.

*Method 2: Bayes' Formula* Because $U_1$ and $U_2$ partition the sample space, by Bayes' formula we have

$$P(U_2|R) = \frac{P(U_2)P(R|U_2)}{P(U_1)P(R|U_1) + P(U_2)P(R|U_2)}$$

$$= \frac{\left(\dfrac{1}{2}\right)(1)}{\left(\dfrac{1}{2}\right)\left(\dfrac{1}{2}\right) + \left(\dfrac{1}{2}\right)(1)} = \frac{\dfrac{1}{2}}{\dfrac{3}{4}} = \frac{2}{3}. \qquad \blacksquare$$

## ■ Exercise 9.7

1. Suppose events $E$ and $F$ partition a sample space $S$, where $E$ and $F$ have probabilities

$$P(E) = \frac{2}{5}, \qquad P(F) = \frac{3}{5}.$$

If $D$ is an event such that

$$P(D|E) = \frac{1}{10}, \qquad P(D|F) = \frac{1}{5},$$

find the probabilities $P(E|D)$ and $P(F|D')$.

2. A sample space is partitioned by events $E_1$, $E_2$, and $E_3$, whose probabilities are $\frac{1}{5}$, $\frac{3}{10}$, and $\frac{1}{2}$, respectively. Suppose $S$ is an event such that the following conditional probabilities hold:

$$P(S|E_1) = \frac{2}{5}, \qquad P(S|E_2) = \frac{7}{10}, \qquad P(S|E_3) = \frac{1}{2}.$$

Find the probabilities $P(E_1|S)$ and $P(E_3|S')$.

3. **Voting** In a certain precinct, 40% of the eligible voters are registered Democrats, 35% are Republicans, and the remainder are Independents. During the last primary election, 15% of the Democrats, 20% of the Republicans, and 10% of the Independents voted. Find the probability that a person who voted is a Democrat.

4. **Imported vs. Domestic Tires** Out of 3000 tires in the warehouse of a tire distributor, 2000 tires are domestic and 1000 are imported. Among the domestic tires, 40% are whitewalls; of the imported tires, 10% are whitewalls. If a tire is selected at random and it is a whitewall, what is the probability that it is imported?

5. **Disease Testing** A new test was developed for detecting Gamma's disease, which is believed to affect 5% of the population. Results of extensive testing indicate that 98% of persons who have this disease will have a positive reaction to the test, whereas 6% of those who do not have the disease will also have a positive reaction.

   a. What is the probability that a randomly selected person who has a positive reaction will actually have Gamma's disease?

   b. What is the probability that a randomly selected person who has a negative reaction will actually have Gamma's disease?

6. **Earnings and Dividends** Of the companies in a particular sector of the economy, it is believed that one-third will have an increase in quarterly earnings. Of those which do, the percentage that declare a dividend is 60%.

Of those which do not have an increase, the percentage that declare a dividend is 10%. What percentage of companies that declare a dividend will have an increase in quarterly earnings?

7. **Marbles in Urn** An urn contains four red and two green marbles, and a second urn contains two red and three green marbles. An urn is selected at random and a marble is randomly drawn from it. The marble is red. What is the probability that it came from the first urn?

8. **Balls in Urn** Urn I contains two red and three white balls. Urn II contains four red and three white balls. Urn III contains two red, two white, and two green balls. An urn is chosen at random, and then a ball is chosen at random from it. The ball is white. Find the probability that it came from Urn I.

9. **Quality Control** A manufacturing process requires the use of a robotic welder on each of two assembly lines, A and B, which produce 200 and 400 units of product per day, respectively. Based on past experience, it is believed that the welder on A produces 2% defective units, whereas the welder on B produces 5% defective units. At the end of a day, a unit was selected at random from the total production and was found to be defective. What is the probability that it came from line A?

10. **Quality Control** A manufacturer of widgets has four assembly lines: A, B, C, and D. The percentages of total daily output that are produced by the four lines are 35%, 20%, 30%, and 15%, respectively. The percentages of defective units produced by the lines are estimated to be 2%, 4%, 3%, and 4%, respectively. Suppose that a widget is randomly selected from a day's production and is defective. What is the probability that it came from assembly line (a) A? (b) B? (c) C? (d) D? (e) From which assembly line did it most likely come?

11. **Wake-Up Call** Barbara Smith, a sales representative, is staying overnight at a hotel and has a breakfast meeting with an important client the following morning. She asked room service to give her a 7 A.M. wake-up call that morning in order that she be prompt for the meeting. The probability that room service makes the call is 0.9. If the call is made, the probability that she will be on time is 0.9, but if the call is not made, the probability that she will be on time is 0.8. If she is on time for the meeting, what is the probability that the call was made?

12. **Candy Snatcher** On a high shelf are two identical opaque candy jars containing 50 raisin clusters each. The clusters in one of the jars are made with dark chocolate. In the other jar, 20 are made with dark chocolate and 30 are made with milk chocolate. (They are mixed well, however.) Bob Jones, who has a sudden craving for chocolate, reaches up and randomly takes out a raisin cluster from one of the jars. If it is made with dark chocolate, what is the probability that it was drawn from the jar containing only dark chocolate?

13. **Physical Fitness Activity** The week of National Employee Health and Fitness Day, the employees of a large company were asked to exercise a minimum of

three times that week for at least 20 minutes per session. The purpose was to generate "exercise miles." All participants who completed this requirement received a certificate acknowledging their contribution. The activities reported were walking, bicycling, and running. Of all who participated, $\frac{1}{2}$ reported walking, $\frac{1}{4}$ reported bicycling, and $\frac{1}{4}$ reported running. Suppose that the probability that a participant who walks will complete the requirement is $\frac{9}{10}$, and for bicycling and running it is $\frac{4}{5}$ and $\frac{2}{3}$, respectively. What percentage of persons who completed the requirement do you expect reported walking? (Assume that each participant got their exercise from one activity only.)

14. **Battery Reliability** When the weather is extremely frigid, a motorist must charge his car battery during the night in order to improve the likelihood that the car will start early the following morning. If he does not charge it, the probability that the car will not start is $\frac{4}{5}$. If he does charge it, the probability that the car will not start is $\frac{1}{8}$. Past experience shows that the probability that he remembers to charge the battery is $\frac{9}{10}$. One morning, during a cold spell, he cannot start his car. What is the probability that he forgot to charge the battery?

15. **Automobile Satisfaction Survey** In a customer satisfaction survey, $\frac{3}{5}$ of those surveyed had a Japanese-made car, $\frac{1}{10}$ a European-made car, and $\frac{3}{10}$ an American-made car. Of the first group, 85% said they would buy the same make of car again, and for the other two groups the corresponding percentages are 50% and 40%. What is the probability that a person who said they would buy the same make again had a Japanese-made car?

16. **Mineral Test Borings** A geologist believes that the probability that the rare earth mineral junko occurs in a particular region of the country is 0.02. If junko is present in that region, the geologists's test borings will have a positive result 90% of the time. However, if junko is not present, a negative result will occur 95% of the time.

   **a.** If a test is positive on a site in the region, find the probability that junko is there.

   **b.** If a test is negative on such a site, find the probability that junko is there.

17. **Physics Exam** After a physics exam was given, it turned out that only 75% of the class answered every question. Of those who did, 80% passed, but of those who did not, only 50% passed. If a student passed the exam, what is the probability that the student answered every question? (P.S.: The instructor eventually reached the conclusion that the test was too long and curved the exam grades, to be fair and merciful.)

18. **Giving Up Smoking** In a 1990 survey of smokers, 75% predicted that they would still be smoking five years later. Five years later, 70% of those who predicted that they would be smoking did not smoke, and of those who predicted that they would not be smoking, 90% did not smoke. What percentage of those who do not smoke predicted that they would be smoking?

19. **Alien Communication** B.G. Cosmos, a scientist, believes that the probability is $\frac{2}{5}$ that aliens from an advanced civilization on Planet X are trying to communicate with us by sending high-frequency signals to earth. By using sophisticated equipment, Cosmos hopes to pick up these signals. The manufacturer of the equipment, Trekee, Inc., claims that if aliens are indeed sending signals, the probability that the equipment will detect them is $\frac{3}{5}$. However, if aliens are not sending signals, the probability that the equipment will seem to detect such signals is $\frac{1}{10}$. If the equipment detects signals, what is the probability that aliens are actually sending them?

20. **Calculus Grades** In an honors Calculus I class, 60% of students had an A average at midterm. Of these, 70% ended up with a course grade of A, and of those who did not have an A average at midterm, 60% ended up with a course grade of A. If one of the students is selected at random and is found to have received an A for the course, what is the probability that the student did not have an A average at midterm?

21. **Movie Critique** A well-known pair of highly influential movie critics have a popular TV show on which they review new movie releases and recently released videos. Over the past 10 years, they gave a "Two Thumbs Up" to 80% of movies that turned out to be box-office successes; they gave a "Two Thumbs Down" to 90% of movies that proved to be unsuccessful. A new movie, *Math Wizard,* whose release is imminent, is considered favorably by others in the industry who have previewed it; in fact, they give it a prior probability of success of $\frac{7}{10}$. Find the probability that it will be a success, given that the pair of TV critics give it a "Two Thumbs Up" after seeing it.

22. **Balls in Urn** Urn 1 contains four green and three red balls, and Urn 2 contains two green, one white, and two red balls. A ball is randomly drawn from Urn 1 and placed into Urn 2. A ball is then randomly drawn from Urn 2. If the ball is green, find the probability that a green ball was drawn from Urn 1.

23. **Risky Loan** In the loan department of Third National Bank, past experience indicates that 15% of loan requests are considered by bank examiners to fall into the

"substandard" class and should not be approved. However, the bank's loan reviewer, I. M. Risky, is lax at times and concludes that a request is not in the substandard class when it is, and vice versa. Suppose that 20% of requests that are actually substandard are not considered substandard by Risky and that 10% of requests that are not substandard are considered by Risky to be substandard and, hence, not approved.

**a.** Find the probability that a request is considered to be substandard by Risky.

**b.** Find the probability that a request is substandard, given that it is considered to be substandard by Risky.

**c.** Find the probability that Risky makes an error in considering a request. (An error occurs when the request is not substandard, but is considered substandard, or when the request is substandard, but considered to be not substandard.)

**24. Coins in Chests** Each of three identical chests has two drawers. The first chest contains a gold coin in each drawer. The second chest contains a silver coin in each drawer, and the third contains a silver coin in one drawer and a gold coin in the other. A chest is chosen at random and a drawer is opened. There is a gold coin in it. What is the probability that the coin in the other drawer of that chest is silver?

**25. Product Identification after Flood**[6] After a severe flood, a distribution warehouse is stocked with waterproofed boxes of fireworks from which the identification labels have been washed off. There are three types of fireworks: low quality, medium quality, and high quality, each packed in units of 100 in identical boxes. None of the individual fireworks have markings on them, but it is believed that in the entire warehouse, the proportions of boxes with low-, medium-, and high-quality fireworks are 0.25, 0.25, and 0.5, respectively. Because detonating a firework destroys it, extensive testing of the fireworks is impractical. Instead, the distributor decides that two fireworks from each box will be tested. The quality of the fireworks will then be decided on the basis of how many of the two are defective. The manufacturer, on the basis of past experience, estimates the conditional probabilities given in Table 9.14. Suppose two fireworks are selected from a box, and tested, and both are found to detonate satisfactorily. Let the events $L$, $M$, and $H$ be that the box contains low-, medium-, and high-quality fireworks, respectively. Furthermore, let $E$ be the observed event that neither of the fireworks was defective.

**a.** Find $P(L|E)$, the probability that the box contains low-quality fireworks, given $E$.

**b.** Find the probability that the box contains medium-quality fireworks, given $E$.

**TABLE 9.14** Conditional Probabilities of Finding $x$ Defectives, Given that Two Fireworks Were Tested from a Box of Known Quality

| Number of Defectives $x$ | Quality of Box | | |
|---|---|---|---|
| | Low | Medium | High |
| 0 | 0.49 | 0.64 | 0.81 |
| 1 | 0.42 | 0.32 | 0.18 |
| 2 | 0.09 | 0.04 | 0.01 |

**c.** Find the probability that the box contains high-quality fireworks, given $E$.

**d.** What is the most likely quality of the fireworks in the box, given $E$?

**26. Product Identification after Flood**

**a.** Repeat Problem 25 if $E$ is the observed event that exactly one of the tested fireworks is defective.

**b.** Repeat Problem 25 if $E$ is the observed event that both of the tested fireworks are defective.

**27. Weather Forecasting**[7] J. B. Smith, who has lived in the same city many years, assigns a prior probability of 0.2 that today's weather will be inclement. (He thinks that today will be fair with probability 0.8.) Smith listens to an early morning weather forecast to get information on the day's weather. The forecaster makes one of three predictions: fair weather, inclement weather, or uncertain weather. Smith has made estimates of conditional probabilities of the different predictions, given the day's weather, as shown in Table 9.15. For example, Smith believes that, of the fair days, 70% are correctly forecast, 20% are forecast as inclement, and 10% are forecast as uncertain. Suppose that Smith hears the forecaster predict fair weather. What is the posterior probability of fair weather?

**TABLE 9.15** Weather and Forecast

| Day's Weather | Forecast | | |
|---|---|---|---|
| | Fair | Inclement | Uncertain |
| Fair | 0.7 | 0.2 | 0.1 |
| Inclement | 0.3 | 0.6 | 0.1 |

[6]Samuel Goldberg, *Probability, An Introduction* (Prentice-Hall, Inc., 1960, Dover Publications, Inc., 1986), pp. 97–98. Adapted by permission of the author.

[7]Samuel Goldberg, *Probability, An Introduction* (Prentice-Hall, Inc., 1960, Dover Publications, Inc., 1986), pp. 99–100. Adapted by permission of the author.

# 9.8 REVIEW

## IMPORTANT TERMS AND SYMBOLS

**Section 9.1**   tree diagram   basic counting principle   permutation, $_nP_r$

**Section 9.2**   combination, $_nC_r$   permutation with repeated objects   cells

**Section 9.3**   sample space   sample point   finite sample space   event   certain event
impossible event   simple event   Venn diagram   complement, $E'$   union, $\cup$
intersection, $\cap$   mutually exclusive events

**Section 9.4**   equally likely outcomes   trial   relative frequency   equiprobable space
probability of event, $P(E)$   addition law for mutually exclusive events   empirical probability   odds

**Section 9.5**   conditional probability, $P(E|F)$   reduced sample space   general multiplication law (law of
compound probability)   trial (stage)   compound experiment   probability tree

**Section 9.6**   independent events   dependent events   special multiplication law

**Section 9.7**   partition   prior probability   posterior probability   Bayes's formula
Bayes's probability tree

## SUMMARY

It is important to know the number of ways a procedure can occur. Suppose a procedure involves a sequence of $k$ stages. Let $n_1$ be the number of ways the first stage can occur, and let $n_i$ be the the number of ways the $i$th stage can occur after the first $i - 1$ stages have occurred, for $i = 2, 3, \ldots, k$. Then the number of ways the procedure can occur is

$$n_1 \cdot n_2 \cdots n_k.$$

This result is called the basic counting principle.

An ordered arrangement of $r$ objects, without repetition, selected from $n$ distinct objects is called a permutation of the $n$ objects taken $r$ at a time. The number of such permutations is denoted $_nP_r$ and is given by

$$_nP_r = \underbrace{n(n-1)(n-2) \cdots (n-r+1)}_{r \text{ factors}} = \frac{n!}{(n-r)!}.$$

If the arrangement is made without regard to order, then it is called a combination of $n$ objects taken $r$ at a time. The number of such combinations is denoted $_nC_r$ and is given by

$$_nC_r = \frac{n!}{r!(n-r)!}.$$

When some of the objects are repeated, the number of distinguishable permutations of $n$ objects, such that $n_1$ are of one type, $n_2$ are of a second type, . . . , and $n_k$ are of the $k$th type, is

$$\frac{n!}{n_1! n_2! \cdots n_k!}, \tag{1}$$

where $n_1 + n_2 + \cdots + n_k = n$.

The expression in Eq. (1) can also be used to determine the number of assignments of objects to cells. If $n$ distinct objects are placed into $k$ ordered cells, with $n_i$ objects in cell $i$ ($i = 1, 2, \ldots, k$), then the number of such assignments is

$$\frac{n!}{n_1! n_2! \cdots n_k!},$$

where $n_1 + n_2 + \cdots + n_k = n$.

A sample space for an experiment is a set $S$ of all possible outcomes of the experiment. These outcomes are called sample points. A subset $E$ of $S$ is called an event. Two special events are the sample space itself, which is a certain event, and the empty set, which is an impossible event. An event consisting of a single sample point is called a simple event. Two events are said to be mutually exclusive when they have no sample point in common.

A sample space whose outcomes are equally likely is called an equiprobable space. If $E$ is an event for a finite equiprobable space $S$, then the probability that $E$ occurs is given by

$$P(E) = \frac{n(E)}{n(S)}.$$

If $F$ is also an event in $S$, we have

$$P(E \cup F) = P(E) + P(F) - P(E \cap F),$$

$$P(E \cup F) = P(E) + P(F), \quad \text{if } E \text{ and } F \text{ are} \\ \text{mutually exclusive,}$$

$$P(E') = 1 - P(E),$$

$$P(S) = 1,$$

$$P(\varnothing) = 0.$$

For an event $E$, the ratio

$$\frac{P(E)}{P(E')} \left( \text{or equivalently, } \frac{P(E)}{1 - P(E)} \right),$$

gives the odds that $E$ occurs. Conversely, if the odds that $E$ occurs are $a{:}b$, then

$$P(E) = \frac{a}{a + b}.$$

The probability that an event $E$ occurs, given that event $F$ has occurred, is called a conditional probability. It is denoted by $P(E|F)$ and can be computed either by considering a reduced equiprobable sample space and using the formula

$$P(E|F) = \frac{n(E \cap F)}{n(F)},$$

or from the formula

$$P(E|F) = \frac{P(E \cap F)}{P(F)},$$

which involves probabilities with respect to the original sample space.

To find the probability that two events both occur, we may use the general multiplication law:

$$P(E \cap F) = P(E)P(F|E) = P(F)P(E|F).$$

Here we multiply the probability that one of the events occurs by the conditional probability that the other one occurs, given that the first has occurred. For more than two events, the corresponding law is

$$P(E_1 \cap E_2 \cap \cdots \cap E_n)$$
$$= P(E_1)P(E_2|E_1)P(E_3|E_1 \cap E_2) \cdots$$
$$P(E_n|E_1 \cap E_2 \cap \cdots \cap E_{n-1}).$$

The general multiplication law is also called the law of compound probability, because it is useful when applied to a compound experiment—one that can be expressed as a sequence of two or more other experiments, called trials or stages.

When we analyze a compound experiment, a probability tree is extremely useful in keeping track of the possible outcomes for each trial of the experiment. A path is a complete sequence of branches from the start to a tip of the tree. Each path represents an outcome of the compound experiment, and the probability of that path is the product of the probabilities for the branches of the path.

Events $E$ and $F$ are independent when the occurrence of one of them does not affect the probability of the other; that is,

$$P(E|F) = P(E) \quad \text{or} \quad P(F|E) = P(F).$$

Evnts that are not independent are dependent.

If $E$ and $F$ are independent, the general multiplication law simplifies into the special multiplication law:

$$P(E \cap F) = P(E)P(F).$$

Here the probability that $E$ and $F$ both occur is the probability of $E$ times the probability of $F$. The preceding equation forms the basis of an alternative definition of independence: Events $E$ and $F$ are independent if and only if

$$P(E \cap F) = P(E)P(F).$$

Three or more events are independent if and only if, for each set of two or more of the events, the probability of the intersection of the events in that set is equal to the product of the probabilities of those events.

A partition divides a sample space into mutually exclusive events. If $E$ is an event and $F_1, F_2, \ldots, F_n$ is a partition, then, to find the conditional probability of event $F_i$, given $E$, when prior and conditional probabilities are known, we may use Bayes' formula:

$$P(F_i|E) = \frac{P(F_i)P(E|F_i)}{P(F_1)P(E|F_1) + P(F_2)P(E|F_2) + \cdots + P(F_n)P(E|F_n)}.$$

A Bayes'-type problem may also be solved with the aid of a Bayes' probability tree.

# REVIEW PROBLEMS

*In Problems 1–4, determine the values.*

**1.** $_8P_3$.

**2.** $_{20}P_1$.

**3.** $_9C_7$.

**4.** $_{12}C_4$.

**5. License Plate** A five-symbol license plate consists of two letters followed by three numbers, the first of which is not zero. How many different license plates are possible?

**6. Dinner** In a restaurant, a complete dinner consists of one appetizer, one entrée and one dessert. The choices for the appetizer are soup and salad; for the entrée, chicken, steak, lobster, and veal; and for the dessert, ice cream, pie, and pudding. How many complete dinners are possible?

**7. Garage-Door Opener** The transmitter for an electric garage-door opener transmits a coded signal to a receiver. The code is determined by five switches, each of which is either in an "on" or "off" position. Determine the number of different codes that may be transmitted.

**8. Baseball** A baseball manager must determine a batting order for his nine-member team. How many batting orders are possible?

**9. Softball** A softball league has seven teams. In terms of first, second, and third place, in how many ways can the season end? Assume that there are no ties.

**10. Trophies** In a trophy case, nine different trophies are to be placed—two on the top shelf, three on the middle, and four on the bottom. Considering the order of arrangement on each shelf, in how many ways can the trophies be placed in the case?

**11. Elevator** Because of crowding, five of eight people can enter an elevator. How many different groups can enter?

**12. Cards** From a 52-card deck of playing cards, a three-card hand is dealt. In how many ways can two of the cards be of the same face value and the other of a different face value?

**13. Light Bulbs** A carton contains 24 light bulbs, one of which is defective. (a) In how many ways can three bulbs be selected? (b) In how many ways can three bulbs be selected if one is defective?

**14. Multiple-Choice Exam** Each question of a 10-question multiple-choice examination has four choices, only one of which is correct, and is worth 10 points. By guessing, in how many ways is it possible to receive a score of 90 or better?

**15. Letter Arrangement** How many distinguishable horizontal arrangements of the letters in MISSISSIPPI are possible?

**16. Flag Signals** Colored flags arranged vertically on a flagpole indicate a signal (or message). How many different signals are possible if two red and three green flags are all used?

**17. Personnel Agency** A nursing personnel agency provides nurses on a temporary basis to hospitals that are short of staff. The manager has a pool of eight nurses and must send three to hospital A and two to hospital B. In how many ways can the manager make assignments?

**18. Tour Operator** A tour operator has three vans, and each can accommodate eight tourists. Suppose 16 people arrive for a city sightseeing tour and the operator will use only two vans. In how many ways can the operator assign the people to the vans?

**19.** Suppose $S = \{1, 2, 3, 4, 5, 6, 7, 8\}$ is the sample space and $E_1 = \{1, 2, 3, 4, 5, 6\}$ and $E_2 = \{4, 5, 6, 7\}$ are events for an experiment. Find (a) $E_1 \cup E_2$, (b) $E_1 \cap E_2$, (c) $E_1' \cup E_2$, (d) $E_1 \cap E_1'$, and (e) $(E_1 \cap E_2')'$. (f) Are $E_1$ and $E_2$ mutually exclusive?

**20. Die and Coin** A die is rolled and then a coin is tossed. (a) Determine a sample space for this experiment. Determine the events that (b) a 2 shows and (c) a head and an even number show.

**21. Urn** Three urns, labeled 1, 2, and 3, each contain two marbles, one red and the other green. A marble is selected at random from each urn. (a) Determine a sample space for this experiment. Determine the events that (b) exactly two marbles are red and (c) the marbles are the same color.

**22.** Suppose that $E_1$ and $E_2$ are events for an experiment with a finite number of sample points. If $P(E_1) = 0.6$, $P(E_1 \cup E_2) = 0.7$, and $P(E_1 \cap E_2) = 0.2$, find $P(E_2)$.

**23. Quality Control** A manufacturer of computer chips packages 10 chips to a box. For quality control, two chips are selected at random from each box and tested. If any one of the chips is defective, the entire box of chips is rejected for sale. For a box that contains exactly one defective chip, what is the probability that the box is rejected?

**24. Drugs** Each of 80 white mice was injected with one of four drugs, A, B, C, or D. Drug A was given to 25%, B to 20%, and C to 20%. If a mouse is chosen at random, determine the probability that it was injected with either C or D.

**25. Multiple-Choice Exam**   Each question on a five-question multiple-choice examination has four choices, only one of which is correct. If a student answers each question in a random fashion, what is the probability that the student answers exactly two questions incorrectly?

**26. Cola Preference**   To determine the national preference of cola drinkers, an advertising agency conducted a survey of 200 of them. Two cola brands, A and B, were involved. The results of the survey are indicated in Table 9.16. If a cola drinker is selected at random, determine the (empirical) probability that the person

**TABLE 9.16**  Cola Preference

| Like A only | 70 |
| Like B only | 80 |
| Like both A and B | 35 |
| Like neither A nor B | 15 |
| Total | 200 |

**a.** Likes both A and B.
**b.** Likes A, but not B.

**27. Urn**   An urn contains four red and six green marbles.

**a.** If two marbles are randomly selected in succession with replacement, determine the probability that both are red.
**b.** If the selection is made without replacement, determine the probability that both are red.

**28. Dice**   A pair of fair dice is rolled. Determine the probability that the sum of the numbers is (a) 4 or 5, (b) a multiple of 4, and (c) no less than 5.

**29. Cards**   Two cards from a standard deck of 52 playing cards are randomly drawn in succession with replacement. Determine the probability that (a) both cards are red and (b) one card is red and the other is a club.

**30. Cards**   Two cards from a standard deck of 52 playing cards are randomly drawn in succession without replacement. Determine the probability that (a) both are hearts and (b) one is an ace and the other is a red king.

*In Problems 31 and 32, for the given value of P(E), find the odds that E will occur.*

**31.** $P(E) = \dfrac{3}{8}$.        **32.** $P(E) = 0.98$.

*In Problems 33 and 34, the odds that E will occur are given. Find P(E).*

**33.** 6:1.        **34.** 3:4.

**35. Cards**   If a card is randomly drawn from a fair deck of 52 cards, find the probability that it is a face card (a jack, queen, or king), given that it is not a heart.

**36. Dice**   If two fair dice are rolled, find the probability that the sum is greater than 8, given that a 4 shows on at least one of the dice.

**37. Novel and TV Movie**   The probability that a particular novel will be successful is 0.6, and if it is successful, the probability that the rights will be purchased for a made-for-TV movie is 0.7. Find the probability that the novel will be successful and made into a TV movie.

**38. Cards**   Three cards are drawn from a standard deck of cards. Find the probability that the cards are, in order, a queen, a heart, and the ace of clubs if the cards are drawn with replacement.

**39. Dice**   If two dice are thrown, find each of the following.

**a.** The probability of getting a total of 7, given that a 4 occurred on at least one die.
**b.** The probability of getting a total of 7 and that a 4 occurred on at least one die.

**40. Die**   A fair die is tossed two times in succession. Find the probability that the first toss is less than 5, given that the total is greater than 9.

**41. Die**   If a fair die is tossed two times in succession, find the probability that the first number is greater than or equal to the second number, given that the second number is greater than 4.

**42. Cards**   Three cards are drawn without replacement from a standard deck of cards. Find the probability that the second card is a heart.

**43. Seasoning Survey**   A survey of 600 adults was made to determine whether or not they liked the taste of a new seasoning. The results are summarized in Table 9.17.

**TABLE 9.17**  Seasoning Survey

|  | **Like** | **Dislike** | **Total** |
|---|---|---|---|
| Male | 80 | 40 | 120 |
| Female | 320 | 160 | 480 |
| Total | 400 | 200 | 600 |

**a.** If a person in the survey is selected at random, find the probability that the person dislikes the seasoning ($L'$), given that the person is a female ($F$).

**b.** Determine whether the events $L = \{$liking the seasoning$\}$ and $M = \{$being a male$\}$ are independent or dependent.

**44. Chips** An urn contains six chips numbered from 1 to 6. Two chips are randomly drawn with replacement. Let $E$ be the event of getting a 4 on the first draw and $F$ be the event of getting a 4 on the second draw.

**a.** Are $E$ and $F$ mutually exclusive?

**b.** Are $E$ and $F$ independent?

**45. College and Family Income** A survey of 175 students resulted in the data shown in Table 9.18. The table shows the type of college the student attends and the income level of the student's family. If a student is selected at random, determine whether the event of attending a public college and the event of coming from a middle-class family are independent or dependent.

**TABLE 9.18**  Student Survey

| Income | College | | |
|---|---|---|---|
| | Private | Public | Total |
| High | 15 | 10 | 25 |
| Middle | 25 | 55 | 80 |
| Low | 10 | 60 | 70 |
| Total | 50 | 125 | 175 |

**46.** If $P(E) = \dfrac{1}{2}$, $P(F) = \dfrac{7}{12}$, and $P(E|F) = \dfrac{3}{7}$, find $P(E \cup F)$.

**47. Shrubs** When a certain type of shrub is planted, the probability that it will take root is 0.7. If four shrubs are planted, find each of the following. Assume independence.

**a.** The probability that none of them take root.

**b.** The probability that exactly two of them take root.

**c.** The probability that at most two of them take root.

**48. Antibiotic** A certain antibiotic is effective for 80% of the people who take it. Suppose three persons take this drug. What is the probability that it will be effective for at least two of them? Assume independence.

**49. Urns and Marbles** Urn I contains three green and two red marbles, and Urn II contains four red, two green, and two white marbles. A marble is randomly drawn from Urn I and placed into Urn II. If a marble is then randomly drawn from Urn II, find the probability that the marble is red.

**50. Urns and Marbles** Urn I contains 4 red and 2 white marbles. Urn II contains 2 red and 3 white marbles. An urn is chosen at random, and then a marble is randomly drawn from it.

**a.** What is the probability that the marble is white?

**b.** If the marble is white, what is the probability that it was drawn from Urn II?

**51. Grade Distribution** Last semester, the grade distribution for a certain class taking an upper-level college course was analyzed. It was found that the proportion of students receiving a grade of A was 0.4 and the proportion getting an A and being a graduate student was 0.1. If a student is randomly selected from this class and is found to have received an A, find the probability that the student is a graduate student.

**52. Alumni Reunion** At the most recent alumni day at Omega College, 500 persons attended. Of these, 400 lived within the state, and 40 percent of them were attending for the first time. Among the alumni who lived out of the state, 70 percent were attending for the first time. That day a raffle was held, and the person who won had also won it the year before. Find the probability that the winner was from out of state.

**53. Quality Control** A manufacturer produces widgits on two shifts. The first shift produces 300 units of product per day, and the second produces 200. From past experience, it is believed that, of the output produced by the first and second shifts, 1 and 2 percent are defective, respectively. At the end of a day, a unit was selected at random from the total production.

**a.** Find the probability that the unit is defective.

**b.** If the unit is defective, find the probability that it came from the second shift.

**54. Aptitude Test** In the past, a company has hired only experienced personnel for its word-processing department. Because of a shortage in this field, the company has decided to hire inexperienced persons and will provide on-the-job training. It has supplied an employment agency with a new aptitude test that has been designed for applicants who desire such a training position. Of those who recently took the test, 35 percent passed. In order to gauge the effectiveness of the test, everyone who took the test was put in the training program. Of those who passed, 80 percent performed satisfactorily, whereas of those who failed, only 30 percent did satisfactorily. If one of the new trainees is selected at random and is found to be satisfactory, what is the probability that the person passed the exam?

# ADDITIONAL TOPICS IN PROBABILITY

**To develop the probability distribution of a random variable and to represent that distribution geometrically by a graph or a histogram. To compute the mean, variance, and standard deviation of a random variable.**

## 10.1 DISCRETE RANDOM VARIABLES AND EXPECTED VALUE

With some experiments, we are interested in events associated with numbers. For example, if two coins are tossed, our interest may be in the *number* of heads that occur. Thus, we consider the events

$$\{0\}, \quad \{1\}, \quad \{2\}.$$

If we let $X$ be a variable that represents the number of heads that occur, then the only values that $X$ can assume are 0, 1, and 2. The value of $X$ is determined by the outcome of the experiment, and hence by chance. In general, a variable whose values depend on the outcome of a random process is called a **random variable.** Usually, random variables are denoted by capital letters such as $X$, $Y$, or $Z$, and the values that these variables assume may be denoted by corresponding lowercase letters ($x$, $y$, or $z$). Thus, for the number of heads ($X$) that occur in the tossing of two coins, we may indicate the possible values by writing

$$X = x, \quad \text{where } x = 0, 1, 2,$$

or, more simply,

$$X = 0, 1, 2.$$

### EXAMPLE 1 Random Variables

**a.** Suppose a die is rolled and $X$ is the number that turns up. Then $X$ is a random variable and $X = 1, 2, 3, 4, 5, 6$.

**b.** Suppose a coin is successively tossed until a head appears. If $Y$ is the number of such tosses, then $Y$ is a random variable and

$$Y = y, \quad \text{where } y = 1, 2, 3, 4, \ldots.$$

Note that $Y$ may assume infinitely many values.

**c.** A student is taking an exam with a one-hour time limit. If $X$ is the number of minutes it takes to complete the exam, then $X$ is a random variable. The values that $X$ may assume form the interval $(0, 60]$. That is, $0 < X \le 60$. ∎

A random variable is called a **discrete random variable** if it may assume only a finite number of values or if its values can be placed in one-to-one correspondence with the positive integers. In Examples 1(a) and 1(b), $X$ and $Y$ are discrete. A random variable is called a **continuous random variable** if it may assume any value in some interval or intervals, such as $X$ does in Example 1(c). In this chapter, we shall be concerned with discrete random variables; Chapter 18 deals with continuous random variables.

If $X$ is a random variable, the probability of the event that $X$ assumes the value $x$ is denoted $P(X = x)$. Similarly, we can consider the probabilities of events such as $X \leq x$ and $X > x$. If $X$ is discrete, then the function $f$ that assigns the number $P(X = x)$ to each possible value of $X$ is called the **probability function,** the **probability distribution,** or—more simply—the **distribution** of the random variable $X$. Thus,

$$f(x) = P(X = x).$$

### EXAMPLE 2  Distribution of a Random Variable

*Suppose that $X$ is the number of heads that appear on the toss of two well-balanced coins. Determine the distribution of $X$.*

*Solution:* We must find the probabilities of the events $X = 0$, $X = 1$, and $X = 2$. The usual equiprobable sample space is

$$S = \{\text{HH, HT, TH, TT}\}.$$

Hence,

the event $X = 0$ is $\{\text{TT}\}$,

the event $X = 1$ is $\{\text{HT, TH}\}$,

the event $X = 2$ is $\{\text{HH}\}$.

The probability for each of these events is given in the **probability table** in the margin. If $f$ is the distribution for $X$, that is, $f(x) = P(X = x)$, then

$$f(0) = \tfrac{1}{4}, \ f(1) = \tfrac{1}{2}, \ \text{and} \ f(2) = \tfrac{1}{4}. \qquad \blacksquare$$

In Example 2, the distribution $f$ was indicated by the listing
$$f(0) = \tfrac{1}{4}, \ f(1) = \tfrac{1}{2}, \ \text{and} \ f(2) = \tfrac{1}{4}.$$

However, the probability table for $X$ gives the same information and is an acceptable way of expressing the probability distribution. Another way is by the graph of the distribution, as shown in Fig. 10.1. The vertical lines from the $x$-axis to the points on the graph merely emphasize the heights of the points. Another representation of the distribution of $X$ is the rectangle diagram in Fig. 10.2, called the **probability histogram** for $X$. Here a rectangle is centered over each value of $X$. The rectangle above $x$ has width 1 and height $P(X = x)$. Thus, its *area* is the probability $1 \cdot P(X = x) = P(X = x)$. This interpretation of probability as an area is important in Chapter 18.

Note in Example 2 that the sum of $f(0), f(1)$, and $f(2)$ is 1:

$$f(0) + f(1) + f(2) = \tfrac{1}{4} + \tfrac{1}{2} + \tfrac{1}{4} = 1.$$

This must be the case, because the events $X = 0$, $X = 1$, and $X = 2$ are mutually exclusive and the union of all three is the sample space [and $P(S) = 1$].

**Probability Table**

| $x$ | $P(X = x)$ |
|-----|------------|
| 0 | $\tfrac{1}{4}$ |
| 1 | $\tfrac{2}{4} = \tfrac{1}{2}$ |
| 2 | $\tfrac{1}{4}$ |

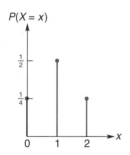

**FIGURE 10.1** Graph of the distribution of $X$.

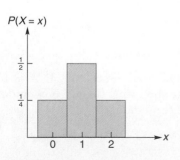

**FIGURE 10.2** Probability histogram for $X$.

We can conveniently indicate the sum $f(0) + f(1) + f(2)$ by the notation

$$\sum_x f(x),$$

which is called *sigma notation*[1] because the Greek letter $\Sigma$ (sigma) is used. Here $\sum_x f(x)$ means that we are to sum all terms of the form $f(x)$ for all values of $x$, which in this case are 0, 1, and 2. Thus,

$$\sum_x f(x) = f(0) + f(1) + f(2).$$

In general, for any probability distribution $f$, we have $0 \le f(x) \le 1$ for all $x$, and the sum of all function values is 1. Therefore,

$$\sum_x f(x) = 1.$$

This means that in any probability histogram, the sum of the areas of the rectangles is 1.

The probability distribution for a random variable $X$ gives the relative frequencies of the values of $X$ in the long run. However, it is often useful to determine the "average" value of $X$ in the long run. In Example 2, for instance, suppose that the two coins were tossed $n$ times, which resulted in $X = 0$ occurring $k_0$ times, $X = 1$ occurring $k_1$ times, and $X = 2$ occurring $k_2$ times. Then the average value of $X$ for these $n$ tosses is

$$\frac{0 \cdot k_0 + 1 \cdot k_1 + 2 \cdot k_2}{n},$$

or, equivalently,

$$0 \cdot \frac{k_0}{n} + 1 \cdot \frac{k_1}{n} + 2 \cdot \frac{k_2}{n}.$$

But the fractions $k_0/n$, $k_1/n$ and $k_2/n$ are the relative frequencies of the events $X = 0$, $X = 1$, and $X = 2$, respectively, that occur in the $n$ tosses. If $n$ is very large, then these relative frequencies approach the probabilities of the events $X = 0$, $X - 1$, and $X = 2$. Thus, it seems reasonable that the average value of $X$ in the long run is

$$0 \cdot f(0) + 1 \cdot f(1) + 2 \cdot f(2) = 0 \cdot \tfrac{1}{4} + 1 \cdot \tfrac{1}{2} + 2 \cdot \tfrac{1}{4} + 1. \tag{1}$$

This means that if we tossed the coins many times, the average number of heads appearing per toss is very close to 1. We define the sum in Eq. (1) to be the *mean, expected value,* or *expectation of $X$* and denote it by $\mu$ (the Greek letter "mu") or $E(X)$. Note that from Eq. (1), $\mu$ has the form $\sum_x x f(x)$. In general, we have the following definition.

**DEFINITION**

*If $X$ is a discrete random variable with probability distribution $f$, then the **mean** (or **expected value** or **expectation**) of $X$, denoted by $\mu$, or E(X), is given by*

$$\mu = E(X) = \sum_x x f(x)$$

The mean of $X$ can be interpreted as the average value of $X$ in the long run. The mean does not necessarily have to be an outcome of the experiment.

---

[1] A more thorough discussion of sigma notation occurs in Sec. 16.5.

**EXAMPLE 3   Expected Gain**

*An insurance company offers an $80,000 catastrophic fire insurance policy to homeowners of a certain type of house. The policy provides protection in the event that such a house is totally destroyed by fire in a one-year period. The company has determined that the probability of such an event is 0.0002. If the annual policy premium is $52, find the expected gain per policy for the company.*

*Strategy:* If an insured house does not suffer a catastrophic fire, the company gains $52. However, if there is such a fire, the company loses $80,000 − $52 (insured value of house minus premium), or $79,948. If $X$ is the gain (in dollars) to the company, then $X$ is a random variable that may assume the values 52 and −79,948. (A loss is considered a negative gain.) The expected gain per policy for the company is the expected value of $X$.

*Solution:* If $f$ is the probability function for $X$, then

$$f(-79{,}948) = P(X = -79{,}948) = 0.0002$$

and

$$f(52) = P(X = 52) = 1 - 0.0002 = 0.9998.$$

The expected value of $X$ is given by

$$E(X) = \sum_x xf(x) = -79{,}948f(-79{,}948) + 52f(52)$$

$$= -79{,}948(0.0002) + 52(0.9998) = 36.$$

Thus, if the company sold many policies, it could expect to gain approximately $36 per policy, which could be applied to such expenses as advertising, overhead, and profit.   ■

Since $E(X)$ is the average value of $X$ in the long run, it is a measure of the central tendency of $X$. However, $E(X)$ does not indicate the dispersion or spread of $X$ from the mean in the long run. For example, Fig. 10.3 shows the graphs of two probability distributions, $f$ and $g$, for the random variables $X$ and $Y$. It can easily be demonstrated that both $X$ and $Y$ have the same mean:

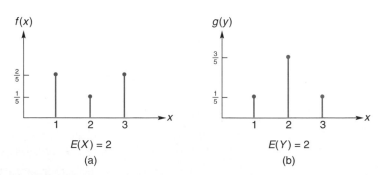

**FIGURE 10.3**   Probability distributions.

$E(X) = 2$ and $E(Y) = 2$. You should verify these results. But from Fig. 10.3, $X$ is more likely to assume the values 1 or 3 than is $Y$, because $f(1)$ and $f(3)$ are $\frac{2}{5}$, whereas $g(1)$ and $g(3)$ are $\frac{1}{5}$. Thus, $X$ has more likelihood of assuming values away from the mean than does $Y$, so there is more dispersion for $X$ in the long run.

There are various ways to measure dispersion for a random variable $X$. One way is to determine the long-run average of the absolute values of the deviations from the mean $\mu$—that is, $E(|X - \mu|)$:[2]

$$E(|X - \mu|) = \sum_x |x - \mu| f(x).$$

However, since absolute value is involved here, $E(|X - \mu|)$ is a mathematically awkward expression.

Although many other measures of dispersion can be considered, two are most widely accepted. One is the *variance,* and the other (which is related to the variance) is the *standard deviation.* The **variance of $X$,** denoted by $\mathrm{Var}(X)$, is the long-run average of the *squares* of the deviations of $X$ from $\mu$:[3]

---

**Variance of $X$**

$$\mathrm{Var}(X) = E[(X - \mu)^2] = \sum_x (x - \mu)^2 f(x). \qquad (2)$$

---

Since $(X - \mu)^2$ is involved in $\mathrm{Var}(X)$, and both $X$ and $\mu$ have the same units of measurement, the units for $\mathrm{Var}(X)$ are those of $X^2$. For instance, in Example 3, $X$ is in dollars; thus, $\mathrm{Var}(X)$ has units of dollars squared. It is convenient to have a measure of dispersion in the same units as $X$. Such a measure is $\sqrt{\mathrm{Var}(X)}$, which is called the **standard deviation of $X$** and is denoted by $\sigma$ (the lowercase Greek letter "sigma"):

---

**Standard Deviation of $X$**

$$\sigma = \sqrt{\mathrm{Var}(X)}.$$

---

Note that $\sigma$ has the property that

$$\sigma^2 = \mathrm{Var}(X).$$

Both $\mathrm{Var}(X)$ [or $\sigma^2$] and $\sigma$ are measures of the dispersion of $X$. The greater the value of $\mathrm{Var}(X)$, or $\sigma$, the greater is the dispersion. One result of a famous theorem, *Chebyshev's inequality,* is that the probability of $X$ falling within two standard deviations of the mean is at least $\frac{3}{4}$.

We can write the formula for variance in Eq. (2) in a different way. Suppose the possible values of $X$ are $x_1$ and $x_2$, and $f$ is the probability function for $X$. Then by Eq. (2),

$$
\begin{aligned}
\mathrm{Var}(X) &= \sum_x (x - \mu)^2 f(x) \\
&= (x_1 - \mu)^2 f(x_1) + (x_2 - \mu)^2 f(x_2) \\
&= (x_1^2 - 2x_1\mu + \mu^2) f(x_1) + (x_2^2 - 2x_2\mu + \mu^2) f(x_2) \\
&= x_1^2 f(x_1) - 2x_1\mu f(x_1) + \mu^2 f(x_1) + x_2^2 f(x_2) - 2x_2\mu f(x_2) + \mu^2 f(x_2).
\end{aligned}
$$

Regrouping and factoring, we have

$$
\begin{aligned}
\mathrm{Var}(X) &= [x_1^2 f(x_1) + x_2^2 f(x_2)] - 2\mu[x_1 f(x_1) + x_2 f(x_2)] + \mu^2[f(x_1) + f(x_2)] \\
&= \sum_x x^2 f(x) - 2\mu \sum_x x f(x) + \mu^2 \sum_x f(x).
\end{aligned}
$$

---

[2]It can be shown that if $Y = g(X)$, then $E(Y) = \sum_x g(x) f(x)$, where $f$ is the probability function for $X$. For example, if $Y = |X - \mu|$, then $E(|X - \mu|) = \sum_x |x - \mu| f(x)$.

[3]By the preceding footnote, if $Y = (X - \mu)^2$, then $E[(X - \mu)^2] = \sum_x (x - \mu)^2 f(x)$.

But $\sum_x x^2 f(x) = E(X^2),$[4] $\sum_x xf(x) = \mu$ and $\sum_x f(x) = 1.$ Therefore,

$$\text{Var}(X) = \sigma^2 = E(X^2) - 2\mu^2 + \mu^2,$$

or

$$\text{Var}(X) = \sigma^2 = E(X^2) - \mu^2$$
$$= \sum_x x^2 f(x) - \mu^2. \tag{3}$$

This formula for variance is quite useful, since it usually simplifies computations.

### EXAMPLE 4   Mean, Variance, and Standard Deviation

*An urn contains 10 marbles, each of which shows a number. Five marbles show 1, two show 2, and three show 3. A marble is drawn at random. If X is the number that shows, determine $\mu$, Var(X), and $\sigma$.*

***Solution:*** The sample space consists of 10 equally likely outcomes (the marbles). The values that $X$ can assume are 1, 2, and 3. The events $X = 1$, $X = 2$, and $X = 3$ contain 5, 2, and 3 sample points, respectively. Thus, if $f$ is the probability function for $X$,

$$f(1) = P(X = 1) = \tfrac{5}{10} = \tfrac{1}{2},$$

$$f(2) = P(X = 2) = \tfrac{2}{10} = \tfrac{1}{5},$$

$$f(3) = P(X = 3) = \tfrac{3}{10}.$$

Calculating the mean gives

$$\mu = \sum_x xf(x) = 1 \cdot f(1) + 2 \cdot f(2) + 3 \cdot f(3)$$

$$= 1 \cdot \tfrac{5}{10} + 2 \cdot \tfrac{2}{10} + 3 \cdot \tfrac{3}{10} = \tfrac{18}{10} = \tfrac{9}{5}.$$

To find $\text{Var}(X)$, either Eq. (2) or Eq. (3) can be used. Both will be used here, so that we can compare the arithmetical computations involved. By Eq. (2),

$$\text{Var}(X) = \sum_x (x - \mu)^2 f(x)$$

$$= \left(1 - \frac{9}{5}\right)^2 f(1) + \left(2 - \frac{9}{5}\right)^2 f(2) + \left(3 - \frac{9}{5}\right)^2 f(3)$$

$$= \left(-\frac{4}{5}\right)^2 \cdot \frac{5}{10} + \left(\frac{1}{5}\right)^2 \cdot \frac{2}{10} + \left(\frac{6}{5}\right)^2 \cdot \frac{3}{10}$$

$$= \frac{16}{25} \cdot \frac{5}{10} + \frac{1}{25} \cdot \frac{2}{10} + \frac{36}{25} \cdot \frac{3}{10}$$

$$= \frac{80 + 2 + 108}{250} = \frac{190}{250} = \frac{19}{25}.$$

---

[4]This is a consequence of a preceding footnote.

By Eq. (3),

$$Var(X) = \sum_x x^2 f(x) - \mu^2$$

$$= 1^2 \cdot f(1) + 2^2 \cdot f(2) + 3^2 \cdot f(3) - \left(\frac{9}{5}\right)^2$$

$$= 1 \cdot \frac{5}{10} + 4 \cdot \frac{2}{10} + 9 \cdot \frac{3}{10} - \frac{81}{25}$$

$$= \frac{5 + 8 + 27}{10} - \frac{81}{25} = \frac{40}{10} - \frac{81}{25}$$

$$= 4 - \frac{81}{25} = \frac{19}{25}.$$

Notice that Eq. (2) involves $(x - \mu)^2$, but Eq. (3) involves $x^2$. Because of this, it is often easier to compute variances by Eq. (3) than by Eq. (2).

Since $\sigma^2 = Var(X) = \frac{19}{25}$, the standard deviation is

$$\sigma = \sqrt{Var(X)} = \sqrt{\frac{19}{25}} = \frac{\sqrt{19}}{5}.$$

■

## ■ Exercise 10.1

*In Problems 1–4, the distribution of the random variable X is given. Determine μ, Var(X), and σ. In Problem 1, construct the probability histogram. In Problem 2, graph the distribution.*

**1.** $f(0) = 0.1, f(1) = 0.4, f(2) = 0.2, f(3) = 0.3$.

**2.** $f(4) = 0.4, f(5) = 0.6$.

**3.** See Fig. 10.4

**4.** See Fig. 10.5

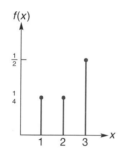

**FIGURE 10.4**
Distribution for
Problem 3.

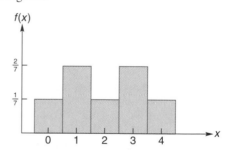

**FIGURE 10.5** Distribution for
Problem 4.

**5.** The random variable $X$ has the following distribution:

| x | P(X = x) |
|---|---|
| 2 | |
| 4 | 0.5 |
| 7 | 0.4 |

**a.** Find $P(X = 2)$.
**b.** Find $\mu$.
**c.** Find $\sigma^2$.

**6.** The random variable $X$ has the following distribution:

| x | P(X = x) |
|---|---|
| 2 | 2n |
| 4 | n |
| 6 | 0.4 |

**a.** Find $P(X = 2)$ and $P(X = 4)$.
**b.** Find $\mu$.

*In Problems 7–10, determine E(X), σ², and σ for the random variable X.*

7. **Coin Toss** Three fair coins are tossed. Let $X$ be the number of heads that occur.

8. **Marbles in Urn** An urn contains six marbles, each of which shows a number. Four marbles show a 1 and two show a 2. A marble is randomly selected and the number that shows, $X$, is observed.

9. **Committee** From a group of two women and three men, two persons are selected at random to form a committee. Let $X$ be the number of women on the committee.

10. **Marbles in Urn** An urn contains two red and three green marbles. Two marbles are randomly drawn in succession with replacement, and the number of red marbles, $X$, is observed.

11. **Marbles in Urn** An urn contains three red and two white marbles. Two marbles are randomly drawn in succession without replacement. Let $X$ = the number of red marbles drawn. Find the distribution $f$ for $X$.

12. **Subcommittee** From a student council committee consisting of two juniors and five seniors, a subcommittee of two students is to be randomly selected. Let $X$ be the number of juniors in the subcommittee. Find a general formula, in terms of combinations, that gives $P(X = x)$, where $x = 0, 1, 2$.

13. **Raffle** A charitable organization is having a raffle for a single prize of $8500. Each raffle ticket costs $1 and 10,000 tickets have been sold.

   a. Find the expected gain for the purchaser of a single ticket.

   b. Find the expected gain for the purchaser of two tickets.

14. **Coin Game** Consider the following game. You are to toss three fair coins. If three heads or three tails turn up, your friend pays you $10. If either one or two heads turn up, you must pay your friend $6. What are your expected winnings or losses per game?

15. **Earnings** A landscaper earns $200 per day when working and loses $30 per day when not working. If the probability of working on any day is $\frac{4}{7}$, find the landscaper's expected daily earnings.

16. **Fast-Food Restaurant** A fast-food chain estimates that if it opens a restaurant in a shopping center, the probability that the restaurant is successful is 0.65. A successful restaurant earns an annual profit of $75,000; a restaurant that is not successful loses $20,000. What is the expected gain to the chain if it opens a restaurant in a shopping center?

17. **Insurance** An insurance company offers a hospitalization policy to individuals in a certain group. For a one-year period, the company will pay $100 per day, up to a maximum of five days, for each day the policyholder is hospitalized. The company estimates that the probability that any person in this group is hospitalized for exactly one day is 0.001; for exactly two days, 0.002; for exactly three days, 0.003; for exactly four days, 0.004; and for five

or more days, 0.008. Find the expected gain per policy to the company if the annual premium is $10.

18. **Demand** The following table gives the probability that $X$ units of a company's product are demanded weekly:

| $x$ | 0 | 1 | 2 | 3 | 4 | 5 |
|---|---|---|---|---|---|---|
| $P(X = x)$ | 0.05 | 0.20 | 0.40 | 0.24 | 0.10 | 0.01 |

Determine the expected weekly demand.

19. **Insurance Premium** In Example 3, if the company wants an expected gain of $50 per policy, determine the annual premium.

20. **Roulette** In the game of roulette, there is a wheel with 37 slots numbered with the integers from 0 to 36, inclusive. A player bets $1 (for example) and chooses a number. The wheel is spun and a ball rolls on the wheel. If the ball lands in the slot showing the chosen number, the player receives the $1 bet plus $35. Otherwise, the player loses the $1 bet. Assume that all numbers are equally likely, and determine the expected gain or loss per play.

21. **Coin Game** Suppose that you pay $1.25 to play a game in which two fair coins are tossed. You receive the number of dollars equal to the number of heads that occur. What is your expected gain (or loss) on each play? The game is said to be *fair* to you when your expected gain is $0. What should you pay to play if this is to be a fair game?

**To develop the binomial distribution and relate it to the binomial theorem.**

## 10.2 THE BINOMIAL DISTRIBUTION

### Binomial Theorem

Later in this section you will see that the terms in the expansion of a power of a binomial are useful in describing the probability distributions of certain random variables. It is worthwhile, therefore, to first discuss the *binomial theorem,* which is a formula for expanding $(a + b)^n$, where $n$ is a positive integer.

Regardless of $n$, there are patterns in the expansion of $(a + b)^n$. To illustrate, we shall consider the *cube* of the binomial $a + b$. By successively applying the distributive law, we have

$$(a + b)^3 = [(a + b)(a + b)](a + b)$$
$$= [a(a + b) + b(a + b)](a + b)$$
$$= [aa + ab + ba + bb](a + b)$$
$$= aa(a + b) + ab(a + b) + ba(a + b) + bb(a + b)$$
$$= aaa + aab + aba + abb + baa + bab + bba + bbb, \quad (1)$$

so that

$$(a + b)^3 = a^3 + 3a^2b + 3ab^2 + b^3. \quad (2)$$

Three observations can be made about the right side of Eq. (2). First, notice that the number of terms is four, which is one more than the power to which $a + b$ is raised (3). Second, the first and last terms are the *cubes* of $a$ and $b$; the powers of $a$ *decrease* from left to right (from 3 to 0), and the powers of $b$ *increase* (from 0 to 3). Third, for each term, the sum of the exponents of $a$ and $b$ is 3, which is the power to which $a + b$ is raised.

Let us now focus on the coefficients of the terms in Eq. (2). Consider the coefficient of the $ab^2$-term. It is the number of terms in Eq. (1) that involve exactly two $b$'s, namely, 3. But let us see *why* there are three terms that involve two $b$'s. Notice in Eq. (1) that each term is the product of three numbers, each of which is either $a$ or $b$. Because of the distributive law, each of the three $a + b$ factors in $(a + b)^3$ contributes either an $a$ or $b$ to the term. Thus, the number of terms involving one $a$ and two $b$'s is equal to the number of ways of choosing two of the three factors to supply a $b$, namely, ${}_3C_2 = \dfrac{3!}{2!1!} = 3$. Similarly,

the coefficient of the $a^3$-term is ${}_3C_0$,

the coefficient of the $a^2b$-term is ${}_3C_1$,

and

the coefficient of the $b^3$-term is ${}_3C_3$.

Generalizing our results, we obtain a formula for expanding $(a + b)^n$, called the *binomial theorem.*

---

**Binomial Theorem**

If $n$ is a positive integer, then

$$(a + b)^n = {}_nC_0a^n + {}_nC_1a^{n-1}b + {}_nC_2a^{n-2}b^2$$
$$+ \cdots + {}_nC_{n-1}ab^{n-1} + {}_nC_nb^n.$$

The coefficients ${}_nC_r$ are called **binomial coefficients.**

**EXAMPLE 1  Binomial Theorem**

*Use the binomial theorem to expand* $(q + p)^4$.

*Solution:* Here $n = 4$, $a = q$, and $b = p$. Thus,

$$(q + p)^4 = {}_4C_0 q^4 + {}_4C_1 q^3 p + {}_4C_2 q^2 p^2 + {}_4C_3 q p^3 + {}_4C_4 p^4$$

$$= \frac{4!}{0!4!}q^4 + \frac{4!}{1!3!}q^3 p + \frac{4!}{2!2!}q^2 p^2 + \frac{4!}{3!1!}q p^3 + \frac{4!}{4!0!}p^4.$$

Recalling that $0! = 1$, we have

$$(q + p)^4 = q^4 + 4q^3 p + 6q^2 p^2 + 4q p^3 + p^4. \qquad \blacksquare$$

## Binomial Distribution

We now turn our attention to repeated trials of an experiment in which the outcome of any trial does not affect the outcome of any other trial. These are referred to as **independent trials.** For example, when a fair die is rolled five times, the outcome on one roll does not affect the outcome on any other roll. Here we have five independent trials of rolling a die. Together, these five trials can be considered as a five-stage compound experiment involving independent events, so we can use the special multiplication law of Sec. 9.6 to determine the probability of obtaining specific outcomes of the trials.

To illustrate, let us find the probability of getting exactly two 4's in the five rolls of the die. We shall consider getting a 4 as a *success* (S) and getting any of the other five numbers as a *failure* (F). For example, the sequence

SSFFF

denotes getting

4, 4, followed by three other numbers.

This sequence can be considered as the intersection of five independent events: success on the first trial, success on the second, failure on the third, and so on. Since the probability of success on any trial is $\frac{1}{6}$, and the probability of failure is $1 - \frac{1}{6} = \frac{5}{6}$, by the special multiplication law for the intersection of independent events, the probability of the sequence SSFFF occurring is

$$\frac{1}{6} \cdot \frac{1}{6} \cdot \frac{5}{6} \cdot \frac{5}{6} \cdot \frac{5}{6} = \left(\frac{1}{6}\right)^2 \left(\frac{5}{6}\right)^3.$$

In fact, this is the probability for *any* particular order of the two S's and three F's. Let us determine how many ways a sequence of two S's and three F's can be formed. Out of five trials, the number of ways of choosing the two trials for success is ${}_5C_2$. So the probability of getting exactly two 4's in the five rolls is

$${}_5C_2 \left(\frac{1}{6}\right)^2 \left(\frac{5}{6}\right)^3. \qquad (3)$$

If we denote the probability of success by $p$ and the probability of failure by $q \, (= 1 - p)$, then (3) takes the form

$${}_5C_2 p^2 q^3,$$

which is the term involving $p^2$ in the expansion of $(q + p)^5$.

More generally, consider the probability of getting exactly $x$ 4's in $n$ rolls of the die. Then $n - x$ of the rolls must be some other number. For a particular order, the probability is

$$p^x q^{n-x}.$$

Since the number of possible orders is $_nC_x$,

$$P(X = x) = {_nC_x}p^x q^{n-x},$$

which is a general expression for the terms in $(q + p)^n$. In summary, the distribution for $X$ (the number of 4's that occur in $n$ rolls) is given by the terms in $(q + p)^n$.

Whenever we have $n$ independent trials of an experiment in which each trial has only two possible outcomes (success and failure) and the probability of success in each trial remains the same, the trials are called **Bernoulli trials.** Because the distribution of the number of successes corresponds to the expansion of a power of a binomial, the experiment is called a **binomial experiment,** and the distribution of the number of successes is called a **binomial distribution.**

---

**Binomial Distribution**

If $X$ is the number of successes in $n$ independent trials of a binomial experiment with probability $p$ of success and $q$ of failure on any trial, then the distribution $f$ for $X$ is given by

$$f(x) = P(X = x) = {_nC_x}p^x q^{n-x},$$

where $x$ is an integer such that $0 \le x \le n$ and $q = 1 - p$. Any random variable with this distribution is called a **binomial random variable** and is said to have a **binomial distribution.** The mean and standard deviation of $X$ are given, respectively, by

$$\mu = np, \qquad \sigma = \sqrt{npq}.$$

---

**EXAMPLE 2   Binomial Distribution**

*Suppose $X$ is a binomial random variable with $n = 4$ and $p = \frac{1}{3}$. Find the distribution for $X$.*

***Solution:*** Here $q = 1 - p = 1 - \frac{1}{3} = \frac{2}{3}$. So we have

$$P(X = x) = {_nC_x}p^x q^{n-x}, \qquad x = 0, 1, 2, 3, 4.$$

Thus,

$$P(X = 0) = {_4C_0}\left(\frac{1}{3}\right)^0\left(\frac{2}{3}\right)^4 = \frac{4!}{0!4!} \cdot 1 \cdot \frac{16}{81} = 1 \cdot 1 \cdot \frac{16}{81} = \frac{16}{81},$$

$$P(X = 1) = {_4C_1}\left(\frac{1}{3}\right)^1\left(\frac{2}{3}\right)^3 = \frac{4!}{1!3!} \cdot \frac{1}{3} \cdot \frac{8}{27} = 4 \cdot \frac{1}{3} \cdot \frac{8}{27} = \frac{32}{81},$$

$$P(X = 2) = {_4C_2}\left(\frac{1}{3}\right)^2\left(\frac{2}{3}\right)^2 = \frac{4!}{2!2!} \cdot \frac{1}{9} \cdot \frac{4}{9} = 6 \cdot \frac{1}{9} \cdot \frac{4}{9} = \frac{8}{27},$$

$$P(X = 3) = {_4C_3}\left(\frac{1}{3}\right)^3\left(\frac{2}{3}\right)^1 = \frac{4!}{3!1!} \cdot \frac{1}{27} \cdot \frac{2}{3} = 4 \cdot \frac{1}{27} \cdot \frac{2}{3} = \frac{8}{81},$$

$$P(X = 4) = {_4C_4}\left(\frac{1}{3}\right)^4\left(\frac{2}{3}\right)^0 = \frac{4!}{4!0!} \cdot \frac{1}{81} \cdot 1 = 1 \cdot \frac{1}{81} \cdot 1 = \frac{1}{81}.$$

**Principles in Practice 1**
**Binomial Distribution**

Let $X$ be the number of persons out of four job applicants who are hired. If the probability of any one applicant being hired is 0.3, find the distribution of $X$.

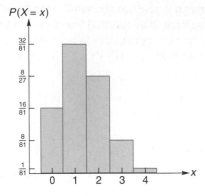

**FIGURE 10.6** Binomial distribution, $n = 4$, $p = \frac{1}{3}$.

The probability histogram for $X$ is given in Fig. 10.6. Note that the mean $\mu$ for $X$ is $np = 4(\frac{1}{3}) = \frac{4}{3}$, and the standard deviation is

$$\sigma = \sqrt{npq} = \sqrt{4 \cdot \frac{1}{3} \cdot \frac{2}{3}} = \sqrt{\frac{8}{9}} = \frac{2\sqrt{2}}{3}.$$

**EXAMPLE 3** **At Least Two Heads in Eight Coin Tosses**

*A fair coin is tossed eight times. Find the probability of getting at least two heads.*

*Solution:* If $X$ is the number of heads that occur, then $X$ has a binomial distribution with $n = 8$, $p = \frac{1}{2}$, and $q = \frac{1}{2}$. To simplify our work, we use the fact that

$$P(X \geq 2) = 1 - P(X < 2).$$

Now,

$$P(X < 2) = P(X = 0) + P(X = 1)$$

$$= {}_8C_0\left(\frac{1}{2}\right)^0\left(\frac{1}{2}\right)^8 + {}_8C_1\left(\frac{1}{2}\right)^1\left(\frac{1}{2}\right)^7$$

$$= \frac{8!}{0!8!} \cdot 1 \cdot \frac{1}{256} + \frac{8!}{1!7!} \cdot \frac{1}{2} \cdot \frac{1}{128}$$

$$= 1 \cdot 1 \cdot \frac{1}{256} + 8 \cdot \frac{1}{2} \cdot \frac{1}{128} = \frac{9}{256}.$$

Thus,

$$P(X \geq 2) = 1 - \frac{9}{256} = \frac{247}{256}.$$

A probability histogram for $X$ is given in Fig. 10.7.

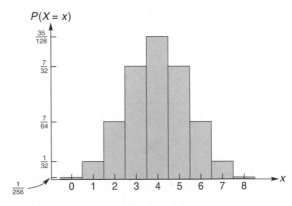

**FIGURE 10.7** Binomial distribution, $n = 8$, $p = \frac{1}{2}$.

**EXAMPLE 4** **Income Tax Audit**

*For a particular group of individuals, 20% of their income tax returns are audited each year. Of five randomly chosen individuals, what is the probability that exactly two will have their returns audited?*

*Solution:* We shall consider this to be a binomial experiment with five trials (selecting an individual). Actually, the experiment is not truly binomial, because selecting an individual from this group affects the probability that another individual's return will be audited. For example, if there are 5000 individuals, then 20%, or 1000, will be audited. The probability that the first individual selected will be audited is $\frac{1000}{5000}$. If that event occurs, the probability that the second individual selected will be audited is $\frac{999}{4999}$. Thus, the trials are not independent. However, we assume that the number of individuals is large, so that for all practical purposes, the probability of auditing an individual remains constant from trial to trial.

For each trial, the two outcomes are *being audited* and *not being audited*. Here we shall define a success as being audited. Letting $X$ be the number of returns audited, $p = 0.2$, and $q = 1 - 0.2 = 0.8$, we have

$$P(X = 2) = {}_5C_2(0.2)^2(0.8)^3 = \frac{5!}{2!3!}(0.04)(0.512)$$

$$= 10(0.04)(0.512) = 0.2048. \qquad \blacksquare$$

## ■ Exercise 10.2

*In Problems 1–4, determine the distribution f for the binomial random variable X if the number of trials is n and the probability of success on any trial is p. Also, find μ and σ.*

**1.** $n = 2, \quad p = \frac{1}{4}$.

**2.** $n = 3, \quad p = \frac{1}{2}$.

**3.** $n = 3, \quad p = \frac{2}{3}$.

**4.** $n = 4, \quad p = 0.4$.

*In Problems 5–10, determine the given probability if X is a binomial random variable, n is the number of trials, and p is the probability of success on any trial.*

**5.** $P(X = 5); \quad n = 6, p = 0.2$.

**6.** $P(X = 3); \quad n = 5, p = \frac{2}{3}$.

**7.** $P(X = 2); \quad n = 4, p = \frac{4}{5}$.

**8.** $P(X = 6); \quad n = 8, p = 0.1$.

**9.** $P(X < 2); \quad n = 5, p = \frac{1}{2}$.

**10.** $P(X \geq 2); \quad n = 6, p = \frac{2}{3}$.

**11. Coin**   A fair coin is tossed 10 times. What is the probability that exactly eight heads occur?

**12. Multiple-Choice Quiz**   Each question in a five-question multiple-choice quiz has four choices, only one of which is correct. If a student guesses at all five questions, find the probability that exactly three will be correct.

**13. Marbles**   An urn contains four red and six green marbles, and four marbles are randomly drawn in succession with replacement. Determine the probability that exactly one marble is green.

**14. Cards**   From a deck of 52 playing cards, three cards are randomly selected in succession with replacement. Determine the probability that exactly two cards are aces.

**15. Quality Control**   A manufacturer produces electrical switches, of which 2% are defective. From a production run of 50,000 switches, four are randomly selected and each is tested. Determine the probability that the sample contains exactly two defective switches. Round your answer to three decimal places. Assume that the four trials are independent and that the number of defective switches in the sample has a binomial distribution.

**16. Coin**   A coin is biased so that $P(H) = 0.2$ and $P(T) = 0.8$. If $X$ is the number of heads in three tosses, determine a formula for $P(X = x)$.

**17. Coin**   A biased coin is tossed three times in succession. The probability of heads on any toss is $\frac{1}{4}$. Find the probability that (a) exactly two heads occur and (b) two or three heads occur.

**18. Cards**   From an ordinary deck of 52 playing cards, 6 cards are randomly drawn in succession with replacement. Find the probability that there are (a) exactly four hearts and (b) at least four hearts.

**19. Quality Control**   In a large production lot of electronic devices, it is believed that one-third are defective. If a sample of four is randomly selected, find the probability that no more than one will be defective.

**20. Computer**   For a certain large population, the probability that a randomly selected person has a computer is 0.6. If four persons are selected at random, find the probability that at least three have a computer.

**21. Baseball**   The probability that a certain baseball player gets a hit is 0.300. Find the probability that if he goes to bat four times, he will get at least one hit.

**22. Stocks**   A financial advisor claims that 60% of the stocks that he recommends for purchase increase in value. From a list of 200 recommended stocks, a client selects 4 at random. Determine the probability, rounded to two decimal places, that at least 2 of the chosen stocks increase in value. Assume that the selections of the stocks are independent trials and that the number of stocks that increase in value has a binomial distribution.

**23. Sexes of Children** If a family has five children, find the probability that at least two are girls. (Assume that the probability that a child is a girl is $\frac{1}{2}$.)

**24.** If $X$ is a binomially distributed random variable with $n = 50$ and $p = \frac{2}{5}$, find $\sigma^2$.

**25.** Suppose $X$ is a binomially distributed random variable such that $\mu = 2$ and $\sigma^2 = \frac{3}{2}$. Find $P(X = 1)$.

**26. Quality Control** In a production process, the probability of a defective unit is 0.04. Suppose a sample of 20 units is selected at random. Let $X$ be the number of defectives.

**a.** Find the expected number of defective units.

**b.** Find Var$(X)$.

**c.** Find $P(X \le 1)$. Round your answer to two decimal places.

---

**To develop the notions of a Markov chain and the associated transition matrix. To find state vectors and the steady-state vector.**

## 10.3 MARKOV CHAINS

We conclude this chapter with a discussion of a special type of stochastic process called a *Markov*[5]*chain.*

> **Markov Chain**
>
> A **Markov chain** is a sequence of trials of an experiment in which the possible outcomes of each trial remain the same from trial to trial, are finite in number, and have probabilities that depend only upon the outcome of the previous trial.

To illustrate a Markov chain, we consider the following situation. Imagine that a small town has only two service stations—say, stations 1 and 2—that handle the servicing needs of the town's automobile owners. (These customers form the population under consideration.) Each time a customer needs car servicing, he or she must make a *choice* of which station to use.

Thus, each customer can be placed into a category according to which of the two stations he or she most recently chose. We can view a customer and the service stations as a *system*. If a customer most recently chose station 1, we shall refer to this as *state* 1 of the system. Similarly, if a customer most recently chose station 2, we say that the system is currently in state 2. Hence, at any given time, the system is in one of its two states. Of course, over a period of time, the system may move from one state to the other. For example, the sequence 1, 2, 2, 1 indicates that in four successive car servicings, the system changed from state 1 to state 2, remained at state 2, and then changed to state 1.

This situation can be thought of as a sequence of trials of an experiment (choosing a service station) in which the possible outcomes for each trial are the two states (station 1 and station 2). Each trial involves observing the state of the system at that time.

If we know the current state of the system, we realize that we cannot be sure of its state at the next observation. However, we may know the *likelihood* of its being in a particular state. For example, suppose that if a customer most recently used station 1, then the probability that the customer uses station 1 the next time is 0.7. (This means that, of those customers who used station 1 most recently, 70 percent continued to use station 1 the next time and 30 percent changed to station 2.) Assume also that if a customer used station 2 most recently, the probability is 0.8 that the customer also uses station 2 the next time. You may recognize these probabilities as being *conditional* probabilities.

---

[5]After the Russian mathematician Andrei Markov (1856–1922).

That is,

$$P(\text{remaining in state } 1 \mid \text{currently in state } 1) = 0.7,$$

$$P(\text{changing to state } 2 \mid \text{currently in state } 1) = 0.3,$$

$$P(\text{remaining in state } 2 \mid \text{currently in state } 2) = 0.8,$$

$$P(\text{changing to state } 1 \mid \text{currently in state } 2) = 0.2.$$

These four probabilities can be organized in a square matrix $\mathbf{T} = [t_{ij}]$ by letting entry $t_{ij}$ be the probability that a customer currently in state $i$ will be in state $j$ at the next observation:

$$
\mathbf{T} = \begin{array}{c} \text{State 1} \\ \text{State 2} \end{array} \begin{bmatrix} 0.7 & 0.3 \\ 0.2 & 0.8 \end{bmatrix}.
$$

Current State; Next State: State 1, State 2

Matrix $\mathbf{T}$ is called a *transition matrix* because it indicates the probabilities of transition from one state to another in *one step*—that is, as we go from one observation period to the next. The entries are called *transition probabilities*. We emphasize that *the transition matrix remains the same at every stage of the sequence of observations*. Note that all entries of the matrix are nonnegative, because they are probabilities. Moreover, the sum of the entries in each row must be 1, because, for each current state, the probabilities account for all possible transitions.

Let us summarize our service station situation up to now. We have a sequence of trials in which the possible outcomes (or states) are the same from trial to trial and are finite in number (two). The probability that the system is in a particular state for a given trial depends only on the state for the preceding trial. Thus, we have a so-called *two-state Markov chain*. A Markov chain determines a square matrix $\mathbf{T}$, called a transition matrix.

> **Transition Matrix**
>
> A **transition matrix** for a $k$-state Markov chain is a $k \times k$ matrix $\mathbf{T} = [t_{ij}]$ in which the entry $t_{ij}$ is the probability of moving from state $i$ to state $j$ from one trial to the next. All entries are nonnegative, and the sum of the entries in each row is 1.

Suppose that when observations are initially made, 60 percent of all customers used station 1 most recently and 40 percent used station 2. This means that, before any additional trials (car servicings) are considered, the probabilities that a customer is in state 1 or 2 are 0.6 and 0.4, respectively. These probabilities are called *initial state probabilities* and are collectively referred to as being the *initial distribution*. They can be represented by a row vector, called an **initial state vector**, which is denoted by $\mathbf{X_0}$. In this case,

$$\mathbf{X_0} = [0.6 \quad 0.4].$$

We would like to find the vector that gives the state probabilities for a customer's *next* visit to a service station. This state vector is denoted by $\mathbf{X_1}$. More generally, a state vector is defined as follows:

> **State Vector**
>
> The **state vector** $\mathbf{X_n}$ for a $k$-state Markov chain is a $k$-entry row vector in which the entry $x_{1j}$ is the probability of being in state $j$ after the $n$th trial.

*For example, the sum of the entries in row 1 of $\mathbf{T}$ is $0.7 + 0.3 = 1$.*

*A subscript of 0 is used for the initial state vector.*

We can find the entries for $\mathbf{X_1}$ from the probability tree in Figure 10.8. We see that the probability of being in state 1 after the next visit is the sum

$$(0.6)(0.7) + (0.4)(0.2) = 0.5, \tag{1}$$

and the probability of being in state 2 is

$$(0.6)(0.3) + (0.4)(0.8) = 0.5. \tag{2}$$

Thus,

$$\mathbf{X_1} = [0.5 \quad 0.5].$$

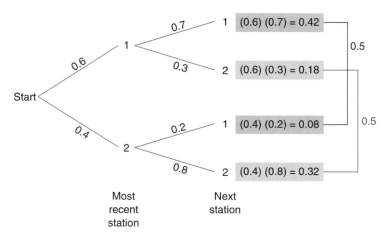

**FIGURE 10.8** Probability tree for two-state Markov chain.

The sums of products on the left sides of Eqs. (1) and (2) remind us of matrix multiplication. In fact, they are the entries in the matrix $\mathbf{X_0 T}$ obtained by multiplying the initial state vector by the transition matrix:

$$\mathbf{X_1} = \mathbf{X_0 T} = [0.6 \quad 0.4]\begin{bmatrix} 0.7 & 0.3 \\ 0.2 & 0.8 \end{bmatrix} = [0.5 \quad 0.5].$$

This pattern of taking the product of a state vector and the transition matrix to get the next state vector continues, allowing us to find state probabilities for future observations. For example, to find $\mathbf{X_2}$, the state vector that gives the probabilities for each state after two trials (following the initial observation), we have

$$\mathbf{X_2} = \mathbf{X_1 T} = [0.5 \quad 0.5]\begin{bmatrix} 0.7 & 0.3 \\ 0.2 & 0.8 \end{bmatrix} = [0.45 \quad 0.55].$$

Thus, the probability of being in state 1 after two car servicings is 0.45. Note that, since $\mathbf{X_1} = \mathbf{X_0 T}$, we can write

$$\mathbf{X_2} = (\mathbf{X_0 T})\mathbf{T},$$

or, more simply,

$$\mathbf{X_2} = \mathbf{X_0 T}^2.$$

In general, the $n$th state vector can be found by multiplying the previous state vector, $\mathbf{X_{n-1}}$, by $\mathbf{T}$:

If **T** is the transition matrix for a Markov chain, then the state vector $\mathbf{X}_n$ for the $n$th trial is given by

$$\mathbf{X}_n = \mathbf{X}_{n-1}\mathbf{T}.$$

Equivalently, we can find $\mathbf{X}_n$ by using only the initial state matrix $\mathbf{X}_0$ and the transition matrix **T:**

*Here we find $\mathbf{X}_n$ by using powers of* **T.**

$$\mathbf{X}_n = \mathbf{X}_0\mathbf{T}^n. \tag{3}$$

Let us now consider the situation in which we know the initial state of the system. For example, take the case of observing initially that a customer has most recently chosen station 1. This means the probability that the system is in state 1 is 1, so the initial state vector must be

$$\mathbf{X}_0 = \begin{bmatrix} 1 & 0 \end{bmatrix}.$$

Suppose we determine $\mathbf{X}_2$, the state vector that gives the state probabilities after the next two visits. This is given by

$$\mathbf{X}_2 = \mathbf{X}_0\mathbf{T}^2 = \begin{bmatrix} 1 & 0 \end{bmatrix}\begin{bmatrix} 0.7 & 0.3 \\ 0.2 & 0.8 \end{bmatrix}^2$$

$$= \begin{bmatrix} 1 & 0 \end{bmatrix}\begin{bmatrix} 0.55 & 0.45 \\ 0.30 & 0.70 \end{bmatrix} = \begin{bmatrix} 0.55 & 0.45 \end{bmatrix}.$$

Thus, for this customer, the probabilities of using station 1 or station 2 after two steps are 0.55 and 0.45, respectively. Observe that these probabilities form the *first row* of $\mathbf{T}^2$. On the other hand, if the system were initially in state 2, then the state vector after two steps would be

$$\begin{bmatrix} 0 & 1 \end{bmatrix}\mathbf{T}^2 = \begin{bmatrix} 0 & 1 \end{bmatrix}\begin{bmatrix} 0.55 & 0.45 \\ 0.30 & 0.70 \end{bmatrix} = \begin{bmatrix} 0.30 & 0.70 \end{bmatrix}.$$

Hence, for this customer, the probabilities of using station 1 or station 2 after two steps are 0.30 and 0.70, respectively. Observe that these probabilities form the *second row* of $\mathbf{T}^2$. Based on our observations, we now have a way of interpreting $\mathbf{T}^2$: The entries in

$$\mathbf{T}^2 = \begin{array}{c} 1 \\ 2 \end{array}\begin{array}{cc} \phantom{x}1 & \phantom{x}2 \\ \begin{bmatrix} 0.55 & 0.45 \\ 0.30 & 0.70 \end{bmatrix} \end{array}$$

give the probabilities of moving from one state to another in *two* steps. In general, we have the following:

*This gives the significance of the entries in* $\mathbf{T}^n$.

If **T** is a transition matrix, then for $\mathbf{T}^n$, the entry in row $i$ and column $j$ gives the probability of moving from state $i$ to state $j$ in $n$ steps.

### EXAMPLE 1   Demography

*A certain county is divided into three demographic regions. Research indicates that each year 20 percent of the residents in region 1 move to region 2 and 10 percent move to region 3. (The others remain in region 1.) Of the residents in region 2, 10 percent move to region 1 and 10 percent move to region 3. Of the residents in region 3, 20 percent move to region 1 and 10 percent move to region 2.*

**a.** *Find the transition matrix* **T** *for this situation.*

*Solution:* We have

$$
\begin{array}{c}
\textit{From} \\
\textit{Region}
\end{array}
\quad
\begin{array}{ccc}
\textit{To Region} \\
1 \quad 2 \quad 3
\end{array}
$$

$$
\mathbf{T} =
\begin{array}{c} 1 \\ 2 \\ 3 \end{array}
\begin{bmatrix}
0.7 & 0.2 & 0.1 \\
0.1 & 0.8 & 0.1 \\
0.2 & 0.1 & 0.7
\end{bmatrix}.
$$

Note that to find $t_{11}$, we subtracted the sum of the other two entries in the first row from 1. The entries $t_{22}$ and $t_{33}$ are found similarly.

**b.** *Find the probability that a resident of region 1 this year is a resident of region 1 next year; in two years.*

*Solution:* From entry $t_{11}$ in transition matrix **T**, the probability that a resident of region 1 remains in region 1 after one year is 0.7. The probabilities of moving from one region to another in two steps are given by $\mathbf{T}^2$:

Powers of **T** can be conveniently found with a graphics calculator.

$$
\mathbf{T}^2 =
\begin{array}{c} 1 \\ 2 \\ 3 \end{array}
\begin{array}{ccc}
1 & 2 & 3
\end{array}
\begin{bmatrix}
0.53 & 0.31 & 0.16 \\
0.17 & 0.67 & 0.16 \\
0.29 & 0.19 & 0.52
\end{bmatrix}
$$

Thus, the probability that a resident of region 1 is in region 1 after two years is 0.53.

**c.** *This year, suppose 40 percent of county residents live in region 1, 30 percent live in region 2, and 30 percent live in region 3. Find the probability that a resident of the county lives in region 2 after three years.*

*Solution:* The initial state vector is

$$
\mathbf{X_0} = [0.40 \quad 0.30 \quad 0.30].
$$

The distribution of the population after three years is given by state vector $\mathbf{X_3}$. From Eq. (3) with $n = 3$, we have

$$
\mathbf{X_3} = \mathbf{X_0}\mathbf{T}^3 = \mathbf{X_0}\mathbf{T}^2\mathbf{T}
$$

Of course, $\mathbf{X_3}$ can be easily obtained with a graphics calculator: Enter $\mathbf{X_0}$ and **T**, and then evaluate $\mathbf{X_0}\mathbf{T}^3$ directly.

$$
= [0.40 \quad 0.30 \quad 0.30]
\begin{bmatrix}
0.53 & 0.31 & 0.16 \\
0.17 & 0.67 & 0.16 \\
0.29 & 0.19 & 0.52
\end{bmatrix}
\begin{bmatrix}
0.7 & 0.2 & 0.1 \\
0.1 & 0.8 & 0.1 \\
0.2 & 0.1 & 0.7
\end{bmatrix}
$$

$$
= [0.3368 \quad 0.4024 \quad 0.2608].
$$

This result means that in three years, 33.68 percent of the county residents live in region 1, 40.24 percent live in region 2, and 26.08 percent live in region 3. Thus, the probability that a resident lives in region 2 in three years is 0.4024. ∎

## Steady-State Vectors

Let us now return to our service station problem. Recall that if the initial state vector is

$$
\mathbf{X_0} = [0.6 \quad 0.4],
$$

then

$$
\mathbf{X_1} = [0.5 \quad 0.5],
$$

$$
\mathbf{X_2} = [0.45 \quad 0.55].
$$

Some state vectors beyond the second are

$$\mathbf{X_3} = \mathbf{X_2 T} = \begin{bmatrix} 0.45 & 0.55 \end{bmatrix} \begin{bmatrix} 0.7 & 0.3 \\ 0.2 & 0.8 \end{bmatrix} = \begin{bmatrix} 0.425 & 0.575 \end{bmatrix},$$

$$\mathbf{X_4} = \mathbf{X_3 T} = \begin{bmatrix} 0.425 & 0.575 \end{bmatrix} \begin{bmatrix} 0.7 & 0.3 \\ 0.2 & 0.8 \end{bmatrix} = \begin{bmatrix} 0.4125 & 0.5875 \end{bmatrix},$$

$$\mathbf{X_5} = \mathbf{X_4 T} = \begin{bmatrix} 0.4125 & 0.5875 \end{bmatrix} \begin{bmatrix} 0.7 & 0.3 \\ 0.2 & 0.8 \end{bmatrix} = \begin{bmatrix} 0.40625 & 0.59375 \end{bmatrix},$$

$$\vdots$$

$$\mathbf{X_{10}} = \mathbf{X_9 T} \approx \begin{bmatrix} 0.40020 & 0.59980 \end{bmatrix}.$$

These results strongly suggest, and it is indeed the case, that, as the number of trials increases, the entries in the state vectors tend to get closer and closer to the corresponding entries in the vector

$$\mathbf{Q} = \begin{bmatrix} 0.40 & 0.60 \end{bmatrix}.$$

(Equivalently, it can be shown that the entries in each row of $\mathbf{T}^n$ approach the corresponding entries in those of $\mathbf{Q}$ as $n$ increases.) Vector $\mathbf{Q}$ has a special property. Observe the result of multiplying $\mathbf{Q}$ by the transition matrix $\mathbf{T}$:

$$\mathbf{QT} = \begin{bmatrix} 0.40 & 0.60 \end{bmatrix} \begin{bmatrix} 0.7 & 0.3 \\ 0.2 & 0.8 \end{bmatrix} = \begin{bmatrix} 0.40 & 0.60 \end{bmatrix} = \mathbf{Q}.$$

We thus have

$$\mathbf{QT} = \mathbf{Q},$$

which indicates that $\mathbf{Q}$ *remains unchanged from trial to trial.*

In summary, as the number of trials increases, the state vectors get closer and closer to $\mathbf{Q}$, which remains unchanged from trial to trial. Essentially, the distribution of the population between the service stations is stabilizing. That is, in the long run, approximately 40 percent of the population will have their cars serviced at station 1 and 60 percent at station 2. To describe this, we say that $\mathbf{Q}$ is the **steady-state vector** of this process. It can be shown that the steady-state vector is unique. (There is only one such vector.) Moreover, $\mathbf{Q}$ does not depend on the initial state vector $\mathbf{X_0}$, but depends only on the transition matrix $\mathbf{T}$. For this reason, we say that $\mathbf{Q}$ is the *steady-state vector for* $\mathbf{T}$.

*The steady-state vector is unique and does not depend on the initial distribution.*

What we need now is a procedure for finding the steady-state vector $\mathbf{Q}$ without having to compute state vectors for large values of $n$. Fortunately, the previously stated property that $\mathbf{QT} = \mathbf{Q}$ can be used to find $\mathbf{Q}$. If we let $\mathbf{Q} = \begin{bmatrix} q_1 & q_2 \end{bmatrix}$, we have

$$\mathbf{QT} = \mathbf{Q},$$

$$\begin{bmatrix} q_1 & q_2 \end{bmatrix} \begin{bmatrix} 0.7 & 0.3 \\ 0.2 & 0.8 \end{bmatrix} = \begin{bmatrix} q_1 & q_2 \end{bmatrix},$$

$$\begin{bmatrix} 0.7q_1 + 0.2q_2 & 0.3q_1 + 0.8q_2 \end{bmatrix} = \begin{bmatrix} q_1 & q_2 \end{bmatrix}.$$

By matrix equality, this implies that

$$\begin{cases} 0.7q_1 + 0.2q_2 = q_1, \\ 0.3q_1 + 0.8q_2 = q_2. \end{cases}$$

Simplifying gives

$$\begin{cases} -0.3q_1 + 0.2q_2 = 0, \\ \phantom{-}0.3q_1 - 0.2q_2 = 0. \end{cases}$$

These equations are equivalent, so there are infinitely many solutions. However, it can be shown that $q_1 + q_2$ must be 1. So we now have the system

$$\begin{cases} -0.3q_1 + 0.2q_2 = 0, \\ \phantom{-}0.3q_1 - 0.2q_2 = 0, \\ \phantom{-0.3}q_1 + q_2 = 1. \end{cases}$$

Solving this system by elimination or matrix techniques gives $q_1 = 0.40$ and $q_2 = 0.60$. Thus, the steady-state vector is

$$\mathbf{Q} = [0.40 \quad 0.60],$$

which confirms our previous assumption.

We must point out that for Markov chains in general, the state vectors do not always approach a steady-state vector. However, it can be shown that a steady-state vector for **T** does exist, provided that **T** is *regular*:

> A transition matrix **T** is **regular** if there exists an integer power of **T** for which all entries are positive.

Only regular transition matrices will be considered in this section. A Markov chain whose transition matrix is regular is called a **regular Markov chain.**

In summary, we have the following:

> Suppose **T** is the transition matrix for a regular Markov chain. Then the steady-state row vector
>
> $$\mathbf{Q} = [q_1 \quad q_2 \quad \cdots \quad q_k]$$
>
> is the solution to the matrix equation
>
> $$\mathbf{QT} = \mathbf{Q}, \tag{4}$$
>
> subject to the condition that
>
> $$q_1 + q_2 + \cdots + q_k = 1. \tag{5}$$

Eqs. (4) and (5) are at the heart of finding the steady-state vector.

### EXAMPLE 2 Steady-State Vector

*For the demography problem of Example 1, in the long run what percentage of county residents will live in each region?*

*Solution:* The population distribution in the long run is given by the steady-state vector **Q,** which we now proceed to find. From Eq. (4), $\mathbf{QT} = \mathbf{Q}$; that is,

$$[q_1 \quad q_2 \quad q_3] \begin{bmatrix} 0.7 & 0.2 & 0.1 \\ 0.1 & 0.8 & 0.1 \\ 0.2 & 0.1 & 0.7 \end{bmatrix} = [q_1 \quad q_2 \quad q_3].$$

Multiplying the matrices on the left side gives a row matrix. By equating the entries of this matrix with the corresponding entries of the row matrix on the right side, we obtain the system

$$\begin{cases} 0.7q_1 + 0.1q_2 + 0.2q_3 = q_1, \\ 0.2q_1 + 0.8q_2 + 0.1q_3 = q_2, \\ 0.1q_1 + 0.1q_2 + 0.7q_3 = q_3. \end{cases}$$

Combining terms gives

$$\begin{cases} -0.3q_1 + 0.1q_2 + 0.2q_3 = 0, \\ \phantom{-}0.2q_1 - 0.2q_2 + 0.1q_3 = 0, \\ \phantom{-}0.1q_1 + 0.1q_2 - 0.3q_3 = 0. \end{cases}$$

Solving this system, together with the condition that

$$q_1 + q_2 + q_3 = 1,$$

gives

$$\mathbf{Q} = [0.3125 \quad 0.4375 \quad 0.2500].$$

Thus, in the long run, the percentages of county residents living in regions 1, 2, and 3 are 31.25%, 43.75%, and 25%, respectively. ■

We conclude this section by developing a matrix procedure for finding the steady-state vector

$$\mathbf{Q} = [q_1 \; q_2 \; \cdots \; q_k]$$

for the transition matrix $\mathbf{T}$ of a $k$-state regular Markov chain. Since

$$\mathbf{QT} = \mathbf{Q},$$

we have

$$\mathbf{QT} - \mathbf{Q} = \mathbf{O},$$
$$\mathbf{QT} - \mathbf{QI} = \mathbf{O},$$
$$\mathbf{Q}(\mathbf{T} - \mathbf{I}) = \mathbf{O}.$$

Taking transposes gives

Here we make use of a number of properties of the transpose of a matrix. These properties were discussed in Chapter 6.

$$[\mathbf{Q}(\mathbf{T} - \mathbf{I})]^{\mathsf{T}} = \mathbf{O}^{\mathsf{T}},$$
$$(\mathbf{T} - \mathbf{I})^{\mathsf{T}}\mathbf{Q}^{\mathsf{T}} = \mathbf{O},$$
$$(\mathbf{T}^{\mathsf{T}} - \mathbf{I}^{\mathsf{T}})\mathbf{Q}^{\mathsf{T}} = \mathbf{O},$$
$$(\mathbf{T}^{\mathsf{T}} - \mathbf{I})\mathbf{Q}^{\mathsf{T}} = \mathbf{O}.$$

The last equation corresponds to a homogeneous system of linear equations. This system, along with the condition

$$q_1 + q_2 + \cdots + q_k = 1,$$

form a nonhomogeneous system that can be solved by reducing a matrix of the form

$$\left[\begin{array}{c|c} \mathbf{T}^{\mathsf{T}} - \mathbf{I} & \mathbf{O} \\ \hline 1 \ldots 1 & 1 \end{array}\right].$$

To illustrate, for the situation in Example 2 with

$$\mathbf{T} = \begin{bmatrix} 0.7 & 0.2 & 0.1 \\ 0.1 & 0.8 & 0.1 \\ 0.2 & 0.1 & 0.7 \end{bmatrix},$$

we have

$$\mathbf{T}^{\mathsf{T}} - \mathbf{I} = \begin{bmatrix} 0.7 & 0.1 & 0.2 \\ 0.2 & 0.8 & 0.1 \\ 0.1 & 0.1 & 0.7 \end{bmatrix} - \begin{bmatrix} 1 & 0 & 0 \\ 0 & 1 & 0 \\ 0 & 0 & 1 \end{bmatrix}$$

$$= \begin{bmatrix} -0.3 & 0.1 & 0.2 \\ 0.2 & -0.2 & 0.1 \\ 0.1 & 0.1 & -0.3 \end{bmatrix}.$$

We now reduce the augmented matrix

$$\begin{bmatrix} -0.3 & 0.1 & 0.2 & | & 0 \\ 0.2 & -0.2 & 0.1 & | & 0 \\ 0.1 & 0.1 & -0.3 & | & 0 \\ 1 & 1 & 1 & | & 1 \end{bmatrix},$$

which gives

$$\begin{bmatrix} 1 & 0 & 0 & | & 0.3125 \\ 0 & 1 & 0 & | & 0.4375 \\ 0 & 0 & 1 & | & 0.2500 \\ 0 & 0 & 0 & | & 0 \end{bmatrix}.$$

Thus, $q_1 = 0.3125$, $q_2 = 0.4375$, and $q_3 = 0.2500$, as before.

## ▪ Exercise 10.3

*In Problems 1–6, can the given matrix be a transition matrix for a Markov chain?*

**1.** $\begin{bmatrix} \frac{1}{2} & 0 \\ \frac{2}{3} & \frac{1}{3} \end{bmatrix}$.

**2.** $\begin{bmatrix} 0.1 & 0.9 \\ 1 & 0 \end{bmatrix}$.

**3.** $\begin{bmatrix} \frac{1}{2} & -\frac{1}{4} & \frac{3}{4} \\ \frac{1}{8} & \frac{5}{8} & \frac{1}{4} \\ \frac{1}{3} & \frac{1}{3} & \frac{1}{3} \end{bmatrix}$.

**4.** $\begin{bmatrix} 0.2 & 0.7 & 0.1 \\ 0.6 & 0.2 & 0.2 \\ 0 & 0 & 0 \end{bmatrix}$.

**5.** $\begin{bmatrix} 0.4 & 0.2 & 0.4 \\ 0 & 0.1 & 0.9 \\ 0.5 & 0.3 & 0.2 \end{bmatrix}$.

**6.** $\begin{bmatrix} 0.5 & 0.4 & 0.6 \\ 0.1 & 0.3 & 0.6 \\ 0.3 & 0.3 & 0.4 \end{bmatrix}$.

*In Problems 7–10, a transition matrix for a Markov chain is given. Determine the values of the letter entries.*

**7.** $\begin{bmatrix} \frac{2}{3} & a \\ b & \frac{1}{4} \end{bmatrix}$.

**8.** $\begin{bmatrix} a & \frac{1}{8} \\ b & a \end{bmatrix}$.

**9.** $\begin{bmatrix} 0.4 & a & 0.2 \\ a & 0.1 & b \\ a & b & c \end{bmatrix}$.

**10.** $\begin{bmatrix} a & a & a \\ a & b & \frac{1}{4} \\ a & b & c \end{bmatrix}$.

*In Problems 11–14, determine whether the given vector could be a state vector for a Markov chain.*

**11.** $[0.4 \quad 0.6]$.

**12.** $[1 \quad 0]$.

**13.** $[0.2 \quad 0.7 \quad 0.5]$.

**14.** $[0.7 \quad -0.3 \quad 0.6]$.

*In Problems 15–20, a transition matrix $\mathbf{T}$ and an initial state vector $\mathbf{X}_0$ are given. Compute the state vectors $\mathbf{X}_1$, $\mathbf{X}_2$, and $\mathbf{X}_3$.*

**15.** $\mathbf{T} = \begin{bmatrix} \frac{2}{3} & \frac{1}{3} \\ 1 & 0 \end{bmatrix}$,

$\mathbf{X}_0 = [\frac{1}{4} \quad \frac{3}{4}]$.

**16.** $\mathbf{T} = \begin{bmatrix} \frac{1}{2} & \frac{1}{2} \\ \frac{1}{4} & \frac{3}{4} \end{bmatrix}$,

$\mathbf{X}_0 = [\frac{1}{2} \quad \frac{1}{2}]$.

**17.** $\mathbf{T} = \begin{bmatrix} 0.5 & 0.5 \\ 0.5 & 0.5 \end{bmatrix}$,

$\mathbf{X}_0 = [0.4 \quad 0.6]$.

**18.** $\mathbf{T} = \begin{bmatrix} 0.1 & 0.9 \\ 0.9 & 0.1 \end{bmatrix}$,

$\mathbf{X}_0 = \begin{bmatrix} 0.2 & 0.8 \end{bmatrix}$.

**19.** $\mathbf{T} = \begin{bmatrix} 0.1 & 0.2 & 0.7 \\ 0 & 0.4 & 0.6 \\ 0.3 & 0.3 & 0.4 \end{bmatrix}$,

$\mathbf{X}_0 = \begin{bmatrix} 0.2 & 0 & 0.8 \end{bmatrix}$.

**20.** $\mathbf{T} = \begin{bmatrix} 0.5 & 0 & 0.5 \\ 0.1 & 0.1 & 0.8 \\ 0.2 & 0.3 & 0.5 \end{bmatrix}$,

$\mathbf{X}_0 = \begin{bmatrix} 0.1 & 0.3 & 0.6 \end{bmatrix}$.

*In Problems 21–24, a transition matrix* $\mathbf{T}$ *is given.*

**a.** Compute $\mathbf{T}^2$ and $\mathbf{T}^3$.

**b.** What is the probability of going from state 1 to state 2 after two steps?

**c.** What is the probability of going from state 2 to state 1 after three steps?

**21.** $\begin{bmatrix} \frac{1}{4} & \frac{3}{4} \\ \frac{3}{4} & \frac{1}{4} \end{bmatrix}$.

**22.** $\begin{bmatrix} \frac{1}{3} & \frac{2}{3} \\ \frac{1}{2} & \frac{1}{2} \end{bmatrix}$.

**23.** $\begin{bmatrix} 0 & 1 & 0 \\ 0.5 & 0.4 & 0.1 \\ 0.3 & 0.3 & 0.4 \end{bmatrix}$.

**24.** $\begin{bmatrix} 0.1 & 0.2 & 0.7 \\ 0.1 & 0.1 & 0.8 \\ 0.1 & 0.1 & 0.8 \end{bmatrix}$.

*In Problems 25–30, find the steady-state vector for the given transition matrix.*

**25.** $\begin{bmatrix} \frac{1}{2} & \frac{1}{2} \\ \frac{3}{4} & \frac{1}{4} \end{bmatrix}$.

**26.** $\begin{bmatrix} \frac{1}{2} & \frac{1}{2} \\ \frac{1}{4} & \frac{3}{4} \end{bmatrix}$.

**27.** $\begin{bmatrix} \frac{1}{5} & \frac{4}{5} \\ \frac{3}{5} & \frac{2}{5} \end{bmatrix}$.

**28.** $\begin{bmatrix} \frac{1}{4} & \frac{3}{4} \\ \frac{1}{3} & \frac{2}{3} \end{bmatrix}$.

**29.** $\begin{bmatrix} 0.4 & 0.3 & 0.3 \\ 0.6 & 0.3 & 0.1 \\ 0.6 & 0.1 & 0.3 \end{bmatrix}$.

**30.** $\begin{bmatrix} 0.1 & 0.2 & 0.7 \\ 0.4 & 0.2 & 0.4 \\ 0.3 & 0.3 & 0.4 \end{bmatrix}$.

**31. Spread of Flu** A flu has attacked a college dorm that has 200 students. Suppose the probability that a student having the flu will still have it 4 days later is 0.1. However, for a student who does not have the flu, the probability of having the flu 4 days later is 0.2.

**a.** Find a transition matrix for this situation.

**b.** If 120 students now have the flu, how many students (to the nearest integer) can be expected to have the flu 8 days from now? 12 days from now?

**32. Physical Fitness** A physical-fitness center has found that, of those members who perform high-impact exercising on one visit, 60 percent will do the same on the next visit and 40 percent will do low-impact exercising. Of those who perform low-impact exercising on one visit, 70 percent will do the same on the next visit and 30 percent will do high-impact exercising. On the last visit, suppose that 60 percent of members did high-impact exercising and 40 percent did low-impact exercising. After two more visits, what percentage of members will be performing high-impact exercising?

**33. Newspapers** In a certain area, two daily newspapers are available. It has been found that if a customer buys newspaper A on one day, then the probability is 0.1 that he or she will change to the other newspaper the next day. If a customer buys newspaper B on one day, then the probability is 0.7 that he or she will buy the same newspaper the next day.

**a.** Find the transition matrix for this situation.

**b.** Find the probability that a person who buys A on Monday will buy A on Thursday.

**34. Video Rentals** A video rental store has three locations in a city. A video can be rented from any of the three locations and returned to any of them. Studies show that videos are rented from one location and returned to a location according to the probabilities given by the following matrix:

|         | Returned to |     |     |
|---------|-----|-----|-----|
| Rented From | 1 | 2 | 3 |
| 1 | 0.7 | 0.1 | 0.2 |
| 2 | 0.2 | 0.8 | 0 |
| 3 | 0.2 | 0.2 | 0.6 |

Suppose that 20 percent of the videos are initially rented from location 1, 50 percent from 2, and 30 percent from 3. Find the percentages of videos that can be expected to be returned to each location:

**a.** After this rental.   **b.** After the next rental.

**35. Voting** In a certain region, voter registration was analyzed according to party affiliation: Democratic, Republican, and other. It was found that on a year-to-year basis, the probability that a voter switches registration from Democratic to Republican is 0.1; from Democratic to other, 0.1; from Republican to Democratic, 0.1; from Republican to other, 0.1; and from other to Democratic, 0.3; and from other to Republican, 0.1.

**a.** Find a transition matrix for this situation.

**b.** What is the probability that a presently registered Republican voter will be registered Democratic two years from now?

**c.** If 40 percent of the present voters are Democratic and 40 percent are Republican, what percentage can be expected to be Republican one year from now?

**36. Demography** The residents of a certain region are classified as urban (U), suburban (S), or rural (R). A marketing firm has found that over successive 5-year periods, residents shift from one classification to another according to the probabilities given by the following matrix:

$$\begin{array}{c} \\ U \\ S \\ R \end{array} \begin{array}{c} U \qquad S \qquad R \\ \begin{bmatrix} 0.7 & 0.2 & 0.1 \\ 0.1 & 0.8 & 0.1 \\ 0.1 & 0.1 & 0.8 \end{bmatrix} \end{array}.$$

**a.** Find the probability that a suburban resident will be a rural resident in 15 years.

**b.** Suppose the initial population of the region is 50 percent urban, 25 percent suburban, and 25 percent rural. Determine the expected population distribution in 15 years.

**37. Long-Distance Telephone Service** A major long-distance telephone company (company A) has studied the tendency of telephone users to switch from one carrier to another. The company believes that over successive six-month periods, the probability that a customer who uses A's service will switch

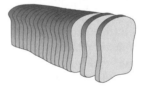

to a competing service is 0.2 and the probability that a customer of any competing service will switch to A is 0.3.

**a.** Find a transition matrix for this situation.

**b.** If A presently controls 70 percent of the market, what percentage can it expect to control six months from now?

**c.** What percentage of the market can A expect to control in the long run?

**38. Automobile Purchases** In a certain region, a study of car ownership was made. It was determined that if a person presently owns a Ford, then the probability that the next car the person buys is also a Ford is 0.6. If a person does not presently own a Ford, then the probability that the person will buy a Ford on the next car purchase is 0.3.

**a.** Find the transition matrix for this situation.

**b.** In the long run, what proportion of car purchases in the region can be expected to be Fords?

**39. Laboratory Mice** Suppose 100 mice are in a two-compartment cage and are free to move between the compartments. At regular time intervals, the number of mice in each compartment is observed. It has been found that if a mouse is in compartment 1 at one observation, then the probability that the mouse will be in

compartment 1 at the next observation is $\frac{5}{7}$. If a mouse is in compartment 2 at one observation, then the probability that the mouse will be in compartment 2 at the next observation is $\frac{4}{7}$. Initially, suppose that 50 mice are placed into each compartment.

**a.** Find the transition matrix for this situation.

**b.** After two observations, what percentage of mice (rounded to two decimal places) can be expected to be in each compartment?

**c.** In the long run, what percentage of mice can be expected in each compartment?

**40. Vending Machines** A typical cry of college students is "Don't put your money into that soda machine; I tried it and the machine didn't work!" Suppose that if a vending machine is working properly one time, then the probability that it will work properly the next time is 0.7. On the other hand, suppose that if the machine is not working properly one time, then the probability that it will not work properly the next time is 0.9.

**a.** Find a transition matrix for this situation.

**b.** Suppose that you want four sodas from such a vending machine and initially you get your first soda. What is the probability that you get the three other sodas in three tries?

**c.** If there are 40 such vending machines on a college campus, in the long run how many machines do you expect to work properly?

**41. Advertising** A supermarket chain sells bread from bakeries A and B. Presently, A accounts for 50 percent of the chain's daily bread sales. To increase sales, A launches a new advertising campaign. The bakery

believes that the change in bread sales at the chain will be based on the following transition matrix:

$$\begin{array}{c} \\ A \\ B \end{array} \begin{array}{c} A \qquad B \\ \begin{bmatrix} \frac{3}{4} & \frac{1}{4} \\ \frac{1}{2} & \frac{1}{2} \end{bmatrix} \end{array}.$$

**a.** Find the steady-state vector.

**b.** In the long run, by what percentage can A expect to increase present sales at the chain? Assume that the total daily sales of bread at the chain remain the same.

**42. Bank Branches** A bank with three branches, A, B, and C, finds that customers usually return to the same branch for their banking needs. However, at times a customer may go to a different branch because of a changed circumstance. For example, a person who usually goes to branch A may sometimes deviate and go to branch B because the person has business to conduct in the vicinity of branch B. For customers of branch A, suppose that 80 percent return to A on their next visit, 10 percent to go B, and 10 percent go to C. For customers of branch B, suppose that 70 percent return to B on their next visit, 20 percent go to A, and 10 percent go to C. For customers of branch C, suppose that 70 percent return to C on their next visit, 20 percent go to A, and 10 percent go to B.

a. Find a transition matrix for this situation.

b. If a customer most recently went to branch B, what is the probability that the customer returns to B on the second bank visit from now?

c. Initially, suppose 200 customers go to A, 200 go to B, and 100 go to C. On their next visit, how many can be expected to go to A? To B? To C?

d. Of the initial 500 customers, in the long run how many can be expected to go to A? To B? To C?

43. Show that the transition matrix $\mathbf{T} = \begin{bmatrix} \frac{1}{2} & \frac{1}{2} \\ 1 & 0 \end{bmatrix}$ is regular.

[*Hint:* Examine the entries in $\mathbf{T}^2$.]

44. Show that the transition matrix $\begin{bmatrix} 1 & 0 \\ 0 & 1 \end{bmatrix}$ is not regular.

# 10.4 REVIEW

## IMPORTANT TERMS AND SYMBOLS

**Section 10.1**   discrete random variable   continuous random variable   probability function   probability distribution   probability histogram   sigma notation, $\sum_x f(x)$   mean, $\mu$   expected value, $E(X)$   variance, $\text{Var}(X)$, $\sigma^2$   standard deviation, $\sigma$

**Section 10.2**   binomial theorem   binomial coefficients   independent trials   Bernoulli trials   binomial experiment   binomial distribution

**Section 10.3**   Markov chain   states of system   transition matrix, $\mathbf{T}$   state vector, $\mathbf{X}_n$   regular transition matrix   steady-state vector, $\mathbf{Q}$

## SUMMARY

If $X$ is a discrete random variable and $f$ is the function such that $f(x) = P(X = x)$, then $f$ is called the probability function, or probability distribution, of $X$. In general,

$$\sum_x f(x) = 1.$$

The mean, or expected value, of $X$ is the long-run average of $X$ and is denoted $\mu$ or $E(X)$:

$$\mu = E(X) = \sum_x xf(x).$$

The mean can be interpreted as a measure of the central tendency of $X$ in the long run. A measure of the dispersion of $X$ is the variance, denoted $\text{Var}(X)$ and is given by

$$\text{Var}(X) = \sum_x (x - \mu)^2 f(x),$$

or, equivalently, by

$$\text{Var}(X) = \sum_x x^2 f(x) - \mu^2.$$

Another measure of dispersion of $X$ is the standard deviation $\sigma$:

$$\sigma = \sqrt{\text{Var}(X)}.$$

If an experiment is repeated several times, then each performance of the experiment is called a trial. The trials are independent when the outcome of any single trial does not affect the outcome of any other. If there are only two possible outcomes (success and failure) for each independent trial, and the probabilities of success and failure do not change from trial to trial, then the experiment is called a binomial experiment. For such an experiment, if $X$ is the number of successes in $n$ trials, then the distribution $f$ of $X$ is called a binomial distribution, and

$$f(x) = P(X = x) = {}_nC_x p^x q^{n-x},$$

where $p$ is the probability of success on any trial and $q = 1 - p$ is the probability of failure. The mean $\mu$ and standard deviation $\sigma$ of $X$ are given by

$$\mu = np \quad \text{and} \quad \sigma = \sqrt{npq}.$$

A binomial distribution is intimately connected with the binomial theorem, which is a formula for expanding the $n$th power of a binomial, namely,

$$(a + b)^n = {_nC_0}a^n + {_nC_1}a^{n-1}b + \cdots + {_nC_{n-1}}ab^{n-1} + {_nC_n}b^n,$$

where $n$ is a positive integer.

A Markov chain is a sequence of trials of an experiment in which the possible outcomes of each trial, which are called states, remain the same from trial to trial, are finite in number, and have probabilities that depend only upon the outcome of the previous trial. For a $k$-state Markov chain, if the probability of moving from state $i$ to state $j$ from one trial to the next is represented by $t_{ij}$, then the $k \times k$ matrix $\mathbf{T} = [t_{ij}]$ is called the transition matrix for the chain. The entries in the $n$th power of $\mathbf{T}$ also represent probabilities; the entry in the $i$th row and $j$th column of $\mathbf{T}^n$ gives the probability of moving from state $i$ to state $j$ in $n$ steps. A $k$-entry row vector in which the entry $x_{1j}$ is the probability of being in state $j$ after the $n$th trial is called a state vector and is denoted $\mathbf{X}_n$. The initial state probabilities are represented by the initial state vector $\mathbf{X}_0$. The state vector $\mathbf{X}_n$ can be found by multiplying the previous state vector $\mathbf{X}_{n-1}$ by the transition matrix $\mathbf{T}$:

$$\mathbf{X}_n = \mathbf{X}_{n-1}\mathbf{T}.$$

Alternatively, $\mathbf{X}_n$ can be found by multiplying the initial state vector $\mathbf{X}_0$ by $\mathbf{T}^n$:

$$\mathbf{X}_n = \mathbf{X}_0\mathbf{T}^n.$$

If the transition matrix $\mathbf{T}$ is regular (that is, if there is a positive integer power of $\mathbf{T}$ for which all entries are positive), then, as the number $n$ of trials increases, $\mathbf{X}_n$ gets closer and closer to the vector $\mathbf{Q}$, called the steady-state vector of $\mathbf{T}$. If

$$\mathbf{Q} = [q_1 \ q_2 \ \cdots \ q_k],$$

then the entries of $\mathbf{Q}$ indicate the long-run probability distribution of the states. The vector $\mathbf{Q}$ can be found by solving the matrix equation

$$\mathbf{QT} = \mathbf{Q},$$

subject to the condition that

$$q_1 + q_2 + \cdots + q_k = 1.$$

From a procedural point of view, the entries of $\mathbf{Q}$ can be found by reducing a matrix of the form

$$\left[ \begin{array}{c|c} \mathbf{T}^{\mathrm{T}} - \mathbf{I} & \mathbf{O} \\ \hline 1 \ldots 1 & 1 \end{array} \right].$$

## REVIEW PROBLEMS

*In Problems 1 and 2, the distribution for the random variable X is given. Construct the probability histogram and determine $\mu$, Var(X), and $\sigma$.*

**1.** $f(1) = 0.7, f(2) = 0.1, f(3) = 0.2$.

**2.** $f(0) = \frac{1}{2}, f(1) = \frac{1}{8}, f(2) = \frac{3}{8}$.

**3. Coin and Die** A fair coin and a fair die are tossed. Let $X$ be the sum of the number of heads and the number of dots that show. Determine (a) the distribution $f$ for $X$ and (b) $E(X)$.

**4. Cards** Two cards from a standard deck of 52 playing cards are randomly drawn in succession without replacement, and the number of aces, $X$, is observed. Determine (a) the distribution $f$ for $X$ and (b) $E(X)$.

**5. Card Game** In a game, a player pays $0.25 to randomly draw 2 cards, with replacement, from a standard deck of 52 playing cards. For each 10 that appears, the player receives $1. What is the player's expected gain or loss? Give your answer to the nearest cent.

**6. Gas Station Profits** An oil company determines that the probability that a gas station located along an interstate highway is successful is 0.45. A successful station earns an annual profit of $40,000; a station that is not

successful loses $10,000 annually. What is the expected gain to the company if it locates a station along an interstate highway?

**7. Mail-Order Computers** A mail-order computer company offers a 30-day money-back guarantee to any customer who is not completely satisfied with its product. The company realizes a profit of $200 for each computer sold, but assumes a loss of $100 for shipping and handling for each unit returned. The probability that a unit is returned is 0.08.

**a.** What is the expected gain for each unit shipped?

**b.** If the distributor ships 4000 units per year, what is the expected annual profit?

**8. Lottery** In a lottery, you pay $0.50 to choose a number (integer) between 0 and 9999, inclusive. If that number is drawn, you win $2500. What is your expected gain (or loss) per play?

*In Problems 9 and 10, determine the distribution f for the binomial random variable X if the number of trials is n and the probability of success on any trial is p. Also, find μ and σ.*

**9.** $n = 3, p = 0.1$.

**10.** $n = 4, p = \dfrac{1}{4}$.

*In Problems 11 and 12, determine the given probability if X is a binomial random variable, n is the number of trials, and p is the probability of success on any trial.*

**11.** $P(X \le 1)$; $n = 5, p = \dfrac{3}{4}$.

**12.** $P(X > 2)$; $n = 6, p = \dfrac{2}{3}$.

**13. Die**  A fair die is rolled four times. Find the probability that exactly three of the rolls result in a 2 or a 3.

**14. Seed Germination**  The probability that a certain type of seed germinates is 0.8. If five seeds are planted, what is the probability that none will germinate?

**15. Coin**  A biased coin is tossed four times. The probability that a head occurs on any toss is $\frac{1}{3}$. Find the probability that at least two heads occur.

**16. Marbles**  An urn contains two red, three green, and five black marbles. Five marbles are randomly drawn in succession with replacement. Find the probability that at most two of the marbles are red.

*In Problems 17 and 18, a transition matrix for a Markov chain is given. Determine the values of a, b, and c.*

**17.** $\begin{bmatrix} 0.1 & a & 0.6 \\ 2a & b & b \\ a & b & c \end{bmatrix}$.

**18.** $\begin{bmatrix} a & a & 0.2 \\ a & b & c \\ c & c & a \end{bmatrix}$.

*In Problems 19 and 20, a transition matrix T and an initial state vector $X_0$ for a Markov chain are given. Compute the state vectors $X_1$ and $X_2$.*

**19.**  $T = \begin{bmatrix} 0.1 & 0.2 & 0.7 \\ 0.3 & 0.4 & 0.3 \\ 0.1 & 0.1 & 0.8 \end{bmatrix}$,

$X_0 = [0.5 \quad 0 \quad 0.5]$.

**20.**  $T = \begin{bmatrix} 0.4 & 0.2 & 0.4 \\ 0.1 & 0.6 & 0.3 \\ 0.1 & 0.5 & 0.4 \end{bmatrix}$,

$X_0 = [0.2 \quad 0.2 \quad 0.6]$.

*In Problems 21 and 22, a transition matrix T for a Markov chain is given.*

 **a.** Compute $T^2$ and $T^3$.

 **b.** What is the probability of going from state 1 to state 2 after two steps?

 **c.** What is the probability of going from state 2 to state 1 after three steps?

**21.** $\begin{bmatrix} \frac{1}{7} & \frac{6}{7} \\ \frac{3}{7} & \frac{4}{7} \end{bmatrix}$.

**22.** $\begin{bmatrix} 0 & 1 & 0 \\ 0.5 & 0.1 & 0.4 \\ 0.1 & 0.1 & 0.8 \end{bmatrix}$.

*In Problems 23 and 24, find the steady-state vector for the given transition matrix for a Markov chain.*

**23.** $\begin{bmatrix} \frac{1}{3} & \frac{2}{3} \\ \frac{2}{3} & \frac{1}{3} \end{bmatrix}$.

**24.** $\begin{bmatrix} 0.4 & 0.3 & 0.3 \\ 0.4 & 0.2 & 0.4 \\ 0.3 & 0.3 & 0.4 \end{bmatrix}$.

**25. Automobile Market**  For a particular segment of the automobile market, the results of a survey indicate that 80 percent of Japanese-car owners would buy a Japanese car the next time and 20 percent would buy a non-Japanese car. Of non-Japanese-car owners, 40 percent would buy a non-Japanese car the next time and 60 percent would buy a Japanese car.

 **a.** Of those who currently own a Japanese car, what percentage will buy a Japanese car two cars later?

 **b.** If 60 percent of this segment currently own a Japanese car and 40 percent own a non-Japanese car, what will be the distribution for this segment of the market two cars from now?

 **c.** How will this segment be distributed in the long run?

26. **Voting** Suppose that the probabilities of voting for particular parties in a future election depend on the voting patterns in the previous election. For a certain region where there is a three-party political system, assume that these probabilities are contained in the matrix

$$\mathbf{T} = [t_{ij}] = \begin{bmatrix} 0.6 & 0.2 & 0.2 \\ 0.3 & 0.6 & 0.1 \\ 0.3 & 0.2 & 0.5 \end{bmatrix},$$

where $t_{ij}$ represents the probability that a voter who voted for party $i$ in the last election will vote for party $j$ in the next election.

a. At the last election, 40 percent of the electorate voted for party 1, 30% for party 2, and 30% for party 3. What is the expected percentage distribution of votes for the next election?

b. In the long run, what is the percentage distribution of votes? Give your answers to the nearest percent.

# MATHEMATICAL *SNAPSHOT*

In this snapshot you will determine the probability of winning a game that you might have seen on the popular television show *The Price Is Right*. For the moment think of yourself as being a contestant.

The game consists of two parts. First, four consumer products are placed before you: item 1, item 2, and so on. Accompanying each item is a fictitious price, which is not the true price of the item (see Fig. 10.9). For each item you must predict whether its

**FIGURE 10.9**   Four items with fictitious prices.

true price is higher or lower than the price shown. The number of your correct predictions is used in the second part of the game, as follows. Four shells rest on a table (see Fig. 10.10), and hidden under one of them is a small ball. If you identify the shell that covers the

**FIGURE 10.10**   Ball is under a shell.

ball, then you win the grand prize. You are allowed to randomly select as many shells as there are correct predictions from the first part of the game. For example, making three correct predictions in the first part allows you to select three of the shells. The problem is

to determine the probability that you will win the grand prize.

To solve the problem, consider taking an informal approach. That is, use intuition to make an educated guess about the solution. As a first step, simplify the situation a bit. For each of the four items, assume the probability of the event that you make a correct prediction is $p$. Then for each item the probability of making an incorrect prediction is $q = 1 - p$. You can think of making these predictions as being four independent trials of a binomial experiment in which $p$ is the probability of success and $q$ is the probability of failure in any trial.

Let $X$ be the number of correct predictions. Then $X$ has a binomial distribution with $n = 4$. The expected value of $X$ (that is, the average number of correct predictions) is given by

$$E(X) = np = 4p.$$

Thus the average number of shells that you will choose is also $4p$. Let $B$ be the event that you choose the shell with the ball under it. Then it seems reasonable to guess that

$$P(B) = \frac{\text{number of shells selected}}{4}$$

$$= \frac{4p}{4} = p.$$

......................................
[6]Adapted from D. W. Turner, D. M. Young, and V. R. Marco, "Maybe the Price Doesn't Have to Be Right: Analysis of a Popular TV Game Show," *The College Mathematics Journal,* 19, no. 5 (1988), 419–21. By permission of the Mathematical Association of America.

As you know, this result assumes that for each item the probability that you will make a correct prediction is $p$. To handle the more general situation in which this is not necessarily true, you can proceed as follows. Let $p_i$ be the probability of the event that you make a correct prediction for item $i$, where $i = 1, 2, 3, 4$. Then $q_i = 1 - p_i$. Also, let

$$I_i = \begin{bmatrix} 1, & \text{if your prediction for item } i \text{ is correct,} \\ 0, & \text{otherwise.} \end{bmatrix}$$

Then

$$X = I_1 + I_2 + I_3 + I_4,$$

so

$$E(X) = E(I_1 + I_2 + I_3 + I_4).$$

It can be proved that the expected value of a sum of random variables is equal to the sum of their expected values. Thus you can write

$$E(X) = E(I_1) + E(I_2) + E(I_3) + E(I_4).$$

Computing $E(I_1)$ gives $E(I_1) = 1 \cdot p_1 + 0 \cdot q_1 = p_1$, and similar results are obtained for $E(I_2)$, and so on. Hence

$$E(X) = p_1 + p_2 + p_3 + p_4.$$

Again, dividing the average number of correct predictions by 4 gives your guess as to the probability of winning the grand prize:

$$P(B) = \frac{p_1 + p_2 + p_3 + p_4}{4} \tag{1}$$

[In particular, for the case that $p_1 = p_2 = p_3 = p_4 = p$, then this formula gives $P(B) = p$, which agrees with your previous result for the simplified situation.] Although $P(B)$ was intuitively obtained here, Turner, Young, and Marco give a rigorous derivation of $P(B)$ and it confirms your intuitive result.

If your knowledge of prices is perfect, then $p_i = 1$ for all $i$, so Eq. (1) gives

$$P(B) = \frac{1 + 1 + 1 + 1}{4} = 1.$$

On the other hand, just guessing gives $p_i = \frac{1}{2}$ for all $i$, so

$$P - (B) = \frac{\frac{1}{2} + \frac{1}{2} + \frac{1}{2} + \frac{1}{2}}{4} = \frac{1}{2}.$$

Thus, provided you do not try to make incorrect predictions on purpose, you have

$$\frac{1}{2} \le P(B) \le 1.$$

As Turner, Young, and Marco state, "this result is rather surprising since people intuitively believe that when the items are unfamiliar to the contestant, $P(B)$ will be below $\frac{1}{2}$." In a class experiment that they performed involving 69 student "contestants," approximately 51% won the game. As a final comment, Turner, Young, and Marco remark: "We wonder if these results were known by the game's designers."

### ■ Exercises

1. Suppose that the game is revised so that there are only three consumer items, but the number of shells remains at four. If, for each item, you merely guess whether the price is higher or lower than the fictitious price, find the probability of choosing the shell with the ball under it.

2. Suppose that the game is revised so that there are five shells, but the number of consumer items remains at four. If, for each item, you merely guess whether the price is higher or lower than the fictitious price, find the probability of choosing the shell with the ball under it.

# Limits and Continuity

**To study limits and their basic properties.**

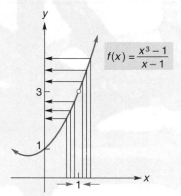

**FIGURE 11.1** $\displaystyle\lim_{x \to 1} \frac{x^3 - 1}{x - 1} = 3.$

## 11.1 LIMITS

Perhaps you have been in a parking-lot situation in which you must "inch up" to the car in front, but yet you do not want to bump or touch it. This notion of getting closer and closer to something, but yet not touching it, is very important in mathematics and is involved in the concept of *limits,* which lies at the foundation of calculus. Basically, we will let a variable "inch up" to a particular value and examine the effect it has on the values of a function.

For example, consider the function

$$f(x) = \frac{x^3 - 1}{x - 1}.$$

Although this function is not defined at $x = 1$, we may be curious about the behavior of the function values as $x$ gets very close to 1. Table 11.1 gives some values of $x$ that are slightly less than 1 and some that are slightly greater, as well as their corresponding function values. Notice that as $x$ takes on values closer and closer to 1, regardless of whether $x$ approaches it *from the left* $(x < 1)$ or *from the right* $(x > 1)$, the corresponding values of $f(x)$ get closer and closer to one and only one number, 3. This is also clear from the graph of $f$ in Figure 11.1. Notice there that even though the function is

**TABLE 11.1**

| x < 1 | | x > 1 | |
|---|---|---|---|
| **x** | **f(x)** | **x** | **f(x)** |
| 0.8 | 2.44 | 1.2 | 3.64 |
| 0.9 | 2.71 | 1.1 | 3.31 |
| 0.95 | 2.8525 | 1.05 | 3.1525 |
| 0.99 | 2.9701 | 1.01 | 3.0301 |
| 0.995 | 2.985025 | 1.005 | 3.015025 |
| 0.999 | 2.997001 | 1.001 | 3.003001 |

not defined at $x = 1$ (as indicated by the hollow dot), the function values get closer and closer to 3 as $x$ gets closer and closer to 1. To express this, we say that the **limit** of $f(x)$ as $x$ approaches 1 is 3 and write

$$\lim_{x \to 1} \frac{x^3 - 1}{x - 1} = 3.$$

We can make $f(x)$ as close to 3 as we wish by taking $x$ sufficiently close to, but not equal to, 1. The limit exists at 1, even though 1 is not in the domain of $f$.

We can also consider the limit of a function as $x$ approaches a number that is in the domain. Let us examine the limit of $f(x) = x + 3$ as $x$ approaches 2 $(x \to 2)$:

$$\lim_{x \to 2} (x + 3).$$

Obviously, if $x$ is close to 2 (but not equal to 2), then $x + 3$ is close to 5. This is also apparent from the table and graph in Fig. 11.2. Thus,

$$\lim_{x \to 2} (x + 3) = 5.$$

| $x < 2$ | | $x > 2$ | |
|---|---|---|---|
| $x$ | $f(x)$ | $x$ | $f(x)$ |
| 1.5 | 4.5 | 2.5 | 5.5 |
| 1.9 | 4.9 | 2.1 | 5.1 |
| 1.95 | 4.95 | 2.05 | 5.05 |
| 1.99 | 4.99 | 2.01 | 5.01 |
| 1.999 | 4.999 | 2.001 | 5.001 |

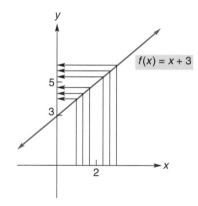

**FIGURE 11.2**   $\lim_{x \to 2} (x + 3) = 5.$

Although $x + 3$ is 5 when $x = 2$, this has no bearing on the existence of a limit.

In general, for any function $f$, we have the following definition of a limit.

**DEFINITION**

*The **limit** of $f(x)$ as $x$ approaches $a$ is the number $L$, written*

$$\lim_{x \to a} f(x) = L,$$

*provided that $f(x)$ is arbitrarily close to $L$ for all $x$ sufficiently close to, but not equal to, $a$.*

We emphasize that, when finding a limit, we are concerned not with what happens to $f(x)$ when $x$ *equals a,* but only with what happens to $f(x)$ when $x$ is *close to a.* Moreover, a limit must be independent of the way in which $x$ approaches $a$. That is, the limit must be the same whether $x$ approaches $a$ from the left or from the right (for $x < a$ or $x > a$, respectively).

**EXAMPLE 1   Estimating a Limit from a Graph**

**a.** *Estimate* $\lim_{x \to 1} f(x)$, *where the graph of f is given in Fig. 11.3(a).*

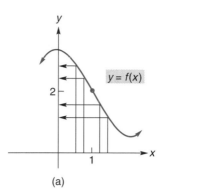 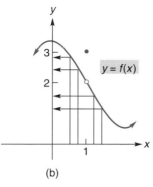

**FIGURE 11.3** Investigation of $\lim\limits_{x \to 1} f(x)$.

*Solution:* If we look at the graph for values of $x$ near 1, we see that $f(x)$ is near 2. Moreover, as $x$ gets closer and closer to 1, $f(x)$ appears to get closer and closer to 2. Thus, we estimate that

$$\lim\limits_{x \to 1} f(x) \text{ is } 2.$$

**b.** *Estimate* $\lim\limits_{x \to 1} f(x)$, *where the graph of f is given in Fig.* 11.3(b).

*Solution:* Although $f(1) = 3$, this fact has no bearing whatsoever on the *limit* of $f(x)$ as $x$ approaches 1. We see that as $x$ gets closer and closer to 1, $f(x)$ appears to get closer and closer to 2. Thus, we estimate that

$$\lim\limits_{x \to 1} f(x) \text{ is } 2. \qquad \blacksquare$$

Up to now, all of the limits that we have considered did indeed exist. Next we shall look at some situations in which a limit does not exist.

**EXAMPLE 2** Limits That Do Not Exist

**a.** *Estimate* $\lim\limits_{x \to -2} f(x)$ *if it exists, where the graph of f is given in Fig.* 11.4.

*Solution:* As $x$ approaches $-2$ from the left $(x < -2)$, the values of $f(x)$ appear to get closer to 1. But as $x$ approaches $-2$ from the right $(x > -2)$, $f(x)$ appears to get closer to 3. Hence, as $x$ approaches $-2$, the

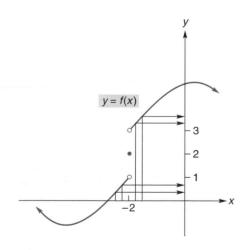

**FIGURE 11.4** $\lim\limits_{x \to -2} f(x)$ does not exist.

function values do not settle down to one and only one number. We conclude that

$$\lim_{x \to -2} f(x) \text{ does not exist.}$$

Note that the limit does not exist even though the function is defined at $x = -2$.

**b.** *Estimate* $\lim_{x \to 0} \dfrac{1}{x^2}$ *if it exists.*

*Solution:* Let $f(x) = 1/x^2$. The table in Figure 11.5 gives values of $f(x)$ for some values of $x$ near 0. As $x$ gets closer and closer to 0, the values of $f(x)$

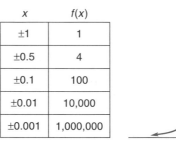
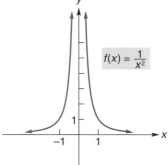

| x | f(x) |
|---|---|
| ±1 | 1 |
| ±0.5 | 4 |
| ±0.1 | 100 |
| ±0.01 | 10,000 |
| ±0.001 | 1,000,000 |

$$f(x) = \frac{1}{x^2}$$

**FIGURE 11.5** $\lim_{x \to 0} \dfrac{1}{x^2}$ does not exist.

get larger and larger without bound. This is also clear from the graph. Since the values of $f(x)$ do not approach a number as $x$ approaches 0,

$$\lim_{x \to 0} \frac{1}{x^2} \text{ does not exist.} \qquad \blacksquare$$

## TECHNOLOGY

**Problem:** Estimate $\lim_{x \to 2} f(x)$ if

$$f(x) = \frac{x^3 + 2.1x^2 - 10.2x + 4}{x^2 + 2.5x - 9}.$$

**Solution:** One method of finding the limit is by constructing a table of function values $f(x)$ when $x$ is close to 2. From Figure 11.6, we estimate the limit to be 1.57. Alternatively, we can estimate the limit from the graph of $f$. Figure 11.7 shows the graph of $f$ with the standard window of $[-10, 10] \times [-10, 10]$ First we zoom in several times around $x = 2$ and obtain Figure 11.8. After tracing around $x = 2$, we estimate the limit to be 1.57.

| X | Y1 |
|---|---|
| 1.9 | 1.4688 |
| 1.99 | 1.5592 |
| 1.999 | 1.5682 |
| 1.9999 | 1.5691 |
| 2.01 | 1.5793 |
| 2.001 | 1.5702 |
| 2.0001 | 1.5693 |

X=2.0001

**FIGURE 11.6** $\lim_{x \to 2} f(x) \approx 1.57$.

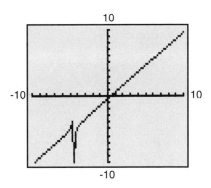

**FIGURE 11.7** Graph of $f(x)$ in standard window.

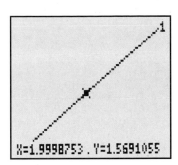

**FIGURE 11.8** Zooming and tracing around $x = 2$ gives $\lim_{x \to 2} f(x) \approx 1.57$.

### Properties of Limits

To determine limits, we do not always want to compute function values or sketch a graph. Alternatively, there are several properties of limits that we may be able to employ. The following properties may seem reasonable to you:

**1.** If $f(x) = c$ is a constant function, then

$$\lim_{x \to a} f(x) = \lim_{x \to a} c = c$$

**2.** $\lim_{x \to a} x^n = a^n$, for any positive integer $n$.

**EXAMPLE 3    Applying Limit Properties 1 and 2**

**a.** $\lim_{x \to 2} 7 = 7;$    $\lim_{x \to -5} 7 = 7.$

**b.** $\lim_{x \to 6} x^2 = 6^2 = 36.$

**c.** $\lim_{t \to -2} t^4 = (-2)^4 = 16.$ ∎

Some other properties of limits are as follows:

If $\lim_{x \to a} f(x)$ and $\lim_{x \to a} g(x)$ exist, then:

**3.** $\lim_{x \to a} [f(x) \pm g(x)] = \lim_{x \to a} f(x) \pm \lim_{x \to a} g(x).$

*That is, the limit of a sum or difference is the sum or difference, respectively, of the limits.*

**4.** $\lim_{x \to a} [f(x) \cdot g(x)] = \lim_{x \to a} f(x) \cdot \lim_{x \to a} g(x).$

*That is, the limit of a product is the product of the limits.*

**5.** $\lim_{x \to a} [cf(x)] = c \cdot \lim_{x \to a} f(x),$    where $c$ is a constant.

*That is, the limit of a constant times a function is the constant times the limit of the function.*

**Principles in Practice 2**

**Applying Limit Properties**

The volume of helium in a spherical balloon (in cubic centimeters), as a function of the radius $r$ in centimeters, is given by $V(r) = \dfrac{4}{3}\pi r^3$. Find $\lim_{r\to 1} V(r)$.

**EXAMPLE 4   Applying Limit Properties**

**a.** $\lim_{x\to 2} (x^2 + x) = \lim_{x\to 2} x^2 + \lim_{x\to 2} x$   (Property 3)

$$= 2^2 + 2 = 6 \qquad \text{(Property 2).}$$

**b.** Property 3 can be extended to the limit of a finite number of sums and differences. For example,

$$\lim_{q\to -1} (q^3 - q + 1) = \lim_{q\to -1} q^3 - \lim_{q\to -1} q + \lim_{q\to -1} 1$$

$$= (-1)^3 - (-1) + 1 = 1.$$

**c.** $\lim_{x\to 2} [(x+1)(x-3)] = \lim_{x\to 2} (x+1)\cdot \lim_{x\to 2} (x-3)$   (Property 4)

$$= [\lim_{x\to 2} x + \lim_{x\to 2} 1]\cdot[\lim_{x\to 2} x - \lim_{x\to 2} 3]$$

$$= (2+1)\cdot(2-3) = 3(-1) = -3.$$

**d.** $\lim_{x\to -2} 3x^3 = 3\cdot \lim_{x\to -2} x^3$   (Property 5)

$$= 3(-2)^3 = -24. \qquad \blacksquare$$

**Principles in Practice 3**

**Limit of a Polynomial**

The revenue function for a certain product is given by $R(x) = 500x - 6x^2$. Find $\lim_{x\to 8} R(x)$.

**EXAMPLE 5   Limit of a Polynomial Function**

Let $f(x) = c_n x^n + c_{n-1}x^{n-1} + \cdots + c_1 x + c_0$ define a polynomial function. Then

$$\lim_{x\to a} f(x) = \lim_{x\to a} (c_n x^n + c_{n-1}x^{n-1} + \cdots + c_1 x + c_0)$$

$$= c_n\cdot \lim_{x\to a} x^n + c_{n-1}\cdot \lim_{x\to a} x^{n-1} + \cdots c_1\cdot \lim_{x\to a} x + \lim_{x\to a} c_0$$

$$= c_n a^n + c_{n-1} a^{n-1} + \cdots + c_1 a + c_0 = f(a).$$

Thus, we have the following property:

> If $f$ is a polynomial function, then
>
> $$\lim_{x\to a} f(x) = f(a).$$

That is, the limit of a polynomial function as $x$ approaches $a$ is merely the value of the function at $a$. $\qquad \blacksquare$

The result of Example 5 allows us to find many limits as $x \to a$ by just substituting $a$ for $x$. For example, we can find

$$\lim_{x\to -3} (x^3 + 4x^2 - 7)$$

by substituting $-3$ for $x$ because $x^3 + 4x^2 - 7$ is a polynomial function:

$$\lim_{x\to -3} (x^3 + 4x^2 - 7) = (-3)^3 + 4(-3)^2 - 7 = 2.$$

Similarly,

$$\lim_{h\to 3} [2(h - 1)] = 2(3 - 1) = 4.$$

We want to stress that one does not evaluate limits simply by "plugging in," unless one is relying on a rule that covers the situation. We were able to find the previous two limits by direct substitution because we have a rule that applies to limits of polynomial functions. However, indiscriminate use of substitution can lead to erroneous results. To illustrate, in Example 1(b) we have

$f(1) = 3$, which is not the limit as $x \to 1$; in Example 2(a), $f(-2) = 2$, which is not the limit as $x \to -2$.

The next two limit properties concern quotients and roots.

If $\lim\limits_{x \to a} f(x)$ and $\lim\limits_{x \to a} g(x)$ exist, then:

**6.** $\lim\limits_{x \to a} \dfrac{f(x)}{g(x)} = \dfrac{\lim\limits_{x \to a} f(x)}{\lim\limits_{x \to a} g(x)},$  if  $\lim\limits_{x \to a} g(x) \neq 0.$

*That is, the limit of a quotient is the quotient of limits, provided that the denominator does not have a limit of 0.*

**7.** $\lim\limits_{x \to a} \sqrt[n]{f(x)} = \sqrt[n]{\lim\limits_{x \to a} f(x)}.$[1]

### EXAMPLE 6  Applying Limit Properties 6 and 7

**a.** $\lim\limits_{x \to 1} \dfrac{2x^2 + x - 3}{x^3 + 4} = \dfrac{\lim\limits_{x \to 1} (2x^2 + x - 3)}{\lim\limits_{x \to 1} (x^3 + 4)} = \dfrac{2 + 1 - 3}{1 + 4} = \dfrac{0}{5} = 0.$

Note that in Example 6(a) the numerator and denominator of the function are polynomials. In general, we can determine the limit of a rational function as $x \to a$ by direct substitution, provided that the denominator is not 0 at $a$.

**b.** $\lim\limits_{t \to 4} \sqrt{t^2 + 1} = \sqrt{\lim\limits_{t \to 4} (t^2 + 1)} = \sqrt{17}.$

**c.** $\lim\limits_{x \to 3} \sqrt[3]{x^2 + 7} = \sqrt[3]{\lim\limits_{x \to 3} (x^2 + 7)} = \sqrt[3]{16} = \sqrt[3]{8 \cdot 2} = 2\sqrt[3]{2}.$  ∎

## Limits and Algebraic Manipulation

We now consider limits to which our limit properties do not apply and which cannot be evaluated by direct substitution. Our technique will be to algebraically manipulate $f(x)$ so as to obtain a form to which our limit properties will apply.

### Principles in Practice 4
**Applying Limit Property**

The rate of change of productivity $p$ (in number of units produced per hour) increases with time on the job by the function $p = \dfrac{50(t^2 + 4t)}{t^2 + 3t + 20}$. Find $\lim\limits_{t \to 2} p$.

### EXAMPLE 7  Finding a Limit by Factoring and Cancellation

*Find* $\lim\limits_{x \to -1} \dfrac{x^2 - 1}{x + 1}.$

*Solution:* As $x \to -1$, both numerator and denominator approach zero. Because the limit of the denominator is 0, we *cannot* use Property 6. However, since what happens to the quotient when $x$ equals $-1$ is of no concern, we can assume that $x \neq -1$ and simplify the fraction:

$$\frac{x^2 - 1}{x + 1} = \frac{(x + 1)(x - 1)}{x + 1} = x - 1.$$

This algebraic manipulation (factoring and cancellation) on the original function $\dfrac{x^2 - 1}{x + 1}$ yields a new function $x - 1$, which is the same as the original function for $x \neq -1$. Thus,

$$\lim_{x \to -1} \frac{x^2 - 1}{x + 1} = \lim_{x \to -1} \frac{(x + 1)(x - 1)}{x + 1} = \lim_{x \to -1} (x - 1) = -1 - 1 = -2.$$

---

[1]If $n$ is even, we require that $\lim\limits_{x \to a} f(x)$ be positive.

Notice that, although the original function is not defined at $-1$, it *does* have a limit as $x \to -1$. ■

In Example 7, the method of finding a limit by direct substitution does not work. Replacing $x$ by $-1$ gives 0/0, which has no meaning. When the meaningless form 0/0 arises, algebraic manipulation (as in Example 7) may result in a form for which the limit *can* be determined.

In the beginning of this section, we found

$$\lim_{x \to 1} \frac{x^3 - 1}{x - 1}$$

by examining a table of function values of $f(x) = (x^3 - 1)/(x - 1)$ and also by considering the graph of $f$. This limit has the form 0/0. Now we shall determine the limit by using the technique described in Example 7 (the technique of factoring and cancellation).

When both $f(x)$ and $g(x)$ approach 0 as $x \to a$, then the limit

$$\lim_{x \to a} \frac{f(x)}{g(x)}$$

is said to have the *form* 0/0.

### EXAMPLE 8  Form 0/0

*Find* $\lim\limits_{x \to 1} \dfrac{x^3 - 1}{x - 1}$.

**Solution:** As $x \to 1$, both the numerator and denominator approach 0. Thus, we shall try to express the quotient in a different form for $x \neq 1$. By factoring, we have

$$\frac{x^3 - 1}{x - 1} = \frac{(x - 1)(x^2 + x + 1)}{x - 1} = x^2 + x + 1.$$

(Alternatively, long division would give the same result.) Therefore,

$$\lim_{x \to 1} \frac{x^3 - 1}{x - 1} = \lim_{x \to 1}(x^2 + x + 1) = 1^2 + 1 + 1 = 3,$$

as we showed before. ■

### EXAMPLE 9  Form 0/0

*If* $f(x) = x^2 + 1$, *find* $\lim\limits_{h \to 0} \dfrac{f(x + h) - f(x)}{h}$.

**Solution:**

$$\lim_{h \to 0} \frac{f(x + h) - f(x)}{h} = \lim_{h \to 0} \frac{[(x + h)^2 + 1] - (x^2 + 1)}{h}.$$

The expression

$$\frac{f(x + h) - f(x)}{h}$$

is called a *difference quotient*. The limit of the difference quotient lies at the heart of differential calculus. You will encounter such limits in Chapter 12.

Here we treat $x$ as a constant because $h$, not $x$, is changing. As $h \to 0$, both the numerator and denominator approach 0. Therefore, we shall try to express the quotient in a different form for $h \neq 0$. We have

$$\lim_{h \to 0} \frac{[(x + h)^2 + 1] - (x^2 + 1)}{h} = \lim_{h \to 0} \frac{[x^2 + 2xh + h^2 + 1] - x^2 - 1}{h}$$

$$= \lim_{h \to 0} \frac{2xh + h^2}{h} = \lim_{h \to 0} \frac{h(2x + h)}{h}$$

$$= \lim_{h \to 0}(2x + h) = 2x. \quad ■$$

### A Special Limit

We conclude this section with a note concerning a most important limit, namely,

$$\lim_{x\to 0}(1 + x)^{1/x}.$$

Figure 11.9 shows the graph of $f(x) = (1 + x)^{1/x}$. Although $f(0)$ does not exist, as $x \to 0$ it is clear that the limit of $(1 + x)^{1/x}$ exists. It is approximately

### Principles in Practice 5

#### Form 0/0

The length of a material increases as it is heated up according to the equation $l = 125 + 2x$. The rate at which the length is increasing is given by

$$\lim_{h\to 0}\frac{125 + 2(x + h) - (125 + 2x)}{h}.$$

Calculate this limit.

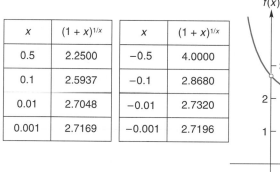

| x | $(1 + x)^{1/x}$ | x | $(1 + x)^{1/x}$ |
|------|--------|--------|--------|
| 0.5 | 2.2500 | −0.5 | 4.0000 |
| 0.1 | 2.5937 | −0.1 | 2.8680 |
| 0.01 | 2.7048 | −0.01 | 2.7320 |
| 0.001 | 2.7169 | −0.001 | 2.7196 |

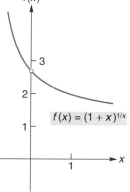

**FIGURE 11.9** $\lim_{x\to 0}(1 + x)^{1/x} = e$.

2.71828 and is denoted by the letter $e$. This, you may recall, is the base of the system of natural logarithms. The limit

This limit will be used in Chapter 13.

$$\lim_{x\to 0}(1 + x)^{1/x} = e$$

can actually be considered the definition of $e$.

## ■ Exercise 11.1

*In Problems 1–4, use the graph of f to estimate each limit if it exists.*

**1.** Graph of $f$ appears in Fig. 11.10.

   **a.** $\lim_{x\to 0} f(x)$.    **b.** $\lim_{x\to 1} f(x)$.    **c.** $\lim_{x\to 2} f(x)$.

**2.** Graph of $f$ appears in Fig. 11.11.

   **a.** $\lim_{x\to -1} f(x)$.    **b.** $\lim_{x\to 0} f(x)$.    **c.** $\lim_{x\to 1} f(x)$.

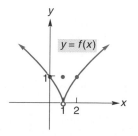

**FIGURE 11.10**
Diagram for Problem 1.

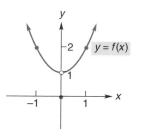

**FIGURE 11.11**
Diagram for Problem 2.

**3.** Graph of $f$ appears in Fig. 11.12.

   **a.** $\lim\limits_{x \to -1} f(x)$.
   **b.** $\lim\limits_{x \to 1} f(x)$.
   **c.** $\lim\limits_{x \to 2} f(x)$.

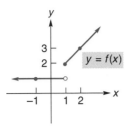

**FIGURE 11.12**
Diagram for Problem 3.

**4.** Graph of $f$ appears in Fig. 11.13.

   **a.** $\lim\limits_{x \to -1} f(x)$.
   **b.** $\lim\limits_{x \to 0} f(x)$.
   **c.** $\lim\limits_{x \to 1} f(x)$.

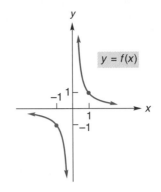

**FIGURE 11.13**
Diagram for Problem 4.

*In Problems 5–8, use your calculator to complete the table, and use your results to estimate the given limit.*

**5.** $\lim\limits_{x \to 1} \dfrac{2x^2 - x - 1}{x - 1}$.

| $x$ | 0.9 | 0.99 | 0.999 | 1.001 | 1.01 | 1.1 |
|-----|-----|------|-------|-------|------|-----|
| $f(x)$ | | | | | | |

**6.** $\lim\limits_{x \to -2} \dfrac{x^2 - 4}{x + 2}$.

| $x$ | $-2.1$ | $-2.01$ | $-2.001$ | $-1.999$ | $-1.99$ | $-1.9$ |
|-----|--------|---------|----------|----------|---------|--------|
| $f(x)$ | | | | | | |

**7.** $\lim\limits_{x \to 0} \dfrac{e^x - 1}{x}$.

| $x$ | $-0.1$ | $-0.01$ | $-0.001$ | 0.001 | 0.01 | 0.1 |
|-----|--------|---------|----------|-------|------|-----|
| $f(x)$ | | | | | | |

**8.** $\lim\limits_{h \to h} \dfrac{\sqrt{1 + h} - 1}{h}$.

| $h$ | $-0.1$ | $-0.01$ | $-0.001$ | 0.001 | 0.01 | 0.1 |
|-----|--------|---------|----------|-------|------|-----|
| $f(x)$ | | | | | | |

*In Problems 9–34, find the limits.*

**9.** $\lim\limits_{x \to 2} 16$.

**10.** $\lim\limits_{x \to 3} 2x$.

**11.** $\lim\limits_{t \to -5} (t^2 - 5)$.

**12.** $\lim\limits_{t \to 1/2} (3t - 5)$.

**13.** $\lim\limits_{x \to -1} (x^3 - 3x^2 - 2x + 1)$.

**14.** $\lim\limits_{r \to 9} \dfrac{4r - 3}{11}$.

**15.** $\lim\limits_{t \to -3} \dfrac{t - 2}{t + 5}$.

**16.** $\lim\limits_{x \to -6} \dfrac{x^2 + 6}{x - 6}$.

**17.** $\lim\limits_{h \to 0} \dfrac{h}{h^2 - 7h + 1}$.

**18.** $\lim\limits_{h \to 0} \dfrac{h^2 - 2h - 4}{h^3 - 1}$.

**19.** $\lim\limits_{p \to 4} \sqrt{p^2 + p + 5}$

**20.** $\lim\limits_{y \to 9} \sqrt{y + 3}$.

**21.** $\lim\limits_{x \to -2} \dfrac{x^2 + 2x}{x + 2}$.

**22.** $\lim\limits_{x \to -1} \dfrac{x + 1}{x + 1}$.

**23.** $\lim\limits_{x \to 2} \dfrac{x^2 - x - 2}{x - 2}$.

**24.** $\lim\limits_{t \to 0} \dfrac{t^2 + 2t}{t^2 - 2t}$.

**25.** $\lim\limits_{x \to -1} \dfrac{x^2 + 2x + 1}{x + 1}$.

**26.** $\lim\limits_{t \to 1} \dfrac{t^2 - 1}{t - 1}$.

**27.** $\lim\limits_{x \to 3} \dfrac{x - 3}{x^2 - 9}$.

**28.** $\lim\limits_{x \to 0} \dfrac{x^2 - 2x}{x}$.

**29.** $\lim\limits_{x \to 4} \dfrac{x^2 - 9x + 20}{x^2 - 3x - 4}$.

**30.** $\lim\limits_{x \to 3} \dfrac{x^4 - 81}{x^2 - x - 6}$.

**31.** $\lim\limits_{x \to 2} \dfrac{3x^2 - x - 10}{x^2 + 5x - 14}$.

**32.** $\lim\limits_{x \to -4} \dfrac{x^2 + 2x - 8}{x^2 + 5x + 4}$.

**33.** $\lim\limits_{h\to0}\dfrac{(2+h)^2-2^2}{h}$.

**34.** $\lim\limits_{x\to0}\dfrac{(x+2)^2-4}{x}$.

**35.** Find $\lim\limits_{h\to0}\dfrac{(x+h)^2-x^2}{h}$ by treating $x$ as a constant.

**36.** Find $\lim\limits_{h\to0}\dfrac{2(x+h)^2+5(x+h)-2x^2-5x}{h}$ by treating $x$ as a constant.

*In Problems 37–42, find $\lim\limits_{h\to0}\dfrac{f(x+h)-f(x)}{h}$.*

**37.** $f(x)=4-x$.

**38.** $f(x)=2x+3$.

**39.** $f(x)=x^2-3$.

**40.** $f(x)=x^2+x+1$.

**41.** $f(x)=2x^2-3x$.

**42.** $f(x)=5-2x-x^2$.

**43.** Find $\lim\limits_{x\to6}\dfrac{\sqrt{x-2}-2}{x-6}$ [*Hint:* First rationalize the numerator by multiplying both the numerator and denominator by $\sqrt{x-2}+2$.]

**44.** Find the constant $c$ so that $\lim\limits_{x\to3}\dfrac{x^2+x+c}{x^2-5x+6}$ exists. For that value of $c$, determine the limit. [*Hint:* Find the value of $c$ for which $x-3$ is a factor of the numerator.]

**45. Power Plant** The maximum theoretical efficiency of a power plant is given by

$$E=\frac{T_h-T_c}{T_h},$$

where $T_h$ and $T_c$ are the respective absolute temperatures of the hotter and colder reservoirs. Find (a) $\lim\limits_{T_c\to0}E$ and (b) $\lim\limits_{T_c\to T_h}E$.

**46. Satellite** When a 3200-lb satellite revolves about the earth in a circular orbit of radius $r$ ft, the total mechanical energy $E$ of the earth-satellite system is given by

$$E=-\frac{7.0\times10^{17}}{r}\ \text{ft}=\text{lb}.$$

Find the limit of $E$ as $r\to2.2\times10^7$ ft.

*In Problems 47–50, use a graphics calculator to graph the functions, and then estimate the limits. **Round your answers to two decimal places.***

 **47.** $\lim\limits_{x\to2}\dfrac{x^4+x^3-24}{x^2-4}$.

 **48.** $\lim\limits_{x\to4}\dfrac{\sqrt{x}-2}{\sqrt{2x-8}}$.

 **49.** $\lim\limits_{x\to16}\dfrac{x-\sqrt{x}-12}{4-\sqrt{x}}$.

**50.** $\lim\limits_{x\to1}\dfrac{x^3+x^2-5x+3}{x^3+2x^2-7x+4}$.

 **51. Water Purification** The cost of purifying water is given by $C=\dfrac{50{,}000}{p}-6500$, where $p$ is the percent of impurities remaining after purification. Graph this function on your graphics calculator, and determine $\lim\limits_{p\to0}C$. Discuss what this means.

**52. Profit Function** The profit function for a certain business is given by $P(x)=225x-3x^2-800$. Graph this function on your graphics calculator, and use the evaluation function to determine $\lim\limits_{x\to52.7}P(x)$, utilizing the rule about the limit of a polynomial function.

---

**OBJECTIVE**

To study one-sided limits, infinite limits, and limits at infinity.

## 11.2 LIMITS (CONTINUED)

### One-Sided Limits

Figure 11.14 shows the graph of a function $f$. Notice that $f(x)$ is not defined when $x=0$. As $x$ approaches 0 *from the right*, $f(x)$ approaches 1. We write this as

$$\lim\limits_{x\to0^+}f(x)=1.$$

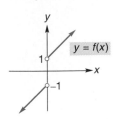

**FIGURE 11.14**
$\lim\limits_{x \to 0} f(x)$ does not exist.

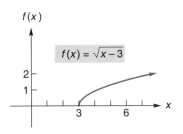

**FIGURE 11.15**
$\lim\limits_{x \to 3^+} \sqrt{x - 3} = 0.$

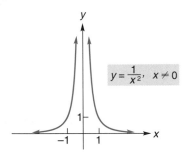

| $x$ | $f(x)$ |
|---|---|
| $\pm 1$ | 1 |
| $\pm 0.5$ | 4 |
| $\pm 0.1$ | 100 |
| $\pm 0.01$ | 10,000 |
| $\pm 0.001$ | 1,000,000 |

**FIGURE 11.16** $\lim\limits_{x \to 0} \dfrac{1}{x^2} = \infty.$

This Pitfall is extremely important.

On the other hand, as $x$ approaches 0 *from the left*, $f(x)$ approaches $-1$, and we write

$$\lim_{x \to 0^-} f(x) = -1.$$

Limits like these are called **one-sided limits**. From the preceding section, we know that the limit of a function as $x \to a$ is independent of the way $x$ approaches $a$. Thus, the limit will exist if and only if both one-sided limits exist and are equal. We therefore conclude that

$$\lim_{x \to 0^-} f(x) = -1 \text{ does not exist.}$$

As another example of a one-sided limit, consider $f(x) = \sqrt{x - 3}$ as $x$ approaches 3. Since $f$ is defined only when $x \geq 3$, we may speak of the limit as $x$ approaches 3 from the right. If $x$ is slightly greater than 3, then $x - 3$ is a positive number that is close to 0, so $\sqrt{x - 3}$ is close to 0. We conclude that

$$\lim_{x \to 3^+} \sqrt{x - 3} = 0.$$

This limit is also evident from Figure 11.15.

## Infinite Limits

In the previous section, we considered limits of the form 0/0—that is, limits where both the numerator and denominator approach 0. Now we shall examine limits where the denominator approaches 0, but the numerator approaches a number different from 0. For example, consider

$$\lim_{x \to 0} \frac{1}{x^2}.$$

Here, as $x$ approaches 0, the denominator approaches 0 and the numerator approaches 1. Let us investigate the behavior of $f(x) = 1/x^2$ when $x$ is close to 0. The number $x^2$ is positive and also close to 0. Thus, dividing 1 by such a number results in a very large number. In fact, the closer $x$ is to 0, the larger the value of $f(x)$. For example, see the table of values in Fig. 11.16, which also shows the graph of $f$. Clearly, as $x \to 0$ both from the left and from the right, $f(x)$ increases without bound. Hence, no limit exists at 0. We say that as $x \to 0$, $f(x)$ becomes positively infinite, and symbolically, we express this "infinite limit" by writing

$$\lim_{x \to 0} \frac{1}{x^2} = \infty.$$

*Pitfall* ▼ The use of the "equals" sign in this situation does not mean that the limit exists. On the contrary, the symbolism here ($\infty$) is a way of saying specifically that there is no limit, and it indicates why there is no limit.

Consider now the graph of $y = f(x) = 1/x$ for $x \neq 0$. (See Fig. 11.17.) As $x$ approaches 0 from the right, $1/x$ becomes positively infinite; as $x$ approaches 0 from the left, $1/x$ becomes negatively infinite. Symbolically, these infinite limits are written

$$\lim_{x \to 0^+} \frac{1}{x} = \infty \qquad \text{and} \qquad \lim_{x \to 0^-} \frac{1}{x} = -\infty.$$

Either one of these facts implies that

$$\lim_{x \to 0} \frac{1}{x} \text{ does not exist.}$$

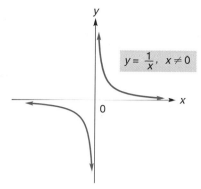

| x | f(x) |
|---|---|
| 0.01 | 100 |
| 0.001 | 1,000 |
| 0.0001 | 10,000 |
| −0.01 | −100 |
| −0.001 | −1,000 |
| −0.0001 | −10,000 |

$y = \dfrac{1}{x}, \; x \neq 0$

**FIGURE 11.17**  $\displaystyle\lim_{x \to 0} \frac{1}{x}$ does not exist.

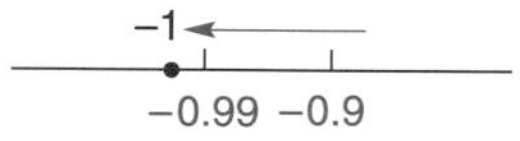

**FIGURE 11.18**  $x \to -1^+$.

### EXAMPLE 1  Infinite Limits

*Find the limit (if it exists).*

**a.** $\displaystyle\lim_{x \to -1^+} \frac{2}{x + 1}$.

*Solution:* As $x$ approaches $-1$ from the right (think of values of $x$ such as $-0.9, -0.99$, and so on, as shown in Fig. 11.18), $x + 1$ approaches 0, but is always positive. Since we are dividing 2 by positive numbers approaching 0, the results, $2/(x + 1)$, are positive numbers that are becoming arbitrarily large. Thus,

$$\lim_{x \to -1^+} \frac{2}{x + 1} = \infty,$$

and the limit does not exist. By a similar analysis, you should be able to show that

$$\lim_{x \to -1^-} \frac{2}{x + 1} = -\infty.$$

**b.** $\displaystyle\lim_{x \to 2} \frac{x + 2}{x^2 - 4}$.

*Solution:* As $x \to 2$, the numerator approaches 4 and the denominator approaches 0. Hence, we are dividing numbers near 4 by numbers near 0. The results are numbers that become arbitrarily large in magnitude. At this stage, we can write

$$\lim_{x \to 2} \frac{x + 2}{x^2 - 4} \text{ does not exist.}$$

However, let us see if we can use the symbol $\infty$ or $-\infty$ to be more specific about "does not exist." Notice that

$$\lim_{x \to 2} \frac{x + 2}{x^2 - 4} = \lim_{x \to 2} \frac{x + 2}{(x + 2)(x - 2)} = \lim_{x \to 2} \frac{1}{x - 2}.$$

Since

$$\lim_{x \to 2^+} \frac{1}{x - 2} = \infty \quad \text{and} \quad \lim_{x \to 2^-} \frac{1}{x - 2} = -\infty,$$

then $\lim\limits_{x \to 2} \dfrac{x+2}{x^2-4}$ is neither $\infty$ nor $-\infty$. ∎

Example 1 considered limits of the form $k/0$, where $k \neq 0$. It is important that you distinguish the form $k/0$ from the form $0/0$, which was discussed in Sec. 11.1. These two forms are handled quite differently.

### EXAMPLE 2  Finding a Limit

*Find* $\lim\limits_{t \to 2} \dfrac{t-2}{t^2-4}$.

**Solution:** As $t \to 2$, *both* numerator and denominator approach 0 (form 0/0). Thus, we first simplify the fraction, as we did in Sec. 11.1, and then take the limit:

$$\lim_{t \to 2} \frac{t-2}{t^2-4} = \lim_{t \to 2} \frac{t-2}{(t+2)(t-2)} = \lim_{t \to 2} \frac{1}{t+2} = \frac{1}{4}.$$ ∎

### Limits at Infinity

Now let us examine the function

$$f(x) = \frac{1}{x}$$

as $x$ becomes infinite, first in a positive sense and then in a negative sense. From Table 11.2, you can see that as $x$ increases without bound through positive values, the values of $f(x)$ approach 0. Likewise, as $x$ decreases without bound through negative values, the values of $f(x)$ also approach 0. These observations

| **TABLE 11.2** | Behavior of $f(x)$ as $x \to \pm\infty$, | | |
|---|---|---|---|
| $x$ | $f(x)$ | $x$ | $f(x)$ |
| 1,000 | 0.001 | $-1,000$ | $-0.001$ |
| 10,000 | 0.0001 | $-10,000$ | $-0.0001$ |
| 100,000 | 0.00001 | $-100,000$ | $-0.00001$ |
| 1,000,000 | 0.000001 | $-1,000,000$ | $-0.000001$ |

are also apparent from the graph in Fig. 11.17. There, as you move to the right along the curve through positive $x$-values, the corresponding $y$-values approach 0. Similarly, as you move to the left along the curve through negative $x$-values, the corresponding $y$-values approach 0. Symbolically, we write

$$\lim_{x \to \infty} \frac{1}{x} = 0 \quad \text{and} \quad \lim_{x \to -\infty} \frac{1}{x} = 0.$$

Both of these limits are called *limits at infinity*.

You should be able to obtain

$$\lim_{x \to \infty} \frac{1}{x} \quad \text{and} \quad \lim_{x \to -\infty} \frac{1}{x}$$

without the benefit of a graph or a table. Conceptually, for $x \to \infty$, increasing the value of the denominator in $1/x$ means that the fraction is getting smaller and smaller in magnitude—that is, closer to 0. Alternatively, dividing 1 by a large positive number results in a number close to 0. A similar argument can be made for the limit as $x \to -\infty$.

### EXAMPLE 3  Limits at Infinity

*Find the limit (if it exists).*

**a.** $\lim\limits_{x \to \infty} \dfrac{4}{(x-5)^3}$.

## Principles in Practice 1
## Limits at Infinity

The demand function for a certain product is given by

$$p(x) = \frac{10,000}{(x+1)^2},$$

where $p$ is the price in dollars and $x$ is the quantity sold. Graph this function on your graphics calculator in the window $[0, 10] \times [0, 10,000]$. Use the TRACE function to find $\lim_{x \to \infty} p(x)$. Determine what is happening to the graph and what this means about the demand function.

**Solution:** As $x$ becomes very large, so does $x - 5$. Since the cube of a large number is also large, $(x - 5)^3 \to \infty$. Dividing 4 by very large numbers results in numbers near 0. Thus,

$$\lim_{x \to \infty} \frac{4}{(x-5)^3} = 0.$$

**b.** $\lim_{x \to -\infty} \sqrt{4 - x}.$

**Solution:** As $x$ gets negatively infinite, $4 - x$ becomes positively infinite. Because square roots of large numbers are large numbers, we conclude that

$$\lim_{x \to -\infty} \sqrt{4 - x} = \infty.$$  ∎

In our next discussion we shall need a certain limit, namely, $\lim_{x \to \infty} 1/x^p$, where $p > 0$. As $x$ becomes very large, so does $x^p$. Dividing 1 by very large numbers results in numbers near 0. Thus, $\lim_{x \to \infty} 1/x^p = 0$. In general,

$$\lim_{x \to \infty} \frac{1}{x^p} = 0 \quad \text{and} \quad \lim_{x \to -\infty} \frac{1}{x^p} = 0,$$

where $p > 0$.[2] For example,

$$\lim_{x \to \infty} \frac{1}{\sqrt[3]{x}} = \lim_{x \to \infty} \frac{1}{x^{1/3}} = 0.$$

Let us now find the limit of the rational function

$$f(x) = \frac{4x^2 + 5}{2x^2 + 1}$$

as $x \to \infty$. (Recall from Sec. 3.2 that a rational function is a quotient of polynomials.) As $x$ gets larger and larger, *both* the numerator and denominator of $f(x)$ become infinite. However, the form of the quotient can be changed, so that we can draw a conclusion as to whether or not it has a limit. To do this, we divide both the numerator and denominator by the greatest power of $x$ that occurs in the denominator. Here it is $x^2$. This gives

$$\lim_{x \to \infty} \frac{4x^2 + 5}{2x^2 + 1} = \lim_{x \to \infty} \frac{\dfrac{4x^2 + 5}{x^2}}{\dfrac{2x^2 + 1}{x^2}} = \lim_{x \to \infty} \frac{\dfrac{4x^2}{x^2} + \dfrac{5}{x^2}}{\dfrac{2x^2}{x^2} + \dfrac{1}{x^2}}$$

$$= \lim_{x \to \infty} \frac{4 + \dfrac{5}{x^2}}{2 + \dfrac{1}{x^2}} = \frac{\lim_{x \to \infty} 4 + 5 \cdot \lim_{x \to \infty} \dfrac{1}{x^2}}{\lim_{x \to \infty} 2 + \lim_{x \to \infty} \dfrac{1}{x^2}}.$$

Since $\lim_{x \to \infty} 1/x^p = 0$ for $p > 0$,

$$\lim_{x \to \infty} \frac{4x^2 + 5}{2x^2 + 1} = \frac{4 + 5(0)}{2 + 0} = \frac{4}{2} = 2.$$

[2]For $\lim_{x \to -\infty} 1/x^p$, we assume that $p$ is such that $1/x^p$ is defined for $x < 0$.

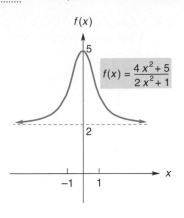

**FIGURE 11.19** $\lim_{x \to \infty} f(x) = 2$ and $\lim_{x \to -\infty} f(x) = 2.$

Similarly, the limit as $x \to -\infty$ is 2. These limits are clear from the graph of $f$ in Fig. 11.19.

For the preceding function, there is an easier way to find $\lim_{x \to \infty} f(x)$. For *large* values of $x$, in the numerator the term involving the greatest power of $x$, namely, $4x^2$, dominates the sum $4x^2 + 5$, and the dominant term in the denominator, $2x^2 + 1$, is $2x^2$. Thus, as $x \to \infty$, $f(x)$ can be approximated by $(4x^2)/(2x^2)$. As a result, to determine the limit of $f(x)$, it suffices to determine the limit of $(4x^2)/(2x^2)$. That is,

$$\lim_{x \to \infty} \frac{4x^2 + 5}{2x^2 + 1} = \lim_{x \to \infty} \frac{4x^2}{2x^2} = \lim_{x \to \infty} 2 = 2,$$

as we saw before. In general, we have the following rule:

---

**Limits at Infinity for Rational Functions**

If $f(x)$ is a *rational function* and $a_n x^n$ and $b_m x^m$ are the terms in the numerator and denominator, respectively, with the greatest powers of $x$, then

$$\lim_{x \to \infty} f(x) = \lim_{x \to \infty} \frac{a_n x^n}{b_m x^m}$$

and

$$\lim_{x \to -\infty} f(x) = \lim_{x \to -\infty} \frac{a_n x^n}{b_m x^m}.$$

---

Let us apply this rule to the situation where the degree of the numerator is greater than the degree of the denominator. For example,

$$\lim_{x \to -\infty} \frac{x^4 - 3x}{5 - 2x} = \lim_{x \to -\infty} \frac{x^4}{-2x} = \lim_{x \to -\infty} \left( -\frac{1}{2} x^3 \right) = \infty.$$

(Note that in the next-to-last step, as $x$ becomes very negative, so does $x^3$; moreover, $-\frac{1}{2}$ times a very negative number is very positive.) Similarly,

$$\lim_{x \to \infty} \frac{x^4 - 3x}{5 - 2x} = \lim_{x \to \infty} \left( -\frac{1}{2} x^3 \right) = -\infty.$$

From this illustration, we make the following conclusion:

---

If the degree of the numerator of a *rational function* is greater than the degree of the denominator, then the function has no limit as $x \to \infty$ or as $x \to -\infty$.

---

**Principles in Practice 2**

**Limits at Infinity for Rational Functions**

The yearly amount of sales $y$ of a certain company (in thousands of dollars) is related to the amount the company spends on advertising, $x$ (in thousands of dollars), according to the equation $y(x) = \dfrac{500x}{x + 20}$. Graph this function on your graphics calculator in the window $[0,1000] \times [0, 550]$. Use TRACE to explore $\lim_{x \to \infty} y(x)$, and determine what this means to the company.

**EXAMPLE 4   Limits at Infinity for Rational Functions**

*Find the limit (if it exists).*

**a.** $\lim_{x \to \infty} \dfrac{x^2 - 1}{7 - 2x + 8x^2}.$

*Solution:*

$$\lim_{x \to \infty} \frac{x^2 - 1}{7 - 2x + 8x^2} = \lim_{x \to \infty} \frac{x^2}{8x^2} = \lim_{x \to \infty} \frac{1}{8} = \frac{1}{8}.$$

**b.** $\lim_{x \to -\infty} \dfrac{x}{(3x - 1)^2}.$

*Solution:*

$$\lim_{x \to \infty} \frac{x}{(3x - 1)^2} = \lim_{x \to -\infty} \frac{x}{9x^2 - 6x + 1} = \lim_{x \to -\infty} \frac{x}{9x^2}$$

$$= \lim_{x \to -\infty} \frac{1}{9x} = \frac{1}{9} \cdot \lim_{x \to -\infty} \frac{1}{x} = \frac{1}{9}(0) = 0.$$

**c.** $\lim\limits_{x \to \infty} \dfrac{x^5 - x^4}{x^4 - x^3 + 2}.$

*Solution:* Since the degree of the numerator is greater than that of the denominator, there is no limit. More precisely,

$$\lim_{x \to \infty} \frac{x^5 - x^4}{x^4 - x^3 + 2} = \lim_{x \to \infty} \frac{x^5}{x^4} = \lim_{x \to \infty} x = \infty. \qquad \blacksquare$$

*Pitfall* ▼ Our preceding technique applies only to limits of rational functions at *infinity*. To find $\lim\limits_{x \to 0} \dfrac{x^2 - 1}{7 - 2x + 8x^2}$, we do not determine the limit of $\dfrac{x^2}{8x^2}$, because $x$ does not approach $\infty$ or $-\infty$. Instead, we have

$$\lim_{x \to 0} \frac{x^2 - 1}{7 - 2x + 8x^2} = \frac{0 - 1}{7 - 0 + 0} = -\frac{1}{7}.$$

Let us now consider the limit of the polynomial function $f(x) = 8x^2 - 2x$ as $x \to \infty$:

$$\lim_{x \to \infty} (8x^2 - 2x).$$

Because a polynomial is a rational function with denominator 1, we have

$$\lim_{x \to \infty} (8x^2 - 2x) = \lim_{x \to \infty} \frac{8x^2 - 2x}{1} = \lim_{x \to \infty} \frac{8x^2}{1} = \lim_{x \to \infty} 8x^2.$$

That is, the limit of $8x^2 - 2x$ as $x \to \infty$ is the same as the limit of the term involving the greatest power of $x$, namely, $8x^2$. As $x$ becomes very large, so does $8x^2$. Thus,

$$\lim_{x \to \infty} (8x^2 - 2x) = \lim_{x \to \infty} 8x^2 = \infty.$$

In general, we have the following:

> As $x \to \infty$ (or $x \to -\infty$), the limit of a *polynomial function* is the same as the limit of its term that involves the greatest power of $x$.

**Principles in Practice 3**
**Limits at Infinity for Polynomial Functions**

The cost $C$ of producing $x$ units of a certain product is given by $C(x) = 50,000 + 200x + 0.3x^2$. Use your graphics calculator to explore $\lim\limits_{x \to \infty} C(x)$ and determine what this means.

**EXAMPLE 5   Limits at Infinity for Polynomial Functions**

**a.** $\lim\limits_{x \to -\infty} (x^3 - x^2 + x - 2) = \lim\limits_{x \to -\infty} x^3$. As $x$ becomes very negative, so does $x^3$. Thus,

$$\lim_{x \to -\infty} (x^3 - x^2 + x - 2) = \lim_{x \to -\infty} x^3 = -\infty.$$

**b.** $\lim\limits_{x \to -\infty} (-2x^3 + 9x) = \lim\limits_{x \to -\infty} -2x^3 = \infty$, because $-2$ times a very negative number is very positive. $\qquad \blacksquare$

Do not use dominant terms when a function is not rational.

The technique of focusing on dominant terms to find limits as $x \to \infty$ or $x \to -\infty$ is valid for *rational functions,* but it is not necessarily valid for other types of functions. For example, consider

$$\lim_{x \to \infty} (\sqrt{x^2 + x} - x). \tag{1}$$

Notice that $\sqrt{x^2 + x} - x$ is not a rational function. It is *incorrect* to infer that because $x^2$ dominates in $x^2 + x$, the limit in (1) is the same as

$$\lim_{x \to \infty} (\sqrt{x^2} - x) = \lim_{x \to \infty} (x - x) = \lim_{x \to \infty} 0 = 0.$$

It can be shown (see Problem 62) that the limit in (1) is not 0, but is $\frac{1}{2}$.

The ideas discussed in this section will now be applied to a compound function.

---

**Principles in Practice 4**

**Limits for a Compound Function**

A plumber charges $100 for the first hour of work at your house and $75 for every hour (or fraction thereof) afterwards. The function for what an $x$-hour visit will cost you is

$$f(x) = \begin{cases} \$100 & 0 < x \leq 1 \\ \$175 & 1 < x \leq 2 \\ \$250 & 2 < x \leq 3 \\ \$325 & 3 < x \leq 4 \end{cases}.$$

Find $\lim_{x \to 1} f(x)$ and $\lim_{x \to 2.5} f(x)$.

---

**EXAMPLE 6    Limts for a Compound Function**

If $f(x) = \begin{cases} x^2 + 1, & \text{if } x \geq 1 \\ 3, & \text{if } x < 1 \end{cases}$, *find the limit (if it exists).*

**a.** $\lim_{x \to 1^+} f(x).$

*Solution:* Here $x$ gets close to 1 from the right. For $x > 1$, we have $f(x) = x^2 + 1$. Thus,

$$\lim_{x \to 1^+} f(x) = \lim_{x \to 1^+} (x^2 + 1).$$

If $x$ is greater than 1, but close to 1, then $x^2 + 1$ is close to 2. Therefore,

$$\lim_{x \to 1^+} f(x) = \lim_{x \to 1^+} (x^2 + 1) = 2.$$

**b.** $\lim_{x \to 1^-} f(x).$

*Solution:* Here $x$ gets close to 1 from the left. For $x < 1$, $f(x) = 3$. Hence,

$$\lim_{x \to 1^-} f(x) = \lim_{x \to 1^-} 3 = 3.$$

**c.** $\lim_{x \to 1} f(x).$

*Solution:* We want the limit as $x$ approaches 1. However, the rule of the function depends on whether $x \geq 1$ or $x < 1$. Thus, we must consider one-sided limits. The limit as $x$ approaches 1 will exist if and only if both one-sided limits exist and are the same. From parts (a) and (b),

$$\lim_{x \to 1^+} f(x) \neq \lim_{x \to 1^-} f(x), \quad \text{since } 2 \neq 3.$$

Therefore,

$$\lim_{x \to 1} f(x) \quad \text{does not exist.}$$

**d.** $\lim_{x \to \infty} f(x).$

*Solution:* For very large values of $x$, we have $x \geq 1$, so $f(x) = x^2 + 1$. Thus,

$$\lim_{x \to \infty} f(x) = \lim_{x \to \infty} (x^2 + 1) = \lim_{x \to \infty} x^2 = \infty.$$

**e.** $\lim_{x \to -\infty} f(x).$

***Solution:*** For very negative values of $x$, we have $x < 1$, so $f(x) = 3$. Hence,

$$\lim_{x \to -\infty} f(x) = \lim_{x \to -\infty} 3 = 3.$$

All the limits in parts (a) through (c) should be obvious from the graph of $f$ in Fig. 11.20.

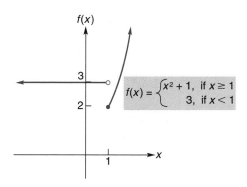

**FIGURE 11.20** Graph of compound function.

## ■ Exercise 11.2

**1.** For the function $f$ given in Fig. 11.21, find the following limits. If the limit does not exist, so state that, or use the symbol $\infty$ or $-\infty$ where appropriate.

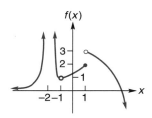

**FIGURE 11.21**
Diagram for Problem 1.

**2.** For the function $f$ given in Fig. 11.22, find the following limits. If the limit does not exist, so state that, or use the symbol $\infty$ or $-\infty$ where appropriate.

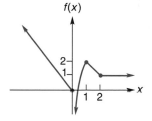

**FIGURE 11.22**
Diagram for Problem 2.

**a.** $\lim\limits_{x \to 1^-} f(x)$.　　**b.** $\lim\limits_{x \to 1^+} f(x)$.

**c.** $\lim\limits_{x \to 1} f(x)$.　　**d.** $\lim\limits_{x \to \infty} f(x)$.

**e.** $\lim\limits_{x \to -2^-} f(x)$.　　**f.** $\lim\limits_{x \to 2^+} f(x)$.

**g.** $\lim\limits_{x \to -2} f(x)$.　　**h.** $\lim\limits_{x \to -\infty} f(x)$.

**i.** $\lim\limits_{x \to -1} f(x)$.　　**j.** $\lim\limits_{x \to -1^+} f(x)$.

**k.** $\lim\limits_{x \to -1} f(x)$.

**a.** $\lim\limits_{x \to 0} f(x)$.　　**b.** $\lim\limits_{x \to 0^+} f(x)$.

**c.** $\lim\limits_{x \to 0} f(x)$.　　**d.** $\lim\limits_{x \to -\infty} f(x)$.

**e.** $\lim\limits_{x \to 1} f(x)$.　　**f.** $\lim\limits_{x \to \infty} f(x)$.

**g.** $\lim\limits_{x \to 2^+} f(x)$.　　**h.** $\lim\limits_{x \to \infty} f(x)$.

*In each of Problems 3–54, find the limit. If the limit does not exist, so state that, or use the symbol $\infty$ or $-\infty$ where appropriate.*

**3.** $\lim\limits_{x \to 3^+} (x - 2)$.　　**4.** $\lim\limits_{x \to -1^+} (1 - x^2)$.　　**5.** $\lim\limits_{x \to -\infty} 5x$.　　**6.** $\lim\limits_{x \to \infty} 3$.

**7.** $\lim_{x \to 0^-} \dfrac{6x}{x^4}.$

**8.** $\lim_{x \to 0} \dfrac{5}{x-1}.$

**9.** $\lim_{x \to -\infty} x^2.$

**10.** $\lim_{t \to \infty} (t-1)^3.$

**11.** $\lim_{h \to 0^+} \sqrt{h}.$

**12.** $\lim_{h \to 5^-} \sqrt{5-h}.$

**13.** $\lim_{x \to 5} \dfrac{3}{x-5}.$

**14.** $\lim_{x \to 0^-} 2^{1/2}.$

**15.** $\lim_{x \to 1^+} (4\sqrt{x} - 1).$

**16.** $\lim_{x \to 2^+} (x\sqrt{x^2 - 4}).$

**17.** $\lim_{x \to \infty} \sqrt{x+10}.$

**18.** $\lim_{x \to -\infty} \sqrt{2-3x}.$

**19.** $\lim_{x \to \infty} \dfrac{3}{\sqrt{x}}.$

**20.** $\lim_{x \to \infty} \dfrac{3}{2x\sqrt{x}}.$

**21.** $\lim_{x \to \infty} \dfrac{x+2}{x+3}.$

**22.** $\lim_{x \to \infty} \dfrac{2x-4}{3-2x}.$

**23.** $\lim_{x \to -\infty} \dfrac{x^2-1}{x^3+4x-3}.$

**24.** $\lim_{r \to \infty} \dfrac{r^3}{r^2+1}.$

**25.** $\lim_{t \to \infty} \dfrac{5t^2+2t+1}{4t+7}.$

**26.** $\lim_{x \to -\infty} \dfrac{2x}{3x^6-x+4}$

**27.** $\lim_{x \to \infty} \dfrac{7}{2x+1}.$

**28.** $\lim_{x \to -\infty} \dfrac{1}{(4x-1)^3}.$

**29.** $\lim_{x \to \infty} \dfrac{3-4x-2x^3}{5x^3-8x+1}.$

**30.** $\lim_{x \to \infty} \dfrac{7-2x-x^4}{9-3x^4+2x^2}.$

**31.** $\lim_{x \to 3^-} \dfrac{x+3}{x^2-9}.$

**32.** $\lim_{x \to -2^+} \dfrac{2x}{4-x^2}.$

**33.** $\lim_{w \to \infty} \dfrac{2w^2-3w+4}{5w^2+7w-1}.$

**34.** $\lim_{x \to \infty} \dfrac{4-3x^3}{x^3-1}.$

**35.** $\lim_{x \to \infty} \dfrac{6-4x^2+x^3}{4+5x-7x^2}.$

**36.** $\lim_{x \to -\infty} \dfrac{3x-x^3}{x^3+x+1}.$

**37.** $\lim_{x \to -5} \dfrac{2x^2+9x-5}{x^2+5x}.$

**38.** $\lim_{t \to 2} \dfrac{t^2+2t-8}{2t^2-5t+2}.$

**39.** $\lim_{x \to 1} \dfrac{x^2-3x+1}{x^2+1}.$

**40.** $\lim_{x \to -1} \dfrac{3x^3-x^2}{2x+1}.$

**41.** $\lim_{x \to 1^+} \left[1 + \dfrac{1}{x-1}\right].$

**42.** $\lim_{x \to -\infty} \dfrac{x^3+2x^2+1}{x^3-4}.$

**43.** $\lim_{x \to -7^-} \dfrac{x^2+1}{\sqrt{x^2-49}}.$

**44.** $\lim_{x \to -3^+} \dfrac{x}{\sqrt{9-x^2}}.$

**45.** $\lim_{x \to 0^+} \dfrac{2}{x+x^2}.$

**46.** $\lim_{x \to \infty} \left(x + \dfrac{1}{x}\right).$

**47.** $\lim_{x \to 1} x(x-1)^{-1}.$

**48.** $\lim_{x \to 1/2} \dfrac{1}{2x-1}.$

**49.** $\lim_{x \to 0^+} \left(-\dfrac{3}{x}\right).$

**50.** $\lim_{x \to 0} \left(-\dfrac{3}{x}\right).$

**51.** $\lim_{x \to 0} |x|.$

**52.** $\lim_{x \to 0} \left|\dfrac{1}{x}\right|.$

**53.** $\lim_{x \to -\infty} \dfrac{x+1}{x}.$

**54.** $\lim_{x \to \infty} \left[\dfrac{2}{x} - \dfrac{x^2}{x^2-1}\right].$

*In Problems 55–58, find the indicated limits. If the limit does not exist, so state that, or use the symbol $\infty$ or $-\infty$ where appropriate.*

**55.** $f(x) = \begin{cases} 2, & \text{if } x \le 2 \\ 1, & \text{if } x > 2 \end{cases};$

   **a.** $\lim_{x \to 2^+} f(x).$   **b.** $\lim_{x \to 2^-} f(x).$   **c.** $\lim_{x \to 2} f(x).$

   **d.** $\lim_{x \to \infty} f(x).$   **e.** $\lim_{x \to -\infty} f(x).$

**56.** $f(x) = \begin{cases} x, & \text{if } x \le 1 \\ 2, & \text{if } x > 1 \end{cases};$

   **a.** $\lim_{x \to 1^+} f(x).$   **b.** $\lim_{x \to 1^-} f(x).$   **c.** $\lim_{x \to 1} f(x).$

   **d.** $\lim_{x \to \infty} f(x).$   **e.** $\lim_{x \to -\infty} f(x).$

**57.** $g(x) = \begin{cases} x, & \text{if } x < 0 \\ -x, & \text{if } x > 0 \end{cases};$

   **a.** $\lim_{x \to 0^+} g(x).$   **b.** $\lim_{x \to 0} g(x).$   **c.** $\lim_{x \to 0} g(x).$

   **d.** $\lim_{x \to \infty} g(x).$   **e.** $\lim_{x \to -\infty} g(x).$

**58.** $g(x) = \begin{cases} x^2, & \text{if } x < 0 \\ -x, & \text{if } x > 0 \end{cases};$

   **a.** $\lim_{x \to 0^+} g(x).$   **b.** $\lim_{x \to 0^-} g(x).$   **c.** $\lim_{x \to 0} g(x).$

   **d.** $\lim_{x \to \infty} g(x).$   **e.** $\lim_{x \to -\infty} g(x).$

**59. Average Cost** If $c$ is the total cost in dollars to produce $q$ units of a product, then the average cost per unit for an output of $q$ units is given by $\bar{c} = c/q$. Thus, if the total cost equation is $c = 5000 + 6q$, then

$$\bar{c} = \dfrac{5000}{q} + 6.$$

For example, the total cost of an output of 5 units is $5030, and the average cost per unit at this level of production is $1006. By finding $\lim_{q \to \infty} \bar{c}$, show that the average cost approaches a level of stability if the producer continually increases output. What is the limiting value of

the average cost? Sketch the graph of the average-cost function.

**60. Average Cost** Repeat Problem 59, given that the fixed cost is $12,000 and the variable cost is given by the function $c_v = 7q$.

**61. Population** The population of a certain small city $t$ years from now is predicted to be

$$N = 20,000 + \dfrac{10,000}{(t+2)^2}.$$

Find the population in the long run; that is, find $\lim_{t \to \infty} N.$

**62.** Show that

$$\lim_{x \to \infty} (\sqrt{x^2 + x} - x) = \frac{1}{2}.$$

[*Hint*: Rationalize the numerator by multiplying the expression $\sqrt{x^2 + x} - x$ by

$$\frac{\sqrt{x^2 + x} + x}{\sqrt{x^2 + x} + x}.$$

Then express the denominator in a form such that $x$ is a factor.]

**63. Host–Parasite Relationship**   For a particular host–parasite relationship, it was determined that when the host density (number of hosts per unit of area) is $x$, the number of hosts parasitized over a period of time is

$$y = \frac{900x}{10 + 45x}.$$

If the host density were to increase without bound, what value would $y$ approach?

**64.** If $f(x) = \begin{cases} \sqrt{2 - x}, & \text{if } x < 2, \\ x^3 + k(x + 1), & \text{if } x \ge 2, \end{cases}$ determine the value of the constant $k$ for which $\lim_{x \to 2} f(x)$ exists.

*In Problems 65 and 66, use a calculator to evaluate the given function when $x = 1, 0.5, 0.2, 0.1, 0.01, 0.001,$ and $0.0001$. From your results, draw a conclusion about $\lim_{x \to 0^+} f(x)$*

**65.** $f(x) = x^{2x}$.

**66.** $f(x) = x \ln x$.

 **67.** Graph $f(x) = \sqrt{4x^2 - 1}$. Use the graph to estimate $\lim_{x \to 1/2^+} f(x)$.

 **68.** Graph $f(x) = \dfrac{\sqrt{x^2 - 4}}{2 - x}$. Use the graph to estimate $\lim_{x \to 2^+} f(x)$ if it exists. Use the symbol $\infty$ or $-\infty$ if appropriate.

 **69.** Graph $f(x) = \begin{cases} 2x^2 + 3, & \text{if } x < 2, \\ 2x + 5, & \text{if } x \ge 2. \end{cases}$ Use the graph to estimate each of the following limits if it exists:

a. $\lim_{x \to 2^-} f(x)$,   b. $\lim_{x \to 2^+} f(x)$,   c. $\lim_{x \to 2} f(x)$.

---

**To extend the notion of compound interest discussed in Chapters 5 and 8 to the situation where interest is compounded continuously. To develop formulas for compound amount and present value.**

## 11.3  Interest Compounded Continuously

When money is invested at a given annual rate, the interest earned each year depends on how frequently interest is compounded. For example, more interest is earned if it is compounded monthly rather than semiannually. We can successively get more interest by compounding it weekly, daily, per hour, and so on. However, there is a maximum interest that can be earned, which we now examine.

Suppose a principal of $P$ dollars is invested for $t$ years at an annual rate of $r$. If interest is compounded $k$ times a year, then the rate per conversion period is $r/k$, and there are $kt$ periods. From Sec. 5.1, the compound amount is given by

$$S = P\left(1 + \frac{r}{k}\right)^{kt}.$$

If $k \to \infty$, the number of conversion periods increases indefinitely, and the length of each period approaches 0. In this case, we say that interest is **compounded continuously**—that is, at every instant of time. The compound amount is

$$\lim_{k \to \infty} P\left(1 + \frac{r}{k}\right)^{kt},$$

which may be written

$$P\left[\lim_{k \to \infty} \left(1 + \frac{r}{k}\right)^{k/r}\right]^{rt}.$$

By making a substitution, we get the limit to have a familiar form.

By letting $x = r/k$, then as $k \to \infty$, we have $x \to 0$. Thus, the limit inside the brackets has the form $\lim_{x \to 0} (1 + x)^{1/x}$ which, as we saw in Sec. 11.1, is $e$. Therefore, we have the following formula:

> **Compound Amount under Continuous Interest**
>
> The formula
> $$S = Pe^{rt}$$
> gives the compound amount $S$ of a principal of $P$ dollars after $t$ years at an annual interest rate $r$ compounded continuously.

### EXAMPLE 1   Compound Amount

*If $100 is invested at an annual rate of 5% compounded continuously, find the compound amount at the end of*

**a.** 1 *year.*          **b.** 5 *years.*

*Solution:*

The interest of $5.13 is the maximum amount of compound interest that can be earned at an annual rate of 5%.

**a.** Here $P = 100$, $r = 0.05$, and $t = 1$, so
$$S = Pe^{rt} = 100e^{(0.05)(1)} \approx \$105.13.$$

We can compare this value with the value after one year of a $100 investment at an annual rate of 5% compounded semiannually—namely, $100(1.025)^2 \approx \$105.06$. The difference is not significant.

**b.** Here $P = 100$, $r = 0.05$, and $t = 5$, so
$$S = 100e^{(0.05)(5)} = 100e^{0.25} \approx \$128.40. \blacksquare$$

We can find an expression that gives the effective rate that corresponds to an annual rate of $r$ compounded continuously. (From Chapter 8, the effective rate is the equivalent rate compounded annually.) If $i$ is the corresponding effective rate, then after one year, a principal $P$ accumulates to $P(1 + i)$. This must equal the accumulated amount under continuous interest, $Pe^r$. Thus, $P(1 + i) = Pe^r$, from which it follows that $1 + i = e^r$, so $i = e^r - 1$.

> **Effective Rate under Continuous Interest**
>
> The effective rate corresponding to an annual rate of $r$ compounded continuously is
> $$e^r - 1.$$

### EXAMPLE 2   Effective Rate

*Find the effective rate that corresponds to an annual rate of 5% compounded continuously.*

*Solution:* The effective rate is
$$e^r - 1 = e^{0.05} - 1 \approx 0.0513,$$

or 5.13%. $\blacksquare$

If we solve $S = Pe^{rt}$ for $P$, we get $P = S/e^{rt}$. In this formula, $P$ is the principal that must be invested now at an annual rate of $r$ compounded continuously so that at the end of $t$ years the compound amount is $S$. We call $P$ the **present value** of $S$.

---

### Present Value under Continuous Interest

The formula

$$P = Se^{-rt}$$

gives the present value $P$ of $S$ dollars due at the end of $t$ years at an annual rate of $r$ compounded continuously.

---

**EXAMPLE 3**   **Trust Fund**

*A trust fund is being set up by a single payment so that at the end of 20 years there will be $25,000 in the fund. If interest is compounded continuously at an annual rate of 7%, how much money (to the nearest dollar) should be paid into the fund initially?*

*Solution:*  We want the present value of $25,000 due in 20 years. Therefore,

$$P = Se^{-rt} = 25{,}000e^{-(0.07)(20)}$$
$$= 25{,}000e^{-1.4} \approx 6165.$$

Thus, $6165 should be paid initially.   ■

## ■ Exercise 11.3

*In Problems 1 and 2, find the compound amount and compound interest if $4000 is invested for six years and interest is compounded continuously at the given annual rate.*

**1.** $5\frac{1}{2}\%$.      **2.** $9\%$.

*In Problems 3 and 4, find the present value of $2500 due eight years from now if interest is compounded continuously at the given annual rate.*

**3.** $6\frac{3}{4}\%$.      **4.** $8\%$.

*In Problems 5–8, find the effective rate of interest that corresponds to the given annual rate compounded continuously.*

**5.** $4\%$.      **6.** $7\%$.      **7.** $10\%$.      **8.** $9\%$.

**9. Investment**  If $100 is deposited in a savings account that earns interest at an annual rate of $5\frac{1}{2}\%$ compounded continuously, what is the value of the account at the end of two years?

**10. Investment**  If $1000 is invested at an annual rate of 6% compounded continuously, find the compound amount at the end of eight years.

**11. Stock Redemption**  The board of directors of a corporation agrees to redeem some of its callable preferred stock in five years. At that time, $1,000,000 will be required. If the corporation can invest money at an annual interest rate of 8% compounded continuously, how much should it presently invest so that the future value is sufficient to redeem the shares?

**12. Trust Fund**  A trust fund is being set up by a single payment so that at the end of 30 years there will be $50,000 in the fund. If interest is compounded continuously at an annual rate of 9%, how much money should be paid into the fund initially?

**13. Trust Fund**  As a gift for their newly born daughter's 20th birthday, the Smiths want to give her at that time a sum of money which has the same buying power as does $10,000 on the date of her birth. To accomplish this, they will make a single initial payment into a trust fund set up specifically for the purpose.

**a.** Assume that the annual effective rate of inflation is 10%. In 20 years, what sum will have the same buying power as does $10,000 at the date of the Smith's daughter's birth? Round your answer to the nearest dollar.

**b.** What should be the amount of the single initial payment into the fund if interest is compounded continuously at an annual rate of 12%? Round your answer to the nearest dollar.

**14. Investment**   Presently, the Smiths have $50,000 to invest for 18 months. They have two options open to them:

**a.** Invest the money in a certificate paying interest at the nominal rate of 9% compounded quarterly;

**b.** Invest the money in a savings account earning interest at the annual rate of 8.5% compounded continuously.

How much money will they have in 18 months with each option?

**15.** What annual rate compounded continuously is equivalent to an effective rate of 5%?

**16.** What annual rate $r$ compounded continuously is equivalent to a nominal rate of 6% compounded semiannually? [*Hint:* First show that $r = 2 \ln(1.03)$.]

**17. Continuous Annuity**   An annuity in which $R$ dollars are paid each year by uniform payments that are payable continuously is called a *continuous annuity* or a *continuous income stream*. The present value of a continuous annuity for $t$ years is

$$R \cdot \frac{1 - e^{-n}}{r},$$

where $r$ is the annual rate of interest compounded continuously. Find the present value of a continuous annuity of $100 a year for 20 years at 9% compounded continuously. Give your answer to the nearest dollar.

**18. Profit**   Suppose a business has an annual profit of $40,000 for the next five years and the profits are earned continuously throughout each year. Then the profits can be thought of as a continuous annuity. (See Problem 17.) If money is worth 5% compounded continuously, find the present value of the profits.

**19.** If interest is compounded continuously at an annual rate of 0.07, how many years would it take for a principal $P$ to triple? Give your answer to the nearest year.

**20.** If interest is compounded continuously, at what annual rate will a principal of $P$ double in 10 years? Give your answer to the nearest percent.

**21. Savings Options**   On July 1, 1998, Mr. Green had $1000 in a savings account at the First National Bank. This account earns interest at an annual rate of 10% compounded continuously. A competing bank was attempting to attract new customers by offering to add $50 immediately to any new account opened with a minimum $1000 deposit, and the new account would earn interest at the annual rate of 10% compounded semiannually. Mr. Green decided to choose one of the following three options on July 1, 1998:

**a.** Leave the money at the First National Bank;

**b.** Move the money to the competing bank;

**c.** Leave half the money at the First National Bank and move the other half to the competing bank.

For each of these three options, find Mr. Green's accumulated amount on July 1, 2000.

**22. Investment**

**a.** On November 1, 1985, Ms. Rodgers invested $10,000 in a 10-year certificate of deposit that paid interest at the annual rate of 8% compounded continuously. When the certificate matured on November 1, 1995, she reinvested the entire accumulated amount in corporate bonds, which earn interest at the rate of 10% compounded annually. To the nearest dollar, what will be Ms. Rodgers's accumulated amount on November 1, 2000?

**b.** If Ms. Rodgers had made a single investment of $10,000 in 1985 that matured in 2000 and had an effective rate of interest of 9%, would her accumulated amount be more or less than that in part (a) and by how much (to the nearest dollar)?

**23. Investment Strategy**   Suppose that you have $9000 to invest.

**a.** If you invest it with the First National Bank at the nominal rate of 10% compounded quarterly, find the accumulated amount at the end of one year.

**b.** The First National Bank also offers certificates on which it pays 11% compounded continuously. However, a minimum investment of $10,000 is required. Because you have only $9000, the bank is willing to give you a 1-year loan for the extra $1000 that you need. Interest for this loan is at an effective rate of 16%, and both principal and interest are payable at the end of the year. Determine whether or not this strategy of investment is preferable to the strategy in part (a).

---

To study continuity in the context of limits and to find points of discontinuity for a function.

## 11.4 Continuity

Many functions have the property that there is no "break" in their graphs. For example, compare the functions

$$f(x) = x \quad \text{and} \quad g(x) = \begin{cases} x, & \text{if } x \neq 1, \\ 2, & \text{if } x = 1, \end{cases}$$

whose graphs appear in Figs. 11.23 and 11.24, respectively. When $x = 1$, the graph of $f$ is unbroken, but the graph of $g$ has a break. Stated another way, if you were to trace both graphs with a pencil, you would have to lift the pencil

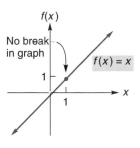

**FIGURE 11.23**
Continuous at 1.

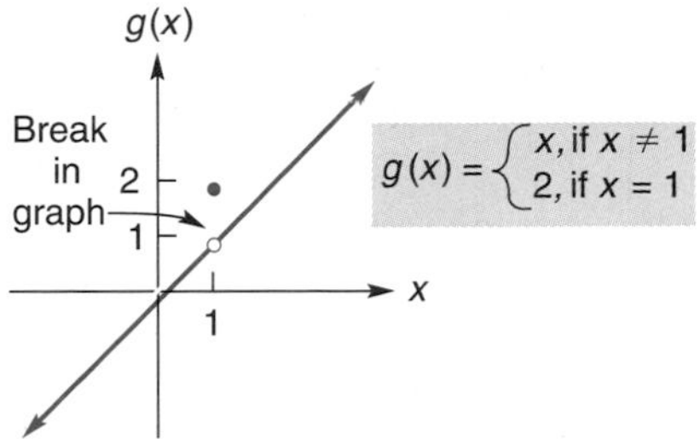

**FIGURE 11.24**
Discontinuous at 1.

off of the graph of $g$ when $x = 1$, but you would not have to lift it off of the graph of $f$. These situations can be expressed by limits. As $x$ approaches 1, compare the limit of each function with the value of the function at $x = 1$:

$$\lim_{x \to 1} f(x) = 1 = f(1),$$

whereas

$$\lim_{x \to 1} g(x) = 1 \neq g(1) = 2.$$

The limit of $f$ as $x \to 1$ is the same as $f(1)$, but the limit of $g$ as $x \to 1$ is *not* the same as $g(1)$. For these reasons, we say that $f$ is *continuous* at 1 and $g$ is *discontinuous* at 1.

**DEFINITION**

*A function f is **continuous** at a if and only if the following three conditions are met:*

    **1.** *$f(x)$ is defined at $x = a$; that is, u is in the domain of f.*

    **2.** *$\lim_{x \to a} f(x)$ exists.*

    **3.** *$\lim_{x \to a} f(x) = f(a)$.*

*If f is defined on an open interval containing a, except perhaps at a itself, and f is not continuous at a, then f is said to be **discontinuous** at a, and a is called a **point of discontinuity**.*

**EXAMPLE 1  Applying the Definition of Continuity**

**a.** *Show that $f(x) = 5$ is continuous at 7.*

    *Solution:* We must verify that the preceding three conditions are met. First, $f(7) = 5$, so $f$ is defined at $x = 7$. Second,

$$\lim_{x \to 7} f(x) = \lim_{x \to 7} 5 = 5.$$

Thus, $f$ has a limit as $x \to 7$. Third,

$$\lim_{x \to 7} f(x) = 5 = f(7).$$

Therefore, $f$ is continuous at 7. (See Fig. 11.25.)

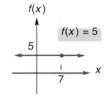

**FIGURE 11.25**  $f$ is continuous at 7.

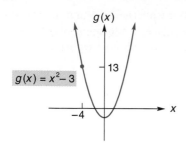

**FIGURE 11.26** *g* is continuous at −4.

**b.** *Show that $g(x) = x^2 - 3$ is continuous at $-4$.*

***Solution:*** The function $g$ is defined at $x = -4$: $g(-4) = 13$. Also,

$$\lim_{x \to -4} g(x) = \lim_{x \to -4} (x^2 - 3) = 13 = g(-4).$$

Therefore, $g$ is continuous at $-4$. (See Fig. 11.26.) ∎

We say that a function is *continuous on an interval* if it is continuous at each point there. In this situation, the graph of the function is connected over the interval. For example, $f(x) = x^2$ is continuous on the interval $[2, 5]$. In fact, in Example 5 of Sec. 11.1, we showed that, for *any* polynomial function $f$, $\lim_{x \to a} f(x) = f(a)$. This means that:

> A polynomial function is continuous at every point.

It follows that such a function is continuous on every interval. We say that polynomial functions are **continuous everywhere** or, more simply, that they are continuous.

**EXAMPLE 2** **Continuity of Polynomial Functions**

The functions $f(x) = 7$ and $g(x) = x^2 - 9x + 3$ are polynomial functions. Therefore, they are continuous. For example, they are continuous at 3. ∎

When is a function discontinuous? We can say that a function $f$ defined on an open interval containing $a$ is discontinuous at $a$ if

**1.** $f$ has no limit as $x \to a$,

or

**2.** as $x \to a$, $f$ has a limit that is different from $f(a)$.

Obviously, if $f$ is not defined at $a$, it is not continuous there. However, if $f$ is not defined at $a$, but *is* defined for all nearby values, then not only is $f$ not continuous at $a$, but also, it is discontinuous there. In Fig. 11.27, we can find points of discontinuity by inspection.

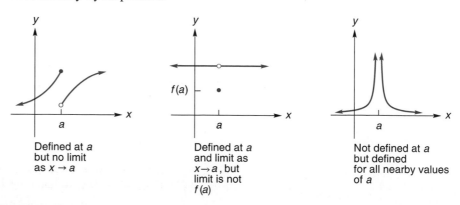

**FIGURE 11.27** Discontinuities at $a$.

**EXAMPLE 3** **Discontinuities**

**a.** Let $f(x) = 1/x$. (See Fig. 11.28.) Note that $f$ is not defined at $x = 0$, but it is defined for all other $x$ nearby. Thus, $f$ is discontinuous at 0. Moreover,

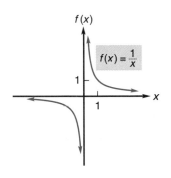

**FIGURE 11.28** Infinite discontinuity at 0.

$\lim\limits_{x \to 0^+} f(x) = \infty$ and $\lim\limits_{x \to 0^-} f(x) = -\infty$. A function is said to have an **infinite discontinuity** at $a$ when at least one of the one-sided limits is either $\infty$ or $-\infty$ as $x \to a$. Hence, $f$ has an *infinite discontinuity* at $x = 0$.

**b.** Let $f(x) = \begin{cases} 1, & \text{if } x > 0, \\ 0, & \text{if } x = 0, \\ -1, & \text{if } x < 0. \end{cases}$ (See Fig. 11.29.) Although $f$ is defined at $x = 0$,

$\lim\limits_{x \to 0} f(x)$ does not exist. Thus, $f$ is discontinuous at 0. ∎

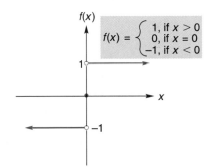

**FIGURE 11.29** Discontinuous compound function.

The following property indicates where the discontinuities of a rational function occur:

**Discontinuities of a Rational Function**

A rational function is discontinuous at points where the denominator is 0 and is continuous otherwise.

Is the function $f(x) = \dfrac{x + 1}{x + 1}$ continuous everywhere? No, it is discontinuous at $-1$. Some students think that $f(x)$ is identically 1, but it is not. Actually,

$$f(x) = 1 \text{ for } x \neq -1,$$

and $f$ is not defined for $x = -1$.

**EXAMPLE 4  Locating Discontinuities in Rational Functions**

*For each of the following functions, find all points of discontinuity.*

**a.** $f(x) = \dfrac{x^2 - 3}{x^2 + 2x - 8}$.

*Solution:* This rational function has denominator

$$x^2 + 2x - 8 = (x + 4)(x - 2),$$

which is 0 when $x = -4$ or $x = 2$. Thus, $f$ is discontinuous only at $-4$ and 2.

**b.** $h(x) = \dfrac{x + 4}{x^2 + 4}$.

*Solution:* For this rational function, the denominator is never 0. (It is always positive.) Therefore, $h$ has no discontinuity. ∎

**EXAMPLE 5  Locating Discontinuities in Compound Functions**

*For each of the following functions, find all points of discontinuity.*

**a.** $f(x) = \begin{cases} x + 6, & \text{if } x \geq 3, \\ x^2, & \text{if } x < 3. \end{cases}$

*Solution:* The only possible trouble may occur when $x = 3$, because this is the only place at which the graph of $f$ could be disconnected. We know that $f(3) = 3 + 6 = 9$. So because

$$\lim_{x \to 3^+} f(x) = \lim_{x \to 3^+} (x + 6) = 9$$

and

$$\lim_{x \to 3^-} f(x) = \lim_{x \to 3^-} x^2 = 9,$$

we can conclude that $\lim_{x \to 3} f(x) = 9 = f(3)$. Thus, the function is continuous at 3, as well as at all other $x$. We can reach the same conclusion by inspecting the graph of $f$ in Fig. 11.30.

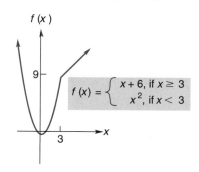

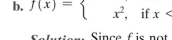

**FIGURE 11.30** Continuous compound function.

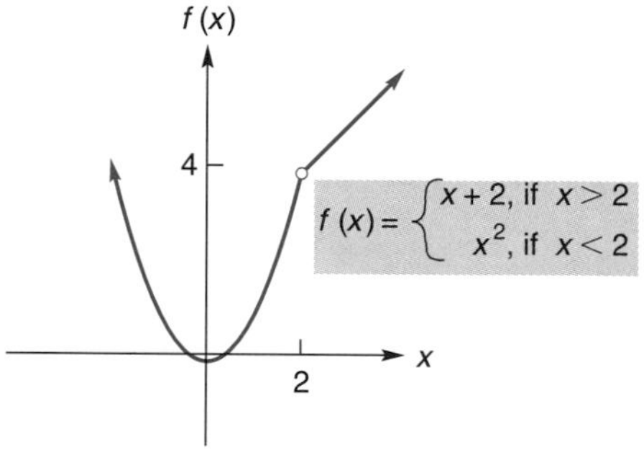

**FIGURE 11.31** Discontinuous at 2.

**b.** $f(x) = \begin{cases} x + 2, & \text{if } x > 2, \\ x^2, & \text{if } x < 2. \end{cases}$

*Solution:* Since $f$ is not defined at $x = 2$, but it is defined for all other $x$ nearby, $f$ is discontinuous at 2. It is continuous for all other $x$. (See Fig. 11.31.) ∎

**EXAMPLE 6**  **"Post-Office" Function**

The "post-office" function

$$c = f(x) = \begin{cases} 32, & \text{if } 0 < x \le 1, \\ 55, & \text{if } 1 < x \le 2, \\ 78, & \text{if } 2 < x \le 3, \\ 101, & \text{if } 3 < x \le 4, \end{cases}$$

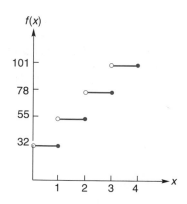

**FIGURE 11.32**  Post-office function.

gives the cost $c$ (in cents) of mailing a parcel of weight $x$ (ounces), for $0 < x \le 4$, in January 1998. It is clear from its graph in Fig. 11.32 that $f$ has discontinuities at 1, 2, and 3 and is constant for values of $x$ between successive discontinuities. Such a function is called a *step function* because of the appearance of its graph. ∎

There is another way to express continuity besides that given in the definition. If we take the statement

$$\lim_{x \to a} f(x) = f(a)$$

and replace $x$ by $a + h$, then as $x \to a$, we have $h \to 0$ (Fig. 11.34). Thus, the statement

$$\lim_{h \to 0} f(a + h) = f(a)$$

defines continuity at $a$.

This method of expressing continuity at $a$ is used frequently in mathematical proofs.

## TECHNOLOGY

By observing the graph of a function, we may be able to determine where a discontinuity occurs. However, we can be fooled. For example, the function

$$f(x) = \frac{x - 1}{x^2 - 1}$$

is discontinuous at $\pm 1$, but the discontinuity at 1 is not obvious from the graph of $f$ in Fig. 11.33. On the other hand, the discontinuity at $-1$ is obvious.

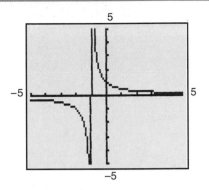

**FIGURE 11.33**   Discontinuity at 1 is not apparent from graph of $f(x) = \dfrac{x - 1}{x^2 - 1}$.

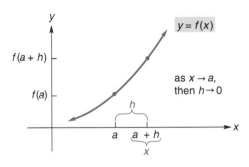

**FIGURE 11.34**   Diagram for continuity at $a$.

Often, it is helpful to describe a situation by a continuous function. For example, the demand schedule in Table 11.3 indicates the number of units of a particular product that consumers will demand per week at various prices. This

**TABLE 11.3**   Demand Schedule

| Price/Unit, $p$ | Quantity/Week, $q$ |
|---|---|
| $20 | 0 |
| 10 | 5 |
| 5 | 15 |
| 4 | 20 |
| 2 | 45 |
| 1 | 95 |

information can be given graphically, as in Fig. 11.35(a), by plotting each quantity-price pair as a point. Clearly, the graph does not represent a continuous function. Furthermore, it gives us no information as to the price at which, say, 35 units would be demanded. However, if we connect the points in Fig. 11.35(a) by a smooth curve [see Fig. 11.35(b)], we get a so-called demand curve. From it, we could guess that at about $2.50 per unit, 35 units would be demanded.

Frequently, it is possible and useful to describe a graph, as in Fig. 11.35(b), by means of an equation that defines a continuous function $f$. Such a

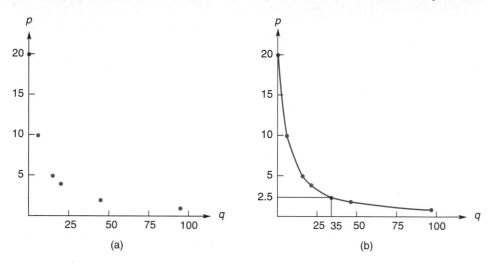

**FIGURE 11.35** Viewing data via a continuous function.

function not only gives us a demand equation, $p = f(q)$, which allows us to anticipate corresponding prices and quantities demanded, but also permits a convenient mathematical analysis of the nature and basic properties of demand. Of course, some care must be used in working with equations such as $p = f(q)$. Mathematically, $f$ may be defined when $q = \sqrt{37}$, but from a practical standpoint, a demand of $\sqrt{37}$ units could be meaningless to our particular situation. For example, if a unit is an egg, then a demand of $\sqrt{37}$ eggs make no sense.

In general, it will be our desire to view practical situations in terms of continuous functions whenever possible, so that we may be better able to analyze their nature.

## ■ Exercise 11.4

*In Problems 1–6, use the definition of continuity to show that the given function is continuous at the indicated point.*

**1.** $f(x) = x^3 - 5x$; $x = 2$.

**2.** $f(x) = \dfrac{x - 3}{9x}$; $x = -3$.

**3.** $g(x) = \sqrt{2 - 3x}$; $x = 0$.

**4.** $f(x) = \dfrac{1}{8}$; $x = 2$.

**5.** $h(x) = \dfrac{x - 4}{x + 4}$; $x = 4$.

**6.** $f(x) = \sqrt[3]{x}$; $x = -1$

*In Problems 7–12, determine whether the function is continuous at the given points.*

**7.** $f(x) = \dfrac{x + 4}{x - 2}$; $-2, 0$.

**8.** $f(x) = \dfrac{x^2 - 4x + 4}{6}$; $2, -2$.

**9.** $g(x) = \dfrac{x - 3}{x^2 - 9}$; $3, -3$.

**10.** $h(x) = \dfrac{3}{x^2 + 4}$; $2, -2$.

**11.** $f(x) = \begin{cases} x + 2, & \text{if } x \ge 2 \\ x^2 & \text{if } x < 2 \end{cases}$; $2, 0$.

**12.** $f(x) = \begin{cases} \frac{1}{x}, & \text{if } x \ne 0 \\ 0, & \text{if } x \ne 0 \end{cases}$; $0, -1$.

*In Problems 13–16, give a reason why the function is continuous everywhere.*

**13.** $f(x) = 2x^2 - 3$.

**14.** $f(x) = \dfrac{x + 2}{5}$.

**15.** $f(x) = \dfrac{x - 1}{x^2 + 4}$.

**16.** $f(x) = x(1 - x)$.

*In Problems 17–34, find all points of discontinuity.*

**17.** $f(x) = 3x^2 - 3$.

**18.** $h(x) = x - 2$.

**19.** $f(x) = \dfrac{3}{x - 4}$.

**20.** $f(x) = \dfrac{x^2 + 3x - 4}{x + 4}$.

**21.** $g(x) = \dfrac{(x^2 - 1)^2}{5}$.

**22.** $f(x) = 0$.

**23.** $f(x) = \dfrac{x^2 + 6x + 9}{x^2 + 2x - 15}$.

**24.** $g(x) = \dfrac{x - 3}{x^2 + x}$.

**25.** $h(x) = \dfrac{x - 7}{x^3 - x}$.

**26.** $f(x) = \dfrac{x}{x}$.

**27.** $p(x) = \dfrac{x}{x^2 + 1}$.

**28.** $f(x) = \dfrac{x^4}{x^4 - 1}$.

**29.** $f(x) = \begin{cases} 1, & \text{if } x \ge 0, \\ -1, & \text{if } x < 0. \end{cases}$

**30.** $f(x) = \begin{cases} 2x + 1, & \text{if } x \ge -1, \\ 1, & \text{if } x < -1. \end{cases}$

**31.** $f(x) = \begin{cases} 0, & \text{if } x \le 1, \\ x - 1, & \text{if } x > 1. \end{cases}$

**32.** $f(x) = \begin{cases} x - 3, & \text{if } x > 2, \\ 3 - 2x, & \text{if } x < 2. \end{cases}$

**33.** $f(x) = \begin{cases} x^2, & \text{if } x > 2, \\ x - 1, & \text{if } x < 2. \end{cases}$

**34.** $f(x) = \begin{cases} 5, & \text{if } x \ge 3, \\ 2x - 1, & \text{if } x < 3. \end{cases}$

**35. Telephone Rates** Suppose the long-distance rate for a telephone call from Hazleton, Pennsylvania, to Los Angeles, California, is $0.20 for the first minute and $0.20 for each additional minute or fraction thereof. If $y = f(t)$ is a function that indicates the total charge $y$ for a call of $t$ minutes' duration, sketch the graph of $f$ for $0 < t \le 4\frac{1}{2}$. Use your graph to determine the values of $t$, where $0 < t \le 4\frac{1}{2}$, at which discontinuities occur.

**36.** The *greatest integer function*, $f(x) = [x]$, is defined to be the greatest integer less than or equal to $x$, where $x$ is any real number. For example, $[3] = 3, [1.999] = 1, [\frac{1}{4}] = 0$ and $[-4.5] = -5$. Sketch the graph of this function for $-3.5 \le x \le 3.5$. Use your sketch to determine the values of $x$ at which discontinuities occur.

**37. Inventory** Sketch the graph of

$$y = f(x) = \begin{cases} -100x + 600, & \text{if } 0 \le x < 5, \\ -100x + 1100, & \text{if } 5 \le x < 10, \\ -100x + 1600, & \text{if } 10 \le x < 15. \end{cases}$$

A function such as this might describe the inventory $y$ of a company at time $x$. Is $f$ continuous at 2? At 5? At 10?

**38.** Graph $g(x) = e^{-1/x^2}$. Because $g$ is not defined at $x = 0$, but is defined for all other nearby $x$, $g$ is discontinuous at 0. Based on the graph of $g$, is

$$f(x) = \begin{cases} e^{-1/x^2}, & \text{if } x \ne 0, \\ 0, & \text{if } x = 0, \end{cases}$$

continuous at 0?

---

**OBJECTIVE**

To develop techniques for solving nonlinear inequalities.

## 11.5 CONTINUITY APPLIED TO INEQUALITIES

In Sec. 2.2, we solved linear inequalities. We now turn our attention to showing how the notion of continuity can be applied to solving a nonlinear inequality such as $x^2 + 3x - 4 > 0$.

Recall (from Sec. 3.4) that there is a relationship between the $x$-intercepts of the graph of a function $g$ (that is, the points where the graph meets the $x$-axis) and the roots of the equation $g(x) = 0$, which are called zeros of $g$. If the graph of $g$ has an $x$-intercept $(r, 0)$, then $g(r) = 0$, so $r$ is a root of the equation $g(x) = 0$, and thus, $r$ is a zero of $g$. Hence, from the graph of $y = g(x)$ in Fig. 11.36, we conclude that $r_1, r_2$, and $r_3$ are zeros of $g$. On the

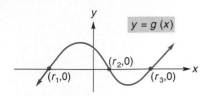

**FIGURE 11.36** $r_1, r_2,$ and $r_3$ are zeros of $g$.

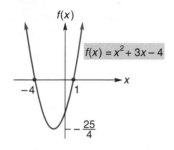

**FIGURE 11.37** $-4$ and 1 are zeros of $f$.

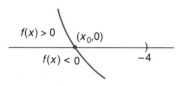

**FIGURE 11.38** Change of sign for continuous function.

Notice the importance of continuity in the solution of inequalities.

other hand, if $r$ is any real root of the equation $g(x) = 0$, then $g(r) = 0$, and hence, $(r, 0)$ lies on the graph of $g$. This means that all real zeros of $g$ can be represented by the points where the graph of $g$ meets the $x$-axis. Note also in Fig. 11.36 that the three zeros determine four open intervals on the $x$-axis:

$$(-\infty, r_1), \quad (r_1, r_2), \quad (r_2, r_3), \quad \text{and} \quad (r_3, \infty).$$

To solve $x^2 + 3x - 4 > 0$, we shall let

$$f(x) = x^2 + 3x - 4 = (x + 4)(x - 1).$$

Because $f$ is a polynomial function, it is continuous everywhere. The zeros of $f$ are $-4$ and 1; hence, the graph of $f$ has $x$-intercepts $(-4, 0)$ and $(1, 0)$. (See Fig. 11.37.) The zeros—or, to be more precise, the $x$-intercepts—determine three intervals on the $x$-axis:

$$(-\infty, -4), \quad (-4, 1), \quad \text{and} \quad (1, \infty).$$

Consider the interval $(-\infty, -4)$. Since $f$ is continuous on this interval, we claim that either $f(x) > 0$ or $f(x) < 0$ *throughout* the interval. If this were not the case, then $f(x)$ would indeed change sign on the interval. By the continuity of $f$, there would be a point where the graph intersects the $x$-axis—for example, at $(x_0, 0)$. (See Fig. 11.38.) But then $x_0$ would be a zero of $f$. However, this cannot be, because there is no zero of $f$ that is less than $-4$. Hence, $f(x)$ must be strictly positive or strictly negative on $(-\infty, -4)$. A similar argument can be made for each of the other intervals.

To determine the sign of $f(x)$ on any one of the three intervals, it suffices to determine its sign at any point in the interval. For instance, $-5$ is in $(-\infty, -4)$ and

$$f(-5) = 6 > 0. \qquad \text{Thus, } f(x) > 0 \text{ on } (-\infty, -4).$$

Similarly, 0 is in $(-4, 1)$, and

$$f(0) = -4 < 0. \qquad \text{Thus, } f(x) < 0 \text{ on } (-4, 1).$$

Finally, 3 is in $(1, \infty)$, and

$$f(3) = 14 > 0. \qquad \text{Thus, } f(x) > 0 \text{ on } (1, \infty).$$

(See the "sign chart" in Fig. 11.39.) Therefore,

$$x^2 + 3x - 4 > 0 \quad \text{on} \quad (-\infty, -4) \text{ and } (1, \infty).$$

so we have solved the inequality. These results are obvious from the graph in Fig. 11.37. The graph lies above the $x$-axis [that is, $f(x) > 0$] on $(-\infty, -4)$ and $(1, \infty)$.

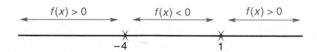

**FIGURE 11.39** Sign chart for $x^2 + 3x - 4$.

### EXAMPLE 1 Solving a Quadratic Inequality

*Solve $x^2 - 3x - 10 < 0$.*

**Solution:** If $f(x) = x^2 - 3x - 10$, then $f$ is a polynomial (quadratic) function and thus is continuous everywhere. To find the (real) zeros of $f$,

we have

$$x^2 - 3x - 10 = 0,$$

$$(x + 2)(x - 5) = 0,$$

$$x = -2, 5.$$

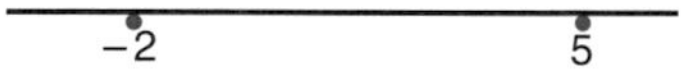

**FIGURE 11.40**   Zeros of
$x^2 - 3x - 10$.

The zeros $-2$ and $5$ are shown in Fig. 11.40. These zeros determine three intervals:

$$(-\infty, -2), \quad (-2, 5), \quad \text{and} \quad (5, \infty).$$

To determine the sign of $f(x)$ on $(-\infty, -2)$, we choose a point in that interval, say, $-3$. The sign of $f(x)$ throughout $(-\infty, -2)$ is the same as that of $f(-3)$. Because

$$f(x) = (x + 2)(x - 5) \qquad \text{[factored form of } f(x)\text{]}.$$

we have

$$f(-3) = (-3 + 2)(-3 - 5) = (-)(-) = \ + \ .$$

[Note how we conveniently found the sign of $f(-3)$ by using the signs of the factors of $f(x)$.] Thus, $f(x) > 0$ on $(-\infty, -2)$. To test the other two intervals, we shall choose the points $0$ and $6$. We find that

$$f(0) = (0 + 2)(0 - 5) = (\ +\ )(-) = -,$$

so $f(x) < 0$ on $(-2, 5)$, and

$$f(6) = (6 + 2)(6 - 5) = (\ +\ )(\ +\ ) = +,$$

so $f(x) > 0$ on $(5, \infty)$. A summary of our results is in the sign chart in Fig. 11.41. Therefore, $x^2 - 3x - 10 < 0$ on $(-2, 5)$.

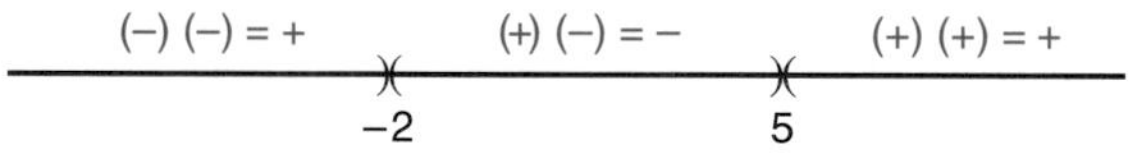

$$
\begin{array}{ccccc}
(-)\,(-) = + & & (+)\,(-) = - & & (+)\,(+) = + \\
\hline
& \!\!\times\!\! & & \!\!\times\!\! & \\
& -2 & & 5 &
\end{array}
$$

**FIGURE 11.41**   Sign chart for $(x + 2)(x - 5)$.   ■

**EXAMPLE 2**   **Solving a Polynomial Inequality**

*Solve $x(x - 1)(x + 4) \le 0$.*

***Solution:*** If $f(x) = x(x - 1)(x + 4)$, then $f$ is a polynomial function and is continuous everywhere. The zeros of $f$ are $0, 1$, and $-4$, which are shown in Fig. 11.42. These zeros determine four intervals:

$$(-\infty, -4), \quad (-4, 0), \quad (0, 1), \quad \text{and} \quad (1, \infty).$$

**FIGURE 11.42**   Zeros of
$x(x - 1)(x + 4)$.

Now, at a test point in each interval, we find the sign of $f(x)$:

$$f(-5) = (-)(-)(-) = -, \qquad \text{so } f(x) < 0 \text{ on } (-\infty, -4);$$

$$f(-2) = (-)(-)(+) = +, \qquad \text{so } f(x) > 0 \text{ on } (-4, 0);$$

$$f(\tfrac{1}{2}) = (+)(-)(+) = -, \qquad \text{so } f(x) < 0 \text{ on } (0, 1);$$

and

$$f(2) = (+)(+)(+) = +, \qquad \text{so } f(x) > 0 \text{ on } (1, \infty).$$

**Principles in Practice 1**
Solving a Polynomial Inequality

An open box is formed by cutting a square piece out of each corner of an 8-inch–by–10-inch piece of metal. If each side of the cut out squares is $x$ inches long, the volume of the box is given by $V(x) = x(8 - 2x)(10 - 2x)$. This problem makes sense only when this volume is positive. Find the values of $x$ for which the volume is positive.

Figure 11.43 shows the sign chart for $f(x)$. Thus, $x(x - 1)(x + 4) \leq 0$ on $(-\infty, -4]$ and $[0, 1]$. Note that $-4, 0$, and $1$ are included in the solution because these values satisfy the equality $(=)$ part of the inequality $(\leq)$.

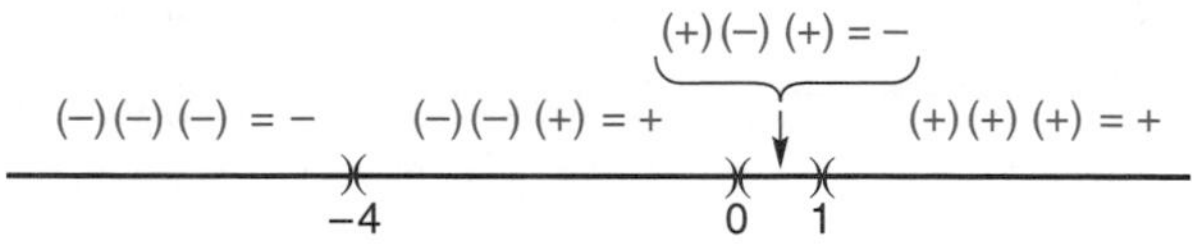

**FIGURE 11.43** Sign chart for $x(x - 1)(x + 4)$.

**EXAMPLE 3** **Solving a Rational Function Inequality**

*Solve* $\dfrac{x^2 - 6x + 5}{x} \geq 0$.

*Solution:* Let $f(x) = \dfrac{x^2 - 6x + 5}{x} = \dfrac{(x - 1)(x - 5)}{x}$. For a rational function $f$, we solve the inequality by considering the intervals determined by the zeros of $f$ and the points where $f$ is discontinuous, for it is around such points that the sign of $f(x)$ may change. Here the zeros are 1 and 5. The function is discontinuous at 0 and continuous otherwise. In Fig. 11.44, we have placed a hollow dot at 0 to indicate that $f$ is not defined there. We thus consider the intervals

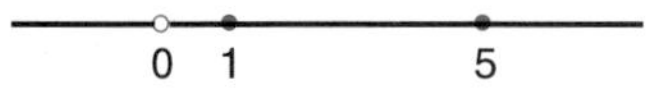

$$(-\infty, 0), \quad (0, 1), \quad (1, 5), \quad \text{and} \quad (5, \infty).$$

Determining the sign of $f(x)$ at a test point in each interval, we find that

$$f(-1) = \frac{(-)(-)}{(-)} = -, \qquad \text{so } f(x) < 0 \text{ on } (-\infty, 0);$$

$$f\left(\frac{1}{2}\right) = \frac{(-)(-)}{(+)} = +, \qquad \text{so } f(x) > 0 \text{ on } (0, 1);$$

$$f(2) = \frac{(+)(-)}{(+)} = -, \qquad \text{so } f(x) < 0 \text{ on } (1, 5);$$

and

$$f(6) = \frac{(+)(+)}{(+)} = +, \qquad \text{so } f(x) > 0 \text{ on } (5, \infty).$$

The sign chart for $f(x)$ is shown in Fig. 11.45. Therefore, $f(x) \geq 0$ on $(0, 1]$

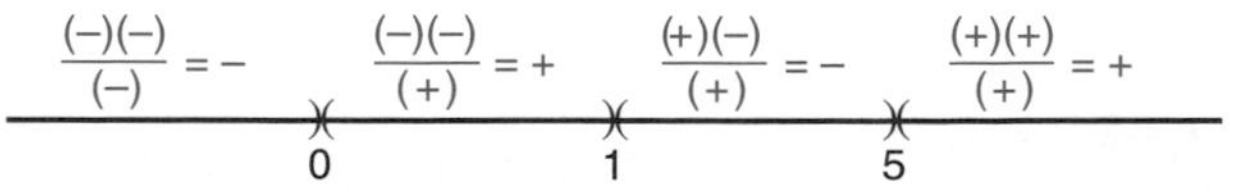

**FIGURE 11.44** Zeros and points of discontinuity for $\dfrac{(x - 1)(x - 5)}{x}$.

**FIGURE 11.45** Sign chart for $\dfrac{(x - 1)(x - 5)}{x}$.

and $[5, \infty)$. Why are 1 and 5 included, but 0 excluded? Figure 11.46 shows the graph of $f$. Notice that the solution of $f(x) \geq 0$ consists of all $x$ values for which the graph lies on or above the $x$-axis.

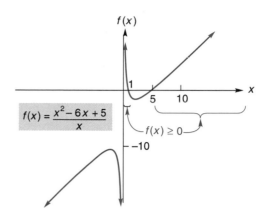

**FIGURE 11.46**   Graph of $f(x) = \dfrac{x^2 - 6x + 5}{x}$.

**EXAMPLE 4**   **Solving Nonlinear Inequalities**

**a.** *Solve $x^2 + 1 > 0$.*

*Solution:* The equation $x^2 + 1 = 0$ has no real roots. Thus, $f(x) = x^2 + 1$ has no real zero. Also, $f$ is continuous everywhere. Therefore, $f(x)$ is always positive or is always negative. But $x^2$ is always positive or zero, so $x^2 + 1$ is always positive. Hence, the solution of $x^2 + 1 > 0$ is $(-\infty, \infty)$.

**b.** *Solve $x^2 + 1 < 0$.*

*Solution:* From part (a), $x^2 + 1$ is always positive, so $x^2 + 1 < 0$ has no solution. ■

## ▪ Exercise 11.5

*In Problems 1–26, solve the inequalities by the technique discussed in this section.*

**1.** $x^2 - 3x - 4 > 0$.

**2.** $x^2 - 8x + 15 > 0$.

**3.** $x^2 - 5x + 6 \le 0$.

**4.** $14 - 5x - x^2 \le 0$.

**5.** $2x^2 + 11x + 14 < 0$.

**6.** $x^2 - 4 < 0$.

**7.** $x^2 + 4 < 0$.

**8.** $2x^2 - x - 2 \le 0$.

**9.** $(x + 2)(x - 3)(x + 6) \le 0$.

**10.** $(x - 5)(x - 2)(x + 3) \ge 0$.

**11.** $-x(x - 5)(x + 4) > 0$.

**12.** $(x + 2)^2 > 0$.

**13.** $x^3 + 4x \ge 0$.

**14.** $(x + 2)^2(x^2 - 1) < 0$.

**15.** $x^3 + 2x^2 - 3x > 0$.

**16.** $x^3 - 4x^2 + 4x > 0$.

**17.** $\dfrac{x}{x^2 - 1} < 0$.

**18.** $\dfrac{x^2 - 1}{x} < 0$.

**19.** $\dfrac{4}{x - 1} \ge 0$.

**20.** $\dfrac{3}{x^2 - 5x + 6} > 0$.

**21.** $\dfrac{x^2 - x - 6}{x^2 + 4x - 5} \ge 0$.

**22.** $\dfrac{x^2 + 2x - 8}{x^2 + 3x + 2} \ge 0$.

**23.** $\dfrac{3}{x^2 + 6x + 8} \le 0$.

**24.** $\dfrac{2x + 1}{x^2} \le 0$.

**25.** $x^2 + 2x \ge 2$.

**26.** $x^4 - 16 \ge 0$.

27. **Revenue** Suppose that consumers will purchase $q$ units of a product when the price of *each* unit is $20 - 0.1q$ dollars. How many units must be sold in order that sales revenue will be no less than $750?

28. **Forest Management** A lumber company owns a forest that is of rectangular shape, 1 mi $\times$ 2 mi. The company wants to cut a uniform strip of trees along the outer edges of the forest. At most, how wide can the strip be if the company wants at least $\frac{3}{4}$ mi$^2$ of forest to remain?

29. **Container Design** A container manufacturer wishes to make an open box by cutting a 4-in. square from each corner of a square sheet of aluminum and then turning up the sides. The box is to contain at least 324 in.$^3$. Find the dimensions of the smallest sheet of aluminum that can be used.

30. **Workshop Participation** Imperial Education Services (I.E.S.) is offering a workshop in data processing to key

personnel at Zeta Corporation. The price per person is $50, and Zeta Corporation guarantees that at least 50 people will attend. Suppose I.E.S. offers to reduce the charge for *everybody* by $0.50 for each person over the 50 who attends. How should I.E.S. limit the size of the group so that the total revenue it receives will never be less than that received for 50 persons?

 31. Graph $f(x) = x^3 + 7x^2 - 5x + 4$. Use the graph to determine the solution of

$$x^3 + 7x^2 - 5x + 4 \leq 0.$$

 32. Graph $f(x) = \dfrac{3x^2 - 0.5x + 2}{6.2 - 4.1x}$. Use the graph to determine the solution of

$$\frac{3x^2 - 0.5x + 2}{6.2 - 4.1x} > 0.$$

 *A novel way of solving a nonlinear inequality like $f(x) > 0$ is by examining the graph of $g(x) = f(x)/|f(x)|$, whose range consists only of 1 and $-1$:*

$$g(x) = \frac{f(x)}{|f(x)|} = \begin{cases} 1, & \text{if } f(x) > 0. \\ -1, & \text{if } f(x) < 0. \end{cases},$$

*The solution of $f(x) > 0$ consists of all intervals for which $g(x) = 1$. Using this technique, solve the inequalities in Problems 33 and 34.*

33. $6x^2 - x - 2 > 0$.

34. $\dfrac{x^2 - x - 2}{x^2 - 4x + 3} < 0$.

## 11.6  REVIEW

## IMPORTANT TERMS AND SYMBOLS

**Section 11.1**    $\lim_{x \to a} f(x) = L$

**Section 11.2**    one-sided limits    $\lim_{x \to a^-} f(x) = L$    $\lim_{x \to a^+} f(x) = L$    $\lim_{x \to a} f(x) = \infty$

$\lim_{x \to -\infty} f(x) = L$    $\lim_{x \to \infty} f(x) = L$

**Section 11.3**    compounding continuously

**Section 11.4**    continuous    discontinuous    continuous on an interval    continuous everywhere

## SUMMARY

The notion of a limit lies at the foundation of calculus. To say that $\lim_{x \to a} f(x) = L$ means that the values of $f(x)$ can be made as close to the number $L$ as we choose by taking $x$ sufficiently close to $a$. If $\lim_{x \to a} f(x)$ and $\lim_{x \to a} g(x)$ exist and $c$ is a constant, then:

1. $\lim_{x \to a} c = c$,

2. $\lim_{x \to a} x^n = a^n$,

3. $\lim_{x \to a} [f(x) \pm g(x)] = \lim_{x \to a} f(x) \pm \lim_{x \to a} g(x)$,

4. $\lim_{x \to a} [f(x) \cdot g(x)] = \lim_{x \to a} f(x) \cdot \lim_{x \to a} g(x)$,

5. $\lim_{x \to a} [cf(x)] = c \cdot \lim_{x \to a} f(x)$,

6. $\lim_{x \to a} \dfrac{f(x)}{g(x)} = \dfrac{\lim_{x \to a} f(x)}{\lim_{x \to a} g(x)}$ if $\lim_{x \to a} g(x) \neq 0$,

7. $\lim_{x \to a} \sqrt[n]{f(x)} = \sqrt[n]{\lim_{x \to a} f(x)}$,

8. If $f$ is a polynomial function, then $\lim_{x \to a} f(x) = f(a)$.

Property 8 implies that the limit of a polynomial function as $x \to a$ can be found by simply substituting $a$ for $x$. However, with other functions, substitution may lead to the meaningless form $0/0$. In such cases, algebraic manipulation such as factoring and cancellation may give a form from which the limit can be determined.

If $f(x)$ approaches $L$ as $x$ approaches $a$ from the right, then we write $\lim_{x \to a^+} f(x) = L$. If $f(x)$ approaches $L$ as $x$ approaches $a$ from the left, we write $\lim_{x \to a^-} f(x) = L$. These limits are called one-sided limits.

The infinity symbol $\infty$, which does not represent a number, is used in describing limits. The statement

$$\lim_{x \to \infty} f(x) = L$$

means that as $x$ increases without bound, the values of $f(x)$ approach the number $L$. A similar statement applies for the situation when $x \to -\infty$, which means that $x$ is decreasing without bound. In general, if $p > 0$, then

$$\lim_{x \to \infty} \frac{1}{x^p} = 0 \quad \text{and} \quad \lim_{x \to -\infty} \frac{1}{x^p} = 0.$$

If $f(x)$ increases without bound as $x \to a$, then we write $\lim_{x \to a} f(x) = \infty$. Similarly, if $f(x)$ decreases bound, we have $\lim_{x \to a} f(x) = -\infty$. To say that the limit of a function is $\infty$ (or $-\infty$) does not mean that the limit exists. Rather, it is a way of saying that the limit does not exist and tells *why* there is no limit.

There is a rule for evaluating the limit of a rational function (quotient of polynomials) as $x \to \infty$ or $-\infty$. If $f(x)$ is a rational function and $a_n x^n$ and $b_m x^m$ are the terms in the numerator and denominator, respectively, with the greatest powers of $x$, then

$$\lim_{x \to \infty} f(x) = \lim_{x \to \infty} \frac{a_n x^n}{b_m x^m}$$

and

$$\lim_{x \to -\infty} f(x) = \lim_{x \to -\infty} \frac{a_n x^n}{b_m x^m}.$$

In particular, as $x \to \infty$ or $-\infty$, the limit of a polynomial is the same as the limit of the term that involves the greatest power of $x$. This means that, for a nonconstant polynomial, the limit as $x \to \infty$ or $-\infty$ is either $\infty$ or $-\infty$.

When interest is compounded at every instant of time, we say that it is compounded continuously. Under continuous compounding at an annual rate $r$ for $t$ years, the formula $S = Pe^{rt}$ gives the compound amount $S$ of a principal of $P$ dollars. The formula $P = Se^{-rt}$ gives the present value $P$ of $S$ dollars. The effective rate corresponding to an annual rate $r$ compounded continuously is $e^r - 1$.

A function $f$ is continuous at $a$ if and only if

> **1.** $f(x)$ is defined at $x = a$,
>
> **2.** $\lim_{x \to a} f(x)$ exists,
>
> **3.** $\lim_{x \to a} f(x) = f(a)$.

Geometrically this means that the graph of $f$ has no break when $x = a$. If a function is not continuous at $a$ and is defined on an open interval containing $a$, except perhaps at $a$ itself, then the function is said to be discontinuous at $a$. Polynomial functions are continuous everywhere, and rational functions are discontinuous only at points where the denominator is zero.

To solve the inequality $f(x) > 0$ [or $f(x) < 0$], we first find the real zeros of $f$ and the values of $x$ for which $f$ is discontinuous. These values determine intervals, and on each interval, $f(x)$ is either always positive or always negative. To find the sign on any one of these intervals, it suffices to find the sign of $f(x)$ at any point there. After the signs are determined for all intervals, it is easy to give the solution of $f(x) > 0$ [or $f(x) < 0$].

## REVIEW PROBLEMS

*In Problems 1–28, find the limits if they exist. If the limit does not exist, so state that, or use the symbol $\infty$ or $-\infty$ where appropriate.*

**1.** $\lim_{x \to -1} (2x^2 + 6x - 1)$.

**2.** $\lim_{x \to 0} \dfrac{2x^2 - 3x + 1}{2x^2 - 2}$.

**3.** $\lim\limits_{x\to 3} \dfrac{x^2 - 9}{x^2 - 3x}$.

**4.** $\lim\limits_{x\to -2} \dfrac{x + 1}{x^2 - 2}$.

**5.** $\lim\limits_{h\to 0} (x + h)$.

**6.** $\lim\limits_{x\to 2} \dfrac{x^2 - 4}{x^2 - 3x + 2}$.

**7.** $\lim\limits_{x\to -4} \dfrac{x^3 + 4x^2}{x^2 + 2x - 8}$.

**8.** $\lim\limits_{x\to 1} \dfrac{x^2 + x - 2}{x^2 + 4x - 5}$.

**9.** $\lim\limits_{x\to \infty} \dfrac{2}{x + 1}$.

**10.** $\lim\limits_{x\to \infty} \dfrac{x^2 + 1}{x^2}$.

**11.** $\lim\limits_{x\to \infty} \dfrac{3x - 2}{5x + 3}$.

**12.** $\lim\limits_{x\to -\infty} \dfrac{1}{x^4}$.

**13.** $\lim\limits_{t\to 3} \dfrac{2t - 3}{t - 3}$.

**14.** $\lim\limits_{x\to -\infty} \dfrac{x^6}{x^5}$.

**15.** $\lim\limits_{x\to -\infty} \dfrac{x + 3}{1 - x}$.

**16.** $\lim\limits_{x\to 4} \sqrt{4}$.

**17.** $\lim\limits_{x\to \infty} \dfrac{x^2 - 1}{(3x + 2)^2}$.

**18.** $\lim\limits_{x\to 1} \dfrac{x^2 + x - 2}{x - 1}$.

**19.** $\lim\limits_{x\to 3^-} \dfrac{x + 3}{x^2 - 9}$.

**20.** $\lim\limits_{x\to 2} \dfrac{2 - x}{x - 2}$.

**21.** $\lim\limits_{x\to \infty} \sqrt{3x}$.

**22.** $\lim\limits_{y\to 5^+} \sqrt{y - 5}$.

**23.** $\lim\limits_{x\to \infty} \dfrac{x^{100} + (1/x^2)}{e - x^{98}}$.

**24.** $\lim\limits_{x\to -\infty} \dfrac{\pi x^2 - x^5}{23x - 3x^4}$.

**25.** $\lim\limits_{x\to 1} f(x)$ if $f(x) = \begin{cases} x^2, & \text{if } 0 \le x < 1, \\ x, & \text{if } > 1. \end{cases}$

**26.** $\lim\limits_{x\to 3} f(x)$ if $f(x) = \begin{cases} x + 5, & \text{if } x < 3, \\ 6, & \text{if } x \ge 3. \end{cases}$

**27.** $\lim\limits_{x\to 4^+} \dfrac{\sqrt{x^2 - 16}}{4 - x}$. [*Hint:* For $x > 4$,
$$\sqrt{x^2 - 16} = \sqrt{x - 4}\,\sqrt{x + 4}.]$$

**28.** $\lim\limits_{x\to 2^+} \dfrac{x^2 + x - 6}{\sqrt{x - 2}}$. [*Hint:* For $x > 2$,
$$\dfrac{x - 2}{\sqrt{x - 2}} = \sqrt{x - 2}.]$$

**29.** If $f(x) = 8x - 2$, find $\lim\limits_{h\to 0} \dfrac{f(x + h) - f(x)}{h}$.

**30.** If $f(x) = 2x^2 - 3$, find $\lim\limits_{h\to 0} \dfrac{f(x + h) - f(x)}{h}$.

**31. Host-Parasite Relationship** For a particular host-parasite relationship, it was determined that when the host density (number of hosts per unit of area) is $x$, then the number of hosts parasitized over a certain period of time is

$$y = 11\left(1 - \dfrac{1}{1 + 2x}\right).$$

If the host density were to increase without bound, what value would $y$ approach?

**32. Predator-Prey Relationship** For a particular predator-prey relationship, it was determined that the number $y$ of prey consumed by an individual predator over a period of time was a function of the prey density $x$ (the number of prey per unit of area). Suppose

$$y = f(x) = \dfrac{10x}{1 + 0.1x}.$$

If the prey density were to increase without bound, what value would $y$ approach?

**33.** For an annual interest of 7% compounded continuously, find:

**a.** the compound amount of $2500 after 14 years.
**b.** the present value of $2500 due in 14 years.

**34.** For an annual interest rate of 6% compounded continuously, find:

**a.** the compound amount of $800 after 9 years.
**b.** the present value of $800 due in 9 years.

**35.** Find the effective rate equivalent to an annual rate of 6% compounded continuously.

**36.** Find the effective rate equivalent to an annual rate of 1% compounded continuously.

**37.** If interest is earned at the annual rate of 10% compounded continuously, determine the exact number of years required for a principal of $P$ to double.

**38. Investment** The Smiths have just made two investments: $400 in account $A$, which earns interest at the rate of 12% compounded semiannually, and $400 in account $B$, which earns interest at the rate of 11% compounded continuously.

**a.** Calculate the effective interest rate for each investment.
**b.** Which investment is worth more at the end of 5 years and by how much?

**39.** Using the definition of continuity, show that the function $f(x) = x + 5$ is continuous at $x = 7$.

**40.** Using the definition of continuity, show that the function $f(x) = \dfrac{x - 3}{x^2 + 4}$ is continuous at $x = 3$.

**41.** State whether $f(x) = x/4$ is continuous everywhere. Give a reason for your answer.

**42.** State whether $f(x) = x^2 - 2$ is continuous everywhere. Give a reason for your answer.

*In Problems 43–50, find the points of discontinuity (if any) for each function.*

**43.** $f(x) = \dfrac{x^2}{x + 3}.$

**44.** $f(x) = \dfrac{0}{x^3}.$

**45.** $f(x) = \dfrac{x - 1}{2x^2 + 3}.$

**46.** $f(x) = (3 - 2x)^2.$

**47.** $f(x) = \dfrac{4 - x^2}{x^2 + 3x - 4}.$

**48.** $f(x) = \dfrac{2x + 6}{x^3 + x}.$

**49.** $f(x) = \begin{cases} x + 4, & \text{if } x > -2, \\ 3x + 6, & \text{if } x \le -2. \end{cases}$

**50.** $f(x) = \begin{cases} 1/x, & \text{if } x < 1, \\ 1, & \text{if } x \ge 1. \end{cases}$

*In Problems 51–58, solve the given inequalities.*

**51.** $x^2 + 4x - 12 > 0.$

**52.** $2x^2 - 6x + 4 \le 0.$

**53.** $x^3 \ge 2x^2.$

**54.** $x^3 + 8x^2 + 15x \ge 0.$

**55.** $\dfrac{x + 5}{x^2 - 1} < 0.$

**56.** $\dfrac{x(x + 5)(x + 8)}{3} < 0.$

**57.** $\dfrac{x^2 + 3x}{x^2 + 2x - 8} \ge 0.$

**58.** $\dfrac{x^2 - 4}{x^2 + 2x + 1} \le 0.$

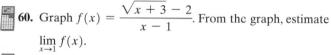 **59.** Graph $f(x) = \dfrac{x^3 + x^2 - 6x}{x^3 - 2x^2 + 6x - 12}.$ Use the graph to estimate $\lim\limits_{x \to 2} f(x).$

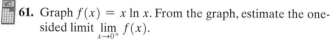 **60.** Graph $f(x) = \dfrac{\sqrt{x + 3} - 2}{x - 1}.$ From the graph, estimate $\lim\limits_{x \to 1} f(x).$

**61.** Graph $f(x) = x \ln x.$ From the graph, estimate the one-sided limit $\lim\limits_{x \to 0^+} f(x).$

**62.** Graph $f(x) = \dfrac{e^x - 1}{(e^x + 1)(e^{2x} - e^x)}.$ Use the graph to estimate $\lim\limits_{x \to 0} f(x).$

**63.** Graph $f(x) = x^3 - x^2 + x - 6.$ Use the graph to determine the solution of

$$x^3 - x^2 + x - 6 \ge 0.$$

**64.** Graph $f(x) = \dfrac{x^5 + 4}{x^3 - 1}.$ Use the graph to determine the solution of

$$\dfrac{x^5 + 4}{x^3 - 1} \le 0.$$

# MATHEMATICAL *SNAPSHOT*

## NATIONAL DEBT ....................................................

The size of the U.S. national debt is of great concern to many people and is frequently a topic in the news. The magnitude of the debt affects the confidence in the U.S. economy of both domestic and foreign investors, corporate officials, and political leaders. There are those who believe that to reduce the debt there must be cuts in government spending, which could affect government programs, or there must be an increase in revenues, possibly through tax increases.

Suppose that it is possible for the debt to be reduced continuously at an annual fixed rate. This is similar to compounding interest continuously, except that instead of adding interest to an amount at each instant of time, you would be subtracting from the debt at each instant. Let us see how you could model this situation.

Suppose the debt $D_0$ at time $t = 0$ is reduced at an annual rate $r$. Furthermore, assume that there are $k$ time periods of equal length in a year. At the end of the first period, the original debt is reduced by $D_0\left(\dfrac{r}{k}\right)$, so the new debt is

$$D_0 - D_0\left(\frac{r}{k}\right) = D_0\left(1 - \frac{r}{k}\right).$$

At the end of the second period, this debt is reduced by $D_0\left(1 - \dfrac{r}{k}\right)\dfrac{r}{k}$, so the new debt is

$$D_0\left(1 - \frac{r}{k}\right) - D_0\left(1 - \frac{r}{k}\right)\frac{r}{k}$$

$$= D_0\left(1 - \frac{r}{k}\right)\left(1 - \frac{r}{k}\right)$$

$$= D_0\left(1 - \frac{r}{k}\right)^2.$$

The pattern continues. At the end of the third period the debt is $D_0\left(1 - \dfrac{r}{k}\right)^3$, and so on. At the end of $t$ years the number of periods is $kt$ and the debt is $D_0\left(1 - \dfrac{r}{k}\right)^{kt}$. If the debt is to be reduced at each instant of time, then $k \to \infty$. Thus you want to find

$$\lim_{k \to \infty} D_0\left(1 - \frac{r}{k}\right)^{kt},$$

which may be rewritten as

$$D_0\left[\lim_{k \to \infty}\left(1 - \frac{r}{k}\right)^{-k/r}\right]^{-rt}.$$

If you let $x = -r/k$, then the condition $k \to \infty$ implies that $x \to 0$. Hence the limit inside the brackets has the form $\lim_{x \to 0}(1 + x)^{1/x}$, which is well known to be $e$. Therefore if the debt $D_0$ at time $t = 0$ is reduced continuously at an annual rate $r$, then $t$ years later the debt $D$ is given by

$$D = D_0 e^{-rt}.$$

For example, assume a current debt of \$5345 billion and a continuous reduction rate of 6% annually. Then the debt $t$ years from now is given by

$$D = 5345 e^{-0.06t},$$

where $D$ is in billions of dollars. This means that in

10 years, the debt will be $5345e^{-0.6} \approx \$2933$ billion. Figure 11.47 shows the graph of $D = 5345e^{-rt}$ for various rates $r$. Of course the greater the value of $r$, the faster the debt reduction. Notice that for $r = 0.06$, the debt at the end of 30 years is still considerable (approximately $884 billion).

It is interesting to note that decaying radioactive elements also follow the model of continuous debt reduction, $D = D_0 e^{-rt}$.

## ■ Exercises

*In the following problems, assume a current national debt of $5345 billion.*

1. If the debt were reduced to $4500 billion a year from now, what annual rate of continuous debt reduction would be involved? Give your answer to the nearest percent.

2. For a continuous debt reduction at an annual rate of 6%, determine the number of years from now required for the debt to be reduced by one-half. Give your answer to the nearest year.

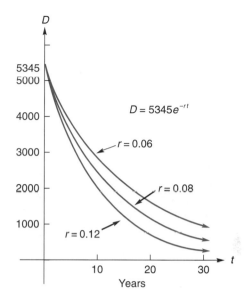

**FIGURE 11.47**    Budget debt is reduced continuously.

# Differentiation

Now we begin our study of calculus. The ideas involved in calculus are completely different from those of algebra and geometry. The power and importance of these ideas and their applications will be clear to you later in the book. In this chapter we introduce the so-called derivative of a function, and you will learn important rules for finding derivatives. You will also see how the derivative is used to analyze the rate of change of a quantity, such as the rate at which the position of a body is changing.

OBJECTIVE

To develop the idea of a tangent
line to a curve, to define the slope
of a curve, and to define a
derivative and give it a geometric
interpretation. To compute
derivatives by using the limit
definition.

## 12.1 THE DERIVATIVE

One of the main problems with which calculus deals is finding the slope of the *tangent line* at a point on a curve. In geometry you probably thought of a tangent line, or *tangent,* to a circle as a line that meets the circle at exactly one point (Fig. 12.1). However, this idea of a tangent is not very useful for other kinds of curves. For example, in Fig. 12.2(a), the lines $L_1$ and $L_2$ intersect the curve at exactly one point $P$. Although we would not think of $L_2$ as the tangent at this point, it seems natural that $L_1$ is. In Fig. 12.2(b) we intuitively would consider $L_3$ to be the tangent at point $P$, even though $L_3$ intersects the curve at other points.

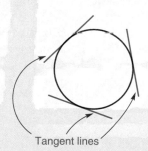

**FIGURE 12.1** Tangent lines to a circle.

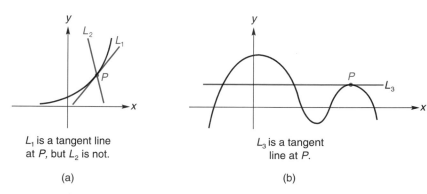

$L_1$ is a tangent line at $P$, but $L_2$ is not.

(a)

$L_3$ is a tangent line at $P$.

(b)

**FIGURE 12.2** Tangent line at a point.

From the previous examples, you can see that we must drop the idea that a tangent is simply a line that intersects a curve at only one point. To obtain a suitable definition of tangent line, we use the limit concept and the

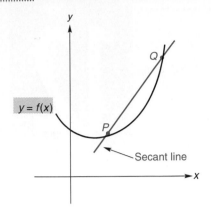

**FIGURE 12.3** Secant line $PQ$.

geometrical notion of a *secant line*. A **secant line** is a line that intersects a curve at two or more points.

Look at the graph of the function $y = f(x)$ in Fig. 12.3. We wish to define the tangent line at point $P$. If $Q$ is a different point on the curve, the line $PQ$ is a secant line. If $Q$ moves along the curve and approaches $P$ from the right (see Fig. 12.4), typical secant lines are $PQ'$, $PQ''$, and so on. As $Q$ approaches $P$ from the left, they are $PQ_1$, $PQ_2$, and so on. *In both cases, the secant lines approach the same limiting position.* This common limiting position of the secant lines is defined to be the **tangent line** to the curve at $P$. This definition seems reasonable and applies to curves in general, not just circles.

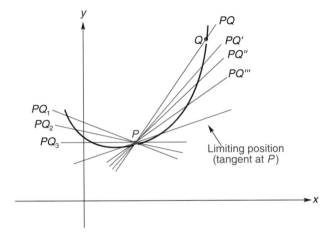

**FIGURE 12.4** The tangent line is a limiting position of secant lines.

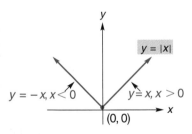

**FIGURE 12.5** No tangent line to graph of $y = |x|$ at $(0, 0)$.

A curve does not necessarily have a tangent line at each of its points. For example, the curve $y = |x|$ does not have a tangent at $(0, 0)$. As you can see in Fig. 12.5, a secant line through $(0, 0)$ and a nearby point to its right on the curve must always be the line $y = x$. Thus the limiting position of such secant lines is also the line $y = x$. However, a secant line through $(0, 0)$ and a nearby point to its left on the curve must always be the line $y = -x$. Hence, the limiting position of such secant lines is also the line $y = -x$. Since there is no common limiting position, there is no tangent line at $(0, 0)$.

Now that we have a suitable definition of a tangent to a curve at a point, we can define the *slope of a curve* at a point.

**DEFINITION**

*The **slope of a curve** at a point $P$ is the slope, if it exists, of the tangent line at $P$.*

Since the tangent at $P$ is a limiting position of secant lines $PQ$, we consider the slope of the tangent to be the limiting value of the slopes of the secant lines as $Q$ approaches $P$. For example, let us consider the curve $f(x) = x^2$ and the slopes of some secant lines $PQ$, where $P = (1, 1)$. For the point $Q = (2.5, 6.25)$, the slope of $PQ$ (see Fig. 12.6) is

$$m_{PQ} = \frac{y_2 - y_1}{x_2 - x_1} = \frac{6.25 - 1}{2.5 - 1} = 3.5.$$

Table 12.1 includes other points $Q$ on the curve, as well as the corresponding slopes of $PQ$. Notice that as $Q$ approaches $P$, the slopes of the secant lines seem to approach 2. Thus, we expect the slope of the indicated tangent line at $(1, 1)$ to be 2. This will be confirmed later, in Example 1. But first, we wish to generalize our procedure.

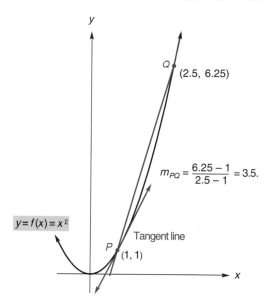

**FIGURE 12.6**  Secant lines to $f(x) = x^2$ through $(1, 1)$ and $(2.5, 6.25)$.

**TABLE 12.1**   Slopes of secant lines to the curve $f(x) = x^2$ at $P = (1,1)$

| $Q$ | Slope of $PQ$ |
|---|---|
| $(2.5, 6.25)$ | $(6.25 - 1)/(2.5 - 1) = 3.5$ |
| $(2, 4)$ | $(4 - 1)/(2 - 1) = 3$ |
| $(1.5, 2.25)$ | $(2.25 - 1)/(1.5 - 1) = 2.5$ |
| $(1.25, 1.5625)$ | $(1.5625 - 1)/(1.25 - 1) = 2.25$ |
| $(1.1, 1.21)$ | $(1.21 - 1)/(1.1 - 1) = 2.1$ |
| $(1.01, 1.0201)$ | $(1.021 - 1)/(1.01 - 1) = 2.01$ |

For the curve $y = f(x)$ in Fig. 12.7, we shall find an expression for the slope at the point $P = (x_1, f(x_1))$. If $Q = (x_2, f(x_2))$, the slope of the secant line $PQ$ is

$$m_{PQ} = \frac{f(x_2) - f(x_1)}{x_2 - x_1}.$$

If the difference $x_2 - x_1$ is called $h$, then we can write $x_2$ as $x_1 + h$. Here we must have $h \neq 0$, for if $h = 0$, then $x_2 = x_1$, and no secant line exists. Accordingly,

$$m_{PQ} = \frac{f(x_1 + h) - f(x_1)}{(x_1 + h) - x_1} = \frac{f(x_1 + h) - f(x_1)}{h}.$$

As $Q$ moves along the curve toward $P$, $x_2$ approaches $x_1$. This means that $h$ approaches zero. The limiting value of the slopes of the secant lines—which is the slope of the tangent line at $(x_1, f(x_1))$—is

$$m_{\text{tan}} = \lim_{h \to 0} m_{\text{sec}},$$

or, more precisely,

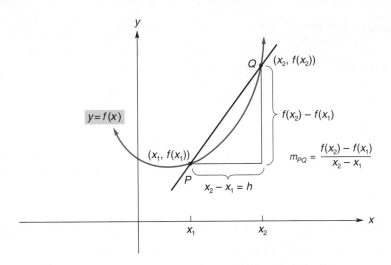

**FIGURE 12.7** Secant line through $P$ and $Q$.

$$m_{\text{tan}} = \lim_{h \to 0} \frac{f(x_1 + h) - f(x_1)}{h}. \tag{1}$$

In Example 1, we shall use this limit to confirm our previous conclusion that the slope of the tangent line to the curve $f(x) = x^2$ at $(1, 1)$ is 2.

**EXAMPLE 1   Finding the Slope of a Tangent Line**

*Find the slope of the tangent line to the curve $y = f(x) = x^2$ at the point $(1, 1)$.*

*Solution:* The slope is the limit in Eq. (1) with $f(x) = x^2$ and $x_1 = 1$:

$$\lim_{h \to 0} \frac{f(1 + h) - f(1)}{h} = \lim_{h \to 0} \frac{(1 + h)^2 - (1)^2}{h}$$

$$= \lim_{h \to 0} \frac{1 + 2h + h^2 - 1}{h} = \lim_{h \to 0} \frac{2h + h^2}{h}$$

$$= \lim_{h \to 0} \frac{h(2 + h)}{h} = \lim_{h \to 0} (2 + h) = 2.$$

Therefore, the tangent line to $y = x^2$ at $(1, 1)$ has slope 2. (Refer back to Fig. 12.6.)   ■

We can generalize Eq. (1) so that it applies to any point $(x, f(x))$ on a curve. Replacing $x_1$ by $x$ gives a function, called the *derivative* of $f$, whose input is $x$ and whose output is the slope of the tangent line to the curve at $(x, f(x))$, provided that the tangent line *has* a slope (that is, provided that the tangent line is nonvertical). We thus have the following definition, which forms the basis of differential calculus:

**DEFINITION**

*The **derivative** of a function $f$ is the function denoted $f'$ (read "$f$ prime") and defined by*

$$f'(x) = \lim_{h \to 0} \frac{f(x + h) - f(x)}{h},$$

*provided that this limit exists. If $f'(x)$ can be found, $f$ is said to be **differentiable**, and $f'(x)$ is called the derivative of $f$ at $x$ or the derivative of $f$ with respect to $x$. The process of finding the derivative is called **differentiation**.*

In the definition of the derivative, the expression

$$\frac{f(x + h) - f(x)}{h}$$

is called a **difference quotient.** Thus $f'(x)$ is the limit of a difference quotient as $h \to 0$.

### EXAMPLE 2 Using the Definition to Find the Derivative

*If $f(x) = x^2$, find the derivative of $f$.*

*Solution:* Applying the definition of a derivative gives

Don't be sloppy when applying the limit definition of a derivative. Write $\lim_{h\to0}$ at each step before the limit is actually taken. Unfortunately, some students neglect to take the final limit, and $h$ appears in their answer. This is a quick way to lose points on an examination.

$$f'(x) = \lim_{h\to0} \frac{f(x + h) - f(x)}{h}$$

$$= \lim_{h\to0} \frac{(x + h)^2 - x^2}{h} = \lim_{h\to0} \frac{x^2 + 2xh + h^2 - x^2}{h}$$

$$= \lim_{h\to0} \frac{2xh + h^2}{h} = \lim_{h\to0} \frac{h(2x + h)}{h} = \lim_{h\to0} (2x + h) = 2x.$$

*Observe that, in taking the limit, we treated $x$ as a constant, because it was $h$, not $x$, that was changing.* Also, note that $f'(x) = 2x$ defines a function of $x$, which we can interpret as giving the slope of the tangent line to the graph of $f$ at $(x, f(x))$. For example, if $x = 1$, then the slope is $f'(1) = 2(1) = 2$, which confirms the result in Example 1. ∎

Besides the notation $f'(x)$, other common ways to denote the derivative of $y = f(x)$ at $x$ are

$$\frac{dy}{dx} \qquad \text{(pronounced "dee } y, \text{dee } x\text{"),}$$

$$\frac{d}{dx}[f(x)] \qquad [\text{dee } f(x), \text{dee } x],$$

$$y' \qquad (y \text{ prime}),$$

$$D_x y \qquad (\text{dee } x \text{ of } y),$$

$$D_x[f(x)] \qquad [\text{dee } x \text{ of } f(x)].$$

*Pitfall* ▼ The notation $\dfrac{dy}{dx}$, which is called *Leibniz notation*, should **not** be thought of as a fraction, although it looks like one. It is a single symbol for a derivative. We have not yet attached any meaning to individual symbols such as $dy$ and $dx$.

If the derivative $f'(x)$ can be evaluated at $x = x_1$, the resulting *number* $f'(x_1)$ is called the **derivative of $f$ at $x_1$**, and $f$ is said to be *differentiable* at $x_1$. Because the derivative gives the slope of the tangent line,

$f'(x_1)$ is the slope of the tangent line to the graph of $y = f(x)$ at $(x_1, f(x_1))$.

Two other notations for the derivative of $f$ at $x_1$ are

$$\left.\frac{dy}{dx}\right|_{x=x_1} \quad \text{and} \quad y'(x_1).$$

**EXAMPLE 3   Finding an Equation of a Tangent Line**

*If $f(x) = 2x^2 + 2x + 3$, find an equation of the tangent line to the graph of $f$ at $(1, 7)$.*

*Solution:*

> *Strategy:* We shall first determine the slope of the tangent line by computing the derivative and evaluating it at $x = 1$. Using this result and the point $(1, 7)$ in a point–slope form gives an equation of the tangent line.

We have

$$f'(x) = \lim_{h \to 0} \frac{f(x + h) - f(x)}{h}$$

$$= \lim_{h \to 0} \frac{[2(x + h)^2 + 2(x + h) + 3] - (2x^2 + 2x + 3)}{h}$$

$$= \lim_{h \to 0} \frac{2x^2 + 4xh + 2h^2 + 2x + 2h + 3 - 2x^2 - 2x - 3}{h}$$

$$= \lim_{h \to 0} \frac{4xh + 2h^2 + 2h}{h} = \lim_{h \to 0} (4x + 2h + 2).$$

So $\qquad f'(x) = 4x + 2$

and $\qquad f'(1) = 4(1) + 2 = 6.$

Thus, the tangent line to the graph at $(1, 7)$ has slope 6. A point–slope form of this tangent is

$$y - 7 = 6(x - 1).$$

It will be our custom to express an equation of the tangent line in slope–intercept form:

$$y - 7 = 6x - 6,$$

$$y = 6x + 1. \qquad \blacksquare$$

***Pitfall*** ▼ In Example 3 it is ***not*** correct to say that, since the derivative is $4x + 2$, the tangent line at $(1, 7)$ is $y - 7 = (4x + 2)(x - 1)$. The derivative must be **evaluated** at the point of tangency to determine the slope of the tangent line.

**EXAMPLE 4   Finding the Slope of a Curve at a Point**

*Find the slope of the curve $y = 2x + 3$ at the point where $x = 6$.*

*Solution:* The slope of the curve is the slope of the tangent line. Letting $y = f(x) = 2x + 3$, we have

$$\frac{dy}{dx} = \lim_{h \to 0} \frac{f(x + h) - f(x)}{h} = \lim_{h \to 0} \frac{[2(x + h) + 3] - (2x + 3)}{h}$$

$$= \lim_{h \to 0} \frac{2h}{h} = \lim_{h \to 0} 2 = 2.$$

Since $dy/dx = 2$, the slope when $x = 6$, or in fact at any point, is 2. Note that the curve is a straight line and thus has the same slope at each point. ∎

**EXAMPLE 5   A Function with a Vertical Tangent Line**

*Find* $\frac{d}{dx}(\sqrt{x})$.

*Solution:* Letting $f(x) = \sqrt{x}$, we have

$$\frac{d}{dx}(\sqrt{x}) = \lim_{h \to 0} \frac{f(x + h) - f(x)}{h} = \lim_{h \to 0} \frac{\sqrt{x + h} - \sqrt{x}}{h}.$$

As $h \to 0$, both the numerator and denominator approach zero. This can be avoided by rationalizing the *numerator:*

You should become familiar with the procedure of rationalizing the *numerator.*

$$\frac{\sqrt{x + h} - \sqrt{x}}{h} = \frac{\sqrt{x + h} - \sqrt{x}}{h} \cdot \frac{\sqrt{x + h} + \sqrt{x}}{\sqrt{x + h} + \sqrt{x}}$$

$$= \frac{(x + h) - x}{h(\sqrt{x + h} + \sqrt{x})} = \frac{h}{h(\sqrt{x + h} + \sqrt{x})}$$

$$= \frac{1}{\sqrt{x + h} + \sqrt{x}}.$$

Therefore,

$$\frac{d}{dx}(\sqrt{x}) = \lim_{h \to 0} \frac{1}{\sqrt{x + h} + \sqrt{x}} = \frac{1}{\sqrt{x} + \sqrt{x}} = \frac{1}{2\sqrt{x}}.$$

Note that the original function, $\sqrt{x}$, is defined for $x \geq 0$, but its derivative, $1/(2\sqrt{x})$, is defined only when $x > 0$. The reason for this is clear from the graph of $y = \sqrt{x}$ in Fig. 12.8. When $x = 0$, the tangent is a vertical line, so its slope is not defined. ∎

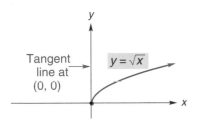

**FIGURE 12.8**   Vertical tangent line at $(0, 0)$.

In Example 5 we saw that the function $y = \sqrt{x}$ is not differentiable when $x = 0$, because the tangent line is vertical at that point. It is worthwhile mentioning that $y = |x|$ also is not differentiable when $x = 0$, but for a different reason: There is *no* tangent line at all at that point. (Refer back to Fig. 12.5.)

It is wise for you to see variables other than $x$ and $y$ involved in a problem. Example 6 illustrates the use of other variables.

To indicate a derivative, Leibniz notation is often useful because it makes it convenient to emphasize the independent and dependent variables involved. For example, if the variable $p$ is a function of the variable $q$, we speak of the derivative of $p$ with respect to $q$, written $dp/dq$.

**EXAMPLE 6   Finding the Derivative of $p$ with Respect to $q$**

*If* $p = f(q) = \dfrac{1}{2q}$, *find* $\dfrac{dp}{dq}$.

**Principles in Practice 1**

**Finding the Derivative of *H* with Respect to *t***

If a ball is thrown upward at a speed of 40 feet/s from a height of 6 feet, its height $H$ in feet after $t$ seconds is given by $H = 6 + 40t - 16t^2$. Find $\dfrac{dH}{dt}$.

*Solution:*

$$\frac{dp}{dq} = \frac{d}{dq}\left(\frac{1}{2q}\right) = \lim_{h \to 0} \frac{f(q + h) - f(q)}{h}$$

$$= \lim_{h \to 0} \frac{\dfrac{1}{2(q + h)} - \dfrac{1}{2q}}{h} = \lim_{h \to 0} \frac{\dfrac{q - (q + h)}{2q(q + h)}}{h}$$

$$= \lim_{h \to 0} \frac{q - (q + h)}{h[2q(q + h)]} = \lim_{h \to 0} \frac{-h}{h[2q(q + h)]}$$

$$= \lim_{h \to 0} \frac{-1}{2q(q + h)} = -\frac{1}{2q^2}.$$

Note that when $q = 0$, neither the function nor its derivative exists. ∎

Keep in mind that the derivative of $y = f(x)$ at $x$ is nothing more than a limit, namely

$$\lim_{h \to 0} \frac{f(x + h) - f(x)}{h}.$$

Although we can interpret the derivative as a function that gives the slope of the tangent line to the curve $y = f(x)$ at the point $(x, f(x))$, this interpretation is simply a geometric convenience that assists our understanding. The preceding limit may exist, aside from any geometric considerations at all. As you will see later, there are other useful interpretations of the derivative.

## TECHNOLOGY

Many graphics calculators have a numerical derivative feature that estimates the derivative of a function at a point. With the TI-82 we would use the "nDeriv" command, in which we must enter the function, the variable, and the value of the variable (separated by commas) in the format

nDeriv(function, variable, value of variable).

For example, the derivative of $f(x) = \sqrt{x^3 + 2}$ at $x = 1$ is estimated in Fig. l2.9. Thus, $f'(1) \approx 0.866$.

On the other hand, we can take the "limit of a difference quotient" approach to estimate this derivative. To make use of the table feature of a graphics calculator, we can enter $f(x)$ as $Y_1$. Then, for $Y_2$, we enter the following form of the difference quotient:

$$(Y_1(1 + X) - Y_1(1))/X.$$

(Here, $x$ plays the role of $h$.) Figure 12.10 shows a table for $Y_2$ as $x$ approaches 0 from both the left and right. This table strongly suggests that $f'(1) \approx 0.866$.

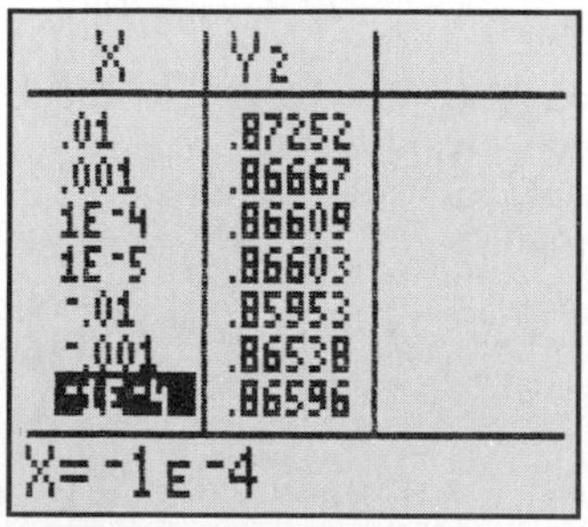

**FIGURE 12.9** Numerical derivative.

**FIGURE 12.10** Limit of a difference quotient as $x \to 0$.

## ■ Exercise 12.1

*In Problems 1 and 2, a function f and a point P on its graph are given.*

    **a. Find the slope of the secant line PQ for each point**
    $Q = (x, f(x))$ *whose x-value is given in the table.*
    *Round your answers to four decimal places.*

    **b. Use your results from part a to estimate the slope of**
    **the tangent line at P.**

**1.** $f(x) = x^3 + 3$, $P = (2, 11)$.

| x-value of $Q$ | 3 | 2.5 | 2.2 | 2.1 | 2.01 | 2.001 |
|---|---|---|---|---|---|---|
| $m_{PQ}$ | | | | | | |

**2.** $f(x) = e^{2x}$, $P = (0, 1)$.

| x-value of $Q$ | 1 | 0.5 | 0.2 | 0.1 | 0.01 | 0.001 |
|---|---|---|---|---|---|---|
| $m_{PQ}$ | | | | | | |

*In Problems 3–18, use the definition of the derivative to find each of the following.*

**3.** $f'(x)$ if $f(x) = x$.

**4.** $f'(x)$ if $f(x) = 4x - 1$.

**5.** $\dfrac{dy}{dx}$ if $y = 3x + 7$.

**6.** $\dfrac{dy}{dx}$ if $y = -5x$.

**7.** $\dfrac{d}{dx}(5 - 4x)$.

**8.** $\dfrac{d}{dx}\left(2 - \dfrac{x}{4}\right)$.

**9.** $f'(x)$ if $f(x) = 3$.

**10.** $f'(x)$ if $f(x) = 7.01$.

**11.** $\dfrac{d}{dx}(x^2 + 4x - 8)$.

**12.** $y'$ if $y = x^2 + 5$.

**13.** $\dfrac{dp}{dq}$ if $p = 2q^2 + 5q - 1$.

**14.** $\dfrac{d}{dx}(x^2 - x - 3)$.

**15.** $y'$ if $y = \dfrac{1}{x}$.

**16.** $\dfrac{dC}{dq}$ if $C = 7 + 2q - 3q^2$.

**17.** $f'(x)$ if $f(x) = \sqrt{x + 2}$.

**18.** $g'(x)$ if $g(x) = \dfrac{2}{x - 3}$.

**19.** Find the slope of the curve $y = x^2 + 4$ at the point $(-2, 8)$.

**20.** Find the slope of the curve $y = 2 - 3x^2$ at the point $(1, -1)$.

**21.** Find the slope of the curve $y = 4x^2 - 5$ when $x = 0$.

**22.** Find the slope of the curve $y = \sqrt{x}$ when $x = 1$.

*In Problems 23–28, find an equation of the tangent line to the curve at the given point.*

**23.** $y = x + 4$; $(3, 7)$.

**24.** $y = 2x^2 - 5$; $(-2, 3)$.

**25.** $y = 3x^2 + 3x - 4$; $(-1, -4)$.

**26.** $y = (x - 1)^2$; $(0, 1)$.

**27.** $y = \dfrac{3}{x + 1}$; $(2, 1)$.

**28.** $y = \dfrac{5}{1 - 3x}$; $(2, -1)$.

**29. Banking** Equations may involve derivatives of functions. In an article on interest rate deregulation, Christofi and Agapos[1] solve the equation

$$r = \left(\frac{\eta}{1 + \eta}\right)\left(r_L - \frac{dC}{dD}\right)$$

for $\eta$ (the Greek letter "eta"). Here $r$ is the deposit rate paid by commercial banks, $r_L$ is the rate earned by commercial banks, $C$ is the administrative cost of transforming deposits into return-earning assets, $D$ is the savings deposits level, and $\eta$ is the deposit elasticity with respect to the deposit rate. Find $\eta$.

*In Problems 30 and 31, use the numerical derivative feature of your graphics calculator to estimate the derivatives of the functions at the indicated values. Round your answers to three decimal places.*

 **30.** $f(x) = \sqrt{3x^2 + 2x}$; $x = 1$, $x = -1$.

 **31.** $f(x) = e^x(4x - 7)$; $x = 0$, $x = 1.5$.

*In Problems 32 and 33, use the "limit of a difference quotient" approach to estimate f'(x) at the indicated values of x. Round your answers to three decimal places.*

 **32.** $f(x) = \dfrac{\ln x}{x + 3}$; $x = 0.5$, $x = 10$.

 **33.** $f(x) = \dfrac{x^2 + 4x + 2}{x^3 - 3}$; $x = 2$, $x = -4$.

[1]A. Christofi and A. Agapos, "Interest Rate Deregulation: An Empirical Justification," *Review of Business and Economic Research*, XX, no. 1 (1984), 39–49.

**34.** Find an equation of the tangent line to the curve $f(x) = x^2 + x$ at the point $(-2, 2)$ Graph both the curve and the tangent line. Notice that the tangent line is a good approximation to the curve near the point of tangency.

**35.** The derivative of $f(x) = x^3 - x + 2$ is $f'(x) = 3x^2 - 1$. Graph both the function $f$ and its derivative $f'$. Observe that there are two points on the graph of $f$ where the tangent line is horizontal. For the $x$-values of these points, what are the corresponding values of $f'(x)$? Why are these results expected? Observe the intervals where $f'(x)$ is positive. Notice that tangent lines to the graph of $f$ have positive slopes over these intervals. Observe the interval where $f'(x)$ is negative. Notice that tangent lines to the graph of $f$ have negative slopes over this interval.

---

**To develop basic differentiation rules, namely, formulas for the derivative of a constant, of $x^n$, of a constant times a function, and of sums and differences of functions.**

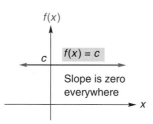

**FIGURE 12.11** The slope of a constant function is 0.

## 12.2 RULES FOR DIFFERENTIATION

You would probably agree that differentiating a function by direct use of the definition of a derivative can be tedious. Fortunately, there are rules that give us completely mechanical and efficient procedures for differentiation. They also avoid the direct use of limits. We shall look at some of these rules in this section.

We begin by showing that the derivative of a constant function is zero. Recall that the graph of the constant function $f(x) = c$ is a horizontal line (see Fig. 12.11), which has a slope of zero at each point. This means that $f'(x) = 0$ regardless of $x$. As a formal proof of this result, we apply the definition of the derivative to $f(x) = c$:

$$f'(x) = \lim_{h \to 0} \frac{f(x + h) - f(x)}{h} = \lim_{h \to 0} \frac{c - c}{h}$$

$$= \lim_{h \to 0} \frac{0}{h} = \lim_{h \to 0} 0 = 0.$$

Thus, we have our first rule:

---

**Rule 1   Derivative of a Constant**

*If c is a constant then*

$$\frac{d}{dx}(c) = 0.$$

*That is, the derivative of a constant function is zero*

---

**EXAMPLE 1   Derivatives of Constant Functions**

**a.** $\frac{d}{dx}(3) = 0$ because 3 is a constant function.

**b.** If $g(x) = \sqrt{5}$, then $g'(x) = 0$ because $g$ is a constant function. For example, the derivative of $g$ when $x = 4$ is $g'(4) = 0$.

**c.** If $s(t) = (1{,}938{,}623)^{807.4}$, then $ds/dt = 0$. ■

The next rule gives a formula for the derivative of $x$ raised to a constant power— that is, the derivative of $f(x) = x^n$. A function of this form is called a **power function.** For example, $f(x) = x^2$ is a power function. To prove the rule, we must expand $(x + h)^n$. Recall that

$$(x + h)^2 = x^2 + 2xh + h^2$$

and

$$(x + h)^3 = x^3 + 3x^2h + 3xh^2 + h^3.$$

In both expansions, the exponents of $x$ decrease from left to right while those of $h$ increase. This is true for the general case $(x + h)^n$, where $n$ is a positive integer. It can be shown that

$$(x + h)^n = x^n + nx^{n-1}h + (\ )x^{n-2}h^2 + \cdots + (\ )xh^{n-1} + h^n,$$

where the missing numbers inside the parentheses are certain constants. This formula is used to prove the following rule:

---

**Rule 2   Derivative of $x^n$**

*If n is any real number, then*

$$\frac{d}{dx}(x^n) = nx^{n-1},$$

*provided that $x^{n-1}$ is defined. That is, the derivative of a constant power of x is the exponent times x raised to a power one less than the given power.*

---

Although the rule states that $n$ can be any real number, our proof is for a positive integer only. The general proof is not given in this text.

*Proof.* We shall give a proof for the case where $n$ is a positive integer. If $f(x) = x^n$, applying the definition of the derivative gives

$$f'(x) = \lim_{h \to 0} \frac{f(x + h) - f(x)}{h} = \lim_{h \to 0} \frac{(x + h)^n - x^n}{h}.$$

By our previous discussion on expanding $(x + h)^n$,

$$f'(x) = \lim_{h \to 0} \frac{x^n + nx^{n-1}h + (\ )x^{n-2}h^2 + \cdots + h^n - x^n}{h}.$$

In the numerator, the sum of the first and last terms is 0. Dividing each of the remaining terms by $h$ gives

$$f'(x) = \lim_{h \to 0} [nx^{n-1} + (\ )x^{n-2}h + \cdots + h^{n-1}].$$

Each term after the first has $h$ as a factor and must approach 0 as $h \to 0$. Hence, $f'(x) = nx^{n-1}$.

**EXAMPLE 2   Derivatives of Powers of $x$**

a.  By Rule 2, $\dfrac{d}{dx}(x^2) = 2x^{2-1} = 2x$.

b.  If $F(x) = x = x^1$, then $F'(x) = 1 \cdot x^{1-1} = 1 \cdot x^0 = 1$. Thus, the derivative of $x$ with respect to $x$ is 1.

c.  If $f(x) = x^{-10}$ then $f'(x) = -10x^{-10-1} = -10x^{-11}$.   ∎

When we apply a differentiation rule to a function, sometimes the function must first be rewritten so that it has the proper form for that rule. For example, to differentiate $f(x) = \dfrac{1}{x^{10}}$ we would first rewrite $f$ as $f(x) = x^{-10}$ and then proceed as in Example 2(c).

**EXAMPLE 3**  **Rewriting Functions in the Form $x^n$**

**a.** To differentiate $y = \sqrt{x}$, we rewrite $\sqrt{x}$ as $x^{1/2}$ so that it has the form $x^n$. Thus,

$$\frac{dy}{dx} = \frac{1}{2}x^{(1/2)-1} = \frac{1}{2}x^{-1/2} = \frac{1}{2\sqrt{x}}.$$

**b.** Let $h(x) = \dfrac{1}{x\sqrt{x}}$. To apply Rule 2, we *must* rewrite $h(x)$ as $h(x) = x^{-3/2}$ so that it has the form $x^n$. We have

$$h'(x) = \frac{d}{dx}(x^{-3/2}) = -\frac{3}{2}x^{(-3/2)-1} = -\frac{3}{2}x^{-5/2}.$$

***Pitfall*** ▼ In Example 3(b), do not rewrite $\dfrac{1}{x\sqrt{x}}$ as $\dfrac{1}{x^{3/2}}$ and then merely differentiate the denominator; that is

$$\frac{d}{dx}\left(\frac{1}{x^{3/2}}\right) \neq \frac{1}{\frac{3}{2}x^{1/2}}.$$

Now that we can say immediately that the derivative of $x^3$ is $3x^2$, the question arises as to what we could say about the derivative of a *multiple* of $x^3$, such as $5x^3$. Our next rule will handle this situation of differentiating a constant times a function.

---

**Rule 3**  **Constant Factor Rule**

*If $f$ is a differentiable function and $c$ is a constant, then $cf(x)$ is differentiable, and*

$$\frac{d}{dx}[cf(x)] = cf'(x).$$

*That is, the derivative of a constant times a function is the constant times the derivative of the function.*

---

*Proof.* If $g(x) = cf(x)$, applying the definition of the derivative of $g$ gives

$$g'(x) = \lim_{h \to 0} \frac{g(x+h) - g(x)}{h} = \lim_{h \to 0} \frac{cf(x+h) - cf(x)}{h}$$

$$= \lim_{h \to 0}\left[c \cdot \frac{f(x+h) - f(x)}{h}\right] = c \cdot \lim_{h \to 0}\frac{f(x+h) - f(x)}{h}.$$

But $\displaystyle\lim_{h \to 0}\frac{f(x+h) - f(x)}{h}$ is $f'(x)$; so $g'(x) = cf'(x)$.

**EXAMPLE 4**  **Differentiating a Constant Times a Function**

*Differentiate the following functions.*

**a.** $g(x) = 5x^3$.

*Solution:* Here $g$ is a constant (5) times a function ($x^3$). So

$$\frac{d}{dx}(5x^3) = 5\frac{d}{dx}(x^3) \qquad \text{(Rule 3)}$$

$$= 5(3x^{3-1}) = 15x^2 \quad \text{(Rule 2)}.$$

**b.** $f(q) = \dfrac{13q}{5}$.

*Solution:*

> *Strategy:* We first rewrite $f$ as a constant times a function and then apply Rule 2.

Because $\dfrac{13q}{5} = \dfrac{13}{5}q$, $f$ is the constant $\dfrac{13}{5}$ times the function $q$. Thus,

$$f'(q) = \frac{13}{5}\frac{d}{dq}(q) \qquad \text{(Rule 3)}$$

$$= \frac{13}{5}\cdot 1 = \frac{13}{5} \qquad \text{(Rule 2)}.$$

**c.** $y = \dfrac{0.25}{\sqrt[5]{x^2}}$.

*Solution:* We can express $y$ as a constant times a function:

$$y = 0.25\cdot\frac{1}{\sqrt[5]{x^2}} = 0.25x^{-2/5}.$$

Hence,

$$y' = 0.25\frac{d}{dx}(x^{-2/5}) \qquad\qquad \text{(Rule 3)}$$

$$= 0.25\left(-\frac{2}{5}x^{-7/5}\right) = -0.1x^{-7/5} \quad \text{(Rule 2)}. \qquad \blacksquare$$

*Pitfall* ▼ To differentiate $f(x) = (4x)^3$, you may be tempted to write $f'(x) = 3(4x)^2$. This is **incorrect!** Do you see why? The reason is that Rule 2 applies to a power of the variable $x$, **not** a power of an expression involving $x$, such as $4x$. To apply our rules, we must get a suitable form for $f(x)$. We can rewrite $(4x)^3$ as $4^3x^3$ or $64x^3$. Thus,

$$f'(x) = 64\frac{d}{dx}(x^3) = 64(3x^2) = 192x^2.$$

The next rule involves derivatives of sums and differences of functions.

> **Rule 4   Derivative of a Sum or Difference**
>
> *If $f$ and $g$ are differentiable functions, then $f + g$ and $f - g$ are differentiable, and*
>
> $$\frac{d}{dx}[f(x) + g(x)] = f'(x) + g'(x)$$
>
> *and*

$$\frac{d}{dx}[f(x) - g(x)] = f'(x) - g'(x).$$

*That is, the derivative of the sum (difference) of two functions is the sum (difference) of their derivatives.*

*Proof.* For the case of a sum, if $F(x) = f(x) + g(x)$, applying the definition of the derivative of $F$ gives

$$F'(x) = \lim_{h \to 0} \frac{F(x + h) - F(x)}{h}$$

$$= \lim_{h \to 0} \frac{[f(x + h) + g(x + h)] - [f(x) + g(x)]}{h}$$

$$= \lim_{h \to 0} \frac{[f(x + h) - f(x)] + [g(x + h) - g(x)]}{h} \qquad \text{(regrouping)}$$

$$= \lim_{h \to 0} \left[ \frac{f(x + h) - f(x)}{h} + \frac{g(x + h) - g(x)}{h} \right].$$

Because the limit of a sum is the sum of the limits,

$$F'(x) = \lim_{h \to 0} \frac{f(x + h) - f(x)}{h} + \lim_{h \to 0} \frac{g(x + h) - g(x)}{h}.$$

But these two limits are $f'(x)$ and $g'(x)$. Thus,

$$F'(x) = f'(x) + g'(x).$$

The proof for the derivative of a difference of two functions is similar.

Rule 4 can be extended to the derivative of any number of sums and differences of functions. For example,

$$\frac{d}{dx}[f(x) - g(x) + h(x) + k(x)] = f'(x) - g'(x) + h'(x) + k'(x).$$

**Principles in Practice 1**

**Differentiating Sums and Differences of Functions**

If the revenue function for a certain product is $r(q) = 50q - 0.3q^2$, find the derivative of this function, also known as the marginal revenue.

**EXAMPLE 5  Differentiating Sums and Differences of Functions**

*Differentiate the following functions.*

**a.** $F(x) = 3x^5 + \sqrt{x}.$

**Solution:** Here $F$ is the sum of two functions, $3x^5$ and $\sqrt{x}$. Therefore,

$$F'(x) = \frac{d}{dx}(3x^5) + \frac{d}{dx}(x^{1/2}) \qquad \text{(Rule 4)}$$

$$= 3\frac{d}{dx}(x^5) + \frac{d}{dx}(x^{1/2}) \qquad \text{(Rule 3)}$$

$$= 3(5x^4) + \frac{1}{2}x^{-1/2} = 15x^4 + \frac{1}{2\sqrt{x}} \qquad \text{(Rule 2).}$$

**b.** $f(z) = \frac{z^4}{4} - \frac{5}{z^{1/3}}.$

*Solution:* To apply our rules, we shall rewrite $f(z)$ in the form $f(z) = \frac{1}{4}z^4 - 5z^{-1/3}$. Since $f$ is the difference of two functions,

$$f'(z) = \frac{d}{dz}\left(\frac{1}{4}z^4\right) - \frac{d}{dz}(5z^{-1/3}) \qquad \text{(Rule 4)}$$

$$= \frac{1}{4}\frac{d}{dz}(z^4) - 5\frac{d}{dz}(z^{-1/3}) \qquad \text{(Rule 3)}$$

$$= \frac{1}{4}(4z^3) - 5\left(-\frac{1}{3}z^{-4/3}\right) \qquad \text{(Rule 2)}$$

$$= z^3 + \frac{5}{3}z^{-4/3}.$$

**c.** $y = 6x^3 - 2x^2 + 7x - 8$.

*Solution:*

$$\frac{dy}{dx} = \frac{d}{dx}(6x^3) - \frac{d}{dx}(2x^2) + \frac{d}{dx}(7x) - \frac{d}{dx}(8)$$

$$= 6\frac{d}{dx}(x^3) - 2\frac{d}{dx}(x^2) + 7\frac{d}{dx}(x) - \frac{d}{dx}(8)$$

$$= 6(3x^2) - 2(2x) + 7(1) - 0$$

$$= 18x^2 - 4x + 7.$$ ∎

In Examples 6 and 7, we need to rewrite the given function in a form to which our rules apply.

**EXAMPLE 6    Evaluating a Derivative**

*Find the derivative of $f(x) = 2x(x^2 - 5x + 2)$ when $x = 2$.*

*Solution:*  We multiply and then differentiate each term:

$$f(x) = 2x^3 - 10x^2 + 4x.$$
$$f'(x) = 2(3x^2) - 10(2x) + 4(1)$$
$$= 6x^2 - 20x + 4,$$
$$f'(2) = 6(2)^2 - 20(2) + 4 = -12.$$ ∎

**EXAMPLE 7    Finding an Equation of a Tangent Line**

*Find an equation of the tangent line to the curve*

$$y = \frac{3x^2 - 2}{x}$$

*when $x = 1$.*

*Solution:*

*Strategy:*  First we find $\dfrac{dy}{dx}$, which gives the slope of the tangent line at any point. Evaluating $\dfrac{dy}{dx}$ when $x = 1$ gives the slope of the required tangent line. We then determine the $y$-coordinate of the point on the curve when $x = 1$. Finally, we substitute the slope and both of the coordinates of the point in a point-slope form to obtain an equation of the tangent line.

Rewriting $y$ as a difference of two functions, we have

$$y = \frac{3x^2}{x} - \frac{2}{x} = 3x - 2x^{-1}.$$

Thus,

$$\frac{dy}{dx} = 3(1) - 2[(-1)x^{-2}] = 3 + \frac{2}{x^2}.$$

The slope of the tangent line to the curve when $x = 1$ is

$$\left.\frac{dy}{dx}\right|_{x=1} = 3 + \frac{2}{1^2} = 5.$$

To find the $y$-coordinate of the point on the curve where $x = 1$, we substitute this value of $x$ into the equation of the *curve*. This gives

$$y = \frac{3(1)^2 - 2}{1} = 1.$$

Hence, the point $(1, 1)$ lies on both the curve and the tangent line. Therefore, an equation of the tangent line is

$$y - 1 = 5(x - 1).$$

In slope–intercept form, we have

$$y = 5x - 4.$$

**Pitfall** ▼ To obtain the $y$-value of the point on the curve when $x = 1$, substitute into the equation of the *curve,* not into the formula for the derivative! ■

## ■ Exercise 12.2

*In Problems **1–74,** differentiate the functions.*

**1.** $f(x) = 5$.

**2.** $f(x) = (\frac{11}{13})^{4/5}$.

**3.** $y = x^6$.

**4.** $f(x) = x^{21}$.

**5.** $y = x^{80}$.

**6.** $y = x^{6.1}$.

**7.** $f(x) = 9x^2$.

**8.** $y = 4x^4$.

**9.** $g(w) = 4w^5$.

**10.** $v(x) = x^e$.

**11.** $y = \frac{2}{3}x^4$.

**12.** $f(p) = \sqrt{3}p^4$.

**13.** $f(t) = \frac{t^9}{18}$.

**14.** $y = \frac{x^3}{3}$.

**15.** $f(x) = x + 3$.

**16.** $f(x) = 3x - 2$.

**17.** $f(x) = 3x^2 - 2x + 3$.

**18.** $f(x) = 7x^2 - 5x$.

**19.** $g(p) = p^4 - 3p^3 - 1$.

**20.** $f(t) = -13t^2 + 14t + 1$.

**21.** $y = -x^8 + x^5$.

**22.** $y = -8x^4 + \ln 2$.

**23.** $y = -13x^3 + 14x^2 - 2x + 3$.

**24.** $V(r) = r^8 - 7r^6 + 3r^2 + 1$.

**25.** $f(x) = 2(13 - x^4)$.

**26.** $f(s) = 5(s^4 - 3)$.

**27.** $g(x) = \frac{13 - x^4}{3}$.

**28.** $f(x) = \frac{5(x^4 - 3)}{2}$.

**29.** $h(x) = 4x^4 + x^3 - \frac{9x^2}{2} + 9x$.

**30.** $f(x) = -3x^2 + \frac{9}{2}x + 2$.

**31.** $f(x) = \frac{3x^4}{2} + \frac{7}{3}x^3$.

**32.** $p(x) = \frac{x^7}{7} + \frac{2x}{3}$.

**33.** $f(x) = x^{7/2}$.

**34.** $f(x) = 2x^{-14/5}$.

**35.** $y = x^{3/4} + x^{5/3}$.

**36.** $y = 5x^3 - x^{-2/5}$.

**37.** $y = \sqrt{x}$.

**38.** $y = \sqrt[3]{x^2}$.

**39.** $f(r) = 6\sqrt[3]{r}$.

**40.** $y = 4\sqrt[8]{x^2}$.

**41.** $f(x) = x^{-4}$.

**42.** $f(s) = 3s^{-2}$.

**43.** $f(x) = x^{-3} + x^{-5} - 2x^{-6}$.

**44.** $f(x) = 100x^{-3} + 10x^{1/2}$.

**45.** $y = \dfrac{1}{x}$.

**46.** $f(x) = \dfrac{2}{x^3}$.

**47.** $y = \dfrac{3}{x^5}$.

**48.** $y = \dfrac{1}{4x^5}$.

**49.** $g(x) = \dfrac{4}{3x^3}$.

**50.** $y = \dfrac{3}{2x^6}$.

**51.** $f(t) = \dfrac{1}{2t}$.

**52.** $g(x) = \dfrac{7}{3x}$.

**53.** $f(x) = \dfrac{x}{7} + \dfrac{7}{x}$.

**54.** $H(x) = \dfrac{x^2}{2} - \dfrac{2}{x^2}$.

**55.** $f(x) = -9x^{1/3} + 5x^{-2/5}$.

**56.** $f(z) = 3z^{1/4} - 12^2 - 8z^{-3/4}$.

**57.** $q(x) = \dfrac{1}{\sqrt[5]{x}}$.

**58.** $f(x) = \dfrac{3}{\sqrt[4]{x^3}}$

**59.** $y = \dfrac{2}{\sqrt{x}}$.

**60.** $y = \dfrac{1}{2\sqrt{x}}$.

**61.** $y = x^2\sqrt{x}$.

**62.** $f(x) = x^3(3x^2)$.

**63.** $f(x) = x(3x^2 - 7x + 7)$.

**64.** $f(x) = x^3(3x^6 - 5x^2 + 4)$.

**65.** $f(x) = x^3(3x)^2$.

**66.** $f(x) = \sqrt{x}(5 - 6x + 3\sqrt[4]{x})$.

**67.** $v(x) = x^{-2/3}(x + 5)$.

**68.** $f(x) = x^{3/5}(x^2 + 7x + 1)$.

**69.** $f(q) = \dfrac{4q^3 + 7q - 4}{q}$.

**70.** $f(w) = \dfrac{w - 5}{w^5}$.

**71.** $f(x) = (x + 1)(x + 3)$.

**72.** $f(x) = x^2(x - 2)(x + 4)$.

**73.** $w(x) = \dfrac{x^2 + x^3}{x^2}$.

**74.** $f(x) = \dfrac{7x^3 + x}{2\sqrt{x}}$.

*For each curve in Problems 75–78, find the slopes at the indicated points.*

**75.** $y = 3x^2 + 4x - 8$; $(0, ^-8), (2, 12), (-3, 7)$.

**76.** $y = 5 - 6x - 2x^3$; $(0, 5), (\tfrac{3}{2}, -\tfrac{43}{4}), (-3, 77)$.

**77.** $y = 4$; when $x = -4, x = 7, x = 22$.

**78.** $y = 2x - 3\sqrt{x}$; when $x = 1, x = 16, x = 25$.

*In Problems 79–82, find an equation of the tangent line to the curve at the indicated point.*

**79.** $y = 4x^2 + 5x + 2$; $(1, 11)$.

**80.** $y = \dfrac{1 - x^2}{5}$; $(4, -3)$.

**81.** $y = \dfrac{2}{x^2}$; $(1, 2)$.

**82.** $y = \sqrt[3]{x}$; $(8, 2)$.

**83.** Find an equation of the tangent line to the curve

$$y = 3 + x - 5x^2 + x^4$$

when $x = 0$.

**84.** Repeat Problem 83 for the curve

$$y = \dfrac{\sqrt{x}(2 - x^2)}{x}$$

when $x = 4$.

**85.** Find all points on the curve

$$y = \tfrac{1}{3}x^3 - x^2$$

where the tangent line is horizontal.

**86.** Repeat Problem 85 for the curve

$$y = \dfrac{x^5}{5} - x + 1$$

**87.** Find all points on the curve

$$y = x^2 - 5x + 3$$

where the slope is 1.

**88.** Repeat Problem 87 for the curve

$$y = \dfrac{x^3}{3} - 3x + 4$$

**89.** If $f(x) = \sqrt{x} + \dfrac{1}{\sqrt{x}}$, evaluate the expression

$$\dfrac{x - 1}{2x\sqrt{x}} - f'(x).$$

**90. Economics** Eswaran and Kotwal[2] consider agrarian economies in which there are two types of workers, permanent and casual. Permanent workers are employed on long-term contracts and may receive benefits such as holiday gifts and emergency aid. Casual workers are hired on a daily basis and perform routine and menial tasks such as weeding, harvesting, and threshing. The difference $z$ in the present-value cost of hiring a permanent worker over that of hiring a casual worker is given by

$$z = (1 + b)w_p - bw_c,$$

where $w_p$ and $w_c$ are wage rates for permanent labor and

casual labor, respectively, $b$ is a constant, and $w_p$ is a function of $w_c$. Eswaran and Kotwal claim that

$$\frac{dz}{dw_c} = (1 + b)\left[\frac{dw_p}{dw_c} - \frac{b}{1 + b}\right].$$

Verify this.

**91.** Find an equation of the tangent line to the graph of $y = x^3 - 3x$ at the point $(2, 2)$. Graph both the function and the tangent line on the same screen. Notice that the line passes through $(2, 2)$ and the line appears to be tangent to the curve.

**92.** Find an equation of the tangent line to the graph of $y = \sqrt{x}$, at the point $(1, 1)$. Graph both the function and the tangent line on the same screen. Notice that the line passes through $(1, 1)$ and the line appears to be tangent to the curve.

[2]M. Eswaran and A. Kotwal, "A Theory of Two-Tier Labor Markets in Agrarian Economies," *The American Economic Review,* 75, no 1 (1985), 162–77.

---

## OBJECTIVE

To motivate the instantaneous rate of change of a function by means of velocity and to interpret the derivative as an instantaneous rate of change. To develop the "marginal" concept, which is frequently used in business and economics.

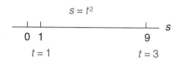

**FIGURE 12.12** Motion along a number line.

## 12.3 THE DERIVATIVE AS A RATE OF CHANGE

We have given a geometrical interpretation of the derivative as being the slope of the tangent line to a curve at a point. Historically, an important application of the derivative involves the motion of an object traveling in a straight line. This gives us a convenient way to interpret the derivative as a *rate of change.*

To denote the change in a variable such as $x$, the symbol $\Delta x$ (read "delta $x$") is commonly used. For example, if $x$ changes from 1 to 3, then the change in $x$ is $\Delta x = 3 - 1 = 2$. The new value of $x( = 3)$ is the old value plus the change, or $1 + \Delta x$. Similarly, if $t$ increases by $\Delta t$, the new value is $t + \Delta t$. We shall use $\Delta$-notation in the discussion that follows.

Suppose an object moves along the number line in Fig. 12.12 according to the equation

$$s = f(t) = t^2,$$

where $s$ is the position of the object at time $t$. This equation is called an *equation of motion,* and $f$ is called a **position function.** Assume that $t$ is in seconds and $s$ is in meters. At $t = 1$ the position is $s = f(1) = 1^2 = 1$, and at $t = 3$ the position is $s = f(3) = 3^2 = 9$. Over this two-second time interval, the object has a change in position, or a *displacement,* of $9 - 1 = 8$ m, and the *average velocity* of the object is defined as

$$v_{\text{ave}} = \frac{\text{displacement}}{\text{length of time interval}} \quad (1)$$

$$= \frac{8}{2} = 4 \text{ m/s}$$

To say that the average velocity is 4 m/s from $t = 1$ to $t = 3$ means that, *on the average,* the position of the object changed by 4 m to the right each second during that time interval. Let us denote the changes in $s$-values and $t$-values by $\Delta s$ and $\Delta t$, respectively. Then the average velocity is given by

$$v_{\text{ave}} = \frac{\Delta s}{\Delta t} = 4 \text{ m/s} \qquad (\text{for the interval } t = 1 \text{ to } t = 3 ).$$

The ratio $\Delta s/\Delta t$ is also called the **average rate of change of $s$ with respect to $t$** over the interval from $t = 1$ to $t = 3$.

Now, let the time interval be only 1 second long (that is, $\Delta t = 1$). Then, for the *shorter* interval from $t = 1$ to $t = 1 + \Delta t = 2$, we have $f(2) = 2^2 = 4$, so

$$v_{\text{ave}} = \frac{\Delta s}{\Delta t} = \frac{f(2) - f(1)}{\Delta t} = \frac{4 - 1}{1} = 3 \text{m/s}.$$

More generally, over the time interval from $t = 1$ to $t = 1 + \Delta t$, the object moves from position $f(1)$ to position $f(1 + \Delta t)$. Thus, its displacement is

$$\Delta s = f(1 + \Delta t) - f(1).$$

Since the time interval has length $\Delta t$, the object's average velocity is given by

$$v_{\text{ave}} = \frac{\Delta s}{\Delta t} = \frac{f(1 + \Delta t) - f(1)}{\Delta t}.$$

If $\Delta t$ were to become smaller and smaller, the average velocity over the interval from $t = 1$ to $t = 1 + \Delta t$ would be close to what we might call the *instantaneous velocity* at time $t = 1$, that is, the velocity at a *point* in time $(t = 1)$, as opposed to the velocity over an *interval* of time. For some typical values of $\Delta t$ between 0.1 and 0.001, we get the average velocities in Table 12.2, which you may verify.

**TABLE 12.2**

| Length of Time Interval $\Delta t$ | Time Interval, $t = 1$ to $t = 1 + \Delta t$ | Average Velocity, $\dfrac{\Delta s}{\Delta t} = \dfrac{f(1 + \Delta t) - f(1)}{\Delta t}$ | |
|---|---|---|---|
| 0.1 | $t = 1$ to $t = 1.1$ | 2.1 | m/s |
| 0.07 | $t = 1$ to $t = 1.07$ | 2.07 | m/s |
| 0.05 | $t = 1$ to $t = 1.05$ | 2.05 | m/s |
| 0.03 | $t = 1$ to $t = 1.03$ | 2.03 | m/s |
| 0.01 | $t = 1$ to $t = 1.01$ | 2.01 | m/s |
| 0.001 | $t = 1$ to $t = 1.001$ | 2.001 | m/s |

The table suggests that as the length of the time interval approaches zero, the average velocity approaches the value 2 m/s. In other words, as $\Delta t$ approaches 0, $\Delta s/\Delta t$ approaches 2 m/s. We define the limit of the average velocity as $\Delta t \to 0$ to be the **instantaneous velocity** (or simply the **velocity**), $v$, at time $t = 1$. This limit is also called the **instantaneous rate of change** of $s$ with respect to $t$ at $t = 1$:

$$v = \lim_{\Delta t \to 0} v_{\text{ave}} = \lim_{\Delta t \to 0} \frac{\Delta s}{\Delta t} = \lim_{\Delta t \to 0} \frac{f(1 + \Delta t) - f(1)}{\Delta t}.$$

If we think of $\Delta t$ as $h$, then the limit on the right is simply the derivative of $s$ with respect to $t$ at $t = 1$. Thus, the instantaneous velocity of the object at $t = 1$ is just $ds/dt$ at $t = 1$. Because $s = t^2$ and

$$\frac{ds}{dt} = 2t,$$

the velocity at $t = 1$ is

$$v = \frac{ds}{dt}\bigg|_{t=1} = 2(1) = 2 \text{ m/s,}$$

which confirms our previous conclusion.

In summary, if $s = f(t)$ is the position function of an object moving in a straight line, then the average velocity of the object over the time interval $[t, t + \Delta t]$ is given by

$$v_{\text{ave}} = \frac{\Delta s}{\Delta t} = \frac{f(t + \Delta t) - f(t)}{\Delta t},$$

and the velocity at time $t$ is given by

$$v = \lim_{\Delta t \to 0} \frac{f(t + \Delta t) - f(t)}{\Delta t} = \frac{ds}{dt}.$$

**EXAMPLE 1  Finding Average Velocity and Velocity**

*Suppose the position function of an object moving along a number line is given by $s = f(t) = 3t^2 + 5$, where t is in seconds and s is in meters.*

**a.** *Find the average velocity over the interval $[10, 10.1]$.*

**b.** *Find the velocity when $t = 10$.*

*Solution:*

**a.** Here $t = 10$ and $\Delta t = 10.1 - 10 = 0.1$. So we have

$$v_{\text{ave}} = \frac{\Delta s}{\Delta t} = \frac{f(t + \Delta t) - f(t)}{\Delta t}$$

$$= \frac{f(10 + 0.1) - f(10)}{0.1}$$

$$= \frac{f(10.1) - f(10)}{0.1}$$

$$= \frac{311.03 - 305}{0.1} = \frac{6.03}{0.1} = 60.3 \text{ m/s.}$$

**b.** The velocity at time $t$ is given by

$$v = \frac{ds}{dt} = 6t.$$

When $t = 10$, the velocity is

$$\frac{ds}{dt}\bigg|_{t=10} = 6(10) = 60 \text{ m/s.}$$

Notice that the average velocity over the interval $[10, 10.1]$ is close to the velocity at $t = 10$. This is to be expected because the length of the interval is small. ∎

Our discussion of the rate of change of $s$ with respect to $t$ applies equally well to *any* function $y = f(x)$. This means that we have the following:

If $y = f(x)$, then

$$\frac{\Delta y}{\Delta x} = \frac{f(x + \Delta x) - f(x)}{\Delta x} = \begin{cases} \text{average rate of change} \\ \text{of } y \text{ with respect to } x \\ \text{over the interval from} \\ x \text{ to } x + \Delta x \end{cases}$$

and

$$\frac{dy}{dx} = \lim_{\Delta x \to 0} \frac{\Delta y}{\Delta x} = \begin{cases} \text{instantaneous rate of change} \\ \text{of } y \text{ with respect to } x. \end{cases} \qquad (2)$$

Because the instantaneous rate of change of $y = f(x)$ at a point is a derivative, it is also the *slope of the tangent line* to the graph of $y = f(x)$ at that point. For convenience, we usually refer to the instantaneous rate of change simply as the **rate of change.** The interpretation of a derivative as a rate of change is extremely important.

Let us now consider the significance of the rate of change of $y$ with respect to $x$. From Eq. (2), if $\Delta x$ (a change in $x$) is close to 0, then $\Delta y / \Delta x$ is close to $dy/dx$. That is,

$$\frac{\Delta y}{\Delta x} \approx \frac{dy}{dx}.$$

Therefore,

$$\Delta y \approx \frac{dy}{dx} \Delta x. \qquad (3)$$

That is, if $x$ changes by $\Delta x$, then the change in $y$, $\Delta y$, is approximately $dy/dx$ times the change in $x$. In particular,

if $x$ changes by 1, an estimate of the change in $y$ is $\dfrac{dy}{dx}$.

## Principles in Practice 1
### Estimating $\Delta P$ by Using $dP/dp$

Suppose that the profit $P$ made by selling a certain product at a price $p$ per unit is given by $P = f(p)$ and the rate of change of that profit with respect to change in price is $\dfrac{dP}{dp} = 5$ at $p = 25$. Estimate the change in the profit $P$ if the price changes from 25 to 25.5.

**EXAMPLE 2  Estimating $\Delta y$ by Using $dy/dx$**

*Suppose that $y = f(x)$ and $\dfrac{dy}{dx} = 8$ when $x = 3$. Estimate the change in $y$ if $x$ changes from 3 to 3.5.*

**Solution:** We have $dy/dx = 8$ and $\Delta x = 3.5 - 3 = 0.5$. The change in $y$ is given by $\Delta y$, and, from (3),

$$\Delta y \approx \frac{dy}{dx} \Delta x = 8(0.5) = 4.$$

We remark that, since $\Delta y = f(3.5) - f(3)$, we have $f(3.5) = f(3) + \Delta y$. For example, if $f(3) = 5$, then $f(3.5)$ can be estimated by $5 + 4 = 9$. ■

**EXAMPLE 3  Finding a Rate of Change**

*Find the rate of change of $y = x^4$ with respect to $x$, and evaluate it when $x = 2$ and when $x = -1$. Interpret your results.*

**Principles in Practice 2**
**Finding a Rate of Change**

The position of an object thrown upward at a speed of 16 feet/s from a height of 0 feet is given by $y(t) = 16t - 16t^2$. Find the rate of change of $y$ with respect to $t$, and evaluate it when $t = 0.5$. Use your graphics calculator to graph $y(t)$. Use the graph to interpret the behavior of the object when $t = 0.5$.

*Solution:* The rate of change is

$$\frac{dy}{dx} = 4x^3.$$

When $x = 2$, $dy/dx = 4(2)^3 = 32$. This means that if $x$ increases by a small amount, then $y$ increases approximately 32 times as much. More simply, we say that $y$ is increasing 32 times as fast as $x$ does. When $x = -1$, $dy/dx = 4(-1)^3 = -4$. The significance of the minus sign on $-4$ is that $y$ is *decreasing* 4 times as fast as $x$ increases. ∎

**EXAMPLE 4  Rate of Change of Price with Respect to Quantity**

*Let $p = 100 - q^2$ be the demand function for a manufacturer's product. Find the rate of change of price $p$ per unit with respect to quantity $q$. How fast is the price changing with respect to $q$ when $q = 5$? Assume that $p$ is in dollars.*

*Solution:* The rate of change of $p$ with respect to $q$ is

$$\frac{dp}{dq} = \frac{d}{dq}(100 - q^2) = -2q.$$

Thus,

$$\left.\frac{dp}{dq}\right|_{q=5} = -2(5) = -10.$$

This means that when five units are demanded, an *increase* of one extra unit demanded corresponds to a decrease of approximately \$10 in the price per unit that consumers are willing to pay. ∎

**EXAMPLE 5  Rate of Change of Volume**

*A spherical balloon is being filled with air. Find the rate of change of the volume of the balloon with respect to its radius. Evalute this rate of change when the radius is 2 ft.*

*Solution:* The formula for the volume $V$ of a sphere of radius $r$ is $V = \frac{4}{3}\pi r^3$. The rate of change of $V$ with respect to $r$ is

$$\frac{dV}{dr} = \frac{4}{3}\pi(3r^2) = 4\pi r^2.$$

When $r = 2$ ft, the rate of change is

$$\left.\frac{dV}{dr}\right|_{r=2} = 4\pi(2)^2 = 16\pi \frac{ft^3}{ft}.$$

This means that when the radius is 2 ft, changing the radius by 1 ft will change the volume by approximately $16\pi$ ft$^3$. ∎

**EXAMPLE 6  Rate of Change of Enrollment**

*A sociologist is studying various suggested programs that can aid in the education of preschool-age children in a certain city. The sociologist believes that $x$ years after the beginning of a particular program, $f(x)$ thousand preschoolers will be enrolled, where*

$$f(x) = \frac{10}{9}(12x - x^2), \qquad 0 \le x \le 12.$$

*At what rate would enrollment change (a) after three years from the start of this program and (b) after nine years?*

**Solution:**  The rate of change of $f(x)$ is

$$f'(x) = \frac{10}{9}(12 - 2x).$$

**a.**  After three years, the rate of change is

$$f'(3) = \frac{10}{9}[12 - 2(3)] = \frac{10}{9} \cdot 6 = \frac{20}{3} = 6\frac{2}{3}.$$

Thus, enrollment would be increasing at the rate of $6\frac{2}{3}$ thousand preschoolers per year.

**b.**  After nine years, the rate is

$$f'(9) = \frac{10}{9}[12 - 2(9)] = \frac{10}{9}[-6] = -\frac{20}{3} = -6\frac{2}{3}.$$

Thus, enrollment would be *decreasing* at the rate of $6\frac{2}{3}$ thousand preschoolers per year.

### Applications of Rate of Change to Economics

A manufacturer's **total-cost function,** $c = f(q)$, gives the total cost $c$ of producing and marketing $q$ units of a product. The rate of change of $c$ with respect to $q$ is called the **marginal cost.** Thus,

$$\text{marginal cost} = \frac{dc}{dq}.$$

For example, suppose $c = f(q) = 0.1q^2 + 3$ is a cost function, where $c$ is in dollars and $q$ is in pounds. Then

$$\frac{dc}{dq} = 0.2q.$$

The marginal cost when 4 lb are produced is $dc/dq$, evaluated when $q = 4$:

$$\left.\frac{dc}{dq}\right|_{q=4} = 0.2(4) = 0.80.$$

This means that if production is increased by 1 lb, from 4 lb to 5 lb, then the change in cost is approximately \$0.80. That is, the additional pound costs about \$0.80. In general, *we interpret marginal cost as the approximate cost of one additional unit of output.* [The actual cost of producing one more pound beyond 4 lb is $f(5) - f(4) = 5.5 - 4.6 = \$0.90$.]

  If $c$ is the total cost of producing $q$ units of a product, then the **average cost per unit,** $\bar{c}$, is

$$\bar{c} = \frac{c}{q}. \qquad (4)$$

For example, if the total cost of 20 units is \$100, then the average cost per unit is $\bar{c} = 100/20 = \$5$. By multiplying both sides of Eq. (4) by $q$, we have

$$c = q\bar{c}.$$

That is, total cost is the product of the number of units produced and the average cost per unit.

### EXAMPLE 7  Marginal Cost

*If a manufacturer's average-cost equation is*

$$\bar{c} = 0.0001q^2 - 0.02q + 5 + \frac{5000}{q},$$

*find the marginal-cost function. What is the marginal cost when 50 units are produced?*

*Solution:*

> *Strategy:* The marginal-cost function is the derivative of the total-cost function $c$. Thus, we first find $c$ by multiplying $\bar{c}$ by $q$.

We have

$$c = q\bar{c}$$

$$= q\left[ 0.0001q^2 - 0.02q + 5 + \frac{5000}{q} \right].$$

$$c = 0.0001q^3 - 0.02q^2 + 5q + 5000.$$

Differentiating $c$, we have the marginal-cost function:

$$\frac{dc}{dq} = 0.0001(3q^2) - 0.02(2q) + 5(1) + 0$$

$$= 0.0003q^2 - 0.04q + 5.$$

The marginal cost when 50 units are produced is

$$\left. \frac{dc}{dq} \right|_{q=50} = 0.0003(50)^2 - 0.04(50) + 5 = 3.75.$$

If $c$ is in dollars and production is increased by one unit, from $q = 50$ to $q = 51$, then the cost of the additional unit is approximately \$3.75. If production is increased by $\frac{1}{3}$ unit, from $q = 50$, then the cost of the additional output is approximately $(\frac{1}{3})(3.75) = \$1.25$. ∎

Suppose $r = f(q)$ is the **total-revenue function** for a manufacturer. The equation $r = f(q)$ states that the total dollar value received for selling $q$ units of a product is $r$. The **marginal revenue** is defined as the rate of change of the total dollar value received with respect to the total number of units sold. Hence, marginal revenue is merely the derivative of $r$ with respect to $q$:

$$\text{marginal revenue} = \frac{dr}{dq}.$$

Marginal revenue indicates the rate at which revenue changes with respect to units sold. We interpret it as *the approximate revenue received from selling one additional unit of output.*

**EXAMPLE 8**  **Marginal Revenue**

Suppose a manufacturer sells a product at $2 per unit. If $q$ units are sold, the total revenue is given by

$$r = 2q.$$

The marginal-revenue function is

$$\frac{dr}{dq} = \frac{d}{dq}(2q) = 2,$$

which is a constant function. Thus, the marginal revenue is 2 regardless of the number of units sold. This is what we would expect, because the manufacturer receives $2 for each unit sold. ∎

## Relative and Percentage Rates of Change

For the total-revenue function in Example 8, namely, $r = f(q) = 2q$, we have

$$\frac{dr}{dq} = 2.$$

This means that revenue is changing at the rate of $2 per unit, regardless of the number of units sold. Although this is valuable information, it may be more significant when compared to $r$ itself. For example, if $q = 50$, then $r = 2(50) = \$100$. Thus, the rate of change of revenue is $2/100 = 0.02$ **of $r$.** On the other hand, if $q = 5000$, then $r = 2(5000) = \$10,000$, so the rate of change of $r$ is $2/10,000 = 0.0002$ **of $r$.** Although $r$ changes at the same rate at each level, compared to $r$ itself, this rate is relatively smaller when $r = 10,000$ than when $r = 100$. By considering the ratio

$$\frac{dr/dq}{r},$$

we have a means of comparing the rate of change of $r$ with $r$ itself. This ratio is called the *relative rate of change* of $r$. We have shown that the relative rate of change when $q = 50$ is

$$\frac{dr/dq}{r} = \frac{2}{100} = 0.02,$$

and when $q = 5000$, it is

$$\frac{dr/dq}{r} = \frac{2}{10,000} = 0.0002.$$

By multiplying relative rates by 100, we obtain so-called *percentage rates of change*. The percentage rate of change when $q = 50$ is $(0.02)(100) = 2\%$; when $q = 5000$, it is $(0.0002)(100) = 0.02\%$. For example, if an additional unit beyond 50 is sold, then revenue increases by approximately 2%.

In general, for any function $f$, we have the following definition:

**DEFINITION**

*The **relative rate of change** of f(x) is*

$$\frac{f'(x)}{f(x)}.$$

The **percentage rate of change** of $f(x)$ is

$$\frac{f'(x)}{f(x)} \cdot 100.$$

---

**Principles in Practice 3**

Relative and Percentage Rates of Change

The volume $V$ of a capsule-shaped container with a cylindrical height of 4 feet and radius $r$ is given by

$$V(r) = \frac{4}{3}\pi r^3 + 4\pi r^2.$$

Determine the relative and percentage rates of change of volume with respect to the radius when the radius is 2 feet.

---

**EXAMPLE 9   Relative and Percentage Rates of Change**

*Determine the relative and percentage rates of change of*

$$y = f(x) = 3x^2 - 5x + 25$$

*when $x = 5$.*

*Solution:* Here

$$f'(x) = 6x - 5.$$

Since $f'(5) = 6(5) - 5 = 25$ and $f(5) = 3(5)^2 - 5(5) + 25 = 75$, the relative rate of change of $y$ when $x = 5$ is

$$\frac{f'(5)}{f(5)} = \frac{25}{75} \approx 0.333.$$

Multiplying 0.333 by 100 gives the percentage rate of change: $(0.333)(100) = 33.3\%$. ∎

---

## ▪ Exercise 12.3

**1.** Suppose that the position function of an object moving along a straight line is $s = f(t) = t^3 + t$, in seconds and $s$ is in meters. Find the average velocity $\Delta s/\Delta t$ over the interval $[1, 1 + \Delta t]$, where $\Delta t$ is given in the following table:

| $\Delta t$ | 1 | 0.5 | 0.2 | 0.1 | 0.01 | 0.001 |
|---|---|---|---|---|---|---|
| $\Delta s/\Delta t$ | | | | | | |

From your results, estimate the velocity when $t = 1$. Verify your estimate by using differentiation.

**2.** If $y = f(x) = \sqrt{2x + 3}$, find the average rate of change of $y$ with respect to $x$ over the interval $[3, 3 + \Delta x]$, where $\Delta x$ is given in the following table:

| $\Delta x$ | 1 | 0.5 | 0.2 | 0.1 | 0.01 | 0.001 |
|---|---|---|---|---|---|---|
| $\Delta y/\Delta x$ | | | | | | |

From your result, estimate the rate of change of $y$ with respect to $x$ when $x = 3$.

*In each of Problems 3–8, a position function is given, where t is in seconds and s is in meters.*
*a. Find the position at the given t-value.*
*b. Find the average velocity over the given interval.*
*c. Find the velocity at the given t-value.*

**3.** $s = t^2 - 3t$; $[4, 4.5]$; $t = 4$.

**4.** $s = \frac{1}{2}t + 1$; $[2, 2.1]$; $t = 2$.

**5.** $s = 2t^3 + 6$; $[1, 1.02]$; $t = 1$.

**6.** $s = -3t^2 + 2t + 1$; $[1, 1.25]$; $t = 1$.

**7.** $s = t^4 - 2t^3 + t$; $[2, 2.1]$; $t = 2$.

**8.** $s = t^4 - t^{5/2}$; $[0, \frac{1}{4}]$; $t = 0$.

**9. Electricity** The current $i$ in a certain resistor as a function of the power $P$ developed in the resistor is given by $i = \sqrt{P}$. Find the rate of change of $i$ with respect to $P$ when $P = 4$.

**10. Physics** The volume $V$ of a certain gas varies with pressure $p$ according to the equation $p = 150/V$. Find the rate of change of $p$ with respect to $V$ when $V = 5$,

**11. Income-Education** Sociologists studied the relation between income and number of years of education for members of a particular urban group. They found that a person with $x$ years of education before seeking regular employment can expect to receive an average yearly income of $y$ dollars per year, where

$$y = 5x^{5/2} + 5900, \qquad 4 \le x \le 16.$$

Find the rate of change of income with respect to number of years of education. Evaluation the expression when $x = 9$.

12. Find the rate of change of the area $A$ of a circle with respect to its radius $r$ if

$$A = \pi r^2.$$

Evaluate the expression when $r = 3$ in.

13. **Skin Temperature**    The approximate temperature $T$ of the skin in terms of the temperature $T_e$ of the environment is given by

$$T = 32.8 + 0.27(T_e - 20),$$

where $T$ and $T_e$ are in degrees Celsius.[3] Find the rate of change of $T$ with respect to $T_e$.

14. **Biology**    The volume $V$ of a spherical cell is given by $V = \frac{4}{3}\pi r^3$, where $r$ is the radius. Find the rate of change of volume with respect to the radius when $r = 6.5 \times 10^{-4}$ cm.

*In Problems 15–20, cost functions are given, where c is the cost of producing q units of a product. In each case, find the marginal-cost function. What is the marginal cost at the given value(s) of q?*

15. $c = 500 + 10q;\ q = 100.$

16. $c = 5000 + 6q;\ q = 36.$

17. $c = 0.3q^2 + 2q + 850;\ q = 3.$

18. $c = 0.1q^2 + 3q + 2;\ q = 3.$

19. $c = q^2 + 50q + 1000;\ q = 15, q = 16, q = 17.$

20. $c = 0.03q^3 - 0.6q^2 + 4.5q + 7700;\ q = 10, q = 20, q = 100.$

*In Problems 21–24, $\bar{c}$ represents average cost per unit, which is a function of the number q of units produced. Find the marginal-cost function and the marginal cost for the indicated values of q.*

21. $\bar{c} = 0.01q + 5 + \dfrac{500}{q};\ q = 50,\ q = 100.$

22. $\bar{c} = 2 + \dfrac{1000}{q};\ q = 25,\ q = 235.$

23. $\bar{c} = 0.00002q^2 - 0.01q + 6 + \dfrac{20,000}{q};\ q = 100, q = 500.$

24. $\bar{c} = 0.001q^2 - 0.3q + 40 + \dfrac{7000}{q};\ q = 10,\ q = 20.$

*In Problems 25–28, r represents total revenue and is a function of the number q of units sold. Find the marginal–revenue function and the marginal revenue for the indicated values of q.*

25. $r = 0.7q;\ q = 8, q = 100, q = 200.$

26. $r = q(15 - \frac{1}{30}q);\ q = 5,\ q = 15,\ q = 150.$

27. $r = 250q + 45q^2 - q^3;\ q = 5, q = 10, q = 25.$

28. $r = 2q(30 - 0.1q);\ q = 10,\ q = 20.$

29. **Hosiery Mill**    The total-cost function for a hosiery mill is estimated by Dean[4] to be

$$c = -10,484.69 + 6.750q - 0.000328q^2,$$

where $q$ is output in dozens of pairs and $c$ is total cost in dollars. Find the marginal-cost function and evaluate it when $q = 5000$.

30. **Light and Power Plant**    The total-cost function for an electric light and power plant is estimated by Nordin[5] to be

$$c = 32.07 - 0.79q + 0.02142q^2 - 0.0001q^3,$$

$$20 \le q \le 90,$$

where $q$ is the eight-hour total output (as a percentage of capacity) and $c$ is the total fuel cost in dollars. Find the marginal–cost function and evaluate it when $q = 70$.

31. **Urban Concentration**    Suppose the 100 largest cities in the United States in 1920 are ranked according to magnitude (areas of cities). From Lotka,[6] the following relation holds approximately:

$$PR^{0.93} = 5,000,000.$$

Here, $P$ is the population of the city having respective rank $R$. This relation is called the *law of urban concentration* for 1920. Solve for $P$ in terms of $R$, and then find how fast the population is changing with respect to rank.

32. **Depreciation**    Under the straight-line method of depreciation, the value $v$ of a certain machine after $t$ years have elapsed given by

$$v = 50,000 - 5000t,$$

where $0 \le t \le 10$. How fast is $v$ changing with respect to $t$ when $t = 2$? $t = 3$? at any time?

[3]R. W. Stacy et al., *Essentials of Biological and Medical Physics* (New York: McGraw-Hill Book Company, 1955).

[4]J. Dean, "Statistical Cost Functions of a Hosiery Mill," *Studies in Business Administration*, XI, no. 4 (Chicago: University of Chicago Press, 1941).

[5]J. A. Nordin, "Note on a Light Plant's Cost Curves," *Econometrica*, 15 (1947), 231–35.

[6]A. J. Lotka, *Elements of Mathematical Biology* (New York: Dover Publications, Inc., 1956).

33. **Winter Moth**   A study of the winter moth was made in Nova Scotia (adapted from Embree.)[7] The prepupae of the moth fall onto the ground from host trees. At a distance of $x$ ft from the base of a host tree, the prepupal density (number of prepupae per square foot of soil) was $y$, where

$$y = 59.3 - 1.5x - 0.5x^2, \quad 1 \le x \le 9.$$

   **a.** At what rate is the prepupal density changing with respect to distance from the base of the tree when $x = 6$?

   **b.** For what value of $x$ is the prepupal density decreasing at the rate of 6 prepupae per square foot per foot?

34. **Cost Function**   For the cost function

$$c = 0.4q^2 + 4q + 5,$$

   find the rate of change of $c$ with respect to $q$ when $q = 2$. Also, what is $\Delta c / \Delta q$ over the interval $[2, 3]$?

*In Problems 35–40, find (a) the rate of change of y with respect x and (b) the relative rate of change of y. At the given value of x, find (c) the rate of change of y, (d) the relative rate of change of y, and (e) the percentage rate of change of y.*

35. $y = f(x) = x + 4;\ x = 5.$

36. $y = f(x) = 4 - 2x;\ x = 3.$

37. $y = 3x^2 + 6;\ x = 2.$

38. $y = 2 - x^2;\ x = 0.$

39. $y = 8 - x^3;\ x = 1.$

40. $y = x^2 + 3x - 4;\ x = -1.$

41. **Cost Function**   For the cost function

$$c = 0.2q^2 + 1.2q + 4,$$

   how fast does $c$ change with respect to $q$ when $q = 5$? Determine the percentage rate of change of $c$ with respect to $q$ when $q = 5$.

42. **Organic Matters/Species Diversity**   In a discussion of contemporary waters of shallow seas, Odum[8] claims that in such waters the total organic matter $y$ (in milligrams per liter) is a function of species diversity $x$ (in number of species per thousand individuals). If $y = 100/x$, at what rate is the total organic matter changing with respect to species diversity when $x = 10$? What is the percentage rate of change when $x = 10$?

43. **Revenue**   For a certain manufacturer, the revenue obtained from the sale of $q$ units of a product is given by

$$r = 30q - 0.3q^2.$$

   (a) How fast does $r$ change with respect to $q$? When $q = 10$, (b) find the relative rate of change of $r$, and (c) to the nearest percent, find the percentage rate of change of $r$.

44. **Revenue**   Repeat Problem 43 for the revenue function given by $r = 20q - 0.1q^2$ and $q = 100$.

45. **Weight of Limb**   The weight of a limb of a tree is given by $W = 2t^{0.432}$, where $t$ is time. Find the relative rate of change of $W$ with respect to $t$.

46. **Response to Shock**   A psychological experiment[9] was conducted to analyze human responses to electrical shocks (stimuli). The subjects received shocks of various intensities. The response $R$ to a shock of intensity $I$ (in

microamperes) was to be a number that indicated the perceived magnitude relative to that of a "standard" shock. The standard shock was assigned a magnitude of 10. Two groups of subjects were tested under slightly different conditions. The responses $R_1$ and $R_2$ of the first and second groups to a shock of intensity $I$ were given by

$$R_1 = \frac{I^{1.3}}{1855.24}, \quad 800 \le I \le 3500,$$

and

$$R_2 = \frac{I^{1.3}}{1101.29}, \quad 800 \le I \le 3500,$$

   **a.** For each group, determine the relative rate of change of response with respect to intensity.

   **b.** How do these changes compare with each other?

   **c.** In general, if $f(x) = C_1 x^n$ and $g(x) = C_2 x^n$, where $C_1$ and $C_2$ are constants, how do the relative rates of change of $f$ and $g$ compare?

47. **Cost**   A manufacturer of mountain bikes has found that when 20 bikes are produced per day, the average cost is $150 and the marginal cost is $125. Based on that information, approximate the total cost of producing 21 bikes per day.

48. **Marginal and Average Costs**   Suppose that the cost function for a certain product is $c = f(q)$. If the relative rate of change of $c$ (with respect to $q$) is $\dfrac{1}{q}$, prove that the marginal-cost function and the average-cost function are equal.

---

[7]D. G. Embree, "The Population Dynamics of the Winter Moth in Nova Scotia, 1954–1962," *Memoirs of the Entomological Society of Canada*, no. 46 (1965).

[8]H. T. Odum, "Biological Circuits and the Marine Systems of Texas," in *Pollution and Marine Biology*, ed. T. A. Olsen and F. J. Burgess (New York: Interscience Publishers, 1967).

[9]H. Babkoff, "Magnitude Estimation of Short Electrocutaneous Pulses," *Psychological Research*, 39, no. 1 (1976), 39–49.

*In Problems 49 and 50, use the numerical derivative feature of your graphics calculator.*

 **49.** If the total-cost function for a manufacturer is given by

$$c = \frac{5q^2}{\sqrt{q^2 + 3}} + 5000,$$

where $c$ is in dollars, find the marginal cost when 10 units are produced. Round your answer to the nearest cent.

 **50.** The population of a city $t$ years from now is given by

$$P = 20,000e^{0.03t}.$$

Find the rate of change of population with respect to time $t$ four years from now. Round your answer to the nearest integer.

---

**O B J E C T I V E**

To relate differentiability to continuity.

## 12.4 DIFFERENTIABILITY AND CONTINUITY

In the next section, we shall make use of the following important relationship between differentiability and continuity:

> If $f$ is differentiable at $a$, then $f$ is continuous at $a$.

To establish this result, we shall assume that $f$ is differentiable at $a$. Then $f'(a)$ exists, and

$$\lim_{h \to 0} \frac{f(a + h) - f(a)}{h} = f'(a) \quad .$$

Consider the numerator $f(a + h) - f(a)$ as $h \to 0$. We have

$$\lim_{h \to 0} [f(a + h) - f(a)] = \lim_{h \to 0} \left[ \frac{f(a + h) - f(a)}{h} \cdot h \right]$$

$$= \lim_{h \to 0} \frac{f(a + h) - f(a)}{h} \cdot \lim_{h \to 0} h$$

$$= f'(a) \cdot 0 = 0.$$

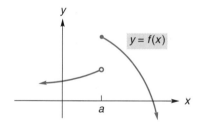

Thus, $\lim_{h \to 0} [f(a + h) - f(a)] = 0$. This means that $f(a + h) - f(a)$ approaches 0 as $h \to 0$. Consequently,

$$\lim_{h \to 0} f(a + h) = f(a).$$

As stated in Sec. 11.4, this condition means that $f$ is continuous at $a$. The foregoing, then, proves that $f$ is continuous at $a$ when $f$ is differentiable there. More simply, we say that **differentiability at a point implies continuity at that point.**

If a function is not continuous at a point, then it cannot have a derivative there. For example, the function in Fig. 12.13 is discontinuous at $a$. The curve has no tangent at that point, so the function is not differentiable there.

**FIGURE 12.13** $f$ is not continuous at $a$, so $f$ is not differentiable at $a$.

**EXAMPLE 1  Continuity and Differentiability**

**a.** Let $f(x) = x^2$. The derivative, $2x$, is defined for all values of $x$, so $f(x) = x^2$ must be continuous for all values of $x$.

**b.** The function $f(p) = \dfrac{1}{2p}$ is not continuous at $p = 0$ because $f$ is not defined there. Thus, the derivative does not exist at $p = 0$.  ■

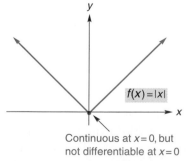

Continuous at $x = 0$, but not differentiable at $x = 0$

**FIGURE 12.14** Continuity does not imply differentiability.

The converse of the statement that differentiability implies continuity is *false.* That is, it is false that continuity implies differentiability. In Example 2, you will see a function that is continuous at a point, but not differentiable there.

**EXAMPLE 2    Continuity Does Not Imply Differentiability**

The function $y = f(x) = |x|$ is continuous at $x = 0$. (See Fig. 12.14 on the previous page.) As we mentioned in Sec. 12.1, there is no tangent line at $x = 0$. Thus, the derivative does not exist there. This shows that continuity does *not* imply differentiability.    ■

OBJECTIVE

To find derivatives by applying the product and quotient rules, and to develop the concepts of marginal propensity to consume and marginal propensity to save.

## 12.5  Product and Quotient Rules

The equation $F(x) = (x^2 + 3x)(4x + 5)$ expresses $F(x)$ as a product of two functions: $x^2 + 3x$ and $4x + 5$. To find $F'(x)$ by using only our previous rules, we first multiply the functions. Then we differentiate the result, term by term:

$$F(x) = (x^2 + 3x)(4x + 5) = 4x^3 + 17x^2 + 15x,$$

$$F'(x) = 12x^2 + 34x + 15. \tag{1}$$

However, in many problems that involve differentiating a product of functions, the multiplication is not as simple as it is here. At times, it is not even practical to attempt it. Fortunately, there is a rule for differentiating a product, and the rule avoids such multiplications. Since the derivative of a sum of functions is the sum of their derivatives, you might think that the derivative of a product of two functions is the product of their derivatives. This is **not** the case, as the next rule shows.

> **Rule 5 Product Rule**
>
> *If f and g are differentiable functions, then the product fg is differentiable, and*
>
> $$\frac{d}{dx}[f(x)g(x)] = f(x)g'(x) + g(x)f'(x).$$
>
> *That is, the derivative of the product of two functions is the first function times the derivative of the second, plus the second function times the derivative of the first.*
>
> $$\frac{d}{dx}(\text{product}) = (\text{first})\left(\begin{array}{c}\text{derivative}\\\text{of second}\end{array}\right) + (\text{second})\left(\begin{array}{c}\text{derivative}\\\text{of first}\end{array}\right).$$

*Proof.* If $F(x) = f(x)g(x)$, then, by the definition of the derivative of $F$,

$$F'(x) = \lim_{h\to 0}\frac{F(x + h) - F(x)}{h}$$

$$= \lim_{h\to 0}\frac{f(x + h)g(x + h) - f(x)g(x)}{h}.$$

Now we use a "trick." Adding and subtracting $f(x + h)g(x)$ in the numerator, we have

$$F'(x)$$

$$= \lim_{h\to 0}\frac{f(x + h)g(x + h) - f(x)g(x) + f(x + h)g(x) - f(x + h)g(x)}{h}.$$

Regrouping gives

$$F'(x)$$
$$= \lim_{h \to 0} \frac{[f(x+h)g(x+h) - f(x+h)g(x)] + [f(x+h)g(x) - f(x)g(x)]}{h}$$

$$= \lim_{h \to 0} \frac{f(x+h)[g(x+h) - g(x)] + g(x)[f(x+h) - f(x)]}{h}$$

$$= \lim_{h \to 0} \frac{f(x+h)[g(x+h) - g(x)]}{h} + \lim_{h \to 0} \frac{g(x)[f(x+h) - f(x)]}{h}$$

$$= \lim_{h \to 0} f(x+h) \cdot \lim_{h \to 0} \frac{g(x+h) - g(x)}{h} + \lim_{h \to 0} g(x) \cdot \lim_{h \to 0} \frac{f(x+h) - f(x)}{h}.$$

Since we assumed that $f$ and $g$ are differentiable, then

$$\lim_{h \to 0} \frac{f(x+h) - f(x)}{h} = f'(x)$$

and

$$\lim_{h \to 0} \frac{g(x+h) - g(x)}{h} = g'(x).$$

The differentiability of $f$ implies that $f$ is continuous, and from Sec. 11.4,

$$\lim_{h \to 0} f(x+h) = f(x).$$

Thus,

$$F'(x) = f(x)g'(x) + g(x)f'(x).$$

## EXAMPLE 1  Applying the Product Rule

*If $F(x) = (x^2 + 3x)(4x + 5)$, find $F'(x)$.*

*Solution:* We shall consider $F$ as a product of two functions:

$$F(x) = \underbrace{(x^2 + 3x)}_{f(x)} \underbrace{(4x + 5)}_{g(x)}$$

Therefore, we can apply the product rule:

$$F'(x) = f(x)g'(x) + g(x)f'(x).$$

$$= \underbrace{(x^2 + 3x)}_{\text{First}} \underbrace{\frac{d}{dx}(4x + 5)}_{\substack{\text{Derivative} \\ \text{of} \\ \text{second}}} + \underbrace{(4x + 5)}_{\text{Second}} \underbrace{\frac{d}{dx}(x^2 + 3x)}_{\substack{\text{Derivative} \\ \text{of} \\ \text{first}}}$$

$$= (x^2 + 3x)(4) + (4x + 5)(2x + 3)$$

$$= 12x^2 + 34x + 15 \qquad \text{(simplifying)}.$$

This agrees with our previous result. [See Eq. (1).] Although there doesn't seem to be much advantage to using the product rule here, you will see that there are times when it is practical to use.  ∎

*Pitfall* ▼ It is worthwhile to repeat that the derivative of the product of two functions is **not** the product of their derivatives. For example, $\frac{d}{dx}(x^2 + 3x) = 2x + 3$ and $\frac{d}{dx}(4x + 5) = 4$, but from Example 1,

$$\frac{d}{dx}[(x^2 + 3x)(4x + 5)] = 12x^2 + 34x + 15 \neq (2x + 3)4.$$

---

**Principles in Practice 1**

**Applying the Product Rule**

A taco stand usually sells 225 tacos per day at \$2 each. A business student's research tells him that for every \$0.15 decrease in the price, the stand will sell 20 more tacos per day. The revenue function for the taco stand is $R(x) = (2 - 0.15x)(225 + 20x)$, where $x$ is the number of \$0.15 reductions in price. Find $\frac{dR}{dx}$.

---

**EXAMPLE 2    Applying the Product Rule**

If $y = (x^{2/3} + 3)(x^{-1/3} + 5x)$, *find dy/dx*.

*Solution:* Applying the product rule gives

$$\frac{dy}{dx} = (x^{2/3} + 3)\frac{d}{dx}(x^{-1/3} + 5x) + (x^{-1/3} + 5x)\frac{d}{dx}(x^{2/3} + 3)$$

$$= (x^{2/3} + 3)(-\tfrac{1}{3}x^{-4/3} + 5) + (x^{-1/3} + 5x)(\tfrac{2}{3}x^{-1/3})$$

$$= \tfrac{25}{3}x^{2/3} + \tfrac{1}{3}x^{-2/3} - x^{-4/3} + 15.$$

Alternatively, we could have found the derivative without the product rule by first finding the product $(x^{2/3} + 3)(x^{-1/3} + 5x)$ and then differentiating the result, term by term. ∎

**EXAMPLE 3    Differentiating a Product of Three Factors**

If $y = (x + 2)(x + 3)(x + 4)$, *find y'*.

*Solution:*

> *Strategy:*   We would like to use the product rule, but it applies only to *two* factors. By treating the first two factors as a single factor, we can consider $y$ to be a product of two functions:
>
> $$y = [(x + 2)(x + 3)](x + 4).$$

The product rule gives

$$y' = [(x + 2)(x + 3)]\frac{d}{dx}(x + 4) + (x + 4)\frac{d}{dx}[(x + 2)(x + 3)]$$

$$= [(x + 2)(x + 3)](1) + (x + 4)\frac{d}{dx}[(x + 2)(x + 3)].$$

Applying the product rule again, we have

$$y'$$
$$= (x + 2)(x + 3) + (x + 4)\left[(x + 2)\frac{d}{dx}(x + 3) + (x + 3)\frac{d}{dx}(x + 2)\right]$$

$$= (x + 2)(x + 3) + (x + 4)[(x + 2)(1) + (x + 3)(1)].$$

After simplifying, we obtain

$$y' = 3x^2 + 18x + 26.$$

Two other ways of finding the derivative are as follows:

**1.** Multiply the first two factors of $y$ to obtain

$$y = (x^2 + 5x + 6)(x + 4),$$

and then apply the product rule.

**2.** Multiply all three factors to obtain

$$y = x^3 + 9x^2 + 26x + 24,$$

and then differentiate, term by term. ∎

**EXAMPLE 4  Using the Product Rule to Find Slope**

*Find the slope of the graph of* $f(x) = (7x^3 - 5x + 2)(2x^4 + 7)$ *when* $x = 1$.

*Solution:*

> *Strategy:* We find the slope by evaluating the derivative when $x = 1$. Because $f$ is a product of two functions, we can find the derivative by using the product rule.

We have

$$f'(x) = (7x^3 - 5x + 2)\frac{d}{dx}(2x^4 + 7) + (2x^4 + 7)\frac{d}{dx}(7x^3 - 5x + 2)$$

$$= (7x^3 - 5x + 2)(8x^3) + (2x^4 + 7)(21x^2 - 5).$$

Since we must compute $f'(x)$ when $x = 1$, *there is no need to simplify* $f'(x)$ *before evaluating it.* Substituting into $f'(x)$, we obtain

$$f'(1) = 4(8) + 9(16) = 176.$$ ∎

Usually, we do not use the product rule when simpler ways are obvious. For example, if $f(x) = 2x(x + 3)$, then it is quicker to write $f(x) = 2x^2 + 6x$, from which $f'(x) = 4x + 6$. Similarly, we do not usually use the product rule to differentiate $y = 4(x^2 - 3)$. Since the 4 is a constant factor, by the constant-factor rule we have $y' = 4(2x) = 8x$.

The next rule is used for differentiating a *quotient* of two functions.

The product rule (and quotient rule that follows) should not be applied when a more direct and efficient method is available.

**Principles in Practice 2**
**Derivative of a Product without the Product Rule**

One hour after $x$ milligrams of a particular drug are given to a person, the change in body temperature $T(x)$, in degrees Fahrenheit, is given approximately by $T(x) = x^2\left(1 - \dfrac{x}{3}\right)$. The rate at which $T$ changes with respect to the size of the dosage $x$, $T'(x)$, is called the *sensitivity* of the body to the dosage. Find the sensitivity when the dosage is 1 milligram. Do not use the product rule.

**Rule 6 Quotient Rule**

*If $f$ and $g$ are differentiable functions and $g(x) \neq 0$, then the quotient $f/g$ is also differentiable, and*

$$\frac{d}{dx}\left[\frac{f(x)}{g(x)}\right] = \frac{g(x)f'(x) - f(x)g'(x)}{[g(x)]^2}.$$

*That is, the derivative of the quotient of two functions is the denominator times the derivative of the numerator, minus the numerator times the derivative of the denominator, all divided by the square of the denominator.*

$$\frac{d}{dx}(\text{quotient})$$

$$= \frac{(\text{denominator})\left(\begin{array}{c}\text{derivative}\\\text{of numerator}\end{array}\right) - (\text{numerator})\left(\begin{array}{c}\text{derivative}\\\text{of denominator}\end{array}\right)}{(\text{denominator})^2}$$

*Proof of formula.* If $F(x) = \dfrac{f(x)}{g(x)}$, then

$$F(x)g(x) = f(x).$$

By the product rule,

$$F(x)g'(x) + g(x)F'(x) = f'(x).$$

Solving for $F'(x)$, we have

$$F'(x) = \frac{f'(x) - F(x)g'(x)}{g(x)}.$$

But $F(x) = f(x)/g(x)$. Thus,

$$F'(x) = \frac{f'(x) - \dfrac{f(x)g'(x)}{g(x)}}{g(x)}.$$

Simplifying gives[10]

$$F'(x) = \frac{g(x)f'(x) - f(x)g'(x)}{[g(x)]^2}.$$

**EXAMPLE 5   Applying the Quotient Rule**

*If* $F(x) = \dfrac{4x^2 + 3}{2x - 1}$, *find* $F'(x)$.

*Solution:*

> *Strategy:*   We recognize $F$ as a quotient, so we can apply the quotient rule.

Let $f(x) = 4x^2 + 3$ and $g(x) = 2x - 1$. Then

$$F'(x) = \frac{g(x)f'(x) - f(x)g'(x)}{[g(x)]^2}$$

$$= \frac{\overbrace{(2x - 1)}^{\text{Denominator}} \overbrace{\dfrac{d}{dx}(4x^2 + 3)}^{\substack{\text{Derivative} \\ \text{of} \\ \text{numerator}}} - \overbrace{(4x^2 + 3)}^{\text{Numerator}} \overbrace{\dfrac{d}{dx}(2x - 1)}^{\substack{\text{Derivative} \\ \text{of} \\ \text{denominator}}}}{\underbrace{(2x - 1)^2}_{\substack{\text{Square} \\ \text{of} \\ \text{denominator}}}}$$

$$= \frac{(2x - 1)(8x) - (4x^2 + 3)(2)}{(2x - 1)^2}$$

$$= \frac{8x^2 - 8x - 6}{(2x - 1)^2} = \frac{2(4x^2 - 4x - 3)}{(2x - 1)^2}.$$

**Pitfall** ▼ The derivative of the quotient of two functions is **not** the quotient of their derivatives. For example,

$$\frac{d}{dx}\left(\frac{4x^2 + 3}{2x - 1}\right) \neq \frac{\dfrac{d}{dx}(4x^2 + 3)}{\dfrac{d}{dx}(2x - 1)} = \frac{8x}{2}.$$

----

[10]You may have observed that this proof assumes the existence of $F'(x)$. However, the rule can be proven without this assumption.

**EXAMPLE 6** **Rewriting before Differentiating**

*Differentiate* $y = \dfrac{1}{x + \dfrac{1}{x + 1}}$.

*Solution:*

> *Strategy:* To simplify the differentiation, we shall rewrite the function so that no fraction appears in the denominator.

We have

$$y = \frac{1}{x + \dfrac{1}{x+1}} = \frac{1}{\dfrac{x(x+1)+1}{x+1}} = \frac{x+1}{x^2 + x + 1}.$$

$$\frac{dy}{dx} = \frac{(x^2 + x + 1)(1) - (x+1)(2x+1)}{(x^2 + x + 1)^2} \qquad \text{(quotient rule)}$$

$$= \frac{(x^2 + x + 1) - (2x^2 + 3x + 1)}{(x^2 + x + 1)^2}$$

$$= \frac{-x^2 - 2x}{(x^2 + x + 1)^2} = -\frac{x^2 + 2x}{(x^2 + x + 1)^2}. \qquad ∎$$

Although a function may have the form of a quotient, this does not necessarily mean that the quotient rule must be used to find the derivative. The next example illustrates some typical situations in which, although the quotient rule can be used, a simpler and more efficient method is available.

**EXAMPLE 7** **Differentiating Quotients without Using the Quotient Rule**

*Differentiate the following functions.*

**a.** $f(x) = \dfrac{2x^3}{5}$.

*Solution:* Rewriting, we have $f(x) = \frac{2}{5}x^3$. By the constant-factor rule,

$$f'(x) = \frac{2}{5}(3x^2) = \frac{6x^2}{5}.$$

**b.** $f(x) = \dfrac{4}{7x^3}$.

*Solution:* Rewriting, we have $f(x) = \frac{4}{7}(x^{-3})$. Thus,

$$f'(x) = \frac{4}{7}(-3x^{-4}) = -\frac{12}{7x^4}.$$

**c.** $f(x) = \dfrac{5x^2 - 3x}{4x}$.

*Solution:* Rewriting, we have $f(x) = \frac{1}{4}\left(\dfrac{5x^2 - 3x}{x}\right) = \frac{1}{4}(5x - 3)$. Thus,

$$f'(x) = \frac{1}{4}(5) = \frac{5}{4}. \qquad ∎$$

***Pitfall*** ▼ To differentiate $f(x) = \dfrac{1}{x^2 - 2}$, you might he tempted to first rewrite the quotient as $(x^2 - 2)^{-1}$. It would be a mistake to do this because we presently have no rule for differentiating that form. In short, we have no choice but to use the quotient rule. However, in the next section we shall develop a rule that allows us to differentiate $(x^2 - 2)^{-1}$ in a direct and efficient way.

### EXAMPLE 8 Marginal Revenue

*If the demand equation for a manufacturer's product is*

$$p = \frac{1000}{q + 5},$$

*where p is in dollars, find the marginal-revenue function and evaluate it when $q = 45$.*

*Solution:*

> *Strategy:* First we must find the revenue function. The revenue *r* received for selling *q* units when the price per unit is *p* is given by
>
> **revenue = (price)(quantity),** or **$r = pq$.**
>
> Using the demand equation we shall express *r* in terms of *q* only. Then we shall differentiate to find the marginal-revenue function, $dr/dq$.

The revenue function is

$$r = \left(\frac{1000}{q + 5}\right)q = \frac{1000q}{q + 5}.$$

Thus, the marginal-revenue function is given by

$$\frac{dr}{dq} = \frac{(q + 5)\dfrac{d}{dq}(1000q) - (1000q)\dfrac{d}{dq}(q + 5)}{(q + 5)^2}$$

$$= \frac{(q + 5)(1000) - (1000q)(1)}{(q + 5)^2} = \frac{5000}{(q + 5)^2},$$

and

$$\left.\frac{dr}{dq}\right|_{q=45} = \frac{5000}{(45 + 5)^2} = \frac{5000}{2500} = 2.$$

This means that selling one additional unit beyond 45 results in approximately $2 more in revenue. ▬

### Consumption Function

A function that plays an important role in economic analysis is the **consumption function.** The consumption function $C = f(I)$ expresses a relationship between the total national income *I* and the total national consumption *C*. Usually, both *I* and *C* are expressed in billions of dollars and *I* is restricted to some interval. The *marginal propensity to consume* is defined as the rate of change of consumption with respect to income. It is merely the derivative of *C* with respect to *I*:

> Marginal propensity to consume $= \dfrac{dC}{dI}.$

If we assume that the difference between income $I$ and consumption $C$ is savings $S$, then

$$S = I - C.$$

Differentiating both sides with respect to $I$ gives

$$\frac{dS}{dI} = \frac{d}{dI}(I) - \frac{d}{dI}(C) = 1 - \frac{dC}{dI}.$$

We define $dS/dI$ as the **marginal propensity to save.** Thus, the marginal propensity to save indicates how fast savings change with respect to income, and

$$\frac{\text{Marginal propensity}}{\text{to save}} = 1 - \frac{\text{Marginal propensity}}{\text{to consume}}.$$

**EXAMPLE 9**   **Finding Marginal Propensities to Consume and to Save**

*If the consumption function is given by*

$$C = \frac{5(2\sqrt{I^3} + 3)}{I + 10},$$

*determine the marginal propensity to consume and the marginal propensity to save when $I = 100$.*

***Solution:***

$$\frac{dC}{dI} = 5\left[\frac{(I + 10)\dfrac{d}{dI}(2I^{3/2} + 3) - (2\sqrt{I^3} + 3)\dfrac{d}{dI}(I + 10)}{(I + 10)^2}\right]$$

$$= 5\left[\frac{(I + 10)(3I^{1/2}) - (2\sqrt{I^3} + 3)(1)}{(I + 10)^2}\right].$$

When $I = 100$, the marginal propensity to consume is

$$\frac{dC}{dI}\bigg|_{I=100} = 5\left[\frac{1297}{12{,}100}\right] \approx 0.536.$$

The marginal propensity to save when $I = 100$ is $1 - 0.536 = 0.464$. This means that if a current income of \$100 billion increases by \$1 billion, the nation consumes approximately 53.6% (536/1000) and saves 46.4% (464/1000) of that increase.   ■

## ■ Exercise 12.5

*In Problems 1–48, differentiate the functions.*

**1.** $f(x) = (4x + 1)(6x + 3)$.

**2.** $f(x) = (3x - 1)(7x + 2)$.

**3.** $s(t) = (8 - 7t)(t^2 - 2)$.

**4.** $Q(x) = (5 - 2x)(x^2 + 1)$.

**5.** $f(r) = (3r^2 - 4)(r^2 - 5r + 1)$.

**6.** $C(I) = (2I^2 - 3)(3I^2 - 4I + 1)$.

**7.** $f(x) = x^2(x^2 - 5)$.

**8.** $f(x) = 3x^3(x^2 - 2x + 2)$.

**9.** $y = (x^2 + 3x - 2)(2x^2 - x - 3)$.

**10.** $y = (2 - 3x + 4x^2)(1 + 2x - 3x^2)$.

**11.** $f(w) = (8w^2 + 2w - 3)(5w^3 + 2)$.

**12.** $f(x) = (3x - x^2)(3 - x - x^2)$.

**13.** $y = (x^2 - 1)(3x^3 - 6x + 5) - 4(4x^2 + 2x + 1)$.

**14.** $h(x) = 4(x^5 - 3) + 3(8x^2 - 5)(3x + 2)$.

**15.** $f(p) = \frac{3}{2}(\sqrt{p} - 4)(4p - 5)$.

**16.** $g(x) = (\sqrt{x} - 3x + 1)(\sqrt[4]{x} - 2\sqrt{x})$.

**17.** $y = 7 \cdot \frac{2}{3}$.

**18.** $y = (x - 1)(x - 2)(x - 3)$.

**19.** $y = (2x - 1)(3x + 4)(x + 7)$.

**20.** $y = \dfrac{2x - 3}{4x + 1}$.

**21.** $f(x) = \dfrac{x}{x - 1}$.

**22.** $f(x) = \dfrac{-2x}{1 - x}$.

**23.** $f(x) = \dfrac{3}{2x^6}$.

**24.** $f(x) = \dfrac{5(x^2 - 2)}{7}$.

**25.** $y = \dfrac{x + 2}{x - 1}$.

**26.** $h(w) = \dfrac{3w^2 + 5w - 1}{w - 3}$.

**27.** $h(z) = \dfrac{5 - 2z}{z^2 - 4}$.

**28.** $y = \dfrac{x^2 - 4x + 2}{x^2 + x + 1}$.

**29.** $y = \dfrac{8x^2 - 2x + 1}{x^2 - 5x}$.

**30.** $f(x) = \dfrac{x^3 - x^2 + 1}{x^2 + 1}$.

**31.** $y = \dfrac{x^2 - 4x + 3}{2x^2 - 3x + 2}$.

**32.** $F(z) = \dfrac{z^4 + 4}{3z}$.

**33.** $g(x) = \dfrac{1}{x^{100} + 1}$.

**34.** $y = \dfrac{3}{7x^3}$.

**35.** $u(v) = \dfrac{v^5 - 8}{v}$.

**36.** $y = \dfrac{x - 5}{2\sqrt{x}}$.

**37.** $y = \dfrac{3x^2 - x - 1}{\sqrt[3]{x}}$.

**38.** $y = \dfrac{x^{0.3} - 2}{2x^{2.1} + 1}$.

**39.** $y = 7 - \dfrac{4}{x - 8} + \dfrac{2x}{3x + 1}$.

**40.** $q(x) = 13x^2 + \dfrac{x - 1}{2x + 3} - \dfrac{4}{x}$.

**41.** $y = \dfrac{x - 5}{(x + 2)(x - 4)}$.

**42.** $y = \dfrac{(2x - 1)(3x + 2)}{4 - 5x}$.

**43.** $s(t) = \dfrac{t^2 + 3t}{(t^2 - 1)(t^3 + 7)}$.

**44.** $f(s) = \dfrac{17}{s(5s^2 - 10s + 4)}$.

**45.** $y = 3x - \dfrac{\frac{2}{x} - \frac{3}{x - 1}}{x - 2}$.

**46.** $y = 7 - 10x^2 + \dfrac{1 - \frac{7}{x^2 + 3}}{x + 2}$.

**47.** $f(x) = \dfrac{a - x}{a + x}$, where $a$ is a constant.

**48.** $f(x) = \dfrac{x^{-1} + a^{-1}}{x^{-1} - a^{-1}}$, where $a$ is a constant.

**49.** Find the slope of the curve

$$y = (4x^2 + 2x - 5)(x^3 + 7x + 4)$$

at $(-1, 12)$.

**50.** Find the slope of the curve $y = \dfrac{x^3}{x^4 + 1}$ at $(1, \frac{1}{2})$.

*In Problems 51–54, find an equation of the tangent line to the curve at the given point.*

**51.** $y = \dfrac{6}{x - 1}$; $(3, 3)$.

**52.** $y = \dfrac{4x + 5}{x^2}$; $(-1, 1)$.

**53.** $y = (2x + 3)[2(x^4 - 5x^2 + 4)]$; $(0, 24)$.

**54.** $y = \dfrac{x + 1}{x^2(x - 4)}$; $(2, -\frac{3}{8})$.

*In Problems 55 and 56, determine the relative rate of change of y with respect to x for the given value of x.*

**55.** $y = \dfrac{x}{2x - 6};\ x = 1.$

**56.** $y = \dfrac{1 - x}{1 + x};\ x = 5.$

**57. Motion** The position function for an object moving in a straight line is

$$s = \frac{2}{t^3 + 1},$$

where $t$ is in seconds and $s$ is in meters. Find the position and velocity of the object at $t = 1$.

**58. Motion** The position function for an object moving in a straight-line path is

$$s = \frac{t + 2}{t^2 + 12},$$

Where $t$ is in seconds and $s$ is in meters. Find the positive value(s) of $t$ for which the velocity of the object is 0.

*In Problems 59–62, each equation represents a demand function for a certain product, where p denotes the price per unit for q units. Find the marginal-revenue function in each case. Recall that revenue = pq.*

**59.** $p = 25 - 0.02q.$

**60.** $p = 500/q.$

**61.** $p = \dfrac{108}{q + 2} - 3.$

**62.** $p = \dfrac{q + 750}{q + 50}.$

**63. Consumption Function** For the United States (1922–1942), the consumption function is estimated by[11]

$$C = 0.672I + 113.1.$$

Find the marginal propensity to consume.

**64. Consumption Function** Repeat Problem 63 if $C = 0.712I + 95.05$ for the United States for 1929–1941.[12]

*In Problems 65–68, each equation represents a consumption function. Find the marginal propensity to consume and the marginal propensity to save for the given value of I.*

**65.** $C = 2 + 2\sqrt{I};\ I = 9.$

**66.** $C = 6 + \dfrac{3I}{4} - \dfrac{\sqrt{I}}{3};\ I = 25.$

**67.** $C = \dfrac{16\sqrt{I} + 0.8\sqrt{I^3} - 0.2I}{\sqrt{I} + 4};\ I = 36.$

**68.** $C = \dfrac{20\sqrt{I} + 0.5\sqrt{I^3} - 0.4I}{\sqrt{I} + 5};\ I = 100.$

**69. Consumption Function** Suppose that a country's consumption function is given by

$$C = \frac{10\sqrt{I} + 0.7\sqrt{I^3} - 0.2I}{\sqrt{I}},$$

where $C$ and $I$ are expressed in billions of dollars.

**a.** Find the marginal propensity to save when income is 25 billion dollars.

**b.** Determine the relative rate of change of $C$ with respect to $I$ when income is 25 billion dollars.

**70. Marginal Propensities to Consume and to Save** Suppose that the savings function of a country is

$$S = \frac{I - \sqrt{I} - 6}{\sqrt{I} + 2},$$

where the national income ($I$) and the national savings ($S$) are measured in billions of dollars. Find the country's marginal propensity to consume and its marginal propensity to save when the national income is $16 billion. [*Hint:* It may be helpful to first factor the numerator.]

**71. Marginal Cost** If the total-cost function for a manufacturer is given by

$$c = \frac{5q^2}{q + 3} + 5000,$$

find the marginal-cost function.

**72. Marginal and Average Costs** Given the cost function $c = f(q)$, show that if $\dfrac{d}{dq}(\bar{c}) = 0$, then the marginal-cost function and average-cost function are equal.

**73. Host-Parasite Relation** For a particular host-parasite relationship, it is determined that when the host density (number of hosts per unit of area) is $x$, the number of hosts that are parasitized is $y$, where

$$y = \frac{900x}{10 + 45x}.$$

At what rate is the number of hosts parasitized changing with respect to host density when $x = 2$?

**74. Acoustics** The persistence of sound in a room after the source of the sound is turned off is called *reverberation*. The *reverberation time* RT of the room is the time it takes for the intensity level of the sound to fall 60 deci-

[11]T. Haavelmo, "Methods of Measuring the Marginal Propensity to Consume," *Journal of the American Statistical Association*, XLII (1947), 105-22.

[12]*Ibid.*

bels. In the acoustical design of an auditorium, the following formula may be used to compute the RT of the room:[13]

$$\text{RT} = \frac{0.05V}{A + xV}.$$

Here $V$ is the room volume, $A$ is the total room absorption, and $x$ is the air absorption coefficient. Assuming that $A$ and $x$ are positive constants, show that the rate of change of RT with respect to $V$ is always positive. If the total room volume increases by one unit, does the reverberation time increase or decrease?

**75. Predator-Prey** In a predator-prey experiment,[14] it was statistically determined that the number of prey consumed, $y$, by an individual predator was a function of the prey density $x$ (the number of prey per unit of area), where

$$y = \frac{0.7355x}{1 + 0.02744x}.$$

Determine the rate of change of prey consumed with respect to prey density.

**76. Social Security Benefits** In a discussion of social security benefits, Feldstein[15] differentiates a function of the form

$$f(x) = \frac{a(1 + x) - b(2 + n)x}{a(2 + n)(1 + x) - b(2 + n)x},$$

where $a$, $b$, and $n$ are constants. He determines that

$$f'(x) = \frac{-1(1 + n)ab}{[a(1 + x) - bx]^2(2 + n)}.$$

Verify this. [*Hint:* For convenience, let $2 + n = c$.]

**77. Business** The manufacturer of a product has found that when 20 units are produced per day, the average cost is \$150 and the marginal cost is \$125. What is the relative rate of change of average cost with respect to quantity when $q = 20$?

**78.** If $y$ is the product of three differentiable functions, that is,

$$y = f(x)g(x)h(x),$$

prove that $dy/dx$ is given by

$$f(x)g(x)h'(x) + f(x)g'(x)h(x) + f'(x)g(x)h(x).$$

**79.** Use the result of Problem 78 to find $dy/dx$ if

$$y = (2x - 1)(x - 2)(x + 3).$$

[13]L.L. Doelle, *Environmental Acoustics* (New York: McGraw-Hill Book Company, 1972).

[14]C. S. Holling, "Some Characteristics of Simple Types of Predation and Parasitism," *The Canadian Entomologist,* XCI, no. 7 (1959), 385–98.

[15]M. Feldstein, "The Optimal Level of Social Security Benefits," *The Quarterly Journal of Economics,* C, no. 2 (1985), 303–20.

## OBJECTIVE

To introduce and apply the chain rule, to derive the power rule as a special case of the chain rule, and to develop the concept of the marginal-revenue product as an application of the chain rule.

## 12.6 THE CHAIN RULE AND THE POWER RULE

Our next rule, the *chain rule,* is one of the most important rules for finding derivatives. It involves a situation in which $y$ is a function of the variable $u$, but $u$ is a function of $x$, and we want to find the derivative of $y$ with respect to $x$. For example, the equations

$$y = u^2 \qquad \text{and} \qquad u = 2x + 1$$

define $y$ as a function of $u$ and $u$ as a function of $x$. If we substitute $2x + 1$ for $u$ in the first equation, we can consider $y$ to be a function of $x$:

$$y = (2x + 1)^2.$$

To find $dy/dx$, we first expand $(2x + 1)^2$:

$$y = 4x^2 + 4x + 1.$$

Then

$$\frac{dy}{dx} = 8x + 4.$$

From this example, you can see that finding $dy/dx$ by first performing a substitution could be quite involved. For instance, if originally we had been given $y = u^{100}$ instead of $y = u^2$, we wouldn't even want to try substituting. Fortunately, the chain rule will allow us to handle such situations with ease.

> **Rule 7   Chain Rule**
>
> If $y$ is a differentiable function of $u$ and $u$ is a differentiable function of $x$, then $y$ is a differentiable function of $x$ and
>
> $$\frac{dy}{dx} = \frac{dy}{du} \cdot \frac{du}{dx}.$$

We can show you why the chain rule is reasonable by considering rates of change. Suppose

$$y = 8u + 5 \quad \text{and} \quad u = 2x - 3.$$

Let $x$ change by one unit. How does $u$ change? To answer this question, we differentiate and find $du/dx = 2$. But for *each* one-unit change in $u$, there is a change in $y$ of $dy/du = 8$. Therefore, what is the change in $y$ if $x$ changes by one unit; that is, what is $dy/dx$? The answer is $8 \cdot 2$, which is $\dfrac{dy}{du} \cdot \dfrac{du}{dx}$. Thus, $\dfrac{dy}{dx} = \dfrac{dy}{du} \cdot \dfrac{du}{dx}$.

We shall now use the chain rule to redo the problem at the beginning of this section. If

$$y = u^2 \quad \text{and} \quad u = 2x + 1,$$

then

$$\frac{dy}{dx} = \frac{dy}{du} \cdot \frac{du}{dx} = \frac{d}{du}(u^2) \cdot \frac{d}{dx}(2x + 1)$$

$$= (2u)2 = 4u.$$

Replacing $u$ by $2x + 1$ gives

$$\frac{dy}{dx} = 4(2x + 1) = 8x + 4,$$

which agrees with our previous result.

**Principles in Practice 1**

**Using the Chain Rule**

If an object moves horizontally according to $x = 6t$, where $t$ is in seconds, and vertically according to $y = 4x^2$, find its vertical velocity $\dfrac{dy}{dt}$.

**EXAMPLE 1   Using the Chain Rule**

**a.** *If $y = 2u^2 - 3u - 2$ and $u = x^2 + 4$, find $dy/dx$.*

*Solution:* By the chain rule,

$$\frac{dy}{dx} = \frac{dy}{du} \cdot \frac{du}{dx} = \frac{d}{du}(2u^2 - 3u - 2) \cdot \frac{d}{dx}(x^2 + 4)$$

$$= (4u - 3)(2x).$$

We can write our answer in terms of $x$ alone by replacing $u$ by $x^2 + 4$.

$$\frac{dy}{dx} = [4(x^2 + 4) - 3](2x) = [4x^2 + 13](2x) = 8x^3 + 26x.$$

**b.** *If $y = \sqrt{w}$ and $w = 7 - t^3$, find $dy/dt$.*

*Solution:* Here, $y$ is a function of $w$ and $w$ is a function of $t$, so we can view $y$ as a function of $t$. By the chain rule,

$$\frac{dy}{dt} = \frac{dy}{dw} \cdot \frac{dw}{dt} = \frac{d}{dw}(\sqrt{w}) \cdot \frac{d}{dt}(7 - t^3)$$

$$= \left(\frac{1}{2} w^{-1/2}\right)(-3t^2) = \frac{1}{2\sqrt{w}}(-3t^2)$$

$$= -\frac{3t^2}{2\sqrt{w}} = -\frac{3t^2}{2\sqrt{7 - t^3}}.$$

■

**EXAMPLE 2** **Using the Chain Rule**

*If* $y = 4u^3 + 10u^2 - 3u - 7$ *and* $u = 4/(3x - 5)$, *find dy/dx when* $x = 1$.

*Solution:* By the chain rule,

$$\frac{dy}{dx} = \frac{dy}{du} \cdot \frac{du}{dx} = \frac{d}{du}(4u^3 + 10u^2 - 3u - 7) \cdot \frac{d}{dx}\left(\frac{4}{3x - 5}\right)$$

$$= (12u^2 + 20u - 3) \cdot \frac{(3x - 5)\dfrac{d}{dx}(4) - 4\dfrac{d}{dx}(3x - 5)}{(3x - 5)^2}$$

$$= (12u^2 + 20u - 3) \cdot \frac{-12}{(3x - 5)^2}.$$

Do not simply replace *x* by 1 and leave your answer in terms of *u*.

Even though $dy/dx$ is in terms of $x$'s and $u$'s, we can evaluate it when $x = 1$ if we determine the corresponding value of $u$. When $x = 1$,

$$u = \frac{4}{3(1) - 5} = -2.$$

Thus,

$$\left.\frac{dy}{dx}\right|_{x=1} = [12(-2)^2 + 20(-2) - 3] \cdot \frac{-12}{[3(1) - 5]^2}$$

$$= 5 \cdot (-3) = -15. \qquad \blacksquare$$

The chain rules states that if $y = f(u)$ and $u = g(x)$, then

$$\frac{dy}{dx} = \frac{dy}{du} \cdot \frac{du}{dx}.$$

Actually, the chain rule applies to a composition function, because

$$y = f(u) = f(g(x)) = (f \circ g)(x).$$

Thus $y$, as a function of $x$, is $f \circ g$. This means that we can use the chain rule to differentiate a function when we recognize the function as a composition. However, we must first break down the function into composite parts.

For example, to differentiate

$$y = (x^3 - x^2 + 6)^{100},$$

we think of the function as a composition. Let

$$y = f(u) = u^{100} \qquad \text{and} \qquad u = g(x) = x^3 - x^2 + 6.$$

Then $y = (x^3 - x^2 + 6)^{100} = [g(x)]^{100} = f(g(x))$. Now that we have a composition, we differentiate. Since $y = u^{100}$ and $u = x^3 - x^2 + 6$, by the chain rule we have

$$\frac{dy}{dx} = \frac{dy}{du} \cdot \frac{du}{dx}$$

$$= (100u^{99})(3x^2 - 2x)$$

$$= 100(x^3 - x^2 + 6)^{99}(3x^2 - 2x).$$

We have just used the chain rule to differentiate $y = (x^3 - x^2 + 6)^{100}$, which is a power of a *function* of *x*, not simply a power of *x*. The following rule, called the *power rule*, generalizes our result and is a special case of the chain rule:

**Rule 8   Power Rule**

*If u is a differentiable function of x and n is any real number, then*

$$\frac{d}{dx}(u^n) = nu^{n-1}\frac{du}{dx}.$$

*Proof.* Let $y = u^n$. Since $y$ is a differentiable function of $u$ and $u$ is a differentiable function of $x$, the chain rule gives

$$\frac{dy}{dx} = \frac{dy}{du} \cdot \frac{du}{dx}.$$

But $dy/du = nu^{n-1}$. Thus,

$$\frac{dy}{dx} = nu^{n-1}\frac{du}{dx},$$

which is the power rule.

Another way of writing the power-rule formula is

$$\frac{d}{dx}([u(x)]^n) = n[u(x)]^{n-1}u'(x).$$

**EXAMPLE 3   Using the Power Rule**

*If $y = (x^3 - 1)^7$, find $y'$.*

*Solution:* Since $y$ is a power of a *function* of $x$, the power rule applies. Letting $u(x) = x^3 - 1$ and $n = 7$, we have

$$y' = n[u(x)]^{n-1}u'(x)$$

$$= 7(x^3 - 1)^{7-1}\frac{d}{dx}(x^3 - 1)$$

$$= 7(x^3 - 1)^6(3x^2) = 21x^2(x^3 - 1)^6. \quad\blacksquare$$

**EXAMPLE 4   Using the Power Rule**

*If $y = \sqrt[3]{(4x^2 + 3x - 2)^2}$, find $dy/dx$ when $x = -2$.*

*Solution:* Since $y = (4x^2 + 3x - 2)^{2/3}$, we use the power rule with

$$u = 4x^2 + 3x - 2$$

and $n = \frac{2}{3}$. We have

$$\frac{dy}{dx} = \frac{2}{3}(4x^2 + 3x - 2)^{(2/3)-1}\frac{d}{dx}(4x^2 + 3x - 2)$$

$$= \frac{2}{3}(4x^2 + 3x - 2)^{-1/3}(8x + 3)$$

$$= \frac{2(8x + 3)}{3\sqrt[3]{4x^2 + 3x - 2}}.$$

Thus,

$$\frac{dy}{dx}\bigg|_{x=-2} = \frac{2(-13)}{3\sqrt[3]{8}} = -\frac{13}{3}. \quad\blacksquare$$

**EXAMPLE 5** Using the Power Rule

If $y = \dfrac{1}{x^2 - 2}$, find $\dfrac{dy}{dx}$.

The technique used in Example 5 is frequently used when the numerator of a quotient is a constant and the denominator is not.

*Solution:* Although the quotient rule can be used here, a more efficient approach is to treat the right side as the power $(x^2 - 2)^{-1}$ and use the power rule. Let $u = x^2 - 2$. Then $y = u^{-1}$, and

$$\frac{dy}{dx} = nu^{n-1}\frac{du}{dx}$$

$$= (-1)(x^2 - 2)^{-1-1}\frac{d}{dx}(x^2 - 2)$$

$$= (-1)(x^2 - 2)^{-2}(2x)$$

$$= -\frac{2x}{(x^2 - 2)^2}.$$ ∎

In Example 5, the technique of moving the denominator to the numerator is not usually used when the numerator and denominator of a quotient involve variables. For example, we would not rewrite

$$\frac{x^2 + 1}{x^2 - 2} \quad \text{as} \quad (x^2 + 1)(x^2 - 2)^{-1},$$

for the following reason: Using the quotient rule on the first form results in an expression that is easily simplified; but using the product rule on the second form results in a sum of terms involving negative exponents, which is not as easily simplified.

**EXAMPLE 6** Differentiating a Power of a Quotient

If $z = \left(\dfrac{2s + 5}{s^2 + 1}\right)^4$, find $\dfrac{dz}{ds}$.

The problem here is to recognize the basic form of the function to be differentiated. In this case it is a power, not a quotient.

*Solution:* Since $z$ is a power of a function, we first use the power rule:

$$\frac{dz}{ds} = 4\left(\frac{2s + 5}{s^2 + 1}\right)^{4-1}\frac{d}{ds}\left(\frac{2s + 5}{s^2 + 1}\right).$$

Now we use the quotient rule:

$$\frac{dz}{ds} = 4\left(\frac{2s + 5}{s^2 + 1}\right)^3\left[\frac{(s^2 + 1)(2) - (2s + 5)(2s)}{(s^2 + 1)^2}\right].$$

Simplifying, we have

$$\frac{dz}{ds} = 4\cdot\frac{(2s + 5)^3}{(s^2 + 1)^3}\left[\frac{-2s^2 - 10s + 2}{(s^2 + 1)^2}\right]$$

$$= -\frac{8(s^2 + 5s - 1)(2s + 5)^3}{(s^2 + 1)^5}.$$ ∎

**EXAMPLE 7** Differentiating a Product of Powers

If $y = (x^2 - 4)^5(3x + 5)^4$, find $y'$.

*Solution:* Since $y$ is a product, we first apply the product rule:

$$y' = (x^2 - 4)^5\frac{d}{dx}[(3x + 5)^4] + (3x + 5)^4\frac{d}{dx}[(x^2 - 4)^5].$$

Now we use the power rule:

$$y' = (x^2 - 4)^5[4(3x + 5)^3(3)] + (3x + 5)^4[5(x^2 - 4)^4(2x)]$$
$$= 12(x^2 - 4)^5(3x + 5)^3 + 10x(3x + 5)^4(x^2 - 4)^4.$$

In differentiating a product in which at least one factor is a power, simplifying the derivative usually involves factoring.

To simplify, we first remove common factors:

$$y' = 2(x^2 - 4)^4(3x + 5)^3[6(x^2 - 4) + 5x(3x + 5)]$$
$$= 2(x^2 - 4)^4(3x + 5)^3(21x^2 + 25x - 24). \qquad \blacksquare$$

Usually, the power rule should be used to differentiate $y = [u(x)]^n$. Although a function such as $y = (x^2 + 2)^2$ may be written $y = x^4 + 4x^2 + 4$ and differentiated easily, this method is impractical for a function such as $y = (x^2 + 2)^{1000}$. Since $y = (x^2 + 2)^{1000}$ is of the form $y = [u(x)]^n$, we have

$$y' = 1000(x^2 + 2)^{999}(2x).$$

## Marginal-Revenue Product

Let us now use our knowledge of calculus to develop a concept relevant to economic studies. Suppose a manufacturer hires $m$ employees who produce a total of $q$ units of a product per day. We can think of $q$ as a function of $m$. If $r$ is the total revenue the manufacturer receives for selling these units, then $r$ can also be considered a function of $m$. Thus, we can look at $dr/dm$, the rate of change of revenue with respect to the number of employees. The derivative $dr/dm$ is called the **marginal-revenue product.** It is approximately the change in revenue that results when a manufacturer hires an extra employee.

### EXAMPLE 8   Marginal-Revenue Product

*A manufacturer determines that m employees will produce a total of q units of a product per day, where*

$$q = \frac{10m^2}{\sqrt{m^2 + 19}}. \qquad (1)$$

*If the demand equation for the product is $p = 900/(q + 9)$, determine the marginal-revenue product when $m = 9$.*

**Solution:** We must find $dr/dm$, where $r$ is revenue. Note that, by the chain rule,

$$\frac{dr}{dm} = \frac{dr}{dq} \cdot \frac{dq}{dm}.$$

Thus, we must find both $dr/dq$ and $dq/dm$ when $m = 9$. We begin with $dr/dq$. The revenue function is given by

$$r = pq = \left(\frac{900}{q + 9}\right)q = \frac{900q}{q + 9}, \qquad (2)$$

so, by the quotient rule,

$$\frac{dr}{dq} = \frac{(q + 9)(900) - 900q(1)}{(q + 9)^2} = \frac{8100}{(q + 9)^2}.$$

In order to evaluate this expression when $m = 9$, we first use the given equation $q = 10m^2/\sqrt{m^2 + 19}$ to find the corresponding value of $q$:

$$q = \frac{10(9)^2}{\sqrt{9^2 + 19}} = 81.$$

Hence,

$$\frac{dr}{dq}\bigg|_{m=9} = \frac{dr}{dq}\bigg|_{q=81} = \frac{8100}{(81+9)^2} = 1.$$

Now we turn to $dq/dm$. From the quotient and power rules, we have

$$\frac{dq}{dm} = \frac{d}{dm}\left(\frac{10m^2}{\sqrt{m^2+19}}\right)$$

$$= \frac{(m^2+19)^{1/2}\frac{d}{dm}(10m^2) - (10m^2)\frac{d}{dm}[(m^2+19)^{1/2}]}{[(m^2+19)^{1/2}]^2}$$

$$= \frac{(m^2+19)^{1/2}(20m) - (10m^2)[\frac{1}{2}(m^2+19)^{-1/2}(2m)]}{m^2+19},$$

so

$$\frac{dq}{dm}\bigg|_{m=9} = \frac{(81+19)^{1/2}(20\cdot9) - (10\cdot81)[\frac{1}{2}(81+19)^{-1/2}(2\cdot9)]}{81+19}$$

$$= 10.71.$$

A direct formula for the marginal-revenue product is

$$\frac{dr}{dm} = \frac{dq}{dm}\left(p + q\frac{dp}{dq}\right).$$

Therefore, from the chain rule,

$$\frac{dr}{dm}\bigg|_{m=9} = (1)(10.71) = 10.71.$$

This means that if a tenth employee is hired, revenue will increase by approximately \$10.71 per day.

---

## TECHNOLOGY

In Example 8 the marginal-revenue product, $dr/dm$, was found by using the chain rule. Another method, which is well suited to a graphics calculator, is to use substitution to express $r$ as a function of $m$ and then differentiate directly. First we take Eq. (1) and substitute for $q$ in the revenue function, Eq. (2); this gives $r$ as a function of $m$. Here are the details: In our function menu, we enter

$$Y_1 = 10X^2/\sqrt{(X^2 + 19)},$$
$$Y_2 = 900Y_1/(Y_1 + 9).$$

$Y_2$ expresses revenue as a function of the number of employees. Finally, to find the marginal-revenue product when $m = 9$, we compute nDeriv($Y_2$,X,9). You should verify that this method gives (approximately) 10.71.

---

## ■ Exercise 12.6

*In Problems 1–8, use the chain rule.*

**1.** If $y = u^2 - 2u$ and $u = x^2 - x$, find $dy/dx$.

**2.** If $y = 2u^3 - 8u$ and $u = 7x - x^3$, find $dy/dx$.

**3.** If $y = \dfrac{1}{w^2}$ and $w = 2 - x$, find $dy/dx$.

**4.** If $y = \sqrt[3]{z}$ and $z = x^6 - x^2 + 1$, find $dy/dx$.

**5.** If $w = u^2$ and $u = \dfrac{t+1}{t-1}$, find $dw/dt$ when $t = 3$.

**6.** If $z = u^2 + \sqrt{u} + 9$ and $u = 2s^2 - 1$, find $dz/ds$ when $s = -1$.

**7.** If $y = 3w^2 - 8w + 4$ and $w = 3x^2 + 1$, find $dy/dx$ when $x = 0$.

**8.** If $y = 3u^3 - u^2 + 7u - 2$ and $u = 3x - 2$, find $dy/dx$ when $x = 1$.

*In Problems 9–52, find y′.*

**9.** $y = (3x + 2)^6$.

**10.** $y = (x^2 - 4)^4$.

**11.** $y = (5 - x^2)^3$.

**12.** $y = (x^2 - x)^3$.

**13.** $y = 3(x^3 - 8x^2 + x)^{100}$.

**14.** $y = \dfrac{(2x^2 + 1)^4}{2}$.

**15.** $y = (x^2 - 2)^{-3}$.

**16.** $y = (3x^2 - 5x)^{-10}$.

**17.** $y = (2x^2 - 3x - 1)^{-10/3}$.

**18.** $y = (7x - x^4)^{-3/2}$.

**19.** $y = \sqrt{5x^2 - x}$.

**20.** $y = \sqrt{3x^2 - 7}$.

**21.** $y = \sqrt[4]{2x - 1}$.

**22.** $y = \sqrt[3]{8x^2 - 1}$.

**23.** $y = 2\sqrt[5]{(x^3 + 1)^2}$.

**24.** $y = 3\sqrt[3]{(x^2 + 1)^2}$.

**25.** $y = \dfrac{6}{2x^2 - x + 1}$.

**26.** $y = \dfrac{2}{x^4 + 2}$.

**27.** $y = \dfrac{1}{(x^2 - 3x)^2}$.

**28.** $y = \dfrac{1}{(1 - x)^3}$.

**29.** $y = \dfrac{2}{\sqrt{8x - 1}}$.

**30.** $y = \dfrac{1}{(3x^2 - x)^{2/3}}$.

**31.** $y = \sqrt[3]{7x} + \sqrt[3]{7}x$.

**32.** $y = \sqrt{2x} + \dfrac{1}{\sqrt{2x}}$.

**33.** $y = x^2(x - 4)^5$.

**34.** $y = 2x(x + 4)^4$.

**35.** $y = 2x\sqrt{6x - 1}$.

**36.** $y = x\sqrt{1 - x}$.

**37.** $y = (x^2 + 2x - 1)^3(5x)$.

**38.** $y = x^2(x^3 - 1)^4$.

**39.** $y = (8x - 1)^3(2x + 1)^4$.

**40.** $y = (6x + 1)^7(2x - 3)^3$.

**41.** $y = \left(\dfrac{x - 7}{x + 4}\right)^{10}$.

**42.** $y = \left(\dfrac{2x}{x + 2}\right)^4$.

**43.** $y = \sqrt{\dfrac{x - 2}{x + 3}}$.

**44.** $y = \sqrt[3]{\dfrac{8x^2 - 3}{x^2 + 2}}$.

**45.** $y = \dfrac{2x - 5}{(x^2 + 4)^3}$.

**46.** $y = \dfrac{(2x + 3)^3}{x^2 + 4}$.

**47.** $y = \dfrac{(8x - 1)^5}{(3x - 1)^3}$.

**48.** $y = \sqrt{(x - 1)(x + 2)^3}$.

**49.** $y = 6(5x^2 + 2)\sqrt{x^4 + 5}$.

**50.** $y = 6 + 3x - 4x(7x + 1)^2$.

**51.** $y = 8t + \dfrac{t - 1}{t + 4} - \left(\dfrac{8t - 7}{4}\right)^2$.

**52.** $y = \dfrac{(4x^2 - 2)(8x - 1)}{(3x - 1)^2}$.

*In Problems 53 and 54, use the quotient rule and power rule to find y′. Do not simplify your answer.*

**53.** $y = \dfrac{(2x + 1)(3x - 5)^2}{(x^2 - 7)^4}$.

**54.** $y = \dfrac{\sqrt{x + 2}(4x^2 - 1)^2}{9x - 3}$.

**55.** If $y = (5u + 6)^3$ and $u = (x^2 + 1)^4$, find $dy/dx$ when $x = 0$.

**56.** If $z = 2y^2 - 4y + 5$, $y = 6x - 5$, and $x = 2t$, find $dz/dt$ when $t = 1$.

**57.** Find the slope of the curve $y = (x^2 - 7x - 8)^3$ at the point $(8, 0)$.

**58.** Find the slope of the curve $y = \sqrt{x + 1}$ at the point $(8, 3)$.

*In Problems 59–62, find an equation of the tangent line to the curve at the given point.*

**59.** $y = \sqrt[3]{(x^2 - 8)^2}$; $(3, 1)$.

**60.** $y = (2x + 3)^2$; $(-2, 1)$.

**61.** $y = \dfrac{\sqrt{7x + 2}}{x + 1}$; $(1, \frac{3}{2})$.

**62.** $y = \dfrac{-3}{(3x^2 + 1)^3}$; $(0, -3)$.

*In Problems 63 and 64, determine the percentage rate of change of y with respect to x for the given value of x.*

**63.** $y = (x^2 + 9)^3$; $x = 4$.

**64.** $y = \dfrac{1}{(x^2 + 1)^2}$; $x = -3$.

*In Problems 65–68, q is the total number of units produced per day by m employees of a manufacturer, and p is the price per unit at which the q units are sold. In each case, find the marginal-revenue product for the given value of m.*

**65.** $q = 2m$, $p = -0.5q + 20$; $m = 5$.

**66.** $q = (200m - m^2)/20$, $p = -0.1q + 70$; $m = 40$.

**67.** $q = 10m^2/\sqrt{m^2 + 9}$, $p = 525/(q + 3)$; $m = 4$.

**68.** $q = 100m/\sqrt{m^2 + 19}$, $p = 4500/(q + 10)$; $m = 9$.

**69. Demand Equation** Suppose $p = 100 - \sqrt{q^2 + 20}$ is a demand equation for a manufacturer's product. (a) Find the rate of change of $p$ with respect to $q$. (b) Find the relative rate of change of $p$ with respect to $q$. (c) Find the marginal-revenue function.

**70. Marginal-Revenue Product** If $p = k/q$, where $k$ is a constant, is the demand equation for a manufacturer's product and $q = f(m)$ defines a function that gives the total number of units produced per day by $m$ employees, show that the marginal-revenue product is always zero.

**71. Cost Function** The cost $c$ of producing $q$ units of a product is given by

$$c = 4000 + 10q + 0.1q^2.$$

If the price per unit $p$ is given by the equation

$$q = 800 - 2.5p,$$

use the chain rule to find the rate of change of cost with respect to price per unit when $p = 80$.

**72. Hospital Discharges** A governmental health agency examined the records of a group of individuals who were hospitalized with a particular illness. It was found that the total proportion that had been discharged at the end of $t$ days of hospitalization was given by

$$f(t) = 1 - \left( \frac{300}{300 + t} \right)^3.$$

Find $f'(300)$ and interpret your answer.

**73. Marginal Cost.** If the total-cost function for a manufacturer is given by

$$c = \frac{5q^2}{\sqrt{q^2 + 3}} + 5000,$$

find the marginal-cost function.

**74. Status/Education** For a certain population, if $E$ is the number of years of a person's education and $S$ represents a numerical value of the person's status based on that educational level, then

$$S = 4\left( \frac{E}{4} + 1 \right)^2.$$

(a) How fast is status changing with respect to education when $E = 16$? (b) At what level of education does the rate of change of status equal 8?

**75. Biology** The volume of a spherical cell is given by $V = \frac{4}{3}\pi r^3$, where $r$ is the radius. At time $t$ seconds, the radius (in centimeters) is given by

$$r = 10^{-8}t^2 + 10^{-7}t.$$

Use the chain rule to find $dV/dt$ when $t = 10$.

**76. Pressure in Body Tissue** Under certain conditions, the pressure $p$ developed in body tissue by ultrasonic beams is given as a function of the beam's intensity via the equation[16]

$$p = (2\rho V I)^{1/2},$$

where $\rho$ (a Greek letter read "rho") is density of the affected tissue and $V$ is the velocity of propagation of the beam. Here $\rho$ and $V$ are constants. (a) Find the rate of change of $p$ with respect to $I$. (b) Find the relative rate of change of $p$ with respect to $I$.

**77. Demography** Suppose that, for a certain group of 20,000 births, the number of people surviving to age $x$ years is

$$l_x = 2000\sqrt{100 - x}, \qquad 0 \le x \le 100.$$

(a) Find the rate of change of $l_x$ with respect to $x$, and evaluate your answer for $x = 36$. (b) Find the relative rate of change of $l_x$ when $x = 36$.

**78. Muscle Contraction** A muscle has the ability to shorten when a load, such as a weight, is imposed on it. The equation

$$(P + a)(v + b) = k$$

is called the "fundamental equation of muscle contraction."[17] Here $P$ is the load imposed on the muscle, $v$ is the velocity of the shortening of the muscle fibers, and $a$, $b$, and $k$ are positive constants. Express $v$ as a function of $P$. Use your result to find $dv/dP$.

**79. Economics** Suppose $pq = 100$ is the demand equation for a manufacturer's product. Let $c$ be the total cost, and assume that the marginal cost is 0.01 when $q = 200$. Use the chain rule to find $dc/dp$ when $q = 200$.

**80. Marginal-Revenue Product** A monopolist who employs $m$ workers finds that they produce

$$q = 2m(2m + 1)^{3/2}$$

units of product per day. The total revenue $r$ (in dollars) is given by

$$r = \frac{50q}{\sqrt{1000 + 3q}}.$$

**a.** What is the price per unit (to the nearest cent) when there are 12 workers?

**b.** Determine the marginal revenue when there are 12 workers.

**c.** Determine the marginal-revenue product when $m = 12$.

**81.** Suppose $y = f(x)$, where $x = g(t)$. Given that $g(2) = 3$, $g'(2) = 4$, $f(2) = 5$, $f'(2) = 6$, $g(3) = 7$, $g'(3) = 8$, $f(3) = 9$, and $f'(3) = 10$, determine the value of $\dfrac{dy}{dt}\bigg|_{t=2}$.

**82. Business** A manufacturer has determined that, for his product, the daily average cost (in hundreds of dollars) is given by

$$\bar{c} = \frac{324}{\sqrt{q^2 + 35}} + \frac{5}{q} + \frac{19}{18}.$$

**a.** As daily production increases, the average cost approaches a constant dollar amount. What is this amount?

**b.** Determine the manufacturer's marginal cost when 17 units are produced per day.

---

[16]R. W. Stacy et al., *Essentials of Biological and Medical Physics* (New York: McGraw-Hill Book Company, 1955).

[17]*Ibid.*

**c.** The manufacturer determines that if production (and sales) were increased to 18 units per day, revenue would increase by $275. Should this move be made? Why?

 **83.** If

$$y = (u + 1)\sqrt{u + 5}$$

and

$$u = x(x^2 + 5)^5,$$

find $dy/dx$ when $x = 0.1$. Round your answer to two decimal places.

 **84.** If

$$y = \frac{9u - 4}{3u^2 + 2}$$

and

$$u = \frac{4x - 1}{(2x - 5)^3},$$

find $dy/dx$ when $x = 5$. Round your answer to two decimal places.

## 12.7 REVIEW

### IMPORTANT TERMS AND SYMBOLS

**Section 12.1**   secant line    tangent line    slope of a curve    derivative

$$\lim_{h \to 0} \frac{f(x + h) - f(x)}{h}$$    difference quotient    $f'(x)$    $y'$    $\dfrac{dy}{dx}$    $\dfrac{d}{dx}[f(x)]$

**Section 12.3**   $\Delta x$    position function    velocity    rate of change    total-cost function    marginal cost    average cost    total-revenue function    marginal revenue    relative rate of change    percentage rate of change

**Section 12.5**   product rule    quotient rule    consumption function    marginal propensity to consume    marginal propensity to save

**Section 12.6**   chain rule    power rule    marginal-revenue product

### SUMMARY

The tangent line (or tangent) to a curve at point $P$ is the limiting position of secant lines $PQ$ as $Q$ approaches $P$ along the curve. The slope of the tangent at $P$ is called the slope of the curve at $P$.

If $y = f(x)$, the derivative of $f$ at $x$ is the function $f'(x)$ defined by the limit in the equation

$$f'(x) = \lim_{h \to 0} \frac{f(x + h) - f(x)}{h}.$$

Geometrically, the derivative gives the slope of the curve $y = f(x)$ at the point $(x, f(x))$. An equation of the tangent at a particular point $(x_1, y_1)$ is obtained by evaluating $f'(x_1)$, which is the slope $m$ of the tangent, and substituting into the point–slope form of a line, $y - y_1 = m(x - x_1)$. Any function that is differentiable at a point must also be continuous there.

The basic rules for finding derivatives are as follows, where we assume that all functions are differentiable:

$$\frac{d}{dx}(c) = 0, \text{ where } c \text{ is any constant.}$$

$$\frac{d}{dx}(x^n) = nx^{n-1}, \text{ where } n \text{ is any real number.}$$

$$\frac{d}{dx}[cf(x)] = cf'(x), \text{ where } c \text{ is a constant.}$$

$$\frac{d}{dx}[f(x) + g(x)] = f'(x) + g'(x).$$

$$\frac{d}{dx}[f(x) - g(x)] = f'(x) - g'(x).$$

$$\frac{d}{dx}[f(x)g(x)] = f(x)g'(x) + g(x)f'(x).$$

$$\frac{d}{dx}\left[\frac{f(x)}{g(x)}\right] = \frac{g(x)f'(x) - f(x)g'(x)}{[g(x)]^2}.$$

$\dfrac{dy}{dx} = \dfrac{dy}{du} \cdot \dfrac{du}{dx}$, where $y$ is a function of $u$ and $u$ is a function of $x$.

$\dfrac{d}{dx}(u^n) = nu^{n-1}\dfrac{du}{dx}$, where $u$ is a function of $x$ and $n$ is any real number.

The derivative $dy/dx$ can also be interpreted as giving the (instantaneous) rate of change of $y$ with respect to $x$:

$$\frac{dy}{dx} = \lim_{\Delta x \to 0} \frac{\Delta y}{\Delta x} = \lim_{\Delta x \to 0} \frac{\text{change in } y}{\text{change in } x}.$$

In particular, if $s = f(t)$ is a position function, where $s$ is position at time $t$, then

$$\frac{ds}{dt} = \text{velocity at time } t.$$

In economics, the term *marginal* is used to describe derivatives of specific types of functions. If $c = f(q)$ is a total-cost function ($c$ is the total cost of $q$ units of a product), then the rate of change

$$\frac{dc}{dq} \text{ is called marginal cost.}$$

We interpret marginal cost as the approximate cost of one additional unit of output. (Average cost per unit, $\bar{c}$, is related to total cost $c$ by $\bar{c} = c/q$, or $c = \bar{c}q$.)

A total-revenue function $r = f(q)$ gives a manufacturer's revenue $r$ for selling $q$ units of product. (Revenue $r$ and price $p$ are related by $r = pq$.) The rate of change

$$\frac{dr}{dq} \text{ is called marginal revenue,}$$

which is interpreted as the approximate revenue obtained from selling one additional unit of output.

If $r$ is the revenue that a manufacturer receives when the total output of $m$ employees is sold, then the derivative $dr/dm$ is called the marginal-revenue product and gives the approximate change in revenue that results when the manufacturer hires an extra employee.

If $C = f(I)$ is a consumption function, where $I$ is national income and $C$ is national consumption, then

$$\frac{dC}{dI} \text{ is marginal propensity to consume}$$

and

$$1 - \frac{dC}{dI} \text{ is marginal propensity to save.}$$

For any function, the relative rate of change of $f(x)$ is

$$\frac{f'(x)}{f(x)},$$

which compares the rate of change of $f(x)$ with $f(x)$ itself. The percentage of rate of change is

$$\frac{f'(x)}{f(x)} \cdot 100.$$

## REVIEW PROBLEMS

*In Problems 1–4, use the definition of the derivative to find $f'(x)$.*

**1.** $f(x) = 2 - x^2$.

**2.** $f(x) = 2x^2 - 3x + 1$.

**3.** $f(x) = \sqrt{3x}$.

**4.** $f(x) = \dfrac{2}{1 + 4x}$.

*In Problems 5–38, differentiate.*

**5.** $y = 6^3$.

**6.** $y = x$.

**7.** $y = 7x^4 - 6x^3 + 5x^2 + 1$.

**8.** $y = 4(x^2 + 5) - 7x$.

**9.** $f(s) = s^2(s^2 + 2)$.

**10.** $y = \sqrt{x + 3}$.

**11.** $y = \dfrac{x^2 + 3}{5}$.

**12.** $y = \dfrac{1}{x^3}$.

**13.** $y = (x^2 + 6x)(x^3 - 6x^2 + 4)$.

**14.** $y = (x^2 + 1)^{100}(x - 6)$.

**15.** $f(x) = (2x^2 + 4x)^{100}$.

**16.** $f(w) = w\sqrt{w} + w^2$.

**17.** $y = \dfrac{1}{2x + 1}$.

**18.** $y = \dfrac{x^4 + 4x^2}{2x}$.

**19.** $y = (8 + 2x)(x^2 + 1)^4$.

**20.** $g(z) = (2z)^{3/5} + 5$.

**21.** $f(z) = \dfrac{z^2 - 1}{z^2 + 1}$.

**22.** $y = \dfrac{x - 5}{(x + 2)^2}$.

**23.** $y = \sqrt[3]{4x - 1}$.

**24.** $h(t) = (1 + 2^2)^0$.

**25.** $y = \dfrac{1}{\sqrt{1 - x}}$.

**26.** $y = \dfrac{x(x + 1)}{2x^2 + 3}$.

**27.** $h(x) = (x - 6)^4(x + 5)^3$.

**28.** $y = \dfrac{(x + 3)^5}{x}$.

**29.** $y = \dfrac{5x - 4}{x + 1}$.

**30.** $f(x) = 5x\sqrt{1 - 2x}$.

**31.** $y = 2x^{-3/8} + (2x)^{-3/8}$.

**32.** $y = \sqrt{\dfrac{x}{2}} + \sqrt{\dfrac{2}{x}}$.

**33.** $y = \dfrac{x^2 + 6}{\sqrt{x^2 + 5}}$.

**34.** $y = \sqrt[3]{(1 - 3x^2)^2}$.

**35.** $y = (x^3 + 6x^2 + 9)^{3/5}$.

**36.** $y = 0.5x(x + 1)^{-2} + 0.3$.

**37.** $g(z) = \dfrac{-7z}{(z - 1)^{-1}}$.

**38.** $g(z) = \dfrac{-3}{4(z^5 + 2z - 5)^4}$.

*In Problems 39–42, find an equation of the tangent line to the curve at the point corresponding to the given value of x.*

**39.** $y = x^2 - 6x + 4$, $x = 1$.

**40.** $y = -2x^3 + 6x + 1$, $x = 2$.

**41.** $y = \sqrt[3]{x}$, $x = 8$.

**42.** $y = \dfrac{x}{1 - x}$, $x = 3$.

**43.** If $f(x) = 4x^2 + 2x + 8$ find the relative and percentage rates of change of $f(x)$ when $x = 1$.

**44.** If $f(x) = x/(x + 4)$, find the relative and percentage rates of change of $f(x)$ when $x = 1$.

**45. Marginal Revenue** If $r = q(20 - 0.1q)$ is a total-revenue function, find the marginal-revenue function.

**46. Marginal Cost** If

$$c = 0.0001q^3 - 0.02q^2 + 3q + 6000$$

is a total-cost function, find the marginal cost when $q = 100$.

**47. Consumption Function** If

$$C = 7 + 0.6I - 0.25\sqrt{I}$$

is a consumption function, find the marginal propensity to consume and the marginal propensity to save when $I = 16$.

**48. Demand Equation** If $p = \dfrac{q + 14}{q + 4}$, is a demand equation, find the rate of change of price $p$ with respect to quantity $q$.

**49. Demand Equation** If $p = -0.5q + 450$ is a demand equation, find the marginal-revenue function.

**50. Average Cost** If $\bar{c} = 0.03q + 1.2 + \dfrac{3}{q}$ is an average-cost function, find the marginal cost when $q = 100$.

**51. Power-Plant Cost Function** The total-cost function of an electric light and power plant is estimated by[18]

$$c = 16.68 + 0.125q + 0.00439q^2, \quad 20 \le q \le 90,$$

where $q$ is the eight-hour total output (as a percentage of capacity) and $c$ is the total fuel cost in dollars. Find the marginal-cost function and evaluate it when $q = 70$.

**52. Marginal-Revenue Product** A manufacturer has determined that $m$ employees will produce a total of $q$ units of product per day, where

$$q = m(50 - m).$$

If the demand function is given by

$$p = -0.01q + 9,$$

find the marginal-revenue product when $m = 10$.

**53. Winter Moth** In a study of the winter moth in Nova Scotia,[19] it was determined that the average number of eggs, $y$, in a female moth was a function of the female's abdominal width $x$ (in millimeters), where

$$y = f(x) = 14x^3 - 17x^2 - 16x + 34$$

and $1.5 \le x \le 3.5$. At what rate does the number of eggs change with respect to abdominal width when $x = 2$?

[18]J. A. Nordin, "Note on a Light Plant's Cost Curves," *Econometrica*, 15 (1947), 231–55.

[19]D. G. Embree, "The Population Dynamics of the Winter Moth in Nova Scotia, 1954–1962," *Memoirs of the Entomological Society of Canada*, no. 46 (1965).

**54. Host-Parasite Relation** For a particular host-parasite relationship, it is found that when the host density (number of hosts per unit of area) is $x$, the number of hosts that are parasitized is

$$y = 10\left(1 - \frac{1}{1 + 2x}\right), \qquad x \geq 0.$$

For what value of $x$ does $dy/dx$ equal $\frac{1}{5}$?

**55. Bacteria Growth** Bacteria are growing in a culture. The time $t$ (in hours) for the number of bacteria to double in number (the generation time) is a function of the temperature $T$ (in degrees Celsius) of the culture and is given by

$$t = f(T) = \begin{cases} \frac{1}{24}T + \frac{11}{4}, & \text{if } 30 \leq T \leq 36, \\ \frac{4}{3}T - \frac{175}{4}, & \text{if } 36 < T \leq 39. \end{cases}$$

Find $dt/dT$ when (a) $T = 38$ and (b) $T = 35$.

**56. Motion** The position function of a particle moving in a straight line is

$$s = \frac{5}{2t^2 + 3},$$

where $t$ is in seconds and $s$ is in meters. Find the velocity of the particle at $t = 1$.

**57. Rate of Change** The volume of a sphere is given by $V = \frac{1}{6}\pi d^3$, where $d$ is the diameter. Find the rate of change of $V$ with respect to $d$ when $d = 4$ ft.

**58. Motion** The position function for a ball thrown vertically upward from the ground is

$$s = 218t - 16t^2,$$

where $s$ is the height in feet above the ground after $t$ seconds. For what value(s) of $t$ is the velocity 64 ft/s?

**59.** Find the marginal-cost function if the average-cost function is

$$\bar{c} = 2q + \frac{10,000}{q^2}.$$

**60.** Find an equation of the tangent line to the curve

$$y = \frac{(x^4 + 1)\sqrt{3x + 1}}{x^5 + x},$$

at the point on the curve where $x = 1$.

**61.** A manufacturer has found that when $m$ employees are working, the number of units of product produced per day is

$$q = 10\sqrt{m^2 + 3600} - 600.$$

The demand equation for the product is

$$9q + p^2 - 7200 = 0,$$

where $p$ is the selling price when the demand for the product is $q$ units per day.

**a.** Determine the manufacturer's marginal-revenue product when $m = 80$.

**b.** Find the relative rate of change of revenue with respect to the number of employees when $m = 80$.

**c.** Suppose it would cost the manufacturer $300 more per day to hire an additional employee. Would you advise the manufacturer to hire the 81st employee? Why?

**62.** If $f(x) = x^2 \ln x$, use the "limit of a difference quotient" approach to estimate $f'(5)$. Round your answer to three decimal places.

**63.** If $f(x) = \sqrt[3]{x^2 + 3x - 4}$, use the numerical derivative feature of your graphics calculator to estimate the derivative when $x = 5$. Round your answer to three decimal places.

**64.** The total-cost function for a manufacturer is given by

$$c = \frac{6q^2 + 4}{\sqrt{q^2 + 8}} + 2000,$$

where $c$ is in dollars. Use the numerical derivative feature of your graphics calculator to estimate the marginal cost when 12 units are produced. Round your answer to the nearest cent.

**65.** If

$$y = (u + 3)\sqrt{u + 6},$$

and

$$u = \frac{x + 4}{x + 3},$$

find $dy/dx$ when $x = 0.3$. Round your answer to two decimal places.

# Additional Differentiation Topics

**To develop a differentiation formula for $y = \ln u$, to apply the formula, and to use it to differentiate a logarithmic function to a base other than $e$.**

## 13.1 Derivatives of Logarithmic Functions

In this section, we develop formulas for differentiating logarithmic functions. We begin with the derivative of $f(x) = \ln x$, where $x > 0$. By the definition of the derivative,

$$\frac{d}{dx}(\ln x) = \lim_{h \to 0} \frac{f(x+h) - f(x)}{h} = \lim_{h \to 0} \frac{\ln(x+h) - \ln x}{h}.$$

Using the property of logarithms that $\ln m - \ln n = \ln(m/n)$, we have

$$\frac{d}{dx}(\ln x) = \lim_{h \to 0} \frac{\ln\left(\dfrac{x+h}{x}\right)}{h}$$

$$= \lim_{h \to 0}\left[\frac{1}{h}\ln\left(\frac{x+h}{x}\right)\right] = \lim_{h \to 0}\left[\frac{1}{h}\ln\left(1 + \frac{h}{x}\right)\right].$$

Writing $\dfrac{1}{h}$ as $\dfrac{1}{x} \cdot \dfrac{x}{h}$ gives

$$\frac{d}{dx}(\ln x) = \lim_{h \to 0}\left[\frac{1}{x} \cdot \frac{x}{h}\ln\left(1 + \frac{h}{x}\right)\right]$$

$$= \lim_{h \to 0}\left[\frac{1}{x}\ln\left(1 + \frac{h}{x}\right)^{x/h}\right] \quad (\text{since } r \ln m = \ln m^r)$$

$$= \frac{1}{x} \cdot \lim_{h \to 0}\left[\ln\left(1 + \frac{h}{x}\right)^{x/h}\right].$$

Because it can be shown that the limit of a logarithm is the logarithm of the limit ($\lim \ln u = \ln \lim u$), we have

$$\frac{d}{dx}(\ln x) = \frac{1}{x}\ln\left[\lim_{h \to 0}\left(1 + \frac{h}{x}\right)^{x/h}\right]. \tag{1}$$

To evaluate $\lim\limits_{h \to 0}\left(1 + \dfrac{h}{x}\right)^{x/h}$, first note that as $h \to 0$, then $\dfrac{h}{x} \to 0$. Thus, if we replace $\dfrac{h}{x}$ by $k$, the limit has the form

$$\lim_{k \to 0}(1 + k)^{1/k}.$$

As stated in Sec. 11.1, this limit is $e$. Hence, Eq. (1) becomes

$$\frac{d}{dx}(\ln x) = \frac{1}{x} \ln e = \frac{1}{x}(1) = \frac{1}{x}.$$

Therefore,

$$\boxed{\frac{d}{dx}(\ln x) = \frac{1}{x}.} \tag{2}$$

**EXAMPLE 1   Differentiating Functions Involving ln $x$**

**a.** *Differentiate* $f(x) = 5 \ln x$.

*Solution:* Here $f$ is a constant (5) times a function $(\ln x)$, so by Eq. (2), we have

$$f'(x) = 5 \frac{d}{dx}(\ln x) = 5 \cdot \frac{1}{x} = \frac{5}{x}.$$

**b.** *Differentiate* $y = \dfrac{\ln x}{x^2}$.

*Solution:* By the quotient rule and Eq. (2),

$$y' = \frac{x^2 \dfrac{d}{dx}(\ln x) - (\ln x)\dfrac{d}{dx}(x^2)}{(x^2)^2}$$

$$= \frac{x^2\left(\dfrac{1}{x}\right) - (\ln x)(2x)}{x^4} = \frac{x - 2x \ln x}{x^4} = \frac{1 - 2 \ln x}{x^3}. \qquad \blacksquare$$

We shall now extend Eq. (2) to cover a broader class of functions. Let $y = \ln u$, where $u$ is a positive differentiable function of $x$. By the chain rule,

The chain rule is invaluable in developing the differentation formula for ln $u$.

$$\frac{d}{dx}(\ln u) = \frac{dy}{du} \cdot \frac{du}{dx} = \frac{d}{du}(\ln u) \cdot \frac{du}{dx} = \frac{1}{u} \cdot \frac{du}{dx}.$$

Thus,

$$\boxed{\frac{d}{dx}(\ln u) = \frac{1}{u} \cdot \frac{du}{dx}.} \tag{3}$$

**EXAMPLE 2   Differentiating Functions Involving ln $u$**

**a.** *Differentiate* $y = \ln(x^2 + 1)$.

*Solution:* This function has the form $\ln u$ with $u = x^2 + 1$. Using Eq. (3) gives

Don't forget about $\dfrac{d}{dx}(x^2 + 1)$.

$$\frac{dy}{dx} = \frac{1}{x^2 + 1}\frac{d}{dx}(x^2 + 1) = \frac{1}{x^2 + 1}(2x) = \frac{2x}{x^2 + 1}.$$

**b.** *Differentiate* $y = x^2 \ln(4x + 2)$.

*Solution:* Using the product rule gives

$$\frac{dy}{dx} = x^2 \frac{d}{dx}[\ln(4x + 2)] + [\ln(4x + 2)]\frac{d}{dx}(x^2).$$

By Eq. (3) with $u = 4x + 2$,

$$\frac{dy}{dx} = x^2\left(\frac{1}{4x + 2}\right)(4) + [\ln(4x + 2)](2x)$$

$$= \frac{2x^2}{2x + 1} + 2x\ln(4x + 2).$$

**c.** *Differentiate* $y = \ln(\ln x)$.

*Solution:* This has the form $y = \ln u$ with $u = \ln x$. Using Eq. (3), we obtain

$$y' = \frac{1}{\ln x}\frac{d}{dx}(\ln x) = \frac{1}{\ln x}\left(\frac{1}{x}\right) = \frac{1}{x\ln x}. \quad \blacksquare$$

Frequently, we can reduce the work involved in differentiating the logarithm of a product, quotient, or power by using properties of logarithms to rewrite the logarithm *before* differentiating. The next example will illustrate.

**EXAMPLE 3  Rewriting Logarithmic Functions before Differentiating**

**a.** *Find* $\dfrac{dy}{dx}$ *if* $y = \ln(2x + 5)^3$.

*Solution:* Here we have the logarithm of a power. First we simplify the right side by using properties of logarithms. Then we differentiate. We have

$$y = \ln(2x + 5)^3 = 3\ln(2x + 5),$$

$$\frac{dy}{dx} = 3\left(\frac{1}{2x + 5}\right)(2) = \frac{6}{2x + 5}.$$

Alternatively, if the simplification were not performed first, we would write

$$\frac{dy}{dx} = \frac{1}{(2x + 5)^3}\frac{d}{dx}[(2x + 5)^3]$$

$$= \frac{1}{(2x + 5)^3}(3)(2x + 5)^2(2) = \frac{6}{2x + 5}.$$

**b.** *Find* $f'(p)$ *if* $f(p) = \ln[(p + 1)^2(p + 2)^3(p + 3)^4]$.

*Solution:* We simplify the right side and then differentiate:

$$f(p) = 2\ln(p + 1) + 3\ln(p + 2) + 4\ln(p + 3),$$

$$f'(p) = 2\left(\frac{1}{p + 1}\right)(1) + 3\left(\frac{1}{p + 2}\right)(1) + 4\left(\frac{1}{p + 3}\right)(1)$$

$$= \frac{2}{p + 1} + \frac{3}{p + 2} + \frac{4}{p + 3}. \quad \blacksquare$$

**Principles in Practice 1**
**Differentiating Functions Involving ln $u$**

The supply of $q$ units of a product at a price of $p$ dollars per unit is given by $q(p) = 25 + 2\ln(3p^2 + 4)$. Find the rate of change of supply with respect to price, $\dfrac{dq}{dp}$.

Comparing both methods, we note that the easier one is to simplify first and then differentiate.

**EXAMPLE 4** **Differentiating Functions Involving Logarithms**

**a.** *Find $f'(w)$ if $f(w) = \ln\sqrt{\dfrac{1 + w^2}{w^2 - 1}}$.*

*Solution:* We simplify by using properties of logarithms and then differentiate:

$$f(w) = \frac{1}{2}\left[\ln(1 + w^2) - \ln(w^2 - 1)\right],$$

$$f'(w) = \frac{1}{2}\left[\frac{1}{1 + w^2}(2w) - \frac{1}{w^2 - 1}(2w)\right]$$

$$= \frac{w}{1 + w^2} - \frac{w}{w^2 - 1} = -\frac{2w}{w^4 - 1}.$$

**b.** *Find $f'(x)$ if $f(x) = \ln^3(2x + 5)$.*

*Solution:* The exponent 3 refers to the cubing of $\ln(2x + 5)$. That is,

$$f(x) = \ln^3(2x + 5) = [\ln(2x + 5)]^3.$$

By the power rule,

$$f'(x) = 3[\ln(2x + 5)]^2 \frac{d}{dx}[\ln(2x + 5)]$$

$$= 3[\ln(2x + 5)]^2\left[\frac{1}{2x + 5}(2)\right]$$

$$= \frac{6}{2x + 5}[\ln(2x + 5)]^2 = \frac{6}{2x + 5}\ln^2(2x + 5).$$

*Pitfall* ▼ Do not confuse $\ln^3(2x + 5)$ with $\ln(2x + 5)^3$, which occurred in Example 3(a). ▬

## Derivatives of Logarithmic Functions to the Base $b$

To differentiate a logarithmic function to a base different from $e$, we can first convert the logarithm to natural logarithms via the change-of-base formula and then differentiate the resulting expression. For example, consider $y = \log_b u$, where $u$ is a differentiable function of $x$. By the change-of-base formula,

$$y = \log_b u = \frac{\ln u}{\ln b}.$$

Differentiating, we have

$$\frac{d}{dx}(\log_b u) = \frac{d}{dx}\left(\frac{\ln u}{\ln b}\right) = \frac{1}{\ln b}\frac{d}{dx}(\ln u) = \frac{1}{\ln b} \cdot \frac{1}{u}\frac{du}{dx}.$$

Thus,

$$\frac{d}{dx}(\log_b u) = \frac{1}{(\ln b)u} \cdot \frac{du}{dx}.$$

Rather than memorize this rule, we suggest that you remember the procedure used to obtain it.

> **Procedure to Differentiate $\log_b u$**
>
> Convert $\log_b u$ to natural logarithms to obtain $\dfrac{\ln u}{\ln b}$, and then differentiate.

**EXAMPLE 5** Differentiating a Logarithmic Function to the Base 2

*Differentiate* $y = \log_2 x$.

*Solution:* Following the foregoing procedure, we have

$$\frac{d}{dx}(\log_2 x) = \frac{d}{dx}\left(\frac{\ln x}{\ln 2}\right) = \frac{1}{\ln 2}\frac{d}{dx}(\ln x) = \frac{1}{x \ln 2}.$$

It is worth mentioning that we can write our answer in terms of the original base. Because

$$\frac{1}{\ln b} = \frac{1}{\dfrac{\log_b b}{\log_b e}} = \frac{\log_b e}{1} = \log_b e,$$

we can express $\dfrac{1}{x \ln 2}$ as $\dfrac{1}{x}\log_2 e$. ∎

---

**Principles in Practice 2**
Differentiating a Logarithmic Function to the Base 10

The intensity of an earthquake is measured on the Richter scale. The reading is given by $R = \log \dfrac{I}{I_0}$, where $I$ is the intensity and $I_0$ is a standard minimum intensity. If $I_0 = 1$, find $\dfrac{dR}{dI}$, the rate of change of the Richter-scale reading with respect to the intensity.

---

**EXAMPLE 6** Differentiating a Logarithmic Function to the Base 10

*If* $y = \log(2x + 1)$, *find the rate of change of* $y$ *with respect to* $x$.

*Solution:* The rate of change is $dy/dx$, and the base involved is 10. Therefore, we have

$$\frac{dy}{dx} = \frac{d}{dx}[\log(2x + 1)] = \frac{d}{dx}\left[\frac{\ln(2x + 1)}{\ln 10}\right]$$

$$= \frac{1}{\ln 10}\cdot\frac{1}{2x + 1}(2) = \frac{2}{(2x + 1)\ln 10}.$$ ∎

## ▪ Exercise 13.1

*In Problems 1–44, differentiate the functions. If possible, first use properties of logarithms to simplify the given function.*

**1.** $y = 4\ln x$.

**2.** $y = \dfrac{\ln x}{14}$.

**3.** $y = \ln(3x - 4)$.

**4.** $y = \ln(5x - 6)$.

**5.** $y = \ln x^2$.

**6.** $y = \ln(ax^2 + b)$.

**7.** $y = \ln(1 - x^2)$.

**8.** $y = \ln(-x^2 + 6x)$.

**9.** $f(p) = \ln(2p^3 + 3p)$.

**10.** $f(r) = \ln(2r^4 - 3r^2 + 2r + 1)$.

**11.** $f(t) = t\ln t$.

**12.** $y = x^2\ln x$.

**13.** $y = x^2\ln(4x + 3)$.

**14.** $y = (2x + 5)^2\ln(2x + 5)$.

**15.** $y = \log_3(2x - 1)$.

**16.** $f(w) = \log(w^2 + w)$.

**17.** $y = x^2 + \log_2(x^2 + 4)$.

**18.** $y = x^3\log_4 x$.

**19.** $f(z) = \dfrac{\ln z}{z}$.

**20.** $y = \dfrac{x^2}{\ln x}$.

**21.** $y = \dfrac{x^2 - 1}{\ln x}$.

**22.** $y = \ln x^{100}$.

**23.** $y = \ln(x^2 + 4x + 5)^3$.

**24.** $y = 6\ln\sqrt[3]{x}$.

**25.** $y = \ln\sqrt{1 + x^2}$.

**26.** $f(s) = \ln\left(\dfrac{s^2}{1 + s^2}\right)$.

**27.** $f(l) = \ln\left(\dfrac{1 + l}{1 - l}\right)$.

**28.** $y = \ln\left(\dfrac{2x + 3}{3x - 4}\right)$.

**29.** $y = \ln\sqrt[4]{\dfrac{1 + x^2}{1 - x^2}}$.

**30.** $y = \ln\sqrt{\dfrac{x^4 - 1}{x^4 + 1}}$.

**31.** $y = \ln[(x^2 + 2)^2(x^3 + x - 1)]$.

**32.** $y = \ln[(5x + 2)^4(8x - 3)^6]$.

**33.** $y = \ln(x\sqrt{2x + 1})$.

**34.** $y = \ln \dfrac{x}{\sqrt{2x + 1}}$.

**35.** $y = (x^2 + 1)\ln(2x + 1)$.

**36.** $y = (ax + b)\ln(ax)$.

**37.** $y = \ln x^3 + \ln^3 x$.

**38.** $y = x^{\ln 3}$.

**39.** $y = \ln^4(ax)$.

**40.** $y = \ln^2(2x + 3)$.

**41.** $y = x \ln\sqrt{x - 1}$.

**42.** $y = \ln(x^2\sqrt{3x - 2})$.

**43.** $y = \sqrt{4 + \ln x}$.

**44.** $y = \ln(x + \sqrt{1 + x^2})$.

**45.** Find an equation of the tangent line to the curve
$$y = \ln(x^2 - 2x - 2)$$
when $x = 3$.

**46.** Find an equation of the tangent line to the curve
$$y = x[\ln(x) - 1]$$
at the point where $x = e$.

**47.** Find the slope of the curve $y = \dfrac{x}{\ln x}$ when $x = 2$.

**48.** **Marginal Revenue** Find the marginal-revenue function if the demand function is $p = 25/\ln(q + 2)$.

**49.** **Marginal Cost** A total-cost function is given by
$$c = 25 \ln(q + 1) + 12.$$
Find the marginal cost when $q = 6$.

**50.** **Marginal Cost** A manufacturer's average-cost function, in dollars, is given by
$$\overline{c} = \frac{400}{\ln(q + 5)}.$$
Find the marginal cost (rounded to two decimal places) when $q = 45$.

**51.** **Supply Change** The supply of $q$ units of a product at a price of $p$ dollars per unit is given by $q(p) = 25 + 10\ln(2p + 1)$. Find the rate of change of supply with respect to price, $\dfrac{dq}{dp}$.

**52.** **Sound Perception** The loudness of sound ($L$, measured in decibels) perceived by the human ear depends upon intensity levels ($I$) according to $L = 10\log\dfrac{I}{I_0}$, where $I_0$ is the standard threshold of audibility. If $I_0 = 17$, find $\dfrac{dL}{dI}$, the rate of change of the loudness with respect to the intensity.

**53.** **Biology** In a certain experiment with bacteria, it is observed that the relative activeness of a given bacteria colony is described by
$$A = 6\ln\left(\frac{T}{a - T} - a\right),$$
where $a$ is a constant and $T$ is the surrounding temperature. Find the rate of change of $A$ with respect to $T$.

**54.** Show that the relative rate of change of $y = f(x)$ with respect to $x$ is equal to the derivative of $y = \ln f(x)$.

**55.** Show that $\dfrac{d}{dx}(\log_b u) = \dfrac{1}{u}(\log_b e)\dfrac{du}{dx}$.

In Problems 56 and 57, use differentiation rules to find $f'(x)$. Then use your graphics calculator to find all zeros of $f'(x)$. Round your answers to two decimal places.

**56.** $f(x) = x^2 \ln x$.

**57.** $f(x) = \dfrac{\ln(2x)}{x}$.

---

**OBJECTIVE**

To develop a differentiation formula for $y = e^u$, to apply the formula, and to use it to differentiate an exponential function with a base other than $e$.

## 13.2 DERIVATIVES OF EXPONENTIAL FUNCTIONS

We now obtain a formula for the derivative of the exponential function
$$y = e^u,$$
where $u$ is a differentiable function of $x$. In logarithmic form, we have
$$u = \ln y.$$

Differentiating both sides with respect to $x$ gives
$$\frac{d}{dx}(u) = \frac{d}{dx}(\ln y),$$
$$\frac{du}{dx} = \frac{1}{y}\frac{dy}{dx}.$$

Solving for $dy/dx$ and then replacing $y$ by $e^u$ gives

$$\frac{dy}{dx} = y\frac{du}{dx} = e^u\frac{du}{dx}.$$

Thus,

$$\frac{d}{dx}(e^u) = e^u\frac{du}{dx}. \tag{1}$$

As a special case, let $u = x$. Then $du/dx = 1$, and

$$\frac{d}{dx}(e^x) = e^x. \tag{2}$$

Note that the function and its derivative are the same.

*Pitfall* ▼ The power rule does not apply to $e^x$. That is, $\dfrac{d}{dx}(e^x) \neq xe^{x-1}$.

### EXAMPLE 1  Differentiating Functions Involving $e^x$

**a.**  *Find* $\dfrac{d}{dx}(3e^x)$.

*Solution:*  Since 3 is a constant factor,

$$\frac{d}{dx}(3e^x) = 3\frac{d}{dx}(e^x)$$

$$= 3e^x \quad [\text{by Eq. (2)}].$$

**b.**  *If* $y = \dfrac{x}{e^x}$, *find* $\dfrac{dy}{dx}$.

*Solution:*  We *first* use the quotient rule and then use Eq. (2):

$$\frac{dy}{dx} = \frac{e^x\dfrac{d}{dx}(x) - x\dfrac{d}{dx}(e^x)}{(e^x)^2} = \frac{e^x(1) - x(e^x)}{(e^x)^2} = \frac{e^x(1-x)}{e^{2x}} = \frac{1-x}{e^x}.$$

**c.**  *If* $y = e^2 + e^x + \ln 3$, *find* $y'$.

*Solution:*  Since $e^2$ and $\ln 3$ are constants, $y' = 0 + e^x + 0 = e^x$.  ■

### EXAMPLE 2  Differentiating Functions Involving $e^u$

**a.**  *Find* $\dfrac{d}{dx}(e^{x^3+3x})$.

*Solution:*  The function has the form $e^u$ with $u = x^3 + 3x$. From Eq. (1),

$\dfrac{d}{dx}(e^u) = e^u\dfrac{du}{dx}$. Don't forget the $\dfrac{du}{dx}$.

$$\frac{d}{dx}(e^{x^3+3x}) = e^{x^3+3x}\frac{d}{dx}(x^3 + 3x) = e^{x^3+3x}(3x^2 + 3)$$

$$= 3(x^2 + 1)e^{x^3+3x}.$$

**b.**  *Find* $\dfrac{d}{dx}[e^{x+1}\ln(x^2 + 1)]$

## Principles in Practice 1

**Differentiating Functions Involving** $e^u$

When an object is moved from one environment to another, the change in temperature of the object is given by $T = Ce^{kt}$, where $C$ is the temperature difference between the two environments, $t$ is the time in the new environment, and $k$ is a constant. Find the rate of change of temperature with respect to time.

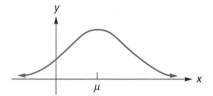

**FIGURE 13.1** The normal-distribution density function.

*Solution:* By the product rule,

$$\frac{d}{dx}[e^{x+1}\ln(x^2+1)] = e^{x+1}\frac{d}{dx}[\ln(x^2+1)] + [\ln(x^2+1)]\frac{d}{dx}(e^{x+1})$$

$$= e^{x+1}\left(\frac{1}{x^2+1}\right)(2x) + [\ln(x^2+1)]e^{x+1}(1)$$

$$= e^{x+1}\left[\frac{2x}{x^2+1} + \ln(x^2+1)\right].$$ ∎

### EXAMPLE 3   The Normal-Distribution Density Function

*An important function used in the social sciences is the **normal-distribution density function***

$$y = f(x) = \frac{1}{\sigma\sqrt{2\pi}}e^{-(1/2)[(x-\mu)/\sigma]^2}$$

*where $\sigma$ (a Greek letter read "sigma") and $\mu$ (a Greek letter read "mu") are constants. The graph of this function, called the normal curve, is "bell shaped." (See Fig. 13.1.) Determine the rate of change of $y$ with respect to $x$ when $x = \mu$.*

*Solution:* The rate of change of $y$ with respect to $x$ is $dy/dx$. We note that the factor $\dfrac{1}{\sigma\sqrt{2\pi}}$ is a constant and the second factor has the form $e^u$, where

$$u = -\frac{1}{2}\left(\frac{x-\mu}{\sigma}\right)^2.$$

Thus,

$$\frac{dy}{dx} = \frac{1}{\sigma\sqrt{2\pi}}\left[e^{-(1/2)[(x-\mu)/\sigma]^2}\right]\left[-\frac{1}{2}(2)\left(\frac{x-\mu}{\sigma}\right)\left(\frac{1}{\sigma}\right)\right].$$

Evaluating $dy/dx$ when $x = \mu$, we obtain

$$\left.\frac{dy}{dx}\right|_{x=\mu} = 0.$$

Geometrically, this means that the slope of the tangent line to the graph of $y = f(x)$ at $x = \mu$ is horizontal. (Refer back to Fig. 13.1.) ∎

### Differentiating Exponential Functions to the Base $a$

Now that we are familiar with the derivative of $e^u$, we consider the derivative of the more general exponential function $a^u$. By replacing $a$ with the equivalent form $e^{\ln a}$, we can express $a^u$ as an exponential function with the base $e$, a form we can differentiate. We have

$$\frac{d}{dx}(a^u) = \frac{d}{dx}[(e^{\ln a})^u] = \frac{d}{dx}(e^{u\ln a})$$

$$= e^{u\ln a}\frac{d}{dx}(u\ln a)$$

$$= e^{u\ln a}\left(\frac{du}{dx}\right)\ln a$$

$$= a^u(\ln a)\frac{du}{dx} \quad (\text{since } e^{u\ln a} = a^u).$$

Therefore,

$$\frac{d}{dx}(a^u) = a^u(\ln a)\frac{du}{dx}. \tag{3}$$

Note that if $a = e$, then the factor $\ln a$ in Eq. (3) is 1. Thus, if exponential functions to the base $e$ are used, we have a simpler differentiation formula with which to work. This is the reason natural exponential functions are used extensively in calculus. Rather than memorizing Eq. (3), we suggest that you remember the procedure for obtaining it.

**Procedure to Differentiate $a^u$**

Convert $a^u$ to a natural exponential function by using the property that $a = e^{\ln a}$, and then differentiate.

The next example will illustrate this procedure.

**EXAMPLE 4** **Differentiating an Exponential Function to the Base 4**

*Find* $\dfrac{d}{dx}(4^x)$.

*Solution:* Using the preceding procedure, we have

$$\frac{d}{dx}(4^x) = \frac{d}{dx}\left[(e^{\ln 4})^x\right]$$

$$= \frac{d}{dx}\left[e^{(\ln 4)x}\right] \quad \left[\text{form: } \frac{d}{dx}(e^u)\right]$$

$$= e^{(\ln 4)x}(\ln 4) \quad \left[\text{by Eq. (1)}\right]$$

Verify the result by using Eq. (3) directly.

$$= 4^x(\ln 4). \qquad\blacksquare$$

**EXAMPLE 5** **Differentiating Different Forms**

*Find* $\dfrac{d}{dx}(e^2 + x^e + 2^{\sqrt{x}})$.

*Solution:* Here we must differentiate three different forms; do not confuse them! The first $(e^2)$ is a constant base to a constant power, so it is a constant itself. Thus, its derivative is zero. The second $(x^e)$ is a variable base to a constant power, so the power rule applies. The third $(2^{\sqrt{x}})$ is a constant base to a variable power, so we must differentiate an exponential function. Taken all together, we have

$$\frac{d}{dx}(e^2 + x^e + 2^{\sqrt{x}}) = 0 + ex^{e-1} + \frac{d}{dx}\left[e^{(\ln 2)\sqrt{x}}\right]$$

$$= ex^{e-1} + \left[e^{(\ln 2)\sqrt{x}}\right](\ln 2)\left(\frac{1}{2\sqrt{x}}\right)$$

$$= ex^{e-1} + \frac{2^{\sqrt{x}}\ln 2}{2\sqrt{x}}. \qquad\blacksquare$$

## ▪ Exercise 13.2

*In Problems 1–28, differentiate the functions.*

**1.** $y = 7e^x$.

**2.** $y = \dfrac{2e^x}{5}$.

**3.** $y = e^{x^2+1}$.

**4.** $y = e^{2x^2+5}$.

**5.** $y = e^{3-5x}$.

**6.** $f(q) = e^{-q^3+6q-1}$.

**7.** $f(r) = e^{3r^2+4r+4}$.

**8.** $y = e^{9x^2+5x^3-6}$.

**9.** $y = xe^x$.

**10.** $y = x^2e^{-x}$.

**11.** $y = x^2e^{-x^2}$.

**12.** $y = xe^{2x}$.

**13.** $y = \dfrac{e^x + e^{-x}}{2}$.

**14.** $y = \dfrac{e^x - e^{-x}}{2}$.

**15.** $y = 4^{3x^2}$.

**16.** $y = 2^x x^2$.

**17.** $f(w) = \dfrac{e^{2w}}{w^2}$.

**18.** $y = e^{x-\sqrt{x}}$.

**19.** $y = e^{1+\sqrt{x}}$.

**20.** $y = (e^{3x} + 1)^4$.

**21.** $y = x^3 - 3^x$.

**22.** $f(z) = e^{1/z}$.

**23.** $y = \dfrac{e^x - 1}{e^x + 1}$.

**24.** $y = e^{2x}(x + 1)$.

**25.** $y = e^{\ln x}$.

**26.** $y = e^{-x} \ln x$.

**27.** $y = e^{x \ln x}$.

**28.** $y = \ln e^{4x+1}$.

**29.** If $f(x) = ee^x e^{x^2}$, find $f'(-1)$.

**30.** If $f(x) = 5^{x^2 \ln x}$, find $f'(1)$.

**31.** Find an equation of the tangent line to the curve $y = e^x$ when $x = 2$.

**32.** Find an equation of the tangent line to the curve $y = x^e e^x$ at the point $(1, e)$.

*For each of the demand equations in Problems 33 and 34, find the rate of change of price p with respect to quantity q. What is the rate of change for the indicated value of q?*

**33.** $p = 15e^{-0.001q}$; $\quad q = 500$.

**34.** $p = 8e^{-3q/800}$; $\quad q = 400$.

*In Problems 35 and 36, $\overline{c}$ is the average cost of producing q units of a product. Find the marginal-cost function and the marginal cost for the given values of q.*

**35.** $\overline{c} = \dfrac{7000e^{q/700}}{q}$; $\quad q = 350, \; q = 700$.

**36.** $\overline{c} = \dfrac{850}{q} + 4000\dfrac{e^{(2q+6)/800}}{q}$; $\quad q = 97, \; q = 197$.

**37.** If $w = e^{x^3-4x} + x \ln(x - 1)$ and $x = \dfrac{t + 1}{t - 1}$, find $\dfrac{dw}{dt}$ when $t = 3$.

**38.** If $f'(x) = e^{-2x}$ and $u = \ln x^2$, show that

$$\frac{d}{dx}[f(u)] = \frac{2}{x^5}.$$

**39.** Determine the value of the positive constant $c$ if

$$\frac{d}{dx}(c^x - x^c) = 0$$

when $x = 1$.

**40.** Calculate the relative rate of change of

$$f(x) = 10^{-x} + \ln(8 + x) + 0.01e^{x-2}$$

when $x = 2$. Round your answer to four decimal places.

**41. Production Run** For a firm, the daily output on the $t$th day of a production run is given by

$$q = 500(1 - e^{-0.2t}).$$

Find the rate of change of output $q$ with respect to $t$ on the tenth day.

**42. Normal-Density Function** For the normal-density function

$$f(x) = \frac{1}{\sqrt{2\pi}} e^{-x^2/2},$$

find $f'(0)$.

**43. Population** The population of a city $t$ years from now is given by $P = 20{,}000e^{0.03t}$. Show that $dP/dt = kP$, where $k$ is a constant. This means that the rate of change of population at any time is proportional to the population at that time.

**44. Market Penetration** In a discussion of diffusion of a new process into a market, Hurter and Rubenstein[1] refer to an equation of the form

$$Y = k\alpha^{\beta^t},$$

where $Y$ is the cumulative level of diffusion of the new process at time $t$ and $k$, $\alpha$, and $\beta$ are positive constants. Verify their claim that

$$\frac{dY}{dt} = k\alpha^{\beta^t}(\beta^t \ln \alpha) \ln \beta.$$

[1]A. P. Hurter, Jr., A. H. Rubenstein, et al., "Market Penetration by New Innovations: The Technological Literature," *Technological Forecasting and Social Change,* 11 (1978), 197–221.

**45. Finance**   After $t$ years, the value $S$ of a principal of $P$ dollars invested at the annual rate of $r$ compounded continuously is given by $S = Pe^{rt}$. Show that the relative rate of change of $S$ with respect to $t$ is $r$.

**46. Predator–Prey Relationship**   In an article concerning predators and prey, Holling[2] refers to an equation of the form

$$y = K(1 - e^{-ax}),$$

where $x$ is the prey density, $y$ is the number of prey attacked, and $K$ and $a$ are constants. Verify his statement that

$$\frac{dy}{dx} = a(K - y).$$

**47. Earthquakes**   According to Richter,[3] the number of earthquakes of magnitude $M$ or greater per unit of time is given by $N = 10^A 10^{-bM}$, where $A$ and $b$ are constants. Find $dN/dM$.

**48. Psychology**   Short-term retention was studied by Peterson and Peterson.[4] The two researchers analyzed a procedure in which an experimenter verbally gave a subject a three-letter consonant syllable, such as CHJ, followed by a three-digit number, such as 309. The subject then repeated the number and counted backwards by 3's, such as 309, 306, 303, . . . . After a period of time, the subject was signaled by a light to recite the three-letter consonant syllable. The time between the experimenter's completion of the last consonant to the onset of the light was called the *recall interval*. The time between the onset of the light and the completion of a response was referred to as *latency*. After many trials, it was determined that, for a recall interval of $t$ seconds, the approximate proportion of correct recalls with latency below 2.83 seconds was

$$p = 0.89[0.01 + 0.99(0.85)^t].$$

**a.** Find $dp/dt$ and interpret your result.

**b.** Evaluate $dp/dt$ when $t = 2$. Round your answer to two decimal places.

**49. Medicine**   Suppose a tracer, such as a colored dye, is injected instantly into the heart at time $t = 0$ and mixes uniformly with blood inside the heart. Let the initial concentration of the tracer in the heart be $C_0$, and assume that the heart has constant volume $V$. Also assume that, as fresh blood flows into the heart, the diluted mixture of blood and tracer flows out at the constant positive rate $r$. Then the concentration of the tracer in the heart at time $t$ is given by

$$C(t) = C_0 e^{-(r/V)t}.$$

Show that $dC/dt = (-r/V)C(t)$.

**50. Medicine**   In Problem 49, suppose the tracer is injected at a constant rate $R$. Then the concentration at time $t$ is

$$C(t) = \frac{R}{r}[1 - e^{-(r/V)t}].$$

**a.** Find $C(0)$.

**b.** Show that $\dfrac{dC}{dt} = \dfrac{R}{V} - \dfrac{r}{V}C(t)$.

**51. Schizophrenia**   Several models have been used to analyze the length of stay in a hospital. For a particular group of schizophrenics, one such model is[5]

$$f(t) = 1 - e^{-0.008t},$$

Where $f(t)$ is the proportion of the group that was discharged at the end of $t$ days of hospitalization. Find the rate of discharge (the proportion discharged per day) at the end of 100 days. Round your answer to four decimal places.

**52. Savings and Consumption**   A country's savings $S$ (in billions of dollars) is related to its national income $I$ (in billions of dollars) by the equation

$$S = \ln \frac{3}{2 + e^{-I}}.$$

**a.** Show that the marginal propensity to consume as a function of income is $\dfrac{2}{2 + e^{-I}}$.

**b.** To the nearest million dollars, what is the national income when the marginal propensity to save is $\frac{1}{10}$? [*Hint*: 1 billion = 1000 million].

 *In Problems 53 and 54, use differentiation rules to find $f'(x)$. Then use your graphics calculator to find all real zeros of $f'(x)$. Round your answers to two decimal places.*

**53.** $f(x) = e^{x^4 + 2x^2 - 4x}$.

**54.** $f(x) = \dfrac{x^2}{2} + e^{-x}$.

[2] C. S. Holling, "Some Characteristics of Simple Types of Predation and Parasitism," *The Canadian Entomologist*, XCI, no.7 (1959), 385–98.

[3] C. F. Richter, *Elementary Seismology* (San Francisco: W. H. Freeman and Company, Publishers, 1958).

[4] L. R. Peterson and M.J. Peterson, "Short-Term Retention of Individual Verbal Items," *Journal of Experimental Psychology*, 58 (1959), 193–98.

[5] W. W. Eaton and G. A. Whitmore, "Length of Stay as a Stochastic Process: A General Approach and Application to Hospitalization for Schizophrenia," *Journal of Mathematical Sociology*, 5 (1977), 273–92.

**To discuss the notion of a
function defined implicitly and
to determine derivatives by
means of implicit differentiation.**

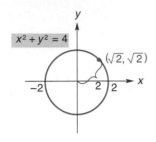

**FIGURE 13.2**   The circle
$x^2 + y^2 = 4$.

# 13.3  IMPLICIT DIFFERENTIATION

Implicit differentiation is a technique for differentiating functions that are not
given in the usual form $y = f(x)$. To introduce this technique, we shall find
the slope of a tangent line to a circle. Let us take the circle of radius 2 whose
center is at the origin (Fig. 13.2). Its equation is

$$x^2 + y^2 = 4,$$

$$x^2 + y^2 - 4 = 0. \tag{1}$$

The point $(\sqrt{2}, \sqrt{2})$ lies on the circle. To find the slope at this point, we
need to find $dy/dx$ there. Until now, we have always had $y$ given explicitly
(directly) in terms of $x$ before determining $y'$; that is, our equation was always
in the form $y = f(x)$. In Eq. (1), this is not so. We say that Eq. (1) has the form
$F(x, y) = 0$, where $F(x, y)$ denotes a function of two variables. The obvious
thing to do is solve Eq. (1) for $y$ in terms of $x$:

$$x^2 + y^2 - 4 = 0,$$

$$y^2 = 4 - x^2,$$

$$y = \pm\sqrt{4 - x^2}. \tag{2}$$

A problem now occurs: Equation (2) may give two values of $y$ for a value
of $x$. It does not define $y$ explicitly as a function of $x$. We can, however, suppose
that Eq. (1) defines $y$ as one of two different functions of $x$,

$$y = +\sqrt{4 - x^2} \quad \text{and} \quad y = -\sqrt{4 - x^2},$$

whose graphs are given in Fig. 13.3. Since the point $(\sqrt{2}, \sqrt{2})$ lies on the graph

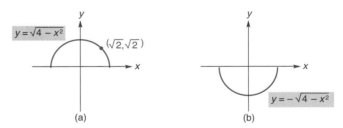

**FIGURE 13.3**    $x^2 + y^2 = 4$ gives rise to two different
functions of $x$.

of $y = \sqrt{4 - x^2}$, we should differentiate that function:

$$y = \sqrt{4 - x^2},$$

$$\frac{dy}{dx} = \frac{1}{2}(4 - x^2)^{-1/2}(-2x)$$

So

$$= -\frac{x}{\sqrt{4 - x^2}}.$$

$$\left.\frac{dy}{dx}\right|_{x=\sqrt{2}} = -\frac{\sqrt{2}}{\sqrt{4 - 2}} = -1.$$

Thus, the slope of the circle $x^2 + y^2 - 4 = 0$ at the point $(\sqrt{2}, \sqrt{2})$ is $-1$.

Let us summarize the difficulties we had. First, $y$ was not originally given explicitly in terms of $x$. Second, after we tried to find such a relation, we ended up with more than one function of $x$. In fact, depending on the equation given, it may be very complicated or even impossible to find an explicit expression for $y$. For example, it would be difficult to solve $ye^x + \ln(x + y) = 0$ for $y$. We shall now consider a method that avoids such difficulties.

An equation of the form $F(x, y) = 0$, such as we had originally, is said to express $y$ *implicitly* as a function of $x$. The word "implicitly" is used, since $y$ is not given explicitly as a function of $x$. However, it is assumed or *implied* that the equation defines $y$ as at least one differentiable function of $x$. Thus, we assume that Eq. (1), $x^2 + y^2 - 4 = 0$, defines some differentiable function of $x$, say, $y = f(x)$. Next, we treat $y$ as a function of $x$ and differentiate both sides of Eq. (1) with respect to $x$. Finally, we solve the result for $dy/dx$. Applying this procedure, we obtain

$$\frac{d}{dx}(x^2 + y^2 - 4) = \frac{d}{dx}(0),$$

$$\frac{d}{dx}(x^2) + \frac{d}{dx}(y^2) - \frac{d}{dx}(4) = \frac{d}{dx}(0). \qquad (3)$$

We know that $\dfrac{d}{dx}(x^2) = 2x$ and that both $\dfrac{d}{dx}(4)$ and $\dfrac{d}{dx}(0)$ are 0. But $\dfrac{d}{dx}(y^2)$ is **not** $2y$, because we are differentiating with respect to $x$, not $y$. That is, $y$ is not the independent variable. Since $y$ is assumed to be a function of $x$, $y^2$ has the form $u^n$, where $y$ plays the role of $u$. Just as the power rule states that $\dfrac{d}{dx}(u^2) = 2u\dfrac{du}{dx}$, we have $\dfrac{d}{dx}(y^2) = 2y\dfrac{dy}{dx}$. Hence, Eq. (3) becomes

$$2x + 2y\frac{dy}{dx} = 0.$$

Solving for $dy/dx$ gives

$$2y\frac{dy}{dx} = -2x,$$

$$\frac{dy}{dx} = -\frac{x}{y}. \qquad (4)$$

Notice that the expression for $dy/dx$ involves the variable $y$ as well as $x$. This means that to find $dy/dx$ at a point, both coordinates of the point must be substituted into $dy/dx$. Thus,

$$\frac{dy}{dx}\bigg|_{(\sqrt{2},\,\sqrt{2})} = -\frac{\sqrt{2}}{\sqrt{2}} = -1, \text{ as before.}$$

This method of finding $dy/dx$ is called **implicit differentiation.** We note that Eq. (4) is not defined when $y = 0$. Geometrically, this is clear, since the tangent line to the circle at either $(2, 0)$ or $(-2, 0)$ is vertical, and the slope is not defined.

Here are the steps to follow when differentiating implicitly:

**Implicit Differentiation Procedure**

For an equation that we assume defines $y$ implicitly as a differentiable function of $x$, the derivative $\dfrac{dy}{dx}$ can be found as follows:

1. Differentiate both sides of the equation with respect to $x$.

2. Collect all terms involving $\dfrac{dy}{dx}$ on one side of the equation, and collect all other terms on the other side.

3. Factor $\dfrac{dy}{dx}$ from the side involving the $\dfrac{dy}{dx}$ terms.

4. Solve for $\dfrac{dy}{dx}$.

**EXAMPLE 1   Implicit Differentiation**

*Find* $\dfrac{dy}{dx}$ *by implicit differentiation if* $y + y^3 - x = 7$.

**Solution:** Here $y$ is not given as an explicit function of $x$ [that is, not in the form $y = f(x)$]. Thus, we assume that $y$ is an implicit (differentiable) function of $x$ and apply the preceding four-step procedure:

1. Differentiating both sides with respect to $x$, we have

$$\frac{d}{dx}(y + y^3 - x) = \frac{d}{dx}(7),$$

$$\frac{d}{dx}(y) + \frac{d}{dx}(y^3) - \frac{d}{dx}(x) = \frac{d}{dx}(7).$$

Now, $\dfrac{d}{dx}(y)$ can be written $\dfrac{dy}{dx}$, and $\dfrac{d}{dx}(x) = 1$. By the power rule,

$$\frac{d}{dx}(y^3) = 3y^2\frac{dy}{dx}.$$

The derivative of $y^3$ with respect to $x$ is $3y^2\dfrac{dy}{dx}$, not $3y^2$.

Hence, we obtain

$$\frac{dy}{dx} + 3y^2\frac{dy}{dx} - 1 = 0.$$

2. Collecting all $\dfrac{dy}{dx}$ terms on the left side and all other terms on the right side gives

$$\frac{dy}{dx} + 3y^2\frac{dy}{dx} = 1.$$

3. Factoring $\dfrac{dy}{dx}$ from the left side, we have

$$\frac{dy}{dx}(1 + 3y^2) = 1.$$

In an implicit-differentiation problem, we are able to find the derivative of a function without knowing the function.

**4.** We solve for $\dfrac{dy}{dx}$ by dividing both sides by $1 + 3y^2$:

$$\frac{dy}{dx} = \frac{1}{1 + 3y^2}.$$ ∎

## Principles in Practice 1
### Implicit Differentiation

Suppose that $P$, the proportion of people affected by a certain disease, is described by $\ln\left(\dfrac{P}{1 - P}\right) = 0.5t$, where $t$ is the time in months. Find $\dfrac{dP}{dt}$, the rate at which $P$ grows with respect to time.

### EXAMPLE 2   Implicit Differentiation

*Find* $\dfrac{dy}{dx}$ *if* $x^3 + 4xy^2 - 27 = y^4$.

**Solution:** Since $y$ is not given explicitly in terms of $x$, we shall use the method of implicit differentiation:

**1.** Assuming that $y$ is a function of $x$ and differentiating both sides with respect to $x$, we get

$$\frac{d}{dx}(x^3 + 4xy^2 - 27) = \frac{d}{dx}(y^4),$$

$$\frac{d}{dx}(x^3) + 4\frac{d}{dx}(xy^2) - \frac{d}{dx}(27) = \frac{d}{dx}(y^4).$$

To find $\dfrac{d}{dx}(xy^2)$, we use the product rule:

$$3x^2 + 4\left[x\frac{d}{dx}(y^2) + y^2\frac{d}{dx}(x)\right] - 0 = 4y^3\frac{dy}{dx},$$

$$3x^2 + 4\left[x\left(2y\frac{dy}{dx}\right) + y^2(1)\right] = 4y^3\frac{dy}{dx},$$

$$3x^2 + 8xy\frac{dy}{dx} + 4y^2 = 4y^3\frac{dy}{dx}.$$

**2.** Collecting $\dfrac{dy}{dx}$ terms on the left side and other terms on the right gives

$$8xy\frac{dy}{dx} - 4y^3\frac{dy}{dx} = -3x^2 - 4y^2.$$

**3.** Factoring $\dfrac{dy}{dx}$ from the left side yields

$$\frac{dy}{dx}(8xy - 4y^3) = -3x^2 - 4y^2.$$

## Principles in Practice 2
### Implicit Differentiation

The volume $V$ of a spherical balloon is given by the equation $V = \dfrac{4}{3}\pi r^3$, where $r$ is the radius of the balloon. If the radius is increasing at a rate of 5 inches/minute $\left(\text{that is, } \dfrac{dr}{dt} = 5\right)$, then find $\dfrac{dV}{dt}$, the rate of increase of the volume of the balloon, when the radius is 12 inches.

**4.** Solving for $\dfrac{dy}{dx}$, we have

$$\frac{dy}{dx} = \frac{-3x^2 - 4y^2}{8xy - 4y^3} = \frac{3x^2 + 4y^2}{4y^3 - 8xy}.$$ ∎

### EXAMPLE 3   Implicit Differentiation

*Find the slope of the curve* $x^3 = (y - x^2)^2$ *at* $(1, 2)$.

**Solution:** The slope at $(1, 2)$ is the value of $dy/dx$ at that point. Finding $dy/dx$ by implicit differentiation, we have

$$\frac{d}{dx}(x^3) = \frac{d}{dx}[(y - x^2)^2],$$

$$3x^2 = 2(y - x^2)\left(\frac{dy}{dx} - 2x\right),$$

$$3x^2 = 2\left(y\frac{dy}{dx} - 2xy - x^2\frac{dy}{dx} + 2x^3\right),$$

$$3x^2 = 2y\frac{dy}{dx} - 4xy - 2x^2\frac{dy}{dx} + 4x^3,$$

$$3x^2 + 4xy - 4x^3 = 2y\frac{dy}{dx} - 2x^2\frac{dy}{dx},$$

$$3x^2 + 4xy - 4x^3 = 2\frac{dy}{dx}(y - x^2),$$

$$\frac{dy}{dx} = \frac{3x^2 + 4xy - 4x^3}{2(y - x^2)}.$$

Thus, the slope of the curve at $(1, 2)$ is

$$\left.\frac{dy}{dx}\right|_{(1,2)} = \frac{3(1)^2 + 4(1)(2) - 4(1)^3}{2[2 - (1)^2]} = \frac{7}{2}.$$

■

## Principles in Practice 3
### Implicit Differentiation

A 10-foot ladder is placed against a vertical wall. Suppose the bottom of the ladder slides away from the wall at a constant rate of 3 feet/s. $\left(\text{That is, } \dfrac{dx}{dt} = 3.\right)$ How fast is the top of the ladder sliding down the wall when the top of the ladder is 8 feet (that is, when $y = 8$) from the ground? $\left(\text{That is, what is } \dfrac{dy}{dt}?\right)$ (Use the Pythagorean theorem for right triangles, $x^2 + y^2 = z^2$, where $x$ and $y$ are the legs of the triangle and $z$ is the hypotenuse.)

**EXAMPLE 4  Implicit Differentiation**

If $q - p = \ln q + \ln p$, find $dq/dp$.

*Solution:* We assume that $q$ is a function of $p$ and differentiate both sides with respect to $p$:

$$\frac{d}{dp}(q) - \frac{d}{dp}(p) = \frac{d}{dp}(\ln q) + \frac{d}{dp}(\ln p),$$

$$\frac{dq}{dp} - 1 = \frac{1}{q}\frac{dq}{dp} + \frac{1}{p},$$

$$\frac{dq}{dp} - \frac{1}{q}\frac{dq}{dp} = \frac{1}{p} + 1,$$

$$\frac{dq}{dp}\left(1 - \frac{1}{q}\right) = \frac{1}{p} + 1,$$

$$\frac{dq}{dp}\left(\frac{q - 1}{q}\right) = \frac{1 + p}{p} \quad \text{(simplifying)},$$

$$\frac{dq}{dp} = \frac{(1 + p)q}{p(q - 1)}.$$

■

## ■ Exercise 13.3

*In Problems 1–24, find the dy/dx by implicit differentiation.*

**1.** $x^2 + 4y^2 = 4.$

**2.** $3x^2 + 6y^2 = 1.$

**3.** $3y^4 - 5x = 0.$

**4.** $2x^2 - 3y^2 = 4.$

**5.** $\sqrt{x} + \sqrt{y} = 3.$

**6.** $x^{1/5} + y^{1/5} = 4.$

**7.** $x^{3/4} + y^{3/4} = 7.$

**8.** $y^3 = 4x.$

9. $xy = 4$.

10. $x + xy - 2 = 0$.

11. $xy - y - 4x = 5$.

12. $x^2 + y^2 = 2xy + 3$.

13. $x^3 + y^3 - 12xy = 0$.

14. $2x^3 + 3xy + y^3 = 0$.

15. $x = \sqrt{y} + \sqrt[3]{y}$.

16. $x^3y^3 + x = 9$

17. $3x^2y^3 - x + y = 25$.

18. $y^2 + y = \ln x$.

19. $y \ln x = xe^y$.

20. $\ln(xy) + x = 4$.

21. $xe^y + y = 4$.

22. $ax^2 - by^2 = c$.

23. $(1 + e^{3x})^2 = 3 + \ln(x + y)$.

24. $y^2 = \ln(x + y)$.

25. If $x + xy + y^2 = 7$, find $dy/dx$ at $(1, 2)$.

26. If $x\sqrt{y + 1} = y\sqrt{x + 1}$, find $dy/dx$ at $(3, 3)$.

27. Find the slope of the curve $4x^2 + 9y^2 = 1$ at the point $\left(0, \dfrac{1}{3}\right)$; at the point $(x_0, y_0)$.

28. Find the slope of the curve $(x^2 + y^2)^3 = 16y^2$ at the point $(0, 2)$.

29. Find an equation of the tangent line to the curve of

$$x^3 + y^2 = 3$$

at the point $(-1, 2)$.

30. Repeat Problem 29 for the curve

$$y^2 + xy - x^2 = 5$$

at the point $(4, 3)$.

*For the demand equations in Problems 31–34, find the rate of change of q with respect to p.*

31. $p = 100 - q^2$.

32. $p = 400 - \sqrt{q}$.

33. $p = \dfrac{20}{(q + 5)^2}$.

34. $p = \dfrac{20}{q^2 + 5}$.

35. **Radioactivity**  The relative activity $I/I_0$ of a radioactive element varies with elapsed time according to the equation

$$\ln\left(\frac{I}{I_0}\right) = -\lambda t,$$

where $\lambda$ (a Greek letter read "lambda") is the disintegration *constant* and $I_0$ is the initial intensity (a constant). Find the rate of change of the intensity $I$ with respect to the elapsed time $t$.

36. **Physical Science**  For a star whose brightness is not much different from that of our sun, the relation between the star's mass $m$ and its luminosity $L$ is given by

$$\log m = 0.06 + 0.26 \log L.$$

    **a.** Find $dm/dL$ by implicit differentiation.

    **b.** Suppose now that the mass of a certain star is changing with respect to time $t$ at the rate $dm/dt$. Find an expression for $dL/dt$, the corresponding rate of change in luminosity.

37. **Earthquakes**  The magnitude $M$ of an earthquake and its energy $E$ are related by the equation[6]

$$1.5M = \log\left(\frac{E}{2.5 \times 10^{11}}\right).$$

Here $M$ is given in terms of Richter's preferred scale of 1958 and $E$ is in ergs. Determine the rate of change of energy with respect to magnitude.

38. **Physical Scale**  The relationship between the speed ($v$), frequency ($f$), and wavelength ($\lambda$) of any wave is given by

$$v = f\lambda.$$

Find $df/d\lambda$ by differentiating implicitly. (Treat $v$ as a constant.) Then show that the same result is obtained if you first solve the equation for $f$ and then differentiate with respect to $\lambda$.

39. **Physical Science**  The relation between temperature $T$ and volume $V$ when a certain gas undergoes an adiabatic process is given by

$$TV^{0.4} = 1500.$$

Find the rate of change of volume with respect to temperature.

40. **Biology**  The equation $(P + a)(v + b) = k$ is called the "fundamental equation of muscle contraction."[7] Here $P$ is the load imposed on the muscle, $v$ is the velocity of the shortening of the muscle fibres, and $a$, $b$, and $k$ are positive constants. Use implicit differentiation to show that, in terms of $P$,

$$\frac{dv}{dP} = -\frac{k}{(P + a)^2}.$$

41. **Marginal Propensity to Consume**  A country's savings $S$ is defined implicitly in terms of its national income $I$ by the equation

$$S^2 + \frac{1}{4}I^2 = SI + I,$$

where both $S$ and $I$ are in billions of dollars. Find the marginal propensity to *consume* when $I = 16$ and $S = 12$.

42. **Technological Substitution**  New products or technologies often tend to replace old ones. For example, today most commercial airlines use jet engines rather than prop engines. In discussing the forecasting of techno-

---

[6]K. E. Bullen, *An Introduction to the Theory of Seismology* (Cambridge, U. K.: Cambridge at the University Press, 1963).

[7]R. W. Stacy et al., *Esssentials of Biological and Medical Physics* (New York: McGraw-Hill Book Company, 1955).

logical substitution, Hurter and Rubenstein[8] refer to the equation

$$\ln \frac{f(t)}{1 - f(t)} + \sigma \frac{1}{1 - f(t)} = C_1 + C_2 t,$$

where $f(t)$ is the market share of the substitute over time $t$ and $C_1$, $C_2$, and $\sigma$ (a Greek letter read "sigma") are constants. Verify their claim that the rate of substitution is

$$f'(t) = \frac{C_2 f(t)[1 - f(t)]^2}{\sigma f(t) + [1 - f(t)]}.$$

[8]A. P. Hurter, Jr., A. H. Rubenstein, et al., "Market Penetration by New Innovations: The Technological Literature," *Technological Forecasting and Social Change*, 11 (1978), 197-221.

O B J E C T I V E

**To describe the method of logarithmic differentiation and to show how to differentiate a function of the form $u^v$.**

# 13.4 LOGARITHMIC DIFFERENTIATION

A technique called **logarithmic differentiation** often simplifies the differentiation of $y = f(x)$ when $f(x)$ involves products, quotients, or powers. The procedure is as follows:

> **Logarithmic Differentiation**
>
> To differentiate $y = f(x)$,
>
> 1. Take the natural logarithm of both sides. This results in
>
> $$\ln y = \ln[f(x)].$$
>
> 2. Simplify $\ln[f(x)]$ by using properties of logarithms.
>
> 3. Differentiate both sides with respect to $x$.
>
> 4. Solve for $\dfrac{dy}{dx}$.
>
> 5. Express the answer in terms of $x$ only. This requires substituting $f(x)$ for $y$.

The next example illustrates this procedure.

**EXAMPLE 1    Logarithmic Differentiation**

*Find $y'$ if $y = \dfrac{(2x - 5)^3}{x^2 \sqrt[4]{x^2 + 1}}$.*

**Solution:** Differentiating this function in the usual way is messy because it involves the quotient, power, and product rules. Logarithmic differentiation makes the work less of a chore.

1. We take the natural logarithm of both sides:

$$\ln y = \ln \frac{(2x - 5)^3}{x^2 \sqrt[4]{x^2 + 1}}.$$

2. Simplifying by using properties of logarithms, we have

$$\ln y = \ln(2x - 5)^3 - \ln[x^2 \sqrt[4]{x^2 + 1}]$$
$$= \ln(2x - 5)^3 - [\ln x^2 + \ln(x^2 + 1)^{1/4}],$$
$$\ln y = 3 \ln(2x - 5) - 2 \ln x - \frac{1}{4} \ln(x^2 + 1).$$

Since $y$ is a function of $x$, differentiating $\ln y$ with respect to $x$ gives $\frac{1}{y} y'$.

3. Differentiating with respect to $x$ gives

$$\frac{1}{y} y' = 3 \left( \frac{1}{2x - 5} \right)(2) - 2 \left( \frac{1}{x} \right) - \frac{1}{4} \left( \frac{1}{x^2 + 1} \right)(2x),$$

$$\frac{y'}{y} = \frac{6}{2x - 5} - \frac{2}{x} - \frac{x}{2(x^2 + 1)}.$$

**4.** Solving for $y'$ yields

$$y' = y\left[\frac{6}{2x - 5} - \frac{2}{x} - \frac{x}{2(x^2 + 1)}\right].$$

**5.** Substituting the original expression for $y$ gives $y'$ in terms of $x$ only:

$$y' = \frac{(2x - 5)^3}{x^2\sqrt[4]{x^2 + 1}}\left[\frac{6}{2x - 5} - \frac{2}{x} - \frac{x}{2(x^2 + 1)}\right]. \qquad \blacksquare$$

Logarithmic differentiation can also be used to differentiate a function of the form $\boldsymbol{y = u^v}$, where both $u$ and $v$ are differentiable functions of $x$. Because neither the base nor the exponent is necessarily a constant, the differentiation techniques for $u^n$ and $a^u$ do not apply here.

**EXAMPLE 2**  *Differentiating the Form $u^v$*

*Differentiate $y = x^x$ by using logarithmic differentiation.*

*Solution:* This equation has the form $y = u^v$, where $u$ and $v$ are functions of $x$. Taking the natural logarithm of both sides gives $\ln y = \ln x^x$ or

$$\ln y = x \ln x.$$

Differentiating both sides with respect to $x$ yields

$$\frac{1}{y}\, y' = x\left(\frac{1}{x}\right) + (\ln x)(1),$$

$$\frac{y'}{y} = 1 + \ln x.$$

Solving for $y'$ and substituting $x^x$ for $y$, we obtain

$$y' = y(1 + \ln x) = x^x(1 + \ln x). \qquad \blacksquare$$

It is worthwhile mentioning that an alternative technique for differentiating a function of the form $y = u^v$ is to convert it to an exponential function to the base $e$. To illustrate, for the function in Example 2, we have

$$y = x^x = (e^{\ln x})^x = e^{x \ln x},$$

$$y' = e^{x \ln x}\left(x \cdot \frac{1}{x} + \ln x\right)$$

$$= x^x(1 + \ln x).$$

**EXAMPLE 3**  Differentiating the Form $u^v$

*Find the derivative of $y = (1 + e^x)^{\ln x}$.*

*Solution:* This has the form $y = u^v$, where $u = 1 + e^x$ and $v = \ln x$. Using logarithmic differentiation, we have

$$\ln y = \ln\left[(1 + e^x)^{\ln x}\right],$$

$$\ln y = (\ln x) \ln(1 + e^x),$$

$$\frac{1}{y}\, y' = (\ln x)\left[\frac{1}{1 + e^x} \cdot e^x\right] + [\ln(1 + e^x)]\left(\frac{1}{x}\right),$$

$$\frac{1}{y} y' = \frac{e^x \ln x}{1 + e^x} + \frac{\ln(1 + e^x)}{x},$$

$$y' = y \left[ \frac{e^x \ln x}{1 + e^x} + \frac{\ln(1 + e^x)}{x} \right],$$

$$y' = (1 + e^x)^{\ln x} \left[ \frac{e^x \ln x}{1 + e^x} + \frac{\ln(1 + e^x)}{x} \right]. \qquad \blacksquare$$

After completing this section, you should understand how to differentiate each of the following forms:

$$y = \begin{cases} [f(x)]^n, & \text{(a)} \\ a^{f(x)}, & \text{(b)} \\ [f(x)]^{g(x)}. & \text{(c)} \end{cases}$$

For type (a), you may utilize the power rule. For type (b), use the differentiation formula for exponential functions. [If $a \neq e$, you can first convert $a^{f(x)}$ to an $e^u$ function.] For type (c), use logarithmic differentiation or first convert to an $e^u$ function. Do not apply a rule in a situation where the rule does not apply. For example, the derivative of $x^x$ is not $x \cdot x^{x-1}$.

## ■ Exercise 13.4

*In Problems 1–12, find $y'$ by using logarithmic differentiation.*

**1.** $y = (x + 1)^2(x - 1)(x^2 + 3)$.

**2.** $y = (3x + 4)(8x - 1)^2(3x^2 + 1)^4$.

**3.** $y = (3x^3 - 1)^2(2x + 5)^3$.

**4.** $y = (3x + 1)\sqrt{8x - 1}$.

**5.** $y = \sqrt{x + 1}\sqrt{x^2 - 2}\sqrt{x + 4}$.

**6.** $y = (x + 2)\sqrt{x^2 + 9}\sqrt[3]{2x + 1}$.

**7.** $y = \dfrac{\sqrt{1 - x^2}}{1 - 2x}$.

**8.** $y = \sqrt{\dfrac{x^2 + 5}{x + 9}}$.

**9.** $y = \dfrac{(2x^2 + 2)^2}{(x + 1)^2(3x + 2)}$.

**10.** $y = \dfrac{x(1 + x^2)^2}{\sqrt{2 + x^2}}$.

**11.** $y = \sqrt{\dfrac{(x - 1)(x + 1)}{3x - 4}}$.

**12.** $y = \sqrt[3]{\dfrac{6(x^3 + 1)^2}{x^6 e^{-4x}}}$.

*In Problems 13–20, find $y'$.*

**13.** $y = x^{2x+1}$.

**14.** $y = (2x)^{\sqrt{x}}$.

**15.** $y = x^{1/x}$.

**16.** $y = \left(\dfrac{2}{x}\right)^x$.

**17.** $y = (3x + 1)^{2x}$.

**18.** $y = x^{x^2}$.

**19.** $y = e^x x^{3x}$.

**20.** $y = (\ln x)^{e^x}$.

**21.** If $y = (4x - 3)^{2x+1}$, find $dy/dx$ when $x = 1$.

**22.** If $y = (\ln x)^{\ln x}$, find $dy/dx$ when $x = e$.

**23.** Find an equation of the tangent line to
$$y = (x + 1)(x + 2)^2(x + 3)^2$$
at the point where $x = 0$.

**24.** Find an equation of the tangent line to the graph of
$$y = e^x(-x)^x$$
at the point where $x = -1$.

**25.** Find an equation of the tangent line to the graph of
$$y = 3e^x(x^2 - x + 1)^x$$
at the point where $x = 1$.

**26.** If $y = x^{2x}$, find the relative rate of change of $y$ with respect to $x$ when $x = 2$.

**27.** If $y = (3x)^{-2x}$, find the value of $x$ for which the *percentage* rate of change of $y$ with respect to $x$ is 60.

**28.** Suppose $f(x)$ and $g(x)$ are positive differentiable functions and $y = [f(x)]^{g(x)}$. Use logarithmic differentiation to show that
$$\frac{dy}{dx} = [f(x)]^{g(x)}\left( f'(x)\frac{g(x)}{f(x)} + g'(x)\ln[f(x)] \right).$$

**To find higher-order derivatives both directly and implicitly.**

# 13.5  HIGHER-ORDER DERIVATIVES

We know that the derivative of a function $y = f(x)$ is itself a function, $f'(x)$. If we differentiate $f'(x)$, the resulting function is called the **second derivative** of $f$ at $x$. It is denoted $f''(x)$, which is read "$f$ double prime of $x$." Similarly, the derivative of the second derivative is called the **third derivative,** written $f'''(x)$. Continuing in this way, we get *higher-order derivatives*. Some notations for higher-order derivatives are given in Table 13.1. To avoid clumsy notation, primes are not used beyond the third derivative.

**TABLE 13.1**

| First derivative: | $y'$, | $f'(x)$, | $\dfrac{dy}{dx}$, | $\dfrac{d}{dx}[f(x)]$, | $D_x y$ |
|---|---|---|---|---|---|
| Second derivative: | $y''$, | $f''(x)$, | $\dfrac{d^2 y}{dx^2}$, | $\dfrac{d^2}{dx^2}[f(x)]$, | $D_x^2 y$ |
| Third derivative: | $y'''$, | $f'''(x)$, | $\dfrac{d^3 y}{dx^3}$, | $\dfrac{d^3}{dx^3}[f(x)]$, | $D_x^3 y$ |
| Fourth derivative: | $y^{(4)}$, | $f^{(4)}(x)$, | $\dfrac{d^4 y}{dx^4}$, | $\dfrac{d^4}{dx^4}[f(x)]$, | $D_x^4 y$ |

*Pitfall* ▼ The symbol $d^2 y/dx^2$ represents the second derivative of $y$. It is not the same as $(dy/dx)^2$, the square of the first derivative of $y$. That is,

$$\frac{d^2 y}{dx^2} \neq \left(\frac{dy}{dx}\right)^2.$$

**EXAMPLE 1   Finding Higher-Order Derivatives**

**a.** If $f(x) = 6x^3 - 12x^2 + 6x - 2$, *find all higher-order derivatives.*

*Solution:* Differentiating $f(x)$ gives

$$f'(x) = 18x^2 - 24x + 6.$$

Differentiating $f'(x)$ yields

$$f''(x) = 36x - 24.$$

Similarly,

$$f'''(x) = 36,$$
$$f^{(4)}(x) = 0.$$

All successive derivatives are also 0: $f^{(5)}(x) = 0$, and so on.

**b.** If $f(x) = 7$, *find $f''(x)$.*

*Solution:*

$$f'(x) = 0,$$
$$f''(x) = 0.$$ ■

## Principles in Practice 1

**Finding a Second-Order Derivative**

The height $h(t)$ of a rock dropped off of a 200-foot building is given by $h(t) = 200 - 16t^2$, where $t$ is the time measured in seconds. Find $\dfrac{d^2 h}{dt^2}$, the acceleration of the rock at time $t$.

**EXAMPLE 2   Finding a Second-Order Derivative**

*If $y = e^{x^2}$, find $\dfrac{d^2 y}{dx^2}$.*

*Solution:*

$$\frac{dy}{dx} = e^{x^2}(2x) = 2xe^{x^2}.$$

By the product rule,

$$\frac{d^2y}{dx^2} = 2[x(e^{x^2})(2x) + e^{x^2}(1)] = 2e^{x^2}(2x^2 + 1). \qquad \blacksquare$$

---

**Principles in Practice 2**

**Evaluating a Second-Order Derivative**

If the cost to produce $q$ units of a product is $c(q) = 7q^2 + 11q + 19$ and the marginal cost function is $c'(q)$, find the rate of change of the marginal cost function with respect to $q$ when $q = 3$.

---

**EXAMPLE 3    Evaluating a Second-Order Derivative**

If $y = f(x) = \dfrac{16}{x+4}$, *find* $\dfrac{d^2y}{dx^2}$ *and evaluate it when* $x = 4$.

*Solution:* Since $y = 16(x+4)^{-1}$, the power rule gives

$$\frac{dy}{dx} = -16(x+4)^{-2},$$

$$\frac{d^2y}{dx^2} = 32(x+4)^{-3} = \frac{32}{(x+4)^3}.$$

Evaluating when $x = 4$, we obtain

$$\left.\frac{d^2y}{dx^2}\right|_{x=4} = \frac{32}{8^3} = \frac{1}{16}.$$

The second derivative evaluated at $x = 4$ is also denoted $f''(4)$ or $y''(4)$.    $\blacksquare$

**EXAMPLE 4    Finding the Rate of Change of $f''(x)$**

If $f(x) = x\ln x$, *find the rate of change of* $f''(x)$.

*Solution:* To find the rate of change of any function, we must find its derivative. Thus, we want the derivative of $f''(x)$, which is $f'''(x)$. Accordingly,

The rate of change of $f''(x)$ is $f'''(x)$.

$$f'(x) = x\left(\frac{1}{x}\right) + (\ln x)(1) = 1 + \ln x,$$

$$f''(x) = 0 + \frac{1}{x} = \frac{1}{x},$$

$$f'''(x) = \frac{d}{dx}(x^{-1}) = (-1)x^{-2} = -\frac{1}{x^2}. \qquad \blacksquare$$

## Higher-Order Implicit Differentiation

We shall now find a higher-order derivative by means of implicit differentiation. Keep in mind that we shall assume $y$ to be a function of $x$.

**EXAMPLE 5    Higher-Order Implicit Differentiation**

*Find* $\dfrac{d^2y}{dx^2}$ *if* $x^2 + 4y^2 = 4$.

*Solution:* Differentiating both sides with respect to $x$, we obtain

$$2x + 8y\frac{dy}{dx} = 0,$$

$$\frac{dy}{dx} = \frac{-x}{4y}, \qquad (1)$$

$$\frac{d^2y}{dx^2} = \frac{4y\dfrac{d}{dx}(-x) - (-x)\dfrac{d}{dx}(4y)}{(4y)^2}$$

$$= \frac{4y(-1) - (-x)\left(4\dfrac{dy}{dx}\right)}{16y^2}$$

$$= \frac{-4y + 4x\dfrac{dy}{dx}}{16y^2}.$$

$$\frac{d^2y}{dx^2} = \frac{-y + x\dfrac{dy}{dx}}{4y^2} \qquad (2)$$

Although we have found an expression for $d^2y/dx^2$, our answer involves the derivative $dy/dx$. It is customary to express the answer without the derivative —that is, in terms of $x$ and $y$ only. This is easy to do. From Eq. (1), $\dfrac{dy}{dx} = \dfrac{-x}{4y}$, so by substituting into Eq. (2), we have

$$\frac{d^2y}{dx^2} = \frac{-y + x\left(\dfrac{-x}{4y}\right)}{4y^2} = \frac{-4y^2 - x^2}{16y^3} = -\frac{4y^2 + x^2}{16y^3}.$$

We can further simplify the answer. Since $x^2 + 4y^2 = 4$ (the original equation),

$$\frac{d^2y}{dx^2} = -\frac{4}{16y^3} = -\frac{1}{4y^3}. \qquad \blacksquare$$

In Example 5, the simplification of $d^2y/dx^2$ by making use of the original equation is not unusual.

### EXAMPLE 6 Higher-Order Implicit Differentiation

*Find* $\dfrac{d^2y}{dx^2}$ *if* $y^2 = e^{x+y}$.

*Solution:* Differentiating both sides with respect to $x$ gives

$$2y\frac{dy}{dx} = e^{x+y}\left(1 + \frac{dy}{dx}\right).$$

Solving for $dy/dx$, we obtain

$$2y\frac{dy}{dx} = e^{x+y} + e^{x+y}\frac{dy}{dx},$$

$$2y\frac{dy}{dx} - e^{x+y}\frac{dy}{dx} = e^{x+y},$$

$$(2y - e^{x+y})\frac{dy}{dx} = e^{x+y},$$

$$\frac{dy}{dx} = \frac{e^{x+y}}{2y - e^{x+y}}.$$

Since $y^2 = e^{x+y}$ (the original equation),

$$\frac{dy}{dx} = \frac{y^2}{2y - y^2} = \frac{y}{2 - y}.$$

$$\frac{d^2y}{dx^2} = \frac{(2 - y)\dfrac{dy}{dx} - y\left(-\dfrac{dy}{dx}\right)}{(2 - y)^2} = \frac{2\dfrac{dy}{dx}}{(2 - y)^2}.$$

Now we express our answer without $dy/dx$. Since $\dfrac{dy}{dx} = \dfrac{y}{2 - y}$,

$$\frac{d^2y}{dx^2} = \frac{2\left(\dfrac{y}{2 - y}\right)}{(2 - y)^2} = \frac{2y}{(2 - y)^3}.$$ ∎

■ **Exercise 13.5**

*In Problems 1–20, find the indicated derivatives.*

**1.** $y = 4x^3 - 12x^2 + 6x + 2$, $y'''$.

**2.** $y = 2x^4 - 6x^2 + 7x - 2$, $y'''$.

**3.** $y = 7 - x$, $\dfrac{d^2y}{dx^2}$.

**4.** $y = -x - x^2$, $\dfrac{d^2y}{dx^2}$.

**5.** $y = x^3 + e^x$, $y^{(4)}$.

**6.** $f(q) = \ln q$, $f'''(q)$.

**7.** $f(x) = x^2 \ln x$, $f''(x)$.

**8.** $y = \dfrac{1}{x}$, $y'''$.

**9.** $f(p) = \dfrac{1}{6p^3}$, $f'''(p)$.

**10.** $f(x) = \sqrt{x}$, $f''(x)$.

**11.** $f(r) = \sqrt{1 - r}$, $f''(r)$.

**12.** $y = e^{-4x^2}$, $y''$.

**13.** $y = \dfrac{1}{5x - 6}$, $\dfrac{d^2y}{dx^2}$.

**14.** $y = (2x + 1)^4$, $y''$.

**15.** $y = \dfrac{x + 1}{x - 1}$, $y''$.

**16.** $y = 2x^{1/2} + (2x)^{1/2}$, $y''$.

**17.** $y = \ln[x(x + 1)]$, $y''$.

**18.** $y = \ln\dfrac{(2x - 3)(4x - 5)}{x + 3}$, $y''$.

**19.** $f(z) = z^2 e^z$, $f''(z)$.

**20.** $y = \dfrac{x}{e^x}$, $\dfrac{d^2y}{dx^2}$.

**21.** If $y = e^{2x}$, find $\dfrac{d^5y}{dx^5}\bigg|_{x=0}$.

**22.** If $y = e^{2\ln(x^3+1)}$, find $y''$ when $x = 1$.

*In Problems 23–32, find $y''$.*

**23.** $x^2 + 4y^2 - 16 = 0$.   **24.** $x^2 - y^2 = 16$.

**25.** $y^2 = 4x$.   **26.** $4x^2 + 3y^2 = 4$.

**27.** $\sqrt{x} + 4\sqrt{y} = 4$.   **28.** $y^2 - 6xy = 4$.

**29.** $xy + y - x = 4$.   **30.** $xy + y^2 = 1$.

**31.** $y = e^{x+y}$.   **32.** $e^x - e^y = x^2 + y^2$.

**33.** If $x^2 + 8y = y^2$, find $d^2y/dx^2$ when $x = 3$ and $y = -1$.

**34.** Show that the equation

$$f''(x) + 4f'(x) + 4f(x) = 0$$

is satisfied if $f(x) = (3x - 5)e^{-2x}$.

**35.** Find the rate of change of $f'(x)$ if $f(x) = (5x - 3)^4$.

**36.** Find the rate of change of $f''(x)$ if

$$f(x) = 6\sqrt{x} + \frac{1}{6\sqrt{x}}.$$

**37. Marginal Cost** If $c = 0.3q^2 + 2q + 850$ is a cost function, how fast is marginal cost changing when $q = 100$?

**38. Marginal Revenue** If $p = 1000 - 45q - q^2$ is a demand equation, how fast is marginal revenue changing when $q = 10$?

**39.** If $f(x) = x^4 - 6x^2 + 5x - 6$, determine the values of $x$ for which $f''(x) = 0$.

**40.** Suppose that $e^y = y^2e^x$. (a) Determine $dy/dx$, and express your answer in terms of $y$ only. (b) Determine $d^2y/dx^2$, and express your answer in terms of $y$ only.

*In Problems **41** and **42**, determine $f''(x)$. Then use your graphics calculator to find all real zeros of $f''(x)$. Round your answers to two decimal places.*

**41.** $f(x) = 6e^x - x^3 - 15x^2$.

**42.** $f(x) = \dfrac{x^5}{20} - \dfrac{x^4}{4} + \dfrac{5x^3}{6} + \dfrac{x^2}{2}$.

## 13.6 REVIEW

### IMPORTANT TERMS AND SYMBOLS

**Section 13.3**    implicit differentiation

**Section 13.4**    logarithmic differentiation

**Section 13.5**    higher-order derivatives, $f''(x)$, $\dfrac{d^3y}{dx^3}$, $\dfrac{d^4}{dx^4}[f(x)]$, and so on

### SUMMARY

If an equation implicitly defines $y$ as a function of $x$ [rather than defining it explicitly in the form $y = f(x)$], then $dy/dx$ can be found by implicit differentiation. With this method, we treat $y$ as a function of $x$ and differentiate both sides of the equation with respect to $x$. When doing this, remember that

$$\frac{d}{dx}(y^n) = ny^{n-1}\frac{dy}{dx}.$$

Finally, we solve the resulting equation for $dy/dx$.

The derivative formulas for natural logarithmic and exponential functions are

$$\frac{d}{dx}(\ln u) = \frac{1}{u}\frac{du}{dx}$$

and

$$\frac{d}{dx}(e^u) = e^u\frac{du}{dx}.$$

To differentiate logarithmic and exponential functions in bases other than $e$, you can first transform the function to base $e$ and then differentiate the result. Alternatively, differentiation formulas can be applied:

$$\frac{d}{dx}(\log_b u) = \frac{1}{(\ln b)u}\cdot\frac{du}{dx},$$

$$\frac{d}{dx}(a^u) = a^u(\ln a)\frac{du}{dx}.$$

Suppose that $f(x)$ consists of products, quotients, or powers. To differentiate $y = \log_b[f(x)]$, it may be helpful to use properties of logarithms to rewrite $\log_b[f(x)]$ in terms of simpler logarithms and then differentiate that form. To differentiate $y = f(x)$, where $f(x)$ consists of products, quotients, or powers, the method of logarithmic differentiation may be used. In that method, we take the natural logarithm of both sides of $y = f(x)$ to obtain $\ln y = \ln[f(x)]$. After simplifying $\ln[f(x)]$ by using properties of logarithms, we differentiate both sides of $\ln y = \ln[f(x)]$ with respect to $x$ and then solve for $y'$. Logarithmic differentiation is also used to differentiate $y = u^v$, where both $u$ and $v$ are functions of $x$.

Because the derivative $f'(x)$ of a function $y = f(x)$ is itself a function, it can be successively differentiated to obtain the second derivative $f''(x)$, the third derivative $f'''(x)$, and other higher-order derivatives.

### REVIEW PROBLEMS

*In Problems **1–30**, differentiate.*

**1.** $y = 2e^x + e^2 + e^{x^2}$.

**2.** $f(w) = we^w + w^2$.

**3.** $f(r) = \ln(r^2 + 5r)$.

**4.** $y = e^{\ln x}$.

**5.** $y = e^{x^2+4x+5}$.

**6.** $f(t) = \log_6 \sqrt{t^2 + 1}$.

**7.** $y = e^x(x^2 + 2)$.

**8.** $y = 2^{7x^2}$.

**9.** $y = \sqrt{(x - 6)(x + 5)(9 - x)}$.

**10.** $f(t) = e^{\sqrt{t}}$.

**11.** $y = \dfrac{\ln x}{e^x}$.

**12.** $y = \dfrac{e^x + e^{-x}}{x^2}$.

**13.** $f(q) = \ln[(q + 1)^2(q + 2)^3]$.

**14.** $y = (x - 6)^4(x + 4)^3(6 - x)^2$.

**15.** $y = 10^{2-7x}$.

**16.** $y = (e + e^2)^0$.

**17.** $y = \dfrac{4e^{3x}}{xe^{x-1}}$.

**18.** $y = \dfrac{e^x}{\ln x}$.

**19.** $y = \log_2(8x + 5)^2$.

**20.** $y = \ln\left(\dfrac{1}{x}\right)$.

**21.** $f(l) = \ln(1 + l + l^2 + l^3)$.  **22.** $y = x^{x^3}$.

**23.** $y = (x + 1)^{x+1}$.

**24.** $y = \dfrac{1 + e^x}{1 - e^x}$.

**25.** $f(t) = \ln(t^2\sqrt{1 - t})$.  **26.** $y = (x + 2)^{\ln x}$.

**27.** $y = \dfrac{(x^2 + 2)^{3/2}(x^2 + 9)^{4/9}}{(x^3 + 6x)^{4/11}}$  **28.** $y = \dfrac{\ln x}{\sqrt{x}}$.

**29.** $y = (x^x)^x$.

**30.** $y = x^{(x^x)}$.

*In Problems 31–34, evaluate $y'$ at the given value of $x$.*

**31.** $y = (x + 1)\ln x^2$,  $x = 1$.

**32.** $y = \dfrac{e^{x^2 - 4}}{\sqrt{x + 7}}$,  $x = 2$.

**33.** $y = e^{e + x\ln(1/x)}$,  $x = e$.

**34.** $y = \left[\dfrac{4^{3x}(x^3 - x + 1)^{1/5}}{(x^2 + x + 1)^4}\right]^{-2}$,  $x = 0$.

*In Problems 35 and 36, find an equation of the tangent line to the curve at the point corresponding to the given value of $x$.*

**35.** $y = e^x$,  $x = \ln 2$.

**36.** $y = x + x^2 \ln x$,  $x = 1$.

**37.** Find the $y$-intercept of the tangent line to the graph of $y = x(2^{2-x^2})$ at the point where $x = 1$.

**38.** If $w = 5^{x+1} + \ln(1 - x^2)$ and

$$x = \log_5(t^2 + 4) - e^{(t-1)^2},$$

find $w$ and $dw/dt$ when $t = 1$. Simplify your answers.

*In Problems 39–42, find the indicated derivative at the given point.*

**39.** $y = e^{x^2 - 4}$,  $y''$,  $(2, 1)$.
**41.** $y = \ln(2x)$,  $y'''$,  $(1, \ln 2)$.

**40.** $y = x^2 e^x$,  $y'''$,  $(1, e)$.
**42.** $y = x \ln x$,  $y''$,  $(1, 0)$.

*In Problems 43–46, find $dy/dx$.*

**43.** $2xy + y^2 = 6$.

**44.** $x^2 y^2 = 1$.

**45.** $\ln(xy^2) = xy$.

**46.** $(\ln y)e^{y\ln x} = e^2$.

*In Problems 47 and 48, find $d^2y/dx^2$ at the given point.*

**47.** $x + xy + y = 5$,  $(2, 1)$.

**48.** $xy + y^2 = 2$,  $(1, 1)$.

**49.** If $y$ is defined implicitly by $e^y = (y + 1)e^x$, determine both $dy/dx$ and $d^2y/dx^2$ as explicit functions of $y$ only.

**50.** If $\sqrt{x} + \sqrt{y} = 4$, show that

$$\frac{dy}{dx} = -\frac{\sqrt{y}}{\sqrt{x}} \quad \text{and} \quad \frac{d^2y}{dx^2} = 2x^{-3/2}.$$

**51. Schizophrenia** Several models have been used to analyze the length of stay in a hospital. For a particular group of schizophrenics, one such model is[9]

$$f(t) = 1 - (0.8e^{-0.01t} + 0.2e^{-0.0002t}),$$

where $f(t)$ is the proportion of the group that was discharged at the end of $t$ days of hospitalization. Determine the discharge rate (proportion discharged per day) at the end of $t$ days.

**52. Earthquakes** According to Richter,[10] the number $N$ of earthquakes of magnitude $M$ or greater per unit of time is given by $\log N = A - bM$, where $A$ and $b$ are constants. He claims that

$$\log\left(-\frac{dN}{dM}\right) = A + \log\left(\frac{b}{q}\right) - bM,$$

where $q = \log e$. Verify this statement.

 **53.** If $f(x) = e^{9x^4 + 4x^3 - 36x}$, find all real zeros of $f'(x)$. Round your answers to two decimal places.

 **54.** If $f(x) = \dfrac{x^5}{10} + \dfrac{x^4}{12} + \dfrac{2x^3}{3} + x^2 + 3$, find all zeros of $f''(x)$. Round your answers to two decimal places.

[9]Adapted from W. W. Eaton and G. A. Whitmore, "Length of Stay as a Stochastic Process: A General Approach and Application to Hospitalization for Schizophrenia," *Journal of Mathematical Sociology,* 5 (1977) 273-92.

[10]C. F. Richter, *Elementary Seismology* (San Francisco: W. H. Freeman and Company, Publishers, 1958).

# Curve Sketching

**To find when a function is increasing or decreasing, to find critical values, to locate relative maxima and relative minima, and to state the first-derivative test. Also, to sketch the graph of a function by using the information obtained from the first derivative.**

## 14.1  RELATIVE EXTREMA

### Increasing or Decreasing Nature of a Function

Examining the graphical behavior of equations is a basic part of mathematics and has applications to many areas of study. When we sketch a curve, just plotting points may not give enough information about its shape. For example, the points $(-1, 0), (0, -1)$, and $(1, 0)$ satisfy the equation given by $y = (x + 1)^3(x - 1)$. On the basis of these points, you might hastily conclude that the graph should appear as in Fig. 14.1(a), but in fact the true shape is given in Fig. 14.1(b). In this chapter we shall explore the powerful role that differentiation plays in analyzing a function, so that we may determine the true shape and behavior of its graph.

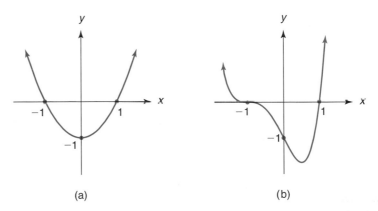

(a)                                      (b)

**FIGURE 14.1**  Curves passing through $(-1, 0), (0, -1)$, and $(1, 0)$.

We begin by analyzing the graph of the function $y = f(x)$ in Fig. 14.2. Notice that as $x$ increases (goes from left to right) on the interval $I_1$, between $a$ and $b$, the values of $f(x)$ increase and the curve is rising. Mathematically, this observation means that if $x_1$ and $x_2$ are any two points in $I_1$ such that $x_1 < x_2$, then $f(x_1) < f(x_2)$. Here $f$ is said to be an *increasing function* on $I_1$. On the other hand, as $x$ increases on the interval $I_2$ between $c$ and $d$, the curve is falling. On this interval, $x_3 < x_4$ implies that $f(x_3) > f(x_4)$, and $f$

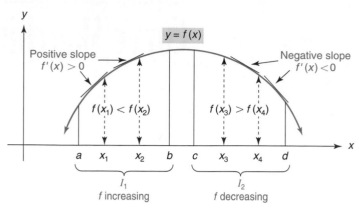

**FIGURE 14.2** Increasing or decreasing nature of function.

is said to be a *decreasing function* on $I_2$. We summarize these observations in the following definition.

### DEFINITION

*A function f is said to be **increasing** on the interval I if, for any two numbers $x_1$, $x_2$ in I, where $x_1 < x_2$, then $f(x_1) < f(x_2)$. A function f is **decreasing** on the interval I if, for any two numbers $x_1$, $x_2$ in I, where $x_1 < x_2$, then $f(x_1) > f(x_2)$.*

Turning again to Fig. 14.2, we note that over the interval $I_1$, tangent lines to the curve have positive slopes, so $f'(x)$ must be positive for all $x$ in $I_1$. Basically, a positive derivative implies that the curve is rising. Over the interval $I_2$, the tangent lines have negative slopes, so $f'(x) < 0$ for all $x$ in $I_2$. Basically, the curve is falling where the derivative is negative. We thus have the following rule, which allows us to use the derivative to determine when a function is increasing or decreasing:

> **Rule 1   Criteria for Increasing or Decreasing Function**
>
> Let $f$ be differentiable on the interval $(a, b)$. If $f'(x) > 0$ for all $x$ in $(a, b)$, then $f$ is increasing on $(a, b)$. If $f'(x) < 0$ for all $x$ in $(a, b)$, then $f$ is decreasing on $(a, b)$

To illustrate these ideas, we shall use Rule 1 to find the intervals on which $y = 18x - \frac{2}{3}x^3$ is increasing or decreasing. Letting $y = f(x)$, we must determine when $f'(x)$ is positive and when $f'(x)$ is negative. We have

$$f'(x) = 18 - 2x^2 = 2(9 - x^2) = 2(3 + x)(3 - x).$$

Using the technique of Sec. 11.5, we can find the sign of $f'(x)$ by testing the intervals determined by the roots of $2(3 + x)(3 - x) = 0$, namely, 3 and $-3$. (See Fig. 14.3.) In each interval, the sign of $f'(x)$ is determined by the signs of its factors:

if $x < -3$, then $f'(x) = 2(\;-\;)(\;+\;) = \;-\;$, so $f$ is decreasing;

if $-3 < x < 3$, then $f'(x) = 2(\;+\;)(\;+\;) = \;+\;$, so $f$ is increasing;

if $x > 3$, then $f'(x) = 2(\;+\;)(\;-\;) = \;-\;$, so $f$ is decreasing.

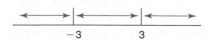

**FIGURE 14.3** Intervals determined by roots of $f'(x) = 0$.

These results are indicated in Fig. 14.4(a). Thus, $f$ is decreasing on $(-\infty, -3)$ and $(3, \infty)$, and is increasing on $(-3, 3)$. This corresponds to the rising and falling nature of the graph of $f$ shown in Fig. 14.4(b). These results could be sharpened by noting that, by definition, $f$ is decreasing on $(-\infty, -3]$ and

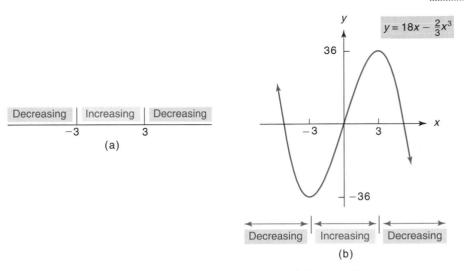

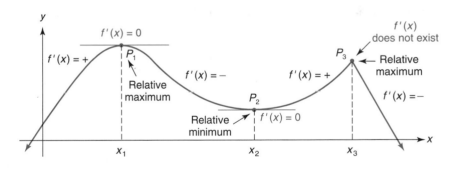

**FIGURE 14.4** Increasing/decreasing for $y = 18x - \frac{2}{3}x^3$.

$[3, \infty)$, and increasing on $[-3, 3]$. However, for our purposes, open intervals are sufficient. *It will be our practice to determine* **open** *intervals on which a function is increasing or decreasing.*

### Extrema

Look now at the graph of $y = f(x)$ in Fig. 14.5. Some observations can be made. First, there is something special about the points $P_1$, $P_2$, and $P_3$. Notice

**FIGURE 14.5** Relative maxima and relative minima.

that $P_1$ is *higher* than any other "nearby" point on the curve—and likewise for $P_3$. The point $P_2$ is *lower* than any other "nearby" point on the curve. Since $P_1$, $P_2$, and $P_3$ may not necessarily be the highest or lowest points on the *entire* curve, we simply say that the graph of $f$ has a *relative maximum (point)* when $x = x_1$ (or at $x_1$) and when $x = x_3$, and has a *relative minimum (point)* when $x = x_2$. On the function level, when $x = x_1$ and $x = x_3$, $f$ has *relative maximum values* of $f(x_1)$ and $f(x_3)$. Similarly, when $x = x_2$, $f$ has a *relative minimum value* of $f(x_2)$. When we refer to a relative maximum or minimum, it is understood that we are referring to a point or value, depending on the context. Turning back to the graph, we see that there is an *absolute maximum* (highest point on the entire curve) when $x = x_1$, but there is no *absolute minimum* (lowest point on the entire curve) because the curve is assumed to extend downward indefinitely. More precisely, we define these new terms as follows:

**DEFINITION**

*A function f has a **relative maximum** when $x = x_0$ if there is an open interval containing $x_0$ on which $f(x_0) \geq f(x)$ for all x in the interval. The relative maximum is $f(x_0)$. A function f has a **relative minimum** when $x = x_0$ if there is an open interval containing $x_0$ on which $f(x_0) \leq f(x)$ for all x in the interval. The relative minimum is $f(x_0)$.*

**DEFINITION**

*A function f has an **absolute maximum** when $x = x_0$ if $f(x_0) \geq f(x)$ for all x in the domain of f. The absolute maximum is $f(x_0)$. A function f has an **absolute minimum** when $x = x_0$ if $f(x_0) \leq f(x)$ for all x in the domain of f. The absolute minimum is $f(x_0)$.*

If it exists, an absolute maximum is unique; however, it may occur for more than one value of x. A similar statement is true for an absolute minimum.

We refer to either a relative maximum or a relative minimum as a **relative extremum** (plural: *relative extrema*). Similarly, we speak of **absolute extrema.**

When dealing with relative extrema, we compare the function value at a point with values of nearby points; however, when dealing with absolute extrema, we compare the function value at a point with all other values determined by the domain. Thus, relative extrema are "local" in nature, whereas absolute extrema are "global" in nature.

Referring back to Fig. 14.5, we notice that at a relative extremum the derivative may not be defined (as when $x = x_3$). But whenever it is defined at a relative extremum, it is 0 (as when $x = x_1$ and $x = x_2$), and hence, the tangent line is horizontal. We may state:

---

**Rule 2    A Necessary Condition for Relative Extrema**

*If f has a relative extremum when $x = x_0$, then $f'(x_0) = 0$ or $f'(x_0)$ is not defined.*

---

The implication in Rule 2 goes in only one direction:

$$\left.\begin{array}{c} \text{relative extremum} \\ \text{at } x_0 \end{array}\right\} \rightarrow \left\{\begin{array}{c} f'(x_0) = 0 \\ \text{or} \\ f'(x_0) \text{ not defined.} \end{array}\right.$$

Rule 2 does *not* say that if $f'(x_0)$ is 0 or is not defined, then there must be a relative extremum at $x_0$. In fact, there may not be one at all. For example, in Fig. 14.6, $f'(x_0)$ is 0 because the tangent line is horizontal at $x_0$, but yet there is no relative extremum there.

In general, the best we can say is that relative extrema *may* occur at points on the graph of f where $f'(x) = 0$ or where $f'(x)$ is not defined. Because these points are so important for locating relative extrema, they are called *critical points*, and their x-coordinates are called *critical values*. Thus, in Fig. 14.5, the numbers $x_1$, $x_2$, and $x_3$ are critical values, and $P_1$, $P_2$, and $P_3$ are critical points.

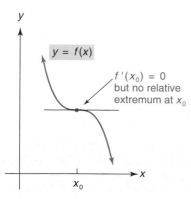

**FIGURE 14.6**    No relative extremum at $x_0$.

**DEFINITION**

*If $x_0$ is in the domain of f and either $f'(x_0) = 0$ or $f'(x_0)$ is not defined, then $x_0$ is called a **critical value** of f. If $x_0$ is a critical value, then the point $(x_0, f(x_0))$ is called a **critical point.***

At a critical point, there may be a relative maximum, a relative minimum, or neither. Moreover, from Fig. 14.5, we observe that each relative extremum occurs at a point around which the sign of $f'(x)$ is changing. For the relative maximum when $x = x_1$, $f'(x)$ goes from $+$ for $x < x_1$ to $-$ for $x > x_1$, as *long as x is near* $x_1$. At the relative minimum when $x = x_2$, $f'(x)$ goes from $-$ to $+$, and at the relative maximum when $x = x_3$, it again goes from $+$ to $-$. Thus, *around relative maxima, f is increasing and then decreasing, and the reverse holds for relative minima.* More precisely, we have the following rule:

### Rule 3   Criteria for Relative Extrema

*Suppose f is continuous on an open interval I that contains the critical value $x_0$ and f is differentiable on I, except possibly at $x_0$.*

**a.** *If $f'(x)$ changes from positive to negative as x increases through $x_0$, then f has a relative maximum when $x = x_0$.*

**b.** *If $f'(x)$ changes from negative to positive as x increases through $x_0$, then f has a relative minimum when $x = x_0$.*

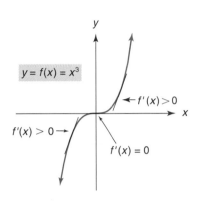

**FIGURE 14.7**   Zero is a critical value, but does not give a relative extremum.

***Pitfall*** ▼   We remind you again that not every critical value corresponds to a relative extremum. For example, if $y = f(x) = x^3$, then $f'(x) = 3x^2$. Since $f'(0) = 0$ and $f(0)$ is defined, 0 is a critical value. Now if $x < 0$, then $3x^2 > 0$. If $x > 0$, then $3x^2 > 0$. Since $f'(x)$ does not change sign, no relative maximum or minimum exists. Indeed, since $f'(x) \geq 0$ for all $x$, the graph of $f$ never falls, and $f$ is said to be *nondecreasing*. (See Fig. 14.7).

From our discussions and the preceding pitfall, you should realize that a critical value is only a "candidate" for a relative extremum. It may correspond to a relative maximum, a relative minimum, or neither.

It is important to understand that not every value of $x$ where $f'(x)$ does not exist is a critical value. For example, if

$$y = f(x) = \frac{1}{x^2}, \quad \text{then} \quad f'(x) = -\frac{2}{x^3}.$$

Although $f'(x)$ is not defined when $x = 0$, 0 is not a critical value, because 0 is not in the domain of $f$. That is, no $y$-value corresponds to $x = 0$. Thus, a relative extremum cannot occur when $x = 0$. Nevertheless, the derivative may change sign around any $x$-value where $f'(x)$ is not defined, so such values are important in determining intervals over which $f$ is increasing or decreasing.

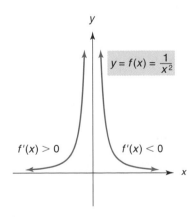

**FIGURE 14.8**   $f'(0)$ is not defined, but 0 is not a critical value because 0 is not in the domain of $f$.

$$\text{If } x < 0, \quad \text{then} \quad f'(x) = -\frac{2}{x^3} > 0.$$

$$\text{If } x > 0, \quad \text{then} \quad f'(x) = -\frac{2}{x^3} < 0.$$

Hence, $f$ is increasing on $(-\infty, 0)$ and decreasing on $(0, \infty)$. (See Fig. 14.8.)

In Rule 3 the hypotheses must be satisfied, or the conclusion need not hold. For example, take the case of the compound function

$$f(x) = \begin{cases} \dfrac{1}{x^2}, & \text{if } x \neq 0 \\[2mm] 0, & \text{if } x = 0. \end{cases}$$

As you can see in Fig. 14.9, 0 is in the domain and $f'(0)$ does not exist, so 0 is a critical value. Although $f'(x) = +$ for $x < 0$ and $f'(x) = -$ for $x > 0$, $f$ does not have a relative maximum at 0. Rule 3 does not apply because $f$ is not

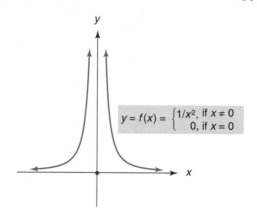

$$y = f(x) = \begin{cases} 1/x^2, & \text{if } x \neq 0 \\ 0, & \text{if } x = 0 \end{cases}$$

**FIGURE 14.9**  Zero is a critical value, but not a relatative extremum.

continuous at 0. Actually, 0 is an absolute minimum, according to the definition. This shows that if $f'(x_0)$ does not exist and $f$ is not continuous at $x_0$, you need to examine carefully what happens around $x_0$.

Summarizing the results of this section, we have the *first-derivative test* for the relative extrema of $y = f(x)$:

**First-Derivative Test for Relative Extrema**

**Step 1.**  Find $f'(x)$.

**Step 2.**  Determine all values of $x$ where $f'(x) = 0$ or $f'(x)$ is not defined. (These values include critical values and points of discontinuity.)

**Step 3.**  On the intervals suggested by the values in step 2, determine whether $f$ is increasing $[f'(x) > 0]$ or decreasing $[f'(x) < 0]$.

**Step 4.**  For each critical value $x_0$ at which $f$ is continuous, determine whether $f'(x)$ changes sign as $x$ increases through $x_0$. There is a relative maximum when $x = x_0$ if $f'(x)$ changes from $+$ to $-$ going from left to right and a relative minimum if $f'(x)$ changes from $-$ to $+$ going from left to right. If $f'(x)$ does not change sign, there is no relative extremum when $x = x_0$.

**Principles in Practice 1**

**First-Derivative Test**

The cost equation for a hot dog stand is given by $c(q) = 2q^3 - 21q^2 + 60q + 500$, where $q$ is the number of hot dogs sold, and $c(q)$ is the cost in dollars. Use the first-derivative test to find when relative extrema occur.

**EXAMPLE 1   First-Derivative Test**

*If $y = f(x) = x + \dfrac{4}{x + 1}$, use the first-derivative test to find when relative extrema occur.*

**Solution:**

**Step 1.**  $f(x) = x + 4(x + 1)^{-1}$, so

$$f'(x) = 1 + 4(-1)(x + 1)^{-2} = 1 - \frac{4}{(x + 1)^2}$$

$$= \frac{(x + 1)^2 - 4}{(x + 1)^2} = \frac{x^2 + 2x - 3}{(x + 1)^2} = \frac{(x + 3)(x - 1)}{(x + 1)^2}.$$

Note that we expressed $f'(x)$ as a quotient with numerator and denominator factored. This enables us in step 2 to easily determine when $f'(x)$ is 0 or not defined.

**Step 2.** Setting $f'(x) = 0$ gives $x = -3, 1$. The denominator of $f'(x)$ is 0 when $x$ is $-1$, so $f'(-1)$ does not exist. The values $-3$ and 1 are critical values, but $-1$ is not because $f(-1)$ is not defined ($f$ is discontinuous at $x = -1$).

**Step 3.** The three values in step 2 lead us to test four intervals (Fig. 14.10). (On each of these intervals, $f$ is differentiable and is not zero.) Note that we have boxed the value $-1$ to indicate that it cannot correspond to a relative extremum. However, it is essential that $-1$ be considered in our analysis of increasing/decreasing.

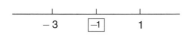

**FIGURE 14.10** Four intervals to test for increasing/decreasing.

If $x < -3$, then $f'(x) = \dfrac{(-)(-)}{(+)} = +$, so $f$ is increasing;

if $-3 < x < -1$, then $f'(x) = \dfrac{(+)(-)}{(+)} = -$, so $f$ is decreasing;

if $-1 < x < 1$, then $f'(x) = \dfrac{(+)(-)}{(+)} = -$, so $f$ is decreasing;

if $x > 1$, then $f'(x) = \dfrac{(+)(+)}{(+)} = +$, so $f$ is increasing (Fig. 14.11).

Thus, $f$ is increasing on the intervals $(-\infty, -3)$ and $(1, \infty)$ and is decreasing on $(-3, -1)$ and $(-1, 1)$.

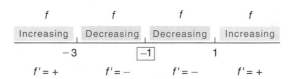

**FIGURE 14.11** Sign chart for
$$f'(x) = \frac{(x+3)(x-1)}{(x+1)^2}.$$

**Step 4.** From Fig. 14.11, we conclude that when $x = -3$, there is a relative maximum, since $f'(x)$ changes from $+$ to $-$. [This relative maximum value is $f(-3) = -3 + (4/-2) = -5$.] When $x = 1$, there is a relative minimum, because $f'(x)$ changes from $-$ to $+$. We ignore $x = -1$ since $-1$ is not a critical value. The graph is shown in Fig. 14.12. ∎

**EXAMPLE 2    A Relative Extremum where $f'(x)$ Does Not Exist**

*Test $y = f(x) = x^{2/3}$ for relative extrema.*

*Solution:* We have

$$f'(x) = \frac{2}{3} x^{-1/3}$$

$$= \frac{2}{3\sqrt[3]{x}}.$$

When $x = 0$, $f'(x)$ is not defined, but $f(x)$ is defined. Thus, 0 is a critical value and there are no others. If $x < 0$, then $f'(x) < 0$. If $x > 0$, then

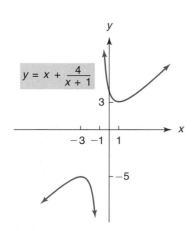

**FIGURE 14.12** Graph of
$y = x + \dfrac{4}{x+1}.$

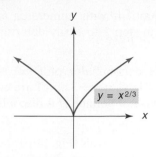

**FIGURE 14.13** Derivative does not exist at 0 and there is a minimum at 0.

---

**Principles in Practice 2**

**Finding Relative Extrema**

A drug is injected into a patient's bloodstream. The concentration of the drug in the bloodstream $t$ hours after the injection is approximated by $C(t) = \dfrac{0.14t}{t^2 + 4t + 4}$. Find the relative extrema for $t > 0$, and use them to determine when the drug is at its greatest concentration.

---

$f'(x) > 0$. Therefore, there is a relative (as well as an absolute) minimum when $x = 0$. (See Fig. 14.13.) Note that when $x = 0$, the tangent line exists and is vertical. ∎

**EXAMPLE 3   Finding Relative Extrema**

*Test $y = f(x) = x^2 e^x$ for relative extrema.*

**Solution:** By the product rule,

$$f'(x) = x^2 e^x + e^x(2x) = xe^x(x + 2).$$

Noting that $e^x$ is always positive, we obtain the critical values 0 and $-2$. From the signs of $f'(x)$ given in Fig. 14.14, we conclude that there is a relative maximum when $x = -2$ and a relative minimum when $x = 0$.

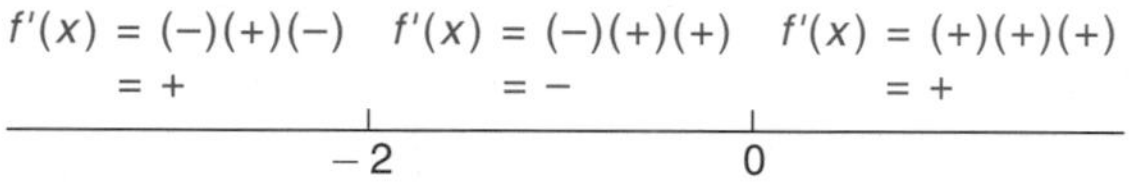

$$
\begin{array}{ccc}
f'(x) = (-)(+)(-) & f'(x) = (-)(+)(+) & f'(x) = (+)(+)(+) \\
= + & = - & = + \\
\end{array}
$$

$$\underset{-2}{\rule{0pt}{0pt}} \qquad \underset{0}{\rule{0pt}{0pt}}$$

**FIGURE 14.14** Sign chart for $f'(x) = xe^x(x + 2)$. ∎

## Curve Sketching

In the next example we show how the first-derivative test, in conjunction with the notions of intercepts and symmetry, can be used as an aid in sketching the graph of a function.

**EXAMPLE 4   Curve Sketching**

*Sketch the graph of $y = f(x) = 2x^2 - x^4$ with the aid of intercepts, symmetry, and the first-derivative test.*

**Solution:**

***Intercepts*** If $x = 0$, then $y = 0$. If $y = 0$, then

$$0 = 2x^2 - x^4 = x^2(2 - x^2) = x^2(\sqrt{2} + x)(\sqrt{2} - x),$$

so $x = 0, \pm\sqrt{2}$. Thus, the intercepts are $(0, 0)$, $(\sqrt{2}, 0)$, and $(-\sqrt{2}, 0)$.

***Symmetry*** Testing for $y$-axis symmetry, we have

$$y = 2(-x)^2 - (-x)^4, \qquad \text{or} \qquad y = 2x^2 - x^4.$$

Since this is the original equation, there is $y$-axis symmetry. Because $y$ is a function (and not the zero function), there is no $x$-axis symmetry and, hence, no symmetry about the origin.

***First-Derivative Test***

**Step 1.** $y' = 4x - 4x^3 = 4x(1 - x^2) = 4x(1 + x)(1 - x)$.

**Step 2.** Setting $y' = 0$ gives the critical values $x = 0, \pm 1$. Since we are interested in a graph, the critical *points* are important to us. By substituting the critical values into the *original* equation, $y = 2x^2 - x^4$, we obtain the $y$-coordinates of these points. We find the critical points to be $(-1, 1)$, $(0, 0)$, and $(1, 1)$.

**FIGURE 14.15**
Intervals for sign chart of
$y' = 4x(1 + x)(1 - x)$.

$$\frac{y'>0 \quad y'<0 \quad y'>0 \quad y'<0}{-1 \qquad 0 \qquad 1}$$

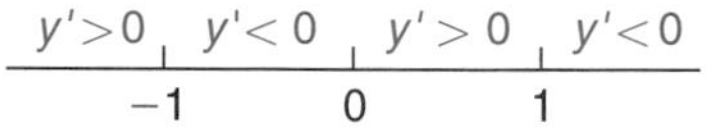

**FIGURE 14.16**
Sign chart of
$y' = 4x(1 + x)(1 - x)$.

**Step 3.** There are four intervals to consider in Fig. 14.15:

if $x < -1$, then $y' = 4(-)(-)(+) = +$, so $f$ is increasing;

if $-1 < x < 0$, then $y' = 4(-)(+)(+) = -$, so $f$ is decreasing;

if $0 < x < 1$, then $y' = 4(+)(+)(+) = +$, so $f$ is increasing;

if $x > 1$, then $y' = 4(+)(+)(-) = -$, so $f$ is decreasing (Fig. 14.16).

**Step 4.** From Fig. 14.16, we conclude that relative maxima occur at $(-1, 1)$ and $(1, 1)$; a relative minimum occurs at $(0, 0)$.

***Discussion*** In Fig. 14.17(a), we have indicated the horizontal tangents at the relative maximum and minimum points. We know the curve rises from the left, has a relative maximum, then falls, has a relative minimum, then rises to a relative maximum, and falls thereafter. A sketch is shown in Fig. 14.17(b).

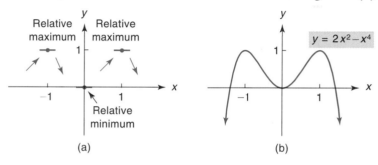

(a)                                    (b)

**FIGURE 14.17**  Putting together the graph of $y = 2x^2 - x^4$.

As a passing comment, we note that in Example 4 *absolute* maxima occur at $x = \pm 1$. [See Fig. 14.17(b).] There is no absolute minimum.

## TECHNOLOGY

A graphics calculator is a powerful tool for investigating relative extrema. For example, consider the function

$$f(x) = 3x^4 - 4x^3 + 4,$$

whose graph is shown in Fig. 14.18. It appears that there is a relative minimum near $x = 1$. We can locate this minimum by either using "trace and zoom" or (on a T1-82) using the "minimum" feature. Figure 14.19 shows the latter approach. The relative minimum point is estimated to be $(1.00, 3)$.

Now let us see how the graph of $f'$ indicates when extrema occur. We have

$$f'(x) = 12x^3 - 12x^2,$$

whose graph is shown in Fig. 14.20. It appears that $f'(x)$ is 0 at two points. Using "trace and zoom" or the "root" feature, we estimate the zeros of $f'$ (the critical values of $f$) to be 1 and 0. Around $x = 1$, we see

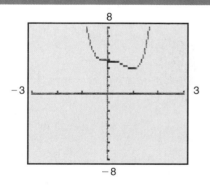

**FIGURE 14.18**  Graph of
$f(x) = 3x^4 - 4x^3 + 4$.

that $f'(x)$ goes from negative values to positive values. (That is, the graph of $f'$ goes from below the $x$-axis to above it.) Thus, we conclude that $f$ has a relative minimum at $x = 1$, which confirms our previous result. Around the critical value $x = 0$, the values of $f'(x)$ are negative. Since $f'(x)$ does not change sign, we conclude that there is no relative extremum at $x = 0$. This is also apparent from the graph in Fig. 14.18.

It is worthwhile to note that we can approximate the graph of $f'$ without determining $f'(x)$ itself. We make use of the "nDeriv" feature. First we enter the function $f$ as $Y_1$. Then we set

$$Y_2 = \text{nDeriv}(Y_1, X, X).$$

The graph of $Y_2$ approximates the graph of $f'(x)$.

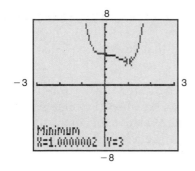

**FIGURE 14.19** Relative minimum at $(1.00, 3)$.

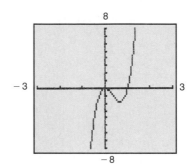

**FIGURE 14.20** Graph of $f'(x) = 12x^3 - 12x^2$.

## ■ Exercise 14.1

*In Problems 1–4, the graph of a function is given. Find the open intervals on which the function is increasing or decreasing, and find the coordinates of all relative extrema.*

**1.**

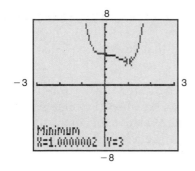

**FIGURE 14.21** Graph for Problem 1.

**2.**

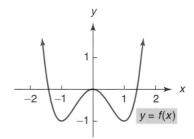

**FIGURE 14.22** Graph for Problem 2.

**3.**

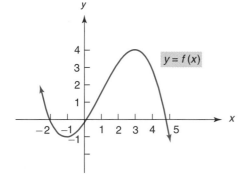

**FIGURE 14.23** Graph for Problem 3.

**4.**

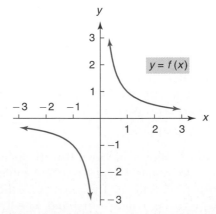

**FIGURE 14.24** Graph for Problem 4.

*In Problems 5–8, the derivative of the continuous function f is given. Find open intervals on which f is increasing or decreasing, and find the x-values of all relative extrema.*

**5.** $f'(x) = (x + 1)(x - 3)$.

**6.** $f'(x) = 2x(x - 1)^3$.

**7.** $f'(x) = (x + 1)(x - 3)^2$.

**8.** $f'(x) = \dfrac{x(x + 2)}{x^2 + 1}$.

*In Problems 9–52, determine when the function is increasing or decreasing, and determine when relative maxima and minima occur. Do not sketch the graph.*

**9.** $y = x^2 + 2$.

**10.** $y = x^2 + 4x + 3$.

**11.** $y = x - x^2 + 2$.

**12.** $y = 4x - x^2$.

**13.** $y = -\dfrac{x^3}{3} - 2x^2 + 5x - 2$.

**14.** $y = 4x^3 - 3x^4$.

**15.** $y = x^4 - 2x^2$.

**16.** $y = -2 + 12x - x^3$.

**17.** $y = x^3 - 6x^2 + 9x$.

**18.** $y = x^3 - 6x^2 + 12x - 6$.

**19.** $y = 2x^3 - \dfrac{11}{2}x^2 - 10x + 2$.

**20.** $y = -5x^3 + x^2 + x - 1$.

**21.** $y = x^3 + 2x^2 - x - 1$.

**22.** $y = \dfrac{9}{5}x^5 - \dfrac{47}{3}x^3 + 10x$.

**23.** $y = 3x^5 - 5x^3$.

**24.** $y = 5x - x^5$.

**25.** $y = -x^5 - 5x^4 + 200$.

**26.** $y = 3x^4 - 4x^3 + 1$.

**27.** $y = 8x^4 - x^8$.

**28.** $y = \dfrac{4}{5}x^5 - \dfrac{13}{3}x^3 + 3x + 4$.

**29.** $y = (x^3 + 1)^3$.

**30.** $y = \sqrt[3]{x}(x - 2)$.

**31.** $y = \dfrac{1}{x - 1}$.

**32.** $y = \dfrac{3}{x}$.

**33.** $y = \dfrac{10}{\sqrt{x}}$.

**34.** $y = \dfrac{x}{x + 1}$.

**35.** $y = \dfrac{x^2}{1 - x}$.

**36.** $y = x + \dfrac{4}{x}$.

**37.** $y = \dfrac{x^2 - 3}{x + 2}$.

**38.** $y = \dfrac{x^2}{x^2 - 9}$.

**39.** $y = \dfrac{5x + 3}{x^2 + 1}$.

**40.** $y = \sqrt[3]{x^3 - 9x}$.

**41.** $y = (x + 2)^3(x - 5)^2$.

**42.** $y = x^2(x + 3)^4$.

**43.** $y = x^3(x - 4)^4$.

**44.** $y = x(1 - x)^{2/5}$.

**45.** $y = e^{-2x}$.

**46.** $y = x \ln x$.

**47.** $y - x^2 - 2 \ln x$.

**48.** $y = xe^x$.

**49.** $y = e^x + e^{-x}$.

**50.** $y = e^{-x^2}$.

**51.** $y = x \ln x - x$.

**52.** $y = (x^2 + 1)e^{-x}$.

*In Problems 53–64, determine intervals on which the function is increasing or decreasing, relative maxima and minima, symmetry, and those intercepts that can be obtained conveniently. Then sketch the graph.*

**53.** $y = x^2 - 6x - 7$.

**54.** $y = 2x^2 - 5x - 12$.

**55.** $y = 3x - x^3$.

**56.** $y = x^4 - 16$.

**57.** $y = 2x^3 - 9x^2 + 12x$.

**58.** $y = x^3 - 9x^2 + 24x - 19$.

**59.** $y = x^4 + 4x^3 + 4x^2$.

**60.** $y = x^5 - \frac{5}{4}x^4$.

**61.** $y = (x - 1)^2(x + 2)^2$.

**62.** $y = (3 - x)\sqrt{x}$.

**63.** $y = 2\sqrt{x} - x$.

**64.** $y = x^{5/3} + 5x^{2/3}$.

**65.** Sketch the graph of a continuous function $f$ such that $f(1) = 2$, $f(3) = 1$, $f'(1) = f'(3) = 0$, $f'(x) > 0$ for $x < 1$, $f'(x) < 0$ for $1 < x < 3$, and $f$ has a relative minimum when $x = 3$.

**66.** Sketch the graph of a continuous function $f$ such that $f(1) = 2$, $f(4) = 5$, $f'(1) = 0$, $f'(x) \geq 0$ for $x < 4$, $f$ has a relative maximum when $x = 4$, and there is a vertical tangent line when $x = 4$.

**67. Average Cost** If $c_f = 25{,}000$ is a fixed-cost function,

show that the average fixed-cost function $\overline{c}f = c_f/q$ is a decreasing function for $q > 0$. Thus, as output $q$ increases, each unit's portion of fixed cost declines.

**68. Marginal Cost** If $c = 4q - q^2 + 2q^3$ is a cost function, when is marginal cost increasing?

**69. Marginal Revenue** Given the demand function

$$p = 400 - 2q,$$

find when marginal revenue is increasing.

70. **Cost Function** For the cost function $c = \sqrt{q}$, show that marginal and average costs are always decreasing for $q > 0$.

71. **Revenue** For a manufacturer's product, the revenue function is given by $r = 240q + 57q^2 - q^3$. Determine the output for maximum revenue.

72. **Labor Markets** Eswaran and Kotwal[1] consider agrarian economies in which there are two types of workers, permanent and casual. Permanent workers are employed on long-term contracts and may receive benefits such as holiday gifts and emergency aid. Casual workers are hired on a daily basis and perform routine and menial tasks such as weeding, harvesting, and threshing. The difference $z$ in the present-value cost of hiring a permanent worker over that of hiring a casual worker is given by

$$z = (1 + b)w_p - bw_c,$$

where $w_p$ and $w_c$ are wage rates for permanent labor and casual labor, respectively, $b$ is a positive constant, and $w_p$ is a function of $w_c$.
(a) Show that

$$\frac{dz}{dw_c} = (1 + b)\left[\frac{dw_p}{dw_c} - \frac{b}{1 + b}\right].$$

(b) If $dw_p/dw_c < b/(1 + b)$, show that $z$ is a decreasing function of $w_c$.

73. **Thermal Pollution** In Shonle's discussion of thermal pollution,[2] the efficiency of a power plant is given by

$$E = 0.71\left(1 - \frac{T_c}{T_h}\right),$$

where $T_h$ and $T_c$ are the respective absolute temperatures of the hotter and colder reservoirs. Assume that $T_c$ is a positive constant and that $T_h$ is positive. Using calculus, show that as $T_h$ increases, the efficiency increases.

74. **Telephone Service** In a discussion of the pricing of local telephone service, Renshaw[3] determines that total revenue $r$ is given by

$$r = 2F + \left(1 - \frac{a}{b}\right)p - p^2 + \frac{a^2}{b},$$

where $p$ is an indexed price per call, and $a$, $b$, and $F$ are constants. Determine the value of $p$ that maximizes revenue.

75. **Storage and Shipping Costs** In his model for storage and shipping costs of materials for a manufacturing process, Lancaster[4] derives the cost function

$$C(k) = 100\left(100 + 9k + \frac{144}{k}\right), \quad 1 \le k \le 100,$$

where $C(k)$ is the total cost (in dollars) of storage and transportation for 100 days of operation if a load of $k$ tons of material is moved every $k$ days. (a) Find $C(1)$. (b) For what value of $k$ does $C(k)$ have a minimum? (c) What is the minimum value?

76. **Physiology—The Bends** When a deep-sea diver undergoes decompression or a pilot climbs to a high altitude, nitrogen may bubble out of the blood, causing what is commonly called *the bends*. Suppose the percentage $P$ of people who suffer effects of the bends at an altitude of $h$ thousand feet is given by[5]

$$P = \frac{100}{1 + 100,000e^{-0.36h}}.$$

Is $P$ an increasing function of $h$?

*In Problems 77–80, from the graph of the function, find the coordinates of all relative extrema. Round your answers to two decimal places.*

77. $y = 0.5x^2 + 4.1x + 5.$    78. $y = 2x^4 - 3x^3 - 4x + 5.$    79. $y = \dfrac{5.2x}{0.4x^2 + 3}.$    80. $y = \dfrac{e^x(5 - x)}{7x^2 + 1}.$

81. Graph the function

$$f(x) = [x(x - 2)(2x - 3)]^2$$

in the window $-1 \le x \le 3$, $-1 \le y \le 3$. Upon first glance, it may appear that this function has two relative minimum points and one relative maximum point. However, in reality, it has three relative minimum points and two relative maximum points. Determine the $x$-values of all these points. Round answers to two decimal places.

82. If $f(x) = 3x^3 - 7x^2 + 4x + 2$, display the graphs of $f$ and $f'$ on the same screen. Notice that $f'(x) = 0$ where relative extrema of $f$ occur.

83. Let $f(x) = 6 + 4x - 3x^2 - x^3$. (a) Find $f'(x)$. (b) Graph $f'(x)$. (c) Observe where $f'(x)$ is positive and where it is negative. Give the intervals (rounded to two decimal places) where $f$ is increasing and where $f$ is decreasing. (d) Graph $f$ and $f'$ on the same screen, and verify your results to part (c).

84. If $f(x) = x^4 - 3x^2 - (2x - 1)^2$, find $f'(x)$. Determine the critical values of $f$. Round your answers to two decimal places.

[1]M. Eswaran and A. Kotwal, "A Theory of Two-Tier Labor Markets in Agrarian Economics," *The American Economic Review*, 75, no. 1 (1985), 162-77.

[2]J. I. Shonle, *Environmental Applications of General Physics* (Reading, MA: Addition-Wesley Publishing Company, Inc., 1975).

[3]E. Renshaw, "A Note of Equity and Efficiency in the Pricing of Local Telephone Services," *The American Economic Review*, 75, no. 3 (1985), 515-18.

[4]P. Lancaster, *Mathematics: Models of the Real World* (Englewood Cliffs, NJ: Prentice-Hall, Inc., 1976).

[5]Adapted from G. E. Folk, Jr., *Textbook of Environmental Physiology*. 2nd ed. (Philadlephia: Lea & Febiger, 1974).

**To find extreme values on a closed interval.**

# 14.2 ABSOLUTE EXTREMA ON A CLOSED INTERVAL

If a function $f$ is *continuous* on a *closed* interval $[a, b]$, it can be shown that of *all* the function values $f(x)$ for $x$ in $[a, b]$, there must be an (absolute) maximum value and an (absolute) minimum value. These two values are called **extreme values** of $f$ on that interval. This important property of continuous functions is called the *extreme-value theorem*.

> **Extreme-Value Theorem**
>
> If a function is continuous on a closed interval, then the function has *both* a maximum value and a minimum value on that interval.

For example, each function in Fig. 14.25 is continuous on the closed interval $[1, 3]$. Geometrically, the extreme-value theorem assures us that over this interval each graph has a highest point and a lowest point.

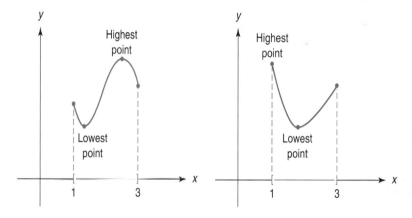

**FIGURE 14.25**   Illustrating the extreme-value theorem.

In the extreme-value theorem, you must realize that we are dealing with

**1.** a closed interval

and

**2.** a function continuous on that interval.

If either condition (1) or condition (2) is not met, then extreme values are not guaranteed. For example, Fig. 14.26(a) shows the graph of the continuous function $f(x) = x^2$ on the *open* interval $(-1, 1)$. You can see that $f$ has no maximum value on the interval (although $f$ has a minimum value there). Now consider the function $f(x) = 1/x^2$ on the closed interval $[-1, 1]$. Here $f$ is *not continuous* at 0. From the graph of $f$ in Fig. 14.26(b), you can see that $f$ has no maximum value (although there is a minimum value).

In the previous section, our emphasis was on relative extrema. Now we shall focus our attention on absolute extrema and make use of the extreme-value theorem where possible. If the domain of a function is a closed interval, to determine *absolute* extrema we must examine the function not only at critical values, but also at the endpoints. For example, Fig. 14.27 shows the graph of the continuous function $y = f(x)$ over $[a, b]$. The extreme-value theorem guarantees absolute extrema over the interval. Clearly, the important points on the graph occur at $x = a, b, c,$ and $d$, which correspond to endpoints or

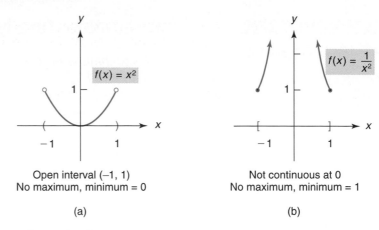

Open interval $(-1, 1)$
No maximum, minimum $= 0$

(a)

Not continuous at 0
No maximum, minimum $= 1$

(b)

**FIGURE 14.26** Extreme-value theorem does not apply.

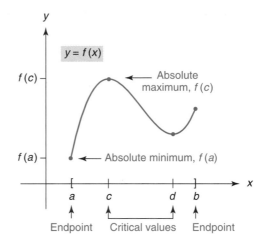

**FIGURE 14.27** Absolute extrema.

critical values. Notice that the absolute maximum occurs at the critical value $c$ and the absolute minimum occurs at the endpoint $a$. These results suggest the following procedure:

---

**Procedure to Find Absolute Extrema for a Function $f$ That Is Continuous on $[a, b]$**

**Step 1.** Find the critical values of $f$.

**Step 2.** Evaluate $f(x)$ at the endpoints $a$ and $b$ and at the critical values on $(a, b)$.

**Step 3.** The maximum value of $f$ is the greatest of the values found in step 2. The minimum value of $f$ is the least of the values found in step 2.

---

**EXAMPLE 1    Finding Extreme Values on a Closed Interval**

*Find absolute extrema for $f(x) = x^2 - 4x + 5$ over the closed interval* $[1, 4]$.

*Solution:* Since $f$ is continuous on $[1, 4]$, the foregoing procedure applies.

**Step 1.** To find the critical values of $f$, we first find $f'$:

$$f'(x) = 2x - 4 = 2(x - 2).$$

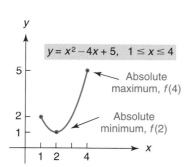

**FIGURE 14.28**  Extreme values for Example 1.

This gives the critical value $x = 2$.

**Step 2.** Evaluating $f(x)$ at the endpoints 1 and 4 and at the critical value 2, we have

$$f(1) = 2,$$
$$f(4) = 5,$$

values of $f$ at endpoints

and

$$f(2) = 1,$$    value of $f$ at critical value in $(1, 4)$.

**Step 3.** From the function values in step 2, we conclude that the maximum is $f(4) = 5$ and the minimum is $f(2) = 1$. (See Fig. 14.28.)  ∎

## ▪ Exercise 14.2

*In Problems* **1–14,** *find the absolute extrema of the given function on the given interval.*

**1.** $f(x) = x^2 - 2x + 3, \quad [-1, 2]$.

**2.** $f(x) = -2x^2 - 6x + 5, \quad [-2, 3]$.

**3.** $f(x) = \frac{1}{3}x^3 - x^2 - 3x + 1, \quad [0, 2]$.

**4.** $f(x) = \frac{1}{4}x^4 - \frac{3}{2}x^2, \quad [0, 1]$.

**5.** $f(x) = 4x^3 + 3x^2 - 18x + 3, \quad [\frac{1}{2}, 3]$.

**6.** $f(x) = x^{4/3}, \quad [-8, 8]$.

**7.** $f(x) = -3x^5 + 5x^3, \quad [-2, 0]$.

**8.** $f(x) = \frac{7}{3}x^3 + 2x^2 - 3x + 1, \quad [0, 3]$.

**9.** $f(x) = 3x^4 - x^6, \quad [-1, 2]$.

**10.** $f(x) = 2 + \frac{1}{3}x^3 - \frac{3}{2}x^4, \quad [-1, 1]$.

**11.** $f(x) = x^4 - 9x^2 + 2, \quad [-1, 3]$.

**12.** $f(x) = \dfrac{x}{x^2 + 1}, \quad [0, 2]$.

 **13.** $f(x) = x^{2/3}, \quad [-2, 3]$.

**14.** $f(x) = 0.3x^3 - 4.2x + 5, \quad [-1, 4]$.

**15.** Consider the function

$$f(x) = x^4 + 8x^3 + 21x^2 + 20x + 9$$

over the interval $[-4, 9]$.

**a.** Determine the value(s) (rounded to two decimal places) of $x$ at which $f$ attains a minimum value.

**b.** What is the minimum value (rounded to two decimal places) of $f$?

**c.** Determine the value(s) of $x$ at which $f$ attains a maximum value.

**d.** What is the maximum value of $f$?

---

### OBJECTIVE

To test a function for concavity and inflection points. Also, to sketch curves with the aid of the information obtained from the first and second derivatives.

## 14.3 CONCAVITY

You have seen that the first derivative provides much information for sketching curves. It is used to determine when a function is increasing or decreasing and to locate relative maxima and minima. However, to be sure we know the true shape of a curve, we may need more information. For example, consider the curve $y = f(x) = x^2$. Since $f'(x) = 2x$, $x = 0$ is a critical value. If $x < 0$, then $f'(x) < 0$, and $f$ is decreasing; if $x > 0$, then $f'(x) > 0$, and $f$ is increasing. Thus, there is a relative minimum when $x = 0$. In Figure 14.29, both curves meet the preceeding conditions. But which one truly describes the curve $y = x^2$? This question will easily be settled by using the second derivative and the notion of *concavity*.

In Fig. 14.30, note that each curve $y = f(x)$ "bends" (or opens) upward. This means that if tangent lines are drawn to each curve, the curves lie *above* them. Moreover, the slopes of the tangent lines *increase* in value as $x$ increases: In part (a), the slopes go from small positive values to larger values; in part (b), they are negative and approaching zero (and thus increasing); in part (c), they pass from negative values to positive values. Since $f'(x)$ gives the slope at a

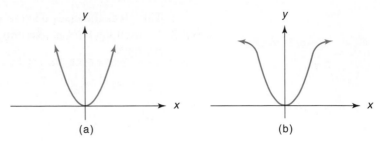

**FIGURE 14.29**   Two functions with $f'(x) < 0$ for $x < 0$ and $f'(x) > 0$ for $x > 0$.

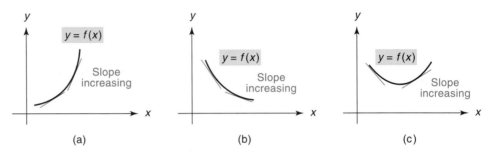

**FIGURE 14.30**   Each curve is concave up.

point, an increasing slope means that $f'$ must be an increasing function. To describe this property, each curve (or function $f$) is said to be *concave up*.

In Fig. 14.31, it can be seen that each curve lies *below* the tangent lines and the curves are bending downward. As $x$ increases, the slopes of the tangent lines are *decreasing*. Thus, $f'$ must be a decreasing function here, and we say that $f$ is *concave down*.

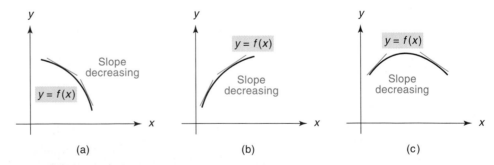

**FIGURE 14.31**   Each curve is concave down.

### DEFINITION

*Let f be differentiable on the interval $(a, b)$. Then f is said to be **concave up** [**concave down**] on $(a, b)$ if f' is increasing [decreasing] on $(a, b)$.*

**Pitfall** ▼   Concavity relates to whether $f'$, not $f$, is increasing or decreasing. In Fig. 14.30(b), note that $f$ is concave up and decreasing; however, in Fig. 14.31(a), $f$ is concave down and decreasing.

*Remember:* If $f$ is concave up on an interval, then, geometrically, its graph is bending upward there. If $f$ is concave down, then its graph is bending downward.

Since $f'$ is increasing when its derivative $f''(x)$ is positive, and $f'$ is decreasing when $f''(x)$ is negative, we can state the following rule:

**Rule 4   Criteria for Concavity**

*Let $f'$ be differentiable on the interval $(a, b)$. If $f''(x) > 0$ for all $x$ in $(a, b)$, then $f$ is concave up on $(a, b)$. If $f''(x) < 0$ for all $x$ in $(a, b)$, then $f$ is concave down on $(a, b)$.*

A function $f$ is also said to be concave up at a point $x_0$ if there exists an open interval around $x_0$ on which $f$ is concave up. In fact, for the functions that we shall consider, if $f''(x_0) > 0$, then $f$ is concave up at $x_0$. Similarly, $f$ is concave down at $x_0$ if $f''(x_0) < 0$.

**EXAMPLE 1   Testing for Concavity**

*Determine where the given function is concave up and where it is concave down.*

**a.** $y = f(x) = (x - 1)^3 + 1$.

*Solution:* To apply Rule 4, we must examine the signs of $y''$. Now, $y' = 3(x - 1)^2$, so

$$y'' = 6(x - 1).$$

Thus, $f$ is concave up when $6(x - 1) > 0$, that is, when $x > 1$. And $f$ is concave down when $6(x - 1) < 0$, that is, when $x < 1$. (See Fig. 14.32.)

**b.** $y = x^2$.

*Solution:* We have $y' = 2x$ and $y'' = 2$. Because $y''$ is always positive, the graph of $y = x^2$ must always be concave up, as in Fig. 14.29(a). The graph cannot appear as in Fig. 14.29(b), for that curve is sometimes concave down.

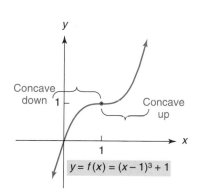

**FIGURE 14.32**  Concavity for $f(x) = (x - 1)^3 + 1$.

A point on a graph where concavity changes from concave down to concave up, or vice versa, such as $(1, 1)$ in Fig. 14.32, is called an *inflection point* or a *point of inflection*. Around such a point, the sign of $f''(x)$ must go from $-$ to $+$ or from $+$ to $-$. More precisely, we have the following definition:

The definition of an inflection point implies that $x_0$ is in the domain of $f$.

**DEFINITION**

*A function $f$ has an **inflection point** when $x = x_0$ if and only if $f$ is continuous at $x_0$ and $f$ changes concavity at $x_0$.*

To test a function for concavity and inflection points, first find the values of $x$ where $f''(x)$ is 0 or not defined. These values of $x$ determine intervals. On each interval, determine whether $f''(x) > 0$ ($f$ is concave up) or $f''(x) < 0$ ($f$ is concave down). If concavity changes around one of these $x$-values and $f$ is continuous there, then $f$ has an inflection point at this $x$-value. The continuity requirement implies that the $x$-value must be in the domain of the function. In brief, a *candidate* for an inflection point must satisfy two conditions:

**1.** $f''$ must be 0 or not defined at that point.

**2.** $f$ must be continuous at that point.

The candidate *will be* an inflection point if concavity changes around it. For example, if $f(x) = x^{1/3}$, then $f'(x) = \frac{1}{3}x^{-2/3}$ and

$$f''(x) = -\frac{2}{9}x^{-5/3} = -\frac{2}{9x^{5/3}}.$$

Because $f''$ is not defined at 0, but $f$ is continuous at 0, there is a candidate for an inflection point when $x = 0$. If $x > 0$, then $f''(x) < 0$, so $f$ is concave down for $x > 0$; if $x < 0$, then $f''(x) > 0$, so $f$ is concave up for $x < 0$. Because concavity changes at $x = 0$, there is an inflection point there. (See Fig. 14.33.)

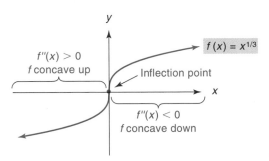

**FIGURE 14.33** Inflection point for $f(x) = x^{1/3}$.

### EXAMPLE 2 Concavity and Inflection Points

*Test $y = 6x^4 - 8x^3 + 1$ for concavity and inflection points.*

*Solution:* We have

$$y' = 24x^3 - 24x^2,$$
$$y'' = 72x^2 - 48x = 24x(3x - 2).$$

To find where $y'' = 0$, we set each factor in $y''$ equal to 0. This gives $x = 0, \frac{2}{3}$. We also note that $y''$ is never undefined. Thus, there are three intervals to consider. (See Fig. 14.34.) Since $y$ is continuous at 0 and $\frac{2}{3}$, these points are candidates for inflection points.

If $x < 0$, then $y'' = 24(-)(-) = +$, so the curve is concave up;

if $0 < x < \frac{2}{3}$, then $y'' = 24(+)(-) = -$, so the curve is concave down;

if $x > \frac{2}{3}$, then $y'' = 24(+)(+) = +$, so the curve is concave up. (See Fig. 14.35.)

Since concavity changes at the points where $x = 0$ and $\frac{2}{3}$, these candidates are indeed inflection points. (See Fig. 14.36.) In summary, the curve is concave up on $(-\infty, 0)$ and $(\frac{2}{3}, \infty)$, and is concave down on $(0, \frac{2}{3})$. Inflection points occur when $x = 0$ or $x = \frac{2}{3}$. These points are $(0, 1)$ and $(\frac{2}{3}, -\frac{5}{27})$.  ∎

**FIGURE 14.34** Intervals to consider for concavity of $y = 6x^4 - 8x^3 + 1$.

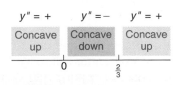

**FIGURE 14.35** Sign chart of $y'' = 24x(3x - 2)$.

### EXAMPLE 3 A Change in Concavity with No Inflection Point

*Discuss concavity and find all inflection points for $f(x) = \frac{1}{x}$.*

*Solution:* Since $f(x) = x^{-1}$,

$$f'(x) = -x^{-2},$$
$$f''(x) = 2x^{-3} = \frac{2}{x^3}.$$

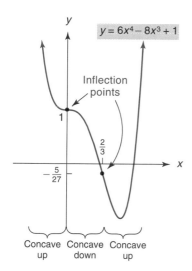

**FIGURE 14.36** Graph of $y = 6x^4 - 8x^3 + 1$.

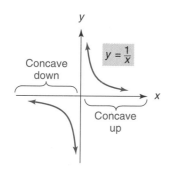

**FIGURE 14.38** Graph of $y = \frac{1}{x}$.

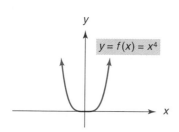

**FIGURE 14.39** Graph of $f(x) = x^4$.

We see that $f''(x)$ is never 0, but it is not defined when $x = 0$. Since $f$ is not continuous at 0, we conclude that 0 is not a candidate for an inflection point. Thus, the given function has no inflection point. However, 0 must be considered in an analysis of concavity. (See the number line in Fig. 14.37; note that we have boxed the value 0 to indicate that it cannot correspond to an inflection point.) If $x > 0$, then $f''(x) > 0$; if $x < 0$, then $f''(x) < 0$. Hence, $f$ is

$$\boxed{0}$$

**FIGURE 14.37** Intervals to consider for concavity analysis.

concave up on $(0, \infty)$ and concave down on $(-\infty, 0)$. (See Fig. 14.38.) Although concavity changes around $x = 0$, there is no inflection point there because $f$ is not continuous at 0 (nor is it even defined there). ∎

*Pitfall* ▼ A candidate for an inflection point may not necessarily be an inflection point. For example, if $f(x) = x^4$, then $f''(x) = 12x^2$ and $f''(0) = 0$. But $x < 0$ implies that $f''(x) > 0$, and $x > 0$ also implies that $f''(x) > 0$. Thus, concavity does not change, and there are no inflection points. (See Fig. 14.39.)

**Curve Sketching**

**EXAMPLE 4   Curve Sketching**

*Sketch the graph of $y = 2x^3 - 9x^2 + 12x$.*

*Solution:*

*Intercepts* If $x = 0$, then $y = 0$. Setting $y = 0$ gives $0 = x(2x^2 - 9x + 12)$. Clearly, $x = 0$, and using the quadratic formula on $2x^2 - 9x + 12 = 0$ gives no real roots. Thus, the only intercept is $(0, 0)$.

*Symmetry* None.

*Maxima and Minima* Letting $y = f(x)$, we have

$$f'(x) = 6x^2 - 18x + 12 = 6(x^2 - 3x + 2) = 6(x - 1)(x - 2).$$

The critical values are $x = 1, 2$. (See Fig. 14.40.)

If $x < 1$, then $f'(x) = 6(-)(-) = \ +\ $, so $f$ is increasing;

if $1 < x < 2$, then $f'(x) = 6(+)(-) = -$, so $f$ is decreasing;

if $x > 2$, then $f'(x) = 6(+)(+) = +$, so $f$ is increasing. (See Fig. 14.41.)

**FIGURE 14.40** Critical values of $y = 2x^3 - 9x^2 + 12x$.

**FIGURE 14.41** Sign chart of $f'(x)$.

There is a relative maximum when $x = 1$ and a relative minimum when $x = 2$.

*Concavity*   $f''(x) = 12x - 18 = 6(2x - 3)$.

Setting $f''(x) = 0$ gives a possible inflection point at $x = \frac{3}{2}$.

If $x < \frac{3}{2}$, then $f''(x) < 0$, so $f$ is concave down;

if $x > \frac{3}{2}$, then $f''(x) > 0$, so $f$ is concave up. (See Fig. 14.42.)

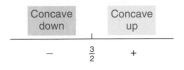

**FIGURE 14.42** Sign chart of $f''(x)$.

Since concavity changes around $x = \frac{3}{2}$ and $f$ is continuous there, $f$ has an inflection point when $x = \frac{3}{2}$.

***Discussion*** We now find the coordinates of the important points on the graph (and any other points if there is doubt as to the behavior of the curve). We have the following table:

| $x$ | 0 | 1 | $\frac{3}{2}$ | 2 |
|---|---|---|---|---|
| $y$ | 0 | 5 | $\frac{9}{2}$ | 4 |

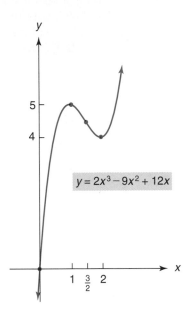

**FIGURE 14.43** Graph of $y = 2x^3 - 9x^2 + 12x$.

As $x$ increases, the function is first concave down and increases to a relative maximum at $(1, 5)$; it then decreases to $(\frac{3}{2}, \frac{9}{2})$, after which it becomes concave up, but continues to decrease until it reaches a relative minimum at $(2, 4)$; thereafter, it increases and is still concave up. (See Fig. 14.43.) ∎

## TECHNOLOGY

Suppose that you need to find the inflection points for

$$f(x) = \frac{1}{20}x^5 - \frac{17}{16}x^4 + \frac{273}{32}x^3 - \frac{4225}{128}x^2 + \frac{750}{4}.$$

The second derivative of $f$ is given by

$$f''(x) = x^3 - \frac{51}{4}x^2 + \frac{819}{16}x - \frac{4225}{64}.$$

Here the zeros of $f''$ are not obvious. Thus, we shall graph $f''$ using a graphics calculator. (See Fig. 14.44.) We find that the zeros of $f''$ are approximately 3.25 and 6.25. Around $x = 6.25$, $f''(x)$ goes from negative to positive values. Therefore, at $x = 6.25$, there is an inflection point. Around $x = 3.25$, $f''(x)$ does not change sign, so no inflection point exists at $x = 3.25$. Comparing our results with the graph of $f$ in Fig. 14.45, we see that everything checks out.

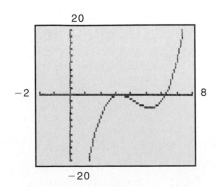

**FIGURE 14.44** Graph of $f''$; zeros of $f''$ are 3.25 and 6.25.

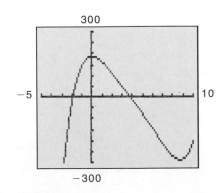

**FIGURE 14.45** Graph of $f$; inflection point at $x = 6.25$, but not at $x = 3.25$.

## ▪ Exercise 14.3

*In Problems 1–6, a function and its second derivative are given. Determine the concavity of f and x-values where points of inflection occur.*

**1.** $f(x) = x^4 - 3x^3 + 7x - 5$;  $f''(x) = 6x(2x - 3)$.

**2.** $f(x) = \dfrac{x^5}{20} + \dfrac{x^4}{4} - 2x^2$;  $f''(x) = (x - 1)(x + 2)^2$.

**3.** $f(x) = \dfrac{2 + x - x^2}{x^2 - 2x + 1}$;  $f''(x) = \dfrac{2(7 - x)}{(x - 1)^4}$.

**4.** $f(x) = -\dfrac{x^2}{(2 - x)^2}$;  $f''(x) = -\dfrac{8(x + 1)}{(2 - x)^4}$.

**5.** $f(x) = \dfrac{x^2 + 1}{x^2 - 2}$;  $f''(x) = \dfrac{6(3x^2 + 2)}{(x^2 - 2)^3}$.

**6.** $f(x) = x\sqrt{4 - x^2}$;  $f''(x) = \dfrac{2x(x^2 - 6)}{(4 - x^2)^{3/2}}$.

*In Problems 7–34, determine concavity and the x-values where points of inflection occur. Do not sketch the graphs.*

**7.** $y = -2x^2 + 4x$.

**8.** $y = 3x^2 - 6x + 5$.

**9.** $y = 4x^3 + 12x^2 - 12x$.

**10.** $y = x^3 - 6x^2 + 9x + 1$.

**11.** $y = 4x^3 - 21x^2 + 5x$.

**12.** $y = x^4 - 8x^2 - 6$.

**13.** $y = x^4 - 6x^2 + 5x - 6$.

**14.** $y = -\dfrac{x^4}{4} + \dfrac{9x^2}{2} + 2x$.

**15.** $y = x^{1/5}$.

**16.** $y = \dfrac{1}{x^5}$.

**17.** $y = \dfrac{x^4}{2} + \dfrac{19x^3}{6} - \dfrac{7x^2}{2} + x + 5$.

**18.** $y = -\dfrac{5}{2}x^4 - \dfrac{1}{6}x^3 + \dfrac{1}{2}x^2 + \dfrac{1}{3}x - \dfrac{2}{5}$

**19.** $y = \dfrac{1}{20}x^5 - \dfrac{1}{4}x^4 + \dfrac{1}{6}x^3 - \dfrac{1}{2}x - \dfrac{2}{3}$.

**20.** $y = \dfrac{9}{5}x^5 - \dfrac{32}{3}x^3 + 10x - 2$.

**21.** $y = \dfrac{1}{30}x^6 - \dfrac{7}{12}x^4 + 5x^2 + 2x - 1$

**22.** $y = x^6 - 3x^4$.

**23.** $y = \dfrac{x + 1}{x - 1}$.

**24.** $y = x + \dfrac{1}{x}$.

**25.** $y = \dfrac{x^2}{x^2 + 1}$.

**26.** $y = \dfrac{x^2}{x + 3}$.

**27.** $y = \dfrac{21x + 40}{6(x + 3)^2}$.

**28.** $y = (x^2 - 4)^2$.

**29.** $y = e^x$.

**30.** $y = e^x - e^{-x}$.

**31.** $y = xe^x$.

**32.** $y = xe^{-x}$.

**33.** $y = \dfrac{\ln x}{x}$.

**34.** $y = \dfrac{x^2 + 1}{e^x}$.

*In Problems 35–62, sketch each curve. Then determine intervals on which the function is increasing, decreasing, concave up, and concave down; relative maxima and minima; inflection points; symmetry; and those intercepts that can be obtained conveniently.*

**35.** $y = x^2 + 4x + 3$.

**36.** $y = x^2 + 2$.

**37.** $y = 4x - x^2$.

**38.** $y = x - x^2 + 2$.

**39.** $y = x^3 - 9x^2 + 24x - 19$.

**40.** $y = 3x - x^3$.

**41.** $y = \dfrac{x^3}{3} - 4x$.

**42.** $y = x^3 - 6x^2 + 9x$.

**43.** $y = x^3 - 3x^2 + 3x - 3$.

**44.** $y = 2x^3 - 9x^2 + 12x$.

**45.** $y = 4x^3 - 3x^4$.

**46.** $y = -\dfrac{x^3}{3} - 2x^2 + 5x - 2$.

**47.** $y = -2 + 12x - x^3$.

**48.** $y = (3 + 2x)^3$.

**49.** $y = x^3 - 6x^2 + 12x - 6$.

**50.** $y = \dfrac{x^5}{100} - \dfrac{x^4}{20}$.

**51.** $y = 5x - x^5$.

**52.** $y = x(1 - x)^3$.

**53.** $y = 3x^4 - 4x^3 + 1$.

**54.** $y = 3x^5 - 5x^3$.

**55.** $y = 4x^2 - x^4$.

**56.** $y = x^4 - 2x^2$.

**57.** $y = x^{1/3}(x - 8)$.

**58.** $y = (x - 1)^2(x + 2)^2$.

**59.** $y = 4x^{1/3} + x^{4/3}$.

**60.** $y = 2x\sqrt{x + 3}$.

**61.** $y = 2x + 3x^{2/3}$.

**62.** $y = 5x^{2/3} - x^{5/3}$.

**63.** Sketch the graph of a continuous function $f$ such that $f(2) = 4, f'(2) = 0, f'(x) < 0$ if $x < 2$, and $f''(x) > 0$ if $x > 2$.

**64.** Sketch the graph of a continuous function $f$ such that $f(3) = 2, f'(3) = 0, f''(x) > 0$ for $x < 3$, and $f''(x) < 0$ for $x > 3$.

**65.** Sketch the graph of a continuous function $f$ such that $f(1) = 1, f'(1) = 0$, and $f''(x) < 0$ for all $x$.

**66.** Sketch the graph of a continuous function $f$ such that $f(3) = 4$, both $f'(x) > 0$ and $f''(x) > 0$ for $x < 3$, and both $f'(x) < 0$ and $f''(x) > 0$ for $x > 3$.

**67. Demand Equation** Show that the graph of the demand equation $p = \dfrac{100}{q+2}$ is decreasing and concave up for $q > 0$.

**68. Average Cost** For the cost function

$$c = 3q^2 + 5q + 6,$$

show that the graph of the average-cost function $\bar{c}$ is always concave up for $q > 0$.

**69. Species of Plants** The number of species of plants on a plot may depend on the size of the plot. For example, in Fig. 14.46, we see that on 1-m² plots there are three species (A, B, and C on the left plot, A, B, and D on the right plot), and on a 2-m² plot there are four species (A, B, C, and D).

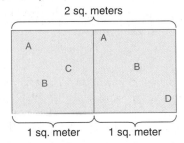

**FIGURE 14.46** Species of plants.

In a study of rooted plants in a certain geographic region,[6] it was determined that the average number of species, S, occuring on plots of size $A$ (in square meters) is given by

$$S = f(A) = 12\sqrt[4]{A}, \qquad 0 \le A \le 625.$$

Sketch the graph of $f$. (*Note:* Your graph should be rising and concave down. Thus, the number of species is increasing with respect to area, but at a decreasing rate.)

**70. Inferior Good** In a discussion of an inferior good, Persky[7] considers a function of the form

$$g(x) = e^{(U_0/A)}e^{-x^2/(2A)},$$

where $x$ is a quantity of a good, $U_0$ is a constant that represents utility, and $A$ is a positive constant. Persky claims that the graph of $g$ is concave down for $x < \sqrt{A}$ and concave up for $x > \sqrt{A}$. Verify this.

**71. Psychology** In a psychological experiment involving conditioned response,[8] subjects listened to four tones, denoted 0, 1, 2, and 3. Initially, the subjects were conditioned to tone 0 by receiving a shock whenever this tone was heard. Later, when each of the four tones (stimuli)

were heard without shocks, the subjects' responses were recorded by means of a tracking device that measures galvanic skin reaction. The average response to each stimulus (without shock) was determined, and the results were plotted on a coordinate plane where the $x$- and $y$-axes represent the stimuli (0, 1, 2, 3) and the average galvanic responses, respectively. It was determined that the points fit a curve that is approximated by the graph of

$$y = 12.5 + 5.8(0.42)^x.$$

Show that this function is decreasing and concave up.

**72. Entomology** In a study of the effects of food deprivation on hunger,[9] an insect was fed until its appetite was completely satisfied. Then it was deprived of food for $t$ hours (the deprivation period). At the end of this period, the insect was refed until its appetite was again completely satisfied. The weight $H$ (in grams) of the food that was consumed at this time was statistically found to be a function of $t$, where

$$H = 1.00[1 - e^{-(0.0464t + 0.0670)}].$$

Here $H$ is a measure of hunger. Show that $H$ is increasing with respect to $t$ and is concave down.

**73. Insect Dispersal** In an experiment on the dispersal of a particular insect,[10] a large number of insects are placed at a release point in an open field. Surrounding this point are traps that are placed in a concentric circular arrangement at a distance of 1m, 2m, 3m, and so on, from the release point. Twenty-four hours after the insects are released, the number of insects in each trap is counted. It is determined that a distance of $r$ meters from the release point, the average number of insects contained in a trap is

$$n = f(r) = 0.1 \ln(r) + \frac{7}{r} - 0.8, \qquad 1 \le r \le 10.$$

(a) Show that the graph of $f$ is always falling and concave up. (b) Sketch the graph of $f$. (c) When $r = 5$, at what rate is the average number of insects in a trap decreasing with respect to distance?

**74.** Graph $y = 0.25x^3 - 3.1x^2 + 9.9x - 6.1$, and from the graph, determine the number of (a) relative maximum points, (b) relative minimum points, and (c) inflection points.

**75.** Graph $y = x^5(x - 2.3)$, and from the graph, determine the number of inflection points.

**76.** Graph $y = 1 - 2^{-x^2}$, and from the graph, determine the number of inflection points.

[6]Adapted from R.W. Poole, *An Introduction to Quantitative Ecology* (New York: McGraw-Hill Book Company, 1974).

[7]A. L. Persky, "An Inferior Good and a Novel Indifference Map," *The American Economist* XXIX, no. 1 (1985), 67–69.

[8]Adapted from C. I. Hovland, "The Generalization of Conditioned Responses: I. The Sensory Generalization of Conditioned Responses with Varying Frequencies of Tone," *Journal of General Psychology*, 17 (1937), 125–48.

[9]C. S. Holling, "The Functional Response of Invertebrate Predators to Prey Density," *Memoirs of the Entomological Society of Canada*, no. 48 (1966).

[10] Adapted from Poole, op. cit.

**77.** Graph the curve $y = x^3 - 2x^2 + x + 3$, and also graph the tangent line to the curve at $x = 2$. Around $x = 2$, does the curve lie above or below the tangent line? From your observation determine the concavity at $x = 2$.

 **78.** If $f(x) = x^3 + 2x^2 - 3x + 2$, find $f'(x)$ and $f''(x)$. Note that where $f'$ has a relative minimum, $f$ changes its direction of bending. Why?

**79.** If $f(x) = x^6 + 3x^5 - 4x^4 + 2x^2 + 1$, find the $x$-values (rounded to two decimal places) of the inflection points of $f$.

**80.** If $f(x) = \dfrac{x + 3}{x^2 + 1}$, find the $x$-values (rounded to two decimal places) of the inflection points of $f$.

---

**O B J E C T I V E**

To locate relative extrema by applying the second-derivative test.

## 14.4 The Second-Derivative Test

The second derivative may be used to test certain critical values for relative extrema. Observe in Fig. 14.47 that when $x = x_0$, there is a horizontal tangent; that is, $f'(x_0) = 0$. Furthermore, around $x_0$, the function is concave up. [That is, $f''(x_0) > 0$.] This leads us to conclude that there is a relative minimum at

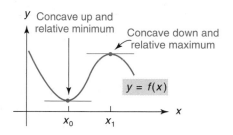

**FIGURE 14.47** Relating concavity to relative extrema.

$x_0$. On the other hand, around $x_1$, the function is concave down. [That is, $f''(x_1) < 0$.] Because the tangent line is horizontal at $x_1$, we conclude that a relative maximum exists there. This technique of examining the second derivative at points where the first derivative is 0 is called the *second-derivative test* for relative extrema.

> **Second-Derivative Test for Relative Extrema**
>
> Suppose $f'(x_0) = 0$.
>
> If $f''(x_0) < 0$, then $f$ has a relative maximum at $x_0$.
>
> If $f''(x_0) > 0$, then $f$ has a relative minimum at $x_0$.

We want to emphasize that **the second-derivative test does *not* apply when $f'(x_0) = 0$ and $f''(x_0) = 0$.** Under these conditions, there may be a relative maximum, a relative minimum, or neither at $x_0$. In such cases, the first-derivative test should be used to analyze what is happening at $x_0$. Also, the second-derivative test does not apply when $f'(x_0)$ is not defined.

**EXAMPLE 1** Second-Derivative Test

*Test the following for relative maxima and minima. Use the second-derivative test if possible.*

**a.** $y = 18x - \frac{2}{3}x^3$.

*Solution:*

$$y' = 18 - 2x^2 = 2(9 - x^2) = 2(3 + x)(3 - x),$$

$$y'' = -4x.$$

Solving $y' = 0$ gives the critical values $x = \pm 3$.

$$\text{If } x = 3, \text{ then } y'' = -4(3) = -12 < 0,$$

so there is a relative maximum when $x = 3$.

$$\text{If } x = -3, \text{ then } y'' = -4(-3) = 12 > 0,$$

so there is a relative minimum when $x = -3$. (Refer back to Fig. 14.4.)

Although the second-derivative test is very useful, do not depend entirely on it. Not only may the test not apply, but at times it may be awkward to find the second derivative.

**b.** $y = 6x^4 - 8x^3 + 1$.

*Solution:*

$$y' = 24x^3 - 24x^2 = 24x^2(x - 1),$$

$$y'' = 72x^2 - 48x.$$

Solving $y' = 0$ gives the critical values $x = 0, 1$. We see that

$$\text{if } x = 0, \text{ then } y'' = 0,$$

and

$$\text{if } x = 1, \text{ then } y'' > 0.$$

By the second-derivative test, there is a relative minimum when $x = 1$. We cannot apply the test when $x = 0$ because $y'' = 0$ there. To analyze what is happening at 0, we turn to the first-derivative test:

$$\text{If } x < 0, \text{ then } y' < 0;$$

$$\text{if } 0 < x < 1, \text{ then } y' < 0.$$

Thus, no maximum or minimum exists when $x = 0$. (Refer back to Fig. 14.36.) ∎

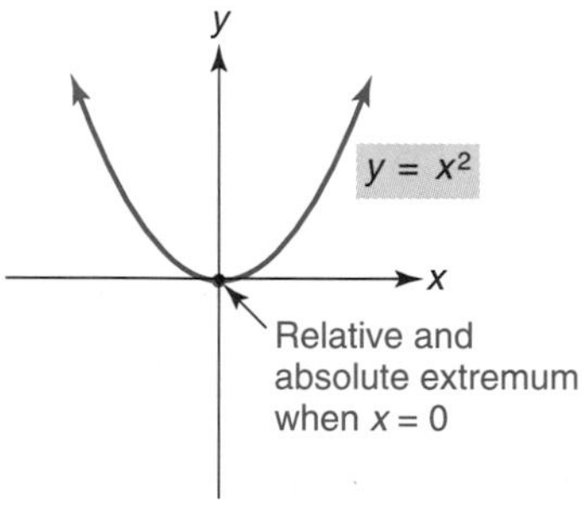

**FIGURE 14.48** Exactly one relative extremum implies an absolute extremum.

If a continuous function has *exactly one* relative extremum on an interval, it can be shown that the relative extremum must also be an *absolute* extremum on the interval. To illustrate, in Fig. 14.48 the function $y = x^2$ has a relative minimum when $x = 0$, and there are no other relative extrema. Since $y = x^2$ is continuous, this relative minimum is also an absolute minimum for the function.

### EXAMPLE 2  Absolute Extrema

*If $y = f(x) = x^3 - 3x^2 - 9x + 5$, determine when absolute extrema occur on the interval $(0, \infty)$.*

*Solution:* We have

$$f'(x) = 3x^2 - 6x - 9 = 3(x^2 - 2x - 3)$$

$$= 3(x + 1)(x - 3).$$

The only critical value on the interval $(0, \infty)$ is 3. Applying the second-derivative test at this point gives

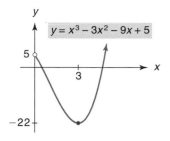

**FIGURE 14.49** On $(0, \infty)$, there is an absolute minimum when $x = 3$.

$$f''(x) = 6x - 6,$$
$$f''(3) = 6(3) - 6 = 12 > 0.$$

Thus, there is a relative minimum when $x = 3$. Since this is the only relative extremum on $(0, \infty)$ and $f$ is continuous there, we conclude by our previous discussion that there is an *absolute* minimum value when $x = 3$; this value is $f(3) = -22$. (See Fig. 14.49.) ∎

### ■ Exercise 14.4

*In Problems* **1–14,** *test for relative maxima and minima. Use the second-derivative test if possible. In Problems* **1-4,** *state whether the relative extrema are also absolute extrema.*

**1.** $y = x^2 - 5x + 6$.

**2.** $y = -2x^2 + 6x + 12$.

**3.** $y = -4x^2 + 2x - 8$.

**4.** $y = 3x^2 - 5x + 6$.

**5.** $y = x^3 - 27x + 1$.

**6.** $y = x^3 - 12x + 1$.

**7.** $y = -x^3 + 3x^2 + 1$.

**8.** $y = x^4 - 2x^2 + 4$.

**9.** $y = 2x^4 + 2$.

**10.** $y = -x^7$.

**11.** $y = 81x^5 - 5x$.

**12.** $y = \dfrac{13}{3}x^3 + \dfrac{21}{2}x^2 - 10x - 7$.

**13.** $y = (x^2 + 7x + 10)^2$.

**14.** $y = \dfrac{x^3}{3} + 2x^2 + x - 5$.

---

**OBJECTIVE**

To determine horizontal and vertical asymptotes for a curve and to sketch the graphs of functions having asymptotes.

## 14.5 ASYMPTOTES

### Vertical Asymptotes

In this section, we conclude our discussion of curve-sketching techniques by investigating functions having *asymptotes*. Basically, an asymptote is a line that a curve approaches arbitrarily closely. For example, in each part of Fig. 14.50, the dashed line $x = a$ is an asymptote. But to be precise about it, we need to make use of infinite limits. In Fig. 14.50(a), notice that as $x \to a^+$, $f(x)$ becomes positively infinite:

$$\lim_{x \to a^+} f(x) = \infty.$$

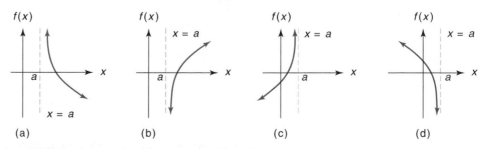

(a)  (b)  (c)  (d)

**FIGURE 14.50** Vertical asymptotes $x = a$.

In Fig. 14.50(b), as $x \to a^+$, $f(x)$ becomes negatively infinite:

$$\lim_{x \to a^+} f(x) = -\infty.$$

In Figs. 14.50(c) and (d), we have

$$\lim_{x \to a^-} f(x) = \infty \qquad \text{and} \qquad \lim_{x \to a^-} f(x) = -\infty,$$

respectively.

Loosely speaking, we can say that each graph in Fig. 14.50 has an "explosion" around the dashed vertical line $x = a$, in the sense that a one-sided limit of $f(x)$ at $a$ is either $\infty$ or $-\infty$. The line $x = a$ is called a *vertical asymptote* for the graph. A vertical asymptote is not part of the graph, but is a useful aid in sketching it because part of the graph approaches the asymptote. Because of the explosion around $x = a$, the function is *not* continuous at $a$.

**DEFINITION**

*The line $x = a$ is a **vertical asymptote** for the graph of the function f if and only if at least one of the following is true:*

$$\lim_{x \to a^+} f(x) = \infty \ (or\ -\infty)$$

*or*

$$\lim_{x \to a^-} f(x) = \infty \ (or\ -\infty).$$

To determine vertical asymptotes, we must find values of $x$ around which $f(x)$ increases or decreases without bound. For a rational function (a quotient of two polynomials), these $x$-values are precisely those for which the denominator is zero but the numerator is not zero. For example, consider the rational function

$$f(x) = \frac{3x - 5}{x - 2}.$$

When $x$ is 2, the denominator is 0, but the numerator is not. If $x$ is slightly larger than 2, then $x - 2$ is both close to 0 and positive, and $3x - 5$ is close to 1. Thus, $(3x - 5)/(x - 2)$ is very large, so

$$\lim_{x \to 2^+} \frac{3x - 5}{x - 2} = \infty.$$

This limit is sufficient to conclude that the line $x = 2$ is a vertical asymptote. Because we are ultimately interested in the behavior of a function around a vertical asymptote, it is worthwhile to examine what happens to this function as $x$ approaches 2 from the left. If $x$ is slightly less than 2, then $x - 2$ is very close to 0 but negative, and $3x - 5$ is close to 1. Hence, $(3x - 5)/(x - 2)$ is "very negative," so

$$\lim_{x \to 2^-} \frac{3x - 5}{x - 2} = -\infty.$$

We conclude that the function increases without bound as $x \to 2^+$ and decreases without bound as $x \to 2^-$. The graph appears in Fig. 14.51.

In summary, we have a rule for vertical asymptotes.

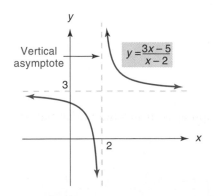

**FIGURE 14.51**   Graph of
$y = \dfrac{3x - 5}{x - 2}.$

> **Vertical-Asymptote Rule for Rational Functions**
>
> Suppose that
>
> $$f(x) = \frac{P(x)}{Q(x)},$$
>
> where $P$ and $Q$ are polynomial functions. The line $x = a$ is a vertical asymptote for the graph of $f$ if and only if $Q(a) = 0$ and $P(a) \neq 0$.

### EXAMPLE 1 Finding Vertical Asymptotes

*Determine vertical asymptotes for the graph of*

$$f(x) = \frac{x^2 - 4x}{x^2 - 4x + 3}.$$

*Solution:* Since $f$ is a rational function, the vertical-asymptote rule applies. Writing

$$f(x) = \frac{x(x - 4)}{(x - 3)(x - 1)}$$

makes it clear that the denominator is 0 when $x$ is 3 or 1. Neither of these values makes the numerator 0. Thus, the lines $x = 3$ and $x = 1$ are vertical asymptotes. (See Fig. 14.52.)

Although the vertical-asymptote rule guarantees that the lines $x = 3$ and $x = 1$ are vertical asymptotes, it does not indicate the precise nature of the "explosion" around these lines. A precise analysis requires the use of one-sided limits.

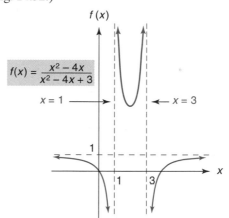

**FIGURE 14.52** Graph of $f(x) = \dfrac{x^2 - 4x}{x^2 - 4x + 3}$.

## Horizontal Asymptotes

A curve $y = f(x)$ may have another kind of asymptote. In Fig. 14.53(a), as $x$ increases without bound ($x \to \infty$), the graph approaches the horizontal line $y = b$. That is,

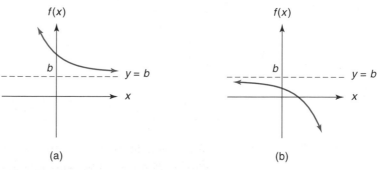

(a)                    (b)

**FIGURE 14.53** Horizontal asymptotes $y = b$.

$$\lim_{x \to \infty} f(x) = b.$$

In Fig. 14.53(b), as $x$ becomes negatively infinite, the graph approaches the horizontal line $y = b$. That is

$$\lim_{x \to -\infty} f(x) = b.$$

In each case, the dashed line $y = b$ is called a *horizontal asymptote* for the graph. It is a horizontal line around which the graph "settles" either as $x \to \infty$ or as $x \to -\infty$.

Although the graph of a horizontal line settles around itself as $x \to \infty$ or as $x \to -\infty$, a line is not considered to have asymptotes. In summary, we have the following definition:

**DEFINITION**

*Let $f$ be a nonlinear function. The line $y = b$ is a **horizontal asymptote** for the graph of $f$ if and only if at least one of the following is true:*

$$\lim_{x \to \infty} f(x) = b \qquad or \qquad \lim_{x \to -\infty} f(x) = b.$$

To test for horizontal asymptotes, we must find the limits of $f(x)$ as $x \to \infty$ and as $x \to -\infty$. To illustrate, we again consider

$$f(x) = \frac{3x - 5}{x - 2}.$$

Since this is a rational function, we can use the procedures of Sec. 11.2 to find the limits. Because the dominant term in the numerator is $3x$ and the dominant term in the denominator is $x$, we have

$$\lim_{x \to \infty} \frac{3x - 5}{x - 2} = \lim_{x \to \infty} \frac{3x}{x} = \lim_{x \to \infty} 3 = 3.$$

Thus, the line $y = 3$ is a horizontal asymptote. (See Fig. 14.54.) Also,

$$\lim_{x \to -\infty} \frac{3x - 5}{x - 2} = \lim_{x \to -\infty} \frac{3x}{x} = \lim_{x \to -\infty} 3 = 3.$$

Hence, the graph settles down near the horizontal line $y = 3$ both as $x \to \infty$ and as $x \to -\infty$.

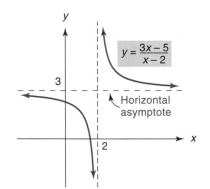

$y = \dfrac{3x - 5}{x - 2}$

Horizontal asymptote

**FIGURE 14.54**  Graph of $f(x) = \dfrac{3x - 5}{x - 2}$.

**EXAMPLE 2  Finding Horizontal Asymptotes**

*Find horizontal asymptotes for the graph of*

$$f(x) = \frac{x^2 - 4x}{x^2 - 4x + 3}.$$

*Solution:* We have

$$\lim_{x \to \infty} \frac{x^2 - 4x}{x^2 - 4x + 3} = \lim_{x \to \infty} \frac{x^2}{x^2} = \lim_{x \to \infty} 1 = 1.$$

Therefore, the line $y = 1$ is a horizontal asymptote. The same result is obtained as $x \to -\infty$. (Refer back to Fig. 14.52.) ■

A few remarks about asymptotes are appropriate now. With vertical asymptotes, we are examining the behavior of a graph around specific $x$-values.

However, with horizontal asymptotes we are examining the graph as $x$ becomes unbounded. Although a graph may have numerous vertical asymptotes, it can have at most two different horizontal asymptotes.

From Sec. 11.2, when the numerator of a rational function has degree greater than that of the denominator, no limit exists as $x \to \infty$ or $x \to -\infty$. From this observation, we conclude that *whenever the degree of the numerator of a rational function is greater than the degree of the denominator, the graph of the function cannot have a horizontal asymptote.*

### EXAMPLE 3   Finding Horizontal and Vertical Asymptotes

*Find vertical and horizontal asymptotes for the graph of the polynomial function*

$$y = f(x) = x^3 + 2x.$$

*Solution:* We begin with vertical asymptotes. This is a rational function with denominator 1, which is never zero. By the vertical-asymptote rule, there are no vertical asymptotes. Because the degree of the numerator (3) is greater than the degree of the denominator (0), there are no horizontal asymptotes. However, let us examine the behavior of the graph of $f$ as $x \to \infty$ and $x \to -\infty$. We have

$$\lim_{x \to \infty} (x^3 + 2x) = \lim_{x \to \infty} x^3 = \infty$$

and $\quad \lim_{x \to -\infty} (x^3 + 2x) = \lim_{x \to -\infty} x^3 = -\infty.$

Thus, as $x \to \infty$, the graph must extend indefinitely upward, and as $x \to -\infty$, the graph must extend indefinitely downward. (See Fig. 14.55.) ■

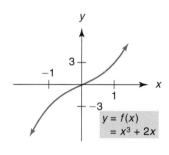

**FIGURE 14.55**   Graph of $y = x^3 + 2x$ has neither horizontal nor vertical asymptotes.

The results in Example 3 can be generalized to any polynomial function:

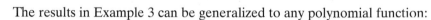

A polynomial function has neither a horizontal nor a vertical asymptote.

### EXAMPLE 4   Finding Horizontal and Vertical Asymptotes

*Find horizontal and vertical asymptotes for the graph of $y = e^x - 1$.*

*Solution:* Testing for horizontal asymptotes, we let $x \to \infty$. Then $e^x$ increases without bound, so

$$\lim_{x \to \infty} (e^x - 1) = \infty.$$

Thus, the graph does not settle down as $x \to \infty$. However, as $x \to -\infty$, we have $e^x \to 0$, so

$$\lim_{x \to -\infty} (e^x - 1) = \lim_{x \to -\infty} e^x - \lim_{x \to -\infty} 1 = 0 - 1 = -1.$$

Therefore, the line $y = -1$ is a horizontal asymptote. The graph has no vertical asymptotes because $e^x - 1$ neither increases nor decreases without bound around any fixed value of $x$. (See Fig. 14.56.) ■

## Curve Sketching

We conclude this chapter by showing you how to graph a function by making use of all the curve-sketching tools that we have developed.

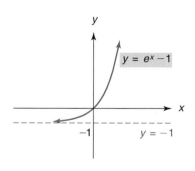

**FIGURE 14.56**   Graph of $y = e^x - 1$ has a horizontal asymptote.

### EXAMPLE 5   Curve Sketching

*Sketch the graph of $y = \dfrac{1}{4 - x^2}$.*

*Solution:*

*Intercepts* When $x = 0$, $y = \frac{1}{4}$. If $y = 0$, then $0 = 1/(4 - x^2)$, which has no solution. Thus $(0, \frac{1}{4})$ is the only intercept.

*Symmetry* There is symmetry about the $y$-axis: Replacing $x$ by $-x$ gives

$$y = \frac{1}{4 - (-x)^2}, \qquad \text{or} \qquad y = \frac{1}{4 - x^2},$$

which is the same as the original equation. It can be shown that there is no other symmetry.

*Asymptotes* Testing for horizontal asymptotes, we have

$$\lim_{x \to \infty} \frac{1}{4 - x^2} = \lim_{x \to \infty} \frac{1}{-x^2} = -\lim_{x \to \infty} \frac{1}{x^2} = 0.$$

Similarly,

$$\lim_{x \to -\infty} \frac{1}{4 - x^2} = 0.$$

Thus, $y = 0$ (the $x$-axis) is a horizontal asymptote. Because the denominator of $1/(4 - x^2)$ is 0 when $x = \pm 2$, and the numerator is not 0 for these values of $x$, the lines $x = 2$ and $x = -2$ are vertical asymptotes.

*Maxima and Minima* Since $y = (4 - x^2)^{-1}$,

$$y' = -1(4 - x^2)^{-2}(-2x) = \frac{2x}{(4 - x^2)^2}.$$

We see that $y'$ is 0 when $x = 0$ and $y'$ is undefined when $x = \pm 2$. However, only 0 is a critical value, because $y$ is not defined at $\pm 2$. There are four intervals to test for increasing/decreasing:

| | | |
|---|---|---|
| If $x < -2$, | then | $y' < 0$; |
| if $-2 < x < 0$, | then | $y' < 0$; |
| if $0 < x < 2$, | then | $y' > 0$; |
| if $x > 2$, | then | $y' > 0$. |

The function is decreasing on $(-\infty, -2)$ and $(-2, 0)$ and increasing on $(0, 2)$ and $(2, \infty)$. (See Fig. 14.57.) There is a relative minimum when $x = 0$.

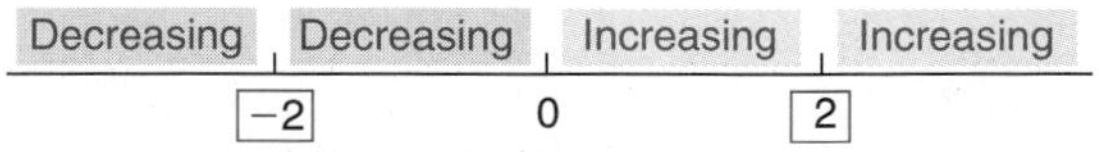

| Decreasing | Decreasing | Increasing | Increasing |
|---|---|---|---|
| | $-2$ | $0$ | $2$ |

**FIGURE 14.57** Increasing/decreasing analysis.

*Concavity*

$$y'' = \frac{(4 - x^2)^2(2) - (2x)2(4 - x^2)(-2x)}{(4 - x^2)^4}$$

$$= \frac{2(4 - x^2)[(4 - x^2) - (2x)(-2x)]}{(4 - x^2)^4} = \frac{2(4 + 3x^2)}{(4 - x^2)^3}.$$

Setting $y'' = 0$, we get no real roots. However, $y''$ is undefined when $x = \pm 2$. Although concavity may change around these values of $x$, the values cannot correspond to inflection points because they are not in the domain of the function. There are three intervals to test for concavity:

$$\text{If } x < -2, \qquad \text{then} \qquad y'' < 0;$$

$$\text{if } -2 < x < 2, \qquad \text{then} \qquad y'' > 0;$$

$$\text{if } x > 2, \qquad \text{then} \qquad y'' < 0.$$

The graph is concave up on $(-2, 2)$ and concave down on $(-\infty, -2)$ and $(2, \infty)$. (See Fig. 14.58.) Although concavity changes around $x = \pm 2$, as we said before, these values cannot correspond to inflection points.

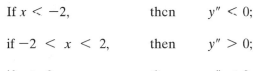

**FIGURE 14.58** Concavity analysis.

**Discussion** Plotting the points in the table in Fig. 14.59, some arbitrarily chosen, and using the preceding information, we get the indicated graph. Due to symmetry about the $y$-axis, our table has only $x \geq 0$.

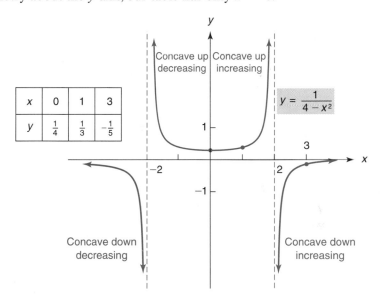

**FIGURE 14.59** Graph of $y = \dfrac{1}{4 - x^2}$.

## EXAMPLE 6    Curve Sketching

*Sketch the graph of $y = \dfrac{4x}{x^2 + 1}$.*

**Solution:**

**Intercepts** When $x = 0$, $y = 0$; when $y = 0$, $x = 0$. Thus, $(0, 0)$ is the only intercept.

**Symmetry** There is symmetry about the origin: Replacing $x$ by $-x$ and $y$ by $-y$ gives

$$-y = \frac{4(-x)}{(-x)^2 + 1}, \qquad \text{or} \qquad y = \frac{4x}{x^2 + 1},$$

which is the same as the original equation. No other symmetry exists.

**Asymptotes** Testing for horizontal asymptotes, we have

$$\lim_{x \to \infty} \frac{4x}{x^2 + 1} = \lim_{x \to \infty} \frac{4x}{x^2} = \lim_{x \to \infty} \frac{4}{x} = 0,$$

and similarly,

$$\lim_{x \to -\infty} \frac{4x}{x^2 + 1} = 0.$$

Thus, $y = 0$ (the $x$-axis) is a horizontal asymptote. Because the denominator of $4x/(x^2 + 1)$ is never 0, there is no vertical asymptote.

**_Maxima and Minima_** Letting $y = f(x)$, we have

$$f'(x) = \frac{(x^2 + 1)(4) - 4x(2x)}{(x^2 + 1)^2} = \frac{4 - 4x^2}{(x^2 + 1)^2} = \frac{4(1 + x)(1 - x)}{(x^2 + 1)^2}.$$

The critical values are $x = \pm 1$, so there are three intervals to consider:

If $x < -1$, then $f'(x) = \dfrac{4(-)(+)}{(+)} = -$, so $f$ is decreasing;

if $-1 < x < 1$, then $f'(x) = \dfrac{4(+)(+)}{(+)} = +$, so $f$ is increasing;

if $x > 1$, then $f'(x) = \dfrac{4(+)(-)}{(+)} = -$, so $f$ is decreasing. (See Fig. 14.60.)

There is a relative minimum when $x = -1$ and a relative maximum when $x = 1$.

| Decreasing | Increasing | Decreasing |
|:---:|:---:|:---:|
| | $-1$ | $1$ |

**FIGURE 14.60** Increasing/decreasing analysis.

**_Concavity_** Since $f'(x) = \dfrac{4 - 4x^2}{(x^2 + 1)^2}$,

$$f''(x) = \frac{(x^2 + 1)^2(-8x) - (4 - 4x^2)(2)(x^2 + 1)(2x)}{(x^2 + 1)^4}$$

$$= \frac{8x(x^2 + 1)(x^2 - 3)}{(x^2 + 1)^4} = \frac{8x(x + \sqrt{3})(x - \sqrt{3})}{(x^2 + 1)^3}.$$

Setting $f''(x) = 0$, we conclude that the possible points of inflection are when $x = \pm\sqrt{3}, 0$. There are four intervals to consider:

If $x < -\sqrt{3}$, then $f''(x) = \dfrac{8(-)(-)(-)}{(+)} = -$, so $f$ is concave down;

if $-\sqrt{3} < x < 0, f''(x) = \dfrac{8(-)(+)(-)}{(+)} = +$, so $f$ is concave up;

if $0 < x < \sqrt{3}$, then $f''(x) = \dfrac{8(+)(+)(-)}{(+)} = -$, so $f$ is concave down;

if $x > \sqrt{3}$, then $f''(x) = \dfrac{8(+)(+)(+)}{(+)} = +$, so $f$ is concave up.

(See Fig. 14.61.)

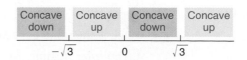

| Concave down | Concave up | Concave down | Concave up |
|:---:|:---:|:---:|:---:|
| | $-\sqrt{3}$ | $0$ | $\sqrt{3}$ |

**FIGURE 14.61** Concavity analysis.

Inflection points occur when $x = 0, \pm\sqrt{3}$.

***Discussion*** After consideration of all of the preceeding information, the graph of $y = 4x/(x^2 + 1)$ is given in Fig. 14.62, together with a table of important points.

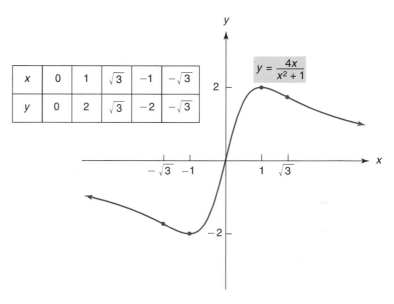

| $x$ | 0 | 1 | $\sqrt{3}$ | $-1$ | $-\sqrt{3}$ |
|---|---|---|---|---|---|
| $y$ | 0 | 2 | $\sqrt{3}$ | $-2$ | $-\sqrt{3}$ |

**FIGURE 14.62**   Graph of $y = \dfrac{4x}{x^2 + 1}$.

## ■ Exercise 14.5

*In Problems 1–24, find the horizontal and vertical asymptotes for the graphs of the functions. Do not sketch the graphs.*

**1.** $y = \dfrac{x}{x + 1}$.

**2.** $y = \dfrac{x + 1}{x}$.

**3.** $f(x) = \dfrac{x - 1}{2x + 3}$.

**4.** $y = \dfrac{2x + 1}{2x - 1}$.

**5.** $y = \dfrac{4}{x}$.

**6.** $y = -\dfrac{4}{x^2}$.

**7.** $y = \dfrac{1}{x^2 - 1}$.

**8.** $y = \dfrac{x}{x^2 - 4}$.

**9.** $y = x^2 - 5x + 8$.

**10.** $y = \dfrac{x^3}{x^2 - 9}$.

**11.** $f(x) = \dfrac{2x^2}{x^2 + x - 6}$.

**12.** $f(x) = \dfrac{x^2}{5}$.

**13.** $y = \dfrac{4 + 2x - 7x^2}{x^2 - 5}$.

**14.** $y = \dfrac{2x^3 + 1}{3x(2x - 1)(4x - 3)}$.

**15.** $y = \dfrac{4}{x - 6} + 4$.

**16.** $f(x) = \dfrac{x^2 - 1}{2x^2 - 9x + 4}$.

**17.** $f(x) = \dfrac{3 - x^4}{x^3 + x^2}$.

**18.** $y = \dfrac{x^2 + x}{x}$.

**19.** $y = \dfrac{x^2 - 3x - 4}{1 + 4x + 4x^2}$.

**20.** $y = \dfrac{x^4 + 1}{1 - x^4}$.

**21.** $y = \dfrac{9x^2 - 16}{(3x + 4)^2}$.

**22.** $y = \dfrac{2}{9} + \dfrac{3x}{14x^2 + x - 3}$.

**23.** $y = 2e^{x+2} + 4$.

**24.** $f(x) = e^{-x}$.

*In Problems 25–46, sketch each curve. Determine intervals on which the function is increasing, decreasing, concave up, and concave down; relative maxima and minima; inflection points; symmetry; horizontal and vertical asymptotes; and those intercepts that can be obtained conveniently.*

**25.** $y = \dfrac{3}{x}$.

**26.** $y = \dfrac{1}{x - 1}$.

**27.** $y = \dfrac{x}{x + 1}$.

**28.** $y = \dfrac{10}{\sqrt{x}}$.

**29.** $y = x^2 + \dfrac{1}{x^2}$.

**30.** $y = \dfrac{x^2}{1 - x}$.

**31.** $y = \dfrac{1}{x^2 - 1}$.

**32.** $y = \dfrac{1}{x^2 + 1}$.

**33.** $y = \dfrac{1 + x}{1 - x}$.

**34.** $y = \dfrac{1 - x}{x^2}$.

**35.** $y = \dfrac{x^2}{7x + 4}$.

**36.** $y = \dfrac{x^3 + 1}{x}$.

**37.** $y = \dfrac{9}{9x^2 - 6x - 8}$.

**38.** $y = \dfrac{37x^2 + 36x + 9}{9x^2}$.

**39.** $y = \dfrac{2x - 3}{(2x - 9)^2}$.

**40.** $y = \dfrac{3x + 1}{(6x + 5)^2}$.

**41.** $y = \dfrac{x^2 - 1}{x^3}.$

**42.** $y = \dfrac{x}{(x + 1)^2}.$

**43.** $y = x + \dfrac{1}{x + 1}.$

**44.** $y = \dfrac{3x^4 + 1}{x^3}.$

**45.** $y = \dfrac{-3x^2 + 2x - 5}{3x^2 - 2x - 1}.$

**46.** $y = 2x + 1 + \dfrac{4}{2x + 1}.$

**47.** Sketch the graph of a function $f$ such that $f(0) = 0$, there is a horizontal asymptote $y = 1$ for $x \to \pm\infty$, there is a vertical asymptote $x = 2$, both $f'(x) < 0$ and $f''(x) < 0$ for $x < 2$, and both $f'(x) < 0$ and $f''(x) > 0$ for $x > 2$.

**48.** Sketch the graph of a function $f$ such that $f(0) = 0$, there is a horizontal asymptote $y = 2$ for $x \to \pm\infty$, there is a vertical asymptote $x = -1$, both $f'(x) > 0$ and $f''(x) > 0$ for $x < -1$, and both $f'(x) > 0$ and $f''(x) < 0$ for $x > -1$.

**49.** Sketch the graph of a function $f$ such that $f(0) = 0$, there is a horizontal asymptote $y = 0$ for $x \to \pm\infty$, there are vertical asymptotes $x = -1$ and $x = 2$, $f'(x) < 0$ for $x < -1$ and $-1 < x < 2$, and $f''(x) < 0$ for $x > 2$.

**50.** Sketch the graph of a function $f$ such that $f(-1) = 0$, $f(0) = 1$, $f(3) = 0$, there is a horizontal asymptote $y = 0$ for $x \to \pm\infty$, there are vertical asymptotes $x = -2$ and $x = 1$, $f''(x) < 0$ for $x < -2$, and $f'(x) > 0$ for $-2 < x < 1$ and $1 < x < 3$.

**51. Purchasing Power** In discussing the time pattern of purchasing, Mantell and Sing[11] use the curve

$$y = \frac{x}{a + bx},$$

as a mathematical model. They claim that $y = 1/b$ is an asymptote. Verify this.

**52.** Sketch the graphs of $y = 6 - 3e^{-x}$ and $y = 6 + 3e^{-x}$. Show that they are asymptotic to the same line. What is the equation of this line?

[11]L. H. Mantell and F. P. Sing, *Economics for Business Decisions* (New York: McGraw-Hill Book Company, 1972), p. 107.

**53. Market for Product** For a new product, the yearly number of thousand packages sold, $y$, after $t$ years from its introduction is predicted to be given by

$$y = f(t) = 150 - 76e^{-t}.$$

Show that $y = 150$ is a horizontal asymptote for the graph. This reveals that after the product is established with consumers, the market tends to be constant.

 **54.** Graph $y = \dfrac{x^2 - 5}{x^3 + 4x^2 + 13x + 1}.$ From the graph, locate any horizontal or vertical asymptotes.

 **55.** Graph $y = \dfrac{6x^3 - 2x^2 + 6x - 1}{3x^3 - 2x^2 - 18x + 12}.$ From the graph, locate any horizontal or vertical asymptotes.

 **56.** Graph $y = \dfrac{\ln(x + 4)}{x^2 - 8x + 5}$, in the standard window. The graph suggests that there are two vertical asymptotes of the form $x = k$, where $k > 0$. Also, it appears that the graph "begins" near $x = -4$. As $x \to -4^+$,

$$\ln(x + 4) \to -\infty \quad \text{and} \quad x^2 - 8x + 5 \to 53 .$$

Thus, $\lim\limits_{x \to 4^+} y = -\infty$, $y = -\infty$. This gives the vertical asymptote $x = -4$. So in reality, there are *three* vertical asymptotes. Use the zoom feature to make the asymptote $x = -4$ apparent from the display.

 **57.** Graph $y = \dfrac{0.34e^{0.7x}}{4.2 + 0.71e^{0.7x}}$, where $x > 0$. From the graph, determine an equation of the horizontal asymptote by examining the $y$-values as $x \to \infty$. To confirm this equation algebraically, find $\lim\limits_{x \to \infty} y$ by first dividing both the numerator and denominator by $e^{0.7x}$.

# 14.6 REVIEW

## IMPORTANT TERMS

**Section 14.1**  increasing function   decreasing function   relative maximum   relative minimum
relative extrema   absolute extrema   critical value   critical point
first-derivative test

**Section 14.2**  extreme-value theorem

**Section 14.3**  concave up   concave down   inflection point

**Section 14.4**  second-derivative test

**Section 14.5**  vertical asymptote   horizontal asymptote   vertical-asymptote rule for rational functions

# SUMMARY

Calculus is a great aid in sketching the graph of a function. The first derivative is used to determine when a function is increasing or decreasing and to locate relative maxima and minima. If $f'(x)$ is positive throughout an interval, then over that interval, $f$ is increasing and its graph rises (from left to right). If $f'(x)$ is negative throughout an interval, then over that interval, $f$ is decreasing and its graph is falling.

A point $(x_0, y_0)$ on the graph at which $f'(x)$ is 0 or is not defined is a candidate for a relative extremum, and $x_0$ is called a critical value. For a relative extremum to occur at $x_0$, the first derivative must change sign around $x_0$. The following procedure is the first-derivative test for the relative extrema of $y = f(x)$:

**First-Derivative Test for Relative Extrema**

**Step 1.** Find $f'(x)$.

**Step 2.** Determine all values of $x$ where $f'(x) = 0$ or $f'(x)$ is not defined.

**Step 3.** On the intervals suggested by the values in step 2, determine whether $f$ is increasing $[f'(x) > 0]$ or decreasing $[f'(x) < 0]$.

**Step 4.** For each critical value $x_0$ at which $f$ is continuous, determine whether $f'(x)$ changes sign as $x$ increases through $x_0$. There is a relative maximum when $x = x_0$ if $f'(x)$ changes from $+$ to $-$, and a relative minimum if $f'(x)$ changes from $-$ to $+$. If $f'(x)$ does not change sign, there is no relative extremum when $x = x_0$.

Under certain conditions, a function is guaranteed to have absolute extrema. The extreme-value theorem states that if $f$ is continuous on a closed interval, then $f$ has an absolute maximum value and an absolute minimum value over the interval. To locate absolute extrema, the following procedure may be used:

**Procedure to Find Absolute Extrema for a Function $f$ Continuous on $[a, b]$**

**Step 1.** Find the critical values of $f$.

**Step 2.** Evaluate $f(x)$ at the endpoints $a$ and $b$ and at the critical values on $(a, b)$.

**Step 3.** The maximum value of $f$ is the greatest of the values found in step 2. The minimum value of $f$ is the least of the values found in step 2.

The second derivative is used to determine concavity and points of inflection. If $f''(x) > 0$ throughout an interval, then $f$ is concave up over that interval, and its graph bends upward. If $f''(x) < 0$ over an interval, then $f$ is concave down throughout that interval, and its graph bends downward. A point on the graph where $f$ is continuous and its concavity changes is an inflection point. The point $(x_0, y_0)$ on the graph is a possible point of inflection if $f''(x_0)$ is 0 or is not defined and $f$ is continuous at $x_0$.

The second derivative also provides a means for testing certain critical values for relative extrema:

**Second-Derivative Test for Relative Extrema**

Suppose $f'(x_0) = 0$. Then

If $f''(x_0) < 0$, $f$ has a relative maximum at $x_0$;

if $f''(x_0) > 0$, $f$ has a relative minimum at $x_0$.

Asymptotes are also aids in curve sketching. Graphs "explode" near vertical asymptotes, and they "settle" near horizontal asymptotes. The line $x = a$ is a vertical asymptote for the graph of a function $f$ if $\lim f(x) = \infty$ or $-\infty$ as $x$ approaches $a$ from the right $(x \to a^+)$ or the left $(x \to a^-)$. For the case of a rational function, $f(x) = P(x)/Q(x)$, we can find vertical asymptotes without evaluating limits. If $Q(a) = 0$, but $P(a) \neq 0$, then the line $x = a$ is a vertical asymptote.

The line $y = b$ is a horizontal asymptote for the graph of a nonlinear function $f$ if at least one of the following is true:

$$\lim_{x \to \infty} f(x) = b \qquad \text{or} \qquad \lim_{x \to -\infty} f(x) = b.$$

In particular, a polynomial function has neither a horizontal nor a vertical asymptote. Moreover, a rational function whose numerator has degree greater than that of the denominator does not have a horizontal asymptote.

# REVIEW PROBLEMS

*In Problems 1–4, find horizontal and vertical asymptotes.*

**1.** $y = \dfrac{3x^2}{x^2 - 16}$.

**2.** $y = \dfrac{x + 1}{4x - 2x^2}$.

**3.** $y = \dfrac{5x^2 - 3}{(3x + 2)^2}$.

**4.** $y = \dfrac{4x + 1}{3x - 5} - \dfrac{3x + 1}{2x - 11}$

*In Problems 5–8, find critical values.*

**5.** $f(x) = \dfrac{x^2}{2 - x^2}.$      **6.** $f(x) = (x - 1)^2(x + 6)^4.$      **7.** $f(x) = \dfrac{\sqrt[3]{x + 1}}{3 - 4x}.$      **8.** $f(x) = \dfrac{6xe^{-5x/6}}{6x + 5}.$

*In Problems 9–12, find intervals on which the function is increasing or decreasing.*

**9.** $f(x) = -x^3 + 6x^2 - 9x$   **10.** $f(x) = \dfrac{x^2}{(x + 1)^2}.$      **11.** $f(x) = \dfrac{x^4}{x^2 - 3}.$      **12.** $f(x) = \sqrt[3]{3x^3 - 4x}.$

*In Problems 13–18, find intervals on which the function is concave up or concave down.*

**13.** $f(x) = x^4 - x^3 - 14$      **14.** $f(x) = \dfrac{x + 1}{x - 1}$      **15.** $f(x) = \dfrac{1}{2x - 1}.$

**16.** $f(x) = x^3 + 2x^2 - 5x + 2.$      **17.** $f(x) = (4x + 1)^3(4x + 9).$      **18.** $f(x) = (x^2 - x - 1)^2.$

*In Problems 19–24, test for relative extrema.*

**19.** $f(x) = 2x^3 - 9x^2 + 12x + 7.$      **20.** $f(x) = \dfrac{2x + 1}{x^2}.$      **21.** $f(x) = \dfrac{x^6}{6} + \dfrac{x^3}{3}.$

**22.** $f(x) = \dfrac{x^2}{x^2 - 4}.$      **23.** $f(x) = x^{2/3}(x + 1).$      **24.** $f(x) = x^3(2x - 1)^4.$

*In Problems 25–30, find the x-values where inflection points occur.*

**25.** $y = x^5 - 5x^4 + 3x.$      **26.** $y = \dfrac{x^2 + 1}{x}.$      **27.** $y = (3x - 5)(x^4 + 2).$

**28.** $y = x^2 + 2\ln(-x).$      **29.** $y = \dfrac{x^2}{e^x}.$      **30.** $y = (x^2 - 4)^3.$

*In Problems 31–34, test for absolute extrema on the given interval.*

**31.** $f(x) = 3x^4 - 4x^3, \quad [0, 2].$      **32.** $f(x) = 2x^3 - 15x^2 + 36x, \quad [0, 3].$

**33.** $f(x) = \dfrac{x}{(5x - 6)^2}, \quad [-2, 0].$      **34.** $f(x) = (x - 2)^2(x + 1)^{2/3}, \quad [-1, 1].$

**35.** Let $f(x) = (x^2 + 1)e^{-x}.$

   **a.** Determine the values of $x$ at which relative maxima and relative minima, if anym, occur.

   **b.** Determine the interval(s) on which the graph of $f$ is concave down, and find the coordinates of all points of inflection.

**36.** Let $f(x) = \dfrac{x}{x^2 + 1}.$ It can be shown that

$$f'(x) = \dfrac{1 - x^2}{(x^2 + 1)^2}.$$

   **a.** Determine whether the graph of $f$ is symmetric about the $x$-axis, $y$-axis, or origin.

   **b.** Find the interval(s) on which $f$ is increasing.

   **c.** Find the coordinates of all relative extrema of $f.$

   **d.** Determine $\lim\limits_{x \to -\infty} f(x)$ and $\lim\limits_{x \to \infty} f(x).$

   **e.** Sketch the graph of $f.$

   **f.** State the absolute minimum and absolute maximum values of $f(x)$ (if they exist).

*In Problems 37–48, sketch the graphs of the functions. Indicate intervals on which the function is increasing, decreasing, concave up, or concave down; indicate relative maximum points, relative minimum points, points of inflection, horizontal asymptotes, vertical asymptotes, symmetry, and those intercepts that can be obtained conveniently.*

**37.** $y = x^2 - 2x - 24.$      **38.** $y = x^3 - 27x.$      **39.** $y = x^3 - 12x + 20.$

**40.** $y = x^4 - 4x^3 - 20x^2 + 150.$      **41.** $y = x^3 + x.$      **42.** $y = \dfrac{x + 2}{x - 3}.$

**43.** $f(x) = \dfrac{100(x + 5)}{x^2}.$      **44.** $y = \dfrac{x^2 - 4}{x^2 - 1}.$      **45.** $y = \dfrac{x}{(2x - 1)^3}.$

**46.** $y = 6x^{1/3}(2x - 1).$      **47.** $f(x) = \dfrac{e^x + e^{-x}}{2}.$      **48.** $f(x) = 1 + \ln(x^2).$

**49.** Are the following statements true or false?

**a.** If $f'(x_0) = 0$, then $f$ must have a relative extremum at $x_0$.

**b.** Since the function $f(x) = 1/x$ is decreasing on the intervals $(-\infty, 0)$ and $(0, \infty)$, it is impossible to find $x_1$ and $x_2$ in the domain of $f$ such that $x_1 < x_2$ and $f(x_1) < f(x_2)$.

**c.** On the interval $(-1, 1]$, the function $f(x) = x^4$ has an absolute maximum and an absolute minimum.

**d.** If $f''(x_0) = 0$, then $(x_0, f(x_0))$ must be a point of inflection.

**e.** A function $f$ defined on the interval $(-2, 2)$ with exactly one relative maximum must have an absolute maximum.

**50.** An important function in probability theory is the standard normal-density function

$$f(x) = \frac{1}{\sqrt{2\pi}} e^{-x^2/2}.$$

**a.** Determine whether the graph of $f$ is symmetric about the $x$-axis, $y$-axis, or origin.

**b.** Find the intervals on which $f$ is increasing and those on which it is decreasing.

**c.** Find the coordinates of all relative extrema of $f$.

**d.** Find $\lim\limits_{x \to -\infty} f(x)$ and $\lim\limits_{x \to \infty} f(x)$.

**e.** Find the intervals on which the graph of $f$ is concave up and those on which it is concave down.

**f.** Find the coordinates of all points of inflection.

**g.** Sketch the graph of $f$.

**h.** Find all absolute extrema.

**51. Marginal Cost**  If $c = q^3 - 6q^2 + 12q + 18$ is a total-cost function, for what values of $q$ is marginal cost increasing?

**52. Marginal Revenue**  If $r = 160q^{3/2} - 3q^2$ is the revenue function for a manufacturer's product, determine the intervals on which the marginal revenue function is increasing.

**53. Revenue Function**  The demand equation for a manufacturer's product is

$$p = 150 - \frac{\sqrt{q}}{10}, \qquad \text{where } q > 0.$$

Show that the graph of the revenue function is concave down wherever it is defined.

**54. Contraception**  In a model of the effect of contraception on birthrate,[12] the equation,

$$R = f(x) = \frac{x}{4.4 - 3.4x}, \qquad 0 \le x \le 1,$$

gives the proportional reduction $R$ in the birthrate as a function of the efficiency $x$ of a contraception method. An efficiency of 0.2 (or 20%) means that the probability of becoming pregnant is 80% of the probability of be-

coming pregnant without the contraceptive. Find the reduction (as as percentage) when efficiency is (a) 0, (b) 0.5, and (c) 1. Find $dR/dx$ and $d^2R/dx^2$, and sketch the graph of the equation.

**55. Learning and Memory**  If you were to recite members of a category, such as four-legged animals, the words that you utter would probably occur in "chunks," with distinct pauses between such chunks. For example, you might say the following for the category of four-legged animals:

dog, cat, mouse, rat,

(pause)

horse, donkey, mule,

(pause)

cow, pig, goat, lamb

etc.

The pauses may occur because one has to mentally search for subcategories (animals around the house, beasts of burden, farm animals, etc.).

The elapsed time between onsets of successive words is called *interresponse time*. A function has been used to analyze the length of time for pauses and the chunk size (number of words in a chunk).[13] This function $f$ is such that

$$f(t) = \begin{cases} \text{the average number of words} \\ \text{that occur in succession with} \\ \text{interresponse times less than } t. \end{cases}$$

The graph of $f$ has a shape similar to that in Fig. 14.63 and is best fit by a third-degree polynomial, such as

$$f(t) = At^3 + Bt^2 + Ct + D.$$

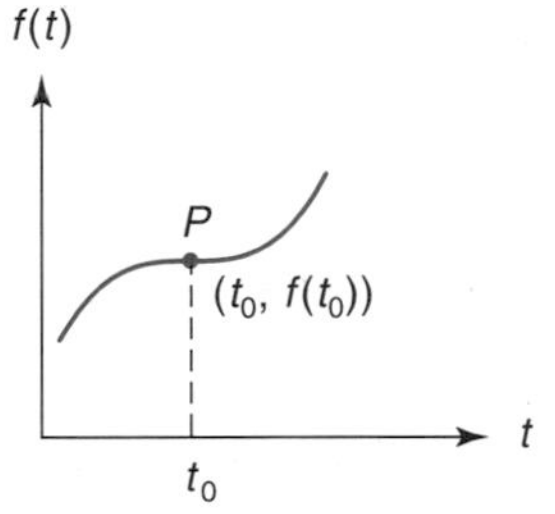

**FIGURE 14.63**  Diagram for Problem 55.

The point $P$ has special meaning. It is such that the value $t_0$ separates interresponse times *within* chunks from those *between* chunks. Mathematically, $P$ is a critical point that is also a point of inflection. Assume these two conditions, and show that (a) $t_0 = -B/(3A)$ and (b) $B^2 = 3AC$.

---

[12]R. K. Leik and B. F. Meeker, *Mathematical Sociology* (Englewood Cliffs, NJ: Prentice-Hall, Inc., 1975).

[13]A. Graesser and G. Mandler, "Limited Processing Capacity Constrains the Storage of Unrelated Sets of Words and Retrieval from Natural Categories," *Human Learning and Memory*, 4, no. 1 (1978), 86–100.

56. **Market Penetration** In a model for the market penetration of a new product, sales $S$ of the product at time $t$ are given by[14]

$$S = g(t) = \frac{m(p + q)^2}{p} \left[ \frac{e^{-(p+q)t}}{\left( \frac{q}{p} e^{-(p+q)t} + 1 \right)^2} \right].$$

where $p$, $q$, and $m$ are nonzero constants.

a. Show that

$$\frac{dS}{dt} = \frac{\frac{m}{p}(p + q)^3 e^{-(p+q)t} \left[ \frac{q}{p} e^{-(p+q)t} - 1 \right]}{\left( \frac{q}{p} e^{-(p+q)t} + 1 \right)^3}.$$

b. Determine the value of $t$ for which maximum sales occur. You may assume that $S$ attains a maximum when $dS/dt = 0$.

[14]A. P. Hurter, Jr., A. H. Rubenstein, et. al., "Market Penetration by New Innovations: The Technological Literature," *Technological Forecasting and Social Change*, vol. 11 (1978), 197–221.

*In Problems 57–60, where appropriate, round your answers to two decimal places.*

 57. From the graph of $y = 4x^3 + 5.3x^2 - 7x + 3$, find the coordinates of all relative extrema.

 58. From the graph of $f(x) = 2x^4 - 3x^3 + 8x - 4$, determine the absolute extrema of $f$ over the interval $[-1, 1]$.

 59. The graph of a function $f$ has exactly one inflection point. If

$$f''(x) = \frac{x^3 + 3x + 2}{5x^2 - 2x + 4},$$

use the graph of $f''$ to determine the $x$-value of the inflection point of $f$.

 60. Graph $y = \frac{4x - 2x^2}{x^3 + 4x + 4}$. From the graph, locate any horizontal or vertical asymptotes.

# Applications of Differentiation

**OBJECTIVE**

**To model situations involving maximizing or minimizing a quantity.**

The aim of this example is to set up a cost function from which cost is minimized.

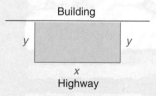

Building

$y$ $y$

$x$

Highway

**FIGURE 15.1** Fencing problem of Example 1.

## 15.1 Applied Maxima and Minima

By using techniques from the previous chapter, we can solve problems that involve maximizing or minimizing a quantity. For example, we might want to maximize profit or minimize cost. The crucial part is expressing the quantity to be maximized or minimized as a function of some variable in the problem. Then we differentiate and test the resulting critical values. For this, the first-derivative test or the second-derivative test may be used, although it may be obvious from the nature of the problem whether or not a critical value represents an appropriate answer. Because our interest is in *absolute* maxima and minima, sometimes we must examine endpoints of the domain of the function.

**EXAMPLE 1** Minimizing the Cost of a Fence

*For insurance purposes, a manufacturer plans to fence in a 10,800-ft$^2$ rectangular storage area adjacent to a building by using the building as one side of the enclosed area. The fencing parallel to the building faces a highway and will cost $3 per foot installed, whereas the fencing for the other two sides costs $2 per foot installed. Find the amount of each type of fence so that the total cost of the fence will be a minimum. What is the minimum cost?*

**Solution:** As a first step in a problem like this, it is a good idea to draw a diagram that reflects the situation. In Fig. 15.1, we have labeled the length of the side parallel to the building as $x$ and the lengths of the other two sides as $y$, where $x$ and $y$ are in feet.

Since we want to minimize cost, our next step is to determine a function that gives cost. The cost obviously depends on how much fencing is along the highway and how much is along the other two sides. Along the highway the cost per foot is 3 (dollars), so the total cost of that fencing is $3x$. Similarly, along *each* of the other two sides, the cost is $2y$. Thus, the total cost of the fencing is given by the cost function

$$C = 3x + 2y + 2y,$$

or

$$C = 3x + 4y. \tag{1}$$

We need to find the absolute minimum value of $C$. To do this, we use the techniques discussed in Chapter 14; that is, we examine $C$ at critical values (and any endpoints) in the domain. But in order to differentiate, we need to first express $C$ as a function of one variable only. [Equation (1) gives $C$ as a function of *two* variables, $x$ and $y$]. We can accomplish this by first finding a relationship between $x$ and $y$. From the statement of the problem, we are told that the storage area, which is $xy$, must be 10,800:

$$xy = 10{,}800. \tag{2}$$

With this equation, we can express one variable (say, $y$) in terms of the other $(x)$. Then, substitution into Eq. (1) will give $C$ as a function of one variable only. Solving Eq. (2) for $y$ gives

$$y = \frac{10{,}800}{x}. \tag{3}$$

Substituting into Eq. (1), we have

$$C = 3x + 4\left(\frac{10{,}800}{x}\right),$$

$$C = 3x + \frac{43{,}200}{x}. \tag{4}$$

From the physical nature of the problem, the domain of $C$ is $x > 0$.

We now find $dC/dx$, set it equal to 0, and solve for $x$. We have

$$\frac{dC}{dx} = 3 - \frac{43{,}200}{x^2},$$

$$3 - \frac{43{,}200}{x^2} = 0,$$

$$3 = \frac{43{,}200}{x^2},$$

from which it follows that

$$x^2 = \frac{43{,}200}{3} = 14{,}400,$$

$$x = 120 \quad (\text{since } x > 0).$$

Thus, 120 is the *only* critical value, and there are no endpoints to consider. To test this value, we shall use the second-derivative test.

$$\frac{d^2C}{dx^2} = \frac{86{,}400}{x^3}.$$

When $x = 120$, $d^2C/dx^2 > 0$, so we conclude that $x = 120$ gives a relative minimum. However, since 120 is the only critical value on the open interval $(0, \infty)$ and $C$ is continuous on that interval, this relative minimum must also be an absolute minimum.

We are not done yet! The questions posed in the problem must be answered. For minimum cost, the number of feet of fencing along the highway is 120. When $x = 120$, we have, from Eq. (3), $y = 10{,}800/120 = 90$. Therefore, the number of feet of fencing for the other two sides is $2y = 180$. It follows that 120 ft of the \$3 fencing and 180 ft of the \$2 fencing are needed. The minimum cost can be obtained from the cost function, Eq. (4), and is

$$C = 3x + \frac{43{,}200}{x} = 3(120) + \frac{43{,}200}{120} = \$720. \quad \blacksquare$$

Based on Example 1, the following guide may be helpful in solving an applied maximum or minimum problem:

---

**Guide for Solving Applied Max–Min Problems**

**Step 1.** When appropriate, draw a diagram that reflects the information in the problem.

**Step 2.** Set up a function for the quantity that you want to maximize or minimize.

**Step 3.** Express the function in step 2 as a function of one variable only, and note the domain of this function. The domain may be implied by the nature of the problem itself.

**Step 4.** Find the critical values of the function. After testing each critical value, determine which one gives the absolute extreme value you are seeking. If the domain of the function includes endpoints, be sure to also examine function values at these endpoints.

**Step 5.** Based on the results of step 4, answer the question(s) posed in the problem.

---

This example involves maximizing revenue when a demand equation is known.

**EXAMPLE 2   Maximizing Revenue**

*The demand equation for a manufacturer's product is*

$$p = \frac{80 - q}{4}, \qquad 0 \le q \le 80,$$

*where q is the number of units and p is the price per unit. At what value of q will there be maximum revenue? What is the maximum revenue?*

*Solution:* Let $r$ represent total revenue, which is the quantity to be maximized. Since

$$\text{revenue} = (\text{price})(\text{quantity}),$$

we have

$$r = pq = \frac{80 - q}{4} \cdot q = \frac{80q - q^2}{4},$$

where $0 \le q \le 80$. Setting $dr/dq = 0$, we obtain

$$\frac{dr}{dq} = \frac{80 - 2q}{4} = 0,$$

$$80 - 2q = 0,$$

$$q = 40.$$

Thus, 40 is the only critical value. Now we see whether this gives a maximum. Examining the first derivative for $0 \le q < 40$, we have $dr/dq > 0$, so $r$ is increasing. If $q > 40$, then $dr/dq < 0$, so $r$ is decreasing. Because to the left of 40 we have $r$ increasing, and to the right $r$ is decreasing, we conclude that $q = 40$ gives the *absolute* maximum revenue, namely, $[80(40) - (40)^2]/4 = 400$. ∎

### EXAMPLE 3   Minimizing Average Cost

This example involves minimizing average cost when the cost function is known

*A manufacturer's total-cost function is given by*

$$c = \frac{q^2}{4} + 3q + 400,$$

*where c is the total cost of producing q units. At what level of output will average cost per unit be a minimum? What is this minimum?*

**Solution:**  The quantity to be minimized is the average cost $\bar{c}$. The average-cost function is

$$\bar{c} = \frac{c}{q} = \frac{\dfrac{q^2}{4} + 3q + 400}{q} = \frac{q}{4} + 3 + \frac{400}{q}. \tag{5}$$

Here $q$ must be positive. To minimize $\bar{c}$, we differentiate:

$$\frac{d\bar{c}}{dq} = \frac{1}{4} - \frac{400}{q^2} = \frac{q^2 - 1600}{4q^2}.$$

To get the critical values, we solve $d\bar{c}/dq = 0$:

$$q^2 - 1600 = 0,$$
$$(q - 40)(q + 40) = 0,$$
$$q = 40 \qquad (\text{since } q > 0).$$

To determine whether this level of output gives a relative minimum, we shall use the second-derivative test. We have

$$\frac{d^2\bar{c}}{dq^2} = \frac{800}{q^3},$$

which is positive for $q = 40$. Thus, $\bar{c}$ has a relative minimum when $q = 40$. We note that $\bar{c}$ is continuous for $q > 0$. Since $q = 40$ is the only relative extremum, we conclude that this relative minimum is indeed an absolute minimum. Substituting $q = 40$ in Eq. (5) gives the minimum average cost $\bar{c} = 23$. ∎

### EXAMPLE 4   Maximization Applied to Enzymes

This example is a biological application involving maximizing the rate at which an enzyme is formed. The equation involved is a literal equation.

*An enzyme is a protein that acts as a catalyst for increasing the rate of a chemical reaction that occurs in cells. In a certain reaction, an enzyme is converted to another enzyme called the product. The product acts as a catalyst for its own formation. The rate R at which the product is formed (with respect to time) is given by*

$$R = kp(l - p),$$

*where l is the total initial amount of both enzymes, p is the amount of the product enzyme, and k is a positive constant. For what value of p will R be a maximum?*

**Solution:**  We can write $R = k(pl - p^2)$. Setting $dR/dp = 0$ and solving for $p$ gives

$$\frac{dR}{dp} = k(l - 2p) = 0,$$

$$p = \frac{l}{2}.$$

Now, $d^2R/dp^2 = -2k$. Since $k > 0$, the second derivative is always negative. Hence, $p = l/2$ gives a relative maximum. Moreover, since $R$ is a continuous function of $p$, we conclude that we indeed have an absolute maximum when $p = l/2$ ∎

Calculus can be applied to inventory decisions, as the following example shows.

### EXAMPLE 5  Economic Lot Size

This example involves determining the number of units in a production run in order to minimize certain costs.

*A company annually produces and sells 10,000 units of a product. Sales are uniformly distributed throughout the year. The company wishes to determine the number of units to be manufactured in each production run in order to minimize total annual setup costs and carrying costs. The same number of units is produced in each run. This number is referred to as the **economic lot size** or **economic order quantity**. The production cost of each unit is $20, and carrying costs (insurance, interest, storage, etc.) are estimated to be 10% of the value of the average inventory. Setup costs per production run are $40. Find the economic lot size.*

**Solution:** Let $q$ be the number of units in a production run. Since sales are distributed at a uniform rate, we shall assume that inventory varies uniformly from $q$ to 0 between production runs. Thus, we take the average inventory to be $q/2$ units. The production costs are $20 per unit, so the value of the average inventory is $20(q/2)$. Carrying costs are 10% of this value:

$$0.10(20)\left(\frac{q}{2}\right).$$

The number of production runs per year is $10{,}000/q$. Hence, the total setup costs are

$$40\left(\frac{10{,}000}{q}\right).$$

Therefore, the total of the annual carrying costs and setup costs is given by

$$C = 0.10(20)\left(\frac{q}{2}\right) + 40\left(\frac{10{,}000}{q}\right)$$

$$= q + \frac{400{,}000}{q}, \qquad (q > 0).$$

$$\frac{dC}{dq} = 1 - \frac{400{,}000}{q^2} = \frac{q^2 - 400{,}000}{q^2}.$$

Setting $dC/dq = 0$, we get

$$q^2 = 400{,}000.$$

Since $q > 0$, we choose

$$q = \sqrt{400{,}000} = 200\sqrt{10} \approx 632.5.$$

To determine whether this value of $q$ minimizes $C$, we shall examine the first derivative. If $0 < q < \sqrt{400{,}000}$, then $dC/dq < 0$. If $q > \sqrt{400{,}000}$, then $dC/dq > 0$. We conclude that there is an *absolute* minimum at $q = 632.5$. The number of production runs is $10{,}000/632.5 \approx 15.8$. For practical purposes, there would be 16 lots, each having an economic lot size of 625 units. ■

### EXAMPLE 6 Maximizing TV Cable Company Revenue

The aim of this example is to set up a revenue function from which revenue is maximized over a closed interval.

*The Vista TV Cable Co. currently has* 2000 *subscribers who are each paying a monthly rate of* $20. *A survey reveals that there will be* 50 *more subscribers for each* $0.25 *decrease in the rate. At what rate will maximum revenue be obtained, and how many subscribers will there be at this rate?*

*Solution:* Let $x$ be the number of $0.25 decreases. The monthly rate is then $20 - 0.25x$, where $0 \le x \le 80$ (the rate cannot be negative), and the number of *new* subscribers is $50x$. Thus, the total number of subscribers is $2000 + 50x$. We want to maximize the revenue, which is given by

$$r = (\text{number of subscribers})(\text{rate per subscriber})$$
$$= (2000 + 50x)(20 - 0.25x)$$
$$= 50(40 + x)(20 - 0.25x)$$
$$= 50(800 + 10x - 0.25x^2).$$

Setting $r' = 0$ and solving for $x$, we have

$$r' = 50(10 - 0.5x) = 0,$$
$$x = 20.$$

Since the domain of $r$ is the closed interval $[0, 80]$, the absolute maximum value of $r$ must occur at $x = 20$ or at one of the endpoints of the interval. We now compute $r$ at these three points:

$$\text{If } x = 0, \quad \text{then } r = 40{,}000;$$
$$\text{if } x = 20, \text{ then } r = 45{,}000;$$
$$\text{if } x = 80, \text{ then } r = 0.$$

Accordingly, the maximum revenue occurs when $x = 20$. This corresponds to twenty $0.25 decreases, for a total decrease of $5; that is, the monthly rate is $15. The number of subscribers at that rate is $2000 + 50(20) = 3000$. ■

### EXAMPLE 7 Maximizing the Number of Recipients of Health-Care Benefits

Here we maximize a function over a closed interval.

*An article in a sociology journal stated that if a particular health-care program for the elderly were initiated, then t years after its start, n thousand elderly people would receive direct benefits, where*

$$n = \frac{t^3}{3} - 6t^2 + 32t, \qquad 0 \le t \le 12.$$

*For what value of t does the maximum number receive benefits?*

*Solution:* Setting $dn/dt = 0$, we have

$$\frac{dn}{dt} = t^2 - 12t + 32 = 0,$$
$$(t - 4)(t - 8) = 0,$$
$$t = 4 \quad \text{or} \quad t = 8.$$

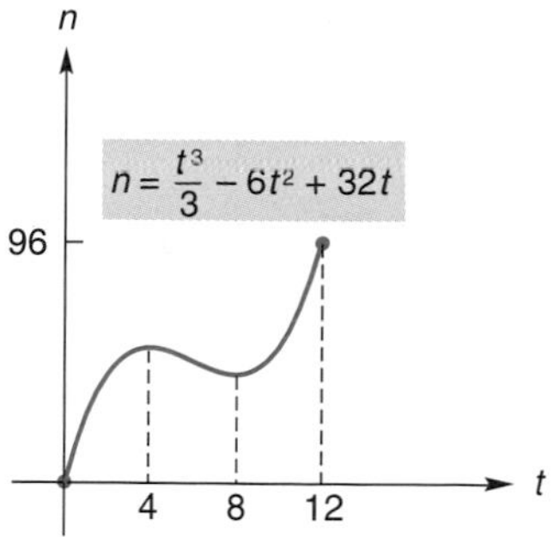

**FIGURE 15.2** Graph of
$n = \dfrac{t^3}{3} - 6t^2 + 32t$ on $[0, 12]$.

Since the domain of $n$ is the closed interval $[0, 12]$, the absolute maximum value of $n$ must occur at $t = 0, 4, 8$, or 12:

If $t = 0$, then $n = 0$,

if $t = 4$, then $n = \dfrac{160}{3} = 53\dfrac{1}{3}$,

if $t = 8$, then $n = \dfrac{128}{3} = 42\dfrac{2}{3}$,

if $t = 12$, then $n = 96$.

Thus, an absolute maximum occurs when $t = 12$. A graph of the function is given in Fig. 15.2. ■

*Pitfall* ▼ The preceding example illustrates that you should not ignore endpoints when finding absolute extrema on a closed interval.

In the next example, we use the word *monopolist*. Under a situation of monopoly, there is only one seller of a product for which there are no similar substitutes, and the seller—that is, the monopolist— controls the market. By considering the demand equation for the product, the monopolist may set the price (or volume of output) so that maximum profit will be obtained.

### EXAMPLE 8 Profit Maximization

*Suppose that the demand equation for a monopolist's product is $p = 400 - 2q$ and the average-cost function is $\bar{c} = 0.2q + 4 + (400/q)$, where $q$ is number of units, and both $p$ and $\bar{c}$ are expressed in dollars per unit.*

This example involves maximizing profit when the demand and average-cost functions are known. In part (d), a tax is imposed on the monopolist, and a new profit function is analyzed.

**a.** *Determine the level of output at which profit is maximized.*

**b.** *Determine the price at which maximum profit occurs.*

**c.** *Determine the maximum profit.*

**d.** *If, as a regulatory device, the government imposes a tax of $22 per unit on the monopolist, what is the new price for profit maximization?*

*Solution:* We know that

$$\text{profit} = \text{total revenue} - \text{total cost}.$$

Since total revenue $r$ and total cost $c$ are given by

$$r = pq = 400q - 2q^2$$

and

$$c = q\bar{c} = 0.2q^2 + 4q + 400,$$

the profit is

$$P = r - c = 400q - 2q^2 - (0.2q^2 + 4q + 400),$$

or

$$P = 396q - 2.2q^2 - 400, \tag{6}$$

where $q > 0$.

**a.** To maximize profit, we set $dP/dq = 0$:

$$\frac{dp}{dq} = 396 - 4.4q = 0,$$

$$q = 90.$$

Now, $d^2P/dq^2 = -4.4$ is always negative, so it is negative at the critical value $q = 90$. By the second-derivative test then, there is a relative maximum there. Since $q = 90$ is the only critical value on $(0, \infty)$, we must have an absolute maximum there.

**b.** The price at which maximum profit occurs is obtained by setting $q = 90$ in the demand equation:

$$p = 400 - 2(90) = \$220.$$

**c.** The maximum profit is obtained by replacing $q$ by 90 in Eq. (6), which gives

$$P = \$17,420.$$

**d.** The tax of \$22 per unit means that for $q$ units, the total cost increases by $22q$. The new cost function is $c_1 = 0.2q^2 + 4q + 400 + 22q$, and the new profit is given by

$$P_1 = 400q - 2q^2 - (0.2q^2 + 4q + 400 + 22q)$$
$$= 374q - 2.2q^2 - 400.$$

Setting $dP_1/dq = 0$ gives

$$\frac{dP_1}{dq} = 374 - 4.4q = 0,$$

$$q = 85.$$

Since $d^2P_1/dq^2 = -4.4 < 0$, we conclude that, to maximize profit, the monopolist must restrict output to 85 units at a higher price of $p_1 = 400 - 2(85) = \$230$. Since this price is only \$10 more than before, only part of the tax has been shifted to the consumer, and the monopolist must bear the cost of the balance. The profit now is \$15,495, which is less than the former profit. ∎

This discussion leads to the economic principle that when profit is maximum, marginal revenue is equal to marginal cost.

We conclude this section by using calculus to develop an important principle in economics. Suppose $p = f(q)$ is the demand function for a firm's product, where $p$ is price per unit and $q$ is the number of units produced and sold. Then the total revenue is given by $r = qp = qf(q)$, which is a function of $q$. Let the total cost of producing $q$ units be given by the cost function $c = g(q)$. Thus, the total profit, which is total revenue minus total cost, is also a function of $q$, namely,

$$P = r - c = qf(q) - g(q).$$

Let us consider the most profitable output for the firm. Ignoring special cases, we know that profit is maximized when $dP/dq = 0$ and $d^2P/dq^2 < 0$. We have

$$\frac{dP}{dq} = \frac{d}{dq}(r - c) = \frac{dr}{dq} - \frac{dc}{dq}.$$

Consequently, $dP/dq = 0$ when

$$\frac{dr}{dq} = \frac{dc}{dq}.$$

That is, at the level of maximum profit, the slope of the tangent to the total-revenue curve must equal the slope of the tangent to the total-cost curve

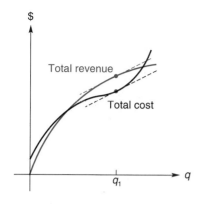

**FIGURE 15.3** At maximum profit, marginal revenue equals marginal cost.

(Fig 15.3). But $dr/dq$ is the marginal revenue MR, and $dc/dq$ is the marginal cost MC. Thus, under typical conditions,

> to maximize profit, it is necessary that
>
> $$MR = MC.$$

For this to indeed correspond to a maximum, it is necessary that $d^2P/dq^2 < 0$:

$$\frac{d^2P}{dq^2} = \frac{d^2}{dq^2}(r - c) = \frac{d^2r}{dq^2} - \frac{d^2c}{dq^2} < 0 \quad \text{or} \quad \frac{d^2r}{dq^2} < \frac{d^2c}{dq^2}.$$

That is, when MR = MC, in order to ensure maximum profit, the slope of the marginal-revenue curve must be less than the slope of the marginal-cost curve.

The condition that $d^2P/dq^2 < 0$ when $dP/dq = 0$ can be viewed another way. Equivalently, to have MR = MC correspond to a maximum, $dP/dq$ must go from $+$ to $-$; that is, it must go from $dr/dq - dc/dq > 0$ to $dr/dq - dc/dq < 0$. Hence, as output increases, we must have MR > MC and then MR < MC. This means that at the point $q_1$ of maximum profit, *the marginal-cost curve must cut the marginal-revenue curve from below* (Fig. 15.4). For production up to $q_1$, the revenue from additional output would be greater than the cost of such output, and the total profit would increase. For output beyond $q_1$, MC > MR, and each unit of output would add more to total costs than to total revenue. Hence, total profits would decline.

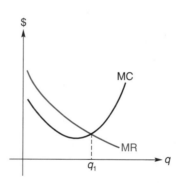

**FIGURE 15.4** At maximum profit, the marginal-cost curve cuts the marginal-revenue curve from below.

## ■ Exercise 15.1

*In this set of problems, unless otherwise specified, p is price per unit (in dollars) and q is output per unit of time. Fixed costs refer to costs that remain constant at all levels of production during a given time period. (An example is rent.)*

**1.** Find two numbers whose sum is 40 and whose product is a maximum.

**2.** Find two nonnegative numbers whose sum is 20 and which are such that the product of twice one number and the square of the other number will be a maximum.

**3. Fencing** A company has set aside $3000 to fence in a rectangular portion of land adjacent to a stream by using the stream for one side of the enclosed area. The cost of the fencing parallel to the stream is $5 per foot

installed, and the fencing for the remaining two sides costs $3 per foot installed. Find the dimensions of the maximum enclosed area.

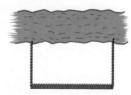

4. **Fencing** The owner of the Laurel Nursery Garden Center wants to fence in 1000 ft$^2$ of land in a rectangular plot to be used for different types of shrubs. The plot is to be divided into four equal plots with three fences parallel to the same pair of sides, as shown in Fig. 15.5. What is the least number of feet of fence needed?

**FIGURE 15.5** Diagram for Problem 4.

5. **Average Cost** A manufacturer finds that the total cost $c$ of producing a product is given by the cost function
$$c = 0.05q^2 + 5q + 500.$$
At what level of output will average cost per unit be a minimum?

6. **Automobile Expense** The cost per hour (in dollars) of operating an automobile is given by
$$C = 0.12s - 0.0012s^2 + 0.08, \quad 0 \le s \le 60,$$
where $s$ is the speed in miles per hour. At what speed is the cost per hour a minimum?

7. **Revenue** The demand equation for a monopolist's product is
$$p = -5q + 30.$$
At what price will revenue be maximized?

8. **Revenue** For a monopolist's product, the demand function is
$$q = 10{,}000e^{-0.02p}.$$
Find the value of $p$ for which maximum revenue is obtained.

9. **Weight Gain** A group of biologists studied the nutritional effects on rats that were fed a diet containing 10% protein.[1] The protein consisted of yeast and cottonseed

flour. By varying the percent $p$ of yeast in the protein mix. the group found that the (average) weight gain (in grams) of a rat over a period of time was
$$f(p) = 160 - p - \frac{900}{p + 10}, \quad 0 \le p \le 100.$$
Find (a) the maximum weight gain and (b) the minimum weight gain.

10. **Drug Dose** The severity of the reaction of the human body to an initial dose $D$ of a drug is given by[2]
$$R = f(D) = D^2\left(\frac{C}{2} - \frac{D}{3}\right),$$
where the constant $C$ denotes the maximum amount of the drug that may be given. Show that $R$ has a maximum *rate of change* when $D = C/2$.

11. **Profit** For a monopolist's product, the demand function is
$$p = 72 - 0.04q$$
and the cost function is
$$c = 500 + 30q.$$
At what level of output will profit be maximized? At what price does this occur, and what is the profit?

12. **Profit** For a monopolist, the cost per unit of producing a product is $3, and the demand equation is
$$p = \frac{10}{\sqrt{q}}.$$
What price will give the greatest profit?

13. **Profit** For a monopolist's product, the demand equation is
$$p = 42 - 4q,$$
and the average-cost function is
$$\bar{c} = 2 + \frac{80}{q}.$$
Find the profit-maximizing price.

14. **Profit** For a monopolist's product, the demand function is
$$p = \frac{50}{\sqrt{q}},$$
and the average-cost function is
$$\bar{c} = 0.50 + \frac{1000}{q}.$$
Find the profit-maximizing price and output. At this level, show that marginal revenue is equal to marginal cost.

15. **Profit** A manufacturer can produce at most 120 units of a certain product each year. The demand equation for the product is
$$p = q^2 - 100q + 3200,$$
and the manufacturer's average-cost function is
$$\bar{c} = \frac{2}{3}q^2 - 40q + \frac{10{,}000}{q}.$$

[1]Adapted from R. Bressani, "The Use of Yeast in Human Foods," in *Single-Cell Protein,* ed. R.I. Mateles and S. R. Tannenbaum (Cambridge, MA: MIT Press, 1968).

[2]R. M. Thrall, J. A. Mortimer. K. R. Rebman. and R. F. Baum, eds., *Some Mathematical Models in Biology,* rev. ed., Report No. 40241-R-7. Prepared at University of Michigan, 1967.

Determine the profit-maximizing output $q$ and the corresponding maximum profit.

**16. Cost**  A manufacturer has determined that, for a certain product, the average cost (in dollars per unit) is given by

$$\bar{c} = 2q^2 - 36q + 210 - \frac{200}{q},$$

where $2 \leq q \leq 10$.

**a.** At what level within the interval [2, 10] should production be fixed in order to minimize total cost? What is the minimum total cost?

**b.** If production were required to lie within the interval [5, 10], what value of $q$ would minimize total cost?

**17. Profit**  For XYZ Manufacturing Co., total fixed costs are $1200, material and labor costs combined are $2 per unit, and the demand equation is

$$p = \frac{100}{\sqrt{q}}.$$

What level of output will maximize profit? Show that this occurs when marginal revenue is equal to marginal cost. What is the price at profit maximization?

**18. Revenue**  A real-estate firm owns 100 garden-type apartments. At $400 per month, each apartment can be rented. However, for each $10-per-month increase, there will be two vacancies with no possibility of filling them. What rent per apartment will maximize monthly revenue?

**19. Revenue**  A TV cable company has 4800 subscribers who are each paying $18 per month. It can get 150 more subscribers for each $0.50 decrease in the monthly fee. What rate will yield maximum revenue, and what will this revenue be?

**20. Profit**  A manufacturer of a product finds that, for the first 500 units that are produced and sold, the profit is $50 per unit. The profit on each of the units beyond 500 is decreased by $0.10 times the number of additional units produced. For example, the total profit when 502 units are produced and sold is $500(50) + 2(49.80)$. What level of output will maximize profit?

**21. Container Design**  A container manufacturer is designing a rectangular box, open at the top and with a square base, that is to have a volume of 32 ft.³ If the box is to require the least amount of material, what must be its dimensions? (See Fig. 15.6.)

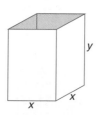

**FIGURE 15.6**  Open-top box for Problems 21 and 22.

**22. Container Design**  An open-top box with a square base is to be constructed from 192 ft² of material. What should be the dimensions of the box if the volume is to be a maximum? What is the maximum volume? (See Fig. 15.6.)

**23. Container Design**  An open box is to be made by cutting equal squares from each corner of a 12-inch-square piece of cardboard and then folding up the sides. Find the length of the side of the square that must be cut out if the volume of the box is to be maximized. What is the maximum volume? (See Fig. 15.7.)

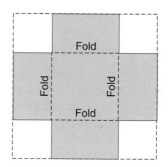

**FIGURE 15.7**  Box for Problem 23.

**24. Poster Design**  A rectangular cardboard poster is to have 150 in.² for printed matter. It is to have a 3-in. margin at the top and bottom and a 2-in. margin on each side. Find the dimensions of the poster so that the amount of cardboard used is minimized. (See Fig. 15.8.) [*Hint:* First find the values of $x$ and $y$ in Fig. 15.8 that minimize the amount of cardboard.]

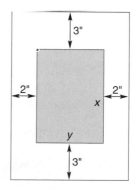

**FIGURE 15.8**  Poster for Problem 24.

**25. Container Design**  A cylindrical can, open at the top, is to have a fixed volume of $K$. Show that if the least amount of material is to be used, then both the radius and height are equal to $\sqrt[3]{K/\pi}$. (See Fig.15.9)

Volume = $\pi r^2 h$
Surface area = $2\pi rh + \pi r^2$

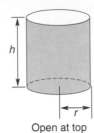

Open at top

**FIGURE 15.9** Can for Problems 25 and 26.

26. **Container Design** A cylindrical can, open at the top, is to be made from a fixed amount of material, $K$. If the volume is to be a maximum, show that both the radius and height are equal to $\sqrt{K/(3\pi)}$. (See Fig. 15.9.)

27. **Profit** The demand equation for a monopolist's product is

$$p = 600 - 2q,$$

and the total-cost function is

$$c = 0.2q^2 + 28q + 200.$$

Find the profit-maximizing output and price, and determine the corresponding profit. If the government were to impose a tax of $22 per unit on the manufacturer, what would be the new profit-maximizing output and price? What is the profit now?

28. **Profit** Use the *original* data in Problem 27, and assume that the government imposes a license fee of $100 on the manufacturer. This is a lump-sum amount without regard to output. Show that the profit-maximizing price and ouput remain the same. Show, however, that there will be less profit.

29. **Economic Lot Size** A manufacturer has to produce annually 1000 units of a product that is sold at a uniform rate during the year. The production cost of each unit is $10, and carrying costs (insurance, interest, storage, etc.) are estimated to be 12.8% of the value of average inventory. Setup costs per production run are $40. Find the economic lot size.

30. **Profit** For a monopolist's product, the cost function is

$$c = 0.004q^3 + 20q + 5000,$$

and the demand function is

$$p = 450 - 4q.$$

Find the profit-maximizing output.

31. **Workshop Attendance** Imperial Educational Services (I.E.S.) is considering offering a workshop in resource allocation to key personnel at Acme Corp. To make the offering economically feasible, I.E.S. feels that at least 30 persons must attend at a cost of $50 each. Moreover, I.E.S. will agree to reduce the charge for *everybody* by

$1.25 for each person over the 30 who attends. How many people should be in the group for I.E.S. to maximize revenue? Assume that the maximum allowable number in the group is 40.

32. **Cost of Leasing Motor** The Kiddie Toy Company plans to lease an electric motor that will be used 90,000 horsepower-hours per year in manufacturing. One horsepower-hour is the work done in 1 hour by a 1-horsepower motor. The annual cost to lease a suitable motor is $150, plus $0.60 per horsepower. The cost per horsepower-hour of operating the motor is $0.006/N$, where $N$ is the horsepower. What size motor, in horsepower, should be leased in order to minimize cost?

33. **Transportation Cost** The cost of operating a truck on a thruway (excluding the salary of the driver) is

$$0.11 + \frac{s}{300}$$

dollars per mile, where $s$ is the (steady) speed of the truck in miles per hour. The truck driver's salary is $12 per hour. At what speed should the truck driver operate the truck to make a 700-mile trip most economical?

34. **Cost** For a manufacturer, the cost of making a part is $3 per unit for labor and $1 per unit for materials; overhead is fixed at $2000 per week. If more than 5000 units are made each week, labor is $4.50 per unit for those units in excess of 5000. At what level of production will average cost per unit be a minimum?

35. **Profit** Ms. Jones owns a small insurance agency that sells policies for a large insurance company. For each policy sold, Ms. Jones, who does not sell policies herself, is paid a commission of $50 by the insurance company. From previous experience, Ms. Jones has determined that, when she employs $m$ salespeople,

$$q = m^3 - 12m^2 + 60m$$

policies can be sold per week. She pays each of the $m$ salespeople a salary of $750 per week, and her weekly fixed cost is $2500. Current office facilities can accommodate at most seven salespeople. Determine the number of salespeople that Ms. Jones should hire to maximize her weekly profit. What is the corresponding maximum profit?

**36. Profit** A manufacturing company sells high-quality jackets through a chain of specialty shops. The demand equation for these jackets is

$$p = 400 - 50q.$$

where $p$ is the selling price (in dollars per jacket) and $q$ is the demand (in hundreds of jackets). If this company's marginal-cost function is given by

$$\frac{dc}{dq} = \frac{800}{q + 5}$$

show that there is a maximum profit, and determine the number of jackets that must be sold to obtain this maximum profit.

**37. Chemical Production** Each day, a firm makes $x$ tons of chemical A ($x \geq 4$) and

$$y = \frac{24 - 6x}{5 - x}$$

tons of chemical B. The profit on chemical A is $2000 per ton, and on B it is $1000 per ton. How much of chemical A should be produced per day to maximize profit? Answer the same question if the profit on A is $P$ per ton and that on B is $P/2$ per ton.

**38. Rate of Return** To erect an office building, fixed costs are $250,000 and include land, architect's fees, a basement, a foundation, etc. If $x$ floors are constructed, the cost (excluding fixed costs) is

$$c = \frac{x}{2}[100,000 + 5000(x - 1)].$$

The revenue per month is $5000 per floor. How many floors will yield a maximum rate of return on investment? (Rate of return = total revenue/total cost.)

**39. Gait and Power Output of an Animal** In a model by Smith,[3] the power output of an animal at a given speed as a function of its movement or *gait j*, is found to be

$$P(j) = Aj\frac{L^4}{V} + B\frac{V^3L^2}{1 + j},$$

where $A$ and $B$ are constants, $j$ is a measure of the "jumpiness" of the gait, $L$ is a constant representing linear dimension, and $V$ is a constant forward speed.

Assume that $P$ is a minimum when $dP/dj = 0$. Show that when this occurs,

$$(1 + j)^2 = \frac{BV^4}{AL^2}.$$

As a passing comment, Smith indicates that "at top speed, $j$ is zero for an elephant, 0.3 for a horse, and 1 for a greyhound, approximately."

**40. Traffic Flow** In a model of traffic flow on a lane of a freeway, the number of cars the lane can carry per unit time is given by[4]

$$N = \frac{-2a}{-2at_r + v - \frac{2al}{v}},$$

where $a$ is the acceleration of a car when stopping ($a < 0$), $t_r$ is the reaction time to begin braking, $v$ is the average speed of the cars, and $l$ is the length of a car. Assume that $a$, $t_r$, and $l$ are constant. To find how many cars a lane can carry at most, we want to find the speed $v$ that maximizes $N$. To maximize $N$, it suffices to minimize the denominator

$$-2at_r + v - \frac{2al}{v}.$$

**a.** Find the value of $v$ that minimizes the denominator.

**b.** Evaluate your answer in (a) when $a = -19.6$ (ft/sec$^2$), $l = 20$ (ft), and $t_r = 0.5$ (sec). Your answer will be in feet per second.

**c.** Find the corresponding value of $N$ to one decimal place. Your answer will be in cars per second. Convert your answer to cars per hour.

**41. Average Cost** During the Christmas season, a promotional company purchases cheap red felt stockings, glues fake white fur and sequins onto them, and packages them for distribution. The total cost of producing $q$ cases of stockings is given by

$$c = 3q^2 + 50q - 18q \ln q + 120.$$

Find the number of cases that should be processed in order to minimize the average cost per case. Determine (to two decimal places) this minimum average cost.

[3]J. M. Smith, *Mathematical Ideas in Biology* (London: Cambridge University Press, 1968).

[4]J. I. Shonle, *Environmental Application* Addison-Wesley Publishing Co.

**42. Profit** A monopolist's demand equation is given by

$$p = q^2 - 21q + 164,$$

where $p$ is the selling price (in thousands of dollars) per ton when $q$ tons of product are sold. Suppose that fixed cost is $100 thousand and that each ton costs $20 thousand to produce. If current equipment has a maximum production capacity of 10 tons, use the graph of the profit function to determine at what production level the maximum profit occurs. Find the corresponding maximum profit and selling price per ton.

---

O B J E C T I V E

To define the differential, interpret it geometrically, and use it in approximations. Also, to state the reciprocal relationship between $dx/dy$ and $dy/dx$.

## 15.2 DIFFERENTIALS

We shall soon give you a reason for using the symbol $dy/dx$ to denote the derivative of $y$ with respect to $x$. To do this, we introduce the notion of the *differential* of a function.

### DEFINITION

*Let $y = f(x)$ be a differentiable function of $x$, and let $\Delta x$ denote a change in $x$, where $\Delta x$ can be any real number. Then the **differential** of $y$, denoted $dy$ or $d[f(x)]$, is given by*

$$dy = f'(x)\Delta x.$$

Note that $dy$ is a function of two variables, namely, $x$ and $\Delta x$.

### EXAMPLE 1    Computing a Differential

*Find the differential of $y = x^3 - 2x^2 + 3x - 4$, and evaluate it when $x = 1$ and $\Delta x = 0.04$.*

*Solution:* The differential is

$$dy = \frac{d}{dx}(x^3 - 2x^2 + 3x - 4)\Delta x$$

$$= (3x^2 - 4x + 3)\Delta x.$$

When $x = 1$ and $\Delta x = 0.04$,

$$dy = [3(1)^2 - 4(1) + 3](0.04) = 0.08. \qquad \blacksquare$$

If $y = x$, then $dy = d(x) = 1\Delta x = \Delta x$. Hence, the differential of $x$ is $\Delta x$. We abbreviate $d(x)$ by $dx$. Thus, $dx = \Delta x$. From now on, it will be our practice to write $dx$ for $\Delta x$ when finding a differential. For example,

$$d(x^2 + 5) = \frac{d}{dx}(x^2 + 5)dx = 2x\,dx.$$

Summarizing, we say that if $y = f(x)$ defines a differentiable function of $x$, then

$$dy = f'(x)\,dx,$$

where $dx$ is any real number. Provided that $dx \neq 0$, we can divide both sides by $dx$:

$$\frac{dy}{dx} = f'(x).$$

That is, *dy*/*dx* can be viewed either as the quotient of two differentials, namely, *dy* divided by *dx*, or as one symbol for the derivative of *f* at *x*. It is for this reason that we introduced the symbol *dy*/*dx* to denote the derivative.

**EXAMPLE 2**  **Finding a Differential in Terms of *dx***

**a.** If $f(x) = \sqrt{x}$, then

$$d(\sqrt{x}) = \frac{d}{dx}(\sqrt{x})\,dx = \frac{1}{2}x^{-1/2}\,dx = \frac{1}{2\sqrt{x}}\,dx.$$

**b.** If $u = (x^2 + 3)^5$, then $du = 5(x^2 + 3)^4(2x)\,dx = 10x(x^2 + 3)^4\,dx.$  ■

The differential can be interpreted geometrically. In Fig. 15.10, the point $P(x, f(x))$ is on the curve $y = f(x)$. Suppose *x* changes by *dx*, a real number, to the new value $x + dx$. Then the new function value is $f(x + dx)$, and the

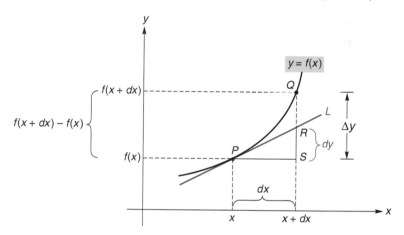

**FIGURE 15.10**  Geometric interpretation of *dy* and $\Delta x$.

corresponding point on the curve is $Q(x + dx, f(x + dx))$. Passing through *P* and *Q* are horizontal and vertical lines, respectively, that intersect at *S*. A line *L* tangent to the curve at *P* intersects segment *QS* at *R*, forming the right triangle *PRS*. Observe that the graph of *f* near *P* is approximated by the tangent line at *P*. The slope of *L* is $f'(x)$, or, equivalently, it is $\overline{SR}/\overline{PS}$:

$$f'(x) = \frac{\overline{SR}}{\overline{PS}}.$$

Since $dy = f'(x)\,dx$ and $dx = \overline{PS}$,

$$dy = f'(x)\,dx = \frac{\overline{SR}}{\overline{PS}} \cdot \overline{PS} = \overline{SR}.$$

Thus, if *dx* is a change in *x* at *P*, then *dy* is the corresponding vertical change along the **tangent line** at *P*. Note that for the same *dx*, the vertical change along the **curve** is $\Delta y = \overline{SQ} = f(x + dx) - f(x)$. Do not confuse $\Delta y$ with *dy*. However, from Fig. 15.10, the following is apparent:

When *dx* is close to 0, *dy* is an approximation to $\Delta y$. Therefore,

$$\Delta y \approx dy.$$

This fact is useful in estimating $\Delta y$, a change in *y*, as Example 3 shows.

**EXAMPLE 3** Using the Differential to Estimate a Change in a Quantity

*A governmental health agency examined the records of a group of individuals who were hospitalized with a particular illness. It was found that the total proportion P that was discharged at the end of t days of hospitalization is given by*

$$P = P(t) = 1 - \left(\frac{300}{300 + t}\right)^3.$$

*Use differentials to approximate the change in the proportion discharged if t changes from 300 to 305.*

**Solution:** The change in $t$ from 300 to 305 is $\Delta t = dt = 305 - 300 = 5$. The change in $P$ is $\Delta P = P(305) - P(300)$. We approximate $\Delta P$ by $dP$:

$$\Delta P \approx dP = P'dt = -3\left(\frac{300}{300 + t}\right)^2\left[-\frac{300}{(300 + t)^2}\right]dt.$$

When $t = 300$ and $dt = 5$,

$$dP = -3\left(\frac{300}{600}\right)^2\left[-\frac{300}{(600)^2}\right]5$$

$$= -3\left(\frac{1}{2}\right)^2\left[-\frac{1}{2(600)}\right]5 = \frac{1}{320} \approx 0.0031.$$

For a comparison, the true value of $\Delta P$ is $P(305) - P(300) = 0.87807 - 0.87500 = 0.00307$ (to five decimal places). ∎

We said that if $y = f(x)$, then $\Delta y \approx dy$ if $dx$ is close to zero. Thus,

$$\Delta y = f(x + dx) - f(x) \approx dy$$

or

$$f(x + dx) \approx f(x) + dy. \tag{1}$$

This formula gives us a way of estimating a function value $f(x + dx)$. For example, suppose we estimate $\ln(1.06)$. Letting $y = f(x) = \ln x$, we need to estimate $f(1.06)$. Since $d(\ln x) = (1/x)\,dx$, we have, from (1),

$$f(x + dx) \approx f(x) + dy,$$

$$\ln(x + dx) \approx \ln x + \frac{1}{x}dx.$$

We know the exact value of $\ln 1$, so we shall let $x = 1$ and $dx = 0.06$. Then $x + dx = 1.06$, and $dx$ is close to zero. Therefore,

$$\ln(1 + 0.06) \approx \ln(1) + \frac{1}{1}(0.06),$$

$$\ln(1.06) \approx 0 + 0.06 = 0.06.$$

The true value of $\ln(1.06)$ to five decimal places is 0.05827.

**EXAMPLE 4** Using the Differential to Estimate a Function Value

*The demand function for a product is given by*

$$p = f(q) = 20 - \sqrt{q},$$

---

Formula (1) is used to approximate a function value, whereas the formula $\Delta y \approx dy$ is used to approximate a change in function values.

*where p is the price per unit in dollars for q units. By using differentials, approximate the price when* 99 *units are demanded.*

***Solution:*** We want to approximate $f(99)$. By formula (1),

$$f(q + dq) \approx f(q) + dp,$$

where

$$dp = -\frac{1}{2\sqrt{q}} \, dq.$$

We choose $q = 100$ and $dq = -1$ because $q + dq = 99$, $dq$ is small, and it is easy to compute $f(100) = 20 - \sqrt{100} = 10$. We thus have

$$f(99) = f[100 + (-1)] \approx f(100) - \frac{1}{2\sqrt{100}}(-1),$$

$$f(99) \approx 10 + 0.05 = 10.05.$$

Hence, the price per unit when 99 units are demanded is approximately \$10.05.

⬛

The equation $y = x^3 + 4x + 5$ defines $y$ as a function of $x$. However, it also defines $x$ implicitly as a function of $y$. Accordingly, we can look at the derivative of $x$ with respect to $y$, $dx/dy$. Since $dx/dy$ can be considered a quotient of differentials, we are motivated to write (and it is indeed true that)

$$\frac{dx}{dy} = \frac{1}{\dfrac{dy}{dx}}, \qquad \text{provided that } dy/dx \neq 0.$$

But $dy/dx$ is the derivative of $y$ with respect to $x$ and equals $3x^2 + 4$. Thus,

$$\frac{dx}{dy} = \frac{1}{3x^2 + 4}.$$

This is the *reciprocal* of $dy/dx$.

**EXAMPLE 5**   **Finding** *dp/dq* **from** *dq/dp*

*Find* $\dfrac{dp}{dq}$ *if* $q = \sqrt{2500 - p^2}$.

***Solution:***

*Strategy:* There are a number of ways to find *dp/dq*. One approach is to solve the given equation for *p* explicitly in terms of *q* and then differentiate directly. Another approach to find *dp/dq* is to use implicit differentiation. However, since *q* is given explicitly as a function of *p*, we can easily find *dq/dp* and then use the preceding reciprocal relation to find *dp/dq*. We shall take this approach.

We have

$$\frac{dq}{dp} = \frac{1}{2}(2500 - p^2)^{-1/2}(-2p) = -\frac{p}{\sqrt{2500 - p^2}}.$$

Hence,

$$\frac{dp}{dq} = \frac{1}{\dfrac{dq}{dp}} = -\frac{\sqrt{2500 - p^2}}{p}.$$

■

## ■ Exercise 15.2

*In Problems 1–10, find the differential of the function in terms of x and dx.*

**1.** $y = 3x - 4$.

**2.** $y = 2$.

**3.** $f(x) = \sqrt{x^4 + 2}$.

**4.** $f(x) = (4x^2 - 5x + 2)^3$.

**5.** $u = \dfrac{1}{x^2}$.

**6.** $u = \dfrac{1}{\sqrt{x}}$.

**7.** $p = \ln(x^2 + 7)$.

**8.** $p = e^{x^3+5}$.

**9.** $y = (4x + 3)e^{2x^2+3}$.

**10.** $y = \ln\sqrt{x^4 + 1}$.

*In Problems 11–16, find Δy and dy for the given values of x and dx.*

**11.** $y = 4 - 7x$; $x = 3, dx = 0.02$.

**12.** $y = 5x^2$; $x = -1, dx = -0.02$.

**13.** $y = 4x^2 - 3x + 10$; $x = -1, dx = 0.25$.

**14.** $y = (3x + 2)^2$; $x = -1, dx = -0.03$.

**15.** $y = \sqrt{25 - x^2}$; $x = 3, dx = -0.1$. Round your answer to three decimal places.

**16.** $y = \ln(-x)$; $x = -5, dx = 0.1$.

**17.** Let $f(x) = \dfrac{x + 5}{x + 1}$.

   **a.** Evaluate $f'(1)$.

   **b.** Use differentials to estimate the value of $f(1.1)$.

**18.** Let $f(x) = x^{3x}$.

   **a.** Evaluate $f'(1)$.

   **b.** Use differentials to estimate the value of $f(0.98)$.

*In Problems 19–26, approximate each expression by using differentials.*

**19.** $\sqrt{101}$.

**20.** $\sqrt{120}$.

**21.** $\sqrt[3]{63}$.

**22.** $\sqrt[4]{17}$.

**23.** $\ln 0.97$.

**24.** $\ln 1.01$.

**25.** $e^{0.01}$.

**26.** $e^{-0.01}$.

*In Problems 27–32, find dx/dy or dp/dq.*

**27.** $y = 2x - 1$.

**28.** $y = 5x^2 + 3x + 2$.

**29.** $q = (p^2 + 5)^3$.

**30.** $q = \sqrt{p + 5}$.

**31.** $q = \dfrac{1}{p}$.

**32.** $q = e^{5-p}$.

**33.** If $y = 5x^2 + 3x + 2$, find the value of $dx/dy$ when $x = 1$.

**34.** If $y = \ln x$, find the value of $dx/dy$ when $x = 2$.

*In Problems 35 and 36, find the rate of change of q with respect to p for the indicated value of q.*

**35.** $p = \dfrac{500}{q + 2}$; $q = 18$.

**36.** $p = 50 - \sqrt{q}$; $q = 100$.

**37. Profit** Suppose that the profit (in dollars) of producing $q$ units of a product is

$$P = 396q - 2.2q^2 - 400.$$

Using differentials, find the approximate change in profit if the level of production changes from $q = 80$ to $q = 81$. Find the true change.

**38. Revenue** Given the revenue function

$$r = 250q + 45q^2 - q^3,$$

use differentials to find the approximate change in revenue if the number of units increases from $q = 40$ to $q = 41$. Find the true change.

**39. Demand** The demand equation for a product is

$$p = \frac{10}{\sqrt{q}}.$$

Using differentials, approximate the price when 24 units are demanded.

**40. Demand**   Given the demand function

$$p = \frac{100}{\sqrt{q + 4}},$$

use differentials to estimate the price per unit when 12.5 units are demanded.

**41.** If $y = f(x)$, then the *proportional change in y* is defined to be $\Delta y/y$, which can be approximated with differentials by $dy/y$. Use this last form to approximate the proportional change in the cost function

$$c = f(q) = \frac{q^4}{2} + 3q + 400$$

when $q = 10$ and $dq = 2$. Round your answer to one decimal place.

**42. Status/Income**   Suppose that $S$ is a numerical value of status based on a person's annual income $I$ (in thousands of dollars). For a certain population, suppose $S = 20\sqrt{I}$. Use differentials to approximate the change in $S$ if annual income decreases from \$25,000 to \$24,500.

**43. Biology**   The volume of a spherical cell is given by $V = \frac{4}{3}\pi r^3$, where $r$ is the radius. Estimate the change in volume when the radius changes from $6.5 \times 10^{-4}$ cm to $6.6 \times 10^{-4}$ cm.

**44. Muscle Contraction**   The equation

$$(P + a)(v + b) = k$$

[5]R. W. Stacy et al., *Essentials of Biological and Medical Physics* (New York: McGraw-Hill Book Company, 1955).

is called the "fundamental equation of muscle contraction."[5] Here $P$ is the load imposed on the muscle, $v$ is the velocity of the shortening of the muscle fibers, and $a$, $b$, and $k$ are positive constants. Find $v$ in terms of $P$, and then use the differential to approximate the change in $v$ due to a small change in $P$.

**45. Demand**   The demand $q$, for a monopolist's product is related to the price per unit, $p$, according to the equation

$$2 + \frac{q^2}{200} = \frac{4000}{p^2}.$$

a. Verify that 40 units will be demanded when the price per unit is \$20.

b. Show that $\dfrac{dq}{dp} = -2.5$ when the price per unit is \$20.

c. Use differentials and the results of parts (a) and (b) to approximate the number of units that will be demanded if the price per unit is reduced to \$19.20.

**46. Profit**   The demand equation for a monopolist's product is

$$p = \frac{1}{3}q^2 - 76q + 6000,$$

and the average-cost function is

$$\overline{c} = 600 - q + \frac{100,000}{3q}.$$

a. Verify that the profit is \$90,000 when 100 units are demanded.

b. Use differentials and the result of part (a) to estimate the profit when 97 units are demanded.

---

**To give a mathematical analysis of the economic concept of elasticity.**

## 15.3  ELASTICITY OF DEMAND

*Elasticity of demand* is a means by which economists measure how a change in the price of a product will affect the quantity demanded. That is, it refers to consumer response to price changes. Loosely speaking, elasticity of demand is the ratio of the resulting percentage change in quantity demanded to a given percentage change in price:

$$\frac{\text{percentage change in quantity}}{\text{percentage change in price}}.$$

For example, if, for a price increase of 5%, quantity demanded were to decrease by 2%, we would loosely say that elasticity of demand is $-2/5$.

To be more general, suppose $p = f(q)$ is the demand function for a product. Consumers will demand $q$ units at a price of $f(q)$ per unit and will demand $q + h$ units at a price of $f(q + h)$ per unit (Fig. 15.11). The *percentage change in quantity demanded* from $q$ to $q + h$ is

$$\frac{(q + h) - q}{q} \cdot 100 = \frac{h}{q} \cdot 100.$$

The corresponding percentage change in price per unit is

$$\frac{f(q + h) - f(q)}{f(q)} \cdot 100.$$

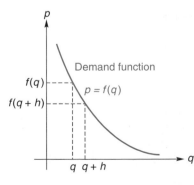

**FIGURE 15.11**   Change in demand.

The ratio of these percentage changes is

$$\frac{\dfrac{h}{q} \cdot 100}{\dfrac{f(q+h)-f(q)}{f(q)} \cdot 100} = \frac{h}{q} \cdot \frac{f(q)}{f(q+h)-f(q)}$$

$$= \frac{f(q)}{q} \cdot \frac{h}{f(q+h)-f(q)}$$

$$= \frac{\dfrac{f(q)}{q}}{\dfrac{f(q+h)-f(q)}{h}}. \qquad (1)$$

If $f$ is differentiable, then as $h \to 0$, the limit of $[f(q+h)-f(q)]/h$ is $f'(q) = dp/dq$. Thus, the limit of (1) is

$$\frac{\dfrac{f(q)}{q}}{\dfrac{dp}{dq}} \quad \text{or} \quad \frac{\dfrac{p}{q}}{\dfrac{dp}{dq}},$$

which is called the *point elasticity of demand.*

**DEFINITION**

*If $p = f(q)$ is a differentiable demand function, the **point elasticity of demand,** denoted by the Greek letter $\eta$ (eta), at $(q, p)$ is given by*

$$\eta = \frac{\dfrac{p}{q}}{\dfrac{dp}{dq}}.$$

To illustrate, let us find the point elasticity of demand for the demand function $p = 1200 - q^2$. We have

$$\eta = \frac{\dfrac{p}{q}}{\dfrac{dp}{dq}} = \frac{\dfrac{1200-q^2}{q}}{-2q} = -\frac{1200-q^2}{2q^2} = -\left[\frac{600}{q^2} - \frac{1}{2}\right]. \qquad (2)$$

For example, if $q = 10$, then $\eta = -[(600/10^2) - \frac{1}{2}] = -5\frac{1}{2}$. Since

$$\eta \approx \frac{\% \text{ change in demand}}{\% \text{ change in price}},$$

we have

$$(\% \text{ change in price})(\eta) \approx \% \text{ change in demand}.$$

Thus, if price were increased by 1% when $q = 10$, then quantity demanded would change by approximately

$$(1\%)\left(-5\frac{1}{2}\right) = -5\frac{1}{2}\%.$$

That is, demand would decrease $5\frac{1}{2}\%$. Similarly, decreasing price by $\frac{1}{2}\%$ when $q = 10$ results in a change in demand of approximately

$$\left(-\frac{1}{2}\%\right)\left(-5\frac{1}{2}\right) = 2\frac{3}{4}\%.$$

Hence, demand increases by $2\frac{3}{4}\%$.

Note that when elasticity is evaluted, no units are attached to it—it is nothing more than a real number. For normal behavior of demand, an increase (decrease) in price corresponds to a decrease (increase) in quantity. Thus, $dp/dq$ will always be negative or 0, and $\eta$ (where defined) will always be negative or 0. Some economists disregard the minus sign; in the preceding situation, they would consider the elasticity to be $5\frac{1}{2}$. We shall not adopt this practice.

There are three categories of elasticity:

1. When $|\eta| > 1$, demand is *elastic.*

2. When $|\eta| = 1$, demand has *unit elasticity.*

3. When $|\eta| < 1$, demand is *inelastic.*

For example, in Eq. (2), since $|\eta| = 5\frac{1}{2}$ when $q = 10$, demand is elastic. If $q = 20$, then $|\eta| = |-[(600/20^2) - \frac{1}{2}]| = 1$ so demand has unit elasticity. If $q = 25$, then $|\eta| = |-\frac{23}{50}|$, and demand is inelastic.

Loosely speaking, for a given percentage change in price, there is a greater percentage change in quantity demanded if demand is elastic, a smaller percentage change if demand is inelastic, and an equal percentage change if demand has unit elasticity.

**EXAMPLE 1** **Finding Point Elasticity of Demand**

*Determine the point elasticity of the demand equation*

$$p = \frac{k}{q}, \qquad where \ k > 0 \ and \ q > 0.$$

*Solution:* From the definition, we have

$$\eta = \frac{\dfrac{p}{q}}{\dfrac{dp}{dq}} = \frac{\dfrac{k}{q^2}}{\dfrac{-k}{q^2}} = -1.$$

Thus, the demand has unit elasticity for all $q > 0$. The graph of $p = k/q$ is called an *equilateral hyperbola* and is often found in economics texts in discussions of elasticity. (See Fig. 3.14 for a graph of such a curve.) ∎

**EXAMPLE 2** **Finding Point Elasticity of Demand**

*Determine the point elasticity of the demand equation*

$$q = p^2 - 40p + 400, \qquad where \ q > 0.$$

*Solution:* To compute $\eta$, we need to know $dp/dq$. In the given demand equation, $p$ is not an explicit function of $q$, so we cannot find $dp/dq$ directly. However, from Sec. 15.2,

$$\frac{dp}{dq} = \frac{1}{\frac{dq}{dp}}.$$

Thus,

$$\eta = \frac{\frac{p}{q}}{\frac{dp}{dq}} = \frac{p}{q} \cdot \frac{1}{\frac{dp}{dq}} = \frac{p}{q} \frac{dq}{dp}.$$

Since $dq/dp = 2p - 40$, we have

$$\eta = \frac{p}{q}(2p - 40).$$

For example, if $p = 15$, then $q = 25$; hence, $\eta = [15(-10)]/25 = -6$, so demand is elastic.  ■

*Here we analyze elasticity for linear demand.*

Point elasticity for a *linear* demand equation is quite interesting. Suppose the equation has the form

$$p = mq + b, \quad \text{where } m < 0 \text{ and } b > 0.$$

(See Fig. 15.12.) We assume that $q > 0$; thus, $p < b$. The point elasticity of demand is

$$\eta = \frac{\frac{p}{q}}{\frac{dp}{dq}} = \frac{\frac{p}{q}}{m} = \frac{p}{mq} = \frac{p}{p - b}.$$

By considering $d\eta/dp$, we shall show that $\eta$ is a decreasing function of $p$. By the quotient rule,

$$\frac{d\eta}{dp} = \frac{(p - b) - p}{(p - b)^2} = -\frac{b}{(p - b)^2}.$$

Since $b > 0$ and $(p - b)^2 > 0$, it follows that $d\eta/dp < 0$, so $\eta$ is a decreasing function of $p$: As $p$ increases, $\eta$ must decrease. However, $p$ ranges between 0 and $b$, and at the midpoint of this range, $b/2$,

$$\eta = \frac{\frac{b}{2}}{\frac{b}{2} - b} = \frac{\frac{b}{2}}{-\frac{b}{2}} = -1.$$

Therefore, if $p < b/2$, then $\eta > -1$; if $p > b/2$, then $\eta < -1$. Because we must have $\eta \le 0$, we can state these facts another way: When $p < b/2$, $|\eta| < 1$, and demand is inelastic; when $p = b/2$, $|\eta| = 1$, and demand has unit elasticity; when $p > b/2$, $|\eta| > 1$, and demand is elastic. This shows that the slope of a demand curve is not a measure of elasticity. The slope of the line in Fig. 15.12 is $m$ everywhere, but elasticity varies with the point on the line.

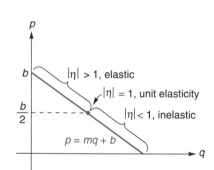

**FIGURE 15.12**  Elasticity for linear demand.

### Elasticity and Revenue

*Here we analyze the relationship between elasticity and the rate of change of revenue.*

Turning to a different situation, we can relate how elasticity of demand affects changes in revenue (marginal revenue). If $p = f(q)$ is a manufacturer's demand function, the total revenue is given by

$$r = pq.$$

To find the marginal revenue, $dr/dq$, we differentiate $r$ by using the product rule:

$$\frac{dr}{dq} = p + q\frac{dp}{dq}. \tag{3}$$

Factoring the right side of Eq. (3), we have

$$\frac{dr}{dq} = p\left(1 + \frac{q}{p}\frac{dp}{dq}\right).$$

But

$$\frac{q}{p}\frac{dp}{dq} = \frac{\dfrac{dp}{dq}}{\dfrac{p}{q}} = \frac{1}{\eta}.$$

Thus,

$$\boxed{\frac{dr}{dq} = p\left(1 + \frac{1}{\eta}\right).} \tag{4}$$

If demand is elastic, then $\eta < -1$, so $1 + \dfrac{1}{\eta} > 0$. If demand is inelastic, then $\eta > -1$, so $1 + \dfrac{1}{\eta} < 0$. Let us assume that $p > 0$. From Eq. (4) we can conclude that $dr/dq > 0$ on intervals for which demand is elastic; hence, total revenue $r$ is increasing there. On the other hand, marginal revenue is negative on intervals for which demand is inelastic; consequently, total revenue is decreasing there.

Thus, we conclude from the preceeding argument that as more units are sold, a manufacturer's total revenue increases if demand is elastic, but decreases if demand is inelastic. That is, if demand is elastic, a lower price will increase revenue. This means that a lower price will cause a large enough increase in demand to actually increase revenue. If demand is inelastic, a lower price will decrease revenue. For unit elasticity, a lower price leaves total revenue unchanged.

## ▪ Exercise 15.3

*In Problems 1–14, find the point elasticity of the demand equations for the indicated values of q or p, and determine whether demand is elastic, is inelastic, or has unit elasticity.*

**1.** $p = 40 - 2q$; $q = 5$.

**2.** $p = 12 - 0.03q$; $q = 300$.

**3.** $p = \dfrac{1000}{q}$; $q = 288$.

**4.** $p = \dfrac{1000}{q^2}$; $q = 156$.

**5.** $p = \dfrac{500}{q + 2}$; $q = 100$.

**6.** $p = \dfrac{800}{2q + 1}$; $q = 25$.

**7.** $p = 150 - e^{q/100}$; $q = 100$.

**8.** $p = 100e^{-q/200}$; $q = 200$.

**9.** $q = 600 - 100p$; $p = 3$.

**10.** $q = 100 - p$; $p = 50$

**11.** $q = \sqrt{2500 - p}$;  $p = 900$.

**13.** $q = \dfrac{(p - 100)^2}{2}$;  $p = 20$.

**12.** $q = \sqrt{2500 - p^2}$;  $p = 20$.

**14.** $q = p^2 - 60p + 898$;  $p = 10$.

**15.** For the linear demand equation $p = 13 - 0.05q$, verify that demand is elastic when $p = 10$, is inelastic when $p = 3$, and has unit elasticity when $p = 6.50$.

**16.** For what value (or values) of $q$ do the following demand equations have unit elasticity?

  **a.** $p = 26 - 0.10q$.

  **b.** $p = 1200 - q^2$.

**17.** The demand equation for a product is

$$q = 500 - 40p + p^2,$$

where $p$ is the price per unit (in dollars) and $q$ is the quantity of units demanded (in thousands). Find the point elasticity of demand when $p = 15$. If this price of 15 is increased by $\frac{1}{2}\%$, what is the approximate change in demand?

**18.** The demand equation for a certain product is

$$q = \sqrt{2500 - p^2},$$

where $p$ is in dollars. Find the point elasticity of demand when $p = 30$, and use this value to compute the approximate percentage change in demand if the price of $30 is decreased to $28.50.

**19.** For the demand equation $p = 500 - 2q$, verify that demand is elastic and total revenue is increasing for $0 < q < 125$. Verify that demand is inelastic and total revenue is decreasing for $125 < q < 250$.

**20.** Verify that $\dfrac{dr}{dq} = p\left(1 + \dfrac{1}{\eta}\right)$ if $p = 40 - 2q$.

**21.** Repeat Problem 20 for $p = \dfrac{1000}{q^2}$.

**22.** Suppose $p = mq + b$ is a linear demand equation, where $m \neq 0$ and $b > 0$.

  **a.** Show that $\lim\limits_{p \to b^-} \eta = -\infty$.

  **b.** Show that $\eta = 0$ when $p = 0$.

**23.** The demand equation for a manufacturer's product is

$$p = \dfrac{200}{\sqrt{6000 + 10q^2}}.$$

  **a.** Verify that $q = 20$ when $p = 2$.

  **b.** Determine the point elasticity of demand when $p = 2$. Is demand elastic, is it inelastic, or does it have unit elasticity at this point?

  **c.** If the price when $p = 2$ is decreased by 2%, what is the approximate number of units by which demand changes?

  **d.** If the price when $p = 2$ is decreased by 2%, will the total revenue increase, decrease, or remain constant? Give a reason for your answer.

**24.** Given the demand equation $q^2(1 + p)^2 = p$, determine the point elasticity of demand when $p = 9$.

**25.** The demand equation for a product is

$$q = \dfrac{60}{p} + \ln(65 - p^3).$$

  **a.** Determine the point elasticity of demand when $p = 4$, and classify the demand as elastic, inelastic, or of unit elasticity at this price level.

  **b.** If the price is lowered by 2% (from $4.00 to $3.92), use the answer to part (a) to estimate the corresponding percentage change in quantity sold.

  **c.** Will the changes in part (b) result in an increase or decrease in revenue? Explain.

**26.** The demand equation for a manufacturer's product is

$$p = 50(201 - q)^{0.001\sqrt{q+25}}.$$

  **a.** Show that $dp/dq = -0.75$ when 200 units are demanded. Use logarithmic differentiation.

  **b.** Using the result in part (a), determine the point elasticity of demand when 200 units are demanded. At this level, is demand elastic, inelastic, or of unit elasticity?

  **c.** Use the result in part (b) to approximate the price per unit if demand decreases from 200 to 188 units.

  **d.** If the current demand is 200 units, should the manufacturer increase or decrease price in order to increase revenue? (Justify your answer.)

**27.** A manufacturer of aluminum doors currently is able to sell 500 doors per week at a price of $80 each. If the price were lowered to $75 each, an additional 50 doors per week could be sold. Estimate the current elasticity of demand for the doors, and also estimate the current value of the manufacturer's marginal-revenue function.

**28.** Given the demand equation

$$p = 1000 - q^2,$$

where $5 \leq q \leq 30$, for what value of $q$ is $|\eta|$ a maximum? For what value is it a minimum?

**29.** Repeat Problem 28 for

$$p = \dfrac{200}{q + 5}$$

such that $5 \leq q \leq 95$.

**To approximate real roots of an equation by using calculus. The method shown is suitable for calculators.**

## 15.4 NEWTON'S METHOD

It is quite easy to solve equations of the form $f(x) = 0$ when $f$ is a linear or quadratic function. For example, we can solve $x^2 + 3x - 2 = 0$ by the quadratic formula. However, if $f(x)$ has degree greater than two (or is not a polynomial), it may be difficult, or even impossible, to find solutions (or roots) of $f(x) = 0$ by the methods to which you are accustomed. For this reason, we may settle for approximate solutions, which can be obtained in a variety of efficient ways. For example, a graphics calculator may be used to estimate the real roots of $f(x) = 0$. In this section, you will learn how the derivative may be so used (provided that $f$ is differentiable). The procedure we shall develop, called *Newton's method,* is well suited to a calculator or computer.

Newton's method requires an initial estimate for a root of $f(x) = 0$. One way of obtaining this estimate is by making a rough sketch of the graph of $y = f(x)$ and estimating the root from the graph. A point on the graph where $y = 0$ is an $x$-intercept, and the $x$-value of this point is a root of $f(x) = 0$. Another way of locating a root is based on the following fact:

> If $f$ is continuous on the interval $[a, b]$ and $f(a)$ and $f(b)$ have opposite signs, then the equation $f(x) = 0$ has at least one real root between $a$ and $b$.

Figure 15.13 depicts this situation. The $x$-intercept between $a$ and $b$ corresponds to a root of $f(x) = 0$, and we can use either $a$ or $b$ to approximate this root.

Assuming that we have an estimated (but incorrect) value for a root, we turn to a way of getting a better approximation. In Fig. 15.14, you can see that

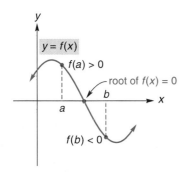

**FIGURE 15.13** Root of $f(x) = 0$ between $a$ and $b$, where $f(a)$ and $f(b)$ have opposite signs.

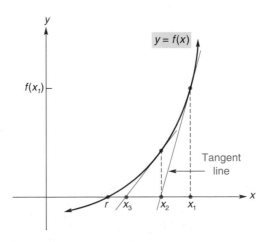

**FIGURE 15.14** Improving approximation of root via tangent line.

$f(r) = 0$, so $r$ is a root of the equation $f(x) = 0$. Suppose $x_1$ is an initial approximation to $r$ (and one that is close to $r$). Observe that the tangent line to the curve at $(x_1, f(x_1))$ intersects the $x$-axis at the point $(x_2, 0)$, and $x_2$ is a better approximation to $r$ than is $x_1$.

We can find $x_2$ from the equation of the tangent line. The slope of the tangent line is $f'(x_1)$, so a point-slope form for this line is

$$y - f(x_1) = f'(x_1)(x - x_1). \qquad (1)$$

Since $(x_2, 0)$ is on the tangent line, its coordinates must satisfy Eq. (1). This gives

$$0 - f(x_1) = f'(x_1)(x_2 - x_1),$$

$$-\frac{f(x_1)}{f'(x_1)} = x_2 - x_1 \qquad [\text{if } f'(x_1) \neq 0].$$

Thus,

$$x_2 = x_1 - \frac{f(x_1)}{f'(x_1)}. \tag{2}$$

To get a better approximation to $r$, we again perform the procedure described, but this time we use $x_2$ as our starting point. This gives the approximation

$$x_3 = x_2 - \frac{f(x_2)}{f'(x_2)}. \tag{3}$$

Repeating (or *iterating*) this computation over and over, we hope to obtain better approximations, in the sense that the sequence of values

$$x_1, x_2, x_3, \ldots$$

will approach $r$. In practice, we terminate the process when we have reached a desired degree of accuracy.

If you analyze Eqs. (2) and (3), you can see how $x_2$ is obtained from $x_1$ and how $x_3$ is obtained from $x_2$. In general, $x_{n+1}$ is obtained from $x_n$ by means of the following general formula, called **Newton's method:**

**Newton's Method**

$$x_{n+1} = x_n - \frac{f(x_n)}{f'(x_n)}, \quad n = 1, 2, 3, \ldots. \tag{4}$$

A formula, like Eq. (4), that indicates how one number in a sequence is obtained from the preceding one is called a **recursion formula,** or an *iteration equation.*

### EXAMPLE 1  Approximating a Root by Newton's Method

*Approximate the root of $x^4 - 4x + 1 = 0$ that lies between 0 and 1. Continue the approximation procedure until two successive approximations differ by less than 0.0001.*

*Solution:* Letting $f(x) = x^4 - 4x + 1$, we have

$$f(0) = 0 - 0 + 1 = 1$$

and

$$f(1) = 1 - 4 + 1 = -2.$$

In the event that a root lies between $a$ and $b$, and $f(a)$ and $f(b)$ are equally close to 0, choose either $a$ or $b$ as the first approximation.

(Note the change in sign.) Since $f(0)$ is closer to 0 than is $f(1)$, we choose 0 to be our first approximation, $x_1$. Now,

$$f'(x) = 4x^3 - 4,$$

so

$$f(x_n) = x_n^4 - 4x_n + 1 \qquad \text{and} \qquad f'(x_n) = 4x_n^3 - 4.$$

## Principles in Practice 1

**Approximating a Root by Newton's Method**

If the total profit (in dollars) from the sale of $x$ televisions is $P(x) = 20x - 0.01x^2 - 850 + 3\ln(x)$, use Newton's method to approximate the break-even quantities. (*Note:* There are two break-even quantities; one is between 10 and 50, and the other is between 1900 and 2000.) Give the $x$-value to the nearest integer.

**TABLE 15.1**

| $n$ | $x_n$ | $x_{n+1}$ |
|---|---|---|
| 1 | 0.00000 | 0.25000 |
| 2 | 0.25000 | 0.25099 |
| 3 | 0.25099 | 0.25099 |

Substituting into Eq. (4) gives the recursion formula

$$x_{n+1} = x_n - \frac{f(x_n)}{f'(x_n)} = x_n - \frac{x_n^4 - 4x_n + 1}{4x_n^3 - 4}$$

$$= \frac{4x_n^4 - 4x_n - x_n^4 + 4x_n - 1}{4x_n^3 - 4},$$

so

$$x_{n+1} = \frac{3x_n^4 - 1}{4x_n^3 - 4}. \tag{5}$$

Since $x_1 = 0$, letting $n = 1$ in Eq. (5) gives

$$x_2 = \frac{3x_1^4 - 1}{4x_1^3 - 4} = \frac{3(0)^4 - 1}{4(0)^3 - 4} = 0.25.$$

Letting $n = 2$ in Eq. (5) gives

$$x_3 = \frac{3x_2^4 - 1}{4x_2^3 - 4} = \frac{3(0.25)^4 - 1}{4(0.25)^3 - 4} \approx 0.25099.$$

Letting $n = 3$ in Eq. (5) gives

$$x_4 = \frac{3x_3^4 - 1}{4x_3^3 - 4} = \frac{3(0.25099)^4 - 1}{4(0.25099)^3 - 4} \approx 0.25099.$$

The data obtained thus far are displayed in Table 15.1. Since the values of $x_3$ and $x_4$ differ by less than 0.0001, we take the root to be 0.25099 (that is, $x_4$). ∎

**EXAMPLE 2  Approximating a Root by Newton's Method**

*Approximate the root of $x^3 = 3x - 1$ that lies between $-1$ and $-2$. Continue the approximation procedure until two successive approximations differ by less than 0.0001.*

**Solution:** Letting $f(x) = x^3 - 3x + 1$ [we need the form $f(x) = 0$], we find that

$$f(-1) = (-1)^3 - 3(-1) + 1 = 3$$

and

$$f(-2) = (-2)^3 - 3(-2) + 1 = -1.$$

(Note the change in sign.) Since $f(-2)$ is closer to 0 than is $f(-1)$, we choose $-2$ to be our first approximation, $x_1$. Now,

$$f'(x) = 3x^2 - 3,$$

so

$$f(x_n) = x_n^3 - 3x_n + 1 \quad \text{and} \quad f'(x_n) = 3x_n^2 - 3.$$

Substituting into Eq. (4) gives the recursion formula

$$x_{n+1} = x_n - \frac{f(x_n)}{f'(x_n)} = x_n - \frac{x_n^3 - 3x_n + 1}{3x_n^2 - 3},$$

so

$$x_{n+1} = \frac{2x_n^3 - 1}{3x_n^2 - 3}. \tag{6}$$

**TABLE 15.2**

| n | $x_n$ | $x_{n+1}$ |
|---|---|---|
| 1 | −2.00000 | −1.88889 |
| 2 | −1.88889 | −1.87945 |
| 3 | −1.87945 | −1.87939 |

[[M 670]]The situation where $x_1$ leads to a derivative of 0 occurs in Problems 2 and 8 of Exercise 15.4. [[E 670]]

Since $x_1 = -2$, letting $n = 1$ in Eq. (6) gives

$$x_2 = \frac{2x_1^3 - 1}{3x_1^2 - 3} = \frac{2(-2)^3 - 1}{3(-2)^2 - 3} \approx -1.88889.$$

Continuing in this way, we obtain Table 15.2. Because the values of $x_3$ and $x_4$ differ by 0.00006, which is less than 0.0001, we take the root to be −1.87939 (that is, $x_4$). ∎

In case your choice for the initial approximation, $x_1$, gives the derivative a value of zero, choose a different number that is close to the desired root. A graph of $f$ could be helpful in this situation. Finally, we remark that there are times when the sequence of approximations does not approach the root. A discussion of such situations is beyond the scope of this book.

## TECHNOLOGY

Figure 15.15 gives a short TI-82 program for Newton's method. Before the program is executed, the first approximation to the root of $f(x) = 0$ is stored as X, and $f(x)$ and $f'(x)$ are stored as $Y_1$ and $Y_2$, respectively. When executed, the program computes the first iteration and pauses. Successive iterations are obtained by successively pressing ENTER. Figure 15.16 shows the iterations for the problem in Example 2.

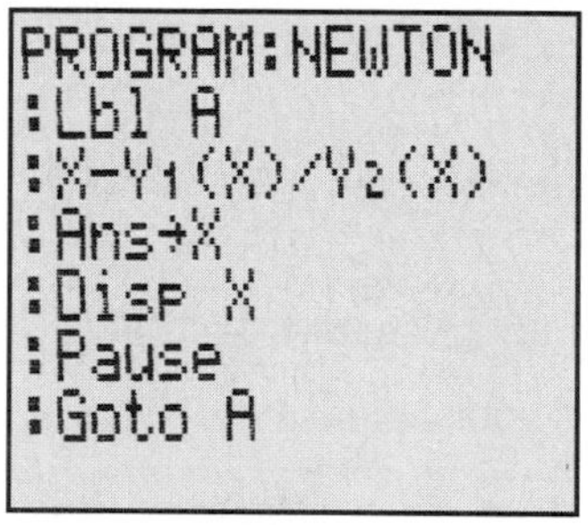

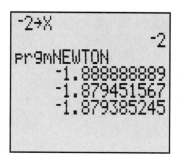

**FIGURE 15.15** Calculator program for Newton's method.

**FIGURE 15.16** Iterations for problem in Example 2.

## ■ Exercise 15.4

*In Problems 1–10, use Newton's method to approximate the indicated root of the given equation. Continue the approximation procedure until the difference of two successive approximations is less than 0.0001.*

1. $x^3 - 4x + 1 = 0$; root between 0 and 1.
2. $x^3 + 2x^2 - 1 = 0$; root between 0 and 1.
3. $x^3 - x - 1 = 0$; root between 1 and 2.
4. $x^3 - 9x + 6 = 0$; root between 2 and 3.
5. $x^3 + x + 16 = 0$; root between −3 and −2.
6. $x^3 = 2x + 5$; root between 2 and 3.
7. $x^4 = 3x - 1$; root between 0 and 1.
8. $x^4 + 4x - 1 = 0$; root between −2 and −1.
9. $x^4 - 2x^3 + x^2 - 3 = 0$; root between 1 and 2.
10. $x^4 - x^3 + x - 2 = 0$; root between 1 and 2.

11. Estimate, to three-decimal-place accuracy, the cube root of 70. [*Hint:* Show that the problem is equivalent to finding a root of $f(x) = x^3 - 70 = 0$. Choose 4 as the initial estimate. Continue the iteration until two successive approximations, rounded to three decimal places, are the same.]

12. Estimate $\sqrt[5]{40}$, to two-decimal-place accuracy. Use 2 as your initial estimate.

13. Find, to two-decimal-place accuracy, all real solutions of the equation $e^x = x + 5$. [*Hint:* A rough sketch of the graphs of $y = e^x$ and $y = x + 5$ makes it clear how many solutions there are. Use nearby integer values for your initial estimates.]

14. Find, to three-decimal-place accuracy, all real solutions of the equation $\ln x = 5 - x$.

15. **Break-Even Quantity** The cost of manufacturing $q$ tons of a certain product is given by

$$c = 250 + 2q - 0.1q^3,$$

and the revenue obtained by selling the $q$ tons is given by

$$r = 3q.$$

Approximate, to two-decimal-place accuracy, the break-even quantity. [*Hint:* Approximate a root of $r - c = 0$ by choosing 13 as your initial estimate.]

16. **Break-Even Quantity** The total cost of manufacturing $q$ hundred pencils is $c$ dollars, where

$$c = 40 + 3q + \frac{q^2}{1000} + \frac{1}{q}.$$

Pencils are sold for $7 per hundred.

**a.** Show that the break-even quantity is a solution of the equation

$$f(q) = \frac{q^3}{1000} - 4q^2 + 40q + 1 = 0.$$

**b.** Use Newton's method to approximate the solution of $f(q) = 0$, where $f(q)$ is given in part (a). Use 10 as your initial approximation, and give your answer to two-decimal-place accuracy.

17. **Equilibrium** Given the supply equation $p = 2q + 5$ and the demand equation $p = \frac{100}{q^2 + 1}$, use Newton's method to estimate the market equilibrium quantity. Give your answer to three-decimal-place accuracy.

18. **Equilibrium** Given the supply equation

$$p = 0.1q^3 + 0.6q + 2$$

and the demand equation $p = 30 - q$, use Newton's method to estimate the market equilibrium quantity, and find the corresponding equilibrium price. Use 5 as an initial estimate for the required value of $q$, and give your answer to two-decimal-place accuracy.

19. Use Newton's method to approximate (to two-decimal-place accuracy) a critical value of the function

$$f(x) = \frac{x^3}{3} - x^2 - 5x + 1$$

on the interval $[3, 4]$.

# 15.5 REVIEW

## IMPORTANT TERMS AND SYMBOLS

**Section 15.1**     economic lot size

**Section 15.2**     differential, $dy$, $dx$

**Section 15.3**     point elasticity of demand     elastic     inelastic     unit elasticity

**Section 15.4**     Newton's method

## SUMMARY

In a practical sense, the power of calculus is that it allows us to maximize or minimize quantities. For example, in the area of economics, we can maximize profit or minimize cost. Some important relationships that are used in economics problems are the following:

$$\bar{c} = \frac{c}{q}, \quad \text{average cost per unit} = \frac{\text{total cost}}{\text{quantity}},$$

$$r = pq, \quad \text{revenue} = (\text{price})(\text{quantity}),$$

$$P = r - c, \quad \text{profit} = \text{total revenue} - \text{total cost}.$$

If $y = f(x)$ is a differentiable function of $x$, we define the differential $dy$ by

$$dy = f'(x)\, dx,$$

where $dx$ (or $\Delta x$) is a change in $x$ and can be any real number. If $dx$ is close to zero, then $dy$ is an approximation to $\Delta y$, a change in $y$:

$$\Delta y \approx dy.$$

Moreover, $dy$ can be used to estimate a function value. We use the relationship

$$f(x + dx) \approx f(x) + dy.$$

Here $f(x + dx)$ is the value to be estimated; $x$ and $dx$ are chosen so that $f(x)$ is easy to compute and $dx$ is small.

If an equation defines $y$ as a function of $x$, then the derivative of $x$ with respect to $y$ is given by

$$\frac{dx}{dy} = \frac{1}{\frac{dy}{dx}}, \quad dy/dx \neq 0.$$

Point elasticity of demand is a number that measures how consumer demand is affected by a change in price. It is given by

$$\eta = \frac{p/q}{dp/dq},$$

where $p$ is the price per unit at which $q$ units are demanded. The three categories of elasticity are as follows:

$$|\eta| > 1, \quad \text{demand is elastic,}$$

$$|\eta| = 1, \quad \text{unit elasticity,}$$

$$|\eta| < 1, \quad \text{demand is inelastic.}$$

To put it simply, for a given percentage change in price, there is a greater percentage change in quantity demanded if demand is elastic, a smaller percentage change if demand is inelastic, and an equal percentage change if demand has unit elasticity.

The relationship between elasticity and the rate of change of revenue is given by

$$\frac{dr}{dq} = p\left(1 + \frac{1}{\eta}\right).$$

Newton's method is the name given to the following formula, which is used to approximate the roots of the equation $f(x) = 0$, provided that $f$ is differentiable:

$$x_{n+1} = x_n - \frac{f(x_n)}{f'(x_n)}, \quad n = 1, 2, 3, \ldots$$

## REVIEW PROBLEMS

1. **Maximization of Production** A manufacturer determined that $m$ employees on a certain production line will produce $q$ units per month, where

$$q = 80m^2 - 0.1m^4.$$

To obtain maximum monthly production, how many employees should be assigned to the production line?

2. **Revenue** The demand function for a manufacturer's product is given by $p = 100e^{-0.1q}$. For what value of $q$ does the manufacturer maximize total revenue?

3. **Revenue** The demand function for a monopolist's product is

$$p = \sqrt{600 - q}.$$

If the monopolist wants to produce at least 100 units, but not more than 300 units, how many units should be produced to maximize total revenue?

4. **Average Cost** If $c = 0.01q^2 + 5q + 100$ is a cost function, find the average-cost function. At what level of production $q$ is there a minimum average cost?

5. **Profit** The demand function for a monopolist's product is

$$p = 400 - 2q,$$

and the average cost per unit for producing $q$ units is

$$\bar{c} = q + 160 + \frac{2000}{q},$$

where $p$ and $\bar{c}$ are in dollars per unit. Find the maximum profit that the monopolist can achieve.

6. **Container Design** A rectangular box is to be made by cutting out equal squares from each corner of a piece of cardboard 10 in. by 16 in. and then folding up the sides. What must be the length of the side of the square cut out if the volume of the box is to be maximum?

7. **Fencing** A rectangular field is to be enclosed by a fence and divided equally into three parts by two fences parallel to one pair of the sides. If a total of 800 ft of fencing is to be used, find the dimensions of the field if its area is to be maximized.

8. **Poster Design** A rectangular poster having an area of 500 in.$^2$ is to have a 4-in. margin at each side and at the bottom and a 6-in. margin at the top. The remainder of the poster is for printed matter. Find the dimensions of the poster so that the area for the printed matter is maximized.

9. **Cost** A furniture company makes personal-computer stands. For a certain model, the total cost (in thousands of dollars) when $q$ *hundred* stands are produced is given by

$$c = 2q^3 - 9q^2 + 12q + 20.$$

a. The company is currently capable of manufacturing between 75 and 600 stands (inclusive) per week. Determine the number of stands that should be produced per week to minimize the total cost, and find the corresponding average cost per stand.

**b.** Suppose that between 300 and 600 stands must be produced. How many should the company now produce in order to minimize total cost?

**10. Bacteria**   In a laboratory, an experimental antibacterial agent is applied to a population of 100 bacteria. Data indicate that the number of bacteria $t$ hours after the agent is introduced is given by

$$N = \frac{14,400 + 120t + 100t^2}{144 + t^2}.$$

For what value of $t$ does the maximum number of bacteria in the population occur? What is this maximum number?

*In Problems 11 and 12, determine the differentials of the functions in terms of $x$ and $dx$.*

**11.** $f(x) = x^2 \ln(x + 5)$.

**12.** $f(x) = \dfrac{x^2 + 5}{x - 7}$.

**13. Temperature**   Fahrenheit temperature F and Celsius temperature C are related by $F = \frac{9}{5}C + 32$. Using differentials, find how much F would change due to a change in C of $\frac{1}{2}°$.

**14.** If $p = q^2 + 8q$, use differentials to estimate $\Delta p$ if $q$ changes from 4 to 4.02.

*In Problems 15 and 16, approximate the expressions by using differentials.*

**15.** $e^{-0.01}$.

**16.** $\sqrt{25.5}$.

**17.** If $x = 4y^2 + 7y - 3$, find $dy/dx$.

*For the demand equations in Problems 18–20, determine whether demand is elastic, is inelastic, or has unit elasticity for the indicated value of $q$.*

**18.** $p = \dfrac{500}{q}$;   $q = 200$.

**19.** $p = 900 - q^2$;   $q = 10$.

**20.** $p = 18 - 0.02q$;   $q = 600$.

**21.** The demand equation for a product is

$$p = 30 - \sqrt{q}.$$

**a.** Find the point elasticity of demand when $p = 10$.

**b.** Verify that demand is inelastic if $0 < p < 10$.

**22.** The demand equation of a product is

$$q = \sqrt{2500 - p^2}.$$

Find the point elasticity of demand when $p = 30$. If the price of 30 decreases $\frac{2}{3}\%$, what is the approximate change in demand?

**23.** The demand equation for a product is

$$q = \sqrt{100 - p}, \qquad \text{where } 0 < p < 100.$$

**a.** Find all prices that correspond to elastic demand.

**b.** Compute the point elasticity of demand when $p = 40$. Use your answer to estimate the percentage increase or decrease in demand when price is increased by 5% to $p = 42$.

**24.** The equation $x^3 - 2x - 2 = 0$ has a root between 1 and 2. Use Newton's method to estimate the root. Continue the approximation procedure until the difference of two successive approximations is less than 0.0001. Round your answer to four decimal places.

**25.** Find, to an accuracy of three decimal places, all real solutions of the equation $e^x = 3x$.

# Integration

Chapters 12–15 dealt with differential calculus. We differentiated a function and obtained another function, its derivative. *Integral calculus* is concerned with the reverse process: We are given the derivative of a function and must find the original function. The need for doing this arises in a natural way. For example, we may have a marginal-revenue function and want to find the revenue function from it. Integral calculus also involves a concept that allows us to take the limit of a special kind of sum as the number of terms in the sum becomes infinite. This is the real power of integral calculus! With such a notion, we may find the area of a region that cannot be found by any other convenient method.

## 16.1 THE INDEFINITE INTEGRAL

Given a function $f$, if $F$ is a function such that

$$F'(x) = f(x), \tag{1}$$

then $F$ is called an *antiderivative* of $f$. Thus,

an antiderivative of $f$ is simply a function whose derivative is $f$.

Multiplying both sides of Eq. (1) by the differential $dx$ gives $F'(x)\,dx = f(x)\,dx$. However, because $F'(x)\,dx$ is the differential of $F$, we have $dF = f(x)\,dx$. Hence, we can think of an antiderivative of $f$ as a function whose differential is $f(x)\,dx$.

**DEFINITION**

*An* **antiderivative** *of a function $f$ is a function $F$ such that*

$$F'(x) = f(x),$$

*or equivalently, in differential notation,*

$$dF = f(x)\,dx.$$

For example, because the derivative of $x^2$ is $2x$, $x^2$ is an antiderivative of $2x$. However, it is not the only antiderivative of $2x$: Since

$$\frac{d}{dx}(x^2 + 1) = 2x \quad \text{and} \quad \frac{d}{dx}(x^2 - 5) = 2x,$$

both $x^2 + 1$ and $x^2 - 5$ are also antiderivatives of $2x$. In fact, it is obvious that because the derivative of a constant is zero, $x^2 + C$ is also an antiderivative of $2x$ for *any* constant $C$. Thus, $2x$ has infinitely many antiderivatives. More importantly, *all* antiderivatives of $2x$ must be functions of the form $x^2 + C$, because of the following fact:

> Any two antiderivatives of a function differ only by a constant.

Since $x^2 + C$ describes all antiderivatives of $2x$, we can refer to it as being the *most general antiderivative* of $2x$, denoted by $\int 2x\, dx$, which is read "the *indefinite integral* of $2x$ with respect to $x$." Thus, we write

$$\int 2x\, dx = x^2 + C.$$

The symbol $\int$ is called the **integral sign,** $2x$ is the **integrand,** and $C$ is the **constant of integration.** The $dx$ is part of the integral notation and indicates the variable involved. Here $x$ is the **variable of integration.**

More generally, the **indefinite integral** of any function $f$ with respect to $x$ is written $\int f(x)\, dx$ and denotes the most general antiderivative of $f$. Since all antiderivatives of $f$ differ only by a constant, if $F$ is any antiderivative of $f$, then

$$\int f(x)\, dx = F(x) + C, \quad \text{where } C \text{ is a constant.}$$

To *integrate* $f$ means to find $\int f(x)\, dx$. In summary,

> $$\int f(x)\, dx = F(x) + C \quad \text{if and only if} \quad F'(x) = f(x).$$

---

**Principles in Practice 1**

**Finding an Indefinite Integral**

If the marginal cost for a company is $f(q) = 28.3$, find $\int 28.3\, dq$, which gives the form of the cost function.

---

**EXAMPLE 1**    **Finding an Indefinite Integral**

*Find* $\int 5\, dx$.

*Solution:*

> *Strategy:* First we must find (perhaps better words are "guess at") a function whose derivative is 5. Then we add the constant of integration.

Since we know that the derivative of $5x$ is 5, $5x$ is an antiderivative of 5. Therefore,

$$\int 5\, dx = 5x + C. \qquad \blacksquare$$

**Pitfall** ▼ It is **incorrect** to write

$$\int 5dx = 5x.$$

**Do not forget the constant of integration.**

Using differentiation formulas from Chapters 12 and 13, we have compiled a list of basic integration formulas in Table 16.1. These formulas are easily verified.

**TABLE 16.1**   Basic Integration Formulas

**1.** $\int k\, dx = kx + C,\quad k$ is a constant.

**2.** $\int x^n\, dx = \dfrac{x^{n+1}}{n+1} + C,\quad n \neq -1.$

**3.** $\int e^x\, dx = e^x + C.$

**4.** $\int kf(x)\, dx = k \int f(x)\, dx,\quad k$ is a constant.

**5.** $\int [f(x) \pm g(x)]\, dx = \int f(x)\, dx \pm \int g(x)\, dx.$

For example, Formula 2 is true because the derivative of $x^{n+1}/(n+1)$ is $x^n$ for $n \neq -1$. (We must have $n \neq -1$ because the denominator is 0 when $n = -1$.) Formula 2 states that the indefinite integral of a power of $x$ (except $x^{-1}$) is obtained by increasing the exponent of $x$ by one, dividing by the new exponent, and adding on the constant of integration. The indefinite integral of $x^{-1}$ will be discussed in Sec. 16.3.

To verify Formula 4, we must show that the derivative of $k \int f(x)\, dx$ is $kf(x)$. Since the derivative of $k \int f(x)\, dx$ is simply $k$ times the derivative of $\int f(x)\, dx$, which is $f(x)$, Formula 4 is verified. You should verify the other formulas. Formula 5 can be extended to any number of sums or differences.

**EXAMPLE 2**   Indefinite Integrals of a Constant and of a Power of $x$

**a.** *Find* $\int 1\, dx.$

*Solution:* By Formula 1 with $k = 1$,
$$\int 1\, dx = 1x + C = x + C.$$
Usually, we write $\int 1\, dx$ as $\int dx$. Thus, $\int dx = x + C.$

**b.** *Find* $\int x^5\, dx.$

*Solution:* By Formula 2 with $n = 5$,
$$\int x^5\, dx = \frac{x^{5+1}}{5+1} + C = \frac{x^6}{6} + C.$$

**EXAMPLE 3**   Indefinite Integral of a Constant Times a Function of $x$

*Find* $\int 7x\, dx.$

*Solution:* By Formula 4 with $k = 7$ and $f(x) = x$,
$$\int 7x\, dx = 7 \int x\, dx.$$
Since $x$ is $x^1$, by Formula 2 we have
$$\int x^1\, dx = \frac{x^{1+1}}{1+1} + C_1 = \frac{x^2}{2} + C_1,$$
where $C_1$ is the constant of integration. Therefore,

**Principles in Practice 2**

Indefinite Integral of a Constant Times a Function of $t$

If the rate of change of a company's revenues can be modeled by $\dfrac{dR}{dt} = 0.12t^2$, then find $\int 0.12t^2\, dt$, which gives the form of the company's revenue function.

$$\int 7x \, dx = 7 \int x \, dx = 7 \left[ \frac{x^2}{2} + C_1 \right] = \frac{7}{2}x^2 + 7C_1.$$

Since $7C_1$ is just an arbitrary constant, we shall replace it by $C$ for simplicity. Thus,

$$\int 7x \, dx = \frac{7}{2}x^2 + C.$$

It is not necessary to write all intermediate steps when integrating. More simply, we write

$$\int 7x \, dx = (7)\frac{x^2}{2} + C = \frac{7}{2}x^2 + C.$$ ∎

**Pitfall** ▼ Only a *constant* factor of the integrand can "jump" in front of an integral sign. Because $x$ is not a constant, $\int 7x \, dx \neq 7x \int dx = (7x)(x + C) = 7x^2 + 7Cx$.

**EXAMPLE 4** Indefinite Integral of a Constant Times a Function of $x$

*Find* $\int -\frac{3}{5}e^x \, dx$.

*Solution:*

$$\int -\frac{3}{5}e^x \, dx = -\frac{3}{5}\int e^x \, dx \qquad \text{(Formula 4).}$$

$$= -\frac{3}{5}e^x + C \qquad \text{(Formula 3).}$$ ∎

---

**Principles in Practice 3**

**Finding Indefinite Integrals**

Due to new competition, the number of subscriptions to a certain magazine is declining at a rate of $-\frac{480}{t^3}$, subscriptions per month, where $t$ is the number of months since the competition entered the market. Find the form of the equation for the number of subscribers to the magazine.

---

**EXAMPLE 5** Finding Indefinite Integrals

**a.** *Find* $\int \frac{1}{\sqrt{t}} \, dt$.

*Solution:* Here $t$ is the variable of integration. We rewrite the integrand so that a basic formula can be used. Since $1/\sqrt{t} = t^{-1/2}$, applying Formula 2 gives

$$\int \frac{1}{\sqrt{t}} \, dt = \int t^{-1/2} \, dt = \frac{t^{(-1/2)+1}}{-\frac{1}{2} + 1} + C = \frac{t^{1/2}}{\frac{1}{2}} + C = 2\sqrt{t} + C.$$

**b.** *Find* $\int \frac{1}{6x^3} \, dx$.

*Solution:*

$$\int \frac{1}{6x^3} \, dx = \frac{1}{6}\int x^{-3} \, dx = \left(\frac{1}{6}\right)\frac{x^{-3+1}}{-3 + 1} + C$$

$$= -\frac{x^{-2}}{12} + C = -\frac{1}{12x^2} + C.$$ ∎

**EXAMPLE 6** Indefinite Integral of a Sum

*Find* $\int (x^2 + 2x) \, dx$.

The rate of growth of the population of a new city is estimated by $\frac{dN}{dt} = 500 + 300\sqrt{t}$, where $t$ is in years. Find $\int (500 + 300\sqrt{t})\, dt$.

When integrating an expression involving more than one term, one needs only one constant of integration.

**Solution:** By Formula 5,

$$\int (x^2 + 2x)\, dx = \int x^2\, dx + \int 2x\, dx.$$

Now,

$$\int x^2\, dx = \frac{x^{2+1}}{2+1} + C_1 - \frac{x^3}{3} + C_1,$$

and

$$\int 2x\, dx = 2\int x\, dx = (2)\frac{x^{1+1}}{1+1} + C_2 = x^2 + C_2.$$

Thus,

$$\int (x^2 + 2x)\, dx = \frac{x^3}{3} + x^2 + C_1 + C_2.$$

For convenience, we shall replace the constant $C_1 + C_2$ by $C$. We then have

$$\int (x^2 + 2x)\, dx = \frac{x^3}{3} + x^2 + C.$$

Omitting intermediate steps, we simply integrate term by term and write

$$\int (x^2 + 2x)\, dx + \frac{x^3}{3} + (2)\frac{x^2}{2} + C = \frac{x^3}{3} + x^2 + C. \qquad \blacksquare$$

Suppose the rate of savings in the United States is given by $\frac{dS}{dt} = 2.1t^2 - 65.4t + 491.6$, where $t$ is the time in years and $S$ is the amount of money saved in billions of dollars. Find the form of the equation for the amount of money saved.

**EXAMPLE 7   Indefinite Integral of a Sum and Difference**

*Find* $\int (2\sqrt[5]{x^4} - 7x^3 + 10e^x - 1)\, dx$.

**Solution:**

$$\int (2\sqrt[5]{x^4} - 7x^3 + 10e^x - 1)\, dx$$

$$= 2\int x^{4/5}\, dx - 7\int x^3\, dx + 10\int e^x\, dx - \int 1\, dx$$

$$= (2)\frac{x^{9/5}}{\frac{9}{5}} - (7)\frac{x^4}{4} + 10e^x - x + C$$

$$= \frac{10}{9}x^{9/5} - \frac{7}{4}x^4 + 10e^x - x + C. \qquad \blacksquare$$

Sometimes, in order to apply the basic integration formulas, it is necessary to first perform algebraic manipulations on the integrand, as Example 8 shows.

**EXAMPLE 8   Using Algebraic Manipulation to Find an Indefinite Integral**

*Find* $\int y^2(y + \frac{2}{3})\, dy$.

**Solution:** The integrand does not fit a familiar integration form. However, by multiplying the integrand we get

$$\int y^2(y + \tfrac{2}{3})\, dy = \int (y^3 + \tfrac{2}{3}y^2)\, dy$$

$$= \frac{y^4}{4} + \left(\frac{2}{3}\right)\frac{y^3}{3} + C = \frac{y^4}{4} + \frac{2y^3}{9} + C. \qquad \blacksquare$$

**Pitfall** ▼ In Example 8, we first multiplied the factors in the integrand. We point out that

The integral of a product is *not* the product of integrals.

$$\int y^2(y + \tfrac{2}{3})\, dy \neq \left[\int y^2\, dy\right]\left[\int (y + \tfrac{2}{3})\, dy\right].$$

More generally,

$$\int f(x)g(x)\, dx \neq \int f(x)\, dx \cdot \int g(x)\, dx.$$

**EXAMPLE 9**  Using Algebraic Manipulation to Find an Indefinite Integral

a. *Find* $\displaystyle \int \frac{(2x - 1)(x + 3)}{6}\, dx.$

**Solution:** By factoring out the constant $\frac{1}{6}$ and multiplying the binomials, we get

$$\int \frac{(2x - 1)(x + 3)}{6}\, dx = \frac{1}{6}\int (2x^2 + 5x - 3)\, dx$$

$$= \frac{1}{6}\left[(2)\frac{x^3}{3} + (5)\frac{x^2}{2} - 3x\right] + C$$

$$= \frac{x^3}{9} + \frac{5x^2}{12} - \frac{x}{2} + C.$$

b. *Find* $\displaystyle \int \frac{x^3 - 1}{x^2}\, dx.$

Another algebraic approach to (b) is

$$\int \frac{x^3 - 1}{x^2}\, dx$$

$$= \int (x^3 - 1)x^{-2}\, dx$$

$$= \int (x - x^{-2})\, dx,$$

and so on.

**Solution:** We can break up the integrand into fractions by dividing each term in the numerator by the denominator:

$$\int \frac{x^3 - 1}{x^2}\, dx = \int \left(\frac{x^3}{x^2} - \frac{1}{x^2}\right) dx = \int (x - x^{-2})\, dx$$

$$= \frac{x^2}{2} - \frac{x^{-1}}{-1} + C = \frac{x^2}{2} + \frac{1}{x} + C.$$ ■

## ▪ Exercise 16.1

*In Problems 1–52, find the indefinite integrals.*

**1.** $\displaystyle \int 5\, dx.$

**2.** $\displaystyle \int \tfrac{1}{2}\, dx.$

**3.** $\displaystyle \int x^8\, dx.$

**4.** $\displaystyle \int 2x^{25}\, dx.$

**5.** $\displaystyle \int 5x^{-7}\, dx.$

**6.** $\displaystyle \int \frac{z^{-3}}{3}\, dz.$

**7.** $\displaystyle \int \frac{1}{x^{10}}\, dx.$

**8.** $\displaystyle \int \frac{7}{x^4 dx}.$

**9.** $\displaystyle \int \frac{1}{y^{11/5}}\, dy.$

**10.** $\displaystyle \int \frac{7}{2x^{9/4}}\, dx.$

**11.** $\displaystyle \int (8 + u)\, du.$

**12.** $\displaystyle \int (r^3 + 2r)\, dr.$

**13.** $\displaystyle \int (y^5 - 5y)\, dy.$

**14.** $\displaystyle \int (7 - 3w - 2w^2)\, dw.$

**15.** $\displaystyle \int (3t^2 - 4t + 5)\, dt.$

**16.** $\displaystyle \int (1 + u + u^2 + u^3)\, du.$

**17.** $\displaystyle \int (7 + e)\, dx.$

**18.** $\displaystyle \int (5 - 2^{-1})\, dx.$

**19.** $\displaystyle \int \left(\frac{x}{7} - \frac{3}{4}x^4\right) dx.$

**20.** $\displaystyle \int \left(\frac{2x^2}{7} - \frac{8}{3}x^4\right) dx.$

**21.** $\displaystyle \int 3e^x\, dx.$

**22.** $\displaystyle \int \left(\frac{e^x}{3} + 2x\right) dx.$

**23.** $\int (x^{8.3} - 9x^6 + 3x^{-4} + x^{-3})\, dx.$

**24.** $\int (0.3y^4 - 8y^{-3} + 2)\, dy.$

**25.** $\int \dfrac{-2\sqrt{x}}{3}\, dx.$

**26.** $\int dw.$

**27.** $\int \dfrac{1}{4\sqrt[8]{x^2}}\, dx.$

**28.** $\int \dfrac{-3}{(2x)^2}\, dx.$

**29.** $\int \left( \dfrac{x^3}{3} - \dfrac{3}{x^3} \right) dx.$

**30.** $\int \left( \dfrac{1}{2x^3} - \dfrac{1}{x^4} \right) dx.$

**31.** $\int \left( \dfrac{3w^2}{2} - \dfrac{2}{3w^2} \right) dw.$

**32.** $\int \dfrac{2}{e^{-s}}\, ds.$

**33.** $\int \dfrac{2z - 5}{7}\, dz.$

**34.** $\int \dfrac{1}{12}\left( \dfrac{1}{3}e^x \right) dx.$

**35.** $\int (x^e + e^x)\, dx.$

**36.** $\int \left( 3y^3 - 2y^2 + \dfrac{e^y}{6} \right) dy.$

**37.** $\int (2\sqrt{x} - 3\sqrt[4]{x})\, dx.$

**38.** $\int 0\, dx.$

**39.** $\int \left( -\dfrac{\sqrt[3]{x^2}}{5} - \dfrac{7}{2\sqrt{x}} + 6x \right) dx.$

**40.** $\int \left( \sqrt{x} - \dfrac{1}{\sqrt[3]{x}} \right) dx.$

**41.** $\int (x^2 + 5)(x - 3)\, dx.$

**42.** $\int x^4(x^3 + 3x^2 + 7)\, dx.$

**43.** $\int \sqrt{x}(x + 3)\, dx.$

**44.** $\int (z + 2)^2\, dz.$

**45.** $\int (2u + 1)^2\, du.$

**46.** $\int \left( \dfrac{1}{\sqrt[3]{x}} + 1 \right)^2 dx.$

**47.** $\int v^{-2}(2v^4 + 3v^2 - 2v^{-3})\, dv.$

**48.** $\int [6e^u - u^3(\sqrt{u} + 1)]\, du.$

**49.** $\int \dfrac{z^4 + 10z^3}{5z^2}\, dz.$

**50.** $\int \dfrac{x^4 - 3x^2 + 4x}{x}\, dx.$

**51.** $\int \dfrac{e^x + e^{2x}}{e^x}\, dx.$

**52.** $\int \dfrac{(x^4 + 1)^2}{x^3}\, dx.$

**53.** If $F(x)$ and $G(x)$ are such that $F'(x) = G'(x)$, is it true that $F(x) - G(x)$ must be zero?

**54. a.** Find a function $F$ such that $\int F(x)\, dx = xe^x + C.$

**b.** Is there only one function $F$ satisfying the equation given in part (a), or are there many such functions?

**55.** Find $\int \dfrac{d}{dx}\left( \dfrac{1}{\sqrt{x^2 + 1}} \right) dx.$

---

OBJECTIVE

To find a particular antiderivative of a function that satisfies certain conditions. This involves evaluating a constant of integration.

## 16.2  INTEGRATION WITH INITIAL CONDITIONS

If we know the rate of change, $f'$, of the function $f$, then the function $f$ itself is an antiderivative of $f'$ (since the derivative of $f$ is $f'$). Of course, there are many antiderivatives of $f'$, and the most general one is denoted by the indefinite integral. For example, if

$$f'(x) = 2x,$$

then,

$$f(x) = \int f'(x)\, dx = \int 2x\, dx = x^2 + C. \qquad (1)$$

That is, *any* function of the form $f(x) = x^2 + C$ has its derivative equal to $2x$. Because of the constant of integration, notice that we do not know $f(x)$ specifically. However, if $f$ must assume a certain function value for a particular value of $x$, then we can determine the value of $C$ and thus determine $f(x)$ specifically. For instance, if $f(1) = 4$, then from Eq. (1),

$$f(1) = 1^2 + C,$$

$$4 = 1 + C,$$

$$C = 3.$$

Thus,

$$f(x) = x^2 + 3.$$

That is, we now know the particular function $f(x)$ for which $f'(x) = 2x$ and $f(1) = 4$. The condition $f(1) = 4$, which gives a function value of $f$ for a specific value of $x$, is called an *initial condition* (or *boundary value*).

---

**Principles in Practice 1**

**Initial-Condition Problem**

The rate of growth of a species of bacteria is estimated by $\dfrac{dN}{dt} = 800 + 200e^t$, where $N$ is the number of bacteria (in thousands) after $t$ hours. If $N(5) = 40{,}000$, find $N(t)$.

---

**EXAMPLE 1**  **Initial-Condition Problem**

*If $y$ is a function of $x$ such that $y' = 8x - 4$ and $y(2) = 5$, find $y$. [Note: $y(2) = 5$ means that $y = 5$ when $x = 2$.] Also, find $y(4)$.*

*Solution:* Here $y(2) = 5$ is the initial condition. Since $y' = 8x - 4$, $y$ is an antiderivative of $8x - 4$:

$$y = \int (8x - 4)\, dx = 8 \cdot \frac{x^2}{2} - 4x + C = 4x^2 - 4x + C. \qquad (2)$$

We can determine the value of $C$ by using the initial condition. Because $y = 5$ when $x = 2$, from Eq. (2), we have,

$$5 = 4(2)^2 - 4(2) + C,$$
$$5 = 16 - 8 + C,$$
$$C = -3.$$

Replacing $C$ by $-3$ in Eq. (2) gives the function that we seek:

$$y = 4x^2 - 4x - 3. \qquad (3)$$

To find $y(4)$, we let $x = 4$ in Eq. (3):

$$y(4) = 4(4)^2 - 4(4) - 3 = 64 - 16 - 3 = 45. \qquad \blacksquare$$

---

**Principles in Practice 2**

**Initial-Conditions Problem Involving $y''$**

The acceleration of an object after $t$ seconds is given by $y'' = 84t + 24$, the velocity at 8 seconds is given by $y'(8) = 2891$ feet/sec, and the position at 2 seconds is given by $y(2) = 185$ feet. Find $y(t)$.

---

**EXAMPLE 2**  **Initial-Conditions Problem Involving $y''$**

*Given that $y'' = x^2 - 6$, $y'(0) = 2$, and $y(1) = -1$, find $y$.*

*Solution:*

> *Strategy:* To go from $y''$ to $y$, two integrations are needed: the first to take us from $y''$ to $y'$ and the other to take us from $y'$ to $y$. Hence, there will be two constants of integration, which we shall denote by $C_1$ and $C_2$.

Since $y'' = \dfrac{d}{dx}(y') = x^2 - 6$, $y'$ is an antiderivative of $x^2 - 6$. Thus,

$$y' = \int (x^2 - 6)\, dx = \frac{x^3}{3} - 6x + C_1. \qquad (4)$$

Now, $y'(0) = 2$ means that $y' = 2$ when $x = 0$; therefore, from Eq. (4), we have

$$2 = \frac{0^3}{3} - 6(0) + C_1.$$

Hence, $C_1 = 2$, so

$$y' = \frac{x^3}{3} - 6x + 2.$$

By integration, we can find $y$:

$$y = \int \left( \frac{x^3}{3} - 6x + 2 \right) dx$$

$$= \left( \frac{1}{3} \right) \frac{x^4}{4} - (6)\frac{x^2}{2} + 2x + C_2,$$

so

$$y = \frac{x^4}{12} - 3x^2 + 2x + C_2. \tag{5}$$

Now, since $y = -1$ when $x = 1$, we have, from Eq. (5),

$$-1 = \frac{1^4}{12} - 3(1)^2 + 2(1) + C_2.$$

Thus, $C_2 = -\frac{1}{12}$, so

$$y = \frac{x^4}{12} - 3x^2 + 2x - \frac{1}{12}. \qquad \blacksquare$$

Integration with initial conditions is applicable to many applied situations, as the next three examples illustrate.

### EXAMPLE 3  Income and Education

*For a particular urban group, sociologists studied the current average yearly income y (in dollars) that a person can expect to receive with x years of education before seeking regular employment. They estimated that the rate at which income changes with respect to education is given by*

$$\frac{dy}{dx} = 10x^{3/2}, \qquad 4 \le x \le 16,$$

*where y = 9872 when x = 9. Find y.*

***Solution:*** Here $y$ is an antiderivative of $10x^{3/2}$. Thus,

$$y = \int 10x^{3/2}\,dx = 10\int x^{3/2}\,dx$$

$$= (10)\frac{x^{5/2}}{\frac{5}{2}} + C.$$

$$y = 4x^{5/2} + C. \tag{6}$$

The initial condition is that $y = 9872$ when $x = 9$. By substituting these values into Eq. (6), we can determine the value of $C$:

$$9872 = 4(9)^{5/2} + C$$

$$= 4(243) + C.$$

$$9872 = 972 + C.$$

Therefore, $C = 8900$, and

$$y = 4x^{5/2} + 8900. \qquad \blacksquare$$

**EXAMPLE 4   Finding the Demand Function from Marginal Revenue**

*If the marginal-revenue function for a manufacturer's product is*

$$\frac{dr}{dq} = 2000 - 20q - 3q^2,$$

*find the demand function.*

*Solution:*

> *Strategy:* By integrating $dr/dq$ and using an initial condition, we can find the revenue function $r$. But revenue is also given by the general relationship $r = pq$, where $p$ is the price per unit. Thus, $p = r/q$. Replacing $r$ in this equation by the revenue function yields the demand function.

Since $dr/dq$ is the derivative of total revenue $r$,

$$r = \int (2000 - 20q - 3q^2)\, dq$$

$$= 2000q - (20)\frac{q^2}{2} - (3)\frac{q^3}{3} + C,$$

or

$$r = 2000q - 10q^2 - q^3 + C. \tag{7}$$

Revenue is 0 when $q$ is 0.

We assume that **when no units are sold, total revenue is 0;** that is, $r = 0$ when $q = 0$. This is our initial condition. Substituting these values into Eq. (7) gives

$$0 = 2000(0) - 10(0)^2 - 0^3 + C.$$

Although $q = 0$ gives $C = 0$, this is not true in general. It occurs in this section because the revenue functions are polynomials. In later sections, evaluating when $q = 0$ may produce a nonzero value for $C$.

Hence, $C = 0$, and

$$r = 2000q - 10q^2 - q^3.$$

To find the demand function, we use the fact that $p = r/q$ and substitute for $r$:

$$p = \frac{r}{q} = \frac{2000q - 10q^2 - q^3}{q}.$$

$$p = 2000 - 10q - q^2. \qquad\blacksquare$$

**EXAMPLE 5   Finding Cost from Marginal Cost**

*In the manufacture of a product, fixed costs per week are $4000. Fixed costs are costs, such as rent and insurance, that remain constant at all levels of production during a given time period. If the marginal-cost function is*

$$\frac{dc}{dq} = 0.000001(0.002q^2 - 25q) + 0.2,$$

*where c is the total cost (in dollars) of producing q pounds of product per week, find the cost of producing 10,000 lb in 1 week.*

*Solution:* Since $dc/dq$ is the derivative of the total cost $c$,

$$c = \int \left[ 0.000001(0.002q^2 - 25q) + 0.2 \right] dq$$

$$= 0.000001 \int (0.002q^2 - 25q)\, dq + \int 0.2\, dq.$$

$$c = 0.000001 \left( \frac{0.002q^3}{3} - \frac{25q^2}{2} \right) + 0.2q + C.$$

When $q$ is 0, total cost is equal to fixed cost.

Although $q = 0$ gives $C$ a value equal to fixed costs, this is not true in general. It occurs in this section because the cost functions are polynomials. In later sections, evaluating when $q = 0$ may produce a value for $C$ that is different from fixed cost.

Fixed costs are constant regardless of output. Therefore, when $q = 0$, $c = 4000$, which is our initial condition. By substitution, we find that $C = 4000$, so

$$c = 0.000001\left(\frac{0.002q^3}{3} - \frac{25q^2}{2}\right) + 0.2q + 4000. \tag{8}$$

From Eq. (8), when $q = 10,000, c = 5416\frac{2}{3}$. Thus, the total cost for producing 10,000 pounds of product in 1 week is \$5416.67. ∎

## ▪ Exercise 16.2

*In Problems 1 and 2, find y, subject to the given conditions.*

**1.** $dy/dx = 3x - 4$;  $y(-1) = \frac{13}{2}$.

**2.** $dy/dx = x^2 - x$;  $y(3) = 4$.

*In Problems 3 and 4, if y satisfies the given conditions, find y(x) for the given value of x.*

**3.** $y' = 4/\sqrt{x}$,  $y(4) = 10$;  $x = 9$.

**4.** $y' = -x^2 + 2x$,  $y(2) = 1$;  $x = 1$.

*In Problems 5–8, find y, subject to the given conditions.*

**5.** $y'' = -x^2 - 2x$;  $y'(1) = 0$, $y(1) = 1$.

**6.** $y'' = x + 1$;  $y'(0) = 0$, $y(0) = 5$.

**7.** $y''' = 2x$;  $y''(-1) = 3$, $y'(3) = 10$, $y(0) = 2$.

**8.** $y''' = e^x + 1$;  $y''(0) = 1$, $y'(0) = 2$, $y(0) = 3$.

*In Problems 9–12, dr/dq is a marginal-revenue function. Find the demand function.*

**9.** $dr/dq = 0.7$.

**10.** $dr/dq = 15 - \frac{1}{15}q$.

**11.** $dr/dq = 275 - q - 0.3q^2$.

**12.** $dr/dq = 10,000 - 2(2q + q^3)$.

*In Problems 13–16, dc/dq is a marginal-cost function, and fixed costs are indicated in braces. For Problems 13 and 14, find the total-cost function. For Problems 15 and 16, find the total cost for the indicated value of q.*

**13.** $dc/dq = 1.35$;  $\{200\}$.

**14.** $dc/dq = 2q + 50$;  $\{1000\}$.

**15.** $dc/dq = 0.09q^2 - 1.2q + 4.5$;  $\{7700\}$;  $q = 10$.

**16.** $dc/dq = 0.000102q^2 - 0.034q + 5$; $\{10,000\}$; $q = 100$

**17. Diet for Rats** A group of biologists studied the nutritional effects on rats that were fed a diet containing 10% protein.[1] The protein consisted of yeast and corn flour.

Over a period of time, the group found that the (approximate) rate of change of the average weight gain $G$ (in grams) of a rat with respect to the percentage $P$ of yeast in the protein mix was

$$\frac{dG}{dP} = -\frac{P}{25} + 2, \qquad 0 \le P \le 100.$$

If $G = 38$ when $P = 10$, find $G$.

**18. Winter Moth** A study of the winter moth was made in Nova Scotia.[2] The prepupae of the moth fall onto the ground from host trees. It was found that the (approximate) rate at which prepupal density $y$ (the number of prepupae per square foot of soil) changes with respect to distance $x$ (in feet) from the base of a host tree is

$$\frac{dy}{dx} = -1.5 - x, \qquad 1 \le x \le 9.$$

If $y = 57.3$ when $x = 1$, find $y$.

**19. Fluid Flow** In the study of the flow of fluid in a tube of constant radius $R$, such as blood flow in portions of the body, one can think of the tube as consisting of concen-

[1]Adapted from R. Bressani, "The Use of Yeast in Human Foods," in *Single-Cell Protein*, ed. R. I, Mateles and S. R. Tannenbaum (Cambridge, MA: MIT Press, 1968).

[2]Adapted from D. G. Embree, "The Population Dynamics of the Winter Moth in Nova Scotia, 1954–1962," *Memoirs of the Entomological Society of Canada*, no. 46 (1965).

tric tubes of radius $r$, where $0 \geq r \geq R$. The velocity $v$ of the fluid is a function of $r$ and is given by[3]

$$v = \int - \frac{(P_1 - P_2)r}{2l\eta} \, dr,$$

where $P_1$ and $P_2$ are pressures at the ends of the tube, $\eta$ (a Greek letter read "eta") is fluid viscosity, and $l$ is the length of the tube. If $v = 0$ when $r = R$, show that

$$v = \frac{(P_1 - P_2)(R^2 - r^2)}{4l\eta}.$$

20. **Elasticity of Demand** The sole producer of a product has determined that the marginal-revenue function is

$$\frac{dr}{dq} = 100 - 3q^2.$$

[3] R.W. Stacy et al., *Essentials of Biological and Medical Physics* (New York: McGraw-Hill, 1955).

Determine the point elasticity of demand for the product when $q = 5$. [*Hint:* First find the demand function.]

21. **Average Cost** A manufacturer has determined that the marginal-cost function is

$$\frac{dc}{dq} = 0.003q^2 - 0.4q + 40,$$

where $q$ is the number of units produced. If marginal cost is \$27.50 when $q = 50$, and fixed costs are \$5000, what is the *average* cost of producing 100 units?

22. If $f''(x) = 6x + 2$ and $f'(-1) = 5$, evaluate

$$f(1) - f(-1).$$

---

## 16.3 More Integration Formulas

### Power Rule for Integration

The formula

$$\int x^n \, dx = \frac{x^{n+1}}{n + 1} + C, \quad \text{if } n \neq -1,$$

which applies to a power of $x$, can be generalized to handle a power of a *function* of $x$. Let $u$ be a differentiable function of $x$. By the power rule for differentiation, if $n \neq -1$, then

$$\frac{d}{dx}\left(\frac{[u(x)]^{n+1}}{n + 1}\right) = \frac{(n + 1)[u(x)]^n \cdot u'(x)}{n + 1} = [u(x)]^n \cdot u'(x).$$

Thus,

$$\int [u(x)]^n \cdot u'(x) \, dx = \frac{[u(x)]^{n+1}}{n + 1} + C, \quad n \neq -1.$$

We call this the *power rule for integration*. Note that $u'(x) \, dx$ is the differential of $u$, namely $du$. In mathematical shorthand, we can replace $u(x)$ by $u$ and $u'(x) \, dx$ by $du$:

> **Power Rule for Integration**
>
> If $u$ is differentiable, then
>
> $$\int u^n \, du = \frac{u^{n+1}}{n + 1} + C, \quad \text{if } n \neq -1.$$

It is essential that you realize the difference between the power rule for integration and the formula for $\int x^n \, dx$. In the power rule, $u$ represents a function, whereas in $\int x^n \, dx$, $x$ is a variable.

**EXAMPLE 1 Applying the Power Rule for Integration**

**a.** *Find* $\int (x + 1)^{20} \, dx.$

**Solution:** Since the integrand is a power of the function $x + 1$, we shall set $u = x + 1$. Then $du = dx$, and $\int (x + 1)^{20} \, dx$ has the form $\int u^{20} \, du$. By the power rule for integration,

$$\int (x + 1)^{20} \, dx = \int u^{20} \, du = \frac{u^{21}}{21} + C = \frac{(x + 1)^{21}}{21} + C.$$

Note that we give our answer not in terms of $u$, but explicitly in terms of $x$.

**b.** *Find* $\int 3x^2(x^3 + 7)^3 \, dx.$

**Solution:** We observe that the integrand contains a power of the function $x^3 + 7$. Let $u = x^3 + 7$. Then $du = 3x^2 \, dx$. Fortunately, $3x^2$ appears as a factor in the integrand and can be used as part of $du$. We thus have

$$\int 3x^2(x^3 + 7)^3 \, dx = \int (x^3 + 7)^3 [3x^2 dx] = \int u^3 \, du$$

$$= \frac{u^4}{4} + C = \frac{(x^3 + 7)^4}{4} + C. \qquad \blacksquare$$

After integrating, you may wonder what happened to $3x^2$; $3x^2$, together with $dx$, forms the differential of $u$ in the power rule.

In order to apply the power rule for integration, sometimes an adjustment must be made to obtain $du$ in the integrand, as Example 2 illustrates.

**EXAMPLE 2 Adjusting for *du***

*Find* $\int x\sqrt{x^2 + 5} \, dx.$

**Solution:** We can write this as $\int x(x^2 + 5)^{1/2} \, dx$. Notice that the integrand contains a power of the function $x^2 + 5$. If $u = x^2 + 5$, then $du = 2x \, dx$. Since the *constant* factor 2 in $du$ does *not* appear in the integrand, this integral does not have the form $\int u^n \, du$. However, we can put the given integral in this form by first multiplying and dividing the integrand by 2. This does not change its value. Thus,

$$\int x(x^2 + 5)^{1/2} \, dx = \int \frac{2}{2}x(x^2 + 5)^{1/2} \, dx = \int \frac{1}{2}(x^2 + 5)^{1/2}[2x \, dx].$$

Moving the *constant* factor $\frac{1}{2}$ in front of the integral sign, we have

$$\int x(x^2 + 5)^{1/2} \, dx = \frac{1}{2}\int (x^2 + 5)^{1/2}[2x \, dx] \qquad (1)$$

$$= \frac{1}{2}\int u^{1/2} \, du = \frac{1}{2}\left[\frac{u^{3/2}}{\frac{3}{2}}\right] + C.$$

Going back to $x$ gives

$$\int x\sqrt{x^2 + 5} \, dx = \frac{(x^2 + 5)^{3/2}}{3} + C. \qquad \blacksquare$$

In Example 2, we needed the factor 2 in the integrand. In Eq. (1), it was inserted, and the integral was simultaneously multiplied by $\frac{1}{2}$. More generally, if $c$ is a nonzero constant, then

$$\int f(x) \, dx = \int \frac{c}{c}f(x) \, dx = \frac{1}{c}\int cf(x) \, dx.$$

We can adjust for constant factors, but not variable factors.

In effect, we can multiply the integrand by a nonzero constant $c$, as long as we compensate for this by multiplying the entire integral by $1/c$. Such a manipulation **cannot** be done with *variable* factors.

**Pitfall** ▼ When using the form $\int u^n \, du$, do not neglect $du$. For example,

$$\int (4x + 1)^2 \, dx \neq \frac{(4x + 1)^3}{3} + C.$$

The proper way to do this problem is as follows. Letting $u = 4x + 1$, we have $du = 4 \, dx$. Thus,

$$\int (4x + 1)^2 \, dx = \frac{1}{4} \int (4x + 1)^2 [4 \, dx] = \frac{1}{4} \int u^2 \, du$$

$$= \frac{1}{4} \cdot \frac{u^3}{3} + C = \frac{(4x + 1)^3}{12} + C.$$

**EXAMPLE 3    Applying the Power Rule for Integration**

**a.** *Find* $\int \sqrt[3]{6y} \, dy$.

*Solution:* The integrand is $(6y)^{1/3}$, a power of a function. Let us try to use the power rule for integration. If we set $u = 6y$, then $du = 6 \, dy$. Since the factor 6 does not appear in the integrand, we insert a factor of 6 and adjust for it with a factor of $\frac{1}{6}$ in front of the integral. We then have

$$\int \sqrt[3]{6y} \, dy = \int (6y)^{1/3} \, dy = \frac{1}{6} \int (6y)^{1/3} [6 \, dy] = \frac{1}{6} \int u^{1/3} \, du$$

$$= \left(\frac{1}{6}\right) \frac{u^{4/3}}{\frac{4}{3}} + C = \frac{(6y)^{4/3}}{8} + C.$$

Example 3(a) can be handled without the use of the power rule by rewriting the integrand as $\sqrt[3]{6} y^{1/3}$. This gives the equivalent answer $\dfrac{3\sqrt[3]{6}}{4} y^{4/3} + C$.

**b.** *Find* $\displaystyle \int \frac{2x^3 + 3x}{(x^4 + 3x^2 + 7)^4} \, dx$.

*Solution:* We can write this as $\int (x^4 + 3x^2 + 7)^{-4}(2x^3 + 3x) \, dx$. Let us try to use the power rule for integration. If $u = x^4 + 3x^2 + 7$, then $du = (4x^3 + 6x) \, dx$, which is two times the quantity $(2x^3 + 3x) \, dx$ in the integral. Thus, we insert a factor of 2 and adjust for it with a factor of $\frac{1}{2}$ in front of the integral, as follows:

$$\int (x^4 + 3x^2 + 7)^{-4}(2x^3 + 3x) \, dx$$

$$= \frac{1}{2} \int (x^4 + 3x^2 + 7)^{-4} [2(2x^3 + 3x) \, dx]$$

$$= \frac{1}{2} \int (x^4 + 3x^2 + 7)^{-4} [(4x^3 + 6x) \, dx]$$

$$= \frac{1}{2} \int u^{-4} \, du = \frac{1}{2} \cdot \frac{u^{-3}}{-3} + C = -\frac{1}{6u^3} + C$$

$$= -\frac{1}{6(x^4 + 3x^2 + 7)^3} + C.$$

In using the power rule for integration, take care when making your choice for *u*. In Example 3(b), you would *not* be able to proceed very far if, for instance, you let $u = 2x^3 + 3x$. At times you may find it necessary to try many different choices. So don't just sit and look at the integral. Try something even if it is wrong, because it may give you a hint as to what might work. **Skill at integration comes only after many hours of practice and conscientious study.**

**EXAMPLE 4  An Integral to Which the Power Rule Does Not Apply**

*Find* $\int 4x^2(x^4 + 1)^2 \, dx$.

*Solution:* If we set $u = x^4 + 1$, then $du = 4x^3 \, dx$. To get *du* in the integral, we need an additional factor of the *variable x*. However, we can adjust only for **constant** factors. Thus, we cannot use the power rule. Instead, to find the integral, we shall first expand $(x^4 + 1)^2$:

$$\int 4x^2(x^4 + 1)^2 \, dx = 4 \int x^2(x^8 + 2x^4 + 1) \, dx$$

$$= 4 \int (x^{10} + 2x^6 + x^2) \, dx$$

$$= 4\left( \frac{x^{11}}{11} + \frac{2x^7}{7} + \frac{x^3}{3} \right) + C. \qquad \blacksquare$$

**Integrating Natural Exponential Functions**

We now turn our attention to integrating exponential functions. If *u* is a differentiable function of *x*, then

$$\frac{d}{dx}(e^u) = e^u \frac{du}{dx}.$$

Corresponding to this differentiation formula is the integration formula

$$\int e^u \frac{du}{dx} \, dx = e^u + C.$$

But $\frac{du}{dx} \, dx$ is the differential of *u*, namely, *du*. Thus,

$$\int e^u \, du = e^u + C. \qquad (2)$$

**EXAMPLE 5  Integrals Involving Exponential Functions**

**a.** *Find* $\int 2xe^{x^2} \, dx$

*Solution:* Let $u = x^2$. Then $du = 2x \, dx$, and by Eq. (2),

$$\int 2xe^{x^2} \, dx = \int e^{x^2}[2x \, dx] = \int e^u \, du$$

$$= e^u + C = e^{x^2} + C.$$

**b.** *Find* $\int (x^2 + 1)e^{x^3+3x} \, dx$.

**Principles in Practice 1**
**Integrals Involving Exponential Functions**

When an object is moved from one environment to another, its temperature $T$ changes at a rate given by $\dfrac{dT}{dt} = kCe^{kt}$, where $t$ is the time (in hours) after changing environments, $C$ is the temperature difference (original minus new) between the environments, and $k$ is a constant. If the original environment is 70°, the new environment is 60°, and $k = -0.5$, find the general form of $T(t)$.

**Solution:** If $u = x^3 + 3x$, then $du = (3x^2 + 3)\,dx = 3(x^2 + 1)\,dx$. If the integrand contained a factor of 3, the integral would have the form $\int e^u\,du$. Thus, we write

$$\int (x^2 + 1)e^{x^3+3x}\,dx = \frac{1}{3}\int e^{x^3+3x}[3(x^2 + 1)\,dx]$$

$$= \frac{1}{3}\int e^u\,du = \frac{1}{3}e^u + C$$

$$= \frac{1}{3}e^{x^3+3x} + C.$$ ∎

**Pitfall** ▼ Do not apply the power-rule formula for $\int u^n\,du$ to $\int e^u\,du$. For example,

$$\int e^x\,dx \neq \frac{e^{x+1}}{x + 1} + C.$$

### Integrals Involving Logarithmic Functions

As you know, the power-rule formula $\int u^n\,du = u^{n+1}/(n + 1) + C$ does not apply when $n = -1$. To handle that situation, namely, $\int u^{-1}\,du = \int \dfrac{1}{u}\,du$, we first recall that

$$\frac{d}{dx}(\ln u) = \frac{1}{u}\frac{du}{dx}.$$

It would seem that $\int \dfrac{1}{u}\dfrac{du}{dx}\,dx = \int \dfrac{1}{u}\,du = \ln u + C$. However, the logarithm of $u$ is defined if and only if $u$ is positive. If $u < 0$, then $\ln u$ is not defined. Thus,

$$\int \frac{1}{u}\,du = \ln u + C, \qquad \text{provided that } u > 0.$$

On the other hand, if $u < 0$, then $-u > 0$, and $\ln(-u)$ is defined. Moreover,

$$\frac{d}{dx}[\ln(-u)] = \frac{1}{-u}(-1)\frac{du}{dx} = \frac{1}{u}\frac{du}{dx}.$$

In this case ($u < 0$),

$$\int \frac{1}{u}\frac{du}{dx}\,dx = \int \frac{1}{u}\,du = \ln(-u) + C.$$

In summary, if $u > 0$, then $\int \dfrac{1}{u}\,du = \ln u + C$; if $u < 0$, then $\int \dfrac{1}{u}\,du = \ln(-u) + C$. Combining these cases, we have

$$\int \frac{1}{u}\,du = \ln |u| + C. \tag{3}$$

In particular, if $u = x$, then $du = dx$, and

$$\int \frac{1}{x}\,dx = \ln |x| + C. \tag{4}$$

**Principles in Practice 2**

**Integrals Involving $\frac{1}{u}\, du$**

If the rate of vocabulary memorization of the average student in a foreign language is given by $\frac{dv}{dt} = \frac{35}{t+1}$, where $v$ is the number of vocabulary words memorized in $t$ hours of study, how many words would the student memorize in two hours?

**EXAMPLE 6**  Integrals Involving $\frac{1}{u}\, du$

**a.** *Find* $\int \frac{7}{x}\, dx.$

*Solution:* From Eq. (4),

$$\int \frac{7}{x}\, dx = 7\int \frac{1}{x}\, dx = 7\ln|x| + C.$$

Using properties of logarithms, we can write this answer another way:

$$\int \frac{7}{x}\, dx = \ln|x^7| + C.$$

**b.** *Find* $\int \frac{2x}{x^2 + 5}\, dx.$

*Solution:* Let $u = x^2 + 5$. Then $du = 2x\, dx$. From Eq. (3),

$$\int \frac{2x}{x^2 + 5}\, dx = \int \frac{1}{x^2 + 5}[2x\, dx] = \int \frac{1}{u}\, du$$

$$= \ln|u| + C = \ln|x^2 + 5| + C.$$

Since $x^2 + 5$ is always positive, we can omit the absolute-value bars:

$$\int \frac{2x}{x^2 + 5}\, dx = \ln(x^2 + 5) + C. \qquad \blacksquare$$

**EXAMPLE 7**  An Integral Involving $\frac{1}{u}\, du$

*Find* $\int \frac{(2x^3 + 3x)\, dx}{x^4 + 3x^2 + 7}.$

*Solution:* If $u = x^4 + 3x^2 + 7$, then $du = (4x^3 + 6x)\, dx$, which is two times the numerator. To apply Eq. (3), we insert a factor of 2 and adjust for it with a factor of $\frac{1}{2}$, as follows:

$$\int \frac{2x^3 + 3x}{x^4 + 3x^2 + 7}\, dx = \frac{1}{2}\int \frac{2(2x^3 + 3x)}{x^4 + 3x^2 + 7}\, dx$$

$$= \frac{1}{2}\int \frac{1}{x^4 + 3x^2 + 7}[(4x^3 + 6x)\, dx]$$

$$= \frac{1}{2}\int \frac{1}{u}\, du = \frac{1}{2}\ln|u| + C$$

$$= \frac{1}{2}\ln|x^4 + 3x^2 + 7| + C$$

$$= \ln\sqrt{x^4 + 3x^2 + 7} + C. \qquad \blacksquare$$

**EXAMPLE 8**  An Integral Involving Two Forms

*Find* $\int \left[ \frac{1}{(1-w)^2} + \frac{1}{w-1} \right] dw.$

*Solution:*

$$\int \left[ \frac{1}{(1-w)^2} + \frac{1}{w-1} \right] dw = \int (1-w)^{-2} \, dw + \int \frac{1}{w-1} \, dw$$

$$= -1 \int (1-w)^{-2}[-dw] + \int \frac{1}{w-1} dw.$$

The first integral has the form $\int u^{-2} du$, and the second has the form $\int \frac{1}{v} \, dv$.

Thus,

$$\int \left[ \frac{1}{(1-w)^2} + \frac{1}{w-1} \right] dw = -\frac{(1-w)^{-1}}{-1} + \ln |w-1| + C$$

$$= \frac{1}{1-w} + \ln |w-1| + C. \qquad \blacksquare$$

For your convenience, we list in Table 16.2 the basic integration formulas so far discussed. We assume that $u$ is a function of $x$.

---

**TABLE 16.2**   Basic Integration Formulas

**1.** $\int k \, du = ku + C,$   $k$  a constant.

**2.** $\int u^n du = \dfrac{u^{n+1}}{n+1} + C,$   $n \neq -1.$

**3.** $\int e^u du = e^u + C.$

**4.** $\int \dfrac{1}{u} \, du = \ln |u| + C,$   $u \neq 0.$

**5.** $\int kf(x) \, dx = k \int f(x) \, dx.$

**6.** $\int [f(x) \pm g(x)] dx = \int f(x) dx \pm \int g(x) dx.$

---

■ **Exercise 16.3**

*In Problems 1–80, find the indefinite integrals.*

**1.** $\int (x+5)^7 \, dx.$

**2.** $\int 15(x+2)^4 \, dx.$

**3.** $\int 2x(x^2+3)^5 \, dx.$

**4.** $\int (3x^2+14x)(x^3+7x^2+1) \, dx.$

**5.** $\int (3y^2+6y)(y^3+3y^2+1)^{2/3} \, dy.$

**6.** $\int (-12z^2-12z+1)(-4z^3-6z^2+z)^{18} \, dx.$

**7.** $\int \dfrac{3}{(3x-1)^3} \, dx.$

**8.** $\int \dfrac{4x}{(2x^2-7)^{10}} \, dx.$

**9.** $\int \sqrt{2x-1} \, dx.$

**10.** $\int \dfrac{1}{\sqrt{x-2}} \, dx.$

**11.** $\int (7x-6)^4 \, dx.$

**12.** $\int x^2(3x^3+7)^3 \, dx.$

**13.** $\int x(x^2+3)^{12} \, dx.$

**14.** $\int x\sqrt{1+2x^2} \, dx.$

**15.** $\int x^4(27+x^5)^{1/3} \, dx$

**16.** $\int (3-2x)^{10} \, dx.$

**17.** $\int 3e^{3x}\,dx.$

**18.** $\int 2e^{2t+5}\,dt.$

**19.** $\int (2t+1)e^{t^2+t}\,dt.$

**20.** $\int -3w^2 e^{-w^3}\,dw.$

**21.** $\int xe^{5x^2}\,dx.$

**22.** $\int x^3 e^{4x^4}\,dx.$

**23.** $\int 6e^{-2x}\,dx.$

**24.** $\int x^4 e^{-6x^5}\,dx.$

**25.** $\int \frac{1}{x+5}\,dx.$

**26.** $\int \frac{2x+1}{x+x^2}\,dx.$

**27.** $\int \frac{3x^2+4x^3}{x^3+x^4}\,dx.$

**28.** $\int \frac{3x^2-2x}{1-x^2+x^3}\,dx.$

**29.** $\int \frac{6z}{(z^2-6)^5}\,dx.$

**30.** $\int \frac{1}{(8y-3)^3}\,dy.$

**31.** $\int \frac{4}{x}\,dx.$

**32.** $\int \frac{3}{1+2y}\,dy.$

**33.** $\int \frac{s^2}{s^3+5}\,ds.$

**34.** $\int \frac{2x^2}{3-4x^3}\,dx.$

**35.** $\int \frac{7}{5-3x}\,dx.$

**36.** $\int \frac{7t}{5t^2-6}\,dt.$

**37.** $\int \sqrt{5x}\,dx.$

**38.** $\int \frac{1}{(4x)^7}\,dx.$

**39.** $\int \frac{x}{\sqrt{x^2-4}}\,dx.$

**40.** $\int \frac{7}{3-2x}\,dx.$

**41.** $\int 2y^3 e^{y^4+1}\,dx.$

**42.** $\int \sqrt{4x-3}\,dx.$

**43.** $\int v^2 e^{-2v^3+1}\,dv.$

**44.** $\int \frac{x^2}{\sqrt[3]{2x^3+9}}\,dx.$

**45.** $\int (e^{-5x}+2e^x)\,dx.$

**46.** $\int 4\sqrt[3]{y+1}\,dy.$

**47.** $\int (x+1)(3-3x^2-6x)^3\,dx.$

**48.** $\int 2ye^{3y^2}\,dy.$

**49.** $\int \frac{x^2+2}{x^3+6x}\,dx.$

**50.** $\int (e^x - e^{-x} + e^{2x})\,dx.$

**51.** $\int \frac{16s-4}{3-2s+4s^2}\,ds.$

**52.** $\int (t^2+4t)(t^3+6t^2)^6\,dt.$

**53.** $\int x(2x^2+1)^{-1}\,dx.$

**54.** $\int (w^3-8w^7+1)(w^4-4w^8+4w)^{-6}\,dw.$

**55.** $\int -(x^2-2x^5)(x^3-x^6)^{-10}\,dx.$

**56.** $\int \frac{3}{7}(v-2)e^{2-4v+v^2}\,dv.$

**57.** $\int (2x^3+x)(x^4+x^2)\,dx.$

**58.** $\int (e^{3.1})^2\,dx.$

**59.** $\int \frac{18+12x}{(4-9x-3x^2)^5}\,dx.$

**60.** $\int (e^x-e^{-x})^2\,dx.$

**61.** $\int x(2x+1)e^{4x^3+3x^2-4}\,dx.$

**62.** $\int (u^2+3-ue^{7-u^2})\,du.$

**63.** $\int x\sqrt{(7-5x^2)^3}\,dx.$

**64.** $\int e^{-x/4}\,dx.$

**65.** $\int \left(\sqrt{2x}-\frac{1}{\sqrt{2x}}\right)\,dx.$

**66.** $\int \frac{x^3}{e^{x^4}}\,dx.$

**67.** $\int (x^2+1)^2\,dx.$

**68.** $\int \left[x(x^2-16)^2-\frac{1}{2x+5}\right]\,dx.$

**69.** $\int \left[\frac{x}{x^2+1}+\frac{x^5}{(x^6+1)^2}\right]\,dx.$

**70.** $\int \left[\frac{1}{x-1}+\frac{1}{(x-1)^2}\right]\,dx.$

**71.** $\int \left[\frac{1}{3x-5}-(x^2-2x^5)(x^3-x^6)^{-10}\right]\,dx.$

**72.** $\int (r^3+5)^2\,dr.$

**73.** $\int \left[\sqrt{3x+1}-\frac{x}{x^2+3}\right]\,dx.$

**74.** $\int \left[\frac{2x}{x^2+3}-\frac{x^3}{(x^4+2)^2}\right]\,dx.$

**75.** $\int \frac{e^{\sqrt{x}}}{\sqrt{x}}\,dx.$

**76.** $\int (e^4-2^e)\,dx.$

**77.** $\int \frac{1+e^{2x}}{e^x}\,dx.$

**78.** $\displaystyle\int \frac{1}{t^2}\sqrt{\frac{1}{t} - 1}\,dt$

**79.** $\displaystyle\int \frac{x+1}{x^2 + 2x}\,\ln(x^2 + 2x)\,dx.$

**80.** $\int \sqrt[3]{x}\,e^{\sqrt[3]{8x^4}}\,dx.$

*In Problems 81–84, find y, subject to the given conditions.*

**81.** $y' = (3 - 2x)^2;\quad y(0) = 1.$

**82.** $y' = \dfrac{x}{x^2 + 4};\quad y(1) = 0.$

**83.** $y'' = \dfrac{1}{x^2};\quad y'(-1) = 1,\ y(1) = 0.$

**84.** $y'' = \sqrt{x + 2};\quad y'(2) = \frac{1}{3},\ y(2) = -\frac{7}{15}.$

**85. Real Estate** The rate of change of the value of a house that cost \$350,000 to build can be modeled by $\dfrac{dV}{dt} = 8e^{0.05t}$, where $t$ is the time in years since the house was built and $V$ is the value (in thousands of dollars) of the house. Find $V(t)$.

**86. Life Span** If the rate of change of the expected life span $l$ at birth of people born in the United States can be modeled by $\dfrac{dl}{dt} = \dfrac{14.1}{t + 20}$, where $t$ is the number of years after 1920 and the expected life span was 43 years in 1920, find the expected life span for people born in 1969.

**87. Oxygen in Capillary** In a discussion of the diffusion of oxygen from capillaries,[4] concentric cylinders of radius $r$ are used as a model for a capillary. The concentration $C$ of oxygen in the capillary is given by

$$C = \int \left( \frac{Rr}{2K} + \frac{B_1}{r} \right) dr,$$

where $R$ is the constant rate at which oxygen diffuses from the capillary, and $K$ and $B_1$ are constants. Find $C$. (Write the constant of integration as $B_2$.)

**88.** Find $f(2)$ if $f(\frac{1}{2}) = 1$ and $f'(x) = e^{2x-1} - 6x.$

[4]W. Simon, *Mathematical Techniques for Physiology and Medicine* (New York: Academic Press, Inc., 1972).

---

## OBJECTIVE

**To discuss techniques of handling more challenging integration problems, namely, by algebraic manipulation and by fitting the integrand to a familiar form. To integrate an exponential function with a base different from e and to find the consumption function, given the marginal propensity to consume.**

# 16.4 Techniques of Integration

Now that you have had some practice in determining indefinite integrals, suppose we consider some problems of a greater degree of difficulty.

When you are integrating fractions, sometimes a preliminary division is needed to get familiar integration forms, as the next example shows.

**EXAMPLE 1 Preliminary Division before Integration**

**a.** *Find* $\displaystyle\int \frac{x^3 + x}{x^2}\,dx.$

*Solution:* A familiar integration form is not apparent. However, we can break up the integrand into two fractions by dividing each term in the numerator by the denominator. We then have

Here we split up the integrand.

$$\int \frac{x^3 + x}{x^2}\,dx = \int \left[ \frac{x^3}{x^2} + \frac{x}{x^2} \right] dx = \int \left[ x + \frac{1}{x} \right] dx$$

$$= \frac{x^2}{2} + \ln|x| + C.$$

**b.** *Find* $\displaystyle\int \frac{2x^3 + 3x^2 + x + 1}{2x + 1}\,dx.$

*Solution:* Here the integrand is a quotient of polynomials in which the degree of the numerator is greater than or equal to that of the denominator and the denominator has more than one term. In such a situation, in order to integrate, we first use long division until the degree of the remainder is less than that of the divisor. We obtain

Here we use long division to rewrite the integrand.

$$\int \frac{2x^3 + 3x^2 + x + 1}{2x + 1} \, dx = \int \left( x^2 + x + \frac{1}{2x + 1} \right) dx$$

$$= \frac{x^3}{3} + \frac{x^2}{2} + \int \frac{1}{2x + 1} \, dx$$

$$= \frac{x^3}{3} + \frac{x^2}{2} + \frac{1}{2} \int \frac{1}{2x + 1} \, [2 \, dx]$$

$$= \frac{x^3}{3} + \frac{x^2}{2} + \frac{1}{2} \ln |2x + 1| + C. \qquad ■$$

### EXAMPLE 2   Indefinite Integrals

**a.** *Find* $\displaystyle\int \frac{1}{\sqrt{x}(\sqrt{x} - 2)^3} \, dx.$

   ***Solution:*** We can write this integral as $\displaystyle\int \frac{(\sqrt{x} - 2)^{-3}}{\sqrt{x}} \, dx.$ Let us try the power rule for integration with $u = \sqrt{x} - 2.$ Then $du = \dfrac{1}{2\sqrt{x}} \, dx,$ and

Here the integral is fit to the form to which the power rule for integration applies.

$$\int \frac{(\sqrt{x} - 2)^{-3}}{\sqrt{x}} \, dx = 2 \int (\sqrt{x} - 2)^{-3} \left[ \frac{1}{2\sqrt{x}} \, dx \right]$$

$$= 2 \int u^{-3} \, du = 2\left( \frac{u^{-2}}{-2} \right) + C$$

$$= -\frac{1}{u^2} + C = -\frac{1}{(\sqrt{x} - 2)^2} + C.$$

**b.** *Find* $\displaystyle\int \frac{1}{x \ln x} \, dx.$

   ***Solution:*** If $u = \ln x,$ then $du = \dfrac{1}{x} \, dx,$ and

Here the integral fits the familiar form $\displaystyle\int \frac{1}{u} \, du.$

$$\int \frac{1}{x \ln x} \, dx = \int \frac{1}{\ln x}\left( \frac{1}{x} \, dx \right) = \int \frac{1}{u} \, du$$

$$= \ln |u| + C = \ln |\ln x| + C.$$

**c.** *Find* $\displaystyle\int \frac{5}{w(\ln w)^{3/2}} \, dw.$

   ***Solution:*** If $u = \ln w,$ then $du = \dfrac{1}{w} \, dw.$ Applying the power rule for integration, we have

Here the integral is fit to the form to which the power rule for integration applies.

$$\int \frac{5}{w(\ln w)^{3/2}} \, dw = 5 \int (\ln w)^{-3/2}\left( \frac{1}{w} \, dw \right)$$

$$= 5 \int u^{-3/2} \, du = 5 \cdot \frac{u^{-1/2}}{-\frac{1}{2}} + C$$

$$= \frac{-10}{u^{1/2}} + C = -\frac{10}{(\ln w)^{1/2}} + C. \qquad ■$$

**Integrating $a^u$**

In Sec. 16.3, we integrated an exponential function to the base $e$:

$$\int e^u\, du = e^u + C.$$

Now let us consider the integral of an exponential function to a base other than $e$:

$$\int a^u\, du.$$

To find this integral, we first convert $a^u$ to an exponential function to the base $e$ by using Property 8 of Sec. 5.3:

$$a = e^{\ln a}. \tag{1}$$

Example 3 will illustrate

**EXAMPLE 3   An Integral Involving $a^u du$**

*Find* $\displaystyle\int 2^{3-x}\, dx.$

**Solution:**

> *Strategy:* We want to integrate an exponential function to the base 2. To do this, we shall first convert from base 2 to base $e$ by using Eq. (1) to write 2 in terms of $e$.

Since $2 = e^{\ln 2}$, we have

$$\int 2^{3-x}\, dx = \int (e^{\ln 2})^{3-x}\, dx = \int e^{(\ln 2)(3-x)}\, dx.$$

The integrand of the last integral is of the form $e^u$, where $u = (\ln 2)(3 - x)$. Since $du = -\ln 2\, dx$, we have

$$\int e^{(\ln 2)(3-x)}\, dx = -\frac{1}{\ln 2} \int e^{(\ln 2)(3-x)}[(-\ln 2)\, dx] \qquad \left(\text{form: } \int e^u\, du\right)$$

$$= -\frac{1}{\ln 2} e^{(\ln 2)(3-x)} + C = -\frac{1}{\ln 2} 2^{3-x} + C.$$

Thus,

$$\int 2^{3-x}\, dx = -\frac{1}{\ln 2} 2^{3-x} + C.$$

Notice that we expressed our answer in terms of an exponential function to the base 2, the base of the original integrand.   ■

Generalizing the procedure described in Example 3, we can obtain a formula for integrating $a^u$:

$$\int a^u\, du = \int (e^{\ln a})^u\, du = \int e^{(\ln a)u}\, du$$

$$= \frac{1}{\ln a} \int e^{(\ln a)u}[(\ln a)\, du]$$

$$= \frac{1}{\ln a} e^{(\ln a)u} + C = \frac{1}{\ln a} (e^{\ln a})^u + C$$

$$= \frac{1}{\ln a} a^u + C.$$

Hence, we have

$$\int a^u \, du = \frac{1}{\ln a} a^u + C.$$

Applying this formula to the integral in Example 3 gives

$$\int 2^{3-x} \, dx \qquad (a = 2, u = 3 - x)$$

$$= -\int 2^{3-x}(-dx) \qquad (du = -dx)$$

$$= -\frac{1}{\ln 2} 2^{3-x} + C,$$

which is the same result that we obtained before.

## Application of Integration

We shall now consider an application of integration that relates a consumption function to the marginal propensity to consume.

**EXAMPLE 4** **Find Consumption Function from Marginal Propensity to Consume**

*For a certain country, the marginal propensity to consume is given by*

$$\frac{dC}{dI} = \frac{3}{4} - \frac{1}{2\sqrt{3I}},$$

*where consumption C is a function of national income I. Here I is expressed in billions of slugs (50 slugs = $0.01). Determine the consumption function for the country if it is known that consumption is 10 billion slugs (C = 10) when I = 12.*

***Solution:*** Since the marginal propensity to consume is the derivative of $C$, we have

$$C = \int \left( \frac{3}{4} - \frac{1}{2\sqrt{3I}} \right) dI = \int \frac{3}{4} \, dI - \frac{1}{2} \int (3I)^{-1/2} \, dI$$

$$= \frac{3}{4} I - \frac{1}{2} \int (3I)^{-1/2} \, dI.$$

If we let $u = 3I$, then $du = 3 \, dI$, and

$$C = \frac{3}{4} I - \left( \frac{1}{2} \right) \frac{1}{3} \int (3I)^{-1/2} [3 \, dI]$$

$$= \frac{3}{4} I - \frac{1}{6} \frac{(3I)^{1/2}}{\frac{1}{2}} + C_1.$$

$$C = \frac{3}{4} I - \frac{\sqrt{3I}}{3} + C_1.$$

This is an example of a initial-value problem.

When $I = 12$, then $C = 10$, so

$$10 = \frac{3}{4}(12) - \frac{\sqrt{3(12)}}{3} + C_1,$$

$$10 = 9 - 2 + C_1.$$

Thus, $C_1 = 3$, and the consumption function is

$$C = \frac{3}{4}I - \frac{\sqrt{3I}}{3} + 3.$$

■

## ▪ Exercise 16.4

*In Problems 1–56, determine the indefinite integrals.*

**1.** $\displaystyle\int \frac{2x^4 + 3x^3 - x^2}{x^3} \, dx.$

**2.** $\displaystyle\int \frac{9x^2 + 5}{3x} \, dx.$

**3.** $\displaystyle\int (3x^2 + 2)\sqrt{2x^3 + 4x + 1} \, dx.$

**4.** $\displaystyle\int \frac{x}{\sqrt[4]{x^2 + 1}} \, dx.$

**5.** $\displaystyle\int \frac{3}{\sqrt{4 - 5x}} \, dx.$

**6.** $\displaystyle\int \frac{xe^{x^2} dx}{e^{x^2} - 2}.$

**7.** $\displaystyle\int 4^{7x} dx.$

**8.** $\displaystyle\int 3^x \, dx.$

**9.** $\displaystyle\int 2x(7 - e^{x^2/4}) \, dx.$

**10.** $\displaystyle\int \left( e^x + x^e + ex + \frac{e}{x} \right) dx.$

**11.** $\displaystyle\int \frac{6x^2 - 11x + 5}{3x - 1} \, dx.$

**12.** $\displaystyle\int \frac{(2x - 1)(x + 3)}{x - 5} \, dx.$

**13.** $\displaystyle\int \frac{3e^{2x}}{e^{2x} + 1} \, dx.$

**14.** $\displaystyle\int (e^{4 - 3x})^2 \, dx.$

**15.** $\displaystyle\int \frac{e^{7/x}}{x^2} \, dx.$

**16.** $\displaystyle\int \frac{2x^4 - 6x^3 + x - 2}{x - 2} \, dx.$

**17.** $\displaystyle\int \frac{2x^3}{x^2 - 4} \, dx.$

**18.** $\displaystyle\int \frac{5 - 4x^2}{3 + 2x} \, dx.$

**19.** $\displaystyle\int \frac{(\sqrt{x} + 2)^2}{3\sqrt{x}} \, dx.$

**20.** $\displaystyle\int \frac{3e^s}{6 + 5e^s} \, ds.$

**21.** $\displaystyle\int \frac{(x^{1/3} + 2)^4}{\sqrt[3]{x^2}} \, dx.$

**22.** $\displaystyle\int \frac{\sqrt{1 + \sqrt{x}}}{\sqrt{x}} \, dx.$

**23.** $\displaystyle\int \frac{\ln x}{x} \, dx.$

**24.** $\displaystyle\int \sqrt{t}(5 - t\sqrt{t})^{0.4} \, dt.$

**25.** $\displaystyle\int \frac{\ln^2(r + 1)}{r + 1} \, dr.$

**26.** $\displaystyle\int \frac{8x^3 - 6x^2 - ex^4}{3x^3} \, dx.$

**27.** $\displaystyle\int \frac{3^{\ln x}}{x} \, dx.$

**28.** $\displaystyle\int \frac{2}{x \ln(2x^2)} \, dx.$

**29.** $\displaystyle\int x\sqrt{e^{x^2 + 3}} \, dx.$

**30.** $\displaystyle\int \frac{x + 3}{x + 6} \, dx$

**31.** $\displaystyle\int \frac{1}{(x + 3)\ln(x + 3)} \, dx.$

**32.** $\displaystyle\int (x^{e^x} + 2x) \, dx.$

**33.** $\displaystyle\int \frac{x^3 + x^2 - x - 3}{x^2 - 3} \, dx.$

**34.** $\displaystyle\int \frac{4x \ln\sqrt{1 + x^2}}{1 + x^2} \, dx.$

**35.** $\displaystyle\int \frac{3x\sqrt{\ln(x^2 + 1)^2}}{x^2 + 1} \, dx.$

**36.** $\displaystyle\int 3(x^2 + 2)^{-1/2}xe^{\sqrt{x^2 + 2}} \, dx.$

**37.** $\displaystyle\int \left( \frac{x^3}{\sqrt{x^4 - 1}} - \ln 4 \right) dx.$

**38.** $\displaystyle\int \frac{x - x^{-2}}{x^2 + 2x^{-1}} \, dx.$

**39.** $\displaystyle\int \frac{2x^4 - 8x^3 - 6x^2 + 4}{x^3} \, dx.$

**40.** $\displaystyle\int \frac{e^x + e^{-x}}{e^x - e^{-x}} \, dx.$

**41.** $\displaystyle\int \frac{x}{x - 1} \, dx.$

**42.** $\displaystyle\int \frac{x}{(x^2 + 1)\ln(x^2 + 1)} \, dx.$

**43.** $\displaystyle\int \frac{xe^{x^2}}{\sqrt{e^{x^2} + 2}} \, dx.$

**44.** $\displaystyle\int \frac{7}{(2x + 1)[1 + \ln(2x + 1)]^2} \, dx.$

**45.** $\displaystyle\int \frac{(e^{-x} + 6)^2}{e^x} \, dx.$

**46.** $\displaystyle\int \left[ \frac{1}{8x + 1} - \frac{1}{e^x(8 + e^{-x})^2} \right] dx.$

**47.** $\displaystyle\int (x^3 + ex)\sqrt{x^2 + e} \, dx.$

**48.** $\displaystyle\int 2^{x \ln x}(1 + \ln x) \, dx.$

**49.** $\int \sqrt{x}\,\sqrt{(8x)^{3/2} + 3}\; dx.$

**50.** $\int \dfrac{3}{x(\ln x)^{1/2}}\; dx.$

**51.** $\int \dfrac{\sqrt{s}}{e^{\sqrt{s^3}}}\; ds.$

**52.** $\int \dfrac{\ln^3 x}{3x}\; dx.$

**53.** $\int e^{\ln(x+2)}\; dx.$

**54.** $\int dx.$

**55.** $\int \dfrac{\ln\,(xe^x)}{x}\; dx.$

**56.** $\int e^{x^2 + \ln x}\; dx.$

*In Problems 57 and 58, dr/dq is a marginal-revenue function. Find the demand function.*

**57.** $\dfrac{dr}{dq} = \dfrac{200}{(q+2)^2}.$

**58.** $\dfrac{dr}{dq} = \dfrac{900}{(2q+3)^3}.$

*In Problems 59 and 60, dc/dq is a marginal-cost function. Find the total-cost function if fixed costs in each case are* **2000.**

**59.** $\dfrac{dc}{dq} = \dfrac{20}{q+5}.$

**60.** $\dfrac{dc}{dq} = 2e^{0.001q}.$

*In Problems 61–63, dC/dI represents the marginal propensity to consume. Find the consumption function, subject to the given condition.*

**61.** $\dfrac{dC}{dI} = \dfrac{1}{\sqrt{I}}; \quad C(9) = 8$

**62.** $\dfrac{dC}{dI} = \dfrac{3}{4} - \dfrac{1}{2\sqrt{3I}}; \quad C(3) = \dfrac{11}{4}.$

**63.** $\dfrac{dC}{dI} = \dfrac{3}{4} - \dfrac{1}{6\sqrt{I}}; \quad C(25) = 23.$

**64. Cost Function** The marginal-cost function for a manufacturer's product is given by

$$\frac{dc}{dq} = 10 - \frac{100}{q+10},$$

where $c$ is the total cost in dollars when $q$ units are produced. When 100 units are produced, the average cost is $50 per unit. To the nearest dollar, determine the manufacturer's fixed cost.

**65. Cost Function** Suppose the marginal-cost function for a manufacturer's product is given by

$$\frac{dc}{dq} = \frac{100q^2 - 4998q + 50}{q^2 - 50q + 1},$$

where $c$ is the total cost in dollars when $q$ units are produced.

**a.** Determine the marginal cost when 50 units are produced.

**b.** If fixed costs are $10,000, find the total cost of producing 50 units.

**c.** Use the results of parts (a) and (b) and differentials to approximate the total cost of producing 52 units.

**66. Cost Function** The marginal-cost function for a manufacturer's product is given by

$$\frac{dc}{dq} = \frac{9}{10}\sqrt{q}\,\sqrt{0.04q^{3/2} + 4},$$

where $c$ is the total cost in dollars when $q$ units are produced. Fixed costs are $360.

**a.** Determine the marginal cost when 25 units are produced.

**b.** Find the total cost of producing 25 units.

**c.** Use the results of parts (a) and (b) and differentials to approximate the total cost of producing 23 units.

**67. Value of Land** It is estimated that $t$ years from now the value $V$ (in dollars) of an acre of land near the ghost town of Cherokee, California, will be increasing at the rate of $\dfrac{8t^3}{\sqrt{0.2t^4 + 8000}}$ dollars per year. If the land is currently worth $500 per acre, how much will it be worth in 10 years? Express your answer to the nearest dollar.

**68. Revenue Function** The marginal-revenue function for a manufacturer's product is given by

$$\frac{dr}{dq} = \frac{3}{e^q + 2},$$

where $r$ is the total revenue received (in dollars) when $q$ units are produced and sold. Find the demand function, and express it in the form $p = f(q)$. [*Hint:* Rewrite $dr/dq$ by multiplying both numerator and denominator by $e^{-q}$.]

**69. Savings** A certain country's marginal propensity to save is given by

$$\frac{dS}{dI} = \frac{5}{(I+2)^2}$$

where $S$ and $I$ represent total national savings and income, respectively, and are measured in billions of dollars. If total national consumption is $7.5 billion when total national income is $8 billion, for what value(s) of $I$ is total national savings equal to zero?

**70. Consumption Function** A certain country's marginal propensity to save is given by

$$\frac{dS}{dI} = \frac{1}{2} - \frac{1.8}{\sqrt[3]{3I^2}},$$

where $S$ and $I$ represent total national savings and income, respectively, and are measured in billions of dollars.

**a.** Determine the marginal propensity to consume when total national income is $81 billion.

**b.** Determine the consumption function, given that savings are $3 billion when total national income is $24 billion.

**c.** Use the result in part (b) to show that consumption is $54.9 billion when total national income is $81 billion.

**d.** Use differentials and the results in parts (a) and (c) to approximate consumption when total national income is $78 billion.

---

OBJECTIVE

To introduce sigma notation and to give summation formulas that will be used in the next section.

## 16.5 SUMMATION

To prepare you for further applications of integration, we need to discuss certain sums.

Consider finding the sum $S$ of the first $n$ positive integers:

$$S = 1 + 2 + \cdots + (n - 1) + n. \qquad (1)$$

Writing the terms on the right side of Eq. (1) in reverse order, we have

$$S = n + (n - 1) + \cdots + 2 + 1. \qquad (2)$$

Adding the corresponding sides of Eqs. (1) and (2) gives

$$
\begin{array}{ccccccccc}
S = & 1 & + & 2 & + \cdots + & (n - 1) & + & n & \\
S = & n & + & (n - 1) & + \cdots + & 2 & + & 1 & \\
\hline
2S = & (n + 1) & + & (n + 1) & + \cdots + & (n + 1) & + & (n + 1). &
\end{array}
$$

On the right side of the last equation, the term $(n + 1)$ occurs $n$ times. Thus, $2S = n(n + 1)$, so

$$S = \frac{n(n + 1)}{2} \qquad \text{(the sum of the first } n \text{ positive integers).} \qquad (3)$$

For example, the sum of the first 100 positive integers corresponds to $n = 100$ and is $100(100 + 1)/2$, or 5050.

For convenience, to indicate a sum we shall introduce *sigma notation*, so named because the Greek letter $\Sigma$ (sigma) is used. For example, the notation

$$\sum_{k=1}^{3} (2k + 5)$$

denotes the sum of those numbers obtained from the expression $2k + 5$ by first replacing $k$ by 1, then by 2, and, finally, by 3. Thus,

$$\sum_{k=1}^{3} (2k + 5) = [2(1) + 5] + [2(2) + 5] + [2(3) + 5]$$

$$= 7 + 9 + 11 = 27.$$

The letter $k$ is called the *index of summation;* the numbers 1 and 3 are the *limits of summation* (1 is the *lower limit* and 3 is the *upper limit*). The values of the index begin at the lower limit and progress through integer values to the upper limit. The symbol used for the index is a "dummy" symbol in the sense that it does not affect the sum of the terms. Any other letter can be used. For example,

$$\sum_{j=1}^{3} (2j + 5) = 7 + 9 + 11 = \sum_{k=1}^{3} (2k + 5).$$

## EXAMPLE 1  Sigma Notation

**a.** *Evaluate* $\displaystyle\sum_{k=4}^{7}\frac{k^2+3}{2}$.

*Solution:* Here the sum begins with $k = 4$. So we have

$$\sum_{k=4}^{7}\frac{k^2+3}{2} = \frac{4^2+3}{2} + \frac{5^2+3}{2} + \frac{6^2+3}{2} + \frac{7^2+3}{2}$$

$$= \frac{19}{2} + \frac{28}{2} + \frac{39}{2} + \frac{52}{2} = 69.$$

**b.** *Evaluate* $\displaystyle\sum_{j=0}^{2}(-1)^{j+1}(j-1)^2$.

*Solution:*

$$\sum_{j=0}^{2}(-1)^{j+1}(j-1)^2$$

$$= (-1)^{0+1}(0-1)^2 + (-1)^{1+1}(1-1)^2 + (-1)^{2+1}(2-1)^2$$

$$= (-1) + 0 + (-1) = -2. \qquad\blacksquare$$

To express the sum of the first $n$ positive integers in sigma notation, we can write

$$\sum_{k=1}^{n}k = 1 + 2 + \cdots + n.$$

By Eq. (3),

$$\sum_{k=1}^{n}k = \frac{n(n+1)}{2}. \tag{4}$$

Note in Eq. (4) that $\displaystyle\sum_{k=1}^{n}k$ is a function of $n$ alone, not of $k$.

## EXAMPLE 2  Applying Formula 4

**a.** *Evaluate* $\displaystyle\sum_{k=1}^{60}k$.

*Solution:* Here we must find the sum of the first 60 positive integers. By Eq. (4) with $n = 60$,

$$\sum_{k=1}^{60}k = \frac{60(60+1)}{2} = 1830.$$

**b.** *Evaluate* $\displaystyle\sum_{k=1}^{n-1}k$.

*Solution:* Here we must add the first $n - 1$ positive integers. Replacing $n$ by $n - 1$ in Eq. (4), we obtain

$$\sum_{k=1}^{n-1}k = \frac{(n-1)[(n-1)+1]}{2} = \frac{(n-1)n}{2}. \qquad\blacksquare$$

Another useful formula is that for the sum of the *squares* of the first $n$ positive integers:

$$\sum_{k=1}^{n} k^2 = \frac{n(n+1)(2n+1)}{6}. \tag{5}$$

**EXAMPLE 3  Applying Formula 5**

*Evaluate* $1 + 4 + 9 + 16 + 25 + 36$.

*Solution:* This sum can be written as $\sum_{k=1}^{6} k^2$. By Eq. (5) with $n = 6$,

$$\sum_{k=1}^{6} k^2 = \frac{6(6+1)[2(6)+1]}{6} = 91.$$

We conclude with a property of sigma. If $x_1, x_2, \cdots, x_n$, are real numbers and $c$ is a constant, then

$$\sum_{i=1}^{n} cx_i = cx_1 + cx_2 + \cdots + cx_n$$

$$= c(x_1 + x_2 + \cdots + x_n) = c \sum_{i=1}^{n} x_i.$$

Thus,

$$\sum_{i=1}^{n} cx_i = c \sum_{i=1}^{n} x_i.$$

This means that a constant factor can "jump" before a sigma. For example,

$$\sum_{i=1}^{5} 3i^2 = 3 \sum_{i=1}^{5} i^2.$$

By Eq. (5), we have

$$\sum_{i=1}^{5} 3i^2 = 3 \sum_{i=1}^{5} i^2 = 3 \left[ \frac{5(6)(11)}{6} \right] = 165.$$

*Pitfall* ▼ Although constant factors can "jump" before a sigma, nothing else can.

## ■ Exercise 16.5

*In Problems 1–10, evaluate the given sum.*

**1.** $\displaystyle\sum_{k=1}^{5} (k + 4)$.

**2.** $\displaystyle\sum_{k=12}^{15} (5 - 2k)$.

**3.** $\displaystyle\sum_{j=1}^{10} (-1)^j$.

**4.** $\displaystyle\sum_{j=0}^{5} 2^j$.

**5.** $\displaystyle\sum_{n=2}^{3} (3n^2 - 7)$.

**6.** $\displaystyle\sum_{n=2}^{4} \frac{n+1}{n-1}$.

**7.** $\displaystyle\sum_{k=3}^{4} \frac{(-1)^k (k+1)}{2^k}$.

**8.** $\displaystyle\sum_{n=1}^{5} 1$.

**9.** $\displaystyle\sum_{k=1}^{3} \frac{(-1)^{k-1}(1 - k^2)}{k}$.

**10.** $\displaystyle\sum_{n=1}^{4} (n^2 + n)$.

*In Problems* **11–16**, *express the given sums in sigma notation.*

**11.** $1 + 2 + 3 + \cdots + 15.$

**12.** $7 + 8 + 9 + 10.$

**13.** $1 + 3 + 5 + 7.$

**14.** $2 + 4 + 6 + 8.$

**15.** $1^2 + 2^2 + 3^2 + \cdots + 12^2.$

**16.** $3 + 6 + 9 + 12.$

*In Problems* **17–22**, *evaluate the sums by using Eqs.* **(4)** *and* **(5).**

**17.** $\displaystyle\sum_{k=1}^{450} k.$

**18.** $\displaystyle\sum_{k=1}^{10} k^2.$

**19.** $\displaystyle\sum_{j=1}^{6} 4j.$

**20.** $\displaystyle\sum_{i=1}^{40} \frac{i}{2}.$

**21.** $\displaystyle\sum_{i=1}^{6} 3i^2.$

**22.** $\displaystyle\sum_{j=1}^{8} \left(\frac{j}{2}\right)^2.$

**23.** A company has an asset whose original value is $3200. The asset has no salvage value. The maintenance cost each year is $100 and increases by $100 each year. Show that the average annual total cost over a period of $n$ years is

$$C = \frac{3200}{n} + 50(n + 1).$$

Find the value of $n$ that minimizes $C$. What is the average annual cost at this value of $n$?

OBJECTIVE

**To motivate, by means of the concept of area, the definite integral as a limit of a special sum; to evaluate simple definite integrals by using a limiting process.**

## 16.6  THE DEFINITE INTEGRAL

Figure 16.1 shows the region $R$ bounded by the lines $y = f(x) = 2x$, $y = 0$ (the $x$-axis), and $x = 1$. The region is simply a right triangle. If $b$ and $h$ are the lengths of the base and the height, respectively, then, from geometry, the area of the triangle is $A = \frac{1}{2}bh = \frac{1}{2}(1)(2) = 1$ square unit. We shall now find this area by another method, which, as you will see later, applies to more complex regions. This method involves the summation of areas of rectangles.

Let us divide the interval $[0, 1]$ on the $x$-axis into four subintervals of equal length by means of the equally spaced points $x_0 = 0$, $x_1 = \frac{1}{4}$, $x_2 = \frac{2}{4}$, $x_3 = \frac{3}{4}$, and $x_4 = \frac{4}{4} = 1$. (See Fig. 16.2.) Each subinterval has length $\Delta x = \frac{1}{4}$. These subintervals determine four subregions of $R$: $R_1, R_2, R_3$, and $R_4$, as indicated.

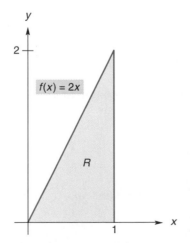

**FIGURE 16.1**   Region bounded by $f(x) = 2x$, $y = 0$, and $x = 1$.

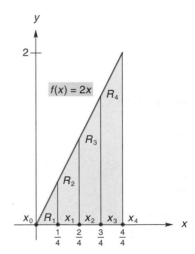

**FIGURE 16.2**   Four subregions of $R$.

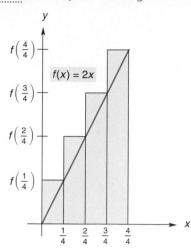

**FIGURE 16.3** Four circumscribed rectangles.

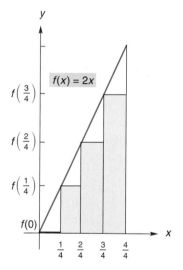

**FIGURE 16.4** Four inscribed rectangles.

With each subregion, we can associate a *circumscribed* rectangle (Fig. 16.3)—that is, a rectangle whose base is the corresponding subinterval and whose height is the *maximum* value of $f(x)$ on that subinterval. Since $f$ is an increasing function, the maximum value of $f(x)$ on each subinterval occurs when $x$ is the right-hand endpoint. Thus, the areas of the circumscribed rectangles associated with regions $R_1$, $R_2$, $R_3$, and $R_4$ are $\frac{1}{4}f(\frac{1}{4})$, $\frac{1}{4}f(\frac{2}{4})$, $\frac{1}{4}f(\frac{3}{4})$, and $\frac{1}{4}f(\frac{4}{4})$, respectively. The area of each rectangle is an approximation to the area of its corresponding subregion. Hence, the sum of the areas of these rectangles, denoted by $\overline{S}_4$ (read "$S$ sub 4 upper bar" or "the fourth upper sum"), approximates the area $A$ of the triangle. We have

$$\overline{S}_4 = \tfrac{1}{4}f(\tfrac{1}{4}) + \tfrac{1}{4}f(\tfrac{2}{4}) + \tfrac{1}{4}f(\tfrac{3}{4}) + \tfrac{1}{4}f(\tfrac{4}{4})$$
$$= \tfrac{1}{4}[2(\tfrac{1}{4}) + 2(\tfrac{2}{4}) + 2(\tfrac{3}{4}) + 2(\tfrac{4}{4})] = \tfrac{5}{4}.$$

You may verify that we can write $\overline{S}_4$ as $\overline{S}_4 = \sum_{i=1}^{4} f(x_i)\,\Delta x$. The fact that $\overline{S}_4$ is greater than the actual area of the triangle might have been expected, since $\overline{S}_4$ includes areas of shaded regions that are not in the triangle. (See Fig. 16.3.)

On the other hand, with each subregion we can also associate an *inscribed* rectangle (Fig. 16.4)—that is, a rectangle whose base is the corresponding subinterval, but whose height is the *minimum* value of $f(x)$ on that subinterval. Since $f$ is an increasing function, the minimum value of $f(x)$ on each subinterval will occur when $x$ is the left-hand endpoint. Thus, the areas of the four inscribed rectangles associated with $R_1$, $R_2$, $R_3$, and $R_4$ are $\frac{1}{4}f(0)$, $\frac{1}{4}f(\frac{1}{4})$, $\frac{1}{4}f(\frac{2}{4})$, and $\frac{1}{4}f(\frac{3}{4})$, respectively. Their sum, denoted $\underline{S}_4$ (read "$S$ sub 4 lower bar" or "the fourth lower sum"), is also an approximation to the area $A$ of the triangle. We have

$$\underline{S}_4 = \tfrac{1}{4}f(0) + \tfrac{1}{4}f(\tfrac{1}{4}) + \tfrac{1}{4}f(\tfrac{2}{4}) + \tfrac{1}{4}f(\tfrac{3}{4})$$
$$= \tfrac{1}{4}[2(0) + 2(\tfrac{1}{4}) + 2(\tfrac{2}{4}) + 2(\tfrac{3}{4})] = \tfrac{3}{4}.$$

Using sigma notation, we can write $\underline{S}_4 = \sum_{i=0}^{3} f(x_i)\Delta x$. Note that $\underline{S}_4$ is less than the area of the triangle, because the rectangles do not account for that portion of the triangle which is not shaded in Fig. 16.4.

Since

$$\tfrac{3}{4} = \underline{S}_4 \le A \le \overline{S}_4 = \tfrac{5}{4},$$

we say that $\underline{S}_4$ is an approximation to $A$ from *below* and $\overline{S}_4$ is an approximation to $A$ from *above*.

If $[0, 1]$ is divided into more subintervals, we expect that better approximations to $A$ will occur. To test this out, let us use six subintervals of equal length $\Delta x = \frac{1}{6}$. Then $\overline{S}_6$, the total area of six circumscribed rectangles (see Fig. 16.5), and $\underline{S}_6$, the total area of six inscribed rectangles (see Fig. 16.6), are

$$\overline{S}_6 = \tfrac{1}{6}f(\tfrac{1}{6}) + \tfrac{1}{6}f(\tfrac{2}{6}) + \tfrac{1}{6}f(\tfrac{3}{6}) + \tfrac{1}{6}f(\tfrac{4}{6}) + \tfrac{1}{6}f(\tfrac{5}{6}) + \tfrac{1}{6}f(\tfrac{6}{6})$$
$$= \tfrac{1}{6}[2(\tfrac{1}{6}) + 2(\tfrac{2}{6}) + 2(\tfrac{3}{6}) + 2(\tfrac{4}{6}) + 2(\tfrac{5}{6}) + 2(\tfrac{6}{6})] = \tfrac{7}{6}$$

and

$$\underline{S}_6 = \tfrac{1}{6}f(0) + \tfrac{1}{6}f(\tfrac{1}{6}) + \tfrac{1}{6}f(\tfrac{2}{6}) + \tfrac{1}{6}f(\tfrac{3}{6}) + \tfrac{1}{6}f(\tfrac{4}{6}) + \tfrac{1}{6}f(\tfrac{5}{6})$$
$$= \tfrac{1}{6}[2(0) + 2(\tfrac{1}{6}) + 2(\tfrac{2}{6}) + 2(\tfrac{3}{6}) + 2(\tfrac{4}{6}) + 2(\tfrac{5}{6})] = \tfrac{5}{6}.$$

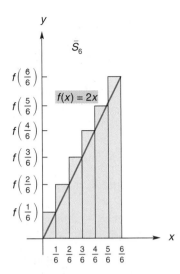

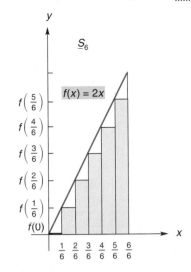

**FIGURE 16.5**   Six circumscribed rectangles.

**FIGURE 16.6**   Six inscribed rectangles.

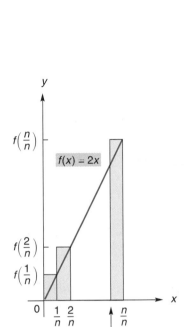

**FIGURE 16.7**   $n$ circumscribed rectangles.

Note that $\underline{S}_6 \leq A \leq \overline{S}_6$, and, with appropriate labeling, both $\overline{S}_6$ and $\underline{S}_6$ will be of the form $\sum f(x)\Delta x$. Clearly, using six subintervals gives better approximations to the area than does four subintervals, as expected.

More generally, if we divide $[0, 1]$ into $n$ subintervals of equal length $\Delta x$, then $\Delta x = 1/n$, and the endpoints of the subintervals are $x = 0$, $1/n$, $2/n$, . . . , $(n - 1)/n$, and $n/n = 1$. (See Fig. 16.7.) The total area of $n$ *circumscribed* rectangles is

$$\overline{S}_n = \frac{1}{n}f\left(\frac{1}{n}\right) + \frac{1}{n}f\left(\frac{2}{n}\right) + \cdots + \frac{1}{n}f\left(\frac{n}{n}\right) \qquad (1)$$

$$= \frac{1}{n}\left[2\left(\frac{1}{n}\right) + 2\left(\frac{2}{n}\right) + \cdots + 2\left(\frac{n}{n}\right)\right]$$

$$= \frac{2}{n^2}[1 + 2 + \cdots + n] \qquad \left(\text{by factoring } \frac{2}{n} \text{ from each term}\right).$$

From Sec. 16.5, the sum of the first $n$ positive integers is $\dfrac{n(n + 1)}{2}$. Thus,

$$\overline{S}_n = \left(\frac{2}{n^2}\right)\frac{n(n + 1)}{2} = \frac{n + 1}{n}.$$

For $n$ *inscribed* rectangles, the total area determined by the subintervals (see Fig. 16.8) is

$$\underline{S}_n = \frac{1}{n}f(0) + \frac{1}{n}f\left(\frac{1}{n}\right) + \cdots + \frac{1}{n}f\left(\frac{n - 1}{n}\right) \qquad (2)$$

$$= \frac{1}{n}\left[2(0) + 2\left(\frac{1}{n}\right) + \cdots + 2\left(\frac{n - 1}{n}\right)\right]$$

$$= \frac{2}{n^2}[1 + \cdots + (n - 1)].$$

Summing the first $n - 1$ positive integers as we did in Example 2(b) of Sec. 16.5, we obtain

$$\underline{S}_n = \left(\frac{2}{n^2}\right)\frac{(n - 1)n}{2} = \frac{n - 1}{n}.$$

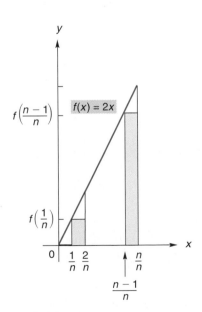

**FIGURE 16.8**   $n$ inscribed rectangles.

From Eqs. (1) and (2), we again see that both $\overline{S}_n$ and $\underline{S}_n$ are sums of the *form*

$$\sum f(x)\,\Delta x, \quad \text{namely,} \quad \overline{S}_n = \sum_{k=1}^{n} f\left(\frac{k}{n}\right)\Delta x \quad \text{and} \quad \underline{S}_n = \sum_{k=0}^{n-1} f\left(\frac{k}{n}\right)\Delta x.$$

From the nature of $\overline{S}_n$ and $\underline{S}_n$, it seems reasonable—and it is indeed true—that

$$\underline{S}_n \leq A \leq \overline{S}_n.$$

As $n$ becomes larger, $\underline{S}_n$ and $\overline{S}_n$ become better approximations to $A$. In fact, let us take the limits of $\underline{S}_n$ and $\overline{S}_n$ as $n$ approaches $\infty$ through positive integral values:

$$\lim_{n\to\infty} \underline{S}_n = \lim_{n\to\infty} \frac{n-1}{n} = \lim_{n\to\infty}\left(1 - \frac{1}{n}\right) = 1,$$

$$\lim_{n\to\infty} \overline{S}_n = \lim_{n\to\infty} \frac{n+1}{n} = \lim_{n\to\infty}\left(1 + \frac{1}{n}\right) = 1.$$

Since $\overline{S}_n$ and $\underline{S}_n$ have the same common limit, namely,

$$\lim_{n\to\infty} \overline{S}_n = \lim_{n\to\infty} \underline{S}_n = 1, \tag{3}$$

and since

$$\underline{S}_n \leq A \leq \overline{S}_n,$$

we shall take this limit to be the area of the triangle. Thus, $A = 1$ square unit, which agrees with our prior finding.

We call the common limit of $\overline{S}_n$ and $\underline{S}_n$, namely, 1, the *definite integral* of $f(x) = 2x$ on the interval from $x = 0$ to $x = 1$, and we denote this quantity by writing

$$\int_0^1 2x\,dx = 1. \tag{4}$$

The reason for using the term "definite integral" and the symbolism in Eq. (4) will become apparent in the next section. The numbers 0 and 1 appearing with the integral sign $\int$ in Eq. (4) are called the *limits of integration;* 0 is the *lower limit* and 1 is the *upper limit.*

In general, for a function $f$ defined on the interval from $x = a$ to $x = b$, where $a < b$, we can form the sums $\overline{S}_n$ and $\underline{S}_n$, which are obtained by considering the maximum and minimum values, respectively, on each of $n$ subintervals of equal length $\Delta x$.[5] We can now state the following:

The common limit of $\overline{S}_n$ and $\underline{S}_n$ as $n \to \infty$, if it exists, is called the **definite integral** of $f$ over $[a, b]$ and is written

$$\int_a^b f(x)\,dx.$$

The numbers $a$ and $b$ are called **limits of integration;** $a$ is the **lower limit** and $b$ is the **upper limit.** The symbol $x$ is called the **variable of integration** and $f(x)$ is the **integrand.**

---

[5]Here we assume that the maximum and minimum values exist.

In terms of a limiting process, we have

$$\sum f(x)\,\Delta x \rightarrow \int_a^b f(x)\,dx.$$

The definite integral is the limit of the form $\sum f(x)\,\Delta x$. This interpretation will be useful in later sections.

Two points must be made about the definite integral. First, the definite integral is the limit of a sum of the form $\sum f(x)\,\Delta x$. In fact, one can think of the integral sign as an elongated "$S$," the first letter of "Summation." Second, for an arbitrary function $f$ defined on an interval, we may be able to calculate the sums $\overline{S}_n$ and $\underline{S}_n$, and determine their common limit if it exists. However, some terms in the sums may be negative if $f(x)$ is negative at points in the interval. These terms are not areas of rectangles (an area is never negative), so the common limit may not represent an area. Thus, **the definite integral is nothing more than a real number; it may or may not represent an area.**

As you saw in Eq. (3), $\lim_{n\to\infty}\underline{S}_n$ is equal to $\lim_{n\to\infty}\overline{S}_n$. For an arbitrary function, this is not always true. However, for the functions that we shall consider, these limits will be equal, and the definite integral will always exist. To save time, we shall just use the **right-hand endpoint** of each subinterval in computing a sum. For the functions in this section, this sum will be denoted $S_n$ and will correspond to either $\underline{S}_n$ or $\overline{S}_n$

### EXAMPLE 1  Computing an Area by Using Right-Hand Endpoints

*Find the area of the region in the first quadrant bounded by $f(x) = 4 - x^2$ and the lines $x = 0$ and $y = 0$.*

In general, over $[a, b]$, we have

$$\Delta x = \frac{b - a}{n}.$$

*Solution:* A sketch of the region appears in Fig. 16.9. The interval over which $x$ varies in this region is seen to be $[0, 2]$, which we divide into $n$ subintervals of equal length $\Delta x$. Since the length of $[0, 2]$ is 2, we take $\Delta x = 2/n$. The endpoints of the subintervals are $x = 0,\ 2/n,\ 2(2/n),\cdots,\ (n-1)(2/n)$, and $n(2/n) = 2$, which are shown in Fig. 16.10. The diagram also shows the corresponding rectangles obtained by using the right-hand endpoint of each subinterval. The area of each rectangle is the product of its width $(2/n)$ and its

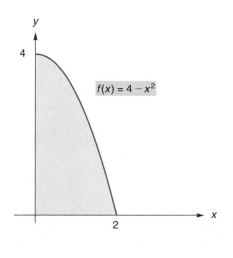

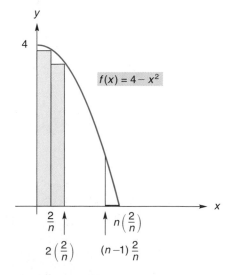

**FIGURE 16.9**  Region of Example 1.

**FIGURE 16.10**  $n$ subintervals and corresponding rectangles for Example 1.

**Principles in Practice 1**

**Computing an Area by Using Right-Hand Endpoints**

A company has determined that its marginal-revenue function is given by $R'(x) = 600 - 0.5x$, where $R$ is the revenue (in dollars) received when $x$ units are sold. Find the total revenue received for selling 10 units by finding the area in the first quadrant bounded by $y = R'(x) = 600 - 0.5x$ and the lines $y = 0$, $x = 0$, and $x = 10$.

height, which is the function value at the right-hand endpoint of its base. Summing these areas, we get

$$S_n = \frac{2}{n}f\left(\frac{2}{n}\right) + \frac{2}{n}f\left[2\left(\frac{2}{n}\right)\right] + \cdots + \frac{2}{n}f\left[n\left(\frac{2}{n}\right)\right]$$

$$= \frac{2}{n}\left[f\left(\frac{2}{n}\right) + f\left[2\left(\frac{2}{n}\right)\right] + \cdots + f\left[n\left(\frac{2}{n}\right)\right]\right]$$

$$= \frac{2}{n}\left[\left\{4 - \left(\frac{2}{n}\right)^2\right\} + \left\{4 - \left[2\left(\frac{2}{n}\right)\right]^2\right\} + \cdots + \left\{4 - \left[n\left(\frac{2}{n}\right)\right]^2\right\}\right].$$

Since the number 4 occurs $n$ times in the sum, we can simplify $S_n$. We get

$$S_n = \frac{2}{n}\left[4n - \left(\frac{2}{n}\right)^2 - 2^2\left(\frac{2}{n}\right)^2 - \cdots - n^2\left(\frac{2}{n}\right)^2\right]$$

$$= \frac{2}{n}\left[4n - \left(\frac{2}{n}\right)^2\{1^2 + 2^2 + \cdots + n^2\}\right].$$

From Sec. 16.5, $\displaystyle\sum_{k=1}^{n} k^2 = \frac{n(n+1)(2n+1)}{6}$, so

$$S_n = \frac{2}{n}\left[4n - \left(\frac{2}{n}\right)^2 \frac{n(n+1)(2n+1)}{6}\right]$$

$$= 8 - \frac{4(n+1)(2n+1)}{3n^2}$$

$$= 8 - \frac{4}{3}\left(\frac{2n^2 + 3n + 1}{n^2}\right).$$

Finally, we take the limit of $S_n$ as $n \to \infty$:

$$\lim_{n\to\infty} S_n = \lim_{n\to\infty}\left[8 - \frac{4}{3}\left(\frac{2n^2 + 3n + 1}{n^2}\right)\right]$$

$$= \lim_{n\to\infty}\left[8 - \frac{4}{3}\left(2 + \frac{3}{n} + \frac{1}{n^2}\right)\right]$$

$$= 8 - \frac{8}{3} = \frac{16}{3}.$$

Thus,

$$\lim_{n\to\infty} S_n = \frac{16}{3}.$$

Hence, the area of the region is $\frac{16}{3}$ square units. ∎

### EXAMPLE 2 Evaluating a Definite Integral

*Evaluate* $\displaystyle\int_0^2 (4 - x^2)\ dx.$

*Solution:* We want to find the definite integral of $f(x) = 4 - x^2$ over the interval $[0, 2]$. Thus, we must compute $\lim\limits_{n \to \infty} S_n$. But this limit is precisely the limit $\frac{16}{3}$ found in Example 1, so we conclude that

$$\int_0^2 (4 - x^2) \, dx = \frac{16}{3}.$$

No units are attached to the answer, since a definite integral is simply a unitless number.

### EXAMPLE 3   Integrating a Function over an Interval

*Integrate $f(x) = x - 5$ from $x = 0$ to $x = 3$; that is, evaluate $\int_0^3 (x - 5) \, dx$.*

*Solution:* We first divide $[0, 3]$ into $n$ subintervals of equal length $\Delta x = 3/n$. The endpoints are $0, 3/n, 2(3/n), \ldots, (n-1)(3/n), n(3/n) = 3$. (See Fig. 16.11.) Using right-hand endpoints, we form the sum

$$S_n = \frac{3}{n}f\left(\frac{3}{n}\right) + \frac{3}{n}f\left[2\left(\frac{3}{n}\right)\right] + \cdots + \frac{3}{n}f\left[n\left(\frac{3}{n}\right)\right].$$

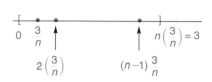

**FIGURE 16.11**   Dividing $[0, 3]$ into $n$ subintervals.

Simplifying, we have

$$S_n = \frac{3}{n}\left[\left\{\frac{3}{n} - 5\right\} + \left\{2\left(\frac{3}{n}\right) - 5\right\} + \cdots + \left\{n\left(\frac{3}{n}\right) - 5\right\}\right]$$

$$= \frac{3}{n}\left[-5n + \frac{3}{n}\{1 + 2 + \cdots + n\}\right]$$

$$= \frac{3}{n}\left[-5n + \left(\frac{3}{n}\right)\frac{n(n+1)}{2}\right]$$

$$= -15 + \frac{9}{2} \cdot \frac{n+1}{n}$$

$$= -15 + \frac{9}{2}\left(1 + \frac{1}{n}\right).$$

Taking the limit, we obtain

$$\lim_{n \to \infty} S_n = \lim_{n \to \infty}\left[-15 + \frac{9}{2}\left(1 + \frac{1}{n}\right)\right] = -15 + \frac{9}{2} = -\frac{21}{2}.$$

Thus,

$$\int_0^3 (x - 5) \, dx = -\frac{21}{2}.$$

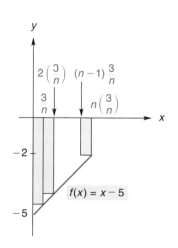

**FIGURE 16.12**   $f(x)$ is negative at each right-hand endpoint.

Note that the definite integral here is a *negative* number. The reason is clear from the graph of $f(x) = x - 5$ over the interval $[0, 3]$. (See Fig. 16.12.) Since the value of $f(x)$ is negative at each right-hand endpoint, each term in $S_n$ must also be negative. Hence, $\lim\limits_{n \to \infty} S_n$, which is the definite integral, is negative. Geometrically, each term in $S_n$ is the negative of the area of a rectangle. (Refer again to Fig. 16.12.) Although the definite integral is simply a number, here we can interpret it as representing the negative of the area of the region bounded by $f(x) = x - 5$, $x = 0$, $x = 3$, and the $x$-axis ($y = 0$). ∎

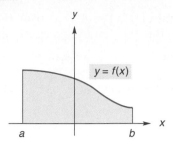

**FIGURE 16.13** If $f$ is continuous and $f(x) \geq 0$ on $[a, b]$, then $\int_a^b f(x)\, dx$ represents area.

In Example 3, it was shown that *the definite integral does not have to represent an area.* In fact, there the definite integral was negative. However, if $f$ is continuous and $f(x) \geq 0$ on $[a, b]$, then $S_n \geq 0$ for all values of $n$. Therefore, $\lim\limits_{n\to\infty} S_n \geq 0$, so $\int_a^b f(x)\, dx \geq 0$. Furthermore, this definite integral gives the area of the region bounded by $y = f(x)$, $y = 0$, $x = a$, and $x = b$. (See Fig. 16.13.)

Although the approach that we took to discuss the definite integral is sufficient for our purposes, it is by no means rigorous. **The important thing to remember about the definite integral is that it is the limit of a special sum.**

## TECHNOLOGY

Here is a program for the TI-82 graphics calculator that will estimate the limit of $S_n$ as $n \to \infty$ for a function $f$ defined on $[a, b]$.

PROGRAM:RIGHTSUM
Lbl 1
Input "SUBINTV",N
$(B - A)/N \to H$
$\emptyset \to S$
$A + H \to X$
$1 \to I$
Lbl 2
$Y_1 + S \to S$
$X + H \to X$
$I + 1 \to I$
If $I \leq N$
Goto 2
$H*S \to S$
Disp S
Pause
Goto 1

```
PrgmRIGHTSUM
SUBINTV100
                -10.455
SUBINTV1000
               -10.4955
SUBINTV2000
              -10.49775
```

**FIGURE 16.14** Values of $S_n$ for $f(x) = x - 5$ on $[0, 3]$.

RIGHTSUM will compute $S_n$ for a given number $n$ of subintervals. Before executing the program,

store $f(x)$, $a$, and $b$ as $Y_1$, A, and B, respectively. Upon execution of the program, you will be prompted to enter the number of subintervals. Then the program proceeds to display the value of $S_n$. Each time ENTER is pressed, the program repeats. In this way, a display of values of $S_n$ for various numbers of subintervals may be obtained. Figure 16.14 shows values of $S_n$ ($n = 100, 1000,$ and $2000$) for the function $f(x) = x - 5$ on the interval $[0, 3]$. As $n \to \infty$, it appears that $S_n \to -10.5$. Thus, we estimate that

$$\lim_{n\to\infty} S_n \approx -10.5,$$

or equivalently,

$$\int_0^3 (x - 5)\, dx \approx -10.5,$$

which agrees with our result in Example 3.

It is interesting to note that the time required for a calculator to compute $S_{2000}$ in Fig. 16.14 was in excess of 1.5 minutes.

## ▪ Exercise 16.6

*In Problems 1–4, sketch the region in the first quadrant that is bounded by the given curves. Approximate the area of the region by the indicated sum. Use the right-hand endpoint of each subinterval.*

**1.** $f(x) = x, y = 0, x = 1;$ $S_3$.

**2.** $f(x) = 3x, y = 0, x = 1;$ $S_5$.

**3.** $f(x) = x^2, y = 0, x = 1;$ $S_3$.

**4.** $f(x) = x^2 + 1, y = 0, x = 0, x = 1;$ $S_2$.

*In Problems 5 and 6, by dividing the indicated interval into n subintervals of equal length, find $S_n$ for the given function. Use the right-hand endpoint of each subinterval. Do not find $\lim\limits_{n\to\infty} S_n$.*

**5.** $f(x) = 4x;$ $[0, 1]$.

**6.** $f(x) = 2x + 1;$ $[0, 2]$.

*In Problems 7 and 8, (a) simplify $S_n$ and (b) find $\lim_{n \to \infty} S_n$.*

**7.** $S_n = \dfrac{1}{n}\left[\left(\dfrac{1}{n} + 1\right) + \left(\dfrac{2}{n} + 1\right) + \cdots + \left(\dfrac{n}{n} + 1\right)\right].$

**8.** $S_n = \dfrac{2}{n}\left[\left(\dfrac{2}{n}\right)^2 + \left(2 \cdot \dfrac{2}{n}\right) + \cdots + \left(n \cdot \dfrac{2}{n}\right)^2\right].$

*In Problems 9–14, sketch the region in the first quadrant that is bounded by the given curves. Determine the exact area of the region by considering the limit of $S_n$ as $n \to \infty$. Use the right-hand endpoint of each subinterval.*

**9.** Region as described in Problem 1.

**10.** Region as described in Problem 2.

**11.** Region as described in Problem 3.

**12.** Region as described in Problem 4.

**13.** $f(x) = 2x^2$, $y = 0$, $x = 2$.

**14.** $f(x) = 9 - x^2$, $y = 0$, $x = 0$.

*In Problems 15–20, evaluate the given definite integral by taking the limit of $S_n$. Use the right-hand endpoint of each subinterval. Sketch the graph, over the given interval, of the function to be integrated.*

**15.** $\displaystyle\int_0^2 3x \, dx.$

**16.** $\displaystyle\int_0^4 9 \, dx.$

**17.** $\displaystyle\int_0^3 -4x \, dx.$

**18.** $\displaystyle\int_0^3 (2x - 9) \, dx.$

**19.** $\displaystyle\int_0^1 (x^2 + x) \, dx.$

**20.** $\displaystyle\int_1^2 (x + 2) \, dx.$

**21.** Find $D_x\left[\displaystyle\int_2^3 \sqrt{x^2 + 1} \, dx\right]$ without the use of limits.

**22.** Find $\displaystyle\int_0^3 f(x) \, dx$ without the use of limits, where

$$f(x) = \begin{cases} 1 - x, & \text{if } 0 \le x \le 1, \\ 2x - 2, & \text{if } 1 \le x \le 2, \\ 2, & \text{if } 2 \le x \le 3. \end{cases}$$

**23.** Find $\displaystyle\int_{-1}^3 f(x) \, dx$ without the use of limits, where

$$f(x) = \begin{cases} 1, & \text{if } x \le 1, \\ 2 - x, & \text{if } 1 \le x \le 2, \\ -1 + \dfrac{x}{2}, & \text{if } x > 2. \end{cases}$$

*In each of Problems 24–26, use a program, such as **RIGHTSUM**, to estimate the area of the region in the first quadrant bounded by the given curves. Round your answer to one decimal place.*

**24.** $f(x) = x^3 + 1$, $y = 0$, $x = 2$, $x = 3.5$.

**25.** $f(x) = \sqrt{x}$, $y = 0$, $x = 0$, $x = 4$.

**26.** $f(x) = e^x$, $y = 0$, $x = 0$, $x = 1$.

*In each of Problems 27–30, use a program, such as **RIGHTSUM**, to estimate the value of the definite integral. Round your answer to one decimal place.*

**27.** $\displaystyle\int_2^5 \dfrac{x + 1}{x + 2} \, dx.$

**28.** $\displaystyle\int_{-2}^{-1} \dfrac{1}{x} \, dx.$

**29.** $\displaystyle\int_{-1}^2 (4x^2 + x - 10) \, dx.$

**30.** $\displaystyle\int_{0.1}^{0.2} \ln x \, dx.$

---

OBJECTIVE

# 16.7 THE FUNDAMENTAL THEOREM OF INTEGRAL CALCULUS

**To informally develop the Fundamental Theorem of Integral Calculus and to use it to compute definite integrals. To obtain a change in function values when the rate of change in the function is known.**

## The Fundamental Theorem

Thus far, the limiting processes of both the derivative and definite integral have been considered separately. We shall now bring these fundamental ideas together and develop the important relationship that exists between them. As a result, we may evaluate definite integrals more efficiently.

The graph of a function $f$ is given in Fig. 16.15. Assume that $f$ is continuous on the interval $[a, b]$ and that its graph does not fall below the $x$-axis. That is, $f(x) \ge 0$. From the preceding section, the area of the region below the graph and above the $x$-axis from $x = a$ to $x = b$ is given by $\displaystyle\int_a^b f(x) \, dx$. We shall now consider another way to determine this area.

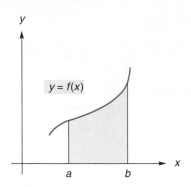

**FIGURE 16.15** On $[a, b]$, $f$ is continuous and $f(x) \geq 0$.

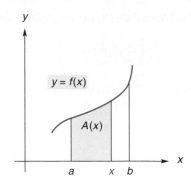

**FIGURE 16.16** $A(x)$ is an area function.

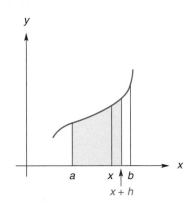

**FIGURE 16.17** $A(x + h)$ gives the area of the shaded region.

Suppose that there is a function $A = A(x)$, which we shall refer to as an area function, that gives the area of the region below the graph of $f$ and above the $x$-axis from $a$ to $x$, where $a \leq x \leq b$. This region is shaded in Fig. 16.16. Do not confuse $A(x)$, which is an area, with $f(x)$, which is the height of the graph at $x$.

From its definition, we can state two properties of $A$ immediately:

**1.** $A(a) = 0$, since there is "no area" from $a$ to $a$;

**2.** $A(b)$ is the area from $a$ to $b$; that is

$$A(b) = \int_a^b f(x) \, dx.$$

If $x$ is increased by $h$ units, then $A(x + h)$ is the area of the shaded region in Fig. 16.17. Hence, $A(x + h) - A(x)$ is the difference of the areas in Figs. 16.17 and 16.16, namely, the area of the shaded region in Fig. 16.18. For $h$ sufficiently close to zero, the area of this region is the same as the area of a rectangle (Fig. 16.19) whose base is $h$ and whose height is some value $\overline{y}$ between

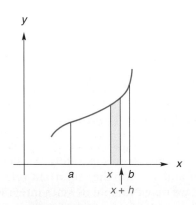

**FIGURE 16.18** Area of shaded region is $A(x + h) - A(x)$.

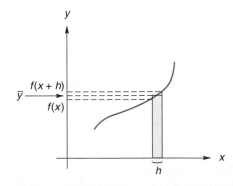

**FIGURE 16.19** Area of rectangle is the same as area of shaded region in Fig. 16.18.

$f(x)$ and $f(x + h)$. Here $\overline{y}$ is a function of $h$. Thus on the one hand, the area of the rectangle is $A(x + h) - A(x)$, and on the other hand it is $h\overline{y}$, so

$$A(x + h) - A(x) = h\overline{y}$$

or

$$\frac{A(x + h) - A(x)}{h} = \overline{y} \qquad \text{(dividing by } h\text{)}.$$

As $h \to 0$, then $\overline{y}$ approaches the number $f(x)$, so

$$\lim_{h \to 0} \frac{A(x + h) - A(x)}{h} = f(x). \qquad (1)$$

But the left side is merely the derivative of $A$. Thus, Eq. (1) becomes

$$A'(x) = f(x).$$

We conclude that the area function $A$ has the additional property that its derivative $A'$ is $f$. That is, $A$ is an antiderivative of $f$. Now, suppose that $F$ is *any* antiderivative of $f$. Then, since both $A$ and $F$ are antiderivatives of the same function, they differ at most by a constant $C$:

$$A(x) = F(x) + C. \qquad (2)$$

Recall that $A(a) = 0$. So, evaluating both sides of Eq. (2) when $x = a$ gives

$$0 = F(a) + C,$$

or

$$C = -F(a).$$

Thus, Eq. (2) becomes

$$A(x) = F(x) - F(a). \qquad (3)$$

If $x = b$, then, from Eq. (3),

$$A(b) = F(b) - F(a). \qquad (4)$$

But recall that

$$A(b) = \int_a^b f(x)\, dx. \qquad (5)$$

From Eqs. (4) and (5), we get

$$\int_a^b f(x)\, dx = F(b) - F(a).$$

A relationship between a definite integral and antidifferentiation has now become clear. To find $\int_a^b f(x)\,dx$, it suffices to find an antiderivative of $f$, say, $F$, and subtract the value of $F$ at the lower limit $a$ from its value at the upper limit $b$. We assumed here that $f$ was continuous and $f(x) \geq 0$ so that we could appeal to the concept of an area. However, our result is true for any continuous function[6] and is known as the *Fundamental Theorem of Integral Calculus.*

**Fundamental Theorem of Integral Calculus**

If $f$ is continuous on the interval $[a, b]$ and $F$ is any antiderivative of $f$ on that interval, then

$$\int_a^b f(x)\, dx = F(b) - F(a).$$

---

[6]If $f$ is continuous on $[a, b]$, it can be shown that $\int_a^b f(x)\, dx$ does indeed exist.

It is important that you understand the difference between a definite integral and an indefinite integral. The **definite integral** $\int_a^b f(x)\,dx$ is a **number** defined to be the limit of a sum. The Fundamental Theorem states that the **indefinite integral** $\int f(x)\,dx$ (an antiderivative of $f$), which is a **function** of $x$ and is related to the differentiation process, can be used to determine this limit.

*The definite integral is a number, and an indefinite integral is a function.*

Suppose we apply the Fundamental Theorem to evaluate $\int_0^2 (4 - x^2)\,dx$. Here $f(x) = 4 - x^2$, $a = 0$, and $b = 2$. Since an antiderivative of $4 - x^2$ is $F(x) = 4x - (x^3/3)$, it follows that

$$\int_0^2 (4 - x^2)\,dx = F(2) - F(0) = \left(8 - \frac{8}{3}\right) - (0) = \frac{16}{3}.$$

This confirms our result in Example 2 of Sec. 16.6. If we had chosen $F(x)$ to be $4x - (x^3/3) + C$, then we would have

$$F(2) - F(0) = \left[\left(8 - \frac{8}{3}\right) + C\right] - [0 + C] = \frac{16}{3},$$

as before. Since the choice of the value of $C$ is immaterial, for convenience we shall always choose it to be 0, as originally done. Usually, $F(b) - F(a)$ is abbreviated by writing

$$F(x)\Big|_a^b.$$

Hence, we have

$$\int_0^2 (4 - x^2)\,dx = \left(4x - \frac{x^3}{3}\right)\Big|_0^2 = \left(8 - \frac{8}{3}\right) - 0 = \frac{16}{3}.$$

**Principles in Practice 1**

**Applying the Fundamental Theorem**

The income (in dollars) from a fast-food chain is increasing at a rate of $f(t) = 10,000e^{0.02t}$, where $t$ is in years. Find $\int_3^6 10,000e^{0.02t}\,dt$, which gives the total income for the chain between the third and sixth years.

**EXAMPLE 1  Applying the Fundamental Theorem**

*Find* $\int_{-1}^3 (3x^2 - x + 6)\,dx$.

***Solution:*** An antiderivative of $3x^2 - x + 6$ is

$$x^3 - \frac{x^2}{2} + 6x.$$

Thus,

$$\int_{-1}^3 (3x^2 - x + 6)\,dx$$

$$= \left(x^3 - \frac{x^2}{2} + 6x\right)\Big|_{-1}^3$$

$$= \left[3^3 - \frac{3^2}{2} + 6(3)\right] - \left[(-1)^3 - \frac{(-1)^2}{2} + 6(-1)\right]$$

$$= \left(\frac{81}{2}\right) - \left(-\frac{15}{2}\right) = 48.$$

### Properties of the Definite Integral

For $\int_a^b f(x)\,dx$, we have assumed that $a < b$. We now define the cases in which $a > b$ or $a = b$. First,

$$\text{if } a > b, \text{ then } \int_a^b f(x)\,dx = -\int_b^a f(x)\,dx.$$

That is, interchanging the limits of integration changes the integral's sign. For example,

$$\int_2^0 (4 - x^2)\,dx = -\int_0^2 (4 - x^2)\,dx.$$

If the limits of integration are equal, we have

$$\int_a^a f(x)\,dx = 0.$$

Some properties of the definite integral deserve mention. The first of the properties that follow restates more formally our comment from the preceding section concerning area.

### Properties of the Definite Integral

**1.** If $f$ is continuous and $f(x) \geq 0$ on $[a, b]$, then $\int_a^b f(x)\,dx$ can be interpreted as the area of the region bounded by the curve $y = f(x)$, the $x$-axis, and the lines $x = a$ and $x = b$.

**2.** $\int_a^b kf(x)\,dx = k \int_a^b f(x)\,dx,$ where $k$ is a constant.

**3.** $\int_a^b [f(x) \pm g(x)]\,dx = \int_a^b f(x)\,dx \pm \int_a^b g(x)\,dx.$

Properties 2 and 3 are similar to rules for indefinite integrals because a definite integral may be evaluated by the Fundamental Theorem in terms of an antiderivative. Two more properties of definite integrals are as follows.

**4.** $\int_a^b f(x)\,dx = \int_a^b f(t)\,dt.$

The variable of integration is a "dummy variable" in the sense that any other variable produces the same result—that is, the same number.

To illustrate property 4, you may verify, for example, that

$$\int_0^2 x^2\,dx = \int_0^2 t^2\,dt.$$

**5.** If $f$ is continuous on an interval $I$ and $a$, $b$, and $c$ are in $I$, then

$$\int_a^c f(x)\,dx = \int_a^b f(x)\,dx + \int_b^c f(x)\,dx.$$

Property 5 means that the definite integral over an interval can be expressed in terms of definite integrals over subintervals. Thus,

$$\int_0^2 (4 - x^2)\,dx = \int_0^1 (4 - x^2)\,dx + \int_1^2 (4 - x^2)\,dx.$$

We shall look at some examples of definite integration now and compute some areas in the next section.

**EXAMPLE 2**   **Using the Fundamental Theorem**

*Find* $\displaystyle\int_0^1 \frac{x^3}{\sqrt{1 + x^4}}\,dx.$

*Solution:* To find an antiderivative of the integrand, we shall apply the power rule for integration:

$$\int_0^1 \frac{x^3}{\sqrt{1 + x^4}}\,dx = \int_0^1 x^3(1 + x^4)^{-1/2}\,dx$$

$$= \frac{1}{4}\int_0^1 (1 + x^4)^{-1/2}[4x^3\,dx] = \left(\frac{1}{4}\right)\frac{(1 + x^4)^{1/2}}{\frac{1}{2}}\Big|_0^1$$

$$= \frac{1}{2}(1 + x^4)^{1/2}\Big|_0^1 = \frac{1}{2}(2)^{1/2} - \frac{1}{2}(1)^{1/2}$$

$$= \frac{1}{2}(\sqrt{2} - 1).$$

$\blacksquare$

**Pitfall** ▼ In Example 2, the value of the antiderivative $\frac{1}{2}(1 + x^4)^{1/2}$ at the lower limit 0 is $\frac{1}{2}(1)^{1/2}$. **Do not** assume that an evaluation at the limit zero will yield 0.

**EXAMPLE 3**   **Evaluating Definite Integrals**

**a.** *Find* $\displaystyle\int_1^2 [4t^{1/3} + t(t^2 + 1)^3]\,dt.$

   *Solution:*

$$\int_1^2 [4t^{1/3} + t(t^2 + 1)^3]\,dt = 4\int_1^2 t^{1/3}\,dt + \frac{1}{2}\int_1^2 (t^2 + 1)^3[2t\,dt]$$

$$= (4)\frac{t^{4/3}}{\frac{4}{3}}\Big|_1^2 + \left(\frac{1}{2}\right)\frac{(t^2 + 1)^4}{4}\Big|_1^2$$

$$= 3(2^{4/3} - 1) + \frac{1}{8}(5^4 - 2^4)$$

$$= 3 \cdot 2^{4/3} - 3 + \frac{609}{8},$$

$$= 6\sqrt[3]{2} + \frac{585}{8}.$$

**b.** *Find* $\displaystyle\int_0^1 e^{3t}\,dt.$

   *Solution:*

$$\int_0^1 e^{3t}\,dt = \frac{1}{3}\int_0^1 e^{3t}[3\,dt]$$

$$= \left(\frac{1}{3}\right)e^{3t}\Big|_0^1 = \frac{1}{3}(e^3 - e^0) = \frac{1}{3}(e^3 - 1).$$

$\blacksquare$

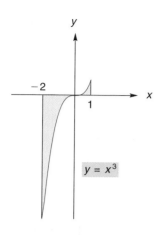

**FIGURE 16.20**  Graph of $y = x^3$ on the interval $[-2, 1]$.

**EXAMPLE 4  Finding and Interpreting a Definite Integral**

*Evaluate* $\int_{-2}^{1} x^3 \, dx$.

*Solution:*

$$\int_{-2}^{1} x^3 \, dx = \frac{x^4}{4} \bigg|_{-2}^{1} = \frac{1^4}{4} - \frac{(-2)^4}{4} = \frac{1}{4} - \frac{16}{4} = -\frac{15}{4}.$$

The reason the result is negative is clear from the graph of $y = x^3$ on the interval $[-2, 1]$. (See Fig. 16.20.) For $-2 \le x < 0$, $f(x)$ is negative. Since a definite integral is a limit of a sum of the form $\Sigma f(x) \, \Delta x$, it follows that $\int_{-2}^{0} x^3 \, dx$ is not only a negative number, but also the negative of the area of the shaded region in the third quadrant. On the other hand, $\int_{0}^{1} x^3 \, dx$ is the area of the shaded region in the first quadrant, since $f(x) \ge 0$ on $[0, 1]$. The definite integral over the entire interval $[-2, 1]$ is the *algebraic* sum of these numbers, because, from property 5,

$$\int_{-2}^{1} x^3 \, dx = \int_{-2}^{0} x^3 \, dx + \int_{0}^{1} x^3 \, dx.$$

Thus, $\int_{-2}^{1} x^3 \, dx$ does not represent the area between the curve and the $x$-axis. However, if area is desired, it can be given by

$$\left| \int_{-2}^{0} x^3 \, dx \right| + \int_{0}^{1} x^3 \, dx. \qquad ■$$

*Pitfall* ▼  Remember that $\int_{a}^{b} f(x) \, dx$ is a limit of a sum. In some cases this limit represents an area. In others it does not. When $f(x) \ge 0$ on $[a, b]$, then the integral represents the area between the graph of $f$ and the $x$-axis from $x = a$ to $x = b$.

## The Definite Integral of a Derivative

Since a function $f$ is an antiderivative of $f'$, by the Fundamental Theorem we have

$$\int_{a}^{b} f'(x) \, dx = f(b) - f(a). \qquad (6)$$

But $f'(x)$ is the rate of change of $f$ with respect to $x$. Hence, if we know the rate of change of $f$ and want to find the difference in function values $f(b) - f(a)$, it suffices to evaluate $\int_{a}^{b} f'(x) \, dx$.

**EXAMPLE 5  Finding a Change in Function Values by Definite Integration**

*A manufacturer's marginal-cost function is*

$$\frac{dc}{dq} = 0.6q + 2.$$

*If production is presently set at $q = 80$ units per week, how much more would it cost to increase production to 100 units per week?*

## Principles in Practice 2
### Change in Function Values

A managerial service determines that the rate of increase in maintenance costs (in dollars per year) for a particular apartment complex is given by $M'(x) = 90x^2 + 5000$, where $x$ is the age of the apartment complex in years and $M(x)$ is the total (accumulated) cost of maintenance for $x$ years. Find the total cost for the first five years.

*Solution:* The total-cost function is $c = c(q)$, and we want to find the difference $c(100) - c(80)$. The rate of change of $c$ is $dc/dq$, so by Eq. (6),

$$c(100) - c(80) = \int_{80}^{100} \frac{dc}{dq}\, dq = \int_{80}^{100} (0.6q + 2)\, dq$$

$$= \left[ \frac{0.6q^2}{2} + 2q \right] \Big|_{80}^{100} = [0.3q^2 + 2q] \Big|_{80}^{100}$$

$$= [0.3(100)^2 + 2(100)] - [0.3(80)^2 + 2(80)]$$

$$= 3200 - 2080 = 1120.$$

If $c$ is in dollars, then the cost of increasing production from 80 units to 100 units is $1120.  ∎

## TECHNOLOGY

Many graphics calculators have the capability to estimate the value of a definite integral. On a TI-82, to estimate

$$\int_{80}^{100} (0.6q + 2)\, dq,$$

we use the "fnInt(" command, as indicated in Fig. 16.21. The parameters that must be entered with this command are

    function to    variable of    lower    upper
  be integrated'   integration'   limit '   limit ·

We see that the value of the definite integral is approximately 1120, which agrees with the result in Example 5.

    Similarly, to estimate

$$\int_{-2}^{1} x^3\, dx$$

we enter

$$\text{fnInt(X}^3\text{,X}, -2,1),$$

or alternatively, if we first store $x^3$ as $Y_1$, we can enter

$$\text{fnInt(Y}_1\text{,X},-2,1).$$

In each case we obtain $-3.75$, which agrees with the result in Example 4.

```
fnInt(.6Q+2,Q,80
,100)
                1120
```

**FIGURE 16.21** Estimating $\int_{80}^{100} (0.6q + 2)\, dq$.

## ■ Exercise 16.7

*In Problems 1–43, evaluate the definite integral.*

**1.** $\int_0^2 5\, dx$.

**2.** $\int_2^4 (1 - e)\, dx$.

**3.** $\int_1^2 7x\, dx$.

**4.** $\int_0^5 -3x\, dx$.

**5.** $\int_{-3}^1 (2x - 3)\, dx$.

**6.** $\int_{-1}^1 (4 - 9y)\, dy$.

**7.** $\int_2^3 (y^2 - 2y + 1)\, dy$.

**8.** $\int_3^2 (2t - t^2)\, dt$.

**9.** $\int_{-2}^{-1} (3w^2 - w - 1)\, dw$.

**10.** $\int_8^9 dt$.

**11.** $\int_1^2 -4t^{-4}\, dt$.

**12.** $\int_1^2 \frac{x^{-2}}{2}\, dx$.

**13.** $\int_{-1}^1 \sqrt[3]{x^5}\, dx$.

**14.** $\int_{1/2}^{3/2} (x^2 + x + 1)\, dx$.

**15.** $\int_{1/2}^3 \frac{1}{x^2}\, dx$.

**16.** $\int_4^9 \left( \frac{1}{\sqrt{x}} - 2 \right) dx$.

**17.** $\int_{-1}^{1}(z+1)^5\,dz.$

**18.** $\int_{1}^{8}(x^{1/3}-x^{-1/3})\,dx.$

**19.** $\int_{0}^{1}2x^2(x^3-1)^3\,dx.$

**20.** $\int_{1}^{3}(x+3)^3\,dx.$

**21.** $\int_{1}^{8}\dfrac{4}{y}\,dy.$

**22.** $\int_{-(e^5)}^{-1}\dfrac{1}{x}\,dx$

**23.** $\int_{0}^{1}e^5\,dx.$

**24.** $\int_{0}^{e}\dfrac{1}{x+1}\,dx.$

**25.** $\int_{0}^{2}x^2e^{x^3}\,dx.$

**26.** $\int_{0}^{1}(3x^2+4x)(x^3+2x^2)^4\,dx.$

**27.** $\int_{4}^{5}\dfrac{2}{(x-3)^3}\,dx.$

**28.** $\int_{0}^{6}\sqrt{2x+4}\,dx.$

**29.** $\int_{1/3}^{2}\sqrt{10-3p}\,dp.$

**30.** $\int_{-1}^{1}q\sqrt{q^2+3}\,dq.$

**31.** $\int_{0}^{1}x^2\sqrt[3]{7x^3+1}\,dx.$

**32.** $\int_{0}^{\sqrt{7}}\left[2x-\dfrac{x}{(x^2+1)^{5/3}}\right]dx.$

**33.** $\int_{0}^{1}\dfrac{2x^3+x}{x^2+x^4+1}\,dx.$

**34.** $\int_{a}^{b}(m+ny)\,dy.$

**35.** $\int_{0}^{1}(e^x-e^{-2x})\,dx.$

**36.** $\int_{-2}^{1}|x|\,dx.$

**37.** $\int_{1}^{e}(x^{-1}+x^{-2}-x^{-3})\,dx.$

**38.** $\int_{1}^{2}\left(6\sqrt{x}-\dfrac{1}{\sqrt{2x}}\right)dx.$

**39.** $\int_{1}^{3}(x+1)e^{x^2+2x}\,dx.$

**40.** $\int_{3}^{4}\dfrac{e^{\ln x}}{x}\,dx.$

**41.** $\int_{0}^{2}\dfrac{x^6+6x^4+x^3+8x^2+x+5}{x^3+5x+1}\,dx.$

**42.** $\int_{-1}^{1}\dfrac{1}{1+e^x}\,dx.$ [*Hint*: Multiply the integrand by $\dfrac{e^{-x}}{e^{-x}}$.]

**43.** $\int_{0}^{2}f(x)\,dx,$ where $f(x)=\begin{cases}4x^2, & \text{if } 0\le x\le\frac{1}{2},\\ 2x, & \text{if }\frac{1}{2}\le x\le 2.\end{cases}$

**44.** Evaluate $\left(\int_{2}^{3}x\,dx\right)^2-\int_{2}^{3}x^2\,dx.$

**45.** Suppose $f(x)=\int_{1}^{x}\dfrac{1}{t^2}\,dt.$ Evaluate $\int_{e}^{1}f(x)\,dx.$

**46.** Evaluate $\int_{10}^{10}e^{x^2}\,dx+\int_{0}^{\ln 1}\dfrac{1}{2\ln 2}\,dx.$

**47.** If $\int_{1}^{3}f(x)\,dx=4$ and $\int_{3}^{2}f(x)\,dx=3,$ find $\int_{1}^{2}f(x)\,dx.$

**48.** If $\int_{1}^{4}f(x)\,dx=6,$ $\int_{2}^{4}f(x)\,dx=5,$

and $\int_{1}^{3}f(x)\,dx=2,$

find $\int_{2}^{3}f(x)\,dx.$

**49.** Evaluate $\int_{1}^{2}\left(\dfrac{d}{dx}\int_{1}^{2}e^{x^2}\,dx\right)dx.$ [*Hint*: it is not necessary to find $\int_{1}^{2}e^{x^2}\,dx.$]

**50.** Suppose that $f(x)=\int_{e}^{x}\dfrac{e^t+e^{-t}}{e^t-e^{-t}}\,dt,$ where $x>e.$ Find $f'(x).$

**51. Severity Index**  In discussing traffic safety, Shonle[7] considers how much acceleration a person can tolerate in a crash so that there is no major injury. The *severity index* is defined as

$$\text{S. I.}=\int_{0}^{T}\alpha^{5/2}\,dt,$$

where $\alpha$ (a Greek letter read "alpha") is considered a constant involved with a weighted average acceleration, and $T$ is the duration of the crash. Find the severity index.

**52. Statistics**  In statistics, the mean $\mu$ (a Greek letter read "mu") of the continuous probability density function $f$ defined on the interval $[a,b]$ is given by

$$\mu=\int_{a}^{b}[x\cdot f(x)]\,dx,$$

and the variance $\sigma^2$ ($\sigma$ is a Greek letter read "sigma") is given by

$$\sigma^2=\int_{a}^{b}(x-\mu)^2 f(x)\,dx.$$

Compute $\mu$ and then $\sigma^2$ if $a=0,b=1,$ and $f(x)=1.$

**53. Distribution of Incomes**  The economist Pareto[8] has stated an empirical law of distribution of higher incomes that gives the number $N$ of persons receiving $x$ or more dollars. If

$$\dfrac{dN}{dx}=-Ax^{-B},$$

where $A$ and $B$ are constants, set up a definite integral that gives the total number of persons with incomes between $a$ and $b,$ where $a<b.$

[7] J. I. Shonle, *Environmental Applications of General Physics* (Reading, MA: Addison-Wesley Publishing Company, Inc., 1975).

[8] G. Tintner, *Methodology of Mathematical Economics and Econometrics* (Chicago: University of Chicago Press, 1967), p. 16

54. **Biology** In a discussion of gene mutation,[9] the following integral occurs:

$$\int_0^{10^{-4}} x^{-1/2}\, dx.$$

Evaluate this integral.

55. **Continuous Income Flow** The present value (in dollars) of a continuous flow of income of \$2000 a year for five years at 6% compounded continuously is given by

$$\int_0^5 2000e^{-0.06t}\, dt.$$

Evaluate the present value to the nearest dollar.

56. **Biology** In biology, problems frequently arise involving the transfer of a substance between compartments. An example is a transfer from the bloodstream to tissue. Evaluate the following integral, which occurs in a two-compartment diffusion problem:[10]

$$\int_0^t (e^{-a\tau} - e^{-b\tau})\, d\tau,$$

Here, $\tau$ (read "tau") is a Greek letter; $a$ and $b$ are constants.

57. **Demography** For a certain population, suppose $l$ is a function such that $l(x)$ is the number or persons who reach the age of $x$ in any year of time. This function is called a *life table function*. Under appropriate conditions, the integral

$$\int_x^{x+n} l(t)\, dt$$

gives the expected number of people in the population between the exact ages of $x$ and $x + n$, inclusive. If

$$l(x) = 10,000\sqrt{100 - x},$$

determine the number of people between the exact ages of 36 and 64, inclusive. Give your answer to the nearest integer, since fractional answers make no sense.

58. **Mineral Consumption** If $c_0$ is the yearly consumption of a mineral at time $t = 0$, then , under continuous consumption, the total amount of the mineral used in the interval $[0, t_1]$ is

$$\int_0^{t_1} c_0 e^{kt}\, dt,$$

where $k$ is the rate of consumption. For a rare-earth mineral, it has been determined that $c_0 = 3000$ units and $k = 0.05$. Evaluate the integral for these data.

59. **Marginal Cost** A manufacturer's marginal-cost function is

$$\frac{dc}{dq} = 0.2q + 3.$$

If $c$ is in dollars, determine the cost involved to increase production from 60 to 70 units.

60. **Marginal Cost** Repeat Problem 59 if

$$\frac{dc}{dq} = 0.003q^2 - 0.6q + 40$$

and production increases from 100 to 200 units.

61. **Marginal Revenue** A manufacturer's marginal-revenue function is

$$\frac{dr}{dq} = \frac{1000}{\sqrt{100q}}.$$

If $r$ is in dollars, find the change in the manufacturer's total revenue if production is increased from 400 to 900 units.

62. **Marginal Revenue** Repeat Problem 61 if

$$\frac{dr}{dq} = 250 + 90q - 3q^2$$

and production is increased from 10 to 20 units.

63. **Crime Rate** A sociologist is studying the crime rate in a certain city. She estimates that $t$ months after the beginning of next year, the total number of crimes committed will increase at the rate of $8t + 10$ crimes per month. Determine the total number of crimes that can be expected to be committed next year. How many crimes can be expected to be committed during the last six months of that year?

64. **Hospital Discharges** For a group of hospitalized individuals, suppose the discharge rate is given by

$$f(t) = \frac{81 \times 10^6}{(300 + t)^4},$$

where $f(t)$ is the proportion of the group discharged per day at the end of $t$ days. What proportion has been discharged by the end of 700 days?

65. **Production** Imagine a one-dimensional country of length $2R$. (See Fig.16.22[11]). Suppose the production of

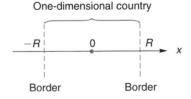

One-dimensional country

**FIGURE 16.22** Diagram for Problem 65.

goods for this country is continuously distributed from border to border. If the amount produced each year per unit of distance is $f(x)$, then the country's total yearly production is given by

[9] W. J. Ewens, *Population Genetics* (London: Methuen & Company Ltd., 1969).

[10] W. Simon, *Mathematical Techniques for Physiology and Medicine* (New York: Academic Press, Inc., 1972).

[11] R. Taagepera, "Why the Trade/GNP Ratio Decreases with Country Size," *Social Science Research*, 5 (1976), 385–404.

$$G = \int_{-R}^{R} f(x)\, dx.$$

Evaluate $G$ if $f(x) = i$, where $i$ is constant.

**66. Exports**   For the one-dimensional country of Problem 65, under certain conditions the amount of the country's exports is given by

$$E = \int_{-R}^{R} \frac{i}{2}\left[e^{-k(R-x)} + e^{-k(R+x)}\right] dx,$$

where $i$ and $k$ are constants ($k \neq 0$). Evaluate $E$.

**67. Average Delivered Price**   In a discussion of a delivered price of a good from a mill to a customer, DeCanio[12] claims that the average delivered price paid by consumers is given by

$$A = \frac{\int_{0}^{R} (m + x)[1 - (m + x)]\, dx}{\int_{0}^{R} [1 - (m + x)]\, dx},$$

where $m$ is mill price, $x$ is the maximum distance to the point of sale. DeCanio determines that

$$A = \frac{m + \dfrac{R}{2} - m^2 - mR - \dfrac{R^2}{3}}{1 - m - \dfrac{R}{2}}.$$

Verify this.

*In each of Problems 68–70, use the Fundamental Theorem of Integral Calculus to determine the value of the definite integral. Confirm your result with your calculator.*

**68.** $\displaystyle\int_{2}^{4} (7x + 3x^2)\, dx.$

**69.** $\displaystyle\int_{0}^{4} \frac{1}{(4x + 4)^2}\, dx.$

**70.** $\displaystyle\int_{0}^{1} e^{3t}\, dt.$ Round your answer to two decimal places.

*In each of Problems 71–74, estimate the value of the definite integral. Round your answer to two decimal places.*

**71.** $\displaystyle\int_{-1}^{5} \frac{x^2 + 1}{x^2 + 4}\, dx.$

**72.** $\displaystyle\int_{1}^{3.4} x \ln^2 x\, dx.$

**73.** $\displaystyle\int_{0}^{4} \sqrt{t^2 + 3}\, dt.$

**74.** $\displaystyle\int_{-1}^{1} \frac{\sqrt{q + 1}}{q + 3}\, dq.$

[12]S. J. DeCanio, "Delivered Pricing and Multiple Basing Point Equilibria: A Reevaluation," *The Quarterly Journal of Economics,* XCIX, no. 2 (1984), 329–49.

## OBJECTIVE

**To use vertical strips and the definite integral to find the area of the region between a curve and the $x$-axis.**

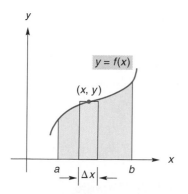

**FIGURE 16.23**   Region with vertical element.

# 16.8  AREA

In Sec. 16.6 we saw that the area of a region could be found by evaluating the limit of a sum of the form $\Sigma f(x)\, \Delta x$, where $f(x)\, \Delta x$ represents the area of a rectangle. This limit is a special case of a definite integral, so it can easily be found by using the Fundamental Theorem.

When using the definite integral to determine area, you should make a rough sketch of the region involved. Let us consider the area of the region bounded by $y = f(x)$ and the $x$-axis from $x = a$ to $x = b$, where $f(x) \geq 0$ on $[a, b]$. (See Fig. 16.23.) To set up the integral, a sample rectangle should be included in the sketch because the area of the region is a limit of sums of areas of rectangles. This will not only help you understand the integration process, but it will also help you find areas of more complicated regions. Such a rectangle (see Fig. 16.23) is called a **vertical element of area** ( or a **vertical strip**). In the diagram, the width of the vertical element is $\Delta x$. The height is the $y$-value of the curve. Hence, the rectangle has area $y\, \Delta x$ or $f(x)\, \Delta x$. The area of the entire region is found by summing the areas of all such elements between $x = a$ and $x = b$ and finding the limit of this sum, which is the definite integral. Symbolically, we have

$$\sum f(x)\Delta x \rightarrow \int_{a}^{b} f(x)\, dx = \text{area}.$$

Example 1 will illustrate.

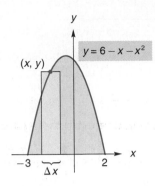

**FIGURE 16.24**  Region
of Example 1 with vertical
element.

### EXAMPLE 1   Using the Definite Integral to Find Area

*Find the area of the region bounded by the curve*

$$y = 6 - x - x^2$$

*and the x-axis.*

**Solution:**  First we must sketch the curve so that we can visualize the region. Since

$$y = -(x^2 + x - 6) = -(x - 2)(x + 3),$$

the *x*-intercepts are $(2, 0)$ and $(-3, 0)$. Using techniques of graphing that were previously discussed, we obtain the graph and region shown in Fig. 16.24. *With this region, it is crucial that the x-intercepts of the curve be found, because they determine the interval over which the areas of the elements must be summed.* That is, *these x-values are the limits of integration.* The vertical element shown has width $\Delta x$ and height $y$. Hence, the area of the element is $y \Delta x$. Summing the areas of all such elements from $x = -3$ to $x = 2$ and taking the limit via the definite integral gives the area:

$$\sum y \,\Delta x \to \int_{-3}^{2} y \, dx = \text{area}.$$

To evaluate the integral, we must express the integrand in terms of the variable of integration, *x*. Since $y = 6 - x - x^2$,

$$\text{area} = \int_{-3}^{2} (6 - x - x^2) \, dx = \left( 6x - \frac{x^2}{2} - \frac{x^3}{3} \right) \Bigg|_{-3}^{2}$$

$$= \left( 12 - \frac{4}{2} - \frac{8}{3} \right) - \left( -18 - \frac{9}{2} - \frac{-27}{3} \right) = \frac{125}{6} \text{ square units.} \quad ■$$

### EXAMPLE 2   Finding the Area of a Region

*Find the area of the region bounded by the curve $y = x^2 + 2x + 2$, the x-axis, and the lines $x = -2$ and $x = 1$.*

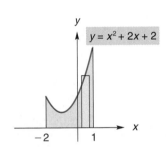

**FIGURE 16.25**  Diagram
for Example 2.

**Solution:**  A sketch of the region is given in Fig. 16.25. We have

$$\text{area} = \int_{-2}^{1} y \, dx = \int_{-2}^{1} (x^2 + 2x + 2) \, dx$$

$$= \left( \frac{x^3}{3} + x^2 + 2x \right) \Bigg|_{-2}^{1} = \left( \frac{1}{3} + 1 + 2 \right) - \left( -\frac{8}{3} + 4 - 4 \right)$$

$$= 6 \text{ square units.} \quad ■$$

### EXAMPLE 3   Finding the Area of a Region

*Find the area of the region between the curve $y = e^x$ and the x-axis from $x = 1$ to $x = 2$.*

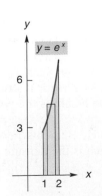

**FIGURE 16.26**
Diagram for
Example 3.

**Solution:**  A sketch of the region is given in Fig. 16.26. We write

$$\text{area} = \int_{1}^{2} y \, dx = \int_{1}^{2} e^x \, dx = e^x \Bigg|_{1}^{2}$$

$$= e^2 - e = e(e - 1) \text{ square units.} \quad ■$$

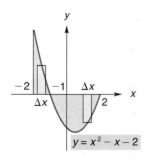

**FIGURE 16.27**
Diagram for Example 4.

**EXAMPLE 4**   **An Area Requiring Two Definite Integrals**

*Find the area of the region bounded by the curve*

$$y = x^2 - x - 2$$

*and the line $y = 0$ (the x-axis) from $x = -2$ to $x = 2$.*

*Solution:* A sketch of the region is given in Fig. 16.27. Notice that the $x$-intercepts are $(-1, 0)$ and $(2, 0)$.

*Pitfall* ▼ It is wrong to write hastily that the area is $\int_{-2}^{2} y \, dx$, for the following reason: For the left rectangle, the height is $y$. However, for the rectangle on the right, $y$ is negative, so its height is the positive number $-y$. This points out the importance of sketching the region.

On the interval $[-2, -1]$, the area of the element is

$$y \, \Delta x = (x^2 - x - 2) \, \Delta x.$$

On $[-1, 2]$ the area is

$$-y \, \Delta x = -(x^2 - x - 2) \, \Delta x.$$

Thus,

$$\text{area} = \int_{-2}^{-1} (x^2 - x - 2) \, dx + \int_{-1}^{2} -(x^2 - x - 2) \, dx$$

$$= \left( \frac{x^3}{3} - \frac{x^2}{2} - 2x \right)\Big|_{-2}^{-1} - \left( \frac{x^3}{3} - \frac{x^2}{2} - 2x \right)\Big|_{-1}^{2}$$

$$= \left[ \left( -\frac{1}{3} - \frac{1}{2} + 2 \right) - \left( -\frac{8}{3} - \frac{4}{2} + 4 \right) \right] -$$

$$\left[ \left( \frac{8}{3} - \frac{4}{2} - 4 \right) - \left( -\frac{1}{3} - \frac{1}{2} + 2 \right) \right]$$

$$= \frac{19}{3} \text{ square units.} \qquad\blacksquare$$

The next example shows the use of area as a probability in statistics.

**EXAMPLE 5**   **Statistics Application**

*In statistics, a (probability)* ***density function*** *$f$ of a variable $x$, where $x$ assumes all values in the interval $[a, b]$, has the following properties:*

1. $f(x) \geq 0$.

2. $\int_a^b f(x) \, dx = 1$.

3. *The probability that $x$ assumes a value between $c$ and $d$, which is written $P(c \leq x \leq d)$, where $a \leq c \leq d \leq b$, is represented by the area of the region bounded by the graph of $f$ and the x-axis between $x = c$ and $x = d$. Hence (see Fig. 16.28),*

$$P(c \leq x \leq d) = \int_c^d f(x) \, dx.$$

*For the density function $f(x) = 6(x - x^2)$, where $0 \leq x \leq 1$, find each of the following probabilities.*

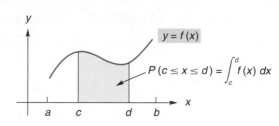

**FIGURE 16.28** Probability as an area.

**a.** $P(0 \le x \le \frac{1}{4})$.

*Solution:* Here $[a, b]$ is $[0, 1]$, $c$ is 0, and $d$ is $\frac{1}{4}$. By property 3, we have

$$P(0 \le x \le \tfrac{1}{4}) = \int_0^{1/4} 6(x - x^2)\,dx = 6\int_0^{1/4}(x - x^2)\,dx$$

$$= 6\left(\frac{x^2}{2} - \frac{x^3}{3}\right)\bigg|_0^{1/4} = (3x^2 - 2x^3)\bigg|_0^{1/4}$$

$$= \left[3\left(\frac{1}{4}\right)^2 - 2\left(\frac{1}{4}\right)^3\right] - 0 = \frac{5}{32}.$$

**b.** $P(x \ge \frac{1}{2})$.

*Solution:* Since the domain of $f$ is $0 \le x \le 1$, to say that $x \ge \frac{1}{2}$ means that $\frac{1}{2} \le x \le 1$. Thus,

$$P(x \ge \tfrac{1}{2}) = \int_{1/2}^1 6(x - x^2)\,dx = 6\int_{1/2}^1(x - x^2)\,dx$$

$$= 6\left(\frac{x^2}{2} - \frac{x^3}{3}\right)\bigg|_{1/2}^1 = (3x^2 - 2x^3)\bigg|_{1/2}^1 = \frac{1}{2}. \quad\blacksquare$$

■ **Exercise 16.8**

*In Problems 1–34, use a definite integral to find the area of the region bounded by the given curve, the x-axis, and the given lines. In each case, first sketch the region. Watch out for areas of regions that are below the x-axis.*

**1.** $y = 4x$, $x = 2$.

**2.** $y = 3x + 1$, $x = 0$, $x = 4$.

**3.** $y = 3x + 2$, $x = 2$, $x = 3$.

**4.** $y = x + 5$, $x = 2$, $x = 4$

**5.** $y = x - 1$, $x = 5$.

**6.** $y = 2x^2$, $x = 1$, $x = 2$.

**7.** $y = x^2$, $x = 2$, $x = 3$.

**8.** $y = 2x^2 - x$, $x = -2$, $x = -1$.

**9.** $y = x^2 + 2$, $x = -1$, $x = 2$.

**10.** $y = 2x + x^3$, $x = 1$.

**11.** $y = x^2 - 2x$, $x = -3$, $x = -1$.

**12.** $y = 3x^2 - 4x$, $x = -2$, $x = -1$.

**13.** $y = 9 - x^2$.

**14.** $y = \dfrac{4}{x}$, $x = 1$, $x = 2$.

**15.** $y = 1 - x - x^3$, $x = -2$, $x = 0$.

**16.** $y = e^x$, $x = 1$, $x = 3$.

**17.** $y = 3 + 2x - x^2$.

**18.** $y = \dfrac{1}{x^2}$, $x = 2$, $x = 3$.

**19.** $y = \dfrac{1}{x}$, $x = 1$, $x = e$.

**20.** $y = \dfrac{1}{x}$, $x = 1$, $x = e^2$.

**21.** $y = \sqrt{x + 9}$, $x = -9$, $x = 0$.

**22.** $y = x^2 - 2x$, $x = 1$, $x = 3$.

**23.** $y = \sqrt{2x - 1}$,   $x = 1$,   $x = 5$.

**24.** $y = x^3 + 3x^2$,   $x = -2$,   $x = 2$.

**25.** $y = \sqrt[3]{x}$,   $x = 2$.

**26.** $y = x^2 - 4$,   $x = -2$,   $x = 2$.

**27.** $y = e^x$,   $x = 0$,   $x = 2$.

**28.** $y = |x|$,   $x = -2$,   $x = 2$.

**29.** $y = x + \dfrac{2}{x}$,   $x = 1$,   $x = 2$.

**30.** $y = 6 - x - x^2$.

**31.** $y = x^3$,   $x = -2$,   $x = 4$.

**32.** $y = \sqrt{x - 2}$,   $x = 2$,   $x = 6$.

**33.** $y = 2x - x^2$,   $x = 1$,   $x = 3$.

**34.** $y = x^2 - x + 1$,   $x = 0$,   $x = 1$.

**35.** Given that

$$f(x) = \begin{cases} 3x^2, & \text{if } 0 \le x \le 2, \\ 16 - 2x, & \text{if } x \ge 2, \end{cases}$$

determine the area of the region bounded by the graph of $y = f(x)$, the $x$-axis, and the line $x = 3$. Include a sketch of the region.

**36.** Under conditions of a continuous uniform distribution, a topic in statistics, the proportion of persons with incomes between $a$ and $t$, where $a \le t \le b$, is the area of the region between the curve $y = 1/(b - a)$ and the $x$-axis from $x = a$ to $x = t$. Sketch the graph of the curve and determine the area of the given region.

**37.** Suppose $f(x) = x/8$, where $0 \le x \le 4$. If $f$ is a density function (refer to Example 5), find each of the following.

  **a.** $P(0 \le x \le 1)$.

  **b.** $P(2 \le x \le 4)$.

  **c.** $P(x \ge 3)$.

**38.** Suppose $f(x) = 3(1 - x)^2$, where $0 \le x \le 1$. If $f$ is a density function (refer to Example 5), find each of the following.

  **a.** $P(\frac{1}{2} \le x \le 1)$.

  **b.** $P(\frac{1}{3} \le x \le \frac{1}{2})$.

  **c.** $P(x \le \frac{1}{3})$.

  **d.** $P(x \ge \frac{1}{3})$, using your result from part (c).

**39.** Suppose $f(x) = 1/x$, where $e \le x \le e^2$. If $f$ is a density function (refer to Example 5), find each of the following.

  **a.** $P(3 < x \le 5)$.

  **b.** $P(x \le 4)$.

  **c.** $P(x \ge 3)$.

  **d.** Verify that $P(e \le x \le e^2) = 1$.

**40. a.** Let $r$ be a real number, where $r > 1$. Evaluate

$$\int_1^r \frac{1}{x^2}\, dx.$$

  **b.** Your answer to part (a) can be interpreted as the area of a certain region of the plane. Sketch this region.

  **c.** Evaluate $\displaystyle\lim_{r \to \infty} \left( \int_1^r \frac{1}{x^2}\, dx \right)$.

  **d.** Your answer to part (c) can be interpreted as the area of a certain region of the plane. Sketch this region.

*In each of Problems 41–44, use definite integration to estimate the area of the region bounded by the given curve, the x-axis, and the given lines. Round your answer to two decimal places.*

**41.** $y = \dfrac{1}{x^2 + 1}$,   $x = -1$,   $x = 1$.

**42.** $y = \dfrac{x}{\sqrt{x + 3}}$,   $x = 1$,   $x = 6$.

**43.** $y = x^4 - 2x^3 - 2$,   $x = 1$,   $x = 3$.

**44.** $y = 1 + 3x - x^4$.

---

**OBJECTIVE**

To find the area of a region bounded by two or more curves by using either vertical or horizontal strips

## 16.9  AREA BETWEEN CURVES

### Vertical Elements

We shall now find the area of a region enclosed by several curves. As before, our procedure will be to draw a sample element of area and use the definite integral to "add together" the areas of all such elements.

For example, consider the area of the region in Fig. 16.29 that is bounded on the top and bottom by the curves $y = f(x)$ and $y = g(x)$ and on the sides by the lines $x = a$ and $x = b$. The width of the indicated vertical element is $\Delta x$, and the height is the $y$-value of the upper curve minus the $y$-value of the lower curve, which we shall write as $y_{\text{upper}} - y_{\text{lower}}$. Thus, the area of the element is

$$[y_{\text{upper}} - y_{\text{lower}}]\, \Delta x,$$

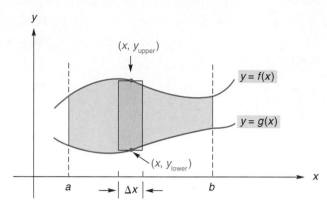

**FIGURE 16.29**   Region between curves.

or,

$$[f(x) - g(x)]\Delta x.$$

Summing the areas of all such elements from $x = a$ to $x = b$ by the definite integral gives the area of the region:

$$\sum [f(x) - g(x)]\Delta x \rightarrow \int_a^b [f(x) - g(x)]\,dx = \text{area}.$$

**EXAMPLE 1**   **Finding an Area between Two Curves**

*Find the area of the region bounded by the curves $y = \sqrt{x}$ and $y = x$.*

*Solution:* A sketch of the region appears in Fig. 16.30. To determine where the curves intersect, we solve the system formed by the equations $y = \sqrt{x}$ and $y = x$. Eliminating $y$ by substitution, we obtain

$$\sqrt{x} = x,$$

$$x = x^2 \qquad \text{(squaring both sides)},$$

$$0 = x^2 - x = x(x - 1).$$

$$x = 0 \quad \text{or} \quad x = 1.$$

If $x = 0$, then $y = 0$; if $x = 1$, then $y = 1$. Thus, the curves intersect at $(0, 0)$ and $(1, 1)$. The width of the indicated element of area is $\Delta x$. The height is the $y$-value on the upper curve minus the $y$-value on the lower curve:

$$y_{\text{upper}} - y_{\text{lower}} = \sqrt{x} - x.$$

Hence, the area of the element is $(\sqrt{x} - x)\Delta x$. Summing the areas of all such elements from $x = 0$ to $x = 1$ by the definite integral, we get the area of the entire region:

$$\text{area} = \int_0^1 (\sqrt{x} - x)\,dx$$

$$= \int_0^1 (x^{1/2} - x)\,dx = \left(\frac{x^{3/2}}{\frac{3}{2}} - \frac{x^2}{2}\right)\Bigg|_0^1$$

$$= \left(\frac{2}{3} - \frac{1}{2}\right) - (0 - 0) = \frac{1}{6} \text{ square unit} \qquad\blacksquare$$

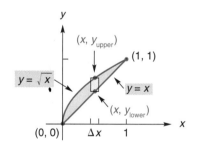

**FIGURE 16.30**   Diagram for Example 1.

It should be obvious to you that knowing the points of intersection is important in determining the limits of integration.

**EXAMPLE 2** Finding an Area between Two Curves

*Find the area of the region bounded by the curves $y = 4x - x^2 + 8$ and $y = x^2 - 2x$.*

*Solution:* A sketch of the region appears in Fig. 16.31. To find where the curves intersect, we solve the system of equations $y = 4x - x^2 + 8$ and $y = x^2 - 2x$:

$$4x - x^2 + 8 = x^2 - 2x,$$
$$-2x^2 + 6x + 8 = 0,$$
$$x^2 - 3x - 4 = 0,$$
$$(x + 1)(x - 4) = 0.$$
$$x = -1 \quad \text{or} \quad x = 4.$$

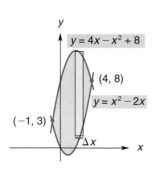

**FIGURE 16.31**
Diagram for Example 2.

When $x = -1$, then $y = 3$; when $x = 4$, then $y = 8$. Thus, the curves intersect at $(-1, 3)$ and $(4, 8)$. The width of the indicated element is $\Delta x$. The height is the $y$-value on the upper curve minus the $y$-value on the lower curve:

$$y_{\text{upper}} - y_{\text{lower}} = (4x - x^2 + 8) - (x^2 - 2x).$$

Therefore, the area of the element is

$$[(4x - x^2 + 8) - (x^2 - 2x)]\Delta x = (-2x^2 + 6x + 8)\Delta x.$$

Summing all such areas from $x = -1$ to $x = 4$, we have

$$\text{area} = \int_{-1}^{4}(-2x^2 + 6x + 8)\, dx = 41\tfrac{2}{3} \text{ square units.} \quad \blacksquare$$

**EXAMPLE 3** Area of a Region Having Two Different Upper Curves

*Find the area of the region between the curves $y = 9 - x^2$ and $y = x^2 + 1$ from $x = 0$ to $x = 3$.*

*Solution:* The region is sketched in Fig. 16.32. The curves intersect when

$$9 - x^2 = x^2 + 1,$$
$$8 = 2x^2,$$
$$4 = x^2,$$
$$x = \pm 2.$$

When $x = \pm 2$, then $y = 5$, so the points of intersection are $(\pm 2, 5)$. Because we are interested in the region from $x = 0$ to $x = 3$, the intersection point that is of concern to us is $(2, 5)$. Notice in Fig. 16.32 that in the region to the *left* of the intersection point $(2, 5)$, an element has

$$y_{\text{upper}} = 9 - x^2 \quad \text{and} \quad y_{\text{lower}} = x^2 + 1,$$

but for an element to the *right* of $(2, 5)$ the reverse is true, namely,

$$y_{\text{upper}} = x^2 + 1 \quad \text{and} \quad y_{\text{lower}} = 9 - x^2.$$

Thus, from $x = 0$ to $x = 2$, the area of an element is

$$(y_{\text{upper}} - y_{\text{lower}})\, \Delta x = [(9 - x^2) - (x^2 + 1]\, \Delta x$$
$$= (8 - 2x^2)\, \Delta x,$$

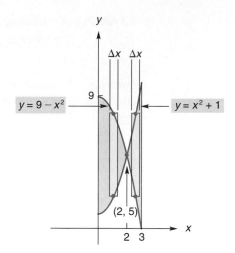

**FIGURE 16.32** $y_{upper}$ is $9 - x^2$ on $[0, 2]$ and is $x^2 + 1$ on $[2, 3]$.

but from $x = 2$ to $x = 3$, it is

$$(y_{upper} - y_{lower})\, \Delta x = [(x^2 + 1) - (9 - x^2)]\, \Delta x$$

$$= (2x^2 - 8)\, \Delta x.$$

Therefore, to find the area of the entire region, we need *two* integrals:

$$\text{area} = \int_0^2 (8 - 2x^2)\, dx + \int_2^3 (2x^2 - 8)\, dx$$

$$= \left(8x - \frac{2x^3}{3}\right)\Big|_0^2 + \left(\frac{2x^3}{3} - 8x\right)\Big|_2^3$$

$$= \left[\left(16 - \frac{16}{3}\right) - 0\right] + \left[(18 - 24) - \left(\frac{16}{3} - 16\right)\right]$$

$$= \frac{46}{3} \text{ square units.} \qquad \blacksquare$$

### Horizontal Elements

Sometimes area can more easily be determined by summing areas of horizontal elements rather than vertical elements. In the following example, an area will be found by both methods. In each case, the element of area determines the form of the integral.

**EXAMPLE 4  The Methods of Vertical Elements and Horizontal Elements**

*Find the area of the region bounded by the curve $y^2 = 4x$ and the lines $y = 3$ and $x = 0$ (the y-axis).*

*Solution:* The region is sketched in Fig. 16.33. When the curves $y = 3$ and $y^2 = 4x$ intersect, $9 = 4x$, so $x = \frac{9}{4}$. Thus, the intersection point is $\left(\frac{9}{4}, 3\right)$. Since the width of the vertical strip is $\Delta x$, we integrate with respect to the variable $x$. Accordingly, $y_{upper}$ and $y_{lower}$ must be expressed as functions of $x$. For the lower curve, $y^2 = 4x$, we have $y = \pm 2\sqrt{x}$. But $y \geq 0$ for the portion

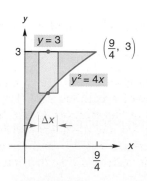

**FIGURE 16.33** Vertical element of area.

of this curve that bounds the region, so we use $y = 2\sqrt{x}$. The upper curve is $y = 3$. Hence, the height of the strip is

$$y_{\text{upper}} - y_{\text{lower}} = 3 - 2\sqrt{x}.$$

Therefore, the strip has an area of $(3 - 2\sqrt{x})\Delta x$, and we wish to sum all such areas from $x = 0$ to $x = \frac{9}{4}$. We have

$$\text{area} = \int_0^{9/4} (3 - 2\sqrt{x}) \, dx = \left( 3x - \frac{4x^{3/2}}{3} \right)\Bigg|_0^{9/4}$$

$$= \left[ 3\left(\frac{9}{4}\right) - \frac{4}{3}\left(\frac{9}{4}\right)^{3/2} \right] - (0)$$

$$= \frac{27}{4} - \frac{4}{3}\left[ \left(\frac{9}{4}\right)^{1/2} \right]^3 = \frac{27}{4} - \frac{4}{3}\left(\frac{3}{2}\right)^3 = \frac{9}{4} \text{ square units.}$$

Let us now approach this problem from the point of view of a **horizontal element of area** (or **horizontal strip**) as shown in Fig. 16.34. The width of the element is $\Delta y$. The length of the element is the *x-value on the rightmost curve minus the x-value on the leftmost curve*. Thus, the area of the element is

$$(x_{\text{right}} - x_{\text{left}}) \, \Delta y.$$

With horizontal elements, the width is $\Delta y$, not $\Delta x$.

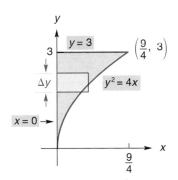

**FIGURE 16.34**  Horizontal element of area.

We wish to sum all such areas from $y = 0$ to $y = 3$:

$$\sum (x_{\text{right}} - x_{\text{left}}) \, \Delta y \rightarrow \int_0^3 (x_{\text{right}} - x_{\text{left}}) \, dy.$$

Since the variable of integration is $y$, we must express $x_{\text{right}}$ and $x_{\text{left}}$ as functions of $y$. The rightmost curve is $y^2 = 4x$ or, equivalently, $x = y^2/4$. The left curve is $x = 0$. Thus,

$$\text{area} = \int_0^3 (x_{\text{right}} - x_{\text{left}}) \, dy$$

$$= \int_0^3 \left( \frac{y^2}{4} - 0 \right) dy = \frac{y^3}{12}\Bigg|_0^3 = \frac{9}{4} \text{ square units.}$$

Note that for this region, horizontal strips make the definite integral easier to evaluate (and set up) than an integral with vertical strips. In any case, remember that **the limits of integration are those limits for the variable of integration.** ∎

### EXAMPLE 5  Advantage of Horizontal Elements

*Find the area of the region bounded by the graphs of $y^2 = x$ and $x - y = 2$.*

*Solution:* The region is sketched in Fig. 16.35. The curves intersect when $y^2 - y = 2$. Thus, $y^2 - y - 2 = 0$, or equivalently, $(y + 1)(y - 2) = 0$, from which it follows that $y = -1$ or $y = 2$. This gives the intersection points $(1, -1)$ and $(4, 2)$. Let us try vertical elements of area. [See Fig. 16.35(a).] Solving $y^2 = x$ for $y$ gives $y = \pm\sqrt{x}$. As seen in Fig. 16.35(a), to the *left* of $x = 1$, the upper end of the element lies on $y = \sqrt{x}$ and the lower end lies on $y = -\sqrt{x}$. To the *right* of $x = 1$, the upper curve is $y = \sqrt{x}$ and the lower curve is $x - y = 2$ (or $y = x - 2$). Thus, with vertical strips, *two* integrals are needed to evaluate the area:

$$\text{area} = \int_0^1 [\sqrt{x} - (-\sqrt{x})] \, dx + \int_1^4 [\sqrt{x} - (x - 2)] \, dx.$$

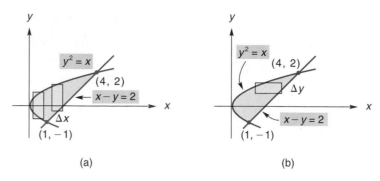

**FIGURE 16.35** Region of Example 5 with vertical and horizontal elements.

Perhaps the use of horizontal strips can simplify our work. In Fig. 16.35(b), the width of the strip is $\Delta y$. The rightmost curve is *always* $x - y = 2$ (or $x = y + 2$), and the leftmost curve is *always* $y^2 = x$ (or $x = y^2$). Therefore, the area of the horizontal strip is $[(y + 2) - y^2]\,\Delta y$, so the total area is

$$\text{area} = \int_{-1}^{2}(y + 2 - y^2)\,dy = \frac{9}{2}\ \text{square units.}$$

Clearly, the use of horizontal strips is the most desirable approach to solving the problem. Only a single integral is needed, and it is much simpler to compute. ∎

## TECHNOLOGY

**Problem:** Estimate the area of the region bounded by the graphs of

$$y = x^4 - 2x^3 - 2 \qquad \text{and} \qquad y = 1 + 2x - 2x^2.$$

***Solution:*** With a TI-82, we enter $x^4 - 2x^3 - 2$ as $Y_1$ and $1 + 2x - 2x^2$ as $Y_2$ and display their graphs. The region in question is shaded in Fig. 16.36; $y_{\text{upper}}$ corresponds to $Y_2$ and $y_{\text{lower}}$ corresponds to $Y_1$. Using vertical strips, we have

$$\text{area} = \int_{A}^{B}(Y_2 - Y_1)\,dx,$$

where A and B are the $x$-values of the intersection points in Quadrants III and IV, respectively. With the intersection feature we find A, as indicated in Fig. 16.37. This value of $x$ is then stored from the home screen as A. (See Fig. 16.38.) Similarly, we find the $x$-value of the intersection point in Quadrant IV, which we store as B. Using the "fnInt(" command (Fig. 16.38), we estimate the area of the region to be 7.54 square units.

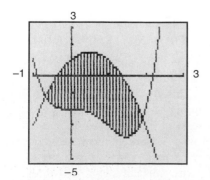

**FIGURE 16.36** Graphs of $Y_1(y_{\text{lower}})$ and $Y_2(y_{\text{upper}})$.

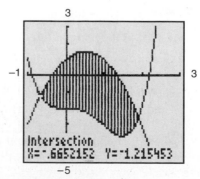

**FIGURE 16.37** Intersection point in Quadrant III.

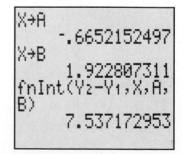

**FIGURE 16.38** Storing $x$-values of intersection points and estimating area.

■ **Exercise 16.9**

*In Problems 1–6, express the area of the shaded region in terms of an integral (or integrals). Do not evaluate your expression.*

**1.** See Fig.16. 39.

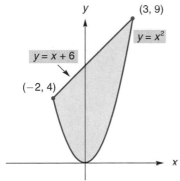

**FIGURE 16.39**   Region for Problem 1.

**2.** See Fig. 16.40.

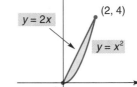

**FIGURE 16.40**   Region for Problem 2.

**3.** See Fig. 16.41.

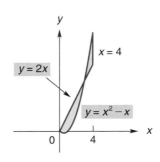

**FIGURE 16.41**   Region for Problem 3.

**4.** See Fig. 16.42.

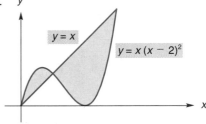

**FIGURE 16.42**   Region for Problem 4.

**5.** See Fig. 16.43.

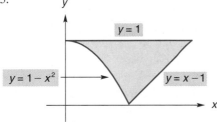

**FIGURE 16.43**   Region for Problem 5.

**6.** See Fig. 16.44.

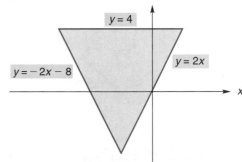

**FIGURE 16.44**   Region for Problem 6.

**7.** Express, in terms of a single integral, the total area of the region to the left of the line $x = 2$ that is between the curves $y = x^2 - 4$ and $y = 11 - 2x^2$. Do *not* evaluate the integral.

**8.** Express, in terms of a single integral, the total area of the region in the *fourth* quadrant bounded by the $x$-axis and the graphs of $y^2 = x$ and $y = 2 - x$. Do *not* evaluate the integral.

*In Problems 9–32, find the area of the region bounded by the graphs of the given equations. Be sure to find any needed points of intersection. Consider whether the use of horizontal strips makes the integral simpler than when vertical strips are used.*

**9.** $y = x^2$,   $y = 2x$.

**10.** $y = x$,   $y = -x + 3$,   $y = 0$.

**11.** $y = x^2$,   $x = 0$,   $y = 4$ $(x \geq 0.)$

**12.** $y = x^2 + 1$,   $y = x + 3$.

13. $y = x^2 + 3$, $y = 9$.

14. $y^2 = x$, $x = 2$.

15. $x = 8 + 2y$, $x = 0$, $y = -1$, $y = 3$.

16. $y = x - 4$, $y^2 = 2x$.

17. $y = 4 - x^2$, $y = -3x$.

18 $x = y^2 + 2$, $x = 6$.

19. $y^2 = x$, $y = x - 2$.

20. $y = x^2$, $y = x + 2$.

21. $2y = 4x - x^2$, $2y = x - 4$.

22. $y = \sqrt{x}$, $y = x^2$.

23. $y^2 = x$, $3x - 2y = 1$.

24. $y = 2 - x^2$, $y = x$.

25. $y = 8 - x^2$, $y = x^2$, $x = -1$, $x = 1$.

26. $y^2 = 4 - x$, $y = x + 2$.

27. $y = x^2$, $y = 2$, $y = 5$.

28. $y = x^3 - x$, x-axis.

29. $y = x^3$, $y = x$.

30. $y = x^3$, $y = \sqrt{x}$.

31. $4x + 4y + 17 = 0$, $y = \dfrac{1}{x}$.

32. $y^2 = -x$, $x - y = 4$, $y = -1$, $y = 2$.

33. Find the area of the region that is between the curves

$$y = x - 1 \quad \text{and} \quad y = 5 - 2x$$

from $x = 0$ to $x = 4$.

34. Find the area of the region that is between the curves

$$y = x^2 - 4x + 4 \quad \text{and} \quad y = 10 - x^2$$

from $x = 2$ to $x = 4$.

35. **Lorentz Curve** A *Lorentz curve* is used in studying income distributions. If $x$ is the cumulative percentage of income recipients, ranked from poorest to richest, and $y$ is the cumulative percentage of income, then equality of income distribution is given by the line $y = x$ in Fig.

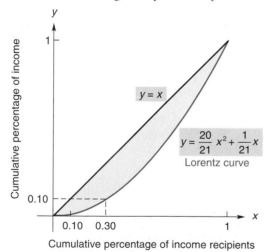

**FIGURE 16.45** Diagram for Problem 35.

16.45, where $x$ and $y$ are expressed as decimals. For example, 10% of the people receive 10% of total income, 20% of the people receive 20% of the income, and so on. Suppose the actual distribution is given by the Lorentz curve defined by

$$y = \tfrac{20}{21}x^2 + \tfrac{1}{21}x.$$

Note, for example, that 30% of the people receive only 10% of total income. The degree of deviation from equality is measured by the *coefficient of inequality*[13] for a Lorentz curve. This coefficient is defined to be the area between the curve and the diagonal, divided by the area under the diagonal:

$$\frac{\text{area between curve and diagonal}}{\text{area under diagonal}}.$$

For example, when all incomes are equal, the coefficient of inequality is zero. Find the coefficient of inequality for the Lorentz curve just defined.

36. **Lorentz curve** Find the coefficient of inequality as in Problem 35 for the Lorentz curve defined by $y = \tfrac{11}{12}x^2 + \tfrac{1}{12}x$.

37. Find the area of the region bounded by the graphs of the equations $y^2 = 4x$ and $y = mx$, where $m$ is a positive constant.

38. **a.** Find the area of the region bounded by the graphs of $y = x^2 - 1$ and $y = 2x + 2$.

 **b.** What percentage of the area in part (a) lies above the *x-axis*?

39. The region bounded by the curve $y = x^2$ and the line $y = 4$ is divided into two parts of equal area by the line $y = k$, where $k$ is a constant. Find the value of $k$.

*In each of Problems 40–44, estimate the area of the region bounded by the graphs of the given equations. Round your answer to two decimal places.*

40. $y = 9x - 15 - x^2$, $y = \dfrac{10}{x}$.

41. $y = \sqrt{25 - x^2}$, $y = 7 - 2x - x^4$.

42. $y = x^3 - 8x + 1$, $y = x^2 - 5$.

43. $y = x^5 - 3x^3 + 2x$, $y = 3x^2 - 4$.

44. $y = x^4 - 3x^3 - 15x^2 + 19x + 30$, $y = x^3 + x^2 - 20x$.

[[F 13]][13]G. Stigler, *The Theory of Price*, 3rd ed. (New York: The Macmillan Company, 1966), pp. 293–94.

**To develop the economic concepts of consumers' surplus and producers' surplus, which are represented by areas.**

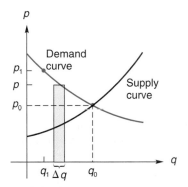

**FIGURE 16.46** Supply and demand curves.

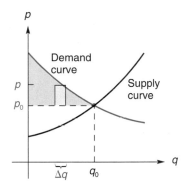

**FIGURE 16.47** Benefit to consumers for $\Delta q$ units.

# 16.10 CONSUMERS' AND PRODUCERS' SURPLUS

Determining the area of a region has applications in economics. Figure 16.46 shows a supply curve for a product. The curve indicates the price $p$ per unit at which the manufacturer will sell (or supply) $q$ units. The diagram also shows a demand curve for the product. This curve indicates the price $p$ per unit at which consumers will purchase (or demand) $q$ units. The point $(q_0, p_0)$ where the two curves intersect is called the *point of equilibrium*. Here $p_0$ is the price per unit at which consumers will purchase the same quantity $q_0$ of a product that producers wish to sell at that price. In short, $p_0$ is the price at which stability in the producer-consumer relationship occurs.

Let us assume that the market is at equilibrium and the price per unit of the product is $p_0$. According to the demand curve, there are consumers who would be willing to pay *more* than $p_0$. For example, at the price per unit of $p_1$, consumers would buy $q_1$ units. These consumers are benefiting from the lower equilibrium price $p_0$.

The vertical strip in Fig. 16.46 has area $p \, \Delta q$. This expression can also be thought of as the total amount of money that consumers would spend by buying $\Delta q$ units of the product if the price per unit were $p$. Since the price is actually $p_0$, these consumers spend only $p_0 \, \Delta q$ for the $\Delta q$ units and thus benefit by the amount $p \, \Delta q - p_0 \, \Delta q$. This expression can be written $(p - p_0) \, \Delta q$, which is the area of a rectangle of width $\Delta q$ and height $p - p_0$. (See Fig. 16.47.) Summing the areas of all such rectangles from $q = 0$ to $q = q_0$ by definite integration, we have

$$\int_0^{q_0} (p - p_0) \, dq.$$

This integral, under certain conditions, represents the total gain to consumers who who are willing to pay more than the equilibrium price. This total gain is called **consumers' surplus,** abbreviated CS. If the demand function is given by $p = f(q)$, then

$$\mathrm{CS} = \int_0^{q_0} [f(q) - p_0] \, dq.$$

Geometrically (see Fig. 16.48), consumers' surplus is represented by the area between the line $p = p_0$ and the demand curve $p = f(q)$ from $q = 0$ to $q = q_0$.

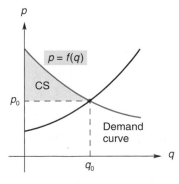

**FIGURE 16.48** Consumers' surplus.

Some of the producers also benefit from the equilibrium price, since they are willing to supply the product at prices *less* than $p_0$. Under certain

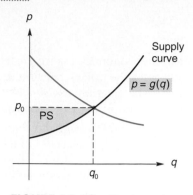

**FIGURE 16.49** Producers' surplus.

conditions, the total gain to the producers is represented geometrically in Fig. 16.49 by the area between the line $p = p_0$ and the supply curve $p = g(q)$ from $q = 0$ to $q = q_0$. This gain, called **producers' surplus** and abbreviated PS, is given by

$$\text{PS} = \int_0^{q_0} [p_0 - g(q)]\, dq.$$

**EXAMPLE 1   Finding Consumers' Surplus and Producers' Surplus**

*The demand function for a product is*

$$p = f(q) = 100 - 0.05q,$$

*where p is the price per unit (in dollars) for q units. The supply function is*

$$p = g(q) = 10 + 0.1q.$$

*Determine consumers' surplus and producers' surplus under market equilibrium.*

**Solution:** First we must find the equilibrium point $(p_0, q_0)$ by solving the system formed by the functions $p = 100 - 0.05q$ and $p = 10 + 0.1q$. We thus equate the two expressions for $p$ and solve:

$$10 + 0.1q = 100 - 0.05q,$$

$$0.15q = 90,$$

$$q = 600.$$

When $q = 600$, then $p = 10 + 0.1(600) = 70$. Hence, $q_0 = 600$ and $p_0 = 70$. Consumers' surplus is

$$\text{CS} = \int_0^{q_0} [f(q) - p_0]\, dq = \int_0^{600} (100 - 0.05q - 70)\, dq$$

$$= \left( 30q - 0.05 \frac{q^2}{2} \right) \Bigg|_0^{600} = 9000.$$

Producers' surplus is

$$\text{PS} = \int_0^{q_0} [p_0 - g(q)]\, dq = \int_0^{600} [70 - (10 + 0.1q)]\, dq$$

$$= \left( 60q - 0.1 \frac{q^2}{2} \right) \Bigg|_0^{600} = 18{,}000.$$

Therefore, consumers' surplus is \$9000 and producers' surplus is \$18,000. ∎

**EXAMPLE 2   Using Horizontal Strips to Find Consumers' Surplus and Producers' Surplus**

*The demand equation for a product is*

$$q = f(p) = \frac{90}{p} - 2$$

*and the supply equation is $q = g(p) = p - 1$. Determine consumers' surplus and producers' surplus when market equilibrium has been established.*

*Solution:* Determining the equilibrium point, we have

$$p - 1 = \frac{90}{p} - 2,$$

$$p^2 + p - 90 = 0,$$

$$(p + 10)(p - 9) = 0.$$

Thus, $p_0 = 9$, so $q_0 = 9 - 1 = 8$. (See Fig. 16.50.) Note that the demand equation expresses $q$ as a function of $p$. Since consumers' surplus can be

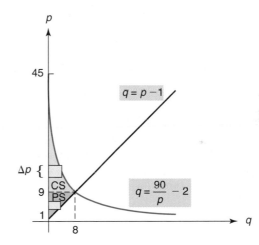

**FIGURE 16.50**   Diagram for Example 2.

considered an area, this area can be determined by means of horizontal strips of width $\Delta p$ and length $q = f(p)$. The areas of these strips are summed from $p = 9$ to $p = 45$ by integrating with respect to $p$:

$$CS - \int_9^{45} \left( \frac{90}{p} - 2 \right) dp = (90 \ln |p| - 2p) \Big|_9^{45}$$

$$= 90 \ln 5 - 72 \approx 72.85.$$

Using horizontal strips for producer's surplus, we have

$$PS = \int_1^9 (p - 1) \, dp = \frac{(p - 1)^2}{2} \Big|_1^9 = 32.$$

■

## ■ Exercise 16.10

*In Problems 1–6, the first equation is a demand equation and the second is a supply equation of a product. In each case, determine consumers' surplus and producers' surplus under market equilibrium.*

**1.** $p = 20 - 0.8q$,
   $p = 4 + 1.2q$.

**2.** $p = 900 - q^2$,
   $p = 100 + q^2$.

**3.** $p = \dfrac{50}{q + 5}$,

   $p = \dfrac{q}{10} + 4.5$.

**4.** $p = 400 - q^2$,
   $p = 20q + 100$.

**5.** $q = 100(10 - p)$,
   $q = 80(p - 1)$.

**6.** $q = \sqrt{100 - p}$,
   $q = \dfrac{p}{2} - 10$.

**7.** The demand equation for a product is

$$q = 10\sqrt{100 - p}.$$

Calculate consumers' surplus under market equilibrium which occurs at a price of $84.

**8.** The demand equation for a product is

$$q = 400 - p^2,$$

and the supply equation is

$$p = \frac{q}{60} + 5.$$

Find producers' surplus and consumers' surplus under market equilibrium.

**9.** The demand equation for a product is $p = 2^{11-q}$, and the supply equation is $p = 2^{q+1}$, where $p$ is the price per unit (in hundreds of dollars) when $q$ units are demanded or supplied. Determine, to the nearest thousand dollars, consumers' surplus under market equilibrium.

**10.** The demand equation for a product is

$$(p + 20)(q + 10) = 800,$$

and the supply equation is

$$q - 2p + 30 = 0.$$

**a.** Verify, by substitution, that market equilibrium occurs when $p = 20$ and $q = 10$.

**b.** Determine consumers' surplus under market equilibrium.

**11.** The demand equation for a product is

$$p = 60 - \frac{50q}{\sqrt{q^2 + 3600}},$$

and the supply equation is

$$p = 10 \ln(q + 20) - 26.$$

Determine consumers' surplus and producers' surplus under market equilibrium. Round your answers to the nearest integer.

## 16.11 REVIEW

### IMPORTANT TERMS AND SYMBOLS

**Section 16.1**  antiderivative   indefinite integral   $\int f(x)\,dx$   integral sign   integrand
variable of integration   constant of integration

**Section 16.2**  initial condition

**Section 16.3**  power rule for integration

**Section 16.5**  $\Sigma$   index of summation   limits of summation

**Section 16.6**  definite integral   $\int_a^b f(x)\,dx$   lower limit of integration   upper limit of integration

**Section 16.7**  Fundamental Theorem of Integral Calculus   $F(x)\big|_a^b$

**Section 16.8**  vertical element of area (vertical strip)

**Section 16.9**  horizontal element of area (horizontal strip)

**Section 16.10**  consumers' surplus   producers' surplus

### SUMMARY

An antiderivative of a function $f$ is a function $F$ such that $F'(x) = f(x)$. Any two antiderivatives of $f$ differ at most by a constant. The most general antiderivative of $f$ is called the indefinite integral of $f$ and is denoted $\int f(x)\,dx$. Thus,

$$\int f(x)\,dx = F(x) + C,$$

where $C$ is called the constant of integration.
Some basic integration formulas are as follows:

$$\int k \, dx = kx + C, \quad k \text{ a constant},$$

$$\int x^n \, dx = \frac{x^{n+1}}{n+1} + C, \quad n \neq -1,$$

$$\int e^x \, dx = e^x + C,$$

$$\int kf(x) \, dx = k \int f(x) \, dx, \quad k \text{ a constant},$$

and $\int [f(x) \pm g(x)] \, dx = \int f(x) \, dx \pm \int g(x) \, dx.$

Another formula is the power rule for integration:

$$\int u^n \, du = \frac{u^{n+1}}{n+1} + C, \quad \text{if } n \neq -1.$$

Here $u$ represents a differentiable function of $x$, and $du$ is its differential. In applying the power rule to a given integral, it is important that the integral be written in a form that precisely matches the power rule. Other integration formulas are

$$\int e^u \, du = e^u + C$$

and $\int \frac{1}{u} \, du = \ln |u| + C, \quad u \neq 0.$

If the rate of change of a function $f$ is known—that is, if $f'$ is known—then $f$ is an antiderivative of $f'$. In addition, if we know that $f$ satisfies an initial condition, then we can find the particular antiderivative. For example, if a marginal-cost function $dc/dq$ is given to us, then by integration, we can find the most general form of $c$. That form involves a constant of integration. However, if we are also given fixed costs (that is, costs involved when $q = 0$), then we can determine the value of the constant of integration and thus find the particular cost function $c$. Similarly, if we are given a marginal-revenue function $dr/dq$, then by integration and by using the fact that $r = 0$ when $q = 0$, we can determine the particular revenue function $r$. Once $r$ is known, the corresponding demand equation can be found by using the equation $p = r/q$.

Sigma notation is convenient for representing sums. This notation is especially useful in determining areas. To find the area of the region bounded by $y = f(x)$ [where $f(x) \geq 0$ and $f$ is continuous] and the $x$-axis from $x = a$ to $x = b$, we divide the interval $[a, b]$ into $n$ subintervals of equal length $\Delta x$. If $x_i$ is the right-hand endpoint of an arbitrary subinterval, then the product $f(x_i)\Delta x$ is the area of a rectangle.

Denoting the sum of all such areas of rectangles for the $n$ subintervals by $S_n$, we define the limit of $S_n$ as $n \to \infty$ as the area of the entire region:

$$\lim_{n \to \infty} S_n = \lim_{n \to \infty} \sum_{i=1}^{n} f(x_i) \, \Delta x = \text{area}.$$

If the restriction that $f(x) \geq 0$ is omitted, this limit is defined as the definite integral of $f$ over $[a, b]$:

$$\lim_{n \to \infty} \sum_{i=1}^{n} f(x_i) \, \Delta x = \int_a^b f(x) \, dx.$$

Instead of evaluating definite integrals by using limits, we may employ the Fundamental Theorem of Integral Calculus. Mathematically,

$$\int_a^b f(x) \, dx = F(x) \Big|_a^b = F(b) - F(a)$$

where $F$ is any antiderivative of $f$.

Some properties of the definite integral are

$$\int_a^b kf(x) \, dx = k \int_a^b f(x) \, dx, \quad k \text{ a constant},$$

$$\int_a^b [f(x) \pm g(x)] \, dx = \int_a^b f(x) \, dx \pm \int_a^b g(x) \, dx,$$

and

$$\int_a^c f(x) \, dx = \int_a^b f(x) \, dx + \int_b^c f(x) \, dx.$$

If the rate of change of a function $f$ is known, then a change in function values of $f$ can easily be found by the formula

$$\int_a^b f'(x) \, dx = f(b) - f(a).$$

If $f(x) \geq 0$ and continuous on $[a, b]$, then the definite integral may be used to find the area of the region bounded by $y = f(x)$ and the $x$-axis from $x = a$ to $x = b$. The definite integral may also be used to find areas of more complicated regions. In these situations, an element of area should be drawn in the region. This will allow you to set up the proper definite integral. In some situations vertical elements should be considered, whereas in others horizontal elements are more advantageous.

One application of finding areas involves consumers' surplus and producers' surplus. Suppose the market for a product is at equilibrium and $(q_0, p_0)$ is the equilibrium point (the point of intersection of the supply and demand curves for the product). Then consumers' surplus, CS, corresponds to the area from $q = 0$ to $q = q_0$, bounded above by the demand curve and below by the line $p = p_0$. Thus,

$$\text{CS} = \int_0^{q_0} [f(q) - p_0]\, dq,$$

where $f$ is the demand function. Producers' surplus, PS, corresponds to the area from $q = 0$ to $q = q_0$, bounded above by the line $p = p_0$ and below by the supply curve. Therefore,

$$\text{PS} = \int_0^{q_0} [p_0 - g(q)]\, dq,$$

where $g$ is the supply function.

## REVIEW PROBLEMS

*In Problems 1–40, determine the integrals.*

**1.** $\int (x^3 + 2x - 7)\, dx.$

**2.** $\int dx.$

**3.** $\int_0^9 (\sqrt{x} + x)\, dx.$

**4.** $\int \dfrac{2}{5 - 3x}\, dx.$

**5.** $\int \dfrac{2}{(x + 5)^3}\, dx.$

**6.** $\int_4^{12} (y - 8)^{501}\, dy.$

**7.** $\int \dfrac{6x^2 - 12}{x^3 - 6x + 1}\, dx.$

**8.** $\int_0^2 xe^{4 - x^2}\, dx.$

**9.** $\int_0^1 \sqrt[3]{3t + 8}\, dt.$

**10.** $\int \dfrac{5 - 3x}{2}\, dx.$

**11.** $\int y(y + 1)^2\, dy.$

**12.** $\int_0^1 10^{-8}\, dx.$

**13.** $\int \dfrac{\sqrt[4]{z} - \sqrt[3]{z}}{\sqrt{z}}\, dz.$

**14.** $\int \dfrac{(0.5x - 0.1)^4}{0.4}\, dx.$

**15.** $\int_1^2 \dfrac{t^2}{2 + t^3}\, dt.$

**16.** $\int \dfrac{4x^2 - x}{x}\, dx.$

**17.** $\int x^2\sqrt{3x^3 + 2}\, dx.$

**18.** $\int (2x^3 + x)(x^4 + x^2)^{3/4}\, dx.$

**19.** $\int (e^{2y} - e^{-2y})\, dy.$

**20.** $\int \dfrac{8x}{3\sqrt[3]{7 - 2x^2}}\, dx.$

**21.** $\int \left( \dfrac{1}{x} + \dfrac{2}{x^2} \right) dx.$

**22.** $\int_0^1 \dfrac{e^{2x}}{1 + e^{2x}}\, dx.$

**23.** $\int_{-2}^1 (y^4 - y + 1)\, dy.$

**24.** $\int_7^{70} dx.$

**25.** $\int_{\sqrt{3}}^2 7x\sqrt{4 - x^2}\, dx.$

**26.** $\int_0^1 (2x + 1)(x^2 + x)^4\, dx.$

**27.** $\int_0^1 \left[ 2x - \dfrac{1}{(x + 1)^{2/3}} \right] dx.$

**28.** $\int_2^8 (\sqrt{2x} - x + 4)\, dx.$

**29.** $\int \dfrac{\sqrt{t} - 3}{t^2}\, dt.$

**30.** $\int \dfrac{z^2}{z - 1}\, dz.$

**31.** $\int_{-1}^0 \dfrac{x^2 + 4x - 1}{x + 2}\, dx.$

**32.** $\int \dfrac{(x^2 + 4)^2}{x^2}\, dx.$

**33.** $\int \sqrt{x}\sqrt{x^{3/2} + 1}\, dx.$

**34.** $\int \dfrac{e^{\sqrt{3x}}}{\sqrt{2x}}\, dx.$

**35.** $\int_1^e \dfrac{e^{\ln x}}{x^2}\, dx.$

**36.** $\int \dfrac{6x^2 + 4}{e^{x^3 + 2x}}\, dx.$

**37.** $\int \dfrac{(1 + e^{3x})^2}{e^{-3x}}\, dx.$

**38.** $\int \dfrac{1}{e^{3x}(6 + e^{-3x})^2}\, dx.$

**39.** $\int \sqrt{10^{3x}}\, dx.$

**40.** $\int \dfrac{3x^3 + 3x^2 + 11x + 1}{x^2 + x + 3}\, dx.$

*In Problems 41 and 42, find y, subject to the given condition.*

**41.** $y' = e^{2x} + 3, \quad y(0) = -\frac{1}{2}.$

**42.** $y' = \dfrac{x + 3}{x}, \quad y(1) = 5.$

*In Problems 43–50, determine the area of the region bounded by the given curve, the x-axis, and the given lines.*

**43.** $y = x^2 - 1, \quad x = 2 \quad (y \geq 0).$

**44.** $y = 4e^x, \quad x = 0, \quad x = 3.$

**45.** $y = \sqrt{x + 4}, \quad x = 0.$

**46.** $y = x^2 - x - 2, \quad x = -2, \quad x = 2.$

**47.** $y = 5x - x^2.$

**48.** $y = \sqrt[4]{x}, \quad x = 1, \quad x = 16.$

**49.** $y = \dfrac{1}{x} + 3$, $x = 1$, $x = 3$.

**50.** $y = x^3 - 1$, $x = -1$.

*In Problems 51–58, find the area of the region bounded by the given curves.*

**51.** $y^2 = 4x$   $x = 0$,   $y = 2$.

**52.** $y = 2x^2$,   $x = 0$,   $y = 2$     $(x \geq 0)$.

**53.** $y = x^2 + 4x - 5$,   $y = 0$.

**54.** $y = 2x^2$,   $y - x^2 + 9$.

**55.** $y = x^2 - 2x$,   $y = 12 - x^2$.

**56.** $y = \sqrt{x}$,   $x = 0$,   $y = 3$.

**57.** $y = \ln x$,   $x = 0$,   $y = 0$,   $y = 1$.

**58.** $y = 1 - x$,   $y = x - 2$,   $y = 0$,   $y = 1$.

**59. Marginal Revenue**   If marginal revenue is given by

$$\frac{dr}{dq} = 100 - \frac{3}{2}\sqrt{2q},$$

determine the corresponding demand equation.

**60. Marginal Cost**   If marginal cost is given by

$$\frac{dc}{dq} = q^2 + 7q + 6,$$

and fixed costs are 2500, determine the total cost of producing six units. Assume that costs are in dollars.

**61. Marginal Revenue**   A manufacturer's marginal-revenue function is

$$\frac{dr}{dq} = 275 - q - 0.3q^2.$$

If $r$ is in dollars, find the increase in the manufacturer's total revenue if production is increased from 10 to 20 units.

**62. Marginal Cost**   A manufacturer's marginal-cost function is

$$\frac{dc}{dq} = \frac{500}{\sqrt{2q + 25}}.$$

If $c$ is in dollars, determine the cost involved to increase production from 100 to 300 units.

**63. Hospital Discharges**   For a group of hospitalized individuals, suppose the discharge rate is given by

$$f(t) = 0.008e^{-0.008t},$$

where $f(t)$ is the proportion discharged per day at the end of $t$ days of hospitalization. What proportion of the group is discharged at the end of 100 days?

**64. Business Expenses**   The total expenditures (in dollars) of a business over the next five years is given by

$$\int_0^5 4000e^{0.05t} \, dt.$$

Evaluate the expenditures.

**65.** Find the area of the region between the curves $y = 9 - 2x$ and $y = x$ from $x = 0$ to $x = 4$.

**66.** Find the area of the region between the curves $y = x^2$ and $y = 4 - 3x$ from $x = -1$ to $x = 2$.

**67. Consumers' and Producers' Surplus**   The demand equation for a product is

$$p = 0.01q^2 - 1.1q + 30,$$

and the supply equation is

$$p = 0.01q^2 + 8.$$

Determine consumers' surplus and producers' surplus when market equilibrium has been established.

**68. Consumers' Surplus**   The demand equation for a product is

$$p = (q - 5)^2,$$

and the supply equation is

$$p = q^2 + q + 3,$$

where $p$ (in thousands of dollars) is the price per 100 units when $q$ hundred units are demanded or supplied. Determine consumers' surplus under market equilibrium.

**69. Biology**   In a discussion of gene mutation,[14] the equation

$$\int_{q_0}^{q_n} \frac{dq}{q - \hat{q}} = -(u + v) \int_0^n dt$$

occurs, where $u$ and $v$ are gene mutation rates, the $q$'s are gene frequencies, and $n$ is the number of generations. Assume that all letters represent constants, except $q$ and $t$. Integrate both sides and then use your result to show that

$$n = \frac{1}{u + v} \ln \left| \frac{q_0 - \hat{q}}{q_n - \hat{q}} \right|.$$

**70. Fluid Flow**   In studying the flow of a fluid in a tube of constant radius $R$, such as blood flow in portions of the body, one can think of the tube as consisting of concentric tubes of radius $r$, where $0 \leq r \leq R$. The velocity $v$ of the fluid is a function of $r$ and is given by[15]

$$v = \frac{(P_1 - P_2)(R^2 - r^2)}{4\eta l},$$

where $P_1$ and $P_2$ are pressures at the ends of the tube, $\eta$ (a Greek letter read "eta") is the fluid viscosity, and $l$ is

[14]W. B. Mather, *Principles of Quantitative Genetics* (Minneapolis: Burgess Publishing Company, 1964).

[15]R. W. Stacy et al. *Essentials of Biological and Medical Physics* (New York: McGraw-Hill Book Company, 1955).

the length of the tube. The volume rate of flow (through the tube is given by

$$Q = \int_0^R 2\pi r v \, dr.$$

Show that $Q = \dfrac{\pi R^4 (P_1 - P_2)}{8\eta l}$. Note that $R$ occurs as a factor to the fourth power. Thus, doubling the radius of the tube has the effect of increasing the flow by a factor of 16. The formula that you derived for the volume rate of flow is called *Poiseuille's law*, after the French physiologist Jean Poiseuille.

**71. Inventory** In a discussion of inventory, Barbosa and Friedman[16] refer to the function

$$g(x) = \frac{1}{k} \int_1^{1/x} k u^r \, du,$$

where $k$ and $r$ are constants, $k > 0$ and $r > -2$, and $x > 0$. Verify the claim that

$$g'(x) = -\frac{1}{x^{r+2}}.$$

[*Hint:* Consider two cases: when $r \neq -1$ and when $r = -1$.]

 *In each of Problems 72–74, estimate the area of the region bounded by the given curves. Round your answer to two decimal places.*

**72.** $y = 3x^3 + 6x^2 - 5x - 4, \quad y = 0.$

**73.** $y = x^3 - 2x - 3, \quad y = 4 + 2x - 3x^2.$

**74.** $y = x^3 - 2x - 3, \quad y = 2x^2 - x^3 - 4.$

 **75.** The demand equation for a product is

$$p = \frac{200}{\sqrt{q + 20}},$$

and the supply equation is

$$p = 2\ln(q + 10) + 5.$$

Determine consumers' surplus and producers' surplus under market equilibrium. Round your answers to the nearest integer.

[16]L. C. Barbosa and M. Friedman, "Deterministic Inventory Lot Size Models—a General Root Law," *Management Science*, 24, no. 8 (1978), 819–26.

# MATHEMATICAL *SNAPSHOT*

## DELIVERED PRICE

Suppose that you are a manufacturer of a product whose sales occur within $R$ miles of your mill. Assume that you charge customers for shipping at the rate $s$, in dollars per mile, for each unit of product sold. If $m$ is the unit price (in dollars) at the mill, then the delivered unit price $p$ to a customer $x$ miles from the mill is the mill price plus the shipping charge $sx$:

$$p = m + sx, \quad 0 \le x \le R. \quad (1)$$

The problem is to determine the average delivered price of the units sold.

Suppose that there is a function $f$ such that $f(t) \ge 0$ on the interval $[0, R]$ and such that the area under the graph of $f$ and above the $t$-axis from $t = 0$ to $t = x$ represents the total number of units $Q$ sold to customers within $x$ miles of the mill. [See Fig. 16.51(a).] You can refer to $f$ as the distribution of demand. Because $Q$ is a function of $x$ and is represented by area,

$$Q(x) = \int_0^x f(t) \, dt.$$

In particular, the total number of units sold within the market area is

$$Q(R) = \int_0^R f(t) \, dt$$

[see Fig. 16.51(b)]. For example, if $f(t) = 10$ and $R = 100$, then the total number of units sold within the market area is

$$Q(100) = \int_0^{100} 10 \, dt = 10t \Big|_0^{100} = 1000 - 0 = 1000.$$

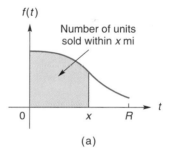

$f(t)$

Number of units sold within $x$ mi

(a)

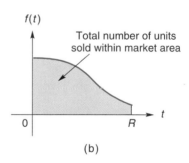

$f(t)$

Total number of units sold within market area

(b)

**FIGURE 16.51** Number of units sold as an area.

The average delivered price $A$ is given by

$$A = \frac{\text{total revenue}}{\text{total number of units sold}}.$$

Because the denominator is $Q(R)$, $A$ can be determined once the total revenue is found.

To find the total revenue, first consider the number of units sold over an interval. If $t_1 < t_2$ [see Fig. 16.52(a)], then the area under the graph of $f$ and above the $t$-axis from $t = 0$ to $t = t_1$ represents the number

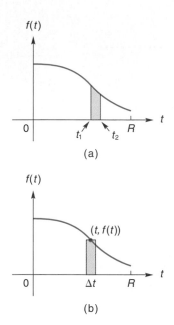

$f(t)$

$0$     $t_1$   $t_2$   $R$   $t$

(a)

$f(t)$

$(t, f(t))$

$0$    $\Delta t$   $R$   $t$

(b)

**FIGURE 16.52**   Number of units sold over an interval.

of units sold within $t_1$ miles of the mill. Similarly, the area under the graph of $f$ and above the $t$-axis from $t = 0$ to $t = t_2$ represents the number of units sold within $t_2$ miles of the mill. Thus the difference in these areas is geometrically the area of the shaded region in Fig. 16.52(a) and represents the number of units sold between $t_1$ and $t_2$ miles of the mill, which is $Q(t_2) - Q(t_1)$. Thus

$$Q(t_2) - Q(t_1) = \int_{t_1}^{t_2} f(t)\, dt.$$

For example, if $f(t) = 10$, then the number of units sold to customers located between 4 and 6 miles of the mill is

$$Q(6) - Q(4) = \int_{4}^{6} 10\, dt = 10t\, \Big|_{4}^{6} = 60 - 40 = 20.$$

The area of the shaded region in Fig. 16.52(a) can be approximated by the area of a rectangle [see Fig. 16.52(b)] whose height is $f(t)$ and whose width is $\Delta t$, where $\Delta t = t_2 - t_1$. Thus the number of units sold over the interval of length $\Delta t$ is approximately $f(t)\,\Delta t$. Because the price of each of these units is [from Eq. (1)] approximately $m + st$, the revenue received is approximately

$$(m + st)f(t)\,\Delta t.$$

The sum of all such products from $t = 0$ to $t = R$ approximates the total revenue. Definite integration gives

$$\sum (m + st)\, f(t)\, \Delta t \rightarrow \int_{0}^{R} (m + st)\, f(t)\, dt.$$

Thus

$$\text{total revenue} = \int_{0}^{R} (m + st)\, f(t)\, dt.$$

Consequently, the average delivered price $A$ is given by

$$A = \frac{\displaystyle\int_{0}^{R} (m + st)\, f(t)\, dt}{Q(R)}$$

or, equivalently,

$$A = \frac{\displaystyle\int_{0}^{R} (m + st)\, f(t)\, dt}{\displaystyle\int_{0}^{R} f(t)\, dt}.$$

For example, if $f(t) = 10$, $m = 200$, $s = 0.25$, and $R = 100$, then

$$\int_{0}^{R} (m + st)\, f(t)\, dt = \int_{0}^{100} (200 + 0.25t) \cdot 10\, dt$$

$$= 10 \int_{0}^{100} (200 + 0.25t)\, dt$$

$$= 10 \left( 200t + \frac{t^2}{8} \right)\Bigg|_{0}^{100}$$

$$= 10 \left[ \left( 20{,}000 + \frac{10{,}000}{8} \right) - 0 \right]$$

$$= 212{,}500.$$

From above,

$$\int_{0}^{R} f(t)\, dt = \int_{0}^{100} 10\, dt = 1000.$$

Thus, the average delivered price is $212{,}500/1000 = \$212.50$.

### ■ Exercises

**1.** If $f(t) = 50 - 2t$, determine the number of units sold to customers located within 5 miles of the mill. Between 10 and 15 miles.

**2.** If $f(t) = 20$, $m = 50$, $s = 0.20$, and $R = 80$, determine (a) the total revenue, (b) the total number of units sold, and (c) the average delivered price.

**3.** If $f(t) = 60 - 2t$, $m = 100$, $s = 1$, and $R = 30$, determine (a) the total revenue, (b) the total number of units sold, and (c) the average delivered price.

# Methods and Applications of Integration

**To develop and apply the formula for integration by parts.**

## 17.1 INTEGRATION BY PARTS[1]

Many integrals cannot be found by our previous methods. However, there are ways of changing certain integrals to forms that are easier to integrate. Of these methods, we shall discuss two: *integration by parts* and (in Sec. 17.2) *integration using partial fractions.*

If $u$ and $v$ are differentiable functions of $x$, we have, by the product rule,

$$(uv)' = uv' + vu'.$$

Rearranging gives

$$uv' = (uv)' - vu'.$$

Integrating both sides with respect to $x$, we get

$$\int uv' dx = \int (uv)' dx - \int vu' dx. \tag{1}$$

For $\int (uv)' \, dx$, we must find a function whose derivative with respect to $x$ is $(uv)'$. Clearly, $uv$ is such a function. Hence $\int (uv)' \, dx = uv + C_1$, and Eq. (1) becomes

$$\int uv' \, dx = uv + C_1 - \int vu' \, dx.$$

Absorbing $C_1$ into the constant of integration for $\int vu' \, dx$ and replacing $v' \, dx$ by $dv$ and $u' \, dx$ by $du$, we have the *formula for integration by parts:*

**Formula for Integration by Parts**

$$\int u \, dv = uv - \int v \, du. \tag{2}$$

This formula expresses an integral, $\int u \, dv$, in terms of another integral, $\int v \, du$, that may be easier to find.

--------

[1] May be omitted without loss of continuity.

To apply the formula to a given integral $\int f(x)\, dx$, we must write $f(x)\, dx$ as the product of two factors (or *parts*) by choosing a function $u$ and a differential $dv$ such that $f(x)\, dx = u\, dv$. However, for the formula to be useful, we must be able to integrate the part chosen for $dv$. To illustrate, consider

$$\int xe^x\, dx.$$

This integral cannot be determined by previous integration formulas. One way to write $xe^x\, dx$ in the form $u\, dv$ is by letting

$$u = x \qquad \text{and} \qquad dv = e^x\, dx.$$

To apply the formula for integration by parts, we must find $du$ and $v$:

$$du = dx \qquad \text{and} \qquad v = \int e^x\, dx = e^x + C_1.$$

Thus,

$$\int \underbrace{x}_{u}\underbrace{e^x\, dx}_{dv} = uv - \int v\, du$$

$$= x(e^x + C_1) - \int (e^x + C_1)\, dx$$

$$= xe^x + C_1 x - e^x - C_1 x + C$$

$$= xe^x - e^x + C = e^x(x - 1) + C.$$

The first constant, $C_1$, does not appear in the final answer. This is a characteristic of integration by parts, and from now on this constant will not be written when finding $v$ from $dv$.

When you are using the formula for integration by parts, sometimes the "best choice" for $u$ and $dv$ may not be obvious. In some cases, one choice may be as good as another; in other cases, only one choice may be suitable. Insight into making a good choice (if any exists) will come only with practice and, of course, trial and error.

**Principles in Practice 1**
Integration by Parts

The monthly sales of a computer keyboard are estimated to decline at the rate of $S'(t) = -4te^{0.1t}$ keyboards per month, where $t$ is time in months and $S(t)$ is the number of keyboards sold each month. If 5000 keyboards are sold now $[S(0) = 5000]$, find $S(t)$.

**EXAMPLE 1  Integration by Parts**

*Find* $\int \dfrac{\ln x}{\sqrt{x}}\, dx$ *by integration by parts.*

*Solution:* We try

$$u = \ln x \qquad \text{and} \qquad dv = \frac{1}{\sqrt{x}}\, dx.$$

Then,

$$du = \frac{1}{x}\, dx \qquad \text{and} \qquad v = \int x^{-1/2}\, dx = 2x^{1/2}.$$

Thus,

$$\int \underbrace{\ln x}_{u} \underbrace{\left(\frac{1}{\sqrt{x}} \, dx\right)}_{dv} = uv - \int v \, du$$

$$= (\ln x)(2\sqrt{x}) - \int (2x^{1/2})\left(\frac{1}{x} \, dx\right)$$

$$= 2\sqrt{x} \ln x - 2 \int x^{-1/2} \, dx$$

$$= 2\sqrt{x} \ln x - 2(2\sqrt{x}) + C$$

$$= 2\sqrt{x} \left[\ln(x) - 2\right] + C. \qquad \blacksquare$$

Example 2 shows how a poor choice for $u$ and $dv$ can be made. If you make a choice that does not work, do not become frustrated. Rather, make other choices until one that works is found (if such a choice exists).

### EXAMPLE 2   Integration by Parts

*Evaluate* $\int_{1}^{2} x \ln x \, dx$.

*Solution:* Since the integral does not fit a familiar form, we shall try integration by parts. Let $u = x$ and $dv = \ln x \, dx$. Then $du = dx$, but $v = \int \ln x \, dx$ is not apparent by inspection. So we shall make a different choice for $u$ and $dv$. Let

$$u = \ln x \qquad \text{and} \qquad dv = x \, dx.$$

Then,

$$du = \frac{1}{x} \, dx \qquad \text{and} \qquad v = \int x \, dx = \frac{x^2}{2}.$$

Therefore,

$$\int_{1}^{2} x \ln x \, dx = (\ln x)\left(\frac{x^2}{2}\right)\Big|_{1}^{2} - \int_{1}^{2} \left(\frac{x^2}{2}\right)\frac{1}{x} \, dx$$

$$= (\ln x)\left(\frac{x^2}{2}\right)\Big|_{1}^{2} - \frac{1}{2}\int_{1}^{2} x \, dx$$

$$= \frac{x^2 \ln x}{2}\Big|_{1}^{2} - \frac{1}{2}\left(\frac{x^2}{2}\right)\Big|_{1}^{2}$$

$$= (2 \ln 2 - 0) - (1 - \tfrac{1}{4}) = 2 \ln 2 - \tfrac{3}{4}. \qquad \blacksquare$$

### EXAMPLE 3   Integration by Parts where $u$ Is the Entire Integrand

*Determine* $\int \ln y \, dy$.

*Solution:* We cannot integrate $\ln y$ by previous methods, so we shall try integration by parts. Let $u = \ln y$ and $dv = dy$. Then $du = (1/y) \, dy$ and $v = y$. So we have:

$$\int \ln y \, dy = (\ln y)(y) - \int y\left(\frac{1}{y} \, dy\right)$$

$$= y \ln y - \int dy = y \ln y - y + C$$

$$= y\left[\ln(y) - 1\right] + C.\qquad \blacksquare$$

Before trying integration by parts, see whether the technique is really needed. Sometimes the integral can be handled by a basic technique, as Example 4 shows.

**EXAMPLE 4   Basic Integration Form**

*Determine* $\displaystyle\int xe^{x^2}\,dx.$

*Solution:* This integral can be fit to the form $\displaystyle\int e^u\,du.$

*Pitfall* ▼ Do not forget about basic integration forms. Integration by parts is not needed here!

$$\int xe^{x^2}\,dx = \frac{1}{2}\int e^{x^2}(2x\,dx)$$

$$= \frac{1}{2}\int e^u\,du \quad (\text{where } u = x^2)$$

$$= \frac{1}{2}e^u + C = \frac{1}{2}e^{x^2} + C.\qquad \blacksquare$$

Sometimes integration by parts must be used more than once, as shown in the following example.

**Principles in Practice 2**

**Applying Integration by Parts Twice**

Suppose a population of bacteria grows at a rate of
$$P'(t) = 0.1t(\ln t)^2.$$
Find the general form of $P(t)$.

**EXAMPLE 5   Applying Integration by Parts Twice**

*Determine* $\displaystyle\int x^2 e^{2x+1}\,dx.$

*Solution:* Let $u = x^2$ and $dv = e^{2x+1}\,dx$. Then $du = 2x\,dx$ and $v = e^{2x+1}/2$.

$$\int x^2 e^{2x+1}\,dx = \frac{x^2 e^{2x+1}}{2} - \int \frac{e^{2x+1}}{2}(2x\,dx)$$

$$= \frac{x^2 e^{2x+1}}{2} - \int xe^{2x+1}\,dx.$$

To find $\displaystyle\int xe^{2x+1}\,dx$, we shall again use integration by parts. Here, let $u = x$ and $dv = e^{2x+1}\,dx$. Then $du = dx$ and $v = e^{2x+1}/2$, and we have

$$\int xe^{2x+1}\,dx = \frac{xe^{2x+1}}{2} - \int \frac{e^{2x+1}}{2}\,dx$$

$$= \frac{xe^{2x+1}}{2} - \frac{e^{2x+1}}{4} + C_1.$$

Thus,

$$\int x^2 e^{2x+1}\,dx = \frac{x^2 e^{2x+1}}{2} - \frac{xe^{2x+1}}{2} + \frac{e^{2x+1}}{4} + C \qquad (\text{where } C = -C_1)$$

$$= \frac{e^{2x+1}}{2}\left(x^2 - x + \frac{1}{2}\right) + C.\qquad \blacksquare$$

## ▪ Exercise 17.1

**1.** In applying integration by parts to

$$\int f(x)\,dx,$$

a student found that $u = x$, $du = dx$, $dv = (x + 5)^{1/2}$, and $v = \frac{2}{3}(x + 5)^{3/2}$. Use this information to find $\int f(x)\,dx$.

**2.** Use integration by parts to find

$$\int xe^{5x+2}\,dx$$

by choosing $u = x$ and $dv = e^{5x+2}\,dx$.

*In Problems 3–29, find the integrals.*

**3.** $\int xe^{-x}\,dx.$

**4.** $\int xe^{2x}\,dx.$

**5.** $\int y^3 \ln y\,dy.$

**6.** $\int x^2 \ln x\,dx.$

**7.** $\int \ln(4x)\,dx.$

**8.** $\int \dfrac{t}{e^t}\,dt.$

**9.** $\int x\sqrt{x+1}\,dx$

**10.** $\int \dfrac{x}{\sqrt{1+4x}}\,dx.$

**11.** $\int \dfrac{x}{(2x+1)^2}\,dx.$

**12.** $\int \dfrac{\ln(x+1)}{2(x+1)}\,dx.$

**13.** $\int \dfrac{\ln x}{x^2}\,dx.$

**14.** $\int \dfrac{x+1}{e^x}\,dx.$

**15.** $\int_1^2 xe^{2x}\,dx.$

**16.** $\int_0^1 xe^{-x}\,dx.$

**17.** $\int_0^1 xe^{-x^2}\,dx.$

**18.** $\int \dfrac{x^3}{\sqrt{4-x^2}}\,dx.$

**19.** $\int_1^2 \dfrac{x}{\sqrt{4-x}}\,dx.$

**20.** $\int (\ln x)^2\,dx.$

**21.** $\int (2x-1)\ln(x-1)\,dx.$

**22.** $\int \dfrac{xe^x}{(x+1)^2}\,dx.$

**23.** $\int x^2 e^x\,dx.$

**24.** $\int_1^e \sqrt{x}\,\ln(x^2)\,dx.$

**25.** $\int (x - e^{-x})^2\,dx.$

**26.** $\int x^2 e^{-2x}\,dx.$

**27.** $\int x^3 e^{x^2}\,dx.$

**28.** $\int x^5 e^{x^3}\,dx.$

**29.** $\int (2^x + x)^2\,dx.$

.............................................................................................

**30.** Find $\int \ln(x + \sqrt{x^2 + 1})\,dx$. *Hint:* Show that

$$\frac{d}{dx}[\ln(x + \sqrt{x^2 + 1})] = \frac{1}{\sqrt{x^2 + 1}}.$$

**31.** Find the area of the region bounded by the $x$-axis, the curve $y = \ln x$, and the line $x = e^3$.

**32.** Find the area of the region bounded by the $x$-axis and the curve $y = xe^{-x}$ between $x = 0$ and $x = 4$.

**33.** Find the area of the region bounded by the $x$-axis and the curve $y = x\sqrt{2x + 1}$ between $x = 0$ and $x = 4$.

**34. Consumers' Surplus** Suppose the demand equation for a manufacturer's product is given by

$$p = 10(q + 10)e^{-(0.1q+1)},$$

where $p$ is the price per unit (in dollars) when $q$ units are demanded. Assume that market equilibrium occurs when $q = 20$. Determine consumers' surplus at market equilibrium.

**35. Revenue** Suppose total revenue $r$ and price per unit $p$ are differentiable functions of output $q$.

**a.** Use integration by parts to show that

$$\int p\,dq = r - \int q\,\frac{dp}{dq}\,dq.$$

**b.** Using part (a), show that

$$r = \int \left(p + q\,\frac{dp}{dq}\right)dq.$$

**c.** Using part (b), prove that

$$r(q_0) = \int_0^{q_0}\left(p + q\,\frac{dp}{dq}\right)dq.$$

[*Hint:* Refer to Sec. 16.7.]

**36.** Suppose $f$ is a differentiable function. Apply integration by parts to $\int f(x)\,e^x\,dx$ to prove that

$$\int f(x)e^x\,dx + \int f'(x)e^x\,dx = f(x)e^x + C.$$

$$\left(\text{Hence, } \int [f(x) + f'(x)]e^x\,dx = f(x)e^x + C.\right)$$

**To show how to integrate a proper rational function by first expressing it as a sum its partial fractions**

## 17.2 Integration by Partial Fractions[2]

### Distinct Linear Factors

We now consider the integral of a rational function (a quotient of two polynomials). Without loss of generality, we may assume that the numerator $N(x)$ and denominator $D(x)$ have no common polynomial factor and that the degree of $N(x)$ is less than the degree of $D(x)$. [That is, $N(x)/D(x)$ defines a *proper rational function*.] For if the numerator were not of lower degree, we could use long division to divide $N(x)$ by $D(x)$:

$$D(x) \overline{\smash{)}N(x)}^{\textstyle P(x)}; \qquad \text{thus,} \qquad \frac{N(x)}{D(x)} = P(x) + \frac{R(x)}{D(x)}.$$

$$\overline{R(x)}$$

Here $P(x)$ would be a polynomial (easily integrable) and $R(x)$ would be a polynomial of lower degree than $D(x)$. Hence, $R(x)/D(x)$ would define a proper rational function. For example,

$$\int \frac{2x^4 - 3x^3 - 4x^2 - 17x - 6}{x^3 - 2x^2 - 3x} \, dx = \int \left( 2x + 1 + \frac{4x^2 - 14x - 6}{x^3 - 2x^2 - 3x} \right) dx$$

$$= x^2 + x + \int \frac{4x^2 - 14x - 6}{x^3 - 2x^2 - 3x} \, dx.$$

Therefore, we shall consider

$$\int \frac{4x^2 - 14x - 6}{x^3 - 2x^2 - 3x} \, dx.$$

It is essential that the denominator be expressed in factored form:

$$\int \frac{4x^2 - 14x - 6}{x(x + 1)(x - 3)} \, dx.$$

Observe that the denominator consists only of **distinct linear factors** and that each factor occurs exactly once. It can be shown that, to each such factor $x - a$, there corresponds a *partial fraction* of the form

$$\frac{A}{x - a} \qquad (A \text{ a constant})$$

such that the integrand is the sum of the partial fractions. If there are $n$ such *distinct* linear factors, there will be $n$ such partial fractions, each of which is easily integrated. Applying these facts, we can write

$$\frac{4x^2 - 14x - 6}{x(x + 1)(x - 3)} = \frac{A}{x} + \frac{B}{x + 1} + \frac{C}{x - 3}. \qquad (1)$$

To determine the constants $A$, $B$, and $C$, we first combine the terms on the right side:

$$\frac{4x^2 - 14x - 6}{x(x + 1)(x - 3)} = \frac{A(x + 1)(x - 3) + Bx(x - 3) + Cx(x + 1)}{x(x + 1)(x - 3)}.$$

---

[2] May be omitted without loss of continuity.

Since the denominators of both sides are equal, we may equate their numerators:

$$4x^2 - 14x - 6 = A(x + 1)(x - 3) + Bx(x - 3) + Cx(x + 1). \quad (2)$$

Although Eq. (1) is not defined for $x = 0$, $x = -1$, and $x = 3$, we want to find values for $A$, $B$, and $C$ that will make Eq. (2) true for all values of $x$. That is, it will be an identity. By successively setting $x$ in Eq. (2) equal to any three different numbers, we can obtain a system of equations that can be solved for $A$, $B$, and $C$. In particular, the work can be simplified by letting $x$ be the roots of $D(x) = 0$; in our case, $x = 0$, $x = -1$, and $x = 3$. Using Eq. (2), we have, for $x = 0$,

$$-6 = A(1)(-3) + B(0) + C(0) = -3A, \quad \text{so } A = 2.$$

If $x = -1$,

$$12 = A(0) + B(-1)(-4) + C(0) = 4B, \quad \text{so } B = 3.$$

If $x = 3$,

$$-12 = A(0) + B(0) + C(3)(4) = 12C, \quad \text{so } C = -1.$$

Thus Eq. (1) becomes

$$\frac{4x^2 - 14x - 6}{x(x + 1)(x - 3)} = \frac{2}{x} + \frac{3}{x + 1} - \frac{1}{x - 3}.$$

Hence,

$$\int \frac{4x^2 - 14x - 6}{x(x + 1)(x - 3)} \, dx$$

$$= \int \left( \frac{2}{x} + \frac{3}{x + 1} - \frac{1}{x - 3} \right) dx$$

$$= 2 \int \frac{dx}{x} + 3 \int \frac{dx}{x + 1} - \int \frac{dx}{x - 3}$$

$$= 2 \ln |x| + 3 \ln |x + 1| - \ln |x - 3| + C$$

$$= \ln \left| \frac{x^2(x + 1)^3}{x - 3} \right| + C \quad \text{(using properties of logarithms)}.$$

For the *original* integral, we can now state that

$$\int \frac{2x^4 - 3x^3 - 4x^2 - 17x - 6}{x^3 - 2x^2 - 3x} \, dx = x^2 + x + \ln \left| \frac{x^2(x + 1)^3}{x - 3} \right| + C.$$

An alternative method of determining $A$, $B$, and $C$ involves expanding the right side of Eq. (2) and combining similar terms:

$$4x^2 - 14x - 6 = A(x^2 - 2x - 3) + B(x^2 - 3x) + C(x^2 + x)$$

$$= Ax^2 - 2Ax - 3A + Bx^2 - 3Bx + Cx^2 + Cx,$$

$$4x^2 - 14x - 6 = (A + B + C)x^2 + (-2A - 3B + C)x + (-3A).$$

For this identity, coefficients of corresponding powers of $x$ on the left and right sides of the equation must be equal:

$$\begin{cases} 4 = A + B + C, \\ -14 = -2A - 3B + C, \\ -6 = -3A. \end{cases}$$

......................

Solving gives $A = 2$, $B = 3$, and $C = -1$ as before.

### **Principles in Practice 1**

**Distinct Linear Factors**

The marginal revenue for a company manufacturing $q$ radios per week is given by $r'(q) = \dfrac{5(q + 4)}{q^2 + 4q + 3}$, where $r(q)$ is the revenue in thousands of dollars. Find the equation for $r(q)$.

### **EXAMPLE 1   Distinct Linear Factors**

*Determine* $\displaystyle\int \frac{2x + 1}{3x^2 - 27}\, dx$ *by using partial fractions.*

**Solution:** Since the degree of the numerator is less than the degree of the denominator, no long division is necessary. The integral can be written as

$$\frac{1}{3}\int \frac{2x + 1}{x^2 - 9}\, dx.$$

Expressing $(2x + 1)/(x^2 - 9)$ as a sum of partial fractions, we have

$$\frac{2x + 1}{x^2 - 9} = \frac{2x + 1}{(x + 3)(x - 3)} = \frac{A}{x + 3} + \frac{B}{x - 3}.$$

Combining terms and equating numerators gives

$$2x + 1 = A(x - 3) + B(x + 3).$$

If $x = 3$, then

$$7 = 6B, \qquad \text{so} \qquad B = \frac{7}{6};$$

If $x = -3$, then

$$-5 = -6A, \qquad \text{so} \qquad A = \frac{5}{6}.$$

Thus,

$$\int \frac{2x + 1}{3x^2 - 27}\, dx = \frac{1}{3}\left[\int \frac{\frac{5}{6}\, dx}{x + 3} + \int \frac{\frac{7}{6}\, dx}{x - 3}\right]$$

$$= \frac{1}{3}\left[\frac{5}{6}\ln|x + 3| + \frac{7}{6}\ln|x - 3|\right] + C$$

$$= \frac{1}{18}\ln\left|(x + 3)^5(x - 3)^7\right| + C. \qquad \blacksquare$$

### **Repeated Linear Factors**

If the denominator of $N(x)/D(x)$ contains only linear factors, some of which are repeated, then, for each factor $(x - a)^k$, where $k$ is the maximum number of times $x - a$ occurs as a factor, there will correspond the sum of $k$ partial fractions:

$$\frac{A}{x - a} + \frac{B}{(x - a)^2} + \cdots + \frac{K}{(x - a)^k}.$$

### **EXAMPLE 2   Repeated Linear Factors**

*Determine* $\displaystyle\int \frac{6x^2 + 13x + 6}{(x + 2)(x + 1)^2}\, dx$ *by using partial fractions.*

**Solution:** Since the degree of the numerator, namely, 2, is less than that of the denominator, namely, 3, no long division is necessary. In the denominator, the linear factor $x + 2$ occurs once and the linear factor $x + 1$ occurs twice. There will thus be three partial fractions and three constants to determine, and we have

$$\frac{6x^2 + 13x + 6}{(x + 2)(x + 1)^2} = \frac{A}{x + 2} + \frac{B}{x + 1} + \frac{C}{(x + 1)^2},$$

$$6x^2 + 13x + 6 = A(x + 1)^2 + B(x + 2)(x + 1) + C(x + 2).$$

Let us choose $x = -2$, $x = -1$, and, for convenience, $x = 0$. For $x = -2$, we have

$$4 = A.$$

If $x = -1$, then

$$-1 = C.$$

If $x = 0$, then

$$6 = A + 2B + 2C = 4 + 2B - 2 = 2 + 2B$$

$$4 = 2B,$$

$$2 = B.$$

Therefore,

$$\int \frac{6x^2 + 13x + 6}{(x + 2)(x + 1)^2} \, dx = 4 \int \frac{dx}{x + 2} + 2 \int \frac{dx}{x + 1} - \int \frac{dx}{(x + 1)^2}$$

$$= 4 \ln |x + 2| + 2 \ln |x + 1| + \frac{1}{x + 1} + C$$

$$= \ln[(x + 2)^4(x + 1)^2] + \frac{1}{x + 1} + C. \quad \blacksquare$$

### Distinct Irreducible Quadratic Factors

Suppose a quadratic factor $x^2 + bx + c$ occurs in $D(x)$ and it cannot be expressed as a product of two linear factors with real coefficients. Such a factor is said to be an *irreducible quadratic factor over the real numbers*. To each distinct irreducible quadratic factor that occurs exactly once in $D(x)$, there will correspond a partial fraction of the form

$$\frac{Ax + B}{x^2 + bx + c}.$$

Note that even after you have expressed a rational function in terms of partial fractions, you may still find it impossible to integrate using only the calculus you have been taught so far.

### EXAMPLE 3    An Integral with a Distinct Irreducible Quadratic Factor

*Determine* $\displaystyle\int \frac{-2x - 4}{x^3 + x^2 + x} \, dx$ *by using partial fractions.*

**Solution:** Since $x^3 + x^2 + x = x(x^2 + x + 1)$, we have the linear factor $x$ and the quadratic factor $x^2 + x + 1$, which does not seem factorable on inspection. If it were factorable into $(x - r_1)(x - r_2)$, with $r_1$ and $r_2$ real, then $r_1$ and $r_2$ would be roots of the equation $x^2 + x + 1 = 0$. By the quadratic formula, the roots are

$$x = \frac{-1 \pm \sqrt{1 - 4}}{2}.$$

Since there are no real roots, we conclude that $x^2 + x + 1$ is irreducible. Thus there will be two partial fractions and *three* constants to determine. We have

$$\frac{-2x - 4}{x(x^2 + x + 1)} = \frac{A}{x} + \frac{Bx + C}{x^2 + x + 1},$$

$$-2x - 4 = A(x^2 + x + 1) + (Bx + C)x$$

$$= Ax^2 + Ax + A + Bx^2 + Cx.$$

$$0x^2 - 2x - 4 = (A + B)x^2 + (A + C)x + A.$$

Equating coefficients of like powers of $x$, we obtain

$$\begin{cases} 0 = A + B, \\ -2 = A + C, \\ -4 = A. \end{cases}$$

Solving gives $A = -4$, $B = 4$, and $C = 2$. Hence,

$$\int \frac{-2x - 4}{x(x^2 + x + 1)} \, dx = \int \left( \frac{-4}{x} + \frac{4x + 2}{x^2 + x + 1} \right) dx$$

$$= -4 \int \frac{dx}{x} + 2 \int \frac{2x + 1}{x^2 + x + 1} \, dx.$$

Both integrals have the form $\int \dfrac{du}{u}$, so

$$\int \frac{-2x - 4}{x(x^2 + x + 1)} \, dx = -4 \ln |x| + 2 \ln |x^2 + x + 1| + C$$

$$= \ln \left[ \frac{(x^2 + x + 1)^2}{x^4} \right] + C. \qquad ■$$

### Repeated Irreducible Quadratic Factors

Suppose $D(x)$ contains factors of the form $(x^2 + bx + c)^k$, where $k$ is the maximum number of times the irreducible factor $x^2 + bx + c$ occurs. Then, to each such factor, there will correspond a sum of $k$ partial fractions of the form

$$\frac{A + Bx}{x^2 + bx + c} + \frac{C + Dx}{(x^2 + bx + c)^2} + \cdots + \frac{M + Nx}{(x^2 + bx + c)^k}.$$

**EXAMPLE 4   Repeated Irreducible Quadratic Factors**

*Determine* $\displaystyle \int \frac{x^5}{(x^2 + 4)^2} \, dx$ *by using partial fractions.*

*Solution:* Since the numerator has degree 5 and the denominator has degree 4, we first use long division, which gives

$$\frac{x^5}{x^4 + 8x^2 + 16} = x - \frac{8x^3 + 16x}{(x^2 + 4)^2}.$$

The quadratic factor $x^2 + 4$ in the denominator of $(8x^3 + 16x)/(x^2 + 4)^2$ is irreducible and occurs as a factor twice. Thus, to $(x^2 + 4)^2$ there correspond two partial fractions and *four* coefficients to be determined. Accordingly we set

$$\frac{8x^3 + 16x}{(x^2 + 4)^2} = \frac{Ax + B}{x^2 + 4} + \frac{Cx + D}{(x^2 + 4)^2}$$

and obtain

$$8x^3 + 16x = (Ax + B)(x^2 + 4) + Cx + D,$$
$$8x^3 + 0x^2 + 16x + 0 = Ax^3 + Bx^2 + (4A + C)x + 4B + D.$$

Equating like powers of $x$ yields

$$\begin{cases} 8 = A, \\ 0 = B, \\ 16 = 4A + C, \\ 0 = 4B + D. \end{cases}$$

Solving gives $A = 8$, $B = 0$, $C = -16$, and $D = 0$. Therefore,

$$\int \frac{x^5}{(x^2 + 4)^2}\, dx = \int \left( x - \left[ \frac{8x}{x^2 + 4} - \frac{16x}{(x^2 + 4)^2} \right] \right) dx$$

$$= \int x\, dx - 4\int \frac{2x}{x^2 + 4}\, dx + 8\int \frac{2x}{(x^2 + 4)^2}\, dx.$$

The second integral on the preceding line has the form $\int \dfrac{du}{u}$, and the third integral has the form $\int \dfrac{du}{u^2}$. So

$$\int \frac{x^5}{(x^2 + 4)^2} = \frac{x^2}{2} - 4\ln(x^2 + 4) - \frac{8}{x^2 + 4} + C. \qquad \blacksquare$$

From our examples, you may have deduced that the number of constants needed to express $N(x)/D(x)$ by partial fractions is equal to the degree of $D(x)$, if it is assumed that $N(x)/D(x)$ defines a proper rational function. This is indeed the case. Note also that the representation of a proper rational function by partial fractions is unique; that is, there is only one choice of constants that can be made. Furthermore, regardless of the complexity of the polynomial $D(x)$, it can always (theoretically) be expressed as a product of linear and irreducible quadratic factors with real coefficients.

**Principles in Practice 2**

**An Integral Not Requiring Partial Fractions**

The rate of change of the voting population of a city with respect to time $t$ (in years) is estimated to be $V'(t) = \dfrac{300t^3}{t^2 + 6}$. Find the general form of $V(t)$.

**EXAMPLE 5   An Integral Not Requiring Partial Fractions**

*Find* $\displaystyle\int \frac{2x + 3}{x^2 + 3x + 1}\, dx.$

***Solution:*** This integral has the form $\displaystyle\int \frac{1}{u}\, du$. Thus,

$$\int \frac{2x + 3}{x^2 + 3x + 1}\, dx = \ln |x^2 + 3x + 1| + C. \qquad \blacksquare$$

*Pitfall* ▼ Do not forget about basic integration forms.

■ **Exercise 17.2**

*In Problems 1–8, express the given rational function in terms of partial fractions. Watch out for any preliminary divisions.*

**1.** $f(x) = \dfrac{10x}{x^2 + 7x + 6}.$

**2.** $f(x) = \dfrac{x + 5}{x^2 - 1}.$

**3.** $f(x) = \dfrac{x^2}{x^2 + 6x + 8}.$

**4.** $f(x) = \dfrac{2x^2 - 15}{x^2 + 5x}.$

**5.** $f(x) = \dfrac{4x - 5}{x^2 + 2x + 1}.$

**6.** $f(x) = \dfrac{2x + 3}{x^2(x - 1)}.$

**7.** $f(x) = \dfrac{x^2 + 3}{x^3 + x}.$

**8.** $f(x) = \dfrac{3x^2 + 5}{(x^2 + 4)^2}.$

*In Problems 9–30, determine the integrals.*

**9.** $\displaystyle\int \frac{5x-2}{x^2-x}\,dx.$

**10.** $\displaystyle\int \frac{3x+8}{x^2+2x}\,dx.$

**11.** $\displaystyle\int \frac{x+10}{x^2-x-2}\,dx.$

**12.** $\displaystyle\int \frac{dx}{x^2-5x+6}.$

**13.** $\displaystyle\int \frac{3x^3-3x+4}{4x^2-4}\,dx.$

**14.** $\displaystyle\int \frac{4-x^2}{(x-4)(x-2)(x+3)}\,dx.$

**15.** $\displaystyle\int \frac{17x-12}{x^3-x^2-12x}\,dx.$

**16.** $\displaystyle\int \frac{4-x}{x^4-x^2}\,dx.$

**17.** $\displaystyle\int \frac{3x^5+4x^3-x}{x^6+2x^4-x^2-2}\,dx.$

**18.** $\displaystyle\int \frac{x^4-3x^3-5x^2+8x-1}{x^3-2x^2-8x}\,dx.$

**19.** $\displaystyle\int \frac{2x^2-5x-2}{(x-2)^2(x-1)}\,dx.$

**20.** $\displaystyle\int \frac{-3x^3+2x-3}{x^2(x^2-1)}\,dx.$

**21.** $\displaystyle\int \frac{x^2+8}{x^3+4x}\,dx.$

**22.** $\displaystyle\int \frac{2x^3-6x^2-10x-6}{x^4-1}\,dx.$

**23.** $\displaystyle\int \frac{-x^3+8x^2-9x+2}{(x^2+1)(x-3)^2}\,dx.$

**24.** $\displaystyle\int \frac{2x^4+9x^2+8}{x(x^2+2)^2}\,dx.$

**25.** $\displaystyle\int \frac{14x^3+24x}{(x^2+1)(x^2+2)}\,dx.$

**26.** $\displaystyle\int \frac{12x^3+20x^2+28x+4}{(x^2+2x+3)(x^2+1)}\,dx.$

**27.** $\displaystyle\int \frac{3x^3+x}{(x^2+1)^2}\,dx.$

**28.** $\displaystyle\int \frac{3x^2-8x+4}{x^3-4x^2+4x-6}\,dx.$

**29.** $\displaystyle\int_0^1 \frac{2-2x}{x^2+7x+12}\,dx.$

**30.** $\displaystyle\int_1^2 \frac{2x^2+1}{(x+3)(x+2)}\,dx.$

**31.** Find the area of the region bounded by the graph of
$$y = \frac{x^2+1}{(x+2)^2}$$
and the x-axis from $x=0$ to $x=1$.

**32. Consumers' Surplus** Suppose the demand equation for a manufacturer's product is given by

$$p = \frac{200(q+3)}{q^2+7q+6},$$

where $p$ is the price per unit (in dollars) when $q$ units are demanded. Assume that market equilibrium occurs at the point $(q, p) = (10, 325/22)$. Determine consumers' surplus at market equilibrium.

---

**OBJECTIVE**

To illustrate the use of the table of integrals in Appendix C.

## 17.3 Integration by Tables

Certain forms of integrals that occur frequently may be found in standard tables of integration formulas.[3] A short table appears in Appendix C, and its use will be illustrated in this section.

A given integral may have to be replaced by an equivalent form before it will fit a formula in the table. The equivalent form must match the formula *exactly*. Consequently, the steps that you perform should *not* be done mentally. *Write them down!* Failure to do this can easily lead to incorrect results. Before proceeding with the exercises, be sure you understand the illustrative examples *thoroughly*.

In the following examples, the formula numbers refer to the Table of Selected Integrals given in Appendix C.

**EXAMPLE 1  Integration by Tables**

*Find* $\displaystyle\int \frac{x\,dx}{(2+3x)^2}.$

*Solution:* Scanning the table, we identify the integrand with Formula 7:

$$\int \frac{u\,du}{(a+bu)^2} = \frac{1}{b^2}\left( \ln|a+bu| + \frac{a}{a+bu} \right) + C.$$

---

[3] See, for example, W. H. Beyer (ed.). *CRC Standard Mathematical Tables and Formulae*, 30th ed. (Boca Raton, FL: CRC Press, 1996.

Now we see if we can exactly match the given integrand with that in the formula. If we replace $x$ by $u$, 2 by $a$, and 3 by $b$, then $du = dx$, and by substitution, we have

$$\int \frac{x\,dx}{(2 + 3x)^2} = \int \frac{u\,du}{(a + bu)^2} = \frac{1}{b^2}\left(\ln|a + bu| + \frac{a}{a + bu}\right) + C.$$

Returning to the variable $x$ and replacing $a$ by 2 and $b$ by 3, we obtain

$$\int \frac{x\,dx}{(2 + 3x)^2} = \frac{1}{9}\left(\ln|2 + 3x| + \frac{2}{2 + 3x}\right) + C.$$

Note that the answer must be given in terms of $x$, the *original* variable of integration. ∎

### EXAMPLE 2　Integration by Tables

*Find* $\int x^2\sqrt{x^2 - 1}\,dx$.

*Solution:* This integral is identified with Formula 24:

$$\int u^2\sqrt{u^2 \pm a^2}\,du = \frac{u}{8}(2u^2 \pm a^2)\sqrt{u^2 \pm a^2} - \frac{a^4}{8}\ln\left|u + \sqrt{u^2 \pm a^2}\right| + C.$$

In the preceding formula, if the bottommost sign in the dual symbol " $\pm$ " on the left side is used, then the bottommost sign in the dual symbols on the right side must also be used. In the original integral, we let $u = x$ and $a = 1$. Then $du = dx$, and by substitution, the integral becomes

$$\int x^2\sqrt{x^2 - 1}\,dx = \int u^2\sqrt{u^2 - a^2}\,du$$

$$= \frac{u}{8}(2u^2 - a^2)\sqrt{u^2 - a^2} - \frac{a^4}{8}\ln\left|u + \sqrt{u^2 - a^2}\right| + C.$$

Since $u = x$ and $a = 1$,

$$\int x^2\sqrt{x^2 - 1}\,dx = \frac{x}{8}(2x^2 - 1)\sqrt{x^2 - 1} - \frac{1}{8}\ln\left|x + \sqrt{x^2 - 1}\right| + C. \quad ∎$$

This example, as well as Examples 4, 5, and 7, shows how to adjust an integral so that it conforms to one in the table.

### EXAMPLE 3　Integration by Tables

*Find* $\int \dfrac{dx}{x\sqrt{16x^2 + 3}}$.

*Solution:* The integrand can be identified with Formula 28:

$$\int \frac{du}{u\sqrt{u^2 + a^2}} = \frac{1}{a}\ln\left|\frac{\sqrt{u^2 + a^2} - a}{u}\right| + C.$$

If we let $u = 4x$ and $a = \sqrt{3}$, then $du = 4\,dx$. Watch closely how, by inserting 4's in the numerator and denominator, we transform the given integral into an equivalent form that matches Formula 28:

$$\int \frac{dx}{x\sqrt{16x^2 + 3}} = \int \frac{(4\,dx)}{(4x)\sqrt{(4x)^2 + (\sqrt{3})^2}} = \int \frac{du}{u\sqrt{u^2 + a^2}}$$

$$= \frac{1}{a}\ln\left|\frac{\sqrt{u^2 + a^2} - a}{u}\right| + C$$

$$= \frac{1}{\sqrt{3}}\ln\left|\frac{\sqrt{16x^2 + 3} - \sqrt{3}}{4x}\right| + C. \quad ∎$$

**EXAMPLE 4 Integration by Tables**

Find $\displaystyle\int \frac{dx}{x^2(2 - 3x^2)^{1/2}}$.

**Solution:** The integrand is identified with Formula 21:

$$\int \frac{du}{u^2\sqrt{a^2 - u^2}} = -\frac{\sqrt{a^2 - u^2}}{a^2 u} + C.$$

Letting $u = \sqrt{3}x$ and $a^2 = 2$, we have $du = \sqrt{3}\,dx$. Hence, by inserting two factors of $\sqrt{3}$ in both the numerator and denominator of the original integral, we have

$$\int \frac{dx}{x^2(2 - 3x^2)^{1/2}} = \sqrt{3}\int \frac{(\sqrt{3}\,dx)}{(\sqrt{3}x)^2[2 - (\sqrt{3}x)^2]^{1/2}} = \sqrt{3}\int \frac{du}{u^2(a^2 - u^2)^{1/2}}$$

$$= \sqrt{3}\left[-\frac{\sqrt{a^2 - u^2}}{a^2 u}\right] + C = \sqrt{3}\left[-\frac{\sqrt{2 - 3x^2}}{2(\sqrt{3}x)}\right] + C$$

$$= -\frac{\sqrt{2 - 3x^2}}{2x} + C. \qquad\blacksquare$$

**EXAMPLE 5 Integration by Tables**

Find $\displaystyle\int 7x^2 \ln(4x)\,dx$.

**Solution:** This is similar to Formula 42 with $n = 2$:

$$\int u^n \ln u\,du = \frac{u^{n+1}\ln u}{n + 1} - \frac{u^{n+1}}{(n + 1)^2} + C.$$

If we let $u = 4x$, then $du = 4\,dx$. Hence,

$$\int 7x^2 \ln(4x)\,dx = \frac{7}{4^3}\int (4x)^2 \ln(4x)(4\,dx)$$

$$= \frac{7}{64}\int u^2 \ln u\,du = \frac{7}{64}\left(\frac{u^3 \ln u}{3} - \frac{u^3}{9}\right) + C$$

$$= \frac{7}{64}\left[\frac{(4x)^3 \ln(4x)}{3} - \frac{(4x)^3}{9}\right] + C$$

$$= 7x^3\left[\frac{\ln(4x)}{3} - \frac{1}{9}\right] + C$$

$$= \frac{7x^3}{9}[3\ln(4x) - 1] + C. \qquad\blacksquare$$

**EXAMPLE 6 Integral Table Not Needed**

Find $\displaystyle\int \frac{e^{2x}\,dx}{7 + e^{2x}}$.

**Solution:** At first glance, we do not identify the integrand with any form in the table. Perhaps rewriting the integral will help. Let $u = 7 + e^{2x}$ then $du = 2e^{2x}\,dx$. So

$$\int \frac{e^{2x}\,dx}{7 + e^{2x}} = \frac{1}{2}\int \frac{(2e^{2x}\,dx)}{7 + e^{2x}} = \frac{1}{2}\int \frac{du}{u} = \frac{1}{2}\ln|u| + C$$

$$= \frac{1}{2}\ln|7 + e^{2x}| + C = \frac{1}{2}\ln(7 + e^{2x}) + C.$$

Thus, we had only to use our knowledge of basic integration forms. Actually, this form appears as Formula 2 in the table. ∎

**EXAMPLE 7**  **Finding a Definite Integral by Using Tables**

*Evaluate* $\displaystyle\int_1^4 \frac{dx}{(4x^2 + 2)^{3/2}}$.

*Solution:* We shall use Formula 32 to get the indefinite integral first:

$$\int \frac{du}{(u^2 \pm a^2)^{3/2}} = \frac{\pm u}{a^2\sqrt{u^2 \pm a^2}} + C.$$

Letting $u = 2x$ and $a^2 = 2$, we have $du = 2\,dx$. Thus,

$$\int \frac{dx}{(4x^2 + 2)^{3/2}} = \frac{1}{2}\int \frac{(2\,dx)}{[(2x)^2 + 2]^{3/2}} = \frac{1}{2}\int \frac{du}{(u^2 + 2)^{3/2}}$$

$$= \frac{1}{2}\left[\frac{u}{2\sqrt{u^2 + 2}}\right] + C.$$

Here we determine the limits of integration with respect to *u*.

Instead of substituting back to $x$ and evaluating from $x = 1$ to $x = 4$, we can determine the corresponding limits of integration with respect to $u$ and then evaluate the last expression between those limits. Since $u = 2x$, when $x = 1$ we have $u = 2$; when $x = 4$ we have $u = 8$. Hence,

$$\int_1^4 \frac{dx}{(4x^2 + 2)^{3/2}} = \frac{1}{2}\int_2^8 \frac{du}{(u^2 + 2)^{3/2}}$$

$$= \frac{1}{2}\left(\frac{u}{2\sqrt{u^2 + 2}}\right)\Big|_2^8 = \frac{2}{\sqrt{66}} - \frac{1}{2\sqrt{6}}.$$  ∎

*Pitfall* ▼  When changing the variable of integration $x$ to the variable of integration $u$, be sure to change the limits of integration so that they agree with $u$. That is, in Example 7,

$$\int_1^4 \frac{dx}{(4x^2 + 2)^{3/2}} \neq \frac{1}{2}\int_1^4 \frac{du}{(u^2 + 2)^{3/2}}.$$

### Integration Applied to Annuities

Tables of integrals are useful when one deals with integrals associated with annuities. Suppose that you must pay out \$100 at the end of each year for the next two years. Recall from Chapter 8 that a series of payments over a period of time, such as this, is called an *annuity*. If you were to pay off the debt now instead, you would pay the present value of the \$100 that is due at the end of the first year, plus the present value of the \$100 that is due at the end of the second year. The sum of these present values is the present value of the annuity. (The present value of an annuity is discussed in Sec. 8.3.) We shall now consider the present value of payments made continuously over the time interval from $t = 0$ to $t = T$, with $t$ in years, when interest is compounded continuously at an annual rate of $r$.

Suppose a payment is made at time $t$ such that on an annual basis this payment is $f(t)$. If we divide the interval $[0, T]$ into subintervals $[t_{i-1}, t_i]$ of length $\Delta t$ (where $\Delta t$ is small), then the total amount of all payments over such a subinterval is approximately $f(t_i)\,\Delta t$. [For example, if $f(t) = 2000$ and $\Delta t$ were one day, the total amount of the payments would be $2000(\frac{1}{365})$.] The present value of these payments is approximately $e^{-rt_i}f(t_i)\,\Delta t$. (See Sec. 11.3.) Over the interval $[0, T]$, the total of all such present values is

$$\sum e^{-rt_i}f(t_i)\,\Delta t.$$

This sum approximates the present value $A$ of the annuity. The smaller $\Delta t$ is, the better is the approximation. That is, as $\Delta t \to 0$, the limit of the sum is the present value. However, this limit is also a definite integral. That is,

$$A = \int_0^T f(t)e^{-rt}\,dt, \tag{1}$$

where $A$ is the **present value of a continuous annuity** at an annual rate $r$ (compounded continuously) for $T$ years if a payment at time $t$ is at the rate of $f(t)$ per year.

We say that Eq. (1) gives the **present value of a continuous income stream.** Equation (1) may also be used to find the present value of future profits of a business. In this situation, $f(t)$ is the annual rate of profit at time $t$.

We can also consider the *future* value of an annuity rather than its present value. If a payment is made at time $t$, then it has a certain value at the *end* of the period of the annuity—that is, $T - t$ years later. This value is

$$\left(\begin{array}{c}\text{amount of}\\\text{payment}\end{array}\right) + \left(\begin{array}{c}\text{interest on this}\\\text{payment for } T - t \text{ years}\end{array}\right).$$

If $S$ is the total of such values for all payments, then $S$ is called the *accumulated amount of a continuous annuity* and is given by the formula

$$S = \int_0^T f(t)e^{r(T-t)}\,dt,$$

where $S$ is the **accumulated amount of a continuous annuity** at the end of $T$ years at an annual rate $r$ (compounded continuously) when a payment at time $t$ is at the rate of $f(t)$ per year.

**EXAMPLE 8** **Present Value of a Continuous Annuity**

*Find the present value (to the nearest dollar) of a continuous annuity at an annual rate of 8% for 10 years if the payment at time t is at the rate of $t^2$ dollars per year.*

*Solution:* The present value is given by

$$A = \int_0^T f(t)e^{-rt}\,dt = \int_0^{10} t^2 e^{-0.08t}\,dt.$$

We shall use Formula 39,

$$\int u^n e^{au}\,du = \frac{u^n e^{au}}{a} - \frac{n}{a}\int u^{n-1}e^{au}\,du.$$

This is called a *reduction formula,* since it reduces one integral to an expression that involves another integral that is easier to determine. If $u = t$, $n = 2$, and $a = -0.08$, then $du = dt$, and we have

$$A = \frac{t^2 e^{-0.08t}}{-0.08}\Big|_0^{10} - \frac{2}{-0.08}\int_0^{10} t e^{-0.08t}\,dt.$$

In the new integral, the exponent of $t$ has been reduced to 1. We can match this integral with Formula 38,

$$\int ue^{au}\, du = \frac{e^{au}}{a^2}(au - 1) + C,$$

by letting $u = t$ and $a = -0.08$. Then $du = dt$, and

$$A = \int_0^{10} t^2 e^{-0.08t}\, dt = \frac{t^2 e^{-0.08t}}{-0.08}\Big|_0^{10} - \frac{2}{-0.08}\left[\frac{e^{-0.08t}}{(-0.08)^2}(-0.08t - 1)\right]\Big|_0^{10}$$

$$= \frac{100e^{-0.8}}{-0.08} - \frac{2}{-0.08}\left[\frac{e^{-0.8}}{(-0.08)^2}(-0.8 - 1) - \frac{1}{(-0.08)^2}(-1)\right]$$

$$\approx 185.$$

The present value is \$185.

## ■ Exercise 17.3

*In Problems 1 and 2, use Formula 19 in Appendix C to determine the integrals.*

**1.** $\displaystyle\int \frac{dx}{(9 - x^2)^{3/2}}.$

**2.** $\displaystyle\int \frac{dx}{(25 - 4x^2)^{3/2}}.$

*In Problems 3 and 4, use Formula 30 in Appendix C to determine the integrals.*

**3.** $\displaystyle\int \frac{dx}{x^2\sqrt{16x^2 + 3}}.$

**4.** $\displaystyle\int \frac{dx}{x^3\sqrt{x^4 - 4}}.$

*In Problems 5–38, find the integrals by using the table in Appendix C.*

**5.** $\displaystyle\int \frac{dx}{x(6 + 7x)}.$

**6.** $\displaystyle\int \frac{x^2\, dx}{(1 + 2x)^2}.$

**7.** $\displaystyle\int \frac{dx}{x\sqrt{x^2 + 9}}.$

**8.** $\displaystyle\int \frac{dx}{(x^2 + 7)^{3/2}}.$

**9.** $\displaystyle\int \frac{x\, dx}{(2 + 3x)(4 + 5x)}.$

**10.** $\displaystyle\int 2^{5x}\, dx.$

**11.** $\displaystyle\int \frac{dx}{4 + 3e^{2x}}.$

**12.** $\displaystyle\int x^2\sqrt{1 + x}\, dx.$

**13.** $\displaystyle\int \frac{2\, dx}{x(1 + x)^2}.$

**14.** $\displaystyle\int \frac{dx}{x\sqrt{5 - 11x^2}}.$

**15.** $\displaystyle\int_0^1 \frac{x\, dx}{2 + x}.$

**16.** $\displaystyle\int \frac{x^2\, dx}{2 + 5x}.$

**17.** $\displaystyle\int \sqrt{x^2 - 3}\, dx.$

**18.** $\displaystyle\int \frac{dx}{(4 + 3x)(4x + 3)}.$

**19.** $\displaystyle\int_0^{1/12} xe^{12x}\, dx.$

**20.** $\displaystyle\int \sqrt{\frac{2 + 3x}{5 + 3x}}\, dx.$

**21.** $\displaystyle\int x^2 e^x\, dx.$

**22.** $\displaystyle\int_1^2 \frac{dx}{x^2(1 + x)}.$

**23.** $\displaystyle\int \frac{\sqrt{4x^2 + 1}}{x^2}\, dx.$

**24.** $\displaystyle\int \frac{dx}{x\sqrt{2 - x}}.$

**25.** $\displaystyle\int \frac{x\, dx}{(1 + 3x)^2}.$

**26.** $\displaystyle\int \frac{dx}{\sqrt{(1 + 2x)(3 + 2x)}}.$

**27.** $\displaystyle\int \frac{dx}{7 - 5x^2}.$

**28.** $\displaystyle\int x^2\sqrt{2x^2 - 9}\, dx.$

**29.** $\displaystyle\int x^5 \ln(3x)\, dx.$

**30.** $\displaystyle\int \frac{dx}{x^2(1 + x)^2}.$

**31.** $\displaystyle\int 2x\sqrt{1 + 3x}\, dx.$

**32.** $\displaystyle\int x^2 \ln x\, dx.$

**33.** $\displaystyle\int \frac{dx}{\sqrt{4x^2 - 13}}.$

**34.** $\displaystyle\int \frac{dx}{x \ln(2x)}.$

**35.** $\displaystyle\int \frac{dx}{x^2\sqrt{9 - 4x^2}}.$

**36.** $\displaystyle\int \frac{\sqrt{2 - 3x^2}}{x}\, dx.$

**37.** $\displaystyle\int \frac{dx}{\sqrt{x}(\pi + 7e^{4\sqrt{x}})}.$

**38.** $\displaystyle\int_0^1 \frac{x^3\, dx}{1 + x^4}.$

*In Problems 39–56, find the integrals by any method.*

**39.** $\displaystyle\int \frac{x\, dx}{x^2 + 1}.$

**40.** $\displaystyle\int \sqrt{x}\, e^{x^{3/2}}\, dx.$

**41.** $\displaystyle\int x\sqrt{2x^2 + 1}\, dx.$

**42.** $\displaystyle\int \frac{4x^2 - \sqrt{x}}{x}\, dx.$

**43.** $\displaystyle\int \frac{dx}{x^2 - 5x + 6}.$

**44.** $\displaystyle\int \frac{e^{2x}}{\sqrt{e^{2x} + 3}}\, dx.$

**45.** $\displaystyle\int x^3 \ln x\, dx.$

**46.** $\displaystyle\int_0^3 xe^{-x}\, dx.$

**47.** $\int xe^{2x}\,dx.$

**48.** $\int_1^2 x^2\sqrt{3+2x}\,dx.$

**49.** $\int \ln^2 x\,dx.$

**50.** $\int_1^e \ln x\,dx.$

**51.** $\int_1^2 \dfrac{x\,dx}{\sqrt{4-x}}.$

**52.** $\int_1^2 x\sqrt{1+2x}\,dx.$

**53.** $\int_0^1 \dfrac{2x\,dx}{\sqrt{8-x^2}}.$

**54.** $\int_0^{\ln 2} x^3 e^{2x}\,dx.$

**55.** $\int_1^2 x\ln(2x)\,dx.$

**56.** $\int_1^2 dx.$

**57. Biology** In a discussion about gene frequency,[4] the integral

$$\int_{q_0}^{q_n} \frac{dq}{q(1-q)},$$

occurs, where the $q$'s represent gene frequencies. Evaluate this integral.

**58. Biology** Under certain conditions, the number $n$ of generations required to change the frequency of a gene from 0.3 to 0.1 is given by[5]

$$n = -\frac{1}{0.4}\int_{0.3}^{0.1} \frac{dq}{q^2(1-q)}.$$

Find $n$ (to the nearest integer).

**59. Continuous Annuity** Find the present value, to the nearest dollar, of a continuous annuity at an annual rate

of $r$ for $T$ years if the payment at time $t$ is at the annual rate of $f(t)$ dollars, given that

**a.** $r = 0.06,\quad T = 10,\quad f(t) = 5000,$

**b.** $r = 0.05,\quad T = 8,\quad f(t) = 200t.$

**60.** If $f(t) = k$, where $k$ is a positive constant, show that the value of the integral in Eq. (1) of this section is

$$k\left(\frac{1-e^{-rT}}{r}\right).$$

**61. Continuous Annuity** Find the accumulated amount, to the nearest dollar, of a continuous annuity at an annual rate of $r$ for $T$ years if the payment at time $t$ is at an annual rate of $f(t)$ dollars, given that

**a.** $r = 0.06,\quad T = 10,\quad f(t) = 400,$

**b.** $r = 0.04,\quad T = 5,\quad f(t) = 40t.$

**62. Value of Business** Over the next five years, the profits of a business at time $t$ are estimated to be $20{,}000t$ dollars per year. The business is to be sold at a price equal to the present value of these future profits. To the nearest 10 dollars, at what price should the business be sold if interest is compounded continuously at the annual rate of 10%?

[4]W. B. Mather, *Principles of Quantitative Genetics* (Minneapolis: Burgess Publishing Company, 1964).

[5]E. O. Wilson and W. H. Bossert, *A Primer of Population Biology* (Stamford, CT: Sinauer Associates, Inc., 1971).

**OBJECTIVE**

To develop the concept of the average value of a function.

## 17.4 Average Value of a Function

If we are given the three numbers 1, 2, and 9, then their average value, or *mean*, is their sum divided by 3. Denoting this average by $\bar{y}$, we have

$$\bar{y} = \frac{1+2+9}{3} = 4.$$

Similarly, suppose we are given a function $f$ defined on the interval $[a, b]$, and the points $x_1, x_2, \ldots, x_n$ are in the interval. Then the average value of the $n$ corresponding function values $f(x_1), f(x_2), \ldots, f(x_n)$ is

$$\bar{y} = \frac{f(x_1) + f(x_2) + \cdots + f(x_n)}{n} = \frac{\displaystyle\sum_{i=1}^{n} f(x_i)}{n}. \tag{1}$$

We can go a step further. Let us divide the interval $[a, b]$ into $n$ subintervals of equal length. We shall choose $x_1$ to be in the first subinterval, $x_2$ to be in the second, etc. Because $[a, b]$ has length $b - a$, each subinterval has length $\dfrac{b-a}{n}$, which we shall call $\Delta x$. Thus, Eq. (1) can be written

We assume that $x_1, x_2$, and so on are the right-hand endpoints of the subintervals.

$$\bar{y} = \frac{\displaystyle\sum_{i=1}^{n} f(x_i)\left(\dfrac{\Delta x}{\Delta x}\right)}{n} = \frac{\dfrac{1}{\Delta x}\displaystyle\sum_{i=1}^{n} f(x_i)\,\Delta x}{n} = \frac{1}{n\,\Delta x}\sum_{i=1}^{n} f(x_i)\,\Delta x. \tag{2}$$

Since $\Delta x = \dfrac{b - a}{n}$, it follows that $n \, \Delta x = b - a$. So the expression $\dfrac{1}{n \, \Delta x}$ in Eq. (2) can be replaced by $\dfrac{1}{b - a}$. Moreover, as $n \to \infty$, the number of function values used in computing $\bar{y}$ increases, and we get the so-called *average value of the function f,* denoted by $\bar{f}$:

$$\bar{f} = \lim_{n \to \infty}\left[\frac{1}{b - a}\sum_{i=1}^{n} f(x_i)\,\Delta x\right] = \frac{1}{b - a}\lim_{n \to \infty}\sum_{i=1}^{n} f(x_i)\,\Delta x.$$

But the limit on the right is just the definite integral $\displaystyle\int_{a}^{b} f(x)\,dx$. Thus, we have the following definition.

### DEFINITION

*The **average** (or **mean**) **value of a function** $y = f(x)$ over the interval $[a, b]$ is denoted $\bar{f}$ (or $\bar{y}$) and is given by*

$$\bar{f} = \frac{1}{b - a}\int_{a}^{b} f(x)\,dx.$$

### EXAMPLE 1   Average Value of a Function

*Find the average value of the function $f(x) = x^2$ over the interval $[1, 2]$.*

**Solution:**

$$\bar{f} = \frac{1}{b - a}\int_{a}^{b} f(x)\,dx$$

$$= \frac{1}{2 - 1}\int_{1}^{2} x^2\,dx = \frac{x^3}{3}\Big|_{1}^{2} = \frac{7}{3}.$$   ∎

In Example 1, we found that the average value of $y = f(x) = x^2$ over the interval $[1, 2]$ is $\frac{7}{3}$. We can interpret this value geometrically. Since

$$\frac{1}{2 - 1}\int_{1}^{2} x^2\,dx = \frac{7}{3},$$

by solving for the integral we have

$$\int_{1}^{2} x^2\,dx = \frac{7}{3}(2 - 1).$$

However, this integral gives the area of the region bounded by $f(x) = x^2$ and the $x$-axis from $x = 1$ to $x = 2$. (See Fig. 17.1.) From the preceding equation, this area is $(\frac{7}{3})(2 - 1)$, which is the area of a rectangle whose height is the average value $\bar{f} = \frac{7}{3}$ and whose width is $b - a = 2 - 1 = 1$.

### EXAMPLE 2   Average Flow of Blood

*Suppose the flow of blood at time t in a system is given by*

$$F(t) = \frac{F_1}{(1 + \alpha t)^2}, \qquad 0 \le t \le T,$$

*where $F_1$ and $\alpha$ (a Greek letter read "alpha") are constants.[6] Find the average flow $\bar{F}$ on the interval $[0, T]$.*

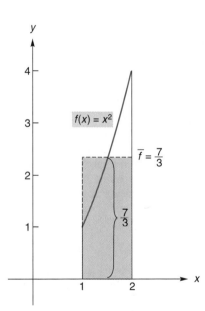

**FIGURE 17.1**
Geometric interpretation of the average value of a function.

[6]W. Simon, *Mathematical Techniques for Physiology and Medicine* (New York: Academic Press, Inc., 1972).

*Solution:*

$$\overline{F} = \frac{1}{T-0} \int_0^T F(t)\, dt$$

$$= \frac{1}{T} \int_0^T \frac{F_1}{(1+\alpha t)^2}\, dt = \frac{F_1}{\alpha T} \int_0^T (1+\alpha t)^{-2} (\alpha\, dt)$$

$$= \frac{F_1}{\alpha T} \left[ \frac{(1+\alpha t)^{-1}}{-1} \right]\Bigg|_0^T = \frac{F_1}{\alpha T} \left[ -\frac{1}{1+\alpha T} + 1 \right]$$

$$= \frac{F_1}{\alpha T} \left[ \frac{-1+1+\alpha T}{1+\alpha T} \right] = \frac{F_1}{\alpha T} \left[ \frac{\alpha T}{1+\alpha T} \right] = \frac{F_1}{1+\alpha T}. \qquad \blacksquare$$

## ■ Exercise 17.4

*In Problems 1–8, find the average value of the function over the given interval.*

**1.** $f(x) = x^2$; $[0, 4]$.

**2.** $f(x) = 3x - 1$; $[1, 2]$.

**3.** $f(x) = 2 - 3x^2$; $[-1, 2]$.

**4.** $f(x) = x^2 + x + 1$; $[1, 3]$.

**5.** $f(t) = 4t^3$; $[-2, 2]$.

**6.** $f(t) = t\sqrt{t^2 + 9}$; $[0, 4]$.

**7.** $f(x) = \sqrt{x}$; $[1, 9]$.

**8.** $f(x) = 1/x$; $[2, 4]$.

**9. Profit** The profit (in dollars) of a business is given by

$$P = P(q) = 396q - 2.1q^2 - 400,$$

where $q$ is the number of units of the product sold. Find the average profit on the interval from $q = 0$ to $q = 100$.

**10. Cost** Suppose the cost (in dollars) of producing $q$ units of a product is given by

$$c = 4000 + 10q + 0.1q^2.$$

Find the average cost on the interval from $q = 100$ to $q = 500$.

**11. Investment** An investment of \$3000 earns interest at an annual rate of 10% compounded continuously. After $t$ years, its value $S$ (in dollars) is given by $S = 3000e^{0.10t}$. Find the average value of a two-year investment.

**12. Medicine** Suppose that colored dye is injected into the bloodstream at a constant rate $R$. At time $t$, let

$$C(t) = \frac{R}{F(t)}$$

be the concentration of dye at a location distant (distal) from the point of injection, where $F(t)$ is as given in Example 2. Show that the average concentration on $[0, T]$ is

$$\overline{C} = \frac{R(1 + \alpha T + \frac{1}{3}\alpha^2 T^2)}{F_1}.$$

**13. Revenue** Suppose a manufacturer receives revenue $r$ from the sale of $q$ units of a product. Show that the average value of the marginal-revenue function over the interval $[0, q_0]$ is the price per unit when $q_0$ units are sold.

**14.** Find the average value of $f(x) = \dfrac{1}{x^2 + 1}$ over the interval $[0, 4]$. Round your answer to two decimal places.

---

**OBJECTIVE**

To estimate the value of a definite integral by using either the trapezoidal rule or Simpson's rule.

## 17.5 APPROXIMATE INTEGRATION

### Trapezoidal Rule

When using the Fundamental Theorem to evaluate $\displaystyle\int_a^b f(x)\, dx$, you may find it extremely difficult, or perhaps impossible, to find an antiderivative of $f$, even with the help of tables. In such a situation, many graphics calculators can estimate the value of the definite integral, provided that the integrand is known. In addition, numerical methods can be used to estimate the integral. These numerical methods use only a finite number of values of $f(x)$. Thus, $f$ does not have to be known on the entire interval $[a, b]$. The numerical methods are especially suitable for computers or calculators. We shall consider two such methods: the *trapezoidal rule* and *Simpson's rule*. In both cases, we assume that $f$ is continuous on $[a, b]$.

In developing the trapezoidal rule, for convenience we shall also assume that $f(x) \geq 0$ on $[a, b]$, so that we can think in terms of area. Basically, this rule involves approximating the graph of $f$ by straight-line segments.

In Fig. 17.2, the interval $[a, b]$ is divided into $n$ subintervals of equal length by the points $a = x_0, x_1, x_2, \ldots,$ and $x_n = b$. Since the length of $[a, b]$ is $b - a$, the length of each subinterval is $(b - a)/n$, which we shall call $h$.

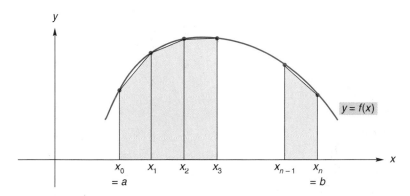

**FIGURE 17.2**   Approximating an area by using trapezoids.

Clearly,

$$x_1 = a + h, \quad x_2 = a + 2h, \quad \ldots, \quad x_n = a + nh = b.$$

With each subinterval, we can associate a trapezoid (a four-sided figure with two parallel sides). The area $A$ of the region bounded by the curve, the $x$-axis, and the lines $x = a$ and $x = b$ is $\int_a^b f(x)\, dx$ and can be approximated by the sum of the areas of the trapezoids determined by the subintervals.

Consider the first trapezoid, which is redrawn in Fig. 17.3. Since the area of a trapezoid is equal to one-half the base times the sum of the lengths of the parallel sides, this trapezoid has area

$$\tfrac{1}{2}h[f(a) + f(a + h)].$$

Similarly, the second trapezoid has area

$$\tfrac{1}{2}h[f(a + h) + f(a + 2h)].$$

The area $A$ under the curve is approximated by the sum of the areas of $n$ trapezoids:

$$A \approx \tfrac{1}{2}h[f(a) + f(a + h)] + \tfrac{1}{2}h[f(a + h) + f(a + 2h)] +$$

$$\tfrac{1}{2}h[f(a + 2h) + f(a + 3h)] + \cdots + \tfrac{1}{2}h[f(a + (n - 1)h) + f(b)].$$

Since $A = \int_a^b f(x)\, dx$, by simplifying the preceding formula we have the trapezoidal rule:

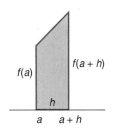

**FIGURE 17.3**
First trapezoid.

---

**The Trapezoidal Rule**

$$\int_a^b f(x)\, dx \approx \frac{h}{2}\{f(a) + 2f(a + h) + 2f(a + 2h) + ,$$

$$\ldots + 2f[a + (n - 1)h] + f(b)\},$$

where $h = (b - a)/n$.

The pattern of the coefficients inside the braces is 1, 2, 2, ..., 2, 1. Usually, the more subintervals, the better is the approximation. In our development, we assumed for convenience that $f(x) \geq 0$ on $[a, b]$. However, the trapezoidal rule is valid without this restriction.

### EXAMPLE 1 Trapezoidal Rule

*Use the trapezoidal rule to estimate the value of*

$$\int_0^1 \frac{1}{1 + x^2} \, dx$$

*for $n = 5$. Compute each term to four decimal places, and round off the answer to three decimal places.*

**Solution:** Here $f(x) = 1/(1 + x^2)$, $n = 5$, $a = 0$, and $b = 1$. Thus,

$$h = \frac{b - a}{n} = \frac{1 - 0}{5} = \frac{1}{5} = 0.2.$$

The terms to be added are

$$
\begin{aligned}
f(a) = \quad f(0) \quad &= 1.0000 \\
2f(a + h) = 2f(0.2) &= 1.9231 \\
2f(a + 2h) = 2f(0.4) &= 1.7241 \\
2f(a + 3h) = 2f(0.6) &= 1.4706 \\
2f(a + 4h) = 2f(0.8) &= 1.2195 \\
f(b) = \quad f(1) \quad &= \underline{0.5000} \\
&\phantom{=} \; 7.8373 = \text{sum.}
\end{aligned}
$$

Hence, our estimate for the integral is

$$\int_0^1 \frac{1}{1 + x^2} \, dx \approx \frac{0.2}{2}(7.8373) \approx 0.784.$$

The actual value of the integral is approximately 0.785. ∎

### Simpson's Rule

Another method for estimating $\int_a^b f(x) \, dx$ is given by Simpson's rule, which involves approximating the graph of $f$ by parabolic segments. We shall omit the derivation.

> **Simpson's Rule**
>
> $$\int_a^b f(x) \, dx \approx \frac{h}{3}\{f(a) + 4f(a + h) + 2f(a + 2h) +$$
>
> $$\cdots + 4f[a + (n - 1)h] + f(b)\},$$
>
> where $h = (b - a)/n$ and $n$ is even.

The pattern of coefficients inside the braces is 1, 4, 2, 4, 2, ..., 2, 4, 1, which requires that **$n$ be even.** Let us use this rule for the integral in Example 1.

### EXAMPLE 2 Simpson's Rule

*Use Simpson's rule to estimate the value of $\int_0^1 \frac{1}{1 + x^2} \, dx$ for $n = 4$. Compute each term to four decimal places, and round off the answer to three decimal places.*

**Principles in Practice 1**
**Trapezoidal Rule**

An oil tanker is losing oil at a rate of $R'(t) = \dfrac{60}{\sqrt{t^2 + 9}}$, where $t$ is the time in minutes and $R(t)$ is the radius of the oil slick in feet. Use the trapezoidal rule with $n = 5$ to approximate $\int_0^5 \dfrac{60}{\sqrt{t^2 + 9}} \, dt$, the size of the radius after five seconds.

**Principles in Practice 2**
**Simpson's Rule**

A yeast culture is growing at the rate of $A'(t) = 0.3e^{0.2t^2}$, where $t$ is the time in hours and $A(t)$ is the amount in grams. Use Simpson's rule with $n = 8$ to approximate $\int_0^4 0.3e^{0.2t^2} \, dt$, the amount the culture grew over the first four hours.

*Solution:* Here $f(x) = 1/(1 + x^2)$, $n = 4$, $a = 0$, and $b = 1$. Thus, $h = (b - a)/n = 1/4 = 0.25$. The terms to be added are:

$$f(a) = \quad f(0) \quad = 1.0000$$
$$4f(a + h) = 4f(0.25) = 3.7647$$
$$2f(a + 2h) = 2f(0.5) \ = 1.6000$$
$$4f(a + 3h) = 4f(0.75) = 2.5600$$
$$f(b) = \quad f(1) \quad = \underline{0.5000}$$
$$9.4247 = \text{sum.}$$

Therefore, by Simpson's rule,

$$\int_0^1 \frac{1}{1 + x^2}\, dx \approx \frac{0.25}{3}\,(9.4247) \approx 0.785.$$

This is a better approximation than that which we obtained in Example 1 by using the trapezoidal rule. ▬

Both Simpson's rule and the trapezoidal rule may be used if we know only $f(a), f(a + h)$, and so on; we need not know $f$ itself. Example 3 will illustrate.

### EXAMPLE 3  Demography

In Example 3, a definite integral is estimated from data points; the function itself is not known.

*A function often used in demography (the study of births, marriages, mortality, etc., in a population) is the **life-table function,** denoted l. In a population having 100,000 births in any year of time, l(x) represents the number of persons who reach the age of x in any year of time. For example, if l(20) = 95,961, then the number of persons who attain age 20 in any year of time is 95,961. Suppose that the function l applies to all people born over an extended period of time. It can be shown that, at any time, the expected number of persons in the population between the exact ages of x and x + m, inclusive, is given by*

$$\int_x^{x+m} l(t)\, dt.$$

*The following table gives values of l(x) for males and females in a certain population.[7] Approximate the number of women in the 20–35 age group by using the trapezoidal rule with n = 3.*

**Life Table**

| Age, $x$ | $l(x)$ Males | $l(x)$ Females | Age, $x$ | $l(x)$ Males | $l(x)$ Females |
|---|---|---|---|---|---|
| 0 | 100,000 | 100,000 | 45 | 89,489 | 93,667 |
| 5 | 97,158 | 97,791 | 50 | 86,195 | 91,726 |
| 10 | 96,921 | 97,618 | 55 | 81,154 | 88,935 |
| 15 | 96,672 | 97,473 | 60 | 73,830 | 84,971 |
| 20 | 95,961 | 97,188 | 65 | 64,108 | 79,445 |
| 25 | 95,000 | 96,839 | 70 | 52,007 | 71,196 |
| 30 | 94,097 | 96,429 | 75 | 38,044 | 59,946 |
| 35 | 93,067 | 95,844 | 80 | 24,900 | 45,662 |
| 40 | 91,628 | 94,961 | | | |

[7]Adapted from *Population: Facts and Methods of Demography,* by Nathan Keyfitz and William Flieger, W. H. Freeman and Company. Copyright © 1971.

*Solution:* We want to estimate

$$\int_{20}^{35} l(t)\, dt.$$

We have $h = \dfrac{b-a}{n} = \dfrac{35-20}{3} = 5$. The terms to be added under the trapezoidal rule are

$$l(20) = 97{,}188$$

$$2l(25) = 2(96{,}839) = 193{,}678$$

$$2l(30) = 2(96{,}429) = 192{,}858$$

$$l(35) = \underline{95{,}844}$$

$$579{,}568 = \text{sum}$$

By the trapezoidal rule,

$$\int_{20}^{35} l(t)\, dt \approx \frac{5}{2}(579{,}568) = 1{,}448{,}920. \qquad \blacksquare$$

Formulas used to determine the accuracy of answers obtained with the trapezoidal or Simpson's rule may be found in standard texts on numerical analysis.

## ■ Exercise 17.5

*In Problems 1 and 2, use the trapezoidal rule or Simpson's rule (as indicated) and the given value of n to estimate the integral.*

**1.** $\displaystyle\int_{-2}^{4} \frac{170}{1+x^2}\, dx$; trapezoidal rule, $n = 6$.

**2.** $\displaystyle\int_{-2}^{4} \frac{170}{1+x^2}\, dx$; Simpson's rule, $n = 6$.

*In Problems 3–8, use the trapezoidal rule or Simpson's rule (as indicated) and the given value of n to estimate the integral. Compute each term to four decimal places, and round the answer to three decimal places. In Problems 3–6, also evaluate the integral by antidifferentiation (the Fundamental Theorem of Integral Calculus).*

**3.** $\displaystyle\int_{0}^{1} x^2\, dx$; trapezoidal rule, $n = 5$.

**4.** $\displaystyle\int_{0}^{1} x^2\, dx$; Simpson's rule, $n = 4$.

**5.** $\displaystyle\int_{1}^{4} \frac{dx}{x}$; Simpson's rule, $n = 6$.

**6.** $\displaystyle\int_{1}^{4} \frac{dx}{x}$; trapezoidal rule, $n = 6$.

**7.** $\displaystyle\int_{0}^{2} \frac{x\, dx}{x+1}$; trapezoidal rule, $n = 4$.

**8.** $\displaystyle\int_{2}^{4} \frac{dx}{x+x^2}$; Simpson's rule, $n = 4$.

*In Problems 9 and 10, use the life table in Example 3 to estimate the given integrals by the trapezoidal rule.*

**9.** $\displaystyle\int_{15}^{40} l(t)\, dt$, males, $n = 5$.

**10.** $\displaystyle\int_{35}^{55} l(t)\, dt$, females, $n = 4$.

*In Problems 11 and 12, suppose the graph of a continuous function f, where f(x) ≥ 0, contains the given points. Use Simpson's rule and all of the points to approximate the area between the graph and the x-axis on the given interval. Round the answer to one decimal place.*

**11.** $(1, 0.4)$, $(2, 0.6)$, $(3, 1.2)$, $(4, 0.8)$, $(5, 0.5)$;   $[1, 5]$.

**12.** $(2, 0)$, $(2.5, 3.6)$, $(3, 10)$, $(3.5, 19.9)$, $(4, 34)$;   $[2, 4]$.

**13.** Using all the information given in Fig. 17.4, estimate $\displaystyle\int_{1}^{3} f(x)\, dx$ by using Simpson's rule. Give your answer in fractional form.

*In Problems 14 and 15, use Simpson's rule and the given value of n to estimate the integral. Compute each term to four decimal places, and round the answer to three decimal places.*

**14.** $\displaystyle\int_{4}^{6} \frac{1}{\sqrt{1+x}}\, dx$; $n = 4$. Also, evaluate the integral by the Fundamental Theorem of Integral Calculus.

**15.** $\displaystyle\int_{0}^{1} \sqrt{1-x^2}\, dx$; $n = 4$.

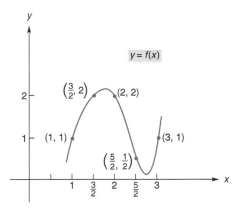

**FIGURE 17.4** Graph of $f$ for Problem 13.

**16. Revenue** Use Simpson's rule to approximate the total revenue received from the production and sale of 80 units of a product if the values of the marginal-revenue function $dr/dq$ are as follows:

| $q$ (units) | 0 | 10 | 20 | 30 | 40 | 50 | 60 | 70 | 80 |
|---|---|---|---|---|---|---|---|---|---|
| $\dfrac{dr}{dq}$ (\$ per unit) | 10 | 9 | 8.5 | 8 | 8.5 | 7.5 | 7 | 6.5 | 7 |

**17. Area of Lake** A straight stretch of highway runs alongside a lake. A surveyor who wishes to know the approximate area of the lake measures the distance from various points along the road to the near and far shores of the lake according to the following table:

| Distance along highway (km) | 0.0 | 0.5 | 1.0 | 1.5 | 2.0 | 2.5 | 3.0 | 3.5 | 4.0 |
|---|---|---|---|---|---|---|---|---|---|
| Distance to near shore (km) | 0.5 | 0.3 | 0.7 | 1.0 | 0.5 | 0.2 | 0.5 | 0.8 | 1.0 |
| Distance to far shore (km) | 0.5 | 2.3 | 2.2 | 3.0 | 2.5 | 2.2 | 1.5 | 1.3 | 1.0 |

Draw a rough sketch of the geographical situation. Then use Simpson's rule to give the best estimate of the lake's area. Give your answer in fractional form.

**18. Producers' Surplus** The supply function for a product is given by the following table, where $p$ is the price per unit (in dollars) at which $q$ units are supplied to the market:

| $q$ | 0 | 10 | 20 | 30 | 40 | 50 |
|---|---|---|---|---|---|---|
| $p$ | 38 | 59 | 69 | 75 | 80 | 84 |

Use the trapezoidal rule to estimate the producers' surplus if the selling price is \$80.

**19. Manufacturing** A manufacturer estimated both marginal cost (MC) and marginal revenue (MR) at various levels of output $(q)$. These estimates are given in the following table:

| $q$ (units) | 0 | 20 | 40 | 60 | 80 | 100 | 120 |
|---|---|---|---|---|---|---|---|
| MC (\$ per unit) | 260 | 255 | 240 | 240 | 245 | 250 | 255 |
| MR (\$ per unit) | 415 | 360 | 320 | 290 | 270 | 260 | 255 |

**a.** Using the trapezoidal rule, estimate the total variable costs of production for 120 units.

**b.** Using Simpson's rule, estimate the total revenue from the sale of 120 units.

**c.** If we assume that maximum profit occurs when MR = MC (that is, when $q = 120$), estimate the maximum profit if fixed costs are \$1500.

---

## 17.6 DIFFERENTIAL EQUATIONS

Occasionally, you may have to solve an equation that involves the derivative of an unknown function. Such an equation is called a **differential equation.** An example is

$$y' = xy^2. \tag{1}$$

More precisely, Eq. (1) is called a **first-order differential equation,** since it involves a derivative of the first order and none of higher order. A solution of Eq. (1) is any function $y = f(x)$ that is defined on an interval and satisfies the equation for all $x$ in the interval.

To solve $y' = xy^2$ or, equivalently,

$$\frac{dy}{dx} = xy^2, \tag{2}$$

we think of $dy/dx$ as a quotient of differentials and algebraically "separate variables" by rewriting the equation so that each side contains only one variable and a differential is not in a denominator:

$$\frac{dy}{y^2} = x\,dx.$$

Integrating both sides and combining the constants of integration, we obtain

$$\int \frac{1}{y^2}\,dy = \int x\,dx,$$

$$-\frac{1}{y} = \frac{x^2}{2} + C_1,$$

$$-\frac{1}{y} = \frac{x^2 + 2C_1}{2}.$$

Since $2C_1$ is an arbitrary constant, we can replace it by $C$.

$$-\frac{1}{y} = \frac{x^2 + C}{2}. \tag{3}$$

Solving Eq. (3) for $y$, we have

$$y = -\frac{2}{x^2 + C}. \tag{4}$$

We can verify by substitution that $y$ is a solution to the differential equation (2):

$$\frac{dy}{dx} = xy^2?$$

$$\frac{4x}{(x^2 + C)^2} = x\left[-\frac{2}{x^2 + C}\right]^2?$$

$$\frac{4x}{(x^2 + C)^2} = \frac{4x}{(x^2 + C)^2}.$$

Note in Eq. (4) that, for *each* value of $C$, a different solution is obtained. We call Eq. (4) the **general solution** of the differential equation. The method that we used to find it is called **separation of variables.**

In the foregoing example, suppose we are given the condition that $y = -\frac{2}{3}$ when $x = 1$ that is, $y(1) = -\frac{2}{3}$. Then the *particular* function that satisfies both Eq. (2) and this condition can be found by substituting the values $x = 1$ and $y = -\frac{2}{3}$ into Eq. (4) and solving for $C$:

$$-\frac{2}{3} = -\frac{2}{1^2 + C},$$

$$C = 2.$$

Therefore, the solution of $dy/dx = xy^2$ such that $y(1) = -\frac{2}{3}$ is

$$y = -\frac{2}{x^2 + 2}. \tag{5}$$

We call Eq. (5) a **particular solution** of the differential equation.

For a clear liquid, light intensity diminishes at a rate of $\dfrac{dI}{dx} = -kI$, where $I$ is the intensity of the light and $x$ is the number of feet below the surface of the liquid. If $k = 0.0085$ and $I = I_0$ when $x = 0$, find $I$ as a function of $x$.

**EXAMPLE 1  Separation of Variables**

*Solve* $y' = -\dfrac{y}{x}$ if $x, y > 0$.

*Solution:* Writing $y'$ as $dy/dx$, separating variables, and integrating, we have

$$\frac{dy}{dx} = -\frac{y}{x},$$

$$\frac{dy}{y} = -\frac{dx}{x}$$

$$\int \frac{1}{y}\,dy = -\int \frac{1}{x}\,dx,$$

$$\ln|y| = C_1 - \ln|x|.$$

Since $x, y > 0$, we can omit the absolute-value bars:

$$\ln y = C_1 - \ln x. \tag{6}$$

To solve for $y$, we convert Eq. (6) to exponential form:

$$y = e^{C_1 - \ln x}.$$

So

$$y = e^{C_1} e^{-\ln x} = \frac{e^{C_1}}{e^{\ln x}}.$$

Replacing $e^{C_1}$ by $C$, where $C > 0$, and rewriting $e^{\ln x}$ as $x$ gives

$$y = \frac{C}{x}, \quad C, x > 0. \qquad \blacksquare$$

In Example 1, note that Eq. (6) expresses the solution implicitly, whereas the final equation ($y = C/x$) states the solution $y$ explicitly in terms of $x$. You will find that the solutions of certain differential equations are often expressed in implicit form for convenience (or necessity because of the difficulty involved in obtaining an explicit form).

## Exponential Growth and Decay

In Sec. 11.3, the notion of interest compounded continuously was developed. Let us now take a different approach to this topic that involves a differential equation. Suppose $P$ dollars are invested at an annual rate $r$ compounded $n$ times a year. Let the function $S = S(t)$ give the compound amount $S$ (or total amount present) after $t$ years from the date of the initial investment. Then the initial principal is $S(0) = P$. Furthermore, since there are $n$ interest periods per year, each period has length $1/n$ years, which we shall denote by $\Delta t$. At the end of the first period, the accrued interest for that period is added to the principal, and the sum acts as the principal for the second period, and so on. Hence, if the beginning of an interest period occurs at time $t$, then the increase in the amount present at the end of a period $\Delta t$ is $S(t + \Delta t) - S(t)$, which we write as $\Delta S$. This increase, $\Delta S$, is also the interest earned for the period. Equivalently, the interest earned is principal times rate times time:

$$\Delta S = S \cdot r \cdot \Delta t.$$

Dividing both sides by $\Delta t$, we obtain

$$\frac{\Delta S}{\Delta t} = rS. \tag{7}$$

As $\Delta t \to 0$, then $n = \frac{1}{\Delta t} \to \infty$, and consequently, interest is being *compounded continuously;* that is, the principal is subject to continuous growth at every instant. However, as $\Delta t \to 0$, then $\Delta S/\Delta t \to dS/dt$, and Eq. (7) takes the form

$$\frac{dS}{dt} = rS. \tag{8}$$

This differential equation means that *when interest is compounded continuously, the rate of change of the amount of money present at time t is proportional to the amount present at time t.* The constant of proportionality is $r$.

To determine the actual function $S$, we solve the differential equation (8) by the method of separation of variables:

$$\frac{dS}{dt} = rS,$$

$$\frac{dS}{S} = r \, dt,$$

$$\int \frac{1}{S} \, dS = \int r \, dt,$$

$$\ln |S| = rt + C_1.$$

We assume that $S > 0$, so $\ln |S| = \ln S$. Thus,

$$\ln S = rt + C_1.$$

To get an explicit form, we can solve for $S$ by converting to exponential form.

$$S = e^{rt + C_1} = e^{C_1} e^{rt}.$$

For simplicity, $e^{C_1}$ can be replaced by $C$ to obtain the general solution

$$S = Ce^{rt}.$$

The condition $S(0) = P$ allows us to find the value of $C$:

$$P = Ce^{r(0)} = C \cdot 1.$$

Hence $C = P$, so

$$S = Pe^{rt}. \tag{9}$$

Equation (9) gives the total value after $t$ years of an initial investment of $P$ dollars compounded continuously at an annual rate $r$. (See Fig. 17.5.)

In our discussion of compound interest, we saw from Eq. (8) that the rate of change in the amount present was proportional to the amount present. There are many natural quantities, such as population, whose rate of growth or decay at any time is considered proportional to the amount of that quantity present. If $N$ denotes the amount of such a quantity at time $t$, then this rate of growth means that

$$\frac{dN}{dt} = kN,$$

where $k$ is a constant. If we separate variables and solve for $N$ as we did for Eq. (8), we get

$$N = N_0 e^{kt}, \tag{10}$$

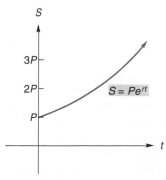

**FIGURE 17.5**
Compounding continuously.

where $N_0$ is a constant. In particular, if $t = 0$, then $N = N_0 e^0 = N_0 \cdot 1 = N_0$. Thus, the constant $N_0$ is simply $N(0)$. Due to the form of Eq. (10), we say that the quantity follows an **exponential law of growth** if $k$ is positive and an **exponential law of decay** if $k$ is negative.

### EXAMPLE 2   Population Growth

*In a certain city, the rate at which the population grows at any time is proportional to the size of the population. If the population was* 125,000 *in* 1970 *and* 140,000 *in* 1990, *what is the expected population in* 2010?

*Solution:* Let $N$ be the size of the population at time $t$. Since the exponential law of growth applies,

$$N = N_0 e^{kt}.$$

To find the population in 2010, we must first find the particular law of growth involved by determining the values of $N_0$ and $k$. Let the year 1970 correspond to $t = 0$. Then $t = 20$ in 1990 and $t = 40$ in 2010. We have

$$N_0 = N(0) = 125{,}000.$$

Thus,

$$N = 125{,}000 e^{kt}.$$

To find $k$, we use the fact that $N = 140{,}000$ when $t = 20$:

$$140{,}000 = 125{,}000 e^{20k}.$$

Hence,

$$e^{20k} = \frac{140{,}000}{125{,}000} = 1.12,$$

$$20k = \ln(1.12) \quad \text{(logarithmic form)},$$

$$k = \tfrac{1}{20} \ln(1.12).$$

Therefore, the law of growth is

$$N = 125{,}000 e^{(t/20)\ln 1.12} \tag{11}$$

$$= 125{,}000 [e^{\ln 1.12}]^{t/20},$$

or

$$N = 125{,}000 (1.12)^{t/20}. \tag{12}$$

Setting $t = 40$ gives the expected population in 2010:

$$N = 125{,}000 (1.12)^2 = 156{,}800.$$

We remark that we can write Eq. (11) in a form different from Eq. (12). Because $\ln(1.12)/20 \approx 0.0057$, we have

$$N \approx 125{,}000 e^{0.0057t}. \qquad \blacksquare$$

In Chapter 5, radioactive decay was discussed. Here we shall consider this topic from the perspective of a differential equation. The rate at which a radioactive element decays at any time is found to be proportional to the amount of that element present. If $N$ is the amount of a radioactive substance at time $t$, then the rate of decay is given by

$$\frac{dN}{dt} = -\lambda N. \tag{13}$$

The positive constant $\lambda$ (a Greek letter read "lambda") is called the **decay constant,** and the minus sign indicates that $N$ is decreasing as $t$ increases. Thus, we have exponential decay. From Eq. (10), the solution of this differential equation is

$$N = N_0 e^{-\lambda t}. \tag{14}$$

If $t = 0$, then $N = N_0 \cdot 1 = N_0$, so $N_0$ represents the amount of the radioactive substance present when $t = 0$.

The time for one-half of the substance to decay is called the **half-life** of the substance. In Sec. 5.2, it was shown that the half-life is given by

$$\text{half-life} = \frac{\ln 2}{\lambda} \approx \frac{0.69315}{\lambda}. \tag{15}$$

Note that the half-life depends on $\lambda$. In Chapter 5, Fig. 5.13 shows the graph of radioactive decay.

### EXAMPLE 3  Finding the Decay Constant and Half-life

*If 60% of a radioactive substance remains after 50 days, find the decay constant and the half-life of the element.*

*Solution:* From Eq. (14),

$$N = N_0 e^{-\lambda t},$$

where $N_0$ is the amount of the element present at $t = 0$. When $t = 50$, then $N = 0.6N_0$, and we have

$$0.6N_0 = N_0 e^{-50\lambda},$$
$$0.6 = e^{-50\lambda},$$
$$-50\lambda = \ln(0.6) \quad \text{(logarithmic form)},$$
$$\lambda = -\frac{\ln(0.6)}{50} \approx 0.01022.$$

Thus, $N \approx N_0 e^{-0.01022t}$. The half-life, from Eq. (15), is

$$\frac{\ln 2}{\lambda} \approx 67.85 \text{ days.} \qquad \blacksquare$$

Radioactivity is useful in dating such things as fossil plant remains and archaeological remains made from organic material. Plants and other living organisms contain a small amount of radioactive carbon 14 ($^{14}C$) in addition to ordinary carbon ($^{12}C$). The $^{12}C$ atoms are stable, but the $^{14}C$ atoms are decaying exponentially. However, $^{14}C$ is formed in the atmosphere due to the effect of cosmic rays. Eventually, this $^{14}C$ is taken up by plants during photosynthesis and replaces what has decayed. As a result, the ratio of $^{14}C$ atoms to $^{12}C$ atoms is considered constant over a long period of time. When a plant dies, it stops absorbing $^{14}C$, and the remaining $^{14}C$ atoms decay. By comparing the proportion of $^{14}C$ to $^{12}C$ in a fossil plant to that of plants found today, we can estimate the age of the fossil. The half-life of $^{14}C$ is approximately 5600 years. Thus, if a fossil is found to have a $^{14}C$-to-$^{12}C$ ratio which is half that of a similar substance found today, we would estimate the fossil to be 5600 years old.

**EXAMPLE 4  Estimating the Age of an Ancient Tool**

*A wood tool found in a Middle East excavation site is found to have a $^{14}C$-to-$^{12}C$ ratio that is 0.6 of the corresponding ratio in a present-day tree. Estimate the age of the tool to the nearest hundred years.*

*Solution:* Let $N$ be the amount of $^{14}C$ present in the wood $t$ years after the tool was made. Then $N = N_0 e^{-\lambda t}$ where $N_0$ is the amount of $^{14}C$ when $t = 0$. Since the ratio of $^{14}C$ to $^{12}C$ is 0.6 of the corresponding ratio in a present-day tree, this means that we want to find the value of $t$ for which $N = 0.6N_0$. Thus, we have

$$0.6N_0 = N_0 e^{-\lambda t},$$

$$0.6 = e^{-\lambda t},$$

$$-\lambda t = \ln(0.6),$$

$$t = -\frac{1}{\lambda}\ln(0.6).$$

From Eq. (15), the half-life is $(\ln 2)/\lambda$, which equals 5600, so $\lambda = (\ln 2)/5600$. Consequently,

$$t = -\frac{1}{(\ln 2)/5600}\ln(0.6)$$

$$= -\frac{5600\ln(0.6)}{\ln 2}$$

$$\approx 4100 \text{ years.} \qquad \blacksquare$$

# ■ Exercise 17.6

*In Problems 1–8, solve the differential equations.*

**1.** $y' = 2xy^2$.

**2.** $y' = x^3y^3$.

**3.** $\dfrac{dy}{dx} - x\sqrt{x^2 + 1} = 0$.

**4.** $\dfrac{dy}{dx} = \dfrac{x}{y}$.

**5.** $\dfrac{dy}{dx} = y,\ y > 0$.

**6.** $y' = e^x y^2$.

**7.** $y' = \dfrac{y}{x},\ x, y > 0$.

**8.** $\dfrac{dy}{dx} + xe^x = 0$.

*In Problems 9–18, solve each of the differential equations, subject to the given conditions.*

**9.** $y' = \dfrac{1}{y};\ y > 0,\ y(2) = 2$.

**10.** $y' = e^{x-y};\ y(0) = 0$. [*Hint:* $e^{x-y} = e^x/e^y$.]

**11.** $e^y y' - x^2 = 0;\ y = 0$ when $x = 0$.

**12.** $x^2 y' + \dfrac{1}{y^2} = 0;\ y(1) = 2$.

**13.** $(4x^2 + 3)^2 y' - 4xy^2 = 0;\ y(0) = \frac{3}{2}$.

**14.** $y' + x^2 y = 0;\ y > 0,\ y = 1$ when $x = 0$.

**15.** $\dfrac{dy}{dx} = \dfrac{3x\sqrt{1 + y^2}}{y};\ y > 0, y(1) = \sqrt{8}$.

**16.** $2y(x^3 + 2x + 1)\dfrac{dy}{dx} = \dfrac{3x^2 + 2}{\sqrt{y^2 + 9}};\ y(0) = 0$.

**17.** $2\dfrac{dy}{dx} = \dfrac{xe^{-y}}{\sqrt{x^2 + 3}};\ y(1) = 0$.

**18.** $x(y^3 + 4)^{3/2}dx = 3e^{x^2}y^2 dy;\ y(0) = 0$.

**19. Cost**  Find the manufacturer's cost function $c = f(q)$ given that

$$(q + 1)^2 \frac{dc}{dq} = cq$$

and fixed cost is $e$.

**20.** Find $f(2)$, given that $f(1) = 0$ and that $y = f(x)$ satisfies the differential equation

$$\frac{dy}{dx} = xe^{x-y}.$$

21. **Circulation of Money** A country has 900 million dollars of paper money in circulation. Each week 45 million dollars is brought into the banks for deposit, and the same amount is paid out. The government decides to issue new paper money; whenever the old money comes into the banks, it is destroyed and replaced by new money. Let $y$ be the amount of old money (in millions of dollars) in circulation at time $t$ (in weeks). Then $y$ satisfies the differential equation

$$\frac{dy}{dt} = -0.05y.$$

How long will it take for 90% of the paper money in circulation to be new? Round your answer to the nearest week. [*Hint:* If money is 90% new, then $y$ is 10% of 900.]

22. **Marginal Revenue and Demand** Suppose that a monopolist's marginal-revenue function is given by the differential equation

$$\frac{dr}{dq} = (50 - 4q)e^{-r/5}.$$

Find the demand equation for the monopolist's product.

23. **Population Growth** In a certain town, the population at any time changes at a rate proportional to the population. If the population in 1985 was 20,000 and in 1995 was 24,000, find an equation for the population at time $t$, where $t$ is the number of years past 1985. Write your answer in two forms, one involving $e$. You may assume that $\ln 1.2 = 0.18$. What is the expected population in 2005?

24. **Population Growth** The population of a town increases by natural growth at a rate proportional to the number $N$ of persons present. If the population at time $t = 0$ is 10,000, find two expressions for the population $N$, $t$ years later, if the population doubles in 50 years. Assume that $\ln 2 = 0.69$. Also, find $N$ for $t = 100$.

25. **Population Growth** Suppose that the population of the world in 1930 was 2 billion and in 1960 was 3 billion. If the exponential law of growth is assumed, what is the expected population in 2000? Give your answer in terms of $e$.

26. **Population Growth** If exponential growth is assumed, in approximately how many years will a population triple if it doubles in 50 years? [*Hint:* Let the population at $t = 0$ be $N_0$.]

27. **Radioactivity** If 30% of the initial amount of a radioactive sample remains after 100 seconds, find the decay constant and the half-life of the element.

28. **Radioactivity** If 30% of the initial amount of a radioactive sample has *decayed* after 100 seconds, find the decay constant and the half-life of the element.

29. **Carbon Dating** An Egyptian scroll was found to have a $^{14}$C-to-$^{12}$C ratio 0.7 of the corresponding ratio in similar present-day material. Estimate the age of the scroll, to the nearest hundred years.

30. **Carbon Dating** A recently discovered archaeological specimen has a $^{14}$C-to-$^{12}$C ratio 0.2 of the corresponding ratio found in present-day organic material. Estimate the age of the specimen, to the nearest hundred years.

31. **Population Growth** Suppose that a population follows exponential growth given by $dN/dt = kN$ for $t \geq t_0$. Suppose also that $N = N_0$ when $t = t_0$. Find $N$, the population size at time $t$.

32. **Radioactivity** Radon has a half-life of 3.82 days. (a) Find the decay constant in terms of $\ln 2$. (b) What fraction of the original amount of radon remains after $2(3.82) = 7.64$ days?

33. **Radioactivity** Radioactive isotopes are used in medical diagnoses as tracers to determine abnormalities that may exist in an organ. For example, if radioactive iodine is swallowed, after some time it is taken up by the thyroid gland. With the use of a detector, the rate at which it is taken up can be measured, and a determination can be made as to whether the uptake is normal. Suppose radioactive technetium-99m, which has a half-life of six hours, is to be used in a brain scan two hours from now. What should be its activity now if the activity when it is used is to be 10 units? Give your answer to one decimal place. [*Hint:* In Eq. (14), let $N =$ activity $t$ hours from now and $N_0 =$ activity now.]

34. **Radioactivity** A radioactive substance that has a half-life of eight days is to be temporarily implanted in a hospital patient until three-fifths of the amount originally present remains. How long should the implant remain in the patient?

35. **Ecology** In a forest, natural litter occurs, such as fallen leaves and branches, dead animals, etc.[8] Let $A = A(t)$ denote the amount of litter present at time $t$, where $A(t)$ is expressed in grams per square meter and $t$ is in years. Suppose that there is no litter at $t = 0$. Thus, $A(0) = 0$. Assume that

   a. Litter falls to the ground continuously at a constant rate of 200 grams per square meter per year.

   b. The accumulated litter decomposes continuously at the rate of 50% of the amount present per year (which is $0.50A$).

   The difference of the two rates is the rate of change of the amount of litter present with respect to time:

$$\left(\begin{array}{c}\text{rate of change}\\\text{of litter present}\end{array}\right) = \left(\begin{array}{c}\text{rate of falling}\\\text{to ground}\end{array}\right) - \left(\begin{array}{c}\text{rate of}\\\text{decomposition}\end{array}\right).$$

   Therefore,

$$\frac{dA}{dt} = 200 - 0.50A.$$

   Solve for $A$. To the nearest gram, determine the amount of litter per square meter after one year.

---

[8] R. W. Poole, *An Introduction to Quantitative Ecology* (New York: McGraw-Hill Book Company, 1974).

36. **Profit and Advertising** A certain company determines that the rate of change of monthly net profit $P$, as a function of monthly advertising expenditure $x$, is proportional to the difference between a fixed amount, \$11,000, and $P$; that is, $dP/dx$ is proportional to \$11,000 $-$ $P$. Furthermore, if no money is spent on monthly advertising, the monthly net profit is \$1000; if \$100 is spent on monthly advertising, the monthly net profit is \$6000. What would the monthly net profit be if \$200 were spent on advertising each month?

37. **Value of Auto** The value of a certain model of automobile depreciates 30% in the first year after the car is purchased. The rate at which the automobile subsequently depreciates is proportional to its value. Suppose that such an automobile was purchased new on July 1, 1997, for \$20,000 and was valued at \$12,600 on January 1, 1999.

   **a.** Determine a formula that expresses the value $V$ of the automobile in terms of $t$, the number of years after July 1, **1998**.

   **b.** Use the formula in part (a) to determine the year and month in which the car has a value of exactly \$7000.

---

**To develop the logistic function as a solution of a differential equation. To model the spread of a rumor. To discuss and apply Newton's law of cooling.**

## 17.7 MORE APPLICATIONS OF DIFFERENTIAL EQUATIONS

### Logistic Growth

In the previous section, we found that if the number $N$ of individuals in a population at time $t$ follows an exponential law of growth, then $N = N_0 e^{kt}$, where $k > 0$ and $N_0$ is the population when $t = 0$. This law assumes that at time $t$ the rate of growth, $dN/dt$, of the population is proportional to the number of individuals in the population. That is, $dN/dt = kN$.

Under exponential growth, a population would get infinitely large as time goes on. In reality, however, when the population gets large enough, environmental factors slow down the rate of growth. Examples are food supply, predators, overcrowding, and so on. These factors cause $dN/dt$ to eventually decrease. It is reasonable to assume that the size of a population is limited to some maximum number $M$, where $0 < N < M$, and as $N \to M$, $dN/dt \to 0$, and the population size tends to be stable.

In summary, we want a population model that has exponential growth initially, but that also includes the effects of environmental resistance to large population growth. Such a model is obtained by multiplying the right side of $dN/dt = kN$ by the factor $(M - N)/M$:

$$\frac{dN}{dt} = kN\left(\frac{M - N}{M}\right).$$

Notice that if $N$ is small, then $(M - N)/M$ is close to 1, and we have growth that is approximately exponential. As $N \to M$, then $M - N \to 0$ and $dN/dt \to 0$, as we wanted in our model. Because $k/M$ is a constant, we can replace it by $K$. Thus,

$$\frac{dN}{dt} = KN(M - N). \tag{1}$$

This states that the rate of growth is proportional to the product of the size of the population and the difference between the maximum size and the actual size of the population. We can solve for $N$ in the differential equation (1) by the method of separation of variables:

$$\frac{dN}{N(M - N)} = K\,dt,$$

$$\int \frac{1}{N(M - N)}\,dN = \int K\,dt. \tag{2}$$

The integral on the left side can be found by using Formula 5 in the table of integrals in Appendix C. Thus, Eq. (2) becomes

$$\frac{1}{M} \ln \left| \frac{N}{M - N} \right| = Kt + C,$$

or

$$\ln \left| \frac{N}{M - N} \right| = MKt + MC.$$

Since $N > 0$ and $M - N > 0$, we can write

$$\ln \frac{N}{M - N} = MKt + MC.$$

In exponential form, we have

$$\frac{N}{M - N} = e^{MKt + MC} = e^{MKt} e^{MC}.$$

Replacing the positive constant $e^{MC}$ by $A$ and solving for $N$ gives

$$\frac{N}{M - N} = Ae^{MKt},$$

$$N = (M - N)Ae^{MKt},$$

$$N = MAe^{MKt} - NAe^{MKt},$$

$$NAe^{MKt} + N = MAe^{MKt},$$

$$N(Ae^{MKt} + 1) = MAe^{MKt},$$

$$N = \frac{MAe^{MKt}}{Ae^{MKt} + 1}.$$

Dividing numerator and denominator by $Ae^{MKt}$, we have

$$N = \frac{M}{1 + \dfrac{1}{Ae^{MKt}}} = \frac{M}{1 + \dfrac{1}{A} e^{-MKt}}.$$

Replacing $1/A$ by $b$ and $MK$ by $c$ yields the so-called *logistic function:*

**Logistic Function**

The function defined by

$$N = \frac{M}{1 + be^{-ct}} \qquad (3)$$

is called the **logistic function** or the **Verhulst–Pearl logistic function.**

The graph of Eq. (3), called a *logistic curve,* is S-shaped and appears in Fig. 17.6. Notice that the line $N = M$ is a horizontal asymptote; that is,

$$\lim_{t \to \infty} \frac{M}{1 + be^{-ct}} = \frac{M}{1 + b(0)} = M.$$

Moreover, from Eq. (1), the rate of growth is

$$KN(M - N),$$

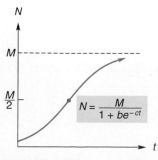

**FIGURE 17.6** Logistic curve.

which can be considered a function of $N$. To find when the maximum rate of growth occurs, we solve $\dfrac{d}{dN}[KN(M-N)]=0$ for $N$:

$$\frac{d}{dN}[KN(M-N)] = \frac{d}{dN}[K(MN-N^2)]$$
$$= K[M-2N] = 0.$$

Thus, $N = M/2$. In other words, the rate of growth increases until the population size is $M/2$ and decreases thereafter. The maximum rate of growth occurs when $N = M/2$ and corresponds to a point of inflection in the graph of $N$. To find the value of $t$ for which this occurs, we substitute $M/2$ for $N$ in Eq. (3) and solve for $t$:

$$\frac{M}{2} = \frac{M}{1+be^{-ct}},$$
$$1+be^{-ct} = 2,$$
$$e^{-ct} = \frac{1}{b},$$
$$e^{ct} = b,$$
$$ct = \ln b \quad \text{(logarithmic form)},$$
$$t = \frac{\ln b}{c}.$$

Therefore, the maximum rate of growth occurs at the point $([\ln b]/c, M/2)$.

We remark that in Eq. (3) we may replace $e^{-c}$ by $C$, and then the logistic function has the following form:

### Alternative Form of Logistic Function

$$N = \frac{M}{1+bC^t}.$$

### EXAMPLE 1  Logistic Growth of Club Membership

*Suppose the membership in a new country club is to be a maximum of* 800 *persons, due to limitations of the physical plant. One year ago the initial membership was* 50 *persons, and now there are* 200. *Provided that enrollment follows a logistic function, how many members will there be three years from now?*

***Solution:*** Let $N$ be the number of members enrolled $t$ years after the formation of the club. Then, from Eq. (3),

$$N = \frac{M}{1+be^{-ct}}.$$

Here $M = 800$, and when $t = 0$, we have $N = 50$. So

$$50 = \frac{800}{1+b},$$
$$1+b = \frac{800}{50} = 16,$$
$$b = 15.$$

Thus,

$$N = \frac{800}{1 + 15e^{-ct}}. \tag{4}$$

When $t = 1$, then $N = 200$, so we have

$$200 = \frac{800}{1 + 15e^{-c}},$$

$$1 + 15e^{-c} = \frac{800}{200} = 4,$$

$$e^{-c} = \frac{3}{15} = \frac{1}{5}.$$

Hence, $c = -\ln\frac{1}{5} = \ln 5$. Rather than substituting this value of $c$ into Eq. (4), it is more convenient to substitute the value of $e^{-c}$ there:

$$N = \frac{800}{1 + 15(\frac{1}{5})^t}.$$

Three years from now, $t$ will be 4. Therefore,

$$N = \frac{800}{1 + 15(\frac{1}{5})^4} \approx 781. \qquad\blacksquare$$

### Modeling the Spread of a Rumor

Let us now consider a simplified model of how a rumor spreads in a population of size $M$. A similar situation would be the spread of an epidemic or new fad.

Let $N = N(t)$ be the number of persons who know the rumor at time $t$. We shall assume that those who know the rumor spread it randomly in the population and that those who are told the rumor become spreaders of the rumor. Furthermore, we shall assume that each knower tells the rumor to $k$ individuals per unit of time. (Some of these $k$ individuals may already know the rumor.) We want an expression for the rate of increase of the knowers of the rumor. Over a unit of time, each of approximately $N$ persons will tell the rumor to $k$ persons. Thus, the total number of persons who are told the rumor over the unit of time is (approximately) $Nk$. However, we are interested only in *new* knowers. The proportion of the population that does not know the rumor is $(M - N)/M$. Hence, the total number of new knowers of the rumor is

$$Nk\left(\frac{M - N}{M}\right),$$

which can be written $(k/M)N(M - N)$. Therefore,

$$\frac{dN}{dt} = \frac{k}{M}N(M - N)$$

$$= KN(M - N), \qquad \text{where } K = \frac{k}{M}.$$

This differential equation has the form of Eq. (1), so its solution, from Eq. (3), is a *logistic function*:

$$N = \frac{M}{1 + be^{-ct}}.$$

**EXAMPLE 2   Campus Rumor**

*In a large university of 45,000 students, a sociology major is researching the spread of a new campus rumor. When she begins her research, she determines that 300 students know the rumor. After one week, she finds that 900 know it. Estimate the number of students who know it four weeks after the research begins by assuming logistic growth. Give the answer to the nearest thousand.*

*Solution:* Let $N$ be the number of students who know the rumor $t$ weeks after the research begins. Then

$$N = \frac{M}{1 + be^{-ct}}.$$

Here $M$, the size of the population, is 45,000, and when $t = 0$, $N = 300$. So we have

$$300 = \frac{45,000}{1 + b},$$

$$1 + b = \frac{45,000}{300} = 150,$$

$$b = 149.$$

Thus,

$$N = \frac{45,000}{1 + 149e^{-ct}}.$$

When $t = 1$, then $N = 900$. Hence,

$$900 = \frac{45,000}{1 + 149e^{-c}},$$

$$1 + 149e^{-c} = \frac{45,000}{900} = 50.$$

Therefore, $e^{-c} = \frac{49}{149}$, so

$$N = \frac{45,000}{1 + 149(\frac{49}{149})^t}.$$

When $t = 4$,

$$N = \frac{45,000}{1 + 149(\frac{49}{149})^4} \approx 16,000.$$

After four weeks, approximately 16,000 students know the rumor.   ■

### Newton's Law of Cooling

We conclude this section with an interesting application of a differential equation. If a homicide is committed, the temperature of the victim's body will gradually decrease from 37°C (normal body temperature) to the temperature of the surroundings (ambient temperature). In general, the temperature of the cooling body changes at a rate proportional to the difference between the temperature of the body and the ambient temperature. This statement is known as **Newton's law of cooling.** Thus, if $T(t)$ is the temperature of the body at time $t$ and the ambient temperature is $a$, then

$$\frac{dT}{dt} = k(T - a),$$

where $k$ is the constant of proportionality. Therefore, Newton's law of cooling is a differential equation. It can be applied to determine the time at which a homicide was committed, as the next example illustrates.

### EXAMPLE 3   Time of Murder

*A wealthy industrialist was found murdered in his home. Police arrived on the scene at 11:00 P.M. The temperature of the body at that time was 31°C, and one hour later it was 30°C. The temperature of the room in which the body was found was 22°C. Estimate the time at which the murder occurred.*

**Solution:**  Let $t$ be the number of hours after the body was discovered and $T(t)$ be the temperature (in degrees Celsius) of the body at time $t$. We want to find the value of $t$ for which $T = 37$ (normal body temperature). This value of $t$ will, of course, be negative. By Newton's law of cooling.

$$\frac{dT}{dt} = k(T - a),$$

where $k$ is a constant and $a$ (the ambient temperature) is 22. Thus,

$$\frac{dT}{dt} = k(T - 22).$$

Separating variables, we have

$$\frac{dT}{T - 22} = k\, dt,$$

$$\int \frac{dT}{T - 22} = \int k\, dt,$$

$$\ln|T - 22| = kt + C.$$

Because $T - 22 > 0$,

$$\ln(T - 22) = kt + C.$$

When $t = 0$, then $T = 31$. Therefore,

$$\ln(31 - 22) = k \cdot 0 + C,$$

$$C = \ln 9.$$

Hence,

$$\ln(T - 22) = kt + \ln 9,$$

$$\ln(T - 22) - \ln 9 = kt,$$

$$\ln \frac{T - 22}{9} = kt.$$

When $t = 1$, then $T = 30$, so

$$\ln \frac{30 - 22}{9} = k \cdot 1,$$

$$k = \ln \frac{8}{9}.$$

Thus,

$$\ln \frac{T - 22}{9} = t \ln \frac{8}{9}.$$

Now we find $t$ when $T = 37$:

$$\ln \frac{37 - 22}{9} = t \ln \frac{8}{9},$$

$$t = \frac{\ln(15/9)}{\ln(8/9)} \approx -4.34.$$

Accordingly, the murder occurred about 4.34 hours *before* the time of discovery of the body (11:00 P.M.). Since 4.34 hours is (approximately) 4 hours and 20 minutes, the industrialist was murdered about 6:40 P.M.    ■

## ■ Exercise 17.7

1. **Population**   The population of a city follows logistic growth and is limited to 40,000. If the population in 1993 was 20,000 and in 1998 was 25,000, what will be the population in 2003? Give your answer to the nearest hundred.

2. **Production**   A company believes that the production of its product in present facilities will follow logistic growth. Presently, 200 units per day are produced, and production will increase to 300 units per day in one year. If production is limited to 500 units per day, what is the anticipated daily production in two years? Give your answer to the nearest unit.

3. **Spread of Rumor**   In a country of 3,000,000 people, the prime minister suffers a heart attack, which the government does not officially publicize. Initially, 50 governmental personnel know of the attack, but are spreading this information as a rumor. At the end of one week, 5000 people know the rumor. Assuming logistic growth, find how many people know the rumor after two weeks. Give your answer to the nearest thousand.

4. **Spread of Fad**   A new fad is sweeping a college campus of 30,000 students. The college newspaper feels that its readers would be interested in a series on the fad. It assigns a reporter when the number of faddists is 400. One week later, there are 1200 faddists. Assuming logistic growth, find a formula for the number $N$ of faddists $t$ weeks after the assignment of the reporter.

5. **Flu Outbreak**   In a city whose population is 100,000, an outbreak of flu occurs. When the city health department begins its record keeping, there are 500 infected persons. One week later, there are 1000 infected persons. Assuming logistic growth, estimate the number of infected persons two weeks after record keeping begins.

6. **Population**   The logistic curve for the U.S. population from 1790 to 1910 is estimated to be[9]

$$N = \frac{197.30}{1 + 35.60e^{-0.031186t}},$$

where $N$ is the population in millions and $t$ is in years counted from 1800. If this logistic function were valid for years after 1910, for what year would the point of inflection occur? Round your answer to one decimal place.

7. **Biology**   In an experiment,[10] five *Paramecia* were placed in a test tube containing a nutritive medium. The number $N$ of *Paramecia* in the tube at the end of $t$ days is given approximately by

$$N = \frac{375}{1 + e^{5.2 - 2.3t}}.$$

   a.   Show that this equation can be written as

$$N = \frac{375}{1 + 181.27e^{-2.3t}}$$

   and hence is a logistic function.

   b.   Find $\lim_{t \to \infty} N$.

8. **Biology**   In a study of the growth of a colony of unicellular organisms,[11] the equation

$$N = \frac{0.2524}{e^{-2.128x} + 0.005125}, \qquad 0 \le x \le 5,$$

was obtained, where $N$ is the estimated area of the growth in square centimeters and $x$ is the age of the colony in days after being first observed.

   a.   Put this equation in the form of a logistic function.

[9] N. Keyfitz, *Introduction to the Mathematics of Population* (Reading, MA: Addison-Wesley Publishing Company, Inc., 1968).

[10] G. F. Gause, *The Struggle for Existence* (New York: Hafner Publishing Co., 1964).

[11] A. J. Lotka, *Elements of Mathematical Biology* (New York: Dover Publications, Inc., 1956).

**b.** Find the area when the age of the colony is 0.

**9. Time of Murder** A waterfront murder was committed, and the victim's body was discovered at 3:15 A.M. by police. At that time, the temperature of the body was 32°C. One hour later, the body temperature was 30°C. After checking with the weather bureau, it was determined that the temperature at the waterfront was 10°C from 10:00 P.M. to 5:00 A.M. About what time did the murder occur?

**10. Enzyme Formation** An enzyme is a protein that acts as a catalyst for increasing the rate of a chemical reaction that occurs in cells. In a certain reaction, an enzyme A is converted to another enzyme B. Enzyme B acts as a catalyst for its own formation. Let $p$ be the amount of enzyme B at time $t$ and $I$ be the total amount of both enzymes when $t = 0$. Suppose the rate of formation of B is proportional to $p(I - p)$. Without directly using calculus, find the value of $p$ for which the rate of formation will be a maximum.

**11. Fund-raising** A small town decides to conduct a fund-raising drive for a fire engine whose cost is $70,000. The initial amount in the fund is $10,000. On the basis of past drives, it is determined that $t$ months after the beginning of the drive, the rate $dx/dt$ at which money is contributed to such a fund is proportional to the difference between the desired goal of $70,000 and the total amount $x$ in the fund at that time. After one month, a total of $40,000 is in the fund. How much will be in the fund after three months?

**12. Birthrate** In a discussion of unexpected properties of mathematical models of population, Bailey[12] considers the case in which the birthrate per *individual* is proportional to the population size $N$ at time $t$. Since the growth rate per individual is $\dfrac{1}{N}\dfrac{dN}{dt}$, this means that

$$\frac{1}{N}\frac{dN}{dt} = kN,$$

or

$$\frac{dN}{dt} = kN^2 \quad \text{(subject to } N = N_0 \text{ at } t = 0\text{),}$$

where $k > 0$. Show that

$$N = \frac{N_0}{1 - kN_0 t}.$$

Use this result to show that

$$\lim N = \infty \quad \text{as} \quad t \to \left(\frac{1}{kN_0}\right)^{-}.$$

This means that over a finite interval of time, there is an infinite amount of growth. Such a model might be useful only for rapid growth over a short interval of time.

**13. Population** Suppose that the rate of growth of a population is proportional to the difference between some maximum size $M$ and the number $N$ of individuals in the population at time $t$. Suppose that when $t = 0$, the size of the population is $N_0$. Find a formula for $N$.

[12]N. T. J. Bailey, *The Mathematical Approach to Biology and Medicine* (New York: John Wiley & Sons, Inc., 1967).

---

**OBJECTIVE**

To define and evaluate improper integrals.

# 17.8 IMPROPER INTEGRALS[13]

Suppose $f(x)$ is continuous and nonnegative for $a \leq x < \infty$. (See Fig. 17.7.) We know that the integral $\displaystyle\int_a^r f(x)\, dx$ is the area of the region between the curve $y = f(x)$ and the $x$-axis from $x = a$ to $x = r$. As $r \to \infty$, we may think of

$$\lim_{r \to \infty} \int_a^r f(x)\, dx$$

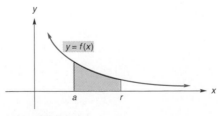

**FIGURE 17.7** Area from $a$ to $r$.

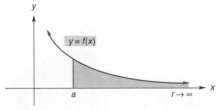

**FIGURE 17.8** Area from $a$ to $r$ as $r \to \infty$.

[13]May be omitted if Chapter 18 is not covered.

as the area of the unbounded region that is shaded in Fig. 17.8. This limit is abbreviated by

$$\int_a^\infty f(x)\,dx, \tag{1}$$

called an **improper integral.** If the limit exists, $\int_a^\infty f(x)\,dx$ is said to be **convergent** or to *converge* to that limit. In this case the unbounded region is considered to have a finite area, and this area is represented by $\int_a^\infty f(x)\,dx$. If the limit does not exist, the improper integral is said to be **divergent,** and the region does not have a finite area.

We can remove the restriction that $f(x) \geq 0$. In general, the improper integral $\int_a^\infty f(x)\,dx$ is defined by

$$\int_a^\infty f(x)\,dx = \lim_{r\to\infty}\int_a^r f(x)\,dx.$$

Other types of improper integrals are

$$\int_{-\infty}^b f(x)\,dx \tag{2}$$

and

$$\int_{-\infty}^\infty f(x)\,dx \tag{3}$$

In each of the three types of improper integrals [(1), (2), and (3)], the interval over which the integral is evaluated has infinite length. The improper integral in (2) is defined by

$$\int_{-\infty}^b f(x)\,dx = \lim_{r\to-\infty}\int_r^b f(x)\,dx.$$

If this limit exists, $\int_{-\infty}^b f(x)\,dx$ is said to be convergent. Otherwise, it is divergent. We shall define the improper integral in (3) after the following example.

## Principles in Practice 1

**Improper Integrals of the Form**

$\int_a^\infty f(x)\,dx$ **and** $\int_{-\infty}^b f(x)\,dx$

The rate at which the human body eliminates a certain drug from its system may be approximated by $R(t) = 3e^{-0.1t} - 3e^{-0.3t}$, where $R(t)$ is in milliliters per minute and $t$ is the time in minutes since the drug was taken. Find $\int_0^\infty (3e^{-0.1t} - 3e^{-0.3t})\,dt$, the total amount of the drug that is eliminated.

**EXAMPLE 1   Improper Integrals of the Form $\int_a^\infty f(x)\,dx$ and $\int_{-\infty}^b f(x)\,dx$**

*Determine whether the following improper integrals are convergent or divergent. For any convergent integral, determine its value.*

**a.** $\int_1^\infty \dfrac{1}{x^3}\,dx.$

*Solution:*

$$\int_1^\infty \frac{1}{x^3}\,dx = \lim_{r\to\infty}\int_1^r x^{-3}\,dx = \lim_{r\to\infty} -\frac{x^{-2}}{2}\Big|_1^r$$

$$= \lim_{r\to\infty}\left[-\frac{1}{2r^2} + \frac{1}{2}\right] = -0 + \frac{1}{2} = \frac{1}{2}.$$

Therefore, $\int_1^\infty \dfrac{1}{x^3}\,dx$ converges to $\dfrac{1}{2}$.

**b.** $\int_{-\infty}^0 e^x\,dx.$

*Solution:*

$$\int_{-\infty}^0 e^x\,dx = \lim_{r\to-\infty}\int_r^0 e^x\,dx = \lim_{r\to-\infty} e^x\Big|_r^0$$

$$= \lim_{r\to-\infty}(1-e^r) = 1 - 0 = 1.$$

(Here we used the fact that as $r \to -\infty$, the graph of $y = e^r$ approaches the $r$-axis, so $e^r \to 0$.) Therefore, $\int_{-\infty}^0 e^x\,dx$ converges to 1.

**c.** $\int_1^\infty \dfrac{1}{\sqrt{x}}\,dx.$

*Solution:*

$$\int_1^\infty \dfrac{1}{\sqrt{x}}\,dx = \lim_{r\to\infty}\int_1^r x^{-1/2}\,dx = \lim_{r\to\infty} 2x^{1/2}\Big|_1^r$$

$$= \lim_{r\to\infty} 2(\sqrt{r}-1) = \infty.$$

Therefore, the improper integral diverges. ■

The improper integral $\int_{-\infty}^\infty f(x)\,dx$ is defined in terms of improper integrals of the forms (1) and (2):

$$\int_{-\infty}^\infty f(x)\,dx = \int_{-\infty}^0 f(x)\,dx + \int_0^\infty f(x)\,dx \qquad (4)$$

If *both* integrals on the right side of Eq. (4) are convergent, then $\int_{-\infty}^\infty f(x)\,dx$ is said to be convergent; otherwise, it is divergent.

**EXAMPLE 2** An Improper Integral of the Form $\int_{-\infty}^\infty f(x)\,dx$

*Determine whether $\int_{-\infty}^\infty e^x\,dx$ is convergent or divergent.*

*Solution:*

$$\int_{-\infty}^\infty e^x\,dx = \int_{-\infty}^0 e^x\,dx + \int_0^\infty e^x\,dx.$$

By Example 1(b), $\int_{-\infty}^0 e^x\,dx = 1$. On the other hand,

$$\int_0^\infty e^x\,dx = \lim_{r\to\infty}\int_0^r e^x\,dx = \lim_{r\to\infty} e^x\Big|_0^r = \lim_{r\to\infty}(e^r-1) = \infty.$$

Since $\int_0^\infty e^x\,dx$ is divergent, $\int_{-\infty}^\infty e^x\,dx$ is also divergent. ■

**EXAMPLE 3** Density Function

*In statistics, a function f is called a density function if $f(x) \geq 0$ and*

$$\int_{-\infty}^{\infty} f(x) \, dx = 1.$$

*Suppose*

$$f(x) = \begin{cases} ke^{-x}, & \text{for } x \geq 0, \\ 0, & \text{elsewhere} \end{cases}$$

*is a density function. Find k.*

*Solution:* We write the equation $\int_{-\infty}^{\infty} f(x) \, dx = 1$ as

$$\int_{-\infty}^{0} f(x) \, dx + \int_{0}^{\infty} f(x) \, dx = 1.$$

Since $f(x) = 0$ for $x < 0$, $\int_{-\infty}^{0} f(x) \, dx = 0$. Thus,

$$\int_{0}^{\infty} ke^{-x} dx = 1,$$

$$\lim_{r \to \infty} \int_{0}^{r} ke^{-x} \, dx = 1,$$

$$\lim_{r \to \infty} -ke^{-x} \Big|_{0}^{r} = 1,$$

$$\lim_{r \to \infty} (-ke^{-r} + k) = 1,$$

$$0 + k = 1,$$

$$k = 1.$$

## ▪ Exercise 17.8

*In Problems 1–12, determine the integrals if they exist. Indicate those that are divergent.*

**1.** $\int_{3}^{\infty} \frac{1}{x^2} \, dx.$

**2.** $\int_{2}^{\infty} \frac{1}{(2x - 1)^3} \, dx.$

**3.** $\int_{1}^{\infty} \frac{1}{x} \, dx.$

**4.** $\int_{1}^{\infty} \frac{1}{\sqrt[3]{x + 1}} \, dx.$

**5.** $\int_{1}^{\infty} e^{-x} \, dx.$

**6.** $\int_{0}^{\infty} (5 + e^{-x}) \, dx.$

**7.** $\int_{1}^{\infty} \frac{1}{\sqrt{x}} \, dx.$

**8.** $\int_{4}^{\infty} \frac{x \, dx}{\sqrt{(x^2 + 9)^3}}.$

**9.** $\int_{-\infty}^{-2} \frac{1}{(x + 1)^3} \, dx.$

**10.** $\int_{-\infty}^{3} \frac{1}{\sqrt{7 - x}} \, dx.$

**11.** $\int_{-\infty}^{\infty} xe^{-x^2} \, dx.$

**12.** $\int_{-\infty}^{\infty} (5 - 3x) \, dx.$

**13. Density Function** The density function for the life $x$, in hours, of an electronic component in a calculator is given by

$$f(x) = \begin{cases} \dfrac{k}{x^2}, & \text{for } x \geq 800, \\ 0, & \text{for } x < 800. \end{cases}$$

**a.** If $k$ satisfies the condition that $\int_{800}^{\infty} f(x) \, dx = 1$, find $k$.

**b.** The probability that the component will last at least 1200 hours is given by $\int_{1200}^{\infty} f(x) \, dx$. Evaluate this integral.

**14. Density Function** Given the density function

$$f(x) = \begin{cases} ke^{-4x}, & \text{for } x \geq 0, \\ 0, & \text{elsewhere,} \end{cases}$$

find $k$. [*Hint:* See Example 3.]

**15. Future Profits** For a business, the present value of all future profits at an annual interest rate $r$ compounded continuously is given by

$$\int_0^\infty p(t)e^{-rt}\,dt,$$

where $p(t)$ is the profit per year in dollars at time $t$. If $p(t) = 240{,}000$ and $r = 0.06$, evaluate this integral.

**16. Psychology** In a psychological model for signal detection,[14] the probability $\alpha$ (a Greek letter read "alpha") of reporting a signal when no signal is present is given by

$$\alpha = \int_{x_c}^\infty e^{-x}\,dx, \quad x \geq 0.$$

The probability $\beta$ (a Greek letter read "beta") of detecting a signal when it is present is

$$\beta = \int_{x_c}^\infty ke^{-kx}\,dx, \quad x \geq 0.$$

In both integrals, $x_c$ is a constant (called a criterion value in this model). Find $\alpha$ and $\beta$ if $k = \frac{1}{8}$.

[14] D. Laming, *Mathematical Psychology* (New York: Academic Press, Inc., 1973).

**17.** Find the area of the region in the first quadrant bounded by the curve $y = e^{-2x}$ and the $x$-axis.

**18. Economics** In discussing entrance of a firm into an industry, Stigler[15] uses the equation

$$V = \pi_0 \int_0^\infty e^{\theta t}e^{-\rho t}\,dt,$$

where $\pi_0, \theta$ (a Greek letter read "theta"), and $\rho$ (a Greek letter read "rho") are constants. Show that $V = \pi_0/(\rho - \theta)$ if $\theta < \rho$.

**19. Population** The predicted rate of growth per year of the population of a certain small city is given by

$$\frac{10{,}000}{(t + 2)^2},$$

where $t$ is the number of years from now. In the long run (that is, as $t \to \infty$), what is the expected change in population from today's level?

[15] G. Stigler, *The Theory of Price,* 3d ed. (New York: Macmillan Publishing Company, 1966), p. 344.

---

# 17.9 REVIEW

## IMPORTANT TERMS AND SYMBOLS

**Section 17.1**  integration by parts

**Section 17.2**  proper rational function  partial fractions

**Section 17.3**  present value of continuous annuity  accumulated amount of continuous annuity

**Section 17.4**  average value of function

**Section 17.5**  trapezoidal rule  Simpson's rule

**Section 17.6**  first-order differential equation  separation of variables  exponential growth  exponential decay  decay constant  half-life

**Section 17.7**  logistic function  Newton's law of cooling

**Section 17.8**  improper integral, $\int_a^\infty f(x)\,dx$, $\int_{-\infty}^b f(x)\,dx$, $\int_{-\infty}^\infty f(x)\,dx$

## SUMMARY

Sometimes we can easily determine an integral whose form is $u\,dv$, where $u$ and $v$ are functions of the same variable, by applying the formula for integration by parts:

$$\int u\,dv = uv - \int v\,du.$$

A proper rational function may be integrated by applying the technique of partial fractions. Here we express the rational function as a sum of fractions, each of which is easier to integrate than the original function was.

To determine an integral that does not have a familiar form, you may be able to match it with a formula in a table of integrals. However, it may be necessary to transform the given integral into an equivalent form before the matching can occur.

An annuity is a series of payments over a period of time. Suppose payments are made continuously for

$T$ years such that a payment at time $t$ is at the rate of $f(t)$ per year. If the annual rate of interest is $r$, compounded continuously, then the present value of the continuous annuity is given by

$$A = \int_0^T f(t)e^{-rt}\, dt,$$

and the accumulated amount is given by

$$S = \int_0^T f(t)e^{r(T-t)}\, dt.$$

The average value $\overline{f}$ of a function $f$ over the interval $[a, b]$ is given by

$$\overline{f} = \frac{1}{b-a} \int_a^b f(x)\, dx.$$

There are formulas that allow us to approximate the value of a definite integral. One formula is the trapezoidal rule:

**Trapezoidal Rule**

$$\int_a^b f(x)dx \approx \frac{h}{2}[f(a) + 2f(a+h) + 2f(a+2h) + \cdots + 2f[a + (n-1)h] + f(b)],$$

where $h = (b-a)/n$.

Another formula is Simpson's rule:

**Simpson's Rule**

$$\int_a^b f(x)dx \approx \frac{h}{3}[f(a) + 4f(a+h) + 2f(a+2h) + \cdots + 4f[a + (n-1)h] + f(b)],$$

where $h = (b-a)/n$ and $n$ is even.

An equation that involves the derivative of an unknown function is called a differential equation. If the highest order derivative that occurs is the first, the equation is called a first-order differential equation. Some first-order differential equations may be solved by the method of separation of variables. In that method, by considering the derivative to be a quotient of differentials, we rewrite the equation so that each side contains only one variable and a differential is not in a denominator. Integrating both sides of the resulting equation gives the solution. This solution involves a constant of integration and is called the general solution of the differential equation. If the unknown function must satisfy the condition that it has a specific function value for a given value of the independent variable, then a particular solution may be found.

Differential equations arise when we know a relation involving the rate of change of a function. For example, if a quantity $N$ at time $t$ is such that it changes at a rate proportional to the amount present, then

$$\frac{dN}{dt} = kN, \quad \text{where } k \text{ is a constant.}$$

The solution of this differential equation is

$$N = N_0 e^{kt},$$

where $N_0$ is the quantity present at $t = 0$. The value of $k$ may be determined when the value of $N$ is known for a given value of $t$ (other than $t = 0$). If $k$ is positive, then $N$ follows an exponential law of growth; if $k$ is negative, $N$ follows an exponential law of decay. If $N$ represents a quantity of a radioactive element, then

$$\frac{dN}{dt} = -\lambda N, \quad \text{where } \lambda \text{ is a positive constant.}$$

Thus, $N$ follows an exponential law of decay, and hence,

$$N = N_0 e^{-\lambda t}.$$

The constant $\lambda$ is called the decay constant. The time for one-half of the element to decay is the half-life of the element:

$$\text{half-life} = \frac{\ln 2}{\lambda} \approx \frac{0.69315}{\lambda}.$$

A quantity $N$ may follow a rate of growth given by

$$\frac{dN}{dt} = KN(M - N), \quad \text{where } K, M \text{ are constants.}$$

Solving this differential equation gives a function of the form

$$N = \frac{M}{1 + be^{-ct}}, \quad \text{where } b, c \text{ are constants,}$$

which is called a logistic function. Many population sizes can be described by a logistic function. In this case, $M$ represents the limit of the size of the population. A logistic function is also used in analyzing the spread of a rumor.

Newton's law of cooling states that the temperature $T$ of a cooling body at time $t$ changes at a rate proportional to the difference $T - a$, where $a$ is the ambient temperature. Thus,

$$\frac{dT}{dt} = k(T - a), \quad \text{where } k \text{ is a constant.}$$

The solution of this differential equation can be used to determine, for example, the time at which a homicide was committed.

An integral of the form

$$\int_a^\infty f(x)\, dx, \quad \int_{-\infty}^b f(x)\, dx, \quad \text{or} \quad \int_{-\infty}^\infty f(x)\, dx$$

is called an improper integral. The first two integrals are defined as follows:

$$\int_a^\infty f(x)\, dx = \lim_{r \to \infty} \int_a^r f(x)\, dx,$$

and

$$\int_{-\infty}^b f(x)\, dx = \lim_{r \to -\infty} \int_r^b f(x)\, dx.$$

If $\int_a^\infty f(x)\, dx$ (or $\int_{-\infty}^b f(x)\, dx$) is a finite number, we say that the integral is convergent; otherwise, it is divergent. The improper integral $\int_{-\infty}^\infty f(x)\, dx$ is defined by

$$\int_{-\infty}^\infty f(x)\, dx = \int_{-\infty}^0 f(x)\, dx + \int_0^\infty f(x)\, dx.$$

If both integrals on the right side are convergent, $\int_{-\infty}^\infty f(x)\, dx$ is said to be convergent; otherwise, it is divergent.

## REVIEW PROBLEMS

*In Problems 1–22, determine the integrals.*

**1.** $\displaystyle\int x \ln x \, dx.$

**2.** $\displaystyle\int \frac{1}{\sqrt{4x^2 + 1}} \, dx.$

**3.** $\displaystyle\int_0^2 \sqrt{4x^2 + 9} \, dx.$

**4.** $\displaystyle\int \frac{2x}{3 - 4x} \, dx.$

**5.** $\displaystyle\int \frac{x \, dx}{(2 + 3x)(3 + x)}.$

**6.** $\displaystyle\int_e^{e^2} \frac{1}{x \ln x} \, dx.$

**7.** $\displaystyle\int \frac{dx}{x(x + 2)^2}.$

**8.** $\displaystyle\int \frac{dx}{x^2 - 1}.$

**9.** $\displaystyle\int \frac{dx}{x^2 \sqrt{9 - 16x^2}}.$

**10.** $\displaystyle\int x^{1/3} \ln \sqrt{x} \, dx.$

**11.** $\displaystyle\int \frac{9 \, dx}{x^2 - 9}.$

**12.** $\displaystyle\int \frac{3x}{\sqrt{1 + 3x}} \, dx.$

**13.** $\displaystyle\int xe^{7x} \, dx.$

**14.** $\displaystyle\int \frac{dx}{2 + 3e^{4x}}.$

**15.** $\displaystyle\int \frac{dx}{2x \ln 2x}.$

**16.** $\displaystyle\int \frac{dx}{x(2 + x)}.$

**17.** $\displaystyle\int \frac{2x}{3 + 2x} \, dx.$

**18.** $\displaystyle\int \frac{dx}{\sqrt{4x^2 - 9}}.$

[16]**19.** $\displaystyle\int \frac{5x^2 + 2}{x^3 + x} \, dx.$

[16]**20.** $\displaystyle\int \frac{3x^3 + 5x^2 + 4x + 3}{x^4 + x^3 + x^2} \, dx.$

[17]**21.** $\displaystyle\int \frac{\ln(x + 1)}{\sqrt{x + 1}} \, dx.$

[17]**22.** $\displaystyle\int (\ln x)^3 \, dx.$

**23.** Find the average value of $f(x) = 3x^2 + 2x$ over the interval $[2, 4]$.

**24.** Find the average value of $f(t) = te^{t^2}$ over the interval $[2, 5]$.

*In Problems 25 and 26, use (a) the trapezoidal rule and (b) Simpson's rule to estimate the integral for the given value of n. Round your answer to three decimal places.*

**25.** $\displaystyle\int_0^3 \frac{1}{x + 1} \, dx, \ n = 6.$

**26.** $\displaystyle\int_0^1 \frac{1}{2 - x^2} \, dx, \ n = 4.$

*In Problems 27 and 28, solve the differential equations.*

**27.** $y' = 3x^2 y + 2xy, \quad y > 0.$

**28.** $y' - 2xe^{x^2 - y + 3} = 0, \quad y(0) = 3.$

*In Problems 29–32, determine the improper integrals if they exist.[18] Indicate those that are divergent.*

**29.** $\displaystyle\int_3^\infty \frac{1}{x^3} \, dx.$

**30.** $\displaystyle\int_{-\infty}^0 e^{3x} \, dx.$

**31.** $\displaystyle\int_1^\infty \frac{1}{2x} \, dx.$

**32.** $\displaystyle\int_{-\infty}^\infty xe^{1 - x^2} \, dx.$

---

[16]Refers to Sec. 17.2.

[17]Refers to Sec. 17.1.

[18]Refers to Sec. 17.8.

**33. Population** The population of a city in 1975 was 100,000 and in 1990 was 120,000. Assuming exponential growth, project the population in 2005.

**34. Population** The population of a city doubles every 10 years due to exponential growth. At a certain time, the population is 10,000. Find an expression for the number of people $N$ at time $t$ years later. Give your answer in terms of ln 2.

**35. Radioactivity** If 95% of a radioactive substance remains after 100 years, find the decay constant, and, to the nearest percent, give the percentage of the original amount present after 200 years.

**36. Medicine** Suppose $q$ is the amount of penicillin in the body at time $t$, and let $q_0$ be the amount at $t = 0$. Assume that the rate of change of $q$ with respect to $t$ is proportional to $q$ and that $q$ decreases as increases. Then we have $dq/dt = -kq$, where $k > 0$. Solve for $q$. What percentage of the original amount present is there when $t = 2/k$?

**37. Biology** Two organisms are initially placed in a medium and begin to multiply. The number $N$ of organisms that are present after $t$ days is recorded on a graph with the horizontal axis labeled $t$ and the vertical axis labeled $N$. It is observed that the points lie on a logistic curve. The number of organisms present after 6 days is 300, and beyond 10 days the number approaches a limit of 450. Find the logistic equation.

**38. College Enrollment** A college believes that enrollment follows logistic growth. Last year enrollment was 1000, and this year it is 1100. If the college can accommodate a maximum of 2000 students, what is the anticipated enrollment next year? Give your answer to the nearest hundred.

**39. Time of Murder** A coroner is called in on a murder case. He arrives at 6:00 P.M. and finds that the victim's temperature is 35°C. One hour later the body temperature is 34°C. The temperature of the room is 25°C. About what time was the murder committed? (Assume that normal body temperature is 37°C.)

**40. Annuity** Find the present value, to the nearest dollar, of a continuous annuity at an annual rate of 5% for 10 years if the payment at time $t$ is at the annual rate of $f(t) = 40t$ dollars.

[19]**41. Hospital Discharges** For a group of hospitalized individuals, suppose the proportion that has been discharged at the end of days $t$ is given by

$$\int_0^t f(x)\, dx,$$

where $f(x) = 0.008e^{-0.01x} + 0.00004e^{-0.0002x}$. Evaluate

$$\int_0^\infty f(x)\, dx.$$

**42. Product Consumption** Suppose that $A(t)$ is the amount of a product that is consumed at time $t$ and that $A$ follows an exponential law of growth. If $t_1 < t_2$ and at time $t_2$ the amount consumed, $A(t_2)$, is double the amount consumed at time $t_1$, $A(t_1)$, then $t_2 - t_1$ is called a doubling period. In a discussion of exponential growth, Shonle[20] states that under exponential growth, "the amount of a product consumed during one doubling period is equal to the total used for all time up to the beginning of the doubling period in question." To justify this statement, reproduce his argument as follows. The amount of the product used up to time $t_1$ is given by

$$\int_{-\infty}^{t_1} A_0 e^{kt}\, dt, \quad k > 0,$$

where $A_0$ is the amount when $t = 0$. Show that this is equal to $(A_0/k)e^{kt_1}$. Next, the amount used during the time interval from $t_1$ to $t_2$ is

$$\int_{t_1}^{t_2} A_0 e^{kt}\, dt.$$

Show that this is equal to

$$\frac{A_0}{k} e^{kt_1}[e^{k(t_2 - t_1)} - 1]. \tag{1}$$

If the interval $[t_1, t_2]$ is a doubling period, then

$$A_0 e^{kt_2} = 2A_0 e^{kt_1}.$$

Show that this relationship implies that $e^{k(t_2 - t_1)} = 2$. Substitute this value into Eq. (1); your result should be the same as the total used during all time up to $t_1$, namely, $(A_0/k)e^{kt_1}$.

**43. Revenue, Cost, and Profit** The following table gives values of a company's marginal-revenue (MR) and marginal-cost (MC) functions:

| $q$ | 0 | 3 | 6 | 9 | 12 | 15 | 18 |
|-----|----|----|----|----|----|----|----|
| MR | 25 | 22 | 18 | 13 | 7 | 3 | 0 |
| MC | 15 | 14 | 12 | 10 | 7 | 4 | 2 |

The company's fixed cost is 25. Assume that profit is a maximum when MR = MC and that this occurs when $q = 12$. Moreover, assume that the output of the company is chosen to maximize the profit. Use the trapezoidal rule and Simpson's rule for each of the following parts.

**a.** Estimate the total revenue by using as many data as possible.

**b.** Estimate the total cost by using as few data as possible.

**c.** Determine how the maximum profit is related to the area enclosed by the line $q = 0$ and the MR and MC curves, and use this relation to estimate the maximum profit as accurately as possible.

---

[19]Refers to Sec. 17.8.

[20]J. I. Shonle, *Environmental Applications of General Physics* (Reading, MA: Addison-Wesley Publishing Company, Inc., 1975).

# MATHEMATICAL *SNAPSHOT*

## DIETING

Today there is great concern about diet and weight loss. Some people want to lose weight in order to "look good." Others lose weight for physical fitness or for health reasons. In fact, some lose weight because of peer pressure. Advertisements for weight control programs frequently appear on television and in newspapers and magazines. In many bookstores, entire sections are devoted to diet and weight control.

Suppose you want to determine a mathematical model of the weight of a person on a restricted caloric diet.[21] A person's weight depends both on the daily rate of energy intake, say $C$ calories per day, and on the daily rate of energy consumption, which is typically between 15 and 20 calories per day for each pound of body weight. Consumption depends on age, sex, metabolic rate, and so on. For an average value of 17.5 calories per pound per day, a person weighing $w$ pounds expends $17.5w$ calories per day. If $C = 17.5w$, then his or her weight remains constant; otherwise weight gain or loss occurs according to whether $C$ is greater or less than $17.5w$.

How fast will weight gain or loss occur? The most plausible physiological assumption is that $dw/dt$ is proportional to the net excess (or deficit) $C - 17.5w$ in the number of calories per day. That is,

$$\frac{dw}{dt} = K(C - 17.5w),$$

where $K$ is a constant. The left side of the equation has units of pounds per day, and $C - 17.5w$ has units of calories per day. Hence the units of $K$ are pounds per calorie. Therefore you need to know how many pounds each excess or deficit calorie puts on or takes off. The commonly used dietetic conversion factor is that 3500 calories is equivalent to one pound. Thus $K = 1/3500$ lb per calorie.

Now, the differential equation modeling weight gain or loss is

$$\frac{dw}{dt} = \frac{1}{3500}(C - 17.5w). \qquad (1)$$

If $C$ is constant, the equation is separable and its solution is

$$w(t) = \frac{C}{17.5} + \left(w_0 - \frac{C}{17.5}\right)e^{-0.005t}, \qquad (2)$$

where $w_0$ is the initial weight and $t$ is in days. In the long run, note that the equilibrium weight (that is, the weight as $t \to \infty$) is $w_{eq} = C/17.5$.

For example, if someone initially weighing 180 lb adopts a diet of 2500 calories per day, then we have $w_{eq} = 2500/17.5 \approx 143$ lb and the weight function is

$$w(t) \approx 143 + (180 - 143)e^{-0.005t}$$
$$= 143 + 37e^{-0.005t}.$$

Figure 17.9 shows the graph of $w(t)$. Notice how long it takes to even get close to the equilibrium weight of 143 lb. The half-life for the process is $(\ln 2)/0.005 \approx 138.6$ days, about 20 weeks. (It would take about 584 days, or 83 weeks, to get to 145 lb.) This may be why so many dieters give up in frustration.

[21]Adapted from A. C. Segal, "A Linear Diet Model," *The College Mathematics Journal,* 18, no. 1 (1987), 44–5. By permission of the Mathematical Association of America.

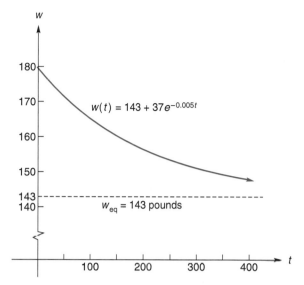

**FIGURE 17.9** Weight as a function of time.

In the figure: $w(t) = 143 + 37e^{-0.005t}$ and $w_{eq} = 143$ pounds.

### ▪ Exercises

**1.** If a person weighing 200 lb adopts a 2000 calorie-per-day diet, determine, to the nearest pound, the equilibrium weight $w_{eq}$. To the nearest day, after how many days will the person reach a weight of 175 lb?

**2.** Show that the solution of Eq. (1) is given by Eq. (2).

**3.** The weight of a person on a restricted caloric diet at time $t$ is given by $w(t)$. [See Eq. (2).] The difference between this weight and the equilibrium weight $w_{eq}$ is $w(t) - w_{eq}$. Suppose it takes $d$ days for the person to lose half of the weight difference. Then

$$w(t + d) = w(t) - \tfrac{1}{2}[w(t) - w_{eq}].$$

By solving this equation for $d$, show that $d = \dfrac{\ln 2}{0.005}$.

# CHAPTER 18

# Continuous Random Variables

## OBJECTIVE

To introduce continuous random variables; to discuss density functions, including uniform and exponential distributions; to discuss cumulative distribution functions; and to compute the mean, variance, and standard deviation for a continuous random variable.

## 18.1 CONTINUOUS RANDOM VARIABLES

### Density Functions

In Chapter 10, the random variables that we considered were primarily discrete. Now we shall concern ourselves with **continuous random variables**. A random variable is continuous if it can assume any value in some interval or intervals. A continuous random variable usually represents data that are measured, such as heights, weights, distances, and periods of time.

For example, the number of hours of life of a calculator battery is a continuous random variable $X$. If the maximum possible life is 1000 hours, then $X$ can assume any value in the interval $[0, 1000]$. In a practical sense, the likelihood that $X$ will assume a single specified value, such as 764.1238, is extremely remote. It is more meaningful to consider the likelihood of $X$ lying within an *interval,* such as that between 764 and 765. (Thus, $764 < X < 765$.) In general, *with a continuous random variable, our concern is the likelihood that it falls within an interval and not that it assumes a particular value.*

As another example, consider an experiment in which a number $X$ is randomly selected from the interval $[0, 2]$. Then $X$ is a continuous random variable. What is the probability that $X$ lies in the interval $[0, 1]$? Because we can loosely think of $[0, 1]$ as being "half" the interval $[0, 2]$, a reasonable (and correct) answer is $\frac{1}{2}$. Similarly, if we think of the interval $[0, \frac{1}{2}]$ as being one-fourth of $[0, 2]$, then $P(0 \le X \le \frac{1}{2}) = \frac{1}{4}$. Actually, each one of these probabilities is simply the length of the given interval divided by the length of $[0, 2]$. For example,

$$P\left(0 \le X \le \frac{1}{2}\right) = \frac{\text{length of } [0, \frac{1}{2}]}{\text{length of } [0, 2]} = \frac{\frac{1}{2}}{2} = \frac{1}{4}.$$

Let us now consider a similar experiment in which $X$ denotes a number chosen at random from the interval $[0, 1]$. As you might expect, the probability that $X$ will assume a value in any given interval within $[0, 1]$ is equal to the length of the given interval divided by the length of $[0, 1]$. Because $[0, 1]$ has length 1, we can simply say that the probability of $X$ falling in an interval is the length of the interval. For example,

$$P(0.2 \le X \le 0.5) = 0.5 - 0.2 = 0.3,$$

and $P(0.2 \leq X \leq 0.2001) = 0.0001$. Clearly, as the length of an interval approaches 0, the probability that $X$ assumes a value in that interval approaches 0. Keeping this in mind, we can think of a single number such as 0.2 as the limiting case of an interval as the length of the interval approaches 0. (Think of $[0.2, 0.2 + x]$ as $x \to 0$.) Thus, $P(X = 0.2) = 0$. In general, *the probability that a continuous random variable $X$ assumes a particular value is 0.* As a result, **the probability that $X$ lies in some interval is not affected by whether or not either of the endpoints of the interval is included or excluded.** For example,

$$
\begin{aligned}
P(X \leq 0.4) &= P(X < 0.4) + P(X = 0.4) \\
&= P(X < 0.4) + 0 \\
&= P(X < 0.4).
\end{aligned}
$$

Similarly, $P(0.2 \leq X \leq 0.5) = P(0.2 < X < 0.5)$.

We can geometrically represent the probabilities associated with a continuous random variable $X$. This is done by means of the graph of a function $y = f(x)$ such that the area under this graph (and above the $x$-axis) between the lines $x = a$ and $x = b$ represents the probability that $X$ assumes a value between $a$ and $b$. (See Fig. 18.1.) Since this area is given by the definite integral $\int_a^b f(x)\,dx$, we have

$$
P(a \leq X \leq b) = \int_a^b f(x)\,dx.
$$

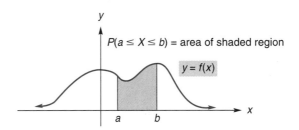

**FIGURE 18.1** Probability density function.

We call the function $f$ the *probability density function* for $X$ (or simply the *density function* for $X$) and say that it defines the *distribution* of $X$. Because probabilities are always nonnegative, it is always true that $f(x) \geq 0$. Also, because the event $-\infty < X < \infty$ must occur, the total area under the density function curve must be 1. That is, $\int_{-\infty}^{\infty} f(x)\,dx = 1$. In summary, we have the following definition.

**DEFINITION**

*If $X$ is a continuous random variable, then a function $y = f(x)$ is called a **(probability) density function** for $X$ if and only if it has the following properties:*

**1.** $f(x) \geq 0$.

**2.** $\int_{-\infty}^{\infty} f(x)\,dx = 1$.

**3.** $P(a \leq X \leq b) = \int_a^b f(x)\,dx$.

To illustrate a density function, we return to the previous experiment in which a number $X$ is chosen at random from the interval $[0, 1]$. Recall that

$$P(a \le X \le b) = \text{length of } [a, b] = b - a, \tag{1}$$

where $a$ and $b$ are in $[0, 1]$. We shall show that the function

$$f(x) = \begin{cases} 1, & \text{if } 0 \le x \le 1, \\ 0, & \text{otherwise,} \end{cases} \tag{2}$$

whose graph appears in Fig. 18.2(a), is a density function for $X$. To do this, we

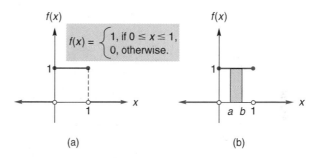

**FIGURE 18.2**   Probability density function.

must verify that $f(x)$ satisfies the three conditions stated in the definition of a density function. First, $f(x)$ is either 0 or 1, so $f(x) \ge 0$. Next, since $f(x) = 0$ for $x$ outside $[0, 1]$,

$$\int_{-\infty}^{\infty} f(x) \, dx = \int_0^1 1 \, dx = x \Big|_0^1 = 1.$$

Finally, to verify that $P(a \le X \le b) = \int_a^b f(x) \, dx$, we compute the area under the graph between $x = a$ and $x = b$ [Fig. 18.2(b)]. We have

$$\int_a^b f(x) \, dx = \int_a^b 1 \, dx = x \Big|_a^b = b - a,$$

which, as stated in Eq. (1), is $P(a \le X \le b)$.

The function in Eq. (2) is called the **uniform density function** over $[0, 1]$, and $X$ is said to have a **uniform distribution.** The word *uniform* is meaningful in the sense that the graph of the density function is horizontal, or "flat," over $[0, 1]$. As a result, $X$ is just as likely to assume a value in one interval within $[0, 1]$ as in another of equal length. A more general uniform distribution is given in Example 1.

### EXAMPLE 1   Uniform Density Function

The uniform density function over $[a, b]$ for the random variable $X$ is given by

$$f(x) = \begin{cases} \dfrac{1}{b - a}, & \text{if } a \le x \le b, \\ 0, & \text{otherwise.} \end{cases}$$

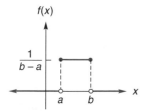

**FIGURE 18.3**   Uniform density function over $[a, b]$.

(See Fig. 18.3.) Note that over $[a, b]$, the region under the graph is a rectangle with height $1/(b - a)$ and width $b - a$. Thus, its area is given by $[1/(b - a)][b - a] = 1$ so $\int_{-\infty}^{\infty} f(x) \, dx = 1$, as must be the case for a density function. If $[c, d]$ is any interval within $[a, b]$, then

$$P(c \le X \le d) = \int_c^d f(x)\, dx = \int_c^d \frac{1}{b-a}\, dx$$

$$= \frac{x}{b-a}\bigg|_c^d = \frac{d-c}{b-a}.$$

For example, suppose $X$ is uniformly distributed over the interval $[1, 4]$ and we need to find $P(2 < X < 3)$. Then $a = 1, b = 4, c = 2,$ and $d = 3$. Therefore,

$$P(2 < X < 3) = \frac{3-2}{4-1} = \frac{1}{3}.$$

■

## Principles in Practice 1

### Density Function

Suppose the time (in minutes) passengers must wait for an airplane is uniformly distributed with density function $f(x) = \frac{1}{60}$, where $0 \le x \le 60$, and $f(x) = 0$ elsewhere. What is the probability that a passenger must wait between 25 and 45 minutes?

## EXAMPLE 2   Density Function

*The density function for a random variable X is given by*

$$f(x) = \begin{cases} kx, & \text{if } 0 \le x \le 2, \\ 0, & \text{otherwise}, \end{cases}$$

*where k is a constant.*

**a.** *Find k.*

*Solution:* Since $\int_{-\infty}^{\infty} f(x)\, dx$ must be 1 and $f(x) = 0$ outside $[0, 2]$, we have

$$\int_{-\infty}^{\infty} f(x)\, dx = \int_0^2 kx\, dx = \frac{kx^2}{2}\bigg|_0^2 = 2k = 1.$$

Thus, $k = \frac{1}{2}$, so $f(x) = \frac{1}{2}x$ on $[0, 2]$.

**b.** *Find $P\left(\frac{1}{2} < X < 1\right)$.*

*Solution:*

$$P\left(\frac{1}{2} < X < 1\right) = \int_{1/2}^1 \frac{1}{2} x\, dx = \frac{x^2}{4}\bigg|_{1/2}^1 = \frac{1}{4} - \frac{1}{16} = \frac{3}{16}.$$

**c.** *Find $P(X < 1)$.*

*Solution:* Since $f(x) = 0$ for $x < 0$, we need only compute the area under the density function between 0 and 1. Thus,

$$P(x < 1) = \int_0^1 \frac{1}{2} x\, dx = \frac{x^2}{4}\bigg|_0^1 = \frac{1}{4}.$$

■

## EXAMPLE 3   Exponential Density Function

*The **exponential density function** is defined by*

$$f(x) = \begin{cases} ke^{-kx}, & \text{if } x \ge 0, \\ 0, & \text{if } x < 0, \end{cases}$$

*where k is a positive constant, called a **parameter,** whose value depends on the experiment under consideration. If X is a random variable with this density function, then X is said to have an **exponential distribution.** Let $k = 1$. Then $f(x) = e^{-x}$ for $x \ge 0$, and $f(x) = 0$ for $x < 0$ (Fig. 18.4).*

**a.** *Find $P(2 < X < 3)$.*

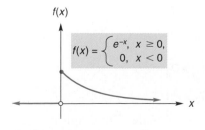

$f(x)$

$$f(x) = \begin{cases} e^{-x}, & x \ge 0, \\ 0, & x < 0 \end{cases}$$

**FIGURE 18.4**   Exponential density function.

*Solution:*

$$P(2 < X < 3) = \int_2^3 e^{-x}\, dx = -e^{-x}\Big|_2^3$$
$$= -e^{-3} - (-e^{-2}) = e^{-2} - e^{-3} \approx 0.086.$$

**b.** *Find* $P(X > 4)$.

*Solution:*

$$P(X > 4) = \int_4^\infty e^{-x}\, dx = \lim_{r\to\infty} \int_4^r e^{-x}\, dx$$

$$= \lim_{r\to\infty} -e^{-x}\Big|_4^r = \lim_{r\to\infty} (-e^{-r} + e^{-4})$$

$$= \lim_{r\to\infty} \left(-\frac{1}{e^r} + e^{-4}\right) = 0 + e^{-4}$$

$$\approx 0.018.$$

Alternatively, we can avoid an improper integral because

$$P(X > 4) = 1 - P(X \le 4) = 1 - \int_0^4 e^{-x}\, dx. \qquad \blacksquare$$

The **cumulative distribution function** $F$ for the continuous random variable $X$ with density function $f$ is defined by

$$F(x) = P(X \le x) = \int_{-\infty}^x f(t)\, dt.$$

For example, $F(2)$ represents the entire area under the density curve that is to the left of the line $x = 2$ (Fig. 18.5). Under certain conditions of continuity, it can be shown that

$$F'(x) = f(x).$$

That is, the derivative of the cumulative distribution function is the density function. Thus, $F$ is an antiderivative of $f$, and by the Fundamental Theorem of Integral Calculus,

$$P(a < X < b) = \int_a^b f(x)\, dx = F(b) - F(a). \qquad (3)$$

This means that the area under the density curve between $a$ and $b$ (Fig. 18.6) is simply the area to the left of $b$ minus the area to the left of $a$.

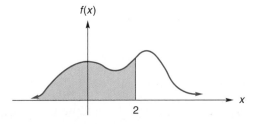

**FIGURE 18.5** $F(2) = P(X \le 2) = $ area of shaded region.

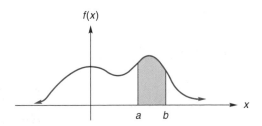

**FIGURE 18.6** $P(a < X < b)$.

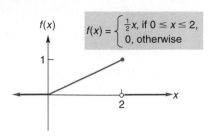

$$f(x) = \begin{cases} \frac{1}{2}x, & \text{if } 0 \le x \le 2, \\ 0, & \text{otherwise} \end{cases}$$

**FIGURE 18.7** Density function for Example 4.

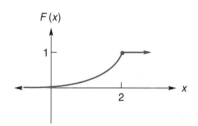

**FIGURE 18.8** Cumulative distribution function for Example 4.

## EXAMPLE 4   Finding and Applying the Cumulative Distribution Function

*Suppose $X$ is a random variable with density function given by*

$$f(x) = \begin{cases} \frac{1}{2}x, & \text{if } 0 \le x \le 2, \\ 0, & \text{otherwise,} \end{cases}$$

*as shown in Fig.* 18.7.

**a.** *Find and sketch the cumulative distribution function.*

*Solution:* Because $f(x) = 0$ if $x < 0$, the area under the density curve to the left of $x = 0$ is 0. Hence, $F(x) = 0$ if $x < 0$. If $0 \le x \le 2$, then

$$F(x) = \int_{-\infty}^{x} f(t)\, dt = \int_{0}^{x} \frac{1}{2} t\, dt = \frac{t^2}{4}\Big|_{0}^{x} = \frac{x^2}{4}.$$

Since $f$ is a density function and $f(x) = 0$ for $x < 0$ and also for $x > 2$, the area under the density curve from $x = 0$ to $x = 2$ is 1. Thus, if $x > 2$, the area to the left of $x$ is 1, so $F(x) = 1$. Hence, the cumulative distribution function is

$$F(x) = \begin{cases} 0, & \text{if } x < 0, \\ \dfrac{x^2}{4}, & \text{if } 0 \le x \le 2, \\ 1, & \text{if } x > 2, \end{cases}$$

which is shown in Fig. 18.8.

**b.** *Find $P(X < 1)$ and $P(1 < X < 1.1)$.*

*Solution:* Using the results of part (a), we have

$$P(X < 1) = F(1) = \frac{1^2}{4} = \frac{1}{4}.$$

From Eq. (3),

$$P(1 < X < 1.1) = F(1.1) - F(1) = \frac{1.1^2}{4} - \frac{1}{4} = 0.0525. \qquad \blacksquare$$

### Mean, Variance, and Standard Deviation

For a random variable $X$ with density function $f$, the **mean** $\mu$ [or **expectation** $E(x)$] is given by

$$\mu = E(X) = \int_{-\infty}^{\infty} x f(x)\, dx$$

and can be thought of as the average value of $X$ in the long run. The **variance** $\sigma^2$ [or Var$(X)$] is given by

$$\sigma^2 = \text{Var}(X) = \int_{-\infty}^{\infty} (x - \mu)^2 f(x)\, dx.$$

You may have noticed that these formulas are similar to the corresponding ones in Chapter 10 for a discrete random variable. It can be shown that an alternative formula for the variance is

$$\sigma^2 = \text{Var}(X) = \int_{-\infty}^{\infty} x^2 f(x)\, dx - \mu^2.$$

The **standard deviation** is

$$\sigma = \sqrt{\operatorname{Var}(X)}.$$

For example, it can be shown that if $X$ is exponentially distributed (see Example 3), $\mu = 1/k$ and $\sigma = 1/k$. As with a discrete random variable, the standard deviation of a continuous random variable $X$ is small if $X$ is likely to assume values close to the mean but unlikely to assume values far from the mean. The standard deviation is large if the reverse is true.

**Principles in Practice 3**

**Finding the Mean and Standard Deviation**

The life expectancy (in years) of patients after they have contracted a certain disease is exponentially distributed with $k = 0.2$. Use the information in the paragraph above Example 5 to find the mean life expectancy and the standard deviation.

**EXAMPLE 5 Finding the Mean and Standard Deviation**

*If $X$ is a random variable with density function given by*

$$f(x) = \begin{cases} \frac{1}{2}x, & \text{if } 0 \le x \le 2, \\ 0, & \text{otherwise,} \end{cases}$$

*find its mean and standard deviation.*

**Solution:** The mean is simply given by

$$\mu = \int_{-\infty}^{\infty} xf(x)\,dx = \int_0^2 x \cdot \frac{1}{2}x\,dx = \left.\frac{x^3}{6}\right|_0^2 = \frac{4}{3}.$$

By the alternative formula for variance, we have

$$\sigma^2 = \int_{-\infty}^{\infty} x^2 f(x)\,dx - \mu^2 = \int_0^2 x^2 \cdot \frac{1}{2}x\,dx - \left(\frac{4}{3}\right)^2$$

$$= \left.\frac{x^4}{8}\right|_0^2 - \frac{16}{9} = 2 - \frac{16}{9} = \frac{2}{9}.$$

Thus, the standard deviation is

$$\sigma = \sqrt{\frac{2}{9}} = \frac{\sqrt{2}}{3}. \qquad \blacksquare$$

We conclude this section by emphasizing that a density function for a continuous random variable must not be confused with a probability distribution function for a discrete random variable. Evaluating such a probability distribution function at a *point* gives a probability. But evaluating a density function at a point does not. Instead, the *area* under the density function curve over an *interval* is interpreted as a probability. That is, probabilities associated with a continuous random variable are given by integrals.

## ■ Exercise 18.1

**1.** Suppose $X$ is a continuous random variable with density function given by

$$f(x) = \begin{cases} \frac{1}{6}(x + 1), & \text{if } 1 < x < 3, \\ 0, & \text{otherwise.} \end{cases}$$

  **a.** Find $P(1 < X < 2)$.

  **b.** Find $P(X < 2.5)$.

  **c.** Find $P(X \ge \frac{3}{2})$.

  **d.** Find $c$ such that $P(X < c) = \frac{1}{2}$. Give your answer in radical form.

**2.** Suppose $X$ is a continuous random variable with density function given by

$$f(x) = \begin{cases} \dfrac{1000}{x^2}, & \text{if } x > 1000, \\ 0, & \text{otherwise.} \end{cases}$$

  **a.** Find $P(3000 < X < 4000)$.

  **b.** Find $P(X > 2000)$.

3. Suppose $X$ is a continuous random variable that is uniformly distributed on $[1, 4]$.

   **a.** What is the formula of the density function for $X$? Sketch its graph.

   **b.** Find $P(2 < X < 3)$.

   **c.** Find $P(0 < X < 1)$.

   **d.** Find $P(X \leq 3.5)$.

   **e.** Find $P(X > 2)$.

   **f.** Find $P(X = 3)$.

   **g.** Find $P(X < 5)$.

   **h.** Find $\mu$.

   **i.** Find $\sigma$.

   **j.** Find the cumulative distribution function $F$ and sketch its graph. Use $F$ to find $P(X < 2)$ and $P(1 < X < 3)$.

4. Suppose $X$ is a continuous random variable that is uniformly distributed on $[0, 5]$.

   **a.** What is the formula of the density function for $X$? Sketch its graph.

   **b.** Find $P(1 < X < 3)$.

   **c.** Find $P(4.5 \leq X < 5)$.

   **d.** Find $P(X = 4)$.

   **e.** Find $P(X > 1)$.

   **f.** Find $P(X < 5)$.

   **g.** Find $P(X > 5)$.

   **h.** Find $\mu$.

   **i.** Find $\sigma$.

   **j.** Find the cumulative distribution function $F$ and sketch its graph. Use $F$ to find $P(1 < X < 3.5)$.

5. Suppose $X$ is uniformly distributed on $[a, b]$.

   **a.** What is the density function for $X$?

   **b.** Find $\mu$.

   **c.** Find $\sigma^2$ and $\sigma$.

6. Suppose $X$ is a continuous random variable with density function given by

   $$f(x) = \begin{cases} k, & \text{if } a \leq x \leq b, \\ 0, & \text{otherwise.} \end{cases}$$

   **a.** Show that $k = \dfrac{1}{b - a}$ and thus $X$ is uniformly distributed.

   **b.** Find the cumulative distribution function $F$.

7. Suppose the random variable $X$ is exponentially distributed with $k = 2$.

   **a.** Find $P(1 < X < 3)$.

   **b.** Find $P(X < 2)$.

   **c.** Find $P(X > 2.5)$.

   **d.** Find $P(\mu - \sigma < X < \mu + \sigma)$.

   **e.** Show that the area under the density function is unity.

8. Suppose the random variable $X$ is exponentially distributed with $k = 0.5$.

   **a.** Find $P(X > 4)$.

   **b.** Find $P(0.5 < X < 2.6)$.

   **c.** Find $P(X < 5)$.

   **d.** Find $P(X = 4)$.

   **e.** Find $c$ such that $P(0 < X < c) = \frac{1}{2}$.

9. The density function for a random variable $X$ is given by

   $$f(x) = \begin{cases} kx, & \text{if } 0 \leq x \leq 4, \\ 0, & \text{otherwise.} \end{cases}$$

   **a.** Find $k$.

   **b.** Find $P(2 < X < 3)$.

   **c.** Find $P(X > 2.5)$.

   **d.** Find $P(X > 0)$.

   **e.** Find $\mu$.

   **f.** Find $\sigma$.

   **g.** Find $c$ such that $P(X < c) = \frac{1}{2}$.

   **h.** Find $P(3 < X < 5)$.

10. The density function for a random variable $X$ is given by

    $$f(x) = \begin{cases} \frac{1}{2}x + k, & \text{if } 2 \leq x \leq 4, \\ 0, & \text{otherwise.} \end{cases}$$

    **a.** Find $k$.

    **b.** Find $P(X \geq 2.5)$.

    **c.** Find $\mu$.

    **d.** Find $P(2 < X < \mu)$.

11. **Waiting Time** At a bus stop, the time $X$ (in minutes) that a randomly arriving person must wait for a bus is uniformly distributed with density function $f(x) = \frac{1}{10}$, where $0 \leq x \leq 10$ and $f(x) = 0$ otherwise. What is the probability that such a person must wait at most seven minutes? What is the average time that a person must wait?

12. **Soft-drink Dispensing** An automatic soft-drink dispenser at a fast-food restaurant dispenses $X$ ounces of cola in a 12-ounce drink. If $X$ is uniformly distributed over $[11.92, 12.08]$, what is the probability that less than 12 ounces will be dispensed? What is the probability that exactly 12 ounces will be dispensed? What is the average amount dispensed?

13. **Emergency Room Arrivals** At a particular hospital, the length of time $X$ (in hours) between successive arrivals at the emergency room is exponentially distributed with $k = 3$. What is the probability that more than one hour passes without an arrival?

14. **Electronic Component Life** The length of life $X$ (in years) of an electronic component has an exponential distribution with $k = \frac{1}{5}$. What is the probability that such a component will fail within four years of use? What is the probability that it will last more than six years?

To discuss the normal distribution, standard units, and the table of areas under the standard normal curve (Appendix D).

## 18.2 THE NORMAL DISTRIBUTION

Quite often, measured data in nature— such as heights of individuals in a population—are represented by a random variable whose density function may be approximated by the bell-shaped curve in Fig. 18.9. The curve extends

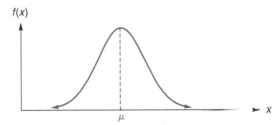

**FIGURE 18.9** Normal curve.

indefinitely to the right and left and never touches the $x$-axis. This curve, called the **normal curve,** is the graph of the most important of all density functions, the *normal density function.*

### DEFINITION

*A continuous random variable $X$ is a **normal random variable** or has a **normal** (or Gaussian)[1] **distribution** if its density function is given by*

$$f(x) = \frac{1}{\sigma\sqrt{2\pi}}\, e^{-(1/2)[(x-\mu)/\sigma]^2}, \qquad -\infty < x < \infty,$$

*called the **normal density function.** The parameters $\mu$ and $\sigma$ are the mean and standard deviation of X, respectively.*

Observe in Fig. 18.9 that $f(x) \to 0$ as $x \to \pm\infty$. That is, the normal curve has the $x$-axis as a horizontal asymptote. Also note that the normal curve is symmetric about the vertical line $x = \mu$. That is, the height of a point on the curve $d$ units to the right of $x = \mu$ is the same as the height of the point on the curve that is $d$ units to the left of $x = \mu$. Because of this symmetry and the fact that the area under the normal curve is 1, the area to the right (or left) of the mean must be $\frac{1}{2}$.

Each choice of values for $\mu$ and $\sigma$ determines a different normal curve. The value of $\mu$ determines where the curve is "centered," and $\sigma$ determines how "spread out" the curve is. The smaller the value of $\sigma$, the less spread out is the area near $\mu$. For example, Fig. 18.10 shows normal curves $C_1$, $C_2$,

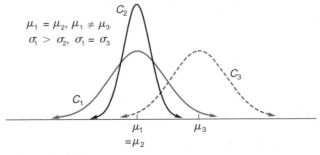

$\mu_1 = \mu_2,\ \mu_1 \neq \mu_3$
$\sigma_1 > \sigma_2,\ \sigma_1 = \sigma_3$

**FIGURE 18.10** Normal curves.

[1]After the German mathematician Carl Friedrich Gauss (1777–1855).

and $C_3$, where $C_1$ has mean $\mu_1$ and standard deviation $\sigma_1$, $C_2$ has mean $\mu_2$, etc. Here $C_1$ and $C_2$ have the same mean but different standard deviations: $\sigma_1 > \sigma_2$. $C_1$ and $C_3$ have the same standard deviation but different means: $\mu_1 < \mu_3$. Curves $C_2$ and $C_3$ have different means and different standard deviations.

The standard deviation plays a significant role in describing probabilities associated with a normal random variable $X$. More precisely, the probability that $X$ will lie within one standard deviation of the mean is approximately 0.68:

$$P(\mu - \sigma < X < \mu + \sigma) = 0.68.$$

In other words, approximately 68% of the area under a normal curve is within one standard deviation of the mean (Fig. 18.11). Between $\mu \pm 2\sigma$ is about

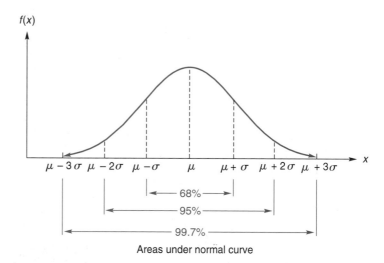

Areas under normal curve

**FIGURE 18.11** Probability and number of standard deviations from $\mu$.

You are encouraged to become familiar with the percentages in Fig. 18.11.

95% of the area, and between $\mu \pm 3\sigma$ is about 99.7%:

$$P(\mu - 2\sigma < X < \mu + 2\sigma) = 0.95,$$
$$P(\mu - 3\sigma < X < \mu + 3\sigma) = 0.997.$$

Thus, it is highly likely that $X$ will lie within three standard deviations of the mean.

**EXAMPLE 1   Analysis of Test Scores**
Let $X$ be a random variable whose values are the scores obtained on a nationwide test given to high school seniors. Suppose, for modeling purposes, that $X$ is normally distributed with mean 600 and standard deviation 90. Then the probability that $X$ lies within $2\sigma = 2(90) = 180$ points of 600 is 0.95. In other words, 95% of the scores lie between 420 and 780. Similarly, 99.7% of the scores are within $3\sigma = 3(90) = 270$ points of 600—that is, between 330 and 870. ∎

If $Z$ is a normally distributed random variable with $\mu = 0$ and $\sigma = 1$, we obtain the normal curve of Fig. 18.12, called the **standard normal curve.**

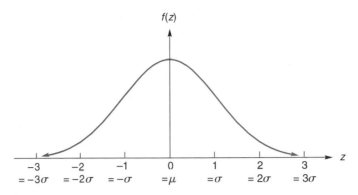

**FIGURE 18.12**   Standard normal curve: $\mu = 0, \sigma = 1$.

## DEFINITION

*A continuous random variable $Z$ is a **standard normal random variable** (or has a **standard normal distribution**) if its density function is given by*

$$f(x) = \frac{1}{\sqrt{2\pi}}e^{-z^2/2},$$

*called the **standard normal density function.** The variable $Z$ has mean $0$ and standard deviation $1$.*

Because a standard normal random variable $Z$ has mean 0 and standard deviation 1, its values are in units of standard deviations from the mean, which are called **standard units.** For example, if $0 < Z < 2.54$, then $Z$ lies within 2.54 standard deviations to the right of 0, the mean. That is, $0 < Z < 2.54\sigma$. To find the probability $P(0 < Z < 2.54)$, we have

$$P(0 < Z < 2.54) = \int_0^{2.54} \frac{e^{-z^2/2}}{\sqrt{2\pi}}\, dz.$$

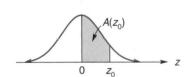

**FIGURE 18.13**
$A(z_0) = P(0 < Z < z_0)$.

The integral on the right cannot be evaluated by elementary methods. However, values for integrals of this kind have been computed and put in tabular form.

One such table is given in Appendix D. The table there gives the area under a standard normal curve between $z = 0$ and $z = z_0$, where $z_0 \geq 0$. This area is shaded in Fig. 18.13 and is denoted by $A(z_0)$. In the left-hand columns of the table are $z$-values to the nearest tenth. The numbers across the top are the hundredths' values. For example, the entry in the row for 2.5 and column under 0.04 corresponds to $z = 2.54$ and is 0.4945. Thus, the area under a standard normal curve between $z = 0$ and $z = 2.54$ is 0.4945:

$$P(0 < Z < 2.54) = A(2.54) = 0.4945.$$

Similarly, you should verify that $A(2) = 0.4772$ and $A(0.33) = 0.1293$.

Using symmetry, we compute an area to the left of $z = 0$ by computing the corresponding area to the right of $z = 0$. For example,

$$P(-z_0 < Z < 0) = P(0 < Z < z_0) = A(z_0),$$

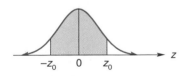

**FIGURE 18.14**
$P(-z_0 < Z < 0) =$
$P(0 < Z < z_0)$.

as shown in Fig. 18.14. Hence, $P(-2.54 < Z < 0) = A(2.54) = 0.4945$.

When computing probabilities for a standard normal variable, you may have to add or subtract areas. A useful aid for doing this properly is a rough sketch of a standard normal curve in which you have shaded the entire area that you want to find, as Example 2 shows.

**EXAMPLE 2** **Probabilities for Standard Normal Variable Z**

**a.** *Find* $P(Z > 1.5)$.

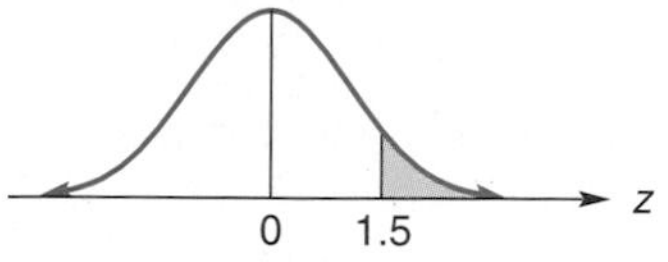

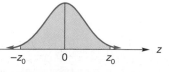

**FIGURE 18.15** $P(Z > 1.5)$.

*Solution:* This probability is the area to the right of $z = 1.5$ (Fig. 18.15). That area is equal to the difference between the total area to the right of $z = 0$, which is 0.5, and the area between $z = 0$ and $z = 1.5$, which is $A(1.5)$. Thus,

$$\begin{aligned} P(Z > 1.5) &= 0.5 - A(1.5) \\ &= 0.5 - 0.4332 = 0.0668. \end{aligned}$$

**b.** *Find* $P(0.5 < Z < 2)$.

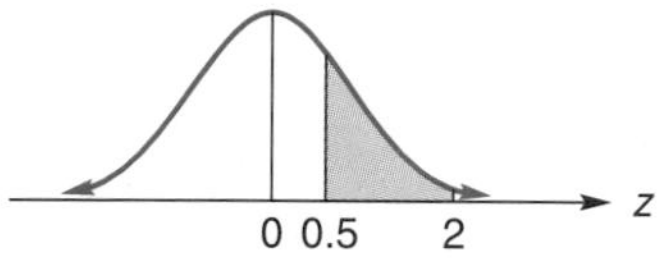

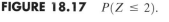

**FIGURE 18.16** $P(0.5 < Z < 2)$.

*Solution:* This probability is the area between $z = 0.5$ and $z = 2$ (Fig. 18.16). That area is the difference of two areas. It is the area between $z = 0$ and $z = 2$, or $A(2)$, minus the area between $z = 0$ and $z = 0.5$, or $A(0.5)$. Thus,

$$\begin{aligned} P(0.5 < Z < 2) &= A(2) - A(0.5) \\ &= 0.4772 - 0.1915 = 0.2857. \end{aligned}$$

**c.** *Find* $P(Z \leq 2)$.

**FIGURE 18.17** $P(Z \leq 2)$.

*Solution:* This probability is the area to the left of $z = 2$ (Fig. 18.17). That area is equal to the sum of the area to the left of $z = 0$, which is 0.5, and the area between $z = 0$ and $z = 2$, or $A(2)$. Thus,

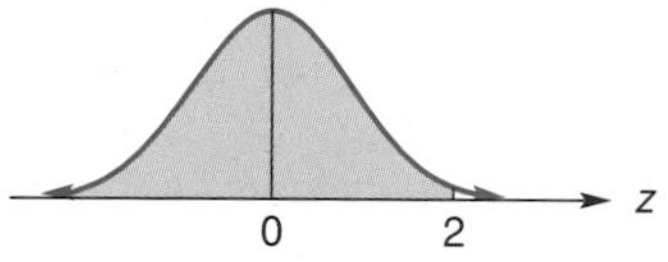

$$\begin{aligned} P(Z \leq 2) &= 0.5 + A(2) \\ &= 0.5 + 0.4772 = 0.9772. \end{aligned}$$

■

**EXAMPLE 3** **Probabilities for Standard Normal Variable Z**

**a.** *Find* $P(-2 < Z < -0.5)$.

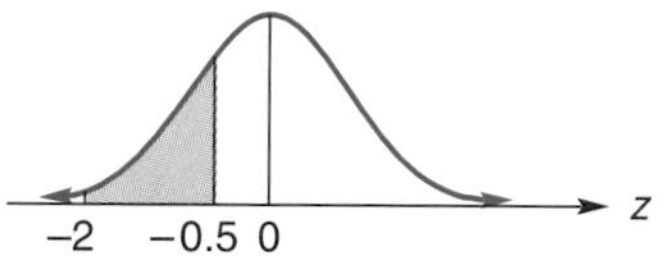

**FIGURE 18.18** $P(-2 < Z < -0.5)$.

*Solution:* This probability is the area between $z = -2$ and $z = -0.5$ (Fig. 18.18). By symmetry, that area is equal to the area between $z = 0.5$ and $z = 2$, which was computed in Example 2(b). We have

$$\begin{aligned} P(-2 < Z < -0.5) &= P(0.5 < Z < 2) \\ &= A(2) - A(0.5) = 0.2857. \end{aligned}$$

**b.** *Find* $z_0$ *such that* $P(-z_0 < Z < z_0) = 0.9642$.

**FIGURE 18.19** $P(-z_0 < Z < z_0) = 0.9642$.

*Solution:* Figure 18.19 shows the corresponding area. Because the total area is 0.9642, by symmetry the area between $z = 0$ and $z = z_0$ is $\frac{1}{2}(0.9642) = 0.4821$, which is $A(z_0)$. Looking at the body of the table in Appendix D, we see that 0.4821 corresponds to a $Z$-value of 2.1. Thus, $z_0 = 2.1$. ■

## Transforming to a Standard Normal Variable Z

If $X$ is normally distributed with mean $\mu$ and standard deviation $\sigma$, you might think that a table of areas is needed for each pair of values of $\mu$ and $\sigma$. Fortunately, this is not the case. Appendix D is still used. But you must first express a given area as an equivalent area under a standard normal curve. This involves transforming $X$ into a standard variable $Z$ (with mean 0 and standard deviation 1) by using the following change-of-variable formula:

$$Z = \frac{X - \mu}{\sigma}. \qquad (1)$$

Here we convert a normal variable to a standard normal variable.

On the right side, subtracting $\mu$ from $X$ gives the distance from $\mu$ to $X$. Dividing by $\sigma$ expresses this distance in terms of units of standard deviation. Thus, $Z$ is the number of standard deviations that $X$ is from $\mu$. That is, formula (1) converts units of $X$ into standard units ($Z$-values). For example, if $X = \mu$, then using formula (1) gives $Z = 0$. Hence, $\mu$ is zero standard deviations from $\mu$.

Suppose $X$ is normally distributed with $\mu = 4$ and $\sigma = 2$. Then to find—for example—$P(0 < X < 6)$, we first use formula (1) to convert the $X$-values 0 and 6 to $Z$-values (standard units):

$$z_1 = \frac{x_1 - \mu}{\sigma} = \frac{0 - 4}{2} = -2,$$

$$z_2 = \frac{x_2 - \mu}{\sigma} = \frac{6 - 4}{2} = 1.$$

It can be shown that

$$P(0 < X < 6) = P(-2 < Z < 1).$$

This means that the area under a normal curve with $\mu = 4$ and $\sigma = 2$ between $x = 0$ and $x = 6$ is equal to the area under a standard normal curve between $z = -2$ and $z = 1$ (Fig. 18.20). This area is the sum of the area $A_1$ between $z = -2$ and $z = 0$ and the area $A_2$ between $z = 0$ and $z = 1$. Using symmetry for $A_1$, we have

$$P(-2 < Z < 1) = A_1 + A_2 = A(2) + A(1)$$

$$= 0.4772 + 0.3413 = 0.8185.$$

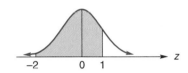

**FIGURE 18.20**
$P(-2 < Z < 1)$.

### EXAMPLE 4  Employees' Salaries

*The weekly salaries of 5000 employees of a large corporation are assumed to be normally distributed with mean $450 and standard deviation $40. How many employees earn less than $400 per week?*

*Solution:* Converting to standard units, we have

$$P(X < 400) = P\left(Z < \frac{400 - 450}{40}\right)$$

$$= P(Z < -1.25).$$

This probability is the area shown in Fig. 18.21(a). By symmetry, that area is equal to the area in Fig. 18.21(b) that corresponds to $P(Z > 1.25)$. This area is

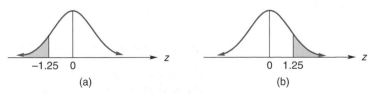

**FIGURE 18.21**  Diagram for Example 4.

the difference between the total area to the right of $z = 0$, which is 0.5, and the area between $z = 0$ and $z = 1.25$, which is $A(1.25)$. Thus,

$$P(X < 400) = P(Z < -1.25) = P(Z > 1.25)$$

$$= 0.5 - A(1.25) = 0.5 - 0.3944$$

$$= 0.1056.$$

That is, 10.56% of the employees have salaries less than $400. This corresponds to $0.1056(5000) = 528$ employees. ■

## ■ Exercise 18.2

1. If $Z$ is a standard normal random variable, find each of the following probabilities.
   a. $P(0 < Z < 1.8)$.
   b. $P(0.45 < Z < 2.81)$.
   c. $P(Z > -1.22)$.
   d. $P(Z \leq 2.93)$.
   e. $P(-2.61 < Z \leq 1.4)$.
   f. $P(Z > 0.07)$.

2. If $Z$ is a standard normal random variable, find each of the following.
   a. $P(-1.96 < Z < 1.96)$.
   b. $P(-2.11 < Z < -1.25)$.
   c. $P(Z < -1.05)$.
   d. $P(Z > 3\sigma)$.
   e. $P(|Z| > 2)$.
   f. $P(|Z| < \frac{1}{2})$.

*In Problems 3–8, find $z_0$ such that the given statement is true. Assume that $Z$ is a standard normal random variable.*

3. $P(Z < z_0) = 0.5517$.

4. $P(Z < z_0) = 0.0668$.

5. $P(Z > z_0) = 0.8599$.

6. $P(Z > z_0) = 0.4960$.

7. $P(-z_0 < Z < z_0) = 0.2662$.

8. $P(|Z| > z_0) = 0.2186$.

9. If $X$ is normally distributed with $\mu = 16$ and $\sigma = 4$, find each of the following probabilities.
   a. $P(X < 22)$.
   b. $P(X < 10)$.
   c. $P(10.8 < X < 12.4)$.

10. If $X$ is normally distributed with $\mu = 200$ and $\sigma = 40$, find each of the following probabilities.
    a. $P(X > 150)$.
    b. $P(210 < X < 250)$.

11. If $X$ is normally distributed with $\mu = -3$ and $\sigma = 2$, find $P(X > -2)$.

12. If $X$ is normally distributed with $\mu = 0$ and $\sigma = 1.5$, find $P(X < 3)$.

13. If $X$ is normally distributed with $\mu = 25$ and $\sigma^2 = 9$, find $P(19 < X \leq 28)$.

14. If $X$ is normally distributed with $\mu = 8$ and $\sigma = 1$, find $P(X > \mu - \sigma)$.

15. If $X$ is normally distributed such that $\mu = 40$ and $P(X > 54) = 0.0401$, find $\sigma$.

16. If $X$ is normally distributed with $\mu = 16$ and $\sigma = 2.25$, find $x_0$ such that the probability that $X$ is between $x_0$ and 16 is 0.4641.

17. **Test Scores** The scores on a national achievement test are normally distributed with mean 500 and standard deviation 100. What percentage of those who took the test had a score greater than 630?

18. **Test Scores** In a test given to a large group of people, the scores were normally distributed with mean 70 and standard deviation 10. What is the least whole-number score that a person could get and yet score in about the top 15%?

19. **Adult Heights** The heights (in inches) of adults in a large population are normally distributed with $\mu = 68$ and $\sigma = 3$. What percentage of the group is under 6 feet tall?

20. **Income** The yearly income for a group of 10,000 professional people is normally distributed with $\mu = \$60,000$ and $\sigma = \$5000$.
    a. What is the probability that a person from this group has a yearly income less than $56,000?
    b. How many of these people have yearly incomes over $70,000?

21. **IQ** The IQs of a large population of children are normally distributed with mean 100.4 and standard deviation 11.6.
    a. What percentage of the children have IQs greater than 125?
    b. About 90% of the children have IQs greater than what value?

22. Suppose $X$ is a random variable with $\mu = 10$ and $\sigma = 2$. If $P(4 < X < 16) = 0.25$, can $X$ be normally distributed?

To show the technique of estimating the binomial distribution by using the normal distribution.

## 18.3 THE NORMAL APPROXIMATION TO THE BINOMIAL DISTRIBUTION

We conclude this chapter by bringing together the notions of a discrete random variable and a continuous random variable. Recall from Chapter 10 that if $X$ is a binomial random variable (which is discrete), and if the probability of success on any trial is $p$, then for $n$ independent trials, the probability of $x$ successes is given by

$$P(X = x) = {}_nC_x p^x q^{n-x},$$

where $q = 1 - p$. You would no doubt agree that calculating probabilities for a binomial random variable can be quite tedious when the number of trials is large. For example, just imagine trying to compute ${}_{100}C_{40}(0.3)^{40}(0.7)^{60}$. To handle expressions like this, we can approximate a binomial distribution by a normal distribution and then use a table of areas.

To show how this is done, let us take a simple example. Figure 18.22 gives a probability histogram for a binomial experiment with $n = 10$ and $p = 0.5$.

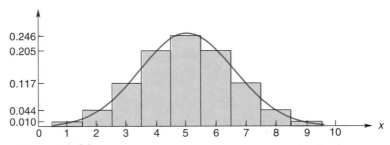

**FIGURE 18.22**   Normal approximation to binomial distribution.

The rectangles centered at $x = 0$ and $x = 10$ are not shown because their heights are very close to 0. Superimposed on the histogram is a normal curve, which approximates it. The approximation would be even better if $n$ were larger. That is, as $n$ gets larger, the width of each unit interval appears to get smaller, and the outline of the histogram tends to take on the appearance of a smooth curve. In fact, *it is not unusual to think of a density curve as the limiting case of a probability histogram.* In spite of the fact that in our case $n$ is only 10, the approximation shown does not seem too bad. The question that now arises is "Which normal distribution approximates the binomial distribution?" Since the mean and standard deviation are measures of central tendency and dispersion of a random variable, we choose the approximating normal distribution to have the same mean and standard deviation as that of the binomial distribution. For this choice, we can estimate the areas of rectangles in the histogram (that is, the binomial probabilities) by finding the corresponding area under the normal curve. In summary, we have the following:

> If $X$ is a binomial random variable and $n$ is sufficiently large, then the distribution of $X$ can be approximated by a normal random variable whose mean and standard deviation are the same as for $X$, which are $np$ and $\sqrt{npq}$, respectively.

Perhaps a word of explanation is appropriate concerning the phrase "$n$ is sufficiently large." Generally speaking, a normal approximation to a binomial distribution is not good if $n$ is small and $p$ is near 0 or 1, because much of the area in the binomial histogram would be concentrated at one end of the distribution (that is, at 0 or $n$). Thus, the distribution would not be fairly symmetric, and a normal curve would not "fit" well. A general rule that you can follow is that the normal approximation to the binomial distribution is reasonable if $np$ and $nq$ are at least 5. This is the case in our example: $np = 10(0.5) = 5$ and $nq = 10(0.5) = 5$.

Let us now use the normal approximation to estimate a binomial probability for $n = 10$ and $p = 0.5$. If $X$ denotes the number of successes, then its mean is

$$np = 10(0.5) = 5$$

and its standard deviation is

$$\sqrt{npq} = \sqrt{10(0.5)(0.5)} = \sqrt{2.5} \approx 1.58.$$

The probability function for $X$ is given by

$$f(x) = {}_{10}C_x(0.5)^x(0.5)^{10-x}.$$

We approximate this distribution by the normal distribution with $\mu = 5$ and $\sigma = \sqrt{2.5}$.

Suppose we estimate the probability that there are between 4 and 7 successes, inclusive, which is given by

$$P(4 \le X \le 7) = P(X = 4) + P(X = 5) + P(X = 6) + P(X = 7)$$

$$= \sum_{x=4}^{7} {}_{10}C_x(0.5)^x(0.5)^{10-x}.$$

This probability is the sum of the areas of the *rectangles* for $X = 4, 5, 6,$ and 7 in Fig. 18.23. Under the normal curve, we have shaded the corresponding area

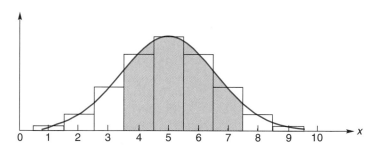

**FIGURE 18.23**    Normal approximation to $P(4 \le X \le 7)$.

that we shall compute as an approximation to this probability. Note that the shading extends not from 4 to 7, but from $4 - \frac{1}{2}$ to $7 + \frac{1}{2}$, that is, from 3.5 to 7.5. This "continuity correction" of 0.5 on each end of the interval allows most of the area in the appropriate rectangles to be included in the approximation, and *such a correction must always be made*. The phrase *continuity correction* is used because $X$ is treated as though it were a continuous random variable. We now convert the $X$-values 3.5 and 7.5 to $Z$-values:

$$z_1 = \frac{3.5 - 5}{\sqrt{2.5}} \approx -0.95,$$

$$z_2 = \frac{7.5 - 5}{\sqrt{2.5}} \approx 1.58.$$

Thus,

$$P(4 \le X \le 7) \approx P(-0.95 \le Z \le 1.58),$$

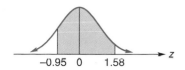

**FIGURE 18.24**
$P(-0.95 < Z \le 1.58)$.

which corresponds to the area under a standard normal curve between $z = -0.95$ and $z = 1.58$ (Fig. 18.24). This area is the sum of the area between $z = -0.95$ and $z = 0$, which, by symmetry, is $A(0.95)$, and the area between $z - 0$ and $z = 1.58$, which is $A(1.58)$. Hence,

$$P(4 \le X \le 7) = P(-0.95 \le Z \le 1.58)$$
$$= A(0.95) + A(1.58)$$
$$= 0.3289 + 0.4429 = 0.7718.$$

This result is close to the true value, 0.7734 (to four decimal places).

### Principles in Practice 1

**Normal Approximation to a Binomial Distribution**

On a game show, the grand prize is hidden behind one of four doors. Assume that the probability of selecting the grand prize is $p = \frac{1}{4}$. There were 20 winners among the last 60 contestants. Suppose that $X$ is the number of contestants that win the grand prize, and $X$ is binomial with $n = 60$. Approximate $P(X = 20)$ by using the normal approximation.

Do not ignore the continuity correction.

**EXAMPLE 1   Normal Approximation to a Binomial Distribution**

*Suppose $X$ is a binomial random variable with $n = 100$ and $p = 0.3$. Estimate $P(X = 40)$ by using the normal approximation.*

**Solution:**  We have

$$P(X = 40) = {}_{100}C_{40}(0.3)^{40}(0.7)^{60},$$

which was mentioned at the beginning of this section. We use a normal distribution with

$$\mu = np = 100(0.3) = 30$$

and

$$\sigma = \sqrt{npq} = \sqrt{100(0.3)(0.7)} = \sqrt{21} \approx 4.58.$$

Converting the corrected $X$-values 39.5 and 40.5 to $Z$-values gives

$$z_1 = \frac{39.5 - 30}{\sqrt{21}} \approx 2.07,$$

$$z_2 = \frac{40.5 - 30}{\sqrt{21}} \approx 2.29.$$

Therefore,

$$P(X = 40) \approx P(2.07 \le Z \le 2.29).$$

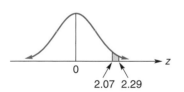

**FIGURE 18.25**
$P(2.07 \le Z \le 2.29)$.

This probability is the area under a standard normal curve between $z = 2.07$ and $z = 2.29$ (Fig. 18.25). That area is the difference of the area between $z = 0$ and $z = 2.29$, which is $A(2.29)$, and the area between $z = 0$ and $z = 2.07$, which is $A(2.07)$. Thus,

$$P(X = 40) \approx P(2.07 \le Z \le 2.29)$$
$$= A(2.29) - A(2.07)$$
$$= 0.4890 - 0.4808 = 0.0082. \quad \blacksquare$$

**EXAMPLE 2   Quality Control**

*In a quality-control experiment, a sample of 500 items is taken from an assembly line. Customarily, 8% of the items produced are defective. What is the probability that more than 50 defective items appear in the sample?*

**Solution:**  If $X$ is the number of defective items in the sample, then we shall consider $X$ to be binomial with $n = 500$ and $p = 0.08$. To find $P(X \ge 51)$, we use the normal approximation to the binomial distribution with

$$\mu = np = 500(0.08) = 40$$

and

$$\sigma = \sqrt{npq} = \sqrt{500(0.08)(0.92)} = \sqrt{36.8} \approx 6.07.$$

Converting the corrected value 50.5 to a $Z$-value gives

$$z = \frac{50.5 - 40}{\sqrt{36.8}} \approx 1.73.$$

Thus,

$$P(X \geq 51) \approx P(Z \geq 1.73).$$

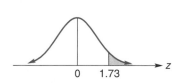

**FIGURE 18.26** $P(Z \geq 1.73)$.

This probability is the area under a standard normal curve to the right of $z = 1.73$ (Fig. 18.26). That area is the difference of the area to the right of $z = 0$, which is 0.5, and the area between $z = 0$ and $z = 1.73$, which is $A(1.73)$. Hence,

$$P(X \geq 51) \approx P(Z \geq 1.73)$$
$$= 0.5 - A(1.73) = 0.5 - 0.4582 = 0.0418. \qquad \blacksquare$$

### ■ Exercise 18.3

*In Problems 1–4, X is a binomial random variable with the given values of n and p. Calculate the indicated probabilities by using the normal approximation.*

**1.** $n = 150$, $p = 0.4$;  $P(X \geq 52)$, $P(X \geq 74)$.

**2.** $n = 50$, $p = 0.3$;  $P(X = 18)$, $P(X \leq 18)$.

**3.** $n = 200$, $p = 0.6$;  $P(X = 125)$, $P(110 \leq X \leq 135)$.

**4.** $n = 25$, $p = 0.25$;  $P(X \geq 5)$.

**5. Die Tossing** Suppose a fair die is tossed 300 times. What is the probability that a 5 turns up between 45 and 60 times, inclusive?

**6. Coin Tossing** For a biased coin, $P(H) = 0.4$ and $P(T) = 0.6$. If the coin is tossed 200 times, what is the probability of getting between 90 and 100 heads, inclusive?

**7. Truck Breakdown** A delivery service has a fleet of 60 trucks. At any given time, the probability of a truck being out of use due to factors such as breakdowns and maintenance is 0.1. What is the probability that 7 or more trucks are out of service at any time?

**8. Quality Control** In a manufacturing plant, a sample of 100 items is taken from the assembly line. For each item in the sample, the probability of being defective is 0.06. What is the probability that there are 3 or more defective items in the sample?

**9. True–False Exam** In a true–false exam with 20 questions, what is the probability of getting at least 12 correct answers by just guessing on all the questions? If there

are 100 questions instead of 20, what is the probability of getting at least 60 correct answers by just guessing?

**10. Multiple-Choice Exam** In a multiple-choice test with 50 questions, each question has four answers, only one of which is correct. If a student guesses on the last 20 questions, what is the probability of getting at least half of them correct?

**11. Poker** In a poker game, the probability of being dealt a hand consisting of three cards of one suit and two cards of another suit (in any order) is about 0.1. In 100 dealt hands, what is the probability that 16 or more of them will be as just described?

**12. Taste Test** A major cola company sponsors a national taste test, in which subjects sample its cola as well as the best-selling brand. Neither cola is identified by brand. The subjects are then asked to choose the cola that tastes better. If each of the 25 subjects in a supermarket actually have no preference and arbitrarily chose one of the colas, what is the probability that 15 or more of them choose the cola from the sponsoring company?

## 18.4 REVIEW

### IMPORTANT TERMS AND SYMBOLS

**Section 18.1**  continuous random variable    density function    uniform density function
uniform distribution    exponential density function    exponential distribution
cumulative distribution function    mean, $\mu$    variance, $\sigma^2$    standard deviation, $\sigma$

**Section 18.2**   normal random variable   normal distribution   normal density function
standard normal curve   standard normal random variable   standard normal distribution
standard normal density function   standard units

**Section 18.3**   continuity correction

## SUMMARY

A continuous random variable $X$ can assume any value in an interval or intervals. A density function for $X$ is a function that has the following properties:

1. $f(x) \geq 0$,

2. $\int_{-\infty}^{\infty} f(x) \, dx = 1$,

3. $P(a \leq X \leq b) = \int_{a}^{b} f(x) \, dx$.

Property 3 means that the area under the graph of $f$ and above the $x$-axis from $x = a$ to $x = b$ is $P(a \leq X \leq b)$. The probability that $X$ assumes a particular value is 0.

The continuous random variable $X$ has a uniform distribution over $[a, b]$ if its density function is given by

$$f(x) = \begin{cases} \dfrac{1}{b - a}, & \text{if } a \leq x \leq b, \\ 0, & \text{otherwise.} \end{cases}$$

$X$ has an exponential density function $f$ if

$$f(x) = \begin{cases} ke^{-kx}, & \text{if } x \geq 0, \\ 0, & \text{if } x < 0, \end{cases}$$

where $k$ is a positive constant.

The cumulative distribution function $F$ for the continuous random variable $X$ with density function $f$ is given by

$$F(x) = P(X \leq x) = \int_{-\infty}^{x} f(t) \, dt.$$

Geometrically, $F(x)$ represents the area under the density curve to the left of $x$. By using $F$, we are able to find $P(a \leq x \leq b)$:

$$P(a \leq x \leq b) = F(b) - F(a).$$

The mean $\mu$ of $X$ [or expectation $E(X)$] is given by

$$\mu = E(X) = \int_{-\infty}^{\infty} xf(x) \, dx;$$

the variance is given by

$$\sigma^2 = \text{Var}(X) = \int_{-\infty}^{\infty} (x - \mu)^2 f(x) \, dx$$
$$= \int_{-\infty}^{\infty} x^2 f(x) \, dx - \mu^2.$$

The standard deviation is given by

$$\sigma = \sqrt{\text{Var}(X)},$$

The graph of the normal density function

$$f(x) = \frac{1}{\sigma\sqrt{2\pi}} e^{-(1/2)[(x-\mu)/\sigma]^2}$$

is called a normal curve and is bell shaped. If $X$ has a normal distribution, then the probability that $X$ lies within one standard deviation of the mean $\mu$ is (approximately) 0.68; within two standard deviations, the probability is 0.95; and within three standard deviations, it is 0.997. If $Z$ is a normal random variable with $\mu = 0$ and $\sigma = 1$, then $Z$ is called a standard normal random variable. The probability $P(0 < Z < z_0)$ is the area under the graph of the standard normal curve from $z = 0$ to $z = z_0$ and is denoted $A(z_0)$. Values of $A(z_0)$ appear in Appendix D.

If $X$ is normally distributed with mean $\mu$ and standard deviation $\sigma$, then $X$ may be transformed into a standard normal random variable by the change-of-variable formula

$$Z = \frac{X - \mu}{\sigma}.$$

With this formula, probabilities for $X$ may be found by using areas under the standard normal curve.

If $X$ is a binomial random variable and the number $n$ of independent trials is large, then the distribution of $X$ may be approximated by using a normal random variable with mean $np$ and standard deviation $\sqrt{npq}$, where $p$ is the probability of success on any trial and $q = 1 - p$. It is important that continuity corrections be considered when one estimates binomial probabilities by a normal random variable.

## REVIEW PROBLEMS

1. Suppose $X$ is a continuous random variable with density function given by

$$f(x) = \begin{cases} \frac{1}{3} + kx^2, & \text{if } 0 \le x \le 1, \\ 0, & \text{otherwise.} \end{cases}$$

   **a.** Find $k$.

   **b.** Find $P\left(\frac{1}{2} < X < \frac{3}{4}\right)$.

   **c.** Find $P\left(X \ge \frac{1}{2}\right)$.

   **d.** Find the cumulative distribution function.

2. Suppose $X$ is exponentially distributed with $k = \frac{1}{4}$. Find $P(X > 1)$.

3. Suppose $X$ is a random variable with density function given by

$$f(x) = \begin{cases} \frac{2}{9}x, & \text{if } 0 \le x \le 3, \\ 0, & \text{otherwise.} \end{cases}$$

   **a.** Find $\mu$.

   **b.** Find $\sigma$.

4. Let $X$ be uniformly distributed over the interval $[2, 6]$. Find $P(X < 5)$.

*Let X be normally distributed with mean 20 and standard deviation 4. In Problems 5–10, determine the given probabilities.*

5. $P(X > 22)$.

6. $P(X < 21)$.

7. $P(12 < X < 18)$.

8. $P(X > 10)$.

9. $P(X < 16)$.

10. $P(22 < X < 32)$.

*In Problems 11 and 12, X is a binomial random variable with $n = 100$ and $p = 0.35$. Find the given probabilities by using the normal approximation.*

11. $P(25 \le X \le 47)$.

12. $P(X = 48)$.

13. **Heights of Individuals**   The heights (in inches) of individuals in a certain group are normally distributed with mean 68 and standard deviation 2. Find the probability that an individual from this group is taller than 6 ft.

14. **Coin Tossing**   If a fair coin is tossed 400 times, use the normal approximation to the binomial distribution to estimate the probability that a head comes up at least 185 times.

# Multivariable Calculus

To discuss functions of several variables and to compute function values. To discuss three-dimensional coordinates and sketch simple surfaces.

## 19.1 FUNCTIONS OF SEVERAL VARIABLES

Suppose a manufacturer produces two products, X and Y. Then the total cost depends on the levels of production of *both* X and Y. Table 19.1 is a

**TABLE 19.1**

| No. of Units of X Produced, $x$ | No. of Units of Y Produced, $y$ | Total Cost of Production, $c$ |
|:---:|:---:|:---:|
| 5 | 6 | 17 |
| 5 | 7 | 19 |
| 6 | 6 | 18 |
| 6 | 7 | 20 |

schedule that indicates total cost at various levels. For example, when 5 units of X and 6 units of Y are produced, the total cost $c$ is 17. In this situation, it seems natural to associate the number 17 with the *ordered pair* (5, 6):

$$(5, 6) \rightarrow 17,$$

The first element of the ordered pair, 5, represents the number of units of X produced, while the second element, 6, represents the number of units of Y produced. Corresponding to the other production situations, we have

$$(5, 7) \rightarrow 19,$$
$$(6, 6) \rightarrow 18,$$

and

$$(6, 7) \rightarrow 20.$$

This correspondence can be considered an input-output relation where the inputs are ordered pairs. With each input, we associate exactly one output. Thus, the correspondence defines a function $f$ whose domain consists of $(5,6), (5,7), (6,6), (6,7)$ and whose range consists of $17, 19, 18, 20$. In function notation,

$$f(5,6) = 17, \qquad f(5,7) = 19,$$
$$f(6,6) = 18, \qquad f(6,7) = 20.$$

We say that the total-cost schedule can be described by $c = f(x, y)$, a function of the two independent variables $x$ and $y$. The letter $c$ is the dependent variable.

Turning to another function of two variables, we see that the equation

$$z = \frac{2}{x^2 + y^2}$$

defines $z$ as a function of $x$ and $y$:

$$z = f(x, y) = \frac{2}{x^2 + y^2}.$$

The domain of $f$ is all ordered pairs of real numbers $(x, y)$ for which the equation has meaning when the first and second elements of $(x, y)$ are substituted for $x$ and $y$, respectively, in the equation. Thus, the domain of $f$ is all ordered pairs except $(0, 0)$. For example, to find $f(2, 3)$, we substitute $x = 2$ and $y = 3$ into the expression $2/(x^2 + y^2)$. We obtain $f(2, 3) = 2/(2^2 + 3^2) = 2/13$.

**Principles in Practice 1**

**Functions of Two Variables**

The cost per day for manufacturing both 12-ounce and 20-ounce coffee mugs is given by $c = 160 + 2x + 3y$, where $x$ is the number of 12-ounce mugs and $y$ is the number of 20-ounce mugs. What is the cost per day of manufacturing

a. 500 12-ounce and 700 20-ounce mugs?

b. 1000 12-ounce and 750 20-ounce mugs?

**EXAMPLE 1   Functions of Two Variables**

a. $f(x, y) = \dfrac{x + 3}{y - 2}$ is a function of two variables. Because the denominator is zero when $y = 2$, the domain of $f$ is all $(x, y)$ such that $y \neq 2$. Some function values are

$$f(0, 3) = \frac{0 + 3}{3 - 2} = 3,$$
$$f(3, 0) = \frac{3 + 3}{0 - 2} = -3.$$

Note that $f(0, 3) \neq f(3, 0)$.

b. $h(x, y) = 4x$ defines $h$ as a function of $x$ and $y$. The domain is all ordered pairs of real numbers. Some function values are

$$h(2, 5) = 4(2) = 8,$$
$$h(2, 6) = 4(2) = 8.$$

Note that the function values are independent of the choice of $y$.

c. If $z^2 = x^2 + y^2$ and $x = 3$ and $y = 4$, then $z^2 = 3^2 + 4^2 = 25$. Consequently, $z = \pm 5$. Thus, with the ordered pair $(3, 4)$, we *cannot* associate exactly one output number. Hence $z$ is *not* a function of $x$ and $y$. ∎

**EXAMPLE 2   Temperature-Humidity Index**

*On hot and humid days, many people tend to feel uncomfortable. The degree of discomfort is numerically given by the temperature-humidity index, THI, which is a function of two variables, $t_d$ and $t_w$:*

$$\text{THI} = f(t_d, t_w) = 15 + 0.4(t_d + t_w),$$

*where $t_d$ is the dry-bulb temperature (in degrees Fahrenheit) and $t_w$ is the wet-bulb temperature (in degrees Fahrenheit) of the air. Evaluate the* THI *when $t_d = 90$ and $t_w = 80$.*

***Solution:*** We want to find $f(90, 80)$:

$$f(90, 80) = 15 + 0.4(90 + 80) = 15 + 68 = 83.$$

When the THI is greater than 75, most people are uncomfortable. In fact, the THI was once called the "discomfort index." Many electric utilities closely follow this index so that they can anticipate the demand that air-conditioning places on their systems.    ■

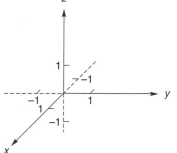

**FIGURE 19.1** Three-dimensional rectangular coordinate system.

If $y = f(x)$ is a function of one variable, the domain of $f$ can be geometrically represented by points on the real-number line. The function itself can be represented by its graph in a coordinate plane, sometimes called a two-dimensional coordinate system. However, for a function of two variables, $z = f(x, y)$, the domain (consisting of ordered pairs of real numbers) can be geometrically represented by a *region* in the plane. The function itself can be geometrically represented in a ***three*-dimensional rectangular coordinate system.** Such a system is formed when three mutually perpendicular real-number lines in space intersect at the origin of each line, as in Fig. 19.1. The three number lines are called the *x-*, *y-*, and *z*-axes, and their point of intersection is called the origin of the system. The arrows indicate the positive directions of the axes, and the negative portions of the axes are shown as dashed lines.

To each point $P$ in space, we can assign a unique ordered triple of numbers, called the *coordinates* of $P$. To do this [see Fig. 19.2(a)], from $P$, we construct a line perpendicular to the *x,y*-plane—that is, the plane determined by the *x*- and *y*-axes. Let $Q$ be the point where the line intersects this plane. From $Q$, we construct perpendiculars to the *x*- and *y*-axes. These lines intersect the *x*- and *y*-axes at $x_0$ and $y_0$, respectively. From $P$, a perpendicular to the *z*-axis is constructed that intersects the axis at $z_0$. Thus, we assign to $P$ the ordered triple $(x_0, y_0, z_0)$. It should also be evident that with each ordered triple of numbers we can assign a unique point in space. Due to this one-to-one correspondence between points in space and ordered triples, an ordered triple may be called a point. In Fig. 19.2(b), points $(2, 0, 0)$, $(2, 3, 0)$, and $(2, 3, 4)$ are shown. Note that the origin corresponds to $(0, 0, 0)$. Typically, the negative portions of the axes are not shown unless required.

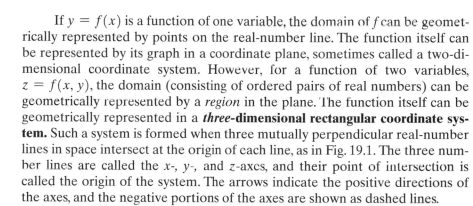

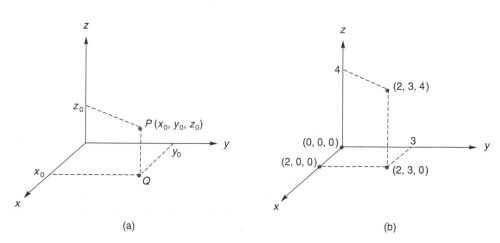

(a)    (b)

**FIGURE 19.2** Points in space.

We can represent a function of two variables, $z = f(x, y)$, geometrically. To each ordered pair $(x, y)$ in the domain of $f$, we assign the point $(x, y, f(x, y))$. The set of all such points is called the *graph* of $f$. Such a graph appears in Fig. 19.3. You can consider $z = f(x, y)$ as representing a surface in space.[1]

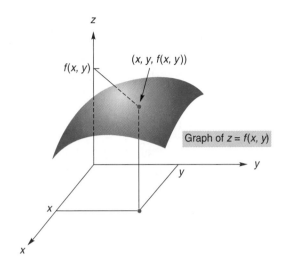

**FIGURE 19.3** Graph of a function of two variables.

In Chapter 11, the continuity of a function of one variable was discussed. If $y = f(x)$ is continuous at $x = x_0$, then points near $x_0$ will have their function values near $f(x_0)$. Extending this concept to a function of two variables, we say that the function $z = f(x, y)$ is continuous at $(x_0, y_0)$ when points near $(x_0, y_0)$ have their function values near $f(x_0, y_0)$. Loosely interpreting this concept, and without delving into it in great depth, we can say that a function of two variables is continuous on its domain (that is, continuous at each point in its domain) if its graph is an "unbroken surface."

Until now, we have considered only functions of either one or two variables. In general, a **function of $n$ variables** is a function whose domain consists of ordered $n$-tuples $(x_1, x_2, \ldots, x_n)$. For example, $f(x, y, z) = 2x + 3y + 4z$ is a function of three variables with a domain consisting of all ordered triples. The function $g(x_1, x_2, x_3, x_4) = x_1 x_2 x_3 x_4$ is a function of four variables with a domain consisting of all ordered 4-tuples. Although functions of several variables are extremely important and useful, we cannot geometrically represent functions of more than two variables.

We now give a brief discussion of sketching surfaces in space. We begin with planes that are parallel to a coordinate plane. By a "coordinate plane" we mean a plane containing two coordinate axes. For example, the plane determined by the $x$- and $y$-axes is the $x,y$-**plane.** Similarly, we speak of the $x,z$-**plane** and the $y,z$-**plane.** The coordinate planes divide space into eight parts, called *octants.* In particular, the part containing all points $(x, y, z)$ such that $x$, $y$, and $z$ are positive is called the **first octant.**

Names are not assigned to the remaining seven octants.

Suppose $S$ is a plane that is parallel to the $x,y$-plane and passes through the point $(0, 0, 5)$. [See Fig. 19.4(a).] Then the point $(x, y, z)$ will lie on $S$ if and only if $z = 5$; that is, $x$ and $y$ can be any real numbers, but $z$ must equal 5. For

[1]We shall freely use the term "surface" in the intuitive sense.

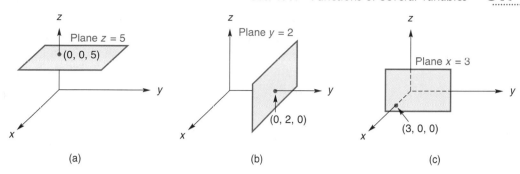

**FIGURE 19.4**    Planes parallel to coordinate planes.

this reason, we say that $z = 5$ is an equation of $S$. Similarly, an equation of the plane parallel to the $x,z$-plane and passing through the point $(0, 2, 0)$ is $y = 2$ [Fig. 19.4(b)]. The equation $x = 3$ is an equation of the plane passing through $(3, 0, 0)$ and parallel to the $y,z$-plane [Fig. 19.4(c)]. Now let us look at planes in general.

In space, the graph of an equation of the form

$$Ax + By + Cz + D = 0,$$

where $D$ is a constant and $A$, $B$, and $C$ are constants that are not all zero, is a plane. Since three distinct points (not lying on the same line) determine a plane, a convenient way to sketch a plane is to first determine the points, if any, where the plane intersects the $x$-, $y$-, and $z$-axes. These points are called *intercepts*.

**EXAMPLE 3    Graphing a Plane**

*Sketch the plane* $2x + 3y + z = 6$.

*Solution:* The plane intersects the $x$-axis when $y = 0$ and $z = 0$. Thus, $2x = 6$, which gives $x = 3$. Similarly, if $x = z = 0$, then $y = 2$; if $x = y = 0$, then $z = 6$. Therefore, the intercepts are $(3, 0, 0)$, $(0, 2, 0)$, and $(0, 0, 6)$. After these points are plotted, a plane is passed through them. The portion of the plane in the first octant is shown in Fig. 19.5(a); however, you should realize that the plane extends indefinitely into space.

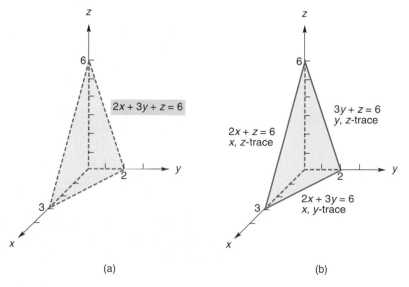

**FIGURE 19.5**    The plane $2x + 3y + z = 6$ and its traces.

A surface can be sketched with the aid of its **traces.** These are the intersections of the surface with the coordinate planes. To illustrate, for the plane $2x + 3y + z = 6$ in Example 3, the trace in the $x,y$-plane is obtained by setting $z = 0$. This gives $2x + 3y = 6$, which is an equation of a *line* in the $x,y$-plane. Similarly, setting $x = 0$ gives the trace in the $y,z$-plane: the line $3y + z = 6$. The $x,z$-trace is the line $2x + z = 6$. [See Fig. 19.5(b).]

Note that this equation places no restriction on $y$.

### EXAMPLE 4   Sketching a Surface

*Sketch the surface $2x + z = 4$.*

*Solution:* This equation has the form of a plane. The $x$- and $z$-intercepts are $(2, 0, 0)$ and $(0, 0, 4)$, and there is no $y$-intercept, since $x$ and $z$ cannot both be zero. Setting $y = 0$ gives the $x,z$-trace $2x + z = 4$, which is a line in the $x,z$-plane. In fact, the intersection of the surface with *any* plane $y = k$ is also $2x + z = 4$. Hence, the plane appears as in Fig. 19.6. ∎

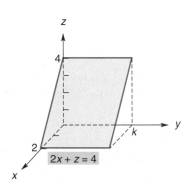

**FIGURE 19.6**   The plane $2x + z = 4$.

Our final examples deal with surfaces that are not planes, but whose graphs can be easily obtained.

### EXAMPLE 5   Sketching a Surface

*Sketch the surface $z = x^2$.*

*Solution:* The $x,z$-trace is the curve $z = x^2$, which is a parabola. In fact, for *any* fixed value of $y$, we get $z = x^2$. Thus, the graph appears as in Fig. 19.7. ∎

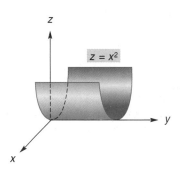

**FIGURE 19.7**   The surface $z = x^2$.

### EXAMPLE 6   Sketching a Surface

*Sketch the surface $x^2 + y^2 + z^2 = 25$.*

*Solution:* Setting $z = 0$ gives the $x,y$-trace $x^2 + y^2 = 25$, which is a circle of radius 5. Similarly, the $y,z$- and $x,z$-traces are the circles $y^2 + z^2 = 25$ and $x^2 + z^2 = 25$, respectively. Note also that since $x^2 + y^2 = 25 - z^2$, the intersection of the surface with the plane $z = k$, where $-5 \le k \le 5$, is a circle. For example, if $z = 3$, the intersection is the circle $x^2 + y^2 = 16$. If $z = 4$, the intersection is $x^2 + y^2 = 9$. That is, cross sections of the surface that are parallel to the $x,y$-plane are circles. The surface appears in Fig. 19.8; it is a sphere.

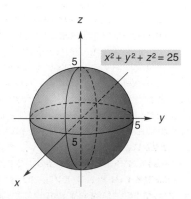

**FIGURE 19.8**   The surface $x^2 + y^2 + z^2 = 25$. ∎

## ■ Exercise 19.1

*In Problems 1–12, determine the indicated function values for the given functions.*

**1.** $f(x, y) = 4x - y^2 + 3$;  $f(2, 1)$.

**2.** $f(x, y) = 3x^2y - 4y$;  $f(1, -2)$.

**3.** $g(x, y, z) = e^x(2y + 3z)$;  $g(0, -1, 2)$.

**4.** $g(x, y, z) = x^2y + xy^2 + yz^2$;  $g(1, -3, 2)$.

**5.** $h(r, s, t, u) = \dfrac{rs}{t^2 - u^2}$;  $h(-3, 3, 5, 4)$.

**6.** $h(r, s, t, u) = \ln(ru)$;  $h(1, 5, 3, 1)$.

**7.** $g(p_A, p_B) = 2p_A(p_A^2 - 5)$;  $g(4, 8)$.

**8.** $g(p_A, p_B) = p_A\sqrt{p_B} + 10$;  $g(8, 4)$.

**9.** $F(x, y, z) = 3$;  $F(2, 0, -1)$.

**10.** $F(x, y, z) = \dfrac{x}{yz}$;  $F(0, 0, 3)$.

**11.** $f(x, y) = 2x - 5y + 4$;  $f(x_0 + h, y_0)$.

**12.** $f(x, y) = x^2y - 3y^3$;  $f(r + t, r)$.

**13. Ecology**  A method of ecological sampling to determine animal populations in a given area involves first marking all the animals obtained in a sample of $R$ animals from the area and then releasing them so that they can mix with unmarked animals. At a later date a second sample is taken of $M$ animals, and the number of these that are marked, $S$, is noted. Based on $R$, $M$, and $S$, an estimate of the total population of animals in the sample area is given by

$$N = f(R, M, S) = \frac{RM}{S}.$$

Find $f(400, 400, 80)$. This method is called the *mark-and-recapture procedure*.[2]

**14. Genetics**  Under certain conditions, if two brown-eyed parents have exactly $k$ children, the probability that there will be exactly $r$ blue-eyed children is given by

$$P(r, k) = \frac{k!\left(\frac{1}{4}\right)^r \left(\frac{3}{4}\right)^{k-r}}{r!(k - r)!}, \qquad r = 0, 1, 2, \ldots, k.$$

Find the probability that, out of a total of four children, exactly three will be blue-eyed.

*In Problems 15–18, find equations of the planes that satisfy the given conditions.*

**15.** Parallel to the $x,z$-plane and also passes through the point $(0, -4, 0)$.

**16.** Parallel to the $y,z$-plane and also passes through the point $(8, 0, 0)$.

**17.** Parallel to the $x,y$-plane and also passes through the point $(2, 7, 6)$.

**18.** Parallel to the $y,z$-plane and also passes through the point $(-4, -2, 7)$.

*In Problems 19–28, sketch the given surfaces.*

**19.** $x + y + z = 1$.

**20.** $2x + y + 2z = 6$.

**21.** $3x + 6y + 2z = 12$.

**22.** $x + 2y + 3z = 4$.

**23.** $x + 2y = 2$.

**24.** $y + z = 1$.

**25.** $z = 4 - x^2$.

**26.** $y = x^2$.

**27.** $x^2 + y^2 + z^2 = 1$.

**28.** $x^2 + y^2 = 1$.

[2]E. P. Odum, *Ecology* (New York: Holt, Rinehart and Winston, 1966).

---

OBJECTIVE

**To compute partial derivatives.**

## 19.2 PARTIAL DERIVATIVES

Figure 19.9 shows the surface $z = f(x, y)$ and a plane that is parallel to the $x,z$-plane and that passes through the point $(x_0, y_0, f(x_0, y_0))$ on the surface. An equation of this plane is $y = y_0$. Hence, any point on the curve that is the intersection of the surface with the plane must have the form $(x, y_0, f(x, y_0))$. Thus, the curve can be described by the equation $z = f(x, y_0)$. Since $y_0$ is constant, $z = f(x, y_0)$ can be considered a function of one variable, $x$. When the derivative of this function is evaluated at $x_0$, it gives the slope of the tangent line to this curve at the point $(x_0, y_0, f(x_0, y_0))$. (See Fig. 19.9.) This slope is called the *partial derivative of f with respect to x* at $(x_0, y_0)$ and is denoted $f_x(x_0, y_0)$. In terms of limits,

$$f_x(x_0, y_0) = \lim_{h \to 0} \frac{f(x_0 + h, y_0) - f(x_0, y_0)}{h}. \tag{1}$$

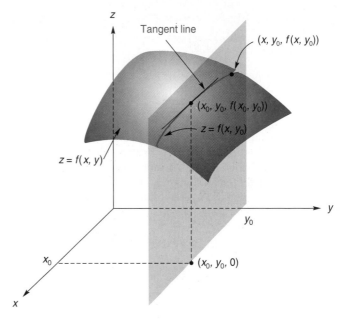

**FIGURE 19.9** Geometric interpretation of $f_x(x_0, y_0)$.

On the other hand, in Fig. 19.10, the plane $x = x_0$ is parallel to the $y,z$-plane and cuts the surface $z = f(x, y)$ in a curve given by $z = f(x_0, y)$, a function of $y$. When the derivative of this function is evaluated at $y_0$, it gives the slope of the tangent line to this curve at the point $(x_0, y_0, f(x_0, y_0))$. This

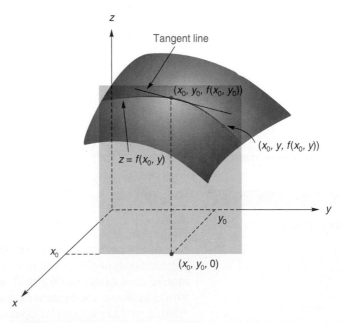

**FIGURE 19.10** Geometric interpretation of $f_y(x_0, y_0)$.

slope is called the *partial derivative of f with respect to y* at $(x_0, y_0)$ and is denoted $f_y(x_0, y_0)$. In terms of limits,

$$f_y(x_0, y_0) = \lim_{h \to 0} \frac{f(x_0, y_0 + h) - f(x_0, y_0)}{h}. \tag{2}$$

*This gives us a geometric interpretation of a partial derivative.*

We say that $f_x(x_0, y_0)$ is the slope of the tangent line to the graph of $f$ at $(x_0, y_0, f(x_0, y_0))$ *in the x-direction;* similarly, $f_y(x_0, y_0)$ is the slope of the tangent line *in the y-direction.*

For generality, by replacing $x_0$ and $y_0$ in Eqs. (1) and (2) by $x$ and $y$, respectively, we get the following definition.

**DEFINITION**

*If $z = f(x, y)$, the* **partial derivative of f with respect to x,** *denoted $f_x$, is the function given by*

$$f_x(x, y) = \lim_{h \to 0} \frac{f(x + h, y) - f(x, y)}{h},$$

*provided that the limit exists.*

*The* **partial derivative of f with respect to y,** *denoted $f_y$, is the function given by*

$$f_y(x, y) = \lim_{h \to 0} \frac{f(x, y + h) - f(x, y)}{h},$$

*provided that the limit exists.*

By analyzing the foregoing definition, we can state the following procedure to find $f_x$ and $f_y$:

*This gives us a mechanical way to find partial derivatives.*

**Procedure to Find $f_x(x, y)$ and $f_y(x, y)$**

To find $f_x$, treat $y$ as a constant, and differentiate $f$ with respect to $x$ in the usual way.

To find $f_y$, treat $x$ as a constant, and differentiate $f$ with respect to $y$ in the usual way.

**EXAMPLE 1    Finding Partial Derivatives**

*If $f(x, y) = xy^2 + x^2y$, find $f_x(x, y)$ and $f_y(x, y)$. Also, find $f_x(3, 4)$ and $f_y(3, 4)$.*

***Solution:*** To find $f_x(x, y)$, we treat $y$ as a constant and differentiate $f$ with respect to $x$:

$$f_x(x, y) = (1)y^2 + (2x)y = y^2 + 2xy.$$

To find $f_y(x, y)$, we treat $x$ as a constant and differentiate with respect to $y$:

$$f_y(x, y) = x(2y) + x^2(1) = 2xy + x^2.$$

Note that $f_x(x, y)$ and $f_y(x, y)$ are each functions of the two variables $x$ and $y$. To find $f_x(3, 4)$ we evaluate $f_x(x, y)$ when $x = 3$ and $y = 4$:

$$f_x(3, 4) = 4^2 + 2(3)(4) = 40.$$

Similarly,

$$f_y(3, 4) = 2(3)(4) + 3^2 = 33.$$  ■

Notations for partial derivatives of $z = f(x, y)$ are in Table 19.2. Table 19.3 gives notations for partial derivatives evaluated at $(x_0, y_0)$. Note that the symbol $\partial$ (not $d$) is used to denote a partial derivative. The symbol $\partial z/\partial x$ is read "the partial derivative of $z$ with respect to $x$."

**TABLE 19.2**

| Partial Derivative of $f$ (or $z$) with Respect to $x$ | Partial Derivative of $f$ (or $z$) with Respect to $y$ |
|---|---|
| $f_x(x, y)$ | $f_y(x, y)$ |
| $\dfrac{\partial}{\partial x}[f(x, y)]$ | $\dfrac{\partial}{\partial y}[f(x, y)]$ |
| $\dfrac{\partial z}{\partial x}$ | $\dfrac{\partial z}{\partial y}$ |

**TABLE 19.3**

| Partial Derivative of $f$ (or $z$) with Respect to $x$ Evaluated at $(x_0, y_0)$ | Partial Derivative of $f$ (or $z$) with Respect to $y$ Evaluated at $(x_0, y_0)$ |
|---|---|
| $f_x(x_0, y_0)$ | $f_y(x_0, y_0)$ |
| $\dfrac{\partial z}{\partial x}\bigg|_{(x_0, y_0)}$ | $\dfrac{\partial z}{\partial y}\bigg|_{(x_0, y_0)}$ |
| $\dfrac{\partial z}{\partial x}\bigg|_{\substack{x=x_0 \\ y=y_0}}$ | $\dfrac{\partial z}{\partial y}\bigg|_{\substack{x=x_0 \\ y=y_0}}$ |

**EXAMPLE 2**   **Finding Partial Derivatives**

**a.** *If* $z = 3x^3y^3 - 9x^2y + xy^2 + 4y$, *find* $\dfrac{\partial z}{\partial x}, \dfrac{\partial z}{\partial y}, \dfrac{\partial z}{\partial x}\bigg|_{(1,0)}$ *and* $\dfrac{\partial z}{\partial y}\bigg|_{(1,0)}$.

*Solution:* To find $\partial z/\partial x$, we differentiate $z$ with respect to $x$ while treating $y$ as a constant:

$$\frac{\partial z}{\partial x} = 3(3x^2)y^3 - 9(2x)y + (1)y^2 + 0$$

$$= 9x^2y^3 - 18xy + y^2.$$

Evaluating the latter equation at $(1, 0)$, we obtain

$$\frac{\partial z}{\partial x}\bigg|_{(1,0)} = 9(1)^2(0)^3 - 18(1)(0) + 0^2 = 0.$$

To find $\partial z/\partial y$, we differentiate $z$ with respect to $y$ while treating $x$ as a constant:

$$\frac{\partial z}{\partial y} = 3x^3(3y^2) - 9x^2(1) + x(2y) + 4(1)$$

$$= 9x^3y^2 - 9x^2 + 2xy + 4.$$

Thus,

$$\frac{\partial z}{\partial y}\bigg|_{(1,0)} = 9(1)^3(0)^2 - 9(1)^2 + 2(1)(0) + 4 = -5.$$

**b.** If $w = x^2 e^{2x+3y}$, find $\partial w/\partial x$ and $\partial w/\partial y$.

*Solution:* To find $\partial w/\partial x$, we treat $y$ as a constant and differentiate with respect to $x$. Since $x^2 e^{2x+3y}$ is a product of two functions, each involving $x$, we use the product rule:

$$\frac{\partial w}{\partial x} = x^2 \frac{\partial}{\partial x}(e^{2x+3y}) + e^{2x+3y}\frac{\partial}{\partial x}(x^2)$$

$$= x^2(2e^{2x+3y}) + e^{2x+3y}(2x)$$

$$= 2x(x+1)e^{2x+3y}.$$

To find $\partial w/\partial y$, we treat $x$ as a constant and differentiate with respect to $y$:

$$\frac{\partial w}{\partial y} = x^2 \frac{\partial}{\partial y}(e^{2x+3y}) = 3x^2 e^{2x+3y}.$$

We have seen that, for a function of two variables, two partial derivatives can be considered. Actually, the concept of partial derivatives can be extended to functions of more than two variables. For example, with $w = f(x, y, z)$ we have three partial derivatives:

the partial with respect to $x$, denoted $f_x(x, y, z)$, $\partial w/\partial x$, etc.;

the partial with respect to $y$, denoted $f_y(x, y, z)$, $\partial w/\partial y$, etc.;

and

the partial with respect to $z$, denoted $f_z(x, y, z)$, $\partial w/\partial z$, etc.

To determine $\partial w/\partial x$, treat $y$ and $z$ as constants, and differentiate $w$ with respect to $x$. For $\partial w/\partial y$, treat $x$ and $z$ as constants, and differentiate with respect to $y$. For $\partial w/\partial z$, treat $x$ and $y$ as constants, and differentiate with respect to $z$. With a function of $n$ variables, we have $n$ partial derivatives, which are determined in the obvious way.

**EXAMPLE 3** **Partial Derivatives of a Function of Three Variables**

If $f(x, y, z) = x^2 + y^2 z + z^3$, find $f_x(x, y, z), f_y(x, y, z),$ and $f_z(x, y, z)$.

*Solution:* To find $f_x(x, y, z)$, we treat $y$ and $z$ as constants and differentiate $f$ with respect to $x$.

$$f_x(x, y, z) = 2x.$$

Treating $x$ and $z$ as constants and differentiating with respect to $y$, we have

$$f_y(x, y, z) = 2yz.$$

Treating $x$ and $y$ as constants and differentiating with respect to $z$, we have

$$f_z(x, y, z) = y^2 + 3z^2.$$

**EXAMPLE 4** **Partial Derivatives of a Function of Four Variables**

If $p = g(r, s, t, u) = \dfrac{rsu}{rt^2 + s^2 t}$, find $\dfrac{\partial p}{\partial s}, \dfrac{\partial p}{\partial t},$ and $\dfrac{\partial p}{\partial t}\Big|_{(0,1,1,1)}$.

*Solution:* To find $\partial p/\partial s$, first note that $p$ is a quotient of two functions, each involving the variable $s$. Thus, we use the quotient rule and treat $r$, $t$, and $u$ as constants:

$$\frac{\partial p}{\partial s} = \frac{(rt^2 + s^2t)\frac{\partial}{\partial s}(rsu) - rsu\frac{\partial}{\partial s}(rt^2 + s^2t)}{(rt^2 + s^2t)^2}$$

$$= \frac{(rt^2 + s^2t)(ru) - (rsu)(2st)}{(rt^2 + s^2t)^2}.$$

Simplifying gives

$$\frac{\partial p}{\partial s} = \frac{ru(rt - s^2)}{t(rt + s^2)^2}.$$

To find $\partial p/\partial t$, we can first write $p$ as

$$p = rsu(rt^2 + s^2t)^{-1}.$$

Next, we use the power rule and treat $r$, $s$, and $u$ as constants:

$$\frac{\partial p}{\partial t} = rsu(-1)(rt^2 + s^2t)^{-2}\frac{\partial}{\partial t}(rt^2 + s^2t)$$

$$= -rsu(rt^2 + s^2t)^{-2}(2rt + s^2),$$

so that

$$\frac{\partial p}{\partial s} = -\frac{rsu(2rt + s^2)}{(rt^2 + s^2t)^2}.$$

Letting $r = 0, s = 1, t = 1$, and $u = 1$ gives

$$\left.\frac{\partial p}{\partial t}\right|_{(0,1,1,1)} = -\frac{0(1)(1)[2(0)(1) + (1)^2]}{[0(1)^2 + (1)^2(1)]^2} = 0. \quad\blacksquare$$

### ■ Exercise 19.2

*In each of Problems 1–26, a function of two or more variables is given. Find the partial derivative of the function with respect to each of the variables.*

**1.** $f(x, y) = 4x^2 + 3y^2 - 6$.

**2.** $f(x, y) = 2x^2 + 3xy$.

**3.** $f(x, y) = 2y + 1$.

**4.** $f(x, y) = e$.

**5.** $g(x, y) = x^3y^2 + 2x^2y - 3xy + 4y$.

**6.** $g(x, y) = (x + 1)^2 + (y - 3)^3 + 5xy^3 - 2$.

**7.** $g(p, q) = \sqrt{pq}$.

**8.** $g(w, z) = \sqrt[3]{w^2 + z^2}$.

**9.** $h(s, t) = \dfrac{s^2 + 4}{t - 3}$.

**10.** $h(u, v) = \dfrac{4uv^2}{u^2 + v^2}$.

**11.** $u(q_1, q_2) = \frac{3}{4}\ln q_1 + \frac{1}{4}\ln q_2$.

**12.** $Q(l, k) = 3l^{0.41}k^{0.59}$.

**13.** $h(x, y) = \dfrac{x^2 + 3xy + y^2}{\sqrt{x^2 + y^2}}$.

**14.** $h(x, y) = \dfrac{\sqrt{x + 4}}{x^2y + y^2x}$.

**15.** $z = e^{5xy}$.

**16.** $z = (x^2 + y)e^{3x + 4y}$.

**17.** $z = 5x\ln(x^2 + y)$.

**18.** $z = \ln(3x^2 + 4y^4)$.

**19.** $f(r, s) = \sqrt{r + 2s}\,(r^3 - 2rs + s^2)$.

**20.** $f(r, s) = \sqrt{rs}\,e^{2+r}$.

**21.** $f(r, s) = e^{3-r} \ln(7 - s)$.

**22.** $f(r, s) = (5r^2 + 3s^3)(2r - 5s)$.

**23.** $g(x, y, z) = 3x^2y + 2xy^2z + 3z^3$.

**24.** $g(x, y, z) = x^2y^3z^5 - 3x^2y^4z^3 + 5xz$.

**25.** $g(r, s, t) = e^{s+t}(r^2 + 7s^3)$.

**26.** $g(r, s, t, u) = rs \ln(2t + 5u)$.

*In Problems 27–34, evaluate the given partial derivatives.*

**27.** $f(x, y) = x^3y + 7x^2y^2$; $f_x(1, -2)$.

**28.** $z = \sqrt{5x^2 + 3xy + 2y}$; $\left.\dfrac{\partial z}{\partial x}\right|_{\substack{x=0 \\ y=2}}$.

**29.** $g(x, y, z) = e^x\sqrt{y + 2z}$; $g_z(0, 1, 4)$.

**30.** $g(x, y, z) = \dfrac{3x^2 + 2y}{xy + xz}$; $g_y(1, 1, 1)$.

**31.** $h(r, s, t, u) = (s^2 + tu) \ln(2r + 7st)$; $h_s(1, 0, 0, 1)$.

**32.** $h(r, s, t, u) = \dfrac{7r + 3s^2u^2}{s}$; $h_t(4, 3, 2, 1)$.

**33.** $f(r, s, t) = rst(r^2 + s^3 + t^4)$; $f_s(1, -1, 2)$.

**34.** $z = \dfrac{x^2 + y^2}{\ln x}$; $\left.\dfrac{\partial z}{\partial x}\right|_{\substack{x=e \\ y=0}}, \left.\dfrac{\partial z}{\partial y}\right|_{\substack{x=e \\ y=0}}$.

**35.** If $z = xe^{x-y} - ye^{y-x}$, show that

$$\frac{\partial z}{\partial x} + \frac{\partial z}{\partial y} = e^{x-y} - e^{y-x}.$$

**36. Stock Prices of a Dividend Cycle** In a discussion of stock prices of a dividend cycle, Palmon and Yaari[3] consider the function $f$ given by

$$u = f(t, r, z) = \frac{(1 + r)^{1-z} \ln(1 + r)}{(1 + r)^{1-z} - t},$$

where $u$ is the instantaneous rate of ask-price appreciation, $r$ is an annual opportunity rate of return, $z$ is the fraction of a dividend cycle over which a share of stock is held by a midcycle seller, and $t$ is the effective rate of capital gains tax. They claim that

$$\frac{\partial u}{\partial z} = \frac{t(1 + r)^{1-z} \ln^2(1 + r)}{[(1 + r)^{1-z} - t]^2}.$$

Verify this.

**37. Money Demand** In a discussion of inventory theory of money demand, Swanson[4] considers the function

$$F(b, C, T, i) = \frac{bT}{C} + \frac{iC}{2}$$

and determines that $\dfrac{\partial F}{\partial C} = -\dfrac{bT}{C^2} + \dfrac{i}{2}$. Verify this partial derivative.

**38. Interest Rate Deregulation** In an article on interest rate deregulation, Christofi and Agapos[5] arrive at the equation

$$r_L = r + D\frac{\partial r}{\partial D} + \frac{dC}{dD}, \tag{3}$$

where $r$ is the deposit rate paid by commercial banks, $r_L$ is the rate earned by commercial banks, $C$ is the administrative cost of transforming deposits into return-earning assets, and $D$ is the savings deposit level. Christofi and Agapos state that

$$r_L = r\left[\frac{1 + \eta}{\eta}\right] + \frac{dC}{dD}, \tag{4}$$

where $\eta = \dfrac{r/D}{\partial r/\partial D}$ is the deposit elasticity with respect to the deposit rate. Express Eq. (3) in terms of $\eta$ to verify Eq. (4).

**39. Advertising and Profitability** In an analysis of advertising and profitability, Swales[6] considers a function $f$ given by

$$R = f(r, a, n) = \frac{r}{1 + a\left(\dfrac{n - 1}{2}\right)},$$

where $R$ is the adjusted rate of profit, $r$ is the accounting rate of profit, $a$ is a measure of advertising expenditures, and $n$ is the number of years that advertising fully depreciates. In the analysis, Swales determines $\partial R/\partial n$. Find this partial derivative.

[3]D. Palmon and U. Yaari, "Taxation of Capital Gains and the Behavior of Stock Prices over the Dividend Cycle," *The American Economist,* XXVII, no. 1 (1983), 13–22.

[4]P. E. Swanson, "Integer Constraints on the Inventory Theory of Money Demand," *Quarterly Journal of Business and Economics,* 23, no. 1 (1984), 32–37.

[5]A. Christofi and A. Agapos, "Interest Rate Deregulation: An Empirical Justification," *Review of Business and Economic Research,* XX (1984), 39–49.

[6]J. K. Swales, "Advertising as an Intangible Asset: Profitability and Entry Barriers: A Comment on Reekie and Bhoyrub," *Applied Economics,* 17, no. 4 (1985), 603–17.

**To develop the notions of partial marginal cost, marginal productivity, and competitive and complementary products.**

Here we have "rate of change" interpretations of partial derivatives.

# 19.3 APPLICATIONS OF PARTIAL DERIVATIVES

From Sec. 19.2, we know that if $z = f(x, y)$, then $\partial z/\partial x$ and $\partial z/\partial y$ can be geometrically interpreted as giving the slopes of the tangent lines to the surface $z = f(x, y)$ in the x- and y-directions, respectively. There are other interpretations: Because $\partial z/\partial x$ is the derivative of $z$ with respect to $x$ when $y$ is held fixed, and because a derivative is a rate of change, we have

$\dfrac{\partial z}{\partial x}$ is the rate of change of $z$ with respect to $x$ when $y$ is held fixed.

Similarly,

$\dfrac{\partial z}{\partial y}$ is the rate of change of $z$ with respect to $y$ when $x$ is held fixed.

We shall now look at some applications in which the "rate of change" notion of a partial derivative is very useful.

Suppose a manufacturer produces $x$ units of product X and $y$ units of product Y. Then the total cost $c$ of these units is a function of $x$ and $y$ and is called a **joint-cost function.** If such a function is $c = f(x, y)$, then $\partial c/\partial x$ is called the **(partial) marginal cost with respect to $x$** and is the rate of change of $c$ with respect to $x$ when $y$ is held fixed. Similarly, $\partial c/\partial y$ is the **(partial) marginal cost with respect to $y$** and is the rate of change of $c$ with respect to $y$ when $x$ is held fixed.

For example, if $c$ is expressed in dollars and $\partial c/\partial y = 2$, then the cost of producing an extra unit of Y when the level of production of X is fixed is approximately two dollars.

If a manufacturer produces $n$ products, the joint-cost function is a function of $n$ variables, and there are $n$ (partial) marginal-cost functions.

### EXAMPLE 1  Marginal Costs

*A company manufactures two types of skis, the Lightning and the Alpine models. Suppose the joint-cost function for producing x pairs of the Lightning model and y pairs of the Alpine model per week is*

$$c = f(x, y) = 0.06x^2 + 65x + 75y + 1000,$$

*where c is expressed in dollars. Determine the marginal costs $\partial c/\partial x$ and $\partial c/\partial y$ when $x = 100$ and $y = 50$, and interpret the results.*

*Solution:* The marginal costs are

$$\frac{\partial c}{\partial x} = 0.12x + 65 \qquad \text{and} \qquad \frac{\partial c}{\partial y} = 75.$$

Thus,

$$\left.\frac{\partial c}{\partial x}\right|_{(100,\,50)} = 0.12(100) + 65 = 77 \tag{1}$$

and

$$\left.\frac{\partial c}{\partial y}\right|_{(100,\,50)} = 75. \tag{2}$$

Equation (1) means that increasing the output of the Lightning model from 100 to 101, while maintaining production of the Alpine model at 50, increases costs by approximately $77. Equation (2) means that increasing the output of the Alpine model from 50 to 51 and holding production of the Lightning model at 100 will increase costs by approximately $75. In fact, since $\partial c/\partial y$ is a constant function, the marginal cost with respect to $y$ is $75 at all levels of production. ∎

**EXAMPLE 2  Loss of Body Heat**

*On a cold day, a person may feel colder when the wind is blowing than when the wind is calm because the rate of heat loss is a function of both temperature and wind speed. The equation*

$$H = (10.45 + 10\sqrt{w} - w)(33 - t)$$

*indicates the rate of heat loss H (in kilocalories per square meter per hour) when the air temperature is t (in degrees Celsius) and the wind speed is w (in meters per second). For H = 2000, exposed flesh will freeze in one minute.*[7]

**a.** *Evaluate H when t = 0 and w = 4.*

*Solution:* When $t = 0$ and $w = 4$,

$$H = (10.45 + 10\sqrt{4} - 4)(33 - 0) = 872.85.$$

**b.** *Evaluate $\partial H/\partial w$ and $\partial H/\partial t$ when $t = 0$ and $w = 4$, and interpret the results.*

*Solution:*

$$\frac{\partial H}{\partial w} = \left(\frac{5}{\sqrt{w}} - 1\right)(33 - t), \qquad \frac{\partial H}{\partial w}\bigg|_{\substack{t=0 \\ w=4}} = 49.5;$$

$$\frac{\partial H}{\partial t} = (10.45 + 10\sqrt{w} - w)(-1), \qquad \frac{\partial H}{\partial t}\bigg|_{\substack{t=0 \\ w=4}} = -26.45.$$

These equations mean that when $t = 0$ and $w = 4$, increasing $w$ by a small amount while keeping $t$ fixed will make $H$ increase approximately 49.5 times as much as $w$ increases. Increasing $t$ by a small amount while keeping $w$ fixed will make $H$ *decrease* approximately 26.45 times as much as $t$ increases.

**c.** *When $t = 0$ and $w = 4$, which has a greater effect on H: a change in wind speed of 1 m/s or a change in temperature of 1°C?*

*Solution:* Since the partial derivative of $H$ with respect to $w$ is greater in magnitude than the partial with respect to $t$ when $t = 0$ and $w = 4$, a change in wind speed of 1 m/s has a greater effect on $H$. ∎

The output of a product depends on many factors of production. Among these may be labor, capital, land, machinery, and so on. For simplicity, let us suppose that output depends only on labor and capital. If the function $P = f(l, k)$ gives the output $P$ when the producer uses $l$ units of labor and $k$ units of capital, then this function is called a **production function.** We define the **marginal productivity with respect to $l$** to be $\partial P/\partial l$. This is the rate of

---

[7]G. E. Folk, Jr., *Textbook of Environmental Physiology,* 2d ed. (Philadelphia: Lea & Febiger, 1974).

change of $P$ with respect to $l$ when $k$ is held fixed. Likewise, the **marginal productivity with respect to $k$** is $\partial P/\partial k$ and is the rate of change of $P$ with respect to $k$ when $l$ is held fixed.

### EXAMPLE 3  Marginal Productivity

*A manufacturer of a popular toy has determined that the production function is $P = \sqrt{lk}$, where $l$ is the number of labor-hours per week and $k$ is the capital (expressed in hundreds of dollars per week) required for a weekly production of $P$ gross of the toy. (One gross is 144 units.) Determine the marginal productivity functions, and evaluate them when $l = 400$ and $k = 16$. Interpret the results.*

*Solution:* Since $P = (lk)^{1/2}$,

$$\frac{\partial P}{\partial l} = \frac{1}{2}(lk)^{-1/2} k = \frac{k}{2\sqrt{lk}}$$

and

$$\frac{\partial P}{\partial k} = \frac{1}{2}(lk)^{-1/2} l = \frac{l}{2\sqrt{lk}}.$$

Evaluating these equations when $l = 400$ and $k = 16$, we obtain

$$\left.\frac{\partial P}{\partial l}\right|_{\substack{l=400 \\ k=16}} = \frac{16}{2\sqrt{400(16)}} = \frac{1}{10}$$

and

$$\left.\frac{\partial P}{\partial k}\right|_{\substack{l=400 \\ k=16}} = \frac{400}{2\sqrt{400(16)}} = \frac{5}{2}.$$

Thus, if $l = 400$ and $k = 16$, increasing $l$ to 401 and holding $k$ at 16 will increase output by approximately $\frac{1}{10}$ gross. But if $k$ is increased to 17 while $l$ is held at 400, the output increases by approximately $\frac{5}{2}$ gross. ∎

### Competitive and Complementary Products

Sometimes two products may be related such that changes in the price of one of them affect the demand for the other. A typical example is that of butter and margarine. If such a relationship exists between products A and B, then the demand for each product is dependent on the prices of both. Suppose $q_A$ and $q_B$ are the quantities demanded for A and B, respectively, and $p_A$ and $p_B$ are their respective prices. Then both $q_A$ and $q_B$ are functions of $p_A$ and $p_B$:

$$q_A = f(p_A, p_B), \qquad \text{demand function for A;}$$
$$q_B = g(p_A, p_B), \qquad \text{demand function for B.}$$

We can find four partial derivatives:

$$\frac{\partial q_A}{\partial p_A}, \quad \textit{the marginal demand for A with respect to } p_A,$$

$$\frac{\partial q_A}{\partial p_B}, \quad \textit{the marginal demand for A with respect to } p_B,$$

$$\frac{\partial q_B}{\partial p_A}, \quad \textit{the marginal demand for B with respect to } p_A,$$

$$\frac{\partial q_B}{\partial p_B}, \quad \textit{the marginal demand for B with respect to } p_B.$$

Under typical conditions, if the price of B is fixed and the price of A increases, then the quantity of A demanded will decrease. Thus, $\partial q_A/\partial p_A < 0$. Similarly, $\partial q_B/\partial p_B < 0$. However, $\partial q_A/\partial p_B$ and $\partial q_B/\partial p_A$ may be either positive or negative. If

$$\frac{\partial q_A}{\partial p_B} > 0 \quad \text{and} \quad \frac{\partial q_B}{\partial p_A} > 0,$$

then A and B are said to be **competitive products** or **substitutes.** In this situation, an increase in the price of B causes an increase in the demand for A, if it is assumed that the price of A does not change. Similarly, an increase in the price of A causes an increase in the demand for B when the price of B is held fixed. Butter and margarine are examples of substitutes.

Proceeding to a different situation, we say that if

$$\frac{\partial q_A}{\partial p_B} < 0 \quad \text{and} \quad \frac{\partial q_B}{\partial p_A} < 0,$$

then A and B are **complementary products.** In this case, an increase in the price of B causes a decrease in the demand for A if the price of A does not change. Similarly, an increase in the price of A causes a decrease in the demand for B when the price of B is held fixed. For example, cameras and film are complementary products. An increase in the price of film will make picture-taking more expensive. Hence, the demand for cameras will decrease.

**EXAMPLE 4  Determining Whether Products Are Competitive or Complementary**

*The demand functions for products* A *and* B *are each a function of the prices of* A *and* B *and are given by*

$$q_A = \frac{50\sqrt[3]{p_B}}{\sqrt{p_A}} \quad \text{and} \quad q_B = \frac{75p_A}{\sqrt[3]{p_B^2}},$$

*respectively. Find the four marginal-demand functions, and also determine whether* A *and* B *are competitive products, complementary products, or neither.*

**Solution:** Writing $q_A = 50p_A^{-1/2}p_B^{1/3}$ and $q_B = 75p_A p_B^{-2/3}$, we have

$$\frac{\partial q_A}{\partial p_A} = 50\left(-\frac{1}{2}\right)p_A^{-3/2}p_B^{1/3} = -25\,p_A^{-3/2}p_B^{1/3},$$

$$\frac{\partial q_A}{\partial p_B} = 50p_A^{-1/2}\left(\frac{1}{3}\right)p_B^{-2/3} = \frac{50}{3}\,p_A^{-1/2}p_B^{-2/3},$$

$$\frac{\partial q_B}{\partial p_A} = 75(1)p_B^{-2/3} = 75p_B^{-2/3},$$

$$\frac{\partial q_B}{\partial p_B} = 75p_A\left(-\frac{2}{3}\right)p_B^{-5/3} = -50p_A p_B^{-5/3}.$$

Since $p_A$ and $p_B$ represent prices, they are both positive. Hence, $\partial q_A/\partial p_B > 0$ and $\partial q_B/\partial p_A > 0$. We conclude that A and B are competitive products. ∎

## ■ Exercise 19.3

*For the joint-cost functions in Problems 1–3, find the indicated marginal cost at the given production level.*

**1.** $c = 4x + 0.3y^2 + 2y + 500$;  $\dfrac{\partial c}{\partial y}$, $x = 20$, $y = 30$.    **2.** $c = x\sqrt{x+y} + 1000$;  $\dfrac{\partial c}{\partial x}$, $x = 40$, $y = 60$.

**3.** $c = 0.03(x + y)^3 - 0.6(x + y)^2 + 4.5(x + y) + 7700;$

$\dfrac{\partial c}{\partial x}, x = 50, \quad y = 50.$

*For the production functions in Problems 4 and 5, find the marginal-production functions $\partial P/\partial k$ and $\partial P/\partial l$.*

**4.** $P = 20lk - 2l^2 - 4k^2 + 800.$

**5.** $P = 1.582l^{0.192}k^{0.764}$

**6. Cobb-Douglas Production Function** In economics, a Cobb-Douglas production function is a production function of the form $P = Al^{\alpha}k^{\beta}$, where $A$, $\alpha$, and $\beta$ are constants and $\alpha + \beta = 1$. For such a function, show that

**a.** $\partial P/\partial l = \alpha P/l.$

**b.** $\partial P/\partial k = \beta P/k.$

**c.** $l\dfrac{\partial P}{\partial l} + k\dfrac{\partial P}{\partial k} = P.$ This means that summing the products of the marginal productivity of each factor and the amount of that factor results in the total product $P$.

*In Problems 7–9, $q_A$ and $q_B$ are demand functions for products A and B, respectively. In each case, find $\partial q_A/\partial p_A$, $\partial q_A/\partial p_B$, $\partial q_B/\partial p_A$, $\partial q_B/\partial p_B$ and determine whether A and B are competitive, complementary, or neither.*

**7.** $q_A = 1000 - 50p_A + 2p_B$;  $q_B = 500 + 4p_A - 20p_B.$

**8.** $q_A = 20 - p_A - 2p_B$;  $q_B = 50 - 2p_A - 3p_B.$

**9.** $q_A = \dfrac{100}{p_A\sqrt{p_B}}$;  $q_B = \dfrac{500}{p_B\sqrt[3]{p_A}}.$

**10. Canadian Manufacturing** The production function for the Canadian manufacturing industries for 1927 is estimated by[8] $P = 33.0l^{0.46}k^{0.52}$, where $P$ is product, $l$ is labor, and $k$ is capital. Find the marginal productivities for labor and capital, and evaluate when $l = 1$ and $k = 1$.

**11. Dairy Farming** An estimate of the production function for dairy farming in Iowa (1939) is given by[9]

$$P = A^{0.27} B^{0.01} C^{0.01} D^{0.23} E^{0.09} F^{0.27},$$

where $P$ is product, $A$ is land, $B$ is labor, $C$ is improvements, $D$ is liquid assets, $E$ is working assets, and $F$ is cash operating expenses. Find the marginal productivities for labor and improvements.

**12. Production Function** Suppose a production function is given by $P = \dfrac{kl}{k + l}.$

**a.** Determine the marginal-productivity functions.

**b.** Show that when $k = l$, the sum of the marginal productivities is a constant.

**13. M.B.A. Compensation** In a study of success among graduates with master's of business administration (M.B.A.) degrees, it was estimated that for staff managers (which include accountants, analysts, etc.), current annual compensation (in dollars) was given by

$$z = 10{,}990 + 1120x + 873y,$$

where $x$ and $y$ are the number of years of work experience before and after receiving the M.B.A. degree, respectively.[10] Find $\partial z/\partial x$ and interpret your result.

**14. Status** A person's general status $S_g$ is believed to be a function of status attributable to education, $S_e$, and status attributable to income, $S_i$, where $S_g$ $S_e$, and $S_i$ are represented numerically. If

$$S_g = 7\sqrt[3]{S_e}\,\sqrt{S_i},$$

determine $\partial S_g/\partial S_e$ and $\partial S_g/\partial S_i$ when $S_e = 125$ and $S_i = 100$, and interpret your results.[11]

**15. Reading Ease** Sometimes we want to evaluate the degree of readability of a piece of writing. Rudolf Flesch[12] developed a function of two variables that will do this, namely,

$$R = f(w, s) = 206.835 - (1.015w + 0.846s),$$

where $R$ is called the *reading ease score*, $w$ is the average number of words per sentence in 100-word samples, and $s$ is the average number of syllables in such samples. Flesch says that an article for which $R = 0$ is "practically unreadable," but one with $R = 100$ is "easy for any literate person." (a) Find $\partial R/\partial w$ and $\partial R/\partial s$. (b) Which is "easier" to read: an article for which $w = w_0$ and $s = s_0$, or one for which $w = w_0 + 1$ and $s = s_0$?

[8]P. Daly and P. Douglas, "The Production Function for Canadian Manufactures," *Journal of the American Statistical Association,* 38 (1943), 178–86.

[9]G. Tintner and O. H. Brownlee, "Production Functions Derived from Farm Records," *American Journal of Agricultural Economics,* 26 (1944), 566–71.

[10]A. G. Weinstein and V. Srinivasen, "Predicting Managerial Success of Master of Business Administration (M.B.A.) Graduates," *Journal of Applied Psychology,* 59, no. 2 (1974), 207–12.

[11]Adapted from R. K. Leik and B. F. Meeker, *Mathematical Sociology* (Englewood Cliffs, NJ Prentice-Hall, Inc., 1975).

[12]R. Flesch, *The Art of Readable Writing* (New York: Harper & Row Publishers, Inc., 1949).

**16. Model for Voice** The study of frequency of vibrations of a taut wire is useful in considering such things as an individual's voice. Suppose

$$\omega = \frac{1}{bL}\sqrt{\frac{\tau}{\pi\rho}},$$

where $\omega$ (a Greek letter read "omega") is frequency, $b$ is diameter, $L$ is length, $\rho$ (a Greek letter read "rho") is density, and $\tau$ (a Greek letter read "tau") is tension.[13] Find $\partial\omega/\partial b$, $\partial\omega/\partial L$, $\partial\omega/\partial\rho$, and $\partial\omega/\partial\tau$.

**17. Traffic Flow** Consider the following traffic-flow situation. On a highway where two lanes of traffic flow in the same direction, there is a maintenance vehicle blocking the left lane. (See Fig. 19.11.) Two vehicles (*lead* and

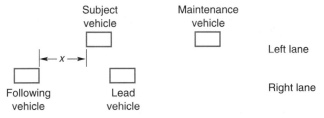

**FIGURE 19.11** Diagram for Problem 17.

*following*) are in the right lane with a gap between them. The *subject* vehicle can choose either to fill or not to fill the gap. That decision may be based not only on the distance $x$ shown in the diagram, but also on other factors (such as the velocity of the *following* vehicle). A *gap index* $g$ has been used in analyzing such a decision.[14,15] The greater the $g$-value, the greater is the propensity for the *subject* vehicle to fill the gap. Suppose

$$g = \frac{x}{V_F} - \left(0.75 + \frac{V_F - V_S}{19.2}\right),$$

where $x$ (in feet) is as before, $V_F$ is the velocity of the *following* vehicle (in feet per second), and $V_S$ is the velocity of the *subject* vehicle (in feet per second). From the diagram, it seems reasonable that if both $V_F$ and $V_S$ are fixed and $x$ increases, then $g$ should increase. Show that this is true by applying calculus to the function $g$. Assume that $x$, $V_F$, and $V_S$ are positive.

**18. Demand** Suppose the demand equations for related products A and B are

$$q_A = e^{-(p_A/p_B)} \quad \text{and} \quad q_B = \frac{16}{p_A p_B^2},$$

where $q_A$ and $q_B$ are the number of units of A and B demanded when the unit prices (in thousands of dollars) are $p_A$ and $p_B$, respectively.

**a.** Classify A and B as competitive, complementary, or neither.

**b.** If the unit prices of A and B are $1000 and $2000, respectively, estimate the change in the demand for A when the price of B is decreased by $40 and the price of A is held constant.

**19. Demand** The demand equations for related products A and B are given by

$$q_A = \frac{30\sqrt{p_B}}{p_A^{2/3}} \quad \text{and} \quad q_B = \frac{50p_A}{p_B^{1/3}},$$

where $q_A$ and $q_B$ are the quantities of A and B demanded and $p_A$ and $p_B$ are the corresponding prices (in dollars) per unit.

**a.** Find the values of the two marginal demands for product A when $p_A = 8$ and $p_B = 64$.

**b.** If $p_B$ were reduced to 60 from 64, with $p_A$ fixed at 8, use part (a) to estimate the corresponding change in demand for product A.

**20. Joint-Cost Function** A manufacturer's joint-cost function for producing $q_A$ units of product A and $q_B$ units of product B is given by

$$c = \frac{q_A^2(q_B^3 + q_A)^{1/2}}{17} + q_A q_B^{1/3} + 600,$$

where $c$ is in dollars.

**a.** Find the marginal-cost functions with respect to $q_A$ and $q_B$.

**b.** Evaluate the marginal-cost function with respect to $q_A$ when $q_A = 17$ and $q_B = 8$. Round your answer to two decimal places.

**c.** Use your answer to part (b) to estimate the change in cost if production of product A is decreased from 17 to 16 units, while production of product B is held constant at 8 units.

**21. Elections** For the congressional elections of 1974, the Republican percentage, $R$, of the Republican–Democratic vote in a district is given (approximately) by[16]

$$R = f(E_r, E_d, I_r, I_d, N)$$

$$= 15.4725 + 2.5945E_r - 0.0804E_r^2 - 2.3648E_d +$$

$$0.0687E_d^2 + 2.1914I_r - 0.0912I_r^2 -$$

$$0.8096I_d + 0.0081I_d^2 - 0.0277E_r I_r +$$

$$0.0493E_d I_d + 0.8579N - 0.0061N^2.$$

[13]R. M. Thrall, J. A. Mortimer, K. R. Rebman, and R. F. Baum, eds., *Some Mathematical Models in Biology,* rev. ed., Report No. 40241-R-7. Prepared at University of Michigan, 1967.

[14]P. M. Hurst, K. Perchonok, and E. L. Seguin, "Vehicle Kinematics and Gap Acceptance," *Journal of Applied Psychology,* 52, no. 4 (1968), 321–24.

[15]K. Perchonok and P. M. Hurst, "Effect of Lane-Closure Signals upon Driver Decision Making and Traffic Flow," *Journal of Applied Psychology,* 52, no. 5 (1968), 410–13.

[16]J. Silberman and G. Yochum, "The Role of Money in Determining Election Outcomes," *Social Science Quarterly,* 58, no. 4 (1978), 671–82.

Here $E_r$ and $E_d$ are the campaign expenditures (in units of \$10,000) by Republicans and Democrats, respectively; $I_r$ and $I_d$ are the number of terms served in Congress, *plus* one, for the Republican and Democratic candidates, respectively, and $N$ is the percentage of the two-party presidential vote that Richard Nixon received in the district for 1968. The variable $N$ gives a measure of Republican strength in the district.

**a.** In the Federal Election Campaign Act of 1974, Congress set a limit of \$188,000 on campaign expenditures. By analyzing $\partial R / \partial E_r$, would you have advised a Republican candidate who served nine terms in Congress to spend \$188,000 on his or her campaign?

**b.** Find the percentage above which the Nixon vote had a negative effect on $R$; that is, find $N$ when $\partial R / \partial N < 0$. Give your answer to the nearest percent.

**22. Sales** After a new product has been launched onto the market, its sales volume (in thousands of units) is given by

$$S = \frac{AT + 450}{\sqrt{A + T^2}},$$

where $T$ is the time (in months) since the product was first introduced and $A$ is the amount (in *hundreds* of dollars) spent each month on advertising.

**a.** Verify that the partial derivative of sales volume with respect to time is given by

$$\frac{\partial S}{\partial T} = \frac{A^2 - 450T}{(A + T^2)^{3/2}}.$$

**b.** Use the result in part (a) to predict the number of months that will elapse before the sales volume begins to decrease if the amount allocated to advertising is held fixed at \$9000 per month.

*Let f be a demand function for product* **A** *and* $q_A = f(p_A, p_B)$, *where* $q_A$ *is the quantity of* **A** *demanded when the price per unit of* **A** *is* $p_A$ *and the price per unit of product* **B** *is* $p_B$. *The partial elasticity of demand for* **A** *with respect to* $p_A$, *denoted* $\eta_{p_A}$, *is defined as* $\eta_{p_A} = (p_A/q_A)(\partial q_A/\partial p_A)$ *The partial elasticity of demand for* **A** *with respect to* $p_B$, *denoted* $\eta_{p_B}$, *is defined as* $\eta_{p_B} = (p_B/q_A)(\partial q_A/\partial p_B)$. *Loosely speaking,* $\eta_{p_A}$ *is the ratio of a percentage change in the quantity of* **A** *demanded to a percentage change in the price of* **A** *when the price of* **B** *is fixed. Similarly,* $\eta_{p_B}$ *can be loosely interpreted as the ratio of a percentage change in the quantity of* **A** *demanded to a percentage change in the price of* **B** *when the price of* **A** *is fixed. In Problems 23–25, find* $\eta_{p_A}$ *and* $\eta_{p_B}$ *for the given values of* $p_A$ *and* $p_B$.

**23.** $q_A = 1000 - 50p_A + 2p_B$;   $p_A = 2, p_B = 10$.

**24.** $q_A = 20 - p_A - 2p_B$;   $p_A = 2, p_B = 2$.

**25.** $q_A = 100/(p_A \sqrt{p_B})$;   $p_A = 1, p_B = 4$.

**OBJECTIVE**

**To find partial derivatives of a function defined implicitly.**

# 19.4 Implicit Partial Differentiation[17]

An equation in $x$, $y$, and $z$ does not necessarily define $z$ as a function of $x$ and $y$. For example, in the equation

$$z^2 - x^2 - y^2 = 0, \tag{1}$$

if $x = 1$ and $y = 1$, then $z^2 - 1 - 1 = 0$, so $z = \pm\sqrt{2}$. Thus, Eq. (1) does not define $z$ as a function of $x$ and $y$. However, solving Eq. (1) for $z$ gives

$$z = \sqrt{x^2 + y^2} \quad \text{or} \quad z = -\sqrt{x^2 + y^2},$$

each of which defines $z$ as a function of $x$ and $y$. Although Eq. (1) does not explicitly express $z$ as a function of $x$ and $y$, it can be thought of as expressing $z$ *implicitly* as one of two different functions of $x$ and $y$. Note that the equation $z^2 - x^2 - y^2 = 0$ has the form $F(x, y, z) = 0$, where $F$ is a function of three variables. Any equation of the form $F(x, y, z) = 0$ can be thought of as expressing $z$ implicitly as one of a set of possible functions of $x$ and $y$. Moreover,

---

[17]May be omitted without loss of continuity.

we can find $\partial z/\partial x$ and $\partial z/\partial y$ directly from the form $F(x, y, z) = 0$.

To find $\partial z/\partial x$ for

$$z^2 - x^2 - y^2 = 0, \tag{2}$$

we first differentiate both sides of Eq. (2) with respect to $x$ while treating $z$ as a function of $x$ and $y$ and treating $y$ as a constant:

$$\frac{\partial}{\partial x}(z^2 - x^2 - y^2) = \frac{\partial}{\partial x}(0),$$

$$\frac{\partial}{\partial x}(z^2) - \frac{\partial}{\partial x}(x^2) - \frac{\partial}{\partial x}(y^2) = 0,$$

$$2z\frac{\partial z}{\partial x} - 2x - 0 = 0.$$

Solving for $\partial z/\partial x$, we obtain

$$2z\frac{\partial z}{\partial x} = 2x,$$

$$\frac{\partial z}{\partial x} = \frac{x}{z}.$$

To find $\partial z/\partial y$, we differentiate both sides of Eq. (2) with respect to $y$ while treating $z$ as a function of $x$ and $y$ and treating $x$ as a constant:

$$\frac{\partial}{\partial y}(z^2 - x^2 - y^2) = \frac{\partial}{\partial y}(0),$$

$$2z\frac{\partial z}{\partial y} - 0 - 2y = 0,$$

$$2z\frac{\partial z}{\partial y} = 2y.$$

Hence,

$$\frac{\partial z}{\partial y} = \frac{y}{z}.$$

The method we used to find $\partial z/\partial x$ and $\partial z/\partial y$ is called *implicit partial differentiation*.

### EXAMPLE 1 Implicit Partial Differentiation

*If* $\dfrac{xz^2}{x + y} + y^2 = 0,$ *evaluate* $\dfrac{\partial z}{\partial x}$ *when* $x = -1, y = 2, and z = 2.$

*Solution:* We treat $z$ as a function of $x$ and $y$ and differentiate both sides of the equation with respect to $x$:

$$\frac{\partial}{\partial x}\left(\frac{xz^2}{x + y}\right) + \frac{\partial}{\partial x}(y^2) = \frac{\partial}{\partial x}(0).$$

Using the quotient rule for the first term on the left, we have

$$\frac{(x + y)\dfrac{\partial}{\partial x}(xz^2) - xz^2\dfrac{\partial}{\partial x}(x + y)}{(x + y)^2} + 0 = 0.$$

Using the product rule for $\frac{\partial}{\partial x}(xz^2)$ gives

$$\frac{(x + y)\left[x\left(2z\frac{\partial z}{\partial x}\right) + z^2(1)\right] - xz^2(1)}{(x + y)^2} = 0.$$

Solving for $\partial z / \partial x$, we obtain

$$2xz(x + y)\frac{\partial z}{\partial x} + z^2(x + y) - xz^2 = 0,$$

$$\frac{\partial z}{\partial x} = \frac{xz^2 - z^2(x + y)}{2xz(x + y)} = -\frac{yz}{2x(x + y)}, \qquad z \neq 0.$$

Thus,

$$\left.\frac{\partial z}{\partial x}\right|_{(-1,2,2)} = 2. \qquad \blacksquare$$

### EXAMPLE 2 Implicit Partial Differentiation

*If $se^{r^2+u^2} = u \ln(t^2 + 1)$, determine $\partial t / \partial u$.*

**Solution:** We consider $t$ as a function of $r$, $s$, and $u$. By differentiating both sides with respect to $u$ while treating $r$ and $s$ as constants, we get

$$\frac{\partial}{\partial u}(se^{r^2+u^2}) = \frac{\partial}{\partial u}[u \ln(t^2 + 1)],$$

$$2sue^{r^2+u^2} = u\frac{\partial}{\partial u}[\ln(t^2 + 1)] + \ln(t^2 + 1)\frac{\partial}{\partial u}(u) \quad \text{(product rule)},$$

$$2sue^{r^2+u^2} = u\frac{2t}{t^2 + 1}\frac{\partial t}{\partial u} + \ln(t^2 + 1).$$

Therefore,

$$\frac{\partial t}{\partial u} = \frac{(t^2 + 1)[2sue^{r^2+u^2} - \ln(t^2 + 1)]}{2ut}. \qquad \blacksquare$$

### ▪ Exercise 19.4

*In Problems 1–11, find the indicated partial derivatives by the method of implicit partial differentiation.*

**1.** $x^2 + y^2 + z^2 = 9$; $\partial z / \partial x$. 

**2.** $z^2 - 3x^2 + y^2 = 0$; $\partial z / \partial x$.

**3.** $2z^3 - x^2 - 4y^2 = 0$; $\partial z / \partial y$. 

**4.** $3x^2 + y^2 + 2z^3 = 9$; $\partial z / \partial y$.

**5.** $x^2 - 2y - z^2 + x^2yz^2 = 20$; $\partial z / \partial x$. 

**6.** $z^3 - xz - y = 0$; $\partial z / \partial x$.

**7.** $e^x + e^y + e^z = 10$; $\partial z / \partial y$. 

**8.** $xyz + 2y^2x - z^3 = 0$; $\partial z / \partial x$.

**9.** $\ln(z) + z - xy = 1$; $\partial z / \partial x$. 

**10.** $\ln x + \ln y - \ln z = e^y$; $\partial z / \partial x$.

**11.** $(z^2 + 6xy)\sqrt{x^3 + 5} = 2$; $\partial z / \partial y$.

*In Problems 12–20, evaluate the indicated partial derivatives for the given values of the variables.*

**12.** $xz + xyz - 5 = 0$; $\partial z / \partial x, x = 1, y = 4, z = 1$. 

**13.** $xz^2 + yz - 12 = 0$; $\partial z / \partial x, x = 2, y = -2, z = 3$.

**14.** $e^{zx} = xyz$; $\partial z / \partial y, x = 1, y = -e^{-1}, z = -1$. 

**15.** $e^{yz} = -xyz$; $\partial z / \partial x, x = -e^2/2, y = 1, z = 2$.

**16.** $\sqrt{xz + y^2} - xy = 0$; $\partial z / \partial y, x = 2, y = 2, z = 6$. 

**17.** $\ln z = x + y$; $\partial z / \partial x, x = 5, y = -5, z = 1$.

**18.** $\dfrac{rs}{s^2 + t^2} = t;$ $\partial r/\partial t, r = 0, s = 1, t = 0.$

**19.** $\dfrac{s^2 + t^2}{rs} = 10;$ $\partial t/\partial r, r = 1, s = 2, t = 4.$

**20.** $\ln(x + z) + xyz = x^2 e^{y+z};$ $\partial z/\partial x, x = 0, y = 1, z = 1.$

**21. Joint-Cost Function** A joint-cost function is defined implicitly by the equation

$$c + \sqrt{c} = 12 + q_A \sqrt{9 + q_B^2},$$

where $c$ denotes the total cost (in dollars) for producing $q_A$ units of product A and $q_B$ units of product B.

**a.** If $q_A = 6$ and $q_B = 4$, find the corresponding value of $c$.

**b.** Determine the marginal costs with respect to $q_A$ and $q_B$ when $q_A = 6$ and $q_B = 4$.

---

**OBJECTIVE**

To compute higher-order partial derivatives.

## 19.5 HIGHER-ORDER PARTIAL DERIVATIVES

If $z = f(x, y)$, then not only is $z$ a function of $x$ and $y$, but also, $f_x$ and $f_y$ are each functions of $x$ and $y$. Hence, we may differentiate $f_x$ and $f_y$ to obtain **second-order partial derivatives** of $f$. Symbolically,

$$f_{xx} \text{ means } (f_x)_x, \qquad f_{xy} \text{ means } (f_x)_y,$$

$$f_{yx} \text{ means } (f_y)_x, \qquad f_{yy} \text{ means } (f_y)_y.$$

In terms of $\partial$-notation,

$$\frac{\partial^2 z}{\partial x^2} \text{ means } \frac{\partial}{\partial x}\left[\frac{\partial z}{\partial x}\right], \qquad \frac{\partial^2 z}{\partial y \partial x} \text{ means } \frac{\partial}{\partial y}\left[\frac{\partial z}{\partial x}\right],$$

$$\frac{\partial^2 z}{\partial x \partial y} \text{ means } \frac{\partial}{\partial x}\left[\frac{\partial z}{\partial y}\right], \qquad \frac{\partial^2 z}{\partial y^2} \text{ means } \frac{\partial}{\partial y}\left[\frac{\partial z}{\partial y}\right].$$

Note that to find $f_{xy}$, we first differentiate $f$ with respect to $x$. For $\partial^2 z / \partial x \, \partial y$, we first differentiate with respect to $y$.

We can extend our notation beyond second-order partial derivatives. For example, $f_{xxy}$ (or $\partial^3 z/\partial y \, \partial x^2$) is a third-order partial derivative of $f$, namely, the partial derivative of $f_{xx}$ (or $\partial^2 z/\partial x^2$) with respect to $y$. A generalization regarding higher-order partial derivatives to functions of more than two variables should be obvious.

**EXAMPLE 1** Second-Order Partial Derivatives

*Find the four second-order partial derivatives of $f(x, y) = x^2 y + x^2 y^2$.*

*Solution:* Since

$$f_x(x, y) = 2xy + 2xy^2,$$

we have

$$f_{xx}(x, y) = \frac{\partial}{\partial x}(2xy + 2xy^2) = 2y + 2y^2$$

and

$$f_{xy}(x, y) = \frac{\partial}{\partial y}(2xy + 2xy^2) = 2x + 4xy.$$

Also, since

$$f_y(x, y) = x^2 + 2x^2 y,$$

we have

$$f_{yy}(x, y) = \frac{\partial}{\partial y}(x^2 + 2x^2y) = 2x^2$$

and

$$f_{yx}(x, y) = \frac{\partial}{\partial x}(x^2 + 2x^2y) = 2x + 4xy.$$ ■

The derivatives $f_{xy}$ and $f_{yx}$ are called **mixed partial derivatives.** Observe in Example 1 that $f_{xy}(x, y) = f_{yx}(x, y)$. Under suitable conditions, mixed partial derivatives of a function are equal; that is, the order of differentiation is of no concern. You may assume that this is the case for all the functions that we consider.

**EXAMPLE 2   Mixed Partial Derivative**

*Find the value of* $\dfrac{\partial^3 w}{\partial z\, \partial y\, \partial x}\bigg|_{(1,2,3)}$ *if* $w = (2x + 3y + 4z)^3$.

*Solution:*

$$\frac{\partial w}{\partial x} = 3(2x + 3y + 4z)^2 \frac{\partial}{\partial x}(2x + 3y + 4z)$$

$$= 6(2x + 3y + 4z)^2,$$

$$\frac{\partial^2 w}{\partial y \partial x} = 6 \cdot 2(2x + 3y + 4z)\frac{\partial}{\partial y}(2x + 3y + 4z)$$

$$= 36(2x + 3y + 4z),$$

$$\frac{\partial^3 w}{\partial z\, \partial y\, \partial x} = 36 \cdot 4 = 144.$$

Thus,

$$\frac{\partial^3 w}{\partial z\, \partial y\, \partial x}\bigg|_{(1,2,3)} = 144.$$ ■

[18]**EXAMPLE 3   Second-Order Partial Derivative of Implicit Function**

*Determine* $\dfrac{\partial^2 z}{\partial x^2}$ *if* $z^2 = xy$.

*Solution:*  By implicit differentiation, we first determine $\partial z/\partial x$:

$$\frac{\partial}{\partial x}(z^2) = \frac{\partial}{\partial x}(xy),$$

$$2z\frac{\partial z}{\partial x} = y,$$

$$\frac{\partial z}{\partial x} = \frac{y}{2z}, \qquad z \neq 0.$$

Differentiating both sides with respect to $x$, we obtain

$$\frac{\partial}{\partial x}\left[\frac{\partial z}{\partial x}\right] = \frac{\partial}{\partial x}\left[\frac{1}{2}yz^{-1}\right],$$

$$\frac{\partial^2 z}{\partial x^2} = -\frac{1}{2}yz^{-2}\frac{\partial z}{\partial x}.$$

_____
[18]Omit if Sec. 19.4 was not covered.

Substituting $y/(2z)$ for $\partial z/\partial x$, we have

$$\frac{\partial^2 z}{\partial x^2} = -\frac{1}{2}\,yz^{-2}\left(\frac{y}{2z}\right) = -\frac{y^2}{4z^3}, \quad z \neq 0.$$ ∎

## ■ Exercise 19.5

*In Problems 1–10, find the indicated partial derivatives.*

**1.** $f(x, y) = 4x^2y;\quad f_x(x, y), f_{xy}(x, y).$

**2.** $f(x, y) = 4x^3 + 5x^2y^3 - 3y;\quad f_x(x, y), f_{xx}(x, y).$

**3.** $f(x, y) = 7x^2 + 3y;\quad f_y(x, y), f_{yy}(x, y), f_{yyx}(x, y).$

**4.** $f(x, y) = (x^2 + xy + y^2)(x^2 + xy + 1);$
$f_x(x, y), f_{xy}(x, y).$

**5.** $f(x, y) = 4e^{2xy};\quad f_y(x, y), f_{yx}(x, y), f_{yxy}(x, y).$

**6.** $f(x, y) = \ln(x^2 + y^2) + 2;\quad f_x(x, y), f_{xx}(x, y),$
$f_{xy}(x, y).$

**7.** $f(x, y) = (x + y)^2(xy);\quad f_x(x, y), f_y(x, y),$
$f_{xx}(x, y), f_{yy}(x, y).$

**8.** $f(x, y, z) = xy^2z^3;\quad f_x(x, y, z), f_{xz}(x, y, z), f_{xy}(x, y, z).$

**9.** $z = \sqrt{x^2 + y^2};\quad \dfrac{\partial z}{\partial x}, \dfrac{\partial^2 z}{\partial x^2}.$

**10.** $z = \dfrac{\ln(x^2 + 5)}{y};\quad \dfrac{\partial z}{\partial x}, \dfrac{\partial^2 z}{\partial y\,\partial x}.$

*In Problems 11–16, find the indicated value.*

**11.** If $f(x, y, z) = 7$, find $f_{yxx}(4, 3, -2).$

**12.** If $f(x, y, z) = z^2(3x^2 - 4xy^3)$, find $f_{xyz}(1, 2, 3).$

**13.** If $f(l, k) = 5l^3k^6 - lk^7$, find $f_{kkl}(2, 1).$

**14.** If $f(x, y) = 2x^2y + xy^2 - x^2y^2$, find $f_{xxy}(0, 1).$

**15.** If $f(x, y) = y^2e^x + \ln(xy)$, find $f_{xyy}(1, 1).$

**16.** If $f(x, y) = x^3 - 3xy^2 + x^2 - y^3$, find $f_{xy}(1, -1)$

**17. Cost Function** Suppose the cost $c$ of producing $q_A$ units of product A and $q_B$ units of product B is given by

$$c = (3q_A^2 + q_B^3 + 4)^{1/3},$$

and the coupled demand functions for the products are given by

$$q_A = 10 - p_A + p_B^2$$

and

$$q_B = 20 + p_A - 11p_B.$$

Find the value of

$$\frac{\partial^2 c}{\partial q_A \partial q_B}$$

when $p_A = 25$ and $p_B = 4$.

**18.** For $f(x, y) = x^4y^4 + 3x^3y^2 - 7x + 4$, show that

$$f_{xyx}(x, y) = f_{xxy}(x, y).$$

**19.** For $f(x, y) = 8x^3 + 2x^2y^2 + 5y^4$, show that

$$f_{xy}(x, y) = f_{yx}(x, y).$$

**20.** For $f(x, y) = xe^{y/x}$, show that

$$xf_{xx}(x, y) + yf_{xy}(x, y) = 0.$$

**21.** For $z = \ln(x^2 + y^2)$, show that $\dfrac{\partial^2 z}{\partial x^2} + \dfrac{\partial^2 z}{\partial y^2} = 0.$

**[19]22.** If $2z^2 - x^2 - 4y^2 = 0$, find $\dfrac{\partial^2 z}{\partial x^2}.$

**[19]23.** If $z^2 - 3x^2 + y^2 = 0$, find $\dfrac{\partial^2 z}{\partial y^2}.$

**[19]24.** If $2z^2 = x^2 + 2xy + xz$, find $\dfrac{\partial^2 z}{\partial x \partial y}.$

--------

[19]Omit if Sec. 19.4 was not covered.

--------

| OBJECTIVE | |
|---|---|

**To show how to find the partial derivative of a function of functions by using the chain rule.**

## 19.6 CHAIN RULE[20]

Suppose a manufacturer of two related products A and B has a joint-cost function given by

$$c = f(q_A, q_B),$$

where $c$ is the total cost of producing quantities $q_A$ and $q_B$ of A and B, respectively. Furthermore, suppose the demand functions for the products are

$$q_A = g(p_A, p_B) \quad \text{and} \quad q_B = h(p_A, p_B),$$

--------

[20]May be omitted without loss of continuity.

where $p_A$ and $p_B$ are the prices per unit of A and B, respectively. Since $c$ is a function of $q_A$ and $q_B$, and since both $q_A$ and $q_B$ are themselves functions of $p_A$ and $p_B$, $c$ can be viewed as a function of $p_A$ and $p_B$. (Appropriately, the variables $q_A$ and $q_B$ are called *intermediate variables* of $c$.) Consequently, we should be able to determine $\partial c / \partial p_A$, the rate of change of total cost with respect to the price of A. One way to do this is to substitute the expressions $g(p_A, p_B)$ and $h(p_A, p_B)$ for $q_A$ and $q_B$, respectively, into $c = f(q_A, q_B)$. Then $c$ is a function of $p_A$ and $p_B$, and we can differentiate $c$ with respect to $p_A$ directly. This approach has some drawbacks—especially when $f$, $g$, or $h$ is given by a complicated expression. Another way to approach the problem would be to use the chain rule (actually *a* chain rule), which we now state without proof.

---

**Chain Rule**

Let $z = f(x, y)$, where both $x$ and $y$ are functions of $r$ and $s$ given by $x = x(r, s)$ and $y = y(r, s)$. If $f$, $x$, and $y$ have continuous partial derivatives, then $z$ is a function of $r$ and $s$, and

$$\frac{\partial z}{\partial r} = \frac{\partial z}{\partial x} \frac{\partial x}{\partial r} + \frac{\partial z}{\partial y} \frac{\partial y}{\partial r}$$

and

$$\frac{\partial z}{\partial s} = \frac{\partial z}{\partial x} \frac{\partial x}{\partial s} + \frac{\partial z}{\partial y} \frac{\partial y}{\partial s}.$$

---

Note that in the chain rule, the number of intermediate variables of $z$ (two) is the same as the number of terms that compose each of $\partial z / \partial r$ and $\partial z / \partial s$.

Returning to the original situation concerning the manufacturer, we see that if $f$, $q_A$, and $q_B$ have continuous partial derivatives, then, by the chain rule,

$$\frac{\partial c}{\partial p_A} = \frac{\partial c}{\partial q_A} \frac{\partial q_A}{\partial p_A} + \frac{\partial c}{\partial q_B} \frac{\partial q_B}{\partial p_A}.$$

**EXAMPLE 1    Rate of Change of Cost**

*For a manufacturer of cameras and film, the total cost $c$ of producing $q_C$ cameras and $q_F$ units of film is given by*

$$c = 30q_C + 0.015q_C q_F + q_F + 900.$$

*The demand functions for the cameras and film are given by*

$$q_C = \frac{9000}{p_C \sqrt{p_F}} \quad and \quad q_F = 2000 - p_C - 400p_F,$$

*where $p_C$ is the price per camera and $p_F$ is the price per unit of film. Find the rate of change of total cost with respect to the price of the camera when $p_C = 50$ and $p_F = 2$.*

*Solution:* We must first determine $\partial c / \partial p_C$. By the chain rule,

$$\frac{\partial c}{\partial p_C} = \frac{\partial c}{\partial q_C} \frac{\partial q_C}{\partial p_C} + \frac{\partial c}{\partial q_F} \frac{\partial q_F}{\partial p_C}$$

$$= (30 + 0.015q_F) \left[ \frac{-9000}{p_C^2 \sqrt{p_F}} \right] + (0.015q_C + 1)(-1).$$

When $p_C = 50$ and $p_F = 2$, then $q_C = 90\sqrt{2}$ and $q_F = 1150$. Substituting these values into $\partial c/\partial p_C$ and simplifying, we have

$$\left.\frac{\partial c}{\partial p_C}\right|_{\substack{p_C=50 \\ p_F=2}} \approx -123.2. \qquad \blacksquare$$

The chain rule can be extended. For example, suppose $z = f(v, w, x, y)$ and $v$, $w$, $x$, and $y$ are all functions of $r$, $s$, and $t$. Then, if certain conditions of continuity are assumed, $z$ is a function of $r$, $s$, and $t$, and we have

$$\frac{\partial z}{\partial r} = \frac{\partial z}{\partial v}\frac{\partial v}{\partial r} + \frac{\partial z}{\partial w}\frac{\partial w}{\partial r} + \frac{\partial z}{\partial x}\frac{\partial x}{\partial r} + \frac{\partial z}{\partial y}\frac{\partial y}{\partial r},$$

$$\frac{\partial z}{\partial s} = \frac{\partial z}{\partial v}\frac{\partial v}{\partial s} + \frac{\partial z}{\partial w}\frac{\partial w}{\partial s} + \frac{\partial z}{\partial x}\frac{\partial x}{\partial s} + \frac{\partial z}{\partial y}\frac{\partial y}{\partial s},$$

and

$$\frac{\partial z}{\partial t} = \frac{\partial z}{\partial v}\frac{\partial v}{\partial t} + \frac{\partial z}{\partial w}\frac{\partial w}{\partial t} + \frac{\partial z}{\partial x}\frac{\partial x}{\partial t} + \frac{\partial z}{\partial y}\frac{\partial y}{\partial t}.$$

Observe that the number of intermediate variables of $z$ (four) is the same as the number of terms that form each of $\partial z/\partial r$, $\partial z/\partial s$, and $\partial z/\partial t$.

Now consider the situation where $z = f(x, y)$ such that $x = x(t)$ and $y = y(t)$. Then

$$\frac{dz}{dt} = \frac{\partial z}{\partial x}\frac{dx}{dt} + \frac{\partial z}{\partial y}\frac{dy}{dt}.$$

**Use the partial derivative symbols and the ordinary derivative symbols appropriately.**

Here we use the symbol $dz/dt$ rather than $\partial z/\partial t$, since $z$ can be considered a function of *one* variable $t$. Likewise, the symbols $dx/dt$ and $dy/dt$ are used rather than $\partial x/\partial t$ and $\partial y/\partial t$. As is typical, the number of terms that compose $dz/dt$ equals the number of intermediate variables of $z$. Other situations would be treated in a similar way.

**EXAMPLE 2  Chain Rule**

**a.** If $w = f(x, y, z) = 3x^2y + xyz - 4y^2z^3$, where

$$x = 2r - 3s, \quad y = 6r + s, \quad and \quad z = r - s,$$

determine $\partial w/\partial r$ and $\partial w/\partial s$.

**Solution:** Since $x$, $y$, and $z$ are functions of $r$ and $s$, then, by the chain rule,

$$\frac{\partial w}{\partial r} = \frac{\partial w}{\partial x}\frac{\partial x}{\partial r} + \frac{\partial w}{\partial y}\frac{\partial y}{\partial r} + \frac{\partial w}{\partial z}\frac{\partial z}{\partial r}$$

$$= (6xy + yz)(2) + (3x^2 + xz - 8yz^3)(6) + (xy - 12y^2z^2)(1)$$

$$= x(18x + 13y + 6z) + 2yz(1 - 24z^2 - 6yz).$$

Also,

$$\frac{\partial w}{\partial s} = \frac{\partial w}{\partial x}\frac{\partial x}{\partial s} + \frac{\partial w}{\partial y}\frac{\partial y}{\partial s} + \frac{\partial w}{\partial z}\frac{\partial z}{\partial s}$$

$$= (6xy + yz)(-3) + (3x^2 + xz - 8yz^3)(1) + (xy - 12y^2z^2)(-1)$$

$$= x(3x - 19y + z) - yz(3 + 8z^2 - 12yz).$$

**b.** If $z = \dfrac{x + e^y}{y}$, where $x = rs + se^{rt}$ and $y = 9 + rt$, evaluate $\partial z/\partial s$ when $r = -2, s = 5, \text{ and } t = 4$.

**Solution:** Since $x$ and $y$ are functions of $r$, $s$, and $t$ (note that we can write $y = 9 + rt + 0 \cdot s$), by the chain rule,

$$\frac{\partial z}{\partial s} = \frac{\partial z}{\partial x}\frac{\partial x}{\partial s} + \frac{\partial z}{\partial y}\frac{\partial y}{\partial s}$$

$$= \left(\frac{1}{y}\right)(r + e^{rt}) + \frac{\partial z}{\partial y} \cdot (0) = \frac{r + e^{rt}}{y}.$$

If $r = -2, s = 5,$ and $t = 4,$ then $y = 1$. Thus,

$$\left.\frac{\partial z}{\partial s}\right|_{\substack{r=-2 \\ s=5 \\ t=4}} = \frac{-2 + e^{-8}}{1} = -2 + e^{-8}. \qquad \blacksquare$$

### EXAMPLE 3 Chain Rule

**a.** Determine $\partial y/\partial r$ if $y = x^2 \ln(x^4 + 6)$ and $x = (r + 3s)^6$.

**Solution:** By the chain rule,

$$\frac{\partial y}{\partial r} = \frac{dy}{dx}\frac{\partial x}{\partial r}$$

$$= \left[x^2 \cdot \frac{4x^3}{x^4 + 6} + 2x \cdot \ln(x^4 + 6)\right][6(r + 3s)^5]$$

$$= 12x(r + 3s)^5\left[\frac{2x^4}{x^4 + 6} + \ln(x^4 + 6)\right].$$

**b.** Given that $z = e^{xy}, x = r - 4s,$ and $y = r - s,$ find $\partial z/\partial r$ in terms of $r$ and $s$.

**Solution:**

$$\frac{\partial z}{\partial r} = \frac{\partial z}{\partial x}\frac{\partial x}{\partial r} + \frac{\partial z}{\partial y}\frac{\partial y}{\partial r}$$

$$= (ye^{xy})(1) + (xe^{xy})(1)$$

$$= (x + y)e^{xy}.$$

Since $x = r - 4s$ and $y = r - s,$

$$\frac{\partial z}{\partial r} = [(r - 4s) + (r - s)]e^{(r-4s)(r-s)}$$

$$= (2r - 5s)e^{r^2 - 5rs + 4s^2}. \qquad \blacksquare$$

## ■ Exercise 19.6

*In Problems 1–12, find the indicated derivatives by using the chain rule.*

**1.** $z = 5x + 3y, x = 2r + 3s, y = r - 2s;$  $\partial z/\partial r, \partial z/\partial s.$

**2.** $z = x^2 + 3xy + 7y^3, x = r^2 - 2s, y = 5s^2;$
  $\partial z/\partial r, \partial z/\partial s.$

**3.** $z = e^{x+y}, x = t^2 + 3, y = \sqrt{t^3};$  $dz/dt.$

**4.** $z = \sqrt{8x + y}, x = t^2 + 3t + 4,$
  $y = t^3 + 4;$  $dz/dt.$

**5.** $w = x^2z^2 + xyz + yz^2, x = 5t,$
  $y = 2t + 3, z = 6 - t;$  $dw/dt.$

**6.** $w = \ln(x^2 + y^2 + z^2)$,
$x = 2 - 3t, y = t^2 + 3, z = 4 - t$; $dw/dt$.

**7.** $z = (x^2 + xy^2)^3, x = r + s + t$,
$y = 2r - 3s + t$; $\partial z/\partial t$.

**8.** $z = \sqrt{x^2 + y^2}, x = r^2 + s - t$,
$y = r - s + t$; $\partial z/\partial r$.

**9.** $w = x^2 + xyz + y^3z^2, x = r - s^2$,
$y = rs, z = 2r - 5s$; $\partial w/\partial s$.

**10.** $w = e^{xyz}, x = r^2s^3, y = r - s, z = rs^2$; $\partial w/\partial r$.

**11.** $y = x^2 - 7x + 5, x = 15rs + 2s^2t^2$; $\partial y/\partial r$.

**12.** $y = 4 - x^2, x = 2r + 3s - 4t$; $\partial y/\partial t$.

**13.** If $z = (4x + 3y)^3$, where $x = r^2s$ and $y = r - 2s$, evaluate $\partial z/\partial r$ when $r = 0$ and $s = 1$.

**14.** If $z = \sqrt{5x + 2y}$, where $x = 4t + 7$ and $y = t^2 - 3t + 4$, evaluate $dz/dt$ when $t = 1$.

**15.** If $w = e^{3x - y}(x^2 + 4z^3)$, where $x = rs, y = 2s - r$, and $z = r + s$, evaluate $\partial w/\partial s$ when $r = 1$ and $s = -1$.

**16.** If $y = x/(x - 5)$, where $x = 2t^2 - 3rs - r^2t$, evaluate $\partial y/\partial t$ when $r = 0, s = 2$, and $t = -1$.

**17. Cost Function** Suppose the cost $c$ of producing $q_A$ units of product A and $q_B$ units of product B is given by

$$c = (3q_A^2 + q_B^3 + 4)^{1/3}$$

and the coupled demand functions for the products are given by

$$q_A = 10 - p_A + p_B^2$$

and

$$q_B = 20 + p_A - 11p_B.$$

Use a chain rule to evaluate $\dfrac{\partial c}{\partial p_A}$ and $\dfrac{\partial c}{\partial p_B}$ when $p_A = 25$ and $p_B = 4$.

**18.** Suppose $w = f(x, y)$, where $x = g(t)$ and $y = h(t)$.
  **a.** State a chain rule that gives $dw/dt$.
  **b.** Suppose $h(t) = t$, so that $w = f(x, t)$ where $x = g(t)$. Use part (a) to find $dw/dt$ and simplify your answer.

**19. a.** Suppose $w$ is a function of $x$ and $y$, where both $x$ and $y$ are functions of $s$ and $t$. State a chain rule that expresses $\partial w/\partial s$ in terms of derivatives of these functions.
  **b.** Let $w = 3x^2 \ln(x - 2y)$, where $x = s\sqrt{t - 2}$ and $y = t - 3e^{1-s}$. Use part (a) to evaluate $\partial w/\partial s$ when $s = 1$ and $t = 3$.

**20. Production Function** In considering a production function $P = f(l, k)$, where $l$ is labor input and $k$ is capital input, Fon, Boulier, and Goldfarb[21] assume that $l = Lg(h)$, where $L$ is the number of workers, $h$ is the number of hours per day per worker, and $g(h)$ is a labor effectiveness function. In maximizing profit $p$ given by

$$p = aP - whL,$$

where $a$ is the price per unit of output and $w$ is the hourly wage per worker, Fon, Boulier, and Goldfarb determine $\partial p/\partial L$ and $\partial p/\partial h$. Assume that $k$ is independent of $L$ and $h$, and determine these partial derivatives.

[21]V. Fon, B. L. Boulier, and R. S. Goldfarb, "The Firm's Demand for Daily Hours of Work: Some Implications," *Atlantic Economic Journal,* XIII, no. 1 (1985), 36–42.

---

**To discuss relative maxima and relative minima, to find critical points, and to apply the second-derivative test for a function of two variables.**

# 19.7 Maxima and Minima for Functions of Two Variables

We now extend the notion of relative maxima and minima (or relative extrema) to functions of two variables.

### DEFINITION

*A function $z = f(x, y)$ is said to have a **relative maximum** at the point $(x_0, y_0)$ —that is, when $x = x_0$ and $y = y_0$— if, for all points $(x, y)$ in the plane that are sufficiently close to $(x_0, y_0)$, we have*

$$f(x_0, y_0) \geq f(x, y). \tag{1}$$

*For a **relative minimum,** we replace $\geq$ by $\leq$ in Eq. (1).*

To say that $z = f(x, y)$ has a relative maximum at $(x_0, y_0)$ means, geometrically, that the point $(x_0, y_0, z_0)$ on the graph of $f$ is higher than (or is as high as) all other points on the surface that are "near" $(x_0, y_0, z_0)$. In

Fig. 19.12(a), $f$ has a relative maximum at $(x_1, y_1)$. Similarly, the function $f$ in Fig. 19.12(b) has a relative minimum when $x = y = 0$, which corresponds to a *low* point on the surface.

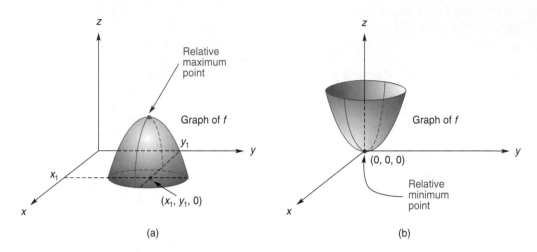

**FIGURE 19.12** Relative extrema.

Recall that in locating extrema for a function $y = f(x)$ of one variable, we examine those values of $x$ in the domain of $f$ for which $f'(x) = 0$ or $f'(x)$ does not exist. For functions of two (or more) variables, a similar procedure is followed. However, for the functions that concern us, extrema will not occur where a derivative does not exist, and such situations will be excluded from consideration.

Suppose $z = f(x, y)$ has a relative maximum at $(x_0, y_0)$, as indicated in Fig. 19.13(a). Then the curve where the plane $y = y_0$ intersects the surface must have a relative maximum when $x = x_0$. Hence, the slope of the tangent line to the surface in the $x$-direction must be 0 at $(x_0, y_0)$. Equivalently, $f_x(x, y) = 0$ at $(x_0, y_0)$. Similarly, on the curve where the plane $x = x_0$ intersects the surface [Fig. 19.13(b)], there must be a relative maximum when $y = y_0$. Thus, in the $y$-direction, the slope of the tangent to the surface must be

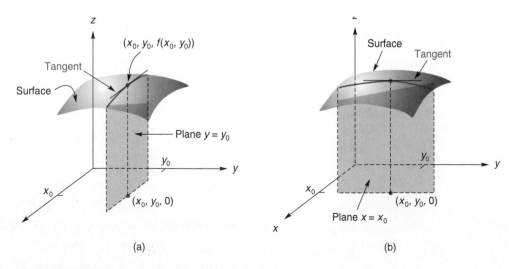

**FIGURE 19.13** At relative extremum, $f_x(x, y) = 0$ and $f_y(x, y) = 0$.

0 at $(x_0, y_0)$. Equivalently, $f_y(x, y) = 0$ at $(x_0, y_0)$. Since a similar discussion applies to a relative minimum, we can combine these results as follows:

---

**Rule 1**

If $z = f(x, y)$ has a relative maximum or minimum at $(x_0, y_0)$, and if both $f_x$ and $f_y$ are defined for all points close to $(x_0, y_0)$, it is necessary that $(x_0, y_0)$ be a solution of the system

$$\begin{cases} f_x(x, y) = 0, \\ f_y(x, y) = 0. \end{cases}$$

---

A point $(x_0, y_0)$ for which $f_x(x, y) = f_y(x, y) = 0$ is called a **critical point** of $f$. Thus, from Rule 1, we infer that, to locate relative extrema for a function, we should examine its critical points.

*Pitfall* ▼ Rule 1 does not imply that there must be an extremum at a critical point. Just as in the case of functions of one variable, a critical point can give rise to a relative maximum, a relative minimum, or neither. A critical point is only a *candidate* for a relative extremum.

Two additional comments are in order: First, Rule 1, as well as the notion of a critical point, can be extended to functions of more than two variables. For example, to locate possible extrema for $w = f(x, y, z)$, we would examine those points for which $w_x = w_y = w_z = 0$. Second, for a function whose domain is restricted, a thorough examination for absolute extrema would include a consideration of boundary points.

### EXAMPLE 1   Finding Critical Points

*Find the critical points of the following functions.*

**a.** $f(x, y) = 2x^2 + y^2 - 2xy + 5x - 3y + 1$.

   **Solution:** Since $f_x(x, y) = 4x - 2y + 5$ and $f_y(x, y) = 2y - 2x - 3$, we solve the system

$$\begin{cases} 4x - 2y + 5 = 0, \\ -2x + 2y - 3 = 0. \end{cases}$$

   This gives $x = -1$ and $y = \frac{1}{2}$. Thus, $(-1, \frac{1}{2})$ is the only critical point.

**b.** $f(l, k) = l^3 + k^3 - lk$.

   **Solution:**

$$\begin{cases} f_l(l, k) = 3l^2 - k = 0, & \text{(2)} \\ f_k(l, k) = 3k^2 - l = 0. & \text{(3)} \end{cases}$$

   From Eq. (2), $k = 3l^2$. Substituting for $k$ in Eq. (3) gives

$$0 = 27l^4 - l = l(27l^3 - 1).$$

   Hence, $l = 0$ or $l = \frac{1}{3}$. If $l = 0$, then $k = 0$; if $l = \frac{1}{3}$, then $k = \frac{1}{3}$. The critical points are therefore $(0, 0)$ and $(\frac{1}{3}, \frac{1}{3})$.

**c.** $f(x, y, z) = 2x^2 + xy + y^2 + 100 - z(x + y - 100)$.

*Solution:* Solving the system

$$\begin{cases} f_x(x, y, z) = 4x + y - z = 0, \\ f_y(x, y, z) = x + 2y - z = 0, \\ f_z(x, y, z) = -x - y + 100 = 0 \end{cases}$$

gives the critical point $(25, 75, 175)$, as you may verify. ∎

## EXAMPLE 2  Finding Critical Points

*Find the critical points of*

$$f(x, y) = x^2 - 4x + 2y^2 + 4y + 7.$$

*Solution:* We have $f_x(x, y) = 2x - 4$ and $f_y(x, y) = 4y + 4$. The system

$$\begin{cases} 2x - 4 = 0, \\ 4y + 4 = 0 \end{cases}$$

gives the critical point $(2, -1)$. Observe that we can write the given function as

$$f(x, y) = x^2 - 4x + 4 + 2(y^2 + 2y + 1) + 1$$

$$= (x - 2)^2 + 2(y + 1)^2 + 1,$$

and $f(2, -1) = 1$. Clearly, if $(x, y) \neq (2, -1)$, then $f(x, y) > 1$. Hence, a relative minimum occurs at $(2, -1)$. Moreover, there is an *absolute minimum* at $(2, -1)$, since $f(x, y) > f(2, -1)$ for *all* $(x, y) \neq (2, -1)$. ∎

Although in Example 2 we were able to show that the critical point gave rise to a relative extremum, in many cases this is not so easy to do. There is, however, a second-derivative test that gives conditions under which a critical point will be a relative maximum or minimum. We state it now, omitting the proof.

**Rule 2**
**Second-Derivative Test for Functions of Two Variables**

Suppose $z = f(x, y)$ has continuous partial derivatives $f_{xx}, f_{yy}$, and $f_{xy}$ at all points $(x, y)$ near the critical point $(x_0, y_0)$. Let $D$ be the function defined by

$$D(x, y) = f_{xx}(x, y) f_{yy}(x, y) - [f_{xy}(x, y)]^2.$$

Then

**a.** if $D(x_0, y_0) > 0$ and $f_{xx}(x_0, y_0) < 0, f$ has a relative maximum at $(x_0, y_0)$;

**b.** if $D(x_0, y_0) > 0$ and $f_{xx}(x_0, y_0) > 0, f$ has a relative minimum at $(x_0, y_0)$;

**c.** if $D(x_0, y_0) < 0, f$ has neither a relative maximum nor a relative minimum at $(x_0, y_0)$;

**d.** if $D(x_0, y_0) = 0$, no conclusion about an extremum at $(x_0, y_0)$ can be drawn, and further analysis is required.

**EXAMPLE 3   Applying the Second-Derivative Test**

*Examine $f(x, y) = x^3 + y^3 - xy$ for relative maxima or minima by using the second-derivative test.*

*Solution:* First we find critical points:

$$f_x(x, y) = 3x^2 - y, \qquad f_y(x, y) = 3y^2 - x.$$

In the same manner as in Example 1(b), solving $f_x(x, y) = f_y(x, y) = 0$ gives the critical points $(0, 0)$ and $(\frac{1}{3}, \frac{1}{3})$.

Now,

$$f_{xx}(x, y) = 6x, \qquad f_{yy}(x, y) = 6y, \qquad f_{xy}(x, y) = -1.$$

Thus,

$$D(x, y) = (6x)(6y) - (-1)^2 = 36xy - 1.$$

Since $D(0, 0) = 36(0)(0) - 1 = -1 < 0$, there is no relative extremum at $(0, 0)$. Also, since $D(\frac{1}{3}, \frac{1}{3}) = 36(\frac{1}{3})(\frac{1}{3}) - 1 = 3 > 0$ and $f_{xx}(\frac{1}{3}, \frac{1}{3}) = 6(\frac{1}{3}) = 2 > 0$, there is a relative minimum at $(\frac{1}{3}, \frac{1}{3})$. At this point, the value of the function is

$$f(\tfrac{1}{3}, \tfrac{1}{3}) = (\tfrac{1}{3})^3 + (\tfrac{1}{3})^3 - (\tfrac{1}{3})(\tfrac{1}{3}) = -\tfrac{1}{27}. \qquad \blacksquare$$

**EXAMPLE 4   A Saddle Point**

*Examine $f(x, y) = y^2 - x^2$ for relative extrema.*

*Solution:* Solving

$$f_x(x, y) = -2x = 0 \qquad \text{and} \qquad f_y(x, y) = 2y = 0,$$

we get the critical point $(0, 0)$. Now we apply the second-derivative test. At $(0, 0)$, and indeed at any point,

$$f_{xx}(x, y) = -2, \qquad f_{yy}(x, y) = 2, \qquad f_{xy}(x, y) = 0.$$

Because $D(0, 0) = (-2)(2) - (0)^2 = -4 < 0$, no relative extremum exists at $(0, 0)$. A sketch of $z = f(x, y) = y^2 - x^2$ appears in Fig. 19.14. Note that, for the surface curve cut by the plane $y = 0$, there is a *maximum* at $(0, 0)$; but for the surface curve cut by the plane $x = 0$, there is a *minimum* at $(0, 0)$. Thus, on the *surface,* no relative extremum can exist at the origin, although $(0, 0)$ is a critical point. Around the origin the curve is saddle shaped, and $(0, 0)$ is called a *saddle point* of $f$. $\qquad \blacksquare$

**EXAMPLE 5   Finding Relative Extrema**

*Examine $f(x, y) = x^4 + (x - y)^4$ for relative extrema.*

*Solution:* If we set

$$f_x(x, y) = 4x^3 + 4(x - y)^3 = 0 \tag{4}$$

and

$$f_y(x, y) = -4(x - y)^3 = 0, \tag{5}$$

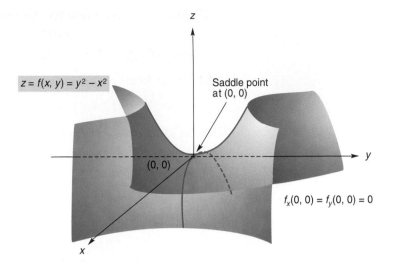

The surface in Fig. 19.14 is called a hyperbolic paraboloid.

$z = f(x, y) = y^2 - x^2$

Saddle point at (0, 0)

(0, 0)

$f_x(0, 0) = f_y(0, 0) = 0$

**FIGURE 19.14**    Saddle point.

then, from Eq. (5), we have $x - y = 0$, or $x = y$. Substituting into Eq. (4) gives $4x^3 = 0$, or $x = 0$. Thus, $x = y = 0$, and $(0, 0)$ is the only critical point. At $(0, 0)$,

$$f_{xx}(x, y) = 12x^2 + 12(x - y)^2 = 0,$$

$$f_{yy}(x, y) = 12(x - y)^2 = 0,$$

$$\text{and} \quad f_{xy}(x, y) = -12(x - y)^2 = 0.$$

Hence, $D(0, 0) = 0$, and the second-derivative test gives no information. However, for all $(x, y) \neq (0, 0)$, we have $f(x, y) > 0$, whereas $f(0, 0) = 0$. Therefore, at $(0, 0)$ the graph of $f$ has a low point, and we conclude that $f$ has a relative (and absolute) minimum at $(0, 0)$.    ∎

### Applications

In many situations involving functions of two variables, and especially in their applications, the nature of the given problem is an indicator of whether a critical point is in fact a relative (or absolute) maximum or a relative (or absolute) minimum. In such cases, the second-derivative test is not needed. Often, in mathematical studies of applied problems, the appropriate second-order conditions are assumed to hold.

### EXAMPLE 6    Maximizing Output

*Let P be a production function given by*

$$P = f(l, k) = 0.54l^2 - 0.02l^3 + 1.89k^2 - 0.09k^3,$$

*where l and k are the amounts of labor and capital, respectively, and P is the quantity of output produced. Find the values of l and k that maximize P.*

***Solution:*** To find the critical points, we solve the system $P_l = 0$ and $P_k = 0$:

$$P_l = 1.08l - 0.06l^2 \qquad P_k = 3.78k - 0.27k^2$$

$$= 0.06l(18 - l) = 0. \qquad = 0.27k(14 - k) = 0.$$

$$l = 0, l = 18. \qquad k = 0, k = 14.$$

There are four critical points: $(0, 0)$, $(0, 14)$, $(18, 0)$, and $(18, 14)$.

Now we apply the second-derivative test to each critical point. We have

$$P_{ll} = 1.08 - 0.12l, \qquad P_{kk} = 3.78 - 0.54k, \qquad P_{lk} = 0.$$

Thus,

$$\begin{aligned} D(l, k) &= P_{ll}P_{kk} - [P_{lk}]^2 \\ &= (1.08 - 0.12l)(3.78 - 0.54k). \end{aligned}$$

At $(0, 0)$,

$$D(0, 0) = 1.08(3.78) > 0.$$

Since $D(0, 0) > 0$ and $P_{ll} = 1.08 > 0$, there is a relative minimum at $(0, 0)$.
At $(0, 14)$,

$$D(0, 14) = 1.08(-3.78) < 0.$$

Because $D(0, 14) < 0$, there is no relative extremum at $(0, 14)$.
At $(18, 0)$,

$$D(18, 0) = (-1.08)(3.78) < 0.$$

Since $D(18, 0) < 0$, there is no relative extremum at $(18, 0)$.
At $(18, 14)$,

$$D(18, 14) = (-1.08)(-3.78) > 0.$$

Because $D(18, 14) > 0$ and $P_{ll} = -1.08 < 0$, there is a relative maximum at $(18, 14)$. Hence, the maximum output is obtained when $l = 18$ and $k = 14$. ∎

### EXAMPLE 7   Profit Maximization

*A candy company produces two types of candy, A and B, for which the average costs of production are constant at $2 and $3 per pound, respectively. The quantities $q_A$, $q_B$ (in pounds) of A and B that can be sold each week are given by the joint-demand functions*

$$q_A = 400(p_B - p_A)$$

*and*

$$q_B = 400(9 + p_A - 2p_B),$$

*where $p_A$ and $p_B$ are the selling prices (in dollars per pound) of A and B, respectively. Determine the selling prices that will maximize the company's profit P.*

*Solution:* The total profit is given by

$$P = \begin{pmatrix} \text{profit} \\ \text{per pound} \\ \text{of A} \end{pmatrix} \begin{pmatrix} \text{pounds} \\ \text{of A} \\ \text{sold} \end{pmatrix} + \begin{pmatrix} \text{profit} \\ \text{per pound} \\ \text{of B} \end{pmatrix} \begin{pmatrix} \text{pounds} \\ \text{of B} \\ \text{sold} \end{pmatrix}.$$

For A and B, the profits per pound are $p_A - 2$ and $p_B - 3$, respectively. Thus,

$$\begin{aligned} P &= (p_A - 2)q_A + (p_B - 3)q_B \\ &= (p_A - 2)[400(p_B - p_A)] + (p_B - 3)[400(9 + p_A - 2p_B)]. \end{aligned}$$

Notice that $P$ is expressed as a function of two variables, $p_A$ and $p_B$. To maximize $P$, we set its partial derivatives equal to 0:

$$\frac{\partial P}{\partial p_A} = (p_A - 2)[400(-1)] + [400(p_B - p_A)](1) + (p_B - 3)[400(1)]$$

$$= 0, \qquad -400 p_A + 800 + 400 p_B - 400 p_A + 400 p_B - 1200$$

$$-800 p_A + 800 p_B - 400$$

$$\frac{\partial P}{\partial p_B} = (p_A - 2)[400(1)] + (p_B - 3)[400(-2)] + [400(9 + p_A - 2p_B)](1)$$

$$= 0.$$

Simplifying the preceding two equations gives

$$\begin{cases} -2p_A + 2p_B - 1 = 0, \\ \phantom{-}2p_A - 4p_B + 13 = 0, \end{cases}$$

whose solution is $p_A = 5.5$ and $p_B = 6$. Moreover, we find that

$$\frac{\partial^2 P}{\partial p_A^2} = -800, \qquad \frac{\partial^2 P}{\partial p_B^2} = -1600, \qquad \frac{\partial^2 P}{\partial p_B \partial p_A} = 800.$$

Therefore,

$$D(5.5, 6) = (-800)(-1600) - (800)^2 > 0.$$

Since $\partial^2 P / \partial p_A^2 < 0$, we indeed have a maximum, and the company should sell candy A at \$5.50 per pound and B at \$6.00 per pound. ∎

## [22]EXAMPLE 8  Profit Maximization for a Monopolist

Suppose a monopolist is practicing price discrimination by selling the same product in two separate markets at different prices. Let $q_A$ be the number of units sold in market A, where the demand function is $p_A = f(q_A)$, and let $q_B$ be the number of units sold in market B, where the demand function is $p_B = g(q_B)$. Then the revenue functions for the two markets are

$$r_A = q_A f(q_A) \quad \text{and} \quad r_B = q_B g(q_B).$$

Assume that all units are produced at one plant, and let the cost function for producing $q \,(= q_A + q_B)$ units be $c = c(q)$. Keep in mind that $r_A$ is a function of $q_A$ and $r_B$ is a function of $q_B$. The monopolist's profit $P$ is

$$P = r_A + r_B - c.$$

To maximize $P$ with respect to outputs $q_A$ and $q_B$, we set its partial derivatives equal to zero. To begin with,

$$\frac{\partial P}{\partial q_A} = \frac{dr_A}{dq_A} + 0 - \frac{\partial c}{\partial q_A}$$

$$= \frac{dr_A}{dq_A} - \frac{dc}{dq}\frac{\partial q}{\partial q_A} = 0 \qquad \text{(chain rule)}.$$

Because

$$\frac{\partial q}{\partial q_A} = \frac{\partial}{\partial q_A}(q_A + q_B) = 1,$$

we have

$$\frac{\partial P}{\partial q_A} = \frac{dr_A}{dq_A} - \frac{dc}{dq} = 0. \tag{6}$$

---

[22]Omit if Sec. 19.6 was not covered.

Similarly,

$$\frac{\partial P}{\partial q_B} = \frac{dr_B}{dq_B} - \frac{dc}{dq} = 0. \tag{7}$$

From Eqs. (6) and (7), we get

$$\frac{dr_A}{dq_A} = \frac{dc}{dq} = \frac{dr_B}{dq_B}.$$

But $dr_A/dq_A$ and $dr_B/dq_B$ are marginal revenues, and $dc/dq$ is marginal cost. Hence, to maximize profit, it is necessary to charge prices (and distribute output) so that the marginal revenues in both markets will be the same and, loosely speaking, will also be equal to the cost of the last unit produced in the plant. ∎

## ▪ Exercise 19.7

*In Problems 1–6, find the critical points of the functions.*

**1.** $f(x, y) = x^2 + y^2 - 5x + 4y + xy$.

**2.** $f(x, y) = x^2 + 4y^2 - 6x + 16y$.

**3.** $f(x, y) = 2x^3 + y^3 - 3x^2 + 1.5y^2 - 12x - 90y$.

**4.** $f(x, y) = xy - \dfrac{1}{x} - \dfrac{1}{y}$.

**5.** $f(x, y, z) = 2x^2 + xy + y^2 + 100 - z(x + y - 200)$.

**6.** $f(x, y, z, w) = x^2 + y^2 + z^2 - w(x - y + 2z - 6)$.

*In Problems 7–20, find the critical points of the functions. For each critical point, determine, by the second-derivative test, whether it corresponds to a relative maximum, to a relative minimum, or to neither, or whether the test gives no information.*

**7.** $f(x, y) = x^2 + 3y^2 + 4x - 9y + 3$.

**8.** $f(x, y) = -2x^2 + 8x - 3y^2 + 24y + 7$.

**9.** $f(x, y) = y - y^2 - 3x - 6x^2$.

**10.** $f(x, y) = x^2 + y^2 + xy - 9x + 1$.

**11.** $f(x, y) = x^3 - 3xy + y^2 + y - 5$.

**12.** $f(x, y) = \dfrac{x^3}{3} + y^2 - 2x + 2y - 2xy$.

**13.** $f(x, y) = \dfrac{1}{3}(x^3 + 8y^3) - 2(x^2 + y^2) + 1$.

**14.** $f(x, y) = x^2 + y^2 - xy + x^3$.

**15.** $f(l, k) = 2lk - l^2 + 264k - 10l - 2k^2$.

**16.** $f(l, k) = l^3 + k^3 - 3lk$.

**17.** $f(p, q) = pq - \dfrac{1}{p} - \dfrac{1}{q}$.

**18.** $f(x, y) = (x - 3)(y - 3)(x + y - 3)$.

**19.** $f(x, y) = (y^2 - 4)(e^x - 1)$.

**20.** $f(x, y) = \ln(xy) + 2x^2 - xy - 6x$.

*In Problems 21–35, unless otherwise indicated, the variables $p_A$ and $p_B$ denote selling prices of products **A** and **B**, respectively. Similarly, $q_A$ and $q_B$ denote quantities of **A** and **B** that are produced and sold during some time period. In all cases, the variables employed will be assumed to be units of output, input, money, etc.*

**21. Maximizing Output** Suppose

$$P = f(l, k) = 1.08l^2 - 0.03l^3 + 1.68k^2 - 0.08k^3$$

is a production function for a firm. Find the quantities of inputs $l$ and $k$ so as to maximize output $P$.

**22. Maximizing Output** In a certain automated manufacturing process, machines M and N are utilized for $m$ and $n$ hours, respectively. If daily output $Q$ is a function of $m$ and $n$, namely,

$$Q = 4.5m + 5n - 0.5m^2 - n^2 - 0.25mn,$$

find the values of $m$ and $n$ that maximize $Q$.

**23. Profit** A candy company produces two varieties of candy, A and B, for which the constant average costs of production are 60 and 70 (cents per lb), respectively. The demand functions for A and B are given by

$$q_A = 5(p_B - p_A) \quad \text{and} \quad q_B = 500 + 5(p_A - 2p_B).$$

Find the selling prices $p_A$ and $p_B$ that maximize the company's profit.

**24. Profit** Repeat Problem 23 if the constant costs of production of A and B are $a$ and $b$ (cents per lb), respectively.

25. **Price Discrimination** Suppose a monopolist is practicing price discrimination in the sale of a product by charging different prices in two separate markets. In market A the demand function is

$$p_A = 100 - q_A,$$

and in B it is

$$p_B = 84 - q_B,$$

where $q_A$ and $q_B$ are the quantities sold per week in A and B, and $p_A$ and $p_B$ are the respective prices per unit. If the monopolist's cost function is

$$c = 600 + 4(q_A + q_B),$$

how much should be sold in each market to maximize profit? What selling prices give this maximum profit? Find the maximum profit.

26. **Profit** A monopolist sells two competitive products, A and B, for which the demand functions are

$$q_A = 1 - 2p_A + 4p_B \quad \text{and} \quad q_B = 11 + 2p_A - 6p_B.$$

If the constant average cost of producing a unit of A is 4 and a unit of B is 1, how many units of A and B should be sold to maximize the monopolist's profit?

27. **Profit** For products A and B, the joint-cost function for a manufacturer is

$$c = 1.5q_A^2 + 4.5q_B^2,$$

and the demand functions are $p_A = 36 - q_A^2$ and $p_B = 30 - q_B^2$. Find the level of production that maximizes profit.

28. **Profit** For a monopolist's products A and B, the joint-cost function is $c = (q_A + q_B)^2$, and the demand functions are $q_A = 26 - p_A$ and $q_B = 10 - 0.25p_B$. Find the values of $p_A$ and $p_B$ that maximize profit. What are the quantities of A and B that correspond to these prices? What is the total profit?

29. **Cost** An open-top rectangular box is to have a volume of 6 ft$^3$. The cost per square foot of materials is $3 for the bottom, $1 for the front and back, and $0.50 for the other two sides. Find the dimensions of the box so that the cost of materials is minimized. (See Fig. 19.15.)

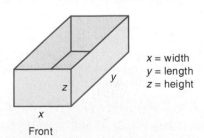

$x$ = width
$y$ = length
$z$ = height

Front

**FIGURE 9.15** Diagram for Problem 29.

30. **Collusion** Suppose A and B are the only two firms in the market selling the same product. (We say that they

are *duopolists.*) The industry demand function for the product is

$$p = 92 - q_A - q_B,$$

where $q_A$ and $q_B$ denote the output produced and sold by A and B, respectively. For A, the cost function is $c_A = 10q_A$; for B, it is $c_B = 0.5q_B^2$. Suppose the firms decide to enter into an agreement on output and price control by jointly acting as a monopoly. In this case, we say they enter into *collusion*. Show that the profit function for the monopoly is given by

$$P = pq_A - c_A + pq_B - c_B.$$

Express $P$ as a function of $q_A$ and $q_B$, and determine how output should be allocated so as to maximize the profit of the monopoly.

31. Suppose $f(x, y) = -2x^2 + 5y^2 + 7$, where $x$ and $y$ must satisfy the equation $3x - 2y = 7$. Find the relative extrema of $f$, subject to the given condition on $x$ and $y$, by first solving the second equation for $y$. Substitute the result for $y$ in the given equation. Thus, $f$ is expressed as a function of one variable, for which extrema may be found in the usual way.

32. Repeat Problem 31 if $f(x, y) = x^2 + 4y^2 + 6$, subject to the condition that $2x - 8y = 20$.

33. Suppose the joint-cost function

$$c = q_A^2 + 3q_B^2 + 2q_A q_B + aq_A + bq_B + d$$

has a relative minimum value of 15 when $q_A = 3$ and $q_B = 1$. Determine the values of the constants $a$, $b$, and $d$.

34. Suppose that the production function $q = f(k, l)$ has a relative maximum value when $k = 35$ and $l = 30$. Suppose also that all the second derivatives of $f$ exist at the point $(35, 30)$ and that $f_{kk}(35, 30) = 0$.
   a. Determine whether
      (i) $f_{kl}(35, 30)$ is a positive number,
      (ii) $f_{kl}(35, 30)$ is a negative number, or
      (iii) $f_{kl}(35, 30)$ is zero, or
      (iv) whether it is impossible to reach one of these conclusions.
   b. Determine whether
      (i) $f_{ll}(35, 30)$ must be positive,
      (ii) $f_{ll}(35, 30)$ must be negative, or
      (iii) $f_{ll}(35, 30)$ must be zero, or
      (iv) whether it is impossible to reach one of these conclusions.

35. **Profit from Competitive Products** A monopolist sells two competitive products, A and B, for which the demand equations are

$$p_A = 35 - 2q_A^2 + q_B$$

and

$$p_B = 20 - q_B + q_A.$$

The joint-cost function is

$$c = -8 - 2q_A^3 + 3q_A q_B + 30q_A + 12q_B + \tfrac{1}{2}q_A^2.$$

**a.** How many units of A and B should be sold to obtain a relative maximum profit for the monopolist? Use the second-derivative test to justify your answer.

**b.** Determine the selling prices required to realize the relative maximum profit. Also, find this relative maximum profit.

**36. Profit and Advertising**   A retailer has determined that the number of TV sets he can sell per week is

$$\frac{4x}{5 + x} + \frac{2y}{10 + y},$$

where $x$ and $y$ represent his weekly expenditures (in dollars) on newspaper and radio advertising, respectively. The profit is $125 per sale, less the cost of advertising, so the weekly profit is given by the formula

$$P = 125\left[\frac{4x}{5 + x} + \frac{2y}{10 + y}\right] - x - y.$$

Find the values of $x$ and $y$ for which the profit is a relative maximum. Use the second-derivative test to verify that your answer corresponds to a relative maximum profit.

**37. Profit from Tomato Crop**   The revenue (in dollars per square meter of ground) obtained from the sale of a crop of tomatoes grown in an artificially heated greenhouse is given by

$$r = 5T(1 - e^{-x}),$$

where $T$ is the temperature (in °C) maintained in the greenhouse and $x$ is the amount of fertilizer applied per square meter. The cost of fertilizer is $20x$ dollars per square meter, and the cost of heating is given by $0.1T^2$ dollars per square meter.

**a.** Find an expression, in terms of $T$ and $x$, for the profit per square meter obtained from the sale of the crop of tomatoes.

**b.** Verify that the pairs

$$(T, x) = (20, \ln 5) \quad \text{and} \quad (T, x) = (5, \ln \tfrac{5}{4})$$

are critical points of the profit function in part (a). [*Note:* You need not derive the pairs.]

**c.** The points in part (b) are the only critical points of the profit function in part (a). Use the second-derivative test to determine whether either of these points corresponds to a relative maximum profit per square meter.

## 19.8 LAGRANGE MULTIPLIERS

We shall now find relative maxima and minima for a function on which certain *constraints* are imposed. Such a situation could arise if a manufacturer wished to minimize a joint-cost function and yet obtain a particular production level. Suppose we want to find the relative extrema of

$$w = x^2 + y^2 + z^2, \tag{1}$$

subject to the constraint that $x$, $y$, and $z$ must satisfy

$$x - y + 2z = 6. \tag{2}$$

We can transform $w$, which is a function of three variables, into a function of two variables such that the new function reflects constraint (2). Solving Eq. (2) for $x$, we get

$$x = y - 2z + 6, \tag{3}$$

which, when substituted for $x$ in Eq. (1), gives

$$w = (y - 2z + 6)^2 + y^2 + z^2. \tag{4}$$

Since $w$ is now expressed as a function of two variables, to find relative extrema we follow the usual procedure of setting the partial derivatives of $w$ equal to 0:

$$\frac{\partial w}{\partial y} = 2(y - 2z + 6) + 2y = 4y - 4z + 12 = 0, \tag{5}$$

$$\frac{\partial w}{\partial z} = -4(y - 2z + 6) + 2z = -4y + 10z - 24 = 0. \tag{6}$$

Solving Eqs. (5) and (6) simultaneously gives $y = -1$ and $z = 2$. Substituting into Eq. (3), we get $x = 1$. Hence, the only critical point of Eq. (1) subject to the constraint represented by Eq. (2) is $(1, -1, 2)$. By using the second-derivative test on Eq. (4) when $y = -1$ and $z = 2$, we have

$$\frac{\partial^2 w}{\partial y^2} = 4, \qquad \frac{\partial^2 w}{\partial z^2} = 10, \qquad \frac{\partial^2 w}{\partial z \, \partial y} = -4,$$

$$D(-1, 2) = 4(10) - (-4)^2 = 24 > 0.$$

Thus $w$, subject to the constraint, has a relative minimum at $(1, -1, 2)$.

This solution was found by using the constraint to express one of the variables in the original function in terms of the other variables. Often this is not practical, but there is another technique, called the method of **Lagrange multipliers,**[23] that avoids this step and yet allows us to obtain critical points.

The method is as follows. Suppose we have a function $f(x, y, z)$ subject to the constraint $g(x, y, z) = 0$. We construct a new function $F$ of *four* variables defined by the following (where $\lambda$ is a Greek letter read "lambda"):

$$F(x, y, z, \lambda) = f(x, y, z) - \lambda g(x, y, z).$$

It can be shown that if $(x_0, y_0, z_0)$ is a critical point of $f$, subject to the constraint $g(x, y, z) = 0$, there exists a value of $\lambda$, say, $\lambda_0$, such that $(x_0, y_0, z_0, \lambda_0)$ is a critical point of $F$. The number $\lambda_0$ is called a **Lagrange multiplier.** Also, if $(x_0, y_0, z_0, \lambda_0)$ is a critical point of $F$, then $(x_0, y_0, z_0)$ is a critical point of $f$, subject to the constraint. Thus, to find critical points of $f$, subject to $g(x, y, z) = 0$, we instead find critical points of $F$. These are obtained by solving the simultaneous equations

$$\begin{cases} F_x(x, y, z, \lambda) = 0, \\ F_y(x, y, z, \lambda) = 0, \\ F_z(x, y, z, \lambda) = 0, \\ F_\lambda(x, y, z, \lambda) = 0. \end{cases}$$

At times, ingenuity must be used to solve the equations. Once we obtain a critical point $(x_0, y_0, z_0, \lambda_0)$ of $F$, we can conclude that $(x_0, y_0, z_0)$ is a critical point of $f$, subject to the constraint $g(x, y, z) = 0$. Although $f$ and $g$ are functions of three variables, the method of Lagrange multipliers can be extended to $n$ variables.

Let us illustrate the method of Lagrange multipliers for the original situation, namely,

$$f(x, y, z) = x^2 + y^2 + z^2, \quad \text{subject to} \quad x - y + 2z = 6.$$

First, we write the constraint as $g(x, y, z) = x - y + 2z - 6 = 0$. Second, we form the function

$$F(x, y, z, \lambda) = f(x, y, z) - \lambda g(x, y, z)$$
$$= x^2 + y^2 + z^2 - \lambda(x - y + 2z - 6).$$

Next, we set each partial derivative of $F$ equal to 0. For convenience, we shall write $F_x(x, y, z, \lambda)$ as $F_x$, and so on:

$$\begin{cases} F_x = 2x - \lambda = 0, & (7) \\ F_y = 2y + \lambda = 0, & (8) \\ F_z = 2z - 2\lambda = 0, & (9) \\ F_\lambda = -x + y - 2z + 6 = 0. & (10) \end{cases}$$

---

[23]After the French mathematician, Joseph-Louis Lagrange (1736–1813).

From Eqs. (7)–(9), we see immediately that

$$x = \frac{\lambda}{2}, \qquad y = -\frac{\lambda}{2}, \qquad \text{and} \qquad z = \lambda. \tag{11}$$

Substituting these values into Eq. (10), we obtain

$$-\frac{\lambda}{2} - \frac{\lambda}{2} - 2\lambda + 6 = 0,$$

$$-3\lambda + 6 = 0,$$

$$\lambda = 2.$$

Thus, from Eq. (11),

$$x = 1, \qquad y = -1, \qquad \text{and} \qquad z = 2.$$

Hence, the only critical point of $f$, subject to the constraint, is $(1, -1, 2)$, at which there may exist a relative maximum, a relative minimum, or neither of these. The method of Lagrange multipliers does not directly indicate which of these possibilities occurs, although from our previous work, we saw that it is indeed a relative minimum. In applied problems, the nature of the problem itself may give a clue as to how a critical point is to be regarded. Often the existence of either a relative minimum or a relative maximum is assumed, and a critical point is treated accordingly. Actually, sufficient second-order conditions for relative extrema are available, but we shall not consider them.

**EXAMPLE 1**  **Method of Lagrange Multipliers**

*Find the critical points for $z = f(x, y) = 3x - y + 6$, subject to the constraint $x^2 + y^2 = 4$.*

*Solution:* We write the constraint as $g(x, y) = x^2 + y^2 - 4 = 0$ and construct the function

$$F(x, y, \lambda) = f(x, y) - \lambda g(x, y) = 3x - y + 6 - \lambda(x^2 + y^2 - 4).$$

Setting $F_x = F_y = F_\lambda = 0$, we have

$$\begin{cases} 3 - 2x\lambda = 0, & (12) \\ -1 - 2y\lambda = 0, & (13) \\ -x^2 - y^2 + 4 = 0. & (14) \end{cases}$$

From Eqs. (12) and (13), we can express $x$ and $y$ in terms of $\lambda$. Then we shall substitute for $x$ and $y$ in Eq. (14) and solve for $\lambda$. Knowing $\lambda$, we can find $x$ and $y$. To begin, from Eqs. (12) and (13), we have

$$x = \frac{3}{2\lambda} \quad \text{and} \quad y = -\frac{1}{2\lambda}.$$

Substituting into Eq. (14), we obtain

$$-\frac{9}{4\lambda^2} - \frac{1}{4\lambda^2} + 4 = 0,$$

$$-\frac{10}{4\lambda^2} + 4 = 0,$$

$$\lambda = \pm \frac{\sqrt{10}}{4}.$$

With these $\lambda$-values, we can find $x$ and $y$. If $\lambda = \sqrt{10}/4$, then

$$x = \frac{3}{2\left(\dfrac{\sqrt{10}}{4}\right)} = \frac{3\sqrt{10}}{5}, \qquad y = -\frac{1}{2\left(\dfrac{\sqrt{10}}{4}\right)} = -\frac{\sqrt{10}}{5}.$$

Similarly, if $\lambda = -\sqrt{10}/4$,

$$x = -\frac{3\sqrt{10}}{5}, \qquad y = \frac{\sqrt{10}}{5}.$$

Thus, the critical points of $f$, subject to the constraint, are $(3\sqrt{10}/5, -\sqrt{10}/5)$ and $(-3\sqrt{10}/5, \sqrt{10}/5)$. Note that the values of $\lambda$ do not appear in the answer; they are simply a means to obtain the solution. ∎

**EXAMPLE 2  Method of Lagrange Multipliers**

*Find critical points for $f(x, y, z) = xyz$, where $xyz \neq 0$, subject to the constraint $x + 2y + 3z = 36$.*

*Solution:* We have

$$F(x, y, z, \lambda) = xyz - \lambda(x + 2y + 3z - 36).$$

Setting $F_x = F_y = F_z = F_\lambda = 0$ gives, respectively,

$$\begin{cases} yz - \lambda = 0, \\ xz - 2\lambda = 0, \\ xy - 3\lambda = 0, \\ -x - 2y - 3z + 36 = 0. \end{cases}$$

Because we cannot directly express $x$, $y$, and $z$ in terms of $\lambda$ only, we cannot follow the procedure in Example 1. However, observe that we can express the products $yz$, $xz$, and $xy$ as multiples of $\lambda$. This suggests that, by looking at quotients of equations, we can obtain a relation between two variables that does not involve $\lambda$. (The $\lambda$'s will cancel.) Proceeding to do this, we write the foregoing system as

$$\begin{cases} yz = \lambda, & (15) \\ xz = 2\lambda, & (16) \\ xy = 3\lambda, & (17) \\ x + 2y + 3z - 36 = 0. & (18) \end{cases}$$

Dividing each side of Eq. (15) by the corresponding side of Eq. (16), we get

$$\frac{yz}{xz} = \frac{\lambda}{2\lambda}, \quad \text{or} \quad y = \frac{x}{2}.$$

This division is valid, since $xyz \neq 0$. Similarly, from Eqs. (15) and (17), we get

$$\frac{yz}{xy} = \frac{\lambda}{3\lambda}, \quad \text{or} \quad z = \frac{x}{3}.$$

Now that we have $y$ and $z$ expressed in terms of $x$ only, we can substitute into Eq. (18) and solve for $x$:

$$x + 2\left(\frac{x}{2}\right) + 3\left(\frac{x}{3}\right) - 36 = 0,$$

$$x = 12.$$

Thus, $y = 6$ and $z = 4$. Hence, $(12, 6, 4)$ is the only critical point satisfying the given conditions. Note that in this situation, we found the critical point without having to find the value for $\lambda$. ∎

**EXAMPLE 3  Minimizing Costs**

*Suppose a firm has an order for 200 units of its product and wishes to distribute its manufacture between two of its plants, plant 1 and plant 2. Let $q_1$ and $q_2$ denote the outputs of plants 1 and 2, respectively, and suppose the total-cost function is given by $c = f(q_1, q_2) = 2q_1^2 + q_1q_2 + q_2^2 + 200$. How should the output be distributed in order to minimize costs?*

*Solution:* We minimize $c = f(q_1, q_2)$, given the constraint $q_1 + q_2 = 200$. We have

$$F(q_1, q_2, \lambda) = 2q_1^2 + q_1q_2 + q_2^2 + 200 - \lambda(q_1 + q_2 - 200),$$

$$\begin{cases} \dfrac{\partial F}{\partial q_1} = 4q_1 + q_2 - \lambda = 0, & (19) \\[2mm] \dfrac{\partial F}{\partial q_2} = q_1 + 2q_2 - \lambda = 0, & (20) \\[2mm] \dfrac{\partial F}{\partial \lambda} = -q_1 - q_2 + 200 = 0. & (21) \end{cases}$$

We can eliminate $\lambda$ from Eqs. (19) and (20) and obtain a relation between $q_1$ and $q_2$. Then, solving this equation for $q_2$ in terms of $q_1$ and substituting into Eq. (21), we can find $q_1$. We begin by subtracting Eq. (20) from Eq. (19), which gives

$$3q_1 - q_2 = 0, \quad \text{so} \quad q_2 = 3q_1.$$

Substituting into Eq. (21), we have

$$-q_1 - 3q_1 + 200 = 0,$$
$$-4q_1 = -200,$$
$$q_1 = 50.$$

Thus, $q_2 = 150$. Accordingly, plant 1 should produce 50 units and plant 2 should produce 150 units in order to minimize costs. ∎

An interesting observation can be made concerning Example 3. From Eq. (19), $\lambda = 4q_1 + q_2 = \partial c / \partial q_1$, the marginal cost of plant 1. From Eq. (20), $\lambda = q_1 + 2q_2 = \partial c / \partial q_2$, the marginal cost of plant 2. Hence, $\partial c / \partial q_1 = \partial c / \partial q_2$, and we conclude that, to minimize cost, it is necessary that the marginal costs of each plant be equal to each other.

**EXAMPLE 4  Least-Cost Input Combination**

*Suppose a firm must produce a given quantity $P_0$ of output in the cheapest possible manner. If there are two input factors $l$ and $k$, and their prices per unit are fixed at $p_l$ and $p_k$, respectively, discuss the economic significance of combining input to achieve least cost. That is, describe the least-cost input combination.*

*Solution:* Let $P = f(l, k)$ be the production function. Then we must minimize the cost function

$$c = lp_l + kp_k,$$

subject to

$$P_0 = f(l, k).$$

We construct

$$F(l, k, \lambda) = lp_l + kp_k - \lambda[f(l, k) - P_0].$$

We have

$$
\begin{cases}
\dfrac{\partial F}{\partial l} = p_l - \lambda \dfrac{\partial}{\partial l}[f(l, k)] = 0, & (22) \\[3mm]
\dfrac{\partial F}{\partial k} = p_k - \lambda \dfrac{\partial}{\partial k}[f(l, k)] = 0, & (23) \\[3mm]
\dfrac{\partial F}{\partial \lambda} = -f(l, k) + P_0 = 0.
\end{cases}
$$

From Eqs. (22) and (23),

$$\lambda = \frac{p_l}{\dfrac{\partial}{\partial l}[f(l, k)]} = \frac{p_k}{\dfrac{\partial}{\partial k}[f(l, k)]}. \qquad (24)$$

Hence,

$$\frac{p_l}{p_k} = \frac{\dfrac{\partial}{\partial l}[f(l, k)]}{\dfrac{\partial}{\partial k}[f(l, k)]}.$$

We conclude that when the least-cost combination of factors is used, the ratio of the marginal products of the input factors must be equal to the ratio of their corresponding unit prices. ∎

## Multiple Constraints

The method of Lagrange multipliers is by no means restricted to problems involving a single constraint. For example, suppose $f(x, y, z, w)$ were subject to constraints $g_1(x, y, z, w) = 0$ and $g_2(x, y, z, w) = 0$. Then there would be two lambdas, $\lambda_1$ and $\lambda_2$ (one corresponding to each constraint), and we would construct the function $F = f - \lambda_1 g_1 - \lambda_2 g_2$. We would then solve the system

$$F_x = F_y = F_z = F_w = F_{\lambda_1} = F_{\lambda_2} = 0.$$

**EXAMPLE 5    Method of Lagrange Multipliers with Two Constraints**

*Find critical points for $f(x, y, z) = xy + yz$, subject to the constraints $x^2 + y^2 = 8$ and $yz = 8$.*

*Solution:* Set

$$F(x, y, z, \lambda_1, \lambda_2) = xy + yz - \lambda_1(x^2 + y^2 - 8) - \lambda_2(yz - 8).$$

Then

$$
\begin{cases}
F_x = y - 2x\lambda_1 = 0, \\
F_y = x + z - 2y\lambda_1 - z\lambda_2 = 0, \\
F_z = y - y\lambda_2 = 0, \\
F_{\lambda_1} = -x^2 - y^2 + 8 = 0, \\
F_{\lambda_2} = -yz + 8 = 0.
\end{cases}
$$

You would probably agree that this appears to be a challenging system to solve. Thus, ingenuity will come into play. Here is one sequence of operations that will allow us to find the critical points. We can write the system as

$$\begin{cases} \dfrac{y}{2x} = \lambda_1, & (25) \\ x + z - 2y\lambda_1 - z\lambda_2 = 0, & (26) \\ \lambda_2 = 1, & (27) \\ x^2 + y^2 = 8, & (28) \\ z = \dfrac{8}{y}. & (29) \end{cases}$$

Substituting $\lambda_2 = 1$ from Eq. (27) into Eq. (26) and simplifying gives the equation $x - 2y\lambda_1 = 0$, so

$$\lambda_1 = \frac{x}{2y}.$$

Substituting into Eq. (25) gives

$$\frac{y}{2x} = \frac{x}{2y},$$

$$y^2 = x^2. \qquad (30)$$

Substituting into Eq. (28) gives $x^2 + x^2 = 8$, from which it follows that $x = \pm 2$. If $x = 2$, then, from Eq. (30), we have $y = \pm 2$. Similarly, if $x = -2$, then $y = \pm 2$. Thus, if $x = 2$ and $y = 2$, then, from Eq. (29), we obtain $z = 4$. Continuing in this manner, we obtain four critical points:

$$(2, 2, 4), (2, -2, -4), (-2, 2, 4), (-2, -2, -4). \qquad \blacksquare$$

### ▪ Exercise 19.8

*In Problems 1–12 find, by the method of Lagrange multipliers, the critical points of the functions, subject to the given constraints.*

**1.** $f(x, y) = x^2 + 4y^2 + 6$;   $2x - 8y = 20$.

**2.** $f(x, y) = -2x^2 + 5y^2 + 7$;   $3x - 2y = 7$.

**3.** $f(x, y, z) = x^2 + y^2 + z^2$;   $2x + y - z = 9$.

**4.** $f(x, y, z) = x + y + z$;   $xyz = 27$.

**5.** $f(x, y, z) = x^2 + xy + 2y^2 + z^2$;   $x - 3y - 4z = 16$.

**6.** $f(x, y, z) = xyz^2$;   $x - y + z = 20$ $(xyz^2 \neq 0)$.

**7.** $f(x, y, z) = xyz$;   $x + 2y + 3z = 18$ $(xyz \neq 0)$.

**8.** $f(x, y, z) = x^2 + y^2 + z^2$;   $x + y + z = 1$.

**9.** $f(x, y, z) = x^2 + 2y - z^2$;   $2x - y = 0, y + z = 0$.

**10.** $f(x, y, z) = x^2 + y^2 + z^2$;   $x + y + z = 1$, $x - y + z = 1$

**11.** $f(x, y, z) = xyz$;   $x + y + z = 12$, $x + y - z = 0$ $(xyz \neq 0)$.

**12.** $f(x, y, z, w) = 2x^2 + 2y^2 + 3z^2 - 4w^2$;   $4x - 8y + 6z + 16w = 6$.

**13. Production Allocation**   To fill an order for 100 units of its product, a firm wishes to distribute production between its two plants, plant 1 and plant 2. The total-cost function is given by

$$c = f(q_1, q_2) = 0.1q_1^2 + 7q_1 + 15q_2 + 1000,$$

where $q_1$ and $q_2$ are the numbers of units produced at plants 1 and 2, respectively. How should the output be distributed in order to minimize costs? (You may assume that the critical point obtained does correspond to the minimum cost.)

**14. Production Allocation**   Repeat Problem 13 if the cost function is

$$c = 3q_1^2 + q_1q_2 + 2q_2^2$$

and a total of 200 units are to be produced.

**15. Maximizing Output**   The production function for a firm is

$$f(l, k) = 12l + 20k - l^2 - 2k^2.$$

The cost to the firm of $l$ and $k$ is 4 and 8 per unit, respectively. If the firm wants the total cost of input to be 88, find the greatest output possible, subject to this budget constraint. (You may assume that the critical point obtained does correspond to the maximum output.)

16. **Maximizing Output** Repeat Problem 15, given that

$$f(l, k) = 60l + 30k - 2l^2 - 3k^2$$

and the budget constraint is $2l + 3k = 30$.

17. **Advertising Budget** A computer company has a monthly advertising budget of $60,000. Its marketing department estimates that if $x$ dollars are spent each month on advertising in newspapers and $y$ dollars per month on television advertising, then the monthly sales will be given by $S = 90x^{1/4}y^{3/4}$ dollars. If the profit is 10% of sales, less the advertising cost, determine how to allocate the advertising budget in order to maximize the monthly profit. (You may assume that the critical point obtained does correspond to the maximum profit.)

18. **Maximizing Production** When $l$ units of labor and $k$ units of capital are invested, a manufacturer's total production $q$ is given by the Cobb-Douglas production function $q = 5l^{1/5}k^{4/5}$. Each unit of labor costs $11 and each unit of capital costs $33. If exactly $11,880 is to be spent on production, determine the numbers of units of labor and capital that should be invested to maximize production. (You may assume that the maximum occurs at the critical point obtained.)

19. **Political Advertising** Newspaper advertisements for political parties always have some negative effects. The recently elected party assumed that the three most important election issues $X$, $Y$, and $Z$, had to be mentioned in each ad, with space $x$, $y$, and $z$ units, respectively, allot-

ted to each. The combined bad effect of this coverage was estimated by the party's back-room boys as

$$B(x, y, z) = x^2 + y^2 + 2z^2.$$

Aesthetics dictated that the total space for $X$ and $Y$ together must be 20, and realism suggested that the total space allotted to $Y$ and $Z$ together must also be 20 units. What values of $x$, $y$, and $z$ in each ad would produce the lowest negative effect? (You may assume that any critical point obtained provides the minimum effect.)

20. **Maximizing Profit** Suppose a manufacturer's production function is given by

$$16q = 65 - 4(l - 4)^2 - 2(k - 5)^2,$$

and the cost to the manufacturer is $4 per unit of labor and $8 per unit of capital, so that the total cost (in dollars) is $4l + 8k$. The selling price of the product is $32 per unit.

a. Express the profit as a function of $l$ and $k$. Give your answer in expanded form.

b. Find all critical points of the profit function obtained in part (a). Apply the second-derivative test at each critical point. If the profit is a relative maximum at a critical point, compute the corresponding relative maximum profit.

c. The profit may be considered a function of $l$, $k$, and $q$ (that is, $P = 32q - 4l - 8k$), subject to the constraint

$$16q = 65 - 4(l - 4)^2 - 2(k - 5)^2.$$

Use the method of Lagrange multipliers to find all critical points of $P = 32q - 8k - 4l$, subject to the constraint.

*Problems 21–24 refer to the following definition. A utility function is a function that attaches a measure to the satisfaction or utility a consumer gets from the consumption of products per unit of time. Suppose $U \, 5 \, f(x, y)$ is such a function, where $x$ and $y$ are the amounts of two products, $X$ and $Y$. The marginal utility of $X$ is $\partial U/\partial x$ and approximately represents the change in total utility resulting from a one-unit change in consumption of product $X$ per unit of time. We define the marginal utility of $Y$ in similar fashion. If the prices of $X$ and $Y$ are $p_x$ and $p_y$, respectively, and the consumer has an income or budget of $I$ to spend, then the budget constraint is*

$$xp_x + yp_y = I.$$

*In Problems 21–23, find the quantities of each product that the consumer should buy, subject to the budget, that will allow maximum satisfaction. That is, in Problems 21 and 22, find values of $x$ and $y$ that maximize $U = f(x, y)$, subject to $xp_x + yp_y = I$. Perform a similar procedure for Problem 23. Assume that such a maximum exists.*

21. $U = x^3y^3$; $p_x = 2$, $p_y = 3$, $I = 48$ ($x^3y^3 \neq 0$).

22. $U = 46x - (5x^2/2) + 34y - 2y^2$; $p_x = 5$, $p_y = 2$, $I = 30$.

23. $U = f(x, y, z) = xyz$; $p_x = 2$, $p_y = 1$, $p_z = 4$, $I = 60$ ($xyz \neq 0$).

24. Let $U = f(x, y)$ be a utility function subject to the budget constraint $xp_x + yp_y = I$, where $p_x$, $p_y$, and $I$ are constants. Show that, to maximize satisfaction, it is necessary that

$$\lambda = \frac{f_x(x, y)}{p_x} = \frac{f_y(x, y)}{p_y},$$

where $f_x(x, y)$ and $f_y(x, y)$ are the marginal utilities of $X$ and $Y$, respectively. Show that $f_x(x, y)/p_x$ is the marginal utility of one dollar's worth of $X$. Hence, maximum satisfaction is obtained when the consumer allocates the budget so that the marginal utility of a dollar's worth of $X$ is equal to the marginal utility per dollar's worth of $Y$. Performing the same procedure as that for $U = f(x, y)$, verify that this is true for $U = f(x, y, z, w)$, subject to the corresponding budget equation. In each case, $\lambda$ is called the *marginal utility of income.*

**To develop the method of least squares and introduce index numbers.**

## 19.9   LINES OF REGRESSION[24]

To study the influence of advertising on sales, a firm compiled the data in Table 19.4. The variable $x$ denotes advertising expenditures in hundreds of dollars,

**TABLE 19.4**

| Expenditures, $x$ | 2 | 3 | 4.5 | 5.5 | 7 |
|---|---|---|---|---|---|
| Revenue, $y$ | 3 | 6 | 8 | 10 | 11 |

and the variable $y$ denotes the resulting sales revenue in thousands of dollars. If each pair $(x, y)$ of data is plotted, the result is called a **scatter diagram** [Fig. 19.16(a)].

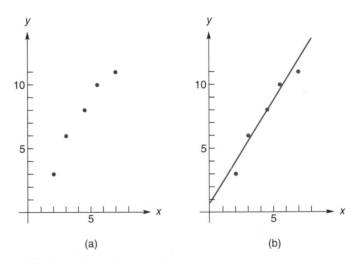

(a)                                      (b)

**FIGURE 19.16**   Scatter diagram and straight-line approximation to data points.

From an observation of the distribution of the points, it is reasonable to assume that a relationship exists between $x$ and $y$ and that it is approximately linear. On this basis, we may fit "by eye" a straight line that approximates the given data [Fig. 19.16(b)] and, from this line, predict a value of $y$ for a given value of $x$. The line seems consistent with the trend of the data, although other lines could be drawn as well. Unfortunately, determining a line "by eye" is not very objective. We want to apply criteria in specifying what we shall call a line of "best fit." A frequently used technique is called the **method of least squares**.

[24]May be omitted without loss of continuity.

To apply the method of least squares to the data in Table 19.4, we first assume that $x$ and $y$ are approximately linearly related and that we can fit a straight line

$$\hat{y} = \hat{a} + \hat{b}x \tag{1}$$

that approximates the given points by a suitable objective choice of the constants $\hat{a}$ and $\hat{b}$ (read "$a$ hat" and "$b$ hat," respectively). For a given value of $x$ in Eq. (1), $\hat{y}$ is the corresponding predicted value of $y$, and $(x, \hat{y})$ will be on the line. Our aim is that $\hat{y}$ be near $y$.

When $x = 2$, the observed value of $y$ is 3. Our predicted value of $y$ is obtained by substituting $x = 2$ in Eq. (1), which yields $\hat{y} = \hat{a} + 2\hat{b}$. The error of estimation, or vertical deviation of the point $(2, 3)$ from the line, is $\hat{y} - y$, or

$$\hat{a} + 2\hat{b} - 3.$$

This vertical deviation is indicated (although exaggerated for clarity) in Fig. 19.17. Similarly, the vertical deviation of $(3, 6)$ from the line is $\hat{a} + 3\hat{b} - 6$, as

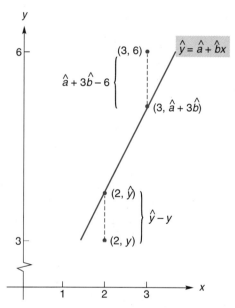

**FIGURE 19.17** Vertical deviation of data points from straight-line approximation.

is also illustrated. To avoid possible difficulties associated with positive and negative deviations, we shall consider the squares of the deviations and shall form the sum $S$ of all such squares for the given data:

$$S = (\hat{a} + 2\hat{b} - 3)^2 + (\hat{a} + 3\hat{b} - 6)^2 + (\hat{a} + 4.5\hat{b} - 8)^2 +$$
$$(\hat{a} + 5.5\hat{b} - 10)^2 + (\hat{a} + 7\hat{b} - 11)^2.$$

The method of least squares requires that we choose as the line of "best fit" the one obtained by selecting $\hat{a}$ and $\hat{b}$ so as to minimize $S$. We can minimize

$S$ with respect to $\hat{a}$ and $\hat{b}$ by solving the system

$$\begin{cases} \dfrac{\partial S}{\partial \hat{a}} = 0, \\[2mm] \dfrac{\partial S}{\partial \hat{b}} = 0. \end{cases}$$

We have

$$\frac{\partial S}{\partial \hat{a}} = 2(\hat{a} + 2\hat{b} - 3) + 2(\hat{a} + 3\hat{b} - 6) + 2(\hat{a} + 4.5\hat{b} - 8) +$$
$$2(\hat{a} + 5.5\hat{b} - 10) + 2(\hat{a} + 7\hat{b} - 11) = 0,$$

$$\frac{\partial S}{\partial \hat{b}} = 4(\hat{a} + 2\hat{b} - 3) + 6(\hat{a} + 3\hat{b} - 6) + 9(\hat{a} + 4.5\hat{b} - 8) +$$
$$11(\hat{a} + 5.5\hat{b} - 10) + 14(\hat{a} + 7\hat{b} - 11) = 0,$$

which, when simplified, give

$$\begin{cases} 5\hat{a} + 22\hat{b} = 38, \\ 44\hat{a} + 225\hat{b} = 384. \end{cases}$$

Solving for $\hat{a}$ and $\hat{b}$, we obtain

$$\hat{a} = \frac{102}{157} \approx 0.65, \quad \hat{b} = \frac{248}{157} \approx 1.58.$$

It can be shown that these values of $\hat{a}$ and $\hat{b}$ lead to a minimum value of $S$. Hence, in the sense of least squares, the line of best fit $\hat{y} = \hat{a} + \hat{b}x$ is

$$\hat{y} = 0.65 + 1.58x. \tag{2}$$

This is, in fact, the line indicated in Fig. 19.16(b). It is called the **least squares line of y on x** or the **linear regression line of y on x**. The constants $\hat{a}$ and $\hat{b}$ are called **linear regression coefficients**. With Eq. (2), we would predict that when $x = 5$, the corresponding value of $y$ is $\hat{y} = 0.65 + 1.58(5) = 8.55$.

More generally, suppose we are given the following $n$ pairs of observations:

$$(x_1, y_1), (x_2, y_2), \ldots, (x_n, y_n).$$

If we assume that $x$ and $y$ are approximately linearly related and that we can fit a straight line

$$\hat{y} = \hat{a} + \hat{b}x$$

that approximates the data, the sum of the squares of the errors $\hat{y} - y$ is

$$S = (\hat{a} + \hat{b}x_1 - y_1)^2 + (\hat{a} + \hat{b}x_2 - y_2)^2 + \cdots + (\hat{a} + \hat{b}x_n - y_n)^2.$$

Since $S$ must be minimized with respect to $\hat{a}$ and $\hat{b}$,

$$\begin{cases} \dfrac{\partial S}{\partial \hat{a}} = 2(\hat{a} + \hat{b}x_1 - y_1) + 2(\hat{a} + \hat{b}x_2 - y_2) + \cdots + 2(\hat{a} + \hat{b}x_n - y_n) = 0, \\[3mm] \dfrac{\partial S}{\partial \hat{b}} = 2x_1(\hat{a} + \hat{b}x_1 - y_1) + 2x_2(\hat{a} + \hat{b}x_2 - y_2) + \cdots + 2x_n(\hat{a} + \hat{b}x_n - y_n) = 0. \end{cases}$$

Dividing both equations by 2 and using sigma notation, we have

$$\begin{cases} \hat{a}n + \hat{b} \sum_{i=1}^{n} x_i - \sum_{i=1}^{n} y_i = 0, \\ \hat{a} \sum_{i=1}^{n} x_i + \hat{b} \sum_{i=1}^{n} x_i^2 - \sum_{i=1}^{n} x_i y_i = 0. \end{cases}$$

Equivalently, we have the system of so-called *normal equations:*

$$\begin{cases} \sum_{i=1}^{n} y_i = \hat{a}n + \hat{b} \sum_{i=1}^{n} x_i, & (3) \\ \sum_{i=1}^{n} x_i y_i = \hat{a} \sum_{i=1}^{n} x_i + \hat{b} \sum_{i=1}^{n} x_i^2. & (4) \end{cases}$$

To solve for $\hat{b}$, we first multiply Eq. (3) by $\sum_{i=1}^{n} x_i$ and Eq. (4) by $n$:

$$\begin{cases} \left( \sum_{i=1}^{n} x_i \right)\left( \sum_{i=1}^{n} y_i \right) = \hat{a}n \sum_{i=1}^{n} x_i + \hat{b}\left( \sum_{i=1}^{n} x_i \right)^2, & (5) \\ n \sum_{i=1}^{n} x_i y_i = \hat{a}n \sum_{i=1}^{n} x_i + \hat{b}n \sum_{i=1}^{n} x_i^2. & (6) \end{cases}$$

Subtracting Eq. (5) from Eq. (6), we obtain

$$n \sum_{i=1}^{n} x_i y_i - \left( \sum_{i=1}^{n} x_i \right)\left( \sum_{i=1}^{n} y_i \right) = \hat{b}n \sum_{i=1}^{n} x_i^2 - \hat{b}\left( \sum_{i=1}^{n} x_i \right)^2$$

$$= \hat{b}\left[ n \sum_{i=1}^{n} x_i^2 - \left( \sum_{i=1}^{n} x_i \right)^2 \right].$$

Thus,

$$\hat{b} = \frac{n \sum_{i=1}^{n} x_i y_i - \left( \sum_{i=1}^{n} x_i \right)\left( \sum_{i=1}^{n} y_i \right)}{n \sum_{i=1}^{n} x_i^2 - \left( \sum_{i=1}^{n} x_i \right)^2}. \qquad (7)$$

Solving Eqs. (3) and (4) for $\hat{a}$ gives

$$\hat{a} = \frac{\left( \sum_{i=1}^{n} x_i^2 \right)\left( \sum_{i=1}^{n} y_i \right) - \left( \sum_{i=1}^{n} x_i \right)\left( \sum_{i=1}^{n} x_i y_i \right)}{n \sum_{i=1}^{n} x_i^2 - \left( \sum_{i=1}^{n} x_i \right)^2}. \qquad (8)$$

It can be shown that these values of $\hat{a}$ and $\hat{b}$ minimize $S$.

Computing the linear regression coefficients $\hat{a}$ and $\hat{b}$ by the formulas of Eqs. (7) and (8) gives the linear regression line of $y$ on $x$, namely, $\hat{y} = \hat{a} + \hat{b}x$, which can be used to estimate $y$ for a given value of $x$.

In the next example, as well as in the exercises, you will encounter **index numbers**. They are used to relate a variable in one period of time to the same variable in another period, called the *base period*. An index number is a *relative* number that describes data that are changing over time. Such data are referred to as *time series*.

**TABLE 19.5**

| Year | Production (in thousands) | Index [1994 = 100] |
|------|--------------------------|--------------------|
| 1993 | 828 | 92 |
| 1994 | 900 | 100 |
| 1995 | 936 | 104 |
| 1996 | 891 | 99 |
| 1997 | 954 | 106 |

For example, consider the time-series data of total production of widgets in the United States for 1993–97, indicated in Table 19.5. If we choose 1994 as the base year and assign the index number 100 to it, then the other index numbers are obtained by dividing each year's production by the 1994 production of 900 and multiplying the result by 100. We can, for example, interpret the index 106 for 1997 as meaning that production for that year was 106% of the production in 1994.

In time-series analysis, index numbers are obviously of great advantage if the data involve numbers of great magnitude. But regardless of the magnitude of the data, index numbers simplify the task of comparing changes in data over periods of time.

**EXAMPLE 1   Determining a Linear-Regression Line**

*By means of the linear-regression line, use the following table to represent the trend for the index of U.S. Government purchases of goods and services from 1982 to 1987 (1982 = 100).*

| Year | 1982 | 1983 | 1984 | 1985 | 1986 | 1987 |
|------|------|------|------|------|------|------|
| Index | 100 | 104 | 108 | 111 | 111 | 114 |

*Source: Economic Report of the President,* 1988, U.S. Government Printing Office, Washington, DC, 1988.

*Solution:* We shall let $x$ denote time and $y$ denote the index and treat $y$ as a linear function of $x$. Also, we shall designate 1982 by $x = 1$, 1983 by $x = 2$, and so on. There are $n = 6$ pairs of measurements. To determine the linear-regression coefficients by using Eqs. (7) and (8), we first perform the arithmetic:

| Year | $x_i$ | $y_i$ | $x_i y_i$ | $x_i^2$ |
|------|-------|-------|-----------|---------|
| 1982 | 1 | 100 | 100 | 1 |
| 1983 | 2 | 104 | 208 | 4 |
| 1984 | 3 | 108 | 324 | 9 |
| 1985 | 4 | 111 | 444 | 16 |
| 1986 | 5 | 111 | 555 | 25 |
| 1987 | 6 | 114 | 684 | 36 |
| Total | $\overline{21}$ | $\overline{648}$ | $\overline{2315}$ | $\overline{91}$ |
| | $= \sum\limits_{i=1}^{6} x_i$ | $= \sum\limits_{i=1}^{6} y_i$ | $= \sum\limits_{i=1}^{6} x_i y_i$ | $= \sum\limits_{i=1}^{6} x_i^2$ |

Hence, by Eq. (8),

$$\hat{a} = \frac{91(648) - 21(2315)}{6(91) - (21)^2} = 98.6,$$

and by Eq. (7),

$$\hat{b} = \frac{6(2315) - 21(648)}{6(91) - (21)^2} \approx 2.69.$$

Thus, the regression line of $y$ on $x$ is

$$\hat{y} = 98.6 + 2.69x,$$

whose graph, as well as a scatter diagram, appears in Fig. 19.18.

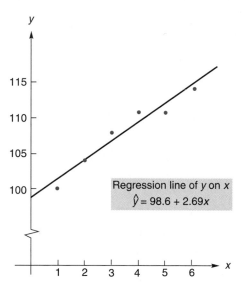

Regression line of $y$ on $x$
$\hat{y} = 98.6 + 2.69x$

**FIGURE 19.18**   Linear-regression line for goods and services.

## TECHNOLOGY

The TI-82 has a utility that computes the equation of the least squares line for a set of data. We shall illustrate by giving the procedure for the six data points $(x_i, y_i)$ of Example 1. After pressing STAT and ENTER, we enter all the $x$-values and then the $y$-values. (See Fig. 19.19.) Next, we press STAT and move to CALC. Finally, pressing 9 and ENTER gives the result shown in Fig. 19.20. [The number $r \approx 0.97057$ is called the *coefficient of correlation* and is a measure of the degree to which the given data are linearly related.]

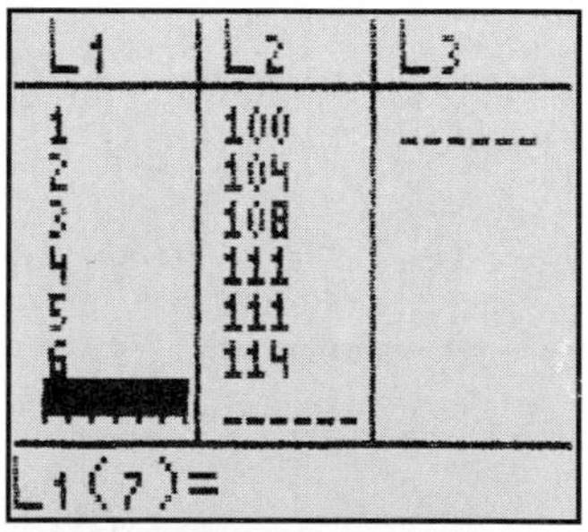

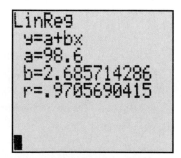

**FIGURE 19.19**   Data of Example 1.

**FIGURE 19.20**   Equation of least squares line.

■ **Exercise 19.9**

*In this exercise set, use a graphics calculator if your instructor permits you to do so.*

*In Problems 1–4, find an equation of the least squares linear-regression line of y on x for the given data, and sketch both the line and the data. Predict the value of y corresponding to x = 3.5.*

1.

| $x$ | 1 | 2 | 3 | 4 | 5 | 6 |
|---|---|---|---|---|---|---|
| $y$ | 1.5 | 2.3 | 2.6 | 3.7 | 4.0 | 4.5 |

2.

| $x$ | 1 | 2 | 3 | 4 | 5 | 6 | 7 |
|---|---|---|---|---|---|---|---|
| $y$ | 1 | 1.8 | 2 | 4 | 4.5 | 7 | 9 |

3.

| $x$ | 2 | 3 | 4.5 | 5.5 | 7 |
|---|---|---|---|---|---|
| $y$ | 3 | 5 | 8 | 10 | 11 |

4.

| $x$ | 2 | 3 | 4 | 5 | 6 | 7 |
|---|---|---|---|---|---|---|
| $y$ | 2.4 | 2.9 | 3.3 | 3.8 | 4.3 | 4.9 |

5. **Demand** A firm finds that when the price of its product is $p$ dollars per unit, the number of units sold is $q$, as indicated in the following table:

| Price, $p$ | 10 | 30 | 40 | 50 | 60 | 70 |
|---|---|---|---|---|---|---|
| Demand, $q$ | 70 | 68 | 63 | 50 | 46 | 32 |

Find an equation of the regression line of $q$ on $p$.

6. **Water and Crop Yield** On a farm, an agronomist finds that the amount of water applied (in inches) and the corresponding yield of a certain crop (in tons per acre) are as given in the following table:

| Water, $x$ | 8 | 16 | 24 | 32 |
|---|---|---|---|---|
| Yield, $y$ | 4.1 | 4.5 | 5.1 | 6.1 |

Find an equation of the regression line of $y$ on $x$. Predict $y$ when $x = 12$.

7. **Virus** A rabbit was injected with a virus, and $x$ hours after the injection the temperature $y$ (in degrees Fahrenheit) of the rabbit was measured.[25] The data are given in the following table:

| Elapsed Time, $x$ | 24 | 32 | 48 | 56 |
|---|---|---|---|---|
| Temperature, $y$ | 102.8 | 104.5 | 106.5 | 107.0 |

Find an equation of the regression line of $y$ on $x$, and estimate the rabbit's temperature 40 hours after the injection.

8. **Psychology** In a psychological experiment, four persons were subjected to a stimulus. Both before and after the stimulus, the systolic blood pressure (in millimeters of mercury) of each subject was measured. The data are given in the following table:

| | Blood Pressure | | | |
|---|---|---|---|---|
| Before Stimulus, $x$ | 130 | 132 | 136 | 140 |
| After Stimulus, $y$ | 139 | 140 | 144 | 146 |

Find an equation of the regression line of $y$ on $x$, where $x$ and $y$ are as defined in the table.

*For the time series in Problems 9 and 10, fit a linear-regression line by least squares; that is, find an equation of the linear-regression line of y on x. In each case, let the first year in the table correspond to x = 1.*

9.

Production of Product A, 1993–1997 (in thousands of units)

| Year | Production |
|---|---|
| 1993 | 10 |
| 1994 | 15 |
| 1995 | 16 |
| 1996 | 18 |
| 1997 | 21 |

10. **Industrial Production** In the following table, let 1975 correspond to $x = 1$, 1977 correspond to $x = 3$, and so on:

Index of Industrial Production—Electrical Machinery (1977 = 100)

| Year | Index |
|---|---|
| 1975 | 77 |
| 1977 | 100 |
| 1979 | 126 |
| 1981 | 134 |

[25]R. R. Sokal and F. J. Rohlf, *Introduction to Biostatistics* (San Francisco: W. H. Freeman & Company, Publishers, 1973).

Source: *Economic Report of the President,* 1988, U.S. Government Printing Office, Washington, D. C. 1988.

**11. Computer Shipments**

a. Find an equation of the least squares line of $y$ on $x$ for the following data (refer to 1994 as year $x = 1$, and so on):

Overseas Shipments
of Computers by
Acme Computer Co.
(in thousands)

| Year | Quantity |
|------|----------|
| 1994 | 35 |
| 1995 | 31 |
| 1996 | 26 |
| 1997 | 24 |
| 1998 | 26 |

b. For the data in part (a), refer to 1994 as year $x = -2$, 1995 as year $x = -1$, 1996 as year $x = 0$ and so on.
Then $\sum_{i=1}^{5} x_i = 0$. Fit a least squares line and observe how the calculation is simplified.

**12. Medical Care** For the following time series, find an equation of the linear-regression line that best fits the data (refer to 1983 as year $x = -2$, 1984 as year $x = -1$, and so on):

Consumer Price Index—
Medical Care, 1983–1987
(1967 = 100)

| Year | Index |
|------|-------|
| 1983 | 357 |
| 1984 | 380 |
| 1985 | 403 |
| 1986 | 434 |
| 1987 | 462 |

*Source: Economic Report of the President*, 1988, U.S. Government Printing Office, Washington, DC, 1988.

OBJECTIVE

**To develop some properties of homogeneous functions, including Euler's theorem.**

# 19.10 A Comment on Homogeneous Functions[26]

Many of the functions that are useful in economic analysis share the property of being homogeneous.

## DEFINITION

*A function $z = f(x, y)$ is said to be **homogeneous of degree n** (n being a constant) if, for **all** positive real values of $\lambda$,*

$$f(\lambda x, \lambda y) = \lambda^n f(x, y).$$

In words, if both $x$ and $y$ are multiplied by the same positive real number, then the resulting function value is a power of the number times the function value $f(x, y)$. For example, if

$$f(x, y) = x^3 - 2xy^2,$$

then

$$f(\lambda x, \lambda y) = (\lambda x)^3 - 2(\lambda x)(\lambda y)^2 = \lambda^3 x^3 - 2\lambda^3 xy^2$$

$$= \lambda^3(x^3 - 2xy^2) = \lambda^3 f(x, y).$$

Thus, $f$ is homogeneous of degree three.

---

[26]This section contains material from Sec. 19.6 and may be omitted without loss of continuity.

An important homogeneous function in economics is the Cobb-Douglas production function:

$$P = f(l, k) = Al^{\alpha}k^{1-\alpha} \quad (\alpha \text{ and } A \text{ constants}).$$

We have

$$f(\lambda l, \lambda k) = A(\lambda l)^{\alpha}(\lambda k)^{1-\alpha} = A\lambda^{\alpha}l^{\alpha}\lambda^{1-\alpha}k^{1-\alpha}$$

$$= \lambda Al^{\alpha}k^{1-\alpha} = \lambda f(l, k).$$

Therefore, $f$ is homogeneous of degree one. For example, $f(l, k) = 2l^{0.3}k^{0.7}$ is a homogeneous function of degree one.

Homogeneous production functions of degree one have an interesting property. If $f$ is such a function, then

$$f(\lambda l, \lambda k) = \lambda f(l, k).$$

For example, if all inputs are doubled, then

$$f(2l, 2k) = 2f(l, k),$$

and output is doubled. Similarly, if all inputs are tripled, output is tripled, etc. In short, the same proportional change in each input factor of production results in the same proportional change in output.

By considering the partial derivatives of a homogeneous function, we can obtain an important result. Let $f(l, k)$ be a homogeneous production function of degree $n$. Then we have the identity

$$f(\lambda l, \lambda k) = \lambda^{n}f(l, k). \tag{1}$$

Consider the left side of Eq. (1). If we set $r = \lambda l$ and $s = \lambda k$, then Eq. (1) becomes

$$f(r, s) = \lambda^{n}f(l, k). \tag{2}$$

Now, for each side, we shall take the partial with respect to $\lambda$. For the left side, $f(r, s)$, we have, by the chain rule,

$$\frac{\partial}{\partial \lambda}[f(r, s)] = \frac{\partial}{\partial r}[f(r, s)]\frac{\partial r}{\partial \lambda} + \frac{\partial}{\partial s}[f(r, s)]\frac{\partial s}{\partial \lambda}$$

$$= \frac{\partial}{\partial r}[f(r, s)]l + \frac{\partial}{\partial s}[f(r, s)]k. \tag{3}$$

For the right side of Eq. (2),

$$\frac{\partial}{\partial \lambda}[\lambda^{n}f(l, k)] = n\lambda^{n-1}f(l, k). \tag{4}$$

Using Eqs. (3) and (4), we set $\dfrac{\partial}{\partial \lambda}[f(r, s)]$ equal to $\dfrac{\partial}{\partial \lambda}[\lambda^{n}f(l, k)]$:

$$l\frac{\partial}{\partial r}[f(r, s)] + k\frac{\partial}{\partial s}[f(r, s)] = n\lambda^{n-1}f(l, k).$$

In particular, if $\lambda = 1$, then $r = l$ and $s = k$, so $f(r, s) = f(l, k)$. Thus, $\dfrac{\partial}{\partial r}[f(r, s)] = \dfrac{\partial}{\partial l}[f(l, k)]$ and $\dfrac{\partial}{\partial s}[f(r, s)] = \dfrac{\partial}{\partial k}[f(l, k)]$. Hence, we have what is called *Euler's theorem* for homogeneous functions:

$$l\frac{\partial}{\partial l}[f(l, k)] + k\frac{\partial}{\partial k}[f(l, k)] = nf(l, k). \tag{5}$$

Now, if $f$ is homogeneous of degree one, such as the Cobb-Douglas function, then $n = 1$, and Eq. (5) becomes

$$l\frac{\partial}{\partial l}[f(l,k)] + k\frac{\partial}{\partial k}[f(l,k)] = f(l,k).$$

We conclude that if we multiply the marginal product of each input by the quantity of the input, the sum is equal to the total product.

OBJECTIVE

To compute double and triple integrals.

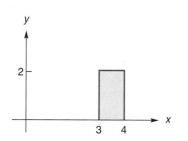

**FIGURE 19.21** Region over which $\int_0^2 \int_3^4 xy\,dx\,dy$ is evaluated.

## 19.11 MULTIPLE INTEGRALS

Recall that the definite integral of a function of one variable is concerned with integration over an *interval*. There are also definite integrals of functions of two variables, called (definite) **double integrals**. These involve integration over a *region* in the plane.

For example, the symbol

$$\int_0^2 \int_3^4 xy\,dx\,dy, \qquad \text{or equivalently,} \qquad \int_0^2 \left[\int_3^4 xy\,dx\right]dy,$$

is the double integral of $f(x,y) = xy$ over a region determined by the limits of integration. That region consists of all points $(x,y)$ in the $x,y$-plane such that $3 \le x \le 4$ and $0 \le y \le 2$. (See Fig. 19.21.) Essentially, a double integral is a limit of a sum of the form $\sum f(x,y)\Delta x\Delta y$, where, in this example, the points $(x,y)$ are in the shaded region. A geometric interpretation of a double integral will be given later.

To evaluate

$$\int_0^2 \int_3^4 xy\,dx\,dy \qquad \text{or} \qquad \int_0^2 \left[\int_3^4 xy\,dx\right]dy,$$

we use successive integrations starting with the innermost integral. First, we evaluate

$$\int_3^4 xy\,dx$$

by treating $y$ as a constant and integrating with respect to $x$ between the limits 3 and 4:

$$\int_3^4 xy\,dx = \left.\frac{x^2 y}{2}\right|_3^4.$$

Substituting the limits for the variable $x$, we have

$$\frac{4^2 \cdot y}{2} - \frac{3^2 \cdot y}{2} = \frac{16y}{2} - \frac{9y}{2} = \frac{7}{2}y.$$

Now we integrate this result with respect to $y$ between the limits 0 and 2:

$$\int_0^2 \frac{7}{2}y\,dy = \left.\frac{7y^2}{4}\right|_0^2 = \frac{7\cdot 2^2}{4} - 0 = 7.$$

Thus,

$$\int_0^2 \int_3^4 xy\,dx\,dy = 7.$$

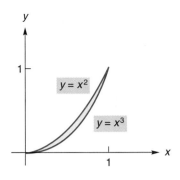

**FIGURE 19.22** Region over which $\int_0^1 \int_{x^3}^{x^2} (x^3 - xy) dy\, dx$ is evaluated.

Now consider the double integral

$$\int_0^1 \int_{x^3}^{x^2} (x^3 - xy)\, dy\, dx, \quad \text{or} \quad \int_0^1 \left[ \int_{x^3}^{x^2} (x^3 - xy) dy \right] dx.$$

Here we integrate first with respect to $y$ and then with respect to $x$. The region over which the integration takes place is all points $(x, y)$ such that $x^3 \le y \le x^2$ and $0 \le x \le 1$. (See Fig. 19.22.) This double integral is evaluated by first treating $x$ as a constant and integrating $x^3 - xy$ with respect to $y$ between $x^3$ and $x^2$, and then integrating the result with respect to $x$ between 0 and 1:

$$\int_0^1 \int_{x^3}^{x^2} (x^3 - xy)\, dy\, dx$$

$$= \int_0^1 \left[ \int_{x^3}^{x^2} (x^3 - xy)\, dy \right] dx = \int_0^1 \left( x^3 y - \frac{xy^2}{2} \right) \Big|_{x^3}^{x^2} dx$$

$$= \int_0^1 \left( \left[ x^3(x^2) - \frac{x(x^2)^2}{2} \right] - \left[ x^3(x^3) - \frac{x(x^3)^2}{2} \right] \right) dx$$

$$= \int_0^1 \left( x^5 - \frac{x^5}{2} - x^6 + \frac{x^7}{2} \right) dx = \int_0^1 \left( \frac{x^5}{2} - x^6 + \frac{x^7}{2} \right) dx$$

$$= \left( \frac{x^6}{12} - \frac{x^7}{7} + \frac{x^8}{16} \right) \Big|_0^1 = \left( \frac{1}{12} - \frac{1}{7} + \frac{1}{16} \right) - 0 = \frac{1}{336}.$$

**EXAMPLE 1** Evaluating a Double Integral

*Find* $\int_{-1}^1 \int_0^{1-x} (2x + 1)\, dy\, dx.$

*Solution:* Here we first integrate with respect to $y$ and then integrate the result with respect to $x$:

$$\int_{-1}^1 \int_0^{1-x} (2x + 1)\, dy\, dx$$

$$= \int_{-1}^1 \left[ \int_0^{1-x} (2x + 1)\, dy \right] dx$$

$$= \int_{-1}^1 (2xy + y) \Big|_0^{1-x} dx = \int_{-1}^1 \{[2x(1 - x) + (1 - x)] - 0\}\, dx$$

$$= \int_{-1}^1 (-2x^2 + x + 1)\, dx = \left( -\frac{2x^3}{3} + \frac{x^2}{2} + x \right) \Big|_{-1}^1$$

$$= \left( -\frac{2}{3} + \frac{1}{2} + 1 \right) - \left( \frac{2}{3} + \frac{1}{2} - 1 \right) = \frac{2}{3}. \qquad \blacksquare$$

**EXAMPLE 2** Evaluating a Double Integral

*Find* $\int_1^{\ln 2} \int_{e^y}^2 dx\, dy.$

*Solution:* Here we first integrate with respect to $x$ and then integrate the result with respect to $y$:

$$\int_1^{\ln 2} \int_{e^y}^2 dx\, dy = \int_1^{\ln 2}\left[\int_{e^y}^2 dx\right] dy = \int_1^{\ln 2} x\Big|_{e^y}^2 dy$$

$$= \int_1^{\ln 2} (2 - e^y)\, dy = (2y - e^y)\Big|_1^{\ln 2}$$

$$= (2\ln 2 - 2) - (2 - e) = 2\ln 2 - 4 + e$$

$$= \ln 4 - 4 + e.$$

A double integral can be interpreted in terms of the volume of a region between the $x,y$-plane and a surface $z = f(x, y)$ if $z \geq 0$. In Fig. 19.23 is a

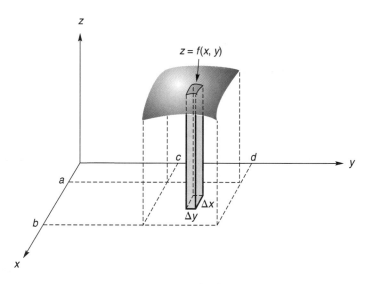

**FIGURE 19.23**   Interpreting $\int_a^b \int_c^d f(x, y)\, dy\, dx$ in terms of volume, where $f(x, y) \geq 0$.

region whose volume we shall consider. The element of volume for this region is a vertical column with height approximately $z = f(x, y)$ and base area $\Delta y\, \Delta x$. Thus, its volume is approximately $f(x, y)\Delta y\, \Delta x$. The volume of the entire region can be found by summing the volumes of all such elements for $a \leq x \leq b$ and $c \leq y \leq d$ via a double integral:

$$\text{volume} = \int_a^b \int_c^d f(x, y)\, dy\, dx.$$

**Triple integrals** are handled by successively evaluating three integrals, as the next example shows.

**EXAMPLE 3**   **Evaluating a Triple Integral**

Find $\displaystyle\int_0^1 \int_0^x \int_0^{x-y} x\, dz\, dy\, dx.$

*Solution:*

$$\int_0^1 \int_0^x \int_0^{x-y} x \, dz \, dy \, dx = \int_0^1 \int_0^x \left[ \int_0^{x-y} x \, dz \right] dy \, dx$$

$$= \int_0^1 \int_0^x (xz) \Big|_0^{x-y} dy \, dx = \int_0^1 \int_0^x [x(x-y) - 0] \, dy \, dx$$

$$= \int_0^1 \int_0^x (x^2 - xy) \, dy \, dx = \int_0^1 \left[ \int_0^x (x^2 - xy) \, dy \right] dx$$

$$= \int_0^1 \left( x^2 y - \frac{xy^2}{2} \right) \Big|_0^x dx = \int_0^1 \left[ \left( x^3 - \frac{x^3}{2} \right) - 0 \right] dx$$

$$= \int_0^1 \frac{x^3}{2} \, dx = \frac{x^4}{8} \Big|_0^1 = \frac{1}{8}. \qquad \blacksquare$$

## ▪ Exercise 19.11

*In Problems 1–22, evaluate the multiple integrals.*

**1.** $\int_0^3 \int_0^4 x \, dy \, dx.$

**2.** $\int_0^2 \int_1^2 y \, dy \, dx.$

**3.** $\int_0^1 \int_0^1 xy \, dx \, dy.$

**4.** $\int_0^2 \int_0^3 x^2 \, dy \, dx.$

**5.** $\int_1^3 \int_1^2 (x^2 - y) \, dx \, dy.$

**6.** $\int_{-1}^2 \int_1^4 (x^2 - 2xy) \, dy \, dx.$

**7.** $\int_0^1 \int_0^2 (x + y) \, dy \, dx.$

**8.** $\int_0^3 \int_0^x (x^2 + y^2) \, dy \, dx.$

**9.** $\int_0^6 \int_0^{3x} y \, dy \, dx.$

**10.** $\int_1^2 \int_0^x \frac{1}{y} \, dy \, dx.$

**11.** $\int_0^1 \int_{3x}^{x^2} 2x^2 y \, dy \, dx.$

**12.** $\int_0^2 \int_0^{x^2} xy \, dy \, dx.$

**13.** $\int_0^2 \int_0^{\sqrt{4-y^2}} x \, dx \, dy.$

**14.** $\int_0^1 \int_y^{\sqrt{y}} y \, dx \, dy.$

**15.** $\int_{-1}^1 \int_x^{1-x} (x + y) \, dy \, dx.$

**16.** $\int_0^3 \int_{y^2}^{3y} x \, dx \, dy.$

**17.** $\int_0^1 \int_0^y e^{x+y} \, dx \, dy.$

**18.** $\int_2^3 \int_0^2 e^{x-y} \, dx \, dy.$

**19.** $\int_{-1}^0 \int_{-1}^2 \int_1^2 6xy^2 z^3 \, dx \, dy \, dz.$

**20.** $\int_0^1 \int_0^x \int_0^{x-y} x \, dz \, dy \, dx.$

**21.** $\int_0^1 \int_{x^2}^x \int_0^{xy} dz \, dy \, dx.$

**22.** $\int_0^2 \int_{y^2}^{3y} \int_0^x dz \, dx \, dy.$

**23. Statistics**  In the study of statistics, a joint density function $z = f(x, y)$ defined on a region in the $x,y$-plane is represented by a surface in space. The probability that

$$a \le x \le b \quad \text{and} \quad c \le y \le d$$

is given by

$$P(a \le x \le b, c \le y \le d) = \int_c^d \int_a^b f(x, y) \, dx \, dy$$

and is represented by the volume between the graph of $f$ and the rectangular region given by

$$a \le x \le b \quad \text{and} \quad c \le y \le d.$$

If $f(x, y) = e^{-(x+y)}$ is a joint density function, where $x \ge 0$ and $y \ge 0$, find

$$P(0 \le x \le 2, 1 \le y \le 2),$$

and give your answer in terms of $e$.

**24. Statistics** In Problem 23, let $f(x, y) = 12e^{-4x-3y}$ for $x, y \geq 0$. Find

$$P(3 \leq x \leq 4, 2 \leq y \leq 6),$$

and give your answer in terms of $e$.

**25. Statistics** In Problem 23, let $f(x, y) = x/8$, where $0 \leq x \leq 2$ and $0 \leq y \leq 4$. Find $P(x \geq 1, y \geq 2)$.

**26. Statistics** In Problem 23, let $f$ be the uniform density function $f(x, y) = 1$ defined over the unit square $0 \leq x \leq 1, 0 \leq y \leq 1$. Determine the probability that $0 \leq x \leq \frac{1}{3}$ and $\frac{1}{4} \leq y \leq \frac{3}{4}$.

# 19.12 REVIEW

## IMPORTANT TERMS AND SYMBOLS

**Section 19.1**  three-dimensional coordinate system  $f(x_1, x_2, \ldots, x_n)$  function of $n$ variables  
$x,y$-plane  $x,z$-plane  $y,z$-plane  octant  traces

**Section 19.2**  partial derivative  $\dfrac{\partial z}{\partial x}$  $f_x(x, y)$  $\dfrac{\partial z}{\partial x}\Big|_{(x_0, y_0)}$  $f_x(x_0, y_0)$

**Section 19.3**  joint-cost function  production function  marginal productivity  competitive products  
complementary products

**Section 19.4**  implicit partial differentiation

**Section 19.5**  $\dfrac{\partial^2 z}{\partial y\, \partial x}$  $\dfrac{\partial^2 z}{\partial x\, \partial y}$  $\dfrac{\partial^2 z}{\partial x^2}$  $\dfrac{\partial^2 z}{\partial y^2}$  $f_{xy}$  $f_{yx}$  $f_{xx}$  $f_{yy}$

**Section 19.6**  chain rule  intermediate variable

**Section 19.7**  relative maxima and minima  critical point  second-derivative test for functions of two variables

**Section 19.8**  Lagrange multipliers

**Section 19.9**  scatter diagram  method of least squares  linear regression line of $y$ on $x$  index numbers

**Section 19.10**  homogeneous function of degree $n$

**Section 19.11**  double integral  triple integral

## SUMMARY

We can extend the concept of a function of one variable to functions of several variables. The inputs for functions of $n$ variables are $n$-tuples. Generally, the graph of a function of two variables is a surface in a three-dimensional coordinate system. Functions of more than two variables cannot be geometrically represented.

For a function of $n$ variables, we can consider $n$ partial derivatives. For example, if $w = f(x, y, z)$, we have the partial derivatives of $f$ with respect to $x$, with respect to $y$, and with respect to $z$, denoted $f_x$, $f_y$, and $f_z$, or $\partial f/\partial x$, $\partial f/\partial y$, and $\partial f/\partial z$, respectively. To find $f_x(x, y, z)$, we treat $y$ and $z$ as constants and differentiate $f$ with respect to $x$ in the usual way. The other partial derivatives are found in a similar fashion. We can interpret $f_x(x, y, z)$ as the approximate change in $w$ that results from a one-unit change in $x$ when $y$ and $z$ are held fixed. There are similar interpretations for the other partial derivatives. A function of several variables may be defined implicitly. In this case, its partial derivatives are found by implicit partial differentiation.

Functions of several variables occur frequently in business and economic analysis, as well as in other areas of study. If a manufacturer produces $x$ units of product X and $y$ units of product Y, then the total cost $c$ of these units is a function of $x$ and $y$ and is called a joint-cost function. The partial derivatives $\partial c/\partial x$ and $\partial c/\partial y$ are called the marginal costs with respect to $x$ and $y$, respectively. We can interpret, for example, $\partial c/\partial x$ as the approximate cost of producing an extra unit of X while the level of production of Y is held fixed.

If $l$ units of labor and $k$ units of capital are used to produce $P$ units of a product, then the function $P = f(l, k)$ is called a production function. The partial derivatives of $P$ are called marginal-productivity functions.

Suppose two products, A and B, are such that the quantity demanded of each is dependent on the prices of both. If $q_A$ and $q_B$ are the quantities of A and B demanded when the prices of A and B are $p_A$ and $p_B$, respectively, then $q_A$ and $q_B$ are each functions of $p_A$ and $p_B$. When $\partial q_A / \partial p_B > 0$ and $\partial q_B / \partial p_A > 0$, then A and B are called competitive products (or substitutes). When $\partial q_A / \partial p_B < 0$ and $\partial q_B / \partial p_A < 0$, then A and B are called complementary products.

If $z = f(x, y)$, where $x = x(r, s)$ and $y = y(r, s)$, then $z$ can be considered as a function of $r$ and $s$. To find, for example, $\partial z / \partial r$, a chain rule may be used:

$$\frac{\partial z}{\partial r} = \frac{\partial z}{\partial x}\frac{\partial x}{\partial r} + \frac{\partial z}{\partial y}\frac{\partial y}{\partial r}.$$

A partial derivative of a function of $n$ variables is itself a function of $n$ variables. By successively taking partial derivatives of partial derivatives, we obtain higher-order partial derivatives. For example, if $f$ is a function of $x$ and $y$, then $f_{xy}$ denotes the partial derivative of $f_x$ with respect to $y$; $f_{xy}$ is called the second-partial derivative of $f$, first with respect to $x$ and then with respect to $y$.

If the function $f(x, y)$ has a relative extremum at $(x_0, y_0)$, then $(x_0, y_0)$ must be a solution of the system

$$f_x(x, y) = 0, \quad f_y(x, y) = 0.$$

Any solution of this system is called a critical point of $f$. Thus, critical points are the candidates at which a relative extremum may occur. The second-derivative test for functions of two variables gives conditions under which a critical point corresponds to a relative maximum or a relative minimum. The test states that if $(x_0, y_0)$ is a critical point of $f$ and

$$D(x, y) = f_{xx}(x, y)f_{yy}(x, y) - [f_{xy}(x, y)]^2,$$

then:

1. if $D(x_0, y_0) > 0$ and $f_{xx}(x_0, y_0) < 0$, $f$ has a relative maximum at $(x_0, y_0)$;

2. if $D(x_0, y_0) > 0$ and $f_{xx}(x_0, y_0) > 0$, $f$ has a relative minimum at $(x_0, y_0)$;

3. if $D(x_0, y_0) < 0$, $f$ has neither a relative maximum nor a relative minimum at $(x_0, y_0)$;

4. if $D(x_0, y_0) = 0$, no conclusion about an extremum at $(x_0, y_0)$ can yet be drawn, and further analysis is required.

To find critical points of a function of several variables, subject to a constraint, we may use the method of Lagrange multipliers. For example, to find the critical points of $f(x, y, z)$, subject to the constraint $g(x, y, z) = 0$, we first form the function

$$F(x, y, z, \lambda) = f(x, y, z) - \lambda g(x, y, z).$$

By solving the system

$$F_x = 0, \quad F_y = 0, \quad F_z = 0, \quad F_\lambda = 0,$$

we obtain the critical points of $F$. If $(x_0, y_0, z_0, \lambda_0)$ is such a critical point, then $(x_0, y_0, z_0)$ is a critical point of $f$, subject to the constraint. It is important to write the constraint in the form $g(x, y, z) = 0$. For example, if the constraint is $2x + 3y - z = 4$, then $g(x, y, z) = 2x + 3y - z - 4$ [or $g(x, y, z) = 4 - 2x - 3y + z$]. If $f(x, y, z)$ is subject to two constraints, $g_1(x, y, z) = 0$ and $g_2(x, y, z) = 0$, then we would form the function $F = f - \lambda_1 g_1 - \lambda_2 g_2$ and solve the system

$$F_x = 0, \quad F_y = 0, \quad F_z = 0, \quad F_{\lambda_1} = 0, \quad F_{\lambda_2} = 0.$$

Sometimes two variables, say, $x$ and $y$, may be related in such a way that the relationship is approximately linear. When the data points $(x_i, y_i)$, where $i = 1, 2, 3, \ldots, n$, are given to us, we can fit a straight line that approximates them. Such a line is the linear-regression line (or least squares line) of $y$ on $x$ and is given by

$$\hat{y} = \hat{a} + \hat{b}x$$

where

$$\hat{a} = \frac{\left(\sum_{i=1}^{n} x_i^2\right)\left(\sum_{i=1}^{n} y_i\right) - \left(\sum_{i=1}^{n} x_i\right)\left(\sum_{i=1}^{n} x_i y_i\right)}{n\sum_{i=1}^{n} x_i^2 - \left(\sum_{i=1}^{n} x_i\right)^2}$$

and

$$\hat{b} = \frac{n\sum_{i=1}^{n} x_i y_i - \left(\sum_{i=1}^{n} x_i\right)\left(\sum_{i=1}^{n} y_i\right)}{n\sum_{i=1}^{n} x_i^2 - \left(\sum_{i=1}^{n} x_i\right)^2}.$$

The $\hat{y}$-values can be used to predict $y$-values for given values of $x$.

When working with functions of several variables, we can consider their multiple integrals. These are determined by successive integration. For example, the double integral

$$\int_1^2 \int_0^y (x + y)\, dx\, dy$$

is determined by first treating $y$ as a constant and integrating $x + y$ with respect to $x$. After evaluating between the limits 0 and $y$, we integrate that result

with respect to $y$ from $y = 1$ to $y = 2$. Thus,

$$\int_1^2 \int_0^y (x + y)\, dx\, dy = \int_1^2 \left[ \int_0^y (x + y)\, dx \right] dy.$$

Triple integrals involve functions of three variables and are also evaluated by successive integration.

## REVIEW PROBLEMS

*In Problems 1–4, sketch the given surfaces.*

**1.** $2x + 3y + z = 9$.

**2.** $z = x$.

**3.** $z = y^2$.

**4.** $x^2 + z^2 = 1$.

*In Problems 5–16, find the indicated partial derivatives.*

**5.** $f(x, y) = 2x^2 + 3xy + y^2 - 1$;  $f_x(x, y), f_y(x, y)$.

**6.** $P = l^3 + k^3 - lk$;  $\partial P/\partial l$,  $\partial P/\partial k$.

**7.** $z = \dfrac{x}{x + y}$;  $\dfrac{\partial z}{\partial x}$,  $\dfrac{\partial z}{\partial y}$.

**8.** $f(p_A, p_B) = 2(p_A - 20) + 3(p_B - 30)$;  $f_{p_A}(p_A, p_B)$.

**9.** $f(x, y) = \ln\sqrt{x^2 + y^2}$;  $\dfrac{\partial}{\partial y}[f(x, y)]$.

**10.** $w = \dfrac{\sqrt{x^2 + y^2}}{y}$;  $\dfrac{\partial w}{\partial x}$.

**11.** $w = e^{x^2 y z}$;  $w_{xy}(x, y, z)$.

**12.** $f(x, y) = xy \ln(xy)$;  $f_{xy}(x, y)$.

**13.** $f(x, y, z) = (x + y)(y + z^2)$;  $\dfrac{\partial^2}{\partial z^2}[f(x, y, z)]$.

**14.** $z = (x^2 - y)(y^2 - 2xy)$;  $\partial^2 z/\partial y^2$.

**15.** $w = xe^{yz} \ln z$;  $\partial w/\partial y$,  $\partial^2 w/\partial x\, \partial z$.

**16.** $P = 100 l^{0.11} k^{0.89}$;  $\partial^2 P/\partial k\, \partial l$.

*In Problems 17 and 18, find the indicated value.*

**17.** If $f(x, y, z) = \dfrac{x + y}{xz}$, find $f_{xyz}(2, 3, 4)$.

**18.** If $f(x, y, z) = (x + 1)e^{y^2 \ln(z+1)}$, find $f_{xyz}(0, 1, 0)$.

[27]**19.** If $w = x^2 + 2xy + 3y^2$, $x = e^r$, and $y = \ln(r + s)$, find $\partial w/\partial r$ and $\partial w/\partial s$.

[27]**20.** If $z = \ln(x/y) + e^y - xy$, $x = r^2 s^2$, and $y = r + s$, find $\partial z/\partial s$.

[27]**21.** If $x^2 + 2xy - 2z^2 + xz + 2 = 0$, find $\partial z/\partial x$.

[27]**22.** If $z^2 - e^{yz} + \ln z + e^{xz} = 0$, find $\partial z/\partial y$.

**23. Production Function** If a manufacturer's production function is defined by $P = 20 l^{0.7} k^{0.3}$, determine the marginal-productivity functions.

**24. Joint-Cost Function** A manufacturer's cost for producing $x$ units of product X and $y$ units of product Y is given by

$$c = 5x + 0.03xy + 7y + 200.$$

Determine the (partial) marginal cost with respect to $x$ when $x = 100$ and $y = 200$.

**25. Competitive/Complementary Products** If $q_A = 200 - 3p_A + p_B$ and $q_B = 50 - 5p_B + p_A$, where $q_A$ and $q_B$ are the number of units demanded of products A and B, respectively, and $p_A$ and $p_B$ are their respective prices per unit, determine whether A and B are competitive or complementary products.

**26. Innovation** For industry, the following model describes the rate $\alpha$ (a Greek letter read "alpha") at which an innovation substitutes for an established process:[28]

$$\alpha = Z + 0.530P - 0.027S.$$

Here, $Z$ is a constant that depends on the particular industry, $P$ is an index of profitability of the innovation, and $S$ is an index of the extent of the investment necessary to make use of the innovation. Find $\partial\alpha/\partial P$ and $\partial\alpha/\partial S$.

**27.** Examine $f(x, y) = x^2 + 2y^2 - 2xy - 4y + 3$ for relative extrema.

**28.** Examine $f(w, z) = 2w^3 + 2z^3 - 6wz + 7$ for relative extrema.

[28]A. P. Hurter, Jr., A. H. Rubenstein, et al., "Market Penetration by New Innovations: The Technological Literature," *Technological Forecasting and Social Change*, 11 (1978), 197–221.

[27]Refers to Sec. 19.4 or Sec. 19.6.

**29. Minimizing Material**   An open-top rectangular cardboard box is to have a volume of 32 cubic feet. Find the dimensions of the box so that the amount of cardboard used is minimized.

**30.** The function

$$f(x, y) = ax^2 + by^2 + cxy - 10x - 20y$$

has a critical point at $(x, y) = (1, 2)$, and the second-derivative test is inconclusive at this point. Determine the values of the constants $a$, $b$, and $c$.

**31. Maximizing Profit**   A dairy produces two types of cheese, A and B, at constant average costs of 50 cents and 60 cents per pound, respectively. When the selling price per pound of A is $p_A$ cents and of B is $p_B$ cents, the demands (in pounds) for A and B, are, respectively,

$$q_A = 250(p_B - p_A)$$

and

$$q_B = 32{,}000 + 250(p_A - 2p_B).$$

Find the selling prices that yield a relative maximum profit. Verify that the profit has a relative maximum at these prices.

**32.** Find all critical points of $f(x, y, z) = xyz$, subject to the condition that

$$3x + 2y + 4z - 120 = 0 \quad (xyz \neq 0).$$

**33.** Find all critical points of $f(x, y, z) = x^2 + y^2 + z^2$, subject to the constraint $3x + 2y + z = 14$.

**34. Surviving Infection**   In an experiment,[29] a group of fish was injected with living bacteria. Of those fish main-

tained at 28°C, the percentage $p$ that survived the infection $t$ hours after the injection is given in the following table:

| $t$ | 8 | 10 | 18 | 20 | 48 |
|---|---|---|---|---|---|
| $p$ | 82 | 79 | 78 | 78 | 64 |

Find the linear-regression line of $p$ on $t$.

**35. Equipment Expenditures**   Find the least squares linear-regression line of $y$ on $x$ for the data given in the following table (refer to year 1993 as year $x = 1$, etc.):

| Equipment Expenditures of Allied Computer Company, 1993–1998 (in millions of dollars) | |
|---|---|
| **Year** | **Expenditures** |
| 1993 | 15 |
| 1994 | 22 |
| 1995 | 21 |
| 1996 | 27 |
| 1997 | 26 |
| 1998 | 34 |

*In Problems 36–39, evaluate the double integrals.*

**36.** $\displaystyle\int_1^2 \int_0^y x^2 y^2 \, dx \, dy.$

**37.** $\displaystyle\int_0^4 \int_{y/2}^2 xy \, dx \, dy.$

**38.** $\displaystyle\int_0^3 \int_{y^2}^{3y} x \, dx \, dy.$

**39.** $\displaystyle\int_0^1 \int_{\sqrt{x}}^{x^2} (x^2 + 2xy - 3y^2) \, dy \, dx.$

[29]J. B. Covert and W. W. Reynolds, "Survival Value of Fever in Fish," *Nature*, 267, no. 5606 (1977), 43–45.

# MATHEMATICAL *SNAPSHOT*

## DATA ANALYSIS TO MODEL COOLING[30]

In Chapter 17 you worked with Newton's law of cooling, which can be used to describe the temperature of a cooling body with respect to time. Here you will determine that relationship in an empirical way via data analysis. This will illustrate how mathematical models are designed in many real-world situations.

Suppose that you want to create a mathematical model of the cooling of hot tea after it is placed in a refrigerator. To do this you place a pitcher containing hot tea and a thermometer in the refrigerator and periodically read and record the temperature of the tea. Table 19.6 gives the collected data, where $T$ is the

Fahrenheit temperature $t$ minutes after the tea is placed in the refrigerator. Initially, that is, at $t = 0$, the temperature is 124°F; when $t = 391$, then $T = 47$°F. After being in the refrigerator overnight, the temperature is 45°F. Figure 19.24 gives a graph of the data points $(t, T)$ for $t = 0$ to $t = 391$.

The pattern of these points strongly suggests that they nearly lie on the graph of a decreasing exponential function, such as the one shown in Fig. 19.24. In particular, because the overnight temperature is 45°F, this exponential function should have $T = 45$ as a horizontal asymptote. Such a function

**TABLE 19.6**

| Time $t$ | Temperature $T$ | Time $t$ | Temperature $T$ |
|---|---|---|---|
| 0 min | 124°F | 128 min | 64°F |
| 5 | 118 | 144 | 62 |
| 10 | 114 | 178 | 59 |
| 16 | 109 | 208 | 55 |
| 20 | 106 | 244 | 51 |
| 35 | 97 | 299 | 50 |
| 50 | 89 | 331 | 49 |
| 65 | 82 | 391 | 47 |
| 85 | 74 | Overnight | 45 |

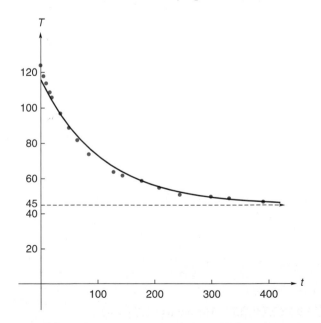

**FIGURE 19.24** Data points and exponential approximation.

[30]Adapted from Gloria Barrett, Dot Doyle, and Dan Teague, "Using Data Analysis in Precalculus to Model Cooling," *The Mathematics Teacher*, 81, no. 8 (November 1988), 680–84. By permission of the National Council of Teachers of Mathematics.

has the form

$$\hat{T} = Ce^{at} + 45, \qquad (1)$$

where $\hat{T}$ gives the predicted temperature at time $t$, and $C$ and $a$ are constants with $a < 0$. (Note that since $a < 0$, then as $t \to \infty$, you have $Ce^{at} \to 0$, so $Ce^{at} + 45 \to 45$.)

Now the problem is to find the values of $C$ and $a$ so that the curve given by Eq. (1) best fits the data. By writing Eq. (1) as

$$\hat{T} - 45 = Ce^{at}$$

and then taking the natural logarithm of each side, you obtain a linear form:

$$\ln(\hat{T} - 45) = \ln(Ce^{at}),$$

$$\ln(\hat{T} - 45) = \ln C + \ln e^{at},$$

$$\ln(\hat{T} - 45) = \ln C + at. \qquad (2)$$

Letting $\hat{T}_l = \ln(\hat{T} - 45)$, Eq. (2) becomes

$$\hat{T}_l = at + \ln C. \qquad (3)$$

Because $a$ and $\ln C$ are constants, Eq. (3) is a linear equation in $\hat{T}_l$ and $t$. This means that for the original data, if you plot the points $(t, \ln(T - 45))$, they should nearly lie on a straight line. These points are shown in Fig. 19.25, where $T_l$ represents $\ln(T - 45)$. Thus for the line given by Eq. (3) that predicts $T_l$, you

can assume that it is the linear regression line of $T_l$ on $t$. That is, $a$ and $\ln C$ are linear regression coefficients. By using the formulas for these coefficients and a calculator, you can determine that

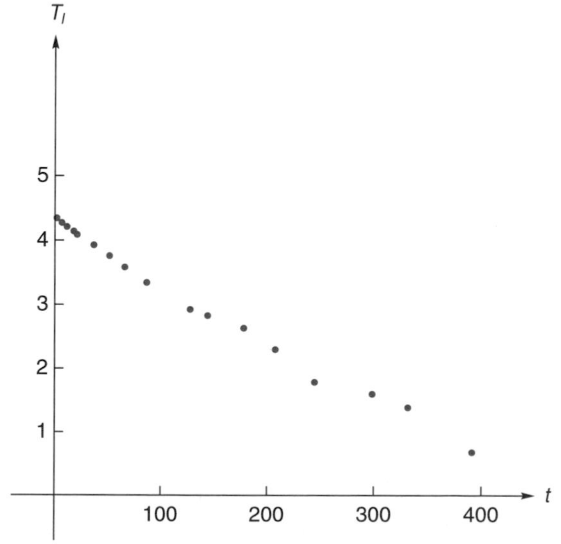

$$a = \frac{17\left(\sum_{i=1}^{17} t_i T_{l_i}\right) - \left(\sum_{i=1}^{17} t_i\right)\left(\sum_{i=1}^{17} T_{l_i}\right)}{17\left(\sum_{i=1}^{17} t_i^2\right) - \left(\sum_{i=1}^{17} t_i\right)^2} \approx -0.00921$$

and

$$\ln C = \frac{\left(\sum_{i=1}^{17} t_i^2\right)\left(\sum_{i=1}^{17} T_{l_i}\right) - \left(\sum_{i=1}^{17} t_i\right)\left(\sum_{i=1}^{17} t_i T_{l_i}\right)}{17\left(\sum_{i=1}^{17} t_i^2\right) - \left(\sum_{i=1}^{17} t_i\right)^2}$$

$$\approx 4.260074.$$

Since $\ln C \approx 4.260074$, then $C \approx e^{4.260074} \approx 70.82$. Thus from Eq. (1),

$$\hat{T} = 70.82e^{-0.00921t} + 45,$$

which is a model that predicts the temperature of the cooling tea. The graph of this function is the curve shown in Fig. 19.24.

■ **Exercises**

1. Plot the following data points on an $x,y$-coordinate plane:

| $x$ | 0 | 1 | 4 | 7 | 10 |
|-----|-----|-----|-----|-----|-----|
| $y$ | 15 | 12 | 9 | 7 | 6 |

Suppose that these points nearly lie on the graph of a decreasing exponential function with horizontal asymptote $y = 5$. Use the technique discussed in this mathematical snapshot to determine the function.

2. Suppose that observed data follow a relation given by $y = C/x^r$, where $x, y, C > 0$. By taking the natural logarithm of each side, show that $\ln x$ and $\ln y$ are linearly related. Thus, the points $(\ln x, \ln y)$ lie on a straight line.

3. Use Newton's law of cooling (see Sec. 17.7) and the data points $(0, 124)$ and $(128, 64)$ to determine the temperature $T$ of the tea discussed in the snapshot at time $t$. Assume that the ambient temperature is $45°F$.

**FIGURE 19.25** The points $(t, T_l)$, where $T_l = \ln(T - 45)$, nearly lie on a straight line.

# Appendix A  Concepts for Calculus

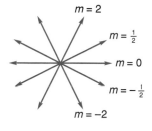

## A.1    REVIEW: SLOPES AND EQUATIONS OF LINES

Slope is a measure of the "steepness" of a line. The **slope of a nonvertical line,** denoted $m$, is given by the formula $m = \dfrac{\text{vertical change}}{\text{horizontal change}} = \dfrac{y_2 - y_1}{x_2 - x_1}$, where $(x_1, y_1)$ and $(x_2, y_2)$ are two distinct points on the line. For example, the line shown in the following diagram has slope $m = \dfrac{y_2 - y_1}{x_2 - x_1} = \dfrac{4 - 1}{1 - (-3)} = \dfrac{3}{4}$:

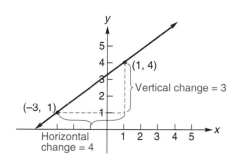

Here are some important observations regarding slopes:

- A line has positive slope if it rises from left to right.
- A line has negative slope if it falls from left to right.
- A horizontal line has slope $m = 0$.
- A vertical line does not have a slope. (This means that the slope is undefined; it does *not* mean that the slope is 0.)
- "Steeper" lines have slopes that are larger in absolute value.
- The slope of a particular line will always be the same, regardless of which two points are chosen to calculate the slope.
- Two lines are parallel if and only if they have the same slope or are both vertical.
- Two lines with slopes $m_1$ and $m_2$ are perpendicular if and only if $m_1 = -\dfrac{1}{m_2}$. Also, a horizontal line and a vertical line are perpendicular.

If you know the slope $m$ of a line and the coordinates $(x_1, y_1)$ of one point on the line, you can easily write an equation of the line in **point-slope form:** $y - y_1 = m(x - x_1)$.

Other useful ways to express an equation of a line include the **slope-intercept form,** $y = mx + b$ (where $m$ is the slope and $b$ is the $y$-intercept), and the **general linear form,** $Ax + By + C = 0$.

**EXAMPLE 1**

*Find a general linear form of the equation of the line that passes through* $(-2, 4)$ *and* $(2, 2)$.

*Solution:* The slope of the line is $m = \dfrac{y_2 - y_1}{x_2 - x_1} = \dfrac{2 - 4}{2 - (-2)} = \dfrac{-2}{4} = -\dfrac{1}{2}$.
Using the known point $(-2, 4)$, we may write the equation in point-slope form and then convert it to general linear form as follows:

$$y - 4 = -\frac{1}{2}[x - (-2)],$$

$$y - 4 = -\frac{1}{2}x - 1,$$

$$2y - 8 = -x - 2,$$

$$x + 2y - 6 = 0.$$

The general linear form is not unique, but one way to write it is $x + 2y - 6 = 0$. ∎

**EXAMPLE 2**

*Determine whether the lines* $y = -\dfrac{2}{3}x - 5$ *and* $2x + 3y = 9$ *are parallel, perpendicular, or neither.*

*Solution:* The first equation is in slope-intercept form with slope $m = -\dfrac{2}{3}$.
We rewrite the second equation in the same form:

$$2x + 3y = 9,$$

$$3y = -2x + 9,$$

$$y = -\frac{2}{3}x + 3.$$

The second line also has slope $m = -\dfrac{2}{3}$. Since the two lines have the same slope, they are parallel. ∎

**EXAMPLE 3**

*Find the slope-intercept form of the equation of the line that passes through* $(0, -2)$ *and is perpendicular to* $y = 4x - 3$.

*Solution:* Since the given line has slope 4, the perpendicular line will have slope $m = -\dfrac{1}{4}$. Also, since the line includes $(0, -2)$, its $y$-intercept is $b = -2$.
Thus, we may directly obtain the slope-intercept form: $y = -\dfrac{1}{4}x + (-2)$, or

$$y = -\frac{1}{4}x - 2.$$ ∎

# ▪ Exercise A.1

**1.** Find the slope of the line that passes through $(-6, 4)$ and $(2, -1)$.

**2.** Find the point-slope form of the equation of the line that contains $(2, 5)$ and has slope 7.

**3.** Find a general linear equation of the line that passes through $(0, 5)$ and has slope $\dfrac{3}{2}$.

**4.** Find a general linear equation of the horizontal line that passes through $(3.7, -2.4)$.

**5.** Find a general linear equation of the line that passes through $(4, -3)$ and $(8, 2)$.

**6.** Find the slope and $y$-intercept of the line determined by $y = \dfrac{2}{3}x + 3$. Sketch the graph.

**7.** If possible, find the slope and $y$-intercept of the line determined by $2 + x = -3$. Sketch the graph.

**8.** Determine whether the lines $2x - 6y = 4$ and $3x + y = 12$ are parallel, perpendicular, or neither.

**9.** Find the slope-intercept form of the equation of the line that passes through $(6, 7)$ and is parallel to $3y = x + 4$.

**10.** Find the slope-intercept form of the equation of the line that passes through $(-2, 4)$ and is perpendicular to $5x - 2y = 10$.

## A.2   SECANT LINES AND AVERAGE RATE OF CHANGE

A line that passes through two points on the graph of a function $f$ is a **secant line** of the graph. The slope of this line gives the **average rate of change** for the interval determined by the first coordinates of the chosen points. Therefore,

$$\text{Average rate of change in } f \text{ from } x_0 \text{ to } x_1 = x_0 + h$$

$$= \text{slope of secant line from } P = (x_0, y_0) = (x_0, f(x_0))$$

$$\text{to } Q = (x_1, y_1) = (x_1, f(x_1)) = (x_0 + h, f(x_0 + h))$$

$$= m$$

$$= \frac{y_1 - y_0}{x_1 - x_0} = \frac{f(x_1) - f(x_0)}{x_1 - x_0} = \frac{f(x_0 + h) - f(x_0)}{(x_0 + h) - x_0}$$

$$= \frac{f(x_0 + h) - f(x_0)}{h}.$$

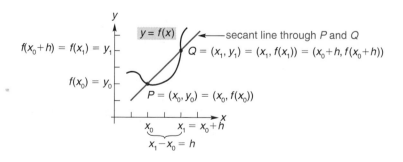

### EXAMPLE 1

*Suppose that Karl is driving to work. Fifteen minutes after he starts he is 7 miles from home, and 20 minutes after he starts he is 11 miles from home. Find his average rate for the interval between 15 and 20 minutes.*

*Solution:* Since $11 - 7 = 4$ and $20 - 15 = 5$, Karl traveled 4 miles in 5 minutes. His average rate is $\dfrac{4 \text{ miles}}{5 \text{ minutes}}$, or 0.8 mile per minute. (This is equivalent to 48 miles per hour.)    ■

The rate of change of distance in Example 1 is called **average velocity,** because its units are distance per unit of time. Unlike a speed, a velocity can be either positive or negative. (Speed is defined as the absolute value of velocity.)

Notice that we did *not* say that Karl was traveling at a constant rate of 48 miles per hour. He may have sped up, slowed down, or even backtracked during the 5-minute interval. If $f(t)$ is a function representing Karl's distance after $t$ minutes, the graph of $f(t)$ might look like any of the following graphs:

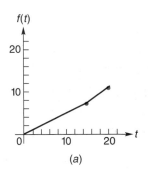
(a)

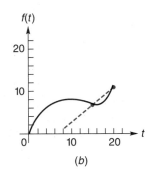
(b)

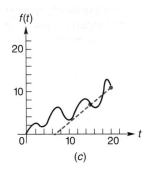
(c)

Each of these graphs contains the points (15, 7) and (20, 11). In each case, the secant line corresponding to the interval [15, 20] is the same, and the average rate of change of distance from home, or average velocity, is the slope of the secant line, which is 0.8 mile per minute.

### EXAMPLE 2

*The following table shows the value of Elizabeth's stock portfolio for several days in November:*

| Date | 1 | 3 | 7 | 10 | 14 | 17 | 21 | 24 | 28 | 31 |
|------|----|----|----|----|----|----|----|----|----|----|
| Value ($ thousands) | 14.2 | 14.6 | 15.2 | 14.7 | 13.9 | 14.1 | 13.7 | 14.3 | 14.7 | 14.5 |

*Find the average rate of change of the value for each of the following intervals:*

**a.** November 3 to 7.   **b.** November 10 to 28.   **c.** November 7 to 21.

*Solution:*

**a.** Average rate $= \dfrac{15{,}200 - 14{,}600}{7 - 3} = \dfrac{600}{4} = \$150$ per day.

**b.** Average rate $= \dfrac{14{,}700 - 14{,}700}{28 - 10} = \dfrac{0}{18} = \$0$ per day.

**c.** Average rate $= \dfrac{13{,}700 - 15{,}200}{21 - 7} = \dfrac{-1500}{14} \approx -\$107.14$ per day.    ■

## EXAMPLE 3

*Sketch the graph of* $f(x) = x^3 - 4x$. *Then sketch the secant line for each interval of x-values, and find each corresponding average rate of change of f.*

**a.** $[-2, -1]$.

**b.** $[-0.5, 1.5]$.

*Solution:* The graph with the secant lines is shown in the following diagram:

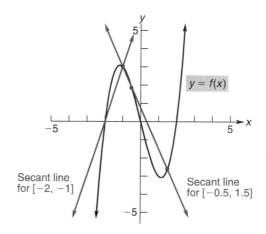

We find the average rates of change by calculating the slope of each line.

**a.** Average rate $= m = \dfrac{y_1 - y_0}{x_1 - x_0} = \dfrac{3 - 0}{-1 - (-2)} = \dfrac{3}{1} = 3.$

**b.** Average rate $= m = \dfrac{y_1 - y_0}{x_1 - x_0} = \dfrac{-2.625 - 1.875}{1.5 - (-0.5)} = \dfrac{-4.5}{2} = -2.25.$ ■

It is not necessary to graph a function in order to calculate an average rate of change. Since the graph of a function $f(x)$ includes the points $(x_0, f(x_0))$ and $(x_1, f(x_1))$, the average rate of change on the interval $[x_0, x_1]$ is given by the expression $\dfrac{f(x_1) - f(x_0)}{x_1 - x_0}$. If you let $h = x_1 - x_0$, then $x_1 = x_0 + h$ and the average rate of change of $f$ on the interval $[x_0, x_0 + h]$ is $\dfrac{f(x_0 + h) - f(x_0)}{h}$.

## EXAMPLE 4

*Let* $g(x) = x^2$. *Find the average rate of change of g for each interval of x-values.*

**a.** $[9, 10]$.    **b.** $[-0.5, -0.4]$.    **c.** $[2, 2 + h]$.    **d.** $[x_0, x_0 + h]$.

*Solution:*

**a.** Average rate $= \dfrac{g(10) - g(9)}{10 - 9} = \dfrac{10^2 - 9^2}{10 - 9} = \dfrac{100 - 81}{10 - 9} = \dfrac{19}{1} = 19.$

**b.** Average rate $= \dfrac{g(-0.4) - g(-0.5)}{-0.4 - (-0.5)} = \dfrac{(-0.4)^2 - (-0.5)^2}{-0.4 + 0.5}$

$= \dfrac{0.16 - 0.25}{0.1} = \dfrac{-0.09}{0.1} = -0.9.$

**c.** Average rate $= \dfrac{g(2 + h) - g(2)}{(2 + h) - 2} = \dfrac{(2 + h)^2 - 2^2}{h}$

$= \dfrac{(4 + 4h + h^2) - 4}{h} = \dfrac{4h + h^2}{h}$

$= 4 + h.$

**d.** Average rate $= \dfrac{g(x_0 + h) - g(x_0)}{(x_0 + h) - x_0} = \dfrac{(x_0 + h)^2 - x_0^2}{h}$

$= \dfrac{(x_0^2 + 2x_0 h + h^2) - x_0^2}{h} = \dfrac{2x_0 h + h^2}{h} = 2x_0 + h.$ ∎

## ▪ Exercise A.2

1. Carmella was 42 inches tall when she was 8 years old, and she was 60 inches tall when she was 12 years old. Find her average rate of growth between 8 and 12 years.

2. A plumber charges $65 for a half-hour job and $128 for a two-hour job. Find the average rate of change, in dollars per hour, between a half hour and two hours.

3. The following table shows the high temperature for several days in March:

| Date | 3 | 5 | 8 | 9 | 11 | 14 | 18 | 21 | 25 | 30 |
|---|---|---|---|---|---|---|---|---|---|---|
| Temperature (°F) | 63 | 57 | 49 | 52 | 61 | 70 | 65 | 73 | 69 | 79 |

Find the average rate of change of temperature for each interval.

**a.** March 8 to 14.     **b.** March 14 to 18.

**c.** March 5 to 21.     **d.** March 3 to 30.

4. Sketch a graph of $f(x) = 0.5x^2$. Then sketch the secant line for each interval of $x$-values, and find each corresponding average rate of change.

**a.** $[0, 3]$.   **b.** $[-4, 1]$.   **c.** $[-2, 2]$.   **d.** $[2, 2.01]$.

5. Sketch a graph of $g(x) = x^3 + 3x$. Then sketch the secant line for each interval of $x$-values, and find each corresponding average rate of change.

**a.** $[1, 3]$.   **b.** $[-2, 2]$.   **c.** $[0, 1]$.   **d.** $[0, 0.1]$.

6. Let $f(x) = x^2 + 2x - 5$. Find the average rate of change of $f$ for each interval of $x$-values.

**a.** $[2, 4]$.   **b.** $[-1, 1]$.   **c.** $[0, h]$.   **d.** $[x_0, x_0 + h]$.

7. Let $g(x) = x^3$. Find the average rate of change of $g$ for each interval of $x$-values.

**a.** $[0, 5]$.   **b.** $[3, 8]$.   **c.** $[3, 3 + h]$.   **d.** $[x_0, x_0 + h]$.

## A.3  SLOPE OF A CURVE AND DERIVATIVE

Suppose you want to study the behavior of the graph of $f(x) = x^2$ at the point $(2, 4)$. If you magnify a tiny section of the graph, including $(2, 4)$, you will find that the graph looks almost like a line. Extending this straight line in both directions and shrinking the graph back to normal size as shown in the accompanying diagram, you will notice that the line intersects the curve in much the same way as a tangent line intersects a circle. For this reason, the line is called the **tangent line** to the curve at $(2, 4)$. The slope of this line is called the **slope of the curve** at the point $(2, 4)$, or the **instantaneous rate of change** of $f(x)$ with respect to $x$ at $x = 2$, or the derivative of $f(x)$ with respect to $x$ at $x = 2$.

How can this slope be calculated? In Example 4c of the previous section, you found that the average rate of change of $g(x)$ for the interval $[2, 2 + h]$ was equal to $4 + h$. As $h$ becomes closer and closer to 0, the value of this expression approaches 4. This number, 4, is called the instantaneous rate of change of $g(x)$ with respect to $x$ when $x = 2$. It can be shown that this instan-

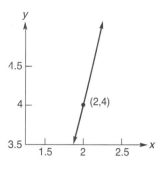
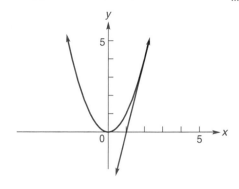

taneous rate of change is the slope of the tangent line. Thus the slope of the curve at $(2, 4)$ is 4.

We have assumed that $h$ is positive, but similar reasoning works for negative values of $h$. If $h$ is negative, the interval is actually $[2 + h, 2]$, but the average rate of change is still given by $4 + h$. Again, as $h$ becomes closer and closer to 0, the value of this expression approaches 4.

In general, if a function $f(x)$ has a "smooth" graph (a graph with no sharp corners or bends), you can calculate the slope of $f(x)$ at a point $(x_0, f(x_0))$ by means of the following procedure:

- Use the expression $\dfrac{f(x_0 + h) - f(x_0)}{h}$ to find the average rate of change of the function between the points $(x_0, f(x_0))$ and $(x_0 + h, f(x_0 + h))$.
- Simplify the expression for average rate of change.
- Determine what value is approached as $h$ approaches zero. (For the functions that we will consider here, you can find this value by substituting the value $h = 0$ into the *simplified* expression for the average rate of change.)

Notice that you cannot substitute $h = 0$ directly into the expression $\dfrac{f(x_0 + h) - f(x_0)}{h}$, because division by zero is undefined. Although we have presented only a brief intuitive outline of instantaneous rates here, these concepts are discussed in great detail in the study of calculus.

The slope of the curve $y = f(x)$ at a point $(x_0, f(x_0))$ is denoted $\dfrac{dy}{dx}(x_0)$ or $\dfrac{df}{dx}(x_0)$ and is called the derivative of the function $f$ with respect to the variable $x$ at the point $(x_0, f(x_0))$. The derivative of $f$ with respect to $x$ varies with $(x_0, f(x_0))$. So the derivative is also a function of $x$, written as $\dfrac{df}{dx}$ or $\dfrac{dy}{dx}$. In short, for $y = f(x)$, a function with a smooth graph, $\dfrac{dy}{dx}$ is the function of $x$ that gives the slope of the curve $y = f(x)$ at the point $P = (x, f(x))$ .

### EXAMPLE 1

*Let $f(x) = x^3$. For each of the following values of x, find the slope of the graph of $f(x)$; in other words, find $\dfrac{df}{dx}(x)$ for the listed values of x.*

**a.** $x = 0$.        **b.** $x = 4$.        **c.** $x = t$.

*Sketch the graph of $y = f(x) = x^3$ and the lines tangent to it at $(0, 0)$ and $(4, 64)$.*

*Solution:*

**a.**  $\dfrac{f(0+h)-f(0)}{h} = \dfrac{h^3-0^3}{h} = \dfrac{h^3}{h} = h^2.$

As $h$ approaches 0, $h^2$ approaches 0. Therefore, the slope is 0, so $\dfrac{df}{dx}(0) = 0.$

**b.**  $\dfrac{f(4+h)-f(4)}{h} = \dfrac{(4+h)^3-4^3}{h} = \dfrac{(64+48h+12h^2+h^3)-64}{h}$

$= \dfrac{48h+12h^2+h^3}{h} = 48+12h+h^2.$

As $h$ approaches 0, $48+12h+h^2$ approaches 48. Hence, the slope is 48, or $\dfrac{df}{dx}(4) = 48.$

**c.**  $\dfrac{f(t+h)-f(t)}{h} = \dfrac{(t+h)^3-t^3}{h} = \dfrac{(t^3+3t^2h+3th^2+h^3)-t^3}{h}$

$= \dfrac{3t^2h+3th^2+h^3}{h} = 3t^2+3th+h^2.$

As $h$ approaches 0, $3t^2+3th+h^2$ approaches $3t^2$. Thus, the slope is $3t^2$, so $\dfrac{df}{dx}(t) = 3t^2$. The diagram below shows the graph of $f(x)=x^3$ and its tangent line at (0,0) and (4,64).

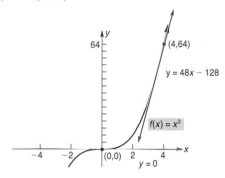

**EXAMPLE 2**

*Let* $f(x) = \dfrac{1}{x+2}$*. For each of the following values of x, find the instantaneous rate of change of* $f(x)$*.*

**a.** $x = 0.$           **b.** $x = 6.$           **c.** $x = x_0.$

*Solution:*

**a.**  $\dfrac{f(0+h)-f(0)}{h} = \dfrac{\dfrac{1}{h+2}-\dfrac{1}{2}}{h} = \dfrac{\dfrac{1}{h+2}-\dfrac{1}{2}}{h}\cdot\dfrac{2(h+2)}{2(h+2)}$

$= \dfrac{2-(h+2)}{2h(h+2)} = \dfrac{-h}{2h(h+2)} = \dfrac{-1}{2(h+2)} = \dfrac{-1}{2h+4}.$

As $h$ approaches 0, the value of $\dfrac{-1}{2h+4}$ approaches $-\dfrac{1}{4}$. Thus, the instantaneous rate of change of $f(x)$ at $x=0$ is $\dfrac{df}{dx}(0) = -\dfrac{1}{4}.$

**b.** $\dfrac{f(6+h)-f(6)}{h} = \dfrac{\dfrac{1}{6+h+2}-\dfrac{1}{6+2}}{h} = \dfrac{\dfrac{1}{8+h}-\dfrac{1}{8}}{h}$

$= \dfrac{8-(8+h)}{8h(8+h)} = \dfrac{-h}{8h(8+h)} = \dfrac{-1}{8(8+h)} = \dfrac{-1}{64+8h}.$

As $h$ approaches 0, the value of $\dfrac{-1}{64+8h}$ approaches $-\dfrac{1}{64}$. Hence, the

instantaneous rate of change of $f(x)$ at $x=6$ is $\dfrac{df}{dx}(6) = -\dfrac{1}{64}$.

**c.** $\dfrac{f(x_0+h)-f(x_0)}{h} = \dfrac{\dfrac{1}{x_0+h+2}-\dfrac{1}{x_0+2}}{h}$

$= \dfrac{(x_0+2)-(x_0+h+2)}{h(x_0+2)(x_0+h+2)}$

$= \dfrac{-h}{h(x_0+2)(x_0+h+2)}$

$= \dfrac{-1}{(x_0+2)(x_0+h+2)}.$

As $h$ approaches 0, the value of $\dfrac{-1}{(x_0+2)(x_0+h+2)}$ approaches

$\dfrac{-1}{(x_0+2)^2}$. Consequently, assuming that $x_0 \neq -2$, the instantaneous rate of

change is $\dfrac{-1}{(x_0+2)^2}$, or $\dfrac{df}{dx}(x_0) = \dfrac{-1}{(x_0+2)^2}.$    ∎

In the study of economics, a **total-cost function,** $c = f(q)$, gives the total cost $c$ of producing and marketing $q$ units of a product. The **marginal cost** is the instantaneous rate of change of $c$, or $f(q)$, with respect to $q$. Thus marginal cost is $\dfrac{dc}{dq}$, and can be interpreted as the approximate cost to produce one additional unit.

### EXAMPLE 3

*Suppose the cost of producing $q$ compact discs is given by $c = 0.001q^2 + 4q + 300$. Find the marginal cost when 1500 compact discs are produced.*

*Solution:* Let $f(q) = 0.001q^2 + 4q + 300$. Then

$\dfrac{f(1500+h)-f(1500)}{h}$

$= \dfrac{[0.001(1500+h)^2+4(1500+h)+300]-[0.001(1500)^2+4(1500)+300]}{h}$

$= \dfrac{(0.001h^2+7h+8550)-8550}{h}$

$= 0.001h + 7.$

As $h$ approaches 0, then $0.001h + 7$ approaches 7. Thus, $\dfrac{df}{dq}(1500) = 7$, or, the marginal cost is \$7 per unit.    ∎

**EXAMPLE 4**

For $y = \sqrt{5x + 11}$, find $\dfrac{dy}{dx}$ and evaluate this derivative at $x = 1$ to find the slope of the curve $y = \sqrt{5x + 11}$ at the point $(1, 4)$.

*Solution:*   Consider $f(x) = \sqrt{5x + 11}$, and compute the slope of the curve $y = f(x)$ at the point $(x, \sqrt{5x + 11})$:

$$\frac{f(x + h) - f(x)}{h} = \frac{\sqrt{5(x + h) + 11} - \sqrt{5x + 11}}{h}$$

$$= \frac{(\sqrt{5(x + h) + 11} - \sqrt{5x + 11})(\sqrt{5(x + h) + 11} + \sqrt{5x + 11})}{h \cdot (\sqrt{5(x + h) + 11} + \sqrt{5x + 11})}$$

$$= \frac{(\sqrt{5(x + h) + 11})^2 - (\sqrt{5x + 11})^2}{h \cdot (\sqrt{5(x + h) + 11} + \sqrt{5x + 11})}$$

$$= \frac{5(x + h) + 11 - (5x + 11)}{h \cdot (\sqrt{5(x + h) + 11} + \sqrt{5x + 11})}$$

$$= \frac{5x + 5h + 11 - 5x - 11}{h(\sqrt{5(x + h) + 11} + \sqrt{5x + 11})}$$

$$= \frac{5h}{h(\sqrt{5(x + h) + 11} + \sqrt{5x + 11})}$$

$$= \frac{5}{\sqrt{5(x + h) + 11} + \sqrt{5x + 11}}.$$

Now set $h = 0$ to get $\dfrac{dy}{dx} = \dfrac{5}{2\sqrt{5x + 11}}$. Letting $x = 1$, we find that the slope of $y = \sqrt{5x + 11}$ at $(1, 4)$ is $\dfrac{5}{8}$.   ■

## ■ Exercise A.3

1. Let $f(x) = x^2$. Find the function $\dfrac{df}{dx}$, the slope of the graph of $f(x)$ at the point $(x, x^2)$. Evaluate the slope at each $x$.
   **a.** $x = 0$.     **b.** $x = -4$.     **c.** $x = x_0$.

2. Let $f(x) = \sqrt{x}$. Find the function $\dfrac{df}{dx}$, the slope of the graph of $f(x)$ at the point $(x, \sqrt{x})$. Evaluate the slope at each $x$.
   **a.** $x = 1$.     **b.** $x = 4$.     **c.** $x = t$ $(t > 0)$.

3. Let $f(x) = \dfrac{1}{x - 4}$. Find the function $\dfrac{df}{dx}$, the instantaneous rate of change of $f(x)$. Evaluate $\dfrac{df}{dx}$ at each $x$.
   **a.** $x = 0$.     **b.** $x = 10$.     **c.** $x = x_0$ $(x_0 \neq 4)$.

4. Let $f(x) = x^3 - 5x$. Find the function $\dfrac{df}{dx}$, the instantaneous rate of change of $f(x)$. Evaluate $\dfrac{df}{dx}$ at each $x$.
   **a.** $x = 0$.     **b.** $x = 5$.     **c.** $x = n$.

5. A ball is thrown straight upward at the rate of 32 feet per second. Its height, in feet, after $t$ seconds is given by the function $h(t) = -16t^2 + 32t$. Find the instantaneous velocity $\dfrac{dh}{dt}$, and evaluate the expression for each value of $t$.
   **a.** $t = 0$ seconds.   **b.** $t = 2$ seconds.   **c.** $t = 3$ seconds.

6. Suppose the cost of producing $q$ copies of a magazine is given by $c = 1.5q + 8000$. Find the marginal cost $\dfrac{dc}{dq}$, and evaluate the expression when $q = 15{,}300$ copies are produced.

7. Suppose the cost of producing $q$ rugs is given by $c = 0.05q^2 + 28q + 5000$. Find the marginal cost $\dfrac{dc}{dq}$, and evaluate the expression when $q = 75$ rugs are produced.

8. Suppose the cost of producing $q$ computers is given by $c = 0.0001q^3 - 0.08q^2 + 1000q + 20{,}000$. Find the marginal cost $\dfrac{dc}{dq}$, and evaluate the expression when $q = 900$ computers are produced.

*For each function $y = f(x)$ in Problems 9–12, find $\dfrac{dy}{dx}$.*

9. **a.** $y = 11$.     **b.** $y = 5x$.     **c.** $y = 5x + 11$.
10. **a.** $y = 10x^2$.          **b.** $y = 10x^2 + 11$.
    **c.** $y = 10x^2 + 5x + 1492$.

**11. a.** $y = \dfrac{1}{5x + 11}$.     **b.** $y = \dfrac{1492}{5x + 11}$.

**c.** $y = \dfrac{3}{12 - 5x}$.     **d.** $y = \dfrac{3}{12 - 5x} + 11$.

**12. a.** $y = \sqrt{5x - 11}$.     **b.** $y = \sqrt{12 - 5x}$.

**c.** $y = \sqrt{12 - 5x} + 10x^2 + 11$.

## A.4   AREAS OF GEOMETRIC SHAPES

In the study of integral calculus, one learns advanced techniques for finding areas of regions formed by graphs in the coordinate plane. Many problems of this sort can be solved using the simple area formulas you learned in the study of geometry. Some of these formulas are as follows:

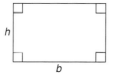

Rectangle
$A = bh$

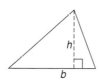

Triangle
$A = \dfrac{1}{2}bh$

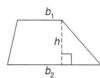

Trapezoid
$A = \dfrac{1}{2}h(b_1 + b_2)$

Circle
$A = \pi r^2$

**EXAMPLE 1**

*Find the area of the preceding figure.*

*Solution:* The figure consists of a trapezoid and two semicircles. The area of the trapezoid is

$$A_{\text{trapezoid}} = \frac{1}{2}h(b_1 + b_2) = \frac{1}{2}(8)(7 + 19) = 104 \text{ cm}^2.$$

The combined area of the semicircles is the same as the area of one circle of radius 5 cm:

$$A_{\text{circle}} = \pi r^2 = \pi(5^2) = 25\pi \text{ cm}^2.$$

The total area of the figure is

$$A = A_{\text{trapezoid}} + A_{\text{circle}} = (104 + 25\pi) \text{ cm}^2, \text{ or about } 182.5 \text{ cm}^2. \quad \blacksquare$$

**EXAMPLE 2**

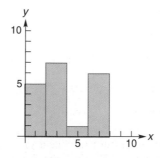

*Find the area of the shaded region in the diagram above.*

*Solution:* The region consists of four rectangles. The area is $(2 \cdot 5) + (2 \cdot 7) + (2 \cdot 1) + (2 \cdot 6) = 2(5 + 7 + 1 + 6) = 2(19) = 38$ square units. ▪

**EXAMPLE 3**

*Find the area of the region bounded by the x-axis and the graph of the function* $y = 9 - |3x|$.

*Solution:*

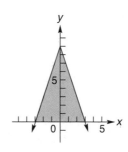

The region is a triangle with base $b = 6$ and height $h = 9$. The area is $\frac{1}{2}bh = \frac{1}{2}(6)(9) = 27$ square units. ▪

## ■ Exercise A.4

*In Problems 1–8, find the area of each figure or region.*

**1.**

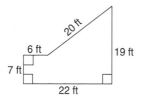

20 ft, 6 ft, 19 ft, 7 ft, 22 ft

**4.**

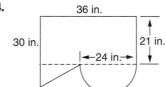

36 in., 30 in., 21 in., 24 in.

**2.**

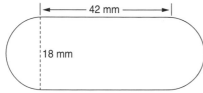

42 mm, 18 mm

**5.**

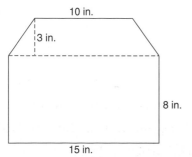

**3.**

10 in., 3 in., 8 in., 15 in.

**6.**

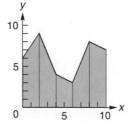

**7.** The region bounded by the graphs of $y = |x| - 4$ and $y = 6 - |x + 3|$.

**8.** The region bounded by the graph of the function $y = \sqrt{25 - x^2}$ and the $x$-axis.

## A.5    Sigma Notation

When many terms are to be added together, it is sometimes convenient to use a special notation to represent the sum. For example, the expression $\displaystyle\sum_{k=5}^{8} 3k$ means the addition of the values of $3k$ for $k = 5$ to 8 ($k$ an integer):

$$\sum_{k=5}^{8} 3k = 3(5) + 3(6) + 3(7) + 3(8) = 15 + 18 + 21 + 24 = 78.$$

This notation is called **sigma notation** because the symbol $\Sigma$ is the capital letter sigma in the Greek alphabet. The constants 5 and 8 are the **lower limit of summation** and **upper limit of summation,** respectively. The variable $k$ is the **index of summation.** Note that it can be replaced by any variable without changing the value of the expression.

### EXAMPLE 1

*Evaluate the given sums.*

**a.** $\displaystyle\sum_{n=3}^{7} (5n - 2).$ 
**b.** $\displaystyle\sum_{j=0}^{6} (j^2 + 1).$

*Solution:*

**a.** $\displaystyle\sum_{n=3}^{7} (5n - 2)$

$= [5(3) - 2] + [5(4) - 2] + [5(5) - 2] + [5(6) - 2] + [5(7) - 2]$

$= 13 + 18 + 23 + 28 + 33$

$= 115.$

**b.** $\displaystyle\sum_{j=0}^{6} (j^2 + 1) = (0^2 + 1) + (1^2 + 1) + (2^2 + 1) + (3^2 + 1) +$

$(4^2 + 1) + (5^2 + 1) + (6^2 + 1)$

$= 1 + 2 + 5 + 10 + 17 + 26 + 37$

$= 98.$ ∎

### EXAMPLE 2

*Express the sum $14 + 16 + 18 + 20 + 22 + \cdots + 100$ in sigma notation.*

*Solution:* There are many ways to express this sum in sigma notation.

*One method:* Notice that the values being added are $2n$, for $n = 7$ to 50. The sum can thus be represented as $\displaystyle\sum_{n=7}^{50} 2n.$

*Another method:* Notice that the values being added are $2k + 12$, for $k = 1$ to 44.

The sum can therefore be represented as $\displaystyle\sum_{k=1}^{44} (2k + 12).$ ∎

Several formulas are useful for computing sums. General formulas include (1) factoring out a common factor of each term in a sum and (2) reordering the terms to be summed. Specifically,

**(1)** $\displaystyle\sum_{k=m}^{n} (c \cdot a_k) = c \cdot \sum_{k=m}^{n} a_k,$

**(2)** $\displaystyle\sum_{k=m}^{n} (a_k + b_k) = \sum_{k=m}^{n} a_k + \sum_{k=m}^{n} b_k.$

On occasion, one may prefer to change the indices, as in the expression

**(3)** $\displaystyle\sum_{k=m}^{n} a_k = \sum_{k=p}^{n+p-m} a_{k+m-p}.$

Also, specific formulas may be used for common sums:

**(4)** $\displaystyle\sum_{k=1}^{n} (1) = n,$

**(5)** $\displaystyle\sum_{k=1}^{n} k = \frac{n(n+1)}{2},$

**(6)** $\displaystyle\sum_{k=1}^{n} k^2 = \frac{n(n+1)(2n+1)}{6}.$

**EXAMPLE 3**

*Evaluate the given sums.*

**a.** $\displaystyle\sum_{k=30}^{100} 4.$     **b.** $\displaystyle\sum_{k=1}^{100} (5k+3).$     **c.** $\displaystyle\sum_{k=1}^{200} 9k^2.$

*Solution:*

**a.** $\displaystyle\sum_{k=30}^{100} 4 = \sum_{k=1}^{71} 4 = 4\left(\sum_{k=1}^{71} 1\right) = 4 \cdot 71 = 284.$

Alternatively,

$\displaystyle\sum_{k=30}^{100} 4 = \sum_{k=1}^{100} 4 - \sum_{k=1}^{29} 4 = 4\left(\sum_{k=1}^{100} 1\right) - 4\left(\sum_{k=1}^{29} 1\right) = 4 \cdot 100 - 4 \cdot 29 = 284.$

**b.** $\displaystyle\sum_{k=1}^{100} (5k+3) = \sum_{k=1}^{100} 5k + \sum_{k=1}^{100} 3$     [by (2)]

$\displaystyle = 5\left(\sum_{k=1}^{100} k\right) + 3\left(\sum_{k=1}^{100} 1\right)$     [by (1)]

$\displaystyle = 5\left(\frac{100 \cdot 101}{2}\right) + 3(100)$     [by (5) and (4)]

$= 25{,}250 + 300 = 25{,}550.$

**c.** $\displaystyle\sum_{k=1}^{200} 9k^2 = 9\sum_{k=1}^{200} k^2$     [by (1)]

$\displaystyle = 9 \cdot \frac{200 \cdot 201 \cdot 401}{6}$     [by (6)]

$= 24{,}180{,}300.$

## ▪ Exercise A.5

*In Problems 1 and 2, give the lower limit of summation, the upper limit of summation, and the index of summation for each expression.*

**1.** $\displaystyle\sum_{t=12}^{17} (8t^2 - 5t + 3).$

**2.** $\displaystyle\sum_{m=3}^{450} (8m - 4).$

*In Problems 3–6, evaluate the given sums.*

**3.** $\displaystyle\sum_{i=1}^{7} 6i.$

**4.** $\displaystyle\sum_{p=0}^{4} 10p.$

**5.** $\displaystyle\sum_{k=3}^{9} (10k + 16).$

**6.** $\displaystyle\sum_{n=10}^{14} (3n - 5).$

*In Problems 7–10, express the given sums in sigma notation.*

**7.** $36 + 37 + 38 + 39 + \cdots + 60.$

**8.** $1 + 4 + 9 + 16 + 25.$

**9.** $5^3 + 5^4 + 5^5 + 5^6 + 5^7 + 5^8.$

**10.** $11 + 15 + 19 + 23 + \cdots + 71.$

*In Problems 11–24, evaluate the given sums.*

**11.** $\displaystyle\sum_{k=1}^{52} 10.$

**12.** $\displaystyle\sum_{k=35}^{135} 2.$

**13.** $\displaystyle\sum_{k=1}^{n} \left(5 \cdot \frac{1}{n}\right).$

**14.** $\displaystyle\sum_{k=1}^{200} (k - 100).$

**15.** $\displaystyle\sum_{k=51}^{100} 10k.$

**16.** $\displaystyle\sum_{k=1}^{n} \frac{n}{n+1} k.$

**17.** $\displaystyle\sum_{k=1}^{20} (5k^2 + 3k).$

**18.** $\displaystyle\sum_{k=1}^{100} \frac{3k^2 - 200k}{101}.$

**19.** $\displaystyle\sum_{k=51}^{100} k^2.$

**20.** $\displaystyle\sum_{k=1}^{50} (k + 50)^2.$

**21.** $\displaystyle\sum_{k=1}^{10} \left\{ \left[ 4 - \left(\frac{2k}{10}\right)^2 \right]\left(\frac{2}{10}\right) \right\}.$

**22.** $\displaystyle\sum_{k=1}^{100} \left\{ \left[ 4 - \left(\frac{2}{100} k\right)^2 \right]\left(\frac{2}{100}\right) \right\}.$

**23.** $\displaystyle\sum_{k=1}^{n} \left\{ \left[ 4 - \left(\frac{2}{n} \cdot k\right)^2 \right]\frac{2}{n} \right\}.$

**24.** $\displaystyle\sum_{k=1}^{n} \frac{k^2}{(n+1)(2n+1)}.$

## A.6   RIEMANN SUMS AND THE DEFINITE INTEGRAL

In the study of calculus, one learns advanced methods for finding the area between a curve and the $x$-axis for a specified interval. For example, you might want to find the area of the region between the graph of $f(x) = x^2$ and the $x$-axis, for $x$-values in the interval $[1, 4]$, as shown in the following diagram:

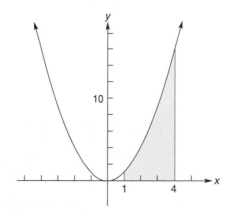

You can use rectangles to estimate the area of the region. To use three rectangles, simply divide the interval into three subintervals, say, [1, 2], [2, 3], and [3, 4]. Then draw a rectangle for each subinterval. The base of each rectangle should be on the $x$-axis, and the top side of each rectangle should intersect the graph of $f(x)$. One way to accomplish this is to use the midpoint of each interval to calculate the height of each rectangle. Thus, we use heights of $f(1.5) = 1.5^2 = 2.25$, $f(2.5) = 2.5^2 = 6.25$, and $f(3.5) = 3.5^2 = 12.25$. The rectangles are shown in the following diagram:

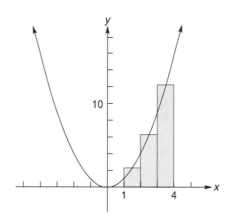

Thus, the desired area, $A$, is approximated by the sum of the areas of the three rectangles:

$$A \approx (1 \cdot 2.25) + (1 \cdot 6.25) + (1 \cdot 12.25) = 20.75 \text{ square units.}$$

A sum of areas of rectangles such as the sum we have just found is called a **Riemann sum.** The subintervals used to create a Riemann sum need not have equal length, but we will continue to use equally spaced subintervals in order to keep the calculations simple. We will use the expression $\Delta x$ to denote the width of each subinterval. (The symbol $\Delta$ is the Greek letter *delta,* which mathematicians often use to represent the change of a quantity.)

Likewise, it is not necessary to use the midpoint of each subinterval, but calculations are usually simplest if you use one of the following conventions for the height of the rectangle in each subinterval $[a, b]$:

- Left-point rule: Use the value at the left endpoint of the subinterval. That is, use $f(a)$.
- Midpoint rule: Use the value at the midpoint of the subinterval. That is, use $f\left(\frac{a + b}{2}\right)$.
- Right-point rule: Use the value at the right endpoint of the subinterval. That is, use $f(b)$.

As you can imagine, using smaller subintervals will result in a more accurate estimate for the area of the region. Of course, small subintervals can also result in very tedious calculations, unless you are using rules for summations or a computer. (Some graphing calculators can be used to compute Riemann sums automatically.)

To write a Riemann sum in summation notation, let $x_k$ represent the chosen $x$-value in the $k$th subinterval, and let $N$ be the number of subintervals; that is, $N = \frac{\text{total interval length}}{\Delta x}$. Then each rectangle has width $\Delta x$ and height $f(x_k)$, so the Riemann sum is $\sum_{k=1}^{N} f(x_k) \cdot \Delta x$, or equivalently, $\Delta x \cdot \left( \sum_{k=1}^{N} f(x_k) \right)$.

Observe that this Riemann sum will approximate the area between the curve $y = f(x)$ and the $x$-axis as long as $f(x_k)$ is the height of a rectangle. Thus, we need $f(x) \geq 0$. For the case where $f(x) \leq 0$, the Riemann sum $\sum_{k=1}^{N} f(x_k) \cdot \Delta x$ would estimate the negative of an area.

For a function $y = f(x)$ and an interval of the $x$-axis $[a, b]$ being subdivided into $N$ subintervals of length $\Delta x$, the Riemann sum $\sum_{k=1}^{N} f(x_k) \cdot \Delta x$ is an estimate (for large $N$) of the definite integral $\int_a^b f(x)\, dx$. For nonnegative continuous functions $f(x)$, $\int_a^b f(x)\, dx$ is the area between $y = f(x)$, the $x$-axis, $x = a$, and $x = b$.

### EXAMPLE 1

*Use each of the following methods to estimate the area of the region bounded by the graph of $f(x) = 8 - x^3$, the line $x = -1$, and the x-axis.*

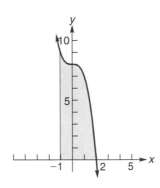

**a.** *Estimate the area $\int_{-1}^{2} (8 - x^3)\, dx$ using a Riemann sum with the right-point rule for $\Delta x = 0.5$.*

**b.** *Write an expression for the Riemann sum with the midpoint rule for $\Delta x = 0.01$. Express your answer using summation notation.*

*Solution:*
**a.** Since the interval length is 3, the number of subintervals is $\frac{3}{\Delta x} = \frac{3}{0.5} = 6$. The subintervals are $[-1, -0.5]$, $[-0.5, 0]$, $[0, 0.5]$, $[0.5, 1]$, $[1, 1.5]$, and $[1.5, 2]$. Using the right-point rule, we must evaluate $f(x)$ at $x = -0.5$, $0, 0.5, 1, 1.5$, and $2$.

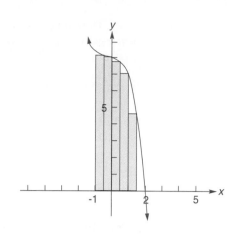

The area of the region is approximated by the Riemann sum:

$$\text{Area} \approx 0.5[f(-0.5) + f(0) + f(0.5) + f(1) + f(1.5) + f(2)]$$

$$= 0.5[(8 - (-0.5)^3) + (8 - 0^3) + (8 - 0.5^3) + (8 - 1^3)$$

$$+ (8 - 1.5^3) + (8 - 2^3)]$$

$$= 0.5[8.125 + 8 + 7.875 + 7 + 4.625 + 0]$$

$$= 0.5[35.625]$$

$$= 17.8125 \text{ square units.}$$

**b.** Since the length of the interval is 0.01, the number of subintervals is $N = \dfrac{3}{\Delta x} = \dfrac{3}{0.01} = 300$. The subintervals are $[-1, -0.99]$, $[-0.99, -0.98]$, and so on. Using the midpoint rule, we must evaluate $f(x)$ at $x = -0.995$, $-0.985$, and so on. Therefore, $x_k = 0.01k - 1.005$. The Riemann sum is

$$\text{Area} \approx \Delta x \cdot \sum_{k=1}^{N} f(x_k) = 0.01 \sum_{k=1}^{300} f(0.01k - 1.005)$$

$$= 0.01 \sum_{k=1}^{300} [8 - (0.01k - 1.005)^3]. \quad \blacksquare$$

In Example 1a, note that the calculated area of 17.8125 is an *underestimate* of the actual area of the region, since the top edge of each rectangle is always on or below the graph of $f(x)$. The actual area of the region is somewhat greater than 17.8125 square units.

Another method of estimating areas that is related to Riemann sums is the **trapezoidal rule,** which uses trapezoids instead of rectangles. For each subinterval $[a, b]$, we use a trapezoid with vertical bases of length $f(a)$ and $f(b)$, respectively, and horizontal "height" $\Delta x = b - a$. The area of this trapezoid is $\dfrac{1}{2} \cdot \Delta x \cdot [f(a) + f(b)]$. The trapezoidal rule is illustrated in Example 2.

**EXAMPLE 2**

*Use the trapezoidal rule to estimate the area $\displaystyle\int_{-2}^{2}(4 - x^2)\, dx$ bounded by the graph of $f(x) = 4 - x^2$ and the x-axis. Use $\Delta x = 1$.*

*Solution:*

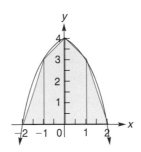

The region corresponds to the interval $[-2, 2]$. (Note that two of the "trapezoids" are actually triangles, since $f(-2) = 0$ and $f(2) = 0$.) Since $f(x)$ is symmetric on this interval, we may first find the trapezoid sum for the interval $[0, 2]$ and then double the result. The subintervals are $[0, 1]$ and $[1, 2]$. The sum for the interval $[0, 2]$ is

$$\int_0^2 (4 - x^2)\, dx = \text{Area} \approx \frac{1}{2} \cdot 1 \cdot [f(0) + f(1)] + \frac{1}{2} \cdot 1 \cdot [f(1) + f(2)]$$

$$= \frac{1}{2} \cdot [(4 - 0^2) + (4 - 1^2)] + \frac{1}{2} \cdot [(4 - 1^2) + (4 - 2^2)]$$

$$= \frac{1}{2} \cdot [4 + 3] + \frac{1}{2} \cdot [3 + 0]$$

$$= 3.5 + 1.5$$

$$= 5 \text{ square units.}$$

This area is estimated using Riemann sums with the right-point rule in Problems 21–23 of Section A.5.

Since an area of 5 square units represents half of the desired region, an estimate for $\int_{-2}^{2} (4 - x^2)\, dx$, the entire area of the region, is $2 \cdot 5 = 10$ square units. ∎

**EXAMPLE 3**

*Evaluate:*   **a.** $\int_0^5 7dx.$   **b.** $\int_3^{10} (2x - 1)\, dx.$

*Solutions:* Each of these is an area of a region depicted:

**a.**
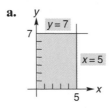

**b.**
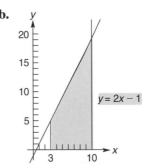

Hence $\int_0^5 7\, dx = 35$ and $\int_3^{10} (2x - 1)\, dx = 84.$ ∎

## ▪ Exercise A.6

1. Estimate the area of the region between the graph of $f(x) = x^3$ and the $x$-axis, for $x$-values in the interval $[0, 6]$. Use a Riemann sum with the left-point rule for $\Delta x = 1$. Is your estimate higher or lower than the actual area?

2. Estimate the area of the region bounded by the graph of $g(x) = 9 - x^2$ and the $x$-axis. Use a Riemann sum with the midpoint rule for $\Delta x = 0.5$.

3. Estimate the area of the region between the graph of $h(x) = x^2 + x$ and the $x$-axis, for $x$-values between 2 and 5. Use a Riemann sum with the right-point rule for $\Delta x = 0.5$.

4. Write an expression for a Riemann-sum estimate of the area of the region described in Problem 1, using the right-point rule for $\Delta x = 0.01$. Express your answer in summation notation.

5. Write an expression for a Riemann-sum estimate of the area of the region described in Problem 2, using the left-point rule for $\Delta x = 0.05$. Express your answer in summation notation.

6. Write an expression for a Riemann-sum estimate of the area of the region described in Problem 3, using the mid-point rule for $\Delta x = 0.02$. Express your answer in summation notation.

7. Estimate the area of the region between the graph of $f(x) = x^2$ and the $x$-axis, for $x$-values in the interval $[1, 4]$. Use the trapezoidal rule with $\Delta x = 1$.

8. Estimate the area of the region bounded by $g(x) = -x^2 + 3x$ and the $x$-axis. Use the trapezoidal rule with $\Delta x = 0.05$.

9. Write an expression for a Riemann-sum estimate of the area of the region described in Problem 7, using the left-point rule for $\Delta x = 0.01$. Express your answer in summation notation.

10. Write an expression for a Riemann-sum estimate of the area of the region described in Problem 8, using the left-point rule for $\Delta x = 0.05$. Express your answer in summation notation.

11. Repeat Problem 3, but this time let $\Delta x = \dfrac{3}{N}$. First write the Riemann sum using summation notation, and then give an answer in terms of $N$, simplified.

12. Using the right-point rule, write a Riemann sum with $N$ rectangles to give an expression (involving $N$) for the area of the region in Problem 7. Simplify your expression to guess a value for $\displaystyle\int_1^4 x^2\, dx$.

13. Using a graph, evaluate each of the following integrals.

   a. $\displaystyle\int_0^3 11\, dx$.

   b. $\displaystyle\int_2^7 \frac{3}{10}\, dx$.

   c. $\displaystyle\int_{10}^{100} \frac{1}{3}\, dx$.

   d. $\displaystyle\int_0^3 8x\, dx$.

   e. $\displaystyle\int_2^4 (8 - 2x)\, dx$.

   f. $\displaystyle\int_2^4 (11 - 2x)\, dx$.

   g. $\displaystyle\int_7^{55} \frac{x + 9}{4}\, dx$.

   h. $\displaystyle\int_{-4}^3 (2x + 10)\, dx$.

14. Evaluate $\displaystyle\int_0^{4000} (6.75 - 0.00065x)\, dx$ and

$$\sum_{k=1}^{4000} (6.75 - 0.00065k).$$

   Explain why these answers should be close to each other.

15. Compare $\displaystyle\int_0^{100} (40 + 0.3x)\, dx$ and $\displaystyle\sum_{k=1}^{100} (40 + 0.3k)$.

## A.7 Area under a Rate-of-Change Curve

When a function $f(x)$ gives a total quantity, its rate of change with respect to $x$ is a measure of the change in quantity per unit of $x$. The area under this rate-of-change curve, given limits on $x$, yields the total accumulation of the quantity between those limits.

### EXAMPLE 1

*Let $y = f(x)$ be the total distance driven by a car one afternoon. The rate of change of $y$ with respect to $x$, $\dfrac{dy}{dx}$, is the velocity at a particular time $x$. How far did the car go between 2:00 P.M. and 5:00 P.M. if the velocity was a constant 55 miles per hour all afternoon?*

*Solution.* Graphing the velocity gives

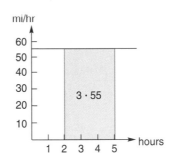

Thus, the total distance is $3 \cdot 55 = 165$ miles. ∎

### EXAMPLE 2

*The marginal cost y in dollars per foot of making x feet of wire is $y = 7 - 0.002x$. A small business has enough funds to make 100 feet of wire, but wants to borrow money to make up to 1000 feet. How much more would it cost to make 1000 feet of wire than 100 feet?*

*Solution:*   Graphing the marginal cost yields

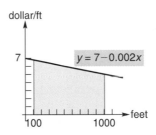

Hence, the area is

$$\int_{100}^{1000} (7 - 0.002x)\, dx = \frac{1}{2}(6.80 + 5.00)(900) = \$5310.$$ ∎

## ▪ Exercise A.7

1. If a train travels at a constant rate of 45 miles per hour, how far would it travel in six hours?

2. If a bank agrees to add funds to a savings account at a constant rate of a penny a minute, how much would be added to the account in a day? in a year?

3. If the marginal revenue for selling an item is $4.95 per item, what would be the total revenue generated from selling 1000 items?

4. In the graph at right, $y$ represents the velocity in miles per hour that a person traveled on a trip, and $x$ measures the time in hours.

   (a) How far did the person travel in the first two hours?

   (b) How far in the second to fourth hour?

   (c) How far for the whole trip?

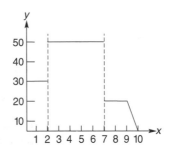

5. In the following graph, $y$ represents the parking rate in cents per minute to park in a lot downtown, and $x$ measures the time parked in minutes.

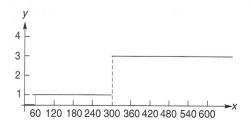

(a) How long can one park for free?

(b) What is the charge to park for 25 minutes?

(c) What is the charge to park for three hours?

(d) What is the charge to park for nine hours?

6. If the marginal cost for making $x$ number of items is

$$y = \begin{cases} 15, & \text{if } 0 \le x \le 3, \\ 6, & \text{if } 3 < x \le 9, \\ 4, & \text{if } 9 < x \le 50, \end{cases}$$

what is the total variable cost to make 30 items?

7. The velocity of a rock falling downward is given as $y = 32x$ feet per second, where $x$ gives time in seconds. Suppose the rock hit the ground after 10 seconds.

(a) How far did the rock fall the first second?

(b) How far did it fall before it hit the ground (i.e., in 10 seconds)?

(c) How far did it fall during the last three seconds before hitting the ground?

8. The rate of change of the area of a semicircle with respect to its radius is given as $y = \pi x$, where $x$ is the radius. Estimate the area when $x = 3$.

9. The marginal cost for a hosiery mill in terms of dollars per dozen pairs of hosiery is given as $y = 6.75 - 0.00065x$. How much more would be the total cost to make 4000 dozen pairs of hosiery than to make no pairs?

10. Over the first five seconds of a carnival ride, the velocity on the ride was recorded as $y = 45x - 9x^2$, where $x$ is time given in seconds and $y$ is the rate in feet per second. Estimate upper and lower bounds on how far this ride travels in five seconds.

11. The rate of change in revenue per day at Joe's Ice Cream Shop, measured from the day he opened in May until he closed at the end of the summer, was recorded (in terms of dollars per day) as $y = 10x - 0.1x^2$. Approximately how much total revenue did Joe collect during the 100 days he was open that summer?

12. The marginal revenue, measured in dollars per foot, from selling $x$ feet of wire was given as $y = 5\sqrt{x}$. Estimate the total revenue from selling 100 feet of wire.

# Appendix B  Compound Interest Tables

$$r = 0.005$$

| $n$ | $(1 + r)^n$ | $(1 + r)^{-n}$ | $a_{\overline{n}|r}$ | $s_{\overline{n}|r}$ |
|---|---|---|---|---|
| 1 | 1.005000 | 0.995025 | 0.995025 | 1.000000 |
| 2 | 1.010025 | 0.990075 | 1.985099 | 2.005000 |
| 3 | 1.015075 | 0.985149 | 2.970248 | 3.015025 |
| 4 | 1.020151 | 0.980248 | 3.950496 | 4.030100 |
| 5 | 1.025251 | 0.975371 | 4.925866 | 5.050251 |
| 6 | 1.030378 | 0.970518 | 5.896384 | 6.075502 |
| 7 | 1.035529 | 0.965690 | 6.862074 | 7.105879 |
| 8 | 1.040707 | 0.960885 | 7.822959 | 8.141409 |
| 9 | 1.045911 | 0.956105 | 8.779064 | 9.182116 |
| 10 | 1.051140 | 0.951348 | 9.730412 | 10.228026 |
| 11 | 1.056396 | 0.946615 | 10.677027 | 11.279167 |
| 12 | 1.061678 | 0.941905 | 11.618932 | 12.335562 |
| 13 | 1.066986 | 0.937219 | 12.556151 | 13.397240 |
| 14 | 1.072321 | 0.932556 | 13.488708 | 14.464226 |
| 15 | 1.077683 | 0.927917 | 14.416625 | 15.536548 |
| 16 | 1.083071 | 0.923300 | 15.339925 | 16.614230 |
| 17 | 1.088487 | 0.918707 | 16.258632 | 17.697301 |
| 18 | 1.093929 | 0.914136 | 17.172768 | 18.785788 |
| 19 | 1.099399 | 0.909588 | 18.082356 | 19.879717 |
| 20 | 1.104896 | 0.905063 | 18.987419 | 20.979115 |
| 21 | 1.110420 | 0.900560 | 19.887979 | 22.084011 |
| 22 | 1.115972 | 0.896080 | 20.784059 | 23.194431 |
| 23 | 1.121552 | 0.891622 | 21.675681 | 24.310403 |
| 24 | 1.127160 | 0.887186 | 22.562866 | 25.431955 |
| 25 | 1.132796 | 0.882772 | 23.445638 | 26.559115 |
| 26 | 1.138460 | 0.878380 | 24.324018 | 27.691911 |
| 27 | 1.144152 | 0.874010 | 25.198028 | 28.830370 |
| 28 | 1.149873 | 0.869662 | 26.067689 | 29.974522 |
| 29 | 1.155622 | 0.865335 | 26.933024 | 31.124395 |
| 30 | 1.161400 | 0.861030 | 27.794054 | 32.280017 |
| 31 | 1.167207 | 0.856746 | 28.650800 | 33.441417 |
| 32 | 1.173043 | 0.852484 | 29.503284 | 34.608624 |
| 33 | 1.178908 | 0.848242 | 30.351526 | 35.781667 |
| 34 | 1.184803 | 0.844022 | 31.195548 | 36.960575 |
| 35 | 1.190727 | 0.839823 | 32.035371 | 38.145378 |
| 36 | 1.196681 | 0.835645 | 32.871016 | 39.336105 |
| 37 | 1.202664 | 0.831487 | 33.702504 | 40.532785 |
| 38 | 1.208677 | 0.827351 | 34.529854 | 41.735449 |
| 39 | 1.214721 | 0.823235 | 35.353089 | 42.944127 |
| 40 | 1.220794 | 0.819139 | 36.172228 | 44.158847 |
| 41 | 1.226898 | 0.815064 | 36.987291 | 45.379642 |
| 42 | 1.233033 | 0.811009 | 37.798300 | 46.606540 |
| 43 | 1.239198 | 0.806974 | 38.605274 | 47.839572 |
| 44 | 1.245394 | 0.802959 | 39.408232 | 49.078770 |
| 45 | 1.251621 | 0.798964 | 40.207196 | 50.324164 |
| 46 | 1.257879 | 0.794989 | 41.002185 | 51.575785 |
| 47 | 1.264168 | 0.791034 | 41.793219 | 52.833664 |
| 48 | 1.270489 | 0.787098 | 42.580318 | 54.097832 |
| 49 | 1.276842 | 0.783182 | 43.363500 | 55.368321 |
| 50 | 1.283226 | 0.779286 | 44.142786 | 56.645163 |

$$r = 0.0075$$

| $n$ | $(1 + r)^n$ | $(1 + r)^{-n}$ | $a_{\overline{n}|r}$ | $s_{\overline{n}|r}$ |
|---|---|---|---|---|
| 1 | 1.007500 | 0.992556 | 0.992556 | 1.000000 |
| 2 | 1.015056 | 0.985167 | 1.977723 | 2.007500 |
| 3 | 1.022669 | 0.977833 | 2.955556 | 3.022556 |
| 4 | 1.030339 | 0.970554 | 3.926110 | 4.045225 |
| 5 | 1.038067 | 0.963329 | 4.889440 | 5.075565 |
| 6 | 1.045852 | 0.956158 | 5.845598 | 6.113631 |
| 7 | 1.053696 | 0.949040 | 6.794638 | 7.159484 |
| 8 | 1.061599 | 0.941975 | 7.736613 | 8.213180 |
| 9 | 1.069561 | 0.934963 | 8.671576 | 9.274779 |
| 10 | 1.077583 | 0.928003 | 9.599580 | 10.344339 |
| 11 | 1.085664 | 0.921095 | 10.520675 | 11.421922 |
| 12 | 1.093807 | 0.914238 | 11.434913 | 12.507586 |
| 13 | 1.102010 | 0.907432 | 12.342345 | 13.601393 |
| 14 | 1.110276 | 0.900677 | 13.243022 | 14.703404 |
| 15 | 1.118603 | 0.893973 | 14.136995 | 15.813679 |
| 16 | 1.126992 | 0.887318 | 15.024313 | 16.932282 |
| 17 | 1.135445 | 0.880712 | 15.905025 | 18.059274 |
| 18 | 1.143960 | 0.874156 | 16.779181 | 19.194718 |
| 19 | 1.152540 | 0.867649 | 17.646830 | 20.338679 |
| 20 | 1.161184 | 0.861190 | 18.508020 | 21.491219 |
| 21 | 1.169893 | 0.854779 | 19.362799 | 22.652403 |
| 22 | 1.178667 | 0.848416 | 20.211215 | 23.822296 |
| 23 | 1.187507 | 0.842100 | 21.053315 | 25.000963 |
| 24 | 1.196414 | 0.835831 | 21.889146 | 26.188471 |
| 25 | 1.205387 | 0.829609 | 22.718755 | 27.384884 |
| 26 | 1.214427 | 0.823434 | 23.542189 | 28.590271 |
| 27 | 1.223535 | 0.817304 | 24.359493 | 29.804698 |
| 28 | 1.232712 | 0.811220 | 25.170713 | 31.028233 |
| 29 | 1.241957 | 0.805181 | 25.975893 | 32.260945 |
| 30 | 1.251272 | 0.799187 | 26.775080 | 33.502902 |
| 31 | 1.260656 | 0.793238 | 27.568318 | 34.754174 |
| 32 | 1.270111 | 0.787333 | 28.355650 | 36.014830 |
| 33 | 1.279637 | 0.781472 | 29.137122 | 37.284941 |
| 34 | 1.289234 | 0.775664 | 29.912776 | 38.564578 |
| 35 | 1.298904 | 0.769880 | 30.682656 | 39.853813 |
| 36 | 1.308645 | 0.764149 | 31.446805 | 41.152716 |
| 37 | 1.318460 | 0.758461 | 32.205266 | 42.461361 |
| 38 | 1.328349 | 0.752814 | 32.958080 | 43.779822 |
| 39 | 1.338311 | 0.747210 | 33.705290 | 45.108170 |
| 40 | 1.348349 | 0.741648 | 34.446938 | 46.446482 |
| 41 | 1.358461 | 0.736127 | 35.183065 | 47.794830 |
| 42 | 1.368650 | 0.730647 | 35.913713 | 49.153291 |
| 43 | 1.378915 | 0.725208 | 36.638921 | 50.521941 |
| 44 | 1.389256 | 0.719810 | 37.358730 | 51.900856 |
| 45 | 1.399676 | 0.714451 | 38.073181 | 53.290112 |
| 46 | 1.410173 | 0.709133 | 38.782314 | 54.689788 |
| 47 | 1.420750 | 0.703854 | 39.486168 | 56.099961 |
| 48 | 1.431405 | 0.698614 | 40.184782 | 57.520711 |
| 49 | 1.442141 | 0.693414 | 40.878195 | 58.952116 |
| 50 | 1.452957 | 0.688252 | 41.566447 | 60.394257 |

$$r = 0.01$$

| $n$ | $(1 + r)^n$ | $(1 + r)^{-n}$ | $a_{\overline{n}|r}$ | $s_{\overline{n}|r}$ |
|---|---|---|---|---|
| 1 | 1.010000 | 0.990099 | 0.990099 | 1.000000 |
| 2 | 1.020100 | 0.980296 | 1.970395 | 2.010000 |
| 3 | 1.030301 | 0.970590 | 2.940985 | 3.030100 |
| 4 | 1.040604 | 0.960980 | 3.901966 | 4.060401 |
| 5 | 1.051010 | 0.951466 | 4.853431 | 5.101005 |
| 6 | 1.061520 | 0.942045 | 5.795476 | 6.152015 |
| 7 | 1.072135 | 0.932718 | 6.728195 | 7.213535 |
| 8 | 1.082857 | 0.923483 | 7.651678 | 8.285671 |
| 9 | 1.093685 | 0.914340 | 8.566018 | 9.368527 |
| 10 | 1.104622 | 0.905287 | 9.471305 | 10.462213 |
| 11 | 1.115668 | 0.896324 | 10.367628 | 11.566835 |
| 12 | 1.126825 | 0.887449 | 11.255077 | 12.682503 |
| 13 | 1.138093 | 0.878663 | 12.133740 | 13.809328 |
| 14 | 1.149474 | 0.869963 | 13.003703 | 14.947421 |
| 15 | 1.160969 | 0.861349 | 13.865053 | 16.096896 |
| 16 | 1.172579 | 0.852821 | 14.717874 | 17.257864 |
| 17 | 1.184304 | 0.844377 | 15.562251 | 18.430443 |
| 18 | 1.196147 | 0.836017 | 16.398269 | 19.614748 |
| 19 | 1.208109 | 0.827740 | 17.226008 | 20.810895 |
| 20 | 1.220190 | 0.819544 | 18.045553 | 22.019004 |
| 21 | 1.232392 | 0.811430 | 18.856983 | 23.239194 |
| 22 | 1.244716 | 0.803396 | 19.660379 | 24.471586 |
| 23 | 1.257163 | 0.795442 | 20.455821 | 25.716302 |
| 24 | 1.269735 | 0.787566 | 21.243387 | 26.973465 |
| 25 | 1.282432 | 0.779768 | 22.023156 | 28.243200 |
| 26 | 1.295256 | 0.772048 | 22.795204 | 29.525631 |
| 27 | 1.308209 | 0.764404 | 23.559608 | 30.820888 |
| 28 | 1.321291 | 0.756836 | 24.316443 | 32.129097 |
| 29 | 1.334504 | 0.749342 | 25.065785 | 33.450388 |
| 30 | 1.347849 | 0.741923 | 25.807708 | 34.784892 |
| 31 | 1.361327 | 0.734577 | 26.542285 | 36.132740 |
| 32 | 1.374941 | 0.727304 | 27.269589 | 37.494068 |
| 33 | 1.388690 | 0.720103 | 27.989693 | 38.869009 |
| 34 | 1.402577 | 0.712973 | 28.702666 | 40.257699 |
| 35 | 1.416603 | 0.705914 | 29.408580 | 41.660276 |
| 36 | 1.430769 | 0.698925 | 30.107505 | 43.076878 |
| 37 | 1.445076 | 0.692005 | 30.799510 | 44.507647 |
| 38 | 1.459527 | 0.685153 | 31.484663 | 45.952724 |
| 39 | 1.474123 | 0.678370 | 32.163033 | 47.412251 |
| 40 | 1.488864 | 0.671653 | 32.834686 | 48.886373 |
| 41 | 1.503752 | 0.665003 | 33.499689 | 50.375237 |
| 42 | 1.518790 | 0.658419 | 34.158108 | 51.878989 |
| 43 | 1.533978 | 0.651900 | 34.810008 | 53.397779 |
| 44 | 1.549318 | 0.645445 | 35.455454 | 54.931757 |
| 45 | 1.564811 | 0.639055 | 36.094508 | 56.481075 |
| 46 | 1.580459 | 0.632728 | 36.727236 | 58.045885 |
| 47 | 1.596263 | 0.626463 | 37.353699 | 59.626344 |
| 48 | 1.612226 | 0.620260 | 37.973959 | 61.222608 |
| 49 | 1.628348 | 0.614119 | 38.588079 | 62.834834 |
| 50 | 1.644632 | 0.608039 | 39.196118 | 64.463182 |

$$r = 0.0125$$

| $n$ | $(1 + r)^n$ | $(1 + r)^{-n}$ | $a_{\overline{n}|r}$ | $s_{\overline{n}|r}$ |
|---|---|---|---|---|
| 1 | 1.012500 | 0.987654 | 0.987654 | 1.000000 |
| 2 | 1.025156 | 0.975461 | 1.963115 | 2.012500 |
| 3 | 1.037971 | 0.963418 | 2.926534 | 3.037656 |
| 4 | 1.050945 | 0.951524 | 3.878058 | 4.075627 |
| 5 | 1.064082 | 0.939777 | 4.817835 | 5.126572 |
| 6 | 1.077383 | 0.928175 | 5.746010 | 6.190654 |
| 7 | 1.090850 | 0.916716 | 6.662726 | 7.268038 |
| 8 | 1.104486 | 0.905398 | 7.568124 | 8.358888 |
| 9 | 1.118292 | 0.894221 | 8.462345 | 9.463374 |
| 10 | 1.132271 | 0.883181 | 9.345526 | 10.581666 |
| 11 | 1.146424 | 0.872277 | 10.217803 | 11.713937 |
| 12 | 1.160755 | 0.861509 | 11.079312 | 12.860361 |
| 13 | 1.175264 | 0.850873 | 11.930185 | 14.021116 |
| 14 | 1.189955 | 0.840368 | 12.770553 | 15.196380 |
| 15 | 1.204829 | 0.829993 | 13.600546 | 16.386335 |
| 16 | 1.219890 | 0.819746 | 14.420292 | 17.591164 |
| 17 | 1.235138 | 0.809626 | 15.229918 | 18.811053 |
| 18 | 1.250577 | 0.799631 | 16.029549 | 20.046192 |
| 19 | 1.266210 | 0.789759 | 16.819308 | 21.296769 |
| 20 | 1.282037 | 0.780009 | 17.599316 | 22.562979 |
| 21 | 1.298063 | 0.770379 | 18.369695 | 23.845016 |
| 22 | 1.314288 | 0.760868 | 19.130563 | 25.143078 |
| 23 | 1.330717 | 0.751475 | 19.882037 | 26.457367 |
| 24 | 1.347351 | 0.742197 | 20.624235 | 27.788084 |
| 25 | 1.364193 | 0.733034 | 21.357269 | 29.135435 |
| 26 | 1.381245 | 0.723984 | 22.081253 | 30.499628 |
| 27 | 1.398511 | 0.715046 | 22.796299 | 31.880873 |
| 28 | 1.415992 | 0.706219 | 23.502518 | 33.279384 |
| 29 | 1.433692 | 0.697500 | 24.200018 | 34.695377 |
| 30 | 1.451613 | 0.688889 | 24.888906 | 36.129069 |
| 31 | 1.469759 | 0.680384 | 25.569290 | 37.580682 |
| 32 | 1.488131 | 0.671984 | 26.241274 | 39.050441 |
| 33 | 1.506732 | 0.663688 | 26.904962 | 40.538571 |
| 34 | 1.525566 | 0.655494 | 27.560456 | 42.045303 |
| 35 | 1.544636 | 0.647402 | 28.207858 | 43.570870 |
| 36 | 1.563944 | 0.639409 | 28.847267 | 45.115505 |
| 37 | 1.583493 | 0.631515 | 29.478783 | 46.679449 |
| 38 | 1.603287 | 0.623719 | 30.102501 | 48.262942 |
| 39 | 1.623328 | 0.616019 | 30.718520 | 49.866229 |
| 40 | 1.643619 | 0.608413 | 31.326933 | 51.489557 |
| 41 | 1.664165 | 0.600902 | 31.927835 | 53.133177 |
| 42 | 1.684967 | 0.593484 | 32.521319 | 54.797341 |
| 43 | 1.706029 | 0.586157 | 33.107475 | 56.482308 |
| 44 | 1.727354 | 0.578920 | 33.686395 | 58.188337 |
| 45 | 1.748946 | 0.571773 | 34.258168 | 59.915691 |
| 46 | 1.770808 | 0.564714 | 34.822882 | 61.664637 |
| 47 | 1.792943 | 0.557742 | 35.380624 | 63.435445 |
| 48 | 1.815355 | 0.550856 | 35.931481 | 65.228388 |
| 49 | 1.838047 | 0.544056 | 36.475537 | 67.043743 |
| 50 | 1.861022 | 0.537339 | 37.012876 | 68.881790 |

$$r = 0.015$$

| $n$ | $(1 + r)^n$ | $(1 + r)^{-n}$ | $a_{\overline{n}|r}$ | $s_{\overline{n}|r}$ |
|---|---|---|---|---|
| 1 | 1.015000 | 0.985222 | 0.985222 | 1.000000 |
| 2 | 1.030225 | 0.970662 | 1.955883 | 2.015000 |
| 3 | 1.045678 | 0.956317 | 2.912200 | 3.045225 |
| 4 | 1.061364 | 0.942184 | 3.854385 | 4.090903 |
| 5 | 1.077284 | 0.928260 | 4.782645 | 5.152267 |
| 6 | 1.093443 | 0.914542 | 5.697187 | 6.229551 |
| 7 | 1.109845 | 0.901027 | 6.598214 | 7.322994 |
| 8 | 1.126493 | 0.887711 | 7.485925 | 8.432839 |
| 9 | 1.143390 | 0.874592 | 8.360517 | 9.559332 |
| 10 | 1.160541 | 0.861667 | 9.222185 | 10.702722 |
| 11 | 1.177949 | 0.848933 | 10.071118 | 11.863262 |
| 12 | 1.195618 | 0.836387 | 10.907505 | 13.041211 |
| 13 | 1.213552 | 0.824027 | 11.731532 | 14.236830 |
| 14 | 1.231756 | 0.811849 | 12.543382 | 15.450382 |
| 15 | 1.250232 | 0.799852 | 13.343233 | 16.682138 |
| 16 | 1.268986 | 0.788031 | 14.131264 | 17.932370 |
| 17 | 1.288020 | 0.776385 | 14.907649 | 19.201355 |
| 18 | 1.307341 | 0.764912 | 15.672561 | 20.489376 |
| 19 | 1.326951 | 0.753607 | 16.426168 | 21.796716 |
| 20 | 1.346855 | 0.742470 | 17.168639 | 23.123667 |
| 21 | 1.367058 | 0.731498 | 17.900137 | 24.470522 |
| 22 | 1.387564 | 0.720688 | 18.620824 | 25.837580 |
| 23 | 1.408377 | 0.710037 | 19.330861 | 27.225144 |
| 24 | 1.429503 | 0.699544 | 20.030405 | 28.633521 |
| 25 | 1.450945 | 0.689206 | 20.719611 | 30.063024 |
| 26 | 1.472710 | 0.679021 | 21.398632 | 31.513969 |
| 27 | 1.494800 | 0.668986 | 22.067617 | 32.986678 |
| 28 | 1.517222 | 0.659099 | 22.726717 | 34.481479 |
| 29 | 1.539981 | 0.649359 | 23.376076 | 35.998701 |
| 30 | 1.563080 | 0.639762 | 24.015838 | 37.538681 |
| 31 | 1.586526 | 0.630308 | 24.646146 | 39.101762 |
| 32 | 1.610324 | 0.620993 | 25.267139 | 40.688288 |
| 33 | 1.634479 | 0.611816 | 25.878954 | 42.298612 |
| 34 | 1.658996 | 0.602774 | 26.481728 | 43.933092 |
| 35 | 1.683881 | 0.593866 | 27.075595 | 45.592088 |
| 36 | 1.709140 | 0.585090 | 27.660684 | 47.275969 |
| 37 | 1.734777 | 0.576443 | 28.237127 | 48.985109 |
| 38 | 1.760798 | 0.567924 | 28.805052 | 50.719885 |
| 39 | 1.787210 | 0.559531 | 29.364583 | 52.480684 |
| 40 | 1.814018 | 0.551262 | 29.915845 | 54.267894 |
| 41 | 1.841229 | 0.543116 | 30.458961 | 56.081912 |
| 42 | 1.868847 | 0.535089 | 30.994050 | 57.923141 |
| 43 | 1.896880 | 0.527182 | 31.521232 | 59.791988 |
| 44 | 1.925333 | 0.519391 | 32.040622 | 61.688868 |
| 45 | 1.954213 | 0.511715 | 32.552337 | 63.614201 |
| 46 | 1.983526 | 0.504153 | 33.056490 | 65.568414 |
| 47 | 2.013279 | 0.496702 | 33.553192 | 67.551940 |
| 48 | 2.043478 | 0.489362 | 34.042554 | 69.565219 |
| 49 | 2.074130 | 0.482130 | 34.524683 | 71.608698 |
| 50 | 2.105242 | 0.475005 | 34.999688 | 73.682828 |

$$r = 0.02$$

| $n$ | $(1 + r)^n$ | $(1 + r)^{-n}$ | $a_{\overline{n}|r}$ | $s_{\overline{n}|r}$ |
|---|---|---|---|---|
| 1 | 1.020000 | 0.980392 | 0.980392 | 1.000000 |
| 2 | 1.040400 | 0.961169 | 1.941561 | 2.020000 |
| 3 | 1.061208 | 0.942322 | 2.883883 | 3.060400 |
| 4 | 1.082432 | 0.923845 | 3.807729 | 4.121608 |
| 5 | 1.104081 | 0.905731 | 4.713460 | 5.204040 |
| 6 | 1.126162 | 0.887971 | 5.601431 | 6.308121 |
| 7 | 1.148686 | 0.870560 | 6.471991 | 7.434283 |
| 8 | 1.171659 | 0.853490 | 7.325481 | 8.582969 |
| 9 | 1.195093 | 0.836755 | 8.162237 | 9.754628 |
| 10 | 1.218994 | 0.820348 | 8.982585 | 10.949721 |
| 11 | 1.243374 | 0.804263 | 9.786848 | 12.168715 |
| 12 | 1.268242 | 0.788493 | 10.575341 | 13.412090 |
| 13 | 1.293607 | 0.773033 | 11.348374 | 14.680332 |
| 14 | 1.319479 | 0.757875 | 12.106249 | 15.973938 |
| 15 | 1.345868 | 0.743015 | 12.849264 | 17.293417 |
| 16 | 1.372786 | 0.728446 | 13.577709 | 18.639285 |
| 17 | 1.400241 | 0.714163 | 14.291872 | 20.012071 |
| 18 | 1.428246 | 0.700159 | 14.992031 | 21.412312 |
| 19 | 1.456811 | 0.686431 | 15.678462 | 22.840559 |
| 20 | 1.485947 | 0.672971 | 16.351433 | 24.297370 |
| 21 | 1.515666 | 0.659776 | 17.011209 | 25.783317 |
| 22 | 1.545980 | 0.646839 | 17.658048 | 27.298984 |
| 23 | 1.576899 | 0.634156 | 18.292204 | 28.844963 |
| 24 | 1.608437 | 0.621721 | 18.913926 | 30.421862 |
| 25 | 1.640606 | 0.609531 | 19.523456 | 32.030300 |
| 26 | 1.673418 | 0.597579 | 20.121036 | 33.670906 |
| 27 | 1.706886 | 0.585862 | 20.706898 | 35.344324 |
| 28 | 1.741024 | 0.574375 | 21.281272 | 37.051210 |
| 29 | 1.775845 | 0.563112 | 21.844385 | 38.792235 |
| 30 | 1.811362 | 0.552071 | 22.396456 | 40.568079 |
| 31 | 1.847589 | 0.541246 | 22.937702 | 42.379441 |
| 32 | 1.884541 | 0.530633 | 23.468335 | 44.227030 |
| 33 | 1.922231 | 0.520229 | 23.988564 | 46.111570 |
| 34 | 1.960676 | 0.510028 | 24.498592 | 48.033802 |
| 35 | 1.999890 | 0.500028 | 24.998619 | 49.994478 |
| 36 | 2.039887 | 0.490223 | 25.488842 | 51.994367 |
| 37 | 2.080685 | 0.480611 | 25.969453 | 54.034255 |
| 38 | 2.122299 | 0.471187 | 26.440641 | 56.114940 |
| 39 | 2.164745 | 0.461948 | 26.902589 | 58.237238 |
| 40 | 2.208040 | 0.452890 | 27.355479 | 60.401983 |
| 41 | 2.252200 | 0.444010 | 27.799489 | 62.610023 |
| 42 | 2.297244 | 0.435304 | 28.234794 | 64.862223 |
| 43 | 2.343189 | 0.426769 | 28.661562 | 67.159468 |
| 44 | 2.390053 | 0.418401 | 29.079963 | 69.502657 |
| 45 | 2.437854 | 0.410197 | 29.490160 | 71.892710 |
| 46 | 2.486611 | 0.402154 | 29.892314 | 74.330564 |
| 47 | 2.536344 | 0.394268 | 30.286582 | 76.817176 |
| 48 | 2.587070 | 0.386538 | 30.673120 | 79.353519 |
| 49 | 2.638812 | 0.378958 | 31.052078 | 81.940590 |
| 50 | 2.691588 | 0.371528 | 31.423606 | 84.579401 |

$$r = 0.025$$

| $n$ | $(1 + r)^n$ | $(1 + r)^{-n}$ | $a_{\overline{n}|r}$ | $s_{\overline{n}|r}$ |
|---|---|---|---|---|
| 1 | 1.025000 | 0.975610 | 0.975610 | 1.000000 |
| 2 | 1.050625 | 0.951814 | 1.927424 | 2.025000 |
| 3 | 1.076891 | 0.928599 | 2.856024 | 3.075625 |
| 4 | 1.103813 | 0.905951 | 3.761974 | 4.152516 |
| 5 | 1.131408 | 0.883854 | 4.645828 | 5.256329 |
| 6 | 1.159693 | 0.862297 | 5.508125 | 6.387737 |
| 7 | 1.188686 | 0.841265 | 6.349391 | 7.547430 |
| 8 | 1.218403 | 0.820747 | 7.170137 | 8.736116 |
| 9 | 1.248863 | 0.800728 | 7.970866 | 9.954519 |
| 10 | 1.280085 | 0.781198 | 8.752064 | 11.203382 |
| 11 | 1.312087 | 0.762145 | 9.514209 | 12.483466 |
| 12 | 1.344889 | 0.743556 | 10.257765 | 13.795553 |
| 13 | 1.378511 | 0.725420 | 10.983185 | 15.140442 |
| 14 | 1.412974 | 0.707727 | 11.690912 | 16.518953 |
| 15 | 1.448298 | 0.690466 | 12.381378 | 17.931927 |
| 16 | 1.484506 | 0.673625 | 13.055003 | 19.380225 |
| 17 | 1.521618 | 0.657195 | 13.712198 | 20.864730 |
| 18 | 1.559659 | 0.641166 | 14.353364 | 22.386349 |
| 19 | 1.598650 | 0.625528 | 14.978891 | 23.946007 |
| 20 | 1.638616 | 0.610271 | 15.589162 | 25.544658 |
| 21 | 1.679582 | 0.595386 | 16.184549 | 27.183274 |
| 22 | 1.721571 | 0.580865 | 16.765413 | 28.862856 |
| 23 | 1.764611 | 0.566697 | 17.332110 | 30.584427 |
| 24 | 1.808726 | 0.552875 | 17.884986 | 32.349038 |
| 25 | 1.853944 | 0.539391 | 18.424376 | 34.157764 |
| 26 | 1.900293 | 0.526235 | 18.950611 | 36.011708 |
| 27 | 1.947800 | 0.513400 | 19.464011 | 37.912001 |
| 28 | 1.996495 | 0.500878 | 19.964889 | 39.859801 |
| 29 | 2.046407 | 0.488661 | 20.453550 | 41.856296 |
| 30 | 2.097568 | 0.476743 | 20.930293 | 43.902703 |
| 31 | 2.150007 | 0.465115 | 21.395407 | 46.000271 |
| 32 | 2.203757 | 0.453771 | 21.849178 | 48.150278 |
| 33 | 2.258851 | 0.442703 | 22.291881 | 30.354034 |
| 34 | 2.315322 | 0.431905 | 22.723786 | 52.612885 |
| 35 | 2.373205 | 0.421371 | 23.145157 | 54.928207 |
| 36 | 2.432535 | 0.411094 | 23.556251 | 57.301413 |
| 37 | 2.493349 | 0.401067 | 23.957318 | 59.733948 |
| 38 | 2.555682 | 0.391285 | 24.348603 | 62.227297 |
| 39 | 2.619574 | 0.381741 | 24.730344 | 64.782979 |
| 40 | 2.685064 | 0.372431 | 25.102775 | 67.402554 |
| 41 | 2.752190 | 0.363347 | 25.466122 | 70.087617 |
| 42 | 2.820995 | 0.354485 | 25.820607 | 72.839808 |
| 43 | 2.891520 | 0.345839 | 26.166446 | 75.660803 |
| 44 | 2.963808 | 0.337404 | 26.503849 | 78.552323 |
| 45 | 3.037903 | 0.329174 | 26.833024 | 81.516131 |
| 46 | 3.113851 | 0.321146 | 27.154170 | 84.554034 |
| 47 | 3.191697 | 0.313313 | 27.467483 | 87.667885 |
| 48 | 3.271490 | 0.305671 | 27.773154 | 90.859582 |
| 49 | 3.353277 | 0.298216 | 28.071369 | 94.131072 |
| 50 | 3.437109 | 0.290942 | 28.362312 | 97.484349 |

$$r = 0.03$$

| $n$ | $(1 + r)^n$ | $(1 + r)^{-n}$ | $a_{\overline{n}|r}$ | $s_{\overline{n}|r}$ |
|---|---|---|---|---|
| 1 | 1.030000 | 0.970874 | 0.970874 | 1.000000 |
| 2 | 1.060900 | 0.942596 | 1.913470 | 2.030000 |
| 3 | 1.092727 | 0.915142 | 2.828611 | 3.090900 |
| 4 | 1.125509 | 0.888487 | 3.717098 | 4.183627 |
| 5 | 1.159274 | 0.862609 | 4.579707 | 5.309136 |
| 6 | 1.194052 | 0.837484 | 5.417191 | 6.468410 |
| 7 | 1.229874 | 0.813092 | 6.230283 | 7.662462 |
| 8 | 1.266770 | 0.789409 | 7.019692 | 8.892336 |
| 9 | 1.304773 | 0.766417 | 7.786109 | 10.159106 |
| 10 | 1.343916 | 0.744094 | 8.530203 | 11.463879 |
| 11 | 1.384234 | 0.722421 | 9.252624 | 12.807796 |
| 12 | 1.425761 | 0.701380 | 9.954004 | 14.192030 |
| 13 | 1.468534 | 0.680951 | 10.634955 | 15.617790 |
| 14 | 1.512590 | 0.661118 | 11.296073 | 17.086324 |
| 15 | 1.557967 | 0.641862 | 11.937935 | 18.598914 |
| 16 | 1.604706 | 0.623167 | 12.561102 | 20.156881 |
| 17 | 1.652848 | 0.605016 | 13.166118 | 21.761588 |
| 18 | 1.702433 | 0.587395 | 13.753513 | 23.414435 |
| 19 | 1.753506 | 0.570286 | 14.323799 | 25.116868 |
| 20 | 1.806111 | 0.553676 | 14.877475 | 26.870374 |
| 21 | 1.860295 | 0.537549 | 15.415024 | 28.676486 |
| 22 | 1.916103 | 0.521893 | 15.936917 | 30.536780 |
| 23 | 1.973587 | 0.506692 | 16.443608 | 32.452884 |
| 24 | 2.032794 | 0.491934 | 16.935542 | 34.426470 |
| 25 | 2.093778 | 0.477606 | 17.413148 | 36.459264 |
| 26 | 2.156591 | 0.463695 | 17.876842 | 38.553042 |
| 27 | 2.221289 | 0.450189 | 18.327031 | 40.709634 |
| 28 | 2.287928 | 0.437077 | 18.764108 | 42.930923 |
| 29 | 2.356566 | 0.424346 | 19.188455 | 45.218850 |
| 30 | 2.427262 | 0.411987 | 19.600441 | 47.575416 |
| 31 | 2.500080 | 0.399987 | 20.000428 | 50.002678 |
| 32 | 2.575083 | 0.388337 | 20.388766 | 52.502759 |
| 33 | 2.652335 | 0.377026 | 20.765792 | 55.077841 |
| 34 | 2.731905 | 0.366045 | 21.131837 | 57.730177 |
| 35 | 2.813862 | 0.355383 | 21.487220 | 60.462082 |
| 36 | 2.898278 | 0.345032 | 21.832252 | 63.275944 |
| 37 | 2.985227 | 0.334983 | 22.167235 | 66.174223 |
| 38 | 3.074783 | 0.325226 | 22.492462 | 69.159449 |
| 39 | 3.167027 | 0.315754 | 22.808215 | 72.234233 |
| 40 | 3.262038 | 0.306557 | 23.114772 | 75.401260 |
| 41 | 3.359899 | 0.297628 | 23.412400 | 78.663298 |
| 42 | 3.460696 | 0.288959 | 23.701359 | 82.023196 |
| 43 | 3.564517 | 0.280543 | 23.981902 | 85.483892 |
| 44 | 3.671452 | 0.272372 | 24.254274 | 89.048409 |
| 45 | 3.781596 | 0.264439 | 24.518713 | 92.719861 |
| 46 | 3.895044 | 0.256737 | 24.775449 | 96.501457 |
| 47 | 4.011895 | 0.249259 | 25.024708 | 100.396501 |
| 48 | 4.132252 | 0.241999 | 25.266707 | 104.408396 |
| 49 | 4.256219 | 0.234950 | 25.501657 | 108.540648 |
| 50 | 4.383906 | 0.228107 | 25.729764 | 112.796867 |

$$r = 0.035$$

| $n$ | $(1 + r)^n$ | $(1 + r)^{-n}$ | $a_{\overline{n}|r}$ | $s_{\overline{n}|r}$ |
|---|---|---|---|---|
| 1 | 1.035000 | 0.966184 | 0.966184 | 1.000000 |
| 2 | 1.071225 | 0.933511 | 1.899694 | 2.035000 |
| 3 | 1.108718 | 0.901943 | 2.801637 | 3.106225 |
| 4 | 1.147523 | 0.871442 | 3.673079 | 4.214943 |
| 5 | 1.187686 | 0.841973 | 4.515052 | 5.362466 |
| 6 | 1.229255 | 0.813501 | 5.328553 | 6.550152 |
| 7 | 1.272279 | 0.785991 | 6.114544 | 7.779408 |
| 8 | 1.316809 | 0.759412 | 6.873956 | 9.051687 |
| 9 | 1.362897 | 0.733731 | 7.607687 | 10.368496 |
| 10 | 1.410599 | 0.708919 | 8.316605 | 11.731393 |
| 11 | 1.459970 | 0.684946 | 9.001551 | 13.141992 |
| 12 | 1.511069 | 0.661783 | 9.663334 | 14.601962 |
| 13 | 1.563956 | 0.639404 | 10.302738 | 16.113030 |
| 14 | 1.618695 | 0.617782 | 10.920520 | 17.676986 |
| 15 | 1.675349 | 0.596891 | 11.517411 | 19.295681 |
| 16 | 1.733986 | 0.576706 | 12.094117 | 20.971030 |
| 17 | 1.794676 | 0.557204 | 12.651321 | 22.705016 |
| 18 | 1.857489 | 0.538361 | 13.189682 | 24.499691 |
| 19 | 1.922501 | 0.520156 | 13.709837 | 26.357180 |
| 20 | 1.989789 | 0.502566 | 14.212403 | 28.279682 |
| 21 | 2.059431 | 0.485571 | 14.697974 | 30.269471 |
| 22 | 2.131512 | 0.469151 | 15.167125 | 32.328902 |
| 23 | 2.206114 | 0.453286 | 15.620410 | 34.460414 |
| 24 | 2.283328 | 0.437957 | 16.058368 | 36.666528 |
| 25 | 2.363245 | 0.423147 | 16.481515 | 38.949857 |
| 26 | 2.445959 | 0.408838 | 16.890352 | 41.313102 |
| 27 | 2.531567 | 0.395012 | 17.285365 | 43.759060 |
| 28 | 2.620172 | 0.381654 | 17.667019 | 46.290627 |
| 29 | 2.711878 | 0.368748 | 18.035767 | 48.910799 |
| 30 | 2.806794 | 0.356278 | 18.392045 | 51.622677 |
| 31 | 2.905031 | 0.344230 | 18.736276 | 54.429471 |
| 32 | 3.006708 | 0.332590 | 19.068865 | 57.334502 |
| 33 | 3.111942 | 0.321343 | 19.390208 | 60.341210 |
| 34 | 3.220860 | 0.310476 | 19.700684 | 63.453152 |
| 35 | 3.333590 | 0.299977 | 20.000661 | 66.674013 |
| 36 | 3.450266 | 0.289833 | 20.290494 | 70.007603 |
| 37 | 3.571025 | 0.280032 | 20.570525 | 73.457869 |
| 38 | 3.696011 | 0.270562 | 20.841087 | 77.028895 |
| 39 | 3.825372 | 0.261413 | 21.102500 | 80.724906 |
| 40 | 3.959260 | 0.252572 | 21.355072 | 84.550278 |
| 41 | 4.097834 | 0.244031 | 21.599104 | 88.509537 |
| 42 | 4.241258 | 0.235779 | 21.834883 | 92.607371 |
| 43 | 4.389702 | 0.227806 | 22.062689 | 96.848629 |
| 44 | 4.543342 | 0.220102 | 22.282791 | 101.238331 |
| 45 | 4.702359 | 0.212659 | 22.495450 | 105.781673 |
| 46 | 4.866941 | 0.205468 | 22.700918 | 110.484031 |
| 47 | 5.037284 | 0.198520 | 22.899438 | 115.350973 |
| 48 | 5.213589 | 0.191806 | 23.091244 | 120.388257 |
| 49 | 5.396065 | 0.185320 | 23.276564 | 125.601846 |
| 50 | 5.584927 | 0.179053 | 23.455618 | 130.997910 |

$$r = 0.04$$

| $n$ | $(1 + r)^n$ | $(1 + r)^{-n}$ | $a_{\overline{n}|r}$ | $s_{\overline{n}|r}$ |
|---|---|---|---|---|
| 1 | 1.040000 | 0.961538 | 0.961538 | 1.000000 |
| 2 | 1.081600 | 0.924556 | 1.886095 | 2.040000 |
| 3 | 1.124864 | 0.888996 | 2.775091 | 3.121600 |
| 4 | 1.169859 | 0.854804 | 3.629895 | 4.246464 |
| 5 | 1.216653 | 0.821927 | 4.451822 | 5.416323 |
| 6 | 1.265319 | 0.790315 | 5.242137 | 6.632975 |
| 7 | 1.315932 | 0.759918 | 6.002055 | 7.898294 |
| 8 | 1.368569 | 0.730690 | 6.732745 | 9.214226 |
| 9 | 1.423312 | 0.702587 | 7.435332 | 10.582795 |
| 10 | 1.480244 | 0.675564 | 8.110896 | 12.006107 |
| 11 | 1.539454 | 0.649581 | 8.760477 | 13.486351 |
| 12 | 1.601032 | 0.624597 | 9.385074 | 15.025805 |
| 13 | 1.665074 | 0.600574 | 9.985648 | 16.626838 |
| 14 | 1.731676 | 0.577475 | 10.563123 | 18.291911 |
| 15 | 1.800944 | 0.555265 | 11.118387 | 20.023588 |
| 16 | 1.872981 | 0.533908 | 11.652296 | 21.824531 |
| 17 | 1.947900 | 0.513373 | 12.165669 | 23.697512 |
| 18 | 2.025817 | 0.493628 | 12.659297 | 25.645413 |
| 19 | 2.106849 | 0.474642 | 13.133939 | 27.671229 |
| 20 | 2.191123 | 0.456387 | 13.590326 | 29.778079 |
| 21 | 2.278768 | 0.438834 | 14.029160 | 31.969202 |
| 22 | 2.369919 | 0.421955 | 14.451115 | 34.247970 |
| 23 | 2.464716 | 0.405726 | 14.856842 | 36.617889 |
| 24 | 2.563304 | 0.390121 | 15.246963 | 39.082604 |
| 25 | 2.665836 | 0.375117 | 15.622080 | 41.645908 |
| 26 | 2.772470 | 0.360689 | 15.982769 | 44.311745 |
| 27 | 2.883369 | 0.346817 | 16.329586 | 47.084214 |
| 28 | 2.998703 | 0.333477 | 16.663063 | 49.967583 |
| 29 | 3.118651 | 0.320651 | 16.983715 | 52.966286 |
| 30 | 3.243398 | 0.308319 | 17.292033 | 56.084938 |
| 31 | 3.373133 | 0.296460 | 17.588494 | 59.328335 |
| 32 | 3.508059 | 0.285058 | 17.873551 | 62.701469 |
| 33 | 3.648381 | 0.274094 | 18.147646 | 66.209527 |
| 34 | 3.794316 | 0.263552 | 18.411198 | 69.857909 |
| 35 | 3.946089 | 0.253415 | 18.664613 | 73.652225 |
| 36 | 4.103933 | 0.243669 | 18.908282 | 77.598314 |
| 37 | 4.268090 | 0.234297 | 19.142579 | 81.702246 |
| 38 | 4.438813 | 0.225285 | 19.367864 | 85.970336 |
| 39 | 4.616366 | 0.216621 | 19.584485 | 90.409150 |
| 40 | 4.801021 | 0.208289 | 19.792774 | 95.025516 |
| 41 | 4.993061 | 0.200278 | 19.993052 | 99.826536 |
| 42 | 5.192784 | 0.192575 | 20.185627 | 104.819598 |
| 43 | 5.400495 | 0.185168 | 20.370795 | 110.012382 |
| 44 | 5.616515 | 0.178046 | 20.548841 | 115.412877 |
| 45 | 5.841176 | 0.171198 | 20.720040 | 121.029392 |
| 46 | 6.074823 | 0.164614 | 20.884654 | 126.870568 |
| 47 | 6.317816 | 0.158283 | 21.042936 | 132.945390 |
| 48 | 6.570528 | 0.152195 | 21.195131 | 139.263206 |
| 49 | 6.833349 | 0.146341 | 21.341472 | 145.833734 |
| 50 | 7.106683 | 0.140713 | 21.482185 | 152.667084 |

$$r = 0.05$$

| $n$ | $(1 + r)^n$ | $(1 + r)^{-n}$ | $a_{\overline{n}|r}$ | $s_{\overline{n}|r}$ |
|---|---|---|---|---|
| 1 | 1.050000 | 0.952381 | 0.952381 | 1.000000 |
| 2 | 1.102500 | 0.907029 | 1.859410 | 2.050000 |
| 3 | 1.157625 | 0.863838 | 2.723248 | 3.152500 |
| 4 | 1.215506 | 0.822702 | 3.545951 | 4.310125 |
| 5 | 1.276282 | 0.783526 | 4.329477 | 5.525631 |
| 6 | 1.340096 | 0.746215 | 5.075692 | 6.801913 |
| 7 | 1.407100 | 0.710681 | 5.786373 | 8.142008 |
| 8 | 1.477455 | 0.676839 | 6.463213 | 3.549109 |
| 9 | 1.551328 | 0.644609 | 7.107822 | 11.026564 |
| 10 | 1.628895 | 0.613913 | 7.721735 | 12.577893 |
| 11 | 1.710339 | 0.584679 | 8.306414 | 14.206787 |
| 12 | 1.795856 | 0.556837 | 8.863252 | 15.917127 |
| 13 | 1.885649 | 0.530321 | 9.393573 | 17.712983 |
| 14 | 1.979932 | 0.505068 | 9.898641 | 19.598632 |
| 15 | 2.078928 | 0.481017 | 10.379658 | 21.578564 |
| 16 | 2.182875 | 0.458112 | 10.837770 | 23.657492 |
| 17 | 2.292018 | 0.436297 | 11.274066 | 25.840366 |
| 18 | 2.406619 | 0.415521 | 11.689587 | 28.132385 |
| 19 | 2.526950 | 0.395734 | 12.085321 | 30.539004 |
| 20 | 2.653298 | 0.376889 | 12.462210 | 33.065954 |
| 21 | 2.785963 | 0.358942 | 12.821153 | 35.719252 |
| 22 | 2.925261 | 0.341850 | 13.163003 | 38.505214 |
| 23 | 3.071524 | 0.325571 | 13.488574 | 41.430475 |
| 24 | 3.225100 | 0.310068 | 13.798642 | 44.501999 |
| 25 | 3.386355 | 0.295303 | 14.093945 | 47.727099 |
| 26 | 3.555673 | 0.281241 | 14.375185 | 51.113454 |
| 27 | 3.733456 | 0.267848 | 14.643034 | 54.669126 |
| 28 | 3.920129 | 0.255094 | 14.898127 | 58.402583 |
| 29 | 4.116136 | 0.242946 | 15.141074 | 62.322712 |
| 30 | 4.321942 | 0.231377 | 15.372451 | 66.438848 |
| 31 | 4.538039 | 0.220359 | 15.592811 | 70.760790 |
| 32 | 4.764941 | 0.209866 | 15.802677 | 75.298829 |
| 33 | 5.003189 | 0.199873 | 16.002549 | 80.063771 |
| 34 | 5.253348 | 0.190355 | 16.192904 | 85.066959 |
| 35 | 5.516015 | 0.181290 | 16.374194 | 90.320307 |
| 36 | 5.791816 | 0.172657 | 16.546852 | 95.836323 |
| 37 | 6.081407 | 0.164436 | 16.711287 | 101.628139 |
| 38 | 6.385477 | 0.156605 | 16.867893 | 107.709546 |
| 39 | 6.704751 | 0.149148 | 17.017041 | 114.095023 |
| 40 | 7.039989 | 0.142046 | 17.159086 | 120.799774 |
| 41 | 7.391988 | 0.135282 | 17.294368 | 127.839763 |
| 42 | 7.761588 | 0.128840 | 17.423208 | 135.231751 |
| 43 | 8.149667 | 0.122704 | 17.545912 | 142.993339 |
| 44 | 8.557150 | 0.116861 | 17.662773 | 151.143006 |
| 45 | 8.985008 | 0.111297 | 17.774070 | 159.700156 |
| 46 | 9.434258 | 0.105997 | 17.880066 | 168.685164 |
| 47 | 9.905971 | 0.100949 | 17.981016 | 178.119422 |
| 48 | 10.401270 | 0.096142 | 18.077158 | 188.025393 |
| 49 | 10.921333 | 0.091564 | 18.168722 | 198.426663 |
| 50 | 11.467400 | 0.087204 | 18.255925 | 209.347996 |

$$r = 0.06$$

| $n$ | $(1 + r)^n$ | $(1 + r)^{-n}$ | $a_{\overline{n}|r}$ | $s_{\overline{n}|r}$ |
|---|---|---|---|---|
| 1 | 1.060000 | 0.943396 | 0.943396 | 1.000000 |
| 2 | 1.123600 | 0.889996 | 1.833393 | 2.060000 |
| 3 | 1.191016 | 0.839619 | 2.673012 | 3.183600 |
| 4 | 1.262477 | 0.792094 | 3.465106 | 4.374616 |
| 5 | 1.338226 | 0.747258 | 4.212364 | 5.637093 |
| 6 | 1.418519 | 0.704961 | 4.917324 | 6.975319 |
| 7 | 1.503630 | 0.665057 | 5.582381 | 8.393838 |
| 8 | 1.593848 | 0.627412 | 6.209794 | 9.897468 |
| 9 | 1.689479 | 0.591898 | 6.801692 | 11.491316 |
| 10 | 1.790848 | 0.558395 | 7.360087 | 13.180795 |
| 11 | 1.898299 | 0.526788 | 7.886875 | 14.971643 |
| 12 | 2.012196 | 0.496969 | 8.383844 | 16.869941 |
| 13 | 2.132928 | 0.468839 | 8.852683 | 18.882138 |
| 14 | 2.260904 | 0.442301 | 9.294984 | 21.015066 |
| 15 | 2.396558 | 0.417265 | 9.712249 | 23.275970 |
| 16 | 2.540352 | 0.393646 | 10.105895 | 25.672528 |
| 17 | 2.692773 | 0.371364 | 10.477260 | 28.212880 |
| 18 | 2.854339 | 0.350344 | 10.827603 | 30.905653 |
| 19 | 3.025600 | 0.330513 | 11.158116 | 33.759992 |
| 20 | 3.207135 | 0.311805 | 11.469921 | 36.785591 |
| 21 | 3.399564 | 0.294155 | 11.764077 | 39.992727 |
| 22 | 3.603537 | 0.277505 | 12.041582 | 43.392290 |
| 23 | 3.819750 | 0.261797 | 12.303379 | 46.995828 |
| 24 | 4.048935 | 0.246979 | 12.550358 | 50.815577 |
| 25 | 4.291871 | 0.232999 | 12.783356 | 54.864512 |
| 26 | 4.549383 | 0.219810 | 13.003166 | 59.156383 |
| 27 | 4.822346 | 0.207368 | 13.210534 | 63.705766 |
| 28 | 5.111687 | 0.195630 | 13.406164 | 68.528112 |
| 29 | 5.418388 | 0.184557 | 13.590721 | 73.639798 |
| 30 | 5.743491 | 0.174110 | 13.764831 | 79.058186 |
| 31 | 6.088101 | 0.164255 | 13.929086 | 84.801677 |
| 32 | 6.453387 | 0.154957 | 14.084043 | 90.889778 |
| 33 | 6.840590 | 0.146186 | 14.230230 | 97.343165 |
| 34 | 7.251025 | 0.137912 | 14.368141 | 104.183755 |
| 35 | 7.686087 | 0.130105 | 14.498246 | 111.434780 |
| 36 | 8.147252 | 0.122741 | 14.620987 | 119.120867 |
| 37 | 8.636087 | 0.115793 | 14.736780 | 127.268119 |
| 38 | 9.154252 | 0.109239 | 14.846019 | 135.904206 |
| 39 | 9.703507 | 0.103056 | 14.949075 | 145.058458 |
| 40 | 10.285718 | 0.097222 | 15.046297 | 154.761966 |
| 41 | 10.902861 | 0.091719 | 15.138016 | 165.047684 |
| 42 | 11.557033 | 0.086527 | 15.224543 | 175.950545 |
| 43 | 12.250455 | 0.081630 | 15.306173 | 187.507577 |
| 44 | 12.985482 | 0.077009 | 15.383182 | 199.758032 |
| 45 | 13.764611 | 0.072650 | 15.455832 | 212.743514 |
| 46 | 14.590487 | 0.068538 | 15.524370 | 226.508125 |
| 47 | 15.465917 | 0.064658 | 15.589028 | 241.098612 |
| 48 | 16.393872 | 0.060998 | 15.650027 | 256.564529 |
| 49 | 17.377504 | 0.057546 | 15.707572 | 272.958401 |
| 50 | 18.420154 | 0.054288 | 15.761861 | 290.335905 |

$$r = 0.07$$

| $n$ | $(1 + r)^n$ | $(1 + r)^{-n}$ | $a_{\overline{n}|r}$ | $s_{\overline{n}|r}$ |
|---|---|---|---|---|
| 1 | 1.070000 | 0.934579 | 0.934579 | 1.000000 |
| 2 | 1.144900 | 0.873439 | 1.808018 | 2.070000 |
| 3 | 1.225043 | 0.816298 | 2.624316 | 3.214900 |
| 4 | 1.310796 | 0.762895 | 3.387211 | 4.439943 |
| 5 | 1.402552 | 0.712986 | 4.100197 | 5.750739 |
| 6 | 1.500730 | 0.666342 | 4.766540 | 7.153291 |
| 7 | 1.605781 | 0.622750 | 5.389289 | 8.654021 |
| 8 | 1.718186 | 0.582009 | 5.971299 | 10.259803 |
| 9 | 1.838459 | 0.543934 | 6.515232 | 11.977989 |
| 10 | 1.967151 | 0.508349 | 7.023582 | 13.816448 |
| 11 | 2.104852 | 0.475093 | 7.498674 | 15.783599 |
| 12 | 2.252192 | 0.444012 | 7.942686 | 17.888451 |
| 13 | 2.409845 | 0.414964 | 8.357651 | 20.140643 |
| 14 | 2.578534 | 0.387817 | 8.745468 | 22.550488 |
| 15 | 2.759032 | 0.362446 | 9.107914 | 25.129022 |
| 16 | 2.952164 | 0.338735 | 9.446649 | 27.888054 |
| 17 | 3.158815 | 0.316574 | 9.763223 | 30.840217 |
| 18 | 3.379932 | 0.295864 | 10.059087 | 33.999033 |
| 19 | 3.616528 | 0.276508 | 10.335595 | 37.378965 |
| 20 | 3.869684 | 0.258419 | 10.594014 | 40.995492 |
| 21 | 4.140562 | 0.241513 | 10.835527 | 44.865177 |
| 22 | 4.430402 | 0.225713 | 11.061240 | 49.005739 |
| 23 | 4.740530 | 0.210947 | 11.272187 | 53.436141 |
| 24 | 5.072367 | 0.197147 | 11.469334 | 58.176671 |
| 25 | 5.427433 | 0.184249 | 11.653583 | 63.249038 |
| 26 | 5.807353 | 0.172195 | 11.825779 | 68.676470 |
| 27 | 6.213868 | 0.160930 | 11.986709 | 74.483823 |
| 28 | 6.648838 | 0.150402 | 12.137111 | 80.697691 |
| 29 | 7.114257 | 0.140563 | 12.277674 | 87.346529 |
| 30 | 7.612255 | 0.131367 | 12.409041 | 94.460786 |
| 31 | 8.145113 | 0.122773 | 12.531814 | 102.073041 |
| 32 | 8.715271 | 0.114741 | 12.646555 | 110.218154 |
| 33 | 9.325340 | 0.107235 | 12.753790 | 118.933425 |
| 34 | 9.978114 | 0.100219 | 12.854009 | 128.258765 |
| 35 | 10.676581 | 0.093663 | 12.947672 | 138.236878 |
| 36 | 11.423942 | 0.087535 | 13.035208 | 148.913460 |
| 37 | 12.223618 | 0.081809 | 13.117017 | 160.337402 |
| 38 | 13.079271 | 0.076457 | 13.193473 | 172.561020 |
| 39 | 13.994820 | 0.071455 | 13.264928 | 185.640292 |
| 40 | 14.974458 | 0.066780 | 13.331709 | 199.635112 |
| 41 | 16.022670 | 0.062412 | 13.394120 | 214.609570 |
| 42 | 17.144257 | 0.058329 | 13.452449 | 230.632240 |
| 43 | 18.344355 | 0.054513 | 13.506962 | 247.776496 |
| 44 | 19.628460 | 0.050946 | 13.557908 | 266.120851 |
| 45 | 21.002452 | 0.047613 | 13.605522 | 285.749311 |
| 46 | 22.472623 | 0.044499 | 13.650020 | 306.751763 |
| 47 | 24.045707 | 0.041587 | 13.691608 | 329.224386 |
| 48 | 25.728907 | 0.038867 | 13.730474 | 353.270093 |
| 49 | 27.529930 | 0.036324 | 13.766799 | 378.999000 |
| 50 | 29.457025 | 0.033948 | 13.800746 | 406.528929 |

$$r = 0.08$$

| $n$ | $(1 + r)^n$ | $(1 + r)^{-n}$ | $a_{\overline{n}|r}$ | $s_{\overline{n}|r}$ |
|---|---|---|---|---|
| 1 | 1.080000 | 0.925926 | 0.925926 | 1.000000 |
| 2 | 1.166400 | 0.857339 | 1.783265 | 2.080000 |
| 3 | 1.259712 | 0.793832 | 2.577097 | 3.246400 |
| 4 | 1.360489 | 0.735030 | 3.312127 | 4.506112 |
| 5 | 1.469328 | 0.680583 | 3.992710 | 5.866601 |
| 6 | 1.586874 | 0.630170 | 4.622880 | 7.335929 |
| 7 | 1.713824 | 0.583490 | 5.206370 | 8.922803 |
| 8 | 1.850930 | 0.540269 | 5.746639 | 10.636628 |
| 9 | 1.999005 | 0.500249 | 6.246888 | 12.487558 |
| 10 | 2.158925 | 0.463193 | 6.710081 | 14.486562 |
| 11 | 2.331639 | 0.428883 | 7.138964 | 16.645487 |
| 12 | 2.518170 | 0.397114 | 7.536078 | 18.977126 |
| 13 | 2.719624 | 0.367698 | 7.903776 | 21.495297 |
| 14 | 2.937194 | 0.340461 | 8.244237 | 24.214920 |
| 15 | 3.172169 | 0.315242 | 8.559479 | 27.152114 |
| 16 | 3.425943 | 0.291890 | 8.851369 | 30.324283 |
| 17 | 3.700018 | 0.270269 | 9.121638 | 33.750226 |
| 18 | 3.996019 | 0.250249 | 9.371887 | 37.450244 |
| 19 | 4.315701 | 0.231712 | 9.603599 | 41.446263 |
| 20 | 4.660957 | 0.214548 | 9.818147 | 45.761964 |
| 21 | 5.033834 | 0.198656 | 10.016803 | 50.422921 |
| 22 | 5.436540 | 0.183941 | 10.200744 | 55.456755 |
| 23 | 5.871464 | 0.170315 | 10.371059 | 60.893296 |
| 24 | 6.341181 | 0.157699 | 10.528758 | 66.764759 |
| 25 | 6.848475 | 0.146018 | 10.674776 | 73.105940 |
| 26 | 7.396353 | 0.135202 | 10.809978 | 79.954415 |
| 27 | 7.988061 | 0.125187 | 10.935165 | 87.350768 |
| 28 | 8.627106 | 0.115914 | 11.051078 | 95.338830 |
| 29 | 9.317275 | 0.107328 | 11.158406 | 103.965936 |
| 30 | 10.062657 | 0.099377 | 11.257783 | 113.283211 |
| 31 | 10.867669 | 0.092016 | 11.349799 | 123.345868 |
| 32 | 11.737083 | 0.085200 | 11.434999 | 134.213537 |
| 33 | 12.676050 | 0.078889 | 11.513888 | 145.950620 |
| 34 | 13.690134 | 0.073045 | 11.586934 | 158.626670 |
| 35 | 14.785344 | 0.067635 | 11.654568 | 172.316804 |
| 36 | 15.968172 | 0.062625 | 11.717193 | 187.102148 |
| 37 | 17.245626 | 0.057986 | 11.775179 | 203.070320 |
| 38 | 18.625276 | 0.053690 | 11.828869 | 220.315945 |
| 39 | 20.115298 | 0.049713 | 11.878582 | 238.941221 |
| 40 | 21.724521 | 0.046031 | 11.924613 | 259.056519 |
| 41 | 23.462483 | 0.042621 | 11.967235 | 280.781040 |
| 42 | 25.339482 | 0.039464 | 12.006699 | 304.243523 |
| 43 | 27.366640 | 0.036541 | 12.043240 | 329.583005 |
| 44 | 29.555972 | 0.033834 | 12.077074 | 356.949646 |
| 45 | 31.920449 | 0.031328 | 12.108402 | 386.505617 |
| 46 | 34.474085 | 0.029007 | 12.137409 | 418.426067 |
| 47 | 37.232012 | 0.026859 | 12.164267 | 452.900152 |
| 48 | 40.210573 | 0.024869 | 12.189136 | 490.132164 |
| 49 | 43.427419 | 0.023027 | 12.212163 | 530.342737 |
| 50 | 46.901613 | 0.021321 | 12.233485 | 573.770156 |

# Appendix C  Table of Selected Integrals

### Rational Forms Containing (a + bu)

1. $\displaystyle \int u^n \, du = \frac{u^{n+1}}{n+1} + C, \quad n \neq -1.$

2. $\displaystyle \int \frac{du}{a+bu} = \frac{1}{b} \ln |a+bu| + C.$

3. $\displaystyle \int \frac{u \, du}{a+bu} = \frac{u}{b} - \frac{a}{b^2} \ln |a+bu| + C.$

4. $\displaystyle \int \frac{u^2 \, du}{a+bu} = \frac{u^2}{2b} - \frac{au}{b^2} + \frac{a^2}{b^3} \ln |a+bu| + C.$

5. $\displaystyle \int \frac{du}{u(a+bu)} = \frac{1}{a} \ln \left| \frac{u}{a+bu} \right| + C.$

6. $\displaystyle \int \frac{du}{u^2(a+bu)} = -\frac{1}{au} + \frac{b}{a^2} \ln \left| \frac{a+bu}{u} \right| + C.$

7. $\displaystyle \int \frac{u \, du}{(a+bu)^2} = \frac{1}{b^2} \left( \ln |a+bu| + \frac{a}{a+bu} \right) + C.$

8. $\displaystyle \int \frac{u^2 \, du}{(a+bu)^2} = \frac{u}{b^2} - \frac{a^2}{b^3(a+bu)} - \frac{2a}{b^3} \ln |a+bu| + C.$

9. $\displaystyle \int \frac{du}{u(a+bu)^2} = \frac{1}{a(a+bu)} + \frac{1}{a^2} \ln \left| \frac{u}{a+bu} \right| + C.$

10. $\displaystyle \int \frac{du}{u^2(a+bu)^2} = -\frac{a+2bu}{a^2 u(a+bu)} + \frac{2b}{a^3} \ln \left| \frac{a+bu}{u} \right| + C.$

**11.** $\displaystyle\int \frac{du}{(a + bu)(c + ku)} = \frac{1}{bc - ak} \ln\left|\frac{a + bu}{c + ku}\right| + C.$

**12.** $\displaystyle\int \frac{u \, du}{(a + bu)(c + ku)} = \frac{1}{bc - ak}\left[\frac{c}{k} \ln|c + ku| - \frac{a}{b} \ln|a + bu|\right] + C.$

## Forms Containing $\sqrt{a + bu}$

**13.** $\displaystyle\int u\sqrt{a + bu} \, du = \frac{2(3bu - 2a)(a + bu)^{3/2}}{15b^2} + C.$

**14.** $\displaystyle\int u^2\sqrt{a + bu} \, du = \frac{2(8a^2 - 12abu + 15b^2u^2)(a + bu)^{3/2}}{105b^3} + C.$

**15.** $\displaystyle\int \frac{u \, du}{\sqrt{a + bu}} = \frac{2(bu - 2a)\sqrt{a + bu}}{3b^2} + C.$

**16.** $\displaystyle\int \frac{u^2 \, du}{\sqrt{a + bu}} = \frac{2(3b^2u^2 - 4abu + 8a^2)\sqrt{a + bu}}{15b^3} + C.$

**17.** $\displaystyle\int \frac{du}{u\sqrt{a + bu}} = \frac{1}{\sqrt{a}} \ln\left|\frac{\sqrt{a + bu} - \sqrt{a}}{\sqrt{a + bu} + \sqrt{a}}\right| + C, \quad a > 0.$

**18.** $\displaystyle\int \frac{\sqrt{a + bu} \, du}{u} = 2\sqrt{a + bu} + a\int \frac{du}{u\sqrt{a + bu}}.$

## Forms Containing $\sqrt{a^2 - u^2}$

**19.** $\displaystyle\int \frac{du}{(a^2 - u^2)^{3/2}} = \frac{u}{a^2\sqrt{a^2 - u^2}} + C.$

**20.** $\displaystyle\int \frac{du}{u\sqrt{a^2 - u^2}} = -\frac{1}{a} \ln\left|\frac{a + \sqrt{a^2 - u^2}}{u}\right| + C.$

**21.** $\displaystyle\int \frac{du}{u^2\sqrt{a^2 - u^2}} = -\frac{\sqrt{a^2 - u^2}}{a^2u} + C.$

**22.** $\displaystyle\int \frac{\sqrt{a^2 - u^2} \, du}{u} = \sqrt{a^2 - u^2} - a \ln\left|\frac{a + \sqrt{a^2 - u^2}}{u}\right| + C, \quad a > 0.$

## Forms Containing $\sqrt{u^2 \pm a^2}$

**23.** $\displaystyle\int \sqrt{u^2 \pm a^2} \, du = \frac{1}{2}\left(u\sqrt{u^2 \pm a^2} \pm a^2 \ln|u + \sqrt{u^2 \pm a^2}|\right) + C.$

**24.** $\displaystyle\int u^2\sqrt{u^2 \pm a^2} \, du = \frac{u}{8}(2u^2 \pm a^2)\sqrt{u^2 \pm a^2} - \frac{a^4}{8} \ln|u + \sqrt{u^2 \pm a^2}| + C.$

25. $\displaystyle\int \frac{\sqrt{u^2 + a^2}\,du}{u} = \sqrt{u^2 + a^2} - a\ln\left|\frac{a + \sqrt{u^2 + a^2}}{u}\right| + C.$

26. $\displaystyle\int \frac{\sqrt{u^2 \pm a^2}\,du}{u^2} = -\frac{\sqrt{u^2 \pm a^2}}{u} + \ln\left|u + \sqrt{u^2 \pm a^2}\right| + C.$

27. $\displaystyle\int \frac{du}{\sqrt{u^2 \pm a^2}} = \ln\left|u + \sqrt{u^2 \pm a^2}\right| + C.$

28. $\displaystyle\int \frac{du}{u\sqrt{u^2 + a^2}} = \frac{1}{a}\ln\left|\frac{\sqrt{u^2 + a^2} - a}{u}\right| + C.$

29. $\displaystyle\int \frac{u^2\,du}{\sqrt{u^2 \pm a^2}} = \frac{1}{2}\left(u\sqrt{u^2 \pm a^2} \mp a^2\ln\left|u + \sqrt{u^2 \pm a^2}\right|\right) + C.$

30. $\displaystyle\int \frac{du}{u^2\sqrt{u^2 \pm a^2}} = -\frac{\pm\sqrt{u^2 \pm a^2}}{a^2 u} + C.$

31. $\displaystyle\int (u^2 \pm a^2)^{3/2}\,du = \frac{u}{8}(2u^2 \pm 5a^2)\sqrt{u^2 \pm a^2} + \frac{3a^4}{8}\ln\left|u + \sqrt{u^2 \pm a^2}\right| + C.$

32. $\displaystyle\int \frac{du}{(u^2 \pm a^2)^{3/2}} = \frac{\pm u}{a^2\sqrt{u^2 \pm a^2}} + C.$

33. $\displaystyle\int \frac{u^2\,du}{(u^2 \pm a^2)^{3/2}} = \frac{-u}{\sqrt{u^2 \pm a^2}} + \ln\left|u + \sqrt{u^2 \pm a^2}\right| + C.$

## Rational Forms Containing $a^2 - u^2$ and $u^2 - a^2$

34. $\displaystyle\int \frac{du}{a^2 - u^2} = \frac{1}{2a}\ln\left|\frac{a + u}{a - u}\right| + C.$

35. $\displaystyle\int \frac{du}{u^2 - a^2} = \frac{1}{2a}\ln\left|\frac{u - a}{u + a}\right| + C.$

## Exponential and Logarithmic Forms

36. $\displaystyle\int e^u\,du = e^u + C.$

37. $\displaystyle\int a^u\,du = \frac{a^u}{\ln a} + C, \quad a > 0, \quad a \ne 1.$

38. $\displaystyle\int ue^{au}\,du = \frac{e^{au}}{a^2}(au - 1) + C.$

39. $\displaystyle\int u^n e^{au}\,du = \frac{u^n e^{au}}{a} - \frac{n}{a}\int u^{n-1}e^{au}\,du.$

**40.** $\displaystyle\int \frac{e^{au}\,du}{u^n} = -\frac{e^{au}}{(n-1)u^{n-1}} + \frac{a}{n-1}\int \frac{e^{au}\,du}{u^{n-1}}.$

**41.** $\displaystyle\int \ln u\,du = u\ln u - u + C.$

**42.** $\displaystyle\int u^n \ln u\,du = \frac{u^{n+1}\ln u}{n+1} - \frac{u^{n+1}}{(n+1)^2} + C, \quad n \neq -1.$

**43.** $\displaystyle\int u^n \ln^m u\,du = \frac{u^{n+1}}{n+1}\ln^m u - \frac{m}{n+1}\int u^n \ln^{m-1} u\,du, \quad m, n \neq -1.$

**44.** $\displaystyle\int \frac{du}{u\ln u} = \ln|\ln u| + C.$

**45.** $\displaystyle\int \frac{du}{a + be^{cu}} = \frac{1}{ac}(cu - \ln|a + be^{cu}|) + C.$

## Miscellaneous Forms

**46.** $\displaystyle\int \sqrt{\frac{a+u}{b+u}}\,du = \sqrt{(a+u)(b+u)} + (a-b)\ln\left(\sqrt{a+u} + \sqrt{b+u}\right) + C.$

**47.** $\displaystyle\int \frac{du}{\sqrt{(a+u)(b+u)}} = \ln\left|\frac{a+b}{2} + u + \sqrt{(a+u)(b+u)}\right| + C.$

**48.** $\displaystyle\int \sqrt{a + bu + cu^2}\,du = \frac{2cu + b}{4c}\sqrt{a + bu + cu^2} -$

$\displaystyle\qquad\qquad \frac{b^2 - 4ac}{8c^{3/2}}\ln\left|2cu + b + 2\sqrt{c}\sqrt{a + bu + cu^2}\right| + C, \quad c > 0.$

# Appendix D   Areas Under the Standard Normal Curve

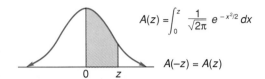

$$A(z) = \int_0^z \frac{1}{\sqrt{2\pi}}\, e^{-x^2/2}\, dx$$

$$A(-z) = A(z)$$

| z | .00 | .01 | .02 | .03 | .04 | .05 | .06 | .07 | .08 | .09 |
|---|-----|-----|-----|-----|-----|-----|-----|-----|-----|-----|
| 0.0 | .0000 | .0040 | .0080 | .0120 | .0160 | .0199 | .0239 | .0279 | .0319 | .0359 |
| 0.1 | .0398 | .0438 | .0478 | .0517 | .0557 | .0596 | .0636 | .0675 | .0714 | .0754 |
| 0.2 | .0793 | .0832 | .0871 | .0910 | .0948 | .0987 | .1026 | .1064 | .1103 | .1141 |
| 0.3 | .1179 | .1217 | .1255 | .1293 | .1331 | .1368 | .1406 | .1443 | .1480 | .1517 |
| 0.4 | .1554 | .1591 | .1628 | .1664 | .1700 | .1736 | .1772 | .1808 | .1844 | .1879 |
| 0.5 | .1915 | .1950 | .1985 | .2019 | .2054 | .2088 | .2123 | .2157 | .2190 | .2224 |
| 0.6 | .2258 | .2291 | .2324 | .2357 | .2389 | .2422 | .2454 | .2486 | .2518 | .2549 |
| 0.7 | .2580 | .2612 | .2642 | .2673 | .2704 | .2734 | .2764 | .2794 | .2823 | .2852 |
| 0.8 | .2881 | .2910 | .2939 | .2967 | .2996 | .3023 | .3051 | .3078 | .3106 | .3133 |
| 0.9 | .3159 | .3186 | .3212 | .3238 | .3264 | .3289 | .3315 | .3340 | .3365 | .3389 |
| 1.0 | .3413 | .3438 | .3461 | .3485 | .3508 | .3531 | .3554 | .3577 | .3599 | .3621 |
| 1.1 | .3643 | .3665 | .3686 | .3708 | .3729 | .3749 | .3770 | .3790 | .3810 | .3820 |
| 1.2 | .3849 | .3869 | .3888 | .3907 | .3925 | .3944 | .3962 | .3980 | .3997 | .4015 |
| 1.3 | .4032 | .4049 | .4066 | .4082 | .4099 | .4115 | .4131 | .4147 | .4162 | .4177 |
| 1.4 | .4192 | .4207 | .4222 | .4236 | .4251 | .4265 | .4279 | .4292 | .4306 | .4319 |
| 1.5 | .4332 | .4345 | .4357 | .4370 | .4382 | .4394 | .4406 | .4418 | .4429 | .4441 |
| 1.6 | .4452 | .4463 | .4474 | .4484 | .4495 | .4504 | .4515 | .4525 | .4535 | .4545 |
| 1.7 | .4554 | .4564 | .4573 | .4582 | .4591 | .4599 | .4608 | .4616 | .4625 | .4633 |
| 1.8 | .4641 | .4649 | .4656 | .4664 | .4671 | .4678 | .4686 | .4693 | .4699 | .4706 |
| 1.9 | .4713 | .4719 | .4726 | .4732 | .4738 | .4744 | .4750 | .4756 | .4761 | .4767 |

| z | .00 | .01 | .02 | .03 | .04 | .05 | .06 | .07 | .08 | .09 |
|---|-----|-----|-----|-----|-----|-----|-----|-----|-----|-----|
| 2.0 | .4772 | .4778 | .4783 | .4788 | .4793 | .4798 | .4803 | .4808 | .4812 | .4817 |
| 2.1 | .4821 | .4826 | .4830 | .4834 | .4838 | .4842 | .4846 | .4850 | .4854 | .4857 |
| 2.2 | .4861 | .4864 | .4868 | .4871 | .4875 | .4878 | .4881 | .4884 | .4887 | .4890 |
| 2.3 | .4893 | .4896 | .4898 | .4901 | .4904 | .4906 | .4909 | .4911 | .4913 | .4916 |
| 2.4 | .4918 | .4920 | .4922 | .4925 | .4927 | .4929 | .4931 | .4932 | .4934 | .4936 |
| 2.5 | .4938 | .4940 | .4941 | .4943 | .4945 | .4946 | .4948 | .4949 | .4951 | .4952 |
| 2.6 | .4953 | .4955 | .4956 | .4957 | .4959 | .4960 | .4961 | .4962 | .4963 | .4964 |
| 2.7 | .4965 | .4966 | .4967 | .4968 | .4969 | .4970 | .4971 | .4972 | .4973 | .4974 |
| 2.8 | .4974 | .4975 | .4976 | .4977 | .4977 | .4978 | .4979 | .4979 | .4980 | .4981 |
| 2.9 | .4981 | .4982 | .4982 | .4983 | .4984 | .4984 | .4985 | .4985 | .4986 | .4986 |
| 3.0 | .4987 | .4987 | .4987 | .4988 | .4988 | .4989 | .4989 | .4989 | .4990 | .4990 |
| 3.1 | .4990 | .4991 | .4991 | .4991 | .4992 | .4992 | .4992 | .4992 | .4993 | .4993 |
| 3.2 | .4993 | .4993 | .4994 | .4994 | .4994 | .4994 | .4994 | .4995 | .4995 | .4995 |
| 3.3 | .4995 | .4995 | .4995 | .4996 | .4996 | .4996 | .4996 | .4996 | .4996 | .4997 |
| 3.4 | .4997 | .4997 | .4997 | .4997 | .4997 | .4997 | .4997 | .4997 | .4997 | .4998 |
| 3.5 | .4998 | .4998 | .4998 | .4998 | .4998 | .4998 | .4998 | .4998 | .4998 | .4998 |

# Answers to Odd-Numbered Problems

## EXERCISE 0.2 (page 3)

**1.** True.  **3.** False; the natural numbers are $1, 2, 3$, and so on.  **5.** True.  **7.** False; $\frac{4}{2} = 2$, a positive integer.
**9.** True.  **11.** True.

## EXERCISE 0.3 (page 6)

**1.** False.  **3.** False.  **5.** True.  **7.** True.  **9.** False.
**11.** Distributive.  **13.** Associaive.  **15.** Commutative.
**17.** Definition of subtraction.  **19.** Distributive.

## EXERCISE 0.4 (page 9)

**1.** $-6$.  **3.** $2$.  **5.** $11$.  **7.** $-2$.  **9.** $-63$.
**11.** $-6$.  **13.** $6 - x$.

**15.** $-12x + 12y$ (or $12y - 12x$).  **17.** $-\frac{1}{3}$.  **19.** $-2$.

**21.** $18$.  **23.** $25$.  **25.** $3x - 12$.  **27.** $-x + 2$.

**29.** $\frac{8}{11}$.  **31.** $-\frac{5x}{7y}$.  **33.** $\frac{2}{3x}$.  **35.** $3$.  **37.** $\frac{7}{xy}$.

**39.** $\frac{5}{6}$.  **41.** $-\frac{1}{6}$.  **43.** $\frac{x - y}{9}$.  **45.** $\frac{1}{24}$.

**47.** $\frac{x}{6y}$.  **49.** Not defined.  **51.** Not defined.

## EXERCISE 0.5 (page 14)

**1.** $2^5 (= 32)$.  **3.** $w^{12}$.  **5.** $\frac{x^8}{x^{17}}$.  **7.** $\frac{x^{10}}{y^{50}}$.

**9.** $8x^6 y^9$.  **11.** $x^6$.  **13.** $x^{14}$.  **15.** $5$.  **17.** $-2$.

**19.** $\frac{1}{2}$.  **21.** $10$.  **23.** $8$.  **25.** $\frac{1}{4}$.  **27.** $\frac{1}{32}$.

**29.** $4\sqrt{2}$.  **31.** $x\sqrt[3]{2}$.  **33.** $4x^2$.  **35.** $-2\sqrt{3} + 4\sqrt[3]{2}$.

**37.** $3z^2$.  **39.** $\frac{9t^2}{4}$.  **41.** $\frac{x^3}{y^2 z^2}$.  **43.** $\frac{2}{x^4}$.  **45.** $\frac{1}{9t^2}$.

**47.** $7^{1/3} s^{2/3}$.  **49.** $x^{1/2} - y^{1/2}$.  **51.** $\frac{x^{9/4} z^{3/4}}{y^{1/2}}$.

**53.** $\sqrt[5]{(8x - y)^4}$.  **55.** $\frac{1}{\sqrt[5]{x^4}}$.  **57.** $\frac{2}{\sqrt[5]{x^4}}$.

**59.** $\frac{3\sqrt{7}}{7}$.  **61.** $\frac{2\sqrt{2x}}{x}$.  **63.** $\frac{\sqrt[3]{9x^2}}{3x}$.  **65.** $4$.

**67.** $\frac{\sqrt[12]{8x^8 y^4}}{xy}$.  **69.** $\frac{2x^6}{y^3}$.  **71.** $t^{2/3}$.  **73.** $\frac{64y^6 x^{1/2}}{x^2}$.

**75.** $xyz$.  **77.** $\frac{1}{9}$.  **79.** $\frac{4y^4}{x^2}$.  **81.** $x^2 y^{5/2}$.  **83.** $\frac{y^{10}}{z^2}$.

**85.** $x^8$.  **87.** $-\frac{4}{s^5}$.  **89.** $\frac{4x^4 z^4}{9y^4}$.

## EXERCISE 0.6 (page 20)

**1.** $11x - 2y - 3$.  **3.** $6t^2 - 2s^2 + 6$.
**5.** $2\sqrt{x} + \sqrt{2y} + \sqrt{3z}$.
**7.** $6x^2 - 9xy - 2z + \sqrt{2} - 4$.
**9.** $\sqrt{2y} - \sqrt{3z}$.  **11.** $-7x + 14y - 19$.
**13.** $x^2 + 9y^2 + xy$.  **15.** $6x^2 + 96$.
**17.** $-6x^2 - 18x - 18$.  **19.** $x^2 + 9x + 20$.
**21.** $x^2 + x - 6$.  **23.** $10x^2 + 19x + 6$.
**25.** $x^2 + 6x + 9$.  **27.** $x^2 - 10x + 25$.
**29.** $2y + 6\sqrt{2}y + 9$.  **31.** $4s - 1$.
**33.** $x^3 + 4x^2 - 3x - 12$.
**35.** $2x^4 + 2x^3 - 5x^2 - 2x + 3$.  **37.** $5x^3 + 5x^2 + 6x$.
**39.** $3x^2 + 2y^2 + 5xy + 2x - 8$.
**41.** $x^3 + 15x^2 + 75x + 125$.
**43.** $8x^3 - 36x^2 + 54x - 27$.  **45.** $z - 4$
**47.** $3x^3 + 2x - \frac{1}{2x^2}$.  **49.** $x + \frac{-1}{x + 3}$.

**51.** $3x^2 - 8x + 17 + \frac{-37}{x + 2}$.

**53.** $t + 8 + \frac{64}{t - 8}$.  **55.** $x - 2 + \frac{7}{3x + 2}$.

## EXERCISE 0.7 (page 23)

**1.** $2(3x + 2)$.  **3.** $5x(2y + z)$.
**5.** $4bc(2a^3 - 3ab^2 d + b^3 cd^2)$.  **7.** $(x + 5)(x - 5)$.
**9.** $(p + 3)(p + 1)$.  **11.** $(4x + 3)(4x - 3)$.
**13.** $(z + 4)(z + 2)$.  **15.** $(x + 3)^2$.
**17.** $2(x + 4)(x + 2)$.  **19.** $3(x - 1)(x + 1)$.
**21.** $(6y + 1)(y + 2)$.  **23.** $2s(3s + 4)(2s - 1)$.
**25.** $x^{2/3} y(1 + 2xy)(1 - 2xy)$.  **27.** $2x(x + 3)(x - 2)$.
**29.** $4(2x + 1)^2$.  **31.** $x(xy - 5)^2$.
**33.** $(x - 2)^2 (x + 2)$.  **35.** $(y + 4)^2 (y + 1)(y - 1)$.
**37.** $(x + 2)(x^2 - 2x + 4)$.
**39.** $(x + 1)(x^2 - x + 1)(x - 1)(x^2 + x + 1)$.
**41.** $2(x + 3)^2 (x + 1)(x - 1)$.  **43.** $P(1 + r)^2$
**45.** $(x^2 + 4)(x + 2)(x - 2)$.
**47.** $(y^4 + 1)(y^2 + 1)(y + 1)(y - 1)$.
**49.** $(x^2 + 2)(x + 1)(x - 1)$.  **51.** $x(x + 1)^2 (x - 1)^2$.

## EXERCISE 0.8 (page 28)

1. $\dfrac{x+2}{x}$.  3. $\dfrac{x-5}{x+5}$.  5. $\dfrac{3x+2}{x+2}$.

7. $-\dfrac{y^2}{(y-3)(y+2)}$.  9. $\dfrac{3-2x}{3+2x}$.

11. $\dfrac{2(x+4)}{(x-4)(x+2)}$.  13. $\dfrac{x}{2}$.  15. $\dfrac{1}{2n}$.  17. $\dfrac{2}{3}$.

19. $-27x^2$.  21. 1.  23. $\dfrac{2x^2}{x-1}$.  25. 1.

27. $-\dfrac{(2x+3)(1+x)}{x+4}$.  29. $x+2$.  31. $\dfrac{5}{3t}$.

33. $\dfrac{1}{1-p^2}$.  35. $\dfrac{2x^2+3x+12}{(2x-1)(x+3)}$.

37. $\dfrac{2x-3}{(x-2)(x+1)(x-1)}$.  39. $\dfrac{35-8x}{(x-1)(x+5)}$.

41. $\dfrac{x^2+2x+1}{x^2}$.  43. $\dfrac{x}{1-xy}$.  45. $\dfrac{x+1}{3x}$.

47. $\dfrac{(x+2)(6x-1)}{2x^2(x+3)}$.  49. $\dfrac{2\sqrt{x}-2\sqrt{x+h}}{\sqrt{x}\sqrt{x+h}}$.

51. $2-\sqrt{3}$.

53. $-\dfrac{\sqrt{6}+2\sqrt{3}}{3}$.  55. $-4-2\sqrt{6}$.  57. $\dfrac{x-\sqrt{5}}{x^2-5}$.

59. $\dfrac{5\sqrt{3}-4\sqrt{2}-13}{2}$.

## PRINCIPLES IN PRACTICE 1.1

1. $P=2(w+2)+2w=2w+4+2w$

$=4w+4$.  2. 200 speciality coffees.

3. 46 weeks; \$1715.  4. $r=\dfrac{d}{t}$.  5. $\sqrt{\dfrac{S}{\pi}}$.

## EXERCISE 1.1 (page 36)

1. 0.  3. $\dfrac{10}{3}$.  5. 0.  7. Adding 5 to both sides; equivalence guaranteed.  9. Squaring both sides; equivalence *not* guaranteed.  11. Dividing both sides by $x$; equivalence *not* guaranteed.  13. Multiplying both sides by $x-1$; equivalence *not* guaranteed.  15. Multiplying both sides by $(x-5)/x$; equivalence *not* guaranteed.

17. $\dfrac{5}{2}$.  19. 0.  21. 1.  23. $\dfrac{12}{5}$.  25. $-1$.

27. 2.  29. $\dfrac{10}{3}$.  31. 90.  33. 8.  35. $-\dfrac{26}{9}$.

37. $-\dfrac{37}{18}$.  39. $\dfrac{60}{17}$.  41. $\dfrac{14}{3}$.  43. 3.  45. $\dfrac{7}{8}$.

47. $P=\dfrac{I}{rt}$.  49. $q=\dfrac{p+1}{8}$.  51. $r=\dfrac{S-P}{Pt}$.

53. $a_1=\dfrac{2S-na_n}{n}$.  55. 120 m.

57. $c=x+0.0825x=1.0825x$.  59. 3.

61. 29 hours.  63. 0.00001.  65. $\dfrac{1}{8}, -\dfrac{1}{14}$.  67. $\dfrac{14}{61}$.

## PRINCIPLES IN PRACTICE 1.2

1. $\dfrac{10}{r+2}=\dfrac{6}{r-2}$; 8 mph.  2. $t=\dfrac{d}{r+w}$; $w=\dfrac{d}{t}-r$.

3. $\sqrt{x^2+16}-x=2$; $x=3$; ramp is 5 feet.

## EXERCISE 1.2 (page 41)

1. $\dfrac{1}{5}$.  3. $\varnothing$.  5. $\dfrac{8}{3}$.  7. $\dfrac{3}{2}$.  9. 0.  11. $\dfrac{5}{3}$.

13. $\dfrac{1}{8}$.  15. 3.  17. $\dfrac{5}{13}$.  19. $\varnothing$.  21. 3.

23. $\dfrac{262}{5}$.  25. $-\dfrac{10}{9}$.  27. 2.  29. 7.  31. $\dfrac{49}{36}$.

33. $-\dfrac{9}{4}$.  35. $d=\dfrac{r}{1+rt}$.  37. $n=\dfrac{2mI}{rB}-1$.

39. 10.  41. $t=\dfrac{d}{r-c}$; $r=\dfrac{d}{t}+c$.

43. Jill is 36 feet above the surface. Jack is 121 feet above the surface.

## PRINCIPLES IN PRACTICE 1.3

1. The number is $-5$ or 6.  2. 50 feet by 60 feet.
3. $1\times1\times5$.  4. 15 items at \$15 per item.
5. 2.5 seconds and 7.5 seconds.  6. \$100
7. Never, they only go up to 400 feet.

## EXERCISE 1.3 (page 47)

1. 2.  3. 4, 3.  5. 3, $-1$.  7. 6.  9. $\pm2$.

11. 0, 8.  13. $\dfrac{1}{2}$.  15. $1, -\dfrac{5}{2}$.  17. 5, $-2$.

19. $0, \dfrac{3}{2}$.  21. 0, 1, $-2$.  23. 0, $\pm8$.  25. $0, \dfrac{1}{2}, -\dfrac{4}{3}$.

27. $-3, -1, 2$.  29. 3, 4.  31. 4, $-6$.  33. $\dfrac{3}{2}$.

35. $\dfrac{5\pm\sqrt{13}}{2}$.  37. No real roots.  39. $\dfrac{1}{2}, -\dfrac{5}{3}$.

41. $4, -\dfrac{5}{2}$.  43. $\dfrac{-2\pm\sqrt{14}}{2}$.  45. $\pm\sqrt{3}, \pm\sqrt{2}$.

47. $2, -\dfrac{1}{2}$.  49. $\pm\dfrac{\sqrt{7}}{7}, \pm\dfrac{\sqrt{2}}{2}$.  51. $-4, 1$.

53. $\dfrac{15}{7}, \dfrac{11}{5}$.  55. $\dfrac{3}{2}, -1$.  57. 6, $-2$.  61. 5, $-2$.

63. $\dfrac{3}{2}$.  65. $-2$.  67. 7, 2.  69. 4, 8.  71. 2.

73. 0, 4.  75. 4.  77. 233.16, 1.84.
79. 6 inches by 8 inches.  83. 1 year and 10 years.
85. 86.8 cm or 33.2 cm.  87. **a.** 9 s, **b.** 3 s or 6 s.
89. 1.5, 0.75.  91. No real root.  93. 1.999, 0.963.

## REVIEW PROBLEMS—CHAPTER 1 (page 50)

**1.** $\frac{1}{4}$. **3.** $-\frac{2}{15}$. **5.** $-\frac{1}{2}$. **7.** $\varnothing$. **9.** $\frac{5}{2}$.

**11.** $\frac{1}{3}$. **13.** $-\frac{9}{7}$. **15.** $-\frac{5}{3}, 1$. **17.** $0, \frac{7}{5}$. **19.** 5.

**21.** $\pm\frac{\sqrt{15}}{3}$. **23.** $\frac{5}{8}, -3$. **25.** $\frac{5 \pm \sqrt{13}}{6}$. **27.** $4, \pm 3$.

**29.** $\frac{1}{2}$. **31.** $\frac{4 \pm \sqrt{13}}{3}$. **33.** 10. **35.** 5.

**37.** No solution. **39.** 10. **41.** $4, 8$. **43.** $-8, 1$.

**45.** $Q = \frac{EA}{4\pi k}$. **47.** $C' = \lambda^2(n - 1 - C)$.

**49.** $T = \pm 2\pi\sqrt{\frac{L}{g}}$. **51.** $\omega = \pm\sqrt{\frac{2mgh - mv^2}{I}}$.

**55.** $6, \frac{5}{4}$. **57.** $-0.709, 0.336$.

## MATHEMATICAL SNAPSHOT—CHAPTER 1 (page 53)

**1. a.** \$111.40; **b.** \$1.14; **c.** 100 lb; **d.** 98.15 lb; **e.** $-1.85\%$
**2.** 4.76%.

## EXERCISE 2.1 (page 59)

**1.** 120. **3.** 48 of $A$, 80 of $B$. **5.** $5\frac{1}{3}$. **7.** 1 m.

**9.** 181,250. **11.** \$4000 at 6%, \$16,000 at $7\frac{1}{2}\%$.

**13.** \$4.25. **15.** 4%. **17.** 80. **19.** \$8000.
**21.** 1600. **23.** \$116.25. **25.** 40. **27.** 46,000.
**29.** Either \$440 or \$460. **31.** \$100. **33.** 77.
**35.** 80 ft by 140 ft. **37.** 9 cm long, 4 cm wide.
**39.** \$112,000. **41.** 60. **43.** Either 125 units of $A$ and 100 units of $B$, or 150 units of $A$ and 125 units of $B$.

## PRINCIPLES IN PRACTICE 2.2

**1.** 5375.
**2.** $150 - x_4 \geq 0; 3x_4 - 210 \geq 0; x_4 + 60 \geq 0; x_4 \geq 0$.

## EXERCISE 2.2 (page 67)

**1.** $(4, \infty)$. **3.** $(-\infty, 5]$. **5.** $\left(-\infty, -\frac{1}{2}\right]$.

**7.** $\left(-\infty, -\frac{2}{5}\right)$. **9.** $(0, \infty)$. **11.** $\left[-\frac{7}{5}, \infty\right)$.

**13.** $\left(-\frac{2}{7}, \infty\right)$. **15.** $\varnothing$. **17.** $\left(-\infty, \frac{\sqrt{3} - 2}{2}\right)$.

**19.** $(-\infty, 6)$. **21.** $(-\infty, -5]$. **23.** $(-\infty, \infty)$.

**25.** $\left(\frac{17}{9}, \infty\right)$. **27.** $\left[-\frac{14}{3}, \infty\right)$. **29.** $(0, \infty)$.

**31.** $(-\infty, 0)$. **33.** $(-\infty, -2]$.

**35.** $444,000 < S < 636,000$. **37.** $x < 70$ degrees.

## EXERCISE 2.3 (page 70)

**1.** 120,001. **3.** 22,500. **5.** 60,000.
**7.** \$25,714.29. **9.** 1000. **11.** $t > 36.5$.
**13.** \$4.50, \$1160.

## PRINCIPLES IN PRACTICE 2.4

**1.** $|w - 22| \leq 0.3$ oz.

## EXERCISE 2.4 (page 75)

**1.** 13. **3.** 6. **5.** 5. **7.** $-3 < x < 3$.
**9.** $\sqrt{5} - 2$.
**11. a.** $|x - 7| < 3$, **b.** $|x - 2| < 3$,
**c.** $|x - 7| \leq 5$, **d.** $|x - 7| = 4$, **e.** $|x + 4| < 2$,
**f.** $|x| < 3$, **g.** $|x| > 6$, **h.** $|x - 6| > 4$, **i.** $|x - 105| < 3$,
**j.** $|x - 850| < 100$.
**13.** $|p_1 - p_2| \leq 2$. **15.** $\pm 7$. **17.** $\pm 6$. **19.** $13, -3$.
**21.** $\frac{2}{5}$. **23.** $\frac{1}{2}, 3$. **25.** $(-4, 4)$.
**27.** $(-\infty, -8) \cup (8, \infty)$. **29.** $(-9, -5)$.
**31.** $(-\infty, 0) \cup (1, \infty)$. **33.** $[2, 3]$.
**35.** $(-\infty, 0] \cup \left[\frac{16}{3}, \infty\right)$. **37.** $|d - 10| \leq 0.01$ m
**39.** $(-\infty, \mu - h\sigma) \cup (\mu + h\sigma, \infty)$.

## REVIEW PROBLEMS—CHAPTER 2 (page 76)

**1.** $(-\infty, 0]$. **3.** $\left(\frac{2}{3}, \infty\right)$. **5.** $\varnothing$.

**7.** $\left(-\infty, \frac{13}{2}\right]$. **9.** $(-\infty, \infty)$. **11.** $-2, 5$.

**13.** $\left(0, \frac{1}{2}\right)$. **15.** $\left(-\infty, -\frac{1}{2}\right] \cup \left[\frac{7}{2}, \infty\right)$.

**17.** 542. **19.** 6000.

## MATHEMATICAL SNAPSHOT—CHAPTER 2 (page 79)

**1.** 1 hour.    **3.** 1 hour.    **5.** 1 hour and 12 minutes.

## PRINCIPLES IN PRACTICE 3.1

**1. a.** $a(r) = \pi r^2$;  **b.** all real numbers;  **c.** $r \geq 0$.

**2. a.** $t(r) = \dfrac{300}{r}$;  **b.** all real numbers except 0;  **c.** $r > 0$;

**d.** $t(x) = \dfrac{300}{x}$; $t\left(\dfrac{x}{2}\right) = \dfrac{600}{x}$; $t\left(\dfrac{x}{4}\right) = \dfrac{1200}{x}$;

**e.** The time is scaled by a factor $c$; $t\left(\dfrac{x}{c}\right) = \dfrac{300c}{x}$.

**3. a.** 300 pizzas;  **b.** \$21.00 per pizza;  **c.** \$16.00 per pizza.

## EXERCISE 3.1 (page 86)

**1.** All real numbers except 0.    **3.** All real numbers $\geq 3$.

**5.** All real numbers.    **7.** All real numbers except $-\dfrac{5}{2}$.

**9.** All real numbers except 0 and 1.

**11.** All real numbers except 4 and $-\dfrac{1}{2}$.

**13.** $1, 7, -7$.    **15.** $-62, 2 - u^2, 2 - u^4$.
**17.** $2, (2v)^2 + 2v = 4v^2 + 2v, (-x^2)^2 + (-x^2) = x^4 - x^2$.
**19.** $4, 0, (x+h)^2 + 2(x+h) + 1$
$\qquad = x^2 + 2xh + h^2 + 2x + 2h + 1.$

**21.** $0, \dfrac{3x-5}{(3x)^2+4} = \dfrac{3x-5}{9x^2+4}, \dfrac{(x+h)-5}{(x+h)^2+4}$
$\qquad = \dfrac{x+h-5}{x^2+2xh+h^2+4}.$

**23.** $0, 256, \dfrac{1}{16}$.    **25. a.** $4x + 4h - 5$;  **b.** 4.

**27. a.** $x^2 + 2hx + h^2 + 2x + 2h$;  **b.** $2x + h + 2$.
**29. a.** $2 - 4x - 4h - 3x^2 - 6hx - 3h^2$;
**b.** $-4 - 6x - 3h$.

**31. a.** $\dfrac{1}{x+h}$;  **b.** $-\dfrac{1}{x(x+h)}$.    **33.** 9.

**35.** $y$ is a function of $x$; $x$ is a function of $y$.
**37.** $y$ is a function of $x$; $x$ is not a function of $y$.
**39.** Yes.    **41.** $V = f(t) = 10{,}000 + 400t$.
**43.** Yes; $P$; $q$.
**45.** 400 pounds per week; 1000 pounds per week; amount supplied increases as the price increases.
**47. a.** 4;  **b.** $8\sqrt[3]{2}$;  **c.** $f(2I_0) = 2\sqrt[3]{2} f(I_0)$; doubling the intensity increases the response by a factor of $2\sqrt[3]{2}$.
**49. a.** $3000, 2900, 2300, 2000; 12, 10$;
**b.** $10, 12, 17, 20; 3000, 2300$.
**51. a.** $-5.13$;  **b.** $2.64$;  **c.** $-17.43$.
**53. a.** $11.33$;  **b.** $50.62$;  **c.** $2.29$.

## PRINCIPLES IN PRACTICE 3.2

**1. a.** $p(n) = \$125$;  **b.** The premiums do not change;
**c.** constant function.
**2. a.** quadratic function;  **b.** 2;  **c.** 3.

**3.** $c(n) = \begin{cases} 3.50n & \text{if } n \leq 5 \\ 3.00n & \text{if } 5 < n \leq 10 \\ 2.75n & \text{if } n > 10 \end{cases}$
**4.** $7! = 5040$.

## EXERCISE 3.2 (page 90)

**1.** Yes.    **3.** No.    **5.** Yes.    **7.** Yes.
**9.** All real numbers.    **11.** All real numbers.
**13. a.** 3;  **b.** 7.    **15. a.** 4;  **b.** $-3$.    **17.** $8, 8, 8$.
**19.** $1, -1, 0, -1$.    **21.** $8, 3, 3, 1$.    **23.** 720.    **25.** 2.
**27.** 5.
**29.** $c(i) = \$3.50$; constant function.
**31. a.** $C = 850 + 3q$;  **b.** 250.

**33.** $c(n) = \begin{cases} 7.50n & \text{if } n < 10 \\ 7.00n & \text{if } n \geq 10 \end{cases}$    **35.** $\dfrac{9}{64}$.

**37. a.** All $T$ such that $30 \leq T \leq 39$;  **b.** $4, \dfrac{17}{4}, \dfrac{33}{4}$.

**39. a.** $237{,}077.34$;  **b.** $-434.97$;  **c.** $52.19$.
**41. a.** $2.21$;  **b.** $9.98$;  **c.** $-14.52$.

## PRINCIPLES IN PRACTICE 3.3

**1.** $c(s(x)) = c(x+3) = 2(x+3) = 2x + 6$.
**2.** Let the length of a side be represented by the function $l(x) = x + 3$ and the area of a square with sides of length $x$ be represented by $a(x) = x^2$. Then $g(x) = (x+3)^2 = [l(x)]^2 = a(l(x))$.

## EXERCISE 3.3 (page 95)

**1. a.** $2x + 8$;  **b.** 8;  **c.** $-2$;  **d.** $x^2 + 8x + 15$;  **e.** 3;
**f.** $\dfrac{x+3}{x+5}$;  **g.** $x + 8$;  **h.** 11;  **i.** $x + 8$.

**3. a.** $2x^2 + x$;  **b.** $-x$;  **c.** $\dfrac{1}{2}$;  **d.** $x^4 + x^3$;
**e.** $\dfrac{x^2}{x^2+x} = \dfrac{x}{x+1}$ (for $x \neq 0$);  **f.** $-1$;
**g.** $(x^2+x)^2 = x^4 + 2x^3 + x^2$;  **h.** $x^4 + x^2$;  **i.** 90.

**5.** $6; -32$.    **7.** $\dfrac{4}{(t-1)^2} + \dfrac{6}{t-1} + 1; \dfrac{2}{t^2+3t}$.

**9.** $\dfrac{1}{v+3}; \sqrt{\dfrac{2w^2+3}{w^2+1}}$.    **11.** $f(x) = x^5, g(x) = 4x - 3$.

**13.** $f(x) = \dfrac{1}{x}, g(x) = x^2 - 2$.

**15.** $f(x) = \sqrt[5]{x}, g(x) = \dfrac{x+1}{3}$.

**17. a.** $r(x) = 9.75x$  **b.** $e(x) = 4.25x + 4500$
**c.** $(r - e)(x) = 5.5x - 4500$.
**19.** $400m - 10m^2$; the total revenue received when the total output of $m$ employees is sold.
**21. a.** $1.52$;  **b.** $1747.24$.    **23. a.** $345.03$;  **b.** $-1.94$.

## PRINCIPLES IN PRACTICE 3.4

**1.** $y = -600x + 7250$; $x$-intercept $\left(12\frac{1}{12}, 0\right)$; $y$-intercept $(0, 7250)$.

**2.** $y = 24.95$; horizontal line; no $x$-intercept; $y$-intercept $(0, 24.95)$.

**3.**

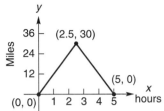

**4.**

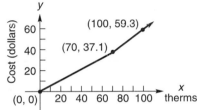

## EXERCISE 3.4 (page 104)

**1.**

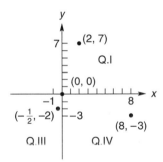

**3. a.** $1, 2, 3, 0$; **b.** all real numbers; **c.** all real numbers; **d.** $-2$.

**5. a.** $0, -1, -1$; **b.** all real numbers; **c.** all nonpositive real numbers; **d.** $0$.

**7.** $(0, 0)$; function; all real numbers; all real numbers.

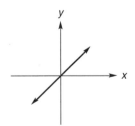

**9.** $(0, -5)$, $\left(\frac{5}{3}, 0\right)$; function; all real numbers; all real numbers.

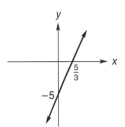

**11.** $(0, 0)$; function; all real numbers; all nonnegative real numbers.

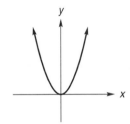

**13.** Every point on $y$-axis; not a function of $x$.

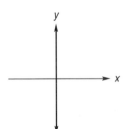

**15.** $(0, 0)$; function; all real numbers; all real numbers.

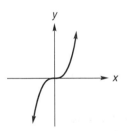

**17.** $(0, 0)$; not a function of $x$.

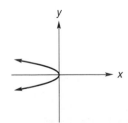

**19.** $(0, 2), (1, 0)$; function; all real numbers; all real numbers.

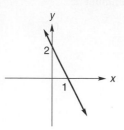

**21.** All real numbers; all real numbers $\leq 4$; $(0, 4), (2, 0), (-2, 0)$.

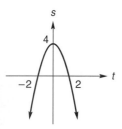

**23.** All real numbers; 2; $(0, 2)$.

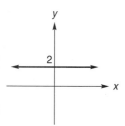

**25.** All real numbers; all real numbers $\geq -3$; $(0, 1), (2 \pm \sqrt{3}, 0)$.

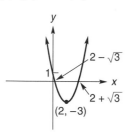

**27.** All real numbers; all real numbers; $(0, 0)$.

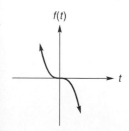

**29.** All real numbers $\geq 5$; all nonnegative real numbers; $(5, 0)$.

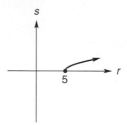

**31.** All real numbers; all nonnegative real numbers; $(0, 1), \left(\dfrac{1}{2}, 0\right)$.

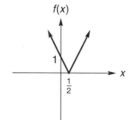

**33.** All nonzero real numbers; all positive real numbers; no intercepts.

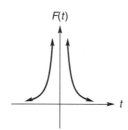

**35.** All nonnegative real numbers; all real numbers $c$ where $0 \leq c \leq 2$.

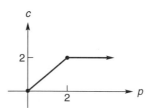

**37.** All real numbers; all nonnegative real numbers.

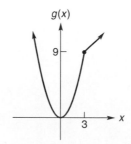

**39.** (a), (b), (d).

**41.**

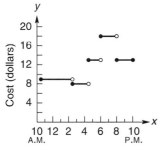

**43.** As price decreases, quantity increaes; $p$ is a function of $q$.

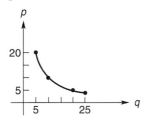

**45.**

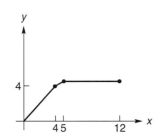

**47.** $-1, -0.35.$   **49.** $1.11, 2.83, 6.06.$
**51.** $-0.84, 2.61.$   **53.** $-0.49, 0.52, 1.25.$
**55. a.** $3.94;$ **b.** $-1.94.$
**57. a.** $(-\infty, \infty);$ **b.** $(-1.73, 0), (0, 4.00).$
**59. a.** $1.93;$ **b.** $[1.93, \infty);$ **c.** $(0, 2.48);$ **d.** no.

## EXERCISE 3.5 (page 111)

**1.** $(0, 0);$ sym. about origin.
**3.** $(\pm 2, 0), (0, 8);$ sym about $y$-axis.
**5.** $(\pm 3, 0);$ sym. about $x$-axis, $y$-axis, origin.
**7.** $(-2, 0);$ sym. about $x$-axis.   **9.** Sym. about $x$-axis.
**11.** $(-21, 0), (0, -7), (0, 3).$
**13.** $(0, 0);$ sym. about origin.   **15.** $(0, 1).$
**17.** $(2, 0), (0, \pm 2);$ sym. about $x$-axis.

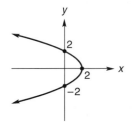

**19.** $(\pm 2, 0), (0, 0);$ sym. about origin.

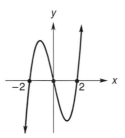

**21.** $(0, 0);$ sym. about $x$-axis, $y$-axis, origin.

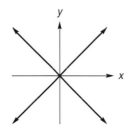

**23.** $(\pm 2, 0), (0, \pm 4);$ sym. about $x$-axis, $y$-axis, origin.

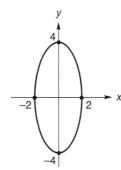

**25. a.** $(\pm 1.18, 0), (0, 2);$ **b.** $2;$ **c.** $(-\infty, 2].$

## EXERCISE 3.6 (page 113)

**1.**

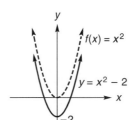

**3.**

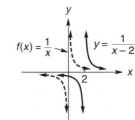

**5.**

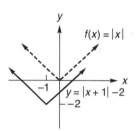

**7.**

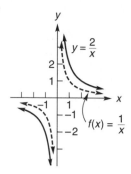

**9.**

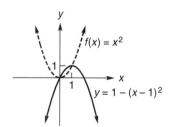

**11.**

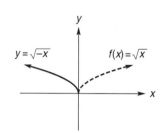

**13.** Translate 4 units to the right and 3 units upward.

**15.** Reflect about the $y$-axis and translate 3 units upward.

## REVIEW PROBLEMS—CHAPTER 3 (page 115)

**1.** All real numbers except 1 and 2.

**3.** All real numbers.

**5.** All nonnegative real numbers except 1.

**7.** $7, 46, 62, 3t^2 - 4t + 7$.   **9.** $0, 3, \sqrt{t}, \sqrt{x^2 - 1}$.

**11.** $\dfrac{3}{5}, 0, \dfrac{\sqrt{x+4}}{x}, \dfrac{\sqrt{u}}{u-4}$.   **13.** $-8, 4, 4, -92$.

**15. a.** $3 - 7x - 7h$; **b.** $-7$.

**17. a.** $4x^2 + 8hx + 4h^2 + 2x + 2h - 5$;

**b.** $8x + 4h + 2$.

**19. a.** $5x + 2$; **b.** 22; **c.** $x - 4$; **d.** $6x^2 + 7x - 3$;

**e.** 10; **f.** $\dfrac{3x - 1}{2x + 3}$; **g.** $3(2x + 3) - 1 = 6x + 8$; **h.** 38;

**i.** $2(3x - 1) + 3 = 6x + 1$.

**21.** $\dfrac{1}{x-1}, \dfrac{1}{x} - 1 = \dfrac{1-x}{x}$.   **23.** $x^3 + 2,\ (x+2)^3$.

**25.** $(0,0), (\pm\sqrt{2/3}, 0)$; sym. about origin.

**27.** $(0,9), (\pm 3, 0)$; sym. about $y$-axis.

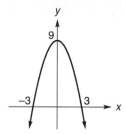

**29.** $(0,2), (-4, 0)$; all $u \geq -4$; all real numbers $\geq 0$.

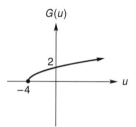

**31.** $\left(0, -\dfrac{1}{2}\right)$; all $t \neq 4$; all nonzero real numbers.

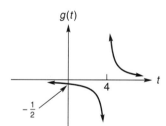

**33.** All real numbers; all real numbers $\geq 1$.

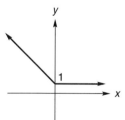

**35.**

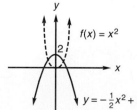

**37.** a, c.

**39.** $-0.67, 0.34, 1.73$.

**41.** $-1.50, -0.88, -0.11, 1.09, 1.40$.
**43. a.** $(-\infty, \infty)$; **b.** $(1.92, 0), (0, 7)$
**45. a.** none; **b.** $1, 3$.

## MATHEMATICAL SNAPSHOT—CHAPTER 3 (page 118)

**1.** $22,656. **3.** $85,190.70.

## PRINCIPLES IN PRACTICE 4.1

**1.** $-2000$; the car depreciated $2000 per year.

**2.** $S = 14T + 8$. **3.** $F = \dfrac{9}{5}C + 32$.

**4.** slope $= \dfrac{125}{3}$; $y$-intercept $= \dfrac{125}{3}$.

**5.** $9C - 5F + 160 = 0$.
**6.**

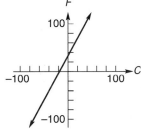

**7.** The slope of $\overline{AB}$ is 0; the slope of $\overline{BC}$ is 7; the slope of $\overline{CA}$ is 1. None of the slopes are negative reciprocals of each other, so the triangle does not have a right angle. The points do not define a right triangle.

## EXERCISE 4.1 (page 126)

**1.** 3. **3.** $-\dfrac{1}{2}$. **5.** Undefined. **7.** 0.

**9.** $6x - y - 4 = 0$. **11.** $x + 4y - 18 - 0$.
**13.** $3x - 7y + 25 = 0$. **15.** $8x - 5y - 29 = 0$.
**17.** $2x - y + 4 = 0$. **19.** $x + 2y + 6 = 0$.
**21.** $y + 2 = 0$. **23.** $x - 2 = 0$. **25.** $2; -1$.

**27.** $-\dfrac{1}{2}; \dfrac{3}{2}$. **29.** Slope undefined; no $y$-intercept.

**31.** $3; 0$. **33.** $0; 1$.

**35.** $2x + 3y - 5 = 0$; $y = -\dfrac{2}{3}x + \dfrac{5}{3}$.

**37.** $4x + 9y - 5 = 0$; $y = -\dfrac{4}{9}x + \dfrac{5}{9}$.

**39.** $3x - 2y + 24 = 0$; $y = \dfrac{3}{2}x + 12$.

**41.** Parallel. **43.** Parallel. **45.** Neither.
**47.** Perpendicular. **49.** Perpendicular.

**51.** $y = 4x + 7$. **53.** $y = 1$. **55.** $y = -\dfrac{1}{3}x + 5$.

**57.** $x = 7$. **59.** $y = -\dfrac{2}{3}x - \dfrac{29}{3}$. **61.** $(5, -4)$.

**63.** $-2$; the stock price dropped an average of $2 per year.
**65.** $y = 3x + 5$. **67.** slope $\approx 0.65$; $y$-intercept $\approx 4.38$

**69. a.** $y = -\dfrac{1}{3}x + \dfrac{1}{6}$; **b.** $y = 3x - \dfrac{3}{2}$.

**71.** $y = -x + 3300$; without modification, the approach angle will cause the plane to crash 700 feet short of the airport.
**73.** $R = 50{,}000T + 80{,}000$.
**75.** The lines are parallel. This is expected because they each have a slope of 1.5.

## PRINCIPLES IN PRACTICE 4.2

**1.** $x =$ number of skis produced;
$y =$ number of boots produced; $8x + 14y = 1000$.

**2.** $p = -\dfrac{3}{8}q + 1025$.

**3.** Answers may vary, but two possible points are $(0, 60)$ and $(2, 140)$.

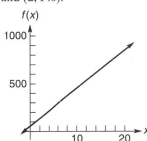

**4.** $f(t) = 2.3t + 32.2$.
**5.** $f(x) = 70x + 150$.

## EXERCISE 4.2 (page 132)

**1.** $-4; 0$.

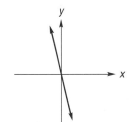

**3.** $2; -4$.

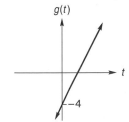

**5.** $-\dfrac{1}{2}; \dfrac{7}{2}$.

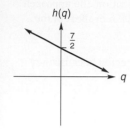

**7.** $f(x) = 4x$.    **9.** $f(x) = -2x + 4$.

**11.** $f(x) = -\dfrac{1}{2}x + \dfrac{15}{4}$.    **13.** $f(x) = x + 1$.

**15.** $p = -\dfrac{2}{5}q + 28$; $16.    **17.** $p = \dfrac{5}{2}q + 90$.

**19.** $c = 3q + 10$; $115.    **21.** $f(x) = 0.115x + 7.95$

**23.** $v = -800t + 8000$; slope $= -800$.

**25.** $f(x) = 11{,}000x + 198{,}000$.    **27.** $f(x) = 65x + 85$.
**29.** $x + 10y = 100$.

**31. a.** $y = \dfrac{5}{11}, x = \dfrac{600}{11}$; **b.** 12.

**33. a.** $p = 0.059t + 0.025$; **b.** 0.556.

**35. a.** $t = \dfrac{1}{4}c + 37$;

**b.** Add 37 to the number of chirps in 15 seconds.

**37.** $P = \dfrac{T}{4} + 80$.    **39. a.** Yes; **b.** 1.8704.

## PRINCIPLES IN PRACTICE 4.3

**1.** Vertex: $(1, 400)$; $y$-intercept: $(0, 399)$;
$x$-intercepts: $(-19, 0), (21, 0)$.

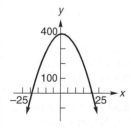

**2.** Vertex: $(1, 24)$; $y$-intercept: $(0, 8)$;
$x$-intercepts: $\left(1 + \dfrac{\sqrt{6}}{2}, 0\right), \left(1 - \dfrac{\sqrt{6}}{2}, 0\right)$.

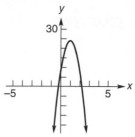

**3.** 1000 units; $3000 maximum revenue.

## EXERCISE 4.3 (page 140)

**1.** Quadratic.    **3.** Not Quadratic.    **5.** Quadratic.
**7.** Quadratic.    **9. a.** $(1, 11)$; **b.** highest.
**11. a.** $-8$; **b.** $-4, 2$; **c.** $(-1, -9)$.
**13.** Vertex: $(3, -4)$; intercepts: $(1, 0), (5, 0), (0, 5)$;
range: all $y \geq -4$.

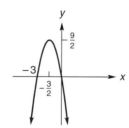

**15.** Vertex: $\left(-\dfrac{3}{2}, \dfrac{9}{2}\right)$; intercepts: $(0, 0), (-3, 0)$;

range: all $y \leq \dfrac{9}{2}$.

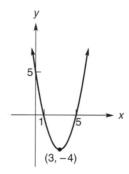

**17.** Vertex: $(-1, 0)$; intercepts: $(-1, 0)$, $(0, 1)$; range: all $s \geq 0$.

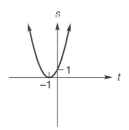

**19.** Vertex: $(2, -1)$; intercept: $(0, -9)$; range: all $y \leq -1$.

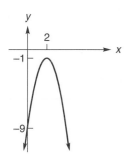

**21.** Vertex: $(4, -3)$; intercepts: $(4 + \sqrt{3}, 0)$, $(4 - \sqrt{3}, 0)$, $(0, 13)$; range: all $t \geq -3$.

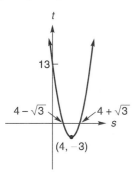

**23.** Minimum; 24.   **25.** Maximum; $-10$.
**27.** $q = 200$; $r = \$120,000$.
**29.** 200 units; $\$240,000$ maximum revenue.
**31.** Vertex: $(9, 225)$; $y$-intercept: $(0, 144)$; $x$-intercepts: $(-6, 0)$, $(24, 0)$.

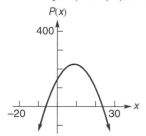

**33.** 70 grams.   **35.** 96 ft; 2 sec.

**37.** Vertex: $\left( \dfrac{5}{2}, 116 \right)$;  $y$-intercept: $(0, 16)$, $x$-intercepts: $\left( \dfrac{5 + \sqrt{29}}{2}, 0 \right)$, $\left( \dfrac{5 - \sqrt{29}}{2}, 0 \right)$.

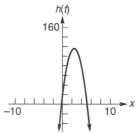

**39. a.** 2.5; **b.** 8.7 m.   **41. a.** $\dfrac{l}{2}$; **b.** $\dfrac{wl^2}{8}$; **c.** 0 and $l$.
**43.** 50 ft $\times$ 100 ft.   **45.** $(1.11, 2.88)$.
**47. a.** 0; **b.** 1; **c.** 2.   **49.** 6.89.

## PRINCIPLES IN PRACTICE 4.4

**1.** $\$120,000$ at 9% and $\$80,000$ at 8%.
**2.** 500 of species A and 1000 of species B.
**3.** Infinitely many solutions of the form
$A = \dfrac{20,000}{3} - \dfrac{4}{3}r$,  $B = r$ where $0 \leq r \leq 5000$.

**4.** $\dfrac{1}{6}$ lb of A; $\dfrac{1}{3}$ lb of B; $\dfrac{1}{2}$ lb of C.

## EXERCISE 4.4 (page 150)

**1.** $x = -1, y = 1$.    **3.** $x = 3, y = -1$.
**5.** $v = 0, w = 18$.    **7.** $x = 2, y = -3$.
**9.** No solution.    **11.** $x = 12, y = -12$.
**13.** $p = \dfrac{3}{2} - 3r$, $q = r$; $r$ is any real number.

**15.** $x = \dfrac{1}{2}, y = \dfrac{1}{2}, z = \dfrac{1}{4}$.    **17.** $x = 1, y = 1, z = 1$

**19.** $x = 1 + 2r, y = 3 - r, z = r$; $r$ is any real number.
**21.** $x = -\dfrac{1}{3}r, y = \dfrac{5}{3}r, z = r$; $r$ is any real number.

**23.** $x = \dfrac{3}{2} - r + \dfrac{1}{2}s, y = r, z = s$; $r$ and $s$ are any real numbers.

**25.** 420 gal of 20% solution, 280 gal of 30% solution.
**27.** 0.5 lb of cotton; 0.25 lb of polyester; 0.25 lb of nylon.
**29.** 275 mi/h (speed of airplane in still air), 25 mi/h (speed of wind).
**31.** 240 units (Early American), 200 units (Contemporary).
**33.** 800 calculators from Exton plant, 700 from Whyton plant.
**35.** 4% on first $\$100,000$, 6% on remainder.
**37.** 60 units of Argon I, 40 units of Argon II.
**39.** 100 chairs, 100 rockers, 200 chaise lounges.
**41.** 40 semiskilled workers, 20 skilled workers, 10 shipping clerks.
**45.** $x = 4, y = 2$.    **47.** $x = 8.3, y = 14.0$.

## EXERCISE 4.5 (page 154)

**1.** $x = 4, y = -12$; $x = -1, y = 3$.
**3.** $p = -3, q = -5$; $p = 2, q = 0$.
**5.** $x = 0, y = 0$; $x = 1, y = 1$.
**7.** $x = 4, y = 8$; $x = -1, y = 3$.
**9.** $p = 0, q = 0$; $p = 1, q = 1$.
**11.** $x = 3\sqrt{2}, y = 2$; $x = -3\sqrt{2}, y = 2$;
$x = \sqrt{15}, y = -1$; $x = -\sqrt{15}, y = -1$.
**13.** $x = 21, y = 15$.      **15.** At $(10, 8.1)$ and $(-10, 7.9)$.
**17.** Three.      **19.** $x = -1.3, y = 5.1$.      **21.** $x = 1.76$.
**23.** $x = -1.46$.

## EXERCISE 4.6 (page 162)

**1.**

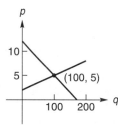

**3.** $(5, 212.50)$.      **5.** $(9, 38)$.      **7.** $(15, 5)$.
**9.**

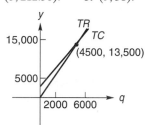

**11.** Cannot break even at any level of production.
**13.** 10 units or 40 units.      **15. a.** $12;  **b.** $12.18.
**17.** 5840 units; 840 units; 1840 units.      **19.** $4.
**21.** Total cost always exceeds total revenue—no break-even point.
**23.** Decreases by $0.70.      **25.** $p_A = 5$; $p_B = 10$.
**27.** 2.4 and 11.3.

## REVIEW PROBLEMS—CHAPTER 4 (page 164)

**1.** 9.      **3.** $y = -x + 1$; $x + y - 1 = 0$.
**5.** $y = \frac{1}{2}x - 1$; $x - 2y - 2 = 0$.
**7.** $y = 4$; $y - 4 = 0$.
**9.** $y = \frac{1}{3}x + 2$; $x - 3y + 6 = 0$.      **11.** Perpendicular.
**13.** Neither.      **15.** Parallel.      **17.** $y = \frac{3}{2}x - 2$; $\frac{3}{2}$.

**19.** $y = \frac{4}{3}$; 0.      **21.** $-2$; $(0, 4)$.

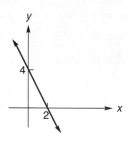

**23.** $(3, 0), (-3, 0), (0, 9)$; $(0, 9)$.

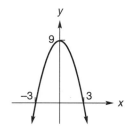

**25.** $(5, 0), (-1, 0), (0, -5)$; $(2, -9)$.

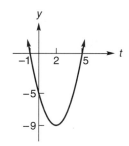

**27.** $3$; $(0, 0)$.

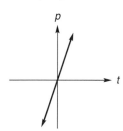

**29.** $(0, -3)$; $(-1, -2)$.

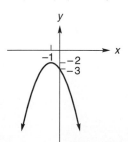

**31.** $x = \frac{17}{7}, y = -\frac{8}{7}$.      **33.** $x = 2, y = -1$.
**35.** $x = 8, y = 4$.      **37.** $x = 0, y = 1, z = 0$.
**39.** $x = -3, y = -4$; $x = 2, y = 1$.

**41.** $x = -2 - 2r$, $y = 7 + r$, $z = r$; $r$ is any real number.
**43.** $x = r$, $y = r$, $z = 0$; $r$ is any real number.

**45.** $a + b - 3 = 0$; $0$.  **47.** $f(x) = -\dfrac{4}{3}x + \dfrac{19}{3}$.

**49.** 50 units; $5000.  **51.** 6.
**53. a.** $R = 75L + 1310$;  **b.** 1385 milliseconds;
**c.** The slope is 75; the time necessary to travel from one level to the next level is 75 milliseconds.
**55.** 300 ft by 250 ft.  **57.** $x = 230$, $y = -130$.
**59.** $x = 0.75$, $y = 1.43$.

## MATHEMATICAL SNAPSHOT—CHAPTER 4 (page 170)

**1. a.** 12;  **b.** $F(x) = \begin{cases} 4, & \text{if } x = 2 \\ 6, & \text{if } x = 3 \\ \left[\dfrac{12}{11}(x - 1) + 4.5\right] & \text{otherwise;} \end{cases}$

**c.** $F(1) = 4$, $F(2) = 4$, $F(3) = 6$, $F(4) = 7$, $F(5) = 8$, $F(6) = 9$, $F(7) = 11$, $F(8) = 12$, $F(9) = 13$, $F(10) = 14$, $F(11) = 15$, $F(12) = 16$.

## PRINCIPLES IN PRACTICE 5.1

**1.** The shape of the graphs are the same. The value of $A$ scales the ordinate of any point by $A$.

**2.**

| Year | Multiplicative Increase | Expression |
|------|------|------|
| 0 | 1 | $1.1^0$ |
| 1 | 1.1 | $1.1^1$ |
| 2 | 1.21 | $1.1^2$ |
| 3 | 1.33 | $1.1^3$ |
| 4 | 1.46 | $1.1^4$ |

1.1; The investment increases by 10% every year $(1 + 1(0.1) = 1 + 0.1 = 1.1)$.

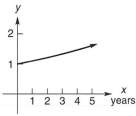

Between 7 and 8 years.

**3.**

| Year | Multiplicative Decrease | Expression |
|------|------|------|
| 0 | 1 | $0.85^0$ |
| 1 | 0.85 | $0.85^1$ |
| 2 | 0.72 | $0.85^2$ |
| 3 | 0.61 | $0.85^3$ |

0.85; The car depreciates by 15% every year $(1 - 1(0.15) = 1 - 0.15 = 0.85)$.

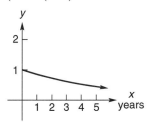

Between 4 and 5 years.
**4.** $y = 1.08^{t-3}$; Shift the graph 3 units to the right.
**5.** $3684.87; $1684.87.
**6.** $2753.79; $753.79.  **7.** 117 employees.
**8.**

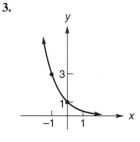

## EXERCISE 5.1 (page 181)

**1.**

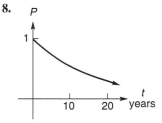

**3.**

**5.**

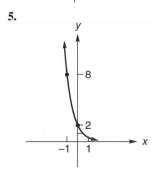

**7.**

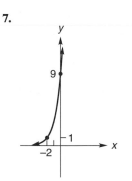

**9.**

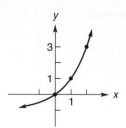

**11.**

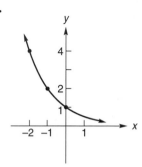

**13.** *B.*   **15.** 140,000.   **17.** $\frac{1}{2}$.

**19. a.** $6014.52; **b.** $2014.52.
**21. a.** $1964.76; **b.** $1264.76.
**23. a.** $14,124.86; **b.** $10,124.86.
**25. a.** $6256.36; **b.** $1256.36.
**27. a.** $9649.69; **b.** $1649.69.   **29.** $10,446.15.
**31. a.** $N = 400(1.05)^t$; **b.** 420; **c.** 486.

**33.**

| Year | Multiplicative Increase | Expression |
|------|-------------------------|------------|
| 0 | 1 | $1.3^0$ |
| 1 | 1.3 | $1.3^1$ |
| 2 | 1.69 | $1.3^2$ |
| 3 | 2.20 | $1.3^3$ |

1.3; The recycling increases by 30% every year
$(1 + 1(0.3) = 1 + 0.3 = 1.3)$.

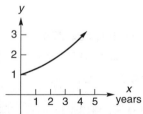

Between 4 and 5 years.
**35.** 97,030.   **37.** 4.4817.   **39.** 0.6703.

**41.**

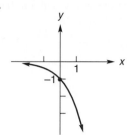

**43.** 0.2240.   **45.** $(e^k)^t$, where $b = e^k$.
**47. a.** 10; **b.** 7.6; **c.** 2.5; **d.** 25 hours.   **49.** 32 years.
**51.** 0.1465.   **55.** 1.58.   **57.** 4.2 min.   **59.** 15.

## PRINCIPLES IN PRACTICE 5.2

**1.** $t = \log_2 16$; $t = $ the number of times the bacteria have
doubled.   **2.** $\dfrac{I}{I_0} = 10^{8.3}$

**3.**

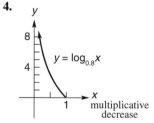

**4.**

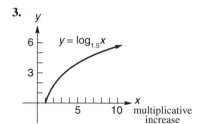

**5.** Approximately 13.9%.   **6.** Approximately 9.2%.

## EXERCISE 5.2 (page 189)

**1.** $\log 10,000 = 4$.   **3.** $2^6 = 64$.   **5.** $\ln 7.3891 = 2$.
**7.** $e^{1.09861} = 3$.

**9.**

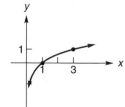

**11.**

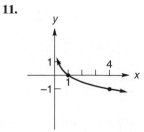

**13.**

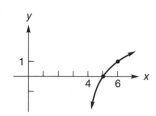

**15.**

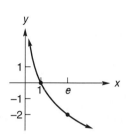

**17.** 2.

**19.** 3.　　**21.** 1.　　**23.** −2.　　**25.** 0.　　**27.** −3.

**29.** 9.　　**31.** 125.　　**33.** $\dfrac{1}{10}$.　　**35.** $e^2$.　　**37.** 2.

**39.** 6.　　**41.** $\dfrac{1}{81}$.　　**43.** 2.　　**45.** $\dfrac{5}{3}$.　　**47.** 4.

**49.** $\dfrac{\ln 2}{3}$.　　**51.** $\dfrac{5 + \ln 3}{2}$.　　**53.** 1.60944.　　**55.** 2.00013.

**57.** $\dfrac{1}{h} = 10^{5.5}$.　　**59.** 41.50.　　**61.** $E = 2.5 \times 10^{11 + 1.5M}$.

**63. a.** 305.2 mm of mercury; **b.** 5.13 km.
**65.** $e^{[u_0 - (x_2^2/2)]/A}$.　　**67.** 21.7 years.
**69.** $y = \dfrac{1}{3} \ln \dfrac{10 - x}{2}$.　　**71.** $(1, 0)$.　　**73.** 7.39.

## PRINCIPLES IN PRACTICE 5.3

**1.** $\log (900{,}000) - \log (9000) = \log \left( \dfrac{900{,}000}{9000} \right)$
$$= \log (100) = 2.$$

**2.** $\log (10{,}000) = \log (10^4) = 4.$

## EXERCISE 5.3 (page 196)

**1.** $b + c$.　　**3.** $a - b$.　　**5.** $3a - b$.　　**7.** $2(a + b)$.

**9.** $\dfrac{b}{a}$.　　**11.** 48.　　**13.** −4.　　**15.** 5.01.　　**17.** −1.

**19.** 4.　　**21.** $\ln x + 2\ln(x + 1)$.
**23.** $2\ln x - 3\ln(x + 1)$.　　**25.** $3[\ln x - \ln(x + 1)]$.
**27.** $\ln x - \ln(x + 1) - \ln(x + 2)$.

**29.** $\dfrac{1}{2} \ln x - 2\ln(x + 1) - 3\ln(x + 2)$.

**31.** $\dfrac{2}{5} \ln x - \dfrac{1}{5} \ln(x + 1) - \ln(x + 2)$.

**33.** $\log 28$.　　**35.** $\log_2 \dfrac{2x}{x + 1}$.　　**37.** $\log[7^9(23)^5]$.

**39.** $\log[100(1.05)^{10}]$.　　**41.** $\dfrac{81}{64}$.　　**43.** 1.　　**45.** $\dfrac{5}{2}$.

**47.** $\pm 2$.　　**49.** $\dfrac{\ln(x + 8)}{\ln 10}$.　　**51.** $\dfrac{\ln(x^2 + 1)}{\ln 3}$.

**53.** $y = \ln \dfrac{z}{7}$.　　**57. a.** 3; **b.** $2 + M_1$.　　**59.** 3.5229.

**61.** 12.4771.
**63.**

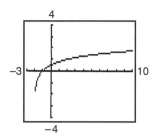

**65.** $\ln 4$.

## PRINCIPLES IN PRACTICE 5.4

**1.** 18.　　**2.** Day 20.　　**3.** The other earthquake is 67.5 times as intense as a zero-level earthquake.

## EXERCISE 5.4 (page 201)

**1.** 5.000.　　**3.** 1.333.　　**5.** −3.000.　　**7.** 2.000.
**9.** 0.083.　　**11.** 0.805.　　**13.** 0.203.　　**15.** 5.140.
**17.** −0.073.　　**19.** 2.322.　　**21.** 3.183.　　**23.** 0.483.
**25.** 2.496.　　**27.** 1.003.000.　　**29.** 6.667.　　**31.** 3.082.
**33.** 3.000.　　**35.** 0.500.　　**37.** $S = 12.4A^{0.26}$.
**39. a.** 100; **b.** 46.　　**41.** 20.5.

**43.** $p = \dfrac{\log (80 - q)}{\log 2}$; 4.32.　　**45.** 7.

**47. a.** 91; **b.** 432; **c.** 8.　　**49.** 0.69.
**51.**

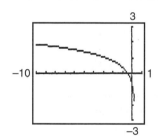

## REVIEW PROBLEMS—CHAPTER 5 (page 204)

**1.** $\log_3 243 = 5$.　　**3.** $16^{1/4} = 2$.　　**5.** $\ln 54.598 = 4$.

**7.** 3.　　**9.** −4.　　**11.** −2.　　**13.** 3.　　**15.** $\dfrac{1}{100}$.

**17.** 3.　　**19.** $3(a + 1)$.　　**21.** $\log \dfrac{25}{27}$.　　**23.** $\ln \dfrac{x^2 y}{z^3}$.

**25.** $\log_2 \dfrac{x^{9/2}}{(x + 1)^3(x + 2)^4}$.　　**27.** $2\ln x + \ln y - 3\ln z$.

**29.** $\dfrac{1}{3}(\ln x + \ln y + \ln z)$.　　**31.** $\dfrac{1}{2}(\ln y - \ln z) - \ln x$.

**33.** $\dfrac{\ln(x + 5)}{\ln 3}$.　　**35.** 1.8295.　　**37.** $2x + \dfrac{1}{2}x$.

**39.** $2x$.　　**41.** $y = e^{x^2 + 2}$.

**43.**

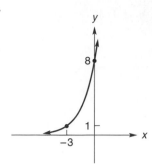

**45.** $\dfrac{1}{3}$. **47.** 1. **49.** 10. **51.** $2^e$. **53.** 0.231.

**55.** $-3.222$. **57.** $-1.596$.

**59. a.** \$3829.04; **b.** \$1229.04. **61.** \$6915.66.

**63. a.** $P = 8000(1.02)^t$; **b.** 8323.

**65. a.** 10 mg; **b.** 4.4; **c.** 0.2; **d.** 1.7; **e.** 5.6.

**67. a.** 6; **b.** 28. **71.** $(-\infty, 0.37]$. **73.** 2.93.

**75.**

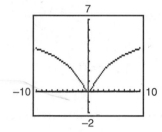

## MATHEMATICAL SNAPSHOT—CHAPTER 5 (page 208)

**1. a.** $P = \dfrac{T(e^{kI} - 1)}{-e^{-dkI}}$;

**b.** $d = \dfrac{1}{kI} \ln\left[\dfrac{P}{P - T(e^{kI} - 1)}\right]$.

**3. a.** 156; **b.** 65.

## PRINCIPLES IN PRACTICE 6.1

**1.** $3 \times 2$ or $2 \times 3$.

**2.** $\begin{bmatrix} 1 & 2 & 4 & 8 & 16 \\ 1 & 2 & 4 & 8 & 16 \\ 1 & 2 & 4 & 8 & 16 \end{bmatrix}$.

## EXERCISE 6.1 (page 214)

**1. a.** $2 \times 3, 3 \times 3, 3 \times 2, 2 \times 2, 4 \times 4, 1 \times 2, 3 \times 1,$
$3 \times 3, 1 \times 1$; **b.** $\mathbf{B}, \mathbf{D}, \mathbf{E}, \mathbf{H}, \mathbf{J}$;
**c.** $\mathbf{H}, \mathbf{J}$ upper triangular; $\mathbf{D}, \mathbf{J}$ lower triangular;
**d.** $\mathbf{F}, \mathbf{J}$; **e.** $\mathbf{G}, \mathbf{J}$.

**3.** 2. **5.** 4. **7.** 6. **9.** 7, 2, 1, 0.

**11.** $\begin{bmatrix} 5 & 8 & 11 & 14 \\ 7 & 10 & 13 & 16 \\ 9 & 12 & 15 & 18 \end{bmatrix}$. **13.** 120 entries, 1, 0, 1, 0.

**15. a.** $\begin{bmatrix} 0 & 0 & 0 & 0 \\ 0 & 0 & 0 & 0 \\ 0 & 0 & 0 & 0 \\ 0 & 0 & 0 & 0 \end{bmatrix}$; **b.** $\begin{bmatrix} 0 & 0 & 0 & 0 & 0 & 0 \\ 0 & 0 & 0 & 0 & 0 & 0 \\ 0 & 0 & 0 & 0 & 0 & 0 \\ 0 & 0 & 0 & 0 & 0 & 0 \\ 0 & 0 & 0 & 0 & 0 & 0 \\ 0 & 0 & 0 & 0 & 0 & 0 \end{bmatrix}$.

**17.** $\begin{bmatrix} 6 & 2 \\ -3 & 4 \end{bmatrix}$. **19.** $\begin{bmatrix} 1 & 3 & -4 \\ 3 & 2 & 2 \\ 2 & -2 & 0 \\ 3 & 0 & 1 \end{bmatrix}$.

**21. a.** $\mathbf{A}$ and $\mathbf{C}$; **b.** all of them.

**25.** $x = 6, y = \dfrac{2}{3}, z = \dfrac{7}{2}$. **27.** $x = 0, y = 0$.

**29. a.** 7; **b.** 3; **c.** February; **d.** deluxe blue; **e.** February;
**f.** February; **g.** 35.

**31.** $-1994$. **33.** $\begin{bmatrix} 3 & 1 & 1 \\ 1 & 7 & 4 \\ 4 & 3 & 1 \\ 2 & 6 & 2 \end{bmatrix}$.

## PRINCIPLES IN PRACTICE 6.2

**1.** $\begin{bmatrix} 230 & 220 \\ 190 & 255 \end{bmatrix}$. **2.** $x_1 = 670, x_2 = 835, x_3 = 1405$.

## EXERCISE 6.2 (page 221)

**1.** $\begin{bmatrix} 4 & -3 & 1 \\ -2 & 10 & 5 \\ 10 & 5 & 3 \end{bmatrix}$. **3.** $\begin{bmatrix} -5 & 5 \\ -9 & 5 \\ 5 & 9 \end{bmatrix}$.

**5.** $\begin{bmatrix} -9 & -7 & 11 \end{bmatrix}$. **7.** Not defined.

**9.** $\begin{bmatrix} -12 & 36 & -42 & -6 \\ -42 & -6 & -36 & 12 \end{bmatrix}$. **11.** $\begin{bmatrix} 5 & -4 & 1 \\ 0 & 7 & -2 \\ -3 & 3 & 13 \end{bmatrix}$.

**13.** $\begin{bmatrix} 6 & 5 \\ -2 & 3 \end{bmatrix}$. **15. O.** **17.** $\begin{bmatrix} 28 & 22 \\ -2 & 6 \end{bmatrix}$.

**19.** Not defined. **21.** $\begin{bmatrix} -22 & -15 \\ -11 & 9 \end{bmatrix}$.

**23.** $\begin{bmatrix} 21 & \frac{29}{2} \\ \frac{19}{2} & -\frac{15}{2} \end{bmatrix}$. **29.** $\begin{bmatrix} 4 & 2 & 5 \\ 7 & -3 & 2 \end{bmatrix}$.

**31.** $\begin{bmatrix} -1 & 5 \\ 6 & -8 \end{bmatrix}$. **33.** Impossible.

**35.** $x = \dfrac{146}{13}, y = -\dfrac{28}{13}$. **37.** $x = 6, y = \dfrac{4}{3}$.

**39.** $x = -6, y = -14, z = 1$. **41.** $\begin{bmatrix} 35 & 65 \\ 75 & 55 \\ 25 & 15 \end{bmatrix}$.

**43.** 1.1. **45.** $\begin{bmatrix} 15 & -4 & 26 \\ 4 & 7 & 30 \end{bmatrix}$. **47.** $\begin{bmatrix} -10 & 22 & 12 \\ 24 & 36 & -44 \end{bmatrix}$.

## PRINCIPLES IN PRACTICE 6.3

**1.** $5780. **2.** $22,843.75.

**3.** $\begin{bmatrix} 1 & \frac{8}{5} \\ 1 & \frac{1}{3} \end{bmatrix} \begin{bmatrix} x \\ y \end{bmatrix} = \begin{bmatrix} \frac{8}{5} \\ \frac{5}{3} \end{bmatrix}.$

## EXERCISE 6.3 (page 233)

**1.** $-12.$ **3.** $19.$ **5.** $7.$ **7.** $2 \times 2; 4.$
**9.** $3 \times 5; 15.$ **11.** $2 \times 1; 2.$ **13.** $3 \times 3; 9.$

**15.** $3 \times 1; 3.$ **17.** $\begin{bmatrix} 1 & 0 & 0 & 0 \\ 0 & 1 & 0 & 0 \\ 0 & 0 & 1 & 0 \\ 0 & 0 & 0 & 1 \end{bmatrix}.$ **19.** $\begin{bmatrix} 10 & -16 \\ 7 & 8 \end{bmatrix}.$

**21.** $\begin{bmatrix} 23 \\ 50 \end{bmatrix}.$ **23.** $\begin{bmatrix} -3 & 4 & 2 \\ 2 & 2 & 4 \\ 5 & 0 & 3 \end{bmatrix}.$

**25.** $\begin{bmatrix} -6 & 16 & 10 & -6 \end{bmatrix}.$ **27.** $\begin{bmatrix} 4 & 6 & -4 & 6 \\ 6 & 9 & -6 & 9 \\ -8 & -12 & 8 & -12 \\ 2 & 3 & -2 & 3 \end{bmatrix}.$

**29.** $\begin{bmatrix} 78 & 84 \\ -21 & -12 \end{bmatrix}.$ **31.** $\begin{bmatrix} -5 & -8 \\ -5 & -20 \end{bmatrix}.$ **33.** $\begin{bmatrix} x \\ y \\ z \end{bmatrix}.$

**35.** $\begin{bmatrix} 2x_1 + x_2 + 3x_3 \\ 4x_1 + 9x_2 + 7x_3 \end{bmatrix}.$ **37.** $\begin{bmatrix} 0 & 0 & 0 \\ 0 & -1 & 1 \\ 1 & 2 & 0 \end{bmatrix}.$

**39.** $\begin{bmatrix} -1 & -20 \\ -2 & 23 \end{bmatrix}.$ **41.** $\begin{bmatrix} \frac{3}{2} & 0 & 0 \\ 0 & \frac{3}{2} & 0 \\ 0 & 0 & \frac{3}{2} \end{bmatrix}.$ **43.** $\begin{bmatrix} -1 & 5 \\ 2 & 17 \\ 1 & 31 \end{bmatrix}.$

**45.** Impossible. **47.** $\begin{bmatrix} 0 & 0 & -4 \\ 2 & -1 & -2 \\ 0 & 0 & 8 \end{bmatrix}.$

**49.** $\begin{bmatrix} 3 & -1 \\ -2 & 2 \end{bmatrix}.$ **51.** $\begin{bmatrix} 0 & 3 & 0 \\ -1 & -1 & 2 \end{bmatrix}.$

**53.** $\begin{bmatrix} 2 & 0 & 0 \\ 0 & 2 & 0 \\ 0 & 0 & 2 \end{bmatrix}.$ **55.** $\begin{bmatrix} 1 & -1 & 0 \\ 0 & 1 & 1 \end{bmatrix}.$ **57.** $\begin{bmatrix} 6 & -7 \\ -7 & 9 \end{bmatrix}.$

**59.** $\begin{bmatrix} 3 & 1 \\ 7 & -2 \end{bmatrix} \begin{bmatrix} x \\ y \end{bmatrix} = \begin{bmatrix} 6 \\ 5 \end{bmatrix}.$

**61.** $\begin{bmatrix} 4 & -1 & 3 \\ 3 & 0 & -1 \\ 0 & 3 & 2 \end{bmatrix} \begin{bmatrix} r \\ s \\ t \end{bmatrix} = \begin{bmatrix} 9 \\ 7 \\ 15 \end{bmatrix}.$

**63.** $2075. **65.** $735,300.

**67. a.** $180,000, $520,000, $400,000, $270,000, $380,000, $640,000; **b.** $390,000, $100,000, $800,000; **c.** $2,390,000;
**d.** $\dfrac{110}{239}, \dfrac{129}{239}.$

**71.** $\begin{bmatrix} 72.82 & -9.8 \\ 51.32 & -36.32 \end{bmatrix}.$ **73.** $\begin{bmatrix} 15.606 & 64.08 \\ -739.428 & 373.056 \end{bmatrix}.$

## PRINCIPLES IN PRACTICE 6.4

**1.** 5 blocks of A, 2 blocks of B, and 1 block of C.
**2.** 3 of X; 4 of Y; 2 of Z.
**3.** $A = 3D; B = 1000 - 2D; C = 500 - D;$
$D = $ any amount $(\leq 500).$

## EXERCISE 6.4 (page 245)

**1.** Not reduced. **3.** Reduced. **5.** Not reduced.

**7.** $\begin{bmatrix} 1 & 0 \\ 0 & 1 \end{bmatrix}.$ **9.** $\begin{bmatrix} 1 & 2 & 3 \\ 0 & 0 & 0 \\ 0 & 0 & 0 \end{bmatrix}.$ **11.** $\begin{bmatrix} 1 & 0 & 0 & 0 \\ 0 & 1 & 0 & 0 \\ 0 & 0 & 1 & 0 \\ 0 & 0 & 0 & 1 \end{bmatrix}.$

**13.** $x = 1, y = 1.$ **15.** No solution.
**17.** $x = -\dfrac{2}{3}r + \dfrac{5}{3}, y = -\dfrac{1}{6}r + \dfrac{7}{6}, z = r,$ where $r$ is any
real number.
**19.** No solution. **21.** $x = -3, y = 1, z = 0.$
**23.** $x = 2, y = -5, z = -1.$
**25.** $x_1 = 0, x_2 = -r, x_3 = -r, x_4 = -r, x_5 = r,$ where $r$ is
any real number.
**27.** Federal, $72,000; state, $24,000.
**29.** A, 2000; B, 4000; C, 5000.
**31. a.** 3 of X, 4 of Z; 2 of X, 1 of Y, 5 of Z; 1 of X; 2 of Y, 6 of
Z; 3 of Y, 7 of Z; **b.** 3 of X, 4 of Z;
**c.** 3 of X, 4 of Z; 3 of Y, 7 of Z.
**33. a.** Let $s, d, g$ represent the numbers of units S, D, G respectively. The six combinations are given by:

| $s$ | 5 | 4 | 3 | 2 | 1 | 0 |
|---|---|---|---|---|---|---|
| $d$ | 8 | 7 | 6 | 5 | 4 | 3 |
| $g$ | 0 | 1 | 2 | 3 | 4 | 5 |

**b.** The combination $s = 0, d = 3, g = 5.$

## PRINCIPLES IN PRACTICE 6.5

**1.** Infinitely many solutions:
$$x + \frac{1}{2}z = 0, y + \frac{1}{2}z = 0;$$

in parametric form: $x = -\dfrac{1}{2}r, y = -\dfrac{1}{2}r, z = r,$

where $r$ is any real number.

## EXERCISE 6.5 (page 251)

**1.** $w = -r - 3s + 2, x = -2r + s - 3, y = r, z = s$
(where $r$ and $s$ are any real numbers).
**3.** $w = -s, x = -3r - 4s + 2, y = r, z = s$
(where $r$ and $s$ are any real numbers).
**5.** $w = -2r + s - 2, x = -r + 4, y = r, z = s$
(where $r$ and $s$ are any real numbers).
**7.** $x_1 = -2r + s - 2t + 1, x_2 = -r - 2s + t + 4,$
$x_3 = r, x_4 = s, x_5 = t$
(where $r, s,$ and $t$ are any real numbers).
**9.** Infinitely many. **11.** Trivial solution.
**13.** Infinitely many. **15.** $x = 0, y = 0.$

**17.** $x = -\dfrac{6}{5}r, y = \dfrac{8}{15}r, z = r.$ **19.** $x = 0, y = 0.$

**21.** $x = r, y = -2r, z = r.$

**23.** $w = -2r, x = -3r, y = r, z = r.$

## PRINCIPLES IN PRACTICE 6.6

**1.** Yes.

**2.** MEET AT NOON FRIDAY.

**3.** $\mathbf{E}^{-1} = \begin{bmatrix} \frac{2}{3} & -\frac{1}{6} & -\frac{1}{3} \\ -\frac{1}{3} & \frac{5}{6} & -\frac{1}{3} \\ -\frac{1}{3} & -\frac{1}{6} & \frac{2}{3} \end{bmatrix}$; $\mathbf{F}$ is not invertible.

**4.** A: 5000 shares; B: 2500 shares; C: 2500 shares

## EXERCISE 6.6 (page 259)

**1.** $\begin{bmatrix} 1 & -1 \\ -5 & 6 \end{bmatrix}$. **3.** Not invertible.

**5.** $\begin{bmatrix} 1 & 0 & 0 \\ 0 & -\frac{1}{3} & 0 \\ 0 & 0 & \frac{1}{4} \end{bmatrix}$. **7.** Not invertible.

**9.** Not invertible (not a square matrix).

**11.** $\begin{bmatrix} 1 & -1 & 0 \\ 0 & 1 & -1 \\ 0 & 0 & 1 \end{bmatrix}$. **13.** $\begin{bmatrix} 1 & 0 & 2 \\ 0 & 1 & 0 \\ 3 & 0 & 7 \end{bmatrix}$.

**15.** $\begin{bmatrix} 1 & -\frac{2}{3} & \frac{5}{3} \\ -1 & \frac{4}{3} & -\frac{10}{3} \\ -1 & 1 & -2 \end{bmatrix}$. **17.** $\begin{bmatrix} \frac{11}{3} & -3 & \frac{1}{3} \\ -\frac{7}{3} & 3 & -\frac{2}{3} \\ \frac{2}{3} & -1 & \frac{1}{3} \end{bmatrix}$.

**19.** $x_1 = 10, x_2 = 6.$ **21.** $x = 17, y = -20.$

**23.** $x = 1, y = 3.$ **25.** $x = -3r + 1, y = r.$

**27.** $x = 0, y = 1, z = 2.$ **29.** $x = 1, y = \dfrac{1}{2}, z = \dfrac{1}{2}.$

**31.** No solution. **33.** $w = 1, x = 3, y = -2, z = 7.$

**35.** $\begin{bmatrix} -\frac{2}{3} & -\frac{1}{3} \\ \frac{1}{3} & -\frac{1}{3} \end{bmatrix}$.

**37. a.** 40 of model A, 60 of model B;

**b.** 45 of model A, 50 of model B.

**39. b.** $\begin{bmatrix} 4 & 6 \\ 7 & 10 \end{bmatrix}$. **41.** Yes.

**43.** D: 5000 shares; E: 1000 shares; F: 4000 shares.

**45. a.** $\begin{bmatrix} 1.46 & 0.56 \\ 0.51 & 1.35 \end{bmatrix}$; **b.** $\begin{bmatrix} \frac{130}{89} & \frac{50}{89} \\ \frac{45}{89} & \frac{120}{89} \end{bmatrix}$.

**47.** $\begin{bmatrix} 1.80 & 1.10 & -0.46 \\ 0.35 & 1.31 & -0.17 \\ 0.44 & 0.42 & 0.59 \end{bmatrix}$.

**49.** $w = 14.44, x = 0.03, y = -0.80, z = 10.33.$

## PRINCIPLES IN PRACTICE 6.7

**1.** 6

## EXERCISE 6.7 (page 268)

**1.** 1. **3.** 0. **5.** $y$. **7.** $-\dfrac{2}{7}$. **9.** 12. **11.** $-12$

**13.** 6. **15.** $\begin{vmatrix} a_{11} & a_{13} & a_{14} \\ a_{21} & a_{23} & a_{24} \\ a_{41} & a_{43} & a_{44} \end{vmatrix}$. **17.** $\begin{vmatrix} a_{21} & a_{22} & a_{24} \\ a_{31} & a_{32} & a_{34} \\ a_{41} & a_{42} & a_{44} \end{vmatrix}$.

**19.** $-16$. **21.** 98. **23.** $-89$. **25.** $-1$. **27.** 2.

**29.** $-90$. **31.** 1. **33.** 24. **35.** 0. **37.** 0.

**39.** 3, 4. **41.** 192. **43. b.** $\dfrac{1}{3}$.

**45.** $c = -1$ or $c = 4$. **47.** $-1630$. **49.** $-3864$.

## EXERCISE 6.8 (page 273)

**1.** $x = \dfrac{9}{5}, y = -\dfrac{2}{5}$. **3.** $x = \dfrac{7}{16}, y = \dfrac{13}{8}$.

**5.** $x = -\dfrac{1}{3}, y = -1$. **7.** $x = \dfrac{6}{5}, z = \dfrac{16}{5}$.

**9.** $x = 4, y = 2, z = 0.$

**11.** $x = \dfrac{2}{3}, y = -\dfrac{28}{15}, z = -\dfrac{26}{15}$.

**13.** $x = 3 - r, y = 0, z = r.$

**15.** $x = 1, y = 3, z = 5.$ **17.** $y = 6, w = 1.$

**19.** Since $\Delta = \begin{vmatrix} 1 & 1 \\ 1 & 1 \end{vmatrix} = 0$, Cramer's rule does not apply.

But the equations in $\begin{cases} x + y = 2, \\ x + y = -3, \end{cases}$ represent distinct parallel lines and hence no solution exists.

**21.** Four games. **23.** $x = 17.85, y = -0.42, z = -24.09.$

## EXERCISE 6.9 (page 277)

**1.** $\begin{bmatrix} 1290 \\ 1425 \end{bmatrix}$; 1405. **3. a.** $\begin{bmatrix} 297.80 \\ 349.54 \\ 443.12 \end{bmatrix}$; **b.** $\begin{bmatrix} 102.17 \\ 125.28 \\ 175.27 \end{bmatrix}$.

**5.** $\begin{bmatrix} 1301 \\ 1215 \\ 1188 \end{bmatrix}$. **7.** $\begin{bmatrix} 1073 \\ 1016 \\ 952 \end{bmatrix}$.

## REVIEW PROBLEMS—CHAPTER 6 (page 279)

**1.** $\begin{bmatrix} 7 & 12 \\ -19 & -5 \end{bmatrix}$. **3.** $\begin{bmatrix} 1 & 35 & 5 \\ 2 & -15 & -7 \\ 1 & 0 & -2 \end{bmatrix}$.

**5.** $\begin{bmatrix} -1 & -2 \\ 5 & 22 \end{bmatrix}$. **7.** $\begin{bmatrix} 6 \\ 32 \end{bmatrix}$. **9.** $\begin{bmatrix} -1 & -2 \\ 2 & 1 \end{bmatrix}$.

**11.** $\begin{bmatrix} 2 & 0 \\ 0 & 17 \end{bmatrix}$. **13.** $x = 3, y = 6.$ **15.** $\begin{bmatrix} 1 & 0 \\ 0 & 1 \end{bmatrix}$.

**17.** $\begin{bmatrix} 1 & 2 & 0 \\ 0 & 0 & 1 \\ 0 & 0 & 0 \end{bmatrix}$. **19.** $x = 0, y = 0.$ **21.** No solution.

**23.** $\begin{bmatrix} -\frac{3}{2} & \frac{5}{6} \\ \frac{1}{2} & -\frac{1}{6} \end{bmatrix}$. **25.** No inverse exists.

**27.** $x = 0, y = 1, z = 0.$ **29.** 18. **31.** 3.

**33.** *rich.*    **35.** $x = 1, y = 2$.    **37.** $-2$.

**39.** $\mathbf{A}^2 = \mathbf{I}_3, \mathbf{A}^{-1} = \mathbf{A}, \mathbf{A}^{1994} = \mathbf{I}_3$.

**41.** $x = 2 - \dfrac{c}{a}, y = \dfrac{c}{a} - 1, z = 1 - \dfrac{a}{c}$.

**43. a.** Let $x, y, z$ represent the weekly doses of capsules of brands I, II, III, respectively. The combinations are given by:

|               | $x$ | $y$ | $z$ |
|---------------|-----|-----|-----|
| combination 1 | 4   | 9   | 0   |
| combination 2 | 3   | 6   | 1   |
| combination 3 | 2   | 3   | 2   |
| combination 4 | 1   | 0   | 3   |

**b.** Combination 4: $x = 1, y = 0, z = 3$.

**45.** $\begin{bmatrix} 215 & 87 \\ 89 & 141 \end{bmatrix}$.    **47.** $\begin{bmatrix} 40.8 \\ 40.56 \end{bmatrix}$.

## MATHEMATICAL SNAPSHOT—CHAPTER 6 (page 283)

**1.** $151.40.

## PRINCIPLES IN PRACTICE 7.1

**1.** $2x + 1.5y > 0.9x + 0.7y + 50$
$y > -1.375x + 62.5$; sketch the dashed line
$y = -1.375x + 62.5$ and shade the half plane above the line. In order to produce a profit, the number of magnets of types A and B produced and sold must be an ordered pair in the region.

**2.** $x \geq 0, y \geq 0, x + y \geq 50, x \geq 2y$
The region consists of points on or above the $x$-axis and on or to the right of the $y$-axis. In addition, the points must be on or above the line $x + y = 50$ and on or below the line $x = 2y$.

## EXERCISE 7.1 (page 289)

**1.**

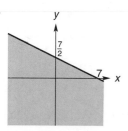

**3.**

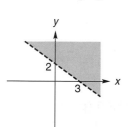

**5.**

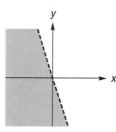

**7.**

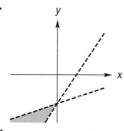

**9.**

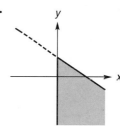

**11.**

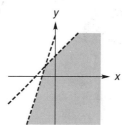

**13.**

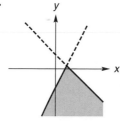

**15.**

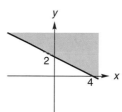

**17.**

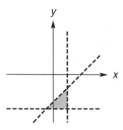

**19.**

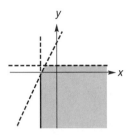

**21.**

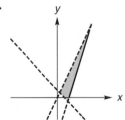

**23.**

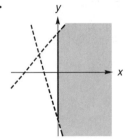

**25.**

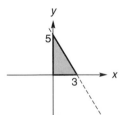

**27.**

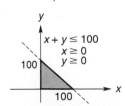

x: number of lb from *A*
y: number of lb from *B*

**29.** $x \geq 0, y \geq 0, 3x + 2y \leq 24, 0.5x + y \leq 8.$

## EXERCISE 7.2 (page 298)

**1.** $P = 640$ when $x = 40, y = 20.$
**3.** $Z = -10$ when $x = 2, y = 3.$
**5.** No optimum solution (empty feasible region).
**7.** $Z = 3$ when $x = 0, y = 1.$
**9.** $C = 2.4$ when $x = \dfrac{3}{5}, y = \dfrac{6}{5}.$
**11.** No optimum solution (unbounded).
**13.** 15 widgets, 25 wadgits; $210.
**15.** 4 units of food A, 4 units of food B; $8.
**17.** 10 tons of ore I, 10 tons of ore II; $1100.
**19.** 6 chambers of type A and 10 chambers of type B.
**21. c.** $x = y = 75.$
**23.** $Z = 15.54$ when $x = 2.56, y = 6.74.$
**25.** $Z = -75.98$ when $x = 9.48, y = 16.67.$

## PRINCIPLES IN PRACTICE 7.3

**1.** Ship $10t + 15$ TV sets from C to A, $-10t + 30$ TV sets from C to B, $-10t + 10$ TV sets from D to A, and $10t$ TV sets from D to B, for $0 \leq t \leq 1$; minimum cost $780.

## EXERCISE 7.3 (page 302)

**1.** $Z = 33$ when $x = (1 - t)(2) + 5t = 2 + 3t,$
$y = (1 - t)(3) + 2t = 3 - t,$ and $0 \leq t \leq 1.$
**3.** $Z = 72$ when $x = (1 - t)(3) + 4t = 3 + t,$
$y = (1 - t)(2) + 0t = 2 - 2t,$ and $0 \leq t \leq 1.$

## PRINCIPLES IN PRACTICE 7.4

**1.** 0 gadgets of Type 1, 72 gadgets of Type 2, 12 gadgets of Type 3; maximum profit of $20,400.

## EXERCISE 7.4 (page 313)

**1.** $Z = 8$ when $x_1 = 0, x_2 = 4.$
**3.** $Z = 14$ when $x_1 = 1, x_2 = 5.$
**5.** $Z = 28$ when $x_1 = 3, x_2 = 2.$
**7.** $Z = 20$ when $x_1 = 0, x_2 = 5, x_3 = 0.$
**9.** $Z = 2$ when $x_1 = 1, x_2 = 0, x_3 = 0.$
**11.** $Z = \dfrac{16}{3}$ when $x_1 = \dfrac{2}{3}, x_2 = \dfrac{14}{3}.$
**13.** $W = 13$ when $x_1 = 1, x_2 = 0, x_3 = 3.$
**15.** $Z = 600$ when $x_1 = 4, x_2 = 1, x_3 = 4, x_4 = 0.$
**17.** 400 from A, 1600 from B; $1100.
**19.** 0 chairs, 300 rockers, 100 chaise lounges; $3600.

## PRINCIPLES IN PRACTICE 7.5

**1.** $35 - 7t$ of device 1, $6t$ of device 2, 0 of device 3, for $0 \leq t \leq 1.$

## EXERCISE 7.5 (page 319)

**1.** Yes; for the tableau, $x_2$ is the entering variable and the quotients $\dfrac{6}{2}$ and $\dfrac{3}{1}$ tie for being the smallest.

**3.** No optimum solution (unbounded).
**5.** $Z = 12$ when $x_1 = 4 + t$, $x_2 = t$, and $0 \le t \le 1$.
**7.** No optimum solution (unbounded).
**9.** $Z = 13$ when $x_1 = \dfrac{3}{2} - \dfrac{3}{2}t$, $x_2 = 6t$, $x_3 = 4 - 3t$, and
$0 \le t \le 1$.
**11.** $3800. If $x_1$, $x_2$, $x_3$ denote the number of chairs, rockers, and chaise lounges produced, respectively, then
$x_1 = 100 - 100t$,
$x_2 = 100 + 150t$,
$x_3 = 200 - 50t$, and
$0 \le t \le 1$.

## PRINCIPLES IN PRACTICE 7.6

**1.** Plant I: 500 standard, 700 deluxe; plant II: 500 standard, 100 deluxe; $89,500 maximum profit.

## EXERCISE 7.6 (page 329)

**1.** $Z = 7$ when $x_1 = 1$, $x_2 = 5$.
**3.** $Z = 4$ when $x_1 = 1$, $x_2 = 2$, $x_3 = 0$.
**5.** $Z = \dfrac{58}{3}$ when $x_1 = \dfrac{14}{3}$, $x_2 = \dfrac{2}{3}$, $x_3 = 0$.
**7.** $Z = -17$ when $x_1 = 3$, $x_2 = 2$.
**9.** No optimum solution (empty feasible region).
**11.** $Z = 2$ when $x_1 = 6$, $x_2 = 10$.
**13.** 255 Standard bookcases, 0 Executive bookcases.
**15.** 30% in A, 0% in AA, 70% in AAA; 6.6%.

## EXERCISE 7.7 (page 333)

**1.** $Z = 54$ when $x_1 = 2$, $x_2 = 8$.
**3.** $Z = 36$ when $x_1 = 9$, $x_2 = 0$, $x_3 = 0$.
**5.** $Z = 4$ when $x_1 = 0$, $x_2 = 0$, $x_3 = 4$.
**7.** $Z = 0$ when $x_1 = 3$, $x_2 = 0$, $x_3 = 1$.
**9.** $Z = 28$ when $x_1 = 3$, $x_2 = 0$, $x_3 = 5$.
**11.** Install device A on kilns producing 700,000 barrels annually, and device B on kilns producing 2,600,000 barrels annually.
**13.** To Exton, 10 from A and 20 from B; to Whyton, 30 from A; $760.
**15. a.** Column 3: 1, 3, 3; column 4: 0, 4, 8.
**b.** $x_1 = 10$, $x_2 = 0$, $x_3 = 20$, $x_4 = 0$.
**c.** 90 in.

## PRINCIPLES IN PRACTICE 7.8

**1.** Minimize $W = 60{,}000y_1 + 2000y_2 + 120y_3$ subject to
$300y_1 + 20y_2 + 3y_3 \ge 300$,
$220y_1 + 40y_2 + y_3 \ge 200$,
$180y_1 + 20y_2 + 2y_3 \ge 200$,
and $y_1, y_2, y_3 \ge 0$.
**2.** Maximize $W = 98y_1 + 80y_2$ subject to
$20y_1 + 8y_2 \le 6$,
$6y_1 + 16y_2 \le 2$,
and $y_1, y_2 \ge 0$.
**3.** 5 device 1, 0 device 2, 15 device 3.

## EXERCISE 7.8 (page 342)

**1.** Minimize $W = 6y_1 + 4y_2$ subject to
$y_1 - y_2 \ge 2$,
$y_1 + y_2 \ge 3$,
$y_1, y_2 \ge 0$.
**3.** Maximize $W = 8y_1 + 2y_2$ subject to
$y_1 - y_2 \le 1$,
$y_1 + 2y_2 \le 8$,
$y_1 + y_2 \le 5$,
$y_1, y_2 \ge 0$.
**5.** Minimize $W = 13y_1 - 3y_2 - 11y_3$ subject to
$-y_1 + y_2 - y_3 \ge 1$,
$2y_1 - y_2 - y_3 \ge -1$,
$y_1, y_2, y_3 \ge 0$.
**7.** Maximize $W = -3y_1 + 3y_2$ subject to
$-y_1 + y_2 \le 4$,
$y_1 - y_2 \le 4$,
$y_1 + y_2 \le 6$,
$y_1, y_2 \ge 0$.
**9.** $Z = 11$ when $x_1 = 0$, $x_2 = \dfrac{1}{2}$, $x_3 = \dfrac{3}{2}$.
**11.** $Z = 26$ when $x_1 = 6$, $x_2 = 1$.
**13.** $Z = 14$ when $x_1 = 1$, $x_2 = 2$.
**15.** $25 on newspaper advertising, $140 on radio advertising; $165.
**17.** 20 shipping clerk apprentices, 40 shipping clerks, 90 semiskilled workers, 0 skilled workers; $600.

## REVIEW PROBLEMS—CHAPTER 7 (page 344)

**1.**

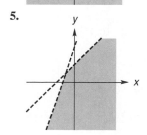

**3.**

**5.**

**7.**

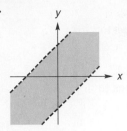

**9.**

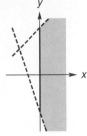

**11.** $Z = 3$ when $x = 3$, $y = 0$.
**13.** $Z = -2$ when $x = 0$, $y = 2$.
**15.** No optimum solution (empty feasible region).
**17.** $Z = 36$ when $x = 2 + 2t$, $y = 3 - 3t$, and $0 \le t \le 1$.
**19.** $Z = 32$ when $x_1 = 8$, $x_2 = 0$.
**21.** $Z = 2$ when $x_1 = 0$, $x_2 = 0$, $x_3 = 2$.
**23.** $Z = 24$ when $x_1 = 0$, $x_2 = 12$.
**25.** $Z = \dfrac{7}{2}$ when $x_1 = \dfrac{5}{4}$, $x_2 = 0$, $x_3 = \dfrac{9}{4}$.
**27.** No optimum solution (unbounded).
**29.** $Z = 70$ when $x_1 = 35$, $x_2 = 0$, $x_3 = 0$.
**31.** 0 units of X, 6 units of Y, 14 units of Z; $398.
**33.** 500,000 gal from A to D, 100,000 gal from A to C, 400,000 gal from B to C; $19,000.
**35.** 100 ounces of food A only.
**37.** $Z = 117.88$ when $x = 7.23$, $y = 3.40$.

## MATHEMATICAL SNAPSHOT—CHAPTER 7 (page 348)

**1.** 2 minutes of radiation.

## PRINCIPLES IN PRACTICE 8.1

**1.** 4.9%.　　**2.** 7 years, 16 days.　　**3.** 7.7208%.
**4.** The $10,000 investment is slightly better over 20 years.

## EXERCISE 8.1 (page 353)

**1. a.** $11,105.58; **b.** $5105.58.
**3.** 8.243%.　　**5.** 8.328%.
**7. a.** 10%; **b.** 10.25%; **c.** 10.381%; **d.** 10.471%;
**e.** 10.516%.
**9.** 8.08%.　　**11.** 9.0 years.　　**13.** $10,282.95.
**15.** $33,130.68.
**17. a.** 18%; **b.** $19.56%.　　**19.** $3198.54.
**21.** 8% compounded annually.
**23. a.** 5.47%; **b.** 5.39%.　　**25.** 11.61%.　　**27.** 11.11%

## EXERCISE 8.2 (page 357)

**1.** $2261.34.　　**3.** $1751.83.　　**5.** $5118.10.
**7.** $4862.31.　　**9.** $6838.95.　　**11.** $8874.49.
**13.** $14,091.10.　　**15.** $1238.58.　　**17.** $1963.28.
**19. a.** $515.62; **b.** profitable.　　**21.** Savings account.
**23.** $103.56.　　**25.** 9.55%.

## PRINCIPLES IN PRACTICE 8.3

**1.** 48 ft, 36 ft, 27 ft, $20\dfrac{1}{4}$ ft, $15\dfrac{3}{16}$ ft.
**2.** 750, 1125, 1688, 2531, 3797, 5695.　　**3.** 35.72 m.
**4.** $176,994.65.　　**5.** 6.20%.　　**6.** $101,925; $121,925.
**7.** $723.03.　　**8.** $11,740.51.　　**9.** $57,355.58.
**10.** $62,345.51.

## EXERCISE 8.3 (page 366)

**1.** 64, 32, 16, 8, 4.　　**3.** 100, 102, 104.04.　　**5.** $\dfrac{422}{243}$.
**7.** 1.11111.　　**9.** 18.664613.　　**11.** 8.213180.
**13.** $2050.10.　　**15.** $29,984.06.　　**17.** $8001.24.
**19.** $90,231.01.　　**21.** $204,977.46.　　**23.** $24,594.36.
**25.** $1937.14.　　**27.** $458.40.
**29. a.** $3048.85; **b.** $648.85.　　**31.** $3474.12.
**33.** $1725.　　**35.** 102.91305.　　**37.** 55,360.30.
**39.** $131.34.　　**41.** $418,288.84.
**43.** $205,073; $142,146.

## EXERCISE 8.4 (page 372)

**1.** $69.33.　　**3.** $1565.56.
**5. a.** $249.11; **b.** $75; **c.** $174.11.

**7.**

| Period | Prin. Outs. at Beginning | Interest for Period | Pmt. at End | Prin. Repaid at End |
|--------|--------------------------|---------------------|-------------|---------------------|
| 1 | 5000.00 | 350.00 | 1476.14 | 1126.14 |
| 2 | 3873.86 | 271.17 | 1476.14 | 1204.97 |
| 3 | 2668.89 | 186.82 | 1476.14 | 1289.32 |
| 4 | 1379.57 | 96.57 | 1476.14 | 1379.57 |
| Total | | 904.56 | 5904.56 | 5000.00 |

**9.**

| Period | Prin. Outs. at Beginning | Interest for Period | Pmt. at End | Prin. Repaid at End |
|---|---|---|---|---|
| 1 | 900.00 | 22.50 | 193.72 | 171.22 |
| 2 | 728.78 | 18.22 | 193.72 | 175.50 |
| 3 | 553.28 | 13.83 | 193.72 | 179.89 |
| 4 | 373.39 | 9.33 | 193.72 | 184.39 |
| 5 | 189.00 | 4.73 | 193.73 | 189.00 |
| Total | | 68.61 | 968.61 | 900.00 |

**11.** 11.    **13.** $1273.
**15. a.** $415.28; **b.** $382.50; **c.** $32.78; **d.** $79,584.
**17.** 23.    **19.** $74,417.    **21.** $38.64.

## REVIEW PROBLEMS—CHAPTER 8 (page 374)

**1.** $\frac{63}{16}$.    **3.** 8.5% compounded annually.
**5.** $586.60.    **7. a.** $1997.13; **b.** $3325.37.
**9.** $936.85.    **11.** $886.98.    **13.** $314.00.
**15.**

| Period | Prin. Outs. at Beginning | Interest for Period | Pmt. at End | Prin. Repaid at End |
|---|---|---|---|---|
| 1 | 15,000.00 | 112.50 | 3067.84 | 2955.34 |
| 2 | 12,044.66 | 90.33 | 3067.84 | 2977.51 |
| 3 | 9067.15 | 68.00 | 3067.84 | 2999.84 |
| 4 | 6067.31 | 45.50 | 3067.84 | 3022.34 |
| 5 | 3044.97 | 22.84 | 3067.81 | 3044.97 |
| Total | | 339.17 | 15,339.17 | 15,000.00 |

**17.** $3296.32.

## MATHEMATICAL SNAPSHOT—CHAPTER 8 (page 378)

**1.** $80.77.    **3.** $7158.28.

## EXERCISE 9.1 (page 384)

**1.**

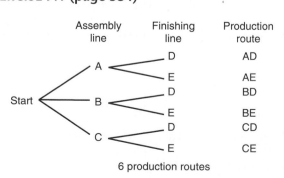

6 production routes

**3.**

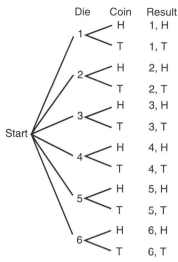

12 possible results

**5.** 20.    **7.** 96.    **9.** 1024.    **11.** 20.    **13.** 720.
**15.** 720.    **17.** 1000; error message is displayed.
**19.** 6.    **21.** 336.    **23.** 216.    **25.** 1320.    **27.** 336.
**29.** 720.    **31.** 2520; 5040.    **33.** 624.    **35.** 24.
**37. a.** 11,880; **b.** 19,008.    **39.** 48.    **41.** 2880.

## EXERCISE 9.2 (page 392)

**1.** 15.    **3.** 1.    **5.** 18.    **9.** 3003.    **11.** 66.
**13.** $\frac{74!}{10! \cdot 64!}$.    **15.** 56.    **17.** 1680.    **19.** 15.
**21.** 720.    **23.** 1680.    **25.** 70.    **27.** 756,756.
**29. a.** 90; **b.** 330.    **31.** 17,325.
**33. a.** 1; **b.** 1; **c.** 18.    **35.** 3744.    **37.** 9,189,180.

## PRINCIPLES IN PRACTICE 9.3

**1.** 10,586,800.

## EXERCISE 9.3 (page 401)

**1.** {9D, 9H, 9C, 9S}.
**3.** {1H, 1T, 2H, 2T, 3H, 3T, 4H, 4T, 5H, 5T, 6H, 6T}.
**5.** {lo, lv, le, ov, oe, ve, ol, vl, el, vo, eo, ev}.
**7. a.** {RR, RW, RB, WR, WW, WB, BR, BW, BB};
**b.** {RW, RB, WR, WB, BR, BW}.
**9.** Sample space consists of ordered sets of six elements and each element is H or T; 64.
**11.** Sample space consists of ordered pairs where first element indicates card drawn and second element indicates number on die; 312.
**13.** Sample space consists of combinations of 52 cards taken 13 at a time; $_{52}C_{13}$.
**15.** {1, 3, 5, 7, 9}.    **17.** {3, 5}.    **19.** {1, 2, 4, 6, 8, 10}.
**21.** $S$.    **23.** $E_1$ and $E_4$, $E_2$ and $E_3$, $E_3$ and $E_4$.
**25.** $E$ and $H$, $G$ and $H$, $H$ and $I$.
**27. a.** {HHH, HHT, HTH, HTT, THH, THT, TTH, TTT};
**b.** {HHH, HHT, HTH, HTT, THH, THT, TTH};
**c.** {HHT, HTH, HTT, THH, THT, TTH, TTT}; **d.** $S$;

**e.** {HHT, HTH, HTT, THH, THT, TTH}; **f.** $\varnothing$;
**g.** {HHH, TTT}.
**29. a.** {ABC, ACB, BAC, BCA, CAB, CBA}.
**b.** {ABC, ACB}; **c.** {BAC, BCA, CAB, CBA}.

## EXERCISE 9.4 (page 412)

**1.** 600. **3. a.** 0.8; **b.** 0.4. **5.** No.

**7. a.** $\frac{5}{36}$; **b.** $\frac{1}{12}$; **c.** $\frac{1}{4}$; **d.** $\frac{1}{36}$; **e.** $\frac{1}{2}$; **f.** $\frac{1}{2}$; **g.** $\frac{5}{6}$.

**9. a.** $\frac{1}{52}$; **b.** $\frac{1}{4}$; **c.** $\frac{1}{13}$; **d.** $\frac{1}{2}$; **e.** $\frac{1}{2}$; **f.** $\frac{1}{52}$; **g.** $\frac{4}{13}$; **h.** $\frac{1}{26}$;
**i.** 0.

**11. a.** $\frac{1}{624}$; **b.** $\frac{4}{624} = \frac{1}{156}$; **c.** $\frac{8}{624} = \frac{1}{78}$; **d.** $\frac{39}{624} = \frac{1}{16}$.

**13. a.** $\frac{12}{2652} = \frac{1}{221}$; **b.** $\frac{338}{2652} = \frac{13}{102}$.

**15. a.** $\frac{1}{8}$; **b.** $\frac{3}{8}$; **c.** $\frac{1}{8}$; **d.** $\frac{7}{8}$. **17. a.** $\frac{4}{5}$; **b.** $\frac{1}{5}$.

**19. a.** 0.1; **b.** 0.35; **c.** 0.7; **d.** 0.95; **e.** 0.1, 0.35, 0.7, 0.95.

**21.** $\frac{1}{10}$. **23. a.** $\frac{1}{2^{10}} = \frac{1}{1024}$; **b.** $\frac{11}{1024}$.

**25.** $\dfrac{13 \cdot {}_4C_3 \cdot 12 \cdot {}_4C_2}{{}_{52}C_5}$.

**27. a.** $\frac{6545}{161,700} \approx 0.040$; **b.** $\frac{4140}{161,700} \approx 0.026$.

**29.** $\frac{1}{9}$. **31. a.** 0.51; **b.** 0.44; **c.** 0.03. **33.** 4:1.

**35.** 3:7. **37.** $\frac{5}{9}$. **39.** $\frac{2}{7}$. **41.** $\frac{3}{4}$.

## EXERCISE 9.5 (page 425)

**1. a.** $\frac{2}{5}$; **b.** $\frac{3}{5}$; **c.** $\frac{1}{3}$; **d.** $\frac{2}{3}$; **e.** $\frac{1}{3}$. **3.** 1. **5.** 0.37.

**7. a.** $\frac{1}{2}$; **b.** $\frac{2}{3}$. **9. a.** $\frac{3}{5}$; **b.** $\frac{2}{5}$; **c.** $\frac{1}{2}$; **d.** $\frac{2}{9}$.

**11. a.** $\frac{5}{8}$; **b.** $\frac{35}{58}$; **c.** $\frac{11}{39}$; **d.** $\frac{8}{25}$; **e.** $\frac{10}{47}$; **f.** $\frac{25}{86}$.

**13. a.** $\frac{1}{2}$; **b.** $\frac{4}{9}$. **15.** $\frac{2}{3}$. **17. a.** $\frac{1}{2}$; **b.** $\frac{1}{4}$. **19.** $\frac{2}{3}$.

**21.** $\frac{1}{11}$. **23.** $\frac{1}{6}$. **25.** $\frac{1}{2}$. **27.** $\frac{1}{13}$. **29.** $\frac{40}{51}$.

**31.** $\frac{8}{16,575}$. **33.** $\frac{11}{850}$. **35.** $\frac{2}{17}$.

**37. a.** $\frac{47}{100}$; **b.** $\frac{27}{47}$. **39. a.** $\frac{3}{4}$; **b.** $\frac{3}{5}$. **41.** $\frac{9}{20}$.

**43.** $\frac{1}{4}$. **45.** $\frac{1}{25}$. **47.** 0.049.

**49. a.** 0.06; **b.** 0.155. **51.** $\frac{1}{8}$.

## EXERCISE 9.6 (page 436)

**1. a.** $\frac{1}{4}$; **b.** $\frac{5}{6}$; **c.** $\frac{1}{3}$; **d.** $\frac{2}{3}$; **e.** $\frac{1}{12}$; **f.** $\frac{1}{2}$; **g.** $\frac{1}{3}$. **3.** $\frac{5}{6}$.

**5.** Independent. **7.** Independent. **9.** Dependent.
**11.** Dependent.
**13. a.** Independent; **b.** independent; **c.** independent;

**d.** no. **15.** Dependent. **17.** $\frac{1}{18}$. **19.** $\frac{1}{25}$.

**21.** $\frac{3}{676}$. **23. a.** $\frac{3}{10}$; **b.** $\frac{1}{40}$; **c.** $\frac{1}{10}$.

**25. a.** $\frac{2}{5}$; **b.** $\frac{1}{5}$; **c.** $\frac{7}{15}$; **d.** $\frac{13}{15}$; **e.** $\frac{2}{15}$.

**27. a.** $\frac{2}{15}$; **b.** $\frac{4}{15}$. **29.** $\frac{7}{18}$. **31.** $\frac{3}{200}$.

**33. a.** $\frac{1}{1728}$; **b.** $\frac{3}{8}$. **35. a.** $\frac{15}{1024}$; **b.** $\frac{1}{64}$; **c.** $\frac{53}{512}$.

**37.** 0.012.

## EXERCISE 9.7 (page 446)

**1.** $P(E \mid D) = \frac{1}{4}, P(F \mid D') = \frac{4}{7}$. **3.** $\frac{12}{31} \approx 0.387$.

**5. a.** $\frac{49}{106} \approx 0.462$; **b.** $\frac{1}{894} \approx 0.001$. **7.** $\frac{5}{8}$. **9.** $\frac{1}{6}$.

**11.** $\frac{81}{89} \approx 0.910$. **13.** $\approx 55.1\%$. **15.** $\frac{3}{4}$.

**17.** $\frac{24}{29} \approx 0.828$. **19.** $\frac{4}{5}$. **21.** $\frac{56}{59} \approx 0.949$.

**23. a.** $\frac{41}{200} = 0.205$; **b.** $\frac{24}{41} \approx 0.585$; **c.** $\frac{23}{200} = 0.115$.

**25. a.** 0.18; **b.** 0.23; **c.** 0.59; **d.** high quality.

**27.** $\frac{28}{31} \approx 0.90$.

## REVIEW PROBLEMS—CHAPTER 9 (page 451)

**1.** 336. **3.** 36. **5.** 608,400. **7.** 32. **9.** 210.
**11.** 56. **13. a.** 2024; **b.** 253. **15.** 34,650.
**17.** 560.
**19. a.** {1, 2, 3, 4, 5, 6, 7}; **b.** {4, 5, 6}; **c.** {4, 5, 6, 7, 8}; **d.** $\varnothing$
**e.** {4, 5, 6, 7, 8}; **f.** no.
**21. a.** {$R_1R_2R_3$, $R_1R_2G_3$, $R_1G_2R_3$, $R_1G_2G_3$, $G_1R_2R_3$,
$G_1R_2G_3$, $G_1G_2R_3$, $G_1G_2G_3$}; **b.** {$R_1R_2G_3$, $R_1G_2R_3$, $G_1R_2R_3$};
**c.** {$R_1R_2R_3$, $G_1G_2G_3$}.

**23.** 0.2. **25.** $\frac{45}{512}$. **27. a.** $\frac{4}{25}$; **b.** $\frac{2}{15}$.

**29. a.** $\frac{1}{4}$; **b.** $\frac{1}{4}$. **31.** 3:5. **33.** $\frac{6}{7}$. **35.** $\frac{3}{13}$.

**37.** 0.42. **39. a.** $\frac{2}{11}$; **b.** $\frac{1}{18}$. **41.** $\frac{1}{4}$.

**43. a.** $\frac{1}{3}$; **b.** independent. **45.** Dependent.

**47. a.** 0.0081; **b.** 0.2646; **c.** 0.3483. **49.** $\frac{22}{45}$.

**51.** $\frac{1}{4}$. **53. a.** 0.014; **b.** $\frac{4}{7} \approx 0.57$.

## EXERCISE 10.1 (page 461)

**1.** $\mu = 1.7$; $\text{Var}(X) = 1.01$; $\sigma \approx 1.00$.

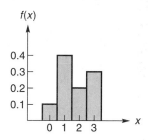

**3.** $\mu = \dfrac{9}{4} = 2.25$; $\text{Var}(X) = \dfrac{11}{16} = 0.6875$; $\sigma \approx 0.83$.

**5. a.** 0.1; **b.** 5; **c.** 3.

**7.** $E(X) = \dfrac{3}{2} = 1.5$; $\sigma^2 = \dfrac{3}{4} = 0.75$; $\sigma \approx 0.87$.

**9.** $E(X) = \dfrac{4}{5} = 0.8$; $\sigma^2 = \dfrac{9}{25} = 0.36$; $\sigma = \dfrac{3}{5} = 0.6$.

**11.** $f(0) = \dfrac{1}{10}, f(1) = \dfrac{3}{5}, f(2) = \dfrac{3}{10}$.

**13. a.** −$0.15 (a loss); **b.** −$0.30 (a loss). **15.** $101.43.
**17.** $3.00. **19.** $66. **21.** Loss of $0.25; $1.

## PRINCIPLES IN PRACTICE 10.2

**1.**

| $x$ | $P(x)$ |
|---|---|
| 0 | $\dfrac{2401}{10,000}$ |
| 1 | $\dfrac{4116}{10,000}$ |
| 2 | $\dfrac{2646}{10,000}$ |
| 3 | $\dfrac{756}{10,000}$ |
| 4 | $\dfrac{81}{10,000}$ |

## EXERCISE 10.2 (page 467)

**1.** $f(0) = \dfrac{9}{16}, f(1) = \dfrac{3}{8}, f(2) = \dfrac{1}{16}$; $\mu = \dfrac{1}{2}$; $\sigma = \dfrac{\sqrt{6}}{4}$.

**3.** $f(0) = \dfrac{1}{27}, f(1) = \dfrac{2}{9}, f(2) = \dfrac{4}{9}, f(3) = \dfrac{8}{27}$; $\mu = 2$;

$\sigma = \dfrac{\sqrt{6}}{3}$. **5.** 0.001536. **7.** $\dfrac{96}{625} = 0.1536$. **9.** $\dfrac{3}{16}$.

**11.** $\dfrac{45}{1024} \approx 0.044$. **13.** $\dfrac{96}{625} = 0.1536$. **15.** 0.002.

**17. a.** $\dfrac{9}{64}$; **b.** $\dfrac{5}{32}$. **19.** $\dfrac{16}{27} \approx 0.593$. **21.** 0.7599.

**23.** $\dfrac{13}{16}$. **25.** $\dfrac{2187}{8192} \approx 0.267$.

## EXERCISE 10.3 (page 476)

**1.** No. **3.** No. **5.** Yes. **7.** $a = \dfrac{1}{3}, b = \dfrac{3}{4}$.

**9.** $a = 0.4, b = 0.5, c = 0.1$. **11.** Yes. **13.** No.

**15.** $\mathbf{X_1} = \begin{bmatrix} \dfrac{11}{12} & \dfrac{1}{12} \end{bmatrix}, \mathbf{X_2} = \begin{bmatrix} \dfrac{25}{36} & \dfrac{11}{36} \end{bmatrix}, \mathbf{X_3} = \begin{bmatrix} \dfrac{83}{108} & \dfrac{25}{108} \end{bmatrix}$.

**17.** $\mathbf{X_1} = [0.5 \quad 0.5], \mathbf{X_2} = [0.5 \quad 0.5], \mathbf{X_3} = [0.5 \quad 0.5]$.

**19.** $\mathbf{X_1} = [0.26 \quad 0.28 \quad 0.46], \mathbf{X_2} = [0.164 \quad 0.302 \quad 0.534]$,
$\mathbf{X_3} = [0.1766 \quad 0.3138 \quad 0.5096]$.

**21. a.** $\mathbf{T}^2 = \begin{bmatrix} \dfrac{5}{8} & \dfrac{3}{8} \\ \dfrac{3}{8} & \dfrac{5}{8} \end{bmatrix}, \mathbf{T}^3 = \begin{bmatrix} \dfrac{7}{16} & \dfrac{9}{16} \\ \dfrac{9}{16} & \dfrac{7}{16} \end{bmatrix}$; **b.** $\dfrac{3}{8}$; **c.** $\dfrac{9}{16}$.

**23. a.** $\mathbf{T}^2 = \begin{bmatrix} 0.50 & 0.40 & 0.10 \\ 0.23 & 0.69 & 0.08 \\ 0.27 & 0.54 & 0.19 \end{bmatrix}$,

$\mathbf{T}^3 = \begin{bmatrix} 0.230 & 0.690 & 0.080 \\ 0.369 & 0.530 & 0.101 \\ 0.327 & 0.543 & 0.130 \end{bmatrix}$; **b.** 0.40; **c.** 0.369.

**25.** $\begin{bmatrix} \dfrac{3}{5} & \dfrac{2}{5} \end{bmatrix}$. **27.** $\begin{bmatrix} \dfrac{3}{7} & \dfrac{4}{7} \end{bmatrix}$. **29.** $[0.5 \quad 0.25 \quad 0.25]$.

**31. a.** $\begin{array}{cc} \text{Flu} & \text{No Flu} \end{array}$
$\begin{bmatrix} 0.1 & 0.9 \\ 0.2 & 0.8 \end{bmatrix}$; **b.** 37, 36.

**33. a.** $\begin{array}{cc} & A \quad B \end{array}$
$\begin{array}{c} A \\ B \end{array}\begin{bmatrix} 0.9 & 0.1 \\ 0.3 & 0.7 \end{bmatrix}$; **b.** 0.804.

**35. a.** $\begin{array}{ccc} D & R & O \end{array}$
$\begin{array}{c} D \\ R \\ O \end{array}\begin{bmatrix} 0.8 & 0.1 & 0.1 \\ 0.1 & 0.8 & 0.1 \\ 0.3 & 0.1 & 0.6 \end{bmatrix}$; **b.** 0.19; **c.** 38%.

**37. a.** $\begin{array}{cc} & A \quad\quad \text{Compet.} \end{array}$
$\begin{array}{c} A \\ \text{Compet.} \end{array}\begin{bmatrix} 0.8 & 0.2 \\ 0.3 & 0.7 \end{bmatrix}$; **b.** 65%; **c.** 60%.

**39. a.** $\begin{array}{cc} & 1 \quad\quad 2 \end{array}$
$\begin{array}{c} 1 \\ 2 \end{array}\begin{bmatrix} \dfrac{5}{7} & \dfrac{2}{7} \\ \dfrac{3}{7} & \dfrac{4}{7} \end{bmatrix}$;

**b.** 59.18% in compartment 1, 40.82% in compartment 2;
**c.** 60% in compartment 1, 40% in compartment 2.

**41. a.** $\begin{bmatrix} \dfrac{2}{3} & \dfrac{1}{3} \end{bmatrix}$; **b.** $33\dfrac{1}{3}\%$.

## REVIEW PROBLEMS—CHAPTER 10 (page 480)

**1.** $\mu = 1.5$, $\text{Var}(X) = 0.65$, $\sigma = 0.81$.

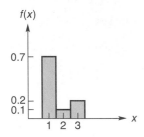

**3. a.** $f(1) = \dfrac{1}{12}, f(2) = f(3) = f(4) = f(5) =$

$f(6) = \dfrac{1}{6}, f(7) = \dfrac{1}{12}$; **b.** 4.    **5.** $0.10 (a loss)

**7. a.** $176; **b.** $704,000.

**9.** $f(0) = 0.729, f(1) = 0.243, f(2) = 0.027$,

$\mu = 0.3; \sigma = \sqrt{0.27} \approx 0.52$.    **11.** $\dfrac{1}{64}$.    **13.** $\dfrac{8}{81}$.

**15.** $\dfrac{11}{27}$.    **17.** $a = 0.3, b = 0.2, c = 0.5$.

**19.** $\mathbf{X_1} = [0.10 \quad 0.15 \quad 0.75], \mathbf{X_2} = [0.130 \quad 0.155 \quad 0.715]$.

**21. a.** $\mathbf{T}^2 = \begin{bmatrix} \dfrac{19}{49} & \dfrac{30}{49} \\ \dfrac{15}{49} & \dfrac{34}{49} \end{bmatrix}, \mathbf{T}^3 = \begin{bmatrix} \dfrac{109}{343} & \dfrac{234}{343} \\ \dfrac{117}{343} & \dfrac{226}{343} \end{bmatrix}$; **b.** $\dfrac{30}{49}$;

**c.** $\dfrac{117}{343}$.    **23.** $\begin{bmatrix} \dfrac{1}{2} & \dfrac{1}{2} \end{bmatrix}$.

**25. a.** 76%; **b.** 74.4% Japanese, 25.6% non-Japanese; **c.** 75% Japanese, 25% non-Japanese.

## MATHEMATICAL SNAPSHOT—CHAPTER 10 (page 484)

**1.** $\dfrac{3}{8}$.

## PRINCIPLES IN PRACTICE 11.1

**1.** The limit as $x \to a$ does not exist if $a$ is an integer, but it exists if $a$ is any other value.
**2.** $36\pi$ cc.    **3.** 3616.    **4.** 20.    **5.** 2.

## EXERCISE 11.1 (page 494)

**1. a.** 1; **b.** 0; **c.** 1.    **3. a.** 1; **b.** does not exist; **c.** 3.
**5.** $f(0.9) = 2.8, f(0.99) = 2.98, f(0.999) = 2.998$,
$f(1.001) = 3.002, f(1.01) = 3.02, f(1.1) = 3.2; 3$.
**7.** $f(-0.1) \approx 0.9516, f(-0.01) \approx 0.9950$,
$f(-0.001) \approx 0.9995, f(0.001) \approx 1.0005, f(0.01) \approx 1.0050$.
$f(0.1) \approx 1.0517; 1$.

**9.** 16.    **11.** 20.    **13.** $-1$.    **15.** $-\dfrac{5}{2}$.    **17.** 0.

**19.** 5.    **21.** $-2$.    **23.** 3.    **25.** 0.    **27.** $\dfrac{1}{6}$.

**29.** $\dfrac{1}{5}$.    **31.** $\dfrac{11}{9}$.    **33.** 4.    **35.** $2x$.    **37.** $-1$.

**39.** $2x$.    **41.** $4x - 3$.    **43.** $\dfrac{1}{4}$.    **45. a.** 1; **b.** 0.

**47.** 11.00.    **49.** $-7.00$.    **51.** Does not exist.

## PRINCIPLES IN PRACTICE 11.2

**1.** $\lim\limits_{x \to \infty} p(x) = 0$. The graph starts out high and quickly goes down toward zero. Accordingly, consumers are willing to purchase large quantities of the product at prices close to 0.
**2.** $\lim\limits_{x \to \infty} y(x) = 500$. The greatest yearly sales they can expect with unlimited advertising is $500,000.
**3.** $\lim\limits_{x \to \infty} C(x) = \infty$. This means that the cost continues to increase without bound as more units are made.
**4.** The limit does not exist; $250.

## EXERCISE 11.2 (page 503)

**1. a.** 2; **b.** 3; **c.** does not exist; **d.** $-\infty$; **e.** $\infty$; **f.** $\infty$; **g.** $\infty$;
**h.** 0; **i.** 1; **j.** 1; **k.** 1.    **3.** 1.    **5.** $-\infty$.    **7.** $-\infty$.
**9.** $\infty$.    **11.** 0.    **13.** Does not exist.    **15.** 0.
**17.** $\infty$.    **19.** 0.    **21.** 1.    **23.** 0.    **25.** $\infty$.

**27.** 0.    **29.** $-\dfrac{2}{5}$.    **31.** $-\infty$.    **33.** $\dfrac{2}{5}$.    **35.** $-\infty$.

**37.** $\dfrac{11}{5}$.    **39.** $-\dfrac{1}{2}$.    **41.** $\infty$.    **43.** $\infty$.    **45.** $\infty$.

**47.** Does not exist.    **49.** $-\infty$.    **51.** 0.    **53.** 1.
**55. a.** 1; **b.** 2; **c.** does not exist; **d.** 1; **e.** 2.
**57. a.** 0; **b.** 0; **c.** 0; **d.** $-\infty$; **e.** $-\infty$.
**59.**                              **61.** 20,000.    **63.** 20.

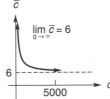

**65.** 1, 0.5, 0.525, 0.631, 0.912, 0.986, 0.998; conclude limit is 1.
**67.** 0.    **69. a.** 11; **b.** 9; **c.** does not exist.

## EXERCISE 11.3 (page 507)

**1.** $5563.87; $1563.87.    **3.** $1456.87.    **5.** 4.08%.
**7.** 10.52%.    **9.** $111.63.    **11.** $670,320.05.
**13. a.** $67,275; **b.** $6103.    **15.** $4.88%.    **17.** $927.
**19.** 16 years.
**21.** Option A: $1221.40; Option B: $1276.28;
Option C: $1218.45.
**23. a.** $9934.32; **b.** This strategy is better by $68.46.

## EXERCISE 11.4 (page 514)

**7.** Continuous at $-2$ and 0.    **9.** Discontinuous at $\pm 3$.
**11.** Continuous at 2 and 0.

**13.** $f$ is a polynomial function.
**15.** $f$ is a rational function and the denominator is never zero. **17.** None. **19.** $x = 4$. **21.** None.
**23.** $x = -5, 3$. **25.** $x = 0, \pm 1$. **27.** None.
**29.** $x = 0$. **31.** None. **33.** $x = 2$.
**35.** Discontinuities at $t = 1, 2, 3, 4$.

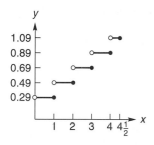

**37.** Yes, no, no.

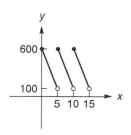

## PRINCIPLES IN PRACTICE 11.5

**1.** $0 < x < 4$.

## EXERCISE 11.5 (page 519)

**1.** $(-\infty, -1), (4, \infty)$. **3.** $[2, 3]$. **5.** $\left(-\dfrac{7}{2}, -2\right)$.
**7.** No solution. **9.** $(-\infty, -6], [-2, 3]$.
**11.** $(-\infty, -4), (0, 5)$. **13.** $[0, \infty)$. **15.** $(-3, 0), (1, \infty)$.
**17.** $(-\infty, -1), (0, 1)$. **19.** $(1, \infty)$.
**21.** $(-\infty, -5), [-2, 1), [3, \infty)$. **23.** $(-4, -2)$.
**25.** $(-\infty, -1 - \sqrt{3}], [-1 + \sqrt{3}, \infty)$.
**27.** Between 50 and 150 inclusive. **29.** 17 in. by 17 in.
**31.** $(-\infty, -7.72]$. **33.** $(-\infty, -0.5), (0.667, \infty)$.

## REVIEW PROBLEMS—CHAPTER 11 (page 521)

**1.** $-5$. **3.** $2$. **5.** $x$. **7.** $-\dfrac{8}{3}$. **9.** $0$. **11.** $\dfrac{3}{5}$.

**13.** Does not exist. **15.** $-1$. **17.** $\dfrac{1}{9}$. **19.** $-\infty$.

**21.** $\infty$. **23.** $-\infty$. **25.** $1$. **27.** $-\infty$. **29.** $8$.
**31.** $11$. **33. a.** \$6661.14; **b.** \$938.28. **35.** 6.18%.
**37.** $10 \ln 2$.
**41.** Continuous everywhere; $f$ is a polynomial function.
**43.** $x = -3$. **45.** None. **47.** $x = -4, 1$.
**49.** $x = -2$. **51.** $(-\infty, -6), (2, \infty)$.
**53.** $[2, \infty), x = 0$. **55.** $(-\infty, -5), (-1, 1)$.
**57.** $(-\infty, -4), [-3, 0], (2, \infty)$. **59.** $1.00$. **61.** $0$.
**63.** $[2.00, \infty)$.

## MATHEMATICAL SNAPSHOT—CHAPTER 11 (page 525)

**1.** 17%.

## PRINCIPLES IN PRACTICE 12.1

**1.** $\dfrac{dH}{dt} = 40 - 32t$.

## EXERCISE 12.1 (page 535)

**1. a.**

| $x$-value of $Q$ | 3 | 2.5 | 2.2 | 2.1 | 2.01 | 2.001 |
|---|---|---|---|---|---|---|
| $m_{PQ}$ | | 19 | 15.25 | 13.24 | 12.61 | 12.0601 | 12.0060 |

**b.** We estimate that $m_{\tan} = 12$.
**3.** $1$. **5.** $3$. **7.** $-4$. **9.** $0$. **11.** $2x + 4$.

**13.** $4q + 5$. **15.** $-\dfrac{1}{x^2}$. **17.** $\dfrac{1}{2\sqrt{x + 2}}$. **19.** $-4$.

**21.** $0$. **23.** $y = x + 4$. **25.** $y = -3x - 7$.

**27.** $y = -\dfrac{1}{3}x + \dfrac{5}{3}$. **29.** $\dfrac{r}{r_L - r - \dfrac{dC}{dD}}$.

**31.** $-3.000, 13.445$. **33.** $-5.120, 0.038$.
**35.** For the $x$-values of the points where the tangent to the graph of $f$ is horizontal, the corresponding values of $f'(x)$ are 0. This is expected because the slope of a horizontal line is zero and the derivative gives the slope of the tangent line.

## PRINCIPLES IN PRACTICE 12.2

**1.** $50 - 0.6q$.

## EXERCISE 12.2 (page 542)

**1.** $0$. **3.** $6x^5$. **5.** $80x^{79}$. **7.** $18x$. **9.** $20w^4$.

**11.** $\dfrac{8}{3}x^3$. **13.** $\dfrac{1}{2}t^8$. **15.** $1$. **17.** $6x - 2$.

**19.** $4p^3 - 9p^2$. **21.** $-8x^7 + 5x^4$.

**23.** $-39x^2 + 28x - 2$. **25.** $-8x^3$. **27.** $-\dfrac{4}{3}x^3$.

**29.** $16x^3 + 3x^2 - 9x + 9$. **31.** $6x^3 + 7x^2$.

**33.** $\dfrac{7}{2}x^{5/2}$. **35.** $\dfrac{3}{4}x^{-1/4} + \dfrac{5}{3}x^{2/3}$.

**37.** $\dfrac{1}{2}x^{-1/2}$ or $\dfrac{1}{2\sqrt{x}}$. **39.** $2r^{-2/3}$. **41.** $-4x^{-5}$.

**43.** $-3x^{-4} - 5x^{-6} + 12x^{-7}$. **45.** $-x^{-2}$ or $-\dfrac{1}{x^2}$.

**47.** $-15x^{-6}$. **49.** $-4x^{-4}$. **51.** $-\dfrac{1}{2}t^{-2}$.

**53.** $\dfrac{1}{7} - 7x^{-2}$. **55.** $-3x^{-2/3} - 2x^{-7/5}$. **57.** $-\dfrac{1}{5}x^{-6/5}$.

**59.** $-x^{-3/2}$. **61.** $\dfrac{5}{2}x^{3/2}$. **63.** $9x^2 - 14x + 7$.

**65.** $45x^4$. **67.** $\dfrac{1}{3}x^{-2/3} - \dfrac{10}{3}x^{-5/3} = \dfrac{1}{3}x^{-5/3}(x - 10)$.

**69.** $8q + \dfrac{4}{q^2}$. **71.** $2(x + 2)$. **73.** 1.

**75.** $4, 16, -14$. **77.** $0, 0, 0$. **79.** $y = 13x - 2$.

**81.** $y = -4x + 6$. **83.** $y = x + 3$.

**85.** $(0, 0), \left(2, -\dfrac{4}{3}\right)$. **87.** $(3, -3)$. **89.** 0.

**91.** The tangent line is $y = 9x - 16$.

## PRINCIPLES IN PRACTICE 12.3

**1.** 2.5 units.

**2.** $\dfrac{dy}{dt} = 16 - 32t$. $\left.\dfrac{dy}{dt}\right|_{t=0.5} = 0$ feet/s. When $t = 0.5$ the object reaches its maximum height.

**3.** 1.2 and 120%.

## EXERCISE 12.3 (page 552)

**1.**

| $\Delta t$ | 1 | 0.5 | 0.2 | 0.1 | 0.01 | 0.001 |
|---|---|---|---|---|---|---|
| $\Delta s / \Delta t$ | 8 | 5.75 | 4.64 | 4.31 | 4.0301 | 4.003001 |

We estimate the velocity $t = 1$ to be 4.0000 m/s. With differentiation the velocity is 4 m/s.

**3. a.** 4 m; **b.** 5.5 m/s; **c.** 5 m/s.

**5. a.** 8 m; **b.** 6.1208 m/s; **c.** 6 m/s.

**7. a.** 2 m; **b.** 10.261 m/s; **c.** 9 m/s. **9.** $\dfrac{1}{4}$.

**11.** $\dfrac{dy}{dx} = \dfrac{25}{2}x^{3/2}$; 337.50. **13.** 0.27.

**15.** $dc/dq = 10$; 10. **17.** $dc/dq = 0.6q + 2$; 3.8.

**19.** $dc/dq = 2q + 50$; 80, 82, 84.

**21.** $dc/dq = 0.02q + 5$; 6, 7.

**23.** $dc/dq = 0.00006q^2 - 0.02q + 6$; 4.6, 11.

**25.** $dr/dq = 0.7$; 0.7, 0.7, 0.7.

**27.** $dr/dq = 250 + 90q - 3q^2$; 625, 850, 625.

**29.** $dc/dq = 6.750 - 0.000656q$; 3.47.

**31.** $dP/dR = -4,650,000R^{-1.93}$. **33. a.** $-7.5$; **b.** 4.5.

**35. a.** 1; **b.** $\dfrac{1}{x+4}$; **c.** 1; **d.** $\dfrac{1}{9} \approx 0.111$; **e.** 11.1%.

**37. a.** $6x$; **b.** $\dfrac{2x}{x^2+2}$; **c.** 12; **d.** $\dfrac{2}{3} \approx 0.667$; **e.** 66.7%.

**39. a.** $-3x^2$; **b.** $-\dfrac{3x^2}{8-x^3}$; **c.** $-3$; **d.** $-\dfrac{3}{7} \approx -0.429$; **e.** $-42.9\%$. **41.** 3.2; 21.3%.

**43. a.** $dr/dq = 30 - 0.6q$; **b.** $\dfrac{4}{45} \approx 0.089$; **c.** 9%.

**45.** $\dfrac{0.432}{t}$. **47.** \$3125. **49.** \$5.07/unit.

## PRINCIPLES IN PRACTICE 12.5

**1.** $\dfrac{dR}{dx} = 6.25 - 6x$.

**2.** $T'(x) = 2x - x^2$; $T'(1) = 1$.

## EXERCISE 12.5 (page 563)

**1.** $(4x + 1)(6) + (6x + 3)(4) = 48x + 18 = 6(8x + 3)$.

**3.** $(8 - 7t)(2t) + (t^2 - 2)(-7) = 14 + 16t - 21t^2$.

**5.** $(3r^2 - 4)(2r - 5) + (r^2 - 5r + 1)(6r)$
$= 12r^3 - 45r^2 - 2r + 20$.

**7.** $4x^3 - 10x$.

**9.** $(x^2 + 3x - 2)(4x - 1) + (2x^2 - x - 3)(2x + 3)$
$= 8x^3 + 15x^2 - 20x - 7$.

**11.** $(8w^2 + 2w - 3)(15w^2) + (5w^3 + 2)(16w + 2)$
$= 200w^4 + 40w^3 - 45w^2 + 32w + 4$.

**13.** $(x^2 - 1)(9x^2 - 6) + (3x^3 - 6x + 5)(2x)$
$- 4(8x + 2)$
$= 15x^4 - 27x^2 - 22x - 2$.

**15.** $\dfrac{3}{2}\left[(p^{1/2} - 4)(4) + (4p - 5)\left(\dfrac{1}{2}p^{-1/2}\right)\right]$
$= \dfrac{3}{4}(12p^{1/2} - 5p^{-1/2} - 32)$.

**17.** 0. **19.** $18x^2 + 94x + 31$.

**21.** $\dfrac{(x - 1)(1) - (x)(1)}{(x - 1)^2} = -\dfrac{1}{(x - 1)^2}$.

**23.** $-\dfrac{9}{x^7}$. **25.** $\dfrac{(x - 1)(1) - (x + 2)(1)}{(x - 1)^2} = -\dfrac{3}{(x - 1)^2}$.

**27.** $\dfrac{(z^2 - 4)(-2) - (5 - 2z)(2z)}{(z^2 - 4)^2} = \dfrac{2(z - 4)(z - 1)}{(z^2 - 4)^2}$.

**29.** $\dfrac{(x^2 - 5x)(16x - 2) - (8x^2 - 2x + 1)(2x - 5)}{(x^2 - 5x)^2}$
$= \dfrac{-38x^2 - 2x + 5}{(x^2 - 5x)^2}$.

**31.** $\dfrac{(2x^2 - 3x + 2)(2x - 4) - (x^2 - 4x + 3)(4x - 3)}{(2x^2 - 3x + 2)^2}$
$= \dfrac{5x^2 - 8x + 1}{(2x^2 - 3x + 2)^2}$.

**33.** $-\dfrac{100x^{99}}{(x^{100} + 1)^2}$. **35.** $\dfrac{4(v^5 + 2)}{v^2}$.

**37.** $\dfrac{15x^2 - 2x + 1}{3x^{4/3}}$. **39.** $\dfrac{4}{(x - 8)^2} + \dfrac{2}{(3x + 1)^2}$.

**41.** $\dfrac{[(x + 2)(x - 4)](1) - (x - 5)(2x - 2)}{[(x + 2)(x - 4)]^2}$
$= \dfrac{-(x^2 - 10x + 18)}{[(x + 2)(x - 4)]^2}$.

**43.**
$\dfrac{[(t^2 - 1)(t^3 + 7)](2t + 3) - (t^2 + 3t)(5t^4 - 3t^2 + 14t)}{[(t^2 - 1)(t^3 + 7)]^2}$
$= \dfrac{-3t^6 - 12t^5 + t^4 + 6t^3 - 21t^2 - 14t - 21}{[(t^2 - 1)(t^3 + 7)]^2}$.

**45.** $3 - \dfrac{2x^3 + 3x^2 - 12x + 4}{[x(x - 1)(x - 2)]^2}$. **47.** $-\dfrac{2a}{(a + x)^2}$.

**49.** $-6$. **51.** $y = -\dfrac{3}{2}x + \dfrac{15}{2}$. **53.** $y = 16x + 24$.

**55.** 1.5. **57.** 1 m, $-1.5$ m/s. **59.** $\dfrac{dr}{dq} = 25 - 0.04q$.

**61.** $\dfrac{dr}{dq} = \dfrac{216}{(q + 2)^2} - 3$. **63.** $\dfrac{dC}{dI} = 0.672$.

**65.** $\dfrac{1}{3}; \dfrac{2}{3}.$  **67.** 0.615; 0.385.   **69. a.** 0.32; **b.** 0.026.

**71.** $\dfrac{dc}{dq} = \dfrac{5q(q + 6)}{(q + 3)^2}.$   **73.** $\dfrac{9}{10}.$   **75.** $\dfrac{0.7355}{(1 + 0.02744x)^2}.$

**77.** $-\dfrac{1}{120}.$   **79.** $6x^2 + 2x - 13.$

## PRINCIPLES IN PRACTICE 12.6

**1.** $288t.$

## EXERCISE 12.6 (page 572)

**1.** $(2u - 2)(2x - 1) = 4x^3 - 6x^2 - 2x + 2.$

**3.** $\left(-\dfrac{2}{w^3}\right)(-1) = \dfrac{2}{(2 - x)^3}.$   **5.** $-2.$   **7.** $0.$

**9.** $18(3x + 2)^5.$   **11.** $-6x(5 - x^2)^2.$

**13.** $300(3x^2 - 16x + 1)(x^3 - 8x^2 + x)^{99}.$

**15.** $-6x(x^2 - x)^{-4}.$

**17.** $-\dfrac{10}{3}(4x - 3)(2x^2 - 3x - 1)^{-13/3}.$

**19.** $\dfrac{1}{2}(10x - 1)(5x^2 - x)^{-1/2}.$   **21.** $\dfrac{1}{2}(2x - 1)^{-3/4}.$

**23.** $\dfrac{12}{5}x^2(x^3 + 1)^{-3/5}.$   **25.** $-6(4x - 1)(2x^2 - x + 1)^{-2}.$

**27.** $-2(2x - 3)(x^2 - 3x)^{-3}.$   **29.** $-8(8x - 1)^{-3/2}.$

**31.** $\dfrac{7}{3}(7x)^{-2/3} + \sqrt[3]{7}.$

**33.** $(x^2)[5(x - 4)^4(1)] + (x - 4)^5(2x)$
$\qquad = x(x - 4)^4(7x - 8).$

**35.** $(2x)\left[\dfrac{1}{2}(6x - 1)^{-1/2}(6)\right] + (\sqrt{6x - 1})(2)$
$\qquad = 6x(6x - 1)^{-1/2} + 2\sqrt{6x - 1}.$

**37.** $(x^2 + 2x - 1)^3(5) + (5x)[3(x^2 + 2x - 1)^2(2x + 2)]$
$\qquad = 5(x^2 + 2x - 1)^2(7x^2 + 8x - 1).$

**39.** $(8x - 1)^3[4(2x + 1)^3(2)]$
$\qquad + (2x + 1)^4[3(8x - 1)^2(8)]$
$\qquad = 16(8x - 1)^2(2x + 1)^3(7x + 1).$

**41.** $10\left(\dfrac{x - 7}{x + 4}\right)^9\left[\dfrac{(x + 4)(1) - (x - 7)(1)}{(x + 4)^2}\right]$
$\qquad = \dfrac{110(x - 7)^9}{(x + 4)^{11}}.$

**43.** $\dfrac{1}{2}\left(\dfrac{x - 2}{x + 3}\right)^{-1/2}\left[\dfrac{(x + 3)(1) - (x - 2)(1)}{(x + 3)^2}\right]$
$\qquad = \dfrac{5}{2(x + 3)^2}\left(\dfrac{x - 2}{x + 3}\right)^{-1/2}.$

**45.** $\dfrac{(x^2 + 4)^3(2) - (2x - 5)[3(x^2 + 4)^2(2x)]}{(x^2 + 4)^6}$
$\qquad = \dfrac{-2(5x^2 - 15x - 4)}{(x^2 + 4)^4}.$

**47.** $\dfrac{(3x - 1)^3[40(8x - 1)^4] - (8x - 1)^5[9(3x - 1)^2]}{(3x - 1)^6}$
$\qquad = \dfrac{(8x - 1)^4(48x - 31)}{(3x - 1)^4}.$

**49.** $6\{(5x^2 + 2)[2x^3(x^4 + 5)^{-1/2}] + (x^4 + 5)^{1/2}(10x)\}$
$\qquad = 12x(x^4 + 5)^{-1/2}(10x^4 + 2x^2 + 25).$

**51.** $8 + \dfrac{5}{(t + 4)^2} - (8t - 7) = 15 - 8t + \dfrac{5}{(t + 4)^2}.$

**53.** $\dfrac{(x^2 - 7)^4[(2x + 1)(2)(3x - 5)(3) + (3x - 5)^2(2)]}{(x^2 - 7)^8} \\ \qquad \dfrac{- (2x + 1)(3x - 5)^2[4(x^2 - 7)^3(2x)]}{}.$

**55.** $0.$   **57.** $0.$   **59.** $y = 4x - 11.$

**61.** $y = -\dfrac{1}{6}x + \dfrac{5}{3}.$   **63.** 96%.   **65.** 20.   **67.** 13.99.

**69. a.** $-\dfrac{q}{\sqrt{q^2 + 20}};$ **b.** $-\dfrac{q}{100\sqrt{q^2 + 20} - q^2 - 20};$

**c.** $100 - \dfrac{q^2}{\sqrt{q^2 + 20}} - \sqrt{q^2 + 20}.$

**71.** $-325.$   **73.** $\dfrac{dc}{dq} = \dfrac{5q(q^2 + 6)}{(q^2 + 3)^{3/2}}.$   **75.** $48\pi(10)^{-19}.$

**77. a.** $-\dfrac{1000}{\sqrt{100 - x}};$ $-125;$ **b.** $-\dfrac{1}{128}.$

**79.** $-4.$   **81.** 40.   **83.** 86,111.22.

## REVIEW PROBLEMS—CHAPTER 12 (page 576)

**1.** $-2x.$   **3.** $\dfrac{\sqrt{3}}{2\sqrt{x}}.$   **5.** $0.$

**7.** $28x^3 - 18x^2 + 10x = 2x(14x^2 - 9x + 5).$

**9.** $4s^3 + 4s = 4s(s^2 + 1).$   **11.** $\dfrac{2x}{5}.$

**13.** $(x^2 + 6x)(3x^2 - 12x) + (x^3 - 6x^2 + 4)(2x + 6)$
$\qquad = 5x^4 \quad 108x^2 + 8x + 24.$

**15.** $100(2x^2 + 4x)^{99}(4x + 4)$
$\qquad = 400(x + 1)[(2x)(x + 2)]^{99}.$

**17.** $-\dfrac{2}{(2x + 1)^2}.$

**19.** $(8 + 2x)(4)(x^2 + 1)^3(2x) + (x^2 + 1)^4(2)$
$\qquad = 2(x^2 + 1)^3(9x^2 + 32x + 1).$

**21.** $\dfrac{(z^2 + 1)(2z) - (z^2 - 1)(2z)}{(z^2 + 1)^2} = \dfrac{4z}{(z^2 + 1)^2}.$

**23.** $\dfrac{4}{3}(4x - 1)^{-2/3}.$

**25.** $-\dfrac{1}{2}(1 - x)^{-3/2}(-1) = \dfrac{1}{2}(1 - x)^{-3/2}.$

**27.** $(x - 6)^4[3(x + 5)^2] + (x + 5)^3[4(x - 6)^3]$
$\qquad = (x - 6)^3(x + 5)^2(7x + 2).$

**29.** $\dfrac{(x + 1)(5) - (5x - 4)(1)}{(x + 1)^2} = \dfrac{9}{(x + 1)^2}.$

**31.** $2\left(-\dfrac{3}{8}\right)x^{-11/8} + \left(-\dfrac{3}{8}\right)(2x)^{-11/8}(2)$
$\qquad = -\dfrac{3}{4}(1 + 2^{-11/8})x^{-11/8}.$

**33.** $\dfrac{\sqrt{x^2 + 5}(2x) - (x^2 + 6)(1/2)(x^2 + 5)^{-1/2}(2x)}{x^2 + 5}$
$\qquad = \dfrac{x(x^2 + 4)}{(x^2 + 5)^{3/2}}.$

**35.** $\left(\dfrac{3}{5}\right)(x^3 + 6x^2 + 9)^{-2/5}(3x^2 + 12x).$

$= \dfrac{9}{5}x(x + 4)(x^3 + 6x^2 + 9)^{-2/5}.$

**37.** $7(1 - 2z).$     **39.** $y = -4x + 3.$

**41.** $y = \dfrac{1}{12}x + \dfrac{4}{3}.$     **43.** $\dfrac{5}{7} \approx 0.714; 71.4\%.$

**45.** $dr/dq = 20 - 0.2q.$     **47.** $0.569, 0.431.$

**49.** $dr/dq = 450 - q.$

**51.** $dc/dq = 0.125 + 0.00878q; 0.7396.$

**53.** 84 eggs/mm.     **55. a.** $\dfrac{4}{3}$; **b.** $\dfrac{1}{24}.$     **57.** $8\pi$ ft³/ft.

**59.** $4q - \dfrac{10{,}000}{q^2}.$

**61. a.** 240; **b.** $\dfrac{1}{100}$; **c.** No, since $dr/dm < 300$ when

$m = 80.$

**63.** 0.397.     **65.** $-0.32.$

## PRINCIPLES IN PRACTICE 13.1

**1.** $\dfrac{dq}{dp} = \dfrac{12p}{3p^2 + 4}.$     **2.** $\dfrac{dR}{dI} = \dfrac{1}{I \ln 10}.$

## EXERCISE 13.1 (page 583)

**1.** $\dfrac{4}{x}.$     **3.** $\dfrac{3}{3x - 4}.$     **5.** $\dfrac{2}{x}.$     **7.** $-\dfrac{2x}{1 - x^2}.$

**9.** $\dfrac{6p^2 + 3}{2p^3 + 3p} = \dfrac{3(2p^2 + 1)}{p(2p^2 + 3)}.$

**11.** $t\left(\dfrac{1}{t}\right) + (\ln t) = 1 + \ln t.$

**13.** $\dfrac{4x^2}{4x + 3} + 2x \ln (4x + 3).$     **15.** $\dfrac{2}{(\ln 3)(2x - 1)}.$

**17.** $2x\left[1 + \dfrac{1}{(\ln 2)(x^2 + 4)}\right].$

**19.** $\dfrac{z\left(\dfrac{1}{z}\right) - (\ln z)(1)}{z^2} = \dfrac{1 - \ln z}{z^2}.$

**21.** $\dfrac{(\ln x)(2x) - (x^2 - 1)\left(\dfrac{1}{x}\right)}{(\ln x)^2} = \dfrac{2x^2 \ln (x) - x^2 + 1}{x \ln^2 x}.$

**23.** $\dfrac{3(2x + 4)}{x^2 + 4x + 5} = \dfrac{6(x + 2)}{x^2 + 4x + 5}.$     **25.** $\dfrac{x}{1 + x^2}.$

**27.** $\dfrac{2}{1 - t^2}.$     **29.** $\dfrac{x}{1 - x^4}.$     **31.** $\dfrac{4x}{x^2 + 2} + \dfrac{3x^2 + 1}{x^3 + x - 1}.$

**33.** $\dfrac{1}{x} + \dfrac{1}{2x + 1}.$     **35.** $\dfrac{2(x^2 + 1)}{2x + 1} + 2x \ln (2x + 1).$

**37.** $\dfrac{3(1 + \ln^2 x)}{x}.$     **39.** $\dfrac{4 \ln^3 (ax)}{x}.$

**41.** $\dfrac{x}{2(x - 1)} + \ln \sqrt{x - 1}.$     **43.** $\dfrac{1}{2x \sqrt{4 + \ln x}}.$

**45.** $y = 4x - 12.$     **47.** $\dfrac{\ln (2) - 1}{\ln^2 2}.$     **49.** $\dfrac{25}{7}.$

**51.** $\dfrac{dq}{dp} = \dfrac{20}{2p + 1}.$     **53.** $\dfrac{6a}{(T - a^2 + aT)(a - T)}.$

**57.** 1.36.

## PRINCIPLES IN PRACTICE 13.2

**1.** $\dfrac{dT}{dt} = Cke^{kt}.$

## EXERCISE 13.2 (page 588)

**1.** $7e^x.$     **3.** $2xe^{x^2+1}.$     **5.** $-5e^{3 - 5x}.$

**7.** $(6r + 4)e^{3r^2+4r+4} = 2(3r + 2)e^{3r^2+4r+4}.$

**9.** $x(e^x) + e^x(1) = e^x(x + 1).$     **11.** $2xe^{-x^2}(1 - x^2).$

**13.** $\dfrac{e^x - e^{-x}}{2}.$     **15.** $(6x)4^{3x^2} \ln 4.$     **17.** $\dfrac{2e^{2w}(w - 1)}{w^3}.$

**19.** $\dfrac{e^{1+\sqrt{x}}}{2\sqrt{x}}.$     **21.** $3x^2 - 3^x \ln 3.$     **23.** $\dfrac{2e^x}{(e^x + 1)^2}.$

**25.** 1.     **27.** $(1 + \ln x)e^{x \ln x}.$     **29.** $-e.$

**31.** $y - e^2 = e^2(x - 2)$ or $y = e^2x - e^2.$

**33.** $dp/dq = -0.015e^{-0.001q}, -0.015e^{-0.5}.$

**35.** $dc/dq = 10e^{q/700}; 10e^{0.5}; 10e.$     **37.** $-5.$

**39.** $e.$     **41.** $100e^{-2}.$     **47.** $-b(10^{A - bM}) \ln 10.$

**51.** 0.0036.     **53.** 0.68.

## PRINCIPLES IN PRACTICE 13.3

**1.** $\dfrac{dP}{dt} = 0.5(P - P^2).$

**2.** $\dfrac{dV}{dt} = 4\pi r^2 \dfrac{dr}{dt}$ and $\dfrac{dV}{dt}\bigg|_{r=12} = 2880\pi$ inches/minute

**3.** The top of the ladder is sliding down at a rate of $\dfrac{9}{4}$ feet/second.

## EXERCISE 13.3 (page 594)

**1.** $-\dfrac{x}{4y}.$     **3.** $\dfrac{5}{12y^3}.$     **5.** $-\dfrac{\sqrt{y}}{\sqrt{x}}.$     **7.** $-\dfrac{y^{1/4}}{x^{1/4}}.$

**9.** $-\dfrac{y}{x}.$     **11.** $\dfrac{4 - y}{x - 1}.$     **13.** $\dfrac{4y - x^2}{y^2 - 4x}.$     **15.** $\dfrac{6y^{2/3}}{3y^{1/6} + 2}.$

**17.** $\dfrac{1 - 6xy^3}{1 + 9x^2y^2}.$     **19.** $\dfrac{xe^y - y}{x(\ln x - xe^y)}.$     **21.** $-\dfrac{e^y}{xe^y + 1}.$

**23.** $6e^{3x}(1 + e^{3x})(x + y) - 1.$     **25.** $-\dfrac{3}{5}.$

**27.** $0; -\dfrac{4x_0}{9y_0}.$     **29.** $y = -\dfrac{3}{4}x + \dfrac{5}{4}.$     **31.** $\dfrac{dq}{dp} = -\dfrac{1}{2q}.$

**33.** $\dfrac{dq}{dp} = -\dfrac{(q + 5)^3}{40}.$     **35.** $-\lambda I.$     **37.** $1.5E \ln 10.$

**39.** $-\dfrac{V}{0.4T} = -2.5\dfrac{V}{T}.$     **41.** $\dfrac{3}{8}.$

## EXERCISE 13.4 (page 598)

**1.** $(x + 1)^2(x - 1)(x^2 + 3)\left[\dfrac{2}{x + 1} + \dfrac{1}{x - 1} + \dfrac{2x}{x^2 + 3}\right]$.

**3.** $(3x^2 - 1)^2(2x + 5)^3\left[\dfrac{18x^2}{3x^3 - 1} + \dfrac{6}{2x + 5}\right]$.

**5.** $\dfrac{\sqrt{x + 1}\,\sqrt{x^2 - 2}\,\sqrt{x + 4}}{2}$.

$\left[\dfrac{1}{x + 1} + \dfrac{2x}{x^2 - 2} + \dfrac{1}{x + 4}\right]$.

**7.** $\dfrac{\sqrt{1 - x^2}}{1 - 2x}\left[\dfrac{x}{x^2 - 1} + \dfrac{2}{1 - 2x}\right]$.

**9.** $\dfrac{(2x^2 + 2)^2}{(x + 1)^2(3x + 2)}\left[\dfrac{4x}{x^2 + 1} - \dfrac{2}{x + 1} - \dfrac{3}{3x + 2}\right]$.

**11.** $\dfrac{1}{2}\sqrt{\dfrac{(x - 1)(x + 1)}{3x - 4}}\left[\dfrac{1}{x - 1} + \dfrac{1}{x + 1} - \dfrac{3}{3x - 4}\right]$.

**13.** $x^{2x+1}\left(\dfrac{2x + 1}{x} + 2\ln x\right)$. **15.** $\dfrac{x^{1/x}(1 - \ln x)}{x^2}$.

**17.** $2(3x + 1)^{2x}\left[\dfrac{3x}{3x + 1} + \ln(3x + 1)\right]$.

**19.** $e^x x^{3x}(4 + 3\ln x)$. **21.** 12. **23.** $y = 96x + 36$.

**25.** $y = 6ex - 3e$. **27.** $\dfrac{1}{3e^{13}}$.

## PRINCIPLES IN PRACTICE 13.5

**1.** $\dfrac{d^2h}{dt^2} = -32$ feet/sec$^2$ (*Note:* Negative values indicate the downward direction.).

**2.** $c''(3) = 14$ dollars/unit$^2$.

## EXERCISE 13.5 (page 602)

**1.** 24. **3.** 0. **5.** $e^x$. **7.** $3 + 2\ln x$. **9.** $-\dfrac{10}{p^6}$.

**11.** $-\dfrac{1}{4(1 - r)^{3/2}}$. **13.** $\dfrac{50}{(5x - 6)^3}$. **15.** $\dfrac{4}{(x - 1)^3}$.

**17.** $-\left[\dfrac{1}{x^2} + \dfrac{1}{(x + 1)^2}\right]$. **19.** $e^z(z^2 + 4z + 2)$.

**21.** 32. **23.** $-\dfrac{1}{y^3}$. **25.** $-\dfrac{4}{y^3}$. **27.** $\dfrac{1}{8x^{3/2}}$.

**29.** $\dfrac{2(y - 1)}{(1 + x)^2}$. **31.** $\dfrac{y}{(1 - y)^3}$. **33.** $-\dfrac{16}{125}$.

**35.** $300(5x - 3)^2$. **37.** 0.6. **39.** $\pm 1$.

**41.** $-4.99$ and $1.94$.

## REVIEW PROBLEMS—CHAPTER 13 (page 603)

**1.** $2e^x + e^x(2x) = 2(e^x + xe^x)$.

**3.** $\dfrac{1}{r^2 + 5r}(2r + 5) = \dfrac{2r + 5}{r(r + 5)}$.

**5.** $e^{x^2+4x+5}(2x + 4) = 2(x + 2)e^{x^2+4x+5}$.

**7.** $e^x(2x) + (x^2 + 2)e^x = e^x(x^2 + 2x + 2)$.

**9.** $\dfrac{\sqrt{(x - 6)(x + 5)(9 - x)}}{2}\left[\dfrac{1}{x - 6} + \dfrac{1}{x + 5} + \dfrac{1}{x - 9}\right]$.

**11.** $\dfrac{e^x\left(\dfrac{1}{x}\right) - (\ln x)(e^x)}{e^{2x}} = \dfrac{1 - x\ln x}{xe^x}$.

**13.** $\dfrac{2}{q + 1} + \dfrac{3}{q + 2}$. **15.** $-7(\ln 10)^{2 - 7x}$.

**17.** $\dfrac{4e^{2x+1}(2x - 1)}{x^2}$. **19.** $\dfrac{16}{(8x + 5)\ln 2}$.

**21.** $\dfrac{1 + 2l + 3l^2}{1 + l + l^2 + l^3}$. **23.** $(x + 1)^{x + 1}[1 + \ln(x + 1)]$.

**25.** $2\left(\dfrac{1}{t}\right) + \dfrac{1}{2}\left(\dfrac{1}{1 - t}\right)(-1) = \dfrac{5t - 4}{2t(t - 1)}$.

**27.** $y\left[\dfrac{3}{2}\left(\dfrac{1}{x^2 + 2}\right)(2x) + \dfrac{4}{9}\left(\dfrac{1}{x^2 + 9}\right)(2x)\right.$

$\left. - \dfrac{4}{11}\left(\dfrac{1}{x^3 + 6x}\right)(3x^2 + 6)\right]$

$= y\left[\dfrac{3x}{x^2 + 2} + \dfrac{8x}{9(x^2 + 9)} - \dfrac{12(x^2 + 2)}{11(x^3 + 6x)}\right]$,

where $y$ is given in problem.

**29.** $(x^x)^x(x + 2x\ln x)$. **31.** 4. **33.** $-2$.

**35.** $y = 2x + 2(1 - \ln 2)$ or $y = 2x + 2 - \ln 4$.

**37.** $(0, 4\ln 2)$. **39.** 18. **41.** 2. **43.** $-\dfrac{y}{x + y}$.

**45.** $\dfrac{xy^2 - y}{2x - x^2y}$. **47.** $\dfrac{4}{9}$.

**49.** $\dfrac{dy}{dx} = \dfrac{y + 1}{y}; \dfrac{d^2y}{dx^2} = -\dfrac{y + 1}{y^3}$.

**51.** $f'(t) = 0.008e^{-0.01t} + 0.00004e^{-0.0002t}$. **53.** 0.90.

## PRINCIPLES IN PRACTICE 14.1

**1.** There is a relative maximum when $x = 1$, and a relative minimum when $x = 3$.

**2.** The drug is at its greatest concentration 2 hours after injection.

## EXERCISE 14.1 (page 614)

**1.** Dec. on $(-\infty, -1)$ and $(3, \infty)$; inc. on $(-1, 3)$; rel. min. $(-1, -1)$; rel. max. $(3, 4)$.

**3.** Dec. on $(-\infty, -2)$ and $(0, 2)$; inc. on $(-2, 0)$ and $(2, \infty)$; rel. min. $(-2, 1)$ and $(2, 1)$; no rel. max.

**5.** Inc. on $(-\infty, -1)$ and $(3, \infty)$; dec. on $(-1, 3)$; rel max. when $x = -1$; rel. min. when $x = 3$.

**7.** Dec. on $(-\infty, -1)$; inc. on $(-1, 3)$ and $(3, \infty)$; rel. min. when $x = -1$.

**9.** Dec. on $(-\infty, 0)$; inc. on $(0, \infty)$; rel. min. when $x = 0$.

**11.** Inc. on $\left(-\infty, \dfrac{1}{2}\right)$; dec. on $\left(\dfrac{1}{2}, \infty\right)$; rel. max. when

$x = \dfrac{1}{2}$.

**13.** Dec. on $(-\infty, -5)$ and $(1, \infty)$; inc. on $(-5, 1)$; rel. min. when $x = -5$; rel. max. when $x = 1$.

**15.** Dec. on $(-\infty, -1)$ and $(0, 1)$; inc. on $(-1, 0)$ and $(1, \infty)$; rel. max. when $x = 0$; rel. min. when $x = \pm 1$.

**17.** Inc. on $(-\infty, 1)$ and $(3, \infty)$; dec. on $(1, 3)$; rel. max. when $x = 1$; rel. min. when $x = 3$.

**19.** Inc. on $\left(-\infty, -\dfrac{2}{3}\right)$ and $\left(\dfrac{5}{2}, \infty\right)$; dec. on $\left(-\dfrac{2}{3}, \dfrac{5}{2}\right)$; rel. max. when $x = -\dfrac{2}{3}$; rel. min. when $x = \dfrac{5}{2}$.

**21.** Inc. on $\left(-\infty, \dfrac{-2 - \sqrt{7}}{3}\right)$ and $\left(\dfrac{-2 + \sqrt{7}}{3}, \infty\right)$; dec. on $\left(\dfrac{-2 - \sqrt{7}}{3}, \dfrac{-2 + \sqrt{7}}{3}\right)$; rel. max. when $x = \dfrac{-2 - \sqrt{7}}{3}$; rel. min. when $x = \dfrac{-2 + \sqrt{7}}{3}$.

**23.** Inc. on $(-\infty, -1)$ and $(1, \infty)$; dec. on $(-1, 0)$ and $(0, 1)$; rel. max. when $x = -1$; rel. min. when $x = 1$.

**25.** Dec. on $(-\infty, -4)$ and $(0, \infty)$; inc. on $(-4, 0)$; rel. min. when $x = -4$; rel. max. when $x = 0$.

**27.** Inc. on $(-\infty, -\sqrt{2})$ and $(0, \sqrt{2})$; dec. on $(-\sqrt{2}, 0)$ and $(\sqrt{2}, \infty)$; rel. max. when $x = \pm\sqrt{2}$; rel. min. when $x = 0$.

**29.** Inc. on $(-\infty, -1)$, $(-1, 0)$, and $(0, \infty)$; never dec.; no rel. extremum.

**31.** Dec. on $(-\infty, 1)$ and $(1, \infty)$; no rel. extremum.

**33.** Dec. on $(0, \infty)$; no rel. extremum.

**35.** Dec. on $(-\infty, 0)$ and $(2, \infty)$; inc. on $(0, 1)$ and $(1, 2)$; rel. min. when $x = 0$; rel. max. when $x = 2$.

**37.** Inc. on $(-\infty, -3)$ and $(-1, \infty)$; dec. on $(-3, -2)$ and $(-2, -1)$; rel. max. when $x = -3$; rel. min. when $x = -1$.

**39.** Dec. on $\left(-\infty, \dfrac{-3 - \sqrt{34}}{5}\right)$ and $\left(\dfrac{-3 + \sqrt{34}}{5}, \infty\right)$; inc. on $\left(\dfrac{-3 - \sqrt{34}}{5}, \dfrac{-3 + \sqrt{34}}{5}\right)$; rel. min. when $x = \dfrac{-3 - \sqrt{34}}{5}$; rel. max. when $x = \dfrac{-3 + \sqrt{34}}{5}$.

**41.** Inc. on $(-\infty, -2)$, $\left(-2, \dfrac{11}{5}\right)$, and $(5, \infty)$; dec. on $\left(\dfrac{11}{5}, 5\right)$; rel. max. when $x = \dfrac{11}{5}$; rel. min. when $x = 5$.

**43.** Inc. on $(-\infty, 0)$, $\left(0, \dfrac{12}{7}\right)$, and $(4, \infty)$; dec. on $\left(\dfrac{12}{7}, 4\right)$; rel. max. when $x = \dfrac{12}{7}$; rel. min. when $x = 4$.

**45.** Dec. on $(-\infty, \infty)$; no rel. extremum.

**47.** Dec. on $(0, 1)$; inc. on $(1, \infty)$; rel. min. when $x = 1$.

**49.** Dec. on $(-\infty, 0)$; inc. on $(0, \infty)$; rel. min. when $x = 0$.

**51.** Dec. on $(0, 1)$; inc. on $(1, \infty)$; rel. min. when $x = 1$; no rel. max.

**53.** Dec. on $(-\infty, 3)$; inc. on $(3, \infty)$; rel. min. when $x = 3$; intercepts: $(7, 0)$, $(-1, 0)$, $(0, -7)$.

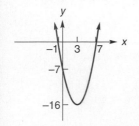

**55.** Dec. on $(-\infty, -1)$ and $(1, \infty)$; inc. on $(-1, 1)$; rel. min. when $x = -1$; rel. max. when $x = 1$; sym. about origin; intercepts: $(\pm\sqrt{3}, 0)$, $(0, 0)$.

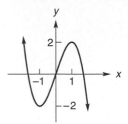

**57.** Inc. on $(-\infty, 1)$ and $(2, \infty)$; dec. on $(1, 2)$; rel. max. when $x = 1$; rel. min. when $x = 2$; intercept: $(0, 0)$.

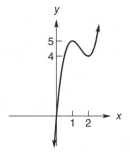

**59.** Inc. on $(-2, -1)$ and $(0, \infty)$; dec. on $(-\infty, -2)$ and $(-1, 0)$; rel. max. when $x = -1$; rel. min. when $x = -2, 0$; intercepts: $(0, 0)$, $(-2, 0)$.

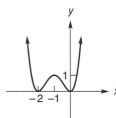

**61.** Dec. on $(-\infty, -2)$ and $\left(-\dfrac{1}{2}, 1\right)$; inc. on $\left(-2, -\dfrac{1}{2}\right)$ and $(1, \infty)$; rel. min. when $x = -2, 1$; rel. max. when $x = -\dfrac{1}{2}$; intercepts: $(1, 0)$, $(-2, 0)$, $(0, 4)$.

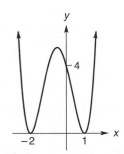

**63.** Dec. on $(1, \infty)$; inc. on $(0, 1)$; rel. max. when $x = 1$; intercepts: $(0, 0), (4, 0)$.

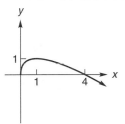

**65.**

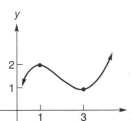

**69.** Never.

**71.** 40.  **75. a.** 25,300; **b.** 4; **c.** 17,200.
**77.** Rel. min.: $(-4.10, -3.41)$.
**79.** Rel. max.: $(2.74, 2.37)$; rel. min.: $(-2.74, -2.37)$.
**81.** Rel. min.: $0, 1.50, 2.00$; rel. max.: $0.57, 1.77$.
**83. a.** $f'(x) = 4 - 6x - 3x^2$
**c.** Dec.: $(-\infty, -2.53), (0.53, \infty)$; inc.: $(-2.53, 0.53)$.

## EXERCISE 14.2 (page 619)

**1.** Maximum: $f(-1) = 6$; minimum: $f(1) = 2$.

**3.** Maximum: $f(0) = 1$; minimum: $f(2) = -\dfrac{19}{3}$.

**5.** Maximum: $f(3) = 84$; minimum: $f(1) = -8$.
**7.** Maximum: $f(-2) = 56$; minimum: $f(-1) = -2$.
**9.** Maximum: $f(\sqrt{2}) = 4$; minimum $f(2) = -16$.

**11.** Maximum: $f(0) = f(3) = 2$; minimum:
$f\left(\dfrac{3\sqrt{2}}{2}\right) = -\dfrac{73}{4}$.
**13.** Maximum: $f(3) \approx 2.08$; minimum: $f(0) = 0$.
**15. a.** $-3.22, -0.78$; **b.** 2.75; **c.** 9; **d.** 14,283.

## EXERCISE 14.3 (page 625)

**1.** Conc. up $(-\infty, 0), \left(\dfrac{3}{2}, \infty\right)$; conc. down $\left(0, \dfrac{3}{2}\right)$; inf. pt.

when $x = 0, \dfrac{3}{2}$.

**3.** Conc. up $(-\infty, 1), (1, 7)$; conc down $(7, \infty)$; inf. pt. when $x = 7$.
**5.** Conc. up $(-\infty, -\sqrt{2}), (\sqrt{2}, \infty)$; conc down $(-\sqrt{2}, \sqrt{2})$; no inf. pt.
**7.** Conc. down. $(-\infty, \infty)$.
**9.** Conc down. $(-\infty, -1)$; conc. up $(-1, \infty)$; inf. pt. when $x = -1$.

**11.** Conc. down. $\left(-\infty, \dfrac{7}{4}\right)$; conc. up $\left(\dfrac{7}{4}, \infty\right)$; inf. pt. when

$x = \dfrac{7}{4}$.

**13.** Conc. up $(-\infty, -1), (1, \infty)$; conc. down $(-1, 1)$; inf. pt. when $x = \pm 1$.
**15.** Conc. up $(-\infty, 0)$; conc. down $(0, \infty)$; inf. pt. when $x = 0$.

**17.** Conc. up $\left(-\infty, -\dfrac{7}{2}\right), \left(\dfrac{1}{3}, \infty\right)$; conc. down $\left(-\dfrac{7}{2}, \dfrac{1}{3}\right)$;

inf. pt. when $x = -\dfrac{7}{2}, \dfrac{1}{3}$.

**19.** Conc. down $(-\infty, 0), \left(\dfrac{3 - \sqrt{5}}{2}, \dfrac{3 + \sqrt{5}}{2}\right)$; conc. up

$\left(0, \dfrac{3 - \sqrt{5}}{2}\right), \left(\dfrac{3 + \sqrt{5}}{2}, \infty\right)$; inf. pt. when

$x = 0, \dfrac{3 \pm \sqrt{5}}{2}$.

**21.** Conc. up $(-\infty, -\sqrt{5}), (-\sqrt{2}, \sqrt{2}), (\sqrt{5}, \infty)$; conc. down $(-\sqrt{5}, -\sqrt{2}), (\sqrt{2}, \sqrt{5})$; inf. pt. when $x = \pm\sqrt{5}, \pm\sqrt{2}$.
**23.** Conc. down $(-\infty, 1)$; conc. up $(1, \infty)$.
**25.** Conc. down. $(-\infty, -1/\sqrt{3}), (1/\sqrt{3}, \infty)$; conc. up $(-1/\sqrt{3}, 1/\sqrt{3})$; inf. pt. when $x = \pm 1/\sqrt{3}$.

**27.** Conc. down. $(-\infty, -3), \left(-3, \dfrac{2}{7}\right)$; conc. up $\left(\dfrac{2}{7}, \infty\right)$; inf.

pt. when $x = \dfrac{2}{7}$.

**29.** Conc. up. $(-\infty, \infty)$.
**31.** Conc. down $(-\infty, -2)$; conc. up $(-2, \infty)$; inf. pt. when $x = -2$.
**33.** Conc. down $(0, e^{3/2})$; conc. up $(e^{3/2}, \infty)$; inf. pt. when $x = e^{3/2}$.
**35.** Int. $(-3, 0), (-1, 0), (0, 3)$; dec. $(-\infty, -2)$; inc. $(-2, \infty)$; rel. min. when $x = -2$; conc. up $(-\infty, \infty)$.

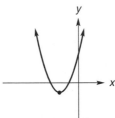

**37.** Int. $(0, 0), (4, 0)$; inc. $(-\infty, 2)$; dec. $(2, \infty)$; rel. max. when $x = 2$; conc. down $(-\infty, \infty)$.

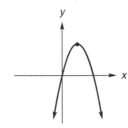

**39.** Int. $(0, -19)$; inc. $(-\infty, 2)$, $(4, \infty)$; dec. $(2, 4)$; rel. max. when $x = 2$; rel. min. when $x = 4$; conc. down $(-\infty, 3)$; conc. up $(3, \infty)$; inf. pt. when $x = 3$.

**41.** Int. $(0, 0)$, $(\pm 2\sqrt{3}, 0)$; inc. $(-\infty, -2)$, $(2, \infty)$; dec. $(-2, 2)$; rel. max. when $x = -2$; rel. min. when $x = 2$; conc. down $(-\infty, 0)$; conc. up $(0, \infty)$; inf. pt. when $x = 0$; sym. about origin.

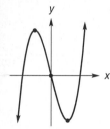

**43.** Int. $(0, -3)$; inc. $(-\infty, 1)$, $(1, \infty)$; no rel. max. or min.; conc. down $(-\infty, 1)$; conc. up $(1, \infty)$; inf. pt. when $x = 1$.

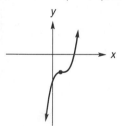

**45.** Int. $(0, 0)$, $(4/3, 0)$; inc. $(-\infty, 0)$, $(0, 1)$; dec. $(1, \infty)$; rel. max. when $x = 1$; conc. up $(0, 2/3)$; conc. down $(-\infty, 0)$, $(2/3, \infty)$; inf. pt. when $x = 0$, $x = 2/3$.

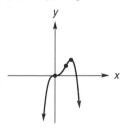

**47.** Int. $(0, -2)$; dec. $(-\infty, -2)$, $(2, \infty)$; inc. $(-2, 2)$; rel. min. when $x = -2$; rel. max. when $x = 2$; conc. up $(-\infty, 0)$; conc. down $(0, \infty)$; inf. pt. when $x = 0$.

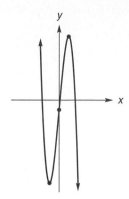

**49.** Int. $(0, -6)$; inc. $(-\infty, 2)$, $(2, \infty)$; conc. down $(-\infty, 2)$; conc. up $(2, \infty)$; inf. pt. when $x = 2$.

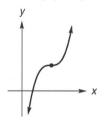

**51.** Int. $(0, 0)$, $(\pm \sqrt[4]{5}, 0)$; dec. $(-\infty, -1)$, $(1, \infty)$; inc. $(-1, 1)$; rel. min. when $x = -1$; rel. max. when $x = 1$; conc. up $(-\infty, 0)$; conc. down $(0, \infty)$; inf. pt. when $x = 0$; sym. about origin.

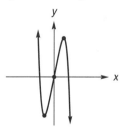

**53.** Int. $(0, 1)$, $(1, 0)$; dec. $(-\infty, 0)$, $(0, 1)$; inc. $(1, \infty)$; rel. min. when $x = 1$; conc. up $(-\infty, 0)$, $(2/3, \infty)$; conc. down $(0, 2/3)$; inf. pt. when $x = 0$, $x = 2/3$.

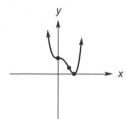

**55.** Int. $(0, 0)$, $(\pm 2, 0)$; inc. $(-\infty, -\sqrt{2})$, $(0, \sqrt{2})$; dec. $(-\sqrt{2}, 0)$, $(\sqrt{2}, \infty)$; rel. max. when $x = \pm\sqrt{2}$; rel. min. when $x = 0$; conc. down $(-\infty, -\sqrt{2/3})$, $(\sqrt{2/3}, \infty)$; conc. up $(-\sqrt{2/3}, \sqrt{2/3})$; inf. pt. when $x = \pm\sqrt{2/3}$; sym. about $y$-axis.

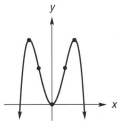

**57.** Int. $(0, 0)$, $(8, 0)$; dec. $(-\infty, 0)$, $(0, 2)$; inc. $(2, \infty)$; rel. min. when $x = 2$; conc. up $(-\infty, -4)$, $(0, \infty)$; conc. down $(-4, 0)$; inf. pt. when $x = -4$, $x = 0$.

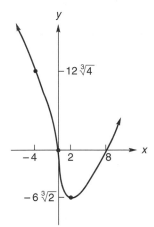

**59.** Int. $(0, 0)$, $(-4, 0)$; dec. $(-\infty, -1)$; inc. $(-1, 0)$, $(0, \infty)$; rel. min. when $x = -1$; conc. up $(-\infty, 0)$, $(2, \infty)$; conc. down $(0, 2)$; inf. pt. when $x = 0$, $x = 2$.

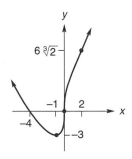

**61.** Int. $(0, 0)$, $\left(-\dfrac{27}{8}, 0\right)$; inc. $(-\infty, -1)$, $(0, \infty)$; dec. $(-1, 0)$; rel. min. when $x = 0$; rel. max. when $x = -1$; conc. down $(-\infty, 0)$, $(0, \infty)$.

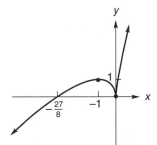

**63.**

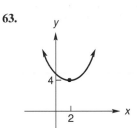

**65.**

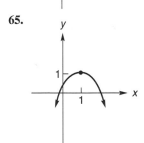

**69.**

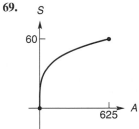

**73. b.** 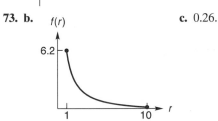  **c.** 0.26.

**75.** Two.  **77.** Above tangent line; concave up.
**79.** $-2.61$, $-0.26$.

## EXERCISE 14.4 (page 629)

**1.** Rel. min. when $x = \dfrac{5}{2}$; abs. min.

**3.** Rel. max. when $x = \dfrac{1}{4}$; abs. max.

**5.** Rel. max. when $x = -3$; rel. min. when $x = 3$.

**7.** Rel. min. when $x = 0$; rel. max. when $x = 2$.

**9.** Test fails, when $x = 0$ there is a rel. min. by first-deriv. test.

**11.** Rel. max. when $x = -\dfrac{1}{3}$; rel. min. when $x = \dfrac{1}{3}$.

**13.** Rel. min. when $x = -5, -2$; rel. max. when $x = -\dfrac{7}{2}$.

## EXERCISE 14.5 (page 637)

**1.** $y = 1, x = -1$.    **3.** $y = \dfrac{1}{2}, x = -\dfrac{3}{2}$.

**5.** $y = 0, x = 0$.    **7.** $y = 0, x = 1, x = -1$.

**9.** None.    **11.** $y = 2, x = 2, x = -3$.

**13.** $y = -7, x = -\sqrt{5}, x = \sqrt{5}$.    **15.** $y = 4, x = 6$.

**17.** $x = 0, x = -1$.    **19.** $y = \dfrac{1}{4}, x = -\dfrac{1}{2}$.

**21.** $y = 1, x = -\dfrac{4}{3}$.    **23.** $y = 4$.

**25.** Dec. $(-\infty, 0), (0, \infty)$; conc. down $(-\infty, 0)$; conc. up $(0, \infty)$; sym. about origin; asymptotes $x = 0, y = 0$.

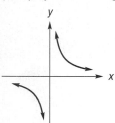

**27.** Int. $(0, 0)$; inc. $(-\infty, -1), (-1, \infty)$; conc. up $(-\infty, -1)$; conc. down $(-1, \infty)$; asymptotes $x = -1, y = 1$.

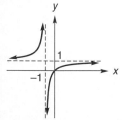

**29.** Dec. $(-\infty, -1), (0, 1)$; inc. $(-1, 0), (1, \infty)$; rel. min. when $x = \pm 1$; conc. up $(-\infty, 0), (0, \infty)$; sym. about $y$-axis; asymptote $x = 0$.

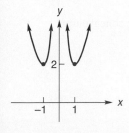

**31.** Int. $(0, -1)$; inc. $(-\infty, -1), (-1, 0)$; dec. $(0, 1), (1, \infty)$; rel. max. when $x = 0$; conc. up $(-\infty, -1), (1, \infty)$; conc. down $(-1, 1)$; asymptotes $x = 1, x = -1, y = 0$; sym. about $y$-axis.

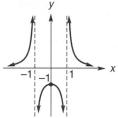

**33.** Int. $(-1, 0), (0, 1)$; inc. $(-\infty, 1), (1, \infty)$; conc. up $(-\infty, 1)$; conc. down $(1, \infty)$; asymptotes $x = 1, y = -1$.

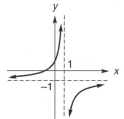

**35.** Int. $(0, 0)$; inc. $\left(-\infty, -\dfrac{8}{7}\right), (0, \infty)$; dec. $\left(-\dfrac{8}{7}, -\dfrac{4}{7}\right)$, $\left(-\dfrac{4}{7}, 0\right)$; rel. max. when $x = -\dfrac{8}{7}$; rel. min. when $x = 0$; conc. down $\left(-\infty, -\dfrac{4}{7}\right)$; conc. up $\left(-\dfrac{4}{7}, \infty\right)$; asymptote $x = -\dfrac{4}{7}$.

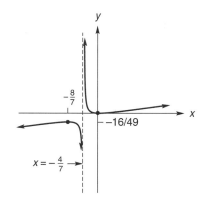

**37.** Int. $\left(0, -\dfrac{9}{8}\right)$; inc. $\left(-\infty, -\dfrac{2}{3}\right)$, $\left(-\dfrac{2}{3}, \dfrac{1}{3}\right)$; dec. $\left(\dfrac{1}{3}, \dfrac{4}{3}\right)$, $\left(\dfrac{4}{3}, \infty\right)$; rel. max. when $x = \dfrac{1}{3}$; conc. up $\left(-\infty, -\dfrac{2}{3}\right)$, $\left(\dfrac{4}{3}, \infty\right)$; conc. down $\left(-\dfrac{2}{3}, \dfrac{4}{3}\right)$; asymptotes $y = 0$, $x = -\dfrac{2}{3}$, $x = \dfrac{4}{3}$.

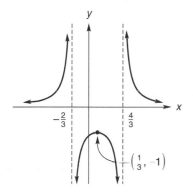

**39.** Int. $\left(\dfrac{3}{2}, 0\right)$, $\left(0, -\dfrac{1}{27}\right)$; dec. $\left(-\infty, -\dfrac{3}{2}\right)$, $\left(\dfrac{9}{2}, \infty\right)$; inc. $\left(-\dfrac{3}{2}, \dfrac{9}{2}\right)$; rel. min. when $x = -\dfrac{3}{2}$; conc. down $\left(-\infty, -\dfrac{9}{2}\right)$; conc. up $\left(-\dfrac{9}{2}, \dfrac{9}{2}\right)$, $\left(\dfrac{9}{2}, \infty\right)$; inf. pt. when $x = -\dfrac{9}{2}$; asymptotes $x = \dfrac{9}{2}$, $y = 0$.

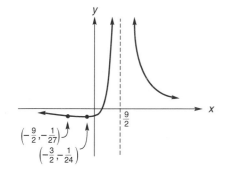

**41.** Int. $(-1, 0)$, $(1, 0)$; inc. $(-\sqrt{3}, 0)$, $(0, \sqrt{3})$; dec. $(-\infty, -\sqrt{3})$, $(\sqrt{3}, \infty)$; rel. max. when $x = \sqrt{3}$; rel. min. when $x = -\sqrt{3}$; conc. down $(-\infty, -\sqrt{6})$, $(0, \sqrt{6})$; conc. up $(-\sqrt{6}, 0)$, $(\sqrt{6}, \infty)$; inf. pt. when $x = \pm\sqrt{6}$; asymptotes $x = 0$, $y = 0$; sym. about origin.

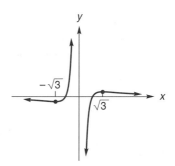

**43.** Int. $(0, 1)$; inc. $(-\infty, -2)$, $(0, \infty)$; dec. $(-2, -1)$, $(-1, 0)$; rel. max. when $x = -2$; rel. min when $x = 0$; conc. down $(-\infty, -1)$; conc. up $(-1, \infty)$; asymptote $x = -1$.

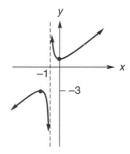

**45.** Int. $(0, 5)$; dec. $\left(-\infty, -\dfrac{1}{3}\right)$, $\left(-\dfrac{1}{3}, \dfrac{1}{3}\right)$; inc. $\left(\dfrac{1}{3}, 1\right)$, $(1, \infty)$; rel. min. when $x = \dfrac{1}{3}$; conc. down $\left(-\infty, -\dfrac{1}{3}\right)$, $(1, \infty)$; conc. up $\left(-\dfrac{1}{3}, 1\right)$; asymptotes $x = -\dfrac{1}{3}$, $x = 1$, $y = -1$.

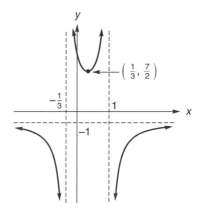

**47.**

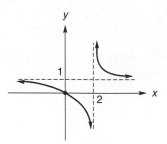

**49.**

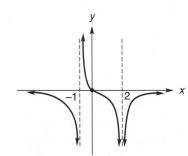

**55.** $x \approx \pm 2.45$, $x \approx 0.67$, $y = 2$.  **57.** $y \approx 0.48$.

## REVIEW PROBLEMS—CHAPTER 14 (page 639)

**1.** $y = 3$, $x = 4$, $x = -4$.  **3.** $y = \dfrac{5}{9}$, $x = -\dfrac{2}{3}$.

**5.** $x = 0, 4$.  **7.** $x = -\dfrac{15}{8}, -1$.

**9.** Inc. $(1, 3)$; dec. on $(-\infty, 1)$ and $(3, \infty)$.
**11.** Dec. on $(-\infty, -\sqrt{6})$, $(0, \sqrt{3})$, $(\sqrt{3}, \sqrt{6})$; inc. on $(-\sqrt{6}, -\sqrt{3})$, $(-\sqrt{3}, 0)$, $(\sqrt{6}, \infty)$.
**13.** Conc. up on $(-\infty, 0)$ and $\left(\dfrac{1}{2}, \infty\right)$; conc. down on $\left(0, \dfrac{1}{2}\right)$.

**15.** Conc. down on $\left(-\infty, \dfrac{1}{2}\right)$; conc. up on $\left(\dfrac{1}{2}, \infty\right)$.

**17.** Conc. up on $\left(-\infty, -\dfrac{5}{4}\right)$, $\left(-\dfrac{1}{4}, \infty\right)$; conc. down on $\left(-\dfrac{5}{4}, -\dfrac{1}{4}\right)$.

**19.** Rel. max. at $x = 1$; rel. min. at $x = 2$.
**21.** Rel. min. at $x = -1$.

**23.** Rel. max. at $x = -\dfrac{2}{5}$; rel. min. at $x = 0$.

**25.** At $x = 3$.  **27.** At $x = 1$.  **29.** At $x = 2 \pm \sqrt{2}$.
**31.** Maximum: $f(2) = 16$; minimum: $f(1) = -1$.

**33.** Maximum: $f(0) = 0$; minimum: $f\left(-\dfrac{6}{5}\right) = -\dfrac{1}{120}$.

**35. a.** $f$ has no relative extrema;
**b.** $f$ is conc. down on $(1, 3)$; inf. pts.: $(1, 2e^{-1})$, $(3, 10e^{-3})$.

**37.** Int. $(-4, 0)$, $(6, 0)$, $(0, -24)$; inc. $(1, \infty)$; dec. $(-\infty, 1)$; rel. min. when $x = 1$; conc. up $(-\infty, \infty)$.

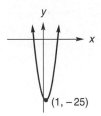

**39.** Int. $(0, 20)$; inc. $(-\infty, -2)$, $(2, \infty)$; dec. $(-2, 2)$; rel. max. when $x = -2$; rel. min. when $x = 2$; conc. up $(0, \infty)$; conc. down $(-\infty, 0)$; inf. pt. when $x = 0$.

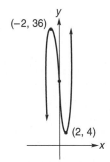

**41.** Int. $(0, 0)$; inc. $(-\infty, \infty)$; conc. down $(-\infty, 0)$; conc. up $(0, \infty)$; inf. pt. when $x = 0$; sym. about origin.

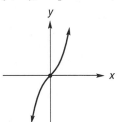

**43.** Int. $(-5, 0)$; inc. $(-10, 0)$; dec. $(-\infty, -10)$, $(0, \infty)$; rel. min. when $x = -10$; conc. up $(-15, 0)$, $(0, \infty)$; conc. down $(-\infty, -15)$; inf. pt. when $x = -15$; horiz. asym. $y = 0$; vert. asym. $x = 0$.

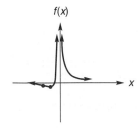

**45.** Int. $(0, 0)$; inc. $\left(-\infty, -\dfrac{1}{4}\right)$; dec. $\left(-\dfrac{1}{4}, \dfrac{1}{2}\right), \left(\dfrac{1}{2}, \infty\right)$; rel.

max. when $x = -\dfrac{1}{4}$; conc. up $\left(-\infty, -\dfrac{1}{2}\right), \left(\dfrac{1}{2}, \infty\right)$; conc.

down $\left(-\dfrac{1}{2}, \dfrac{1}{2}\right)$; inf. pt. when $x = -\dfrac{1}{2}$; horiz. asym. $y = 0$;

vert. asym. $x = \dfrac{1}{2}$.

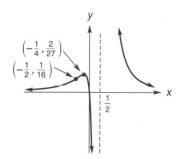

**47.** Int. $(0, 1)$; inc. $(0, \infty)$; dec. $(-\infty, 0)$; rel. min. when $x = 0$; conc. up $(-\infty, \infty)$; sym. about $y$-axis.

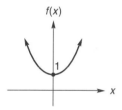

**49. a.** False; **b.** false; **c.** true; **d.** false; **e.** false.
**51.** $q > 2$.
**57.** Rel. max. $(-1.32, 12.28)$; rel. min. $(0.44, 1.29)$.
**59.** $x = -0.60$.

## EXERCISE 15.1 (page 651)

**1.** 20 and 20. **3.** 300 ft by 250 ft. **5.** 100 units.

**7.** $15. **9. a.** 110 grams; **b.** $51\dfrac{9}{11}$ grams.

**11.** 525 units; price $= $51$; profit $= $10,525$. **13.** $22.
**15.** 120 units; $86,000. **17.** 625 units; $4.
**19.** $17; $86,700. **21.** 4 ft by 4 ft by 2 ft.
**23.** 2 in.; 128 in³.
**27.** 130 units, $p = $340, P = $36,980$; 125 units, $p = $350$, $P = $34,175$. **29.** 250 per lot (4 lots). **31.** 35.
**33.** 60 mi/h. **35.** 7; $1000.
**37.** $5 - \sqrt{3}$ tons; $5 - \sqrt{3}$ tons. **41.** 10 cases; $50.55.

## EXERCISE 15.2 (page 660)

**1.** $3\,dx$. **3.** $\dfrac{2x^3}{\sqrt{x^4 + 2}}\,dx$. **5.** $-\dfrac{2}{x^3}\,dx$.

**7.** $\dfrac{2x}{x^2 + 7}\,dx$. **9.** $4e^{2x^2 + 3}(4x^2 + 3x + 1)\,dx$.

**11.** $\Delta y = -0.14, dy = -0.14$.
**13.** $\Delta y = -2.5, dy = -2.75$.

**15.** $\Delta y \approx 0.073, dy = \dfrac{3}{40} = 0.075$. **17. a.** $-1$; **b.** 2.9.

**19.** 10.05. **21.** $3\dfrac{47}{48}$. **23.** $-0.03$. **25.** 1.01.

**27.** $\dfrac{1}{2}$. **29.** $\dfrac{1}{6p(p^2 + 5)^2}$. **31.** $-p^2$. **33.** $\dfrac{1}{13}$.

**35.** $-\dfrac{4}{5}$. **37.** 44; 41.80. **39.** 2.04. **41.** 0.7.

**43.** $(1.69 \times 10^{-11})\pi$ cm³. **45. c.** 42 units.

## EXERCISE 15.3 (page 665)

**1.** $-3$, elastic. **3.** $-1$, unit elasticity.

**5.** $-1.02$, elastic. **7.** $-\left(\dfrac{150}{e} - 1\right)$, elastic.

**9.** $-1$, unit elasticity. **11.** $-\dfrac{9}{32}$, inelastic.

**13.** $-\dfrac{1}{2}$, inelastic.

**15.** $|\eta| = \dfrac{10}{3}$ when $p = 10$, $|\eta| = \dfrac{3}{10}$ when $p = 3$, $|\eta| = 1$
when $p = 6.50$. **17.** $-1.2$, 0.6% decrease.
**23. b.** $\eta = -2.5$, elastic; **c.** 1 unit;
**d.** increase, since demand is elastic.

**25. a.** $\eta = -\dfrac{207}{15} \approx -13.8$, elastic; **b.** 27.6%; **c.** Since

demand is elastic, lowering the price results in an increase

in revenue. **27.** $\eta = -1.6$; $\dfrac{dr}{dq} = 30$.
**29.** Maximum at $q = 5$; minimum at $q = 95$.

## PRINCIPLES IN PRACTICE 15.4

**1.** 43 and 1958.

## EXERCISE 15.4 (page 670)

**1.** 0.25410. **3.** 1.32472. **5.** $-2.38769$. **7.** 0.33767.
**9.** 1.90785. **11.** 4.121. **13.** $-4.99$ and 1.94.
**15.** 13.33. **17.** 2.880. **19.** 3.45.

## REVIEW PROBLEMS—CHAPTER 15 (page 672)

**1.** 20. **3.** 300. **5.** $2800. **7.** 200 ft by 100 ft.
**9. a.** 200, $120; **b.** 300.
**11.** $\left[\dfrac{x^2}{x + 5} + 2x \ln (x + 5)\right]dx$. **13.** $\left(\dfrac{9}{10}\right)^{\circ}$.

**15.** 0.99. **17.** $\dfrac{1}{8y + 7}$. **19.** elastic. **21. a.** $-1$.

**23. a.** $\dfrac{200}{3} < p < 100$;

**b.** $\eta = -\dfrac{1}{3}$; demand decreases by approximately 1.67%.

**25.** 0.619 and 1.512.

## PRINCIPLES IN PRACTICE 16.1

**1.** $\int 28.3 \, dq = 28.3q + C.$   **2.** $\int 0.12t^2 \, dt = 0.04t^3 + C.$

**3.** $\int -\dfrac{480}{t^3} \, dt = \dfrac{240}{t^2} + C.$

**4.** $\int (500 + 300\sqrt{t}) dt = 500t + 200t^{3/2} + C.$

**5.** $S(t) = 0.7t^3 - 32.7t^2 + 491.6t + C.$

## EXERCISE 16.1 (page 680)

**1.** $5x + C.$   **3.** $\dfrac{x^9}{9} + C$   **5.** $-\dfrac{5}{6x^6} + C.$

**7.** $-\dfrac{1}{9x^9} + C.$   **9.** $-\dfrac{5}{6y^{6/5}} + C.$   **11.** $8u + \dfrac{u^2}{2} + C.$

**13.** $\dfrac{y^6}{6} - \dfrac{5y^2}{2} + C$   **15.** $t^3 - 2t^2 + 5t + C.$

**17.** $(7 + e)x + C.$   **19.** $\dfrac{x^2}{14} - \dfrac{3x^5}{20} + C.$

**21.** $3e^x + C.$   **23.** $\dfrac{x^{9.3}}{9.3} - \dfrac{9x^7}{7} - \dfrac{1}{x^3} - \dfrac{1}{2x^2} + C.$

**25.** $-\dfrac{4x^{3/2}}{9} + C.$   **27.** $2\sqrt[8]{x} + C.$

**29.** $\dfrac{x^4}{12} + \dfrac{3}{2x^2} + C.$   **31.** $\dfrac{w^3}{2} + \dfrac{2}{3w} + C.$

**33.** $\dfrac{1}{7}(z^2 - 5z) + C.$   **35.** $\dfrac{x^{e+1}}{e+1} + e^x + C.$

**37.** $\dfrac{4x^{3/2}}{3} - \dfrac{12x^{5/4}}{5} + C.$

**39.** $-\dfrac{3x^{5/3}}{25} - 7x^{1/2} + 3x^2 + C.$

**41.** $\dfrac{x^4}{4} - x^3 + \dfrac{5x^2}{2} - 15x + C.$   **43.** $\dfrac{2x^{5/2}}{5} + 2x^{3/2} + C.$

**45.** $\dfrac{4u^3}{3} + 2u^2 + u + C.$   **47.** $\dfrac{2v^3}{3} + 3v + \dfrac{1}{2v^4} + C.$

**49.** $\dfrac{z^3}{15} + z^2 + C.$   **51.** $x + e^x + C.$

**53.** No, $F(x) - G(x)$ must be a constant.

**55.** $\dfrac{1}{\sqrt{x^2 + 1}} + C.$

## PRINCIPLES IN PRACTICE 16.2

**1.** $N(t) = 800t + 200e^t + 6317.37.$
**2.** $y(t) = 14t^3 + 12t^2 + 11t + 3.$

## EXERCISE 16.2 (page 685)

**1.** $y = \dfrac{3x^2}{2} - 4x + 1.$   **3.** $18.$

**5.** $y = -\dfrac{x^4}{12} - \dfrac{x^3}{3} + \dfrac{4x}{3} + \dfrac{1}{12}.$

**7.** $y = \dfrac{x^4}{12} + x^2 - 5x + 2.$   **9.** $p = 0.7.$

**11.** $p = 275 - 0.5q - 0.1q^2.$   **13.** $c = 1.35q + 200.$

**15.** $7715.$   **17.** $G = -\dfrac{P^2}{50} + 2P + 20.$

**21.** $\$80$ $(dc/dq = 27.50$ when $q = 50$ is not relevant to problem).

## PRINCIPLES IN PRACTICE 16.3

**1.** $T(t) = 10e^{-0.5t} + C.$   **2.** 38 words.

## EXERCISE 16.3 (page 692)

**1.** $\dfrac{(x + 5)^8}{8} + C.$   **3.** $\dfrac{(x^2 + 3)^6}{6} + C.$

**5.** $\dfrac{3}{5}(y^3 + 3y^2 + 1)^{5/3} + C.$   **7.** $-\dfrac{(3x - 1)^{-2}}{2} + C.$

**9.** $\dfrac{1}{3}(2x - 1)^{3/2} + C.$   **11.** $\dfrac{(7x - 6)^5}{35} + C.$

**13.** $\dfrac{(x^2 + 3)^{13}}{26} + C.$   **15.** $\dfrac{3}{20}(27 + x^5)^{4/3} + C.$

**17.** $e^{3x} + C.$   **19.** $e^{t^2 + t} + C.$   **21.** $\dfrac{1}{10}e^{5x^2} + C.$

**23.** $-3e^{-2x} + C.$   **25.** $\ln|x + 5| + C.$

**27.** $\ln|x^3 + x^4| + C.$   **29.** $-\dfrac{3}{4}(z^2 - 6)^{-4} + C.$

**31.** $4\ln|x| + C.$   **33.** $\dfrac{1}{3}\ln|s^3 + 5| + C.$

**35.** $-\dfrac{7}{3}\ln|5 - 3x| + C.$

**37.** $\dfrac{2}{15}(5x)^{3/2} + C = \dfrac{2\sqrt{5}}{3}x^{3/2} + C.$

**39.** $\sqrt{x^2 - 4} + C.$   **41.** $\dfrac{1}{2}e^{y^4 + 1} + C.$

**43.** $-\dfrac{1}{6}e^{-2v^3 + 1} + C.$   **45.** $-\dfrac{1}{5}e^{-5x} + 2e^x + C.$

**47.** $-\dfrac{1}{24}(3 - 3x^2 - 6x)^4 + C.$   **49.** $\dfrac{1}{3}\ln|x^3 + 6x| + C.$

**51.** $2\ln|3 - 2s + 4s^2| + C.$   **53.** $\dfrac{1}{4}\ln(2x^2 + 1) + C.$

**55.** $\dfrac{1}{27}(x^3 - x^6)^{-9} + C.$   **57.** $\dfrac{1}{4}(x^4 + x^2)^2 + C.$

**59.** $\dfrac{1}{2}(4 - 9x - 3x^2)^{-4} + C.$   **61.** $\dfrac{1}{6}e^{4x^3 + 3x^2 - 4} + C.$

**63.** $-\dfrac{1}{25}(7 - 5x^2)^{5/2} + C.$

**65.** $\dfrac{(2x)^{3/2}}{3} - \sqrt{2x} + C = \dfrac{2\sqrt{2}}{3}x^{3/2} - \sqrt{2}x^{1/2} + C.$

**67.** $\dfrac{x^5}{5} + \dfrac{2x^3}{3} + x + C.$

**69.** $\dfrac{1}{2}\ln(x^2 + 1) - \dfrac{1}{6(x^6 + 1)} + C.$

**71.** $\dfrac{1}{3}\ln|3x - 5| + \dfrac{1}{27}(x^3 - x^6)^{-9} + C.$

**73.** $\dfrac{2}{9}(3x + 1)^{3/2} - \ln\sqrt{x^2 + 3} + C.$   **75.** $2e^{\sqrt{x}} + C.$

**77.** $-e^{-x} + e^x + C.$      **79.** $\frac{1}{4}\ln^2(x^2 + 2x) + C.$

**81.** $y = -\frac{1}{6}(3 - 2x)^3 + \frac{11}{2}.$

**83.** $y = -\ln|x| = \ln|1/x|.$      **85.** $160e^{0.05t} + 190.$

**87.** $\frac{Rr^2}{4K} + B_1\ln|r| + B_2.$

## EXERCISE 16.4 (page 698)

**1.** $x^2 + 3x - \ln|x| + C.$      **3.** $\frac{1}{3}(2x^3 + 4x + 1)^{3/2} + C.$

**5.** $-\frac{6}{5}\sqrt{4 - 5x} + C.$      **7.** $\frac{4^{7x}}{7\ln 4} + C.$

**9.** $7x^2 - 4e^{(1/4)x^2} + C.$

**11.** $x^2 - 3x + \frac{2}{3}\ln|3x - 1| + C.$

**13.** $\frac{3}{2}\ln(e^{2x} + 1) + C.$      **15.** $-\frac{1}{7}e^{7/x} + C.$

**17.** $x^2 + 4\ln|x^2 - 4| + C.$      **19.** $\frac{2}{9}(\sqrt{x} + 2)^3 + C.$

**21.** $\frac{3}{5}(x^{1/3} + 2)^5 + C.$      **23.** $\frac{1}{2}(\ln^2 x) + C.$

**25.** $\frac{1}{3}\ln^3(r + 1) + C.$      **27.** $\frac{3^{\ln x}}{\ln 3} + C.$

**29.** $e^{(x^2+3)/2} + C.$      **31.** $\ln|\ln(x + 3)| + C.$

**33.** $\frac{x^2}{2} + x + \ln|x^2 - 3| + C.$

**35.** $\frac{1}{2}\ln^{3/2}[(x^2 + 1)^2] + C.$

**37.** $\frac{1}{2}\sqrt{x^4 - 1} - (\ln 4)x + C.$

**39.** $x^2 - 8x - 6\ln|x| - \frac{2}{x^2} + C.$

**41.** $x + \ln|x - 1| + C.$      **43.** $\sqrt{e^{x^2} + 2} + C.$

**45.** $-\frac{(e^{-x} + 6)^3}{3} + C.$      **47.** $\frac{1}{5}(x^2 + e)^{5/2} + C.$

**49.** $\frac{1}{36\sqrt{2}}[(8x)^{3/2} + 3]^{3/2} + C.$

**51.** $-\frac{2}{3}e^{-\sqrt{s^3}} + C.$      **53.** $\frac{x^2}{2} + 2x + C.$

**55.** $\frac{\ln^2 x}{2} + x + C.$      **57.** $p = \frac{100}{q + 2}.$

**59.** $c = 20\ln|(q + 5)/5| + 2000.$

**61.** $C = 2(\sqrt{I} + 1).$      **63.** $C = \frac{3}{4}I - \frac{1}{3}\sqrt{I} + \frac{71}{12}.$

**65. a.** \$150 per unit; **b.** \$15,000; **c.** \$15,300.

**67.** $2500 - 800\sqrt{5} \approx \$711$ per acre.      **69.** $I = 3.$

## EXERCISE 16.5 (page 702)

**1.** 35.      **3.** 0.      **5.** 25.      **7.** $-\frac{3}{16}.$      **9.** $-\frac{7}{6}.$

**11.** $\sum_{k=1}^{15} k.$      **13.** $\sum_{k=1}^{4}(2k - 1).$      **15.** $\sum_{k=1}^{12} k^2.$

**17.** 101,475.      **19.** 84.      **21.** 273.      **23.** 8; \$850.

## PRINCIPLES IN PRACTICE 16.6

**1.** \$5975.

## EXERCISE 16.6 (page 710)

**1.** $\frac{2}{3}$ square unit.      **3.** $\frac{14}{27}$ square unit.

**5.** $S_n = \frac{1}{n}\left[4\left(\frac{1}{n}\right) + 4\left(\frac{2}{n}\right) + \ldots + 4\left(\frac{n}{n}\right)\right] = \frac{2(n + 1)}{n}.$

**7. a.** $S_n = \frac{n + 1}{2n} + 1;$ **b.** $\frac{3}{2}.$      **9.** $\frac{1}{2}$ square unit.

**11.** $\frac{1}{3}$ square unit.      **13.** $\frac{16}{3}$ square unit.      **15.** 6.

**17.** $-18.$      **19.** $\frac{5}{6}.$      **21.** 0.      **23.** $\frac{11}{4}.$

**25.** 5.3 square units.      **27.** 2.4.      **29.** $-16.5.$

## PRINCIPLES IN PRACTICE 16.7

**1.** \$32,830.      **2.** \$28,750.

## EXERCISE 16.7 (page 718)

**1.** 10.      **3.** $\frac{21}{2}.$      **5.** $-20.$      **7.** $\frac{7}{3}.$      **9.** $\frac{15}{2}.$

**11.** $-\frac{7}{6}.$      **13.** 0.      **15.** $\frac{5}{3}.$      **17.** $\frac{32}{3}.$      **19.** $-\frac{1}{6}.$

**21.** $4\ln 8.$      **23.** $e^5.$      **25.** $\frac{1}{3}(e^8 - 1).$      **27.** $\frac{3}{4}.$

**29.** $\frac{38}{9}.$      **31.** $\frac{15}{28}.$      **33.** $\frac{1}{2}\ln 3.$      **35.** $e + \frac{1}{2e^2} - \frac{3}{2}.$

**37.** $\frac{3}{2} - \frac{1}{e} + \frac{1}{2e^2}.$      **39.** $\frac{e^3}{2}(e^{12} - 1).$      **41.** $6 + \ln 19.$

**43.** $\frac{47}{12}.$      **45.** $2 - e.$      **47.** 7.      **49.** 0.      **51.** $\alpha^{5/2}T.$

**53.** $\int_b^a -Ax^{-B}dx.$      **55.** \$8639.      **57.** 1,973,333.

**59.** \$160.      **61.** \$2000.      **63.** 696; 492.      **65.** $2Ri.$

**69.** 0.05.      **71.** 3.52.      **73.** 11.08.

## EXERCISE 16.8 (page 724)

*In Problems 1–33, answers are assumed to be expressed on square units.*

**1.** 8.      **3.** $\frac{19}{2}.$      **5.** 8.      **7.** $\frac{19}{3}.$      **9.** 9.      **11.** $\frac{50}{3}.$

**13.** 36.      **15.** 8.      **17.** $\frac{32}{3}.$      **19.** 1.      **21.** 18.

**23.** $\dfrac{26}{3}$. **25.** $\dfrac{3}{2}\sqrt[3]{2}$. **27.** $e^2 - 1$.

**29.** $\dfrac{3}{2} + 2\ln 2 = \dfrac{3}{2} + \ln 4$. **31.** 68. **33.** 2.

**35.** 19 square units. **37. a.** $\dfrac{1}{16}$; **b.** $\dfrac{3}{4}$; **c.** $\dfrac{7}{16}$.

**39. a.** $\ln\dfrac{5}{3}$; **b.** $\ln(4) - 1$; **c.** $2 - \ln 3$.

**41.** 1.57 square units. **43.** 11.41 square units.

## EXERCISE 16.9 (page 731)

**1.** Area $= \displaystyle\int_{-2}^{3}[(x+6) - x^2]\,dx$.

**3.** Area $=$
$$\int_{0}^{3}[2x - (x^2 - x)]\,dx + \int_{3}^{4}[(x^2 - x) - 2x]\,dx.$$

**5.** Area $= \displaystyle\int_{0}^{1}[(y+1) - \sqrt{1-y}]\,dy$.

**7.** Area $= \displaystyle\int_{-\sqrt{5}}^{2}[(11 - 2x^2) - (x^2 - 4)]\,dx$.

In Problems **9–33**, answers are assumed to be expressed in square units.

**9.** $\dfrac{4}{3}$. **11.** $\dfrac{16}{3}$. **13.** $8\sqrt{6}$. **15.** 40. **17.** $\dfrac{125}{6}$.

**19.** $\dfrac{9}{2}$. **21.** $\dfrac{125}{12}$. **23.** $\dfrac{32}{81}$. **25.** $\dfrac{44}{3}$.

**27.** $\dfrac{4}{3}(5\sqrt{5} - 2\sqrt{2})$. **29.** $\dfrac{1}{2}$. **31.** $\dfrac{255}{32} - 4\ln 2$.

**33.** 12. **35.** $\dfrac{20}{63}$. **37.** $\dfrac{8}{3m^3}$ square units. **39.** $2^{4/3}$.

**41.** 4.76 square units. **43.** 7.26 square units.

## EXERCISE 16.10 (page 735)

**1.** CS $= 25.6$, PS $= 38.4$.
**3.** CS $= 50\ln(2) - 25$, PS $= 1.25$.
**5.** CS $= 800$, PS $= 1000$. **7.** $426.67. **9.** $254,000.
**11.** CS $\approx 1197$, PS $\approx 477$.

## REVIEW PROBLEMS—CHAPTER 16 (page 738)

**1.** $\dfrac{x^4}{4} + x^2 - 7x + C$. **3.** $\dfrac{117}{2}$.

**5.** $-(x+5)^{-2} + C$. **7.** $2\ln|x^3 - 6x + 1| + C$.

**9.** $\dfrac{11\sqrt[3]{11}}{4} - 4$. **11.** $\dfrac{y^4}{4} + \dfrac{2y^3}{3} + \dfrac{y^2}{2} + C$.

**13.** $\dfrac{4z^{3/4}}{3} - \dfrac{6z^{5/6}}{5} + C$. **15.** $\dfrac{1}{3}\ln\dfrac{10}{3}$.

**17.** $\dfrac{2}{27}(3x^3 + 2)^{3/2} + C$. **19.** $\dfrac{1}{2}(e^{2y} + e^{-2y}) + C$.

**21.** $\ln|x| - \dfrac{2}{x} + C$. **23.** 11.1. **25.** $\dfrac{7}{3}$.

**27.** $4 - 3\sqrt[3]{2}$. **29.** $\dfrac{3}{t} - \dfrac{2}{\sqrt{t}} + C$. **31.** $\dfrac{3}{2} - 5\ln 2$.

**33.** $\dfrac{4}{9}(x^{3/2} + 1)^{3/2} + C$. **35.** 1. **37.** $\dfrac{(1 + e^{3x})^3}{9} + C$.

**39.** $\dfrac{2\sqrt{10^{3x}}}{3\ln 10} + C$. **41.** $y = \dfrac{1}{2}e^{2x} + 3x - 1$.

In Problems **43–57**, answers are assumed to be expressed in square units.

**43.** $\dfrac{4}{3}$. **45.** $\dfrac{16}{3}$. **47.** $\dfrac{125}{6}$. **49.** $6 + \ln 3$. **51.** $\dfrac{2}{3}$.

**53.** 36. **55.** $\dfrac{125}{3}$. **57.** $e - 1$.

**59.** $p = 100 - \sqrt{2q}$. **61.** $1900. **63.** 0.5507.

**65.** 15 square units. **67.** CS $= 166\dfrac{2}{3}$, PS $= 53\dfrac{1}{3}$.

**73.** 24.71 square units. **75.** CS $\approx 1148$, PS $\approx 251$.

## MATHEMATICAL SNAPSHOT—CHAPTER 16 (page 742)

**1.** $225, 125$. **3. a.** $99,000; **b.** 900; **c.** $110.

## PRINCIPLES IN PRACTICE 17.1

**1.** $S(t) = -40te^{0.1t} + 400e^{0.1t} + 4600$.
**2.** $P(t) = 0.025t^2 - 0.05t^2\ln t + 0.05t^2(\ln t)^2 + C$.

## EXERCISE 17.1 (page 747)

**1.** $\dfrac{2}{3}x(x+5)^{3/2} - \dfrac{4}{15}(x+5)^{5/2} + C$.

**3.** $-e^{-x}(x+1) + C$. **5.** $\dfrac{y^4}{4}\left[\ln(y) - \dfrac{1}{4}\right] + C$.

**7.** $x[\ln(4x) - 1] + C$.

**9.** $\dfrac{2x}{3}(x+1)^{3/2} - \dfrac{4}{15}(x+1)^{5/2} + C$
$$= \dfrac{2}{15}(x+1)^{3/2}(3x - 2) + C.$$

**11.** $-\dfrac{x}{2(2x+1)} + \dfrac{1}{4}\ln|2x+1| + C$.

**13.** $-\dfrac{1}{x}(1 + \ln x) + C$. **15.** $\dfrac{1}{4}e^2(3e^2 - 1)$.

**17.** $\dfrac{1}{2}(1 - e^{-1})$, parts not needed.

**19.** $\dfrac{2}{3}(9\sqrt{3} - 10\sqrt{2})$.

**21.** $x(x-1)\ln(x-1) - \dfrac{x^2}{2} + C$.

**23.** $e^x(x^2 - 2x + 2) + C$.

**25.** $\dfrac{x^3}{3} + 2e^{-x}(x+1) - \dfrac{e^{-2x}}{2} + C$.

**27.** $\dfrac{e^{x^2}}{2}(x^2 - 1) + C$.

**29.** $\dfrac{2^{2x-1}}{\ln 2} + \dfrac{2^{x+1}x}{\ln 2} - \dfrac{2^{x+1}}{\ln^2 2} + \dfrac{x^3}{3} + C$.

**31.** $2e^3 + 1$ square units. **33.** $\dfrac{298}{15}$ square units.

## PRINCIPLES IN PRACTICE 17.2

**1.** $r(q) = \frac{5}{2}\ln\left|\frac{3(q+1)^3}{q+3}\right|.$

**2.** $V(t) = 150t^2 - 900\ln(t^2 + 6) + C.$

## EXERCISE 17.2 (page 753)

**1.** $\frac{12}{x+6} - \frac{2}{x+1}.$   **3.** $1 + \frac{2}{x+2} - \frac{8}{x+4}.$

**5.** $\frac{4}{x+1} - \frac{9}{(x+1)^2}.$   **7.** $\frac{3}{x} - \frac{2x}{x^2+1}.$

**9.** $2\ln|x| + 3\ln|x - 1| + C = \ln|x^2(x-1)^3| + C.$

**11.** $-3\ln|x+1| + 4\ln|x-2| + C$
$$= \ln\left|\frac{(x-2)^4}{(x+1)^3}\right| + C.$$

**13.** $\frac{1}{4}\left[\frac{3x^2}{2} + 2\ln|x-1| - 2\ln|x+1|\right] + C$
$$= \frac{1}{4}\left(\frac{3x^2}{2} + \ln\left[\frac{x-1}{x+1}\right]^2\right) + C.$$

**15.** $\ln|x| + 2\ln|x-4| - 3\ln|x+3| + C$
$$= \ln\left|\frac{x(x-4)^2}{(x+3)^3}\right| + C.$$

**17.** $\frac{1}{2}\ln|x^6 + 2x^4 - x^2 - 2| + C$, partial fractions not required.

**19.** $\left[\frac{4}{x-2}\right] - 5\ln|x-1| + 7\ln|x-2| + C$
$$= \left[\frac{4}{x-2}\right] + \ln\left[\frac{(x-2)^7}{(x-1)^5}\right] + C.$$

**21.** $2\ln|x| - \frac{1}{2}\ln(x^2 + 4) + C = \frac{1}{2}\ln\left[\frac{x^4}{x^2+4}\right] + C.$

**23.** $-\frac{1}{2}\ln(x^2 + 1) - \frac{2}{x-3} + C.$

**25.** $5\ln(x^2 + 1) + 2\ln(x^2 + 2) + C$
$$= \ln[(x^2+1)^5(x^2+2)^2] + C.$$

**27.** $\frac{3}{2}\ln(x^2 + 1) + \frac{1}{x^2+1} + C.$

**29.** $18\ln(4) - 10\ln(5) - 8\ln(3).$

**31.** $\frac{11}{6} + 4\ln\frac{2}{3}$ square units.

## EXERCISE 17.3 (page 759)

**1.** $\frac{x}{9\sqrt{9-x^2}} + C.$   **3.** $-\frac{\sqrt{16x^2 + 3}}{3x} + C.$

**5.** $\frac{1}{6}\ln\left|\frac{x}{6+7x}\right| + C.$   **7.** $\frac{1}{3}\ln\left|\frac{\sqrt{x^2+9} - 3}{x}\right| + C.$

**9.** $\frac{1}{2}\left[\frac{4}{5}\ln|4 + 5x| - \frac{2}{3}\ln|2 + 3x|\right] + C.$

**11.** $\frac{1}{8}(2x - \ln[4 + 3e^{2x}]) + C.$

**13.** $2\left[\frac{1}{1+x} + \ln\left|\frac{x}{1+x}\right|\right] + C$   **15.** $1 + \ln\frac{4}{9}.$

**17.** $\frac{1}{2}(x\sqrt{x^2 - 3} - 3\ln|x + \sqrt{x^2 - 3}|) + C.$

**19.** $\frac{1}{144}.$   **21.** $e^x(x^2 - 2x + 2) + C.$

**23.** $2\left(-\frac{\sqrt{4x^2 + 1}}{2x} + \ln|2x + \sqrt{4x^2 + 1}|\right) + C.$

**25.** $\frac{1}{9}\left(\ln|1 + 3x| + \frac{1}{1 + 3x}\right) + C.$

**27.** $\frac{1}{\sqrt{5}}\left(\frac{1}{2\sqrt{7}}\ln\left|\frac{\sqrt{7} + \sqrt{5}x}{\sqrt{7} - \sqrt{5}x}\right|\right) + C.$

**29.** $\frac{1}{3^6}\left[\frac{(3x)^6\ln(3x)}{6} - \frac{(3x)^6}{36}\right] + C$
$$= \frac{x^6}{36}[6\ln(3x) - 1] + C.$$

**31.** $\frac{4(9x - 2)(1 + 3x)^{3/2}}{135} + C.$

**33.** $\frac{1}{2}\ln|2x + \sqrt{4x^2 - 13}| + C.$

**35.** $-\frac{\sqrt{9 - 4x^2}}{9x} + C.$

**37.** $\frac{1}{2\pi}(4\sqrt{x} - \ln|\pi + 7e^{4\sqrt{x}}|) + C.$

**39.** $\frac{1}{2}\ln(x^2 + 1) + C.$   **41.** $\frac{1}{6}(2x^2 + 1)^{3/2} + C.$

**43.** $\ln\left|\frac{x-3}{x-2}\right| + C.$   **45.** $\frac{x^4}{4}\left[\ln(x) - \frac{1}{4}\right] + C.$

**47.** $\frac{e^{2x}}{4}(2x - 1) + C.$

**49.** $x(\ln x)^2 - 2x\ln(x) + 2x + C.$

**51.** $\frac{2}{3}(9\sqrt{3} - 10\sqrt{2}).$   **53.** $2(2\sqrt{2} - \sqrt{7}).$

**55.** $\frac{7}{2}\ln(2) - \frac{3}{4}.$   **57.** $\ln\left|\frac{q_n(1 - q_0)}{q_0(1 - q_n)}\right|.$

**59. a.** \$37,599;  **b.** \$4924.   **61. a.** \$5481;  **b.** \$535.

## EXERCISE 17.4 (page 762)

**1.** $\frac{16}{3}.$   **3.** $-1.$   **5.** $0.$   **7.** $\frac{13}{6}.$   **9.** \$12,400.

**11.** \$3321.

## PRINCIPLES IN PRACTICE 17.5

**1.** 76.90 feet.   **2.** 5.77 grams.

## EXERCISE 17.5 (page 766)

**1.** 413.   **3.** $0.340; \frac{1}{3} \approx 0.333.$   **5.** $1.388; \ln 4 \approx 1.386.$

**7.** 0.883.   **9.** 2,361,375.   **11.** 3.0 square units.

**13.** $\frac{8}{3}.$   **15.** 0.771.   **17.** $\frac{35}{6}$ km².

**19. a.** \$29,750;  **b.** \$36,600;  **c.** \$5350.

## PRINCIPLES IN PRACTICE 17.6

**1.** $I = I_0e^{-0.0085x}$.

## EXERCISE 17.6 (page 773)

**1.** $y = -\dfrac{1}{x^2 + C}$. **3.** $y = \dfrac{1}{3}(x^2 + 1)^{3/2} + C$.

**5.** $y = Ce^x, C > 0$. **7.** $y = Cx, C > 0$.

**9.** $y = \sqrt{2x}$. **11.** $y = \ln\dfrac{x^3 + 3}{3}$.

**13.** $y = \dfrac{4x^2 + 3}{2(x^2 + 1)}$. **15.** $y = \sqrt{\left(\dfrac{3x^2}{2} + \dfrac{3}{2}\right)^2 - 1}$.

**17.** $y = \ln\left(\dfrac{1}{2}\sqrt{x^2 + 3}\right)$. **19.** $c = (q + 1)e^{1/(q + 1)}$.

**21.** 46 weeks.

**23.** $N = 20,000e^{0.018t}$; $N = 20,000(1.2)^{t/10}$; $28,800$.

**25.** $2e^{0.946}$ billion. **27.** $0.01204$; $57.57$ sec.

**29.** 2900 years. **31.** $N = N_0e^{k(t - t_0)}, t \geq t_0$.

**33.** 12.6 units. **35.** $A = 400(1 - e^{-t/2})$, 157 grams.

**37. a.** $V = 14,000e^{(2 \ln 0.9)t}$; **b.** October 2001.

## EXERCISE 17.7 (page 781)

**1.** 29,400. **3.** 430,000. **5.** 1990. **7. b.** 375.

**9.** 1:06 A.M. **11.** \$62,500.

**13.** $N = M - (M - N_0)e^{-kt}$.

## PRINCIPLES IN PRACTICE 17.8

**1.** 20 ml.

## EXERCISE 17.8 (page 785)

**1.** $\dfrac{1}{3}$. **3.** Div. **5.** $\dfrac{1}{e}$. **7.** Div. **9.** $-\dfrac{1}{2}$. **11.** 0.

**13. a.** 800; **b.** $\dfrac{2}{3}$. **15.** 4,000,000. **17.** $\dfrac{1}{2}$ square unit.

**19.** 5000 increase.

## REVIEW PROBLEMS—CHAPTER 17 (page 788)

**1.** $\dfrac{x^2}{4}[2 \ln (x) - 1] + C$. **3.** $5 + \dfrac{9}{4}\ln 3$.

**5.** $\dfrac{1}{21}(9 \ln |3 + x| - 2 \ln |2 + 3x|) + C$.

**7.** $\dfrac{1}{2(x + 2)} + \dfrac{1}{4}\ln\left|\dfrac{x}{x + 2}\right| + C$.

**9.** $-\dfrac{\sqrt{9 - 16x^2}}{9x} + C$. **11.** $\dfrac{3}{2} \ln\left|\dfrac{x - 3}{x + 3}\right| + C$.

**13.** $\dfrac{e^{7x}}{49}(7x - 1) + C$. **15.** $\dfrac{1}{2}\ln |\ln 2x| + C$.

**17.** $x - \dfrac{3}{2}\ln |3 + 2x| + C$.

**19.** $2 \ln |x| + \dfrac{3}{2}\ln (x^2 + 1) + C$.

**21.** $2\sqrt{x + 1}[\ln (x + 1) - 2] + C$. **23.** 34.

**25. a.** 1.405; **b.** 1.388. **27.** $y = Ce^{x^3 + x^2}, C > 0$.

**29.** $\dfrac{1}{18}$. **31.** Div. **33.** 144,000. **35.** 0.0005; 90%.

**37.** $N = \dfrac{450}{1 + 224e^{-1.02t}}$. **39.** 4:16 P.M. **41.** 1.

**43. a.** 207, 208; **b.** 157, 165; **c.** 41, 41.

## MATHEMATICAL SNAPSHOT—CHAPTER 17 (page 791)

**1.** 114; 69.

## PRINCIPLES IN PRACTICE 18.1

**1.** $\dfrac{1}{3}$. **2.** 0.607.

**3.** Mean 5 years, standard deviation 5 years.

## EXERCISE 18.1 (page 799)

**1. a.** $\dfrac{5}{12}$; **b.** $\dfrac{11}{16} = 0.6875$; **c.** $\dfrac{13}{16} = 0.8125$; **d.** $-1 + \sqrt{10}$.

**3. a.** $f(x) = \begin{cases} \dfrac{1}{3} & \text{if } 1 \leq x \leq 4, \\ 0, & \text{otherwise;} \end{cases}$

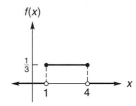

**b.** $\dfrac{1}{3}$; **c.** 0; **d.** $\dfrac{5}{6}$; **e.** $\dfrac{2}{3}$; **f.** 0; **g.** 1; **h.** $\dfrac{5}{2}$; **i.** $\dfrac{\sqrt{3}}{2}$;

**j.** $F(x) = \begin{cases} 0, & \text{if } x < 1, \\ \dfrac{x - 1}{3} & \text{if } 3 \leq x \leq 4, \\ 1, & \text{if } x > 4. \end{cases}$

$P(X < 2) = \dfrac{1}{3}, P(1 < X < 3) = \dfrac{2}{3}$.

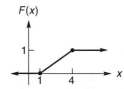

**5. a.** $f(x) = \begin{cases} \dfrac{1}{b - a} & \text{if } a \leq x \leq b, \\ 0, & \text{otherwise;} \end{cases}$

**b.** $\dfrac{a + b}{2}$; **c.** $\sigma^2 = \dfrac{(b - a)^2}{12}, \sigma = \dfrac{b - a}{\sqrt{12}}$.

**7. a.** $e^{-2} - e^{-6} \approx 0.133$; **b.** $1 - e^{-4} \approx 0.982$;

**c.** $e^{-5} \approx 0.007$; **d.** $1 - e^{-2} \approx 0.865$.

**9. a.** $\frac{1}{8}$; **b.** $\frac{5}{16}$; **c.** $\frac{39}{64} \approx 0.609$; **d.** 1; **e.** $\frac{8}{3}$; **f.** $\frac{2\sqrt{2}}{3}$;

**g.** $2\sqrt{2}$; **h.** $\frac{7}{16}$. **11.** $\frac{7}{10}$; 5 min. **13.** $e^{-3} \approx 0.050$.

## EXERCISE 18.2 (page 806)

**1. a.** 0.4641; **b.** 0.3239; **c.** 0.8888; **d.** 0.9983; **e.** 0.9147;
**f.** 0.4721. **3.** 0.13. **5.** −1.08. **7.** 0.34.
**9. a.** 0.9332; **b.** 0.0668; **c.** 0.0873. **11.** 0.3085.
**13.** 0.8185. **15.** 8. **17.** 9.68%. **19.** 90.82%.
**21. a.** 1.7%; **b.** 85.6.

## PRINCIPLES IN PRACTICE 18.3

**1.** 0.0396.

## EXERCISE 18.3 (page 810)

**1.** 0.1056; 0.0122. **3.** 0.0430; 0.9232. **5.** 0.7507.
**7.** 0.4129. **9.** 0.2514; 0.0287. **11.** 0.0336.

## REVIEW PROBLEMS—CHAPTER 18 (page 812)

**1. a.** 2; **b.** $\frac{9}{32}$; **c.** $\frac{3}{4}$;

**d.** $F(x) = \begin{cases} 0, & \text{if } x < 0, \\ \dfrac{x}{3} + \dfrac{2x^3}{3}, & \text{if } 0 \le x \le 1, \\ 1, & \text{if } x > 1. \end{cases}$

**3. a.** 2; **b.** $\frac{1}{\sqrt{2}} \approx 0.71$. **5.** 0.3085. **7.** 0.2857.
**9.** 0.1587. **11.** 0.9817. **13.** 0.0228.

## PRINCIPLES IN PRACTICE 19.1

**1. a.** \$3260; **b.** \$4410.

## EXERCISE 19.1 (page 819)

**1.** 10. **3.** 4. **5.** −1. **7.** 88. **9.** 3.
**11.** $2x_0 + 2h - 5y_0 + 4$. **13.** 2000. **15.** $y = -4$.
**17.** $z = 6$.
**19.**

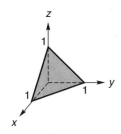

**21.**

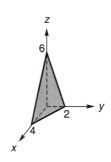

**23.**

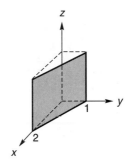

**25.**

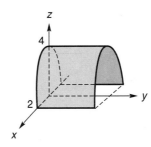

**27.**

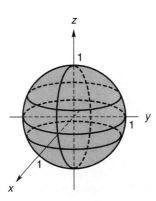

## EXERCISE 19.2 (page 824)

**1.** $f_x(x, y) = 8x$; $f_y(x, y) = 6y$.
**3.** $f_x(x, y) = 0$; $f_y(x, y) = 2$.
**5.** $g_x(x, y) = 3x^2y^2 + 4xy - 3y$;
$\quad g_y(x, y) = 2x^3y + 2x^2 - 3x + 4$.

**7.** $g_p(p, q) = \dfrac{q}{2\sqrt{pq}}$; $g_q(p, q) = \dfrac{p}{2\sqrt{pq}}$.

**9.** $h_s(s, t) = \dfrac{2s}{t - 3}$; $h_t(s, t) = -\dfrac{s^2 + 4}{(t - 3)^2}$.

**11.** $u_{q_1}(q_1, q_2) = \dfrac{3}{4q_1}; u_{q_2}(q_1, q_2) = \dfrac{1}{4q_2}.$

**13.** $h_x(x, y) = (x^3 + xy^2 + 3y^3)(x^2 + y^2)^{-3/2};$
$h_y(x, y) = (3x^3 + x^2y + y^3)(x^2 + y^2)^{-3/2}.$

**15.** $\dfrac{\partial z}{\partial x} = 5ye^{5xy}; \dfrac{\partial z}{\partial y} = 5xe^{5xy}.$

**17.** $\dfrac{\partial z}{\partial x} = 5\left[\dfrac{2x^2}{x^2 + y} + \ln(x^2 + y)\right]; \dfrac{\partial z}{\partial y} = \dfrac{5x}{x^2 + y}.$

**19.** $f_r(r, s) = \sqrt{r + 2s}\,(3r^2 - 2s) + \dfrac{r^3 - 2rs + s^2}{2\sqrt{r + 2s}};$

$f_s(r, s) = 2(s - r)\sqrt{r + 2s} + \dfrac{r^3 - 2rs + s^2}{\sqrt{r + 2s}}.$

**21.** $f_r(r, s) = -e^{3-r}\ln(7 - s); f_s(r, s) = \dfrac{e^{3-r}}{s - 7}.$

**23.** $g_x(x, y, z) = 6xy + 2y^2z; g_y(x, y, z) = 3x^2 + 4xyz;$
$g_z(x, y, z) = 2xy^2 + 9z^2.$

**25.** $g_r(r, s, t) = 2re^{s+t}; g_s(r, s, t)$
$= (7s^3 + 21s^2 + r^2)e^{s+t};$
$g_t(r, s, t) = e^{s+t}(r^2 + 7s^3).$

**27.** 50.  **29.** $\dfrac{1}{3}.$  **31.** 0.  **33.** 26

**39.** $-\dfrac{ra}{2\left[1 + a\dfrac{n-1}{2}\right]^2}.$

## EXERCISE 19.3 (page 829)

**1.** 20.  **3.** 784.5.

**5.** $\dfrac{\partial P}{\partial k} = 1.208648 l^{0.192} k^{-0.236}; \dfrac{\partial P}{\partial l} = 0.303744 l^{-0.808} k^{0.764}.$

**7.** $\dfrac{\partial q_A}{\partial p_A} = -50; \dfrac{\partial q_A}{\partial p_B} = 2; \dfrac{\partial q_B}{\partial p_A} = 4; \dfrac{\partial q_B}{\partial p_B} = -20;$
competitive.

**9.** $\dfrac{\partial q_A}{\partial p_A} = -\dfrac{100}{p_A^2 p_B^{1/2}}; \dfrac{\partial q_A}{\partial p_B} = -\dfrac{50}{p_A p_B^{3/2}};$
$\dfrac{\partial q_B}{\partial p_A} = -\dfrac{500}{3 p_B p_A^{4/3}}; \dfrac{\partial q_B}{\partial p_B} = -\dfrac{500}{p_B^2 p_A^{1/3}};$ complementary.

**11.** $\dfrac{\partial P}{\partial B} = 0.01 A^{0.27} B^{-0.99} C^{0.01} D^{0.23} E^{0.09} F^{0.27};$

$\dfrac{\partial P}{\partial C} = 0.01 A^{0.27} B^{0.01} C^{-0.99} D^{0.23} E^{0.09} F^{0.27}.$

**13.** 1120; if a staff manager with an M.B.A. degree had an extra year of work experience before the degree, the manager would receive $1120 per year in extra compensation.

**15. a.** $-1.015; -0.846;$
**b.** One for which $w = w_0$ and $s = s_0$.

**17.** $\dfrac{\partial g}{\partial x} = \dfrac{1}{V_F} > 0$ for $V_F > 0$. Thus if $x$ increases and $V_F$ and $V_s$ are fixed, then $g$ increases.

**19. a.** When $p_A = 8$ and $p_B = 64, \dfrac{\partial q_A}{\partial p_A} = -5$ and

$\dfrac{\partial q_A}{\partial p_B} = \dfrac{15}{32};$ **b.** Demand for A decreases by approximately

$\dfrac{15}{8}$ units.

**21. a.** No; **b.** 70%.  **23.** $\eta_{p_A} = -\dfrac{5}{46}, \eta_{p_B} = \dfrac{1}{46}$

**25.** $\eta_{p_A} = -1, \eta_{p_B} = -\dfrac{1}{2}$

## EXERCISE 19.4 (page 834)

**1.** $-\dfrac{x}{z}.$  **3.** $\dfrac{4y}{3z^2}.$  **5.** $\dfrac{x(yz^2 + 1)}{z(1 - x^2y)}.$  **7.** $-e^{y-z}.$

**9.** $\dfrac{yz}{1 + z}.$  **11.** $-\dfrac{3x}{z}.$  **13.** $-\dfrac{9}{10}.$  **15.** $-\dfrac{4}{e^2}.$

**17.** 1.  **19.** $\dfrac{5}{2}.$

**21. a.** 36; **b.** With respect to $q_A, \dfrac{60}{13};$ with respect to $q_B, \dfrac{288}{65}.$

## EXERCISE 19.5 (page 837)

**1.** $8xy; 8x.$  **3.** $3; 0; 0.$
**5.** $8xe^{2xy}; 8e^{2xy}(2xy + 1); 32x(1 + xy)e^{2xy}.$
**7.** $3x^2y + 4xy^2 + y^3;$
$3xy^2 + 4x^2y + x^3; 6xy + 4y^2; 6xy + 4x^2.$
**9.** $x(x^2 + y^2)^{-1/2}; y^2(x^2 + y^2)^{-3/2}.$  **11.** 0.  **13.** 1758.

**15.** $2e.$  **17.** $-\dfrac{1}{8}.$  **23.** $-\dfrac{y^2 + z^2}{z^3} = -\dfrac{3x^2}{z^3}.$

## EXERCISE 19.6 (page 840)

**1.** $\dfrac{\partial z}{\partial r} = 13; \dfrac{\partial z}{\partial s} = 9.$  **3.** $\left[2t + \dfrac{3\sqrt{t}}{2}\right]e^{x+y}.$

**5.** $5(2xz^2 + yz) + 2(xz + z^2) - (2x^2z + xy + 2yz).$
**7.** $3(x^2 + xy^2)^2(2x + y^2 + 2xy).$
**9.** $-2s(2x + yz) + r(xz + 3y^2z^2) - 5(xy + 2y^3z).$
**11.** $15s(2x - 7).$  **13.** 324.  **15.** $-1.$

**17.** When $p_A = 25$ and $p_B = 4, \dfrac{\partial c}{\partial p_A} = -\dfrac{1}{4}$ and $\dfrac{\partial c}{\partial p_B} = \dfrac{5}{4}.$

**19. a.** $\dfrac{\partial w}{\partial s} = \dfrac{\partial w}{\partial x}\dfrac{\partial x}{\partial s} + \dfrac{\partial w}{\partial y}\dfrac{\partial y}{\partial s};$ **b.** $-15.$

## EXERCISE 19.7 (page 849)

**1.** $\left(\dfrac{14}{3}, -\dfrac{13}{3}\right).$  **3.** $(2, 5), (2, -6), (-1, 5), (-1, -6).$

**5.** $(50, 150, 350).$  **7.** $\left(-2, \dfrac{3}{2}\right)$, rel. min.

**9.** $\left(-\dfrac{1}{4}, \dfrac{1}{2}\right)$, rel. max.

**11.** $(1, 1)$, rel. min; $\left(\dfrac{1}{2}, \dfrac{1}{4}\right)$, neither.

**13.** $(0, 0)$, rel. max.; $\left(4, \dfrac{1}{2}\right)$, rel. min.; $\left(0, \dfrac{1}{2}\right)$, $(4, 0)$, neither.

**15.** $(122, 127)$, rel. max. **17.** $(-1, -1)$, rel. min.
**19.** $(0, -2)$, $(0, 2)$, neither. **21.** $l = 24$, $k = 14$.
**23.** $p_A = 80$, $p_B = 85$.
**25.** $q_A = 48$, $q_B = 40$, $p_A = 52$, $p_B = 44$, profit $= 3304$.
**27.** $q_A = 3$, $q_B = 2$. **29.** 1 ft by 2 ft by 3 ft.

**31.** $\left(\dfrac{105}{37}, \dfrac{28}{37}\right)$, rel. min. **33.** $a = -8$, $b = -12$, $d = 33$.

**35. a.** 2 units of A and 3 units B;
**b.** Selling price for A is 30 and selling price for B is 19.
Relative maximum profit is 25.
**37. a.** $P = 5T(1 - e^{-x}) - 20x - 0.1T^2$;
**c.** Relative maximum at $(20, \ln 5)$; no relative extremum at
$\left(5, \ln \dfrac{5}{4}\right)$.

## EXERCISE 19.8 (page 857)

**1.** $(2, -2)$. **3.** $\left(3, \dfrac{3}{2}, -\dfrac{3}{2}\right)$. **5.** $\left(\dfrac{4}{3}, -\dfrac{4}{3}, -\dfrac{8}{3}\right)$.

**7.** $(6, 3, 2)$. **9.** $\left(\dfrac{2}{3}, \dfrac{4}{3}, -\dfrac{4}{3}\right)$. **11.** $(3, 3, 6)$.

**13.** Plant 1, 40 units; plant 2, 60 units.
**15.** 74 units (when $l = 8$, $k = 7$).
**17.** \$15,000 on newspaper advertising and \$45,000 on TV advertising.
**19.** $x = 5$, $y = 15$, $z = 5$.
**21.** $x = 12$, $y = 8$. **23.** $x = 10$, $y = 20$, $z = 5$.

## EXERCISE 19.9 (page 865)

**1.** $\hat{y} = 0.98 + 0.61x$; 3.12. **3.** $\hat{y} = 0.057 + 1.67x$; 5.90.
**5.** $\hat{q} = 82.6 - 0.641p$. **7.** $\hat{y} = 100 + 0.13x$; 105.2.
**9.** $\hat{y} = 8.5 + 2.5x$.
**11. a.** $\hat{y} = 35.9 - 2.5x$; **b.** $\hat{y} = 28.4 - 2.5x$.

## EXERCISE 19.11 (page 871)

**1.** 18. **3.** $\dfrac{1}{4}$. **5.** $\dfrac{2}{3}$. **7.** 3. **9.** 324. **11.** $-\dfrac{58}{35}$.

**13.** $\dfrac{8}{3}$. **15.** $-\dfrac{1}{3}$. **17.** $\dfrac{e^2}{2} - e + \dfrac{1}{2}$. **19.** $-\dfrac{27}{4}$.

**21.** $\dfrac{1}{24}$. **23.** $e^{-4} - e^{-2} - e^{-3} + e^{-1}$. **25.** $\dfrac{3}{8}$.

## REVIEW PROBLEMS—CHAPTER 19 (page 874)

**1.**

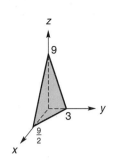

**3.**

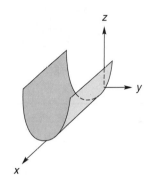

**5.** $4x + 3y$; $3x + 2y$. **7.** $\dfrac{y}{(x + y)^2}$; $-\dfrac{x}{(x + y)^2}$.

**9.** $\dfrac{y}{x^2 + y^2}$. **11.** $2xze^{x^2yz}(1 + x^2yz)$. **13.** $2(x + y)$.

**15.** $xze^{yz} \ln z$; $\dfrac{e^{yz}}{z} + ye^{yz} \ln z = e^{yz}\left(\dfrac{1}{z} + y \ln z\right)$.

**17.** $\dfrac{1}{64}$. **19.** $2(x + y)e^r + 2\left(\dfrac{x + 3y}{r + s}\right)$; $2\left(\dfrac{x + 3y}{r + s}\right)$.

**21.** $\dfrac{2x + 2y + z}{4z - x}$. **23.** $\dfrac{\partial P}{\partial l} = 14l^{-0.3}k^{0.3}$; $\dfrac{\partial P}{\partial k} = 6l^{0.7}k^{-0.7}$.

**25.** Competitive. **27.** $(2, 2)$, rel. min.
**29.** 4 ft by 4 ft by 2 ft.
**31.** A, 89 cents per pound; B, 94 cents per pound.
**33.** $(3, 2, 1)$. **35.** $\hat{y} = 12.87 + 3.23x$

**37.** 8. **39.** $\dfrac{1}{210}$.

## MATHEMATICAL SNAPSHOT—CHAPTER 19 (page 877)

**1.** $y = 9.50e^{-0.22399x} + 5$. **3.** $T = 79e^{-0.01113t} + 45$.

## A.1 Review: Slopes and Equation of Lines

**1.** $-\dfrac{5}{8}$. **3.** $3x - 2y + 10 = 0$.
**5.** $5x - 4y - 32 = 0$.

**7.** slope is undefined. no $y$-intercept.

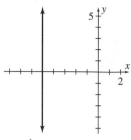

**9.** $y = \dfrac{1}{3}x + 5$.

## A.2 Review: Secant Lines and Average Rate of Change

**1.** 4.5 inches per year.
**3. a.** 3.5 degrees per day; **b.** $-1.25$ degrees per day;
**c.** 1 degree per day; **d.** 0.59 degrees per day.
**5. a.** 16; **b.** 7; **c.** 4; **d.** 3.01.
**7. a.** 25; **b.** 97; **c.** $27 + 9h + h^2$; **d.** $3x_0^2 + 3x_0 h + h^2$.

## A.3 Review: Slope of a Curve and Derivatve

**1. a.** 0; **b.** $-8$; **c.** $2x_0$.
**3. a.** $-\dfrac{1}{16}$; **b.** $-\dfrac{1}{36}$; **c.** $-\dfrac{1}{(x_0 - 4)^2}$.
**5. a.** 32; **b.** $-32$; **c.** $-64$.　　**7.** 35.5.
**9. a.** 0; **b.** 5; **c.** 5.
**11. a.** $-\dfrac{5}{(5x + 11)^2}$; **b.** $-\dfrac{7460}{(5x + 11)^2}$; **c.** $\dfrac{15}{(12 - 5x)^2}$;
**d.** $\dfrac{15}{(12 - 5x)^2}$.

## A.4 Review: Areas of Geometric Shapes

**1.** 250 square feet.　　**3.** 157.5 square inches.
**5.** 51 square units. **7.** 45.5 square units.

## A.5 Review: Sigma Notation

**1.** $12, 17, t$.　　**3.** 168.　　**5.** 532.　　**7.** $\displaystyle\sum_{i=36}^{60} i$.
**9.** $\displaystyle\sum_{j=3}^{8} 5^j$.　　**11.** 520.　　**13.** 5.　　**15.** 37,750.
**17.** 14,980.　　**19.** 295,425.　　**21.** $4\dfrac{23}{25}$.
**23.** $8 - \dfrac{4(n + 1)(2n + 1)}{3n^2}$.

## A.6 Review: Riemann Sums and the Definite Integral

**1.** 225; lower.　　**3.** 55.625.
**5.** $0.05 \displaystyle\sum_{k=1}^{120} [9 - (-3.05 + 0.05k)^2]$.　　**7.** 21.5.

**9.** $0.01 \displaystyle\sum_{k=1}^{300} (0.99 + 0.01k)^2$.
**11.** $\dfrac{3}{N} \displaystyle\sum_{k=1}^{N} \left[ \left( 2 + \dfrac{3}{N}k \right)^2 + 2 + \dfrac{3}{N}k \right] = \dfrac{99}{2} + \dfrac{36}{N} + \dfrac{9}{2N^2}$.
**13. a.** 33; **b.** $\dfrac{3}{2}$; **c.** 30; **d.** 36; **e.** 4; **f.** 10; **g.** 480; **h.** 63.
**15.** $\displaystyle\int_0^{100} (40 + 0.3x)dx = 5500$
　　　$\displaystyle\sum_{k=1}^{100} (40 + 0.3k) = 5515$.

## A.7 Review: Area Under a Rate-of-Change Curve

**1.** 270 miles.　　**3.** $4950$.
**5. a.** 60 minutes; **b.** $0; **c.** $1.20; **d.** $9.60.
**7. a.** 16 feet; **b.** 1600 feet; **c.** 816 feet.
**9.** $21,800.　　**11.** $12,500.

# Index

# Photo Credits

## Business Relations

$$\text{Interest} = (\text{principal})(\text{rate})(\text{time})$$
$$\text{Total cost} = \text{variable cost} + \text{fixed cost}$$
$$\text{Average cost per unit} = \frac{\text{total cost}}{\text{quantity}}$$
$$\text{Total revenue} = (\text{price per unit})(\text{number of units sold})$$
$$\text{Profit} = \text{total revenue} - \text{total cost}$$

## Ordinary Annuity Formulas

$$A = R\frac{1 - (1 + r)^{-n}}{r} = Ra_{\overline{n}|r} \qquad \text{(present value)}$$

$$S = R\frac{(1 + r)^{n} - 1}{r} = Rs_{\overline{n}|r} \qquad \text{(future value)}$$

## Graphs of Elementary Functions

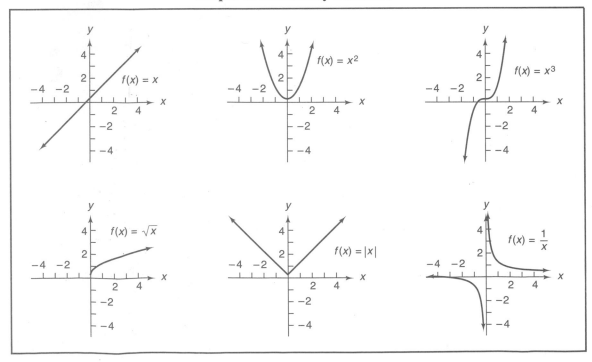